What our readers say about REA's Problem Solvers®...

- *"...the best tools for learn...*
- *"...superb..."*
- *"...taught me more than I imagined..."*

"Your *Problem Solver®* books are the best tools for learning I have ever encountered."

~Instructor, Batavia, Illinois

"I own a large library of your *Problem Solver®* books, which I find to be extremely useful. I use them for reference not only for my own homework and research, but I also tutor undergraduate students and your books help me to do a 'quick refresher' of concepts in certain topics."

~Student, Rochester, New York

"Thank you for the superb work you have done in publishing the *Problem Solvers®*. These books are the best review books on the market."

~Student, New Orleans, Louisiana

"I found your *Problem Solvers®* to be very helpful. I have nine of your books and intend to purchase more."

~Student, Gulfport, Mississippi

"I love your *Problem Solvers®*. The volumes I have, have already taught me more than I imagined."

~Instructor, Atlanta, Georgia

We couldn't have said it any better!

Research & Education Association
Making the world smarter

Visit us online at www.rea.com.
Your comments welcome at info@rea.com.

PROBLEM SOLVERS®

Linear Algebra

Staff of Research & Education Association

Research & Education Association
Visit our website at
www.rea.com

Research & Education Association
61 Ethel Road West
Piscataway, New Jersey 08854
E-mail: info@rea.com

THE LINEAR ALGEBRA
PROBLEM SOLVER®

Printed in the United States of America

Library of Congress Control Number 2006928793

International Standard Book Number 0-87891-518-4

Let REA's Problem Solvers® work for you

REA's *Problem Solvers* are for anyone—from student to seasoned professional—who seeks a thorough, practical resource. Each *Problem Solver* offers hundreds of problems and clear step-by-step solutions not found in any other publication.

Perfect for self-paced study or teacher-directed instruction, from elementary to advanced academic levels, the *Problem Solvers* can guide you toward mastery of your subject.

Whether you are preparing for a test, are seeking to solve a specific problem, or simply need an authoritative resource that will pick up where your textbook left off, the *Problem Solvers* are your best and most trustworthy solution.

Since the problems and solutions in each *Problem Solver* increase in complexity and subject depth, these references are found on the shelves of anyone who requires help, wants to raise the academic bar, needs to verify findings, or seeks a challenge.

For many, *Problem Solvers* are homework helpers. For others, they're great research partners. What will *Problem Solvers* do for you?

- Save countless hours of frustration groping for answers
- Provide a broad range of material
- Offer problems in order of capability and difficulty
- Simplify otherwise complex concepts
- Allow for quick lookup of problem types in the index
- Be a valuable reference for as long as you are learning

Each *Problem Solver* book was created to be a reference for life, spanning a subject's entire breadth with solutions that will be invaluable as you climb the ladder of success in your education or career.

—*Staff of Research & Education Association*

How To Use This Book

The genius of the **Problem Solvers** lies in their simplicity. The problems and solutions are presented in a straightforward manner, the organization of the book is presented so that the subject coverage will easily line up with your coursework, and the writing is clear and instructive.

Each chapter opens with an explanation of principles, problem-solving techniques, and strategies to help you master entire groups of problems for each topic.

The chapters also present progressively more difficult material. Starting with the fundamentals, a chapter builds toward more advanced problems and solutions—just the way one learns a subject in the classroom. The range of problems takes into account critical nuances to help you master the material in a systematic fashion.

Inside, you will find varied methods of presenting problems, as well as different solution methods, all of which take you through a solution in a step-by-step, point-by-point manner.

There are no shortcuts in **Problem Solvers**. You are given no-nonsense information that you can trust and grow with, presented in its simplest form for easy reading and quick comprehension.

As you can see on the facing page, the key features of this book are:

- Clearly labeled chapters

- Solutions presented in a way that will equip you to distinguish key problem types and solve them on your own more efficiently

- Problems numbered and conveniently indexed by the problem number, not by page number

Get smarter....Let **Problem Solvers** go to your head!

Anatomy of a Problem Solver®

CHAPTER 4

Chapter Subject

BASIC MATRIX ARITHMETIC

Problem Topic

Problem Number

ADDITION OF MATRICES

● PROBLEM 4-1

Prove that if $A = (a_{ij})$ and $B = (b_{ij})$ are $m \times n$ matrices over a field K and c is an element of K, then $A + B = (c_{ij})$ where $c_{ij} = a_{ij} + b_{ij}$, $i = 1,2,\ldots,m$, $j = 1,2,\ldots,n$ and $cA = (d_{ij})$ where $d_{ij} = ca_{ij}$, $i = 1,2,\ldots,m$, $j = 1,2,\ldots,n$.

The Full Solution to the Problem

The Problem

<u>Solution</u>: Let e_j represent the vector $(0,\ldots,0,1,0,\ldots,0)$ with the __ __ in the ith place. Letting $C = (c_{ij})$, then the ith row of __ is __ . $c = (c_{i1},c_{i2},\ldots,c_{in})$; $i = 1,2,\ldots,m$. On the other __

__ $B) = e_iA + e_iB = (a_{i1},a_{i2},\ldots,a_{in}) + (b_{i1},b_{i2},\ldots$

__ ponding elements in the ith row,

$$e_i(A+B) = (a_{i1} + b_{i1}, a_{i2} + b_{i2}, \ldots, a_{in} + b_{in}) .$$

But, by definition, $c_{ij} = a_{ij} + b_{ij}$. Therefore, by sybstitution, $e_i(A+B) = $ __ $_{i1},c_{i2},\ldots,c_{in}) = e_iC$.

This is true for all $i = 1,\ldots,m$. Therefore, $A + B = C$.

(b)   Let $D = (d_{ij})$ then,

$$e_i \cdot D = (d_{i1},d_{i2},\ldots,d_{in}) , \ i = 1,2,\ldots,m .$$

On the other hand,

$$e_i(cA) = c(e_iA) = c(a_{i1},a_{i2},\ldots,a_{in}), \ i = 1,2,\ldots,m$$
$$= (ca_{i1}, ca_{i2},\ldots,ca_{in}) .$$

But, by definition, $d_{ij} = ca_{ij}$. Therefore, $e_i(cA) = (d_{i1},\ldots,d_{in})$ by substitution; i.e., $e_i(cA) = e_iD$. This holds for all $i = 1,\ldots,m$. Therefore, $cA = D$.

● PROBLEM 4-2

If $A = \begin{bmatrix} 2 & -2 & 4 \\ -1 & 1 & 1 \end{bmatrix}$ and $B = \begin{bmatrix} 0 & 1 & -3 \\ 1 & 3 & 1 \end{bmatrix}$, find $2A + B$.

120

CONTENTS

CHAPTER 1

VECTOR SPACES

EXAMPLES OF VECTOR SPACES

● **PROBLEM** 1-1

Define a field and give an example of i) an infinite field and ii) a finite field.

<u>Solution</u>:   A field is a set of elements $F = \{a, b, \dots\}$ together with two binary operations in F, called multiplication and addition (not necessarily the multiplication and addition we are familiar with) for which the following postulates hold:

1) Associative Law of Addition: For every triplet (a,b,c) in F, a + (b+c) = (a+b) + c .

2) Commutative Law of Addition: For every pair (a,b) in F, a + b = b + a .

3) Associative Law of Multiplication: For every triplet (a,b,c) , a(bc) = (ab)c .

4) Commutative Law for Multiplication: For every pair (a,b) in F, ab = ba .

5) Distributive Law: For every triplet (a,b,c) in F, a(b+c) = ab + ac .

   The set F must contain an identity element for multiplication and for addition.

6) There exists an element 0 in F such that for any a in F, a + 0 = a .

7) There exists an element in F, call it 1, such that for every a in F, $a \cdot 1 = a$.

   Each non-zero element must have an inverse with respect to each operation.

8) For each element a in F, there is in F an element -a such that a + (-a) = 0   (0 is identity for addition).

9) For each element a in F (a ≠ 0), there is an element a^{-1} in F such that $a^{-1} \cdot a = 1$ (1 is identity for multiplication).

i) The set of real numbers over + and $\times$ is an infinite field. Notice that from the definition of a field, "two binary operations in F", we require that F be closed. That is, each result of an operation between two elements of the field is again in the field. For any three real numbers a,b,c, we know ab and a + b are real numbers. We also know from the properties of real numbers that:

a)   a + (b+c) = (a+b) + c ; b)   a(bc) = (ab)c ; c)   a + b = b + a ;

d) ab = ba ; e) a(b+c) = ab + ac .

We also have identities for multiplication and addition (0,1). We know
that each element a has an additive inverse -a and a multiplicative
one 1/a . Thus, $R(+,x)$ is a field.

ii) We can construct a finite field from the set of integers
$J = \{0, \pm 1, \pm 2, \ldots\}$.

Let p denote a fixed prime and define $a \equiv \pmod{p}$ to mean b is
the remainder when a is div ded by p or, equivalently, b - a is
divisible by p. (i.e., $5 \equiv 2 \pmod{3}$ and $12 \equiv 1 \pmod{11}$). Choosing
p = 5, we can construct a field from the integers mod 5. The modulus
5 (as it is called) relation partitions the integers into 5 mutually
disjoint classes. That is, every integer is 0,1,2,3 or 4 mod 5. This
set of classes (residue classes) is a finite field over addition and
multiplication defined as follows: $a \pmod 5 + b \pmod 5 = (a+b) \pmod 5$,
and $a \pmod 5 \cdot b \pmod 5 = ab \pmod 5$. Since a and b are integers,
we have commutativity, associativity and distribution. We also have
identities 0 and 1 for addition and multiplication, respectively.

If we draw an addition and a multiplication table for integers
mod 5, we see that each element has an additive and a multiplicative
inverse since there is a 0 or a 1 respectively in each row.

+	0	1	2	3	4
0	0	1	2	3	4
1	1	2	3	4	0
2	2	3	4	0	1
3	3	4	0	1	2
4	4	0	1	2	3

.	0	1	2	3	4
0	0	0	0	0	0
1	0	1	2	3	4
2	0	2	4	1	3
3	0	3	1	4	2
4	0	4	3	2	1

Notice 0 has no multiplicative inverse. A similar finite field can be
constructed using any prime integer.

● PROBLEM 1-2

Show that the space R^n (comprised of n-tuples of real numbers
$(x_1, \ldots, x_n)$ is a vector space over the field R of real

numbers. The operations are addition of n-tuples, i.e.,
$(x_1, \ldots, x_n) + (y_1, y_2, \ldots, y_n) = (x_1 + y_1, x_2 + y_2, \ldots, x_n$
$+ y_n)$, and scalar multiplication, $\alpha(x_1, x_2, \ldots, x_n)$
$= (\alpha x_1, \alpha x_2, \ldots, \alpha x_n)$ where $\alpha \in R$.

Solution: Any set that satisfies the axioms for a vector
space over a field is known as a vector space. We must show
that R^n satisfies the vector space axioms. The axioms fall
into two distinct categories:

A) the axioms of addition for elements of a set
B) the axioms involving multiplication of vectors by elements
    from the field .

1) Closure under addition

By definition, $(x_1, x_2, \ldots, x_n) + (y_1, y_2, \ldots, y_n)$

$\quad = (x_1 + y_1, x_2 + y_2, \ldots, x_n + y_n)$.

Now, since $x_1, y_1, x_2, y_2, \ldots, x_n, y_n$ are real numbers, the

sums of $x_1 + y_1, x_2 + y_2, \ldots, x_n + y_n$ are also real numbers.

Therefore, $(x_1 + y_1, x_2 + y_2, \ldots, x_n + y_n)$ is also an n-tuple

of real numbers; hence, it belongs to R^n.

2) Addition is commutative

The numbers $x_1, x_2, \ldots, x_n$ are the coordinates of the vector

$(x_1, x_2, \ldots, x_n)$, and $y_1, y_2, \ldots, y_n$ are the coordinates of

the vector $(y_1, y_2, \ldots, y_n)$.

Show $(x_1, x_2, \ldots, x_n) + (y_1, \ldots, y_n) = (y_1, \ldots, y_n)$

$+ (x_1, \ldots, x_n)$. Now, the coordinates $x_1 + y_1, x_2$

$+ y_2, \ldots, x_n + y_n$ are sums of real numbers. Since real

numbers satisfy the commutativity axiom, $x_1 + y_1 = y_1 + x_1$,

$x_2 + y_2 = y_2 + x_2, \ldots, x_n + y_n = y_n + x_n$.

Thus, $(x_1, x_2, \ldots, x_n) + (y_1, y_2, \ldots, y_n)$

$\qquad\qquad = (x_1 + y_1, x_2 + y_2, \ldots, x_n + y_n)$ (by definition)

$\qquad\qquad = (y_1 + x_1, y_2 + x_2, \ldots, y_n + x_n)$ (by commutati-
$\qquad\qquad\qquad\qquad\qquad\qquad\qquad\qquad\qquad\qquad$vity of real
$\qquad\qquad\qquad\qquad\qquad\qquad\qquad\qquad\qquad\qquad$numbers)

$\qquad\qquad = (y_1, y_2, \ldots, y_n)$

$\qquad\qquad\quad + (x_1, x_2, \ldots, x_n)$ $\qquad\qquad$ (by definition)

We have shown that n-tuples of real numbers satisfy
the commutativity axiom for a vector space.

3) Addition is associative: $(a+b) + c = a + (b+c)$.
Let $(x_1, x_2, \ldots, x_n)$, $(y_1, y_2, \ldots, y_n)$ and $(z_1, z_2, \ldots, z_n)$

be three points in R^n.

Now, $((x_1, x_2, \ldots, x_n) + (y_1, y_2, \ldots, y_n))$

$+ (z_1, z_2, \ldots, z_n)$

$= (x_1 + y_1, x_2 + y_2, \ldots, x_n + y_n) + (z_1, z_2, \ldots, z_n)$

$$= ((x_1 + y_1) + z_1, (x_2 + y_2) + z_2, \ldots, (x_n + y_n) + z_n). (1)$$

The coordinates $(x_i + y_i) + z_i$ $(i = 1, \ldots, n)$ are real numbers. Since real numbers satisfy the associativity axiom, $(x_i + y_i) + z_i = x_i + (y_i + z_i)$. Hence, (1) may be rewritten as

$$(x_1 + (y_1 + z_1), x_2 + (y_2 + z_2), \ldots, x_n + (y_n + z_n))$$

$$= (x_1, x_2, \ldots, x_n) + (y_1 + z_1, y_2 + z_2, \ldots, y_n + z_n)$$

$$= (x_1, x_2, \ldots, x_n) + ((y_1, y_2, \ldots, y_n) + (z_1, z_2, \ldots, z_n)).$$

4) Existence and uniqueness of a zero element

The set R^n should have a member $(a_1, a_2, \ldots, a_n)$ such that for any point $(x_1, \ldots, x_n)$ in R^n, $(x_1, x_2, \ldots, x_n)$ $+ (a_1, a_2, \ldots, a_n) = (x_1, x_2, \ldots, x_n)$. The point $(\underbrace{0, 0, 0, \ldots, 0}_{n \text{ zeros}})$, where 0 is the unique zero of the real number system, satisfies this requirement.

5) Existence and uniqueness of an additive inverse.

Let $(x_1, x_2, \ldots, x_n) \in R^n$. An additive inverse of $(x_1, x_2, \ldots, x_n)$ is an n-tuple $(a_1, a_2, \ldots, a_n)$ such that $(x_1, x_2, \ldots, x_n) + (a_1, a_2, \ldots, a_n) = (0, 0, \ldots, 0)$. Since $x_1, x_2, \ldots, x_n$ belong to the real number system, they have unique additive inverses $(-x_1), (-x_2), \ldots, (-x_n)$. Consider $(-x_1, -x_2, \ldots, -x_n) \in R^n$.

$$(x_1, x_2, \ldots, x_n) + (-x_1, -x_2, \ldots, -x_n)$$

$$= (x_1 + (-x_1), x_2 + (-x_2), \ldots, x_n + (-x_n))$$

$$= (0, 0, \ldots, 0).$$

We now turn to the axioms involving scalar multiplication.

6) Closure under scalar multiplication.

By definition, $\alpha(x_1, x_2, \ldots, x_n) = (\alpha x_1, \alpha x_2, \ldots, \alpha x_n)$ where the coordinates αx_i are real numbers. Hence, $(\alpha x_1, \alpha x_2, \ldots, \alpha x_n) \in R^n$.

7) Associativity of scalar multiplication.

Let α, β be elements of R. We must show that
$(\alpha\ \beta)(x_1,x_2,\ldots,x_n) = \alpha(\beta x_1,\ \beta x_2,\ \ldots,\ \beta x_n)$. But, since α, β
and x_1, x_2, $\ldots$, x_n are real numbers, $(\alpha\ \beta)\ (x_1,\ x_2,\ \ldots,\ x_n)$
$= (\alpha\beta x_1,\ \alpha\beta x_2,\ \ldots,\ \alpha\beta x_n) = \alpha(\beta x_1,\ \beta x_2,\ \ldots,\ \beta x_n)$.

8) The first distributive law

We must show that $\alpha(x+y) = \alpha x + \alpha y$ where x and y are vectors
in R^n and $\alpha \in R$.

$$\alpha[(x_1,\ x_2,\ \ldots,\ x_n) + (y_1,\ y_2,\ \ldots,\ y_n)]$$

$$= \alpha((x_1 + y_1),\ (x_2 + y_2),\ \ldots,\ (x_n + y_n))$$

$$= (\alpha(x_1 + y_1),\ \alpha(x_2 + y_2),\ \ldots,\ \alpha(x_n + y_n))\ (2)$$

(by definition of scalar multiplication).

Since each coordinate is a product of a real number and
the sum of two real numbers, $\alpha(x_i + y_i) = \alpha x_i + \alpha y_i$. Hence,

(2) becomes $[(\ \alpha x_1 + \alpha y_1),\ (\ \alpha x_2 + \alpha y_2),\ \ldots,\ (\alpha x_n + \alpha y_n)]$

$$= (\alpha x_1,\ \alpha x_2,\ \ldots,\ \alpha x_n) + (\alpha y_1,\ \alpha y_2,\ \ldots,\ \alpha y_n)$$

$$= \alpha x + \alpha y.$$

9) The second distributive law

We must show that $(\alpha + \beta)\ x = \alpha x + \beta x$ where α, $\beta \in R$ and x is
a vector in R^n. Since $\alpha + \beta$ is also a scalar,

$$(\alpha + \beta)\ (x_1,\ x_2,\ \ldots,\ x_n)$$

$$= ((\alpha + \beta)\ x_1,\ (\alpha + \beta)\ x_2,\ \ldots,\ (\alpha + \beta)\ x_n)\ (3)$$

Since α, β, x_i are all real numbers, then $(\alpha + \beta)\ x_i$
$= \alpha x_i + \beta x_i$. Therefore, (3) becomes

$((\alpha x_1 + \beta x_1),\ (\alpha x_2 + \beta x_2),\ \ldots,\ (\alpha x_n + \beta x_n))$

$= (\alpha x_1, \alpha x_2,\ \ldots,\ \alpha x_n) + (\beta x_1,\ \beta x_2,\ \ldots,\ \beta x_n)$

$= \alpha(x_1,\ x_2,\ \ldots,\ x_n) + \beta(x_1,\ x_2,\ \ldots,\ x_n)$.

10) The existence of a unit element from the field.

We require that there exist a scalar in the field R, call it
"1", such that $1\ (x_1,\ x_2,\ \ldots,\ x_n) = (x_1,\ x_2,\ \ldots,\ x_n)$.

Now the real number 1 satisfies this requirement.

Since a set defined over a field that satisfies (1) - (10) is a vector space, the set R^n of n-tuples is a vector space when equipped with the given operations of addition and scalar multiplication.

● **PROBLEM** 1-3

Show that V, the set of all functions from a set S ≠ φ to the field R, is a vector space under the following operations: if f(s) and g(s) ∈ V, then (f + g)(s) = f(s) + g(s). If c is a scalar from R, then (cf)(s) = cf(s).

Solution: Since the points of V are functions, V is called a function space. Because our field is R, V is the space of real-valued functions defined on S. Also, since addition of real numbers is commutative, f(s) + g(s) = g(s) + f(s). (Here f(s) and g(s) are real numbers, the values of the functions f and g at the point s ∈ S). Since addition in R is associative, f(s) + [g(s) + h(s)] = [f(s) + g(s)] + h(s) for all s ∈ S. Hence, addition of functions is associative. Next, the unique zero vector is the zero function which assigns to each s ∈ S the value 0 ∈ R. For each f in V, let (-f) be the function given by

$$(-f)(s) = -f(s).$$

Then f + (-f) = 0 as required.

Next, we must verify the scalar axioms. Since multiplication in R is associative,

$$(cd) f(s) = c (df(s)).$$

The two distributive laws are: c (f + g)(s) = cf(s) + cg(s) and (c + d) f(s) = cf(s) + df(s). They hold by the properties of the real numbers. Finally, for 1 ∈ R, 1f(s) = f(s).

Thus, V is a vector space. If, instead of R we had considered a field F, we would have had to verify the above axioms using the general properties of a field (any set that satisfies the axioms for a field).

● **PROBLEM** 1-4

Does the space V of potential functions of the nth degree form a vector space over the field of real numbers? Addition and scalar multiplication are as defined below:

i) f(x, y) + g (x, y) = (f + g) $(x_1 \ y)$ for f, g ∈ V.

ii) c [f (x, y)] = cf (x, y).

Solution: A potential function is any twice differentiable function V (x, y) that satisfies the second-order, partial differential equation

$$\frac{\partial^2 V}{\partial^2 x} + \frac{\partial^2 V}{\partial y^2} = 0 \qquad\qquad (1)$$

Examples of functions that satisfy (1) are 1) $V = k$ (a constant); 2) $V = x$; 3) $V = y$; 4) $V = xy$.

Note that, if V_1 and V_2 are potential functions,

$$\frac{\partial^2 V_1}{\partial x^2} + \frac{\partial^2 V_1}{\partial y^2} + \frac{\partial^2 V_2}{\partial x^2} + \frac{\partial^2 V_2}{\partial y^2} = 0 + 0 = 0,$$

their sum is again a potential function. (This gives us closure under addition.)

If V is a potential function, then so is cV where c is a scalar. To show this, we observe that

$$\frac{\partial^2 cV}{\partial x^2} = c\,\frac{\partial^2 V}{\partial x^2} \quad\text{and}\quad \frac{\partial^2 cV}{\partial y^2} = c\,\frac{\partial^2 V}{\partial y^2}.$$

Then, $\dfrac{\partial^2 cV}{\partial x^2} + \dfrac{\partial^2 cV}{\partial y^2} = c\left[\dfrac{\partial^2 V}{\partial x^2} + \dfrac{\partial^2 V}{\partial x^2}\right] = c(0) = 0.$

Finally, the set of potential functions V is a subset of the vector space of all real valued functions G. Since V is closed under addition and scalar multiplication, it is a subspace of G. Thus, it is a vector space in its own right.

To find the dimension of the nth degree potential function, let

$$V\,(x,\ y) = a_0 x^n + a_1 x^{n-1} y + a_2 x^{n-2} y^2 + \ldots + a_{n-2} x^2 y^{n-2}$$
$$+ a_{n-1}\, xy^{n-1} + a_n y^n.$$

Then, $\dfrac{\partial^2 V}{\partial x^2} + \dfrac{\partial^2 V}{\partial y^2} = 0$ is given by

$$a_0 n(n-1)\, x^{n-2} + a_1 (n-1)(n-2)\, x^{n-3} y + \ldots + 3 \cdot 2 a_{n-3}\, xy^{n-3}$$
$$+ 2a_{n-2} y^{n-2} + 2a_2 x^{n-2} + 3(2) a_3 x^{n-3} y + \ldots + (n-1)(n-2) a_{n-1}\, xy^{n-3}$$
$$+ n(n-1)\, a_n y^{n-2} = 0.$$

$$[n(n-1)a_0 + 2a_2] x^{n-2} + [(n-1)(n-2) a_1 + 3(2) a_3] x^{n-3} y + \ldots$$
$$+ [(n-1)(n-2) a_{n-1} + 3(2) a_{n-3}] xy^{n-3} + [n(n-1)a_n + 2a_{n-2}] y^{n-2} = 0.$$

We see that, from the coefficient of x^{n-2}, $a_2 = \dfrac{-n(n-1)}{2(1)}\, a_0.$ (1)

Similarly, for $x^{n-3} y$, $a_3 = \dfrac{-(n-1)(n-2)}{3(2)}\, a_1.$ $\qquad$ (2)

The general recurrence relationship is, for

$$x^{n-k}y^{(k-2)}, \qquad a_k = \frac{-(n-(k+2))(n-(k+1))a_{k-2}}{k(k-1)}. \qquad (3)$$

Now, $\dfrac{\partial^2 v}{\partial x^2} + \dfrac{\partial^2 v}{\partial y^2} = 0$ only if the coefficients of $x^r y^{n-r+2}$ are

equal to zero. Hence,

$$n(n-1)a_0 + 2a_2 = (n-1)(n-2)a_1 + 3(2)a_3 = \ldots$$

$$= (n-1)(n-2)a_{n-1} + 3(2)a_{n-3} = n(n-1)a_n + 2a_{n-2} = 0.$$

But from (1) and (2), if n and a_0 are given, then a_2, a_4, $\ldots a_{2k}$ can be calculated using the recurrence relation (3). Similarly, if a_1 is given, then a_1, a_3, $\ldots$, a_{2k-1} can be calculated using (3).

Hence, every potential function of degree n can be expressed as a linear combination of two potential functions of degree n, one for a_0 and one for a_1.

Thus, potential functions of degree n form a space of dimension 2.

● **PROBLEM** 1-5

Show that the set of semi-magic squares of order 3 x 3 form a vector space over the field of real numbers with addition defined as: $a_{ij} + b_{ij} = (a+b)_{ij}$ for i, j = 1, ..., 3.

Solution: First, consider an example of a magic square.

$$\begin{bmatrix} 4 & 9 & 2 \\ 3 & 5 & 7 \\ 8 & 1 & 6 \end{bmatrix} \qquad (1)$$

We see that in (1)

i)   every row has the same total, T;

ii)   every column has the same total, T;

iii) the two diagonals have the same total, again T.

A square array of numbers that satisfies i) and ii) but not iii) is called a semi-magic square. For example, from (1)

$$\begin{bmatrix} 6 & 2 & 7 \\ 8 & 4 & 3 \\ 1 & 9 & 5 \end{bmatrix}$$

is semi-magic. Let M be the set of 3 x 3 semi-magic squares. We should notice that when two semi-magic squares are added, the result is a semi-magic square. For example, suppose m_1 and $m_2 \in M$ have row (and column) sums of T_1 and T_2 respectively; then each row and column in $m_1 + m_2$ will have a sum of $T_1 + T_2$. Now, for m_1, m_2 and $m_3 \in M$

i) $m_1 + m_2 = m_2 + m_1$.

This follows from the commutativity property of the real numbers.

ii) $(m_1 + m_2) + m_3 = m_1 + (m_2 + m_3)$

Again, since the elements of m_1, m_2 and m_3 are real numbers and real numbers obey the associative law, semi-magic squares are associative with respect to addition.

iii)   Existence of a zero element

$$\begin{bmatrix} a_{11} & a_{12} & a_{13} \\ a_{21} & a_{22} & a_{23} \\ a_{31} & a_{32} & a_{33} \end{bmatrix} + \begin{bmatrix} 0 & 0 & 0 \\ 0 & 0 & 0 \\ 0 & 0 & 0 \end{bmatrix}$$

$$= \begin{bmatrix} a_{11} & a_{12} & a_{13} \\ a_{21} & a_{22} & a_{23} \\ a_{31} & a_{32} & a_{33} \end{bmatrix}$$

The 3 x 3 array with zeros everywhere is a semi-magic square.

iv)   Existence of an additive inverse.

If we replace every element in a semi-magic square with its negative, the result is a semi-magic square. Adding, we obtain the zero semi-magic square.

Next, let α be a scalar from the field of real numbers. Then, αm_1 is still a semi-magic square. The two distributive laws hold, and $(\alpha\beta)m_1 = \alpha(\beta m_1)$ from the properties of the real numbers. Finally, $1(m_1) = m_1$, i.e., multiplication of a vector in M by the unit element from R leaves the vector unchanged.

Thus, M is a vector space.

9

Show that the set of all arithmetical progressions forms a two-dimensional vector space over the field of real numbers. Addition and scalar multiplication are defined as follows: if $x = (x_1, x_2, \ldots, x_n, \ldots)$ and $y = (y_1, y_2, \ldots, y_n, \ldots)$, then $x + Y = (x_1 + y_1, x_2 + y_2, \ldots, x_n + y_n, \ldots)$ and $\alpha x = \alpha(x_1, x_2, \ldots, x_n, \ldots) = (\alpha x_1, \alpha x_2, \ldots, \alpha x_n, \ldots)$.

Solution: An arithmetic progression of real numbers is a sequence of real numbers such that the difference between any two successive terms is a constant ($|x_{n+1} - x_n| = $ constant).

To make things clearer, consider a numerical example. Let $x = (2, 5, 8, 11, 14, \ldots, 3n-1, \ldots)$ and $y = (6, 11, 16, 21, 26, \ldots, 5n+1, \ldots)$. Then $(x + Y) = (8, 16, 24, 32, 40, \ldots, 8n, \ldots)$. Letting $\alpha = 3$, we have $\alpha x = 3(2, 5, 8, 11, 14, \ldots, 3_{n-1}, \ldots) = (6, 15, 24, 33, 42, \ldots, 9n-3, \ldots)$.

It should be clear that the sum of two progressions is a progression, and that a progression multiplied by a scalar is a progression. That is, the set of progressions is closed with respect to addition and scalar multiplication.

If we view a progression as an infinite tuple of real numbers, we see that they satisfy the axioms for a vector field. Thus,

i) $(x + Y) = (y + x)$;

ii) $(x + y) + z = x + (y + z)$;

iii) $x = (x_1, x_2, \ldots, x_n, \ldots) + \vec{0} = (x_1, x_2, \ldots, x_n), \vec{0}$

$\quad\quad = (0, 0, \ldots)$;

iv) $(x_1, x_2, \ldots, x_n, \ldots) + ((-x_1), (-x_2), \ldots,) = (0, 0, \ldots)$.

Similarly, the axioms of scalar multiplication also hold. Thus, the set of all arithmetic progressions is a vector space.

To find the dimension note that any arithmetical progression can be written as the linear combination of the two vectors: $e_1 = (1, 1, 1, \ldots)$ and $e_2 = (0, 1, 2, 3, \ldots)$.

For example, letting $x = (2, 5, 8, 11, 14, \ldots)$, we can express x as

$$2e_1 + 3e_2 = 2(1, 1, 1, \ldots) + 3(0, 1, 2, 3, \ldots).$$

In general, an arithmetic progression has the form $x = (a, a+b, a+2b, \ldots, a+nb, \ldots)$.

10

This can be decomposed into

(a, a, a, ...) + (0, b, 2b, ..., nb, ...)

= a(1, 1, 1, ...) + b(0, 1, 2, 3, ..., n, ...)

= ae_1 + be_2.

The two progressions e_1 and e_2 form a basis for the space. Since the dimension of a finite dimensional vector space is equal to the number of vectors in the basis, the vector space of arithmetical progressions has dimension 2.

● **PROBLEM** 1-7

Consider the figure shown.

It could, for example, represent an electrical network. Define closed loops to be of the form ab, bcf, cde, etc. What are the appropriate steps whereby one can construct a vector space containing closed loops?

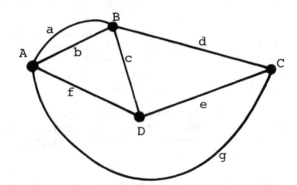

Solution: Examining the figure, we see that the closed loop bdef can be decomposed into the sum of the two loops bcf and cde. This suggests that we define addition of loops using the links a, b, c, ..., g. Let us represent each link by a 7-tuple, since there are 7 segments in the diagram. Thus, a = (1, 0, 0, 0, 0, 0, 0), b = (0, 1, 0, 0, 0, 0, 0), ... g = (0, 0, 0, 0, 0, 0, 1). Now, when we add links, we would like to obtain combinations of links. This can be done by defining addition of links modulo 2. That is, 0 + 0 = 0, 1 + 1 = 0 and 0 + 1 = 1. Then the only values the components can take on are 0 and 1. Thus, bcf + dec

= (0, 1, 1, 0, 0, 1, 0) + (0, 0, 1, 1, 1, 0, 0)

= (0, 1, 0, 1, 1, 1, 0) = bdef.

If we define the empty loop as the loop containing no links, then, using the indicated operation, we can satisfy the vector space axioms.

VECTORS IN 2-SPACE & 3-SPACE

A force of 25 newtons is being opposed by a force of 20 newtons, the acute angle between their lines of action being 60°. Use a scale diagram to approximate the magnitude and direction of the resultant force.

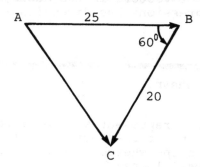

Solution: The situation is shown below where AB and BC represent, respectively, the 25-newton force and the 20-newton force. Using vector addition, we see that AC is the resultant.

The resultant could also be found using Newton's parallelogram law of forces. By actual measurement, we find that the resultant is the vector AC with an approximate magnitude of 23-newtons. Its direction differs from that of the 25-newton force by roughly 50°.

This illustrates that vector addition as it is defined corresponds exactly to Newton's parallelogram law of forces. Force can accurately be represented by vectors.

The water in a river is flowing from west to east at a speed of 5 km/h. If a boat is being propelled across the river at a (water) speed of 20 km/h in a direction 50° north of east, use a scale diagram to approximate the direction and land speed of the boat.

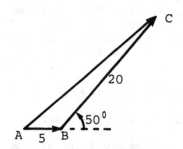

Solution: In physics, vectors are used to represent forces. They have both magnitude and direction. Some common vector forces are velocity, weight, gravitational attraction, etc.

In the given problem, the velocities of the water and boat are represented by AB and BC, respectively, with AC representing the resultant. The resultant can be found using the parallelogram law of forces. In the diagram, the scale is 1 in. = 10 km/hr., and actual measurement reveals that $|AC| \approx 2.4$ in. and $< CAB \approx 40°$. Thus, the land speed of the boat is approximately 25 km/hr. in a direction that is roughly 40° north of east.

● **PROBLEM** 1-10

If $v = (1, - 3, 2)$ and $w = (4, 2, 1)$, find $2v$, $v + w$ and $v - w$.

Solution: v and w represent points in three dimensional space, R^3. If we imagine v connected to the origin (0, 0, 0) by an arrow with base at (0, 0, 0) and tip at (1, - 3, 2), then this arrow may also be thought of as a vector v. (A vector is actually an equivalence class of arrows.)

The vector 2v represents another vector in space, each of whose components is double that of the corresponding components of v. Thus,

$$2v = 2(1, - 3, 2)$$

$$= (2, - 6, 4).$$

The sum of two vectors is clearly another vector. It is found by adding the corresponding coordinates of v and w. Thus,

$$v + w = (1, - 3, 2) + (4, 2, 1) = (5, - 1, 3).$$

Similarly, the difference of two vectors is found by subtracting corresponding coordinates. Thus,

$$v - w = (1, - 3, 2) - (4, 2, 1) = (- 3, - 5, 1).$$

● **PROBLEM** 1-11

Find the norm of the three dimensional vector $u = (- 3, 2, 1)$ and the distance between the points $(- 3, 2, 1)$ and $(4, - 3, 1)$.

Fig. 1

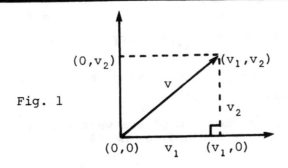

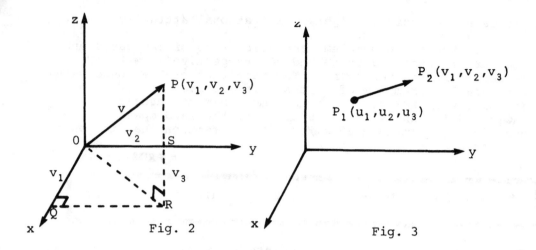

Fig. 2   Fig. 3

Solution: The length of a vector is called its norm and is denoted by $\| v \|$. If we let v be a vector from the origin to the point (v_1, v_2) in R^2, we see that the norm of v is

$$\sqrt{v_1{}^2 + v_2{}^2} \quad :$$

This follows from the theorem of Pythagoras since

$\| v^2 \| = v_1{}^2 + v_2{}^2$, and, hence $\| v \| = \sqrt{v_1{}^2 + v_2{}^2}$. Next, consider a vector from the origin $(0, 0, 0)$ to the point v_1, v_2, v_3 in R^3.

Here, two applications of the Pythagorean theorem are necessary to find $\| v \|$. From the figure, ORP is a right triangle. Hence, $\| v \|^2 = OP^2 = OR^2 + RP^2$.   (1)

But now, the vector OR in the x y plane is itself the hypotenuse of a right triangle whose other sides are OQ and OS. Hence, $OR^2 = OQ^2 + OS^2$ (2)

Substituting (2) into (1),

$$OP^2 = OQ^2 + OS^2 + RP^2$$

$$\| v \|^2 = v_1{}^2 + v_2{}^2 + v_3{}^2$$

$$\| v \| = \sqrt{v_1{}^2 + v_2{}^2 + v_3{}^2}.   (3)$$

The given vector is u = $(- 3, 2, 1)$. Using formula (3),

$$\| u \| = \sqrt{(-3)^2 + (2)^2 + (1)^2} = \sqrt{14}.$$

To find the distance between two points P_1, P_2 in R^3 we reason as follows:

The distance d is the norm of the vector P_1P_2. But P_1P_2 = $(v_1 - u_1, v_2 - u_2, v_3 - u_3)$ and, using (3),

14

$$\| P_1 P_2 \| = \sqrt{(v_1 - u_1)^2 + (v_2 - u_2)^2 + (v_3 - u_3)^2} \quad (4)$$

The given points are u = (- 3, 2, 1) and v = (4, - 3, 1). Hence,

$$\| v - u \| = \sqrt{(4 - (-3)^2 + ((-3) - (2))^2 + (1-1)^2} = \sqrt{74}.$$

Note the following points: i) in n-dimensional space, (n-1) applications of the Pythagorean theorem will yield the norm of a vector v as

$$\| v \| = \sqrt{v_1^2 + v_2^2 + \ldots + v_n^2}$$

where $v = (v_1, v_2, \ldots, v_n)$.

ii) $\| v - u \| = \| u - v \|$, that is, the distance between two points is a symmetric function of its arguments.

• **PROBLEM** 1-12

Find the components of the vector $v = P_1 P_2$ with initial point P_1 (2, -1, 4) and terminal point P_2 (7, 5, -8).

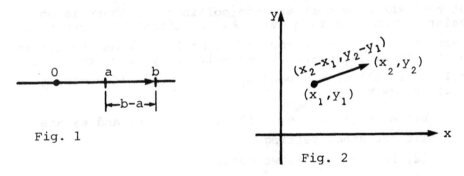

Fig. 1

Fig. 2

Solution: On the real number line a vector with initial point a and terminal point b has length $| b - a |$.

If we take R x R (the Cartesian product of the set of real numbers with itself) we obtain R^2, the Cartesian plane. Here a vector with initial point (x_1, y_1) and terminal point (x_2, y_2) has components $(x_2 - x_1, y_2 - y_1)$.

Forming the Cartesian triple product R^2 x R we obtain the set of all triples. A vector with initial point (x_1, y_1, z_1) and terminal point (x_2, y_2, z_2) has components

$$(x_2 - x_1, y_2 - y_1, z_2 - z_1). \quad (1)$$

In the given problem, $(x_1, y_1, z_1) = (2, -1, 4)$ and $(x_2, y_2, z_2) = (7, 5, -8)$. Hence, $v = P_1 P_2$ has components $(7 - 2, 5 - (-1), - 8 \; -4) = (5, 6, -12).$

15

LINEAR COMBINATIONS

Give an example of a pair of non-collinear vectors in R^2. Then, show that the point $(x_1, x_2) = (8, 7)$ can be expressed as a linear combination of the non-collinear vectors.

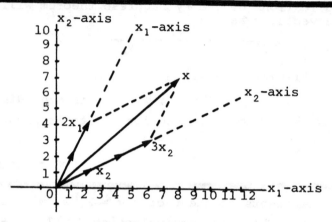

Solution: Two vectors in R^2 are non-collinear when neither is a multiple of the other; that is, v_1, v_2 (v_1 and v_2 are not both the 0 vector) are non-collinear if there is no scalar c such that $v_2 = cv_1$. As justification, note that a vector in R^2 is an arrow (straight line) from the origin to some point (x_1, x_2). If we multiply this vector by a scalar, we will only expand or shorten the old vector to obtain a new vector.

Let $e_1 = (1, 2)$; $e_2 = (2, 1)$. Then e_1 and e_2 are non-collinear since setting

$(2, 1) = c (1, 2)$, we obtain

$(2, 1) = (c, 2c)$

$2 = c$

$1 = 2c$.

There exists no c satisfying these two inconsistent equations.

Two non-collinear vectors in R^2 are sufficient to express any vector in R^2 as a linear combination. Now,

$(8, 7) = c_1 (1, 2) + c_2 (2, 1)$

$(8, 7) = (c_1, 2c_1) + (2c_2, c_2)$

$(8, 7) = (c_1 + 2c_2, 2c_1 + c_2)$.

Setting coordinates equal to each other:

$$8 = c_1 + 2c_2 \qquad\qquad (1)$$

16

$$7 = 2c_1 + c_2. \qquad (2)$$

From (1), $c_1 = 8 - 2c_2$. Then, from (2),

$7 = 2(8-2c_2) + c_2;\ c_2 = 3$. Substituting in (1),

$8 = c_1 + 2(3)$ gives $c_1 = 2$,

and $\quad (8,\ 7) = c_1\ (1,\ 2) + c_2\ (2,\ 1)$

$$= 2\ (1,\ 2) + 3\ (2,\ 1).$$

● **PROBLEM** 1-14

Write the polynomial $v = t^2 + 4t - 3$ over R as a linear combination of the polynomials $e_1 = t^2 - 2t + 5$, $e_2 = 2t^2 - 3t$
and $e_3 = t + 3$.

Solution: The polynomial $v = t^2 + 4t - 3$ is a polynomial
in a three-dimensional vector space. It is possible to see
this because the polynomial is of the form $\alpha_0 + \alpha_1\ \tau + \alpha_2\ \tau^2$,
and, therefore, it takes 3 elements, α_0, α_1 and α_2, to determine a polynomial in this space v_3. Therefore, we need only
3 independent vectors to form a basis. A basis for an n-dimensional vector space consists of n independent vectors. Here
the polynomials $\{e_1,\ e_2,\ e_3\}$ form a basis for V_3. To show
this, set

$c_1\ (t^2 - 2t + 5) + c_2\ (2t^2 - 3t) + c_3\ (t + 3) = 0t^2 + 0t + 0.$

Hence, $\qquad (c_1 + 2c_2)t^2 = 0t^2 \qquad (1)$

$$(-2c_1 - 3c_2 + c_3)t = 0t \qquad (1')$$

$$(5c_1 + 3c_3) = 0 \qquad (1'')$$

From (1), $c_1 = -2c_2$. Substitution into (1') and (1'')
yields

$$-c_2 + c_3 = 0 \qquad (2)$$

$$-10c_2 + 3c_2 = 0 \qquad (2')$$

Multiplying (2) by 3 and subtracting from (2') gives
$c_2 = 0$. From (1), $c_1 = 0$ and, hence, from (1'), $c_3 = 0$.
Thus, $c_1\ e_1 + c_2\ e_2 + c_3\ e_3 = 0$ implies $c_1 = c_2 = c_3 = 0$.
This implies e_1, e_2, e_3 are independent which implies they
form a basis for V_3.

To find v as a linear combination of e_1, e_2 and e_3,
we proceed as follows:

Set v as a linear combination of the e_i using the
unknowns x, y and z: $v = xe_1 + ye_2 + ze_3.$

17

$$t^2 + 4t - 3 = x(t^2 - 2t + 5) + y(2t^2 - 3t) + z(t + 3)$$
$$= xt^2 - 2xt + 5x + 2yt^2 - 3yt + zt + 3z$$
$$= (x + 2y)t^2 + (-2x - 3y + z)t + (5x + 3z)$$

Set coefficients of the same powers of t equal to each other, and reduce the system to echelon form:

$$
\begin{array}{llll}
x + 2y = 1 & & x + 2y = 1 & \\
-2x - 3y + z = 4 & \text{or} & y + z = 6 & \text{or} \\
5x + 3z = -3 & & -10y + 3z = -8 &
\end{array}
$$

$$
\begin{array}{l}
x + 2y = 1 \\
y + z = 6 \\
13z = 52
\end{array}
$$

Note that the system is consistent and so has a solution. Solve for the unknowns to obtain $x = -3$, $y = 2$, $z = 4$. Thus $v = -3e_1 + 2e_2 + 4e_3$.

● **PROBLEM** 1-15

Write the vector $v = (2, -5, 3)$ in R^3 as a linear combination of the vectors $e_1 = (1, -3, 2)$, $e_2 = (2, -4, -1)$ and $e_3 = (1, -5, 7)$.

Solution: To write v as a linear combination of e_1, e_2 and e_3, we set

$$v = c_1 e_1 + c_2 e_2 + c_3 e_3$$

where c_1, c_2 and c_3 are scalars chosen from the field K. Thus,

$$(2, -5, 3) = c_1 (1, -3, 2) + c_2 (2, -4, -1) + c_3 (1, -5, 7)$$

$$(2, -5, 3) = (c_1, -3c_1, 2c_1) + (2c_2, -4c_2, -c_2) + (c_3, -5c_3, 7c_3)$$

$$(2, -5, 3) = (c_1 + 2c_2 + c_3, -3c_1 - 4c_2 - 5c_3, 2c_1 - c_2 + 7c_3)$$

$$
\begin{array}{ll}
L_1: & c_1 + 2c_2 + c_3 = 2 \\
L_2: & -3c_1 - 4c_2 - 5c_3 = -5 \qquad\qquad (1) \\
L_3: & 2c_1 - c_2 + 7c_3 = 3
\end{array}
$$

We may use elementary row operations to reduce this system of equations to echelon form. The system (1) becomes

(Replace L_2 by $+3L_1 + L_2$.

Replace L_3 by $-2L_1 + L_3$).   Replace L_3' by $5L_2 + 2L_3$

$$
\begin{array}{ll}
L_1': & c_1 + 2c_2 + c_3 = 2 \\
L_2': & 0 + 2c_2 - 2c_3 = 1 \\
L_3': & 0 - 5c_2 + 5c_3 = -1
\end{array}
$$

$$
\begin{array}{l}
c_1 + 2c_2 + c_3 = 2 \\
0 + 2c_2 - 2c_3 = 1 \\
0 = 3.
\end{array}
$$

Hence, the system is inconsistent and we conclude that v cannot be written as a combination of e_1, e_2 and e_3.

Checking the set $\{e_1, e_2, e_3\}$ for linear independence we see that

$$c_1 e_1 + c_2 e_2 + c_3 e_3 = (0, 0, 0)$$

$$c_1 (1, -3, 2) + c_2 (2, -4, -1) + c_3 (1, -5, 7) = (0, 0, 0)$$

$$c_1 + 2c_2 + c_3 = 0$$

$$-3c_1 - 4c_2 - 5c_3 = 0$$

$$2c_1 - c_2 + 7c_3 = 0$$

Reducing to echelon form

$$c_1 + 2c_2 + c_3 = 0 \qquad ; \qquad c_1 + 2c_2 + c_3 = 0$$

$$0 + 2c_2 - 2c_3 = 0 \qquad \qquad c_2 - c_3 = 0$$

$$0 - 5c_2 + 5c_3 = 0$$

Hence, $c_2 = c_3$ and $c_1 = -3c_3$ is a general solution of the system, and we see that the vectors e_1, e_2 and e_3 are dependent. Since a basis for R^3 must contain three linearly independent vecotrs, $\{e_1, e_2, e_3\}$ cannot be a basis. This explains why we were unable to write v as a linear combination of e_1, e_2 and e_3.

In fact, setting $e_3 = c_1 e_1 + c_2 e_2$, we have

$$(1, -5, 7) = (c_1 + 2c_2, -3c_1 - 4c_2, 2c_1 - c_2)$$

$$c_1 + 2c_2 = 1 \qquad \qquad c_1 + 2c_2 = 1$$

$$-3c_1 - 4c_2 = -5 \qquad ; \qquad 0 + 2c_2 = -2 .$$

$$2c_1 - c_2 = 7 \qquad \qquad 0 - 5c_2 = 5$$

Hence, $c_2 = -1$ and $c_1 = 3$. Thus,

$$(1, -5, 7) = 3 (1, -3, 2) - (2, -4, -1)$$

and $e_3 = 3e_1 - e_2$.

● PROBLEM 1-16

Can the matrix $E = \begin{bmatrix} 3 & 1 \\ 1 & -1 \end{bmatrix}$ be written as a linear combination

of the matrices

$$m_1 = \begin{bmatrix} 1 & 1 \\ 1 & 0 \end{bmatrix}, \quad m_2 = \begin{bmatrix} 0 & 0 \\ 1 & 1 \end{bmatrix} \quad m_3 = \begin{bmatrix} 0 & 2 \\ 0 & -1 \end{bmatrix} \quad \text{and} \quad m_4 = \begin{bmatrix} 0 & 1 \\ 1 & 0 \end{bmatrix}$$

Solution: One way to proceed would be to attempt to find constants c_1, c_2, c_3 and c_4 such that $E = c_1 m_1 + c_2 m_2 + c_3 m_3 + c_4 m_4$. Another method of solution would be to reason as follows: We are asked whether E can be written as $c_1 m_1 + c_2 m_2 + c_3 m_3 + c_4 m_4$, not actually to find c_1, c_2, c_3 and c_4. Now E is a 2 x 2 matrix. Hence, it has dimension 4. A basis for a four-dimensional vector space must contain four linearly independent vectors. Also, m_1, m_2, m_3, m_4 are four 2 x 2 matrices, i.e., they are elements of the four-dimensional vector space of which E is a member. But are they linearly independent?

A set of vectors $\{v_1, v_2, \ldots, v_n\}$ is linearly independent if

$$c_1 v_1 + c_2 v_2 + \ldots + c_n v_n = 0$$

implies $c_1 = c_2 = \ldots = c_n = 0$.

Letting $\begin{bmatrix} 0 & 0 \\ 0 & 0 \end{bmatrix}$ be the zero element of the vector space of 2 x 2 matrices, we set $c_1 m_1 + c_2 m_2 + c_3 m_3 + c_4 m_4 = 0$. That is,

$$\begin{bmatrix} c_1 & c_1 \\ c_1 & 0 \end{bmatrix} + \begin{bmatrix} 0 & 0 \\ c_2 & c_2 \end{bmatrix} + \begin{bmatrix} 0 & 2c_3 \\ 0 & -c_3 \end{bmatrix} + \begin{bmatrix} 0 & c_4 \\ c_4 & 0 \end{bmatrix}$$

$$= \begin{bmatrix} 0 & 0 \\ 0 & 0 \end{bmatrix} \qquad (1)$$

Two matrices $\{aij\}$ and $\{bij\}$ are equal when $aij = bij$ $(i, j = 1, \ldots, n)$. Thus, from (1)

$$c_1 = 0 \qquad (2)$$

$$c_1 + 2c_3 + c_4 = 0 \qquad (2')$$

$$c_1 + c_2 + c_4 = 0 \qquad (2'')$$

$$c_2 - c_3 = 0 \qquad (2''')$$

From (2), $c_1 = 0$. From (2'''), $c_2 = c_3$. Hence, (2') and (2'') reduce to

$$L_1: \quad 2c_3 + c_4 = 0$$

$$L_2: \quad c_3 + c_4 = 0$$

Now, $-2L_2 + L_1 = -2(c_3 + c_4) + 2c_3 + c_4 = 0$. Hence, $c_4 = 0$.

20

Therefore, $c_3 = c_4 = 0$ can satisfy both of these equations. Further, by (2"'), $c_2 = 0$. Thus, $c_1 = c_2 = c_3 = c_4 = 0$, and we conclude that m_1, m_2, m_3 and m_4 are linearly independent, i.e., they form a basis for the vector space of 2 x 2 matrices.

Therefore, we can say that E can indeed be written as a linear combination of the given matrices m_i (i = 1, ..., 4).

The reader should note the following points: 1) questions about linear independence frequently lead to finding the solution set of a system of linear homogeneous equations; 2) an m✕n matrix is an element of a vector space of dimension mn.

● **PROBLEM** 1-17

Write the vector $v = (1, -2, 5)$ as a linear combination of the vectors $e_1 = (1, 1, 1)$, $e_2 = (1, 2, 3)$ and $e_3 = (2, -1, 1)$.

Solution: We first show that the vectors e_1, e_2, e_3 are linearly independent and form a basis for R^3. Then we find the required constants.

A set of vectors $\{e_1, e_2, ..., e_n\}$ in R^n is linearly independent if, for $c_i \in K$,

$$c_1 e_1 + c_2 e_2 + ..., + c_n e_n = 0$$

implies $c_i = 0$ (i = 1, ..., n). Setting

$$c_1 e_1 + c_2 e_2 + c_3 e_3 = (0, 0, 0),$$

$$c_1 (1, 1, 1) + c_2 (1, 2, 3) + c_3 (2, -1, 1) = (0, 0, 0)$$

$$(c_1, c_1, c_1) + (c_2, 2c_2, 3c_2) + (2c_3, -c_3, c_3) = (0, 0, 0).$$

Adding coordinates,

$$(c_1 + c_2 + 2c_3, c_1 + 2c_2 - c_3, c_1 + 3c_2 + c_3) = (0, 0, 0)$$

Two points in R^3 are equal only if their respective coordinates are equal. Hence,

L_1: $c_1 + c_2 + 2c_3 = 0$ (1)

L_2: $c_1 + 2c_2 - c_3 = 0$

L_3: $c_1 + 3c_2 + c_3 = 0$

We can reduce this system to row echelon form as follows: System (1) becomes

(Replace L_2 by $L_1 - L_2$ and replace L_3 by $L_1 - L_3$)

L_1: $c_1 + c_2 + 2c_3 = 0$ Now replace L_3 by (-2) times

L_2: $0 - c_2 + 3c_3 = 0$. $\qquad$ L_2 added to L_3. We arrive at

L_3: $0 - 2c_2 + c_3 = 0$ $\qquad$ $c_1 + c_2 + 2c_3 = 0$

$$0 - c_2 + 3c_3 = 0$$

$$0 + 0 - 5c_3 = 0$$

Thus, $c_1 = c_2 = c_3 = 0$ is the only solution and $\{e_1, e_2, e_3\}$ is linearly independent. Since a set of n linearly independent n-tuples forms a basis for R^n, the set $\{e_1, e_2, e_3\}$ is a basis for R^3. This means that any triple in R^3 (and in particular the given v) can be written as a linear combination of e_1, e_2 and e_3.

Let $v = c_1 e_1 + c_2 e_2 + c_3 e_3$.

$(1, -2, 5) = c_1 (1, 1, 1) + c_2 (1, 2, 3) + c_3 (2, -1, 1)$

$(1, -2, 5) = (c_1, c_1, c_1) + (c_2, 2c_2 + 3c_2) + (2c_3, - c_3, c_3)$

$(1, -2, 5) = (c_1 + c_2 + 2c_3, c_1 + 2c_2 - c_3, c_1 + 3c_2 + c_3)$.

Setting corresponding coordinates equal to each other,

L_1: $c_1 + c_2 + 2c_3 = 1$

L_2: $c_1 + 2c_2 - c_3 = -2$ $\qquad\qquad\qquad\qquad$ (2)

L_3: $c_1 + 3c_2 + c_3 = 5$.

We can apply row operations on the inhomogeneous system (2) to reduce it to row echelon form. System (2) becomes

L_1: $c_1 + c_2 + 2c_3 = 1$

L_2: $0 + c_2 - 3c_3 = -3$

L_3: $0 + 2c_2 - c_3 = 4$

(multiply L_1 by -1 and add it to L_2. Also multiply L_1 by -1 and add it to L_3).

$$c_1 + c_2 + 2c_3 = 1$$

$$0 + c_2 - 3c_3 = -3$$

$$0 + 0 + 5c_3 = 10$$

Multiply L_2 by -2 and add to L_3, the results of which are given above.

Solving by back-substitution,

$$c_3 = 2; \; c_2 = 3; \; c_1 = -6.$$

22

Hence, $v = -6e_1 + 3e_2 + 2e_3$.

Determine whether or not u and v are linearly dependent if:

(i) $u = (3, 4)$, $v = (1, -3)$

(ii) $u = (2, -3)$, $v = (6, -9)$

(iii) $u = (4, 3, -2)$, $v = (2, -6, 7)$

(iv) $u = (-4, 6, -2)$, $v = (2, -3, 1)$

(v) $u = \begin{bmatrix} 1 & -2 & 4 \\ 3 & 0 & -1 \end{bmatrix}$, $v = \begin{bmatrix} 2 & -4 & 8 \\ 6 & 0 & -2 \end{bmatrix}$

(vi) $u = \begin{bmatrix} 1 & 2 & -3 \\ 6 & -5 & 4 \end{bmatrix}$, $v = \begin{bmatrix} 6 & -5 & 4 \\ 1 & 2 & -3 \end{bmatrix}$

(vii) $u = 2 - 5t + 6t^2 - t^3$, $v = 3 + 2t - 4t^2 + 5t^3$

(viii) $u = 1 - 3t + 2t^2 - 3t^3$, $v = -3 + 9t - 6t^2 + 9t^3$

Solution: Two vectors v_1 and v_2 are linearly dependent if
(1) $c_1 v_1 + c_2 v_2 = 0$ has a solution(s) for c_1 and c_2, and
c_1 and c_2 are not both 0. (1) implies

$$v_2 = -\frac{c_1}{c_2} v_1 \qquad (c_2 \neq 0)$$

That is, if two vectors are dependent then one of the vectors
is a multiple of the other.

i)   If $(3, 4)$ and $(1, -3)$ are dependent, then there exists a
     scalar c_1 such that $(3, 4) = c_1 (1, -3) = (c_1, -3c_1)$. (2)

     In other words,

$$c_1 = 3$$

$$-3c_1 = 4.$$

     These two equations are inconsistent, i.e., there is no
value of c_1 such that (2) holds.

     Thus, u, v are linearly independent.

ii) Using the same reasoning as in i), u and v are dependent
     if there exists a c_1 such that $c_1 u = V$ so,

$$c_1 (2, -3) = (6, -9)$$

23

$$2c_1 = 6 \qquad\qquad (2)$$

$$-3c_1 = -9. \qquad\qquad (3)$$

From (2), $c_1 = 3$ which, when substituted into (3), is true. Thus, $v = 3u$ and the two vectors are linearly dependent.

iii) We wish to find c_1 such that

$$c_1 (4, 3, -2) = (2, -6, 7) \qquad\qquad (4)$$

is true. Setting the corresponding coordinates of the triple of real numbers equal to each other, we obtain

$$4c_1 = 2$$

$$3c_1 = -6 \qquad\qquad (4')$$

$$-2c_1 = 7$$

From the first equation of (4'), $c_1 = \frac{1}{2}$. From the second equation, $c_1 = -2$ and from the third, $c_1 = -\frac{7}{2}$ C_1 is clearly overworked. We conclude that the two vectors are linearly independent since (4) leads to a contradiction.

iv) Setting $c_1 (-4, 6, -2) = (2, -3, 1)$

we have $-4c_1 = 2$

$$6c_1 = -3$$

$$-2c_1 = 1.$$

The value $c_1 = -\frac{1}{2}$ satisfies all three equations. Hence, $v = -\frac{1}{2}u$ and the two vectors are linearly dependent.

v) Here the vectors are 2 x 3 matrices (a m x n matrix is an element of a vector space of dimension mn), but vectors nevertheless. Hence, setting

$$c_1 \begin{pmatrix} 1 & -2 & 4 \\ 3 & 0 & -1 \end{pmatrix} = \begin{pmatrix} 2 & -4 & 8 \\ 6 & 0 & -2 \end{pmatrix},$$

$$\begin{pmatrix} c_1 & -2c_1 & 4c_1 \\ 3c_1 & 0 & -c_1 \end{pmatrix} = \begin{pmatrix} 2 & -4 & 8 \\ 6 & 0 & -2 \end{pmatrix}. \qquad (5)$$

Two matrices $\{a_{ij}, b_{ij}\}$ are equal if and only if, $a_{ij} = b_{ij}$ ($i = 1, \ldots, m$; $j = 1, \ldots, n$). Hence, (5) holds if and only if

$$c_1 = 2 \qquad\qquad 3c_1 = 6$$

$$-2c_1 = -4 \qquad\qquad 0 \;\; = 0$$
$$4c_1 = 8 \qquad\qquad -\,c_1 = -2.$$

Since all of the above equalities are consistent, $c_1 = 2$ and $v = 2u$, thus, we have v is dependent on u (and vice-versa).

vi) Suppose $u = \begin{bmatrix} 1 & 2 & -3 \\ 6 & -5 & 4 \end{bmatrix}$ and $v = \begin{bmatrix} 6 & -5 & 4 \\ 1 & 2 & -3 \end{bmatrix}.$

Then $c\, u = v$ implies

$$\begin{bmatrix} c & 2c & -3c \\ 6c & -5c & 4c \end{bmatrix} = \begin{bmatrix} 6 & -5 & 4 \\ 1 & 2 & -3 \end{bmatrix}$$

and gives $c = 6$ $\qquad\qquad\qquad\qquad\qquad\qquad$ (1)

$\qquad\qquad 2c = -5,$ etc. $\qquad\qquad\qquad\qquad\qquad$ (2)

We do not need to go any further since we already have a contradiction, (2) gives $c = \dfrac{-5}{2}$ but (1) says $c = 6$. Since there exists no such c, v and u are linearly independent.

vii) The set of all polynomials with degree $\le n$ is an $(n + 1)$ dimensional vector space. Here u and v are vectors from v_4. Setting

$$c_1\,(2 - 5t + 6t^2 - t^3) = (3 + 2t - 4t^2 + 5t^3),$$
$$2c_1 - 5c_1 t + 6c_1 t^2 - c_1 t^3 = 3 + 2t - 4t^2 + 5t^3.$$

Two polynomials are equal when coefficients of like powers of x are equal. Thus,

$$2c_1 = 3$$
$$-5c_1 = 2$$
$$6c_1 = -4$$
$$-\,c_1 = 5$$

Since c_1 cannot satisfy all these equations simultaneously, we conclude that it does not exist. Hence, u and v are not multiples of each other, i.e., they are independent.

viii) Again, u and v are two vectors in a 4-dimensional polynomial vector space. Let

$$c_1\,(1 - 3t + 2t^2 - 3t^3) = -\,3 + 9t - 6t^2 + 9t^3.$$

$$c_1 - 3c_1 t + 2ct^2 - 3c_1 t^3 = -\,3 + 9t - 6t^2 + 9t^3.$$

Setting the coefficients of like powers of x equal to each other,

$$c_1 = -3$$

$$-3c_1 = 9$$

$$2c_1 = -6$$

$$-3c_1 = 9.$$

$c_1 = -3$ satisfies all of the above equations. Hence, $v = -3u$ and u and v are dependent.

● PROBLEM 1-19

Determine whether the vectors x, y and z are dependent or independent where

i) $x = (1, 1, -1)$, $y = (2, -3, 1)$, $z = (8, -7, 1)$

ii) $x = (1, -2, -3)$, $y = (2, 3, -1)$, $z = (3, 2, 1)$

iii) $x = (x_1, x_2)$, $y = (y_1, y_2)$, $z = (z_1, z_2)$.

Solution: A set of vectors $\left(v_1, v_2, \ldots, v_n\right)$ is said to be linearly dependent if there exists ci i = 1, 2, ..., n, such that

$$c_1 v_1 + c_2 v_2 + \ldots + c_n v_n = 0 \qquad (1)$$

where at least one of the ci in (1) $\neq$ 0. In the given problems we may proceed as follows:

a) let $c_1 x + c_2 y + c_3 z = 0$ where c_1, c_2, c_3 are unknown scalars;

b) find the equivalent homogeneous system of equations;

c) determine whether the system has a non-zero solution. If the system does, then the vectors are linearly dependent; if the only solution is the trivial solution, they are independent.

i) Let $c_1 x + c_2 y + c_3 z = 0$.

   $c_1 (1, 1, -1) + c_2 (2, -3, 1) + c_3 (8, -7, 1) = (0, 0, 0)$,

$(c_1, c_1, -c_1) + (2c_2, -3c_2, c_2) + (8c_3, -7c_3, c_3) = (0, 0, 0)$

or $c_1 + 2c_2 + 8c_3 = 0$

   $c_1 - 3c_2 - 7c_3 = 0 \qquad (2)$

   $-c_1 + c_2 + c_3 = 0.$

The system (2) may now be reduced to echelon form. Constructing the coefficient matrix from (2),

$$\begin{pmatrix} 1 & 2 & 8 \\ 1 & -3 & -7 \\ -1 & 1 & 1 \end{pmatrix} \underset{T_1}{\leftrightarrow} \begin{pmatrix} 1 & 2 & 8 \\ 0 & -5 & -15 \\ 0 & 3 & 9 \end{pmatrix} \underset{T_2}{\leftrightarrow} \begin{pmatrix} 1 & 2 & 8 \\ 0 & 1 & 3 \\ 0 & 1 & 3 \end{pmatrix} \underset{T_3}{\leftrightarrow} \begin{pmatrix} 1 & 2 & 8 \\ 0 & 1 & 3 \end{pmatrix}.$$

where T_i are the row operations as follows: T_1: replace the third row with the sum of the first and the third rows. Also, replace the second row with the second row minus the first row. T_2: replace the second row with $\left(\frac{-1}{5}\right)$ times the second row and replace the third row with $\left(\frac{1}{3}\right)$ times the third row.

T_3: The third row is the same as the second and so it may be dropped.

Thus, $c_1 + 2c_2 + 8c_3 = 0$ (3)

$$c_2 + 3c_3 = 0.$$

The system (3), equivalent to system (2), has two equations in three unknowns. The second equation in (3) yields $c_2 = -3c_3$. Substituting in the first equation, we get

$c_1 + 2(-3c_3) + 8c_3 = 0$, so $c_1 = -2c_3$. Thus, all of the solutions of this system are $c_1 = -2a$

$$c_2 = -3a \quad \text{where}$$

$$c_3 - a$$

a is any scalar. If $a = 1$, one non-zero solution is $c_1 = -2$, $c_2 = -3$, $c_3 = 1$. Hence, the vectors are dependent.

ii) Let $c_1 x + c_2 y + c_3 z = 0$;
$(c_1, -2c_1 - 3c_1) + (2c_2, 3c_2 - c_2) + (3c_3, 2c_3, c_3) = (0, 0, 0)$

or $c_1 + 2c_2 + 3c_3 = 0$

$$-2c_1 + 3c_2 + 2c_3 = 0 \qquad\qquad\qquad (4)$$

$$-3c_1 - c_2 + c_3 = 0$$

Reducing the coefficient matrix of (4) to echelon form using appropriate row operations,

$$\begin{pmatrix} 1 & 2 & 3 \\ -2 & 3 & 2 \\ -3 & -1 & 1 \end{pmatrix} \underset{T_1}{\leftrightarrow} \begin{pmatrix} 1 & 2 & 3 \\ 0 & 7 & 8 \\ 0 & 5 & 10 \end{pmatrix} \underset{T_2}{\leftrightarrow} \begin{pmatrix} 1 & 2 & 3 \\ 0 & 7 & 8 \\ 0 & 0 & 30 \end{pmatrix}$$

where T_i are elementary row operations: T_1: replace the second row with two times the first row plus the second row;

replace the third row with three times the first row plus the third row.

T_2: replace the third row with the sum of -5 times the second row and 7 times the third row.

The system
$$c_1 + 2c_2 + 3c_3 = 0$$
$$7c_2 + 8c_3 = 0$$
$$30c_3 = 0$$

has only the trivial solution $c_1 = c_2 = c_3 = 0$. Thus, by definition, the vectors are linearly independent.

iii) Note here that x, y, z are arbitrary vectors in R^2.
Let $c_1 (x_1, x_2) + c_2 (y_1, y_2) + c_3 (z_1, z_2) = (0, 0)$.

Then,
$$\begin{pmatrix} x_1 & y_1 & z_1 \\ x_2 & y_2 & z_2 \end{pmatrix} \begin{pmatrix} c_1 \\ c_2 \\ c_3 \end{pmatrix} = \begin{pmatrix} 0 \\ 0 \end{pmatrix}$$

or
$$x_1 c_1 + y_1 c_2 + z_1 c_3 = 0 \tag{5}$$
$$x_2 c_1 + y_2 c_2 + z_2 c_3 = 0.$$

Because there are more unknowns (c_1, c_2, c_3) than equations, the system has a non-zero solution.

Since x, y, z were arbitrary vectors, this shows that any three vectors in R^2 are linearly dependent. In general, any n + 1 vectors in R^n are linearly dependent.

● **PROBLEM 1-20**

Let V be the vector space of 2 x 2 matrices over R. Determine whether the matrices A, B, C ∈ V are dependent where:

(i) $A = \begin{pmatrix} 1 & 1 \\ 1 & 1 \end{pmatrix}$, $B = \begin{pmatrix} 1 & 0 \\ 0 & 1 \end{pmatrix}$, $C = \begin{pmatrix} 1 & 1 \\ 0 & 0 \end{pmatrix}$

(ii) $A = \begin{pmatrix} 1 & 2 \\ 3 & 1 \end{pmatrix}$, $B = \begin{pmatrix} 3 & -1 \\ 2 & 2 \end{pmatrix}$, $C = \begin{pmatrix} 1 & -5 \\ -4 & 0 \end{pmatrix}$

28

Solution: i) $\begin{bmatrix} 0 & 0 \\ 0 & 0 \end{bmatrix}$ is the zero element of the vector space of 2 x 2 matrices. Letting c_1, c_2, c_3 be unknown constants, we set

$$c_1 \begin{bmatrix} 1 & 1 \\ 1 & 1 \end{bmatrix} + c_2 \begin{bmatrix} 1 & 0 \\ 0 & 1 \end{bmatrix} + c_3 \begin{bmatrix} 1 & 1 \\ 0 & 0 \end{bmatrix} = \begin{bmatrix} 0 & 0 \\ 0 & 0 \end{bmatrix}$$

The matrices A, B, C are dependent if in a solution of the above matrix equation at least one c_i , i = 1, 2, 3, is not 0. From the above equation we have

$$\begin{bmatrix} c_1 & c_1 \\ c_1 & c_1 \end{bmatrix} + \begin{bmatrix} c_2 & 0 \\ 0 & c_2 \end{bmatrix} + \begin{bmatrix} c_3 & c_3 \\ 0 & 0 \end{bmatrix} = \begin{bmatrix} 0 & 0 \\ 0 & 0 \end{bmatrix},$$

$$\begin{bmatrix} c_1 + c_2 + c_3 & c_1 + c_3 \\ c_1 & c_1 + c_2 \end{bmatrix} = \begin{bmatrix} 0 & 0 \\ 0 & 0 \end{bmatrix}.$$

Two matrices $\{a_{ij}\}$, $\{b_{ij}\}$ are equal if and only if $a_{ij} = b_{ij}$ (i = 1, ..., m i j = 1, ..., n). Hence,

$$c_1 + c_2 + c_3 = 0$$

$$c_1 \qquad + c_3 = 0$$

$$c_1 \qquad\qquad = 0$$

$$c_1 + c_2 \qquad = 0$$

We see that $c_1 = c_2 = c_3 = c_4 = 0$ is the only solution of the above system. The matrices A, B, C are therefore linearly independent.

ii) Proceeding as in i),

$$c_1 \begin{bmatrix} 1 & 2 \\ 3 & 1 \end{bmatrix} + c_2 \begin{bmatrix} 3 & -1 \\ 2 & 2 \end{bmatrix} + c_3 \begin{bmatrix} 1 & -5 \\ -4 & 0 \end{bmatrix} = \begin{bmatrix} 0 & 0 \\ 0 & 0 \end{bmatrix}$$

so $$\begin{bmatrix} c_1 + 3c_2 + c_3 & 2c_1 - c_2 - 5c_3 \\ 3c_1 + 2c_2 - 4c_3 & c_1 + 2c_2 \end{bmatrix} = \begin{bmatrix} 0 & 0 \\ 0 & 0 \end{bmatrix},$$

which gives $c_1 + 3c_2 + c_3 = 0$

$$2c_1 - c_2 - 5c_3 = 0 \qquad\qquad (1)$$

$$3c_1 + 2c_2 - 4c_3 = 0$$

$$c_1 + 2c_2 \qquad = 0$$

We can reduce the system to row echelon form by applying elementary row operations. First multiply the first equation by (-2) and add it to the second equation. Replace the second equation with the result. Then replace, the third equation with -3 times the first equation plus the third equation. Lastly, replace the fourth equation with the fourth equation minus the first. Thus, (1) becomes

$$c_1 + 3c_2 + c_3 = 0$$

$$- 7c_2 - 7c_3 = 0$$

$$- 7c_2 - 7c_3 = 0$$

$$- c_2 - c_3 = 0$$

or $c_1 + 3c_2 + c_3 = 0$ (2)

$$c_2 + c_3 = 0.$$

$c_1 = 2$; $c_2 = -1$ and $c_3 = 1$ satisfy (2) (among other values). We conclude that the matrices A, B, C are linearly dependent.

● **PROBLEM** 1-21

Determine whether or not the following vectors in R^3 are linearly dependent:

(i) (1, -2, 1), (2, 1, -1), (7, -4, 1)

(ii) (1, -3, 7), (2, 0, -6), (3, -1, -1), (2, 4, -5)

(iii) (1, 2, -3), (1, -3, 2), (2, -1, 5)

(iv) (2, -3, 7), (0, 0, 0), (3, -1, -4)

Solution: All of the given sets of vectors are triples of real numbers. A set of n-tuples $\{v_1, \ldots, v_n\}$ is linearly dependent if there exist c_i (i = 1, ..., n), not all zero, such that

$$c_1 v_1 + c_2 v_2 + \ldots + c_n v_n = 0.$$

If no such c_i exist, $v_1, \ldots, v_n$ are linearly independent.

i)   Set a linear combination of the vectors equal to the zero vector using unknown scalars c_1, c_2 and c_3 from the field K over which the vector space is defined.

$$c_1 (1, -2, 1) + c_2 (2, 1, -1) + c_3 (7, -4, 1) = (0, 0, 0).$$

Setting corresponding coordinates equal to each other, we obtain the following homogeneous system:

30

$$c_1 + 2c_2 + 7c_3 = 0 \quad \text{I}$$

$$-2c_1 + c_2 - 4c_3 = 0 \quad \text{II} \tag{1}$$

$$c_1 - c_2 + c_3 = 0 \quad \text{III}$$

We wish to solve the system (1) for the unknowns c_1, c_2 and c_3. One method is the row reduction method in which we successively reduce the number of unknowns in each equation applying elementary row operations. These are: i) interchanging rows, ii) multiplying a row by a constant, iii) adding a constant multiple of one row to another. Any combination of these operations results in a system of equations which has the same solution(s) as the original system. Applying such operations on (1) we get:

$$c_1 + 2c_2 + 7c_3 = 0 \quad \text{I}$$

$$0 + 5c_2 + 10c_3 = 0 \quad \text{II} \tag{2}$$

$$c_1 - c_2 + c_3 = 0 \quad \text{III}$$

(where we multiplied I of (1) by 2 and added to II of (1) to obtain II of (2)).

Continuing, we get

$$c_1 + 2c_2 + 7c_3 = 0 \quad \text{(same)}$$

$$0 + 5c_2 + 10c_3 = 0 \quad \text{(same)} \tag{3}$$

$$0 - 3c_2 - 6c_3 = 0 \quad (\text{I} \times (-1) + \text{III of (2)}).$$

We could proceed to eliminate c_2 in the third equation, then read off the solution of c_3, and then solve for the other unknowns by back substituting. But examining (3), we see that the second and third equations are equivalent. Hence, (3) becomes

$$c_1 + 2c_2 + 7c_3 = 0$$

$$c_2 + 2c_3 = 0.$$

A solution is $c_1 = 3$, $c_2 = 2$ and $c_3 = -1$. To check, observe that

$$3 (1, -2, 1) + 2 (2, 1, -1) - 1 (7, -4, 1) = (0, 0, 0).$$

Thus, the three vectors are linearly dependent.

ii) A basis for R^3 contains only three vectors. (Note: a basis for R^n contains exactly n vectors). Since a basis also represents the maximal number of independent elements in the space, any set containing more than three vectors must be linearly dependent. Thus, the four vectors in R^3 are dependent.

iii) Form the matrix whose rows are the given vectors:

$$\begin{bmatrix} 1 & 2 & -3 \\ 1 & -3 & 2 \\ 2 & -1 & 5 \end{bmatrix}$$ (4)

We now reduce (4) to echelon form. We find

$$(4) \overset{T_1}{\leftrightarrow} \begin{bmatrix} 1 & 2 & -3 \\ 0 & -5 & 5 \\ 0 & -5 & 11 \end{bmatrix} \overset{T_2}{\leftrightarrow} \begin{bmatrix} 1 & 2 & -3 \\ 0 & -5 & 5 \\ 0 & 0 & 6 \end{bmatrix}.$$

T_1: The first row is unchanged.
The new second row = (row 1) - (row 2).
The new third row = (-2 x row 1) + (row 3).

T_2: First and second rows are unchanged.
The new third row = (row 3) - (row 2)

The echelon matrix has no zero rows. This means the three vectors are independent (elementary row operations are analogous to forming linear combinations - if the three vectors had been dependent, the row operations would have yielded a row of zeros).

iv) Since the set contains the zero vector, it is linearly dependent. To show this consider the following argument: The linear combination

c_1 (2, -3, 7) + c_2 (0, 0, 0) + c_3 (3, -1, -4) = (0, 0, 0)

can always be constructed to have at least one $c_i \neq 0$. Thus, $c_1 = c_3 = 0$ and c_2, being any arbitrary number, satisfies the equation.

● PROBLEM 1-22

Express the point (1, -2) in the xy-plane as a linear combination of (3, 1) and 6, 4).

Solution: Since any three vectors in R^2 are dependent (and any n + 1 vectors in R^n are dependent), we will have

c_1 (1, -2) + c_2 (3, 1) + c_3 (6, 4) = 0 (1)

with not all c_i - 0. If $c_1 \neq 0$ (1) may be written as:

$$\frac{-c_2}{c_1} (3, 1) + \frac{-c_3}{c_1} (6, 4) = (1, -2).$$ (2)

(2) shows explicitly that $(1, -2)$ depends on (or is a linear combination of) $(3, 1)$ and $(6, 4)$.

To find c_1, c_2 and c_3, return to equation (1) and proceed as follows:

(1) yields

A: $c_1 + 3c_2 + 6c_3 = 0$

B: $-2c_1 + c_2 + 4c_3 = 0.$

(3)

Using elementary row operations, we reduce (3) to echelon form.

A = A': $c_1 + 3c_2 + 6c_3 = 0$

2 A + B = B': $7c_2 + 16c_3 = 0$

Now

A' = A": $c_1 + 3c_2 + 6c_3 = 0$ (4)

$\frac{1}{7}$ B' = B": $c_2 + \frac{16}{7}c_3 = 0$

(4) gives: $c_1 = -3c_2 - 6c_3$

$$c_2 = \frac{-16}{7} c_3$$

so $c_1 = -3 \left(\frac{-16}{7} c_3 \right) - 6c_3 = (3 \left| \frac{16}{7} \right| - 6) c_3$

$$c_2 = \frac{-16}{7} c_3$$

Letting $c_3 = 7$, we obtain one non-trivial solution to (4),

$c_1 = 6$ $c_2 = -16$ and $c_3 = 7.$

Since $c_1 = 6 \neq 0$, we may apply (2):

$\frac{8}{3}$ $(3, 1)$ $\frac{-7}{6}$ $(6, 4) = (1, -2).$

● PROBLEM 1-23

Let the functions 1, x and x^2 be defined on the interval $[0, 1]$. Then these functions belong to the vector space of continuous, real-valued functions defined on $[0, 1]$ called C $[0, 1]$. Show that 1, x and x^2 are independent.

Solution: A set of vectors W - $\{w_1, \ldots, w_n\}$ is linearly independent if:

33

$$c_1 w_1 + c_2 w_2 + \ldots + c_n w_n = 0$$

implies $c_i = 0$ $(i = 1, \ldots, n)$.

Thus, we set

$$c_1 (1) + c_2 (x) + c_3 (x^2) = 0. \qquad\qquad (1)$$

If (1) implies $c_1 = c_2 = c_3 = 0$ for all $0 \leq x \leq 1$, then we can conclude that 1, x, and x^2 are linearly independent. Suppose $x = 0$ in (1). Then, $c_1 = 0$. Now, differentiate (1):

$$c_2 + 2c_3 x = 0. \qquad\qquad (2)$$

Setting $x = 0$ in (2) yields $c_2 = 0$. Similarly, differentiating (2),

$$2c_3 = 0.$$

We conclude that the only solution to (1) is $c_1 = c_2 = c_3 = 0$, and, hence, the functions 1, x and x^2 are independent.

● **PROBLEM** 1-24

Show that $p_1(x) = x + 2x^2$, $p_2(x) = 1 + 2x + x^2$ and $p_3(x) = 2 + x$ form a linearly independent set in $V_3(x)$.

<u>Solution</u>:   The set of polynomials with degree ≤ 2 forms a vector space of dimension three over a field K. A set of vectors $\{v_1, v_2, \ldots, v_n\}$ is linearly independent if

$$c_1 v + c_2 v_2 + \ldots + c_n v_n = 0$$

implies $c_1 = c_2 = \ldots = c_n = 0$ where $c_1, c_2, \ldots, c_n$ are scalars from the field K. Thus, let

$$c_1 p_1(x) + c_2 p_2(x) + c_3 p_3(x) = 0 + 0x + 0x^2$$

$$c_1 (x + 2x^2) + c_2 (1 + 2x + x^2) + c_3 (2 + x)$$

$$= 0 + 0x + 0x^2$$

$$c_1 x + 2c_1 x^2 + c_2 + 2c_2 x + c_2 x^2 + 2c_3 + c_3 x$$

$$= 0 + 0x + 0x^2.$$

Gathering like terms,

$$(c_2 + 2c_3) + (c_1 + 2c_2 + c_3)x + (2c_1 + c_2)x^2$$

$$= 0 + 0x + 0x^2$$

Two polynomials are equal when the coefficients of like powers of x are equal. Thus.

$(c_2 + 2c_3) = 0$ (1)

$(c_1 + 2c_2 + c_3) = 0$ (2)

$(2c_1 + c_2) = 0$ (3).

From (3) we get $c_2 = -2c_1$.
Substituting in (2), $c_1 - 4c_1 + c_3 = 0$ or $c_3 = 3c_1$.

Now substituting for c_2 and c_3 in (1), the result is $-2c_1 + 2(3c_1) = 0$. So, $4c_1 = 0$ implies $c_1 = 0$. Then $c_3 = 0$ and $c_2 = 0$.

We conclude that the polynomials are independent.

● **PROBLEM** 1-25

Find the span set of Tx where

a) $T = \{ <2, 1> \}$

b) $T = \{\overline{OP}, \overline{OQ}\}$ where O is the origin and P, Q and O are not collinear.

c) $T = \{1, x, x^2\} \subset F[[0, 1]]$.

Solution: The set of all linear combinations of a set T is called the span set of T and is denoted by SpT.

a) Here (2, 1) is a point in the Cartesian plane. Hence, Sp(T) consists of all vectors of the form $c(2, 1)$ for c, a real number. The graph of T is the line $x = 2y$ in the Cartesian plane.

b) If P, Q are points in space not collinear with O, then

    $SpT = \{a\ \overline{OP} + b\ \overline{OQ}$: a and b are real numbers$\}$.

Thus SpT consists of all $\overline{OR}$ where R is a point in the plane containing $\overline{OP}$ and $\overline{OQ}$. Note: We know $\overline{OP}$ and $\overline{OQ}$ determine a plane since O, P, Q are non-collinear.

c) The set T is defined on the interval [0, 1] of the real line. An arbitrary linear combination of T has the form $a + bx + cx^2$. Thus, SpT consists of all polynomials of degree 2 or less on [0, 1].

● **PROBLEM** 1-26

Let $T \subset R^3$ be $\{(1, 0, 0), (0, 1, 1)\}$. What is the span of T, Sp(T)?

Solution: Let V be a vector space defined over a field K.

If T is a non-empty subset of V, then Sp(T) is the set of all linear combinations of T.

In the given problem $V = R^3$ and we may choose $K = R$, the field of real numbers. Then Sp(T) is the set of all vectors of the form

$$\alpha_1 (1, 0, 0) + \alpha_2 (0, 1, 1) = (\alpha_1, \alpha_2, \alpha_2)$$

for $\alpha_1, \alpha_2 \in R$.

Thus, every element of Sp(T) is of the form $(\alpha_1, \alpha_2, \alpha_2)$. Sp(T) is a subspace of R^3 since i) $(0, 0, 0) \in$ Sp(T); ii) $(\alpha_1^{(1)}, \alpha_2^{(1)}, \alpha_2^{(1)}) + (\alpha_1^{(2)}, \alpha_2^{(2)}, \alpha_2^{(2)})$
$= (\alpha_1^{(1)} + \alpha_1^{(2)}, \alpha_2^{(1)} + \alpha_2^{(2)}, \alpha_2^{(1)} + \alpha_2^{(2)})$ is in

Sp(T). Hence, Sp(T) is closed under addition. iii) $\beta (\alpha_1, \alpha_2, \alpha_2) = (\beta\alpha_1, \beta\alpha_2, \beta\alpha_2) \in$ Sp(T) for $\beta \in R$, and

Sp(T) is closed under scalar multiplication.

● **PROBLEM** 1-27

Find span U when U is each of the following sets of vectors in a vector space V:

a) $U = \{u_1\}$ where u_1 is a non-zero vector of V.

b) $U = \{0\}$, the zero vector for any V.

c) $U = \{(3, -1, 4), (2, 0, 0)\}$ and $V = R^3$.

Solution: Let V be a vector space over the field F and let $u = \{u_1, u_2, \ldots, u_n\}$ be a set of vectors in V. Then the span of u is the set of all linear combinations of $u_i \in U$, i.e., $Sp(u) = c_1 u_1 + c_2 u_2 + \ldots + c_n u_n$ where $c_i \in F$. The span of u is a subspace of V that contains u.

a) Here, we deal with only a single vector. The span of U consists of all the scalar multiples of u_1. To show this, note that the set of such multiples is closed under vector addition and scalar multiplication. That is, $a \cdot u_1 + b \cdot u_1 = (a + b) \cdot u_1$ and $c \cdot (a \cdot u_1) = (c \cdot a) \cdot u_1$. Thus, the set is a subspace of V that contains u_1. Therefore, the span of u_1 is contained in this set of multiples since span U is the smallest subspace of V containing u_1. Conversely, since $u_1 \in$ span U and span U is closed under scalar multiplication, span U contains every scalar multiple of u_1.
    b) Since (0) is a single vector, its span is the set of all scalar multiples of 0. But $K0 = 0$ for all $K \in F$.

36

Hence, $Sp\ (0) = (0)$.

    c) In R^3 the vectors $(3, -1, 4)$ and $(2, 0, 0)$ are non-collinear since they are independent, i.e., $c_1\ (2, 0, 0) + c_2\ (3, -1, 4) = 0$ implies $c_1 = c_2 = 0$. Since $(3, -1, 4)$ and $(2, 0, 0)$ are linearly independent, the span of U is the plane in R^3 determined by $(0, 0, 0)$, $(3, -1, 4)$, $(2, 0, 0)$.

THE BASIS OF A VECTOR SPACE

● **PROBLEM** 1-28

The set $B = \{1, x\}$ is a basis for the vector space P^1 where P^1 is defined to be the vector space of all polynomials of degree less than or equal to 1 over the field of real numbers. Show that the coordinates of an arbitrary function in P^1, using the basis B, are unique.

Solution: An arbitrary element of P^1 has the form: $f(x) = a + bx$ where a, b are elements of R (real numbers). Then $f(x) = a \cdot 1 + b \cdot x$.

    Thus, $f(x)$ has been expressed as a linear combination of vectors in B. To show uniqueness, let $f(x) = c(1) + d(x)$ where c and d, both elements of R, are different from a and b respectively. Certainly, $f(x) - f(x) = 0$ so

$$a(1) + b(x) - [c(1) + d(x)]$$

$$= (a - c)1 + (b - d)x = 0.$$

    But the zero polynomial is the polynomial all of whose coefficients are zero. Therefore, $(a - c) = 0$ and $(b - d) = 0$ which implies that $a = c$ and $b = d$. Hence, $f(x)$ is uniquely represented by $a \cdot 1 + b \cdot x$.

● **PROBLEM** 1-29

Show that the set $T = \left\{ \begin{bmatrix} 1 \\ 0 \\ 1 \\ 0 \end{bmatrix}, \begin{bmatrix} 1 \\ 0 \\ 0 \\ 1 \end{bmatrix}, \begin{bmatrix} 0 \\ 1 \\ 1 \\ 0 \end{bmatrix}, \begin{bmatrix} 0 \\ 1 \\ 0 \\ 0 \end{bmatrix} \right\}$ spans R^4.

Solution: Let B_0 be a basis for a vector space V defined over the field K. Let $B_1 \neq B_0$ be a set containing the same number of elements as B_0, such that every member of B_0 can be expressed as a linear combination of elements from B_1. Then B_1 spans V.

Let us compare T with the standard basis for R^4:

$$\begin{bmatrix} 1 \\ 0 \\ 0 \\ 0 \end{bmatrix} \begin{bmatrix} 0 \\ 1 \\ 0 \\ 0 \end{bmatrix} \begin{bmatrix} 0 \\ 0 \\ 1 \\ 0 \end{bmatrix} \begin{bmatrix} 0 \\ 0 \\ 0 \\ 1 \end{bmatrix}.$$

Now,
$$\begin{bmatrix} 1 \\ 0 \\ 0 \\ 0 \end{bmatrix} = \begin{bmatrix} 1 \\ 0 \\ 1 \\ 0 \end{bmatrix} - \begin{bmatrix} 0 \\ 1 \\ 1 \\ 0 \end{bmatrix} + \begin{bmatrix} 0 \\ 1 \\ 0 \\ 0 \end{bmatrix}$$

$$\begin{bmatrix} 0 \\ 1 \\ 0 \\ 0 \end{bmatrix} = \begin{bmatrix} 0 \\ 1 \\ 0 \\ 0 \end{bmatrix}; \quad \begin{bmatrix} 0 \\ 0 \\ 1 \\ 0 \end{bmatrix} = \begin{bmatrix} 0 \\ 1 \\ 1 \\ 0 \end{bmatrix} - \begin{bmatrix} 0 \\ 1 \\ 0 \\ 0 \end{bmatrix}; \quad \begin{bmatrix} 0 \\ 0 \\ 0 \\ 1 \end{bmatrix} = \begin{bmatrix} 1 \\ 0 \\ 0 \\ 1 \end{bmatrix} - \begin{bmatrix} 1 \\ 0 \\ 1 \\ 0 \end{bmatrix}$$

$$+ \begin{bmatrix} 0 \\ 1 \\ 1 \\ 0 \end{bmatrix} - \begin{bmatrix} 0 \\ 1 \\ 0 \\ 0 \end{bmatrix}.$$

So each element of the standard basis is a linear combination of the elements of T. Therefore, T spans R^4.

● PROBLEM 1-30

The standard basis for R^3 is

$$B_s = \left\{ \begin{bmatrix} 1 \\ 0 \\ 0 \end{bmatrix} \begin{bmatrix} 0 \\ 1 \\ 0 \end{bmatrix} \begin{bmatrix} 0 \\ 0 \\ 1 \end{bmatrix} \right\}.$$

Find another basis for R^3, B_T such that B_T contains the two vectors $\begin{bmatrix} 1 \\ 1 \\ 0 \end{bmatrix}$ and $\begin{bmatrix} 0 \\ 0 \\ 1 \end{bmatrix}$.

Solution: Since $c_1 \begin{bmatrix} 1 \\ 1 \\ 0 \end{bmatrix} + c_2 \begin{bmatrix} 0 \\ 0 \\ 1 \end{bmatrix} = 0$

implies $c_1 = c_2 = 0$, the two vectors are linearly independent and can form part of a basis. Hence, we must find another vector that is linearly independent of the two given vectors since any set of exactly three linearly independent vectors in R^3 forms a basis for R^3.

Select a vector that is linearly independent of $\begin{bmatrix} 0 \\ 0 \\ 1 \end{bmatrix}$

but not equal to $\begin{bmatrix} 1 \\ 1 \\ 0 \end{bmatrix}$. The vector $\begin{bmatrix} 0 \\ 1 \\ 0 \end{bmatrix}$ meets this require-

ment since $\begin{bmatrix} 0 \\ 1 \\ 0 \end{bmatrix} \neq \begin{bmatrix} 1 \\ 1 \\ 0 \end{bmatrix}$ and $c_1 \begin{bmatrix} 0 \\ 0 \\ 1 \end{bmatrix} + c_2 \begin{bmatrix} 0 \\ 1 \\ 0 \end{bmatrix} = \begin{bmatrix} 0 \\ 0 \\ 0 \end{bmatrix}$

implies $c_1 = c_2 = 0$. Now a vector in the span of

$\left\{ \begin{bmatrix} 0 \\ 1 \\ 0 \end{bmatrix}, \begin{bmatrix} 0 \\ 0 \\ 1 \end{bmatrix} \right\}$ has the form

$$\alpha \begin{bmatrix} 0 \\ 1 \\ 0 \end{bmatrix} + \beta \begin{bmatrix} 0 \\ 0 \\ 1 \end{bmatrix} = \begin{bmatrix} 0 \\ \alpha \\ \beta \end{bmatrix}. \tag{1}$$

Hence, $\begin{bmatrix} 1 \\ 1 \\ 0 \end{bmatrix}$ is not in the span of $\begin{bmatrix} 0 \\ 1 \\ 0 \end{bmatrix}$ and $\begin{bmatrix} 0 \\ 0 \\ 1 \end{bmatrix}$.

Thus, $\left\{ \begin{bmatrix} 1 \\ 1 \\ 0 \end{bmatrix}, \begin{bmatrix} 0 \\ 1 \\ 0 \end{bmatrix}, \begin{bmatrix} 0 \\ 0 \\ 1 \end{bmatrix} \right\}$ is a basis for R^3 containing B_s.

● **PROBLEM** 1-31

Show that $(1, 2, 1)$, $(1, -1, 3)$ and $(1, 1, 4)$ form a basis for R^3.

Solution: A basis for a vector space is the smallest set of vectors from the space that spans the space. In the given problem this means that any vector in R^3 (x, y, z) must be uniquely expressible as a linear combination of $(1, 2, 1)$, $(1, -1, 3)$ and $(1, 1, 4)$. Thus, we must show that for any (x, y, z) there exists c_1, c_2, c_3 such that

$$\begin{bmatrix} x \\ y \\ z \end{bmatrix} = c_1 \begin{bmatrix} 1 \\ 2 \\ 1 \end{bmatrix} + c_2 \begin{bmatrix} 1 \\ -1 \\ 3 \end{bmatrix} + c_3 \begin{bmatrix} 1 \\ 1 \\ 4 \end{bmatrix}. \tag{1}$$

Expanding (1), we obtain the three simultaneous equations:

$$x = c_1 + c_2 + c_3$$
$$y = 2c_1 - c_2 + c_3 \quad\quad (2)$$
$$z = c_1 + 3c_2 + 4c_3$$

The system (2) can be rewritten in matrix form as

$$\begin{bmatrix} 1 & 1 & 1 \\ 2 & -1 & 1 \\ 1 & 3 & 4 \end{bmatrix} \begin{bmatrix} c_1 \\ c_2 \\ c_3 \end{bmatrix} = \begin{bmatrix} x \\ y \\ z \end{bmatrix}. \quad\quad (3)$$

If the coefficient matrix of the c-vector is invertible, its inverse premultiplied by $[x\ y\ z]^T$ will yield the unique solution for $[c_1\ c_2\ c_3]^T$. To find the required inverse if it exists, we form the augmented matrix

$$\left[\begin{array}{ccc:ccc} 1 & 1 & 1 & 1 & 0 & 0 \\ 2 & -1 & 1 & 0 & 1 & 0 \\ 1 & 3 & 4 & 0 & 0 & 1 \end{array} \right]. \quad\quad (4)$$

We use elementary row operations to change the left part of (4) to the identity matrix. Simultaneously, the identity matrix on the right will change to the required inverse. Symbolically, $[A : I] \leftrightarrow [I : A^{-1}]$ since changing A to I requires A^{-1} $(AA^{-1} = I)$.

Replace Row 3 with Row 3 - Row 1.
Replace Row 2 with 2 x Row 1 - Row 2.

$$\left[\begin{array}{ccc:ccc} 1 & 1 & 1 & 1 & 0 & 0 \\ 0 & 3 & 1 & 2 & -1 & 0 \\ 0 & 2 & 3 & -1 & 0 & 1 \end{array} \right]$$

Replace Row 2 by 1/3 x Row 2:

$$\left[\begin{array}{ccc:ccc} 1 & 1 & 1 & 1 & 0 & 0 \\ 0 & 1 & \frac{1}{3} & \frac{2}{3} & -\frac{1}{3} & 0 \\ 0 & 2 & 3 & -1 & 0 & 1 \end{array} \right]$$

Replace Row 1 with Row 1 - Row 2.
Replace Row 3 with Row 3 - 2 x Row 2:

$$\left[\begin{array}{ccc:ccc} 1 & 0 & \frac{2}{3} & \frac{1}{3} & \frac{1}{3} & 0 \\ 0 & 1 & \frac{1}{3} & \frac{2}{3} & -\frac{1}{3} & 0 \\ 0 & 0 & \frac{7}{3} & -\frac{7}{3} & \frac{2}{3} & 1 \end{array} \right]$$

Replace Row 3 by $\frac{3}{7}$ x Row 3:

$$\begin{bmatrix} 1 & 0 & \frac{2}{3} & \vdots & \frac{1}{3} & \frac{1}{3} & 0 \\ 0 & 1 & \frac{1}{3} & \vdots & \frac{2}{3} & -\frac{1}{3} & 0 \\ 0 & 0 & 1 & \vdots & -1 & \frac{2}{7} & \frac{3}{7} \end{bmatrix}$$

Replace Row 1 by Row 1 $- \frac{2}{3}$ Row 3.

Replace Row 2 by Row 2 $- \frac{1}{3}$ Row 3.

And, therefore, the matrix (4) becomes

$$\begin{bmatrix} 1 & 0 & 0 & \vdots & 1 & \frac{1}{7} & -\frac{2}{7} \\ 0 & 1 & 0 & \vdots & 1 & -\frac{3}{7} & -\frac{1}{7} \\ 0 & 0 & 1 & \vdots & -1 & \frac{2}{7} & \frac{3}{7} \end{bmatrix} . \tag{5}$$

Premultiplying (3) by the inverse given by (5)

$$c_1 = x + \frac{1}{7}y - \frac{2}{7}z$$

$$c_2 = x - \frac{3}{7}y - \frac{1}{7}z \tag{6}$$

$$c_3 = -x + \frac{2}{7}y + \frac{3}{7}z$$

Any vector in R^3, [x, y, z] can be expressed in the form (1) with coefficients given by (6). To clarify the result, let [x y z]' = [1 2 3]'.

Then $c_1 = 1 + \frac{1}{7}(2) - \frac{2}{7}(3) = \frac{3}{7}$

$c_2 = 1 - \frac{3}{7}(2) - \frac{1}{7}(3) = -\frac{2}{7}$

$c_3 = -1 + \frac{2}{7}(2) + \frac{3}{7}(3) = \frac{6}{7}$,

and $\frac{3}{7} \begin{bmatrix} 1 \\ 2 \\ 1 \end{bmatrix} - \frac{2}{7} \begin{bmatrix} 1 \\ -1 \\ 3 \end{bmatrix} + \frac{6}{7} \begin{bmatrix} 1 \\ 1 \\ 4 \end{bmatrix} = \begin{bmatrix} 1 \\ 2 \\ 3 \end{bmatrix} = $ [x y z]'.

Thus, [1 2 1]' [1 -1 3]   and [1 1 4]' form a basis for R^3 with coordinates given by (6).

● PROBLEM 1-32

What is the usual basis for the vector space V of all real 2 x 3 matrices?

Solution: A basis B for a vector space V is a set of vectors in V such that any element in V can be expressed as a linear combination of vectors in B. Further, B is the smallest such set. (By requiring it to be the smallest such set, we secure the linear independence of its elements.)

An arbitrary, real 2 x 3 matrix has the form

$$A = \begin{bmatrix} a_{11} & a_{12} & a_{13} \\ a_{21} & a_{22} & a_{23} \end{bmatrix} . \quad aij \in \text{Real numbers} \quad (1)$$

Since each element of the matrix is arbitrary, we need at least six matrices to express A as a linear combination. The most natural choice of matrices is:

$$A_1 = \begin{bmatrix} 1 & 0 & 0 \\ 0 & 0 & 0 \end{bmatrix}; \quad A_2 = \begin{bmatrix} 0 & 1 & 0 \\ 0 & 0 & 0 \end{bmatrix}; \quad \dots, \quad A_6 = \begin{bmatrix} 0 & 0 & 0 \\ 0 & 0 & 1 \end{bmatrix}.$$

Then $A = a_{11} A_1 + a_{12} A_2 + a_{13} A_3 + \dots + a_{23} A_6$. The set of matrices $\{A_1, \dots, A_6\}$ is called the usual basis for the vector space of 2 x 3 matrices.

For example, let $A = \begin{bmatrix} 14 & -7 & 6 \\ 3 & 0 & 5 \end{bmatrix}.$

Then $A = 14A_1 - 7A_2 + 6A_3 + 3A_4 + 0A_5 + 5A_6.$

The dimension of a vector space is the number of elements in the basis. We see that $V_{2 \times 3}$ has dimension 6. In general, $V_{m \times n}$ has dimension mn.

● **PROBLEM** 1-33

Define Lagrange interpolation polynomials. Show that they form a basis for the vector space of polynomials with dimension ≤ 2 defined over the field of real numbers.

Solution: The space here is the set of quadratic functions. Thus, we are dealing with a function space. Lagrange interpolation polynomials arise from the problem of determining the quadratic function of x when it takes on the values a, b and c at $x = x_1$, x_2 and x_3 respectively. Suppose we know the values of the quadratic function f(x) when x = 0, 1 or 2. That is, $f(0) = a$, $f(1) = b$ and $f(2) = c$. To find the quadratic function, we must first form three polynomials which we call the Lagrange interpolation polynomials, i.e., three polynomials with the following properties:

$$f_1(0) = 1; \quad f_1(1) = f_1(2) = 0$$

$$f_2(1) = 1; \quad f_2(0) = f_2(2) = 0$$

$$f_3(2) = 1; \quad f_3(0) = f_3(1) = 0.$$

It is relatively simple to find $f_1(x)$. First note that $f_1(x) = 0$ when $x = 1$ or 2, so $(x-1)$ and $(x-2)$ must be factors of $f_1(x)$.

Let $f_1(x) = k_1(x-1)(x-2)$. $f_1(0) = 1$ gives us 1 $= k_1(0-1)(0-2)$ and yields $k_1 = \frac{1}{2}$. Thus, $f_1(x) = \frac{1}{2}(x-1)(x-2)$. Similarly, since $f_2(x) = 0$ at $x = 0$ and $x = 2$, then $f_2(x) = k_2x(x-2)$. But, $f_2(1) = 1$. Therefore, $f_2(1) = k_2(-1) = 1$ implies $k_2 = -1$. Now we have $f_2(x) = -x(x-2)$. Again, $f_3(x) = 0$ when $x = 0$ and $x = 1$. Therefore, $f_3(x) = k_3x(x-1)$. But, $f_3(2) = 1$. We now have $f_3(2) = k_3(2)(1) = 1$ and $k_3 = \frac{1}{2}$. So, $f_3(x) = \frac{1}{2}x(x-1)$.

$$f_1(x) = \frac{1}{2}(x-1)(x-2) = \frac{1}{2}x^2 - \frac{3}{2}x + 1 \qquad (1)$$

$$f_2(x) = -x(x-2) = -x^2 + 2x \qquad (2)$$

$$f_3(x) = \frac{1}{2}x(x-1) = \frac{1}{2}x^2 - \frac{1}{2}x \qquad (3)$$

We can use (1) - (3) to uniquely express any quadratic polynomial which takes values a, b, c at $x = 0$, 1 and 2 respectively. Thus $f(x) = af_1(x) + bf_2(x) + cf_3(x)$.

● PROBLEM 1-34

Which of the following vectors form a basis for R^3?

i)   (1, 1, 1) and (1, -1, 5)

ii)   (1, 2, 3), (1, 0, -1), (3, -1, 0) and (2, 1, -2)

iii)   (1, 1, 1), (1, 2, 3) and (2, -1, 1)

iv)   (1, 1, 2), (1, 2, 5) and (5, 3, 4).

Solution: A basis B for a vector space V has the following properties:

a)   B is the smallest set of vectors that span V:

b)   B is the largest set of linearly independent vectors in V;

c)   the number of vectors in B is the dimension of V.

We may use properties a)-c) to test whether the given sets form a basis for R^3.

i) Since R^3 has dimension 3 (any point in R^3 can be expressed as a unique linear combination of the unit vectors (1, 0, 0)

(0, 1, 0) and (0, 0, 1)$\big)$, (1, 1, 1) and (1, -1, 5) cannot be a basis for R^3 by property c).

ii) Here we have 4 vectors in the alleged basis. Again using property c), these vectors do not form a basis. That is, a basis for R^3 must contain only 3 vectors.

iii) Since there are 3 vectors in this set, it suffices to show that they span R^3. Let (x, y, z) be an arbitrary vector in R^3. Then,

$$(x, y, x) = c_1 (1, 1, 1) + c_2 (1, 2, 3) + c_3 (2, -1, 1)$$

$$(x, y, z) = (c_1, c_1, c_1) + (c_2, 2c_2, 3c_2) + (2c_3, -c_3, c_3)$$

$$(x, y, z) = (c_1 + c_2 + 2c_3, c_1 + 2c_2 - c_3, c_1 + 3c_2 + c_3)$$

$$x = c_1 + c_2 + 2c_3$$
$$y = c_1 + 2c_2 - c_3 \tag{1}$$
$$z = c_1 + 3c_2 + c_3$$

The system (1) may be rewritten in matrix form as

$$\begin{bmatrix} x \\ y \\ z \end{bmatrix} = \begin{bmatrix} 1 & 1 & 2 \\ 1 & 2 & -1 \\ 1 & 3 & 1 \end{bmatrix} \begin{bmatrix} c_1 \\ c_2 \\ c_3 \end{bmatrix} \tag{2}$$

or   X = AC. $\tag{2'}$

Solving (2'), we obtain

$$C = A^{-1} X \tag{2''}$$ as long as A^{-1} exists (That is, A is invertible.)

We must find the inverse of the matrix in (2). Form the augmented matrix

$$\begin{bmatrix} 1 & 1 & 2 & : & 1 & 0 & 0 \\ 1 & 2 & -1 & : & 0 & 1 & 0 \\ 1 & 3 & 1 & : & 0 & 0 & 1 \end{bmatrix} \tag{3}$$
$$\quad\ \ A \qquad\qquad\ \ \ I$$

Apply the following row operations. Start with (3).

i) Replace the first row with 2 times the second row added to the first.

$$\begin{bmatrix} 3 & 5 & 0 & : & 1 & 2 & 0 \\ 1 & 2 & -1 & : & 0 & 1 & 0 \\ 1 & 3 & 1 & : & 0 & 0 & 1 \end{bmatrix} \tag{3'}$$

ii)   Replace the second row of (3') with the second row
plus the third.

$$\left[\begin{array}{ccc:ccc} 3 & 5 & 0 & 1 & 2 & 0 \\ 2 & 5 & 0 & 0 & 1 & 1 \\ 1 & 3 & 1 & 0 & 0 & 1 \end{array}\right]$$

iii)   Replace the first row with the first row minus the
second row.

$$\left[\begin{array}{ccc:ccc} 1 & 0 & 0 & 1 & 1 & -1 \\ 2 & 5 & 0 & 0 & 1 & 1 \\ 1 & 3 & 1 & 0 & 0 & 1 \end{array}\right]$$

iv)   Replace the third row with the third row minus the
first.   Also replace the second row with (-2) times the
first row added to the second.

$$\left[\begin{array}{ccc:ccc} 1 & 0 & 0 & 1 & 1 & -1 \\ 0 & 5 & 0 & -2 & -1 & 3 \\ 0 & 3 & 1 & -1 & -1 & 2 \end{array}\right]$$

v)   In place of the third row, put -3 times the second row
added to 5 times the third row.

$$\left[\begin{array}{ccc:ccc} 1 & 0 & 0 & 1 & 1 & -1 \\ 0 & 5 & 0 & -2 & -1 & 3 \\ 0 & 0 & 5 & 1 & -2 & 1 \end{array}\right]$$

vi)   Multiply the second and third rows by $\frac{1}{5}$.

$$\underset{I}{\left[\begin{array}{ccc} 1 & 0 & 0 \\ 0 & 1 & 0 \\ 0 & 0 & 1 \end{array}\right.}\;\vdots\;\underset{A^{-1}}{\left.\begin{array}{ccc} 1 & 1 & -1 \\ -\frac{2}{5} & -\frac{1}{5} & \frac{3}{5} \\ \frac{1}{5} & -\frac{2}{5} & \frac{1}{5} \end{array}\right]}$$

The right matrix is the required inverse.

Now (2") gives

$$\left[\begin{array}{c} c_1 \\ c_2 \\ c_3 \end{array}\right] = \left[\begin{array}{ccc} 1 & 1 & -1 \\ -\frac{2}{5} & -\frac{1}{5} & \frac{3}{5} \\ \frac{1}{5} & -\frac{2}{5} & \frac{1}{5} \end{array}\right] \left[\begin{array}{c} x \\ y \\ z \end{array}\right]$$

and, hence,

$$c_1 = x + y - z$$
$$c_2 = -\frac{2}{5}x - \frac{1}{5}y + \frac{3}{5}z$$
$$c_3 = \frac{1}{5}x - \frac{2}{5}y + \frac{1}{5}z$$

(5)

Thus, any vector in R^3 can be expressed as a linear combination of the given vectors with the coordinates being given by (5).

iv) If the given vectors are independent, they form a basis for R^3. Form the matrix whose rows are the given vectors, and row reduce to echelon form by applying the required row operations.

$$\begin{bmatrix} 1 & 1 & 2 \\ 1 & 2 & 5 \\ 5 & 3 & 4 \end{bmatrix} \text{ to } \begin{bmatrix} 1 & 1 & 2 \\ 0 & 1 & 3 \\ 0 & -2 & -6 \end{bmatrix} \text{ to } \begin{bmatrix} 1 & 1 & 2 \\ 0 & 1 & 3 \\ 0 & 0 & 0 \end{bmatrix}$$

(6)

From the row echelon matrix (6), we can see that there are only two linearly independent rows, i.e., the given set is not linearly independent. Thus, it cannot be a basis for R^3.

● **PROBLEM** 1-35

One basis for the vector space V of polynomials of degree not exceeding two is the set B = $\{1, x, x^2\}$. Construct another basis for V whose elements are all quadratic functions.

Solution: Since B contains three vectors, we know that any other set containing three linearly independent vectors will be a basis. Consider the three vectors

$$f_1 = \frac{(x-1)(x-2)}{2}; \quad f_2 = \frac{x(x-2)}{-1}; \quad f_3 = \frac{x(x-1)}{2}.$$

(1)

Note that $f_1(1) = f_1(2) = 0; \quad f_1(0) = 1$

(2)

$$f_2(0) = f_2(2) = 0; \quad f_2(1) = 1$$

$$f_3(0) = f_3(1) = 0; \quad f_3(2) = 1.$$

The polynomials f_1, f_2 and f_3 each have degree 2 and thus belong to V.

Let us check for linear independence. Set

$$c_1 f_1(x) + c_2 f_2(x) + c_3 f_3(x) = 0.$$

(3)

Substitute x = 0 in (3); from the relations (2),

$$c_1(1) + c_2(0) + c_3(0) = 0$$

implies $c_1 = 0$.

Now, let $x = 1$ in (3). Then, using (2),

$$0 + c_2(1) + c_3(0) = 0$$

implies $c_2 = 0$. Similarly, letting $x = 2$ in (3) and using (2), we find that $c_3 = 0$. So (3) implies that $c_1 = c_2 = c_3 = 0$. Thus, $f_1(x)$, $f_2(x)$ and $f_3(x)$ are linearly independent and form a basis for V.

To find the coordinates of an arbitrary vector in V, let

$$f(x) = c_1 f_1(x) + c_2 f_2(x) + c_3 f_3(x). \qquad (4)$$

We can use the relations (2) one more time. When $x = 0$, $c_1 = f(0)$. For $x = 1$, $c_2 = f(1)$, and when $x = 2$, $c_3 = f(2)$. Thus, $f(x) = f(0)f_1(x) + f(1)f_2(x) + f(2)f_3(x)$. $\qquad (5)$

The formula (5) is known as the Lagrange interpolation formula.

● **PROBLEM** 1-36

Let F be any field and let $F^n = F \times F \times F \times ... \times F$ (the Cartesian product of F, n times). What is the standard basis of F^n? Give an example.

Solution: We define $e_i^{(n)}$ $(i = 1,...,n)$ as the basic elements of F^n. Then the set $\{e_1^{(n)}, e_2^{(n)}, ..., e_n^{(n)}\} \subset F^n$ is a basis of F^n called the standard basis.

As an example, let $F = R$ and $n = 2$. Then $F^n = R^2$ and the standard basis is $\{(1, 0), (0, 1)\}$. The basis of a vector space plays the role of a coordinate system since every point $(x, y) \in R^2$ can be written as a linear combination of $(1, 0)$ and $(0, 1)$.

SUBSPACES

● **PROBLEM** 1-37

Are the following sets subspaces?

i) $W \subset R^3$ consists of all vectors of the form $(a, b, 1)$ where a and b are any real numbers.

ii) $W \subset R^4$ contains all vectors of the form $(a, b, a - b, a + 2b)$ for $a, b \in R$.

Solution: Let V be a vector space over K and W a non-empty

47

subset of V. Then w is called a subspace if, for all w_1, w_2
$\in$ W and $\alpha \in$ K,

   a) $w_1 + w_2 \in$ W

   b) $\alpha w_1 \in$ W

i) To check whether W is a subspace, let $w_1 = (a_1, b_1, 1)$
and $w_2 = (a_2, b_2, 1)$ be vectors in W.   Then,

$$X + Y = (a_1, b_1, 1) + (a_2, b_2, 1)$$

$$= (a_1 + a_2, b_1 + b_2, 2) \qquad\qquad (1)$$

But $(a_1 + a_2, b_1 + b_2, 2)$ is not of the form $(a, b\ 1)$ since
its third component is not a 1.   So $w_1 + w_2 \notin$ W.

   Hence, W is not closed under addition, i.e., W is not
a subspace.

ii) Let $w_1 = (a_1, b_1, a_1 - b_1, a_1 + 2b_1)$ and

   $w_2 = (a_2, b_2, a_2 - b_2, a_2 + 2b_2)$.

Then, $w_1 + w_2$

$= (a_1 + a_2, b_1 + b_2, a_1 + a_2 - (b_1 + b_2),$

$$a_1 + a_2 +\ (b_1 + b_2))\ (2)$$

   Vectors of the form (2) are again in W.   To see this,
replace $a_1 + a_2$ by a' and $b_1 + b_2$ by b'.

   Thus, $w_1 + w_2 = (a', b', a' - b', a' + 2b')$.

Hence, W is closed under addition.   Next, if α is a scalar,
then

   $\alpha w_1 = \alpha(a_1, b_1, a_1 - b_1, a_1 + 2b_1)$

$$= [\alpha a_1, \alpha b_1, \alpha a_1 - \alpha b_1, \alpha a_1 + 2\alpha b_1] \qquad (3)$$

   Vectors of the form (3) are again in W.   Thus, W is
a subspace of R^4.

● PROBLEM 1-38

Let W be the set consisting of all 2 x 3 matrices of the form

$$\begin{bmatrix} a & b & 0 \\ 0 & c & d \end{bmatrix}$$ where a, b, c, d are   (1)
   real numbers.

Show that W is a subspace of V, the set of all 2 x 3 matrices under the operation of addition over the field of real numbers.

Solution: To show that $W \subseteq V$ is a subspace we must show that, for $w_1, w_2 \in W$ and $\alpha \in R$ (the field),

i) $W \neq \phi$

ii) $w_1 + w_2 \in W$

iii) $\alpha w_1 \in W$.

To show i), let $a = b = c = d = 0$ in the matrix (1). Then W contains at least the zero element and is non-empty.

To show ii), let

$$w_1 = \begin{bmatrix} a_1 & b_1 & 0 \\ 0 & c_1 & d_1 \end{bmatrix} \text{ and } w_2 = \begin{bmatrix} a_2 & b_2 & 0 \\ 0 & c_2 & d_2 \end{bmatrix}$$

be two matrices in W. Then, by the rules of matrix addition,

$$w_1 + w_2 = \begin{bmatrix} a_1 + a_2 & b_1 + b_2 & 0 \\ 0 & c_1 + c_2 & d_1 + d_2 \end{bmatrix}$$

But, by the rules of addition for real numbers, $a_1 + a_2$, $b_1 + b_2$, $c_1 + c_2$ and $d_1 + d_2$ are real numbers. Hence, $w_1 + w_2 \in W$.

Finally, to show iii) let

$$\alpha w_1 = \alpha \begin{bmatrix} a_1 & b_1 & 0 \\ 0 & c_1 & d_1 \end{bmatrix}. \tag{2}$$

By the rules of scalar multiplication for matrices, (2) becomes

$$\begin{bmatrix} \alpha a_1 & \alpha b_1 & 0 \\ 0 & \alpha c_1 & \alpha d_1 \end{bmatrix}.$$

Since α is a real number, by the rules for multiplication of real numbers: αa_1, αb_1, αc_1 and αd_1 are real numbers. Hence, $\alpha w_1 \in W$.

Since W satisfies i), ii) and iii), we conclude that it is a subspace of $V_{2 \times 3}$.

Let $V_{n \times n}$ be the vector space of all square n x n matrices over the field of real numbers. Let W consist of all matrices which commute with a given matrix T, i.e., W = {A $\in$ V: AT = TA}. Show that W is a subspace of V.

Solution: We must show that i) W $\neq$ ϕ; ii) for w_1, w_2 $\in$ W and α, β $\in$ R (the field), $(\alpha w_1 + \beta w_2)$ $\in$ W.

Note that this is a way of combining the requirements of closure for addition and scalar multiplication.

To show i), we note that for given T, if A is the zero matrix of order n, then AT = TA. Hence, W contains the zero vector and is non-empty.

To show ii), we must show that $(\alpha w_1 + \beta w_2)T = T(\alpha w_1 + \beta w_2)$. Now, by the right distributive law,

$$(\alpha w_1 + \beta w_2)T = (\alpha w_1)T + (\beta w_2)T \qquad (1)$$

$$= \alpha(w_1 T) + \beta(w_2 T). \qquad (2)$$

Since w_1 and w_2 $\in$ W, the set of elements that commute with T, (2) becomes $\alpha(T w_1) + \beta(T w_2)$. $\qquad$ (3)

Now we get $T(\alpha w_1) + T(\beta w_2)$, $\qquad$ (4)
since any scalar commutes with a vector in $V_{n \times n}$. Finally,
(4) gives $T(\alpha w_1 + \beta w_2)$ $\qquad$ (5)
by the left distributive law. (1) - (5) show that $(\alpha w_1 + \beta w_2)T = T(\alpha w_1 + \beta w_2)$, so $\alpha w_1 + \beta w_2$ $\in$ W.

Therefore, i) and ii) imply W is a vector space.

Consider the following vector spaces

A) V is any vector space

B) V is the space of all square n x n matrices

C) V is the space of all functions from a set x $\neq$ {0} into the real field R. Give examples of subspaces for these different vector spaces.

Solution: Let W be a subset of a vector space over a field K. Then W is a subspace if it is itself a vector space over K with the operations of addition and scalar multiplication as in the vector space.

To verify that W$\subseteq$V is a subspace, we can show that all

the axioms for V hold when V is restricted to W. But this
is tedious. In fact, W⊆V is a subspace if

i)   W is non-empty

ii)   W is closed under vector addition:
      u,v ∈ W implies u + v ∈ W.

iii) W is closed under scalar multiplication:
      v ∈ W implies kv ∈ w for every k ∈ K.

      We need not verify the other axioms because

i)   the axioms are true for all members of V

ii)   W ⊆ V contains only members of V. Hence, the axioms
      are automatically verified.

A)   Let {0} be the set consisting of the zero vector alone.
      Then, W = {0} is non-empty. Since 0 is the only element
      of W, W is closed under addition. Finally, for every
      k ∈ K, k0 = 0∈ W. Thus, the zero vector is a subspace
      of any vector space.

B)   Let W ⊂ V be the set of matrices (aij) = A such that
      aij = aji. Then W, the set of all n x n symmetric
      matrices, is a subspace of V. To show this, note that
      the nth order zero matrix is symmetric. Hence, W ≠ φ.
      Next, let (aij), (bij) ∈ W. The sum (aij) + (bij) is

$$
\begin{bmatrix} a_{11} \cdots\cdots a_{1n} \\ \vdots \quad a_{22} \quad \vdots \\ a_{n1} \cdots\cdots a_{nn} \end{bmatrix}
+
\begin{bmatrix} b_{11} \cdots\cdots b_{1n} \\ \vdots \quad b_{22} \quad \vdots \\ b_{n1} \cdots\cdots b_{nn} \end{bmatrix}
$$

$$
=
\begin{bmatrix} a_{11}+b_{11} \cdots\cdots a_{1n}+b_{1n} \\ \vdots \quad a_{22}+b_{22} \quad \vdots \\ a_{n1}+b_{n1}, \cdots\cdots a_{nn}+b_{nn} \end{bmatrix} . \qquad (1)
$$

      A matrix is symmetric when it is equal to its transpose,
i.e., interchanging rows and columns leaves the matrix unchanged.
Transposing (1) we obtain

$$
\begin{bmatrix} a_{11}+b_{11} \cdots\cdots a_{n1}+b_{n1} \\ \vdots \quad a_{22}+b_{22} \quad \vdots \\ a_{1n}+b_{1n} \cdots\cdots a_{nn}+b_{nn} \end{bmatrix}
$$

But, since aij = aji and bij = bji, (2) is equal to (1).
Thus, the sum of two symmetric matrices is a symmetric matrix,
i.e., W is closed under addition. Clearly, k(aij) = (kaij)
is a symmetric matrix. Hence, the set of n x n symmetric
matrices is a subset of all n x n matrices.

C) V is a function space. If W is the set of all bounded
functions defined on X (f $\in$ V is bounded if $|f(x)| \leq M$
for all x $\in$ X, where M $\in$ R), then W $\subseteq$ V is a subspace.
This may be shown as follows. Let h $\in$ V be the zero
function that assigns the value 0 to every element of
x. Since $|h(x)| < 1$, h(x) is bounded and belongs to W.
Hence, W $\neq \phi$. Let f,g be any elements of W. Then

$|f(x)| \leq M_1$, $|g(x)| \leq M_2$ where M_1, $M_2 \in$ R. Further,

by the rules of addition of functions, f(x) + g(x)
= (f + g)(x). Now, $|(f + g)(x)| \leq |f(x)| + |g(x)|$
= $M_1 + M_2 = M_3$ (by the triangle inequality). Hence,

the sum of two bounded functions is bounded. Let k $\in$ K.
Then kf(x) is bounded since $|kf(x)| \leq |k|M_1$.

Note: $|kf(x)| = |k||f(x)|$ and $|k||f(x)| \leq |k|M_1$ since

$|f(x)| \leq M_1$. Therefore, $|kf(x)| \leq |k|M_1$. Thus W $\subseteq$ V,

the set of all bounded functions from X to R is a subspace of V.

● **PROBLEM** 1-41

Let V be R^3, i.e., three dimensional space. Show that W
is a subspace of V where

i)   W = {a, b, 0}

ii)   W = {a, b, c : a + b + c = 0}, a, b, c are
      real numbers.

<u>Solution</u>: To show that W is a subspace of V defined over the
field K, we must show that

a)   W $\neq \phi$

b) $k_1 w_1 + k_2 w_2 \in$ W for w_1, $w_2 \in$ W and k_1, $k_2 \in$ K.

i)   Geometrically, W consists of all points in the xy plane.
     Letting a = b = 0, the zero vector (0, 0, 0) lies in the
     xy plane. Hence, W $\neq \phi$.

$k_1 w_1 + k_2 w_2 = k_1 (a_1, b_1, 0) + k_2 (a_2, b_2, 0),$ (1)

and, by the rules of addition of n-tuples of real numbers,
(1) becomes

$(k_1 a_1, k_1 b_1, 0) + (k_2 a_2, k_2 b_2, 0)$

   = $(k_1 a_1 + k_2 a_2, k_1 b_1 + k_2 b_2, 0)$ (2)

Since k_1, k_2, a_1, a_2 and b_1, b_2 are real numbers, all sums and products in (2) are real numbers. Hence, $k_1 w_1 + k_2 w_2 \in W$ and W is a subspace.

ii) Letting a = b = c = 0, a + b + c = 0 and the zero vector (0, 0, 0) is in W. Hence, $W \neq \phi$.

Next, let $w_1 = (a_1, b_1, c_1) \in W$ and
$$w_2 = (a_2, b_2, c_2) \in W.$$

w_1 and w_2 in W implies

$$a_1 + b_1 + c_1 = 0 \qquad\qquad (1)$$
$$a_2 + b_2 + c_2 = 0 \qquad\qquad (2)$$

Now we must show $k_1 w_1 + k_2 w_2 \in W$ $\qquad\qquad$ (3)

$$k_1 w_1 + k_2 w_2 = k_1 (a_1, b_1, c_1) + k_2 (a_2, b_2, c_2)$$
$$= (k_1 a_1 + k_2 a_2, k_1 b_1 + k_2 b_2, k_1 c_1 + k_2 c_2)$$

by the properties of scalar multiplication and vector addition.

To check if $k_1 w_1 + k_2 w_2$ is in W, add up its components; we obtain $k_1 a_1 + k_2 a_2 + k_1 b_1 + k_2 b_2 + k_1 c_1 + k_2 c_2$

$$= k_1 a_1 + k_1 b_1 + k_1 c_1 + k_2 a_2 + k_2 b_2 + k_2 c_2$$

$$= k_1 (a_1 + b_1 + c_1) + k_2 (a_2 + b_2 + c_2)$$

Using (1) and (2), this gives $k_1 (0) + k_2 (0) = 0$. So (3) is verified and W is a subspace.

OPERATIONS IN VECTOR SPACES

● **PROBLEM** 1-42

Let u = (2, -1, 1) and v = (1, 1, 2) be two vectors in R^3. Find the dot product, u•v and the angle between u and v.

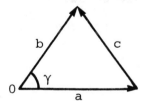

Solution: The dot product in R^3 (or, as it is also known, the inner product) helps us to define the concepts of length of a vector, distance between vectors and the angle between two vectors.

In R^n, the dot product is defined as:

$$x \cdot y = \sum_{i=1}^{n} (x_i \ y_i) \qquad (1)$$

where $x = (x_1, x_2, \ldots, x_n)$ and $y = (y_1, y_2, \ldots, y_n)$. Thus,

in R^3, $u \cdot v = \sum_{i=1}^{3} (u_i \ v_i) = u_1 v_1 + u_2 v_2 + u_3 v_3$ where

$u = (u_1, u_2, u_3)$ and v (v_1, v_2, v_3). For the vectors in the given problem,

$$u \cdot v = (2, -1, 1) \cdot (1, 1, 2)$$
$$= 2(1) + (-1)(1) + 1(2) = 3.$$

Next we show how to find the length of a vector and the angle between two vectors. In (1), if we let $y = x$, we obtain

$$x \cdot x = \sum_{i=1}^{n} (x_i \ x_i) \qquad (2)$$

$$= x_1^2 + x_2^2 + \ldots + x_n^2.$$

The square root of (2) is defined as the norm or length of x. We write

$$\|x\| = (x \cdot x)^{1/2} \qquad (3)$$

For the vectors in the given problem,

$$\|u\| = (u \cdot u)^{1/2} = (u_1^2 + u_2^2 + u_3^2)^{1/2}$$
$$= [(2)^2 + (-1)^2 + (1)^2]^{1/2} = [6]^{1/2} = \sqrt{6}.$$

$$\|v\| = (v \cdot v)^{1/2} = [(1)^2 + (1)^2 + (2)^2]^{1/2} = \sqrt{6}.$$

Consider the triangle below:

The law of cosines states that, for a triangle in the plane,

$$c^2 = a^2 + b^2 - \alpha ab \cos \gamma.$$

In the general case R^n letting $a = u$, $b = v$ and $c = v - u$, we have

$$\|v - u\|^2 = \|u\|^2 + \|v\|^2 - 2 \|u\| \ \|v\| \cos \gamma. \qquad (4)$$

since $(a - b)^2 = a^2 - 2ab + b^2$ and $\|v - u\|^2$

$$= [(v - u) \cdot (v - u)],$$

$[(v - u) \cdot (v - u)] = \|u\|^2 - 2(u \cdot v) + \|v\|^2.$　　(5)

Comparing (4) and (5),

$2 \|u\| \|v\| \cos \gamma = 2 (u \cdot v)$

$$\cos \gamma = \frac{(u \cdot v)}{\|u\| \|v\|}.$$　　(6)

From (6), using the arc cosine relationship, we can find the angle γ.

In the given problem,

$$\cos \gamma = \frac{3}{\sqrt{6}\sqrt{6}} = \frac{1}{2}.$$

The angle whose cosine is $\frac{1}{2}$ is $\gamma = 60°$. Sometimes it is useful to write (6) as $u \cdot v = \|u\| \|v\| \cos \gamma$ which shows that $u \cdot v = 0$ implies u and v are orthogonal.

● **PROBLEM** 1-43

The polynomials $e_1 = 1$, $e_2 = t-1$, $e_3 = (t-1)^2$ form a basis for V_2, the vector space of polynomials with degree ≤ 2. Let $u = 2t^2 - 5t + 6 \in V_2$. Find $[V]e$, the coordinate vector of u relative to the basis $\{e_1, e_2, e_3\}$.

Solution: Let $V = at^2 + bt + c$ be an arbitrary element of V_2 where a, b, c are scalars from the field K. Write V as a combination of the e_i using the unknowns c_1, c_2 and c_3:

$at^2 + bt + c = c_1(1) + c_2(t-1) + c_3(t^2 - 2t +1)$

$\qquad = c_1 + c_2 t - c_2 + c_3 t^2 - 2c_3 t + c_3$

$\qquad = c_3 t^2 + (c_2 - 2c_3)t + (c_1 - c_2 + c_3).$　(1)

Two polynomials are equal when coefficients of like powers are equal. Then, using (1), we solve for the general solution.

$a = c_3; \ b = (c_2 - 2c_3); \ c = (c_1 - c_2 + c_3)$

Solving for c_1, c_2 and c_3: $\quad c_1 = a + b + c;$

$c_2 = 2a+b; \ c_3 = a.$

Hence, $v = at^2 + bt + c = c + bt + at^2$ can be written as:

$c + bt + at^2 = (a + b + c)(1) + (2a + b)(t-1)$

$+ (a)(t^2 - 2t + 1)$　　(2)

55

The given vector in the problem is

$$u = 2t^2 - 5t + 6 = 6 - 5t + 2t^2. \qquad (3)$$

Comparing (2) and (3), $a = 2$, $b = -5$ and $c = 6$.
Hence, $u = (2 - 5 + 6)(1) + (4 - 5)(t-1) + 2(t^2 - 2t + 1)$

$$= 3e_1 - e_2 + 2e_3,$$

and $[V]e = [3, -1, 2]$. In order to complete this problem with confidence, we should verify that e_1, e_2 and e_3 form a basis for V_2. Since it takes only 3 vectors to span V_2 (dim $V_2 = 3$), we need only show that e_1, e_2 and e_3 are linearly independent. If $m_1 e_1 + m_2 e_2 + m_3 e_3 = 0$, by substituting in (1) we get

$$0t^2 + 0t + 0 = c_3 t^2 + (c_2 - 2c_3)t + (c_1 - c_2 + c_3).$$

So $c_3 = 0$ $c_2 - 2c_3 = 0$ and $c_1 - c_2 + c_3 = 0$. This implies $c_1 = c_2 = c_3 = 0$. Hence, e_1, e_2 and e_3 are linearly independent. Thus, we can represent any element in V_2 as a linear combination of them.

● **PROBLEM** 1-44

Let $V_1 = \{t(1, 1) : t \in R\}$ and $V_2 = \{(x_1, x_2) \in R^2 : (1, 1)$
$(x_1, x_2) = 0\}$ be two subsets of R^2. Show that R^2 is the direct sum of V_1 and V_2: $R^2 = V_1 \oplus V_2$.

Solution: We first define what is meant by a direct sum and then show that

$$R^2 = V_1 \oplus V_2.$$

Let V_1, V_2 be subspaces of V, a vector space. The sum of V_1 and V_2, $V_1 + V_2$, is the set of all vectors of the form

$$V_1 + V_2 = \{ v_1 + v_2 : v_1 \in V_1, v_2 \in V_2 \}.$$

If every vector in $V_1 + V_2$ can be uniquely expressed as the sum of a vector from V_1 and a vector from V_2, then $V_1 + V_2$ is called the direct sum of V_1 and V_2, $V_1 \oplus V_2$.

Next, we verify that V_1 and V_2 are indeed subspaces. Consider V_1. For $t = 0$, $t(1, 1) = (0, 0)$, and, hence, V_1 contains the zero vector. (Therefore, $V_1 \neq \phi$.)

The sum of two vectors, $x_1 = t_1(1, 1)$, $y_1 = t_2(1, 1)$, is $x_1 + y_1 = t_1(1, 1) + t_2(1, 1) = (t_1 + t_2)(1, 1) = t(1, 1)$ where $t_1 + t_2 = t$, another real number. This is in V_1. Hence,

V_1 is closed under addition.

Now, let $x = t (1, 1)$ V_2 and $\alpha \in R$. Therefore, $\alpha x = \alpha (t (1, 1)) = \alpha t(1, 1)$. Since α and t are real numbers, αt is again a real number. Thus, $\alpha x \in V_1$ and V_1 is closed under scalar multiplication. Therefore, V_1 is a subspace of R^2.

Consider V_2 which is the set of all vectors in R^2 that are perpendicular to $(1, 1)$. This is true since we know that for any two vectors X and Y, if $X \cdot Y = 0$ then X is perpendicular to Y. Note $(1, -1)$ is perpendicular to $(1, 1)$ since $(1, -1) \cdot (1, 1) = 0$. Therefore, V_2 contains $(1, -1)$, and V_2 is not empty.

Let x, y be two vectors in V_2. By the rules of the dot product,

$$(x + y) \cdot (1, 1) = x \cdot (1, 1) + y \cdot (1, 1) = 0,$$

and, hence, $x + y$ is perpendicular to $(1, 1)$. V_2 is closed under addition. For $\alpha \in R$ and $x \in V_2$, $(\alpha x) \cdot (1, 1) = \alpha (x \cdot (1, 1)) = 0$ and $\alpha x \in V_2$. V_2 is therefore a subspace of R^2.

To find the coordinate form of vectors in V_2, we reason as follows: for $(x_1, x_2) \in V_2$, $(x_1, x_2) \cdot (1, 1) = 0$ implies $x_1 (1) + x_2 (1) = 0$. Thus, $x_2 = - x_1$ and all vectors of the form $t(1, -1)$ for $t \in R$ are in V_2.

The vectors $(1, 1)$ and $(1, -1)$ are linearly independent. To prove this, let $c_1 (1, 1) + c_2 (1, -1) = (0, 0)$ for $c_1, c_2 \in R$. Then,

$$(c_1, c_1) + (c_2, - c_2) = (0, 0)$$

$$c_1 + c_2 = 0$$

$$c_1 - c_2 = 0.$$

Only $c_1 = c_2 = 0$ satisfies both these equations.

Thus $(1, 1)$ and $(1, -1)$ form a basis for R^2. Any vector in R^2 can be uniquely expressed as a linear combination of $(1, 1)$ and $(1, -1)$.

We conclude that $R^2 = V_1 \oplus V_2$.

CHAPTER 2

LINEAR TRANSFORMATIONS

EXAMPLES OF LINEAR TRANSFORMATIONS

● **PROBLEM** 2-1

What are the effects of the following transformations on points (x_1, x_2) in the plane? The transformations are represented by $A = \begin{bmatrix} a_{11} & a_{12} \\ a_{21} & a_{22} \end{bmatrix}$.

i) $A = \begin{bmatrix} 1 & 0 \\ 0 & -1 \end{bmatrix}$ ii) $A = \begin{bmatrix} -1 & 0 \\ 0 & -1 \end{bmatrix}$ iii) $A = \begin{bmatrix} 0 & -1 \\ 1 & 0 \end{bmatrix}$

iv) $A = \begin{bmatrix} 1 & 0 \\ 0 & 0 \end{bmatrix}$ v) $A = \begin{bmatrix} 2 & 0 \\ 0 & 0 \end{bmatrix}$ vi) $A = \begin{bmatrix} 1 & 1 \\ 1 & 1 \end{bmatrix}$

vii) $A = \begin{bmatrix} 2 & 0 \\ 0 & 2 \end{bmatrix}$ viii) $A = \begin{bmatrix} 2 & 0 \\ 0 & 3 \end{bmatrix}$ ix) $A = \begin{bmatrix} 1 & 1 \\ 0 & 1 \end{bmatrix}$.

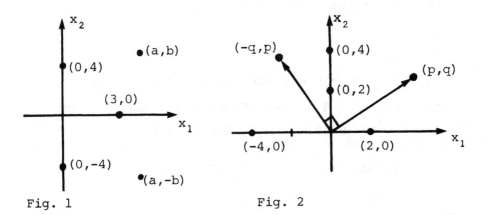

Fig. 1   Fig. 2

Solution: A transformation is a mapping from a vector space V to a vector space W. Any linear transformation can be represented by a matrix. Each of the given transformations is linear. A: $[x_1, x_2] \rightarrow AX$ has the following interpretation.

58

By the rules of matrix multiplication,

$$Ax = \begin{bmatrix} a_{11} & a_{12} \\ a_{21} & a_{22} \end{bmatrix}\begin{bmatrix} x_1 \\ x_2 \end{bmatrix}' = \begin{bmatrix} a_{11}x_1 + {}^2 12x_2 \\ a_{21}x_1 + a_{22}x_2 \end{bmatrix} \qquad (1)$$

i) Using (1), A: $[x_1,x_2] \to \begin{bmatrix} 1 & 0 \\ 0 & -1 \end{bmatrix}\begin{bmatrix} x_1 \\ x_2 \end{bmatrix}' = \begin{bmatrix} x_1 \\ -x_2 \end{bmatrix}.$

Thus, a point with coordinates $[x_1,x_2]$ is sent to a point whose first coordinate is the same, but whose second coordinate is the negative of the original coordinate.

From Fig. 1, we see that only points on the x_1 axis are unaltered by the transformation. All other points are reflected across the x_1 axis.

ii) $Ax = \begin{bmatrix} -1 & 0 \\ 0 & -1 \end{bmatrix}\begin{bmatrix} x_1 \\ x_2 \end{bmatrix} = \begin{bmatrix} -x_1 \\ -x_2 \end{bmatrix}.$

The point (a,b) goes to $(-a,-b)$. The only invariant point is the origin $(0,0)$. The effect on the points geometrically is to shift the vectors they represent in R^2 by an angle of $180°$.

iii) $Ax = \begin{bmatrix} 0 & -1 \\ 1 & 0 \end{bmatrix}\begin{bmatrix} x_1 \\ x_2 \end{bmatrix} = \begin{bmatrix} -x_2 \\ x_1 \end{bmatrix}.$

The transformation is a rotation about the origin through a right angle in the counter-clockwise direction. The origin is the only invariant point.

iv) $Ax = \begin{bmatrix} 1 & 0 \\ 0 & 0 \end{bmatrix}\begin{bmatrix} x_1 \\ x_2 \end{bmatrix} = \begin{bmatrix} x_1 \\ 0 \end{bmatrix}.$

Every point in the plane is projected orthogonally onto the x_1-axis. All points on the x_1 axis are left invariant.

v) $Ax = \begin{bmatrix} 2 & 0 \\ 0 & 0 \end{bmatrix}\begin{bmatrix} x_1 \\ x_2 \end{bmatrix} = \begin{bmatrix} 2x_1 \\ 0 \end{bmatrix}.$

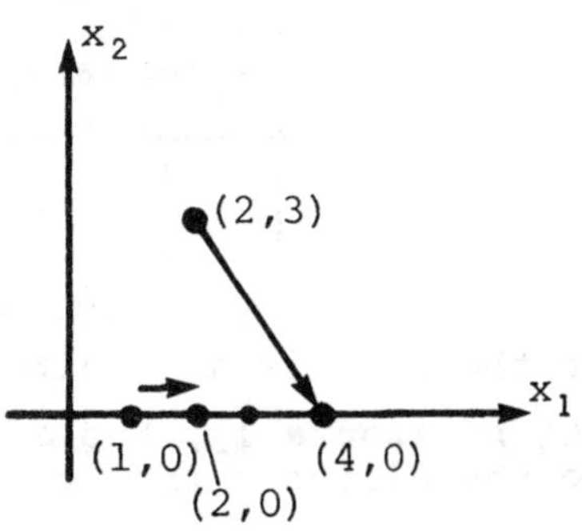

Fig. 3

59

Points on the x_1-axis have their distance from the origin doubled. Other points are projected onto the x_1 axis. The x_1-axis is an invariant line, i.e., the line is unchanged by the transformation although individual points are displaced on it.

vi) $\quad Ax = \begin{bmatrix} 1 & 1 \\ 1 & 1 \end{bmatrix} \begin{bmatrix} x_1 \\ x_2 \end{bmatrix} = \begin{bmatrix} x_1 + x_2 \\ x_1 + x_2 \end{bmatrix}.$

Here, the result of the transformation is that all points are sent to the $x_1 = x_2$ line. In other words, the transformation sends the vector space R^2 into the real line R^1 which, in the x_1, x_2 plane, is represented by $x_1 = x_2$. Vectors on the line $x_1 = x_2$ have their magnitudes doubled.

vii) $\quad Ax = \begin{bmatrix} 2 & 0 \\ 0 & 2 \end{bmatrix} \begin{bmatrix} x_1 \\ x_2 \end{bmatrix} = \begin{bmatrix} 2x_1 \\ 2x_2 \end{bmatrix}.$

The distance of every point from the origin is doubled. The origin is the only invariant point, but every line through the origin is an invariant line.

viii) $\quad Ax = \begin{bmatrix} 2 & 0 \\ 0 & 3 \end{bmatrix} \begin{bmatrix} x_1 \\ x_2 \end{bmatrix} = \begin{bmatrix} 2x_1 \\ 3x_2 \end{bmatrix}.$

The points are stretched; however, the magnitude of the stretch is not the same in all directions. The origin is the only invariant point, but both the x_1 and x_2 axes are invariant lines.

ix) $\quad Ax = \begin{bmatrix} 1 & 1 \\ 0 & 1 \end{bmatrix} \begin{bmatrix} x_1 \\ x_2 \end{bmatrix} = \begin{bmatrix} x_1 + x_2 \\ x_2 \end{bmatrix}.$

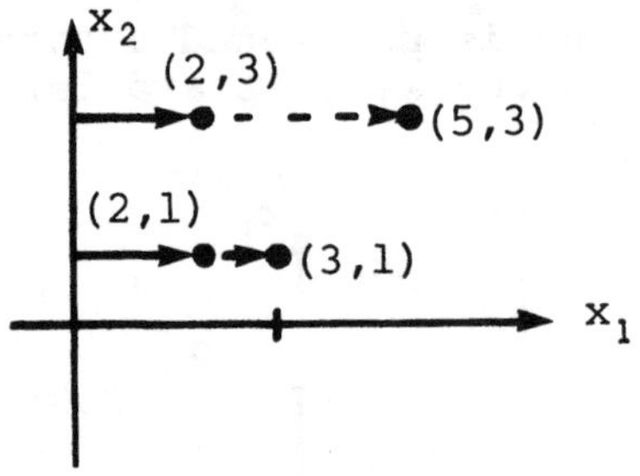

Fig. 4

Points on the x_1 axis are left invariant. Other points are slid along horizontally, the amount of slide being proportional to the distance from the x_1 axis.

Show that the zero mapping and the identity transformation
are linear transformations.

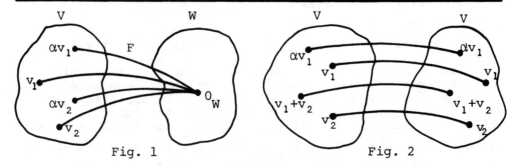

Fig. 1   Fig. 2

Solution: Let V and W be vector spaces over the field K.
Then, the function L: V → W is a linear transformation if:

      i) $L(V_1 + V_2) = L(V_1) + L(V_2)$.

      ii) $L(\alpha V_1) = \alpha L(V_1)$

for $v_1, v_2 \varepsilon V$ and $\alpha \varepsilon K$. An equivalent way of expressing
i) and ii) is $L(\alpha v_1 + v_2) = \alpha L(v_1) + L(v_2)$.

    Let F: V → W be the mapping that sends every element
in V to the zero vector in W. F is the zero mapping. We
see from Fig. 1 that

$$F(v_1 + v_2) = 0_W = F(v_1) + F(v_2).$$

Furthermore,

$$F(\alpha v_1) = 0_W = F(v_1).$$

Thus, the zero mapping is a linear transformation.

    Next, let I: V → W be the function that assigns every
element in V to itself. Thus, W = V. I is the identity
mapping. (That, is, $v_1 \varepsilon V$ gives $I(v_1) = v_1$.) We see from
Fig. 2 that

$$F(v_1 + v_2) = (v_1 + v_2) = v_1 + v_2 = F(v_1) + F(v_2).$$

Also, $F(\alpha v_1) = \alpha v_1 = \alpha F(v_1)$.

Thus, I, the identity mapping, is a linear transformation.

● **PROBLEM** 2-3

Let C[0,1] be the vector space of bounded continuous functions defined on the interval $0 \le x \le 1$. Let M be the subspace of C[0,1] consisting of functions f such that both f
and its derivative f' are defined and continuous for
$0 \le x \le 1$.
    Show that the operations of differentiation and integra-
tion are linear transformation.

Solution: Let V and W be vector spaces over the field R, and let T: V → W be a mapping of elements in V to elements in W. Then T is known as a linear transformation if it satisfies

i) $T(v_1 + v_2) = T(v_1) + T(v_2)$

ii) $T(\alpha v_1) = \alpha T(v_1)$.

for any $v_1, v_2 \in V$ and $\alpha \in R$.

Note that the results of the mapping, $T(v_1)$, $T(v_2)$ are in W.

Let us find the nature of V and W and also the field R for the given problem. Actually, we are more interested in M, the subspace of C[0,1] or V. Let f, g be arbitrary elements in M. Then f', g' are defined and continuous, i.e., they are elements of C[0,1] (but not necessarily of M: the function $\frac{\sin x}{x}$ is continuous and differentiable within [0,1], but its derivative $\frac{\cos x}{x} - \frac{\sin x}{x^2}$ is not differentiable).

Thus, letting D denote the operation of differentiation, we have

$$D: \quad M \rightarrow C[0,1]. \tag{1}$$

Now consider the integration operator. A sufficient condition for f to be integrable is that f be continuous. Hence, the integration operator is defined for all continuous functions over C[0,1], i.e., V = C[0,1]. If f is continuous then

$$I f(x) = \int_0^x f(t)\,dt$$

is also continuous, i.e., W = C[0,1]. Thus,

$$I: \quad C[0,1] \rightarrow C[0,1]. \tag{2}$$

Next, we must show that the functions defined by (1) and (2) are linear transformations.

From the rules of differentiation,

$$\frac{d}{dx}(f + g)(x) = D(f + g)(x) = (f + g)' = f'(x) + g'(x)$$

$$= Df(x) + Dg(x).$$

$$\frac{d}{dx}(cf(x)) = D(cf(x)) = (cf(x))' = cf'(x) = cDf(x).$$

For example, let $f(x) = x^2$; $g(x) = \sin x$. Then,

$$\frac{d}{dx}(x^2 + \sin x) = (2x + \cos x) = D(x^2) + D(\sin x).$$

Thus, D is a linear transformation from M to C[0,1].

Finally, by the rules of integration,

62

$$\int_0^x (f(t) + g(t))dt = \int_0^x f(t)dt + \int_0^x g(t)dt$$

and
$$\int_0^x cf(t)dt = c \int_0^x f(t)dt.$$

Thus, $(f + g)(x) = f(x) + g(x)$

$$(cf(x)) = c\ f(x).$$

For example,

$$I(\sin x + x^2) = \int_0^x \sin t + t^2 dt = -\cos t \Big|_0^x + \frac{t^3}{3}\Big|_0^x$$

$$= 1 - \cos x + \frac{x^3}{3} = I(\sin x) + I(x^2).$$

Thus, I is a linear transformation from the space C[0,1] to C[0,1].

● PROBLEM 2-4

Let the transformation $L: R^3 \to R^3$ be defined by
$$L([x,y,z]) = [x,y] \qquad (1)$$
Show that L is a linear transformation and describe its effect.

Solution: Let $X = [x_1,y_1,z_1]$ and $Y = [x_2,y_2,z_2]$. L is a transformation from the vector space R^3 to the vector space R^2 defined over the field of real numbers.

L is linear if it satisfies the following two conditions:

i) $L(x+y) = L(x) + L(y)$

ii) $L(cx) = cL(x)$ for $c \in R$, the field.

Using (1),

$$L(X+Y) = L([x_1,y_1,z_1] + [x_2,y_2,z_2])$$

$$= L([x_1 + x_2,\ y_1 + y_2,\ z_1 + z_2])$$

$$= [x_1 + x_2,\ y_1 + y_2]$$

$$= [x_1,y_1] + [x_2,y_2] = L(X) + L(Y).$$

$$L(cX) = L([c(x_1,y_1,z_1)])$$

$$= L([cx_1,cy_1,cz_1])$$

$$= [cx_1,cy_1] = c[x_1,y_1] = cL(X) .$$

L is a projection. In n-dimensional Euclidean space (the space of n-tuples of real numbers), the ith projection P_i acts as follows:

$$P[x_1,\ldots,x_{i-1}, x_i, x_{i+1},\ldots,x_n] = [x_1,\ldots,x_{i-1}, x_{i+1},\ldots,x_n] \;,$$

i.e., the ith coordinate is deleted. The projection in the given problem has the following geometric interpretation:

The line PQ is perpendicular to the xy plane, so any v in x y z space (R^3) is sent by L to its "projective image" in the x-y plane (R^2).

• PROBLEM 2-5

Let S be the transformation defined by $S(x,y) = (2x+1,y-2)$. Show that S is non-linear.

Solution: A linear transformation L is a mapping from a vector space V to a vector space W (V and W being defined over a field K) such that, for $v_1,v_2 \in V$ and $k \in K$,

i) $\quad L(v_1 + v_2) = L(v_1) + L(v_2)$

ii) $\quad L(cv_1) = cL(v_1).$

The given transformation is a function from R^2 to R^2 over the field of real numbers. Let $v_1 = (x_1,x_2)$ and $v_2 = (x_2,y_2)$. Then,

$$S(v_1 + v_2) = S[(x_1,y_1) + (x_2,y_2)]$$

$$= S(x_1 + x_2, y_1 + y_2)$$

$$= (2(x_1 + x_2) + 1, y_1 + y_2 - 2)$$

$$= (2x_1 + 2x_2 + 1, y_1 + y_2 - 2). \qquad (1)$$

But, $S(v_1) + S(v_2) = S(x_1,y_1) + S(x_2,y_2)$

$$= (2x_1 + 1, y_1 - 2) + (2x_2 + 1, y_2 - 2)$$

$$= (2x_1 + 2x_2 + 2, y_1 + y_2 - 4). \qquad (2)$$

Since (1) $\neq$ (2), we conclude that $S(x,y)$ is not a linear transformation.

Note, however, that $(2x + 1, y - 2) = (2x,y) + (1,-2)$. Now, $S_1(x,y) = (2x,y)$ is a linear transformation since

$$S_1(v_1 + v_2) = S_1(x_1 + x_2, y_1 + y_2)$$

$$= [2(x_1 + x_2), y_1 + y_2]$$

$$= (2x_1 + y_1, 2x_2 + y_2)$$

$$= S_1(v_1) + S_1(v_2);$$

$$S_1(cv_1) = S_1(cx_1, cy_1) = (2cx_1, cy_1)$$

$$= c(2x_1, y_1) = cS_1(v_1).$$

The remaining part, $(1,-2)$, may be viewed as a constant transformation, i.e., $S_2(x,y) = (1,-2)$ for all (x,y). Then

$$S(x,y) = S_1(x,y) + S_2(x,y) \qquad (3)$$

where $S_1(x,y)$ is a linear transformation and $S_2(x,y)$ is a fixed vector. $S(x,y)$ is an example of an affine transformation, since an affine transformation is any transformation of the form $L(v_1) = T(v_1) + v_0$, where T is a linear transformation and V_0 is a constant vector.

● **PROBLEM** 2-6

Let T_θ denote the operation of rotating a vector in R^2, $V = (x,y)$, counterclockwise through an angle θ. Show that T_θ is a linear transformation.

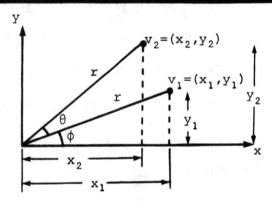

Solution: Let $v_2 = T(v_1)$. From the diagram we see that
A) $|T_\theta v| = |v|$. A rotation does not affect the magnitude of a vector.
B) $x_1 = r \cos \phi$ and $y_1 = r \sin \phi$.
C) $x_2 = r \cos(\theta + \phi)$ and
$y_1 = r \sin(\theta + \phi)$.

Now, using the sine and cosine sum formulae we get:

$$x_2 = r \cos \theta \cos \phi - r \sin \theta \sin \phi = x_1 \cos \theta - y_1 \sin \theta$$

$$y_2 = r \cos \theta \sin \phi + r \sin \theta \cos \phi = x_1 \sin \theta + y_1 \cos \theta.$$

So a rotation by an angle θ, T_θ is

65

$$x_2 = x_1 \cos \theta - y_1 \sin \theta$$

$$y_2 = x_1 \sin \theta + y_1 \cos \theta.$$

(1)

Thus, we know from (1) that

$$T_\theta(x_1, y_1) = (x_1 \cos \theta - y_1 \sin \theta, \ x_1 \sin \theta + y_1 \cos \theta).$$

To demonstrate that an operator T is a linear transformation, it is enough to show that

$$T(Cv_1 + v_2) = CT(v_1) + T(v_2), \quad v_1, v_2 \ \epsilon \ V \text{ and } C \text{ a scalar.}$$

In this case

$$T_\theta(Cv_1 + v_2) = T_\theta(C(x_1, x_2) + (y_1, y_2)) = T_\theta(Cx_1 + y_1, \ Cx_2 + y_2)$$

$$= ((Cx_1 + y_1)\cos \theta - (Cx_2 + y_2)\sin \theta,$$

$$(Cx_1 + y_1)\sin \theta + (Cx_2 + y_2)\cos \theta)$$

$$= (Cx_1 \cos \theta + y_1 \cos \theta - Cx_2 \sin \theta - y_2 \sin \theta,$$

$$Cx_1 \sin \theta + y_1 \sin \theta + Cx_2 \cos \theta + y_2 \cos \theta)$$

$$= (Cx_1 \cos \theta - Cx_2 \sin \theta + y_1 \cos \theta - y_2 \sin \theta,$$

$$Cx_1 \sin \theta + Cy_2 \cos \theta + y_1 \sin \theta + y_2 \cos \theta)$$

$$= (Cx_1 \cos \theta - Cx_2 \sin \theta, \ Cx_1 \sin \theta + Cx_2 \cos \theta)$$

$$+ (y_1 \cos \theta - y_2 \sin \theta, \ y_1 \sin \theta + y_2 \cos \theta)$$

$$= CT(v_1) + v_2.$$

Hence, T_θ is a linear transformation.

● **PROBLEM 2-7**

Let L: $R^3 \to R^3$ be defined by

$$L\left(\begin{bmatrix} x \\ y \\ z \end{bmatrix}\right) = \begin{bmatrix} x+1 \\ 2y \\ z \end{bmatrix}.$$

Show that L is not a linear transformation.

Solution: Let L be a transformation of elements from a vector space V to elements in a vector space W (V and W need not be different) over a field K. Then L is a linear transformation if, for $v_1, v_2 \in V$ and $k \in K$.

$$\text{i)} \quad L(v_1 + v_2) = L(v_1) + L(v_2) \tag{1}$$

$$\text{ii)} \quad L(kv_1) = kL(v_2) \tag{2}$$

The conditions (1) and (2) may be combined into one requirement as follows:

$$L(k_1 v_1 + k_2 v_2) = L(k_1 v_1) + L(k_2 v_2) \quad \text{(by (1))};$$

$$L(k_1 v_1) + L(k_2 v_2) = k_1 L(v_1) + k_2 L(v_2) \quad \text{(by (2))};$$

thus, $L(k_1 v_1 + k_2 v_2) = k_1 L(v_1) + k_2 L(v_2)$ is a defining property of a linear transformation.

Turning to the given function,

$$L(k_1 v_1 + k_2 v_2) = L\left(k_1 \begin{bmatrix} x_1 \\ y_1 \\ z_1 \end{bmatrix} + k_2 \begin{bmatrix} x_2 \\ y_2 \\ z_2 \end{bmatrix} \right)$$

$$= L\left(\begin{bmatrix} k_1 x_1 + k_2 x_2 \\ k_1 y_1 + k_2 y_2 \\ k_1 z_1 + k_2 z_2 \end{bmatrix} \right) = \begin{bmatrix} k_1 x_1 + k_2 x_2 + 1 \\ 2k_1 y_1 + 2k_2 y_2 \\ k_1 z_1 + k_2 z_2 \end{bmatrix} ; \tag{3}$$

however,

$$k_1 L(v_1) + k_2 L(v_2) = k_1 L\begin{bmatrix} x_1 \\ y_1 \\ z_1 \end{bmatrix} + k_2 L\begin{bmatrix} x_2 \\ y_2 \\ z_2 \end{bmatrix}$$

$$= k_1 \begin{bmatrix} x_1 + 1 \\ 2y_2 \\ z_1 \end{bmatrix} + k_2 \begin{bmatrix} x_2 + 1 \\ 2y_2 \\ z_2 \end{bmatrix}$$

$$= \begin{bmatrix} k_1 x_1 + k_1 + k_2 x_2 + k_2 \\ 2k_1 y_1 \qquad + 2k_2 y_2 \\ k_1 z_1 \qquad + k_2 z_2 \end{bmatrix} \tag{4}$$

Since (4) $\neq$ (3), L cannot be a linear transformation. L is, however, an affine transformation, i.e., it can be expressed as $L = L_1 + L_0$ where L_1 is a linear transformation and L_0 is a constant vector. Thus,

$$\begin{bmatrix} x + 1 \\ 2y \\ z \end{bmatrix} = \begin{bmatrix} x \\ 2y \\ z \end{bmatrix} + \begin{bmatrix} 1 \\ 0 \\ 0 \end{bmatrix}$$

and $\quad L\begin{bmatrix} x \\ y \\ z \end{bmatrix} = L_1 + \begin{bmatrix} 1 \\ 0 \\ 0 \end{bmatrix}$ where L_1 is the linear transformation

$$\begin{bmatrix} x \\ 2y \\ z \end{bmatrix}.$$

● **PROBLEM** 2-8

A linear transformation, $T: \quad V \rightarrow W$, is a function defined on a vector space V over a field K that satisfies

     i) $T(v_1 + v_2) = T(v_1) + T(v_2)$

     ii) $T(\alpha v_1) = \alpha T(v_1)$

for $v_1, v_2 \in V$ and $\alpha \in K$.

     Give examples of some non-linear functions by showing that they fail to satisfy either (i) or (ii).

<u>Solution</u>: Let f: $R \rightarrow R$ be given by $f(x) = e^x$. Now,

$$f(x_1 + x_2) = e^{x_1 + x_2} \neq f(x_1) + f(x_2) = e^{x_1} + e^{x_2}$$

In addition, $f(\alpha x_1) = e^{\alpha x_1} \neq \alpha f(x_1) = e^{x_1}$ unless $\alpha = 1$. Thus, in general, $f(x) = e^x$ is a non-linear function.

     Let f: $R \rightarrow R$ be defined by $f(x) = \sin x$. Then, since

$$f(x + y) = \sin(x + y) = \sin x \cos y + \cos x \sin y$$

$$\neq \sin x + \sin y$$

(unless $x = \frac{n\pi}{2}$, $y = \frac{m\pi}{2}$, $n, m = 0, 1, 2, \ldots$), sin x is a non-linear function on any open interval.

     Let f: $R \rightarrow R$ be an nth degree polynomial, i.e.,

$$f(x) = a_0 x^n + a_1 x^{n-1} + \ldots + a_{n-1} x + a_n.$$

Then,

$$f(x_1 + x_2) = a_0(x_1 + x_2)^n + a_1(x_1 + x_2)^{n-1}$$
$$+ \ldots + a_{n-1}(x_1 + x_2) + a_n.$$

Now, by the binomial theorem,

$$(a + b)^n = a^n + na^{n-1}b + \frac{n(n-1)}{2}a^{n-2}b^2$$

$$+ \ldots + \frac{n(n-1)\ldots n-(r-1)}{r!}a^{n-r}b^r + \ldots + b^n$$

$$> a^n + b^n$$

unless a or b = 0.

Hence, $f(x_1 + x_2) \neq f(x_1) + f(x_2)$, and, therefore, any polynomial of degree greater than one is non-linear.

● **PROBLEM** 2-9

Find the projection of the vector X along the vector Y for the points X = (2,3) and Y = (5,-1).

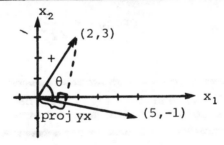

Solution: Let X, Y be two vectors in the plane. The projection of the vector X along the vector Y is obtained by drawing a perpendicular from the point X to the line determined by Y. The vector from the origin to the point of perpendicularity is $\text{proj}_Y X$. It can be seen from the figure that $\text{proj}_Y X = \alpha Y$ where $\alpha < 1$. To find an algebraic expression for $\text{proj}_Y X$ we reason as follows:

From the figure, the length of $\text{proj}_Y X$ is

$$|\text{proj}_Y X| = |X||\cos \theta|.$$

We know that $\cos \theta = \frac{X \cdot Y}{|X||Y|}$. Hence,

$$|\text{proj}_Y X| = |X|\frac{|X \cdot Y|}{|X||Y|} = \frac{|X \cdot Y|}{|Y|},$$

and, therefore,

$$|X \cdot Y| = |proj_Y X||Y|. \tag{1}$$

However, since $|proj_Y X| = |\alpha Y| = |\alpha||Y|$,

$$|\alpha| = \frac{|proj_Y X|}{|Y|}. \tag{2}$$

From (1) and (2), $|proj_Y X| = \frac{|X \cdot Y|}{|Y|} = |\alpha||Y|$, so $|\alpha| = \frac{|X \cdot Y|}{|Y|^2}$. Since $|Y|^2 > 0$ and $X \cdot Y$ and α always have the same sign,

$$\alpha = \frac{X \cdot Y}{|Y|^2} \qquad \text{and, hence}$$

$$proj_Y X = \alpha Y = \frac{X \cdot Y}{|Y|^2} Y . \tag{3}$$

For the given vectors,

$$X \cdot Y = (2,3) \cdot (5,-1) = 7.$$

$$|Y| = \sqrt{Y_1^2 + Y_2^2} = \sqrt{25 + 1} = \sqrt{26}.$$

Thus,
$$proj_Y X = \frac{7}{\sqrt{26}} Y.$$

● **PROBLEM** 2-10

Let $X = [1,0,5,-2]$ and $Y = [3,5,7,1]$ be points in R^4. Find $proj_Y X$ and the cosine of the angle between X and Y.

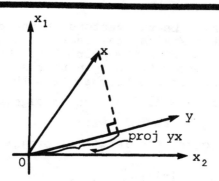

Solution: The projection of X along Y, denoted by $proj_Y X$, is defined as

$$proj_Y X = \frac{X \cdot Y}{|Y|^2} Y. \tag{1}$$

In R^2, the projection can be depicted pictorially. Thus, To find $\cos \theta$, the cosine of the angle between X and Y in R^4, we note that

$$\cos \theta = \frac{X \cdot Y}{|X||Y|}. \tag{2}$$

We are given the points X and Y. Hence, we can find

$$|X| = \left(\sum_{i=1}^{4} x_i^2\right)^{1/2} \quad \text{and} \quad |Y| = \left(\sum_{i=1}^{4} y_i^2\right)^{1/2} \quad \text{and} \quad X \cdot Y.$$

Recall that for two n-tuples in R^n,

$$X = (x_1, x_2, \ldots, x_n) \quad \text{and} \quad Y = (y_1, y_2, \ldots, y_n),$$

$$X \cdot Y = \sum_{i=1}^{n} x_i y_i.$$

Using the given data,

$$X \cdot Y = [1,0,5,-2] \cdot [3,5,7,1] = 36.$$

$$|X| = \sqrt{x_1^2 + x_2^2 + x_3^2 + x_4^2} = \sqrt{1 + 0 + 25 + 4} = \sqrt{30}.$$

$$|Y| = \sqrt{y_1^2 + y_2^2 + y_3^2 + y_4^2} = \sqrt{9 + 25 + 49 + 1} = \sqrt{84}.$$

Thus, from (1),

$$\text{proj}_Y X = \frac{X \cdot Y}{|Y|^2} Y = \frac{36}{84} Y = \frac{3}{7} Y.$$

From (2),

$$\cos \theta = \frac{X \cdot Y}{|X||Y|} = \frac{36}{\sqrt{30}\sqrt{84}} = \frac{6}{\sqrt{70}}.$$

● **PROBLEM** 2-11

Give examples of the following types of linear operators:

    a) two commutative operators
    b) two non-commutative operators.

Solution: Let V be a vector space over a field K. A linear operator T is a function T: $V \rightarrow V$ such that

    i) $T(v_1 + v_2) = T(v_1) + T(v_2)$

    ii) $T(\alpha v_1) = \alpha T(v_1)$

for $v_1, v_2 \in V$ and $\alpha \in K$.

    Let S: $V \rightarrow V$ and T: $V \rightarrow V$ be functions from V into V. Then, for $v \in V$, $T(v)$, $S(v) \in V$, and we can operate on the results, i.e., $S(T(v))$ and $T(S(v))$ are defined.

a)   Let S: V → V and T: V → V be two operators defined on V.   S and T are commutative operators if, for
v ε V,

$$S(T(v)) = T(S(v)). \qquad (1)$$

As an example of commutative operators, let $V = R^2$ and
$K = R$. Let S: $R^2 → R^2$ be a counterclockwise rotation in
R^2 through the angle θ, and let T: $R^2 → R^2$ be a counterclockwise rotation through the angle φ. Now, for a given
vector v ε R^2, S(T(v)) is a counterclockwise rotation of v
from its original position, first through an angle θ and
then through an angle φ. Thus, S(T(v)) is a counterclockwise rotation through φ + θ.
      On the other hand, T(S(v)) is a counterclockwise rota-
tion of v, first through an angle of φ and then through an
angle θ, i.e., the total rotation is θ + φ. Hence,
S(T(v)) = T(S(v)), i.e., S and T commute.

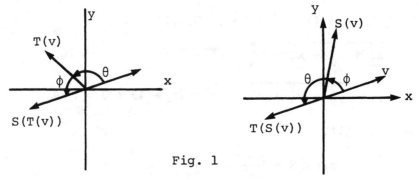

Fig. 1

b)   Let $V = R^2$ and $K = R$.   Let S be a counterclockwise
rotation in R^2 through the angle π/4, and let T be a reflection in the y-axis.   Thus, if v = (x,y) ε V, T(x,y) = (-x,y).

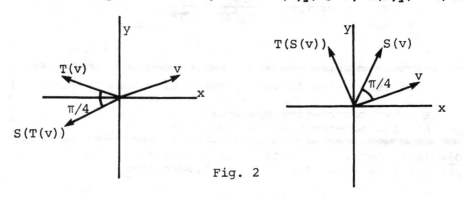

Fig. 2

We see from Fig. 2 above that S(T(v)) ≠ T(S(v)), i.e., S and
T are not commutative.

● PROBLEM 2-12

Give examples of nilpotent operators.

Solution: Let T be a linear transformation from the vector space V to the vector space W. If W = V, then T is known as a linear operator. Since T is a mapping from V into itself, we can operate on the result using T again. That is, $T[T(x)]$ for $x \in V$ is defined. The result is again in V, and we can find $T[T[T(x)]]$ of $T^3(x)$. In general, T^n is defined.

A linear operator T is said to be nilpotent of degree n if T^n equals the zero vector and $0 < m < n$ implies $T^m \neq 0$.

Let $T: V \to W$ be given by $T(v) = 0_W$ where 0 is the zero vector of W. Then T is the zero operator and is nilpotent of degree 1.

Let $T: R^2 \to R^2$ be given by

$$T(x,y) = T\left(x\begin{pmatrix}1\\0\end{pmatrix} + y\begin{pmatrix}0\\1\end{pmatrix}\right) \tag{1}$$

Here $\{(1,0)^T, (0,1)^T\}$ is the standard basis for R^2. Rewriting (1),

$$T(x,y) = T\left(x\begin{pmatrix}1\\0\end{pmatrix} + T\ y\begin{pmatrix}0\\1\end{pmatrix}\right)$$

$$= xT(1,0)^T + yT(0,1)^T. \tag{)}$$

Let $T(1,0)^T = (0,1)^T$ and $T(0,1)^T = (0,0)$. Then the effect of T on an arbitrary vector (x,y) is, from (2),

$$T(x,y) = x(0,1)\ + y(0,0) = x(0,1).$$

Then, $\qquad T^2(x,y) = T(x(0,1)\ )$

$$= xT(0,1)\ = (0,0)$$

and, hence, T is nilpotent of degree two.

THE KERNEL & RANGE OF LINEAR TRANSFORMATIONS

● PROBLEM 2-13

Let $T: R^4 \to R^3$ be a linear transformation defined by

$$T(x,y,z,t) = (x-y+z+t,\ x+2z-t,\ x+y+3z-3t).$$

Find a basis and the dimension of the

    i)   image of T   ii)   kernel of T.

Solution: Let $L: V \longrightarrow W$ be a linear transformation. The image of L is the set of all vectors in W that are the result of L operating on some element in V. The kernel of L is the set of all vectors in V that are mapped to the zero in W.

i) We know that if e_i $i = 1,\ldots,n$ spans R^n and A is any linear operator, then $A(e_i)$ $i = 1,\ldots,n$ spands the image space of A (Im A). So we see first what effect T has on a set of basis vectors in R^4. Choosing the standard basis

$$\left\{(1,0,0,0),\ (0,1,0,0),\ (0,0,1,0),\ (0,0,0,1)\right\},$$

$$T(1,0,0,0) = (1,1,1);\ T(0,1,0,0) = (-1,0,1);$$

$$T(0,0,1,0) = (1,2,3);\ T(0,0,0,1) = (1,-1,-3).$$

To form the matrix representing the linear transformation, we take as rows the vectors that span Im T.

$$\begin{bmatrix} 1 & 1 & 1 \\ -1 & 0 & 1 \\ 1 & 2 & 3 \\ 1 & -1 & -3 \end{bmatrix}.$$

We use elementary row operations on (1) to row reduce it to echelon form:

$$\begin{bmatrix} 1 & 1 & 1 \\ 0 & 1 & 2 \\ 0 & 1 & 2 \\ 0 & -2 & -4 \end{bmatrix} \rightarrow \begin{bmatrix} 1 & 1 & 1 \\ 0 & 1 & 2 \\ 0 & 0 & 0 \\ 0 & 0 & 0 \end{bmatrix}.$$

Thus, there are two independent vectors in the spanning set, and the dimension of Im T is equal to two. A basis for Im T is $((1,1,2),\ (0,1,2))$.

To find a basis for ker T, we argue as follows: ker T, by definition, contains the vectors (x,y,z,t) such that $T(x,y,z,t) = (0,0,0)$. Hence,

$$x - y + z + t = 0$$

$$x \quad + 2z - t = 0 \qquad\qquad (2)$$

$$x + y + 3z - 3t = 0$$

Write the matrix of coefficients of (2) and reduce to echelon form (by row operations).

$$\begin{matrix} R_1: \\ R_2: \\ R_3: \end{matrix} \begin{pmatrix} 1 & -1 & 1 & 1 \\ 1 & 0 & 2 & -1 \\ 1 & 1 & 3 & -3 \end{pmatrix}$$

$$R_1 = R_1': \quad \begin{pmatrix} 1 & -1 & 1 & 1 \\ 0 & 1 & 1 & -2 \\ 0 & 2 & 2 & -4 \end{pmatrix}$$

$R_2 - R_1 = R_2':$

$R_3 - R_1 = R_3':$

$$\begin{aligned} R_2' + R_1' &= R_1'' \\ R_2' &= R_2'' \\ (-2)R_2' + R_3' &= R_3'': \end{aligned} \quad \begin{pmatrix} 1 & 0 & 2 & -1 \\ 0 & 1 & 1 & -2 \\ 0 & 0 & 0 & 0 \end{pmatrix} \qquad (3)$$

(3) has two independent rows so dim(ker t) = 2. In this case, dim(ker T) is equal to dim(Im T). From (3), a basis for ker T is $\{(1,-1,1,1), (0,1,1,-2)\}$.

● **PROBLEM** 2-14

Let T: $R^4 \rightarrow R^4$ be the linear transformation defined by

$$T \begin{bmatrix} x_1 \\ x_2 \\ x_3 \\ x_4 \end{bmatrix} = \begin{bmatrix} 0 \\ x_1 + x_2 \\ x_4 \\ 0 \end{bmatrix}$$

Find the dimensions of Im(T) and ker(T).

Solution: Let T: $V \rightarrow W$ be a linear transformation. Then, according to an important theorem in linear algebra,

$$\dim V = \dim \text{Im}(T) + \dim \ker(T). \qquad (1)$$

The dimension of V for the given transformation is $\dim R^4 = 4$. Hence, we must find either dim(Im (T)) or dim(ker (T)), since then we can solve (1) to find the other. Now, a linear transformation is fully described by its effects on the basis vectors. Taking the standard basis of R^4,

$$T \begin{bmatrix} 1 \\ 0 \\ 0 \\ 0 \end{bmatrix} = \begin{bmatrix} 0 \\ 1 \\ 0 \\ 0 \end{bmatrix} v_1; \quad T \begin{bmatrix} 0 \\ 1 \\ 0 \\ 0 \end{bmatrix} = \begin{bmatrix} 0 \\ 1 \\ 0 \\ 0 \end{bmatrix} v_2; \quad T \begin{bmatrix} 0 \\ 0 \\ 1 \\ 0 \end{bmatrix} = \begin{bmatrix} 0 \\ 0 \\ 0 \\ 0 \end{bmatrix} = v_3; \quad T \begin{bmatrix} 0 \\ 0 \\ 0 \\ 1 \end{bmatrix} = \begin{bmatrix} 0 \\ 0 \\ 1 \\ 0 \end{bmatrix} = v_4$$

From these vectors, we can construct a set that spans the image of T. We know $\{v_i\}$ i = 1,2,3,4 spans Im T so dim(Im T) is the number of independent v_i's. We eliminate

v_3 since the 0-vector is dependent on any vector and v_2 since $v_1 = v_2$ (they are dependent).

Thus,

$$S = \left\{ \begin{bmatrix} 0 \\ 1 \\ 0 \\ 0 \end{bmatrix} \begin{bmatrix} 0 \\ 0 \\ 1 \\ 0 \end{bmatrix} \right\}$$

spans T. Furthermore, the two vectors in S are linearly dependent, i.e. they form a basis. Thus, Im(t) has dimension equal to two. From (1) dim (ker(T)) equals two too and we have

$$\dim \operatorname{Im} (t) = 2$$

$$\dim \ker (t) = 2.$$

● **PROBLEM** 2-15

Let P: $R^3 \rightarrow R^3$ be the mapping $P(x,y,z) = (x,y,0)$. Find the image and kernel of P.

Solution: Let T: $U \rightarrow V$ be a linear transformation from a vector space U to a vector space V over the field K. Then the image of T is the set of all elelments in V that correspond to at least one element in U, i.e.,

$$\operatorname{Im} T = \{v \; \varepsilon \; V: \quad T(u) = v, \text{ for some } u \; \varepsilon \; U\}.$$

The kernel of T is the set of all elements in U that are sent to the zero vector in V. That is,
ker T = $\{u \; \varepsilon \; U: \quad T(u) = 0$, where 0 is the 0-vector in V$\}$.
In the given problem, the image of P consists of all vectors in R^3 with third coordinate equal to zero, i.e., all points (x,y,a) where $a \neq 0$ are not in the image of P. Note that Im P is a subspace of R^3 of dimension 2.

The kernel of P contains all vectors of the form $(0,0,z)$ where z is a real number. That is,

$$\ker P = \{(x,y,z) \; \varepsilon \; R^3: \quad x = 0, \; y = 0, \; z = Z, \; Z \; \varepsilon \; R^1\}$$

since $P(0,0,Z) = (0,0,0)$ $Z \; \varepsilon \; R$. We see that ker P is a subspace of U of dimension 1. A fundamental theorem in linear algebra asserts that

$$\dim \operatorname{Im} P + \dim \ker P = \dim U$$

for any linear map of U into V. In this case, $d(\operatorname{Im} P) = 2$ and $\dim(\ker P) = 1$. Thus, the theorem is verified.
$1 + 2 = 3 = \dim(R^3)$.

Let V and W be vector spaces over the field K, and T: V → W a linear transformation. Find the kernel and range of T if T takes the following forms:

    1) T is the scalar operator αI where I is the identity operator and $\alpha \neq 0$.
    2) T is a rotation of elements in R^2.

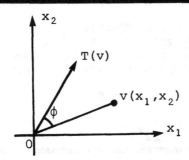

Solution: First, define the kernel and range of a linear transformation; then proceed to answer 1) and 2).

    Let L: V → W be a linear transformation. The kernel of L is the set of all vectors in V that are sent by L to the zero vector in W. Thus,

$$\text{ker } L = \{v \varepsilon V: \quad L(v) = 0_W; \ 0_W \varepsilon W\}. \tag{1}$$

    The range of L is the set of all vectors in W that correspond under L to some vector in V.

$$\text{Ran } L = \{w \varepsilon W: \quad L(v) = w; \ v \varepsilon V\}. \tag{2}$$

    1) As a specific example, let K = R and T = 6I. Thus, T: V → V assigns to every $v_1 \varepsilon V, v_2 = 6v_1$. The only vector that satisfies (1) is the zero vector in V. Hence,

$$\text{ker } T = \{0_V\} \tag{3}$$

Replacing T = 6I by T = I does not affect the set in (3).
    Since every v ε V is the image under T = αI of some v in V, the range of T is the space V.
    If we define the dimension of a vector space as the number of vectors in its basis, the following fact emerges from 1):

    2)   dim V = dim ker T + dim Ran T.

    Let T be a rotation of $v(x_1, x_2)$ through an angle ϕ as shown in Fig. 1. Then only the point (0,0) is sent by T to the point (0,0). Hence, ker T = $\{0_V\}$. We can also see from the figure that every point in the plane could be the result of applying T to some point in the plane. The range of T is R^2. Note once again that

$$\dim V = \dim \ker T + \dim \text{Ran } T. \qquad (4)$$

This is a fundamental thorem of linear algebra which holds for any linear transformation defined on a finite dimensional vector space.

Find the null spaces of the following linear transformations from V to W.

    1)   The zero transformation
    2)   Any isomorphism T: $V \longrightarrow W$.
    3)   The linear transformation associated with the following general system of linear equations:

$$a_{11}x_1 + a_{12}x_2 = 0$$

$$a_{21}x_1 + a_{22}x_2 = 0.$$

Solution: Let L: $V \rightarrow W$ be a linear transformation where V and W are vector spaces defined over the field K. The null space of L is the set of all vectors in V that are sent to the zero vector in W. Thus,

$$N_0 L = \{v \ \varepsilon \ V: \quad L(v) = 0_W; \ 0_W \ \varepsilon \ W\}.$$

The null-space is a subspace of V since, if v_1 and $v_2 \ \varepsilon \ N_0 L$, then

$$L(v_1) = 0; \ L(v_2) = 0.$$

This implies

$$L(v_1 + v_2) = L(v_1) + L(v_2) = 0 + 0 = 0, \qquad \text{and}$$

$$L(\alpha v_1) = \alpha L(v_1) = \alpha \cdot 0 = 0.$$

    1)   The zero transformation assigns every vector in V to the zero vector in W. Thus, the null space of the zero trnasformation is the vector space V itself.
    2)   A linear transformation is an isomorphism if and only if it is 1 - 1 and onto. Since the zero vector in the domain is always sent to the zero vector in the range for any linear transformation, the null space of an isomorphism can consist of only the zero vector.
    3)   The system of linear equations corresponds to the linear transformation $A(x_1, x_2) \rightarrow (0,0)$. The solutions to the system provide the null space of the transformation. Recall that if $a_{11}a_{22} - a_{12}a_{21} \neq 0$, the system has only the unique trivial solution $x_1 = x_2 = 0$. If the equations are dependent, however, there will be an infinite number of solutions.

INVERTIBLE LINEAR TRANSFORMATIONS

Let T: $R^2 \to R^2$ be given by $T(x,y) = (y,2x-y)$. Show that T is nonsingular and find its inverse.

Solution: A transformation T: $V \to U$ is nonsingular if the only vector that maps into $0 \in U$ is $0 \in V$ (That is, ker T = {0}.) If a linear operator L is defined on a finite dimensional vector space, then non-singularity and invertibility are equivalent characteristics of L. Hence, if we can find the inverse of T, we will also have shown that T is non-singular. Let T: $V \to W$ be a linear operator. Then T is defined as

$$T = \{w \in W: \quad T(v) = w; \ v \in V\}, \tag{1}$$

and T^{-1} is defined as

$$T^{-1} = \{v \in V: \quad T^{-1}(w) = v; \ w \in W\}. \tag{2}$$

In the given problem, $V = W = R^2$. Let $T(x,y) = (s,t) \in R^2$. Then, from (1), the image of (x,y) under T is (s,t) and from (2), the image of (s,t) under T^{-1} is (x,y). Now,

$$T(x,y) = (y,2x-y) = (s,t) \tag{3}$$

From (3), setting coordinates equal to each other,

$$s = y \tag{4}$$
$$t = 2x-y$$

Solving (4) for y and s,

$$x = \frac{s + t}{2} ; \quad y = s.$$

Then, using (2),

$$T^{-1}(s,t) = (\frac{s + t}{2}, s) = (x,y). \tag{5}$$

Since T^{-1} is given by (5), we conclude that T is non-singular.

Let T be a linear operator on R^2 defined by

$$T(x_1,x_2) = (x_1 + x_2, x_1).$$

Show that T is non-singular.

Solution: A linear operator is a linear transformation from a vector space V into itself. T is defined as singu-

lar if some non-zero vector in V is mapped into the zero
vector. For example, let $T(x,y) = (6x + 8y, 3x + 4y)$. We
solve $T(x,y) = (0,0)$, $6x + 8y = 0$, $3x + 4y = 0$, which
is equivalent to $3x + 4y = 0$. Thus, $x = -3/4y$. Then,
all vectors of the form $x = t$, $y = -3/4t$ and t a scalar
are mapped by T into the zero vector. Thus, T is singular.
The above definition implies that T is non-singular if and
only if ker T contains only the zero vector.

Let us find ker T for the given transformation.

$$\text{ker } T = \{(x_1, x_2) \ \varepsilon \ V: \ T(x_1, x_2) = (0,0) \ \varepsilon \ V\}. \quad (1)$$

The set (1) is equal to the solution set of the simultaneous
system

$$x_1 + x_2 = 0$$

$$x_1 \qquad = 0$$

which has $(0,0)$ as the unique member.

● **PROBLEM** 2-20

Find the inverses of the following transformations using
the matrices associated with the transformations:

1) T is a counterclockwise rotation in R^2 through
    the angle θ .
2) T is reflection in the y-axis in R^2.

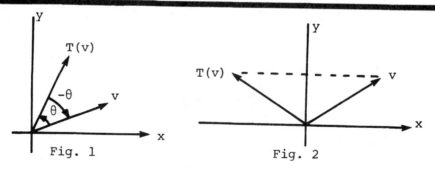

Fig. 1   Fig. 2

Solution: A linear transformation in R^2 is described by
its effects on the basis vectors $B = \big((1,0), (0,1)\big)$.
    Letting $v_1 = (1,0)$, $T(1,0) = (\cos \theta, \sin \theta)$. If
$v_2 = (0,1)$, $T(0,1) = (\cos \theta + \pi/2, \sin \theta + \pi/2)$. Since
$\cos(\theta + \pi/2) = -\sin\theta$ and $\sin(\theta + \pi/2) = \cos \theta$,

$$T(0,1) = (-\sin \theta, \cos \theta).$$

Thus,

$$T(x,y) = \begin{bmatrix} \cos \theta & -\sin \theta \\ \sin \theta & \cos \theta \end{bmatrix} \begin{matrix} x \\ y \end{matrix}. \quad (1)$$

From Fig. 1, the inverse transformation is a clockwise rotation through θ or a counterclockwise rotation through -θ. Hence, its matrix, using (1) is

$$S(x,y) = \begin{bmatrix} \cos(-\theta) & -\sin(-\theta) \\ \sin(-\theta) & \cos(-\theta) \end{bmatrix} \begin{array}{c} x \\ y \end{array} \cdot \qquad (2)$$

To check that (2) is indeed the required inverse, we must show TS(x,y) = ST(x,y) = I. Since the cosine function is even, $\cos(-\theta) = \cos\theta$. Since the sine function is odd, $\sin(-\theta) = -\sin\theta$. Thus, the product of (1) and (2) is:

$$\begin{bmatrix} \cos\theta & -\sin\theta \\ \sin\theta & \cos\theta \end{bmatrix} \begin{bmatrix} \cos\theta & \sin\theta \\ -\sin\theta & \cos\theta \end{bmatrix} \begin{bmatrix} x \\ y \end{bmatrix}$$

$$= \begin{bmatrix} \cos^2\theta + \sin^2\theta & \cos\theta\sin\theta - \cos\theta\sin\theta \\ \sin\cos\theta - \cos\theta\sin\theta & \sin^2\theta + \cos^2\theta \end{bmatrix} \begin{bmatrix} x \\ y \end{bmatrix}$$

$$= \begin{bmatrix} 1 & 0 \\ 0 & 1 \end{bmatrix} \begin{bmatrix} x \\ y \end{bmatrix} = \begin{bmatrix} x \\ y \end{bmatrix} \cdot$$

Similarly, TS(x,y) = I.

From Fig. 2, we see that (x,y) is sent to (-x,y). We can find the matrix associated with this transfromation by describing its effects on the basis vectors B = {(1,0), (0,1)}. Now, T(1,0) = (-1,0) and T(0,1) = (0,1). Hence,

$$T(x,y) = \begin{bmatrix} -1 & 0 \\ 0 & 1 \end{bmatrix} \begin{bmatrix} x \\ y \end{bmatrix} \cdot \qquad (3)$$

Looking again at Fig. 2, we suspect that the inverse of T will be given by another reflection in the y-axis, i.e., the transformation is its own inverse. If so, the matrix in (3) is its own inverse. To check,

$$\begin{bmatrix} -1 & 0 \\ 0 & 1 \end{bmatrix} \begin{bmatrix} -1 & 0 \\ 0 & 1 \end{bmatrix} \begin{bmatrix} x \\ y \end{bmatrix} = \begin{bmatrix} 1 & 0 \\ 0 & 1 \end{bmatrix} \begin{bmatrix} x \\ y \end{bmatrix} = \begin{bmatrix} x \\ y \end{bmatrix} \cdot$$

● PROBLEM 2-21

Find the inverse transformations of the following linear transformations:

 i)   T(x,y) = (2x+y, -x+3y)

 ii) T(x,y,z) = (x+y+z, x-y+z, -x+y+z).

Solution: Let V be a vector space over K. T: V → V is invertible if there is a linear transformation S from V into V such that ST = TS = I, the identity operator on V. A linear transformation is completely specified by its effect on the

basis vectors of V. We use this fact to find the inverses
of i) and ii) above.

    i) $T(x,y) = (2x+y, -x+3y)$.

The standard basis for R^2 is $B = \left\{(1,0), (0,1)\right\}$. Then,

    $T(1,0) = (2,-1) = 2(1,0) - 1(0,1)$

    $T(0,1) = (1,3) = 1(1,0) + 3(0,1)$.

The matrix of the transformation is

$$\begin{bmatrix} 2 & 1 \\ -1 & 3 \end{bmatrix}\begin{bmatrix} x \\ y \end{bmatrix} = \begin{bmatrix} 2x+y \\ -x+3y \end{bmatrix}. \tag{1}$$

To find the inverse transformation, we need only find the
inverse matrix of that given in (1). For a matrix A,

$$A^{-1} = \frac{1}{\det A}[\text{adj. A}]$$

where the adjoint of A is the transpose of the matrix of
cofactors of A.
    If the matrix

$$A = \begin{bmatrix} 2 & 1 \\ -1 & 3 \end{bmatrix},$$

the matrix of cofactors is

$$\begin{bmatrix} 3 & +1 \\ -1 & 2 \end{bmatrix}, \text{ and adj. } A = \begin{bmatrix} 3 & -1 \\ 1 & 2 \end{bmatrix}.$$

Furthermore, det A = 6 + 1 = 7. Hence,

$$A^{-1} = \begin{bmatrix} 3/7 & -1/7 \\ 1/7 & 2/7 \end{bmatrix}$$

and $S(x,y) = [3/7x - 1/7y, 1/7x + 2/7y]$. To show that $S = T^{-1}$,
let a,b ε V. Then, $T(a,b) = (2a+b, -a+3b)$, so $S(T(a,b)) = $
$S(2a+b, -a+3b) = (\frac{3}{7}(2a+b) - \frac{1}{7}(-a+3b), \frac{1}{7}(2a+b) + \frac{2}{7}(-a+3b)) = $
(a,b). Thus, $SoT(a,b) = (a,b)$ as required.
    Similarly, it may be shown that $TS(a,b) = (a,b)$.

    ii) The standard basis for R^3 is

$$B = \{(1,0,0), (0,1,0), (0,0,1)\}$$

$$T(1,0,0) = (1,1,-1)$$

$$T(0,1,0) = (1,-1,1)$$

$$T(0,0,1) = (1,1,1).$$

Thus, the matrix associated with T under the standard basis is

$$A \quad \begin{bmatrix} 1 & 1 & 1 \\ 1 & -1 & 1 \\ -1 & 1 & 1 \end{bmatrix}.$$

To find A^{-1} we can use the Gaussian method for computing inverses. Form the augmented matrix [A: I] and use elementary row operations to convert it to [I: A^{-1}].where I is the identity matrix defined as

$$I = \begin{bmatrix} 1 & 0 & 0 \\ 0 & 1 & 0 \\ 0 & 0 & 1 \end{bmatrix}$$

$$[A: 1] = \begin{bmatrix} 1 & 1 & 1 & | & 1 & 0 & 0 \\ 1 & -1 & 1 & | & 0 & 1 & 0 \\ -1 & 1 & 1 & | & 0 & 0 & 1 \end{bmatrix}.$$

Now perform the following elementary row operations on the above matrix. Add -1 times the first row to the second row, and add the first row to the third row.

$$\begin{bmatrix} 1 & 1 & 1 & | & 1 & 0 & 0 \\ 0 & -2 & 0 & | & -1 & 1 & 0 \\ 0 & 2 & 2 & | & 1 & 0 & 1 \end{bmatrix}.$$

Add the second row to the third row.

$$\begin{bmatrix} 1 & 1 & 1 & | & 1 & 0 & 0 \\ 0 & -2 & 0 & | & -1 & 1 & 0 \\ 0 & 0 & 2 & | & 0 & 1 & 1 \end{bmatrix}.$$

Divide the second row by -2 and the third row by 2.

$$\begin{bmatrix} 1 & 1 & 1 & | & 1 & 0 & 0 \\ 0 & 1 & 0 & | & 1/2 & -1/2 & 0 \\ 0 & 0 & 1 & | & 0 & 1/2 & 1/2 \end{bmatrix}.$$

Add -1 times the second row to the first row.

$$\begin{bmatrix} 1 & 0 & 1 & | & 1/2 & 1/2 & 0 \\ 0 & 1 & 0 & | & 1/2 & -1/2 & 0 \\ 0 & 0 & 1 & | & 0 & 1/2 & 1/2 \end{bmatrix}.$$

Now add -1 times the third row to the first row to obtain
the matrix

$$\begin{bmatrix} 1 & 0 & 0 & \vdots & 1/2 & 0 & -1/2 \\ 0 & 1 & 0 & \vdots & 1/2 & -1/2 & 0 \\ 0 & 0 & 1 & \vdots & 0 & 1/2 & 1/2 \end{bmatrix}$$

Hence,

$$A^{-1} = \begin{bmatrix} 1/2 & 0 & -1/2 \\ 1/2 & -1/2 & 0 \\ 0 & 1/2 & 1/2 \end{bmatrix}, \text{ and}$$

$$S(x,y,z) = A^{-1} \begin{bmatrix} x \\ y \\ z \end{bmatrix} = [1/2x - 1/2z, \ 1/2x - 1/2y, \ 1/2y + 1/2z].$$

$S(x,y,z)$ is the required inverse.

● **PROBLEM** 2-22

A linear transformation $T: \ V \to V$ is said to be invertible
if there exists another linear transformation $T^{-1}: \ V \to V$
such that $T^{-1}T(v) = TT^{-1}(v) = v$ for all $v \ \varepsilon \ V$.
    Necessary and sufficient conditions for T to be inver-
tible are that T be 1 - 1 and onto. Show by example that
the following are possible:

    i)   T is 1 - 1 but not invertible
    ii) T is onto but not invertible.

Solution: It is necessary to use some theorems on linear
transformations to show that if V is a finite dimensional
vector space, then i) and ii) cannot occur. We then investigate linear transformations defined on infinite dimensional vector spaces. Let T: $\ V \to W$ be a linear transformation. The kernel of T is the set of all vectors in V that
are sent to the zero vector in W. In set-theoretic notation

$$\text{ker } T = \{v \ \varepsilon \ V: \ \ T(v) = 0_W; \ 0_W \ \varepsilon \ W\}.$$

The range of T is the set of all vectors in W that are the
image of some v in V. Thus,

$$R(T) = \left(w \ \varepsilon \ W: \ \ T(v) = w; \ v \ \varepsilon \ V\right).$$

Now, T is 1 - 1 if and only if ker T contains only the zero
vector, and T is onto if and only if the range of T is W.
Thus, dim ker T = 0 and dim R(T) = dim W. According to another theorem, in a finite dimensional space dim V = dim ker T
+ dim R(T). Thus, if T is 1-1, then

$$\dim V = 0 + \dim R(T),$$

i.e., T is onto.

Similarly, if T is onto, $\dim R(T) = \dim V$ and, consequently, $\dim \ker T = 0$, i.e., T is 1 - 1.

Thus, if V and W are finite dimensional, i) and ii) cannot occur.

Let $V = \{(x_1, x_2, \ldots) : x_i \in R, i = 1, 2, \ldots\}$. As a result V is the set of all sequences of real numbers. Defining addition of sequences and scalar multiplication as in R^n, V becomes an infinite dimensional vector space over the reals.

Define T: $V \to V$ and S: $V \to V$ by

$$T(x_1, x_2, x_3 \ldots) = (0, x_1, x_2, x_3 \ldots)$$

$$S(x_1, x_2, x_3, \ldots) = (x_2, x_3, \ldots)$$

for all $v \in V$. Both T and S are linear transformations since

$$T(v_1 + v_2) = T((x_1, x_2, \ldots) + (y_1, y_2, \ldots))$$

$$= [(0, x_1, x_2 \ldots) + (0, y_1, y_2, \ldots)]$$

$$= (0, x_1 + y_1, x_2 + y_2, \ldots).$$

On the other hand, $T(v_1) = (0, x_1, x_2, \ldots)$, $T(v_2) = (0, y_1, y_2, \ldots)$ and, hence,

$$T(v_1) + T(v_2) = (0, x_1 + y_1, x_2 + y_2, \ldots) = T(v_1 + v_2).$$

Also, $T(\alpha v_1) = T(x_1, x_2, \ldots) = (0, x_1, x_2) = \alpha T(v_1).$

Similarly, it can be shown that S is a linear transformation.

Further, T is 1 - 1 since, for every $x = (x_1, x_2, \ldots) \neq y$, then $T(x) = (0, x_1, x_2, \ldots) \neq T(y)$. But T is not onto since the vector $(1, 0, 0, \ldots)$, although in V, can never be the image of a vector in V under T. Thus, T is 1 - 1 but not onto.

S is onto since $S(0, x_1, x_2, \ldots) = (x_1, x_2, \ldots)$ for any element $(x_1, x_2, \ldots)$ of V. But S is not 1 - 1 since $S(1, 0, 0, \ldots) = (0, 0, \ldots) = 0$, ie,, the kernel of S contains other elements besides the zero vector. Thus, S is onto but not 1 - 1.

● **PROBLEM 2-23**

Let T: $R^2 \to R^2$ be the linear transformation defined by
$$T \begin{pmatrix} x_1 \\ x_2 \end{pmatrix} = (x_1 x_2).$$
Find the inverse of T.

Solution: A linear transformation L: V → W is said to be invertible if there exists a linear transformation L^{-1}: W → V such that $L^{-1}L(c) = LL^{-1}(v) = v$ for v ε V.

In the given problem, T transposes the set of column vectors in R^2. Hence, it is an isomorphism, i.e., T is 1 - 1 and onto.

If we transpose the row vector $(x_1 x_2)$, we obtain the orginal column vector $\begin{pmatrix} x_1 \\ x_2 \end{pmatrix}$. Thus, $T^{-1}(x_1 x_2) = \begin{pmatrix} x_1 \\ x_2 \end{pmatrix}$ is the required inverse.

● **PROBLEM** 2-24

The vectors $a_1 = (1,2)$ and $a_2 = (3,4)$ form a basis for R^2. Let T: $R^2 → R^3$ be given by $Ta_1 = (3,2,1)$, $Ta_2 = (6,5,4)$. Find $T(1,0)$ in terms of the basis $\{a_1, a_2\}$.

Solution: Since $\left\{a_1, a_2\right\}$ forms a basis, there exist scalars c_1, c_2 such that

$$(1,0) = c_1 a_1 + c_2 a_2. \tag{1}$$

Then, $T(1,0) = T(c_1 a_1 + c_2 a_2) = c_1 T(a_1) + c_2 T(a_2)$ (2)

is the required expression.

To find c_1, c_2, we use (1):

$$(1,0) = c_1(1,2) + c_2(3,4).$$

$$(1,0) = (c_1, 2c_1) + (3c_2, 4c_2)$$

$$1 = c_1 + 3c_2 \tag{3}$$

$$0 = 2c_1 + 4c_2. \tag{4}$$

From (4), we find $c_1 = -2c_2$ and, thus, in (3)

$$-2c_2 + 3c_2 = 1 \qquad \text{or,}$$

$$c_2 = 1 ; \quad \text{hence,} \quad c_1 = -2.$$

As a result,

$$T(1,0) = T(-2a_1 + 1a_2)$$

$$= -2T(1,2) + 1T(3,4)$$

$$= -2(3,2,1) + 1(6,5,4)$$

$$= (-6,-4,-2) + (6,5,4)$$

$$= (0,1,2).$$

Let T: $R^2 \to R^3$ where the action of T on the basis vectors is given by T(1,0) = (1,1,0) and T(0,1) = (0,1,1). If A is the matrix associated with T, show that

$$A \begin{bmatrix} x_1 \\ x_2 \end{bmatrix} = \begin{bmatrix} y_1 \\ y_2 \end{bmatrix}$$

is equivalent to

$$[x_1 x_2]A^T = [y_1 y_2].$$

Solution: It is interesting to generalize the above problem and then obtain the required solution as a particular example. Thus, let L: $R^n \to R^m$ be a linear transformation. Then the matrix associated with the transformation A operates on a vector in R^n to give a vector in R^m. If

$$\begin{bmatrix} x_1 \\ x_2 \\ \cdot \\ \cdot \\ \cdot \\ x_n \end{bmatrix}$$

is a vector in R^n, then

$$\begin{bmatrix} a_{11} & a_{12} & \cdots & a_{1n} \\ a_{21} & \cdot & & \cdot \\ \cdot & \cdot & & \cdot \\ \cdot & \cdot & & \cdot \\ \cdot & \cdot & & \cdot \\ a_{m1} & \cdots & \cdots & a_{mn} \end{bmatrix} \begin{bmatrix} x_1 \\ \cdot \\ \cdot \\ \cdot \\ x_n \end{bmatrix} = \begin{bmatrix} y_1 \\ y_2 \\ \cdot \\ \cdot \\ y_m \end{bmatrix}. \qquad (1)$$

The same result as (1) can be obtained by transposing vectors and the matrix.

$$[x_1 x_2 \cdots x_n] \begin{bmatrix} a_{11} & a_{21} & \cdots & a_{m1} \\ \cdot & & & \\ \cdot & & & \\ \cdot & & & \\ a_{1n} & a_{2n} & \cdots & a_{mn} \end{bmatrix} = [y_1 y_2 \cdots y_m]. \qquad (2)$$

Now apply the above results to the given problem. The

rows of the matrix in (1) yield the coordinates of vectors in R^n while the columns are the coordinates of the transformed vectors in R^m. But A, the matrix associated with T, is the matrix with columns the results of T operating on a basis. Thus,

$$A = \begin{bmatrix} 1 & 0 \\ 1 & 1 \\ 0 & 1 \end{bmatrix}.$$

We have
$T(x,y) = T(x(1,0) + y(0,1)) = xT(1,0) + yT(0,1) = (x,x+y,y)$.
This is the same as

$$\begin{bmatrix} 1 & 0 \\ 1 & 1 \\ 0 & 1 \end{bmatrix} \begin{bmatrix} x \\ y \end{bmatrix} = \begin{bmatrix} x \\ x+y \\ y \end{bmatrix},$$

which is (1), or

$$[x \quad y] \begin{bmatrix} 1 & 1 & 0 \\ 0 & 1 & 1 \end{bmatrix} = [x \quad x+y \quad y],$$

which is (2).

OPERATIONS ON THE SET OF LINEAR TRANSFORMATIONS

● **PROBLEM** 2-26

Let T: $V_1 \rightarrow V_2$; S: $V_2 \rightarrow V_3$; R: $V_3 \rightarrow V_4$ be linear transformations where V_1, V_2, V_3 and V_4 are vector spaces defined over a common field K. If we define multiplication of transformations by

$$S \circ T(v) = S(T(v)),$$

show that multiplication is associative, i.e.,

$$(RS)T(v) = R(ST(v)), \quad \text{where} \quad v \in V_1. \tag{1}$$

Solution: We first identify the domains and co-domains of the various transformations. T transforms V_1, the domain, into V_2. The domain of S is the co-domain of T. Similarly, the domain of R is the co-domain of S. Finally the co-domain of R is V_4.

On the left hand side of (1), since S transforms V_2 into V_3 and R transforms V_3 into V_4, RS transforms V_2 into V_4.

Then, (RS)T transforms V_1 into V_4. On the right hand side, ST transforms V_1 into V_3 and R(ST) transforms V_1 into V_4. Thus, the domains of (RS)T and R(ST) are the same.
Now, let $v \varepsilon V_1$. Then, $((RS)T)v = (RS)(Tv) = R(S(Tv)) = R((ST)v) = (R(ST))v$, i.e., $R(ST) = (RS)T$.
Thus, transformation multiplication is associative.

● **PROBLEM** 2-27

Show that the following linear transformations are isomorphisms:

1) R_θ: $R^2 \rightarrow R^2$. R is rotation of elements in R^2 through the angle θ; $0 \leq \theta \leq \pi/2$.

2) A: $R^3 \rightarrow R^3$ where A is given by

$$3x_1 + x_2 + x_3$$

$$x_2 - x_3$$

$$x_1 - x_2,$$

i.e., $A(x_1, x_2, x_3) = (3x_1 + x_2 + x_3,\ x_2 - x_3,\ x_1 - x_2)$.

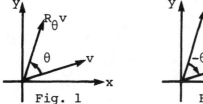

Fig. 1

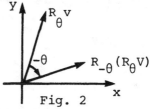

Fig. 2

Solution: A transformation L: $V \rightarrow W$ is said to be an isomorphism if it is 1 - 1 and onto. A function is 1 - 1 if every $v \varepsilon V$ is sent to a different $w \varepsilon W$ and is onto if every $w \varepsilon W$ corresponds to some $v \varepsilon V$. Thus, to answer 1) and 2), we must show that R_θ and A are 1 - 1 and onto.

1) A function is 1 - 1 and onto if and only if it has an inverse.
In Fig. 1, v is sent by R_θ to $R_\theta v$. In Fig. 2, $R_\theta v$ is sent back to v by $R_{-\theta}$. Now, $R_{-\theta}$ is a linear transformation. It is, in fact, the inverse transformation of R_θ, i.e., a vector in R^2 to which R_θ and $R_{-\theta}$ are successively applied remains in its original position. Since R_θ is invertible, it is an isomorphism.

2) To show that $A(x_1, x_2, x_3)$ is an isomorphism we proceed as follows:
Let $(y_1, y_2, y_3) \varepsilon R^3$ be a fixed vector. Then the equa-

tion system

$$3x_1 + x_2 + x_3 = y_1$$

$$x_2 - x_3 = y_2 \tag{1}$$

$$x_1 - x_2 = y_3$$

has the solution $x_1 = \dfrac{y_1 + y_2 + 2y_3}{5}$; $x_2 = \dfrac{y_1 + y_2 - 3y_3}{5}$;

$x_3 = \dfrac{y_1 - 4y_2 - 3y_3}{5}$. Since (y_1, y_3, y_3) was chosen arbitrarily and we obtained solutions for (x_1, x_2, x_3) in terms of (y_1, y_2, y_3), this implies $A(x_1, x_2, x_3)$ is onto.

To show that A is 1 - 1, let (y_1', y_2', y_3') be another solution to (1). Then $(y_1 - y_1', y_2 - y_2', y_3 - y_3')$ is a solution to the system

$$3x_1 + x_2 + x_3 = 0$$

$$x_2 - x_3 = 0 \tag{2}$$

$$x_1 - x_2 = 0.$$

Letting $y_1 - y_1' = x_1$, $y_2 - y_2' = x_2$ and $y_3 - y_3' = x_3$, we obtain

$$3(y_1 - y_1') + (y_2 - y_2') + (y_3 - y_3') = 0$$

$$(y_2 - y_2') - (y_3 - y_3') = 0. \tag{3}$$

$$(y_1 - y_1') - (y_2 - y_2') = 0$$

Comparing (1) and (3), it is seen that

$$(y_1 - y_1') = \tfrac{1}{5}(0 + 0 + 2 \cdot 0) = 0$$

$$(y_2 - y_2') = \tfrac{1}{5}(0 + 0 - 3 \cdot 0) = 0$$

$$(y_3 - y_3') = \tfrac{1}{5}(0 - 4 \cdot 0 - 3 \cdot 0) = 0.$$

Hence, $y_1 = y_1'$; $y_2 = y_2'$; $y_3 = y_3'$ and the solution to system (1) is unique, i.e., $A(x_1, x_2, x_3)$ is 1 - 1. Thus, $A(x_1, x_2, x_3)$ is an isomorphism.

● PROBLEM 2-28

1) Let V be a vector space over a field F and let $\alpha \in F$ be any nonzero scalar. Show that multiplication by α,

M_α: V → V is a vector-space isomorphism.
2) Let V and W be any vector spaces over F and V ε W their Cartesian product. Show that V $\underset{\sim}{}$ V × {0} where 0 ε W.

<u>Solution:</u>   Let T: V → W be a 1 - 1 onto linear transformation. Then T is an isomorphism and the two spaces V and W are said to be isomorphic to each other. If no such operator T exists, V and W are said to be non-isomorphic. When W = V, T is called a linear operator. Let M : V → V denote the scalar operator that sends every v ε V to αv ε V. M is a linear transformation. Further, since α ≠ 0, it has an inverse $\frac{1}{\alpha}$ ε F, and $M_{1/\alpha}$: V → V is a linear operator. Now,

$$(M_\alpha \circ M_{1/\alpha})(v) = M_\alpha((\frac{1}{\alpha})v) = \alpha(\frac{1}{\alpha})v = I_v(v)$$

$$= \frac{1}{\alpha}(\alpha v) = (M_{1/\alpha} \circ M)v \quad \text{for} \quad v \ \varepsilon \ V.$$

Hence, $M_{1/\alpha}$ is the inverse of M_α which implies that M_α is 1 - 1 and onto since an operator has an inverse if and only if it is 1 - 1 and onto.
2)   The set V $\chi\langle 0 \rangle$ is a subspace of V × W. To show that V $\underset{\sim}{}$ V × {0}, we proceed as follows: Define a function of f: $\overline{V}$ → V × {0}, f(v) = (v,0) v ε V. We see that f is 1 - 1 and onto, i.e., the set

{v: v ε V} is isomorphic to the set

{(v,0): v ε V, 0 ε W}.

● **PROBLEM** 2-29

Illustrate by means of an example that isomorphism, although an equivalence relation, is not a congruence relation.

<u>Solution:</u> First, define the following notions: 1) iso-morphism; 2) equivalence relation; 3) congruence relation. Then we can proceed to devise an example.
1)   Let V, W be vector spaces over a field K and T: V → W a linear operator defined on V. Then, T is defined to be an isomorphism if it is 1 - 1 and onto, and the spaces V and W are said to be isomorphic to each other. That is, V and W are said to be isomorphic if there exists an isomorphism which maps V into W (or equivalently W into V) as will be shown.   To denote this, we write V $\underset{\sim}{}$ W.
2)   Let S be a set of three or more objects. Then an equivalence relation N is defined between these objects if, for any a,b,c,ε S ,

i)   a ∿ a
ii)   a ∿ b implies b ∿ a
iii)   a ∿ b and b ∿ c implies a ∿ c.

The relation of isomorphism is an equivalence relation. We have i) V $\underset{\sim}{}$ V (by defining the identity operator T(v) = v, v ε V, which is 1 - 1 and onto)

ii) $V \sim W$ implies $W \sim V$. (Since T is an isomorphism, it is invertible and T^{-1} itself is an isomorphism, $T^{-1}: W \rightarrow V$. Thus, $W \sim V$.)

iii) $V \sim W$ and $W \sim U$ implies $V \sim U$. (Let $T: V \rightarrow W$ and $S: W \rightarrow U$ be isomorphisms. Then $T(v) = w$ and $S(w) = u$ for $v \; \epsilon \; V$, $w \; \epsilon \; W$ and $u \; \epsilon \; U$. Hence, $S(T(v)) = u$, and the product ST is itself 1 - 1 and onto, i.e., $V \sim U$).

3) A congruence relation is a relation between members of a set such that algebraic operations preserve the relation. The equality relation is a congruence relation.

An isomorphism that is not a congruence relation is provided by the following example: let

$$V = R^3 \text{ and let } S = \{(a,0,b): \quad a,b \; \epsilon \; R\},$$

$$T = \{(0,a,b): \quad a,b \; \epsilon \; R\}, \quad R^1 = \{x: \quad x \text{ is real}\}.$$

Note that S, T and $R^1 \subset R^3$, and that $R^1 \cap T = \phi$. Thus, R^3 can be decomposed into the direct sum of two disjoint sub-spaces, $R^3 = R^1 \oplus T$. On the other hand, although $S \sim T$, $R^1 + S$ is not a direct sum. Thus, the two spaces S and T are not congruent although they are isomorphic.

CHAPTER 3

MATRICES OF LINEAR TRANSFORMATIONS

THE REPRESENTATION OF A LINEAR TRANSFORMATION BY A MATRIX

● **PROBLEM** 3-1

Let T: $R^2 \rightarrow R^2$ be a counterclockwise rotation in R^2 through the angle θ . What is the matrix of T with respect to the usual basis {(1,0), (0,1)} ?

__Solution:__ An operation T which is defined for all vectors u in a vector space V resulting in vectors Tu in a vector space W is called a transformation from V into W. If this operation satisfies the conditions

T(u+v) = Tu + Tv and T(au) = a(Tu) ,

it is called a linear transformation or linear operation. The operation of rotation is a linear transformation. Thus T is a linear transformation.

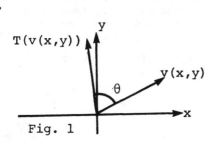

Fig. 1

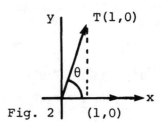

Fig. 2

A linear transformation is specified by its effects on the basis vectors. From Fig. 2 we see that

$$\cos \theta = \frac{x}{\sqrt{x^2 + y^2}} = \frac{x}{1} = x \; ;$$

$$\sin \theta = \frac{y}{\sqrt{x^2 + y^2}} = \frac{y}{1} = y \; .$$

Note that (1,0) is a unit vector and that rotation does not affect its length, i.e., $\sqrt{x^2+y^2} = 1$. Thus, T(1,0) = (cos θ, sin θ).

From Fig. 3: T(0,1) = (cos(θ + π/2), sin(θ + π/2)) .

From Fig. 4, cos(θ + π/2) = –sin θ and sin(θ + π/2) = cos θ . Thus, T(0,1) = (–sin θ, cos θ) . The matrix associated with T is, therefore,

$$A = \begin{bmatrix} \cos \theta & -\sin \theta \\ \sin \theta & \cos \theta \end{bmatrix} \; ,$$

93

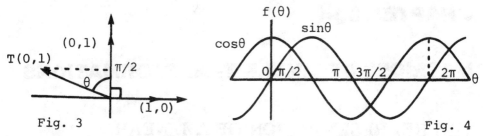

Fig. 3   Fig. 4

since this matrix should have columns $T(0,1)$ and $T(1,0)$. We can now calculate the effect of a rotation through θ using the matrix A.

$$T\begin{bmatrix} x \\ y \end{bmatrix} = A\begin{bmatrix} x \\ y \end{bmatrix} = \begin{bmatrix} \cos\theta & -\sin\theta \\ \sin\theta & \cos\theta \end{bmatrix}\begin{bmatrix} x \\ y \end{bmatrix}$$

$$= ((\cos\theta)x - (\sin\theta)y, (\sin\theta)u + (\cos\theta)y)'.$$

● **PROBLEM** 3-2

Let $T: R^2 \to R^2$ be given by

$$T(x_1, x_2) = (4x_1 - 2x_2, 2x_1 + x_2)$$

and let $\{(1,1), (-1,0)\}$ be a basis for R^2. Compute the matrix of T in the given basis.

Solution: If V is a vector space over K and $T: V \to V$ is a linear transformation of elements of V to elements of V, then T is a linear operator. A linear operator is conveniently described by its effect on the basis vectors. Thus, suppose $\{v_1, v_2, \ldots, v_n\}$ is a basis of V. Then $T(v_1)$, $T(v_2)$, $\ldots$, $T(v_n)$ are vectors in V. In particular,

$$T(v_1) = a_{11}v_1 + a_{12}v_2 + \cdots + a_{1n}v_n$$

$$T(v_2) = a_{21}v_1 + a_{22}v_2 + \cdots + a_{2n}v_n$$

$$\begin{matrix} \cdot & \cdot & \cdot & \cdot \\ \cdot & \cdot & \cdot & \cdot \\ \cdot & \cdot & \cdot & \cdot \end{matrix} \qquad\qquad (1)$$

$$T(v_n) = a_{n1}v_1 + a_{n2}v_2 + \cdots + a_{nn}v_n.$$

Now, suppose u is a vector in V. Then, $u = \alpha_1 v_1 + \alpha_2 v_2 + \cdots + \alpha_n v_n$ and $T(u) = T(\alpha_1 v_1 + \alpha_2 v_2 + \cdots + \alpha_n v_n) = T(\alpha_1 v_1) + T(\alpha_2 v_2) + \cdots + T(\alpha_n v_n)$ $= \alpha_1 T(v_1) + \alpha_2 T(v_2) + \cdots + \alpha_n T(v_n)$.

We see that the transformation T applied to any vector in V can be described in terms of the effect of T on the basis vectors $\{v_1, v_2, \ldots, v_n\}$. We can rewrite (1) in matrix form:

$$T\begin{bmatrix} v_1 \\ v_2 \\ \cdot \\ \cdot \\ \cdot \\ v_n \end{bmatrix} = \begin{bmatrix} a_{11} & a_{12} & \cdots & a_{1n} \\ a_{21} & & & \\ \cdot & & & \\ \cdot & & & \\ \cdot & & & \\ a_{n1} & \cdots & \cdots & a_{nn} \end{bmatrix}\begin{bmatrix} v_1 \\ v_2 \\ \cdot \\ \cdot \\ \cdot \\ v_n \end{bmatrix}$$

94

For computational convenience the above matrix of coefficients is written as its transpose. It is then called the matrix representation of T relative to the basis $\{v_1, v_2, \ldots, v_n\}$. Thus,

$$[T]_v = \begin{bmatrix} a_{11} & a_{21} & \cdots & \cdots & a_{n1} \\ a_{12} & a_{22} & \cdots & \cdots & a_{n2} \\ a_{13} & \cdot & & & \cdot \\ \cdot & \cdot & & & \cdot \\ \cdot & \cdot & & & \cdot \\ a_{1n} & \cdots & \cdots & \cdots & a_{nn} \end{bmatrix} \qquad (2)$$

In this arrangement the effects of T on the basis vectors are conveniently given by the columns of (2).

Now consider the given problem.

$$T(1,1) = (4-2,\ 2+1) = (2,3)$$
$$T(-1,0) = (-4-0,\ -2+0) = (-4,-2)\ .$$

In terms of the basis $\{(1,1),(-1,0)\}$, $(2,3) = a_{11}(1,1) + a_{12}(-1,0)$

or

$$2 = a_{11} - a_{12}$$

$$3 = a_{11}\ .$$

Hence, $a_{11} = 3$, $a_{12} = 1$ and $(2,3) = 3(1,1) + 1(-1,0)$. Similarly, $(-4,-2) = a_{21}(1,1) + a_{22}(-1,0)$; hence, $a_{21} = -2$, $a_{22} = 2$. Thus, $(-4,-2) = -2(1,1) + 2(-1,0)$. The matrix of T relative to the given basis is, therefore,

$$[T]_f = \begin{bmatrix} 3 & -2 \\ 1 & 2 \end{bmatrix}$$

● **PROBLEM** 3-3

Let $\{1,x,x^2\}$ be a basis for the vector space of polynomials of degree not exceeding 2. Let D, D^2 and D^3 denote differentiation operators. Find the matrices of D, D^2 and D^3 relative to the above basis.

Solution: Every linear transformation $T: V \to W$ where V and W are vector spaces has associated with it a matrix. The matrix that represents T depends upon the basis for V on which T operates.

The differentiation operator $D: V \to V$ (an operator is a transformation where the range is the same set as the domain) is a linear transformation. This follows from the rules of calculus since $D(f+g) = Df + Dg$, $D(\alpha f) = \alpha Df$ where f and g are functions and α is a scalar. Similarly, D^2 and D^3 denote the operation of differentiating a function twice and thrice respectively and are both linear operators.

The columns of the matrix representing the operator are the coordinates of each vector in the basis. Thus, since

$$D(1) = 0(1) + 0(x) + 0(x^2)$$

$$D(x) = 1(1) + 0(x) + 0(x^2)$$

$$D(x^2) = 0(1) + 2(x) + 0(x^2)$$

the matrix of D is

$$\begin{bmatrix} 0 & 1 & 0 \\ 0 & 0 & 2 \\ 0 & 0 & 0 \end{bmatrix} .$$

Next, we find the coordinates of the basis vectors under the influence of D^2 :

$$D^2(1) = 0(1) + 0(x) + 0(x^2)$$

$$D^2(x) = 0(1) + 0(x) + 0(x^2)$$

$$D^2(x^2) = 2(1) + 0(x) + 0(x^2) .$$

Thus, D^2 has the matrix

$$\begin{bmatrix} 0 & 0 & 2 \\ 0 & 0 & 0 \\ 0 & 0 & 0 \end{bmatrix} .$$

Finally,

$$D^3(1) = 0(1) + 0(x) + 0(x^2)$$

$$D^3(x) = 0(1) + 0(x) + 0(x^2)$$

$$D^3(x^2) = 0(1) + 0(x) + 0(x^2)$$

and the matrix of D^3 is

$$\begin{bmatrix} 0 & 0 & 0 \\ 0 & 0 & 0 \\ 0 & 0 & 0 \end{bmatrix} .$$

Notice, this shows, as expected, that differentiating a polynomial of degree less than 2 thrice yields the 0 polynomial.

● **PROBLEM** 3-4

Let T be reflection about the y-axis in R^2 .

Find the matrix of T.

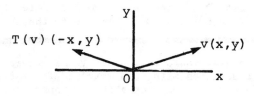

Solution: We must show that T is a linear transformation, find the general matrix of transformations like the one in the figure above and then find the matrix of the given transformation.

Let (x_1,y_1), $(x_2,y_2) \in R^2$. Then,

$$T[(x_1,y_1) + (x_2,y_2)] = T(x_1+x_2, \; y_1+y_2) = (-x_1-x_2, \; y_1+y_2) \tag{1}$$

$$T(x_1,y_1) + T(x_2,y_2) = (-x_1,y_1) + (-x_2,y_2) = (-x_1-x_2,y_1+y_2) \tag{2}$$

Comparing (1) and (2), $T(v_1+v_2) = T(v_1) + T(v_2)$.

$T(\alpha x_1, \alpha y_1) = -\alpha x_1, \ \alpha y_1 = \alpha T(x,y)$. Thus, T is a linear transformation.

Let L: $R^2 \to R^2$ be defined by

$$L[x,y] = [ax + by, \ cx + dy]$$

or

$$L\begin{bmatrix} x \\ y \end{bmatrix} = \begin{bmatrix} ax + by \\ cx + dy \end{bmatrix} = \begin{bmatrix} a & b \\ c & d \end{bmatrix}\begin{bmatrix} x \\ y \end{bmatrix} .$$

Hence, L is a linear transformation from R^2 into R^2 that has the matrix

$$\begin{bmatrix} a & b \\ c & d \end{bmatrix} .$$

Examining the given transformation, we see that T: $R^2 \to R^2$ is given by $T(x,y) = [-x,y] = [-1x+0y, \ 0x+1y]$. Hence, the matrix of T is

$$\begin{bmatrix} -1 & 0 \\ 0 & 1 \end{bmatrix} .$$

● **PROBLEM** 3-5

Let $v = \frac{1}{3}(1,2,2)$ and let Pu be the projection of u onto v. Thus, P is a linear transformation from R^3 into R^3 . Find the matrix of P.

Solution: The projection of u onto v is defined as Pu = (u·v)v . Here (u·v) is the inner product associated with u and v. To find the matrix of P, we have to calculate

$$P\begin{bmatrix} 1 \\ 0 \\ 0 \end{bmatrix}, \quad P\begin{bmatrix} 0 \\ 1 \\ 0 \end{bmatrix}, \quad P\begin{bmatrix} 0 \\ 0 \\ 1 \end{bmatrix}$$

which are the columns of the matrix associated with P. For vectors $x = (x_1, x_2, \ldots, x_n)$ and $y = (y_1, y_2, \ldots, y_n)$ in R^n , the dot product is $\sum_{i=1}^{n} x_i y_i$. Then,

$$P\begin{bmatrix} 1 \\ 0 \\ 0 \end{bmatrix} = \left\{ \begin{bmatrix} 1 \\ 0 \\ 0 \end{bmatrix} \cdot \begin{bmatrix} 1/3 \\ 2/3 \\ 2/3 \end{bmatrix} \right\} \begin{bmatrix} 1/3 \\ 2/3 \\ 2/3 \end{bmatrix} = \frac{1}{9}\begin{bmatrix} 1 \\ 2 \\ 2 \end{bmatrix}$$

$$P\begin{bmatrix} 0 \\ 1 \\ 0 \end{bmatrix} = \left\{ \begin{bmatrix} 0 \\ 1 \\ 0 \end{bmatrix} \cdot \begin{bmatrix} 1/3 \\ 2/3 \\ 2/3 \end{bmatrix} \right\} \begin{bmatrix} 1/3 \\ 2/3 \\ 2/3 \end{bmatrix} = \frac{2}{9}\begin{bmatrix} 1 \\ 2 \\ 2 \end{bmatrix}$$

$$P\begin{bmatrix} 0 \\ 0 \\ 1 \end{bmatrix} = \left\{ \begin{bmatrix} 0 \\ 0 \\ 1 \end{bmatrix} \cdot \begin{bmatrix} 1/3 \\ 2/3 \\ 2/3 \end{bmatrix} \right\} \begin{bmatrix} 1/3 \\ 2/3 \\ 2/3 \end{bmatrix} = \frac{2}{9}\begin{bmatrix} 1 \\ 2 \\ 2 \end{bmatrix}$$

Therefore, the matrix of P is

$$P_0 = \begin{bmatrix} 1/9 & 2/9 & 2/9 \\ 2/9 & 4/9 & 4/9 \\ 2/9 & 4/9 & 4/9 \end{bmatrix}$$

● **PROBLEM** 3-6

Let T: $R^3 \to R^2$ be given by

$$T(x_1, x_2, x_3) = (7x_1 + 2x_2 - 3x_3, \ x_2) .$$

As bases for R^3 and R^2 respectively, let $G = \{g_1, g_2, g_3\} = \{(1,0,0),(0,1,-1),(0,0,1)\}$, $H = \{h_1, h_2\} = \{(1,0),(0,-1)\}$. Compute the matrix representation of T.

Solution: The general linear transformation from R^3 to R^2 has the form $T(x_1, x_2, x_3) = (a_{11}x_1 + a_{12}x_2 + a_{13}x_3, a_{21}x_1 + a_{22}x_2 + a_{23}x_3)$. Its matrix representation is given by its effect on the basis vectors of R^3. Using the basis vectors given in the problem:

$$T(1,0,0) = (a_{11}, a_{21}) \;;$$
$$T(0,1,-1) = (a_{12} - a_{13}, a_{22} - a_{23})$$
$$T(0,0,1) = (a_{13}, a_{23}) \;.$$

These vectors in R^2 are then expressed as linear combinations of the basis vectors of R^2. Using the basis vectors given in the problem,

$$(a_{11}, a_{21}) = b_{11}(1,0) + b_{12}(0,-1)$$
$$= a_{11}(1,0) - a_{21}(0,-1)$$

$$(a_{12} - a_{13}, a_{22} - a_{23}) = b_{21}(1,0) + b_{22}(0,-1)$$
$$= a_{12} - a_{13}(1,0) + a_{23} - a_{22}(0,-1)$$

$$(a_{13}, a_{23}) = b_{31}(1,0) + b_{32}(0,-1)$$
$$= a_{13}(1,0) - a_{23}(0,-1) \;.$$

Thus, the required general matrix is:

$$\begin{bmatrix} a_{11} & a_{12} - a_{13} & a_{13} \\ -a_{21} & a_{23} - a_{22} & -a_{23m} \end{bmatrix}$$

Turning now to the given transformation, $T(1,0,0) = (7,0)$; $T(0,1,0) = (5,1)$ $T(0,0,1) = (-3,0)$. Now,

$$(7,0) = 7(1,0) + 0(0,-1)$$
$$(5,1) = 5(1,0) - 1(0,-1)$$
$$(-3,0) = -3(1,0) + 0(0,-1) \;.$$

Hence, the required matrix is

$$\begin{bmatrix} 7 & 5 & -3 \\ 0 & -1 & 0 \end{bmatrix} \;.$$

● PROBLEM 3-7

Find the linear transformations defined by the following matrices:

$$M = \begin{bmatrix} 2 & 0 \\ 0 & 3 \end{bmatrix} \;; \quad N = \begin{bmatrix} 1 & -1 & 1 \\ -1 & 1 & -1 \end{bmatrix} \;.$$

The linear transformations are of the form $T: R^n \to R^m$, and the bases for R^n and R^m are the usual bases.

Solution: Just as every linear transformation can be represented by a matrix, so, too, every matrix represents some linear transformation with respect to a basis.
The usual basis for R^n is $\{e_1, e_2, \ldots, e_n\}$ where $e_i = (0,0 \ldots 1, 0, \ldots 0, 0)$

Then, $T: R^n \to R^m$ is equivalent to

$$T(e_1) = a_{11}e_1 + a_{12}e_2 + \ldots + a_{1m}e_m$$
$$T(e_2) = a_{21}e_1 + a_{22}e_2 + \ldots + a_{2m}e_m$$
$$\vdots$$
$$T(e_n) = a_{n1}e_1 + a_{n2}e_2 + \ldots + a_{nm}e_m$$

or

$$T \begin{bmatrix} e_1 \\ e_2 \\ \vdots \\ e_n \end{bmatrix} = \begin{bmatrix} a_{11} & a_{12} & \cdots & a_{1m} \\ a_{21} & \cdots & & a_{2m} \\ \vdots & & & \\ a_{n1} & \cdots & & a_{nm} \end{bmatrix} \begin{bmatrix} e_1 \\ e_2 \\ \vdots \\ e_m \end{bmatrix} \qquad (1)$$

The transpose of the matrix in (1) is the matrix of the transformation

$$P = \begin{bmatrix} a_{11} & a_{21} & \cdots & a_{n1} \\ a_{12} & a_{22} & \cdots & a_{n2} \\ \vdots & & & \vdots \\ a_{1m} & a_{2m} & \cdots & a_{nm} \end{bmatrix}.$$

In the matrix P, the components of the ith column are the coordinates of the ith vector in the basis of R^n under the transformation. P is of order $m \times n$. Thus, a linear transformation from R^n to R^m is represented by an $m \times n$ matrix.

Turning to the given matrices, consider M. Since it is a 2×2 matrix, it represents a linear transformation from R^2 into R^2. Thus,

$$T \begin{bmatrix} x \\ y \end{bmatrix} = \begin{bmatrix} 2 & 0 \\ 0 & 3 \end{bmatrix} \begin{bmatrix} x \\ y \end{bmatrix}$$

The first column is $\begin{bmatrix} 2 \\ 0 \end{bmatrix}$ and corresponds to $T\begin{bmatrix} 1 \\ 0 \end{bmatrix}$. Thus, $T\begin{bmatrix} 1 \\ 0 \end{bmatrix} = \begin{bmatrix} 2 \\ 0 \end{bmatrix}$.

Similarly, $T\begin{bmatrix} 0 \\ 1 \end{bmatrix} = \begin{bmatrix} 0 \\ 3 \end{bmatrix}$. Hence, $T\begin{bmatrix} x \\ y \end{bmatrix} = \begin{bmatrix} 2x \\ 3y \end{bmatrix}$.

Now consider N which is a 2×3 matrix and, hence, represents a transformation from R^3 into R^2.

$$T \begin{bmatrix} x \\ y \\ z \end{bmatrix} = \begin{bmatrix} 1 & -1 & 1 \\ -1 & 1 & -1 \end{bmatrix} \begin{bmatrix} x \\ y \\ z \end{bmatrix}$$

Hence, $T\begin{bmatrix} 1 \\ 0 \\ 0 \end{bmatrix} = \begin{bmatrix} 1 \\ -1 \end{bmatrix}$; $T\begin{bmatrix} 0 \\ 1 \\ 0 \end{bmatrix} = \begin{bmatrix} -1 \\ 1 \end{bmatrix}$; $T\begin{bmatrix} 0 \\ 0 \\ 1 \end{bmatrix} = \begin{bmatrix} 1 \\ -1 \end{bmatrix}$, and

$$T \begin{bmatrix} x \\ y \\ z \end{bmatrix} = \begin{bmatrix} x - y + z \\ -x + y - z \end{bmatrix}.$$

That is, $T(x,y,z) = (x-y+z, -x+y-z)$.

● **PROBLEM 3-8**

Let $T: R^2 \to R^2$ be the linear transformation defined by
$$T(x_1, x_2) = (-x_2, x_1).$$

Find the matrix of T relative to the standard basis B = {(1,0),(0,1)} and describe T geometrically.

Solution: The matrix of a linear transformation is found by computing its effects on the basis vectors and arranging the results in columns. Thus,

$$T[1,0] = [0,1] \; ; \; T[0,1] = [-1,0]$$

and

$$A = \begin{bmatrix} 0 & -1 \\ 1 & 0 \end{bmatrix}$$

is the matrix of the transformation.

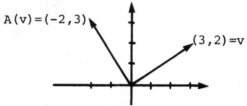

From the figure we see that the effect of T is to rotate (3,2) through $\pi/2$ radians to (-2,3). As this result holds for any point (x,y), we can conclude that any point in the plane is rotated through $\pi/2$ radians by T.

● **PROBLEM** 3-9

a) Let T on R^2 be given by

$$T(1,0) = 3(1,0) + 4(0,1)$$
$$T(0,1) = -5(1,0) + 9(0,1) \; .$$

What is the matrix of T with respect to {(1,0),(0,1)} ?

b) Let the matrix A be given by

$$A = \begin{bmatrix} 2 & 5 & 7 \\ 4 & 9 & 3 \\ -1 & 8 & -5 \end{bmatrix} \; .$$

Find the linear operator T on R^3 that A represents, relative to {(1,0,0),(0,1,0),(0,0,1)} .

Solution: a) Every linear operator from V into V has a matrix representation relative to a given basis. Furthermore, the matrix is constructed by specifying the effects of T on the basis vectors.

We examine the general linear operator $T: R^2 \to R^2$ with respect to the standard basis {(1,0),(0,1)} to see what light it sheds on the given problem. Thus, $T(x,y) = (a_{11}x + a_{21}y, \; a_{12}x + a_{22}y)$,

$$T(1,0) = (a_{11},a_{12}) = a_{11}(1,0) + a_{12}(0,1)$$
$$T(0,1) = (a_{12},a_{22}) = a_{21}(1,0) + a_{22}(0,1) \; .$$

Letting $e_1 = (1,0)$, $e_2 = (0,1)$:

$$T \begin{bmatrix} e_1 \\ e_2 \end{bmatrix} = \begin{bmatrix} a_{11} & a_{12} \\ a_{21} & a_{22} \end{bmatrix} \begin{bmatrix} e_1 \\ e_2 \end{bmatrix} \tag{1}$$

In order to replace T, we must transpose the matrix in (1). Then,

$$\begin{bmatrix} a_{11} & a_{21} \\ a_{12} & a_{22} \end{bmatrix} = A$$

is the matrix of the transformation with respect to the basis $\{e_i\}$.

Applying this to the given problem,

$$T \begin{bmatrix} e_1 \\ e_2 \end{bmatrix} = \begin{bmatrix} 3 & 4 \\ -5 & 9 \end{bmatrix} \begin{bmatrix} e_1 \\ e_2 \end{bmatrix}$$

Thus, the matrix of T with respect to $\{e_i\}$ is:

$$A = \begin{bmatrix} 3 & -5 \\ 4 & 9 \end{bmatrix} .$$

b)   In general, T: $R^3 \to R^3$ is given by:

$$T(x_1, x_2, x_3) = (a_{11}x_1 + a_{21}x_2 + a_{31}x_3,$$

$$a_{12}x_1 + a_{22}x_2 + a_{32}x_3, \ a_{13}x_1 + a_{23}x_2 + a_{33}x_3)$$

The effect of T on the three basis vectors $\{(1,0,0),(0,1,0),(0,0,1)\}$ is:

$$T(1,0,0) = a_{11}(1,0,0) + a_{12}(0,1,0) + a_{13}(0,0,1)$$
$$T(0,1,0) = a_{21}(1,0,0) + a_{22}(0,1,0) + a_{23}(0,0,1)$$
$$T(0,0,1) = a_{31}(1,0,0) + a_{32}(0,1,0) + a_{33}(0,0,1).$$

Letting $e_1 = (1,0,0), \ e_2 = (0,1,0), \ e_3 = (0,0,1)$

$$T \begin{bmatrix} e_1 \\ e_2 \\ e_3 \end{bmatrix} \begin{bmatrix} a_{11} & a_{12} & a_{13} \\ a_{21} & a_{22} & a_{23} \\ a_{31} & a_{32} & a_{33} \end{bmatrix} \begin{bmatrix} e_1 \\ e_2 \\ e_3 \end{bmatrix} \tag{2}$$

The transpose of the matrix in (2) is the required matrix of the transformation.

Since $(A^T)^T = A$, by transposing the given matrix we convert it into the form of (2). Then,

$$T \begin{bmatrix} e_1 \\ e_2 \\ e_3 \end{bmatrix} = \begin{bmatrix} 2 & 4 & -1 \\ 5 & 9 & 8 \\ 7 & 3 & -5 \end{bmatrix} \begin{bmatrix} e_1 \\ e_2 \\ e_3 \end{bmatrix}$$

and so,

$$T(1,0,0) = 2(1,0,0) + 4(0,1,0) + -1(0,0,1)$$

$$T(0,1,0) = 5(1,0,0) + 9(0,1,0) + 8(0,0,1)$$

$$T(0,0,1) = 7(1,0,0) + 3(0,1,0) - 5(0,0,1) .$$

● **PROBLEM** 3-10

Show that the transformation $T(x_1, x_2) = (3x_1, -x_2, -x_1+x_2)$ which has the matrix

$$\begin{bmatrix} 3 & 0 \\ 0 & -1 \\ -1 & 2 \end{bmatrix}$$

can be also considered as a transformation

$$(x_1, x_2)T = (3x_1, -x_2, -x_1 + x_2)$$

with matrix

$$\begin{bmatrix} 3 & 0 & -1 \\ 0 & -1 & 2 \end{bmatrix} .$$

Solution: This problem illustrates the ambiguous notation prevelant in current texts on linear algebra.

One approach to describing the correspondence between linear transformations and matrices is the following:

Let $T: R^n \to R^m$ be a linear transformation. Choose the standard bases for R^n and R^m, i.e., $\{e_i\}$ and $\{e_j\}$, $i = 1,\ldots,n$; $j = 1,\ldots,m$ where e_i and e_j represent the ith and jth n and m-tuples with 1 in the ith and jth coordinate and zeros everywhere else. Then,

$$\begin{aligned}
T(e_1) &= a_{11}e_1 + a_{12}e_2 + \cdots + a_{1m}e_m \\
T(e_2) &= a_{21}e_1 + a_{22}e_2 + \cdots + a_{2m}e_m \\
&\vdots \\
T(e_n) &= a_{n1}e_1 + a_{n2}e_2 + \cdots + a_{nm}e_m
\end{aligned} \tag{1}$$

The system (1) may be expressed in matrix form as:

$$T \begin{bmatrix} e_1 \\ e_2 \\ \vdots \\ e_n \end{bmatrix} = \begin{bmatrix} a_{11} & a_{12} & \cdots & a_{1m} \\ a_{21} & a_{22} & \cdots & a_{2m} \\ \vdots & & & \vdots \\ a_{n1} & a_{n2} & \cdots & a_{nm} \end{bmatrix} \begin{bmatrix} e_1 \\ e_2 \\ \vdots \\ e_m \end{bmatrix} . \tag{2}$$

The transpose of the matrix in (2) is called the matrix of the transformation. Thus,

$$A = \begin{bmatrix} a_{11} & a_{21} & \cdots & a_{n1} \\ a_{12} & a_{22} & & a_{n2} \\ \vdots & \vdots & & \vdots \\ a_{1m} & a_{2m} & \cdots & a_{nm} \end{bmatrix} \tag{3}$$

Why is the matrix in (2) transposed to obtain the matrix (3)? Note that now the operator T can be replaced by the matrix operator A. The operator A and the column of unit vectors of R^n are now conformable for multiplication. In (2) the matrix was of order $n \times m$ and could not be post-multiplied by an $n \times 1$ column vector.

Thus, a transformation $T: R^n \to R^m$ corresponds to an $m \times n$ matrix.

The second approach is as follows: Let $T: R^m \to R^n$ be given by $(x_1, x_2, \ldots)T$. Then, proceeding as in the first approach,

$$e_1 T = a_{11}e_1 + a_{12}e_2 + \cdots + a_{1m}e_m$$

$$e_2 T = a_{21}e_1 + a_{22}e_2 + \cdots + a_{2m}e_m$$

$$\vdots$$

$$e_n T = a_{n1}e_1 + a_{n2}e_2 + \cdots + a_{nm}e_m \quad .$$

Rewriting in matrix form,

$$[e_1 \ e_2 \ldots]T = \begin{bmatrix} a_{11} & a_{12} & \cdots & a_{1m} \\ a_{21} & a_{22} & \cdots & a_{2m} \\ \vdots & \vdots & & \vdots \\ a_{n1} & a_{n2} & \cdots & a_{nm} \end{bmatrix} \begin{bmatrix} e_1 \\ \vdots \\ e_m \end{bmatrix} \tag{4}$$

Now the $n \times m$ matrix in (4) can be used to represent the transformation T. Thus, a transformation $T: R^n \to R^m$ corresponds to an $n \times m$ matrix.

The two approaches are illustrated by the given transformation. First, $T: R^2 \to R^3$ is represented by a 3×2 matrix; then $T: R^2 \to R^3$ is represented by a 2×3 matrix. To check, we see by using matrix multiplication that:

$$T(x_1, x_2) = \begin{bmatrix} 3 & 0 \\ 0 & -1 \\ -1 & 2 \end{bmatrix} \begin{bmatrix} x_1 \\ x_2 \end{bmatrix} = [x_1, x_2] \begin{bmatrix} 3 & 0 & -1 \\ 0 & -1 & 2 \end{bmatrix}$$

$$= (x_1, x_2)T = (3x_1, -x_2, -x_1 + x_2) \quad .$$

● **PROBLEM** 3-11

Find the linear transformation that corresponds to the 2×3 matrix

$$\begin{bmatrix} 1 & 1 & 1 \\ 1 & 0 & 1 \end{bmatrix} \quad .$$

Solution: There is a correspondence between 2×3 matrices and linear transformations from R^3 to R^2 . To see this, let $B_3 = \{(1,0,0),$ $(0,1,0),(0,0,1)\}$ be the standard basis for R^3 , and let

$T(x_1, x_2, x_3) = (a_{11}x_1 + a_{12}x_2 + a_{13}x_3, \ a_{21}x_1 + a_{22}x_2 + a_{23}x_3)$ be a linear transformation from R^3 to R^2 . Then, $T(1,0,0) = a_{11}, a_{21})$, $T(0,1,0) = (a_{12}, a_{22})$, $T(0,0,1) = (a_{13}, a_{23})$.

Let B_2 be the standard basis for R^2 : $B_2 = \{(1,0),(0,1)\}$.

Then $T(1,0,0) = a_{11}(1,0) + a_{21}(0,1)$

$T(0,1,0) = a_{12}(1,0) + a_{22}(0,1)$

$T(0,0,1) = a_{13}(1,0) + a_{23}(0,1)$.

Defining $x = (1,0,0); y = (0,1,0); z = (0,0,1); u = (1,0); v = (0,1)$:

$$T \begin{bmatrix} x \\ y \\ z \end{bmatrix} = \begin{bmatrix} a_{11} & a_{21} \\ a_{12} & a_{22} \\ a_{13} & a_{23} \end{bmatrix} \begin{bmatrix} u \\ v \end{bmatrix} . \tag{1}$$

The transpose of the above matrix is the matrix of the transformation (with respect to the standard basis). Thus, the matrix of the linear transformation is

$$A = \begin{bmatrix} a_{11} & a_{12} & a_{13} \\ a_{21} & a_{22} & a_{23} \end{bmatrix}$$

Now consider the given matrix

$$A = \begin{bmatrix} 1 & 1 & 1 \\ 1 & 0 & 1 \end{bmatrix}$$

Hence, $a_{11} = 1, a_{12} = 1, a_{13} = 1$
$a_{21} = 1, a_{22} = 0, a_{23} = 1$.

Therefore, the linear transformation in R^3 corresponding to the given matrix is $T(x_1, x_2, x_3) = T(x_1 + x_2 + x_3, x_1 + x_3)$.

● PROBLEM 3-12

If $T: R^3 \to R^3$ is defined by the matrix

$$A = \begin{bmatrix} 2 & 3 & 1 \\ 1 & 2 & 3 \\ 3 & 1 & 2 \end{bmatrix}$$

find $(1,1,-1)A$. What is $(x_1, x_2, x_3)T$? Use the standard basis.

Solution: Here T is operating on vectors in R^3 from the right. The matrix A represents T with respect to the standard basis for R^3, $B = \{(1,0,0), (0,1,0), (0,0,1)\}$. Thus, it can replace T wherever T occurs and vice-versa. To find $(1,1,-1)A$ we simply premultiply A by the given row vector.

$$(1,1,-1)A = [1,1,-1] \begin{bmatrix} 2 & 3 & 1 \\ 1 & 2 & 3 \\ 3 & 1 & 2 \end{bmatrix} = [0,4,2] .$$

To find the T that corresponds to A, we premultiply A by (x_1, x_2, x_3), an arbitrary vector in R^3 .

$$[x_1, x_2, x_3] \begin{bmatrix} 2 & 3 & 1 \\ 1 & 2 & 3 \\ 3 & 1 & 2 \end{bmatrix}$$

$$= [2x_1 + x_2 + 3x_3, \ 3x_1 + 2x_2 + x_3, \ x_1 + 3x_2 + 2x_3] .$$

To verify this, notice the following. Let T operate on the basis vectors for R^3 ; then, $T(1,0,0) = (2,3,1); T(0,1,0) = (1,2,3)$ and $T(0,0,1) = (3,1,2)$. Thus, $(x_1, x_2, x_3)T$

$$= [2x_1 + x_2 + 3x_3, \ 3x_1 + 2x_2 + x_3, \ x_1 + 3x_2 + 2x_3] .$$

> Let f be the linear function from R^3 to R^2 such that $f(1,0,0) = (3,2)$, $f(0,1,0) = (1,4)$ and $f(0,0,1) = (2,-5)$. Find $f(2,0,5)$.

Solution: For a linear function f from R^m to R^n, there is a matrix, Mat f, such that $f(u) = (\text{Mat } f)u$ where $u = (u_1, u_2, \ldots, u_n)$ is a vector in R^m. The columns of Mat f are the coordinates of the basis vectors in R^n. They also represent the effect of f on the basis vectors in R^m. Thus, in the given problem,

$$(\text{Mat } f) = \begin{bmatrix} 3 & 1 & 2 \\ 2 & 4 & -5 \end{bmatrix}.$$

Since $f(u) = (\text{Mat } f)u$, letting $u = (2,0,5)$, we obtain

$$\begin{bmatrix} 3 & 1 & 2 \\ 2 & 4 & -5 \end{bmatrix} \begin{bmatrix} 2 \\ 0 \\ 5 \end{bmatrix} \tag{1}$$

Note that (1) is conformable for matrix multiplication, and that the product will be a 2×1 column matrix. Performing the multiplication indicated by (1),

$$f(2,0,5) = \begin{bmatrix} 16 \\ -21 \end{bmatrix}.$$

> Let $B: R^3 \to R^3$ be given by:
> $$(x_1, x_2, x_3)B = (2x_1 + x_3, x_1 + x_2 - x_3, x_1 - x_2 + x_3).$$
> The linear transformation B has a matrix representation with respect to the standard basis for R^3, $\{(1,0,0),(0,1,0),(0,0,1)\}$. What are the row vectors of this matrix?

Solution: A linear transformation $T: R^m \to R^n$ can transform a vector in R^m either from the left $T(x_1, x_2, \ldots, x_m)^T$ or from the right $(x_1, x_2, \ldots, x_m)T$. In the first case, the matrix of the transformation is of order $n \times m$ with the columns representing the effects of T on the basis vectors. In the second case, the matrix is of order $m \times n$ and the coordinates of the transformed basis vectors are the row vectors. We must deal in the given problem with the second case.

Since any vector in R^3 can be written as a linear combination of the basis vectors, we need only study the effect of the transformation on the basis $\{(1,0,0),(0,1,0),(0,0,1)\}$.

$$(1,0,0)B = (2(1) + 0(0), 1(1) + 1(0) - 1(0), 1(1) - 1(0) + 1(0))$$
$$= (2,1,1).$$
$$(0,1,0)B = (0,1,-1).$$
$$(0,0,1)B = (1,-1,1).$$

Thus, the matrix of the transformation is

$$\begin{bmatrix} 2 & 1 & 1 \\ 0 & 1 & -1 \\ 1 & -1 & 1 \end{bmatrix}$$

and the row vectors are $(2,1,1),(0,1,-1)$ and $(1,-1,1)$.

THE TRANSITION MATRIX

Verify, by means of examples, the following theorem:

Let $\{e_1, e_2, \ldots, e_n\}$ be a basis of V, and let T be any operator on V. Let $[T]_e$ and $[v]_e$ be the matrix representations of T and v with respect to the basis. Then, for any vector $v \in V$, $[T]_e [v]_e = [T(v)]_e$.

Solution: An operator is a linear transformation $T: V \to V$. Let V be the vector space of polynomials of degree not greater than three, and let D denote the process of differentiation. Then $D: V \to V$ is an operator. If $a + bt + ct^2 + dt^3$ is an arbitrary vector in V, then

$$D(a + bt + ct^2 + dt^3) = (b + 2ct + 3dt^2) . \tag{1}$$

To find the matrix representation of D, we need a basis. The usual basis for V is $B = \{1, x, x^2, x^3\}$. The coordinates of $a + bt + ct^2 + dt^3$ with respect to B are (a,b,c,d). Thus, $[V]_B = (a,b,c,d)$.

To find the matrix representation of D, we must express its effect on the basis vectors in terms of the basis vectors.

$$D(1) = 0 = 0(1) + 0(t) + 0(t^2) + 0(t^3)$$
$$D(t) = 1 = 1(1) + 0(t) + 0(t^2) + 0(t^3)$$
$$D(t^2) = 2t = 0(1) + 2(t) + 0(t^2) + 0(t^3)$$
$$D(t^3) = 3t^2 = 0(1) + 0(t) + 3(t^2) + 0(t^3) .$$

Hence, $[D]_B =$

$$\begin{bmatrix} 0 & 1 & 0 & 0 \\ 0 & 0 & 2 & 0 \\ 0 & 0 & 0 & 3 \\ 0 & 0 & 0 & 0 \end{bmatrix}$$

Now, $[D]_B [v]_B =$

$$= \begin{bmatrix} 0 & 1 & 0 & 0 \\ 0 & 0 & 2 & 0 \\ 0 & 0 & 0 & 3 \\ 0 & 0 & 0 & 0 \end{bmatrix} \begin{bmatrix} a \\ b \\ c \\ d \end{bmatrix} = \begin{bmatrix} b \\ 2c \\ 3d \\ 0 \end{bmatrix} \tag{2}$$

On the other hand, $D(v)$ from (1) equals $b + 2ct + 3dt^2$. Hence, $[D(v)]_B$ is

$$\begin{bmatrix} b \\ 2c \\ 3d \\ 0 \end{bmatrix} . \tag{3}$$

Since (2) and (3) are equal, this verifies the theorem.

For a second verification, let $T: R^2 \to R^2$ be given by $T(x,y) = (4x - 2y, 2x + y)$. Let $v = (2,3)$ and let $B = \{(1,0),(0,1)\}$ be the usual basis for R^2. Then $[v]_B = (2,3)$. To find $[T]_B$, we need its effect on the basis vectors.

$$T(1,0) = (4,2) = 4(1,0) + 2(0,1).$$
$$T(0,1) = (-2,1) = -2(1,0) + 1(0,1).$$

Hence,

$$[T]_B = \begin{bmatrix} 4 & -2 \\ 2 & 1 \end{bmatrix}.$$

Then $[T]_B[v]_B = \begin{bmatrix} 4 & -2 \\ 2 & 1 \end{bmatrix}\begin{bmatrix} 2 \\ 3 \end{bmatrix} = \begin{bmatrix} 2 \\ 7 \end{bmatrix}.$

On the other hand, since $T(v) = T(2,3) = (2,7) = 2(1,0) + 7(0,1)$, we see that relative to the basis B,

$$[T(v)]_B = \begin{bmatrix} 2 \\ 7 \end{bmatrix}.$$

Since $[T]_B[v]_B = [T(v)]_B$, the theorem is verified.

● PROBLEM 3-16

Let $\{e_i\} = \{(1,0),(0,1)\}$ and $\{f_i\} = \{(1,1),(-1,0)\}$ be two bases for R^2. Find the transition matrix from $\{e_i\}$ to $\{f_i\}$.

Solution: A statement of the general case leads to the solution of the particular problem at hand.

Let V be a vector space over a field K, and let $B_1 = \{u_1,u_2,\ldots,u_n\}$, $B_2 = \{v_1,v_2,\ldots,v_n\}$ be two bases of V. Since B_2 is a basis, every element of B_1 can be written as a linear combination of elements of B_2. Thus,

$$u_1 = a_{11}v_1 + a_{12}v_2 + \ldots + a_{1n}v_n$$
$$u_2 = a_{21}v_1 + a_{22}v_2 + \ldots + a_{2n}v_n$$
$$\cdot$$
$$\cdot \qquad\qquad\qquad\qquad\qquad\qquad\qquad (1)$$
$$\cdot$$
$$u_n = a_{n1}v_1 + a_{n2}v_2 + \ldots + a_{nn}v_n$$

Rewriting the system (1) in matrix form,

$$\begin{bmatrix} u_1 \\ u_2 \\ , \\ \cdot \\ \cdot \\ u_n \end{bmatrix} = \begin{bmatrix} a_{11} & a_{12} & \cdots & a_{1n} \\ a_{21} & a_{22} & \cdots & \\ \cdot & \cdot & & \\ \cdot & \cdot & & \\ a_{n1} & a_{n2} & \cdots & a_{nn} \end{bmatrix}\begin{bmatrix} v_1 \\ v_2 \\ \cdot \\ \cdot \\ \cdot \\ v_n \end{bmatrix}$$

Taking the transpose of the above matrix,

$$P = \begin{bmatrix} a_{11} & a_{21} & \cdots & a_{n1} \\ a_{12} & a_{22} & \cdots & a_{n2} \\ \vdots & \vdots & & \vdots \\ a_{1n} & a_{2n} & \cdots & a_{nn} \end{bmatrix}$$

(2)

The matrix P is the transition matrix from the basis $\{v_1, v_2, \ldots, v_n\}$ to the basis $\{u_1, u_2, \ldots, u_n\}$. Note that the coordinates of u_i are given by the ith column.

We now turn to the given problem.

$$(1,1) = a_{11}(1,0) + a_{12}(0,1)$$
$$(1,1) = (a_{11},0) + (0,a_{12}) .$$

Hence, $a_{11} = 1$, $a_{12} = 1$ and $(1,1) = 1(1,0) + 1(0,1)$. Similarly, $(-1,0) = a_{21}(1,0) + a_{22}(0,1)$ and $a_{21} = -1$; $a_{22} = 0$. Hence, the transition matrix from $\{e_i\}$ to $\{f_i\}$ is

$$P = \begin{bmatrix} 1 & -1 \\ 1 & 0 \end{bmatrix} .$$

We can also find the transition matrix from $\{f_i\}$ to $\{e_i\}$.

$$e_1 = (1,0) = b_{11}(1,1) + b_{12}(-1,0)$$
$$= (1,0) = (b_{11},b_{11}) + (-b_{12},0)$$
$$1 = b_{11} - b_{12}$$

$$0 = b_{11} .$$

Hence, $b_{11} = 0$; $b_{12} = -1$.

$$e_2 = (0,1) = b_{21}(1,1) + b_{22}(-1,0)$$
$$0 = b_{21} - b_{22}$$
$$1 = b_{21} .$$

Hence, $b_{21} = 1$, $b_{22} = 1$ and

$$Q = \begin{bmatrix} 0 & 1 \\ -1 & 1 \end{bmatrix}$$

As a coda to the above composition, observe that $Q = P^{-1}$.

● **PROBLEM** 3-17

Find the matrix representation of each of the following operators T on R^2, first with respect to the usual basis $\{(1,0),(0,1)\}$ and then with respect to the basis $\{(1,3),(2,5)\}$:
i) $T(x,y) = (2y, 3x-y)$,
ii) $T(x,y) = (3x - 4y, x + 5y)$.

Solution: The matrix representation of a linear transformation will change with the basis for the underlying vector space.
i) $T(x,y) = (2y, 3x-y)$.

Let $(a,b) \in R^2$ be an arbitrary vector in R^2. In terms of the usual basis, $(a,b) = a(1,0) + b(0,1)$.

A linear transformation is specified by its effects on the basis vectors.

$$T(1,0) = (0,3) = 0(1,0) + 3(0,1)$$

$$T(0,1) = (2,-1) = 2(1,0) - 1(0,1) .$$

Thus, the matrix of the transformation is

$$A = \begin{bmatrix} 0 & 2 \\ 3 & -1 \end{bmatrix} .$$

Now we must find the coordinates of (a,b) with respect to the basis $\{(1,3),(2,5)\}$.

$$(a,b) = x(1,3) + y(2,5)$$
$$(a,b) = (x+2y, \ 3x+5y) .$$

The solution to the simultaneous equation system
$$x + 2y = a$$
$$3x + 5y = b$$
is $x = -5a + 2b; \ y = 3a - b$. Hence, $(a,b) = [-5a+2b](1,3) + [3a-b](2,5)$. Next, find the effects of T on the basis vectors.

$$T(1,3) = (6,0) = -30(1,3) + 18(2,5)$$

$$T(2,5) = (10,1) = -48(1,3) + 29(2,5) .$$

Hence, the matrix of the transformation with respect to the basis $\{(1,3),(2.5)\}$ is

$$B = \begin{bmatrix} -30 & -48 \\ 18 & 29 \end{bmatrix} .$$

ii) Follow the same route used in i). The matrix of the transformation with respect to the usual basis is:

$$A = \begin{bmatrix} 3 & -4 \\ 1 & 5 \end{bmatrix} .$$

The matrix with respect to the basis $\{(1,3),(2,5)\}$ is a little harder to find.

$$T(1,3) = (-9,16) = 77(1,3) - 43(2,5)$$

$$T(2,5) = (-14,27) = 124(1,3) - 69(2,5).$$

$$B = \begin{bmatrix} 77 & 124 \\ -43 & -69 \end{bmatrix}$$

Thus, the same linear transformation will have different matrix representations in different bases.

● **PROBLEM** 3-18

Let V be the set of arrows (any vector is a class of arrows) in three dimensional space. Then $\{i,j,k\}$ is a basis for V. Another basis for V is $\{i_1,j_1,k_1\}$ where $i_1 = i + j + k$, $j_1 = i - j + k$, $k_1 = i + 2j - k$. Let T be the linear operator on V defined by

$$T(xi + yj + zk) = \frac{x+z}{2} i + (y + z - x)j + \frac{x+z}{2} k \qquad (1)$$

where x,y,z are numbers from the field of real numbers. Find the matrices of T with respect to $\{i,j,k\}$ and $\{i_1,j_1,k_1\}$.

Solution: First, notice that i,j and k are three mutually perpendicular unit vectors as in the figure.

109

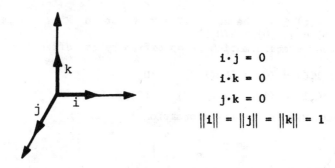

$$i \cdot j = 0$$
$$i \cdot k = 0$$
$$j \cdot k = 0$$
$$\|i\| = \|j\| = \|k\| = 1$$

As any vector in R^3 needs 3 coordinates to specify it and any vector in R^3 can be written as a linear combination of i,j,k ($v = c_1 i + c_2 j + c_3 k$), we know i,j and k span R^3. In fact, any 3 mutually perpendicular vectors form a basis for R^3. Thus, we can use the following theorem from linear algebra: Let T be a linear operator defined on V and let $U = \{u_1, u_2, \ldots, u_n\}$, $V = \{v_1, v_2, \ldots, v_n\}$ be two bases for V. Then if A is the matrix of T with respect to u, $B = QAP$ is the matrix of T with respect to V. Here, P is the transition matrix from U to V and Q is the transition matrix from V to U.

To apply the theorem, we need A, P and Q. From (1), the matrix of T with respect to $\{i,j,k\}$ is:

$$A = \begin{bmatrix} \frac{1}{2} & 0 & \frac{1}{2} \\ -1 & 1 & 1 \\ \frac{1}{2} & 0 & \frac{1}{2} \end{bmatrix}$$

Next, we must find the transition matrix from U to V. Let

$$w = xi + yj + zk = x_1 i_1 + y_1 j_1 + z_1 k_1 .$$

Then, using the given equations relating i,j,k to i_1, j_1, k_1 ,

$$\begin{bmatrix} x \\ y \\ z \end{bmatrix} = \begin{bmatrix} 1 & 1 & 1 \\ 1 & -1 & 2 \\ 1 & 1 & -1 \end{bmatrix} \begin{bmatrix} x_1 \\ y_1 \\ z_1 \end{bmatrix} \tag{2}$$

The transition matrix from $\{i,j,k\}$ to $\{i_1, j_1, k_1\}$ is:

$$P = \begin{bmatrix} 1 & 1 & 1 \\ 1 & -1 & 2 \\ 1 & 1 & -1 \end{bmatrix} \qquad \text{since} \quad i_1 = i + j + k$$
$$j_1 = i - j + k$$
$$k_1 = i + 2j - k .$$

Finally, we require Q. But from (2),

$$\begin{bmatrix} x_1 \\ y_1 \\ z_1 \end{bmatrix} \begin{bmatrix} 1 & 1 & 1 \\ 1 & -1 & 2 \\ 1 & 1 & -1 \end{bmatrix}^{-1} \begin{bmatrix} x \\ y \\ z \end{bmatrix}$$

gives the coordinate transformation from $\begin{bmatrix} x_1 \\ y_1 \\ z_1 \end{bmatrix}$ to $\begin{bmatrix} x \\ y \\ z \end{bmatrix}$.

Hence, $Q = P^{-1}$.

The inverse of a matrix P is given by $P^{-1} = \dfrac{1}{\det P}[\text{adj. }P]$ where $[\text{adj. }P]$ is the transposed matrix of cofactors. The cofactors of the nine elements of P are:

$$P_{11} = + \begin{vmatrix} -1 & 2 \\ 1 & -1 \end{vmatrix} = -1 \ ; \ P_{12} = - \begin{vmatrix} 1 & 2 \\ 1 & -1 \end{vmatrix} = 3 \ ; \ P_{13} = + \begin{vmatrix} 1 & -1 \\ 1 & 1 \end{vmatrix} = 2 \ ;$$

$$P_{21} = - \begin{vmatrix} 1 & 1 \\ 1 & -1 \end{vmatrix} = 2 \ ; \ P_{22} = \begin{vmatrix} 1 & 1 \\ 1 & -1 \end{vmatrix} = -2 \ ; \ P_{23} = - \begin{vmatrix} 1 & 1 \\ 1 & 1 \end{vmatrix} = 0 \ ;$$

$$P_{31} = \begin{vmatrix} 1 & 1 \\ -1 & 2 \end{vmatrix} = 3 \ ; \ P_{32} = - \begin{vmatrix} 1 & 1 \\ 1 & 2 \end{vmatrix} = -1 \ ; \ P_{33} = + \begin{vmatrix} 1 & 1 \\ 1 & -1 \end{vmatrix} = -2 \ .$$

The transpose of the above matrix of cofactors is:

$$\text{adj } P = \begin{bmatrix} -1 & 2 & 3 \\ 3 & -2 & -1 \\ 2 & 0 & -2 \end{bmatrix}$$

By definition of determinant,

$$\det P = (1)(-1)(-1) + (1)(2)(1) + (1)(1)(1) - (1)(-1)(1)$$
$$- (1)(1)(-1) - (1)(2)(1)$$
$$= 4 \ .$$

Thus,

$$P^{-1} = \tfrac{1}{4} \begin{bmatrix} -1 & 2 & 3 \\ 3 & -2 & -1 \\ 2 & 0 & -2 \end{bmatrix} \ .$$

The matrix of T with respect to $\{i_1, j_1, k_1\}$ is, therefore,

$$\begin{bmatrix} -\tfrac{1}{4} & \tfrac{1}{2} & \tfrac{3}{4} \\ \tfrac{3}{4} & -\tfrac{1}{2} & -\tfrac{1}{4} \\ 2/4 & 0 & -\tfrac{1}{2} \end{bmatrix} \begin{bmatrix} \tfrac{1}{2} & 0 & \tfrac{1}{2} \\ -1 & 1 & 1 \\ \tfrac{1}{2} & 0 & \tfrac{1}{2} \end{bmatrix} \begin{bmatrix} 1 & 1 & 1 \\ 1 & -1 & 2 \\ 1 & 1 & -1 \end{bmatrix} = \begin{bmatrix} 1 & 0 & 0 \\ 0 & 1 & 0 \\ 0 & 0 & 1 \end{bmatrix} \ .$$

Thus, $Ti_1 = i_1$, $Tj_1 = j_1$, $Tk_1 = k_1$.

● **PROBLEM** 3-19

Let R^2 be a vector space over R, and let $\{e_i\} = \{(1,0),(0,1)\}$ be the standard basis of R^2. Define $T: R^2 \rightarrow R^2$ by $T(x_1,x_2) = (3x_1 - 2x_2, x_1 + x_2)$. Let $\{f_i\} = \{(-3,0),(1,2)\}$ be another basis for R^2. Find the matrices of T with respect to $\{e_i\}$ and $\{f_i\}$ and the transition matrix between them.

Solution: To find the matrix of T with respect to $\{e_i\}$, evaluate $T(1,0)$ and $T(0,1)$. Thus,

$$T(1,0) = (3-2(0), 1+0) = (3,1)$$
$$T(0,1) = (3(0)-2(1), 0+1) = (-2,1).$$

Hence, the matrix of T with respect to e_i is: $A = \begin{bmatrix} 3 & -2 \\ 1 & 1 \end{bmatrix}$.

Next, to compute the matrix of T with respect to $\{f_i\}$, evaluate $T(-3,0)$ and $T(1,2)$. Then,

$$T(-3,0) = (3(-3) - 2(0),-3+0) = (-9,-3)$$

$$T(1,2) = (3(1) - 2(2),1+2) = (-1,3).$$

We must express $(-9,-3)$ and $(-1,3)$ in terms of the basis vectors $(-3,0)$ and $(1,2)$.

$$(-9,-3) = b_{11}(-3,0) + b_{12}(1,2)$$

$$(-9,-3) = (-3b_{11},0) + (b_{12},2b_{12})$$

$$-9 = -3b_{11} + b_{12}$$

$$-3 = 2b_{12} .$$

Thus, $b_{11} = 5/2$ and $b_{12} = -3/2$. Similarly,

$$(-1,3) = b_{21}(-3,0) + b_{22}(1,2)$$

$$(-1,3) = (-3b_{21},0) + (b_{22},2b_{22})$$

$$-1 = -3b_{21} + b_{22}$$

$$3 = 2b_{22} .$$

Hence, $b_{21} = 5/6$ and $b_{22} = 3/2$. The matrix of T with respect to $\{f_i\}$ is

$$Q = \begin{bmatrix} 5/2 & 5/6 \\ -3/2 & 3/2 \end{bmatrix} .$$

The transition matrix from $\{e_i\}$ to $\{f_i\}$ is given by:

$$(-3,0) = -3(1,0) + 0(0,1)$$

$$(1,2) = 1(1,0) + 2(0,1)$$

$$P = \begin{bmatrix} -3 & 1 \\ 0 & 2 \end{bmatrix}$$

Since $\det P = -6 \neq 0$, P^{-1} exists.

$$P^{-1} = \frac{1}{\det P} \text{ adj. } P = \frac{1}{-6}\begin{bmatrix} 2 & -1 \\ 0 & -3 \end{bmatrix} = \begin{bmatrix} -1/3 & 1/6 \\ 0 & 1/2 \end{bmatrix} .$$

Thus,

$$P^{-1}AP = -\frac{1}{6}\begin{bmatrix} 2 & -1 \\ 0 & -3 \end{bmatrix}\begin{bmatrix} 3 & -2 \\ 1 & 1 \end{bmatrix}\begin{bmatrix} -3 & 1 \\ 0 & 2 \end{bmatrix} = \begin{bmatrix} 5/2 & 5/6 \\ -3/2 & 3/2 \end{bmatrix} = Q .$$

● **PROBLEM** 3-20

Let $T: R^2 \rightarrow R^2$ be given by $T(x,y) = (2x,x+3y)$.
Find the matrix of T with respect to the basis $B = \{(1,1),(0,2)\}$ for R^2. Then, using this matrix, find the matrix of T with respect to the basis $B' = \{(1,2),(0,1)\}$.

Solution: Since T is a linear transformation, it has a matrix representing it with respect to B. We construct the matrix by computing the effect of T on the basis vectors.

$$T(1,1) = (2,4); \quad T(0,2) = (0,6) .$$

Now,

$$(2,4) = a_{11}(1,1) + a_{12}(0,2)$$

$$(2,4) = (a_{11},a_{11} + 2a_{12})$$

Hence, $a_{11} = 2$, $a_{12} = 1$ and $T(1,1) = 2(1,1) + 1(0,2)$. Similarly,
$$(0,6) = a_{21}(1,1) + a_{22}(0,2) .$$
From this we obtain
$$T(0,2) = (0,6) = 0(1,1) + 3(0,2).$$
Thus, the matrix representing T is
$$A = \begin{bmatrix} 2 & 0 \\ 1 & 3 \end{bmatrix} \tag{1}$$

Using (1),
$$\begin{bmatrix} 2 & 0 \\ 1 & 3 \end{bmatrix} \begin{bmatrix} 1 \\ 1 \end{bmatrix} = \begin{bmatrix} 2 \\ 4 \end{bmatrix} ; \quad \begin{bmatrix} 2 & 0 \\ 1 & 3 \end{bmatrix} \begin{bmatrix} 0 \\ 2 \end{bmatrix} = \begin{bmatrix} 0 \\ 6 \end{bmatrix}$$
and
$$\begin{bmatrix} 2 \\ 4 \end{bmatrix} = b_{11} \begin{bmatrix} 1 \\ 2 \end{bmatrix} + b_{12} \begin{bmatrix} 0 \\ 1 \end{bmatrix}$$
$$\begin{bmatrix} 2 \\ 4 \end{bmatrix} = \begin{bmatrix} b_{11} \\ 2b_{11} + b_{12} \end{bmatrix} .$$

Hence, $b_{11} = 2$, $b_{12} = 0$. Also, $\begin{bmatrix} 0 \\ 6 \end{bmatrix} = b_{21} \begin{bmatrix} 1 \\ 2 \end{bmatrix} + b_{22} \begin{bmatrix} 0 \\ 1 \end{bmatrix}$

and, thus, $b_{21} = 0$; $b_{22} = 6$. The matrix of $T(x,y)$ with respect to the
basis $B' = \{(1,2),(0,1)\}$ is, therefore,
$$M = \begin{bmatrix} 2 & 0 \\ 0 & 6 \end{bmatrix} .$$

● **PROBLEM** 3-21

If $A = \begin{bmatrix} 1 & -2 & 3 & 0 \\ 2 & 3 & 0 & 1 \\ 1 & -1 & 1 & 1 \end{bmatrix}$ is the matrix of a linear transforma-

tion $T: V \to W$ relative to bases $\{v_i\}$ and $\{w_i\}$ of V and W
respectively, determine $T(v)$ where $v = 3v_1 + v_2 - 2v_3$.

Solution: Since A represents the linear transformation T with respect
to the basis $\{v_i\}$, we multiply A by the coordinate vector v. Since
A is a 3×4 matrix, realize that T maps a subset of V with dimension 4 into a subset of W with dimension 3. So the coordinate
vector of v is $(3,1,-2,0)$. Then,
$$\begin{bmatrix} 1 & -2 & 3 & 0 \\ 2 & 3 & 0 & 1 \\ 1 & -1 & 1 & 1 \end{bmatrix} \begin{bmatrix} 3 \\ 1 \\ -2 \\ 0 \end{bmatrix} = \begin{bmatrix} -5 \\ 9 \\ 0 \end{bmatrix} .$$
The coordinate vector of Tv in W is $[-5,9,0]$ and, hence, $Tv =$
$-5w_1 + 9w_2 + 0w_3 = -5w_1 + 9w_2$.

● **PROBLEM** 3-22

Let $B_1 = \{(2,1),(1,0)\}$ and $B_2 = \{(-1,2),(3,1)\}$ be two bases for R^2,
and let the matrix of $T(x,y)$ with respect to B_1 be $A = \begin{bmatrix} 2 & 0 \\ 1 & 3 \end{bmatrix}$.

Solution: Since the matrix of a transformation depends on the basis chosen, a change in basis will result in a change in the matrix of the transformation.

Let $(u_1,u_2),(v_1,v_2)$ be two bases for R^2. Then $T: R^2 \to R^2$ is represented by two matrices determined by the relations:

$$T(u_1) = a_{11}u_1 + a_{12}u_2$$
$$T(u_2) = a_{21}u_1 + a_{22}u_2 \tag{1}$$
$$T(v_1) = b_{11}v_1 + b_{12}v_2$$
$$T(v_2) = b_{21}v_1 + b_{22}v_2 \ . \tag{2}$$

From (1) and (2) we obtain the matrices

$$A = \begin{bmatrix} a_{11} & a_{21} \\ a_{12} & a_{22} \end{bmatrix} \qquad B = \begin{bmatrix} b_{11} & b_{21} \\ b_{12} & b_{22} \end{bmatrix} \ .$$

The question is: How can we get from A to B? To do so, the transition matrix from u_i to v_i and the transition matrix from v_i to u_i are needed. Since $v_1, v_2 \in R^2$,

$$v_1 = p_{11}u_1 + p_{12}u_2$$
$$v_2 = p_{21}u_1 + p_{22}u_2 \ . \tag{3}$$

Since $u_1, u_2 \in R^2$,

$$u_1 = q_{11}v_1 + q_{12}v_2$$
$$u_2 = q_{21}v_1 + q_{22}v_2 \ . \tag{4}$$

Thus,

$$P = \begin{bmatrix} p_{11} & p_{21} \\ p_{12} & p_{22} \end{bmatrix} \quad \text{and} \quad Q = \begin{bmatrix} q_{11} & q_{21} \\ q_{12} & q_{33} \end{bmatrix}$$

are the required matrices. According to a theorem in linear algebra, the matrix product $QAP = B$, i.e., it represents T with respect to the basis $\{v_1,v_2\}$. Examining the given problem, we are given A, the matrix of T with respect to B_1. Thus, we must find the transition matrices.

$$(-1,2) = p_{11}(2,1) + p_{12}(1,0) = 2(2,1) - 5(1,0)$$
$$(3,1) = p_{21}(2,1) + p_{22}(1,0) = 1(2,1) + 1(1,0)$$

Hence, $P = \begin{bmatrix} 2 & 1 \\ -5 & 1 \end{bmatrix}$. Similarly,

$$(2,1) = q_{11}(-1,2) + q_{12}(3,1) = \tfrac{1}{7}(-1,2) + \tfrac{5}{7}(3,1)$$
$$(1,0) = q_{21}(-1,2) + q_{22}(3,1) = -\tfrac{1}{7}(-1,2) + \tfrac{2}{7}(3,1) \ .$$

Hence, $Q = \begin{bmatrix} 1/7 & -1/7 \\ 5/7 & 2/7 \end{bmatrix}$. Then,

$$QAP = \begin{bmatrix} 1/7 & -1/7 \\ 5/7 & 2/7 \end{bmatrix} \begin{bmatrix} 2 & 0 \\ 1 & 3 \end{bmatrix} \begin{bmatrix} 2 & 1 \\ -5 & 1 \end{bmatrix}$$

$$= \begin{bmatrix} 1/7 & -1/7 \\ 5/7 & 2/7 \end{bmatrix} \begin{bmatrix} 4 & 2 \\ -13 & 4 \end{bmatrix}$$

$$= \frac{1}{7} \begin{bmatrix} 1 & -1 \\ 5 & 2 \end{bmatrix} \begin{bmatrix} 4 & 2 \\ -13 & 4 \end{bmatrix}$$

$$= \frac{1}{7} \begin{bmatrix} 17 & -2 \\ -6 & 18 \end{bmatrix} = B .$$

Note that $Q = P^{-1}$ or $PQ = I$. Therefore, $B = P^{-1}AP$. Using this equation, the matrix B which represents T with respect to the new basis $\{v_1, v_2\}$ can also be obtained.

ARITHMETIC OF LINEAR TRANSFORMATIONS

● **PROBLEM** 3-23

Let $T: R^4 \rightarrow R^2$ be given by

$$T(x,y,z,t) = (x + 2y + z + t, 2x + 4y - z) .$$

Find the null space of the matrix associated with T under the standard basis:
$$\{(1,0,0,0),(0,1,0,0),(0,0,1,0),(0,0,0,1)\} .$$
The basis for R^2 is also standard, i.e., $\{(1,0),(0,1)\} .$

Solution: Every linear transformation $T: R^m \rightarrow R^n$ has associated with it an $n \times m$ matrix relative to the standard basis. Thus, we expect a 2×4 matrix for the given transformation.

The null space of an $m \times n$ matrix is the solution set to the matrix equation $Au = 0$. This problem has far-reaching ramifications in the theory of linear transformations. To find the matrix associated with T, we compute the effects of T on the basis vectors given.

$$T(1,0,0,0) = (1,2) = 1(1,0) + 2(0,1)$$

$$T(0,1,0,0) = (2,4) = 2(1,0) + 4(0,1)$$

$$T(0,0,1,0) = (1,-1) = 1(1,0) - 1(0,1)$$

$$T(0,0,0,1) = (1,0) = 1(1,0) + 0(0,1) .$$

$$T \begin{bmatrix} e_1 \\ e_2 \\ e_3 \\ e_4 \end{bmatrix} = \begin{bmatrix} 1 & 2 \\ 2 & 4 \\ 1 & -1 \\ 1 & 0 \end{bmatrix} \begin{bmatrix} (1,0) \\ (0,1) \end{bmatrix} \qquad (1)$$

If we wish to replace T by the matrix in (1), the result will not be conformable for multiplication. Hence, we must transpose to obtain

$$\begin{bmatrix} 1 & 2 & 1 & 1 \\ 2 & 4 & -1 & 0 \end{bmatrix} = A . \qquad (2)$$

The matrix (2) is actually the matrix of coefficients to a set of two linear equations in four unknowns. The first step towards solution is to operate on (2) until the leading coefficients have been normalized (set equal to one). This process is called reduction of the coefficient matrix. The matrix (2) reduces to

$$\begin{bmatrix} 1 & 2 & 0 & \frac{1}{3} \\ 0 & 0 & 1 & \frac{2}{3} \end{bmatrix} .$$

Seek the set of solutions to

$$x_1 + 2x_2 + \tfrac{1}{3}x_4 = 0$$

$$x_3 + \tfrac{2}{3}x_4 = 0 .$$

The two leading variables can be set equal to the remaining unknowns:

$$x_1 = -2x_2 - \tfrac{1}{3}x_4 \tag{3}$$
$$x_3 = 0x_2 - \tfrac{2}{3}x_4$$

Two particular solutions to (3) are found by first setting $x_2 = 1$ and $x_4 = 0$ and then setting $x_2 = 0$ and $x_4 = 1$. Then,

$$u_1 = \begin{bmatrix} -2 \\ 1 \\ 0 \\ 0 \end{bmatrix} , \quad u_2 = \begin{bmatrix} -\frac{1}{3} \\ 0 \\ -\frac{2}{3} \\ 1 \end{bmatrix}$$

are solutions to $Au = 0$. The null space of A therefore, consists of all linear combinations of u_1 and u_2, i.e., of the form $c_1u_1 + c_2u_2$ (since u_1 and u_2 are independent and span the null space). The null space of A is nothing more than the kernel of the given linear transformation.

● PROBLEM 3-24

Let $S: R^2 \rightarrow R^2$ and $T: R^2 \rightarrow R^2$ be given by $S(x,y) = (x-2y, 2x+7y)$, $T(x,y) = (3x+2y, x-y)$. Find $(S+T)(x,y)$ and $(3T)(x,y)$, in both functional and matrix form.

Solution: Addition of linear transformations is defined as follows: Let $S,T: V \rightarrow W$ be linear transformations. Let $v \in V$ and $a \in K$, the field over which V,W are defined. Then the sum of S and T at v, $S+T$, is the transformation such that $(S+T)v = S(v) + T(v)$, and the scalar multiple aT is defined by $(aT)v = a(Tv)$.

According to the above definition,

$$(S+T)(x,y) = S(x,y) + T(x,y)$$

$$= (x-2y, 2x+7y) + (3x+2y, x-y)$$

$$= (4x, 3x+6y) \tag{1}$$

and

$$(3T)(x,y) = 3[T(x,y)] = 3(3x+2y, x-y)$$

$$= (9x+6y, 3x-3y) . \tag{2}$$

To find the corresponding matrix forms, first compute the matrices of the linear transformations. Rewriting the vectors $[x,y]$ as a column matrix,

$$S \begin{bmatrix} x \\ y \end{bmatrix} = \begin{bmatrix} x - 2y \\ 2x + 7y \end{bmatrix} = \begin{bmatrix} 1 & -2 \\ 2 & 7 \end{bmatrix} \begin{bmatrix} x \\ y \end{bmatrix}$$

116

$$T \begin{bmatrix} x \\ y \end{bmatrix} = \begin{bmatrix} 3x + 2y \\ x - y \end{bmatrix} = \begin{bmatrix} 3 & 2 \\ 1 & -1 \end{bmatrix} \begin{bmatrix} x \\ y \end{bmatrix}$$

Thus, the matrices of S and T are, respectively,

$$A = \begin{bmatrix} 1 & -2 \\ 2 & 7 \end{bmatrix} \quad \text{and} \quad B = \begin{bmatrix} 3 & 2 \\ 1 & -1 \end{bmatrix}$$

Then, by the rules of matrix addition and scalar multiplication,

$$A + B = \begin{bmatrix} 4 & 0 \\ 3 & 6 \end{bmatrix}, \quad 3B = \begin{bmatrix} 9 & 6 \\ 3 & -3 \end{bmatrix}. \tag{3}$$

Also, $\quad S + T \begin{bmatrix} x \\ y \end{bmatrix} = \begin{bmatrix} 4 & 0 \\ 3 & 6 \end{bmatrix} \begin{bmatrix} x \\ y \end{bmatrix},$ $\qquad\qquad$ (4)

$(S+T)(x,y) = (4x, 3x+6y)$. Thus, (4) is identical to (1).
    Similarly, $3T(x,y) = (9x+6y, 3x-3y)$ and, thus, (3) is the same as (2).

● **PROBLEM** 3-25

Let $T_1: R^4 \rightarrow R^3$ and $T_2: R^4 \rightarrow R^3$ be defined by

$T_1(x_1, x_2, x_3, x_4) = (6x_1 + 2x_2 - 3x_3, x_1 + 4x_3 + 2x_4, 3x_1 - x_2 + 2x_3 - 5x_4)$

$T_2(x_1, x_2, x_3, x_4) = (2x_1 + 4x_2 - 3x_3 + 5x_4, x_1 + 2x_2 + 3x_3 + 4x_4,$
$\qquad\qquad\qquad\qquad\qquad 4x_2 - 2x_3 - x_4)$

Find, relative to the standard basis $\{(1,0,0,0), (0,1,0,0), (0,0,1,0),$
$(0,0,0,1)\}$, the sum $T_1 + T_2$ and the linear combinations $3T_1 + 2T_2$.

Solution: Find the matrices associated with the transformations. Then,
the rules of matrix multiplication and multiplication by a scalar can be
applied to find the required sums.
    To find the matrices of the transformations, compute the effects of
T_1 and T_2 on the basis vectors of R^4 .

$\quad T_1(1,0,0,0) = (6,1,3); \ T(0,1,0,0) = (2,0,-1) \ ; \ T(0,0,1,0) = (-3,4,2);$
$\quad T(0,0,0,1) = (0,2,-5).$

The standard basis for R^3 is $\{(1,0,0), (0,1,0), (0,0,1)\} = \{e_1, e_2, e_3\}$.
Letting $\{f_1, f_2, f_3, f_4\}$ denote the standard basis for R^4 ,

$$T \begin{bmatrix} f_1 \\ f_2 \\ f_3 \\ f_4 \end{bmatrix} = \begin{bmatrix} 6 & 1 & 3 \\ 2 & 0 & -1 \\ -3 & 4 & 2 \\ 0 & 2 & -5 \end{bmatrix} \begin{bmatrix} e_1 \\ e_2 \\ e_3 \end{bmatrix} \tag{1}$$

The transpose of the matrix in (1) is the matrix of the transformation.
Thus,

$$A = \begin{bmatrix} 6 & 2 & -3 & 0 \\ 1 & 0 & 4 & 2 \\ 3 & -1 & 2 & -5 \end{bmatrix}$$

Similarly, the matrix of the transformation T_2 is found to be

$$B = \begin{bmatrix} 2 & 4 & -3 & 5 \\ 1 & 2 & 3 & 4 \\ 0 & 4 & -2 & -1 \end{bmatrix}$$

A and B are conformable for addition. A + B =

$$\begin{bmatrix} 8 & 6 & -6 & 4 \\ 2 & 2 & 7 & 6 \\ 3 & 3 & 0 & -6 \end{bmatrix}$$

We can replace T_1 and T_2 in $2T_1 + 3T_2$ by the matrices A and B respectively. Then, $3A + 2B$

$$3 \begin{bmatrix} 6 & 2 & -3 & 0 \\ 1 & 0 & 4 & 2 \\ 3 & -1 & 2 & -5 \end{bmatrix} + 2 \begin{bmatrix} 2 & 4 & -3 & 5 \\ 1 & 2 & 3 & 4 \\ 0 & 4 & -2 & -1 \end{bmatrix}$$

$$= \begin{bmatrix} 22 & 14 & -15 & 10 \\ 5 & 4 & 18 & 14 \\ 9 & 5 & 2 & -17 \end{bmatrix}$$

If it is required to have the answers in functional form, simply multiply the matrices by an arbitrary column vector of R^4. Thus,

$$T_1 + T_2 = (A+B)(v) \begin{bmatrix} 8 & 6 & -6 & 5 \\ 2 & 2 & 7 & 6 \\ 3 & 3 & 0 & -6 \end{bmatrix} \begin{bmatrix} x_1 \\ x_2 \\ x_3 \\ x_4 \end{bmatrix}$$

$$= (8x_1 + 6x_2 - 6x_3 + tx_4, \; 2x_1 + 2x_2 + 7x_3 + 6x_4, \; 3x_1 + 3x_2 - 6x_4).$$

Similarly, $3A + 2B$

$$= (22x_1 + 14x_2 - 15x_3 + 10x_4, \; 5x_1 + 4x_2 + 18x_3 + 14x_4, \; 9x_1 + 5x_2 + 2x_3 - 17x_4).$$

● **PROBLEM** 3-26

Systems of linear transformations and matrices are isomorphic vector spaces. Show by means of an example that the product of two linear transformations can be represented as the product of two matrices with respect to a given basis.

<u>Solution</u>: Let $T_1: R^3 \rightarrow R^3$, $T_2: R^3 \rightarrow R^3$ be given by $T_1(x_1, x_2, x_3) = (2x_1 - 5x_3, \; 4x_1 + 2x_2 - 2x_3, \; -x_1 + 6x_2 + x_3)$ and $T_2(x_1, x_2, x_3) = (x_1 - x_3, \; 2x_1 - x_2 + 2x_3, \; x_1 + 2x_2 + 3x_3)$. The matrices associated with these transformations are

$$A = \begin{bmatrix} 2 & 4 & -1 \\ 0 & 2 & 6 \\ -5 & -2 & 1 \end{bmatrix} \quad \text{and} \quad B = \begin{bmatrix} 1 & 2 & 1 \\ 0 & -1 & 2 \\ -1 & 2 & 3 \end{bmatrix}$$

respectively.

Let $\{\alpha_1, \alpha_2, \alpha_3\}$ be the basis chosen for R^3. Then, we want to find the matrix representation of $T_1 T_2$ with respect to $\{\alpha_1, \alpha_2, \alpha_3\}$. We must describe the effects of T_1 and T_2 on $\{\alpha_1, \alpha_2, \alpha_3\}$.

118

$$T_1\alpha_1 = 2\alpha_1 - 5\alpha_3 \ ; \ T_1\alpha_2 = 4\alpha_1 + 2\alpha_2 - 2\alpha_3 \ ; \ T_1\alpha_3 = -\alpha_1 + 6\alpha_2 + \alpha_3 \ ;$$
$$T_2\alpha_1 = \alpha_1 - \alpha_3 \ ; \ T_2\alpha_2 - \alpha_2 + 2\alpha_3 \ ; \ T_2\alpha_3 = \alpha_1 + 2\alpha_2 + 3\alpha_3 \ .$$

It follows from these equalities that

$$(T_1 T_2)\alpha_1 = T_1(T_2\alpha_1) = T_1(\alpha_1 - \alpha_3) = T_1\alpha_1 - T_1\alpha_3$$
$$= (2\alpha_1 - 5\alpha_3) - (-\alpha_1 + 6\alpha_2 + \alpha_3) = 3\alpha_1 - 6\alpha_2 - 6\alpha_3$$

$$(T_1 T_2)\alpha_2 = T_1(T_2\alpha_2) = T_1(2\alpha_1 - \alpha_2 + 2\alpha_3) = 2(T_1\alpha_1) - T_1\alpha_2 + 2(T_1\alpha_3)$$
$$= 2(2\alpha_1 - 5\alpha_3) - (4\alpha_1 + 2\alpha_2 - 2\alpha_3) + 2(-\alpha_1 + 6\alpha_2 + \alpha_3)$$
$$= -2\alpha_1 + 10\alpha_2 - 6\alpha_3$$

$$(T_1 T_2)\alpha_3 = T_1(T_2\alpha_3) = T_1(\alpha_1 + 2\alpha_2 + 3\alpha_3) = T_1\alpha_1 + 2(T_1\alpha_2) + 3(T_1\alpha_3)$$
$$= (2\alpha_1 - 5\alpha_3) + 2(4\alpha_1 + 2\alpha_2 - 2\alpha_3) + 3(-\alpha_1 + 6\alpha_2 + \alpha_3)$$
$$= 7\alpha_1 + 22\alpha_2 - 6\alpha_3 \ .$$

Thus, the product $T_1 T_2$ is represented by the matrix C where

$$C = \begin{bmatrix} 3 & -2 & 7 \\ -6 & 10 & 22 \\ -6 & -6 & -6 \end{bmatrix}$$

Similarly, applying the rules of matrix multiplication to the matrices A and B,

$$AB = \begin{bmatrix} 2 & 4 & -1 \\ 0 & 2 & 6 \\ -5 & -2 & 1 \end{bmatrix} \begin{bmatrix} 1 & 2 & 1 \\ 0 & -1 & 2 \\ -1 & 2 & 3 \end{bmatrix} = \begin{bmatrix} 3 & -2 & 7 \\ -6 & 10 & 22 \\ -6 & -6 & -6 \end{bmatrix}$$

The product $T_2 T_1$ would have corresponded to the matrix product BA. Note that the same basis was chosen for both matrix representations.

CHAPTER 4

BASIC MATRIX ARITHMETIC

ADDITION OF MATRICES

● PROBLEM 4-1

Prove that if $A = (a_{ij})$ and $B = (b_{ij})$ are $m \times n$ matrices over a field K and c is an element of K, then $A + B = (c_{ij})$ where $c_{ij} = a_{ij} + b_{ij}$, $i = 1,2,\ldots,m$, $j = 1,2,\ldots,n$ and $cA = (d_{ij})$ where $d_{ij} = ca_{ij}$, $i = 1,2,\ldots,m$, $j = 1,2,\ldots,n$.

<u>Solution</u>: Let e_j represent the vector $(0,\ldots,0,1,0,\ldots,0)$ where the 1 is in the ith place. Letting $C = (c_{ij})$, then the ith row of C is given by $e_i \cdot c = (c_{i1},c_{i2},\ldots,c_{in})$; $i = 1,2,\ldots,m$. On the other hand,

$$e_i(A+B) = e_iA + e_iB = (a_{i1},a_{i2},\ldots,a_{in}) + (b_{i1},b_{i2},\ldots,b_{in})$$

Adding corresponding elements in the ith row,

$$e_i(A+B) = (a_{i1} + b_{i1},a_{i2} + b_{i2},\ldots,a_{in} + b_{in}) .$$

But, by definition, $c_{ij} = a_{ij} + b_{ij}$. Therefore, by sybstitution, $e_i(A+B) = (c_{i1},c_{i2},\ldots,c_{in}) = e_iC$.

This is true for all $i = 1,\ldots,m$. Therefore, $A + B = C$.

(b)   Let $D = (d_{ij})$ then,

$$e_i \cdot D = (d_{i1},d_{i2},\ldots,d_{in}) , \quad i = 1,2,\ldots,m .$$

On the other hand,

$$e_i(cA) = c(e_iA) = c(a_{i1},a_{i2},\ldots,a_{in}), \quad i = 1,2,\ldots,m$$
$$= (ca_{i1}, ca_{i2},\ldots,ca_{in}) .$$

But, by definition, $d_{ij} = ca_{ij}$. Therefore, $e_i(cA) = (d_{i1},\ldots,d_{in})$ by substitution; i.e., $e_i(cA) = e_iD$. This holds for all $i = 1,\ldots,m$. Therefore, $cA = D$.

● PROBLEM 4-2

If $A = \begin{bmatrix} 2 & -2 & 4 \\ -1 & 1 & 1 \end{bmatrix}$ and $B = \begin{bmatrix} 0 & 1 & -3 \\ 1 & 3 & 1 \end{bmatrix}$, find $2A + B$.

Solution: For an $m \times n$ matrix, $A = (a_{ij})$, we know $cA = (ca_{ij})$. Hence,

$$2A = 2 \begin{bmatrix} 2 & -2 & 4 \\ -1 & 1 & 1 \end{bmatrix} = \begin{bmatrix} 2 \cdot 2 & 2 \cdot (-2) & 2 \cdot 4 \\ 2 \cdot (-1) & 2 \cdot 1 & 2 \cdot 1 \end{bmatrix} = \begin{bmatrix} 4 & -4 & 8 \\ -2 & 2 & 2 \end{bmatrix}.$$

For two $m \times n$ matrices, $A = (\alpha_{ij})$ and $B = (\beta_{ij})$, the ith row of the matrix $A + B$ is given by $e_i \cdot (A+B) = (\alpha_{i1} + \beta_{i1}, \ldots, \alpha_{in} + \beta_{in})$. Thus,

$$2A + B = \begin{bmatrix} 4 & -4 & 8 \\ -2 & 2 & 2 \end{bmatrix} + \begin{bmatrix} 0 & 1 & -3 \\ 1 & 3 & 1 \end{bmatrix} = \begin{bmatrix} 4+0 & -4+1 & 8-3 \\ -2+1 & 2+3 & 2+1 \end{bmatrix}$$

$$2A + B = \begin{bmatrix} 4 & -3 & 5 \\ -1 & 5 & 3 \end{bmatrix}.$$

● **PROBLEM 4-3**

Find $A + B$ where

$$A = \begin{bmatrix} 1 & -2 & 4 \\ 2 & -1 & 3 \end{bmatrix}, \quad B = \begin{bmatrix} 0 & 2 & 4 \\ 1 & 3 & 1 \end{bmatrix}.$$

Solution: Definition: If $A = [a_{ij}]$ and $B = [b_{ij}]$ are $m \times n$ matrices, then the sum of A and B is the matrix $C = [c_{ij}]$, defined by

$$c_{ij} = a_{ij} + b_{ij}.$$

That is, C is obtained by adding corresponding elements of A and B. Now,

$$A + B = \begin{bmatrix} 1+0 & -2+2 & 4-4 \\ 2+1 & -1+3 & 3+1 \end{bmatrix} = \begin{bmatrix} 1 & 0 & 0 \\ 3 & 2 & 4 \end{bmatrix}.$$

It should be noted that the sum of the matrices A and B is defined only when A and B have the same number of rows and the same number of columns. That occurs only when A and B are of the same size. We say that the matrices are conformable for addition.

● **PROBLEM 4-4**

Let $A = \begin{bmatrix} 2 & 3 & 7 \\ 4 & m & \sqrt{3} \\ 1 & 5 & a \end{bmatrix}$ $B = \begin{bmatrix} \alpha & \beta & \delta \\ \sqrt{5} & 3 & 1 \\ p & q & 4 \end{bmatrix}$.

Find $A + B$.

Solution: Using the definition of matrix addition, add the (i,j)th element of A to the (i,j)th element of B. Thus,

$$A + B = \begin{bmatrix} 2 & 3 & 7 \\ 4 & m & \sqrt{3} \\ 1 & 5 & a \end{bmatrix} + \begin{bmatrix} \alpha & \beta & \delta \\ \sqrt{5} & 3 & 1 \\ p & q & 4 \end{bmatrix} = \begin{bmatrix} 2+\alpha & 3+\beta & 7+\delta \\ 4+\sqrt{5} & m+3 & \sqrt{3}+1 \\ 1+p & 5+q & a+4 \end{bmatrix}.$$

● **PROBLEM 4-5**

If $A = \begin{bmatrix} 2 & 3 & 4 \\ 1 & 2 & 1 \end{bmatrix}$ and $B = \begin{bmatrix} 0 & 2 & 7 \\ 1 & -3 & 5 \end{bmatrix}$, find $A - B$.

121

<u>Solution</u>: Let A and B be two matrices with the same size, m×n . By definition, $-B = -1 \cdot B$ and $A - B = A + (-B)$. Note that $-1 \cdot B$ is multiplication of a matrix by a scalar which is defined by $-1 \cdot B = (-1 \cdot b_{ij})$ if $B = (b_{ij})$. Also, the $+$ in $A + (-B)$ denotes addition of matrices.

$$B = \begin{bmatrix} 0 & 2 & 7 \\ 1 & -3 & 5 \end{bmatrix} .$$

From the above definitions,

$$-B = (-1)B = \begin{bmatrix} -1 \cdot 0 & -1 \cdot 2 & -1 \cdot 7 \\ -1 \cdot 1 & -1 \cdot -3 & -1 \cdot 5 \end{bmatrix} = \begin{bmatrix} 0 & -2 & -7 \\ 1 & 3 & -5 \end{bmatrix} ,$$

and

$$A - B = A + (-B) = \begin{bmatrix} 2 & 3 & 4 \\ 1 & 2 & 1 \end{bmatrix} + \begin{bmatrix} 0 & -2 & -7 \\ -1 & 3 & -5 \end{bmatrix}$$

$$= \begin{bmatrix} 2+0 & 3-2 & 4-7 \\ 1-1 & 2+3 & 1-5 \end{bmatrix}$$

$$= \begin{bmatrix} 2 & 1 & -3 \\ 0 & 5 & -4 \end{bmatrix} .$$

Observe that $A - B$ can be obtained directly by subtracting the entries of B from the corresponding entries of A.

● **PROBLEM 4-6**

Find $A - B$ if $A = \begin{bmatrix} 3 & -2 & 5 \\ -1 & 2 & 3 \end{bmatrix}$ and $B = \begin{bmatrix} 2 & 3 & 2 \\ -3 & 4 & 6 \end{bmatrix}$.

<u>Solution</u>: Let A and B be two matrices with the same size, i.e., the same number of rows and of columns, say m×n .

$$A = \begin{bmatrix} a_{11} & a_{12} & \cdots & a_{1n} \\ \vdots & \vdots & & \vdots \\ a_{m1} & a_{m2} & \cdots & a_{mn} \end{bmatrix}$$

and

$$B = \begin{bmatrix} b_{11} & b_{12} & \cdots & b_{1n} \\ \vdots & \vdots & & \vdots \\ b_{m1} & b_{m2} & \cdots & b_{mn} \end{bmatrix} .$$

The sum of A and B, written $A + B$, is the matrix obtained by adding corresponding components.

$$A+B = \begin{bmatrix} a_{11}+b_{11} & a_{12}+b_{12} & \cdots & a_{1n}+b_{1n} \\ a_{21}+b_{21} & a_{22}+b_{22} & \cdots & a_{2n}+b_{2n} \\ \cdots & \cdots & \cdots & \cdots \\ a_{m1}+b_{m1} & a_{m2}+b_{m2} & & a_{mn}+b_{mn} \end{bmatrix}$$

Also, define $A - B \equiv A + (-B)$. Here $-B$ represents the matrix $-1 \cdot B$.

In the given problem,

$$A = \begin{bmatrix} 3 & -2 & 5 \\ -1 & 2 & 3 \end{bmatrix} \quad ; \quad B = \begin{bmatrix} 2 & 3 & 2 \\ -3 & 4 & 6 \end{bmatrix}.$$

Then,

$$-1 \cdot B = \begin{bmatrix} -1 \cdot 2 & -1 \cdot 3 & -1 \cdot 2 \\ -1 \cdot (-3) & -1 \cdot 4 & -1 \cdot 6 \end{bmatrix} = \begin{bmatrix} -2 & -3 & -2 \\ 3 & -4 & -6 \end{bmatrix}.$$

Thus,

$$A - B = A + (-B) = \begin{bmatrix} 3 & -2 & 5 \\ -1 & 2 & 3 \end{bmatrix} = \begin{bmatrix} -2 & -3 & -2 \\ 3 & -4 & -6 \end{bmatrix}.$$

● **PROBLEM** 4-7

Show that

a) $A + B = B + A$ where

$$A = \begin{bmatrix} 3 & 1 & 1 \\ 2 & -1 & 1 \end{bmatrix} \quad ; \quad B = \begin{bmatrix} 4 & 2 & -1 \\ 0 & 0 & 2 \end{bmatrix}.$$

b) $(A+B) + C = A + (B+C)$ where

$$A = \begin{bmatrix} -2 & 6 \\ 2 & 1 \end{bmatrix}, \quad B = \begin{bmatrix} 2 & 1 \\ -0 & 3 \end{bmatrix} \quad \text{and} \quad C = \begin{bmatrix} -1 & 0 \\ 7 & 2 \end{bmatrix}.$$

c)   If A and the zero matrix (0_{ij}) have the same size, then
$A + 0 = A$ where

$$A = \begin{bmatrix} 2 & 1 \\ 1 & 2 \end{bmatrix}.$$

d) $A + (-A) = 0$ where

$$A = \begin{bmatrix} 2 & 1 \\ 1 & 2 \end{bmatrix}.$$

e) $(ab)A = a(bA)$ where $a = -5$, $b = 3$ and

$$A = \begin{bmatrix} 6 & -1 & 0 \\ 1 & 2 & 1 \end{bmatrix}.$$

f)   Find B if $2A - 3B + C = 0$ where

$$A = \begin{bmatrix} -1 & 3 \\ 0 & 0 \end{bmatrix} \quad \text{and} \quad C = \begin{bmatrix} -2 & -1 \\ -1 & 1 \end{bmatrix}.$$

<u>Solution</u>: a) By the definition of matrix addition,

$$A + B = \begin{bmatrix} 3 & 1 & 1 \\ 2 & -1 & 1 \end{bmatrix} + \begin{bmatrix} 4 & 2 & -1 \\ 0 & 0 & 2 \end{bmatrix}$$

$$= \begin{bmatrix} 3+4 & 1+2 & 1+(-1) \\ 2+0 & -1+0 & 1+2 \end{bmatrix}$$

$$= \begin{bmatrix} 7 & 3 & 0 \\ 2 & -1 & 3 \end{bmatrix}$$

and

$$B + A = \begin{bmatrix} 4 & 2 & -1 \\ 0 & 0 & 2 \end{bmatrix} + \begin{bmatrix} 3 & 1 & 1 \\ 2 & -1 & 1 \end{bmatrix}$$

$$= \begin{bmatrix} 4+3 & 2+1 & -1+1 \\ 0+2 & 0+(-1) & 2+1 \end{bmatrix} + \begin{bmatrix} 7 & 3 & 0 \\ 2 & -1 & 3 \end{bmatrix}$$

Thus, $A + B = B + A$.

b)
$$A + B = \begin{bmatrix} -2 & 6 \\ 2 & 1 \end{bmatrix} + \begin{bmatrix} 2 & 1 \\ 0 & 3 \end{bmatrix}$$

$$= \begin{bmatrix} -2+2 & 6+1 \\ 2+0 & 1+3 \end{bmatrix} = \begin{bmatrix} 0 & 7 \\ 2 & 4 \end{bmatrix}$$

and
$$(A+B) + C = \begin{bmatrix} 0 & 7 \\ 2 & 4 \end{bmatrix} + \begin{bmatrix} -1 & 0 \\ 7 & 2 \end{bmatrix} = \begin{bmatrix} 0+(-1) & 7+0 \\ 2+7 & 4+2 \end{bmatrix} = \begin{bmatrix} -1 & 7 \\ 9 & 6 \end{bmatrix} .$$

$$B + C = \begin{bmatrix} 2 & 1 \\ 0 & 3 \end{bmatrix} + \begin{bmatrix} -1 & 0 \\ 7 & 2 \end{bmatrix} = \begin{bmatrix} 2+(-1) & 1+0 \\ 0+7 & 3+2 \end{bmatrix} = \begin{bmatrix} 1 & 1 \\ 7 & 5 \end{bmatrix}$$

and
$$A + (B+C) = \begin{bmatrix} -2 & 6 \\ 2 & 1 \end{bmatrix} + \begin{bmatrix} 1 & 1 \\ 7 & 5 \end{bmatrix} = \begin{bmatrix} -2+1 & 6+1 \\ 2+7 & 1+5 \end{bmatrix} = \begin{bmatrix} -1 & 7 \\ 9 & 6 \end{bmatrix} .$$

Thus, $(A+B) + C = A + (B+C)$.

c)   An $m \times n$ matrix all of whose elements are zeros is called a zero matrix and is usually denoted by $0_{m \, n}$.

$$A = \begin{bmatrix} 2 & 1 \\ 1 & 2 \end{bmatrix} \qquad 0 = \begin{bmatrix} 0 & 0 \\ 0 & 0 \end{bmatrix}.$$

Thus,
$$A + 0 = \begin{bmatrix} 2 & 1 \\ 1 & 2 \end{bmatrix} + \begin{bmatrix} 0 & 0 \\ 0 & 0 \end{bmatrix} = \begin{bmatrix} 2+0 & 1+0 \\ 1+0 & 2+0 \end{bmatrix} = \begin{bmatrix} 2 & 1 \\ 1 & 2 \end{bmatrix} .$$

Hence, $A + 0 = A$.

d) $-A = -1 \cdot \begin{bmatrix} 2 & 1 \\ 1 & 2 \end{bmatrix} = \begin{bmatrix} -1 \cdot 2 & -1 \cdot 1 \\ -1 \cdot 1 & -1 \cdot 2 \end{bmatrix} = \begin{bmatrix} -2 & -1 \\ -1 & -2 \end{bmatrix} .$

Thus,
$$A + (-A) = \begin{bmatrix} 2 & 1 \\ 1 & 2 \end{bmatrix} + \begin{bmatrix} -2 & -1 \\ -1 & -2 \end{bmatrix} = \begin{bmatrix} 2+(-2) & 1+(-1) \\ 1+(-1) & 2+(-2) \end{bmatrix} = \begin{bmatrix} 0 & 0 \\ 0 & 0 \end{bmatrix}$$

Therefore, $A + (-A) = 0$.

e)   If $A = \begin{bmatrix} a_1 & b_1 \\ c_1 & d_1 \end{bmatrix}$ and a is any scalar from a field, aA is

defined by
$$aA = \begin{bmatrix} aa_1 & ab_1 \\ ac_1 & ad_1 \end{bmatrix} .$$

So, $bA = 3 \begin{bmatrix} 6 & -1 & 0 \\ 1 & 2 & 1 \end{bmatrix} = \begin{bmatrix} 3 \cdot 6 & 3 \cdot (-1) & 3 \cdot 0 \\ 3 \cdot 1 & 3 \cdot 2 & 3 \cdot 1 \end{bmatrix}$

$$= \begin{bmatrix} 18 & -3 & 0 \\ 3 & 6 & 3 \end{bmatrix}$$

and
$$a(bA) = -5 \begin{bmatrix} 18 & -3 & 0 \\ 3 & 6 & 3 \end{bmatrix} = \begin{bmatrix} -90 & 15 & 0 \\ -15 & -30 & -15 \end{bmatrix}$$

$$(ab)A = ((-5)(3)) \begin{bmatrix} 6 & -1 & 0 \\ 1 & 2 & 1 \end{bmatrix} = -15 \begin{bmatrix} 6 & -1 & 0 \\ 1 & 2 & 1 \end{bmatrix} = \begin{bmatrix} -90 & 15 & 0 \\ -15 & -30 & -15 \end{bmatrix}$$

Thus, $(ab)A = a(bA)$.

f) $2A - 3B + C = 2A + C - 3B = 0$ since matrix addition is commutative.

Now, add 3B to both sides of the equation,

$$2A + C - 3B = 0 ,$$

to obtain $2A + C - 3B + 3B = 0 + 3B$.   (1)

Using the laws we exemplified in parts a) through d), (1) becomes
$2A + C = 3B$.   Now,

$$\tfrac{1}{3}(2A + C) = \tfrac{1}{3}(3B)$$

which implies $B = \tfrac{1}{3}(2A + C)$.

$$2A + C = \begin{bmatrix} 2(-1) & 2(3) \\ 2(0) & 2(0) \end{bmatrix} + \begin{bmatrix} -2 & -1 \\ -1 & 1 \end{bmatrix} = \begin{bmatrix} -4 & 5 \\ -1 & 1 \end{bmatrix}.$$

Thus,

$$\tfrac{1}{3}(2A + C) = \tfrac{1}{3}\begin{bmatrix} -4 & 5 \\ -1 & 1 \end{bmatrix} = \begin{bmatrix} -4/3 & 5/3 \\ -1/3 & 1/3 \end{bmatrix} .$$

MATRIX MULTIPLICATION

a) If $A = (a_{ij})$ is a $p \times q$ matrix and $B = (b_{ij})$ is a $q \times r$ matrix, prove AB is the $p \times r$ matrix (c_{ij}) where

$$c_{ij} = \sum_{k=1}^{q} a_{ik} b_{kj} , \quad \begin{array}{l} i = 1,2,\ldots,p \\ j = 1,2,\ldots,r \end{array} .$$

b) If $A = \begin{bmatrix} 2 & 1 & 1 \\ -1 & 2 & 3 \\ 1 & 0 & 1 \end{bmatrix}$ and $B = \begin{bmatrix} 2 & 1 \\ -1 & 1 \\ 2 & -1 \end{bmatrix}$, find AB .

Solution: Let e_i represent $(0,\ldots,0,1,0,\ldots,0)$ with the 1 in the ith place. Let $C = (c_{ij})$. Consequently, the ith row of C is given by $e_i \cdot (C) = (c_{i1},\ldots,c_{ir}) = \sum_{j=1}^{r} c_{ij} \cdot e_j$, $i = 1,\ldots,p$. On the other hand, since

$$e_i A = \sum_{k=1}^{r} a_{ik} e_k \quad \text{and} \quad e_k B = \sum_{j=1}^{r} b_{kj} e_j ,$$

$$e_i(AB) = (e_i A)B = \sum_{k=1}^{r} (a_{ik} e_k)B = \sum_{k=1}^{r} a_{ik} \left(\sum_{j=1}^{r} b_{kj} e_j \right)$$

$$= \sum_{j=1}^{r} \left(\sum_{k=1}^{r} a_{ik} b_k \right) e_j , \quad i = 1,2,\ldots,p .$$

But, by definition, $\sum_{k=1}^{r} a_{ik} b_{kj} = c_{ij}$. Thus, by substitution,

$$e_i(A \cdot B) = \sum_{j=1}^{r} c_{ij} e_j = e_i C .$$

This holds for all $i = 1,\ldots,p$. Therefore, $AB = C$.

$$AB = \begin{bmatrix} 2 & 1 & 1 \\ -1 & 2 & 3 \\ 1 & 0 & 1 \end{bmatrix}\begin{bmatrix} 2 & 1 \\ -1 & 1 \\ 2 & -1 \end{bmatrix}$$

$$= \begin{bmatrix} (2 \cdot 2 + 1 \cdot (-1) + 1 \cdot 2) & (2 \cdot 1 + 1 \cdot 1 + 1 \cdot (-1)) \\ (-1 \cdot 2 + 2 \cdot (-1) + 3 \cdot 2) & (-1 \cdot 1 + 2 \cdot 1 + 3 \cdot (-1)) \\ (1 \cdot 2 + 0 \cdot (-1) + 1 \cdot 2) & (1 \cdot 1 + 0 \cdot 1 + 1 \cdot (-1)) \end{bmatrix}$$

$$= \begin{bmatrix} 4-1+2 & 2+1-1 \\ -2-2+6 & -1+2-3 \\ 2+0+2 & 1+0-1 \end{bmatrix} = \begin{bmatrix} 5 & 2 \\ 2 & -2 \\ 4 & 0 \end{bmatrix}$$

● **PROBLEM** 4-9

If $A = \begin{bmatrix} 1 & 2 & 4 \\ 2 & 6 & 0 \end{bmatrix}$; $B = \begin{bmatrix} 4 & 1 & 4 & 3 \\ 0 & -1 & 3 & 1 \\ 2 & 7 & 5 & 2 \end{bmatrix}$,

find AB .

Solution: Since A is a 2×3 matrix and B is a 3×4 matrix, the product AB is a 2×4 matrix.

$$AB = \begin{bmatrix} 1 & 2 & 4 \\ 2 & 6 & 0 \end{bmatrix} \begin{bmatrix} 4 & 1 & 4 & 3 \\ 0 & -1 & 3 & 1 \\ 2 & 7 & 5 & 2 \end{bmatrix}$$

$$= \begin{bmatrix} 1 \cdot 4 + 2 \cdot 0 + 4 \cdot 2 & 1 \cdot 1 + 2 \cdot (-1) + 4 \cdot 7 & 1 \cdot 4 + 2 \cdot 3 + 4 \cdot 5 & 1 \cdot 3 + 2 \cdot 1 + 4 \cdot 2 \\ 2 \cdot 2 + 6 \cdot 0 + 0 \cdot 2 & 2 \cdot 1 + 6 \cdot (-1) + 0 \cdot 7 & 2 \cdot 4 + 6 \cdot 3 + 0 \cdot 5 & 2 \cdot 3 + 6 \cdot 1 + 0 \cdot 2 \end{bmatrix}$$

$$= \begin{bmatrix} 4+0+8 & 1-2+28 & 4+6+20 & 3+2+8 \\ 8+0+0 & 2-6+0 & 8+18+0 & 6+6+0 \end{bmatrix}$$

$$AB = \begin{bmatrix} 12 & 27 & 30 & 13 \\ 8 & -4 & 26 & 12 \end{bmatrix}$$

● **PROBLEM** 4-10

If $A = \begin{bmatrix} 1 & 2 \\ -1 & 3 \end{bmatrix}$ and $B = \begin{bmatrix} 2 & 1 \\ 0 & 1 \end{bmatrix}$, show $AB \neq BA$.

Solution: A is 2×2 and B is 2×2 ; the product AB is a 2×2 matrix.

$$AB = \begin{bmatrix} 1 & 2 \\ -1 & 3 \end{bmatrix} \begin{bmatrix} 2 & 1 \\ 0 & 1 \end{bmatrix} = \begin{bmatrix} 1 \cdot 2 + 2 \cdot 0 & 1 \cdot 1 + 2 \cdot 1 \\ -1 \cdot 2 + 3 \cdot 0 & -1 \cdot 1 + 3 \cdot 1 \end{bmatrix} = \begin{bmatrix} 2+0 & 1+2 \\ -2+0 & -1+3 \end{bmatrix}$$

$$= \begin{bmatrix} 2 & 3 \\ -2 & 2 \end{bmatrix} .$$

Now,
$$BA = \begin{bmatrix} 2 & 1 \\ 0 & 1 \end{bmatrix} \begin{bmatrix} 1 & 2 \\ -1 & 3 \end{bmatrix} = \begin{bmatrix} 2 \cdot 1 + 1 \cdot (-1) & 2 \cdot 2 + 1 \cdot 3 \\ 0 \cdot 1 + 1 \cdot (-1) & 0 \cdot 2 + 1 \cdot 3 \end{bmatrix}$$

$$= \begin{bmatrix} 2-1 & 4+3 \\ 0-1 & 0+3 \end{bmatrix} = \begin{bmatrix} 1 & 7 \\ -1 & 3 \end{bmatrix}$$

Thus, $AB \neq BA$.

● **PROBLEM** 4-11

Let $A = \begin{bmatrix} 1 & 1 \\ 3 & 7 \end{bmatrix}$ and $B = \begin{bmatrix} 2 & 5 \\ 4 & 0 \end{bmatrix}$. Show $AB \neq BA$.

<u>Solution</u>:

$$AB = \begin{bmatrix} 1 & 1 \\ 3 & 7 \end{bmatrix} \begin{bmatrix} 2 & 5 \\ 4 & 0 \end{bmatrix} = \begin{bmatrix} 1 \cdot 2 + 1 \cdot 4 & 1 \cdot 5 + 1 \cdot 0 \\ 3 \cdot 2 + 7 \cdot 4 & 3 \cdot 5 + 7 \cdot 0 \end{bmatrix}$$

$$= \begin{bmatrix} 2+4 & 5+0 \\ 6+28 & 15+0 \end{bmatrix} = \begin{bmatrix} 6 & 5 \\ 34 & 15 \end{bmatrix}$$

$$BA = \begin{bmatrix} 2 & 5 \\ 4 & 0 \end{bmatrix} \begin{bmatrix} 1 & 1 \\ 3 & 7 \end{bmatrix} = \begin{bmatrix} 2 \cdot 1 + 5 \cdot 3 & 2 \cdot 1 + 5 \cdot 7 \\ 4 \cdot 1 + 0 \cdot 3 & 4 \cdot 1 + 0 \cdot 7 \end{bmatrix}$$

$$= \begin{bmatrix} 2+15 & 2+35 \\ 4+0 & 4+0 \end{bmatrix} = \begin{bmatrix} 17 & 37 \\ 4 & 4 \end{bmatrix}$$

Therefore, $AB \neq BA$.

● **PROBLEM** 4-12

Let $(r \times s)$ denote a matrix with shape $(r \times s)$. Find the shape of the following products if the product is defined
(i) $(2 \times 3)(3 \times 4)$ (iii) $(1 \times 2)(3 \times 1)$ (v) $(3 \times 4)(3 \times 4)$
(ii) $(4 \times 1)(1 \times 2)$ (iv) $(5 \times 2)(2 \times 3)$ (vi) $(2 \times 2)(2 \times 4)$

<u>Solution</u>: An $m \times p$ matrix and a $q \times n$ matrix are multipliable only when $p = q$. The product is an $m \times n$ matrix. Thus, each of the above products is defined if the inner numbers are equal, in which case the product will have the shape of the outer numbers in the given order.

(i)   The product is a 2×4 matrix.
(ii) The product is a 4×2 matrix.
(iii) The product is not defined since the inner numbers 2 and 3 are not equal.
(iv) The product is a 5×3 matrix.
(v)   The product is not defined even though the matrices have the same shape.
(vi) The product is a 2×4 matrix.

● **PROBLEM** 4-13

Let $A = \begin{bmatrix} 2 & 7 & 3 \\ -1 & 8 & 5 \\ -4 & 9 & -3 \end{bmatrix}$ and $B = \begin{bmatrix} \alpha & \beta & \delta \\ a & b & c \\ \sigma & d & q \end{bmatrix}$.

Find AB.

127

Solution: By the definition of matrix, multiplication,

$$AB = \begin{bmatrix} 2 & 7 & 3 \\ -1 & 8 & 5 \\ -4 & 9 & -3 \end{bmatrix} \begin{bmatrix} \alpha & \beta & \delta \\ a & b & c \\ \sigma & d & q \end{bmatrix}$$

$$= \begin{bmatrix} 2\alpha+7a+3\sigma & 2\beta+7b+3d & 2\delta+7c+3q \\ -\alpha+8a+5\sigma & -\beta+8b+5d & -\delta+8c+5q \\ -4\alpha+9a-3\sigma & -4\beta+9b-3d & -4\delta+9c-3q \end{bmatrix}.$$

● **PROBLEM** 4-14

a) Suppose $A = \begin{bmatrix} 1 & 3 \\ 2 & -1 \end{bmatrix}$ and $B = \begin{bmatrix} 2 & 0 & -4 \\ 3 & -2 & 6 \end{bmatrix}$. Find i) AB and

ii) BA .

b) Suppose $A = [2,1]$ and $B = \begin{bmatrix} 1 & -2 & 0 \\ 4 & 5 & -3 \end{bmatrix}$. Find i) AB, and

ii) BA.

Solution: Suppose $A = (a_{ij})$ and $B = (b_{jk})$ are matrices such that the number of columns of A equals the number of rows of B; also suppose A is an m×n matrix and B is an n×s matrix.

$$A = \begin{bmatrix} a_{11} & \cdots\cdots & a_{1n} \\ \vdots & & \vdots \\ a_{m1} & \cdots\cdots & a_{mn} \end{bmatrix}$$

$$B = \begin{bmatrix} b_{11} & \cdots\cdots & b_{1s} \\ \vdots & & \vdots \\ b_{n1} & \cdots\cdots & b_{ns} \end{bmatrix}.$$

Then the product AB is the m×s matrix whose ik-element is

$$\sum_{j=1}^{n} a_{ij}b_{jk} = a_{i1}b_{1k} + a_{i2}b_{2k} + \cdots + a_{in}b_{nk} .$$ If $A_1,\ldots,A_m$ are the

row vectors of the matrix A, and if $B^1,\ldots,B^s$ are the column vectors of the matrix B, then the ik-element of the product AB is equal to $A_1,\ldots,B^k$. Thus,

$$\begin{matrix} A_1 \cdot B^1 & \cdots\cdots & A_1 \cdot B^s \\ \vdots & & \vdots \\ A_m B^1 & \cdots\cdots & A_m \cdot B^s \end{matrix}$$

a) i) $A = \begin{bmatrix} 1 & 3 \\ 2 & -1 \end{bmatrix}$ and $B = \begin{bmatrix} 2 & 0 & -4 \\ 3 & -2 & 6 \end{bmatrix}$. Here, A is 2×2 and

B is 2×3 . The product AB is a 2×3 matrix. To obtain the components in the first row of AB, multiply the first row (1,3) of A by the columns $\begin{bmatrix} 2 \\ 3 \end{bmatrix}$, $\begin{bmatrix} 0 \\ -2 \end{bmatrix}$ and $\begin{bmatrix} -4 \\ 6 \end{bmatrix}$ of B, respectively.

$$\begin{bmatrix} 1 & 3 \\ 2 & -1 \end{bmatrix} \begin{bmatrix} 2 & 0 & -4 \\ 3 & -2 & 6 \end{bmatrix} = [1 \cdot 2 + 3 \cdot 3 \quad 1 \cdot 0 + 3 \cdot (-2) \quad 1 \cdot (-4) + 3 \cdot 6]$$

$$= [2+0 \quad 0-6 \quad -4+18] = [11 \quad -6 \quad 14]$$

To obtain the components in the second row of AB, multiply the second row (2,01) of A by the columns of B, respectively.

$$\begin{bmatrix} 1 & 3 \\ 2 & -1 \end{bmatrix} \begin{bmatrix} 2 & 0 & -4 \\ 3 & -2 & 6 \end{bmatrix} = [(2 \cdot 2 + (-1) \cdot 3) \quad (2 \cdot 0 + (-1) \cdot (-2)) \quad (2 \cdot (-4) + (-1) \cdot 6)]$$

$$= [4-3 \quad 0+2 \quad -8-6] = [1 \quad 2 \quad -14].$$

Thus,

$$AB = \begin{bmatrix} 11 & -6 & 14 \\ 1 & 2 & -14 \end{bmatrix} .$$

ii) Here B is 2×3 and A is 2×2 . Since the number of columns of B is not equal to the number of rows of A, the product BA is not defined.

b) i) Since A is a 1×2 and B is a 2×3, the product AB is a 1×3 matrix.

$$AB = \begin{bmatrix} 2,1 \end{bmatrix} \begin{bmatrix} 1 & -2 & 0 \\ 4 & 5 & -3 \end{bmatrix} = \begin{bmatrix} (2 \cdot 1 + 1 \cdot 4, & 2 \cdot (-2) + 1 \cdot 5, & 2 \cdot 0 + 1 \cdot (-3) \end{bmatrix}$$

$$= \begin{bmatrix} 6 & 1 & -3 \end{bmatrix}$$

ii) In this case, B is 2×3 and A is 1×2 . Since the number of columns of B is not equal to the number of rows of A, the product BA is not defined.

● PROBLEM 4-15

Let $A = \begin{bmatrix} 2 & -1 \\ 1 & 0 \\ -3 & 4 \end{bmatrix}$ and $B = \begin{bmatrix} 1 & -2 & -5 \\ 3 & 4 & 0 \end{bmatrix}$.

Find (i) AB, (ii) BA .

Solution: Since A is 3×2 and B is 2×3, the product AB is a 3×3 matrix. To obtain the first row of AB, multiply the first row of A by each column of B, respectively. The first row of

$$\begin{bmatrix} 2 & -1 \\ 1 & 0 \\ -3 & 4 \end{bmatrix} \begin{bmatrix} 1 & -2 & -5 \\ 3 & 4 & 0 \end{bmatrix}$$ is $[2 \cdot 1 + (-1) \cdot 3 \quad 2 \cdot (-2) + (-1)(4) \quad 2 \cdot (-5) + -1 \cdot 0]$

$$= [2-3 \quad -4-4 \quad -10+0] = [-1 \quad -8 \quad -10].$$

To obtain the second row of AB, multiply the second row of A by each column of B, respectively.

$$\begin{bmatrix} 2 & -1 \\ 1 & 0 \\ -3 & 4 \end{bmatrix} \begin{bmatrix} 1 & -2 & -5 \\ 3 & 4 & 0 \end{bmatrix} = [1 \cdot 1 + 0 \cdot 3 \quad 1 \cdot (-2) + 0.4 \quad 1 \cdot (-5) + 0.0]$$

$$= [1+0 \quad -2+0 \quad -5+0] = [1 \quad -2 \quad -5]$$

To obtain the third row of AB, multiply the third row of A by each column of B, respectively. Then, AB =

$$\begin{bmatrix} 2 & -1 \\ 1 & 0 \\ -3 & 4 \end{bmatrix} \begin{bmatrix} 1 & -2 & -5 \\ 3 & 4 & 0 \end{bmatrix} = \begin{bmatrix} -1 & -8 & -10 \\ 1 & -2 & -5 \\ (-3)\cdot1+4\cdot3 & (-3)\cdot(-2)+4\cdot4 & (-3)(-5)+4\cdot0 \end{bmatrix}$$

$$= \begin{bmatrix} -1 & -8 & -10 \\ 1 & -2 & -5 \\ -3+12 & 6+16 & 15+0 \end{bmatrix} = \begin{bmatrix} -1 & -8 & -10 \\ 1 & -2 & -5 \\ 9 & 22 & 15 \end{bmatrix}.$$

Thus,

$$AB = \begin{bmatrix} -1 & -8 & -10 \\ 1 & -2 & -5 \\ 9 & 22 & 15 \end{bmatrix}.$$

ii)   Since B is 2×3 and A is 3×2 , BA is defined and is a 2×2 matrix. To obtain the first row of BA, multiply the first row of B by each column of A, respectively.

$$\begin{bmatrix} 1 & -2 & -5 \\ 3 & 4 & 0 \end{bmatrix} \begin{bmatrix} 2 & -1 \\ 1 & 0 \\ -3 & 4 \end{bmatrix}$$

$$= \quad [1\cdot2+(-2)\cdot1+(-5)\cdot(-3) \quad 1\cdot(-1)+(-2)\cdot0+(-5)\cdot4]$$

$$= \quad [2-2+15 \quad -1+0-20] = [15 \quad -21].$$

To obtain the second row of BA, multiply the second row of B by each column of A, respectively. Then, BA =

$$\begin{bmatrix} 1 & -2 & -5 \\ 3 & 4 & 0 \end{bmatrix} \begin{bmatrix} 2 & -1 \\ 1 & 0 \\ -3 & 4 \end{bmatrix}$$

$$= \begin{bmatrix} 15 & -21 \\ 3\cdot2+4\cdot1+0\cdot(-3) & 3\cdot(-1)+4\cdot0+0\cdot4 \end{bmatrix}$$

$$= \begin{bmatrix} 15 & -21 \\ -6+4+0 & -3+0+0 \end{bmatrix} = \begin{bmatrix} 15 & -21 \\ 10 & -3 \end{bmatrix}.$$

Thus,

$$BA = \begin{bmatrix} 15 & -21 \\ 10 & -3 \end{bmatrix}.$$

Remark: In this case, observe that both AB and BA are defined but they are not equal. In fact, they do not even have the same shape.

● PROBLEM 4-16

If $A = \begin{bmatrix} 2 & -1 & 0 \\ 1 & 0 & -3 \end{bmatrix}$ and $B = \begin{bmatrix} 1 & -4 & 0 & 1 \\ 2 & -1 & 3 & -1 \\ 4 & 0 & -2 & 0 \end{bmatrix}$

(1) Determine the shape of AB .
(2) If c_{ij} denotes the element in the ith row and jth column of the

product matrix AB, find c_{23} c_{14} and c_{21}.

Solution: (1) A matrix with m rows and n columns is called an m by n matrix or an $m \times n$ matrix. The pair of numbers (m,n) is called its size or shape. Since A is 2×3 and B is 3×4, the product AB is a 2×4 matrix. Thus, the shape of AB is $(2,4)$.

(2) Now, c_{23} is the element in the 2nd row and 3rd column of the product matrix AB. To obtain c_{23}, multiply the second row of A by the third column of B. Hence,

$$c_{23} = [1 \quad 0 \quad -3] \begin{bmatrix} 0 \\ 3 \\ -2 \end{bmatrix}$$

$$= 1 \cdot 0 + 0 \cdot 3 + (-3) \cdot (-2) = 0 + 0 + 6 = 6 ,$$

and

$$c_{14} = [2 \quad -1 \quad 0] \begin{bmatrix} 1 \\ -1 \\ 0 \end{bmatrix}$$

$$= 2 \cdot 1 + (-1) \cdot (-1) + 0 \cdot 0 = 2 + 1 + 0 = 3 .$$

$$c_{21} = [1 \quad 0 \quad -3] \begin{bmatrix} 1 \\ 2 \\ 4 \end{bmatrix} = 1 \cdot 1 + 0 \cdot 2 + (-3) \cdot 4$$

$$= 1 + 0 - 12 = -11 .$$

● PROBLEM 4-17

Prove $(AB)C = A(BC)$ where $A = \begin{bmatrix} 5 & 2 & 3 \\ 2 & -3 & 4 \end{bmatrix}$, $B = \begin{bmatrix} 2 & -1 & 1 & 0 \\ 0 & 2 & 2 & 2 \\ 3 & 0 & -1 & 3 \end{bmatrix}$

and $C = \begin{bmatrix} 1 & 0 & 2 \\ 2 & -3 & 0 \\ 0 & 0 & 3 \\ 2 & 1 & 0 \end{bmatrix}$

Solution: First, find $(AB)C$ and then $A(BC)$.

$$AB = \begin{bmatrix} 5 & 2 & 3 \\ 2 & -3 & 4 \end{bmatrix} \begin{bmatrix} 2 & -1 & 1 & 0 \\ 0 & 2 & 2 & 2 \\ 3 & 0 & -1 & 3 \end{bmatrix}$$

$$= \begin{bmatrix} 10+0+9 & -5+4+0 & 5+4-3 & 0+4+9 \\ 4+0+12 & -2-6+0 & 2-6-4 & 0-6+12 \end{bmatrix}$$

$$= \begin{bmatrix} 19 & -1 & 6 & 13 \\ 16 & -8 & -8 & 6 \end{bmatrix}$$

Now,

$$(AB)C = \begin{bmatrix} 19 & -1 & 6 & 13 \\ 16 & -8 & -8 & 6 \end{bmatrix} \begin{bmatrix} 1 & 0 & 2 \\ 2 & -3 & 0 \\ 0 & 0 & 3 \\ 2 & 1 & 0 \end{bmatrix}$$

$$= \begin{bmatrix} 19-2+0+26 & 0+3+0+13 & 38+0+18+0 \\ 16-16+0+12 & 0+24+0+6 & 32+0-24+0 \end{bmatrix} = \begin{bmatrix} 43 & 16 & 56 \\ 12 & 30 & 8 \end{bmatrix} .$$

Thus,

$$(AB)C = \begin{bmatrix} 43 & 16 & 56 \\ 12 & 30 & 8 \end{bmatrix} .$$

131

$$BC = \begin{bmatrix} 2 & -1 & 1 & 0 \\ 0 & 2 & 2 & 2 \\ 3 & 0 & -1 & 3 \end{bmatrix} \begin{bmatrix} 1 & 0 & 2 \\ 2 & -3 & 0 \\ 0 & 0 & 3 \\ 2 & 1 & 0 \end{bmatrix}$$

$$= \begin{bmatrix} 2-2+0+0 & 0+3+0+0 & 4+0+3+0 \\ 0+4+0+4 & 0-6+0+2 & 0+0+6+0 \\ 3+0+0+6 & 0+0+0+3 & 6+0-3+0 \end{bmatrix}$$

$$BC = \begin{bmatrix} 0 & 3 & 7 \\ 8 & -4 & 6 \\ 9 & 3 & 3 \end{bmatrix}$$

and

$$A(BC) = \begin{bmatrix} 5 & 2 & 3 \\ 2 & -3 & 4 \end{bmatrix} \begin{bmatrix} 0 & 3 & 7 \\ 8 & -4 & 6 \\ 9 & 3 & 3 \end{bmatrix}$$

$$= \begin{bmatrix} 0+16+27 & 15-8+9 & 35+12+9 \\ 0-24+36 & 6+12+12 & 14-18+12 \end{bmatrix} = \begin{bmatrix} 43 & 16 & 56 \\ 12 & 30 & 8 \end{bmatrix}$$

Thus, $(AB)C = A(BC)$.

● **PROBLEM** 4-18

Find $A(B+C)$ and $AB + AC$ if $A = \begin{bmatrix} 2 & 2 & 3 \\ 3 & -1 & 2 \end{bmatrix}$, $B = \begin{bmatrix} 1 & 0 \\ 2 & 2 \\ 3 & -1 \end{bmatrix}$

and $C = \begin{bmatrix} -1 & 2 \\ 1 & 0 \\ 2 & -2 \end{bmatrix}$.

<u>Solution</u>: $B + C = \begin{bmatrix} 1 & 0 \\ 2 & 2 \\ 3 & -1 \end{bmatrix} + \begin{bmatrix} -1 & 2 \\ 1 & 0 \\ 2 & -2 \end{bmatrix}$

$$\begin{bmatrix} 1+(-1) & 0+2 \\ 2+1 & 2+0 \\ 3+2 & -1+(-2) \end{bmatrix} = \begin{bmatrix} 0 & 2 \\ 3 & 2 \\ 5 & -3 \end{bmatrix}$$

then,

$$A(B+C) = \begin{bmatrix} 2 & 2 & 3 \\ 3 & -1 & 2 \end{bmatrix} \begin{bmatrix} 0 & 2 \\ 3 & 2 \\ 5 & -3 \end{bmatrix}$$

$$= \begin{bmatrix} 2 \cdot 0 + 2 \cdot 3 + 3 \cdot 5 & 2 \cdot 2 + 2 \cdot 2 + 3 \cdot (-3) \\ 3 \cdot 0 + (-1) \cdot 3 + 2 \cdot 5 & 3 \cdot 2 + (-1) \cdot 2 + 2 \cdot (-3) \end{bmatrix}.$$

$$A(B+C) = \begin{bmatrix} 0+6+15 & 4+4-9 \\ 0-3+10 & 6-2-6 \end{bmatrix} = \begin{bmatrix} 21 & -1 \\ 7 & -2 \end{bmatrix}.$$

$$AB = \begin{bmatrix} 2 & 2 & 3 \\ 3 & -1 & 2 \end{bmatrix} \begin{bmatrix} 1 & 0 \\ 2 & 2 \\ 3 & -1 \end{bmatrix}$$

$$= \begin{bmatrix} 2 \cdot 1 + 2 \cdot 2 + 3 \cdot 3 & 2 \cdot 0 + 2 \cdot 2 + 3 \cdot (-1) \\ 3 \cdot 1 + (-1) \cdot 2 + 2 \cdot 3 & 3 \cdot 0 + (-1) \cdot 2 + 2 \cdot (-1) \end{bmatrix}$$

$$= \begin{bmatrix} 2+4+9 & 0+4+(-3) \\ 3-2+6 & 0-2-2 \end{bmatrix} = \begin{bmatrix} 15 & 1 \\ 7 & -4 \end{bmatrix} .$$

$$AC = \begin{bmatrix} 2 & 2 & 3 \\ 3 & -1 & 2 \end{bmatrix} \begin{bmatrix} -1 & 2 \\ 1 & 0 \\ 2 & -2 \end{bmatrix}$$

$$= \begin{bmatrix} 2 \cdot (-1) + 2 \cdot 1 + 3 \cdot 2 & 2 \cdot 2 + 2 \cdot 0 + 3 \cdot (-2) \\ 3 \cdot (-1) + (-1) \cdot 1 + 2 \cdot 2 & 3 \cdot 2 + (-1) \cdot 0 + 2 \cdot (-2) \end{bmatrix}$$

$$AC = \begin{bmatrix} -2+2+6 & 4+0-6 \\ -3-1+4 & 6+0-4 \end{bmatrix} = \begin{bmatrix} 6 & -2 \\ 0 & 2 \end{bmatrix} .$$

Then,

$$AB + AC = \begin{bmatrix} 15 & 1 \\ 7 & -4 \end{bmatrix} + \begin{bmatrix} 6 & -2 \\ 0 & 2 \end{bmatrix} = \begin{bmatrix} 15+6 & 1+(-2) \\ 7+0 & -4+2 \end{bmatrix}$$

$$= \begin{bmatrix} 21 & -1 \\ 7 & -2 \end{bmatrix}$$

Remark: $A(B+C) = \begin{bmatrix} 21 & -1 \\ 7 & -2 \end{bmatrix}$ and $AB + AC = \begin{bmatrix} 21 & -1 \\ 7 & -2 \end{bmatrix} .$

Thus, $A(B+C) = AB + AC$. This is called the left distributive law.

● **PROBLEM** 4-19

Find (i) A^2 (ii) A^3 (iii) A^4 when $A = \begin{bmatrix} 1 & 2 \\ -1 & 1 \end{bmatrix} .$

<u>Solution</u>: $A^2 = AA = \begin{bmatrix} 1 & 2 \\ -1 & 1 \end{bmatrix} \begin{bmatrix} 1 & 2 \\ -1 & 1 \end{bmatrix} = \begin{bmatrix} 1-2 & 2+2 \\ -1-1 & -2+1 \end{bmatrix} = \begin{bmatrix} -1 & 4 \\ -2 & -1 \end{bmatrix} .$

$A^3 = AAA = A^2A = \begin{bmatrix} -1 & 4 \\ -2 & -1 \end{bmatrix} \begin{bmatrix} 1 & 2 \\ -1 & 1 \end{bmatrix} = \begin{bmatrix} -1-4 & -2+4 \\ -2+1 & -4-1 \end{bmatrix} = \begin{bmatrix} -5 & 2 \\ -1 & -5 \end{bmatrix} .$

The usual laws for exponents are $A^m A^n = A^{m+n}$ and $(A^m)^n = A^{mn}$. Thus, $A^4 = A^3A$ or, $A^4 = (A^2)^2 = A^2A^2$.

$$A^4 = A^3A = \begin{bmatrix} -5 & 2 \\ -1 & -5 \end{bmatrix} \begin{bmatrix} 1 & 2 \\ -1 & 1 \end{bmatrix} = \begin{bmatrix} -5-2 & -10+2 \\ -1+5 & -2-5 \end{bmatrix} = \begin{bmatrix} -7 & -8 \\ 4 & -7 \end{bmatrix} .$$

Observe that

$$A^4 = A^2A^2 = \begin{bmatrix} -1 & 4 \\ -2 & -1 \end{bmatrix} \begin{bmatrix} -1 & 4 \\ -2 & -1 \end{bmatrix} = \begin{bmatrix} 1-8 & -4-4 \\ 2+2 & -8+1 \end{bmatrix}$$

$$= \begin{bmatrix} -7 & -8 \\ 4 & -7 \end{bmatrix}$$

● **PROBLEM** 4-20

Suppose S and T are counter-clockwise rotations in R^2 through the angles θ and φ , respectively. The respective matrices of S and R are

$$A = \begin{bmatrix} \cos \theta & -\sin \theta \\ \sin \theta & \cos \theta \end{bmatrix} \text{ and } B = \begin{bmatrix} \cos \varphi & -\sin \varphi \\ \sin \varphi & \cos \varphi \end{bmatrix} .$$

Find AB .

$\quad AB = \begin{bmatrix} \cos\theta & -\sin\theta \\ \sin\theta & \cos\theta \end{bmatrix} \begin{bmatrix} \cos\varphi & -\sin\varphi \\ \sin\varphi & \cos\varphi \end{bmatrix}$

$$= \begin{bmatrix} \cos\theta\,\cos\varphi - \sin\theta\,\sin\varphi & -\cos\theta\,\sin\varphi - \sin\theta\,\cos\varphi \\ \sin\theta\,\cos\varphi + \cos\theta\,\sin\varphi & -\sin\theta\,\sin\varphi + \cos\theta\,\cos\varphi \end{bmatrix}$$

Note that ST is just a counter-clockwise rotation through the angle $\theta + \varphi$ so its matrix is:

$$\begin{bmatrix} \cos(\theta+\varphi) & -\sin(\theta+\varphi) \\ \sin(\theta+\varphi) & \cos(\theta+\varphi) \end{bmatrix} .$$

Since this matrix ST must be equal to the matrix AB, we can write

$$\cos(\theta+\varphi) = \cos\theta\,\cos\varphi - \sin\theta\,\sin\varphi$$
$$\sin(\theta+\varphi) = \cos\theta\,\sin\varphi + \sin\theta\,\cos\varphi .$$

Observe that the familiar formulas for the cosine and sine sums of angles have been derived.

● **PROBLEM** 4-21

---

a)   Show that:

  (i) $A0 = 0$ (ii) $0A = 0$ (iii) $AI = A$ (iv) $IA = A$

where 0 and I denote the zero and identity matrices respectively, and

$$A = \begin{bmatrix} 2 & 1 & 3 \\ 4 & -1 & -1 \end{bmatrix}$$

b)   Give examples of the following rules: (i)   if A has a row of zeros, the same row of AB consists of zeros.   (ii)   if B has a column of zeros, the same column of AB consists of zeros.

Solution: a)   The $m{\times}n$ matrix whose entries are all zero is called the zero matrix and is denoted by 0 . The $n{\times}n$ matrix with 1's on the diagonal and 0's elsewhere, denoted by I, is called the unit or identity matrix; e.g., in R^3 ,

$$I = \begin{bmatrix} 1 & 0 & 0 \\ 0 & 1 & 0 \\ 0 & 0 & 1 \end{bmatrix}$$

(i)
$$A = \begin{bmatrix} 2 & 1 & 3 \\ 4 & -1 & -7 \end{bmatrix}$$

$$0 = \begin{bmatrix} 0 & 0 & 0 & 0 \\ 0 & 0 & 0 & 0 \\ 0 & 0 & 0 & 0 \end{bmatrix}$$

$$A0 = \begin{bmatrix} 2 & 1 & 3 \\ 4 & -1 & -7 \end{bmatrix} \begin{bmatrix} 0 & 0 & 0 & 0 \\ 0 & 0 & 0 & 0 \\ 0 & 0 & 0 & 0 \end{bmatrix}$$

$$= \begin{bmatrix} 0+0+0 & 0+0+0 & 0+0+0 & 0+0+0 \\ 0+0+0 & 0+0+0 & 0+0+0 & 0+0+0 \end{bmatrix} = \begin{bmatrix} 0 & 0 & 0 & 0 \\ 0 & 0 & 0 & 0 \\ 0 & 0 & 0 & 0 \end{bmatrix} .$$

(ii) Likewise, $0A = \begin{bmatrix} 0 & 0 \\ 0 & 0 \end{bmatrix} \begin{bmatrix} 2 & 1 & 3 \\ 4 & -1 & -7 \end{bmatrix}$

$$= \begin{bmatrix} 0+0 & 0+0 & 0+0 \\ 0+0 & 0+0 & 0+0 \end{bmatrix} = \begin{bmatrix} 0 & 0 & 0 \\ 0 & 0 & 0 \end{bmatrix}$$

(iii)
$$I = \begin{bmatrix} 1 & 0 & 0 \\ 0 & 1 & 0 \\ 0 & 0 & 1 \end{bmatrix}$$

$$AI = \begin{bmatrix} 2 & 1 & 3 \\ 4 & -1 & -7 \end{bmatrix} \begin{bmatrix} 1 & 0 & 0 \\ 0 & 1 & 0 \\ 0 & 0 & 1 \end{bmatrix}$$

$$= \begin{bmatrix} 2+0+0 & 0+1+0 & 0+0+3 \\ 4+0+0 & 0-1+0 & 0+0-7 \end{bmatrix}$$

$$= \begin{bmatrix} 2 & 1 & 3 \\ 4 & -1 & -7 \end{bmatrix}.$$

Thus, AI = A.

(iv)
$$I = \begin{bmatrix} 1 & 0 \\ 0 & 1 \end{bmatrix}$$

$$IA = \begin{bmatrix} 1 & 0 \\ 0 & 1 \end{bmatrix} \begin{bmatrix} 2 & 1 & 3 \\ 4 & -1 & -7 \end{bmatrix}$$

$$= \begin{bmatrix} 2+0 & 1+0 & 3+0 \\ 0+4 & 0+(-1) & 0+(-7) \end{bmatrix}$$

$$= \begin{bmatrix} 2 & 1 & 3 \\ 4 & -1 & -7 \end{bmatrix}.$$

Thus, IA = A.

Observe that the zero and identity matrices must have the appropriate size in order for the products to be defined. For example,

$$\begin{bmatrix} 0 & 0 & 0 \\ 0 & 0 & 0 \end{bmatrix} \begin{bmatrix} 2 & 1 & 3 \\ 4 & -1 & -7 \end{bmatrix}$$

is not defined.

b)   Let $A = \begin{bmatrix} -2 & -3 & 1 \\ 0 & 0 & 0 \\ 1 & -1 & 0 \end{bmatrix}$ and $B = \begin{bmatrix} 6 & -2 & 1 \\ 3 & 1 & 2 \\ -1 & 1 & 1 \end{bmatrix}$.

Then,

$$AB = \begin{bmatrix} -2 & -3 & 1 \\ 0 & 0 & 0 \\ 1 & -1 & 0 \end{bmatrix} \begin{bmatrix} 6 & -2 & 1 \\ 3 & 1 & 2 \\ -1 & 1 & 1 \end{bmatrix}$$

$$AB = \begin{bmatrix} -12-9-1 & 4-3+1 & -2-6+1 \\ 0+0+0 & 0+0+0 & 0+0+0 \\ 6-3+0 & -2-1+0 & 1-2+0 \end{bmatrix}$$

$$= \begin{bmatrix} -22 & 2 & -7 \\ 0 & 0 & 0 \\ 3 & -3 & -1 \end{bmatrix}.$$

(ii)   Let $A = \begin{bmatrix} 4 & 1 & 0 \\ -1 & 2 & 3 \\ 1 & 0 & 1 \end{bmatrix}$ and $B = \begin{bmatrix} 1 & 0 & 1 \\ -2 & 0 & 1 \\ 1 & 0 & 1 \end{bmatrix}$.

Then,

$$AB = \begin{bmatrix} 4 & 1 & 0 \\ -1 & 2 & 3 \\ 1 & 0 & 1 \end{bmatrix} \begin{bmatrix} 1 & 0 & 1 \\ -2 & 0 & 1 \\ 1 & 0 & 1 \end{bmatrix}$$

$$= \begin{bmatrix} 4-2+0 & 0+0+0 & 4+1+0 \\ -1-4+3 & 0+0+0 & -1+2+3 \\ 1+0+1 & 0+0+0 & 1+0+1 \end{bmatrix}$$

$$= \begin{bmatrix} 2 & 0 & 5 \\ -2 & 0 & 4 \\ 2 & 0 & 2 \end{bmatrix} .$$

Show that in matrix arithmetic we can have the following:
(a) $AB \neq BA$.
(b) $A \neq 0$, $B \neq 0$ and yet, $AB = 0$.
(c) $A \neq 0$ and $A^2 = 0$.
(d) $A \neq 0$, $A^2 \neq 0$ and $A^3 = 0$.
(e) $A^2 = A$ with $A \neq 0$ and $A \neq I$.
(f) $A^2 = I$ with $A \neq I$ and $A \neq -I$.

Solution: (a) $A = \begin{bmatrix} 2 & 1 \\ -1 & 0 \end{bmatrix}$ $B = \begin{bmatrix} 1 & 0 \\ 3 & 1 \end{bmatrix}$

$$AB = \begin{bmatrix} 2+3 & 0+1 \\ -1+0 & 0 \end{bmatrix} = \begin{bmatrix} 5 & 1 \\ -1 & 0 \end{bmatrix}$$

$$BA = \begin{bmatrix} 1 & 0 \\ 3 & 1 \end{bmatrix} \begin{bmatrix} 2 & 1 \\ -1 & 0 \end{bmatrix} = \begin{bmatrix} 2+0 & 1+0 \\ 6-1 & 3+0 \end{bmatrix} = \begin{bmatrix} 2 & 1 \\ 5 & 3 \end{bmatrix}$$

Thus, $AB \neq BA$.
(b) Let $A = \begin{bmatrix} 0 & 1 \\ 0 & 0 \end{bmatrix}$; $B = \begin{bmatrix} 0 & 4 \\ 0 & 0 \end{bmatrix}$.

$$AB = \begin{bmatrix} 0 & 1 \\ 0 & 0 \end{bmatrix} \begin{bmatrix} 0 & 4 \\ 0 & 0 \end{bmatrix} = \begin{bmatrix} 0+0 & 0+0 \\ 0+0 & 0+0 \end{bmatrix}$$

$$= \begin{bmatrix} 0 & 0 \\ 0 & 0 \end{bmatrix}$$

Thus, $AB = 0$, $A \neq 0$ and $B \neq 0$.
(c) Let $A = \begin{bmatrix} 0 & 1 \\ 0 & 0 \end{bmatrix}$

$$A^2 - AA = \begin{bmatrix} 0 & 1 \\ 0 & 0 \end{bmatrix} \begin{bmatrix} 0 & 1 \\ 0 & 0 \end{bmatrix} = \begin{bmatrix} 0+0 & 0+0 \\ 0+0 & 0+0 \end{bmatrix} = \begin{bmatrix} 0 & 0 \\ 0 & 0 \end{bmatrix} .$$

Let $A = \begin{bmatrix} 1 & -1 \\ 1 & -1 \end{bmatrix}$,

then

$$A^2 = AA = \begin{bmatrix} 1 & -1 \\ 1 & -1 \end{bmatrix} \begin{bmatrix} 1 & -1 \\ 1 & -1 \end{bmatrix} = \begin{bmatrix} 1-1 & -1+1 \\ 1-1 & -1+1 \end{bmatrix} = \begin{bmatrix} 0 & 0 \\ 0 & 0 \end{bmatrix} .$$

Thus, $A \neq 0$ and $A^2 = 0$.
(d) Let $A = \begin{bmatrix} 0 & 1 & 1 \\ 0 & 0 & 1 \\ 0 & 0 & 0 \end{bmatrix}$

$$A^2 = AA = \begin{bmatrix} 0 & 1 & 1 \\ 0 & 0 & 1 \\ 0 & 0 & 0 \end{bmatrix} \begin{bmatrix} 0 & 1 & 1 \\ 0 & 0 & 1 \\ 0 & 0 & 0 \end{bmatrix}$$

$$= \begin{bmatrix} 0+0+0 & 0+0+0 & 0+1+0 \\ 0+0+0 & 0+0+0 & 0+0+0 \\ 0+0+0 & 0+0+0 & 0+0+0 \end{bmatrix} = \begin{bmatrix} 0 & 0 & 1 \\ 0 & 0 & 0 \\ 0 & 0 & 0 \end{bmatrix} .$$

$$A^3 = A^2 A = \begin{bmatrix} 0 & 0 & 1 \\ 0 & 0 & 0 \\ 0 & 0 & 0 \end{bmatrix} \begin{bmatrix} 0 & 1 & 1 \\ 0 & 0 & 1 \\ 0 & 0 & 0 \end{bmatrix}$$

$$= \begin{bmatrix} 0 & 0 & 0 \\ 0 & 0 & 0 \\ 0 & 0 & 0 \end{bmatrix}.$$

(e)   Let $A = \begin{bmatrix} \frac{1}{2} & \frac{1}{2} \\ \frac{1}{2} & \frac{1}{2} \end{bmatrix}$

$$A^2 = AA = \begin{bmatrix} \frac{1}{2} & \frac{1}{2} \\ \frac{1}{2} & \frac{1}{2} \end{bmatrix}\begin{bmatrix} \frac{1}{2} & \frac{1}{2} \\ \frac{1}{2} & \frac{1}{2} \end{bmatrix} = \begin{bmatrix} \frac{1}{4}+\frac{1}{4} & \frac{1}{4}+\frac{1}{4} \\ \frac{1}{4}+\frac{1}{4} & \frac{1}{4}+\frac{1}{4} \end{bmatrix} = \begin{bmatrix} \frac{1}{2} & \frac{1}{2} \\ \frac{1}{2} & \frac{1}{2} \end{bmatrix}.$$

Thus, $A^2 = A$ with $A \neq 0$ and $A \neq I$.

(f)   Let $A = \begin{bmatrix} -1 & 0 \\ 0 & 1 \end{bmatrix}$

$$A^2 = AA = \begin{bmatrix} -1 & 0 \\ 0 & 1 \end{bmatrix}\begin{bmatrix} -1 & 0 \\ 0 & 1 \end{bmatrix}$$

$$A^2 = \begin{bmatrix} 1+0 & 0+0 \\ 0+0 & 0+1 \end{bmatrix} = \begin{bmatrix} 1 & 0 \\ 0 & 1 \end{bmatrix}.$$

Hence, $A^2 = I$ with $A \neq I$ and $A \neq -I$.

● **PROBLEM** 4-23

Compute AB using block multiplication where

$$A = \begin{bmatrix} 1 & 2 & 1 \\ 3 & 4 & 0 \\ \hline 0 & 0 & 2 \end{bmatrix} \quad \text{and} \quad B = \begin{bmatrix} 1 & 2 & 3 & 1 \\ 4 & 5 & 6 & 1 \\ \hline 0 & 0 & 0 & 1 \end{bmatrix} .$$

Solution: Using a system of horizontal and vertical lines, one can partition a matrix A into smaller "submatrices" of A. The matrix A is then called a block matrix. A given matrix may be divided into blocks in different ways. For example,

$$\begin{bmatrix} 1 & -2 & 0 & 1 \\ 2 & 3 & 5 & 7 \\ 3 & 1 & 4 & 5 \end{bmatrix} = \begin{bmatrix} 1 & -2 & 0 & 1 \\ 2 & 3 & 5 & 7 \\ \hline 3 & 1 & 4 & 5 \end{bmatrix}$$

$$= \begin{bmatrix} 1 & -2 & 0 & 1 \\ \hline 2 & 3 & 5 & 7 \\ \hline 3 & 1 & 4 & 5 \end{bmatrix} .$$

$$A = \begin{bmatrix} 1 & 2 & 1 \\ 3 & 4 & 0 \\ \hline 0 & 0 & 2 \end{bmatrix} .$$

Let $E\begin{bmatrix} 1 & 2 \\ 3 & 4 \end{bmatrix}$, $F = \begin{bmatrix} 1 \\ 0 \end{bmatrix}$ and $G[2]$ then,

$$A = \begin{bmatrix} E & F \\ \hline 0 & G \end{bmatrix}$$

$$B = \begin{bmatrix} 1 & 2 & 3 & 1 \\ 4 & 5 & 6 & 1 \\ \hline 0 & 0 & 0 & 1 \end{bmatrix} . \quad \text{Let } R = \begin{bmatrix} 1 & 2 & 3 \\ 4 & 5 & 6 \end{bmatrix} \quad S = \begin{bmatrix} 1 \\ 1 \end{bmatrix} \quad T = [1].$$

137

Then,

$$B = \begin{bmatrix} R & \vdots & S \\ - & - & - \\ 0 & \vdots & T \end{bmatrix} .$$

After partitioning the matrices into block matrices, multiplication of the matrices is the usual matrix multiplication with each entire block considered as a unit entry of the matrix. If two matrices can be multiplied, then they can be multiplied as block matrices if they are each partitioned into blocks similarly; that is, into an equal number of blocks so that corresponding blocks have the same size. Suppose

$$A = \begin{bmatrix} A_1 & \vdots & A_2 \\ - & - & - \\ A_3 & \vdots & A_4 \end{bmatrix} \quad \text{and} \quad B = \begin{bmatrix} B_1 & \vdots & B_2 \\ - & - & - \\ B_3 & \vdots & B_4 \end{bmatrix}$$

where A_1 and B_1, A_2 and B_2, A_3 and B_3, A_4 and B_4 are the same sizes, respectively. Then AB is given by

$$A_1 B_1 + A_2 B_3 \qquad A_1 B_2 + A_2 B_4$$

$$A_3 B_1 + A_3 B_3 \qquad A_3 B_2 + A_4 B_4$$

In the problem,

$$AB = \begin{bmatrix} E & F \\ 0 & G \end{bmatrix} \begin{bmatrix} R & S \\ 0 & T \end{bmatrix} = \begin{bmatrix} ER + F \cdot 0 & ES + FT \\ OR + G \cdot 0 & OS + GT \end{bmatrix}$$

$$= \begin{bmatrix} ER & ES + FT \\ 0 & GT \end{bmatrix} =$$

$$= \begin{bmatrix} \begin{bmatrix} 1 & 2 \\ 3 & 4 \end{bmatrix}\begin{bmatrix} 1 & 2 & 3 \\ 4 & 5 & 6 \end{bmatrix} & \begin{bmatrix} 1 & 2 \\ 3 & 4 \end{bmatrix}\begin{bmatrix} 1 \\ 1 \end{bmatrix} + \begin{bmatrix} 1 \\ 0 \end{bmatrix}[1] \\ 0 & [2][1] \end{bmatrix}$$

$$= \begin{bmatrix} \begin{bmatrix} 1+8 & 2+10 & 3+12 \\ 3+16 & 6+20 & 9+24 \end{bmatrix} & \begin{bmatrix} 1+2 \\ 3+4 \end{bmatrix} + \begin{bmatrix} 1 \\ 0 \end{bmatrix} \\ [0 \quad 0 \quad 0] & [2] \end{bmatrix}$$

$$= \begin{bmatrix} 9 & 12 & 15 & 4 \\ 19 & 26 & 33 & 7 \\ 0 & 0 & 0 & 2 \end{bmatrix}$$

SPECIAL KINDS OF MATRICES

● PROBLEM 4-24

Define (1) An upper triangular matrix.
       (2) A lower triangular matrix.
       (3) A properly triangular matrix.
Give examples.

Solution: A triangular matrix is an n✕n matrix whose non-zero elements lie on the diagonal and all are either above or below the diagonal.

Definition: (1) An upper triangular matrix is an n✕n matrix all of whose non-zero entries lie on its diagonal and above. Example:

$$\begin{bmatrix} 1 & 2 & 3 \\ 0 & 0 & 0 \\ 0 & 0 & 6 \end{bmatrix}$$

(2) A lower triangular matrix is an n✕n matrix all of whose non-zero entries lie on its diagonal and below. Example:

$$\begin{bmatrix} 1 & 0 & 0 \\ 2 & 0 & 0 \\ 3 & 0 & 6 \end{bmatrix} .$$

(3) A properly triangular matrix is an n✕n matrix whose diagonal entries are all zero. Example:

$$\begin{bmatrix} 0 & 1 & 2 \\ 0 & 0 & 3 \\ 0 & 0 & 0 \end{bmatrix}$$

It should be noted that the zero matrix and all diagonal matrices are triangular matrices.

● **PROBLEM** 4-25

If A and B are both diagonal matrices having n rows and n columns, they commute. In other words, show that AB = BA where

$$A = \begin{bmatrix} 2 & 0 & 0 \\ 0 & -1 & 0 \\ 0 & 0 & 3 \end{bmatrix} \quad B = \begin{bmatrix} -2 & 0 & 0 \\ 0 & 4 & 0 \\ 0 & 0 & -6 \end{bmatrix} .$$

Solution: A diagonal matrix is a square matrix whose non-diagonal entries are all zero.

$$AB = \begin{bmatrix} 2 & 0 & 0 \\ 0 & -1 & 0 \\ 0 & 0 & 3 \end{bmatrix} \begin{bmatrix} -2 & 0 & 0 \\ 0 & 4 & 0 \\ 0 & 0 & -6 \end{bmatrix}$$

$$= \begin{bmatrix} 2\cdot(-2) & 0 & 0 \\ 0 & (-1)\cdot 4 & 0 \\ 0 & 0 & 3\cdot(-6) \end{bmatrix} = \begin{bmatrix} -4 & 0 & 0 \\ 0 & -4 & 0 \\ 0 & 0 & -18 \end{bmatrix} .$$

$$BA = \begin{bmatrix} -2 & 0 & 0 \\ 0 & 4 & 0 \\ 0 & 0 & -6 \end{bmatrix} \begin{bmatrix} 2 & 0 & 0 \\ 0 & -1 & 0 \\ 0 & 0 & 3 \end{bmatrix}$$

$$= \begin{bmatrix} (-2)\cdot 2 & 0 & 0 \\ 0 & 4\cdot(-1) & 0 \\ 0 & 0 & (-6)\cdot 3 \end{bmatrix} = \begin{bmatrix} -4 & 0 & 0 \\ 0 & -4 & 0 \\ 0 & 0 & -18 \end{bmatrix}$$

Thus, AB = BA .

A matrix P is called idempotent if $P^2 = P$. Show that the matrices

$$\begin{bmatrix} 25 & -20 \\ 30 & -24 \end{bmatrix}, \quad \begin{bmatrix} -26 & -18 & -27 \\ 21 & 15 & 21 \\ 12 & 8 & 13 \end{bmatrix} \quad \text{and} \quad \begin{bmatrix} 1 & 0 & 0 \\ 0 & 1 & 0 \\ 0 & 0 & 0 \end{bmatrix}$$

are idempotent.

Solution:
$$P^2 = PP = \begin{bmatrix} 25 & -20 \\ 30 & -24 \end{bmatrix} \begin{bmatrix} 25 & -20 \\ 30 & -24 \end{bmatrix}$$

$$= \begin{bmatrix} 25 \cdot 25 + (-20) \cdot 30 & 25 \cdot (-20) + (-20) \cdot (-24) \\ 30 \cdot 25 + (-24) \cdot 30 & 30 \cdot (-20) + (-24) \cdot (-24) \end{bmatrix}$$

$$= \begin{bmatrix} 625 - 600 & -500 + 480 \\ 750 - 720 & -600 + 576 \end{bmatrix}$$

$$= \begin{bmatrix} 25 & -20 \\ 30 & -24 \end{bmatrix} = P.$$

Thus, $P^2 = P$.

$$P^2 = PP = \begin{bmatrix} -26 & -18 & -27 \\ 21 & 15 & 21 \\ 12 & 8 & 13 \end{bmatrix} \begin{bmatrix} -26 & -18 & -27 \\ 21 & 15 & 21 \\ 12 & 8 & 13 \end{bmatrix}$$

$$=$$
$(-26) \cdot (-26) + (-18) \cdot (21) + -27 \cdot 12, \quad (-26) \cdot (-18) + (-18) \cdot 15 + (-27) \cdot 8,$
$$(-26) \cdot (-27) + (-18) \cdot 21 + (-27) \cdot 13$$

$21 \cdot (-26) + 15 \cdot 21 \cdot 12, \; 21 \cdot (-18) + 15 \cdot 15 + 21 \cdot 8, \; 21 \cdot (-27) + 15 \cdot 21 + 21 \cdot 13$

$21 \cdot (-26) + 8 \cdot 21 + 13 \cdot 12, \; 12 \cdot (-18) + 8 \cdot 15 + 13 \cdot 8, \; 12 \cdot (-27) + 8 \cdot 21 + 13 \cdot 13$

$$= \begin{bmatrix} 676 - 378 - 325 & 468 - 270 - 216 & 702 - 378 - 351 \\ -546 + 315 + 252 & -378 + 225 + 168 & -567 + 315 + 273 \\ -312 + 168 + 156 & -216 + 120 + 104 & -324 + 168 + 169 \end{bmatrix}$$

$$= \begin{bmatrix} -26 & -18 & -27 \\ 21 & 15 & 21 \\ 12 & 8 & 13 \end{bmatrix} = P.$$

Thus, the matrix $\begin{bmatrix} -26 & -18 & -27 \\ 21 & 15 & 21 \\ 12 & 8 & 13 \end{bmatrix}$ is idempotent.

$$P^2 = PP = \begin{bmatrix} 1 & 0 & 0 \\ 0 & 1 & 0 \\ 0 & 0 & 0 \end{bmatrix} \begin{bmatrix} 1 & 0 & 0 \\ 0 & 1 & 0 \\ 0 & 0 & 0 \end{bmatrix}$$

$$= \begin{bmatrix} 1 \cdot 1 + 0 \cdot 0 + 0 \cdot 0 & 1 \cdot 0 + 0 \cdot 1 + 0 \cdot 0 & 1 \cdot 0 + 0 \cdot 0 + 0 \cdot 0 \\ 0 \cdot 1 + 1 \cdot 0 + 0 \cdot 0 & 0 \cdot 0 + 1 \cdot 1 + 0 \cdot 0 & 0 \cdot 0 + 1 \cdot 0 + 0 \cdot 0 \\ 0 \cdot 1 + 0 \cdot 0 + 0 \cdot 0 & 0 \cdot 0 + 0 \cdot 1 + 0 ; 0 & 0 \cdot 0 + 0 \cdot 0 + 0 \cdot 0 \end{bmatrix}$$

$$= \begin{bmatrix} 1 & 0 & 0 \\ 0 & 1 & 0 \\ 0 & 0 & 0 \end{bmatrix} = P .$$

Thus, the matrix

$$\begin{bmatrix} 1 & 0 & 0 \\ 0 & 1 & 0 \\ 0 & 0 & 0 \end{bmatrix}$$

is idempotent.

● PROBLEM 4-27

Define the transpose of a matrix. Find the transpose of the following matrices:

$$A = \begin{bmatrix} 4 & -2 & 3 \\ 0 & 5 & -2 \end{bmatrix} \qquad B = \begin{bmatrix} 6 & 2 & -4 \\ 3 & -1 & 2 \\ 0 & 4 & 3 \end{bmatrix}$$

$$C = \begin{bmatrix} 5 & 4 \\ 3 & 2 \\ 2 & -3 \end{bmatrix} \qquad D = \begin{bmatrix} 3 & -5 & 1 \end{bmatrix} \qquad E = \begin{bmatrix} 2 \\ -1 \\ 3 \end{bmatrix}$$

Solution: Definition: If $A = [a_{ij}]$ is an $m \times n$ matrix, then the $n \times m$ matrix $A^t = [a^t_{ij}]$ where

$$a^t_{ij} = a_{ji} \quad [1 \le i \le m . \quad 1 \le j \le n]$$

is called the transpose of A. Thus, the transpose of A is obtained by interchanging the rows and columns of A.

$$A = \begin{bmatrix} 4 & -2 & 3 \\ 0 & 5 & -2 \end{bmatrix} .$$

Then,

$$A^t = \begin{bmatrix} 4 & 0 \\ -2 & 5 \\ 3 & -2 \end{bmatrix}$$

$$B = \begin{bmatrix} 6 & 2 & -4 \\ 3 & -1 & 2 \\ 0 & 4 & 3 \end{bmatrix} .$$

$$B^t = \begin{bmatrix} 6 & 3 & 0 \\ 2 & -1 & 4 \\ -4 & 2 & 3 \end{bmatrix} .$$

$$C = \begin{bmatrix} 5 & 4 \\ -3 & 2 \\ 2 & -3 \end{bmatrix} ; \quad \text{thus,} \quad C^t = \begin{bmatrix} 5 & -3 & 2 \\ 4 & 2 & -3 \end{bmatrix} .$$

$$D = \begin{bmatrix} 3 & -5 & 1 \end{bmatrix} .$$

Then,

$$D^t = \begin{bmatrix} 3 \\ -5 \\ 1 \end{bmatrix} .$$

$$E = \begin{bmatrix} 2 \\ -1 \\ 3 \end{bmatrix} ; \quad \text{hence,} \quad E^t = \begin{bmatrix} 2 & -1 & 3 \end{bmatrix} .$$

Let $A = \begin{bmatrix} 1 & 2 & 0 \\ 3 & -1 & 4 \end{bmatrix}$. Find (i) AA^t, (ii) A^tA.

Solution: The transpose of A, denoted by A^t, is the matrix obtained from A by interchanging the rows and columns of A. For example, if

$$A = \begin{bmatrix} a_{11} & a_{12} & \cdots & a_{1n} \\ a_{21} & a_{22} & \cdots & a_{2n} \\ \vdots & \vdots & \ddots & \vdots \\ a_{m1} & a_{m2} & & a_{mn} \end{bmatrix},$$

then

$$A^t = \begin{bmatrix} a_{11} & a_{21} & \cdots & a_{m1} \\ a_{12} & a_{22} & \cdots & a_{m2} \\ \vdots & \vdots & \ddots & \vdots \\ a_{1n} & a_{2n} & \cdots & a_{mn} \end{bmatrix}$$

Observe that if A is an $m \times n$ matrix, then A^t is an $n \times m$ matrix. Hence, the products AA^t and A^tA are always defined.

$$A = \begin{bmatrix} 1 & 2 & 0 \\ 3 & -1 & 4 \end{bmatrix}$$

then,

$$A^t = \begin{bmatrix} 1 & 3 \\ 2 & -1 \\ 0 & 4 \end{bmatrix}.$$

(i) $AA^t = \begin{bmatrix} 1 & 2 & 0 \\ 3 & -1 & 4 \end{bmatrix} \begin{bmatrix} 1 & 3 \\ 2 & -1 \\ 0 & 4 \end{bmatrix}$

$= \begin{bmatrix} 1 \cdot 1 + 2 \cdot 2 + 0 \cdot 0 & 1 \cdot 3 + 2 \cdot (-1) + 0 \cdot 4 \\ 3 \cdot 1 + (-1) \cdot 2 + (4) \cdot 0 & 3 \cdot 3 + (-1) \cdot (-1) + 4 \cdot 4 \end{bmatrix}$

$= \begin{bmatrix} 1+4+0 & 3-2+0 \\ 3-2+0 & 9+1+16 \end{bmatrix} = \begin{bmatrix} 5 & 1 \\ 1 & 26 \end{bmatrix}.$

(ii) $A^tA = \begin{bmatrix} 1 & 3 \\ 2 & -1 \\ 0 & 4 \end{bmatrix} \begin{bmatrix} 1 & 2 & 0 \\ 3 & -1 & 4 \end{bmatrix}$

$= \begin{bmatrix} 1 \cdot 1 + 3 \cdot 3 & 1 \cdot 2 + 3 \cdot (-1) & 1 \cdot 0 + 3 \cdot 4 \\ 2 \cdot 1 + (-1) \cdot 3 & 2 \cdot 2 + (-1) \cdot (-1) & 2 \cdot 0 + (-1) \cdot 4 \\ 0 \cdot 1 + 4 \cdot 3 & 0 \cdot 2 + 4 \cdot (-1) & 0 \cdot 0 + 4 \cdot 4 \end{bmatrix}$

$= \begin{bmatrix} 1+9 & 2-3 & 0+12 \\ 2-3 & 4+1 & 0-4 \\ 0+12 & 0-4 & 0+16 \end{bmatrix}$

$= \begin{bmatrix} 10 & -1 & 12 \\ -1 & 5 & -4 \\ 12 & -4 & 16 \end{bmatrix}$

CHAPTER 5

DETERMINANTS

PERMUTATIONS

● PROBLEM 5-1

Define permutations. Find the permutations of order 3.

Solution: A permutation of order n, $n = 1, 2, \ldots$, is an ordered set $\langle i_1, i_2, \ldots, i_n \rangle$ of integers in which each of the integers $1, 2, \ldots, n$ occurs exactly once.

We denote permutation σ by:

$$\sigma = \begin{pmatrix} 1 & 2 & \ldots & n \\ i_1 & i_2 & \ldots & i_n \end{pmatrix}$$

or $\sigma = i_1 i_2 \ldots i_n$. The permutations of order 3 are $\langle 123 \rangle$, $\langle 132 \rangle$ $\langle 213 \rangle$, $\langle 231 \rangle$, $\langle 312 \rangle$, $\langle 321 \rangle$. Note that the number of such permutations is $n!$.

This follows from the reasoning below:

Let $\{1, 2, \ldots, n\}$ be the first n integers. There are n values to which 1 can be sent, $(n-1)$ values for 2 and, proceeding, 1 value for $n(n-1)\ldots 1 = n!$

● PROBLEM 5-2

Define an inversion of a permutation. Find the inversion of

$$\langle 3, 1, 4, 2 \rangle .$$

Solution: An inversion of a permutation $\langle i_1, i_2, \ldots, i_n \rangle$ is a pair (p, q) of integers from among $1, 2, \ldots, n$ such that $p > q$ and p occurs before q in the list $i_1, i_2, \ldots, i_n$. $I\langle i_1, i_2, \ldots, i_n \rangle$ will denote the number of inversions of $\langle i_1, i_2, \ldots, i_n \rangle$.

The inversions of $\langle 3, 1, 4, 2 \rangle$ are: $(3,1)$, $(3,2)$ and $(4,2)$. Hence, $I\langle 3, 1, 4, 2 \rangle = 3$. The number,

$$\text{sgn} \langle i_1, i_2, \ldots, i_n \rangle = (-1)^{I\langle i_1, i_2, \ldots, i_n \rangle}$$

is called the sign of the permutation $\langle i_1, i_2, \ldots, i_n \rangle$. Hence, $\text{sgn}\langle 3, 1, 4, 2 \rangle = (-1)^3 = -1$.

Determine the parity of $\sigma = 542163$.

Solution: Consider an arbitrary permutation, σ, in $S_n: \sigma = j_1 j_2 \cdots j_n$.
We say σ is even or odd according as to whether there is an even or odd number of pairs $(i\ K)$ for which

$$i > K \quad \text{but} \quad i \quad \text{precedes} \quad K \quad \text{in} \quad \sigma.$$

Define the sign or parity of σ, written $\text{sgn}\ \sigma$, by:

(a) $\quad \text{sgn}\ \sigma = (-1)^n = +1$ if n is even,

(b) $\quad \text{sgn}\ \sigma = (-1)^n = -1$ if n is odd.

Method 1: $\sigma = 542163$.

It is necessary to obtain the number of pairs (i,j) for which $i > j$ and i precedes j in σ. 5, 4, and 2 precede and are greater than 1. Hence, the number of pairs is 3, $\{(5,1),\ (4,2),\ (2,1)$. 5 and 4 precede and are greater than 2, hence the number of pairs is 2, $\{(5,2),\ (4,2)\}$. 5, 4 and 6 precede, and are greater than 3. Hence, the number of pairs is 3, $\{(5,3),\ (4,3),\ (6,3)\}$. 5 precedes, and is greater than 4. Hence, the number of pairs is 1, $\{(5,4)\}$. Since $3 + 2 + 3 + 1 = 9$ is odd, σ is an odd permutation. Therefore, $\text{sgn}\ \sigma = -1$.

Method 2:

Transpose 1 to the first position as follows:

$$5\ 4\ 2\ 1\ 6\ 3 \quad \text{to} \quad 1\ 5\ 4\ 2\ 6\ 3$$

Transpose 2 to the second position:

$$1\ 5\ 4\ 2\ 6\ 3 \quad \text{to} \quad 1\ 2\ 5\ 4\ 6\ 3$$

Transpose 3 to the third position:

$$1\ 2\ 5\ 4\ 6\ 3 \quad \text{to} \quad 1\ 2\ 3\ 5\ 4\ 6$$

Transpose 4 to the fourth position:

$$1\ 2\ 3\ 5\ 4\ 6 \quad \text{to} \quad 1\ 2\ 3\ 4\ 5\ 6$$

Note that 5 and 6 are in the correct positions. Add the numbers "jumped": $3 + 2 + 3 + 1 = 9$. Since 9 is odd, σ is an odd permutation. Hence, $\text{sgn}\ \sigma = -1$.

Method 3:

An interchange of two numbers in a permutation is equivalent to multiplying the permutation by a transposition. Therefore, transform σ to the identity permutation using transpositions such as:

Thus, the number of transpositions is 5. Since 5 is odd, σ is an odd permutation. Hence, $\text{sgn}\ \sigma = -1$.

Let $\sigma = 24513$, and $\tau = 41352$, be permutations in S_5. Find: (i) the composition permutations $\tau\circ\sigma$ and $\sigma\circ\tau$; (ii) σ^{-1}.

__Solution:__ $\sigma = 24513$ $\tau = 41352$. We can write,

$$\sigma = \begin{pmatrix} 1 & 2 & 3 & 4 & 5 \\ 2 & 4 & 5 & 1 & 3 \end{pmatrix} \qquad \tau = \begin{pmatrix} 1 & 2 & 3 & 4 & 5 \\ 4 & 1 & 3 & 5 & 2 \end{pmatrix}$$

which means,

$\sigma(1) = 2$, $\sigma(2) = 4$, $\sigma(3) = 5$, $\sigma(4)=1$
and
$\sigma(5) = 3$. Also,

$\tau(1) = 4$, $\tau(2) = 1$, $\tau(3) = 3$, $\tau(4) = 5$ and $\tau(5) = 2$.

Now,
$\tau\circ\sigma = \tau\sigma(i) = \tau(\sigma(i))$,
so,
$\tau\sigma(1) = \tau(\sigma(1)) = \tau(2) = 1$
$\tau\sigma(2) = \tau(\sigma(2)) = \tau(4) = 5$
$\tau\sigma(3) = \tau(\sigma(3)) = \tau(5) = 2$
$\tau\sigma(4) = \tau(\sigma(4)) = \tau(1) = 4$
$\tau\sigma(5) = \tau(\sigma(5)) = \tau(3) = 3$.
Thus, $\tau\circ\sigma = 15243$.

$\sigma\circ\tau = \sigma\tau(i) = \sigma(\tau(i))$:
$\sigma(\tau(1)) = \sigma(4) = 1$
$\sigma(\tau(2)) = \sigma(1) = 2$
$\sigma(\tau(3)) = \sigma(3) = 5$
$\sigma(\tau(4)) = \sigma(5) = 3$
$\sigma(\tau(5)) = \sigma(2) = 4$.

Hence, $\sigma\circ\tau = 12534$.
(ii)

$$\sigma = \begin{pmatrix} 1 & 2 & 3 & 4 & 5 \\ 2 & 4 & 5 & 1 & 3 \end{pmatrix}.$$

We know that,
$$\sigma\sigma^{-1} = e$$
where e is the identity permutation,

$$e = \begin{pmatrix} 1 & 2 & 3 & \ldots & n \\ 1 & 2 & 3 & \ldots & n \end{pmatrix}.$$

To obtain e from σ, permute $2\ 4\ 5\ 1\ 3$ to $1\ 2\ 3\ 4\ 5$. Thus,

$$\sigma^{-1} = \begin{pmatrix} 2 & 4 & 5 & 1 & 3 \\ 1 & 2 & 3 & 4 & 5 \end{pmatrix}.$$

Then,
$\sigma^{-1}(1) = 4$, $\sigma^{-1}(2) = 1$, $\sigma^{-1}(3) = 5$, $\sigma^{-1}(4) = 2$
and
$\sigma^{-1}(5) = 3$.

Thus,

$$\sigma^{-1} = \begin{pmatrix} 1 & 2 & 3 & 4 & 5 \\ 4 & 1 & 5 & 2 & 3 \end{pmatrix}$$

So, $\sigma^{-1} = 41523$.

Check the result by showing: $\sigma\sigma^{-1} = e$

$$\sigma\sigma^{-1} = \begin{pmatrix} 1 & 2 & 3 & 4 & 5 \\ 2 & 4 & 5 & 1 & 3 \end{pmatrix} \begin{pmatrix} 1 & 2 & 3 & 4 & 5 \\ 4 & 1 & 5 & 2 & 3 \end{pmatrix} \quad.$$

$$\sigma\sigma^{-1}(1) = \sigma(\sigma^{-1}(1)) = \sigma(4) = 1$$

$$\sigma\sigma^{-1}(2) = 2 \ , \ \sigma\sigma^{-1}(3) = 3 \ , \ \sigma\sigma^{-1}(4) = 4 \ , \ \sigma\sigma^{-1}(5) = 5 \ .$$

Thus,

$$\sigma\sigma^{-1} = \begin{pmatrix} 1 & 2 & 3 & 4 & 5 \\ 1 & 2 & 3 & 4 & 5 \end{pmatrix} = e \quad.$$

● **PROBLEM** 5-5

Find the product of two permutations σ and φ, where,

(i) $\quad \sigma = \begin{pmatrix} 1 & 2 & 3 & 4 & 5 \\ 2 & 4 & 1 & 5 & 3 \end{pmatrix} \qquad \varphi = \begin{pmatrix} 1 & 2 & 3 & 4 & 5 \\ 4 & 1 & 2 & 5 & 3 \end{pmatrix}$

(ii) $\quad \sigma = \begin{pmatrix} 1 & 2 & 3 & 4 & 5 \\ 4 & 1 & 2 & 5 & 3 \end{pmatrix} \qquad \varphi = \begin{pmatrix} 4 & 1 & 2 & 5 & 3 \\ 1 & 2 & 3 & 4 & 5 \end{pmatrix}$

(iii) $\quad \sigma = \begin{pmatrix} 1 & 2 & 3 & 4 & 5 \\ 5 & 2 & 4 & 3 & 1 \end{pmatrix} \qquad \varphi = \begin{pmatrix} 4 & 1 & 2 & 5 & 3 \\ 1 & 2 & 3 & 4 & 5 \end{pmatrix}$

(iv) $\quad \sigma = \begin{pmatrix} 1 & 2 & 3 & 4 & 5 \\ 4 & 1 & 2 & 5 & 3 \end{pmatrix} \qquad \varphi = \begin{pmatrix} 1 & 2 & 3 & 4 & 5 \\ 2 & 4 & 1 & 5 & 3 \end{pmatrix}$

Solution: The product of two permutations σ and φ is defined in terms of function composition:

$$(\sigma\varphi)(i) = \sigma[\varphi(i)] \ , \ i = 1,\ldots,n \ .$$

The set of all permutations on n objects, together with this operation of multiplication is called the symmetric group of degree n, and is denoted by S_n .

(i) $\quad (\sigma\varphi)(i) = \sigma[\varphi(i)] \ , \quad i = 1,\ldots,n;$

then

$$(\sigma\varphi)(1) = \sigma[\varphi(1)] = \sigma(4) = 5$$

$$(\sigma\varphi)(2) = \sigma[\varphi(2)] = \sigma(1) = 2$$

$$(\sigma\varphi)(3) = \sigma[\varphi(3)] = \sigma(2) = 4 \quad.$$

Similarly, we find $\sigma\varphi(4) = 3$ and $\sigma\varphi(5) = 1$. Thus,

(i)

$$\sigma\varphi = \begin{pmatrix} 1 & 2 & 3 & 4 & 5 \\ 2 & 4 & 1 & 5 & 3 \end{pmatrix} \begin{pmatrix} 1 & 2 & 3 & 4 & 5 \\ 4 & 1 & 2 & 5 & 3 \end{pmatrix} = \begin{pmatrix} 1 & 2 & 3 & 4 & 5 \\ 5 & 2 & 4 & 3 & 1 \end{pmatrix}$$

146

(ii)
$$\sigma\varphi = \begin{pmatrix} 1 & 2 & 3 & 4 & 5 \\ 4 & 1 & 2 & 5 & 3 \end{pmatrix} \begin{pmatrix} 4 & 1 & 2 & 5 & 3 \\ 1 & 2 & 3 & 4 & 5 \end{pmatrix} = \begin{pmatrix} 1 & 2 & 3 & 4 & 5 \\ 1 & 2 & 3 & 4 & 5 \end{pmatrix}$$

(iii)
$$\sigma\varphi = \begin{pmatrix} 1 & 2 & 3 & 4 & 5 \\ 5 & 2 & 4 & 3 & 1 \end{pmatrix} \begin{pmatrix} 4 & 1 & 2 & 5 & 3 \\ 1 & 2 & 3 & 4 & 5 \end{pmatrix} = \begin{pmatrix} 1 & 2 & 3 & 4 & 5 \\ 2 & 4 & 1 & 5 & 3 \end{pmatrix}$$

(iv)
$$\sigma\varphi = \begin{pmatrix} 1 & 2 & 3 & 4 & 5 \\ 4 & 1 & 2 & 5 & 3 \end{pmatrix} \begin{pmatrix} 1 & 2 & 3 & 4 & 5 \\ 2 & 4 & 1 & 5 & 3 \end{pmatrix} = \begin{pmatrix} 1 & 2 & 3 & 4 & 5 \\ 1 & 5 & 4 & 3 & 2 \end{pmatrix}$$

Observe that the examples (i) and (iv) show that multiplication of permutations is not always commutative, (i.e., $\sigma\varphi \neq \varphi\sigma$), in general.

EVALUATION OF DETERMINANTS BY MINORS

● PROBLEM 5-6

Define det A and find the determinant of the following matrices:

(a) $[a_{11}]$ (b) $\begin{bmatrix} a_{11} & a_{12} \\ a_{21} & a_{22} \end{bmatrix}$ (c) $\begin{bmatrix} 0 & 0 & 0 \\ 0 & 0 & 0 \\ 0 & 0 & 0 \end{bmatrix}$

(d) $\begin{bmatrix} a_{11} & a_{12} & a_{13} \\ a_{21} & a_{22} & a_{23} \\ a_{31} & a_{32} & a_{33} \end{bmatrix}$

Solution: Determinants are formally defined as:

$$\text{Det}(A) = |A| = \sum_{\sigma} \text{sgn}(\sigma) \prod_{j=1}^{n} a_{\sigma(j)j}$$

$$= \sum_{\sigma} \text{sgn}(\sigma) \, a_{\sigma(1)1} \, a_{\sigma(2)2} \cdots a_{\sigma(n)n}$$

where summation extends over the n!, different permutations, σ, of the n symbols $1, 2, \ldots, n$ and $\text{sgn}(\sigma) = +1$, σ even
-1, σ odd

$|A|$ is also known as an $n \times n$ determinant or a determinant of order n.

(a) $\det A = a_{11}$

(b) $\det A = \begin{vmatrix} a_{11} & a_{12} \\ a_{21} & a_{22} \end{vmatrix} = a_{11}a_{22} - a_{21}a_{12}$

(c) $\det A = |0| = 0$

(d) $\det A = \begin{vmatrix} a_{11} & a_{12} & a_{13} \\ a_{21} & a_{22} & a_{23} \\ a_{31} & a_{32} & a_{33} \end{vmatrix}$

The permutations of S_3 and their signs are:

Permutation	Sign	Permutation	Sign
1 2 3	+	2 1 3	-
1 3 2	-	3 1 2	+
2 3 1	+	3 2 1	-

Then, $\det A = a_{11}a_{22}a_{33} - a_{11}a_{32}a_{23} + a_{21}a_{32}a_{13} - a_{21}a_{12}a_{33} + a_{31}a_{12}a_{23} - a_{31}a_{22}a_{13}$.

● **PROBLEM** 5-7

a) Find the determinant of an arbitrary 3×3 matrix.
b) Find $\det A$ where:

$$A = \begin{bmatrix} -5 & 0 & 2 \\ 6 & 1 & 2 \\ 2 & 3 & 1 \end{bmatrix}$$

Solution: Let

$$A = \begin{bmatrix} b_{11} & b_{12} & b_{13} \\ b_{21} & b_{22} & b_{23} \\ b_{31} & b_{32} & b_{33} \end{bmatrix}.$$

$$\det A = \begin{bmatrix} b_{11} & b_{12} & b_{13} \\ b_{21} & b_{22} & b_{23} \\ b_{31} & b_{32} & b_{33} \end{bmatrix}.$$

Expand the above determinant by minors, using the first column.

$$\det A = +b_{11} \begin{vmatrix} b_{22} & b_{23} \\ b_{32} & b_{33} \end{vmatrix} - b_{21} \begin{vmatrix} b_{12} & b_{13} \\ b_{32} & b_{33} \end{vmatrix} + b_{31} \begin{vmatrix} b_{12} & b_{13} \\ b_{22} & b_{23} \end{vmatrix}$$

$$\det A = b_{11}(b_{22}b_{33} - b_{32}b_{23}) - b_{21}(b_{12}b_{33} - b_{32}b_{13}) + b_{31}(b_{12}b_{23} - b_{22}b_{13}).$$

Now expand the determinant by minors, using the second row:

$$\det A = -b_{21} \begin{vmatrix} b_{12} & b_{13} \\ b_{32} & b_{33} \end{vmatrix} + b_{22} \begin{vmatrix} b_{11} & b_{13} \\ b_{31} & b_{33} \end{vmatrix} - b_{23} \begin{vmatrix} b_{11} & b_{12} \\ b_{31} & b_{32} \end{vmatrix}.$$

$\det A = -b_{21}(b_{12}b_{33} - b_{32}b_{13}) + b_{22}(b_{11}b_{33} - b_{31}b_{13}) - b_{23}(b_{11}b_{32} - b_{31}b_{12})$

$= b_{22}b_{11}b_{33} - b_{22}b_{31}b_{13} - b_{23}b_{11}b_{32} + b_{23}b_{31}b_{12} - b_{21}(b_{12}b_{33} - b_{32}b_{13})$

$= b_{11}(b_{22}b_{33} - b_{32}b_{23}) - b_{21}(b_{12}b_{33} - b_{32}b_{13}) + b_{31}(b_{12}b_{23} - b_{22}b_{13})$

Clearly, this is the same as the first answer. Note, also, that det A can be rearranged algebraically until it can be written as:

$\det A = b_{11}b_{22}b_{33} + b_{12}b_{23}b_{31} + b_{13}b_{32}b_{21}$

$\qquad - [b_{13}b_{22}b_{31} + b_{23}b_{32}b_{11} + b_{33}b_{21}b_{12}]$.

It is easy to remember this result (det of a 3x3 matrix) using the following mnemonic device: Figure 1

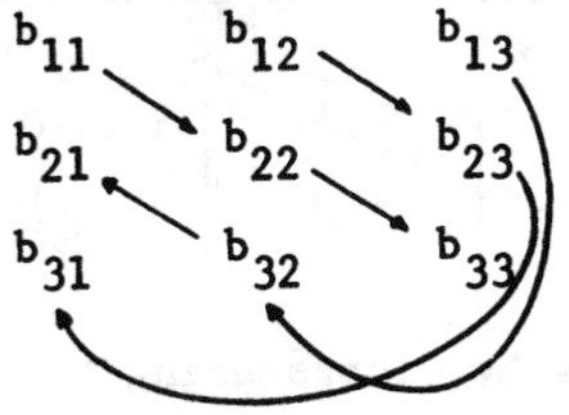

$\left(b_{11}b_{22}b_{33} + b_{12}b_{23}b_{31} + b_{13}b_{32}b_{21}\right)$

Figure 2:

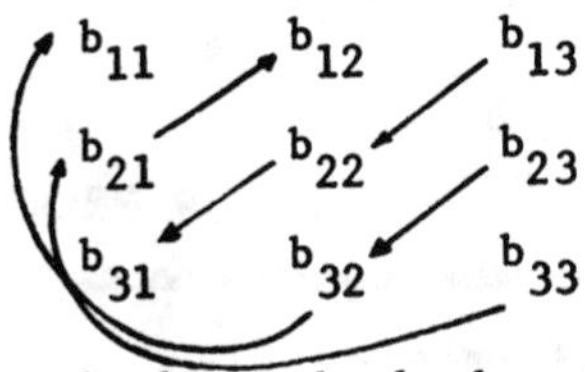

$\left(b_{13}b_{22}b_{31} + b_{23}b_{32}b_{11} + b_{33}b_{21}b_{12}\right)$

$\det A = b_{11}b_{22}b_{33} + b_{12}b_{23}b_{31} + b_{13}b_{32}b_{21}$

$\qquad - [b_{13}b_{22}b_{31} + b_{23}b_{32}b_{11} + b_{33}b_{21}b_{12}]$.

This makes taking 3x3 determinants simpler.

b) Expand the determinant by minors, using the first column.

$$\det A = -5 \begin{vmatrix} 1 & 3 \\ 3 & 1 \end{vmatrix} - 6 \begin{vmatrix} 0 & 2 \\ 3 & 1 \end{vmatrix} + 2 \begin{vmatrix} 0 & 2 \\ 1 & 2 \end{vmatrix}$$

$$= -5(1-6) - 6(0-6) + 2(0-2) = +25 + 36 - 4 = 57.$$

● PROBLEM 5-8

Find the determinant of the following matrix:

$$A = \begin{bmatrix} 2 & 0 & 3 & 0 \\ 2 & 1 & 1 & 2 \\ 3 & -1 & 1 & -2 \\ 2 & 1 & -2 & 1 \end{bmatrix}$$

Solution: Use the method of expansion by minors.

$$A = \begin{bmatrix} 2 & 0 & 3 & 0 \\ 2 & 1 & 1 & 2 \\ 3 & -1 & 1 & -2 \\ 2 & 1 & -2 & 1 \end{bmatrix}$$

Expanding along the first row:

$$\det A = 2 \begin{vmatrix} 1 & 1 & 2 \\ -1 & 1 & -2 \\ 1 & -2 & 1 \end{vmatrix} + 3 \begin{vmatrix} 2 & 1 & 2 \\ 3 & -1 & -2 \\ 2 & 1 & 1 \end{vmatrix}$$

Note that the minors, whose multiplying factors were zero, have been eliminated. This illustrates the general principle that, when evaluating determinants, expansion along the row (or column) containing the most zeros is the optimal procedure.

Add the second row to the first row for each of the 3 by 3 determinants:

$$\det A = 2 \begin{vmatrix} 0 & 2 & 0 \\ -1 & 1 & -2 \\ 1 & -2 & 1 \end{vmatrix} + 3 \begin{vmatrix} 5 & 0 & 0 \\ 3 & -1 & -2 \\ 2 & 1 & 1 \end{vmatrix}$$

Now expand the above determinants by minors using the first row.

$$\det A = 2(-2) \begin{vmatrix} -1 & -2 \\ 1 & 1 \end{vmatrix} + 3(5) \begin{vmatrix} -1 & -2 \\ 1 & 1 \end{vmatrix}$$

$$= (-4)(-1+2) + 15(-1+2)$$

$$= -4+15 = 11 .$$

● PROBLEM 5-9

Evaluate the determinant of the matrix A where

$$A = \begin{bmatrix} 4 & 2 & 1 & 3 \\ -1 & 0 & 2 & 8 \\ 5 & -6 & 0 & -1 \\ 0 & 2 & 2 & 3 \end{bmatrix} .$$

<u>Solution</u>: The value of the determinant of A is not changed if a multiple of one row (column) is added to a multiple of another. Now, add -4 times the third column to the first column and add -2 times the third column to the second column.

$$\begin{bmatrix} 0 & 0 & 1 & 3 \\ -9 & -4 & 2 & 8 \\ 5 & -6 & 0 & -1 \\ -8 & -2 & 2 & 3 \end{bmatrix} .$$

Add -3 times the third column to the fourth column.

$$\begin{bmatrix} 0 & 0 & 1 & 0 \\ -9 & -4 & 2 & 2 \\ 5 & -6 & 0 & -1 \\ -8 & -2 & 2 & -3 \end{bmatrix} .$$

Now,

$$\det A = \begin{vmatrix} 0 & 0 & 1 & 0 \\ -9 & -4 & 2 & 2 \\ 5 & -6 & 0 & -1 \\ -8 & -2 & 2 & -3 \end{vmatrix}.$$

Expand the above determinant by minors using the first row.

$$\det A = +0\begin{vmatrix} -4 & 2 & 2 \\ -6 & 0 & -1 \\ -2 & 2 & -3 \end{vmatrix} - 0\begin{vmatrix} -9 & 2 & 2 \\ 5 & 0 & -1 \\ 8 & 2 & -3 \end{vmatrix}$$

$$+1\begin{vmatrix} -9 & -4 & 2 \\ 5 & -6 & -1 \\ -8 & -2 & -3 \end{vmatrix} - 0\begin{vmatrix} -9 & -4 & 2 \\ 5 & -6 & 0 \\ -8 & -2 & 2 \end{vmatrix}$$

Therefore,

$$\det A = 1\begin{vmatrix} -9 & -4 & 2 \\ 5 & -6 & -1 \\ -8 & -2 & -3 \end{vmatrix}.$$

Add +2 times the second row to the first row and -3 times the second row to the third row.

$$\det A = \begin{vmatrix} 1 & -16 & 0 \\ 5 & -6 & -1 \\ -23 & 16 & 0 \end{vmatrix}$$

Expand the determinant by minors, using column three.

$$\det A = 0\begin{vmatrix} 5 & -6 \\ -23 & 16 \end{vmatrix} + (-1)\begin{vmatrix} 1 & -16 \\ -23 & 16 \end{vmatrix} + 0\begin{vmatrix} 1 & -16 \\ 5 & -16 \end{vmatrix}$$

$$= -1\begin{vmatrix} 1 & -16 \\ -23 & 16 \end{vmatrix}$$

$$= -(16 - 368) = +352.$$

REDUCTION OF DETERMINANTS TO TRIANGULAR FORM

● **PROBLEM** 5-10

Find the determinant of the matrix A where:

$$A = \begin{bmatrix} 2 & 7 & -3 & 8 & 3 \\ 0 & -3 & 7 & 5 & 1 \\ 0 & 0 & 6 & 7 & 6 \\ 0 & 0 & 0 & 9 & 8 \\ 0 & 0 & 0 & 0 & 4 \end{bmatrix}$$

<u>Solution</u>:   A is an upper triangular matrix. As we know, if A is an

nXm triangular matrix (upper or lower), then   det A   is the product of
the entries on the main diagonal.
Hence,

$$\det A = (2) \cdot (-3) \cdot (6) \cdot (9) \cdot (4) = -1296 .$$

● **PROBLEM** 5-11

a)   Compute the determinant of the matrix   A,   where:

$$A = \begin{bmatrix} 2 & 1 & -3 \\ 4 & 1 & 1 \\ 2 & 0 & -2 \end{bmatrix}$$

b)   Prove that the determinant of a lower-triangular matrix   A   is the
product of the diagonal entries of   A.

Solution: a) To find the determinant of   A,   apply row operations until
an upper-triangular matrix is obtained.

$$A = \begin{bmatrix} 2 & 1 & -3 \\ 4 & 1 & 1 \\ 2 & 0 & -2 \end{bmatrix}$$

Now, add   -2 times the first row to the second row and subtract the first
row from the third row.

$$B = \begin{bmatrix} 2 & 1 & -3 \\ 0 & -1 & 7 \\ 0 & -1 & 1 \end{bmatrix}$$

B   has the same determinant as   A.   Subtract the second row from the
third row

$$C = \begin{bmatrix} 2 & 1 & -3 \\ 0 & -1 & 7 \\ 0 & 0 & -6 \end{bmatrix}$$

C   has the same determinant as   B.   C   is an upper triangular matrix and
therefore the determinant of   C   is the product of the diagonal entries
of   C.   Hence,

$$\det A = \det B = \det C = (2) \cdot (-1) \cdot (-6) = 12 .$$

b) Proof:

$$\text{Let } A = \begin{bmatrix} a_{11} & 0 & 0 & \ldots & 0 \\ a_{21} & a_{22} & 0 & \ldots & 0 \\ & & \cdots & & \\ a_{n1} & a_{n2} & a_{n3} & \cdots & a_{nn} \end{bmatrix} .$$

Suppose that $<i_1, i_2, \ldots, i_n>$ is a permutation for which $a_{1i_1} a_{2i_2} \ldots$
$a_{ni_n} \neq 0$. Since $a_{1i_1} \neq 0$, $i_1 = 1$.

Notice that a_{11} is the only non-zero element in the
first row, i.e. $a_{1j} = 0$, $j \neq 1$.

152

Since $a_{2i_2} \neq 0$ and $i_2 \neq i_1 = 1$, $i_2 = 2$. Similarly, $i_3 = 3$, $i_4 = 4, \ldots, i_n = n$. Thus, the only non-zero term in

$$\det A = \Sigma \ \text{sgn} <i_1, \ldots, i_n> \ a_{1i_1} a_{2i_2} \cdots a_{ni_n} \ ,$$

is $a_{11} a_{22} \cdots a_{nn}$. So, $\det A = a_{11} a_{22} \cdots a_{nn}$ equals the product of the diagonal entries.

● **PROBLEM** 5-12

Evaluate $\det A$ where:

$$A = \begin{bmatrix} 0 & 1 & 5 \\ 3 & -6 & 9 \\ 2 & 6 & 1 \end{bmatrix}$$

Solution: Interchange the first and second rows of matrix A, obtaining matrix

$$B = \begin{bmatrix} 3 & -6 & 9 \\ 0 & 1 & 5 \\ 2 & 6 & 1 \end{bmatrix} \ ;$$

and by the properties of the function,

$$\det A = -\det B \ .$$

$$= -\det \begin{bmatrix} 3 & -6 & 9 \\ 0 & 1 & 5 \\ 2 & 6 & 1 \end{bmatrix}$$

or

$$\det A = -3 \det \begin{bmatrix} 1 & -2 & 3 \\ 0 & 1 & 5 \\ 2 & 6 & 1 \end{bmatrix} \ .$$

A common factor of 3 from the first row of the matrix B was taken out. Add -2 times the first row to the third row. The value of the determinant of A will remain the same. Thus,

$$\det A = -3 \det \begin{bmatrix} 1 & -2 & 3 \\ 0 & 1 & 5 \\ 0 & 10 & -5 \end{bmatrix} \ .$$

Add -10 times the second row to the third row. Thus,

$$\det A = -3 \det \begin{bmatrix} 1 & -2 & 3 \\ 0 & 1 & 5 \\ 0 & 0 & -55 \end{bmatrix} \ .$$

As we know, the determinant of a triangular matrix is equal to the product of the diagonal elements. Thus,

$$\det A = (-3) \cdot (1) \cdot (1) \cdot (-55) = 165 \ .$$

Compute the determinant of

$$A = \begin{bmatrix} 1 & 0 & 0 & 3 \\ 2 & 7 & 0 & 6 \\ 0 & 6 & 3 & 0 \\ 7 & 3 & 1 & -5 \end{bmatrix}.$$

Solution: First list the basic properties of the determinant. Let A be a square matrix.
(1) If we interchange two rows (columns) of A, the determinant changes by a factor of (-1).
(2) By adding to one row (column) of A a multiple of another row (column), the determinant is not changed.
(3) If A is triangular, i.e., A has zeros above or below the diagonal, the determinant of A is the product of the diagonal elements of A.
(4) If a row (column) of A is multiplied by a scalar K, but all of the other rows (columns) are left unchanged, then the determinant of the resulting matrix is K times the determinant of A.

Given:

$$A = \begin{bmatrix} 1 & 0 & 0 & 3 \\ 2 & 7 & 0 & 6 \\ 0 & 6 & 3 & 0 \\ 7 & 3 & 1 & -5 \end{bmatrix}.$$

Now add -3 times the first column to the fourth column. But the value of the determinant A will not be changed.
Thus,

$$A = \begin{bmatrix} 1 & 0 & 0 & 0 \\ 2 & 7 & 0 & 0 \\ 0 & 6 & 3 & 0 \\ 7 & 3 & 1 & -26 \end{bmatrix}$$

Now the above matrix A is a lower triangular matrix. Using property (3), yields:

$$\det A = (1) \cdot (3) \cdot (-26) = -546 .$$

THE ADJOINT OF A MATRIX

Find the cofactors of the matrix A where:

$$A = \begin{bmatrix} 1 & 2 & 3 \\ 3 & 2 & 1 \\ 2 & 0 & 2 \end{bmatrix}.$$

Solution: If A is an $n \times n$ matrix, then M_{ij} will denote the $(n-1) \times (n-1)$ matrix obtained from A by deleting its ith row and jth column. The determinant $|M_{ij}|$ is called the minor of the element a_{ij} of A, and we define the cofactor of a_{ij} ; denoted by A_{ij} , as:

$$A_{ij} = (-1)^{i+j} |M_{ij}| .$$

$$A = \begin{bmatrix} 1 & 2 & 3 \\ 3 & 2 & 1 \\ 2 & 0 & 2 \end{bmatrix} .$$

Then

$$A_{11} = (-1)^2 \begin{vmatrix} 2 & 1 \\ 0 & 2 \end{vmatrix} = 4 - 0 = 4$$

$$A_{12} = (-1)^{1+2} \begin{vmatrix} 3 & 1 \\ 2 & 2 \end{vmatrix} = -(6-2) = -4$$

$$A_{13} = (-1)^{1+3} \begin{vmatrix} 3 & 2 \\ 2 & 0 \end{vmatrix} = +(0-4) = -4$$

$$A_{21} = (-1)^{2+1} \begin{vmatrix} 2 & 3 \\ 0 & 2 \end{vmatrix} = -(4-0) = -4$$

$$A_{22} = (-1)^{2+2} \begin{vmatrix} 1 & 3 \\ 2 & 2 \end{vmatrix} = +(2-6) = -4$$

$$A_{23} = (-1)^{2+3} \begin{vmatrix} 1 & 2 \\ 2 & 0 \end{vmatrix} = -(0-4) = 4$$

$$A_{31} = (-1)^{3+1} \begin{vmatrix} 2 & 3 \\ 2 & 1 \end{vmatrix} = +(2-6) = -4$$

$$A_{32} = (-1)^{3+2} \begin{vmatrix} 1 & 3 \\ 3 & 1 \end{vmatrix} = -1(1-9) = 8$$

$$A_{33} = (-1)^{3+3} \begin{vmatrix} 1 & 2 \\ 3 & 2 \end{vmatrix} = +(2-6) = -4$$

Thus, the cofactors of A are:

$$A_{11} = 4 \qquad A_{12} = -4 \qquad A_{13} = -4$$

$$A_{21} = -4 \qquad A_{22} = -4 \qquad A_{23} = 4$$

$$A_{31} = -4 \qquad A_{32} = 8 \qquad A_{33} = -4 .$$

● **PROBLEM** 5-15

a) Given:
$$A = \begin{bmatrix} 5 & 2 & -1 \\ 3 & 1 & 2 \\ 2 & 7 & 4 \end{bmatrix} , \quad \text{find det } A .$$

b) If
$$A = \begin{bmatrix} 2 & 3 & 1 \\ -2 & 4 & 5 \\ 2 & 0 & 7 \end{bmatrix} , \quad \text{find adj } A .$$

Solution: The definition of the determinant in terms of permutations is a mathematical one. For computational purposes, more efficient evaluative procedures are available.

Define the cofactor a_{ij} of an $n \times n$ determinant $|A|$ as $(-1)^{i+j}|C|$ where $|C|$ is the $(n-1) \times (n-1)$ sub-determinant obtained by deleting the ith row and jth column of $|A|$.

Suppose that $A = (a_{ij})$ is an $n \times n$ matrix, and let A_{ij} denote the cofactor of a_{ij}, $i = 1, 2, \ldots, n$; $j = 1, 2, \ldots, n$. Then,

$$\det A = \sum_{k=1}^{n} a_{kj} A_{kj}, \quad j = 1, 2, \ldots, n,$$

and

$$\det A = \sum_{k=1}^{n} a_{ik} A_{ik}, \quad i = 1, 2, \ldots, n.$$

Note that we sum across k to obtain $\det A$. Since $j = 1, 2, \ldots, n$, there are $2n$ ways of computing $\det A$, one for each row (or column) of elements.

$$A = \begin{bmatrix} 5 & 2 & -1 \\ 3 & 1 & 2 \\ 2 & 7 & 4 \end{bmatrix}$$

The cofactors along the first column are:

$$A_{11} = (-1)^{1+1} \begin{vmatrix} 1 & 2 \\ 7 & 4 \end{vmatrix} = + (4-14) = -10$$

$$A_{21} = (-1)^{2+1} \begin{vmatrix} 2 & -1 \\ 7 & 4 \end{vmatrix} = -(8+7) = -15$$

$$A_{31} = (-1)^{3+1} \begin{vmatrix} 2 & -1 \\ 1 & 2 \end{vmatrix} = +(4+1) = 5.$$

Hence,

$$\det A = \sum_{k=1}^{n} a_{k1} A_{k1} = a_{11}A_{11} + a_{21}A_{21} + a_{31}A_{31}$$

$$= 5(-10) + 3(-15) + 2(5)$$
$$= -50 - 45 + 10$$
$$= -85.$$

Also, expanding along the 2nd row,

$$A_{21} = -15, A_{22} = 22, A_{23} = -31.$$

Therefore,

$$\det A = a_{21}A_{21} + a_{22}A_{22} + a_{23}A_{23}$$

$$= 3(-15) + 1(22) + 2(-31)$$
$$= -45 + 22 - 62$$
$$= -85.$$

b) Let A be a $n \times n$ matrix. Then the adjoint of A, adj A, is defined as the matrix of transposed cofactors. Let $C_{11}, C_{12}, \ldots, C_{1n}, C_{21}, \ldots, C_{2n}, \ldots, C_{n1}, \ldots, C_{nn}$ denote the cofactors of $a_{11}, a_{12}, \ldots, a_{nn}$, respectively. Then

$$\text{adj. A} = \begin{bmatrix} C_{11} & C_{21} & \cdots & \cdots & C_{n1} \\ C_{12} & & & & \\ \cdot & & & & \\ \cdot & & & & \\ \cdot & & & & \\ C_{1n} & \cdots & \cdots & \cdots & C_{nn} \end{bmatrix}$$

The adjoint is useful in finding the inverse of a non-singular matrix. The cofactors of A are:

$$A_{11} = + \begin{vmatrix} 4 & 5 \\ 0 & 7 \end{vmatrix} = (28.0) = 28$$

$$A_{12} = - \begin{vmatrix} -2 & 5 \\ 2 & 7 \end{vmatrix} = -(-14-10) = 24$$

$$A_{13} = + \begin{vmatrix} -2 & 4 \\ 2 & 0 \end{vmatrix} = (0-8) = -8$$

$$A_{21} = - \begin{vmatrix} 3 & 1 \\ 0 & 7 \end{vmatrix} = -(21-0) = -21$$

$$A_{22} = + \begin{vmatrix} 2 & 1 \\ 2 & 7 \end{vmatrix} = (14-2) = 12$$

$$A_{23} = - \begin{vmatrix} 2 & 3 \\ 2 & 0 \end{vmatrix} = -(0-6) = 6$$

$$A_{31} = + \begin{vmatrix} 3 & 1 \\ 4 & 5 \end{vmatrix} = (15-4) = 11$$

$$A_{32} = - \begin{vmatrix} 2 & 1 \\ -2 & 5 \end{vmatrix} = -(10+2) = -12$$

$$A_{33} = + \begin{vmatrix} 2 & 3 \\ -2 & 4 \end{vmatrix} = (8+6) = 14 \quad .$$

The matrix of cofactors is:

$$\begin{bmatrix} 28 & 24 & -8 \\ -21 & 12 & 6 \\ 11 & -12 & 14 \end{bmatrix}$$

and the adj of A is:

$$\text{Adj } [A] = \begin{bmatrix} 28 & -21 & 11 \\ 24 & 12 & -12 \\ -8 & 6 & 14 \end{bmatrix} .$$

● PROBLEM 5-16

Find the adjoint of the matrix A, where:

$$A = \begin{bmatrix} 3 & 2 & -1 \\ 1 & 6 & 3 \\ 2 & -4 & 0 \end{bmatrix}.$$

Solution: The transpose of the matrix of cofactors of the elements a_{ij} of A, which is denoted by adj. A, is called the adjoint of A.

The cofactor of a_{ij} of A, denoted by A_{ij} is

$$A_{ij} = (-1)^{i+j} |M_{ij}|,$$

where M_{ij} is the (n-1) square submatrix of A, obtained by deleting its ith row and jth column. The matrix of cofactors is:

$$\begin{bmatrix} A_{11} & A_{12} & \cdots & A_{1n} \\ A_{21} & A_{22} & \cdots & A_{2n} \\ A_{n1} & A_{n2} & \cdots & A_{nn} \end{bmatrix}.$$

Thus,

$$\text{adj. A} = \begin{bmatrix} A_{11} & A_{21} & \cdots & A_{n1} \\ A_{12} & A_{22} & \cdots & A_{n2} \\ \vdots \\ A_{1n} & A_{2n} & \cdots & A_{nn} \end{bmatrix}.$$

Now,

$$A = \begin{bmatrix} 3 & 2 & -1 \\ 1 & 6 & 3 \\ 2 & -4 & 0 \end{bmatrix}.$$

The cofactors of A are:

$$A_{11} = (-1)^{1+1} \begin{vmatrix} 6 & 3 \\ -4 & 0 \end{vmatrix} = + (0+12) = 12$$

$$A_{12} = (-1)^{1+2} \begin{vmatrix} 1 & 3 \\ 2 & 0 \end{vmatrix} = -(0-6) = 6$$

$$A_{13} = (-1)^{1+3} \begin{vmatrix} 1 & 6 \\ 2 & -4 \end{vmatrix} = +(-4 \cdot 12) = -16$$

$$A_{21} = (-1)^{2+1} \begin{vmatrix} 2 & -1 \\ -4 & 0 \end{vmatrix} = -(0-4) = 4$$

$$A_{22} = (-1)^{2+2} \begin{vmatrix} 3 & -1 \\ 2 & 0 \end{vmatrix} = +(0+2) = +2$$

$$A_{23} = (-1)^{2+3} \begin{vmatrix} 3 & 2 \\ 2 & -4 \end{vmatrix} = -(-12-4) = 16$$

$$A_{31} = (-1)^{3+1} \begin{vmatrix} 2 & -1 \\ 6 & 3 \end{vmatrix} = +(6+6) = 12$$

$$A_{32} = (-1)^{3+2} \begin{vmatrix} 3 & -1 \\ 1 & 3 \end{vmatrix} = -(9+1) = -10$$

$$A_{33} = (-1)^{3+3} \begin{vmatrix} 3 & 2 \\ 1 & 6 \end{vmatrix} = +(18-2) = 16.$$

The matrix of cofactors is:

$$\begin{bmatrix} 12 & 6 & -16 \\ 4 & 2 & 16 \\ 12 & -10 & 16 \end{bmatrix} .$$

The adjoint of A is:

$$\text{adj.}(A) = \begin{bmatrix} 12 & 4 & 12 \\ 6 & 2 & -10 \\ -16 & 16 & 16 \end{bmatrix}$$

● PROBLEM 5-17

Find the adjoint A of the following matrices:

(a) $\quad A = \begin{bmatrix} a_{11} & a_{12} \\ a_{21} & a_{22} \end{bmatrix}$ (b) $\quad A = \begin{bmatrix} 1 & 0 & 5 \\ 2 & 1 & 0 \\ 0 & 4 & 0 \end{bmatrix}$

(c) $\quad A = \begin{bmatrix} \lambda_1 & \cdots & 0 \\ & \ddots & \\ 0 & & \lambda_n \end{bmatrix}$, $\quad \lambda_i \neq 0$, $\quad i = 1,2,\ldots,n$

Solution:a)The transpose of the matrix of cofactors of the elements, a_{ij} of A, is called the adjoint of A.

$$A = \begin{bmatrix} a_{11} & a_{12} \\ a_{21} & a_{22} \end{bmatrix} .$$

The cofactors of the four elements are:

$$A_{11} = a_{22} , A_{12} = -a_{21}$$
$$A_{21} = -a_{12}, A_{22} = a_{11} .$$

The matrix of the cofactors is:

$$\begin{bmatrix} a_{22} & -a_{21} \\ -a_{12} & a_{11} \end{bmatrix}$$

Thus,

$$\text{adj } A = \begin{bmatrix} a_{22} & -a_{12} \\ -a_{21} & a_{11} \end{bmatrix} .$$

(b)

$$A = \begin{bmatrix} 1 & 0 & 5 \\ 2 & 1 & 0 \\ 0 & 4 & 0 \end{bmatrix}$$

The cofactors of the nine elements are:

159

$$A_{11} = + \begin{vmatrix} 1 & 0 \\ 4 & 0 \end{vmatrix} = 1(0-0) = 0$$

$$A_{12} = - \begin{vmatrix} 2 & 0 \\ 0 & 0 \end{vmatrix} = 0$$

$$A_{13} = + \begin{vmatrix} 2 & 1 \\ 0 & 4 \end{vmatrix} = (8-0) = +8$$

$$A_{21} = - \begin{vmatrix} 0 & 5 \\ 4 & 0 \end{vmatrix} = -(0-20) = +20$$

$$A_{22} = + \begin{vmatrix} 1 & 5 \\ 0 & 0 \end{vmatrix} = 0$$

$$A_{23} = - \begin{vmatrix} 1 & 0 \\ 0 & 4 \end{vmatrix} = -(4-0) = -4$$

$$A_{31} = + \begin{vmatrix} 0 & 5 \\ 1 & 0 \end{vmatrix} = (0-5) = -5$$

$$A_{32} = - \begin{vmatrix} 1 & 5 \\ 2 & 0 \end{vmatrix} = -(0-10) = +10$$

$$A_{33} = + \begin{vmatrix} 1 & 0 \\ 2 & 1 \end{vmatrix} = (1-0) = 1$$

The matrix of the cofactors is:

$$\begin{bmatrix} 0 & 0 & 8 \\ 20 & 0 & -4 \\ -5 & +10 & 1 \end{bmatrix}$$

Thus,

$$\text{adj } A = \begin{bmatrix} 0 & 20 & -5 \\ 0 & 0 & +10 \\ 8 & -4 & 1 \end{bmatrix}$$

(c)

$$A = \begin{bmatrix} \lambda_1 & & 0 \\ & \ddots & \\ 0 & & \lambda_n \end{bmatrix}$$

Here, the matrix A is a diagonal matrix. We know,

$$A^{-1} = \frac{1}{\det A} (\text{adj } A)$$

or

$$\text{adj } A = |A| A^{-1}.$$

We also know that the inverse of a
diagonal matrix is simply made up of the
inverses of its non-diagonal elements. Now,

$$A = \begin{bmatrix} \lambda_1 & & 0 \\ & \ddots & \\ 0 & & \lambda_n \end{bmatrix}.$$

Then,

$$A^{-1} = \begin{bmatrix} \lambda_1^{-1} & 0 & \cdots & 0 \\ 0 & \lambda_2^{-1} & & \vdots \\ \vdots & & \ddots & \\ 0 & & & \lambda_n^{-1} \end{bmatrix},$$

providing $\lambda_i \neq 0$ $(i = 1,\ldots,n)$. Hence,

$$\text{adj } A = |A| \begin{bmatrix} \lambda_1^{-1} & 0 & \cdots & 0 \\ 0 & \lambda_2^{-1} & \cdots & 0 \\ \vdots & & \ddots & \\ 0 & 0 & & \lambda_n^{-1} \end{bmatrix}$$

Since $|A| = \lambda_1 \lambda_2 \cdots \lambda_n = \prod\limits_{i=1}^{n} \lambda_i$,

$$\text{adj } A = \prod_{i=1}^{n} \lambda_i \begin{bmatrix} \dfrac{1}{\lambda_1} & 0 & \cdots & \cdots & 0 \\ 0 & \dfrac{1}{\lambda_2} & & & \vdots \\ \vdots & & \ddots & & \vdots \\ \vdots & & & \ddots & \dfrac{1}{\lambda_n} \\ 0 & \cdots & \cdots & \cdots & \dfrac{1}{\lambda_n} \end{bmatrix}$$

● **PROBLEM** 5-18

Given:
$$A = \begin{bmatrix} 3 & 1 & 2 \\ 0 & 1 & 1 \\ -1 & 1 & 0 \end{bmatrix}.$$
Show that $(\text{adj } A) \cdot A = (\det A) I$ where I is the identity matrix.

<u>Solution</u>: It is known that the classical adjoint, or adj A, is the transpose of the matrix of cofactors of the elements a_{ij} of A. The cofactors of the nine elements of the given matrix A are:

$$A_{11} = + \begin{vmatrix} 1 & 1 \\ 1 & 0 \end{vmatrix} = (0-1) = -1$$

$$A_{12} = - \begin{vmatrix} 0 & 1 \\ -1 & 0 \end{vmatrix} = -(0+1) = -1$$

$$A_{13} = + \begin{vmatrix} 0 & 1 \\ -1 & 1 \end{vmatrix} = (0+1) = 1$$

$$A_{21} = - \begin{vmatrix} 1 & 2 \\ 1 & 0 \end{vmatrix} = -(0-2) = +2$$

$$A_{22} = + \begin{vmatrix} 3 & 2 \\ -1 & 0 \end{vmatrix} = (0+2) = 2$$

$$A_{23} = - \begin{vmatrix} 3 & 1 \\ -1 & 1 \end{vmatrix} = -(3+1) = -4$$

$$A_{31} = + \begin{vmatrix} 1 & 2 \\ 1 & 1 \end{vmatrix} = (1-2) = -1$$

$$A_{32} = - \begin{vmatrix} 3 & 2 \\ 0 & 1 \end{vmatrix} = -(3-0) = -3$$

$$A_{33} = + \begin{vmatrix} 3 & 1 \\ 0 & 1 \end{vmatrix} = (3-0) = 3 .$$

Then the matrix of the cofactors is:
$$\begin{bmatrix} -1 & -1 & 1 \\ 2 & 2 & -4 \\ -1 & -3 & 3 \end{bmatrix} .$$

161

Hence,

$$\text{adj } A = \begin{bmatrix} -1 & 2 & -1 \\ -1 & 2 & -3 \\ 1 & -4 & 3 \end{bmatrix} .$$

$$(\text{adj } A) \cdot A = \begin{bmatrix} -1 & 2 & -1 \\ -1 & 2 & -3 \\ 1 & -4 & 3 \end{bmatrix} \begin{bmatrix} 3 & 1 & 2 \\ 0 & 1 & 1 \\ -1 & 1 & 0 \end{bmatrix}$$

$$= \begin{bmatrix} -3+0+1 & -1+2-1 & -2+2+0 \\ -3+0+3 & -1+2-3 & -2+2+0 \\ 3+0-3 & 1-4+3 & 2-4+0 \end{bmatrix}$$

$$= \begin{bmatrix} -2 & 0 & 0 \\ 0 & -2 & 0 \\ 0 & 0 & -2 \end{bmatrix} .$$

$$\text{adj } A \cdot = -2 \begin{bmatrix} 1 & 0 & 0 \\ 0 & 1 & 0 \\ 0 & 0 & 1 \end{bmatrix} = -2I$$

$$\det A = \begin{vmatrix} 3 & 1 & 2 \\ 0 & 1 & 1 \\ -1 & 1 & 0 \end{vmatrix} .$$

$$\det A = 3 \begin{vmatrix} 1 & 1 \\ 1 & 0 \end{vmatrix} - 0 \begin{vmatrix} 1 & 2 \\ 1 & 0 \end{vmatrix} - 1 \begin{vmatrix} 1 & 2 \\ 1 & 1 \end{vmatrix}$$

$$= 3(0-1) -1 (1-2)$$

$$= -3 + 1 = -2 .$$

Hence,

$$\text{adj } A \cdot A = -2I = (\det A)I .$$

MISCELLANEOUS RESULTS IN DETERMINANTS

● PROBLEM 5-19

Compute the determinants of each of the following matrices and find which of the matrices are invertible.

(a) $\begin{bmatrix} 3 & 1 & 2 \\ 1 & 0 & 6 \\ -1 & 1 & 1 \end{bmatrix}$ (b) $\begin{bmatrix} -1 & 1 & 3 \\ 2 & 1 & 1 \\ 4 & 2 & 2 \end{bmatrix}$ (c) $\begin{bmatrix} 2 & 1 & 1 \\ 0 & 0 & 0 \\ 4 & 3 & 1 \end{bmatrix}$

<u>Solution</u>: We can evaluate determinants by using the basic properties of the determinant function.

Properties of Determinants:

(1) If each element in a row (or column) is zero, the value of the determinant is zero.

(2) If two rows (or columns) of a determinant are identical, the value of the determinant is zero.

(3) The determinant of a matrix A and its transpose A^t are equal: $|A| = |A^t|$.

(4) The matrix A has an inverse if and only if $\det A \neq 0$.

$$\det A = \begin{vmatrix} 3 & 1 & 2 \\ 1 & 0 & 6 \\ -1 & 1 & 1 \end{vmatrix}$$

$$= 3 \begin{vmatrix} 0 & 6 \\ 1 & 1 \end{vmatrix} - 1 \begin{vmatrix} 1 & 6 \\ -1 & 1 \end{vmatrix} + 2 \begin{vmatrix} 1 & 0 \\ -1 & 1 \end{vmatrix}$$

$$= 3(0-6) - 1(1+6) + 2(1-0)$$

$$= -18-7+2 = -23 .$$

Since $\det A = -23 \neq 0$, this matrix is invertible.

(b)

$$A = \begin{bmatrix} -1 & 1 & 3 \\ 2 & 1 & 1 \\ 4 & 2 & 2 \end{bmatrix}$$

Here $\det A = 0$, since the third row is a multiple of the second row. Since $\det A = 0$, the matrix is not invertible.

(c)

$$A = \begin{bmatrix} 2 & 1 & 1 \\ 0 & 0 & 0 \\ 4 & 3 & 1 \end{bmatrix}$$

$$\det A = \begin{vmatrix} 2 & 1 & 1 \\ 0 & 0 & 0 \\ 4 & 3 & 1 \end{vmatrix} = 0.$$

Here, each element in the second row is zero, therefore, the value of the determinant is zero. Since $\det A = 0$, the matrix is not invertible.

● **PROBLEM** 5-20

Given:
$$A = \begin{bmatrix} 2 & -1 & 1 \\ 4 & 1 & -3 \\ 2 & -1 & 3 \end{bmatrix} ,$$

Evaluate $\det A$ and $\det A^{-1}$. What is the relation between $\det A$ and $\det A^{-1}$?

Solution: First find $\det A$.

$$\det A = \begin{vmatrix} 2 & -1 & 1 \\ 4 & 1 & -3 \\ 2 & -1 & 3 \end{vmatrix}$$

Add the third row to the second row:

$$\det A = \begin{vmatrix} 2 & -1 & 1 \\ 6 & 0 & 0 \\ 2 & -1 & 3 \end{vmatrix}$$

Expand the determinant by minors, using the second row.

$$\det A = -6 \begin{vmatrix} -1 & 1 \\ -1 & 3 \end{vmatrix} + 0 \begin{vmatrix} 2 & 1 \\ 2 & 3 \end{vmatrix} - 0 \begin{vmatrix} 2 & -1 \\ 2 & -1 \end{vmatrix}$$

$$= -6(-3+1) = 12.$$

Thus, $\det A = 12$. Next, find A^{-1}. We know,

$$A^{-1} = \frac{1}{\det A} \cdot \operatorname{adj} A ,$$

where $\operatorname{adj} A$ is the transpose of the matrix of cofactors. Then, the cofactors are:

$$A_{11} = + \begin{vmatrix} 1 & -3 \\ -1 & 3 \end{vmatrix} = (3-3) = 0$$

$$A_{12} = - \begin{vmatrix} 4 & -3 \\ 2 & 3 \end{vmatrix} = -(12+6) = -18$$

$$A_{13} = + \begin{vmatrix} 4 & 1 \\ 2 & -1 \end{vmatrix} = +(-4-2) = -6$$

$$A_{21} = - \begin{vmatrix} -1 & 1 \\ -1 & 3 \end{vmatrix} = -(-3+1) = 2$$

$$A_{22} = + \begin{vmatrix} 2 & 1 \\ 2 & 3 \end{vmatrix} = (6-2) = 4$$

$$A_{23} = - \begin{vmatrix} 2 & -1 \\ 2 & -1 \end{vmatrix} = -(-2+2) = 0$$

$$A_{31} = + \begin{vmatrix} -1 & 1 \\ 1 & -3 \end{vmatrix} = (3-1) = 2$$

$$A_{32} = - \begin{vmatrix} 2 & 1 \\ 4 & -3 \end{vmatrix} = -(-6-4) = 10$$

$$A_{33} = + \begin{vmatrix} 2 & -1 \\ 4 & 1 \end{vmatrix} = (2+4) = 6 .$$

The matrix of the cofactors is:

$$\begin{bmatrix} 0 & -18 & -6 \\ 2 & 4 & 0 \\ 2 & 10 & 6 \end{bmatrix} .$$

Hence,

$$\operatorname{adj} A = \begin{bmatrix} 0 & 2 & 2 \\ -18 & 4 & 10 \\ -6 & 0 & 6 \end{bmatrix} .$$

Then,

$$A^{-1} = \frac{1}{\det A} (\operatorname{adj} A)$$

$$= \frac{1}{12} \begin{bmatrix} 0 & 2 & 2 \\ -18 & 4 & 10 \\ -6 & 0 & 6 \end{bmatrix}$$

$$= \begin{bmatrix} 0 & 2/12 & 2/12 \\ -18/12 & 4/12 & 10/12 \\ -6/12 & 0 & 6/12 \end{bmatrix} = \begin{bmatrix} 0 & 1/6 & 1/6 \\ -3/2 & 1/3 & 5/6 \\ -1/2 & 0 & 1/2 \end{bmatrix} .$$

Now, find $\det A^{-1}$:

$$\det A^{-1} = \begin{vmatrix} 0 & 1/6 & 1/6 \\ -3/2 & 1/3 & 5/6 \\ -1/2 & 0 & 1/2 \end{vmatrix} .$$

Add the third column to the first column.

$$\det A^{-1} = \begin{vmatrix} 1/6 & 1/6 & 1/6 \\ -4/6 & 1/3 & 5/6 \\ 0 & 0 & 1/2 \end{vmatrix} .$$

Expand the determinant by minors, using the third row:

$$\det A^{-1} = 0 \begin{vmatrix} 1/6 & 1/6 \\ 1/3 & 5/6 \end{vmatrix} - 0 \begin{vmatrix} 1/6 & 1/6 \\ -4/6 & 5/6 \end{vmatrix} + 1/2 \begin{vmatrix} 1/6 & 1/6 \\ -4/6 & 1/3 \end{vmatrix}$$

$$= 1/2(1/18 + 4/36)$$

$$= 1/2(1/6) = 1/12 .$$

Thus, $\det A = 12$, while $\det A^{-1} = 1/12$. That is,

$$\det A^{-1} = \frac{1}{\det A} .$$

We emphasize that A has an inverse if and only if $\det A \neq 0$.

● **PROBLEM 5-21**

Compute the determinant of the matrix A, where:

$$A = \begin{bmatrix} 1 & -1 & 0 \\ 3 & 4 & 4 \\ 2 & 1 & 1 \end{bmatrix} .$$

Use the basic definition of the determinant, ie.
$$\det A = \sum_{\sigma} \text{sgn}(\sigma) \, a_{\sigma(j)j} \qquad (j=1,\dots,n)$$
where σ_n is a permutation of the first n integers.

Solution: With the aid of row or column operations, we can simplify the matrix so that the determinant can be more easily computed. Now, add the second column to the first column.

$$A = \begin{bmatrix} 0 & -1 & 0 \\ 7 & 4 & 4 \\ 3 & 1 & 1 \end{bmatrix}$$

Add -4 times the third row to the second row.

$$A = \begin{bmatrix} 0 & -1 & 0 \\ -5 & 0 & 0 \\ 3 & 1 & 1 \end{bmatrix}$$

Add -1 times the third column to the second column and add -3 times the third column to the first column.

$$A = \begin{bmatrix} 0 & -1 & 0 \\ -5 & 0 & 0 \\ 0 & 0 & 1 \end{bmatrix}$$

The only non-zero term in the sum,

$$\sum_{\sigma} \text{sgn}(\sigma) \; a_\sigma$$

is $5 = a_{21}a_{12}a_{33}$. The corresponding permutation, σ, satisfies

$$\sigma(1) = 2, \; \sigma(2) = 1, \; \sigma(3) = 3 \; .$$

Clearly, σ is a (single) transposition, and hence, it is odd. Thus $\text{sgn}(\sigma) = -1$. Therefore,

$$D = -5 \; .$$

● **PROBLEM** 5-22

Given:

$$A = \begin{bmatrix} 1 & 1 & 4 \\ -1 & 2 & -3 \\ 2 & -1 & 3 \end{bmatrix} , \quad B = \begin{bmatrix} -1 & 1 & 2 \\ 0 & 2 & 3 \\ 2 & 3 & -1 \end{bmatrix} .$$

Show that the determinant of the product of these two matrices, A and B, is equal to the product of their determinants, i.e., det AB = det A. det B.

Solution:

$$\det A = \begin{vmatrix} 1 & 1 & 4 \\ -1 & 2 & -3 \\ 2 & -1 & 3 \end{vmatrix}$$

Add the second row to the first row.

$$\det A = \begin{vmatrix} 0 & 3 & 1 \\ -1 & 2 & -3 \\ 2 & -1 & 3 \end{vmatrix}$$

Add -3 times the third column to the second column.

$$\det A = \begin{vmatrix} 0 & 0 & 1 \\ -1 & 11 & -3 \\ 2 & -10 & 3 \end{vmatrix}$$

Expanding along the first row, two terms drop out, since they have factors of 0 .

$$\text{Det } A = 1 \begin{vmatrix} -1 & 11 \\ 2 & -10 \end{vmatrix} = 10 - 22 = -12$$

$$\det B = \begin{vmatrix} -1 & 1 & 2 \\ 0 & 2 & 3 \\ 2 & 3 & -1 \end{vmatrix}$$

Expand the determinant by minors, using the first column.

$$\det B = -1 \begin{vmatrix} 2 & 3 \\ 3 & -1 \end{vmatrix} -0 \begin{vmatrix} 1 & 2 \\ 3 & -1 \end{vmatrix} + 2 \begin{vmatrix} 1 & 2 \\ 2 & 3 \end{vmatrix}$$

$$= -1(-2-9) + 2(3-4)$$

$$= +11 - 2 = 9 \; .$$

$$AB = \begin{bmatrix} 1 & 1 & 4 \\ -1 & 2 & -3 \\ 2 & -1 & 3 \end{bmatrix} \begin{bmatrix} -1 & 1 & 2 \\ 0 & 2 & 3 \\ 2 & 3 & -1 \end{bmatrix}$$

$$= \begin{bmatrix} -1+0+8 & 1+2+12 & 2+3-4 \\ 1+0-6 & -1+4-9 & -2+6+3 \\ -2+0+6 & 2-2+9 & 4-3-3 \end{bmatrix}$$

$$= \begin{bmatrix} 7 & 15 & 1 \\ -5 & -6 & 7 \\ 4 & 9 & -2 \end{bmatrix}$$

Then,

$$\det AB = \begin{vmatrix} 7 & 15 & 1 \\ -5 & -6 & 7 \\ 4 & 9 & -2 \end{vmatrix}$$

$$= 7 \begin{vmatrix} -6 & 7 \\ 9 & -2 \end{vmatrix} - 15 \begin{vmatrix} -5 & 7 \\ 4 & -2 \end{vmatrix} + 1 \begin{vmatrix} -5 & -6 \\ 4 & 9 \end{vmatrix}$$

So,

$$\det AB = 7(12-63) - 15(10-28) + 1(-45+24) = -357 + 270 - 21 = -108$$

and,

$$\det A \cdot \det B = (-12) \times (9)$$
$$= -108 .$$

Hence, $\det AB = \det A \cdot \det B$.

● **PROBLEM** 5-23

Define the absolute determinant. If

(a) $A = \begin{bmatrix} a_{11} \\ a_{21} \end{bmatrix}$, (b) $A = \begin{bmatrix} 1 & 3 \\ 2 & 0 \\ 1 & 4 \end{bmatrix}$,

find the absolute determinant of A .

Solution: Definition: The absolute determinant, A, of a $m \times n$ matrix A, $n \leq m$, is given by the equation:

$$\text{absolute} \quad \det A = (\det A^t A)^{\frac{1}{2}} .$$

The absolute determinant formula is not applied to the case $n > m$, because, then, it would always give a value of zero.

(a) $A = \begin{bmatrix} a_{11} \\ a_{21} \end{bmatrix}$ then $A^t = \begin{bmatrix} a_{11} & a_{21} \end{bmatrix}$ and,

$$A^t A = \begin{bmatrix} a_{11} & a_{21} \end{bmatrix} \begin{bmatrix} a_{11} \\ a_{21} \end{bmatrix} = a_{11}^2 + a_{21}^2 .$$

Hence,

$$\text{absolute } \det A = (\det A^t A)^{\frac{1}{2}}$$
$$= (a_{11}^2 + a_{21}^2)^{\frac{1}{2}} .$$

167

(b)

$$A = \begin{bmatrix} 1 & 3 \\ 2 & 0 \\ 1 & 4 \end{bmatrix} .$$

Then,

$$A^t = \begin{bmatrix} 1 & 2 & 1 \\ 3 & 0 & 4 \end{bmatrix}$$

$$A^t A = \begin{bmatrix} 1 & 2 & 1 \\ 3 & 0 & 4 \end{bmatrix} \begin{bmatrix} 1 & 3 \\ 2 & 0 \\ 1 & 4 \end{bmatrix}$$

$$= \begin{bmatrix} 1+4+1 & 3+0+4 \\ 3+0+4 & 9+0+16 \end{bmatrix}$$

$$= \begin{bmatrix} 6 & 7 \\ 7 & 25 \end{bmatrix} .$$

Hence,

$$\text{absolute det } A = (\det A^t A)^{\frac{1}{2}}$$

$$= \left(\det \begin{vmatrix} 6 & 7 \\ 7 & 25 \end{vmatrix} \right)^{\frac{1}{2}}$$

$$= [150 - 49]^{\frac{1}{2}}$$

$$= (101)^{\frac{1}{2}} = \sqrt{101} .$$

● **PROBLEM** 5-24

Define submatrix and subdeterminant.

Solution: The totality of $m \times n$ matrices (i.e., m rows and n columns) with entries in R will be designated by $M_{m,n}(R)$.

Let r and n be positive integers. Define $\Gamma_{r,n}$ to be the totality of sequences α, $\alpha = (\alpha_1, \ldots, \alpha_r)$, in which each α_1 is an integer satisfying $1 \le \alpha_1 \le n$. Next, let $G_{r,n}$ be the subset of $\Gamma_{r,n}$ consisting of precisely those sequences α for which $1 \le \alpha_1 \le \alpha_2 \le \ldots \le \alpha_r \le n$. Finally, if $r \le n$, then $Q_{r,n}$ will denote the subset of $G_{r,n}$ consisting of those α for which $1 \le \alpha_1 < \ldots < \alpha_r \le n$. Thus, $\Gamma_{r,n}$ is the set of all sequences of length r chosen from $1, \ldots, n$; $G_{r,n}$ is the set of non-decreasing sequences in $\Gamma_{r,n}$ and $Q_{r,n}$ is the set of strictly increasing sequences in $G_{r,n}$.

Let R be a field, and $A \in M_{m,n}(R)$. Let $\alpha \in \Gamma_{r,m}$ and $\beta \in \Gamma_{s,n}$. Then, $A[\alpha|\beta]$ is the matrix in $M_{r,s}(R)$ whose (ij) entry is $a_{\alpha_i \beta_j}$; $i = 1, \ldots, r$, $j = 1, \ldots, s$. If α and β are in $Q_{r,m}$ and $Q_{s,n}$ respectively, then $A[\alpha|\beta]$ is called a submatrix of A. The determinant of a square submatrix is called a subdeterminant. For $\alpha \in Q_{r,m}$ and $\beta \in Q_{s,n}$, let α' and β' denote the sequences in $Q_{m-r,m}$ and $Q_{n-s,n}$ whose integers are complementary to α, and β, respectively.

168

Then, define: $A(\alpha|\beta) = A[\alpha'|\beta']$, $A[\alpha|\beta) = A[\alpha|\beta']$, $A(\alpha|\beta] = A[\alpha'|\beta]$.
If $m = n$, $r = s$ and $\alpha \in Q_{r,n}$ then $A[\alpha|\alpha]$ is called a principal
submatrix of A. The determinant of a principal submatrix is called a
principal subdeterminant. For example, let:

$$A = \begin{bmatrix} a_{11} & a_{12} & a_{13} \\ a_{21} & a_{22} & a_{23} \\ a_{31} & a_{32} & a_{33} \end{bmatrix}$$

and $\alpha = (2,2,3) \in \Gamma_{3,3}$, $\beta = (1,1) \in \Gamma_{2,3}$. Then

$$A[\alpha|\beta] = A \ [2,2,3|1,1] = \begin{bmatrix} a_{21} & a_{21} \\ a_{21} & a_{21} \\ a_{31} & a_{31} \end{bmatrix}$$

Again, if $\alpha = (1,2) \in Q_{2,3}$, $\beta = (2,3) \in Q_{2,3}$, then

$$A[\alpha|\beta] = \begin{bmatrix} a_{12} & a_{13} \\ a_{22} & a_{23} \end{bmatrix},$$

$$A[\alpha|\beta] = A[\alpha'|\beta] = A[3|2,3] = [a_{32} \ a_{33}],$$

and

$$A(\alpha|\beta) = A[\alpha'|\beta'] = A[3|1] = [a_{31}].$$

● **PROBLEM** 5-25

Define the Wronskian.

Solution: The Wronskian of two functions at $x = x_0$ is defined as the
following determinant:

$$W(f,g,x_0) = \det \begin{bmatrix} f(x_0) & g(x_0) \\ f'(x_0) & g'(x_0) \end{bmatrix}$$

If f and g are dependent functions, then there are numbers c_1 and
c_2 not both zero such that

$$c_1 f + c_2 g = 0.$$

Differentiation gives:

$$c_1 f' + c_2 g' = 0.$$

c_1 and c_2 are not both 0 so we know the system of homogeneous equations

$$c_1 f + c_2 g = 0$$
$$c_1 f' + c_2 g' = 0$$

has a non-zero solution.

But a system of homogeneous equations has a non-zero solution if
and only if the determinant of the coefficient matrix is 0. Therefore,

$$W(f,g,x_0) = \det \begin{bmatrix} f(x_0) & g(x_0) \\ f'(x_0) & g'(x_0) \end{bmatrix} = 0$$

so we obtain the following result:

If f and g are dependent, then for each x_0, $W(f,g,x_0) = 0$.
In general, the converse of this is not true. The Wronskian for more
than two functions is defined analogously. For example,

$$w(f,g,h,x_0) = \det \begin{matrix} f(x_0) & g(x_0) & h(x_0) \\ f'(x_0) & g'(x_0) & h'(x_0) \\ f''(x_0) & g''(x_0) & h''(x_0) \end{matrix} \quad .$$

CRAMER'S RULE

● PROBLEM 5-26

Solve the following linear equations by using Cramer's Rule:

$$-2x_1 + 3x_2 - x_3 = 1$$
$$x_1 + 2x_2 - x_3 = 4$$
$$-2x_1 - x_2 + x_3 = -3 \quad .$$

Solution: Consider a system of n linear equations in n unknowns:

$$a_{11}x_1 + a_{12}x_2 + \ldots + a_{1n}x_n = b_1$$

$$a_{21}x_1 + a_{22}x_2 + \ldots + a_{2n}x_n = b_2$$

$$\cdot \cdot \cdot \cdot \cdot \cdot \cdot \cdot \cdot \cdot \cdot \cdot \cdot$$

$$a_{n1}x_1 + a_{n2}x_2 + \ldots + a_{nn}x_n = b_n \quad .$$

Write the above equations in matrix notation.

$$\begin{bmatrix} a_{11} & a_{12} \cdots a_{1n} \\ a_{21} & a_{22} \cdots a_{2n} \\ \vdots & \vdots \\ a_{n1} & a_{n2} \cdots a_{nn} \end{bmatrix} \begin{bmatrix} x_1 \\ x_2 \\ \vdots \\ x_n \end{bmatrix} = \begin{bmatrix} b_1 \\ b_2 \\ \vdots \\ b_n \end{bmatrix}$$

or, $AX = B$.

Let A be an $n \times n$ matrix over the field F such that $\det A \neq 0$.
If $b_1, b_2, \ldots, b_n$ are any scalars in F, the unique solution of the system
of equations $AX = B$ is given by:

$$x_i = \frac{\det A_i}{\det A} \qquad i = 1, 2, \ldots, n \quad ,$$

where A_i is the $n \times n$ matrix obtained from A by replacing the ith
column of A by the column vector

$$\begin{bmatrix} b_1 \\ b_2 \\ \cdot \\ \cdot \\ \cdot \\ b_n \end{bmatrix}$$

The above theorem is known as "Cramer's Rule" for solving systems of linear equations. Cramer's Rule applies only to systems of n linear equations in n unknowns with non-zero determinants.

Consider the given linear system:

$$-2x_1 + 3x_2 - x_3 = 1$$
$$x_1 + 2x_2 - x_3 = 4$$
$$-2x_1 - x_2 + x_3 = -3$$

or,

$$\begin{bmatrix} -2 & 3 & -1 \\ 1 & 2 & --1 \\ -2 & -1 & 1 \end{bmatrix} \begin{bmatrix} x_1 \\ x_2 \\ x_3 \end{bmatrix} = \begin{bmatrix} 1 \\ 4 \\ 3 \end{bmatrix}$$

$$A \qquad\qquad X \;\; = \;\; B$$

$$\text{Det } A = \begin{vmatrix} -2 & 3 & -1 \\ 1 & 2 & -1 \\ -2 & -1 & 1 \end{vmatrix}$$

$$\text{Det } A = -2 \begin{vmatrix} 2 & -1 \\ -1 & 1 \end{vmatrix} - 3 \begin{vmatrix} 1 & -1 \\ -2 & 1 \end{vmatrix} - 1 \begin{vmatrix} 1 & 2 \\ -2 & -1 \end{vmatrix}$$

$$= -2(2-1) - 3(1-2) - (-1+4)$$
$$= -2 + 3 - 3 = -2 \; .$$

Since det A $\neq$ 0, the system has a unique solution. Now,

$$x_1 = \frac{\text{det } A_1}{\text{det } A}, \quad x_2 = \frac{\text{det } A_2}{\text{det } A}, \quad x_3 = \frac{\text{det } A_3}{\text{det } A} \; .$$

Det A_1 is the determinant of the matrix obtained by replacing the 1st column of A by the column of B. Thus,

$$\text{det } A_1 = \begin{vmatrix} 1 & 3 & -1 \\ 4 & 2 & -1 \\ 3 & -1 & 1 \end{vmatrix}$$

$$= -4 \; .$$

Then,

$$x_1 = \frac{-4}{-2} = 2 \; .$$

$$x_2 = \frac{\begin{vmatrix} -2 & 1 & -1 \\ 1 & 4 & -1 \\ -2 & -3 & 1 \end{vmatrix}}{|A|} = \frac{-6}{-2} = 3 \; .$$

$$x_3 = \frac{\det A_3}{\det A} = \frac{\begin{vmatrix} -2 & 3 & 1 \\ 1 & 2 & 4 \\ -2 & -1 & -3 \end{vmatrix}}{-2} = \frac{-8}{-2} = 4 \ .$$

Thus,

$$x_1 = 2 \ , \ x_2 = 3 \ , \ x_3 = 4 \ ,$$

is the unique solution to the given system.

● **PROBLEM** 5-27

Solve the following homogeneous equations:

$$x_1 + 2x_2 + x_3 = 0$$
$$x_2 - 3x_3 = 0$$
$$-x_1 + x_2 - x_3 = 0 \ .$$

Solution: A homogeneous system of linear equations has either a) the unique trivial solution $x_1 = x_2 = \ldots = x_n = 0$ or b) an infinite number of non-trivial solutions, plus the trivial solution.

First, write the above equations in matrix notation:

$$\begin{bmatrix} 1 & 2 & 1 \\ 0 & 1 & -3 \\ -1 & 1 & -1 \end{bmatrix} \begin{bmatrix} x_1 \\ x_2 \\ x_3 \end{bmatrix} = \begin{bmatrix} 0 \\ 0 \\ 0 \end{bmatrix}$$

or, $AX = 0$.

$$\det A = 1 \begin{vmatrix} 1 & -3 \\ 1 & -1 \end{vmatrix} - 2 \begin{vmatrix} 0 & -3 \\ -1 & -1 \end{vmatrix} + 1 \begin{vmatrix} 0 & 1 \\ -1 & 1 \end{vmatrix}$$

$$= [(-1+3) - 2(0-3) + (0+1)]$$

$$= 2 + 6 + 1$$

$$= 9 \ .$$

Hence, $\det A \neq 0$, and according to Cramer's Rule, the above system has a unique solution.

Therefore, $X = 0$, i.e., $x_1 = x_2 = x_3 = 0$ since the homogeneous system $AX = 0$ has a non-zero solution if and only if $\det A = 0$.

● **PROBLEM** 5-28

Solve the system of linear equations:

$$3x + 2y + 4z = 1$$
$$2x - y + z = 0$$
$$x + 2y + 3z = 1 \ .$$

Solution: Use Cramer's Rule to solve this system. Write the above equations in matrix form:

$$\begin{bmatrix} 3 & 2 & 4 \\ 2 & -1 & 1 \\ 1 & 2 & 3 \end{bmatrix} \begin{bmatrix} x \\ y \\ z \end{bmatrix} = \begin{bmatrix} 1 \\ 0 \\ 1 \end{bmatrix} \ .$$

Then

$$A = \begin{bmatrix} 3 & 2 & 4 \\ 2 & -1 & 1 \\ 1 & 2 & 3 \end{bmatrix} , \quad B = \begin{bmatrix} 1 \\ 0 \\ 1 \end{bmatrix} .$$

First, check that $\det A \neq 0$.

$$\det A = \begin{vmatrix} 3 & 2 & 4 \\ 2 & -1 & 1 \\ 1 & 2 & 3 \end{vmatrix}$$

$$\det A = 3 \begin{vmatrix} -1 & 1 \\ 2 & 3 \end{vmatrix} - 2 \begin{vmatrix} 2 & 1 \\ 1 & 3 \end{vmatrix} + 4 \begin{vmatrix} 2 & -1 \\ 1 & 2 \end{vmatrix}$$

$$= 3(-3-2) - 2(6-1) + 4(4+1)$$
$$= -15x - 10 + 20 = x-5.$$

Since $\det A \neq 0$, the system has a unique solution.   Then,

$$x = \frac{\det A_1}{\det A}, \quad y = \frac{\det A_2}{\det A}, \quad z = \frac{\det A_3}{\det A} .$$

$\det A_1$ is the determinant of the matrix obtained by replacing the first column of A by the column vector B.
Thus,

$$\det A_1 = \begin{vmatrix} 1 & 2 & 4 \\ 0 & -1 & 1 \\ 1 & 2 & 3 \end{vmatrix} .$$

Expand the determinant by minors, using the first column.

$$\det A_1 = 1 \begin{vmatrix} -1 & 1 \\ 2 & 3 \end{vmatrix} + 1 \begin{vmatrix} 2 & 4 \\ -1 & 1 \end{vmatrix}$$

$$= 1(-3-2) + 1(2+4)$$

$$= -5 + 6 = + 1 .$$

Now, we have $\det A = -5$.

Thus, $$x = \frac{\det A_1}{\det A} = \frac{1}{-5} = -\frac{1}{5} .$$

$$y = \frac{\det A_2}{\det A} = \frac{\begin{vmatrix} 3 & 1 & 4 \\ 2 & 0 & 1 \\ 1 & 1 & 3 \end{vmatrix}}{-5} .$$

Now, expand $\det A_2$ along the second row:

$$\begin{vmatrix} 3 & 1 & 4 \\ 2 & 0 & 1 \\ 1 & 1 & 3 \end{vmatrix} = -2 \begin{vmatrix} 1 & 4 \\ 1 & 3 \end{vmatrix} - 1 \begin{vmatrix} 3 & 1 \\ 1 & 1 \end{vmatrix}$$

$$= -2(3-4) - 1(3-1)$$

$$= 2 - 2 = 0$$

$$y = \frac{0}{-5} = 0 .$$

$$z = \frac{\det A_3}{\det A} = \frac{\begin{vmatrix} 3 & 2 & 1 \\ 2 & -1 & 0 \\ 1 & 2 & 1 \end{vmatrix}}{-5}$$

Expand determinant A_3 by minors, using the third column.

$$\begin{vmatrix} 3 & 2 & 1 \\ 2 & -1 & 0 \\ 1 & 2 & 1 \end{vmatrix} = + 1 \begin{vmatrix} 2 & -1 \\ 1 & 2 \end{vmatrix} + 1 \begin{vmatrix} 3 & 2 \\ 2 & -1 \end{vmatrix}$$

$$= (4+1) + (-3-4)$$

$$= 5 - 7 = -2 .$$

Then,

$$z = \frac{-2}{-5} = \frac{2}{5} .$$

Thus $x = -1/5$, $y = 0$, $z = 2/5$.

CHAPTER 6

THE INVERSE OF A MATRIX

THE INVERSE & ELEMENTARY ROW OPERATIONS

Find the inverse of

$$\begin{bmatrix} 1 & 2 \\ 3 & 7 \end{bmatrix}$$

Solution: We call B the inverse of A if $A \cdot B = B \cdot A = I$ where I is the identity matrix, and $A \cdot B$ represents matrix multiplication. We also say that A is invertible if A has an inverse, denoted by A^{-1}. Note that

$$A^{-1}A = AA^{-1} = I .$$

We wish to find the inverse of the matrix A.

If A is invertible, there is a procedure for finding its inverse, which we summarize as follows:

1) We first form the $n \times 2n$ matrix (A,I); by applying elementary row operations to (A,I), we

2) reduce the matrix (A,I) to a matrix of the form (I,B) whose first n columns form the identity matrix, and whose last n columns give a matrix we call 'B'.

3) We conclude that $A^{-1} = B$.

We apply the above procedure to the given matrix.

$$A = \begin{bmatrix} 1 & 2 \\ 3 & 7 \end{bmatrix}$$

We first form the matrix (A,I)

$$A = \left[\begin{array}{cc|cc} 1 & 2 & 1 & 0 \\ 3 & 7 & 0 & 1 \end{array}\right]$$

Reduce it so that the left-hand side, "A" becomes the identity matrix:

$$\left[\begin{array}{cc|cc} 1 & 2 & 1 & 0 \\ 3 & 7 & 0 & 1 \end{array}\right] \rightarrow \left[\begin{array}{cc|cc} 1 & 2 & 1 & 0 \\ 0 & 1 & -3 & 1 \end{array}\right]$$

(We added −3 times the first row to the second row)

$$\begin{bmatrix} 1 & 2 & \vdots & 1 & 0 \\ 0 & 1 & \vdots & -3 & 1 \end{bmatrix} \quad \rightarrow \quad \begin{bmatrix} 1 & 0 & \vdots & 7 & -2 \\ 0 & 1 & \vdots & -3 & 1 \end{bmatrix}$$

(We added −2 times the second row to the first row).

Once the 'left−hand side' is reduced to I, the inverse is given by the 'right−hand side', which we called "B". Hence the inverse is

$$\begin{bmatrix} 7 & -2 \\ -3 & 1 \end{bmatrix} .$$

● **PROBLEM** 6-2

Let
$$A = \begin{bmatrix} 1 & 1 & 1 \\ 0 & 1 & 0 \\ 0 & 1 & 1 \end{bmatrix}$$

Find the inverse of A .

Solution: Form the block matrix [A : I] where I is the 3x3 identity matrix. Then apply elementary row operations on [A : I] to reduce it to [I : B]. B is the required inverse.

$$[A : I] \;=\; \begin{bmatrix} 1 & 1 & 1 & \vdots & 1 & 0 & 0 \\ 0 & 1 & 0 & \vdots & 0 & 1 & 0 \\ 0 & 1 & 1 & \vdots & 0 & 0 & 1 \end{bmatrix}$$

Subtract the third row from the first row.

$$\begin{bmatrix} 1 & 0 & 0 & \vdots & 1 & 0 & -1 \\ 0 & 1 & 0 & \vdots & 0 & 1 & 0 \\ 0 & 1 & 1 & \vdots & 0 & 0 & 1 \end{bmatrix}$$

Subtract the second row from the third row

$$\begin{bmatrix} 1 & 0 & 0 & \vdots & 1 & 0 & -1 \\ 0 & 1 & 0 & \vdots & 0 & 1 & 0 \\ 0 & 0 & 1 & \vdots & 0 & -1 & 1 \end{bmatrix}$$

Thus

$$A^{-1} \;=\; \begin{bmatrix} 1 & 0 & -1 \\ 0 & 1 & 0 \\ 0 & -1 & 1 \end{bmatrix}$$

● **PROBLEM** 6-3

Find the inverse of the matrix A   where

$$A = \begin{bmatrix} 1 & 1 & 1 & 1 \\ 0 & 1 & 1 & 1 \\ 0 & 0 & 1 & 1 \\ 0 & 0 & 0 & 1 \end{bmatrix}$$

Show that the inverse of a diagonal matrix is obtained by inverting the diagonal entries.

Solution:

$$[A : I] = \left[\begin{array}{cccc:cccc} 1 & 1 & 1 & 1 & 1 & 0 & 0 & 0 \\ 0 & 1 & 1 & 1 & 0 & 1 & 0 & 0 \\ 0 & 0 & 1 & 1 & 0 & 0 & 1 & 0 \\ 0 & 0 & 0 & 1 & 0 & 0 & 0 & 1 \end{array}\right]$$

Subtract the second row from the first row:

$$\left[\begin{array}{cccc:cccc} 1 & 0 & 0 & 0 & 1 & -1 & 0 & 0 \\ 0 & 1 & 1 & 1 & 0 & 1 & 0 & 0 \\ 0 & 0 & 1 & 1 & 0 & 0 & 1 & 0 \\ 0 & 0 & 0 & 1 & 0 & 0 & 0 & 1 \end{array}\right]$$

Subtract the third row from the second row, and the fourth row from the third row:

$$\left[\begin{array}{cccc:cccc} 1 & 0 & 0 & 0 & 1 & -1 & 0 & 0 \\ 0 & 1 & 0 & 0 & 0 & 1 & -1 & 0 \\ 0 & 0 & 1 & 0 & 0 & 0 & 1 & -1 \\ 0 & 0 & 0 & 1 & 0 & 0 & 0 & 1 \end{array}\right]$$

Hence

$$A^{-1} = \begin{bmatrix} 1 & -1 & 0 & 0 \\ 0 & 1 & -1 & 0 \\ 0 & 0 & 1 & -1 \\ 0 & 0 & 0 & 1 \end{bmatrix}$$

A diagonal matrix is a square matrix whose non-diagonal entries are all zero. Let A be a diagonal matrix whose diagonal entries are all non-zero, and let

$$A = \begin{bmatrix} a_{11} & 0 & \cdots & & 0 \\ 0 & a_{22} & \cdots & & 0 \\ \vdots & & 0 & a_{kk} & \vdots \\ 0 & \cdots & & \cdots & a_{nn} \end{bmatrix}$$

with $a_{ii} \neq 0$, $i = 1,...,n$.

Now apply the procedure for finding the inverse at a matrix.

Then

$$[A : I] = \begin{bmatrix} a_{11} & 0 & \cdots & 0 & \vdots & 1 & 0 & \cdots & 0 \\ 0 & a_{22} & & 0 & \vdots & 0 & 1 & \cdots & 0 \\ \cdot & & & & \vdots & \cdot & & & \\ \cdot & & & & \vdots & \cdot & & & \\ \cdot & & & & \vdots & \cdot & & & \\ 0 & \cdots & \cdot a_{nn} & & 0 & \cdots & \cdots & \cdots & 1 \end{bmatrix}$$

Multiply the first row by $\dfrac{1}{a_{11}}$, the second row by $\dfrac{1}{a_{22}}$ ... and the nth row by $\dfrac{1}{a_{nn}}$, to obtain

$$[I : B] = \begin{bmatrix} 1 & 0 & \cdots & 0 & \vdots & 1/a_{11} & 0 & \cdots & 0 \\ 0 & 1 & \cdots & 0 & \vdots & 0 & 1/a_{22} & \cdots & 0 \\ \cdot & & & & \vdots & \cdot & & & \\ \cdot & & & & \vdots & \cdot & & & \\ \cdot & & & & \vdots & \cdot & & & \\ 0 & \cdots & \cdots & 1 & \vdots & 0 & \cdots & \cdots & 1/a_{nn} \end{bmatrix}$$

Hence

$$A^{-1} = \begin{bmatrix} 1/a_{11} & 0 & \cdots & 0 \\ 0 & 1/a_{22} & \cdots & 0 \\ \cdot & & & \cdot \\ \cdot & & & \cdot \\ \cdot & & & \cdot \\ 0 & \cdots & \cdots & 1/a_{nn} \end{bmatrix}$$

Thus the inverse of a diagonal matrix is obtained by inverting the diagonal entries.

Observe that if one of the diagonal entries is zero, the matrix is not invertible. For example,

$$\begin{bmatrix} 1 & 0 & 0 \\ 0 & 0 & 0 \\ 0 & 0 & 3 \end{bmatrix}$$

is not invertible.

● **PROBLEM** 6-4

Find the inverse of A where

$$A = \begin{bmatrix} 1 & 2 & -1 \\ 2 & 5 & 4 \\ 3 & 7 & 4 \end{bmatrix}$$

<u>Solution</u>:   Form the block matrix [A : I] where I is the 3 x 3 identity matrix. Then apply row operations on A to reduce it to

178

the identity matrix. Simultaneously, I will change into the required inverse.

We have

$$[A,I] = \begin{bmatrix} 1 & 2 & -1 & \vline & 1 & 0 & 0 \\ 2 & 5 & 4 & \vline & 0 & 1 & 0 \\ 3 & 7 & 4 & \vline & 0 & 0 & 1 \end{bmatrix}$$

Now add -2 times the first row to the second row:

$$\begin{bmatrix} 1 & 2 & -1 & \vline & 1 & 0 & 0 \\ 0 & 1 & 6 & \vline & -2 & 1 & 0 \\ 3 & 7 & 4 & \vline & 0 & 0 & 1 \end{bmatrix}$$

Add -3 times the first row to the third row:

$$\begin{bmatrix} 1 & 2 & -1 & \vline & 1 & 0 & 0 \\ 0 & 1 & 6 & \vline & -2 & 1 & 0 \\ 0 & 1 & 7 & \vline & -3 & 0 & 1 \end{bmatrix}$$

Add -1 times the second row to the third row:

$$\begin{bmatrix} 1 & 2 & -1 & \vline & 1 & 0 & 0 \\ 0 & 1 & 6 & \vline & -2 & 1 & 0 \\ 0 & 0 & 1 & \vline & -1 & -1 & 1 \end{bmatrix}$$

Add the third row to the first row:

$$\begin{bmatrix} 1 & 2 & 0 & \vline & 0 & -1 & 1 \\ 0 & 1 & 6 & \vline & -2 & 1 & 0 \\ 0 & 0 & 1 & \vline & -1 & -1 & 1 \end{bmatrix}$$

Add -6 times the third row to the second row:

$$\begin{bmatrix} 1 & 2 & 0 & \vline & 0 & -1 & 1 \\ 0 & 1 & 0 & \vline & 4 & 7 & -6 \\ 0 & 0 & 1 & \vline & -1 & -1 & 1 \end{bmatrix}$$

Add -2 times the second row to the first row:

$$\begin{bmatrix} 1 & 0 & 0 & \vline & -8 & -15 & 13 \\ 0 & 1 & 0 & \vline & 4 & 7 & -6 \\ 0 & 0 & 1 & \vline & -1 & -1 & 1 \end{bmatrix}$$

Hence

$$A^{-1} = \begin{bmatrix} -8 & -15 & 13 \\ 4 & 7 & -6 \\ -1 & -1 & 1 \end{bmatrix}$$

● PROBLEM 6-5

Find the inverse of A where

179

$$A \ = \ \begin{bmatrix} 1 & 2 & -3 \\ 1 & -2 & 1 \\ 5 & -2 & -3 \end{bmatrix}$$

Solution: An $n \times n$ matrix A is said to have an inverse if there exists another matrix, A^{-1} for example, such that $AA^{-1} = I$. A^{-1} is the inverse of A. To find the inverse of A, form the block matrix $[A : I]$. If we convert A to I, we simultaneously convert I to A^{-1}, since $AA^{-1} = I$ and $IA^{-1} = A^{-1}$. We first form the matrix $[A : I]$:

$$\begin{bmatrix} 1 & 2 & -3 & \vdots & 1 & 0 & 0 \\ 1 & -2 & 1 & \vdots & 0 & 1 & 0 \\ 5 & -2 & -3 & \vdots & 0 & 0 & 1 \end{bmatrix}$$

Subtract the first row from the second row:

$$\begin{bmatrix} 1 & 2 & -3 & \vdots & 1 & 0 & 0 \\ 0 & -4 & 4 & \vdots & -1 & 1 & 0 \\ 5 & -2 & -3 & \vdots & 0 & 0 & 1 \end{bmatrix}$$

Add -5 times the first row to the third row:

$$\begin{bmatrix} 1 & 2 & -3 & \vdots & 1 & 0 & 0 \\ 0 & -4 & 4 & \vdots & -1 & 1 & 0 \\ 0 & -12 & 12 & \vdots & -5 & 0 & 1 \end{bmatrix}$$

Add -3 times the second row to the third row:

$$\begin{bmatrix} 1 & 2 & -3 & \vdots & 1 & 0 & 0 \\ 0 & -4 & 4 & \vdots & -1 & 1 & 0 \\ 0 & 0 & 0 & \vdots & -2 & -3 & 1 \end{bmatrix}$$

If the reduced row echelon matrix of A has a row of zeros, then A is singular. Recall these facts:

If A is invertible, it must be nonsingular, and it must be reducible to the identity matrix by elementary row operations. On the other hand, if a matrix is singular, it is not reducible to the identity matrix and it is not invertible. Furthermore, if a matrix A has been transformed by elementary row operations to a matrix F, and F is clearly singular, then A is singular and not reducible to the identity matrix.

Let

$$F \ = \ \begin{bmatrix} 1 & 2 & -3 \\ 0 & -4 & 4 \\ 0 & 0 & 0 \end{bmatrix} \ .$$

Since F has a row of zeros, F is a singular matrix. Therefore, A is a singular and not invertible.

Find the inverse of A where

$$A = \begin{bmatrix} 1 & 1 & 1 \\ 0 & 2 & 3 \\ 5 & 5 & 1 \end{bmatrix}$$

<u>Solution:</u>   We first form the matrix $[A : I]$

$$\begin{bmatrix} 1 & 1 & 1 & \vdots & 1 & 0 & 0 \\ 0 & 2 & 3 & \vdots & 0 & 1 & 0 \\ 5 & 5 & 1 & \vdots & 0 & 0 & 1 \end{bmatrix}$$

and proceed to reduce it to a matrix of the form $[I : B]$, where A
has been reduced to I by elementary row operations. From this
we may conclude that $A^{-1} = B$. We arrange our computations as follows:

$$\begin{bmatrix} 1 & 1 & 1 & \vdots & 1 & 0 & 0 \\ 0 & 2 & 3 & \vdots & 0 & 1 & 0 \\ 5 & 5 & 1 & \vdots & 0 & 0 & 1 \end{bmatrix}$$

Subtract 5 times the first row from the third row to obtain:

$$\begin{bmatrix} 1 & 1 & 1 & \vdots & 1 & 0 & 0 \\ 0 & 2 & 3 & \vdots & 0 & 1 & 0 \\ 0 & 0 & -4 & \vdots & -5 & 0 & 1 \end{bmatrix}$$

Divide the second row by 2 to obtain:

$$\begin{bmatrix} 1 & 1 & 1 & \vdots & 1 & 0 & 0 \\ 0 & 1 & 3/2 & \vdots & 0 & 1/2 & 0 \\ 0 & 0 & -4 & \vdots & -5 & 0 & 1 \end{bmatrix}$$

Subtract the second row from the first row to obtain:

$$\begin{bmatrix} 1 & 0 & -1/2 & \vdots & 1 & -1/2 & 0 \\ 0 & 1 & 3/2 & \vdots & 0 & 1/2 & 0 \\ 0 & 0 & -4 & \vdots & -5 & 0 & 1 \end{bmatrix}$$

Divide the third row by −4 to obtain:

$$\begin{bmatrix} 1 & 0 & -1/2 & \vdots & 1 & -1/2 & 0 \\ 0 & 0 & 3/2 & \vdots & 0 & 1/2 & 0 \\ 0 & 0 & +1 & \vdots & +5/4 & 0 & -1/4 \end{bmatrix}$$

Add −3/2 times the third row to the second row to obtain:

$$\begin{bmatrix} 1 & 0 & -1/2 & \vdots & 1 & -1/2 & 0 \\ 0 & 1 & 0 & \vdots & -15/8 & 1/2 & 3/8 \\ 0 & 0 & 1 & \vdots & 5/4 & 0 & -1/4 \end{bmatrix}$$

Add 1/2 times the third row to the first row to obtain:

$$\begin{bmatrix} 1 & 0 & 0 & \vdots & 13/8 & -1/2 & -1/8 \\ 0 & 1 & 0 & \vdots & -15/8 & 1/2 & 3/8 \\ 0 & 0 & 1 & \vdots & 5/4 & 0 & -1/4 \end{bmatrix} .$$

Hence

$$A^{-1} = \begin{bmatrix} 13/8 & -1/2 & -1/8 \\ -15/8 & 1/2 & 3/8 \\ 5/4 & 0 & -1/4 \end{bmatrix} .$$

● **PROBLEM** 6-7

Find the inverse of

$$A = \begin{bmatrix} 1 & 2 & 3 \\ 2 & 5 & 3 \\ 1 & 0 & 8 \end{bmatrix}$$

Solution: We first form the matrix (A : I)

$$= \begin{bmatrix} 1 & 2 & 3 & \vdots & 1 & 0 & 0 \\ 2 & 5 & 3 & \vdots & 0 & 1 & 0 \\ 1 & 0 & 8 & \vdots & 0 & 0 & 1 \end{bmatrix}$$

and we proceed to reduce it to a matrix of the form (I : B), where A has been reduced to I by elementary row operations. From this we may conclude that $A^{-1} = B$.
Adding -2 times the first row to the second, we obtain:

$$\begin{bmatrix} 1 & 2 & 3 & \vdots & 1 & 0 & 0 \\ 0 & 1 & -3 & \vdots & -2 & 1 & 0 \\ 1 & 0 & 8 & \vdots & 0 & 0 & 1 \end{bmatrix}$$

Adding -1 times the first row to the third, we obtain:

$$\begin{bmatrix} 1 & 2 & 3 & \vdots & 1 & 0 & 0 \\ 0 & 1 & -3 & \vdots & -2 & 1 & 0 \\ 0 & -2 & 5 & \vdots & -1 & 0 & 1 \end{bmatrix}$$

Adding 2 times the second row to the third

$$\begin{bmatrix} 1 & 2 & 3 & \vdots & 1 & 0 & 0 \\ 0 & 1 & -3 & \vdots & -2 & 1 & 0 \\ 0 & 0 & -1 & \vdots & -5 & 2 & 1 \end{bmatrix}$$

Now multiply the third row by -1.

$$\begin{bmatrix} 1 & 2 & 3 & \vdots & 1 & 0 & 0 \\ 0 & 1 & -3 & \vdots & -2 & 1 & 0 \\ 0 & 0 & 1 & \vdots & 5 & -2 & -1 \end{bmatrix} .$$

Add 3 times the third row to the second and -3 times the third row to the first.

$$\begin{bmatrix} 1 & 2 & 0 & | & -14 & 6 & 3 \\ 0 & 1 & 0 & | & 13 & -5 & -3 \\ 0 & 0 & 1 & | & 5 & -2 & -1 \end{bmatrix}$$

Add -2 times the second row to the first.

$$\begin{bmatrix} 1 & 0 & 0 & | & -40 & 16 & 9 \\ 0 & 1 & 0 & | & 13 & -5 & -3 \\ 0 & 0 & 1 & | & 5 & -2 & -1 \end{bmatrix}$$

Thus

$$A^{-1} = \begin{bmatrix} -40 & 16 & 9 \\ 13 & -5 & -3 \\ 5 & -2 & -1 \end{bmatrix}$$

● **PROBLEM** 6-8

Find the inverses of the following matrices.

(1)

$$A = \begin{bmatrix} 3 & 1 \\ -1 & 6 \end{bmatrix}$$

(2)

$$A = \begin{bmatrix} 1 & -7 & -14 \\ 2 & 1 & -1 \\ 1 & 3 & 4 \end{bmatrix}$$

(3)

$$A = \begin{bmatrix} 3 & 1 & 0 \\ 1 & -1 & 2 \\ 1 & 1 & 1 \end{bmatrix}.$$

Solution: The method of solution is the same in all three cases, namely, forming the block matrix $[A : I]$ where I is the $n \times n$ identity matrix, and using elementary row operations to reduce it to $[I : A^{-1}]$.

(1) $A = \begin{bmatrix} 3 & 1 \\ -1 & 6 \end{bmatrix}$.

Now $[A : I] = \begin{bmatrix} 3 & 1 & | & 1 & 0 \\ -1 & 6 & | & 0 & 1 \end{bmatrix}$.

Multiply the first row by 6:

$$\begin{bmatrix} 18 & 6 & | & 6 & 0 \\ -1 & 6 & | & 0 & 1 \end{bmatrix}$$

Subtract the second row from the first row:

$$\begin{bmatrix} 19 & 0 & | & 6 & -1 \\ -1 & 6 & | & 0 & 1 \end{bmatrix}$$

Multiply the second row by 19:

$$\begin{bmatrix} 19 & 0 & \vdots & 6 & -1 \\ -19 & 114 & \vdots & 0 & 19 \end{bmatrix}$$

Add the first row to the second row:

$$\begin{bmatrix} 19 & 0 & \vdots & 6 & -1 \\ 0 & 114 & \vdots & 6 & 18 \end{bmatrix}$$

Divide the first and second rows by 19:

$$\begin{bmatrix} 1 & 0 & \vdots & 6/19 & -1/19 \\ 0 & 6 & \vdots & 6/19 & 18/19 \end{bmatrix}$$

Divide the second row by 6:

$$\begin{bmatrix} 1 & 0 & \vdots & 6/19 & -1/19 \\ 0 & 1 & \vdots & 1/19 & 3/19 \end{bmatrix}$$

Therefore

$$A^{-1} = \begin{bmatrix} 6/19 & -1/19 \\ 1/19 & 3/19 \end{bmatrix}$$

(2)

$$A = \begin{bmatrix} 1 & -7 & -14 \\ 2 & 1 & -1 \\ 1 & 3 & 4 \end{bmatrix}$$

$$[A : I] = \begin{bmatrix} 1 & -7 & -14 & \vdots & 1 & 0 & 0 \\ 2 & 1 & -1 & \vdots & 0 & 1 & 0 \\ 1 & 3 & 4 & \vdots & 0 & 0 & 1 \end{bmatrix}$$

Subtract the first row from the third row:

$$\begin{bmatrix} 1 & -7 & -14 & \vdots & 1 & 0 & 0 \\ 2 & 1 & -1 & \vdots & 0 & 1 & 0 \\ 0 & 10 & 18 & \vdots & -1 & 0 & 1 \end{bmatrix}$$

Divide the third row by 2:

$$\begin{bmatrix} 1 & -7 & -14 & \vdots & 1 & 0 & 0 \\ 2 & 1 & -1 & \vdots & 0 & 1 & 0 \\ 0 & 5 & 9 & \vdots & -1/2 & 0 & 1/2 \end{bmatrix}$$

Add -2 times the first row to the second row:

$$\begin{bmatrix} 1 & -7 & 9 & \vdots & 1 & 0 & 0 \\ 0 & 15 & 27 & \vdots & -2 & 1 & 0 \\ 0 & 5 & 9 & \vdots & -1/2 & 0 & 1/2 \end{bmatrix}$$

Divide the second row by 3:

$$\left[\begin{array}{ccc:ccc} 1 & -7 & -14 & 1 & 0 & 0 \\ 0 & 5 & 9 & -2/3 & 1/3 & 0 \\ 0 & 5 & 9 & -1/2 & 0 & 1/2 \end{array}\right]$$

Subtract the second row from the third row:

$$\left[\begin{array}{ccc:ccc} 1 & -7 & -14 & 1 & 0 & 0 \\ 0 & 5 & 9 & -2/3 & 1/3 & 0 \\ 0 & 0 & 0 & 1/6 & -1/3 & 1/2 \end{array}\right]$$

At this point A is row equivalent to

$$F = \left[\begin{array}{ccc} 1 & -7 & -14 \\ 0 & 5 & 9 \\ 0 & 0 & 0 \end{array}\right]$$

The matrix A is singular and therefore A does not have an inverse.

(3)

$$A = \left[\begin{array}{ccc} 3 & 1 & 0 \\ 1 & -1 & 2 \\ 1 & 1 & 1 \end{array}\right]$$

$$[A : I] = \left[\begin{array}{ccc:ccc} 3 & 1 & 0 & 1 & 0 & 0 \\ 1 & -1 & 2 & 0 & 1 & 0 \\ 1 & 1 & 1 & 0 & 0 & 1 \end{array}\right]$$

Interchange the first and third rows:

$$\left[\begin{array}{ccc:ccc} 1 & 1 & 1 & 0 & 0 & 1 \\ 1 & -1 & 2 & 0 & 1 & 0 \\ 3 & 1 & 0 & 1 & 0 & 0 \end{array}\right]$$

Subtract the first row from the second row and add −3 times the
first row to the third row:

$$\left[\begin{array}{ccc:ccc} 1 & 1 & 1 & 0 & 0 & 1 \\ 0 & -2 & 1 & 0 & 1 & -1 \\ 0 & -2 & -3 & 1 & 0 & -3 \end{array}\right]$$

Divide the second row by −2:

$$\left[\begin{array}{ccc:ccc} 1 & 1 & 1 & 0 & 0 & 1 \\ 0 & 1 & -1/2 & 0 & -1/2 & 1/2 \\ 0 & -2 & -3 & 1 & 0 & -3 \end{array}\right]$$

Subtract the second row from the first row:

$$\begin{bmatrix} 1 & 0 & 3/2 & \vdots & 0 & 1/2 & 1/2 \\ 0 & 1 & -1/2 & \vdots & 0 & -1/2 & 1/2 \\ 0 & -2 & -3 & \vdots & 1 & 0 & -3 \end{bmatrix}$$

Add 2 times the second row to the third row:

$$\begin{bmatrix} 1 & 0 & 3/2 & \vdots & 0 & 1/2 & 1/2 \\ 0 & 1 & -1/2 & \vdots & 0 & -1/2 & 1/2 \\ 0 & 0 & -4 & \vdots & 1 & -1 & -2 \end{bmatrix}$$

Divide the third row by -4:

$$\begin{bmatrix} 1 & 0 & 3/2 & \vdots & 0 & 1/2 & 1/2 \\ 0 & 1 & -1/2 & \vdots & 0 & -1/2 & 1/2 \\ 0 & 0 & 1 & \vdots & -1/4 & 1/4 & +2/4 \end{bmatrix}$$

Add $-3/2$ times the third row to the first row and add $1/2$ times the third row to the second row:

$$\begin{bmatrix} 1 & 0 & 0 & \vdots & 3/8 & +1/8 & -2/8 \\ 0 & 1 & 0 & \vdots & -1/8 & -3/8 & 6/8 \\ 0 & 0 & 1 & \vdots & -1/4 & 1/4 & +2/4 \end{bmatrix}$$

Thus

$$A^{-1} = \begin{bmatrix} 3/8 & 1/8 & -2/8 \\ -1/8 & -3/8 & 6/8 \\ -1/4 & 1/4 & 2/4 \end{bmatrix} .$$

● **PROBLEM** 6-9

Show that A is not invertible where

$$A = \begin{bmatrix} 1 & 6 & 4 \\ 2 & 4 & -1 \\ -1 & 2 & 5 \end{bmatrix}$$

Solution: An $n \times n$ matrix M is said to be invertible if there exists another matrix M^{-1} such that $MM^{-1} = M^{-1}M = I$, the $n \times n$ identity matrix. Matrix inversion corresponds to ordinary division although the rules are quite different.

If a matrix is found to be singular when reduced by elementary row operations, it is not invertible and it cannot be reduced to the identity matrix.

$$[A : I] = \begin{bmatrix} 1 & 6 & 4 & \vdots & 1 & 0 & 0 \\ 2 & 4 & -1 & \vdots & 0 & 1 & 0 \\ -1 & 2 & 5 & \vdots & 0 & 0 & 1 \end{bmatrix}$$

Add -2 times the first row to the second and add the first row to the third:

$$\begin{bmatrix} 1 & 6 & 4 & \vdots & 1 & 0 & 0 \\ 0 & -8 & -9 & \vdots & -2 & 0 & 0 \\ 0 & 8 & 9 & \vdots & 1 & 0 & 1 \end{bmatrix}$$

Add the second row to the third row:

$$\begin{bmatrix} 1 & 6 & 4 & \vdots & 1 & 0 & 0 \\ 0 & -8 & -9 & \vdots & -2 & 0 & 0 \\ 0 & 0 & 0 & \vdots & -1 & 0 & 1 \end{bmatrix}$$

We have obtained a row of zeros on the left side. Therefore, A has been reduced to a singular matrix by elementary row operations, so A is singular and therefore A is not invertible.

THE INVERSE & SYSTEMS OF LINEAR EQUATIONS

● **PROBLEM** 6-10

Let
$$A = \begin{bmatrix} 1 & 2 \\ 3 & 4 \end{bmatrix}$$

Find the inverse of A directly by solving for the entries of the matrix B which satisfies the equation

$$A \cdot B = I ,$$

where A·B is matrix multiplication.

Solution: The problem asks us to solve for the entries a, b, c, and d of the matrix

$$B = \begin{bmatrix} a & b \\ c & d \end{bmatrix} ,$$

given that A·B = I .

Since A·B = I, we have

$$\begin{bmatrix} 1 & 2 \\ 3 & 4 \end{bmatrix} \cdot \begin{bmatrix} a & b \\ c & d \end{bmatrix} = \begin{bmatrix} 1 & 0 \\ 0 & 1 \end{bmatrix}$$

After performing the multiplication of the matrices, we obtain

$$\begin{bmatrix} a+2c & b+2d \\ 3a+4c & 3b+4d \end{bmatrix} = \begin{bmatrix} 1 & 0 \\ 0 & 1 \end{bmatrix} .$$

Recall that two matrices are equal if, and only, if, their corresponding entries are equal. Thus from the last equation we may conclude that $a+2c = 1$, $b+2d = 0$, $3a+4c = 0$, $3b+4d = 1$.

From these four equations, we can obtain two sets of linear equations from which we can solve for each of a, b, c, d. That is, we have the set:

$$a + 2c = 1$$

$$3a + 4c = 0$$

whose solutions are $a = -2$ and $c = 3/2$, and the set

$$b + 2d = 0$$

$$3b + 4d = 1$$

whose solutions are $b = 1$ and $d = -1/2$. Hence,

$$B = \begin{bmatrix} a & b \\ c & d \end{bmatrix} = \begin{bmatrix} -2 & 1 \\ 3/2 & -1/2 \end{bmatrix} .$$

Since B satisfies $AB = I$, $B = A^{-1}$. Hence,

$$A^{-1} = \begin{bmatrix} -2 & 1 \\ 3/2 & -1/2 \end{bmatrix} .$$

The method used consisted of obtaining sets of linear equations whose unique solutions yielded the required inverse.

● **PROBLEM** 6-11

Find the inverse of A where

$$A = \begin{bmatrix} 2 & 3 \\ 3 & 5 \end{bmatrix} .$$

<u>Solution</u>: We know that

$$AA^{-1} = I$$

where I is the identity matrix. Let

$$A^{-1} = \begin{bmatrix} a & b \\ c & d \end{bmatrix} .$$

Since $A A^{-1} = I$, we have

$$\begin{bmatrix} 2 & 3 \\ 3 & 5 \end{bmatrix} \begin{bmatrix} a & b \\ c & d \end{bmatrix} = \begin{bmatrix} 1 & 0 \\ 0 & 1 \end{bmatrix}$$

Performing the matrix multiplication, we obtain:

$$\begin{bmatrix} 2a + 3c & 2b + 3d \\ 3a + 5c & 3b + 5d \end{bmatrix} = \begin{bmatrix} 1 & 0 \\ 0 & 1 \end{bmatrix} .$$

Now this matrix equality is equivalent to the following system of equations to be satisfied by a, b, c, d:

$$2a + 3c = 1 \qquad 2b + 3d = 0$$

$$3a + 5c = 0 \qquad 3b + 5d = 1 .$$

The pair of equations on the left yields $a = 5$ and $c = -3$, while the pair on the right yields $b = -3$ and $d = 2$. Hence

$$A^{-1} = \begin{bmatrix} 5 & -3 \\ -3 & 2 \end{bmatrix} .$$

The method used in this example reduces a matrix inversion problem to one of solving a system of linear equations, and it may be applied to a square matrix of any order.

Find the inverse of the matrix A

$$A = \begin{bmatrix} 2 & 1 & 0 \\ 1 & -1 & 1 \\ 0 & 1 & 3 \end{bmatrix} .$$

Solution: We will apply the method which involves solving directly for the individual entries of the matrix A^{-1} . To do this, we reduce the matrix equation $AXA^{-1} = I$ to systems of linear equations which we can solve for the required entries.

Let

$$A^{-1} = \begin{bmatrix} B_{11} & B_{12} & B_{13} \\ B_{21} & B_{22} & B_{23} \\ B_{31} & B_{32} & B_{33} \end{bmatrix}$$

Since $AA^{-1} = I$, we have

$$\begin{bmatrix} 2 & 1 & 0 \\ 1 & -1 & 1 \\ 0 & 1 & 3 \end{bmatrix} \begin{bmatrix} B_{11} & B_{12} & B_{13} \\ B_{21} & B_{22} & B_{23} \\ B_{31} & B_{32} & B_{33} \end{bmatrix} = \begin{bmatrix} 1 & 0 & 0 \\ 0 & 1 & 0 \\ 0 & 0 & 1 \end{bmatrix}$$

Performing the multiplication of matrices,

$$\begin{bmatrix} 2B_{11}+B_{21} & 2B_{12}+B_{22} & 2B_{13}+B_{23} \\ B_{11}-B_{21}+B_{31} & B_{12}-B_{22}+B_{32} & B_{13}-B_{23}+B_{33} \\ B_{21}+3B_{31} & B_{22}+3B_{32} & B_{23}+3B_{33} \end{bmatrix} = \begin{bmatrix} 1 & 0 & 0 \\ 0 & 1 & 0 \\ 0 & 0 & 1 \end{bmatrix}$$

We now equate corresponding entries in the matrices on either side of the above matrix equation, and obtain:

$2B_{11}+B_{21} = 1$ $\qquad 2B_{12}+B_{22} = 0$ $\qquad 2B_{13}+B_{23} = 0$

$B_{11}-B_{21}+B_{31} = 0$ $\qquad B_{12}-B_{22}+B_{32} = 1$ $\qquad B_{13}-B_{23}+B_{33} = 0$

$B_{21}+3B_{31} = 0$ $\qquad B_{22}+3B_{32} = 0$ $\qquad B_{23}+3B_{33} = 1$

Solving these equations yields:

189

$$B_{11} = 4/11 \qquad B_{12} = 3/11 \qquad B_{13} = -1/11$$

$$B_{21} = 3/11 \qquad B_{22} = -6/11 \qquad B_{23} = 2/11$$

$$B_{31} = -1/11 \qquad B_{32} = 2/11 \qquad B_{33} = 3/11$$

Thus

$$A^{-1} = \begin{bmatrix} 4/11 & 3/11 & -1/11 \\ 3/11 & -6/11 & 2/11 \\ -1/11 & 2/11 & 3/11 \end{bmatrix}.$$

● **PROBLEM** 6-13

Let

$$A = \begin{bmatrix} 1 & 2 & 3 \\ 1 & 3 & 2 \\ 1 & 1 & 5 \end{bmatrix}$$

Show how we obtain the inverse of A by reducing the matrix $[A : I]$ to a matrix of the form $[I : B]$.

Solution: Let e_1, e_2, e_3 be the row vectors of the identity matrix, i.e., $e_1 = (1,0,0)$; $e_2 = (0,1,0)$; $e_3 = (0,0,1)$. Then

$$e_1 A = (1,0,0) \cdot \begin{bmatrix} 1 & 2 & 3 \\ 1 & 3 & 2 \\ 1 & 1 & 5 \end{bmatrix}$$

$$= (1,2,3) = e_1 + 2e_2 + 3e_3$$

$$e_2 A = (0,1,0) \begin{bmatrix} 1 & 2 & 3 \\ 1 & 3 & 2 \\ 1 & 1 & 5 \end{bmatrix}$$

$$= (1,3,2) = e_1 + 3e_2 + 2e_3$$

and

$$e_3 A = (0,0,1) \begin{bmatrix} 1 & 2 & 3 \\ 1 & 3 & 2 \\ 1 & 1 & 5 \end{bmatrix}$$

$$= (1,1,5) = e_1 + e_2 + 5e_3.$$

So we have

$$e_1 + 2e_2 + 3e_3 = e_1 A$$

$$e_1 + 3e_2 + 2e_3 = e_2 A$$

$$e_1 + e_3 + 5e_3 = e_3 A$$

This is equivalent to

$$e_1 + 2e_2 + 3e_3 = e_1 A + 0e_2 A + 0e_3 A$$

$$e_1 + 3e_2 + 2e_3 = 0e_1A + e_2A + 0e_3A$$

$$e_1 + e_2 + 5e_3 = 0e_1A + 0e_2A + e_3A \quad .$$

The coefficients of the above system correspond to the coefficients of the matrix $(A : I)$.

By applying transformations to the equations that correspond to row operations on $(A : I)$, we can transform the system of equations to the system

$$e_1 = 13e_1A - 7e_2A - 5e_3A$$

$$e_2 = -3e_1A + 2e_2A + e_3A$$

$$e_3 = -2e_1A + e_2A + e_3A \quad .$$

This is equivalent to

$$e_1 + 0e_2 + 0e_3 = 13e_1A - 7e_2A - 5e_3A$$

$$0e_1 + e_2 + 0e_3 = -3e_1A + 2e_2A + e_3A$$

$$0e_1 + 0e_2 + e_3 = -2e_1A + e_2A + e_3A \quad .$$

By comparing coefficients, the above equations correspond to the matrix $(I : B)$, and

$$B = \begin{bmatrix} 13 & -7 & -5 \\ -3 & 2 & 1 \\ -2 & 1 & 1 \end{bmatrix} \quad .$$

Multiplying both sides of the equations by A^{-1}, we have

$$e_1A^{-1} = 13e_1 - 7e_2 - 5e_3$$

$$e_2A^{-1} = -3e_1 + 2e_2 + e_3$$

$$e_3A^{-1} = -2e_1 + e_2 + e_3$$

which shows that

$$A^{-1} = \begin{bmatrix} 13 & -7 & -5 \\ -3 & 2 & 1 \\ -2 & 1 & 1 \end{bmatrix} \quad .$$

So $A^{-1} = B$ where B is the matrix obtained by reduction of $(A : I)$.

Let us go through the steps in the transformation of the equations which correspond to row operations on $(A : I)$. Now

$$
\begin{array}{l}
e_1 + 2e_2 + 3e_3 \\
e_1 + 3e_2 + 2e_3 \\
e_1 + e_2 + 5e_3
\end{array}
\longrightarrow
\left[
\begin{array}{ccc:ccc}
1 & 2 & 3 & 1 & 0 & 0 \\
1 & 3 & 2 & 0 & 1 & 0 \\
1 & 1 & 5 & 0 & 0 & 1
\end{array}
\right]
$$

Consider the following three types of operations for transforming linear systems:

T_{ij} : Interchange the ith and jth equations.

$T_i(c)$: Replace the ith equation with c times the ith equation, where $c \neq 0$.

$T_{ij}(c)$: Replace the ith equation with the sum of the ith equation and c times the jth equation (j ≠ i).

Then, $[T_{21}(-1), T_{31}(-1)]$:

$$e_1 + 2e_2 + 3e_3 = e_1A + 0e_2A + 0e_3A$$
$$0e_1 + e_2 - e_3 = -e_1A + e_2A + 0e_3A$$
$$0e_1 - e_2 + 2e_3 = -e_1A + 0e_2A + e_3A$$

$$\longrightarrow \begin{bmatrix} 1 & 2 & 3 & | & 1 & 0 & 0 \\ 0 & 1 & -1 & | & -1 & 1 & 0 \\ 0 & -1 & 2 & | & -1 & 0 & 1 \end{bmatrix}$$

$[T_{32}(1)] =$

$$e_1 + 2e_2 + 3e_3 = e_1A + 0e_2A + 0e_3A$$
$$0e_1 + e_2 - e_3 = -e_1A + e_2A + 0e_3A$$
$$0e_1 + 0e_2 + e_3 = -2e_1A + e_2A + e_3A$$

$$\longrightarrow \begin{bmatrix} 1 & 2 & 3 & | & 1 & 0 & 0 \\ 0 & 1 & -1 & | & -1 & 1 & 0 \\ 0 & 0 & 1 & | & -2 & 1 & 1 \end{bmatrix}$$

$[T_{13}(-3), T_{23}(1)] =$

$$e_1 + 2e_2 + 0e_3 = 7e_1A - 3e_2A - 3e_3A$$
$$0e_1 + e_2 + 0e_3 = -3e_1A + 2e_2A + e_3A$$
$$0e_1 + 0e_2 + e_3 = -2e_1A + e_2A + e_3A$$

$$\longrightarrow \begin{bmatrix} 1 & 2 & 0 & | & 7 & -3 & -3 \\ 0 & 1 & 0 & | & -3 & 2 & 1 \\ 0 & 0 & 1 & | & -2 & 1 & 1 \end{bmatrix}$$

$[T_{12}(-2)] =$

$$e_1 + 0e_2 + 0e_3 = 13e_1A - 7e_2A - 5e_3A$$
$$0e_1 + e_2 + 0e_3 = -3e_1A + 2e_2A + e_3A$$
$$0e_1 + 0e_2 + e_3 = -2e_1A + e_2A + e_3A$$

$$\longrightarrow \begin{bmatrix} 1 & 0 & 0 & | & 13 & -7 & -5 \\ 0 & 1 & 0 & | & -3 & 2 & 1 \\ 0 & 0 & 1 & | & -2 & 1 & 1 \end{bmatrix}$$

Thus

$$A^{-1} = \begin{bmatrix} 13 & -7 & -5 \\ -3 & 2 & 1 \\ -2 & 1 & 1 \end{bmatrix} .$$

● PROBLEM 6-14

Use the classical adjoint to find A^{-1} where

$$A = \begin{bmatrix} 1 & 0 & -1 \\ 0 & 2 & 2 \\ 1 & 1 & -1 \end{bmatrix}$$

Recall some definitions: If $A = (a_{ij})$, then a co-factor of an entry a_{ij} is denoted A_{ij} and is given by $(-1)^{i+j}$ times the determinant of the $(n-1) \times (n-1)$ minor matrix obtained from A by deleting its ith row and jth column.

By the matrix of cofactors, we mean the matrix

$$C = \begin{bmatrix} A_{11} & \cdots\cdots & A_{1n} \\ \cdot & & \cdot \\ \cdot & & \cdot \\ \cdot & & \cdot \\ A_{n1} & \cdots\cdots & A_{nn} \end{bmatrix} \quad .$$

Then the adjoint of A is C^T, i.e.,

$$\text{adj } A = \begin{bmatrix} A_{11} & A_{21} & \cdots & A_{n1} \\ A_{12} & A_{22} & \cdots & \cdot \\ \cdot & \cdot & & \cdot \\ \cdot & \cdot & & \cdot \\ \cdot & \cdot & & \cdot \\ A_{1n} & A_{2n} & \cdots & A_{nn} \end{bmatrix}$$

Recall that A^{-1} exists if and only if $\det A = |A| \neq 0$. The rule for obtaining A^{-1} is then

$$A^{-1} = \frac{1}{|A|} [\text{adj } A]$$

where $|A|$ = determinant of the $n \times n$ square matrix. Let us first compute the determinant of matrix A

$$A = \begin{bmatrix} 1 & 0 & -1 \\ 0 & 2 & 2 \\ 1 & 1 & -1 \end{bmatrix}$$

$$|A| = \begin{vmatrix} 1 & 0 & -1 \\ 0 & 2 & 2 \\ 1 & 1 & -1 \end{vmatrix}$$

$$= 1 \begin{vmatrix} 2 & 2 \\ 1 & -1 \end{vmatrix} - 0 \begin{vmatrix} 0 & 2 \\ 1 & -1 \end{vmatrix} + (-1) \begin{vmatrix} 0 & 2 \\ 1 & 1 \end{vmatrix}$$

$$= 1(-2-2) - 0(0-2) - 1(0-2)$$

$$= -4-0+2 = -2 .$$

We find that $|A| \neq 0$. Therefore A^{-1} exists. The classical adjoint of A is found by replacing each element of A by its cofactor and taking the transpose of the resulting matrix.

Let us now compute the cofactors of the entries of A.

$$A = \begin{bmatrix} 1 & 0 & -1 \\ 0 & 2 & 2 \\ 1 & 1 & -1 \end{bmatrix}$$

To find A_{11}, we delete the first row and first column of A to obtain the matrix

$$\begin{bmatrix} 2 & 2 \\ 1 & -1 \end{bmatrix} .$$

The cofactor A_{11} is then $(-1)^{1+1}$ times the determinant of the above matrix, i.e.,

$$A_{11} = (1)^2 \cdot \begin{vmatrix} 2 & 2 \\ 1 & -1 \end{vmatrix} = \begin{vmatrix} 2 & 2 \\ 1 & -1 \end{vmatrix}$$

$$= (-2-2) = -4 .$$

We find the cofactors of the remaining elements of A by the same method. The cofactors of the nine elements of A are

$$A_{11} = + \begin{vmatrix} 2 & 2 \\ 1 & -1 \end{vmatrix}, \quad A_{12} = - \begin{vmatrix} 0 & 2 \\ 1 & -1 \end{vmatrix}, \quad A_{13} = + \begin{vmatrix} 0 & 2 \\ 1 & 1 \end{vmatrix}$$

$$= (-2-2) \qquad\qquad = -(0-2) \qquad\qquad = (0-2)$$
$$= -4 \qquad\qquad\quad = 2 \qquad\qquad\qquad = -2$$

$$A_{21} = - \begin{vmatrix} 0 & -1 \\ 1 & -1 \end{vmatrix}, \quad A_{22} = + \begin{vmatrix} 1 & -1 \\ 1 & -1 \end{vmatrix}, \quad A_{23} = - \begin{vmatrix} 1 & 0 \\ 1 & 1 \end{vmatrix}$$

$$= -(0+1) \qquad\qquad = (-1+1) \qquad\qquad = -(1-0)$$
$$= -1 \qquad\qquad\quad = 0 \qquad\qquad\qquad = -1$$

$$A_{31} = + \begin{vmatrix} 0 & -1 \\ 2 & 2 \end{vmatrix}, \quad A_{32} = - \begin{vmatrix} 1 & -1 \\ 0 & 2 \end{vmatrix}, \quad A_{33} = + \begin{vmatrix} 1 & 0 \\ 0 & 2 \end{vmatrix}$$

$$= (0+2) \qquad\qquad = -(2-0) \qquad\qquad = (2-0)$$
$$= 2 \qquad\qquad\quad = -2 \qquad\qquad\qquad = 2$$

The matrix of cofactors C is given by

$$C = \begin{bmatrix} -4 & 2 & -2 \\ -1 & 0 & -1 \\ 2 & -2 & 2 \end{bmatrix} .$$

We form the transpose of the matrix of cofactors to obtain the classical adjoint of A:

$$\text{Adj } A = \begin{bmatrix} -4 & -1 & 2 \\ 2 & 0 & -2 \\ -2 & -1 & 2 \end{bmatrix}$$

Now,

$$A^{-1} = \frac{1}{|A|} [\text{adj } A]$$

$$= -\frac{1}{2} \begin{bmatrix} -4 & -1 & 2 \\ 2 & 0 & -2 \\ -2 & -1 & 2 \end{bmatrix} ,$$

So

$$A^{-1} = \begin{bmatrix} 2 & 1/2 & -1 \\ -1 & 0 & 1 \\ 1 & 1/2 & -1 \end{bmatrix} .$$

It is easy to check the computation by verifying that

$$AA^{-1} = I$$

$$\begin{bmatrix} 1 & 0 & -1 \\ 0 & 2 & 2 \\ 1 & 1 & -1 \end{bmatrix} \begin{bmatrix} 2 & 1/2 & -1 \\ -1 & 0 & 1 \\ 1 & 1/2 & -1 \end{bmatrix}$$

$$= \begin{bmatrix} 2 + 0 - 1 & 1/2 + 0 - 1/2 & -1 + 0 + 1 \\ 0 - 2 + 2 & 0 + 0 + 1 & 0 + 2 - 2 \\ 2 - 1 - 1 & 1/2 + 0 - 1/2 & -1 + 1 + 1 \end{bmatrix}$$

$$= \begin{bmatrix} 1 & 0 & 0 \\ 0 & 1 & 0 \\ 0 & 0 & 1 \end{bmatrix} = I \; .$$

CHAPTER 7

THE RANK OF A MATRIX

FINDING THE RANK OF A MATRIX

Find the rank of the matrix A where $A = \begin{bmatrix} 1 & 3 & 2 \\ 2 & 6 & 1 \end{bmatrix}$.

Solution: If A is a matrix, then the rank of A, written $r(A)$, is the maximum number of linearly independent columns or, equivalently, rows. Since A is 2×3, the rank must be two or less. First, check the rows for linear independence. Set

$$c_1(1,3,2) + c_2(2,6,1) = 0$$

to obtain the system of equations:

$L_1 : c_1 + 2c_2 = 0$

$L_2 : 3c_1 + 6c_2 = 0$

$L_3 : 2c_1 + c_2 = 0$.

By solving this sytem of equations, we find that it has only a trivial solution:

$$-3L_1 : -3c_1 - 6c_2 = 0$$

$$+L_2 : \underline{3c_1 + 6c_2 = 0}$$

$$0c_1 + 0c_2 = 0$$

$$-2L_1 : -2c_1 - 4c_2 = 0$$

$$+L_3 : \underline{2c_1 + c_2 = 0}$$

$$0c_1 - 3c_2 = 0$$

$$c_2 = 0$$

From L_1 : $c_1 + 2c_2 = 0$

$$c_1 = 0$$

Solution: $c_1 = c_2 = 0$.

Thus, the two rows are independent and $r(A) = 2$.
We also could have found $r(A)$ by checking the maximum number of

linearly independent columns. The two column vectors

$$\begin{bmatrix} 1 \\ 2 \end{bmatrix} \quad \text{and} \quad \begin{bmatrix} 2 \\ 1 \end{bmatrix}$$

are linearly independent since

$$c_1 \begin{bmatrix} 1 \\ 2 \end{bmatrix} + c_2 \begin{bmatrix} 2 \\ 1 \end{bmatrix} = 0$$

implies $c_1 = c_2 = 0$. Obtaining the system of equations,

$$L_1 : \quad c_1 + 2c_2 = 0$$

$$+L_2 : \quad 2c_1 + c_2 = 0 ,$$

and solving this system of equations, the result is the trivial solution $c_1 = c_2 = 0$.

$$-2L_1 : \quad -2c_1 - 4c_2 = 0$$

$$L_2 : \quad 2c_1 + c_2 = 0$$

$$\overline{\qquad\qquad -3c_2 = 0}$$

$$c_2 = 0$$

$$\text{From } L_1 : \quad c_1 + 2c_2 = 0$$

$$c_1 = 0$$

Solution: $c_1 = c_2 = 0$. Furthermore, since the columns are vectors in
R^2 , the maximum number of linearly independent columns can only equal
two (dim $R^2 = 2$). Thus, again, $r(A) = 2$.

● **PROBLEM** 7-2

Find the rank of AB and BA where

(a) $A = \begin{bmatrix} 1 & 0 \\ 1 & 1 \end{bmatrix}$ and $B = \begin{bmatrix} 0 & 0 \\ 0 & 1 \end{bmatrix}$

(b) $A = \begin{bmatrix} 1 & 1 \\ 1 & 1 \end{bmatrix}$ $B = \begin{bmatrix} 1 & 1 \\ -1 & -1 \end{bmatrix}$

Solution: The following facts are useful in finding the rank of a given
matrix: 1) The rank of a matrix is the dimension of the subspace spanned by the rows. 2) Equivalent matrices have identical ranks. In particular, we can use row operations to obtain an echelon matrix which has
the same rank as the original matrix.

(a) $AB = \begin{bmatrix} 1 & 0 \\ 1 & 1 \end{bmatrix} \begin{bmatrix} 0 & 0 \\ 0 & 1 \end{bmatrix} = \begin{bmatrix} 0 & 0 \\ 0 & 1 \end{bmatrix}$

Hence, $r(AB) = 1$.

$$BA = \begin{bmatrix} 0 & 0 \\ 0 & 1 \end{bmatrix} \begin{bmatrix} 1 & 0 \\ 1 & 1 \end{bmatrix} = \begin{bmatrix} 0 & 0 \\ 1 & 1 \end{bmatrix}$$

and $r(BA) = 1$. Thus, $r(AB) = r(BA) = 1$.

(b) $AB = \begin{bmatrix} 1 & 1 \\ 1 & 1 \end{bmatrix} \begin{bmatrix} 1 & 1 \\ -1 & -1 \end{bmatrix} = \begin{bmatrix} 0 & 0 \\ 0 & 0 \end{bmatrix}$.

Consequently, $r(AB) = 0$.

$BA = \begin{bmatrix} 1 & 1 \\ -1 & -1 \end{bmatrix} \begin{bmatrix} 1 & 1 \\ 1 & 1 \end{bmatrix} = \begin{bmatrix} 2 & 2 \\ -2 & -2 \end{bmatrix}$.

Reduce the matrix BA by elementary row operations. Add the first row to the second row to obtain

$$\begin{bmatrix} 2 & 2 \\ 0 & 0 \end{bmatrix}.$$

Hence, $r(BA) = 1$.

● **PROBLEM** 7-3

Find the rank of the matrix A where:

(i)
$$A = \begin{bmatrix} 1 & 3 & 1 & -2 & -3 \\ 1 & 4 & 3 & -1 & -4 \\ 2 & 3 & -4 & -7 & -3 \\ 3 & 8 & 1 & -7 & -8 \end{bmatrix}$$

(ii)
$$A = \begin{bmatrix} 1 & 2 & -3 \\ 2 & 1 & 0 \\ -2 & -1 & 3 \\ -1 & 4 & -2 \end{bmatrix}$$

(iii)
$$A = \begin{bmatrix} 1 & 3 \\ 0 & -2 \\ 5 & -1 \\ -2 & 3 \end{bmatrix}$$

Solution: (i)   First, reduce the matrix A to echelon form using the elementary row operations.
(a)   Add -1 times the first row to the second row.
(b)   Add -2 times the first row to the third row.
(c)   Add -3 times the first row to the third row.

$$A = \begin{bmatrix} 1 & 3 & 1 & -2 & -3 \\ 0 & 1 & 2 & 1 & -1 \\ 0 & -3 & -6 & -3 & 3 \\ 0 & -1 & -2 & -1 & 1 \end{bmatrix}$$

Add +3 times the second row to the third row.
Add the second row to the fourth row. Then,

$$A = \begin{bmatrix} 1 & 3 & 1 & -2 & -3 \\ 0 & 1 & 2 & 1 & -1 \\ 0 & 0 & 0 & 0 & 0 \\ 0 & 0 & 0 & 0 & 0 \end{bmatrix}.$$

Since the echelon matrix has two nonzero rows, rank (A) = 2.

(ii)   Since row rank equals column rank it is easier to form the transpose of A and then row reduce to echelon form.

$$\begin{bmatrix} 1 & 2 & -2 & -1 \\ 2 & 1 & -1 & 4 \\ -3 & 0 & 3 & -2 \end{bmatrix}$$

Add -2 times the first row to the second row, and add 3 times the first row to the third row.

$$A = \begin{bmatrix} 1 & 2 & -2 & -1 \\ 0 & -3 & 3 & 6 \\ 0 & 6 & -3 & -5 \end{bmatrix}$$

Add 2 times the second row to the third row.

$$A = \begin{bmatrix} 1 & 2 & -2 & -1 \\ 0 & -3 & 3 & 6 \\ 0 & 0 & 3 & 7 \end{bmatrix}$$

Since the echelon matrix has three nonzero rows, rank (A) = 3.

(iii) The two columns are linearly independent since one is not a multiple of the other. Hence, rank [A] = 2 .

● PROBLEM 7-4

Find the rank of the matrix A   where

$$A = \begin{bmatrix} 1 & 0 & 2 & 3 \\ 0 & 0 & 5 & 1 \\ 0 & 0 & 0 & 0 \end{bmatrix}.$$

Solution: The matrix A is an echelon matrix. The rank of an echelon form matrix is the number of non-zero row vectors in the matrix. This follows from the fact that the non-zero rows of an echelon form matrix are linearly independent.

Since the echelon matrix A has two non-zero row vectors, its rank is 2.

THE RANK & BASIS FOR A VECTOR SPACE

● PROBLEM 7-5

Is the set $\{1 + t + t^2, 1 - 3t + 2t^2, 3 - t + 4t^2\}$ independent in   V?

Solution: The coordinates of $1 + t + t^2$, $1 - 3t + 2t^2$, and $3 - t + 4t^2$, with respect to $(1, t, t^2)$, are $(1,1,1)$, $(1,-3,2)$ and $(3,-1,4)$, respectively.

Form the matrix A whose rows are the above coordinate vectors.

$$A = \begin{bmatrix} 1 & 1 & 1 \\ 1 & -3 & 2 \\ 3 & -1 & 4 \end{bmatrix}$$

Now, reduce the above matrix to echelon form. The non-zero rows (or columns) of an echelon matrix are linearly independent. Since the rank of a matrix is defined as the maximum number of linearly independent rows (or columns) and since equivalent matrices have the same rank, the reduction to echelon form seems appropriate.

Add -1 times the first row to the second row, and add -3 times the first row to the third row.

$$\begin{bmatrix} 1 & 1 & 1 \\ 0 & -4 & 1 \\ 0 & -4 & 1 \end{bmatrix}$$

Add -1 times the second row to the third row.

$$\begin{bmatrix} 1 & 1 & 1 \\ 0 & -4 & 1 \\ 0 & 0 & 0 \end{bmatrix}$$

Since the echelon matrix A has two non-zero row vectors, its rank is 2.

The coordinate vectors generate a space of dimension two. We conclude that $\{(1,1,1), (1,-3,2), (3,-1,4)\}$ is independent, and, hence, that $\{1 + t + t^2, 1 - 3t + 2t^2, 3 - t + 4t^2\}$ is dependent.

● **PROBLEM 7-6**

Let V be the subspace of R^4 spanned by $S = \{\alpha_1, \alpha_2, \alpha_3, \alpha_4, \alpha_5\}$ where $\alpha_1 = [1,2,1,2]$, $\alpha_2 = [2,1,2,1]$, $\alpha_3 = [3,2,3,2]$, $\alpha_4 = [3,3,3,3]$ and $\alpha_5 = [5,3,5,3]$. Find a basis for V .

Solution: Form the matrix

$$A = \begin{bmatrix} 1 & 2 & 1 & 2 \\ 2 & 1 & 2 & 1 \\ 3 & 2 & 3 & 2 \\ 3 & 3 & 3 & 3 \\ 5 & 3 & 5 & 3 \end{bmatrix} \tag{1}$$

The rows of A are the vectors $\alpha_1, \alpha_2, \alpha_3, \alpha_4, \alpha_5$. By applying elementary row operations on the matrix (1), we can obtain a row reduced echelon matrix with linearly independent rows. These rows will form a basis for V. Thus, we use the fact that the row spaces of two equivalent matrices are identical. Recall that the row space is the subspace of R^n spanned by the rows of an m×n matrix. Apply the elementary row operations on A.

Replace row 2 with (-row 2) + row 1.
Replace row 3 with (-row 3) + row 1.
Replace row 4 with (-row 4) + row 1.
Replace row 5 with (-row 5) + row 1.

A becomes:

$$\begin{bmatrix} 1 & 2 & 1 & 2 \\ 0 & 3 & 0 & 3 \\ 0 & 4 & 0 & 4 \\ 0 & 3 & 0 & 3 \\ 0 & 7 & 0 & 7 \end{bmatrix}$$

Replace row 2 by 1/3 row 2:

$$\begin{bmatrix} 1 & 2 & 1 & 2 \\ 0 & 1 & 0 & 1 \\ 0 & 4 & 0 & 4 \\ 0 & 3 & 0 & 3 \\ 0 & 7 & 0 & 7 \end{bmatrix}$$

Replace row 1 by row 1 + (-2) $\times$ row 2.
Replace row 3 by -row 3 + (4 $\times$ row 2).
Replace row 4 by -row 4 + (3 $\times$ row 2).
Replace row 5 by -row 5 + (7 $\times$ row 2).
Therefore, A becomes

$$\begin{bmatrix} 1 & 0 & 1 & 0 \\ 0 & 1 & 0 & 1 \\ 0 & 0 & 0 & 0 \\ 0 & 0 & 0 & 0 \\ 0 & 0 & 0 & 0 \end{bmatrix}$$

A basis for the row space of A consists of $\beta_1 = [1,0,1,0]$ and
$\beta_2 = [0,1,0,1]$. $\{\beta_1, \beta_2\}$ form a basis for V.

Note that the above is a good method for finding a basis for a vector
space spanned by a given set $S = \{\alpha_1, \alpha_2, \ldots, \alpha_n\}$ of vectors.

● **PROBLEM 7-7**

Let P_4 denote the vector space of all polynomials of degree at most
equal to four. Let V be the subspace of P_4 spanned by
$S = \{\alpha_1, \alpha_2, \alpha_3, \alpha_4\}$ where $\alpha_1 = t^4 + t^2 + 2t + 1$, $\alpha_2 = t^4 + t^2 + 2t + 2$,
$\alpha_3 = 2t^4 + t^3 + t + 2$ and $\alpha_4 = t^4 + t^3 - t^2 - t$. Find a basis for V.

<u>Solution</u>:   First, find a matrix representation of the vectors in the
given problem. Now P_4 is isomorphic to R^5 . Hence, we can write
the four given polynomials as

$$\begin{bmatrix} 1 & 0 & 1 & 2 & 1 \\ 1 & 0 & 1 & 2 & 2 \\ 2 & 1 & 0 & 1 & 2 \\ 1 & 1 & -1 & -1 & 0 \end{bmatrix} \begin{bmatrix} t^4 \\ t^3 \\ t^2 \\ t_1 \end{bmatrix} = \begin{bmatrix} \alpha_1 \\ \alpha_2 \\ \alpha_3 \\ \alpha_4 \end{bmatrix} \qquad (1)$$

The question is: How can we find a basis for the subspace of R^5 spanned by the rows of the matrix in (1)?

Let V be the row space of the matrix. Then, if B is row equivalent to A, its row space is also V. Now, by applying elementary row operations to the matrix in (1), we can obtain an echelon form matrix from which a basis for V can be found. Thus,

$$A = \begin{bmatrix} 1 & 0 & 1 & 2 & 1 \\ 1 & 0 & 1 & 2 & 2 \\ 2 & 1 & 0 & 1 & 2 \\ 1 & 1 & -1 & -1 & 0 \end{bmatrix}$$

Replace row 2 by row 2 + (-row 1).
Replace row 3 by row 3 + (-2 row 1).
Replace row 4 by row 4 + (-row 1).

$$\begin{bmatrix} 1 & 0 & 1 & 2 & 1 \\ 0 & 0 & 0 & 0 & 1 \\ 0 & 1 & -2 & -3 & 0 \\ 0 & 1 & -2 & -3 & -1 \end{bmatrix}$$

Replace row 4 by row 3 + (-row 4).

$$\begin{bmatrix} 1 & 0 & 1 & 2 & 1 \\ 0 & 0 & 0 & 0 & 1 \\ 0 & 1 & -2 & -3 & 0 \\ 0 & 0 & 0 & 0 & 1 \end{bmatrix}$$

Replace row 2 by row 3.
Replace row 3 by row 2.
Replace row 4 by row 4 + (-row 2).
Therefore, an equivalent matrix is

$$B = \begin{bmatrix} 1 & 0 & 1 & 2 & 0 \\ 0 & 1 & -2 & -3 & 0 \\ 0 & 0 & 0 & 0 & 1 \\ 0 & 0 & 0 & 0 & 0 \end{bmatrix} \qquad (2)$$

Hence, a basis for the row space of A is $\{\beta_1, \beta_2, \beta_3\}$ where $\beta_1 = [1,0,1,2,0]$, $\beta_2 = [0,1,-2,-3,0]$ and $\beta_3 = [0,0,0,0,1]$. Multiplying (2) by the standard basis for P_4,

$$\begin{bmatrix} 1 & 0 & 1 & 2 & 0 \\ 0 & 1 & -2 & -3 & 0 \\ 0 & 0 & 0 & 0 & 1 \\ 0 & 0 & 0 & 0 & 0 \end{bmatrix} \begin{bmatrix} t^4 \\ t^3 \\ t^2 \\ t \\ 1 \end{bmatrix}$$

we see that a basis for the subspace V of P_4 is
$$\{t^4 + t^2 + 2t, \ t^3 - 2t^2 - 3t, 1\} .$$

EXISTENCE OF SOLUTIONS

Let the homogeneous linear system AX = B be given by

$$\begin{bmatrix} 1 & 2 & 0 \\ 0 & 1 & 3 \\ 2 & 1 & 3 \end{bmatrix} \begin{bmatrix} x_1 \\ x_2 \\ x_3 \end{bmatrix} = \begin{bmatrix} 0 \\ 0 \\ 0 \end{bmatrix} . \qquad (1)$$

Show that A has only the trivial solution, $(x_1, x_2, x_3) = (0,0,0)$.

Solution: The system (1) will have only the trivial solution if A^{-1} exists, for then

$$\begin{bmatrix} x_1 \\ x_2 \\ x_3 \end{bmatrix} = \begin{bmatrix} 0 \\ 0 \\ 0 \end{bmatrix} \begin{bmatrix} 1 & 2 & 0 \\ 0 & 1 & 3 \\ 2 & 1 & 3 \end{bmatrix}^{-1} = \begin{bmatrix} 0 \\ 0 \\ 0 \end{bmatrix} .$$

Thus, we must show that A is non-singular. But an n×n matrix is non-singular if and only if rank A = n . To show this, use the folow-ing reasoning: A has full rank (i.e., rank equal to n) if and only if it is equivalent to the n×n identity matrix, I_n . But I_n is non-singular. Hence, A is non-singular. Conversely, if an n×n matrix is non-singular, it is equivalent to an identity matrix, I_n which has rank equal to n. Hence, rank A = n . We must find the rank of the matrix in (1). By applying elementary row operations it can be seen that

$$\begin{bmatrix} 1 & 2 & 0 \\ 0 & 1 & 3 \\ 2 & 1 & 3 \end{bmatrix}$$

is equivalent to

$$\begin{bmatrix} 1 & 0 & -6 \\ 0 & 1 & 3 \\ 0 & 0 & 1 \end{bmatrix} ,$$

i.e., rank A = 3 . Hence, A is non-singular and the system (1) has only the trivial solution.

  The above problem is an illustration of the theorem below. A nec-essary and sufficient condition for a system of n homogeneous linear equations, AX = 0 , to have only the trivial solution, X = 0 , is that rank A = n .

Give an example to illustrate the following theorem: The system of n homogeneous linear equations in n unknowns, AX = 0 , has a non-trivial solution if and only if rank A < n .

Solution: Expanding the matrix equation AX = 0 ,

$$a_{11}x_1 + a_{12}x_2 + \ldots + a_{1n}x_n = 0$$

$$a_{21}x_1 + a_{22}x_2 + \ldots + a_{2n}x_n = 0$$

$$\vdots$$

$$a_{n1}x_1 + a_{n2}x_2 + \ldots + a_{nn}x_n = 0$$

or,

$$\begin{bmatrix} a_{11} & \cdots & a_{1n} \\ a_{21} & \cdots & a_{2n} \\ \cdot & & \cdot \\ \cdot & & \cdot \\ \cdot & & \cdot \\ a_{n1} & \cdots & a_{nn} \end{bmatrix} \begin{bmatrix} x_1 \\ x_2 \\ \cdot \\ \cdot \\ \cdot \\ x_n \end{bmatrix} = \begin{bmatrix} 0 \\ 0 \\ \cdot \\ \cdot \\ \cdot \\ 0 \end{bmatrix} \qquad (1)$$

The rank of the matrix A is the dimension of the subspace of R^n spanned by the rows of A. If we regard A as the matrix representation of a linear operator T on an n-dimensional space, then the rank of A represents the dimension of the range of T. A fundamental theorem of linear algebra is that

$$\dim V = \dim(\text{Range } T) + \dim(\text{kernel } T) .$$

Now, assume that rank $A = n$. Then, $\dim(\text{Range } T) = n$, and, hence, the kernel of T is a zero-dimensional subspace; i.e., the only vector belonging to ker T is the zero vector. But, from (1), ker T is the solution set to $AX = 0$. Hence, for non-trivial solutions, rank $A < n$ must hold. This is the meaning of the above theorem.

Let

$$\begin{bmatrix} x_1 + 2x_2 \\ x_1 + x_2 - 3x_3 \\ x_1 + 3x_2 + 3x_3 \end{bmatrix} = \begin{bmatrix} 0 \\ 0 \\ 0 \end{bmatrix} \qquad (2)$$

Rewriting (2),

$$\begin{bmatrix} 1 & 2 & 0 \\ 1 & 1 & -3 \\ 1 & 3 & 3 \end{bmatrix} \begin{bmatrix} x_1 \\ x_2 \\ x_3 \end{bmatrix} = \begin{bmatrix} 0 \\ 0 \\ 0 \end{bmatrix} \qquad (3)$$

Applying elementary row operations to the matrix in (3), it can be seen that it is equivalent to

$$\begin{bmatrix} 1 & 2 & 0 \\ 0 & 1 & 3 \\ 0 & 0 & 0 \end{bmatrix} .$$

Since the first and second rows are independent, the rows of A span a subspace of R^3 of dimension 2, i.e., rank $A = 2 < 3$. According to the theorem, there exist non-trivial solutions to the system (2). Conversely, we find a non-trivial solution by solving

$$\begin{bmatrix} 1 & 0 & -6 \\ 0 & 1 & 3 \\ 0 & 0 & 0 \end{bmatrix} \begin{bmatrix} x_1 \\ x_2 \\ x_3 \end{bmatrix} = \begin{bmatrix} 0 \\ 0 \\ 0 \end{bmatrix}$$

All vectors of the form

$$t \begin{bmatrix} 6 \\ -3 \\ 1 \end{bmatrix},$$

where t is a scalar, are non trivial solutions of (3). But, we already know that rank A<3. Hence, the necessary condition ("only if") has been illustrated.

● **PROBLEM** 7-10

Consider the system of equations

$$\begin{bmatrix} 2 & 1 & 3 \\ 1 & -2 & 2 \\ 0 & 1 & 3 \end{bmatrix} \begin{bmatrix} x_1 \\ x_2 \\ x_3 \end{bmatrix} = \begin{bmatrix} 1 \\ 2 \\ 3 \end{bmatrix}.$$

Show that the system has a solution without actually computing a solution.

Solution: Let AX = B be a system of m linear equations in n unknowns, where $A = [a_{ij}]$ is an m×n matrix, X is an n-dimensional column vector and B is an m-dimensional column vector. Thus, the system has the form

$$a_{11}x_1 + a_{12}x_2 + \cdots + a_{1n}x_n = b_1$$
$$a_{21}x_1 + a_{22}x_2 + \cdots + a_{2n}x_n = b_2$$
$$\vdots \qquad \vdots \qquad \qquad \vdots \qquad \qquad (1)$$
$$a_{m1}x_1 + a_{n2}x_2 + \cdots + a_{mn}x_n = b_m,$$

The system (1) may also be written as the vector equation

$$x_1 \begin{bmatrix} a_{11} \\ a_{21} \\ \cdot \\ \cdot \\ \cdot \\ a_{m1} \end{bmatrix} + x_2 \begin{bmatrix} a_{12} \\ a_{22} \\ \cdot \\ \cdot \\ \cdot \\ a_{m2} \end{bmatrix} + \cdots + x_n \begin{bmatrix} a_{1n} \\ a_{2n} \\ \cdot \\ \cdot \\ \cdot \\ a_{mn} \end{bmatrix} = \begin{bmatrix} b_1 \\ b_2 \\ \cdot \\ \cdot \\ \cdot \\ b_m \end{bmatrix}$$

Thus, AX = B has a solution when B is a linear combination of the columns of A; that is, if B belongs to the column space of A. But the dimension of the column space of A is the rank of A. Since B is in this space, the rank of A is also the rank of the augmented matrix [A|B]. Thus, a necessary and sufficient condition for AX = B to have a solution is that rank A = rank[A|B].

Applying elementary rwo operations to the given matrix, it can be seen that

$$\begin{bmatrix} 2 & 1 & 3 \\ 1 & -2 & 2 \\ 0 & 1 & 3 \end{bmatrix}$$

is equivalent to

$$\begin{bmatrix} 1 & 0 & 8 \\ 0 & 1 & 3 \\ 0 & 0 & -16 \end{bmatrix} \qquad (2)$$

Since the columns of the matrix (2) form a basis for R^3, rank $A = 3$.
Next, form the augmented matrix

$$\begin{bmatrix} 2 & 1 & 3 & \vdots & 1 \\ 1 & -2 & 2 & \vdots & 2 \\ 0 & 1 & 3 & \vdots & 3 \end{bmatrix}$$

and reduce to echelon form:

$$\begin{bmatrix} 1 & 0 & 0 & \vdots & -1 \\ 0 & 1 & 0 & \vdots & -3/8 \\ 0 & 0 & 1 & \vdots & 9/8 \end{bmatrix}$$

Thus, rank $[A \vdots B] = 3$ and the given system of equations has a solution.

DETERMINANT RANK

● **PROBLEM** 7-11

Let A be the matrix

$$\begin{bmatrix} 0 & 1 & 3 & -2 & -1 & 2 \\ 0 & 2 & 6 & -4 & -2 & 4 \\ 0 & 1 & 3 & -2 & 1 & 4 \\ 0 & 2 & 6 & 1 & -1 & 0 \end{bmatrix}$$

Find the determinant rank of A .

Solution: If A is an m✕n matrix, the determinant rank of A is
defined as follows: The order of the largest non-zero determinant which
is obtainable by the possible deletion of rows and columns from the matrix.
    The standard method of computing the determinant rank is the one shown
below. First, use elementary row operations to reduce the matrix to echelon form (the leading coefficient of each equation equals one). Then,
from the echelon matrix, select the largest upper triangular matrix which
has one's along the main diagonal. The determinant of this matrix is the
product of the diagonal elements, and, hence, the determinant rank is the
order of this determinant. Applying the three elementary row operations
on A, we obtain the equivalent matrix:

$$\begin{bmatrix} 0 & 1 & 3 & -2 & -1 & 2 \\ 0 & 0 & 0 & 1 & 1/5 & -4/5 \\ 0 & 0 & 0 & 0 & 1 & 1 \\ 0 & 0 & 0 & 0 & 0 & 0 \end{bmatrix} \qquad (1)$$

Examining (1), it can be seen that the second, fourth and fifth columns
form the largest possible upper triangular matrix. Thus,

$$\begin{bmatrix} 1 & -2 & -1 \\ 0 & 1 & 1/5 \\ 0 & 0 & 1 \end{bmatrix}$$

has determinant equal to one, and the determinant rank of A is three.
Since (1) contains three non-zero rows, the row-rank of A is also
three, i.e., determinant rank = row rank. The last statement is always
true.

Show that the matrix

$$A = \begin{bmatrix} 0 & 1 & 2 \\ 2 & 3 & 4 \\ 4 & 7 & 10 \end{bmatrix}$$

is equivalent to D_2^{33} where $D_r^{m,n}$ denotes the canonical form under equivalence of A. $D_r^{m,n}$ is the echelon form that has one's along the diagonal and zeros elsewhere, and where all the zero rows are consigned to the depths of the matrix.

Solution: An $m \times n$ matrix A is equivalent to $D_r^{m,n}$ if and only if it has rank r. The matrices $D_r^{m,n}$ are called the canonical forms under equivalence. Each $m \times n$ matrix is equivalent to exactly one canonical form under equivalence.

The given matrix A has rank 2, and may be converted to $D_2^{3,3}$ by the following sequence of matrices, each of which is obtained from the preceding matrix by a row or column operation.

$$A = \begin{bmatrix} 0 & 1 & 2 \\ 2 & 3 & 4 \\ 4 & 7 & 10 \end{bmatrix}$$

Interchange the first and the second rows.

$$A_1 = \begin{bmatrix} 2 & 3 & 4 \\ 0 & 1 & 2 \\ 4 & 7 & 10 \end{bmatrix}$$

Add -2 times the first row to the third row.

$$A_2 = \begin{bmatrix} 2 & 3 & 4 \\ 0 & 1 & 2 \\ 0 & 1 & 2 \end{bmatrix}$$

Add -3 times the second row to the first row, and add -1 times the second row to the third row.

$$A_3 = \begin{bmatrix} 2 & 0 & -2 \\ 0 & 1 & 2 \\ 0 & 0 & 0 \end{bmatrix}$$

Add the first column to the third column.

$$A_4 = \begin{bmatrix} 2 & 0 & 0 \\ 0 & 1 & 2 \\ 0 & 0 & 0 \end{bmatrix}$$

Add -2 times the second column to the third column.

$$A_5 = \begin{bmatrix} 2 & 0 & 0 \\ 0 & 1 & 0 \\ 0 & 0 & 0 \end{bmatrix}$$

Divide the first row by 1/2 . Then,

$$D_2^{3,3} = \begin{bmatrix} 1 & 0 & 0 \\ 0 & 1 & 0 \\ 0 & 0 & 0 \end{bmatrix}$$

● **PROBLEM** 7-13

Let S and T be linear transformations whose matrix representations are given by A and B, respectively. A theorem in linear algebra states that

$$\text{Rank}(A + B) \le \text{Rank } A + \text{Rank } B. \qquad (1)$$

Give examples where a) the strict inequality in (1) holds, and b) the equality in (1) holds.

Solution: The rank of an $m \times n$ matrix, A, is the maximum number of linearly independent rows (or columns) in A. The rank of an operator is defined to be the dimension of its image space. For an operator represented by a matrix, these two definitions are equivalent since the linearly independent rows(columns) of the matrix form a basis for the image space. Let S and T denote the linear transformation of which A and B are the respective matrix representations in some basis. The set of all vectors that are the images of S or T form a subspace of V with dimension less than or equal to dim V. Since $(S+T)u = Su + Tu$, we have that $\text{im}(S+T)$ is a subspace of $[\text{im}S, \text{im}T]$, i.e., the space generated by $\text{im}S$ and $\text{im}T$. Thus,

$$\dim \text{im}(S+T) \le \dim \text{im}S + \dim \text{im}T.$$

But $\dim \text{im}(S+T) = R(S+T)$, so (1) follows.
a) Let A = B be given by the matrix

$$\begin{bmatrix} 2 & 3 \\ 5 & 1 \end{bmatrix}$$

Then, since rank A = rank B = 2 and $\text{rank}(A+B) = \text{rank}\begin{bmatrix} 4 & 6 \\ 10 & 2 \end{bmatrix} = 2$, we have that

$$\text{rank}(A+B) = 2 < \text{rank } A + \text{rank } B = 2 + 2 = 4 .$$

b) Let A be the 2×3 matrix

$$\begin{bmatrix} 3 & 2 & 1 \\ 4 & 5 & 6 \end{bmatrix},$$

and let B be the zero matrix of order 2×3 . The rank of A is 2 (the point (1,6) can be expressed as a linear combination of (3,4) and (2,5)). The rank of B is zero (every vector operated upon by B is sent to the zero vector in V_2 which is a subspace of zero dimension).

The rank $(A+B) = \text{rank} \begin{bmatrix} 3 & 2 & 1 \\ 4 & 5 & 6 \end{bmatrix} = 2$ as previously found. Hence,

$$\text{rank}(A+B) = 2 = \text{rank } A + \text{rank } B = 2 + 0 .$$

CHAPTER 8

SYSTEMS OF LINEAR EQUATIONS

ECHELON FORM MATRICES

Define row-reduced echelon form and give examples.

Solution: A row-reduced echelon form is a matrix such that
(a) Each row that consists entirely of zeros is below each row which contains a non-zero entry.
(b) The first non-zero entry in each row is a 1.
(c) The first non-zero entry in each row is to the right of the first non-zero entry of the preceding row.
(d) Each column that contains the first non-zero entry of some row has zeros everywhere else.

   If a matrix satisfies only properties (a)-(c), it is said to be in row echelon form or just echelon form.

   Note that, together (b) and (d) are equivalent to saying that the column in which the leading entry of a non-zero row "i" occurs is $\vec{e}_i$, i.e.,

$$\begin{bmatrix} 0 \\ \vdots \\ 0 \\ \vdots \\ 1 \\ \vdots \\ 0 \\ \vdots \\ 0 \end{bmatrix}$$

, where the 1 occurs in the ith place.

Examples: The following matrices are in reduced row-echelon form.

(1)
$$\begin{bmatrix} 1 & 0 & 0 & 4 \\ 0 & 1 & 0 & 7 \\ 0 & 0 & 1 & -1 \end{bmatrix}$$
(2)
$$\begin{bmatrix} 1 & 0 & 0 \\ 0 & 1 & 0 \\ 0 & 0 & 1 \end{bmatrix}$$

(3)
$$\begin{bmatrix} 0 & 1 & -2 & 0 & 1 \\ 0 & 0 & 0 & 1 & 3 \\ 0 & 0 & 0 & 0 & 0 \\ 0 & 0 & 0 & 0 & 0 \end{bmatrix}$$
(4)
$$\begin{bmatrix} 0 & 0 \\ 0 & 0 \end{bmatrix}$$

The following matrices are in row-echelon form.

(1)
$$\begin{bmatrix} 1 & 1 & 3 & 7 \\ 0 & 1 & 6 & 2 \\ 0 & 0 & 1 & 5 \end{bmatrix}$$
(2)
$$\begin{bmatrix} 1 & 1 & 0 \\ 0 & 1 & 0 \\ 0 & 0 & 0 \end{bmatrix}$$

(3)
$$\begin{bmatrix} 0 & 1 & 2 & 6 & 0 \\ 0 & 0 & 1 & -1 & 0 \\ 0 & 0 & 0 & 0 & 1 \end{bmatrix}$$

Note that in (1) and (2), although the first non-zero member of the second row occurs in the second column, there are not zeros everywhere else in that column.

● **PROBLEM** 8-2

Define elementary row operations and give an example.

Solution: The three elementary row operations on a matrix A are:
1. Interchange the i-th and the j-th row of A.
2. Add the i-th row of A to the j-th row of A, $i \neq j$.
3. Multiply the i-th row of A by a non-zero scalar k. Let

$$A = \begin{bmatrix} 1 & 6 & 3 & 4 \\ 1 & 2 & 1 & 1 \\ -1 & 2 & 1 & 2 \end{bmatrix}.$$

We perform the following row operations on matrix A.
(1) Interchange the first and the second rows

$$\begin{bmatrix} 1 & 2 & 1 & 1 \\ 1 & 6 & 3 & 4 \\ -1 & 2 & 1 & 2 \end{bmatrix}.$$

(2) Add the first row to the third row and -1 times the first row to the second row. Adding the first row to the third row,

$$\begin{bmatrix} 1 & 2 & 1 & 1 \\ 1 & 6 & 3 & 4 \\ -1+1 & 2+2 & 1+1 & 2+1 \end{bmatrix} = \begin{bmatrix} 1 & 2 & 1 & 1 \\ 1 & 6 & 3 & 4 \\ 0 & 4 & 2 & 3 \end{bmatrix};$$

adding -1 times the first row to the second,

$$\begin{bmatrix} 1 & 2 & 1 & 1 \\ (-1 \cdot 1)+1 & (-1 \cdot 2)+6 & (-1 \cdot 1)+3 & (-1 \cdot 1)+4 \\ 0 & 4 & 2 & 3 \end{bmatrix}$$

$$= \begin{bmatrix} 1 & 2 & 1 & 1 \\ 0 & 4 & 2 & 3 \\ 0 & 4 & 2 & 3 \end{bmatrix}.$$

Add -1 times the second row to the third row.

$$\begin{bmatrix} 1 & 2 & 1 & 1 \\ 0 & 4 & 2 & 3 \\ 0 & 0 & 0 & 0 \end{bmatrix}$$

Divide the second row by 4.

$$\begin{bmatrix} 1 & 2 & 1 & 1 \\ 0 & 1 & \frac{1}{2} & 3/4 \\ 0 & 0 & 0 & 0 \end{bmatrix}$$

Add -2 times the second row to the first row.

$$\begin{bmatrix} (-2\cdot0)+1 & (-2\cdot1)+2 & (-2\cdot\frac{1}{2})+1 & (-2\cdot3/4)+1 \\ 0 & 1 & \frac{1}{2} & 3/4 \\ 0 & 0 & 0 & 0 \end{bmatrix}$$

$$= \begin{bmatrix} 1 & 0 & 0 & -\frac{1}{2} \\ 0 & 1 & \frac{1}{2} & 3/4 \\ 0 & 0 & 0 & 0 \end{bmatrix} .$$

Note that this matrix is in row-reduced echelon form. The elementary row operations can be applied to reduce a matrix to echelon form, and this technique is used in solving systems of linear equations.

● PROBLEM 8-3

Reduce the following matrices to echelon form and then to row reduced echelon form.

(a)
$$A = \begin{bmatrix} 0 & 1 & 3 & -2 \\ 2 & 1 & -4 & 3 \\ 2 & 3 & 2 & -1 \end{bmatrix}$$

(b)
$$A = \begin{bmatrix} 6 & 3 & -4 \\ -4 & 1 & -6 \\ 1 & 2 & -5 \end{bmatrix}$$

Solution: In echelon form, the first non-zero entry of any row is a 1, and any row of zeros lies below the rows with non-zero entries. Furthermore, the first non-zero entry of any row is in a column to the left of the first non-zero entry in the next row. In addition, in reduced echelon form, the entire column containing the first non-zero entry of any row is all zeros except for that entry.

(a) Perform the following row operations: Interchange the first and the second rows.

$$\begin{bmatrix} 2 & 1 & -4 & 3 \\ 0 & 1 & 3 & -2 \\ 2 & 3 & 2 & -1 \end{bmatrix}$$

Add -1 times the first row to the third row.

$$\begin{bmatrix} 2 & 1 & -4 & 3 \\ 0 & 1 & 3 & -2 \\ 0 & 2 & 6 & -4 \end{bmatrix}$$

Add -2 times the second row to the third row.

$$\begin{bmatrix} 2 & 1 & -4 & 3 \\ 0 & 1 & 3 & -2 \\ 0 & 0 & 0 & 0 \end{bmatrix}$$

Finally, to obtain the echelon form, multiply the first column by $\frac{1}{2}$. Hence,

$$\begin{bmatrix} 1 & \frac{1}{2} & 2 & 3/2 \\ 0 & 1 & 3 & -2 \\ 0 & 0 & 0 & 0 \end{bmatrix} .$$

Now add $-\frac{1}{2}$ times the second row to the first row, to obtain the row reduced echelon form

$$\begin{bmatrix} 1 & 0 & -7/2 & 5/2 \\ 0 & 1 & 3 & -2 \\ 0 & 0 & 0 & 0 \end{bmatrix}.$$

(b) First interchange the first and third rows.

$$\begin{bmatrix} 1 & 2 & -5 \\ -4 & 1 & -6 \\ 6 & 3 & -4 \end{bmatrix}.$$

Add 4 times the first row to the second row and -6 times the first row to the third row.

$$\begin{bmatrix} 1 & 2 & -5 \\ 0 & 9 & -26 \\ 0 & -9 & 26 \end{bmatrix}$$

Now add the second row to the third row.

$$\begin{bmatrix} 1 & 2 & -5 \\ 0 & 9 & -26 \\ 0 & 0 & 0 \end{bmatrix}$$

Divide the second row by 9 to obtain the echelon form.

$$\begin{bmatrix} 1 & 2 & -5 \\ 0 & 1 & -26/9 \\ 0 & 0 & 0 \end{bmatrix}$$

Add -2 times the second row to the first row to obtain the row-reduced echelon form.

$$\begin{bmatrix} 1 & 0 & 7/9 \\ 0 & 1 & -26/9 \\ 0 & 0 & 0 \end{bmatrix}.$$

● **PROBLEM 8-4**

Given

$$A = \begin{bmatrix} 1 & -2 & 3 & -1 \\ 2 & -1 & 2 & 2 \\ 3 & 1 & 2 & 3 \end{bmatrix},$$

(i) Reduce A to echelon form.
(ii) Reduce A to row reduced echelon form.

Solution: To obtain the echelon form, the first non-zero entry of any row must be contained in a column to the left of the first non-zero entry in the next row. Also, the first non-zero entry must be a 1, and any row of all zeros lies below all the rows that have non-zero entries.
In reduced echelon form, the column containing the first non-zero entry of the ith row is $\vec{e}_i$.

(i) To reduce A to echelon form, apply the following row operations:

add -2 times the first row to the second row and -3 times the first row to the third row.

$$\begin{bmatrix} 1 & -2 & 3 & -1 \\ 0 & 3 & -4 & 4 \\ 0 & 7 & -7 & 6 \end{bmatrix} .$$

Multiply the second row by 7 and the third row by 3.

$$\begin{bmatrix} 1 & -2 & 3 & -1 \\ 0 & 21 & -28 & 28 \\ 0 & 21 & -21 & 18 \end{bmatrix} .$$

Then, add -1 times the second row to the third row, to obtain

$$\begin{bmatrix} 1 & -2 & 3 & -1 \\ 0 & 3 & -4 & 4 \\ 0 & 0 & 7 & -10 \end{bmatrix} .$$

Divide the second row by 3 and the third row by 7 to obtain the echelon form.

$$\begin{bmatrix} 1 & -2 & 3 & -1 \\ 0 & 1 & -4/3 & 4/3 \\ 0 & 0 & 1 & -10/7 \end{bmatrix} .$$

(ii)   To obtain the reduced echelon form, add 2 times the second row to the first row.

$$\begin{bmatrix} 1 & 0 & 1/3 & 5/3 \\ 0 & 1 & -4/3 & 4/3 \\ 0 & 0 & 1 & -10/7 \end{bmatrix} .$$

Add -1/3 times the third row to the first row and 4/3 times the third row to the second row, resulting in the row reduced echelon form.

$$\begin{bmatrix} 1 & 0 & 0 & 15/7 \\ 0 & 1 & 0 & -4/7 \\ 0 & 0 & 1 & -10/7 \end{bmatrix} .$$

● PROBLEM 8-5

1.   Which of the following matrices are in reduced-row echelon form?

(a) $\begin{bmatrix} 1 & 0 & 0 \\ 0 & 0 & 0 \\ 0 & 0 & 1 \end{bmatrix}$ (b) $\begin{bmatrix} 0 & 1 & 0 \\ 1 & 0 & 0 \\ 0 & 0 & 0 \end{bmatrix}$ (c) $\begin{bmatrix} 1 & 1 & 0 \\ 0 & 1 & 0 \\ 0 & 0 & 0 \end{bmatrix}$

(d) $\begin{bmatrix} 1 & 2 & 0 & 3 & 0 \\ 0 & 0 & 1 & 1 & 0 \\ 0 & 0 & 0 & 0 & 1 \\ 0 & 0 & 0 & 0 & 0 \end{bmatrix}$ (e) $\begin{bmatrix} 1 & 0 & 0 & 5 \\ 0 & 0 & 1 & 3 \\ 0 & 1 & 0 & 4 \end{bmatrix}$

(f) $\begin{bmatrix} 1 & 0 & 3 & 1 \\ 0 & 1 & 2 & 4 \end{bmatrix}$

2. Which of the following matrices are in row-echelon form?

(a) $\begin{bmatrix} 1 & 2 & 3 \\ 0 & 0 & 0 \\ 0 & 0 & 1 \end{bmatrix}$ (b) $\begin{bmatrix} 1 & -7 & 5 & 5 \\ 0 & 1 & 3 & 2 \end{bmatrix}$ (c) $\begin{bmatrix} 1 & 1 & 0 \\ 0 & 1 & 0 \\ 0 & 0 & 0 \end{bmatrix}$

(d) $\begin{bmatrix} 1 & 3 & 0 & 2 & 0 \\ 1 & 0 & 2 & 2 & 0 \\ 0 & 0 & 0 & 0 & 1 \\ 0 & 0 & 0 & 0 & 0 \end{bmatrix}$ (e) $\begin{bmatrix} 2 & 3 & 4 \\ 0 & 1 & 2 \\ 0 & 0 & 3 \end{bmatrix}$

(f) $\begin{bmatrix} 0 & 0 & 0 \\ 0 & 0 & 0 \\ 0 & 0 & 0 \end{bmatrix}$

Solution: A matrix is in row-reduced echelon form if it satisfies the following conditions:
(i) Any zero row lies below the non-zero rows.
(ii) The leading entry of any non-zero row is 1.
(iii) The leading entry of any non-zero row is to the right of the leading entry of each preceding row.
(iv) The column that contains the leading entry of any row has zero for all other entries.
The matrices (d) and (f) satisfy all the necessary conditions; therefore, they are in reduced row-echelon form.

The matrix (a) is not in row-reduced echelon form because it violates condition (i). The matrices (b) and (e) do not satisfy condition (iii) while the matrix (c) does not satisfy condition (iv).

2. If a matrix satisfies the conditions (i), (ii) and (iii), then it is in row echelon form. Only the matrices (b), (c) and (f) satisfy these conditions so they are in row-echelon form.

● **PROBLEM 8-6**

Let
$$A = \begin{bmatrix} 0 & 0 & 0 & 0 & 1 & 1 & 1 \\ 0 & 2 & 6 & 2 & 0 & 0 & 4 \\ 0 & 1 & 3 & 1 & 1 & 0 & 1 \\ 0 & 1 & 3 & 1 & 2 & 1 & 2 \end{bmatrix}.$$

Reduce A to the Hermite normal form.

Solution: The row-reduced echelon matrix is also known as the Hermite normal form. Perform the following elementary row operations to obtain this form. Interchange rows one and three:

$$\begin{bmatrix} 0 & 1 & 3 & 1 & 1 & 0 & 1 \\ 0 & 2 & 6 & 2 & 0 & 0 & 4 \\ 0 & 0 & 0 & 0 & 1 & 1 & 1 \\ 0 & 1 & 3 & 1 & 2 & 1 & 2 \end{bmatrix}.$$

Add -1 times the first row to the fourth row and -2 times the first row to the second row.

$$\begin{bmatrix} 0 & 1 & 3 & 1 & 1 & 0 & 1 \\ 0 & 0 & 0 & 0 & -2 & 0 & 2 \\ 0 & 0 & 0 & 0 & 1 & 1 & 1 \\ 0 & 0 & 0 & 0 & 1 & 1 & 1 \end{bmatrix} .$$

Divide the second row by -2. Add -1 times the third row to the fourth row.

$$\begin{bmatrix} 0 & 1 & 3 & 1 & 1 & 0 & 1 \\ 0 & 0 & 0 & 0 & 1 & 0 & -1 \\ 0 & 0 & 0 & 0 & 1 & 1 & 1 \\ 0 & 0 & 0 & 0 & 0 & 0 & 0 \end{bmatrix} .$$

Add -1 times the second row to the first row and the third row.

$$\begin{bmatrix} 0 & 1 & 3 & 1 & 0 & 0 & 2 \\ 0 & 0 & 0 & 0 & 1 & 0 & -1 \\ 0 & 0 & 0 & 0 & 0 & 1 & 2 \\ 0 & 0 & 0 & 0 & 0 & 0 & 0 \end{bmatrix} .$$

The above matrix is the Hermite normal form of A.

● **PROBLEM** 8-7

1) Define a column-reduced matrix and give an example.
2) Define column-reduced echelon form and give an example.

Solution: The matrix A is said to be column-reduced if
a) The leading entry of each non-zero column is 1.
b) Every row containing the leading entry of some non-zero column has all its other entries zero.
    A non-zero column of a matrix A means a column of A whose entries are not all zero. By the leading entry of a non-zero column of A, we mean the first non-zero entry of that column.
Example: The following matrices are column-reduced:

$$A = \begin{bmatrix} 1 & 0 & 0 \\ 0 & 1 & 0 \\ 0 & 0 & 0 \\ 0 & 0 & 0 \end{bmatrix} , \quad B = \begin{bmatrix} 1 & 0 & 0 \\ 0 & 1 & 0 \\ 0 & 0 & 0 \\ 0 & 0 & 0 \end{bmatrix} .$$

2) A matrix B is said to be in column-reduced echelon form if it satisfies (a) and (b) above and also the following:

c) Each of its zero columns lies to the right of all its non-zero columns.
d) The leading non-zero entry in any column is in the row that lies above the leading entry in the next column.
Note that a matrix in column-reduced echelon form is the transpose of a matrix in row-reduced echelon form.
Example:

$$B = \begin{bmatrix} 1 & 0 & 0 & 0 & 0 \\ 0 & 1 & 0 & 0 & 0 \\ 0 & 2 & 0 & 0 & 0 \\ 0 & 0 & 1 & 0 & 0 \end{bmatrix}$$

215

Given

$$A = \begin{bmatrix} 0 & 0 & 0 & 0 \\ 0 & 0 & 0 & 0 \\ 5 & 7 & 8 & 0 \\ 10 & 9 & 6 & 0 \\ 0 & 10 & 5 & 0 \end{bmatrix} \quad ,$$

reduce A to column-reduced echelon form.

Solution: To reduce a matrix to a form that specifies the arrangement of its columns, it is necessary to perform column operations. These are merely the column analogs of the basic row operations; i.e.:
1. interchanging the ith and jth column.
2. multiplying the ith column by a scalar k.
3. adding the ith column to the jth column.
   To reduce A to column-reduced echelon form, apply the following column operations. Divide the first column by 5.

$$\begin{bmatrix} 0 & 0 & 0 & 0 \\ 0 & 0 & 0 & 0 \\ 1 & 7 & 8 & 0 \\ 2 & 9 & 6 & 0 \\ 0 & 10 & 5 & 0 \end{bmatrix} .$$

Add -7 times the first column to the second column and -8 times the first column to the third column.

$$\begin{bmatrix} 0 & 0 & 0 & 0 \\ 0 & 0 & 0 & 0 \\ 1 & 0 & 0 & 0 \\ 2 & -5 & -10 & 0 \\ 0 & 10 & 5 & 0 \end{bmatrix} .$$

Divide the second column by -5.

$$\begin{bmatrix} 0 & 0 & 0 & 0 \\ 0 & 0 & 0 & 0 \\ 1 & 0 & 0 & 0 \\ 2 & 1 & -10 & 0 \\ 0 & -2 & 5 & 0 \end{bmatrix} .$$

Add -2 times the second column to the first column and 10 times the second column to the third column.

$$\begin{bmatrix} 0 & 0 & 0 & 0 \\ 0 & 0 & 0 & 0 \\ 1 & 0 & 0 & 0 \\ 0 & 1 & 0 & 0 \\ 4 & -2 & -15 & 0 \end{bmatrix} .$$

Divide the third column by -15.

$$\begin{bmatrix} 0 & 0 & 0 & 0 \\ 0 & 0 & 0 & 0 \\ 1 & 0 & 0 & 0 \\ 0 & 1 & 0 & 0 \\ 4 & -2 & 1 & 0 \end{bmatrix} .$$

Add -4 times the third column to the first column and 2 times the third column to the second column to obtain the column-reduced echelon form.

$$\begin{bmatrix} 0 & 0 & 0 & 0 \\ 0 & 0 & 0 & 0 \\ 1 & 0 & 0 & 0 \\ 0 & 1 & 0 & 0 \\ 0 & 0 & 1 & 0 \end{bmatrix} .$$

THE DIMENSION OF THE SOLUTION SPACE

● **PROBLEM** 8-9

Let U and W be the following subspaces of R^4 .

$U = \{(a,b,c,d) : b + c + d = 0\}$,
$W = \{(a,b,c,d) : a + b = 0 , c = 2d\}$.

Find the dimension and a basis of: (i) U, (ii) W, (iii) $U \cap W$.

Solution: A family $(u_1,...,u_k)$ of vectors in a vector space U is a basis for U if: (a) it is linearly independent and (b) it spans U. A vector space U that has a basis consisting of a finite number of vectors is said to be finite-dimensional. The number of vectors in such a basis is called the dimension of U .

(i) We seek a basis for the set of solutions (a,b,c,d) to the equation
$$b + c + d = 0 \quad \text{or} \quad 0 \cdot a + b + c + d = 0 .$$
Choose the free variables to be a,c, and d. Set (1) a = 1, c = 0, d = 0; (2) a = 0, c = 1, d = 0; (3) a = 0, c = 0, d = 1 to obtain the respective solutions.
$$u_1 = (1,0,0,0), \ u_2 = (0,-1,1,0), \ u_3 = (0,-1,0,1).$$
The set $\{u_1,u_2,u_3\}$ is a basis of U, and dim U = 3. Examine more closely what has been done.

The equation b + c + d = 0 is equivalent to the relationship $0 \cdot a + b + c + d = 0$ between the variables a,b,c, and d. We can choose "a" and two others freely and solve for the remaining variable. Choose b to be our dependent variable. Any vector (a,b = c - d, c,d) lies in the subspace, and any set of three linearly independent vectors of this form will span the subspace. The vectors given by (1,b,0,0), (0,b,1,0) and (0,b,0,1) must be linearly independent and, thus, immediately give us a basis for the subspace when we solve for b.

(ii) We seek a basis for the set of solutions (a,b,c,d) to the system
$$\begin{array}{ll} a + b = 0 & a + b = 0 \\ c = 2d & \text{or} \quad c - 2d = 0 . \end{array}$$

Choose the free variables to be b and d. Set (1) b = 1, d = 0; (2) b = 0, d = 1 to obtain the respective solutions

217

$$u_1 = (-1,1,0,0), \quad u_2 = (0,0,2,1) .$$

The set $\{u_1, u_2\}$ is a basis of W and dim W = 2.

(iii) U $\cap$ W consists of those vectors (a,b,c,d) which satisfy the conditions defining U and the conditions defining W, i.e., the three equations

$$b + c + d = 0 \qquad\qquad a + b = 0$$
$$\text{or}$$
$$a + b = 0 \qquad\qquad b + c + d = 0$$
$$c = 2d \qquad\qquad c - 2d = 0 .$$

Choose the free variable to be d. Set d = 1 to obtain the solution u = (3,-3,2,1). Thus, $\{u\}$ is a basis of U $\cap$ W, and dim(U $\cap$ W) = 1.

● **PROBLEM** 8-10

Find the dimension of the vector space spanned by:

(i)   (1,-2,3,-1)   and   (1,1,-2,3)

(ii)   (3,-6,3,-9)   and   (-2,4,-2,6)

(iii) $t^3 + 2t^2 + 3t + 1$ and $2t^3 + 4t^2 + 6t + 2$

(iv) $t^3 - 2t^2 + 5$ and $t^2 + 3t - 4$

(v) $\begin{bmatrix} 1 & 2 \\ 1 & 2 \end{bmatrix}$ and $\begin{bmatrix} 1 & 1 \\ 2 & 2 \end{bmatrix}$

(vi) $\begin{bmatrix} 1 & 1 \\ -1 & -1 \end{bmatrix}$ $\begin{bmatrix} -3 & -3 \\ 3 & 3 \end{bmatrix}$

(vii)   3   and   -3.

Solution: Two non-zero vectors span a space W of dimension 2 if they are independent, and of dimension 1 if they are dependent. Two vectors are dependent if and only if one is a scalar multiple of the other. Now, using the above facts, the dimension of the given vector space can be found. Hence, the dimensions of the subspaces spanned by the given sets of vectors are, respectively:

(1) 2, (ii) 1, (iii) 1, (iv) 2, (v) 2, (vi) 1, and (vii) 1.

Note that in (i) and (ii) the subspace spanned is a subspace of the vector space R^4 ; in (iii) and (iv) it is a subspace of the real vector space of polynomials in T over the field R; in (v) and (vi) it is a subspace of the real vector space of 2 $\times$ 2 matrices with entries in R; and in (vii) it is just R considered as a real vector space.

● **PROBLEM** 8-11

Let V be the real vector space of 2 by 2 symmetric matrices with entries in R. Show that dim V = 3.

Solution: Recall that A = (a_{ij}) is symmetric if A = A^t or, equivalently, $a_{ij} = a_{ji}$. Let A = $\begin{bmatrix} a & b \\ b & c \end{bmatrix}$ be an arbitrary 2 by 2 symmetric matrix where a,b,c $\in$ R. Set (i) a = 1, b = 0, c = 0; (ii) a = 0, b = 1, c = 0; (iii) a = 0, b = 0, c = 1. Then, the respective matrices are:

$$E_1 = \begin{bmatrix} 1 & 0 \\ 0 & 0 \end{bmatrix}, \quad E_2 = \begin{bmatrix} 0 & 1 \\ 1 & 0 \end{bmatrix}, \quad E_3 = \begin{bmatrix} 0 & 0 \\ 0 & 1 \end{bmatrix}.$$

We shall show that $\{E_1, E_2, E_3\}$ is a basis of V, i.e., it (1) spans V and (2) is a linearly independent set.

For the above arbitrary matrix A in V, we have,

$$A = \begin{bmatrix} a & b \\ b & c \end{bmatrix} = aE_1 + bE_2 + cE_3.$$

Thus, $\{E_1, E_2, E_3\}$ generates V, i.e., spans V. (2) By definition, the vectors $v_1, v_2, \ldots, v_m$ are linearly independent if there exist scalars $a_1, a_2, \ldots, a_m \in R$, all of them zero, such that $a_1 v_1 + a_2 v_2 + \ldots + a_m v_m = 0$. If v_1, v_2 and v_3 are linearly independent, the set $\{v_1, v_2, v_3\}$ is called linearly independent. Let $a_1 E_1 + a_2 E_2 + a_3 E_3 = 0$ where a_1, a_2, a_3 are unknown scalars. Note that 0 is the 0 element of V, i.e., the 0 matrix. Then,

$$a_1 \begin{bmatrix} 1 & 0 \\ 0 & 0 \end{bmatrix} + a_2 \begin{bmatrix} 0 & 1 \\ 1 & 0 \end{bmatrix} + a_3 \begin{bmatrix} 0 & 0 \\ 0 & 1 \end{bmatrix} = \begin{bmatrix} 0 & 0 \\ 0 & 0 \end{bmatrix}$$

or,

$$\begin{bmatrix} a_1 & a_2 \\ a_2 & a_3 \end{bmatrix} = \begin{bmatrix} 0 & 0 \\ 0 & 0 \end{bmatrix}.$$

Thus, the result is

$$a_1 = 0, \ a_2 = 0, \ a_3 = 0.$$

Therefore, $\{E_1, E_2, E_3\}$ is independent. Thus, $\{E_1, E_2, E_3\}$ is a basis of V and so the dimension of V is 3.

HOMOGENEOUS SYSTEMS

● **PROBLEM** 8-12

Solve the following system of equations by forming the matrix of coefficients and reducing it to echelon form.

$$\begin{aligned} 3x + 2y - z &= 0 \\ x - y + 2z &= 0 \\ x + y - 6z &= 0 \end{aligned} \tag{1}$$

Solution: The most general linear system of m equations in n unknowns (variables) $x_1, x_2, \ldots, x_n$ is of the form

$$\begin{aligned} a_{11}x_1 + a_{12}x_2 + \ldots + a_{1n}x_n &= c_1 \\ a_{21}x_1 + a_{22}x_2 + \ldots + a_{2n}x_n &= c_2 \\ & \cdots \cdots \cdots \cdots \cdots \cdots \\ a_{m1}x_1 + a_{m2}x_2 + \ldots + a_{mn}x_n &= c_m, \end{aligned} \tag{2}$$

where $a_{11}, a_{12}, \ldots, a_{1n}, \ a_{21}, a_{22}, \ldots, a_{2n}, \ a_{m1}, a_{m2}, \ldots, a_{mn}$ are called the coefficients of the system (2), and $c_1, c_2, \ldots, c_m$ are called the constants of the system (2).

If $c_1 = c_2 = \ldots = c_m = 0$, a system of the form (2) is called a homogeneous system. Let $A = (a_{ij})$ be a matrix of size $m \times n$ where the entries a_{ij} are the same as the coefficients of the system given by (2). Then we say that A is the matrix of coefficients of (2). In order to solve the homogeneous system

$$a_{11}x_1 + a_{12}x_2 + \ldots + a_{1n}x_n = 0$$

$$a_{21}x_1 + a_{22}x_2 + \ldots + a_{2n}x_n = 0 \qquad\qquad (3)$$

$$\bullet$$
$$\bullet$$
$$\bullet$$

$$a_{m1}x_1 + a_{m2}x_2 + \ldots + a_{mn}x_n = 0 \quad,$$

first form the matrix of coefficients $A = (a_{ij})$. Next, obtain a row-echelon matrix $R = (r_{ij})$ by the method of elementary row operations on the matrix A. Since the solutions to (3) are the same as the solutions to

$$r_{11}x_1 + r_{12}x_2 + \ldots + r_{1n}x_n = 0$$

$$r_{21}x_1 + r_{22}x_2 + \ldots + r_{2n}x_n = 0$$

$$\bullet$$
$$\bullet \qquad\qquad (4)$$
$$\bullet$$

$$r_{m1}x_1 + r_{m2}x_2 + \ldots + r_{mn}x_n = 0$$

where the coefficients r_{ij} are identically the entries of R (i.e., R is the coefficient matrix of the homogeneous system (4)) we solve the original system (3) by solving the reduced system (4). Now solve the given homogeneous system (1).

The matrix of coefficients of the system (1) is

$$\begin{bmatrix} 3 & 2 & -1 \\ 1 & -1 & 2 \\ 1 & 1 & -6 \end{bmatrix} \quad .$$

Reduce the above matrix to echelon form. Add -3 times the second row to the first row, and add -1 times the second row to the third row:

$$\begin{bmatrix} 0 & 5 & -7 \\ 1 & -1 & 2 \\ 0 & 2 & -8 \end{bmatrix}$$

Divide the third row by 2.

$$\begin{bmatrix} 0 & 5 & -7 \\ 1 & -1 & 2 \\ 0 & 1 & -4 \end{bmatrix}$$

Add -5 times the third row to the first row, and add the third row to the second row.

$$\begin{bmatrix} 0 & 0 & 13 \\ 1 & 0 & -2 \\ 0 & 1 & -4 \end{bmatrix}$$

Divide row one by 13; then add 2 times the resulting row one to row two. Next, add 4 times the resulting row one to row three:

$$\begin{bmatrix} 0 & 0 & 1 \\ 1 & 0 & 0 \\ 0 & 1 & 0 \end{bmatrix}$$

Interchange rows one and two, then rows two and three:

$$\begin{bmatrix} 1 & 0 & 0 \\ 0 & 1 & 0 \\ 0 & 0 & 1 \end{bmatrix}$$

This matrix is reduced and gives the system

$$x = 0$$
$$y = 0$$
$$z = 0 \quad .$$

Thus, the unique solution to the original system is $x = y = z = 0$, the solution that is called the trivial solution.

● PROBLEM 8-13

For the system
$$2x - y + z = 0$$
$$-7x + 7/2\, y - 7/2\, z = 0$$
$$4x + y - 2z = 0$$

form the matrix of coefficients. Find the form of the solutions to the system by reducing the coefficient matrix to a reduced matrix.

Solution: The coefficient matrix is

$$\begin{bmatrix} 2 & -1 & 1 \\ -7 & 7/2 & -7/2 \\ 4 & 1 & -2 \end{bmatrix} \quad .$$

Add 7/2 times the first row to the second row and add -2 times the first row to the third row.

$$\begin{bmatrix} 2 & -1 & 1 \\ 0 & 0 & 0 \\ 0 & 3 & -4 \end{bmatrix} \qquad (2)$$

Add 1/3 times the third row to the first row.

$$\begin{bmatrix} 2 & 0 & -1/3 \\ 0 & 0 & 0 \\ 0 & 3 & -4 \end{bmatrix}$$

Note the row of zeros which indicates that the second equation has been entirely eliminated, being superfluous. Divide the first row by 2 and the third row by 3. Then interchange the second row and the third row.

$$\begin{bmatrix} 1 & 0 & -1/6 \\ 0 & 1 & -4/3 \\ 0 & 0 & -0 \end{bmatrix}$$

This matrix is reduced. It is the coefficient matrix (after dropping the third row) for the system

$$x - 1/6\ z = 0$$
$$y - 4/3\ z = 0\ .\qquad\qquad (1)$$

The dependence of the solutions on z can be expressed as follows:

$$x = 1/6\ z$$
$$y = 4/3\ z$$
$$z = z\ .$$

To obtain another form for the solution, reduce the matrix (2) in a different manner. Add $1/4$ times the third row of matrix (2) to the first row, obtaining

$$\begin{bmatrix} 2 & -1/4 & 0 \\ 0 & 0 & 0 \\ 0 & 3 & -4 \end{bmatrix}\ .$$

Divide the first row by 2 and the third row by 3, and interchange the second and third rows

$$\begin{bmatrix} 1 & -1/8 & 0 \\ 0 & 1 & -4/3 \\ 0 & 0 & 0 \end{bmatrix}\ .$$

This gives

$$x - 1/8\ y = 0$$
$$y - 4/3\ z = 0 \qquad\qquad (3)$$

The dependence of the solutions on y can be expressed by rewriting system (3) as

$$x = 1/8\ y$$
$$y = y$$
$$z = 3/4\ y\ .$$

● **PROBLEM** 8-14

Let $A = \begin{bmatrix} 2 & 1 & 4 \\ 3 & 0 & 1 \\ 2 & -1 & 1 \end{bmatrix}$ be the coefficient matrix of a homogeneous

system in x,y, and z. Solve this system to illustrate that a homogeneous system of 3 equations in the unknowns, x,y,z has a unique solution.

Solution: The method of solving a homogeneous system is reduction of the coefficient matrix to echelon form. To do this, perform the following row operations on matrix A
Divide the first row by 2

$$\begin{bmatrix} 1 & 1/2 & 2 \\ 3 & 0 & 1 \\ 2 & -1 & 1 \end{bmatrix}\ .$$

Add -3 times the first row to the second row and -2 times the first row to the third row

$$\begin{bmatrix} 1 & 1/2 & 2 \\ 0 & -3/2 & -5 \\ 0 & -2 & -3 \end{bmatrix} \quad .$$

Add $-4/3$ times the second row to the third row,

$$\begin{bmatrix} 1 & 1/2 & 2 \\ 0 & -3/2 & -5 \\ 0 & 0 & 11/3 \end{bmatrix} \quad .$$

Multiply the third row by $3/11$. Next, add -2 times the third row to the first row and 5 times the third row to the second row.

$$\begin{bmatrix} 1 & 1/2 & 0 \\ 0 & -3/2 & 0 \\ 0 & 0 & 1 \end{bmatrix} \quad .$$

Multiply the second row by $-2/3$. Finally, add $-1/2$ times the second row to the first row.

$$\begin{bmatrix} 1 & 0 & 0 \\ 0 & 1 & 0 \\ 0 & 0 & 1 \end{bmatrix} \quad .$$

The above matrix is an identity matrix and the corresponding system of equations is

$$x = 0$$
$$y = 0$$
$$z = 0 \quad .$$

Thus, the system has a unique solution and it is the trivial solution. In general, a homogeneous system of n equations in n unknowns has a unique solution if and only if the coefficient matrix can be reduced to the n by n identity matrix. In this case, the solution is the trivial solution, and the dimension of the solution space is zero. We can look at this in the following way:

Since the coefficient matrix A can be reduced by row operations to the identity matrix, the row vectors of A are linearly independent. Therefore, there cannot exist non-zero real numbers x_1, x_2, x_3 such that

$$a_{j1}x_1 + a_{j2}x_2 + a_{j3}x_3 = 0$$

$$= a_{k1}x_1 + a_{k2}x_2 + a_{k3}x_3 \quad ,$$

where $j \neq k$, $j,k = 1,2,3$; i.e., there is no non-trivial solution.

● **PROBLEM** 8-15

Solve the following system of equations:

$$x + 3y = 0$$
$$2x + 6y + 4z = 0 \tag{1}$$

Solution: To solve the given system of equations, first form the matrix of the coefficients. Then reduce this matrix to echelon form. The matrix of the coefficients is

$$\begin{bmatrix} 1 & 3 & 0 \\ 2 & 6 & 4 \end{bmatrix} \quad .$$

Now add -2 times the first row to the second row

$$\begin{bmatrix} 1 & 3 & 0 \\ 0 & 0 & 4 \end{bmatrix} .$$

Divide the second row by 4

$$\begin{bmatrix} 1 & 3 & 0 \\ 0 & 0 & 1 \end{bmatrix} .$$

The above is the matrix of coefficients for

$$x + 3y = 0$$
$$z = 0 . \qquad (2)$$

This system is easy to solve. We have $z = 0$ and can assign y any value. Then compute x from (2). This gives a solution to (1).

● **PROBLEM** 8-16

Solve the following homogeneous system of linear equations.

$$2x_1 + 2x_2 - x_3 + x_5 = 0$$
$$-x_1 - x_2 + 2x_3 - 3x_4 + x_5 = 0$$
$$x_1 + x_3 - 2x_3 - x_5 = 0 \qquad (1)$$
$$x_3 + x_4 + x_5 = 0$$

Solution: The system (1) has five unknowns but only four equations. We know that a homogeneous system of linear equations with more unknowns than equations has a non-zero (non-trivial) solution. Now, to solve the system (1), form the matrix of the coefficients. Then reduce this matrix to reduced row-echelon form. The coefficient matrix is

$$A = \begin{bmatrix} 2 & 2 & -1 & 0 & 1 \\ -1 & -1 & 2 & -3 & 1 \\ 1 & 1 & -2 & 0 & -1 \\ 0 & 0 & 1 & 1 & 1 \end{bmatrix} . \qquad (1)$$

Add the fourth row to the first row and the third row to the second row.

$$\begin{bmatrix} 2 & 2 & 0 & 1 & 2 \\ 0 & 0 & 0 & -3 & 0 \\ 1 & 1 & -2 & 0 & -1 \\ 0 & 0 & 1 & 1 & 1 \end{bmatrix} . \qquad (2)$$

Divide the second row by -3. Then add -1 times the second row to the first row and to the fourth row.

$$\begin{bmatrix} 2 & 2 & 0 & 0 & 2 \\ 0 & 0 & 0 & 1 & 0 \\ 1 & 1 & -2 & 0 & -1 \\ 0 & 0 & 1 & 0 & 1 \end{bmatrix} \qquad (3)$$

Divide the first row by 2. Then add -1 times the first row to the third row.

$$\begin{bmatrix} 1 & 1 & 0 & 0 & 1 \\ 0 & 0 & 0 & 1 & 0 \\ 0 & 0 & -2 & 0 & -2 \\ 0 & 0 & 1 & 0 & 1 \end{bmatrix} . \qquad (4)$$

Add 2 times the fourth row to the third row

$$\begin{bmatrix} 1 & 1 & 0 & 0 & 1 \\ 0 & 0 & 0 & 1 & 0 \\ 0 & 0 & 0 & 0 & 0 \\ 0 & 0 & 1 & 0 & 1 \end{bmatrix}$$ (5)

Interchange the second and fourth rows. Next, interchange the third and fourth rows.

$$\begin{bmatrix} 1 & 1 & 0 & 0 & 1 \\ 0 & 0 & 1 & 0 & 1 \\ 0 & 0 & 0 & 1 & 0 \\ 0 & 0 & 0 & 0 & 0 \end{bmatrix}$$ (6)

This matrix is in row reduced echelon form. The corresponding system of equations is

$$x_1 + x_2 \qquad + x_5 = 0$$
$$x_3 \qquad + x_5 = 0$$
$$x_4 \qquad = 0 \quad .$$

Solving for the leading variables yields

$$x_1 = -x_2 - x_5$$
$$x_3 = -x_5$$
$$x_4 = 0 \ .$$

The solution set is, therefore, given by

$$x_1 = -s - t, \ x_2 = s, \ x_3 = -t, \ x_4 = 0, \ x_5 = t.$$

That is, we have chosen x_2 and x_5 to be free variables and, hence, set them equal to the parameters s and t, respectively. The dependent variables are then x_1 and x_3 , and their dependence on s and t are given by the reduced form of the system. Thus, any solution vector is of the form $(-s-t, \ s, \ -t, \ 0, \ t)$. Recall that a basis for the subspace of vectors of this form (i.e., the solution space) would be the vectors

(1) $(-0-1,0,-1,0,1) = (-1,0,-1,0,1)$

(2) $(-1-0,1,-0,0,0) = (-1,1,0,0,0,)$

● **PROBLEM** 8-17

[A] Show that each of the following systems has a non-zero solution:

(a) $\begin{aligned} x + 2y - 3z + w &= 0 \\ x - 3y + z - 2w &= 0 \\ 2x + y - 3z + 5w &= 0 \end{aligned}$ (1)

(b) $\begin{aligned} x + y - z &= 0 \\ 2x - 3y + z &= 0 \\ x - 4y + 2z &= 0 \end{aligned}$ (2)

[B] Show that following system has a unique solution:

  $\begin{aligned} x + y - z &= 0 \\ 2x + 4y - z &= 0 \\ 3x + 2y + 2z &= 0 \end{aligned}$ (3)

Solution: [A] For a homogeneous system of linear equations, exactly one of the following is true.
(i) The system has only the trivial solution.
(ii) The system has infinitely many non-trivial solutions.

Any homogeneous linear system that has fewer linearly independent equations than unknowns always has a non-zero solution.

(a) Now, system (1) has a non-zero solution since there are four unknowns but only three equations.

(b) The coefficient matrix for the system (2) is

$$\begin{bmatrix} 1 & 1 & -1 \\ 2 & -3 & 1 \\ 1 & -4 & 2 \end{bmatrix}.$$

Add -2 times the first row to the second row and -1 times the first row to the third row

$$\begin{bmatrix} 1 & 1 & -1 \\ 0 & -5 & 3 \\ 0 & -5 & 3 \end{bmatrix}.$$

Add -1 times the second row to the third row

$$\begin{bmatrix} 1 & 1 & -1 \\ 0 & -5 & 3 \\ 0 & 0 & 0 \end{bmatrix}.$$

This is the matrix of coefficients of the system

$$\begin{aligned} x + y - z &= 0 \\ - 5y + 3z &= 0 \\ 0 &= 0 \,. \end{aligned}$$

The system has a non-zero solution since we obtained only two equations in the three unknowns when we reduced the system to echelon form. For example, let $z = 5$; then $y = 3$ and $x = 2$ solves the system.

[B] The matrix of coefficients for system (3) is

$$\begin{bmatrix} 1 & 1 & -1 \\ 2 & 4 & -1 \\ 3 & 2 & 2 \end{bmatrix}.$$

Reduce it to echelon form. Add -2 times the first row to the second row and -3 times the first row to the third row

$$\begin{bmatrix} 1 & 1 & -1 \\ 0 & 2 & 1 \\ 0 & -1 & 5 \end{bmatrix}.$$

Multiply the third row by 2. Then, add the second row to the third row

$$\begin{bmatrix} 1 & 1 & -1 \\ 0 & 2 & 1 \\ 0 & 0 & 11 \end{bmatrix}.$$

The corresponding system of equations is

$$\begin{aligned} x + y - z &= 0 \\ 2y + z &= 0 \\ 11z &= 0 \,. \end{aligned}$$

Solving this system yields $x = y = z = 0$. In general, a system of homogeneous equations in n unknowns has a zero solution if the corresponding reduced matrix has exactly n rows with non-zero entries.

NON-HOMOGENEOUS SYSTEMS

● PROBLEM 8-18

By forming the augmented matrix and row reducing, determine the solutions of the following system

$$
\begin{aligned}
2x &- y + 3z = 4 \\
3x & + 2z = 5 \\
-2x &+ y + 4z = 6 .
\end{aligned}
\qquad (1)
$$

Solution: The system of equations

$$a_{11}x_1 + a_{12}x_2 + \dots + a_{1n}x_n = c_1$$

$$a_{21}x_1 + a_{22}x_2 + \dots + a_{2n}x_n = c_2$$

$$\vdots$$

$$\qquad\qquad\qquad\qquad\qquad\qquad\qquad (2)$$

$$a_{m1}x_1 + a_{m2}x_2 + \dots + a_{mn}x_n = c_m$$

is called a non-homogeneous linear system if the constants $c_1, c_2, \dots, c_m$ are not all zero.

We form the $m \times (n+1)$ matrix A' defined by

$$
A' = \begin{bmatrix}
a_{11} & a_{12} & \cdots & a_{1n} & c_1 \\
a_{21} & a_{22} & \cdots & a_{2n} & c_2 \\
\vdots & & & & \vdots \\
a_{m1} & a_{m2} & \cdots & a_{mn} & c_m
\end{bmatrix} .
$$

This matrix is called the augmented matrix of the system (2). The first n columns of A' consist of the coefficient matrix of (2), and the last column of A' consists of the corresponding constants.

To solve the non-homogeneous linear system, form the augmented matrix A'. Apply row operations to A' to reduce it to echelon form. Now, the augmented matrix of the system (1) is

$$
\begin{bmatrix}
2 & -1 & 3 & 4 \\
3 & 0 & 2 & 5 \\
-2 & 1 & 4 & 6
\end{bmatrix} .
$$

Add the first row to the third row

$$
\begin{bmatrix}
2 & -1 & 3 & 4 \\
3 & 0 & 2 & 5 \\
0 & 0 & 7 & 10
\end{bmatrix} .
$$

This is the augmented matrix of

$$
\begin{aligned}
2x &- y + 3z = 4 \\
3x & + 2z = 5 \\
&7z = 10 .
\end{aligned}
$$

227

The system has been sufficiently simplified now so that the solution can be found.

From the last equation we have z = 10/7. Substituting this value into the second equation and solving for x gives x = 5/7. Substituting x = 5/7 and z = 10/7 into the first equation and solving for y yields y = 12/7 . The solution to system (1) is, therefore,

$$x = 5/7 \ , \ y = 12/7 \ , \ z = 10/7 \ .$$

Note: We could have further reduced the matrix to row-reduced echelon form and solved the system directly from the reduced matrix. That is, by adding -2/3 times the second row to the first row, we have

$$\begin{bmatrix} 0 & -1 & 5/3 & \vdots & 2/3 \\ 3 & 0 & 2 & \vdots & 5 \\ 0 & 0 & 7 & \vdots & 10 \end{bmatrix} \ .$$

Multiplying the second row by 1/3 and the first row by -1 and interchanging the two, then multiplying the third row by 1/7 results in

$$\begin{bmatrix} 1 & 0 & 2/3 & 5/3 \\ 0 & 1 & -5/3 & -2/3 \\ 0 & 0 & 1 & 10/7 \end{bmatrix} \ .$$

Then, adding 5/3 times the third row to the second and -2/3 times the third row to the first gives

$$\begin{bmatrix} 1 & 0 & 0 & 5/7 \\ 0 & 1 & 0 & 12/7 \\ 0 & 0 & 1 & 10/7 \end{bmatrix} \ .$$

The solution to a non-homogeneous system found in the above manner is called the particular solution. The non-homogeneous system will be satisfied by any sum of the particular solution and a solution to the corresponding homogeneous system. In this case, the only solution to the homogeneous system is the trivial solution. Therefore, the only solution to the non-homogeneous problem is the particular solution.

● **PROBLEM** 8- 19

Solve the following linear system of equations:

$$\begin{array}{rrrrr} 2x & + & 3y & - & 4z & = & 5 \\ -2x & & & + & z & = & 7 \\ 3x & + & 2y & + & 2z & = & 3 \end{array} \qquad (1)$$

Solution: The matrix of coefficients is

$$\begin{bmatrix} 2 & 3 & -4 \\ -2 & 0 & 1 \\ 3 & 2 & 2 \end{bmatrix} \ .$$

Let

$$X = \begin{bmatrix} x \\ y \\ z \end{bmatrix} \quad \text{and} \quad B = \begin{bmatrix} 5 \\ 7 \\ 3 \end{bmatrix} \ .$$

The given linear system can be written in matrix form as

$$AX = B \ .$$

Since the system (1) is non-homogeneous, the solutions to the system are completely determined by its augmented matrix.

228

The augmented matrix for the system (1) is

$$\begin{bmatrix} 2 & 3 & -4 & \vdots & 5 \\ -2 & 0 & 1 & \vdots & 7 \\ 3 & 2 & 2 & \vdots & 3 \end{bmatrix}$$

which can be reduced by using the following sequence of row operations:
Add the first row to the second row

$$\begin{bmatrix} 2 & 3 & -4 & \vdots & 5 \\ 0 & 3 & -3 & \vdots & 12 \\ 3 & 2 & 2 & \vdots & 3 \end{bmatrix} .$$

Divide the first row by 2 and the second row by 3

$$\begin{bmatrix} 1 & 3/2 & -2 & \vdots & 5/2 \\ 0 & 1 & -1 & \vdots & 4 \\ 3 & 2 & 2 & \vdots & 3 \end{bmatrix} .$$

Add -3 times the first row to the third row

$$\begin{bmatrix} 1 & 3/2 & -2 & \vdots & 5/2 \\ 0 & 1 & -1 & \vdots & 4 \\ 0 & -5/2 & 8 & \vdots & -9/2 \end{bmatrix} .$$

Add 5/2 times the second row to the third row

$$\begin{bmatrix} 1 & 3/2 & -2 & \vdots & 5/2 \\ 0 & 1 & -1 & \vdots & 4 \\ 0 & 0 & 11/2 & \vdots & 11/2 \end{bmatrix} .$$

This is the augmented matrix for the system

$$\begin{aligned} x + 3/2\,y - 2z &= 5/2 \\ y - z &= 4 \\ 11/2\,z &= 11/2 . \end{aligned}$$

Now the solution to this system can be easily found. From the last equation we have $z = 1$. Substituting $z = 1$ in the second equation gives $y = 5$. Next, substitute $y = 5$ and $z = 1$ into the first equation. This gives $x = -3$. Therefore, the solution to system (1) is $x = -3$, $y = 5$, $z = 1$.

This is the particular solution. A solution to the corresponding homogeneous problem plus the particular solution also solves the inhomogeneous system and is known as a general solution.

● **PROBLEM** 8-20

Solve the following system

$$\begin{aligned} x + y + 2z &= 9 \\ 2x + 4y - 3z &= 1 \\ 3x + 6y - 5z &= 0 \end{aligned} \qquad (1)$$

Solution: The augmented matrix for the system (1) is

$$\begin{bmatrix} 1 & 1 & 2 & \vdots & 9 \\ 2 & 4 & -3 & \vdots & 1 \\ 3 & 6 & -5 & \vdots & 0 \end{bmatrix} .$$

It can be reduced to echelon form by elementary row operations. Add -2 times the first row to the second row and -3 times the first row to the third row

$$\begin{bmatrix} 1 & 1 & 2 & | & 9 \\ 0 & 2 & -7 & | & -17 \\ 0 & 3 & -11 & | & -27 \end{bmatrix} .$$

Multiply the second row by 1/2

$$\begin{bmatrix} 1 & 1 & 2 & | & 9 \\ 0 & 1 & -7/2 & | & -17/2 \\ 0 & 3 & -11 & | & -27 \end{bmatrix} .$$

Add -3 times the second row to the third row

$$\begin{bmatrix} 1 & 1 & 2 & | & 9 \\ 0 & 1 & -7/2 & | & -17/2 \\ 0 & 0 & -1/2 & | & -3/2 \end{bmatrix} .$$

Multiply the third row by -2 to obtain

$$\begin{bmatrix} 1 & 1 & 2 & | & 9 \\ 0 & 1 & -7/2 & | & -17/2 \\ 0 & 0 & 1 & | & 3 \end{bmatrix} .$$

This is the augmented matrix for the system

$$\begin{aligned} x + y + 2z &= 9 \\ y - 7/2\, z &= -17/2 \\ z &= 3 \end{aligned} .$$

Solving this system gives $x = 1$, $y = 2$, $z = 3$.

● **PROBLEM** 8-21

Solve the following system

$$\begin{aligned} 2x + y - 2z &= 10 \\ 3x + 2y + 2z &= 1 \\ 5x + 4y + 3z &= 4 \end{aligned} \qquad (1)$$

Solution: The augmented matrix is

$$\begin{bmatrix} 2 & 1 & -2 & | & 10 \\ 3 & 2 & 2 & | & 1 \\ 5 & 4 & 3 & | & 4 \end{bmatrix} .$$

Reduce this matrix to echelon form. Add -1 times the second row to the first row

$$\begin{bmatrix} 1 & -1 & -4 & | & 9 \\ 3 & 2 & 2 & | & 1 \\ 5 & 4 & 3 & | & 4 \end{bmatrix} .$$

Add 3 times the first row to the second row and 5 times the first row to the third row.

$$\begin{bmatrix} -1 & -1 & -4 & | & 9 \\ 0 & -1 & -10 & | & 28 \\ 0 & -1 & -17 & | & 49 \end{bmatrix} .$$

Add -1 times the second row to the third row.

$$\begin{bmatrix} -1 & -1 & -4 & \vdots & 9 \\ 0 & -1 & -10 & \vdots & 28 \\ 0 & 0 & -7 & \vdots & 21 \end{bmatrix} .$$

This augmented matrix can be represented in equation form as

$$\begin{aligned} -x - y - 4z &= 9 \\ - y - 10z &= 28 \\ - 7z &= 21 . \end{aligned}$$

Solving this system yields $z = -3$, $y = 2$, $x = 1$. Thus, the solution of the system (1) is $x = 1$, $y = 2$ and $z = 3$.

● **PROBLEM** 8-22

Show that the following non-homogeneous system of linear equations has no solution

$$\begin{aligned} x + 2y - 3z &= -1 \\ 3x - y + 2z &= 7 \\ 5x + 3y - 4z &= 2 \end{aligned} \qquad (1)$$

Solution: The system (1) has no solution if the echelon form of the augmented matrix has a row whose first non-zero entry k appears in the last column. This corresponds in equation form to the statement $0 = k$, which shows that all the equations of the system cannot be satisfied simultaneously. The augmented matrix for the system (1) is

$$\begin{bmatrix} 1 & 2 & -3 & \vdots & -1 \\ 3 & -1 & 2 & \vdots & 7 \\ 5 & 3 & -4 & \vdots & 2 \end{bmatrix}$$

Now, apply row operations to this matrix to reduce it to echelon form. Add -3 times the first row to the second row and -5 times the first row to the third row

$$\begin{bmatrix} 1 & 2 & -3 & \vdots & -1 \\ 0 & -7 & 11 & \vdots & 10 \\ 0 & -7 & 11 & \vdots & 7 \end{bmatrix} .$$

Add -1 times the second row to the third row.

$$\begin{bmatrix} 1 & 2 & -3 & \vdots & -1 \\ 0 & -7 & 11 & \vdots & 10 \\ 0 & 0 & 0 & \vdots & -3 \end{bmatrix} .$$

Thus, this matrix has a row in which the first non-zero entry appears in the last column. Therefore, the system (1) has no solution.

EXISTENCE OF SOLUTIONS

● **PROBLEM** 8-23

Solve the following system by Gauss-Jordan elimination

$$\begin{aligned} x_1 + 3x_2 - 2x_3 + 2x_5 &= 0 \\ 2x_1 + 6x_2 - 5x_3 - 2x_4 + 4x_5 - 3x_6 &= -1 \\ 5x_3 + 10x_4 + 15x_6 &= 5 \\ 2x_1 + 6x_2 + 8x_4 + 4x_5 + 18x_6 &= 6 . \end{aligned}$$

<u>Solution:</u> The augmented matrix for the system is

$$\left[\begin{array}{cccccc|c} 1 & 3 & -2 & 0 & 2 & 0 & 0 \\ 2 & 6 & -5 & -2 & 4 & -3 & -1 \\ 0 & 0 & 5 & 10 & 0 & 15 & 5 \\ 2 & 6 & 0 & 8 & 4 & 18 & 6 \end{array}\right].$$

Reduce this matrix to row-reduced echelon form. Add -2 times the first row to the second and fourth rows

$$\left[\begin{array}{cccccc|c} 1 & 3 & -2 & 0 & 2 & 0 & 0 \\ 0 & 0 & -1 & -2 & 0 & -3 & -1 \\ 0 & 0 & 5 & 10 & 0 & 15 & 5 \\ 0 & 0 & 4 & 8 & 0 & 18 & 6 \end{array}\right].$$

Add 5 times the second row to the third row and 4 times the second row to the fourth row.

$$\left[\begin{array}{cccccc|c} 1 & 3 & -2 & 0 & 2 & 0 & 0 \\ 0 & 0 & -1 & -2 & 0 & -3 & -1 \\ 0 & 0 & 0 & 0 & 0 & 0 & 0 \\ 0 & 0 & 0 & 0 & 0 & 6 & 2 \end{array}\right].$$

Multiply the second row by -1 and the fourth row by 1/6. Then, interchange the third and fourth rows.

$$\left[\begin{array}{cccccc|c} 1 & 3 & -2 & 0 & 2 & 0 & 0 \\ 0 & 0 & 1 & 2 & 0 & 3 & 1 \\ 0 & 0 & 0 & 0 & 0 & 1 & 1/3 \\ 0 & 0 & 0 & 0 & 0 & 0 & 0 \end{array}\right].$$

Add -3 times the third row to the second row. Then add 2 times the second row to the first row.

$$\left[\begin{array}{cccccc|c} 1 & 3 & 0 & 4 & 2 & 0 & 0 \\ 0 & 0 & 1 & 2 & 0 & 0 & 0 \\ 0 & 0 & 0 & 0 & 0 & 1 & 1/3 \\ 0 & 0 & 0 & 0 & 0 & 0 & 0 \end{array}\right].$$

Now the corresponding system of equations is

$$x_1 + 3x_2 + 4x_4 + 2x_5 = 0$$
$$x_3 + 2x_4 = 0$$
$$x_6 = 1/3 .$$

Then, solving for the leading variables results in

$$x_1 = -3x_2 - 4x_4 - 2x_5$$
$$x_3 = -2x_4$$
$$x_6 = 1/3 .$$

If we assign x_2, x_4 and x_5 the arbitrary values r,s, and t, respectively, the solution set is given by the formulas,

$$x_1 = -3r - 4s - 2t, \ x_2 = r, \ x_3 = -2s, \ x_4 = s$$
$$x_5 = t, \ x_6 = 1/3 .$$

For the following system, find the augmented matrix; then, by reducing, determine whether the system has a solution.

$$\begin{aligned} 3x - y + z &= 1 \\ 7x + y - z &= 6 \\ 2x + y - z &= 2 \end{aligned} \tag{1}$$

Solution: The augmented matrix is the matrix of coefficients with an additional column which corresponds to the right hand side of the equalities in the system of equations.

Reduction of the augmented matrix corresponds to reducing the system to a simpler equivalent form. If reduction of the matrix shows a contradiction the system has no solution.

The augmented matrix for the system is

$$\left[\begin{array}{ccc|c} 3 & -1 & 1 & 1 \\ 7 & 1 & -1 & 6 \\ 2 & 1 & -1 & 2 \end{array} \right] .$$

This can be reduced by performing the following row operations. Divide the first row by 3

$$\left[\begin{array}{ccc|c} 1 & -1/3 & 1/3 & 1/3 \\ 7 & 1 & -1 & 6 \\ 2 & 1 & -1 & 2 \end{array} \right] .$$

Now add -7 times the first row to the second row and -2 times the first row to the third row

$$\left[\begin{array}{ccc|c} 1 & -1/3 & 1/3 & 1/3 \\ 0 & 10/3 & -10/3 & 11/3 \\ 0 & 5/3 & -5/3 & 4/3 \end{array} \right] .$$

Divide the second row by $10/3$, and add $-5/3$ times the second row to the third row

$$\left[\begin{array}{ccc|c} 1 & -1/3 & 1/3 & 1/3 \\ 0 & 1 & -1 & 11/10 \\ 0 & 0 & 0 & -1/2 \end{array} \right] .$$

This is the augmented matrix of the system

$$\begin{aligned} x - 1/3\, y + 1/3\, z &= 1/3 \\ y - z &= 11/10 \\ 0 &= -1/2 . \end{aligned} \tag{2}$$

The last equation cannot hold for any choice of x, y, and z. Thus, system (2) has no solution. Therefore, the system (1) has no solution. Observe that a homogeneous system always has at least one solution (namely, the one in which all the variables are equal to zero) while a non-homogeneous system may have no solution.

● **PROBLEM** 8-25

Show that the following system has more than one solution.

$$\begin{aligned} 3x - y + 7z &= 0 \\ 2x - y + 4z &= \tfrac{1}{2} \\ x - y + z &= 1 \\ 6x - 4y + 10z &= 3 . \end{aligned} \tag{1}$$

Solution: The augmented matrix for the system (1) is

$$\begin{bmatrix} 3 & -1 & 7 & \vdots & 0 \\ 2 & -1 & 4 & \vdots & \frac{1}{2} \\ 1 & -1 & 1 & \vdots & 1 \\ 6 & -4 & 10 & \vdots & 3 \end{bmatrix}.$$

Reduce it to row reduced echelon form. Add -3 times the third row to the first row, add -2 times the third row to the second row and add -6 times the third row to the fourth row

$$\begin{bmatrix} 0 & 2 & 4 & \vdots & -3 \\ 0 & 1 & 2 & \vdots & -3/2 \\ 1 & -1 & 1 & \vdots & 1 \\ 0 & 2 & 4 & \vdots & -3 \end{bmatrix}.$$

Add -1 times the first row to the fourth row; add the second row to the third row

$$\begin{bmatrix} 0 & 2 & 4 & \vdots & -3 \\ 0 & 1 & 2 & \vdots & -3/2 \\ 1 & 0 & 3 & \vdots & -1/2 \\ 0 & 0 & 0 & \vdots & 0 \end{bmatrix}.$$

Add -2 times the second row to the first row

$$\begin{bmatrix} 0 & 0 & 0 & \vdots & 0 \\ 0 & 1 & 2 & \vdots & -3/2 \\ 1 & 0 & 3 & \vdots & -1/2 \\ 0 & 0 & 0 & \vdots & 0 \end{bmatrix}.$$

Interchanging row one and row three gives

$$\begin{bmatrix} 1 & 0 & 3 & \vdots & -1/2 \\ 0 & 1 & 2 & \vdots & -3/2 \\ 0 & 0 & 0 & \vdots & 0 \\ 0 & 0 & 0 & \vdots & 0 \end{bmatrix}.$$

This is the augmented matrix for the system

$$
\begin{aligned}
x + 3z &= -\tfrac{1}{2} \\
y + 2z &= -3/2 \\
0 &= 0 \\
0 &= 0 .
\end{aligned}
$$

We can write the above system as

$$
\begin{aligned}
x &= -3z - \tfrac{1}{2} \\
y &= -2z - 3/2 .
\end{aligned}
$$

Since z can be assigned any arbitrary value, there are infinitely many solutions, one for each value of z. Thus, the system (1) has more than one solution.

In general, let R be a row reduced echelon form of the augmented matrix of the given system. Let r be the number of non-zero rows of this R and n be the number of unknowns of the system. If $n > r$, then the system has more than one particular solution.

When $n = r$, the system may or may not have a solution, and there cannot be more than one particular solution. That is, if there is no row whose only non-zero entry is in the last column, then the particular solution exists and is unique.

Suppose that the augmented matrix for a system of linear equations has been reduced by row operations to the given reduced row echelon form. Solve the system.

(a) $\begin{bmatrix} 1 & 0 & 0 & | & 5 \\ 0 & 1 & 0 & | & -2 \\ 0 & 0 & 1 & | & 4 \end{bmatrix}$ (b) $\begin{bmatrix} 1 & 0 & 0 & 4 & | & -1 \\ 0 & 1 & 0 & 2 & | & 6 \\ 0 & 0 & 1 & 3 & | & 2 \end{bmatrix}$

(c) $\begin{bmatrix} 1 & 6 & 0 & 0 & 4 & | & -2 \\ 0 & 0 & 1 & 0 & 3 & | & 1 \\ 0 & 0 & 0 & 1 & 5 & | & 2 \\ 0 & 0 & 0 & 0 & 0 & | & 0 \end{bmatrix}$ (d) $\begin{bmatrix} 1 & 0 & 0 & | & 0 \\ 0 & 1 & 0 & | & 0 \\ 0 & 0 & 0 & | & 1 \end{bmatrix}$

Solution: (a) If the number of non-zero rows, r, of the reduced matrix is equal to the number of unknowns, n, and there is no inconsistent equation $0 = k$, then the system has a unique particular solution.

(a) The corresponding system of equations is

$$x_1 = 5$$
$$x_2 = -2$$
$$x_3 - 4 .$$

Therefore, the solution to the system is

$$x_1 = 5, \; x_2 = -2 \; \text{ and } \; x_3 = 4 .$$

(b) Since $n > r$, we have more than one solution. The corresponding system of equations is

$$x_1 + 4x_4 = -1$$
$$x_2 + 2x_4 = 6$$
$$x_3 + 3x_4 = 2 .$$

The above system can be written as

$$x_1 = -1 - 4x_4$$
$$x_2 = 6 - 2x_4 \qquad\qquad (1)$$
$$x_3 = 2 - 3x_4 .$$

Assign x_4 any value, and then compute x_1, x_2 and x_3 from (1). Thus, we have many solutions, one for each value of x_4.

(c) $n > r$; therefore, the system has many solutions. The corresponding system of equations is

$$x_1 + 6x_2 + 4x_5 = -2$$
$$x_3 + 3x_5 = 1$$
$$x_4 + 5x_5 = 2 .$$

The above equations can be written as

$$x_1 = -2 - 4x_5 - 6x_2$$
$$x_3 = 1 - 3x_5$$
$$x_4 = 2 - 5x_5 .$$

Since x_5 can be assigned an arbitrary value, t, and x_2 can be assigned an arbitrary value, s, there are infinitely many solutions. The solution set is given by the formula

$$x_1 = -2 - 4t - 6s, \ x_2 = s, \ x_3 = 1 - 3t ,$$

$$x_4 = 2 - 5t, \ x_5 = t.$$

(d) This system has no solution since the row-reduced echelon form has a row in which the first non-zero entry is in the last column.

● **PROBLEM 8-27**

Find the solutions of the following systems and describe the solutions in geometric terms.

(a)
$$-2x + y + 3z = 0$$
$$2x - y - 3z = 0 \qquad\qquad (1)$$
$$-6x + 3y + 9z = 0$$

(b)
$$2x + 5y + z = 0$$
$$x - 2y + z = 0 \qquad\qquad (2)$$
$$3x + 3y + 2z = 0$$

(c)
$$x - y + z = 0$$
$$2x - y + z = 0 \qquad\qquad (3)$$
$$x + y + z = 0$$

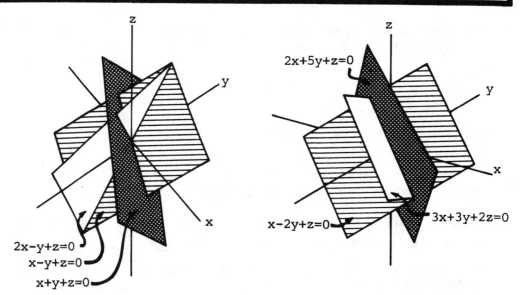

Solution: (a) the coefficient matrix for the system (1) is

$$\begin{bmatrix} -2 & 1 & 3 \\ 2 & -1 & -3 \\ -6 & 3 & 9 \end{bmatrix} .$$

To reduce the matrix, perform the following:
1) Add the first row to the second row.
2) Add 3 times the first row to the third row.
3) Multiply the first row by $-\frac{1}{2}$.
The reduction of the above matrix gives

$$\begin{bmatrix} 1 & -\frac{1}{2} & -3/2 \\ 0 & 0 & 0 \\ 0 & 0 & 0 \end{bmatrix} .$$

Thus, $x - \frac{1}{2}y - 3/2\, z = 0$.

We can choose x and y, and then compute z. We have two free variables and the solution forms a plane which is a two-dimensional linear subspace of R^3 .
(b) The coefficient matrix for the system (2) is

$$\begin{bmatrix} 2 & 5 & 1 \\ 1 & -2 & 1 \\ 3 & 3 & 2 \end{bmatrix} .$$

To reduce the matrix, perform the following:
1) Add -2 times the second row to the first.
2) Add -3 times the second row to the third, obtaining

$$\begin{bmatrix} 0 & 9 & -1 \\ 1 & -2 & 1 \\ 0 & 9 & -1 \end{bmatrix} .$$

3) Add -1 times the first row to the third.
4) Divide the first row by 9, obtaining

$$\begin{bmatrix} 0 & 1 & -1/9 \\ 1 & -2 & 1 \\ 0 & 0 & 0 \end{bmatrix} .$$

5) Add 2 times the first row to the second.
6) Interchange the first two rows. The result is

$$\begin{bmatrix} 1 & 0 & 7/9 \\ 0 & 1 & -1/9 \\ 0 & 0 & 0 \end{bmatrix}$$

which is the coefficient matrix of the system

$$x + 7/9\, z = 0$$
$$y - 1/9\, z = 0 \tag{4}$$

Here we have one free variable, namely z, and the solutions of (4) form a line, which is a one-dimensional subspace of R^3 .
(c) The coefficient matrix for the system (3) is

$$\begin{bmatrix} 1 & -1 & 1 \\ 2 & -1 & 1 \\ 1 & 1 & 1 \end{bmatrix} .$$

To reduce this matrix, perform the following:
1) Add -2 times the first row to the second.
2) Add -1 times the first row to the third, obtaining

$$\begin{bmatrix} 1 & -1 & 1 \\ 0 & 1 & -1 \\ 0 & 2 & 0 \end{bmatrix} .$$

3) Add the second row to the first.
4) Add $-\frac{1}{2}$ times the third row to the second.

$$\begin{bmatrix} 1 & 0 & 0 \\ 0 & 0 & -1 \\ 0 & 2 & 0 \end{bmatrix} .$$

5) Divide the third row by $\frac{1}{2}$, the second row by -1.
6) Interchange the second and third rows. This gives

$$\begin{bmatrix} 1 & 0 & 0 \\ 0 & 1 & 0 \\ 0 & 0 & 1 \end{bmatrix} .$$

Thus, the solution of the system (3) is $x = y = z = 0$.
In this case, we have no choice. Our solution is a single point which is a zero-dimensional space.

● **PROBLEM** 8-28

Find the necessary and sufficient conditions for the existence of a solution to the following system.

$$\begin{aligned} x + y + 2z &= a_1 \\ -2x \quad\quad - z &= a_2 \\ x + 3y + 5z &= a_3 \end{aligned} \tag{1}$$

Solution: The procedure for solving nonhomogeneous systems is to form the augmented matrix of the system and reduce this matrix to a row-echelon matrix. Suppose, during this reduction, we obtain a matrix in which there is a row where the first non-zero entry appears in the last column. This matrix is the augmented matrix of a system with no solution.

If we obtain a reduced matrix with the property that no row has its first non-zero entry in the last column then the system has a solution. Now, the augmented matrix of the system (1) is

$$\left[\begin{array}{ccc|c} 1 & 1 & 2 & a_1 \\ -2 & 0 & -1 & a_2 \\ 1 & 3 & 5 & a_3 \end{array}\right] .$$

Add 2 times the first row to the second row and -1 times the first row to the third row

$$\left[\begin{array}{ccc|c} 1 & 1 & 2 & a_1 \\ 0 & 2 & 3 & a_2+2a_1 \\ 0 & 2 & 3 & a_3-a_1 \end{array}\right] .$$

Add -1 times the second row to the third row

$$\left[\begin{array}{ccc|c} 1 & 1 & 2 & a_1 \\ 0 & 2 & 3 & a_2+2a_1 \\ 0 & 0 & 0 & a_3-a_2-3a_1 \end{array}\right] .$$

Suppose $a_3 - a_2 - 3a_1 \neq 0$. Then the reduced matrix has a row in which the first non-zero entry appears in the last column. Thus, the system has no solution. Therefore, a necessary condition for a solution to exist is that $a_3 - a_2 - 3a_1$ must be equal to zero.

Let $a_3 - a_2 - 3a_1 = 0$. Then, we have

$$\begin{bmatrix} 1 & 1 & 2 & | & a_1 \\ 0 & 2 & 3 & | & a_2 + 2a_1 \\ 0 & 0 & 0 & | & 0 \end{bmatrix}$$

which is the augmented matrix for the system

$$x + y + 2z = a_1$$
$$2y + 3z = a_2 + 2a_1 \quad .$$

The above system can be written as

$$x + \tfrac{1}{2}z = -a_2/2$$
$$y + 3/2z = a_1 + a_2/2 \quad .$$

Assign to z any arbitrary value, then compute x and y from the above equations to obtain a solution to system (1). Thus, we can conclude that $a_3 - a_2 - 3a_1 = 0$ is a necessary and sufficient condition for the existence of a solution to a system (1).

● **PROBLEM 8-29**

Determine the values of a so that the following system of equations has:
(a) no solution, (b) more than one solution, (c) a unique solution.

$$x + y - z = 1$$
$$2x + 3y + az = 3$$
$$x + ay + 3z = 2$$

Solution: First form the augmented matrix for the system and then reduce it to echelon form.

The augmented matrix is

$$\begin{bmatrix} 1 & 1 & -1 & | & 1 \\ 2 & 3 & a & | & 3 \\ 1 & a & 3 & | & 2 \end{bmatrix} \quad .$$

Add -2 times the first row to the second row and -1 times the first row to the third row.

$$\begin{bmatrix} 1 & 1 & -1 & | & 1 \\ 0 & 1 & a+2 & | & 1 \\ 0 & a-1 & 4 & | & 1 \end{bmatrix} \quad .$$

Add $-(a-1)$ times the second row to the third row

$$\begin{bmatrix} 1 & 1 & -1 & | & 1 \\ 0 & 1 & a+2 & | & 1 \\ 0 & 0 & 4-(a+2)(a-1) & | & 1-(a-1) \end{bmatrix} \quad .$$

The above matrix can be written as

$$\begin{bmatrix} 1 & 1 & -1 & | & 1 \\ 0 & 1 & a+2 & | & 1 \\ 0 & 0 & (3+a)(2-a) & | & (2-a) \end{bmatrix} \qquad (1)$$

since $4-(a+2)(a-1) = 4-(a^2+a-2) = 6 - a - a^2 = (3+a)(2-a)$. Suppose $(3+a) = 0$. Then $(2-a) \neq 0$, and we have a reduced matrix with the property that a row has its first non-zero entry in the last column. In this case, the system has no solution. Thus, the system has a solution only if $(3+a) \neq 0$, that is, if $a \neq -3$. If $a = 3$, then the last row of the matrix (1) becomes $[0\ 0\ 0\ 5]$, and the system has no solution. Suppose $a = -2$. Then the last row of the matrix (1) is $[0\ 0\ 0\ 0]$.

In this case, the system has an infinite number of solutions. We can summarize our results as follows: If

(a) $a = -3$, then the system has no solution.
(b) if $a = 2$, the system has more than one solution.
(c) if $a \neq 2$ and $a \neq -3$, the system has a unique solution.

MISCELLANEOUS RESULTS FOR SYSTEMS OF LINEAR EQUATIONS

● **PROBLEM 8-30**

Solve the following system

$$
\begin{aligned}
x_1 - 2x_2 - 3x_3 &= 3 \\
2x_1 - x_2 - 4x_3 &= 7 \\
3x_1 - 3x_2 - 5x_3 &= 8 .
\end{aligned}
\tag{1}
$$

Solution: The matrix of coefficients for the system (1) is

$$
A = \begin{bmatrix} 1 & -2 & -3 \\ 2 & -1 & -4 \\ 3 & -3 & -5 \end{bmatrix} .
$$

The system (1) may be written in matrix form as

$$
\begin{bmatrix} 1 & -2 & -3 \\ 2 & -1 & -4 \\ 3 & -3 & -5 \end{bmatrix} \begin{bmatrix} x_1 \\ x_2 \\ x_3 \end{bmatrix} = \begin{bmatrix} 3 \\ 7 \\ 8 \end{bmatrix} .
\tag{2}
$$

Let

$$
X = \begin{bmatrix} x_1 \\ x_2 \\ x_3 \end{bmatrix}, \quad b = \begin{bmatrix} 3 \\ 7 \\ 8 \end{bmatrix} .
$$

Then equation (2) is written

$$
\vec{AX} = \vec{b} .
\tag{3}
$$

A solution vector $\vec{x}$ can be found by multiplying both sides of equation (3) by A^{-1}. Then we have

$$
A^{-1}AX = A^{-1}b ,
$$

but $A^{-1}A = I$. Hence,

$$
IX = A^{-1}b
$$

or

$$
X = A^{-1}b .
\tag{4}
$$

Thus, the solutions of a system of linear equations can be obtained by finding the inverse matrix of the coefficient matrix of the system and then solving equation (4). To find A^{-1}, first form the matrix $[A : I]$, and reduce this matrix, by applying row operations, to the form $[I : B]$. Then, $B = A^{-1}$. Now,

$$[A : I] = \begin{bmatrix} 1 & -2 & -3 & \vdots & 1 & 0 & 0 \\ 2 & -1 & -4 & \vdots & 0 & 1 & 0 \\ 3 & -3 & -5 & \vdots & 0 & 0 & 1 \end{bmatrix} .$$

Add -2 times the first row to the second row and -3 times the first row to the third row

$$\begin{bmatrix} 1 & -2 & -3 & \vdots & 1 & 0 & 0 \\ 0 & 3 & 2 & \vdots & -2 & 1 & 0 \\ 0 & 3 & 4 & \vdots & -3 & 0 & 1 \end{bmatrix} .$$

Now add -1 times the second row to the third row

$$\begin{bmatrix} 1 & -2 & -3 & \vdots & 1 & 0 & 0 \\ 0 & 3 & 2 & \vdots & -2 & 1 & 0 \\ 0 & 0 & 2 & \vdots & -1 & -1 & 0 \end{bmatrix} .$$

Divide the third row by 2

$$\begin{bmatrix} 1 & -2 & -3 & \vdots & 1 & 0 & 0 \\ 0 & 3 & 2 & \vdots & -2 & 1 & 0 \\ 0 & 0 & 1 & \vdots & -\frac{1}{2} & -\frac{1}{2} & \frac{1}{2} \end{bmatrix} .$$

Add -2 times the third row to the second row and 3 times the third row to the first row

$$\begin{bmatrix} 1 & -2 & 0 & \vdots & -\frac{1}{2} & -3/2 & 3/2 \\ 0 & 3 & 0 & \vdots & -1 & 2 & -1 \\ 0 & 0 & 1 & \vdots & -\frac{1}{2} & -\frac{1}{2} & \frac{1}{2} \end{bmatrix} .$$

Divide the second row by 3; then add 2 times the resulting second row to the first row.

$$\begin{bmatrix} 1 & 0 & 0 & \vdots & -7/6 & -1/6 & 5/6 \\ 0 & 1 & 0 & \vdots & -1/3 & 2/3 & -1/3 \\ 0 & 0 & 1 & \vdots & -\frac{1}{2} & -\frac{1}{2} & \frac{1}{2} \end{bmatrix} .$$

Thus,

$$A^{-1} = \begin{bmatrix} -7/6 & -1/6 & 5/6 \\ -1/3 & 2/3 & -1/3 \\ -\frac{1}{2} & -\frac{1}{2} & \frac{1}{2} \end{bmatrix} .$$

Then equation (4) becomes

$$\begin{bmatrix} x_1 \\ x_2 \\ x_3 \end{bmatrix} = \begin{bmatrix} -7/6 & -1/6 & 5/6 \\ -1/3 & 2/3 & -1/3 \\ -\frac{1}{2} & -\frac{1}{2} & \frac{1}{2} \end{bmatrix} \begin{bmatrix} 3 \\ 7 \\ 8 \end{bmatrix} .$$

Multiplying, we have

$$\begin{bmatrix} x_1 \\ x_2 \\ x_3 \end{bmatrix} = \begin{bmatrix} -21/6 - 7/6 + 40/6 \\ -1 + 14/3 - 8/3 \\ -3/2 - 7/2 + 4 \end{bmatrix} = \begin{bmatrix} 2 \\ 1 \\ -1 \end{bmatrix} .$$

Thus,

$$x_1 = 2, \ x_2 = 1, \ x_3 = -1 .$$

It is interesting to note that the calculation of the inverse matrix is closely related to the solution of simultaneous equations. Indeed, the two processes are essentially the same. To show this, we can use the method of successive elimination. Eliminating x_1 from the second and third equations,

$$x_1 - 2x_2 - 3x_3 = 3$$
$$3x_2 + 2x_3 = 1$$
$$3x_2 + 4x_3 = -1 .$$

Then, eliminating x_2 from the third equation,

$$x_1 - 2x_2 - 3x_3 = 3$$
$$3x_2 + 2x_3 = 1$$
$$2x_3 = -2 .$$

We obtain

$$x_1 - 2x_2 - 3x_3 = 3$$
$$3x_2 + 2x_3 = 1$$
$$x_3 = -1 ,$$

$$x_1 - 2x_2 = 0$$
$$3x_2 = 3$$
$$x_3 = -1$$

and, finally,

$$x_1 = 2$$
$$x_2 = 1$$
$$x_3 = -1 .$$

It can be seen that the solution to the system $A\vec{x} = \vec{b}$, calculated directly, is the same solution we obtained by calculating A^{-1} and finding $A^{-1}\vec{b}$. Observe that the inverse matrix can also be found by using the formula

$$A^{-1} = \frac{1}{\det A} \text{ adj } A .$$

● **PROBLEM** 8-31

Find the solution set of the following system of equations:

$$2x_1 + x_2 - 4x_3 = 8$$
$$3x_1 - x_2 + 2x_3 = -1 .$$

(1)

Solution: If $v = <c_1, c_2, \ldots, c_n>$ is a particular solution of a given system of linear equations and S is the solution set of its associated homogeneous system, then the solution set of the given system is $P = V + S$. The system (1) has more than one particular solution since it has more unknowns than equations. The system (1) has a particular solution, $x = 1$, $x_2 = 2$, $x_3 = -1$, as may be easily verified.

Now, the associated homogeneous system is

$$2x_1 + x_2 - 4x_3 = 0$$
$$3x_1 - x_2 + 2x_3 = 0 ,$$

and its solution set consists of all vectors orthogonal to both $<2,1,-4>$

242

and $<3,-1,2>$. This is the set of scalar multiples of

$$<2,1,-4> \: X \: <3,-1,2> = <-2,-16,-5> \: .$$

Hence, the solution set of the given system is $<1,2,1> + Sp\{<-2,-16,-5>\}$.

● **PROBLEM** 8-32

Consider the following nonhomogeneous system of linear equations.

$$2x + y - 3z = 1 \qquad (1)$$
$$3x + 2y - 2z = 2$$
$$x + y + z = 1 \: .$$

Show that (i) any two solutions to the system (1) differ by a vector which is a solution to the homogeneous system

$$2x + y - 3z = 0$$
$$3x + 2y - 2z = 0 \qquad (2)$$
$$x + y + z = 0 \: .$$

(ii) the sum of a solution to (1) and a solution to (2) gives a solution to (1).

Solution: Consider a system of linear equations

$$a_{11}x_1 + a_{12}x_2 + \ldots + a_{1n}x_n = b_1$$
$$a_{21}x_1 + a_{22}x_2 + \ldots + a_{2n}x_n = b_2 \qquad (3)$$
$$\cdot \: \cdot \: \cdot \: \cdot \: \cdot \: \cdot \: \cdot \: \cdot \: \cdot \: \cdot \: \cdot \: \cdot \: \cdot \: \cdot \: \cdot$$
$$a_{m1}x_1 + a_{m2}x_2 + \ldots + a_{mn}x_n = b_m \: .$$

We say that a vector $v = (c_1, c_2, \ldots, c_n)$ of R^n is a solution vector of (3) if $x_1 = c_1, x_2 = c_2, \ldots, x_n = c_n$ is a solution of (3). Now the augmented matrix for system (1) is

$$\begin{bmatrix} 2 & 1 & -3 & | & 1 \\ 3 & 2 & -2 & | & 2 \\ 1 & 1 & 1 & | & 1 \end{bmatrix} \: .$$

Reduction of the above matrix gives

$$\begin{bmatrix} 1 & 0 & -4 & | & 0 \\ 0 & 1 & 5 & | & 1 \\ 0 & 0 & 0 & | & 0 \end{bmatrix} \: .$$

The corresponding system of equations is

$$x - 4z = 0$$
$$y + 5z = 1 \: .$$

There is more than one solution. Let $z = 1$; then $y = -4$ and $x = 4$. Let $z = 0$; then $y = 1$ and $x = 0$. Thus, $(4,-4,1)$ and $(0,1,0)$ are vector solutions to system (1). The coefficient matrix of the system (2) is

$$\begin{bmatrix} 2 & 1 & -3 \\ 3 & 2 & -2 \\ 1 & 1 & 1 \end{bmatrix} \: .$$

Reduction of this matrix gives

$$\begin{bmatrix} 1 & 0 & -4 \\ 0 & 1 & 5 \\ 0 & 0 & 0 \end{bmatrix}$$

The corresponding system of equations is

$$x - 4z = 0$$
$$y + 5z = 0 .$$

Let $z = 1$; then $y = -5$ and $x = 4$. Now the difference of two solutions of system (1), $(4,-4,1) - (0,1,0) = (4,-5,1)$, is a solution to the homogeneous system (1). Add $(4,-4,1)$ to $(4,-5,1)$. This gives $(8,-9,2)$ which is also a solution to system (1).

(It is easy to check that $(8,-9,2)$ satisfies $x - 4z = 0$ and $y + 5z = 1$.)

Generalize these results as follows: To solve the nonhomogeneous system, first find one solution to this system and all solutions to its associated homogeneous system. Then any other solution to the given nonhomogeneous system is obtained by adding the particular solution of the nonhomogeneous system to the general solution of the homogeneous system.

● PROBLEM 8-33

Solve the following systems of equations.

(1)
$$4x_1 - 3x_2 + x_3 = -1$$
$$x_1 + 5x_2 - 2x_3 = 2$$
$$x_1 + 2x_2 = 0$$

(2)
$$2x_1 + 3x_2 + x_3 - 4x_4 = 0$$
$$x_1 - 5x_2 - 3x_3 + 2x_4 = 0$$
$$5x_1 + 2x_2 - x_4 = 0$$
$$2x_1 - 9x_2 - 5x_3 + 9x_4 = 0$$

(3)
$$8x_1 - 2x_2 + 4x_3 + 3x_4 + x_5 = 2$$
$$x_2 - 4x_3 + x_4 - 2x_5 = -10$$
$$2x_1 + x_2 - 4x_4 = 1$$

Solution: (1) The augmented matrix for system (1) is

$$\begin{bmatrix} 4 & -3 & 1 & | & -1 \\ 1 & 5 & -2 & | & 2 \\ 1 & 2 & 0 & | & 0 \end{bmatrix} .$$

Reduce this matrix to row reduced echelon form. Interchange the first and the third rows

$$\begin{bmatrix} 1 & 2 & 0 & | & 0 \\ 1 & 5 & -2 & | & 2 \\ 4 & -3 & 1 & | & -1 \end{bmatrix} .$$

Add -1 times the first row to the second row and -4 times the first row to the third row.

$$\begin{bmatrix} 1 & 2 & 0 & | & 0 \\ 0 & 3 & -2 & | & 2 \\ 0 & -11 & 1 & | & -1 \end{bmatrix} .$$

Divide the second row by 3 and the third row by -11.

$$\left[\begin{array}{ccc|c} 1 & 2 & 0 & 0 \\ 0 & 1 & -2/3 & 2/3 \\ 0 & 1 & -1/11 & 1/11 \end{array}\right] .$$

Add -1 times the second row to the third row to obtain the row echelon form.

$$\left[\begin{array}{ccc|c} 1 & 2 & 0 & 0 \\ 0 & 1 & -2/3 & 2/3 \\ 0 & 0 & 19/33 & -19/33 \end{array}\right] \qquad (5)$$

It can be seen easily that the system (1) has a unique solution because the system (1) has three unknowns and the row echelon matrix (5) has three non-zero rows. The matrix (5) is the augmented matrix of the system.

$$x_1 + 2x_2 = 0$$

$$x_2 - 2/3\, x_3 = 2/3$$

$$19/33\, x_3 = -19/33 .$$

Solving this system yields $x_1 = 0$, $x_2 = 0$, and $x_3 = -1$.

[2] The matrix of coefficients for the system (2) is

$$\left[\begin{array}{cccc} 2 & 3 & 1 & -4 \\ 1 & -5 & -3 & 2 \\ 5 & 2 & 0 & -1 \\ 2 & -9 & -5 & 9 \end{array}\right] .$$

Interchange the first and the second rows

$$\left[\begin{array}{cccc} 1 & -5 & -3 & 2 \\ 2 & 3 & 1 & -4 \\ 5 & 2 & 0 & -1 \\ 2 & -9 & -5 & 9 \end{array}\right] .$$

Add -2 times the first row to the second row and to the fourth row.
Add -5 times the first row to the third row

$$\left[\begin{array}{cccc} 1 & -5 & -3 & 2 \\ 0 & 13 & 7 & -8 \\ 0 & 27 & 15 & -11 \\ 0 & 1 & 1 & 5 \end{array}\right] .$$

Add 5 times the fourth row to the first row and -13 times the fourth row to the second row. Add -27 times the fourth row to the third row

$$\left[\begin{array}{cccc} 1 & 0 & 2 & 27 \\ 0 & 0 & -6 & -73 \\ 0 & 0 & -12 & -146 \\ 0 & 1 & 1 & 5 \end{array}\right] .$$

Add -2 times the second row to the third row

$$\begin{bmatrix} 1 & 0 & 2 & 27 \\ 0 & 0 & -6 & -73 \\ 0 & 0 & 0 & 0 \\ 0 & 1 & 1 & 5 \end{bmatrix} .$$

Interchange rows two and four, then rows three and four

$$\begin{bmatrix} 1 & 0 & 2 & 27 \\ 0 & 1 & 1 & 5 \\ 0 & 0 & -6 & -73 \\ 0 & 0 & 0 & 0 \end{bmatrix} .$$

Divide the third row by -6. Then add -1 times the third row to the second row and -2 times the third row to the first row

$$\begin{bmatrix} 1 & 0 & 0 & 8/3 \\ 0 & 1 & 0 & -43/6 \\ 0 & 0 & 1 & 73/6 \\ 0 & 0 & 0 & 0 \end{bmatrix} .$$

This is the matrix of coefficients for the system

$$x_1 \qquad + 8/3\ x_4 = 0$$
$$x_2 \quad - 43/6\ x_4 = 0$$
$$x_3 + 73/6\ x_4 = 0 .$$

The above homogeneous system has a non-trivial solution since the system involves more unknowns than equations. Solving for the leading variables yields

$$x_1 = -8/3\ x_4$$
$$x_2 = 43/6\ x_4$$
$$x_3 = -73/6\ x_4 .$$

The solution set is, therefore, given by $x_1 = -8/3\ t$, $x_2 = 43/6\ t$, $x_3 = -73/6\ t$ and $x_4 = t$.

[3] Let V be a particular solution of the system (3), and let $S = \{s_i\}_{i=1},\ldots,n$ be a basis of the solution space of its associated homogeneous system. Note that $n \leq 2$ since the original system has three equations in five unknowns. Then the solution set of the system (3) is

$$p = V + C_1 S_1 + C_2 S_2 .$$

First, find the single solution of system (3). The augmented matrix is

$$\left[\begin{array}{ccccc|c} 8 & -2 & 4 & 3 & 1 & 2 \\ 0 & 1 & -4 & 1 & -2 & -10 \\ 2 & 1 & 0 & -4 & 0 & 1 \end{array}\right] .$$

Interchange rows one and three

$$\left[\begin{array}{ccccc|c} 2 & 1 & 0 & -4 & 0 & 1 \\ 0 & 1 & -4 & 1 & -2 & -10 \\ 8 & -2 & 4 & 3 & 1 & 2 \end{array}\right] .$$

Add -4 times the first row to the third row

$$\begin{bmatrix} 2 & 1 & 0 & -4 & 0 & | & 1 \\ 0 & 1 & -4 & 1 & -2 & | & -10 \\ 0 & -6 & 4 & 19 & 1 & | & -2 \end{bmatrix} .$$

Add -1 times the second row to the first row and 6 times the second
row to the third row

$$\begin{bmatrix} 2 & 0 & 4 & -5 & 2 & | & 11 \\ 0 & 1 & -4 & 1 & -2 & | & -10 \\ 0 & 0 & -20 & 25 & -11 & | & -62 \end{bmatrix} .$$

Divide the third row by -20. Then add -4 times the third row to the
first row and 4 times the third row to the second row

$$\begin{bmatrix} 2 & 0 & 0 & 0 & -1/5 & | & -7/5 \\ 0 & 1 & 0 & -4 & 1/5 & | & 12/5 \\ 0 & 0 & 1 & -25/20 & 11/20 & | & 62/20 \end{bmatrix} .$$

Divide the first row by 2

$$\begin{bmatrix} 1 & 0 & 0 & 0 & -1/10 & | & -7/10 \\ 0 & 1 & 0 & -4 & 1/5 & | & 12/5 \\ 0 & 0 & 1 & -5/4 & 11/20 & | & 62/20 \end{bmatrix} .$$

The corresponding system of equations is

$$x_1 \qquad\qquad - \ 1/10 \ x_5 \ = \ -7/10$$
$$x_2 \quad - \ 4x_4 \ + \ 1/5 \ x_5 \ = \ 12/5$$
$$x_3 \ - \ 5/4 \ x_4 \ + \ 11/20 \ x_5 \ = \ 62/20 .$$

Solving for the leading variables gives

$$x_1 \ = \ -7/10 \ + \qquad + \ 1/10 \ x_5$$
$$x_2 \ = \ 12/5 \ + \ 4x_4 \ - \ 1/5 \ x_5$$
$$x_3 \ + \ 62/20 \ + \ 5/4 \ x_4 \ - \ 11/20 \ x_5 .$$

We obtain a particular solution to the given system by setting $x_4 = x_5 = 0.$
Thus,

$$V \ = \ (-7/10, \ 12/5, \ 62/20, \ 0,0).$$

Now, the associated homogeneous system is

$$8x_1 \ - \ 2x_2 \ + \ 4x_3 \ + \ 3x_4 \ + \ x_5 \ = \ 0$$
$$x_2 \ - \ 4x_3 \ + \ x_4 \ - \ 2x_5 \ = \ 0$$
$$2x_1 \ + \ x_2 \qquad\qquad - \ 4x_4 \qquad\qquad = \ 0$$

Then the matrix of the coefficients is

$$\begin{bmatrix} 8 & -2 & 4 & 3 & 1 \\ 0 & 1 & -4 & 1 & -2 \\ 2 & 1 & 0 & -4 & 0 \end{bmatrix}$$

which can be reduced to the row reduced echelon form,

$$\begin{bmatrix} 1 & 0 & 0 & 0 & -1/10 \\ 0 & 1 & 0 & -4 & 1/5 \\ 0 & 0 & 1 & -5/4 & 11/20 \end{bmatrix} .$$

Thus, the corresponding system of equations is

$$x_1 \qquad\qquad\qquad -1/10\ x_5 = 0$$

$$x_2 \qquad 4x_4 + 1/5\ x_5 = 0$$

$$x_3 - 5/4\ x_4 + 11/20\ x_5 = 0 .$$

Solving for the leading variables yields

$$x_1 = \qquad 1/10\ x_5$$

$$x_2 = 4x_4 - 1/5\ x_5$$

$$x_3 = 5/4\ x_4 - 11/20\ x_5 .$$

Since we can freely choose x_4 and x_5, we set them equal to arbitrary parameters a and b. Then the solution set is given by the formula

$$x_1 \qquad\qquad + 1/10\ b$$

$$x_2 = 4a - 1/5\ b$$

$$x_3 = 5/4\ a - 11/20\ b$$

$$x_4 = a$$

$$x_5 = b .$$

Any element of the solution set satisfies these formulas. Any two linearly independent elements will form a basis. To obtain two such elements, set $a = 1$, $b = 0$; $a = 0$, $b = 1$, respectively. Let $a = 1$ and $b = 0$. Then $x_1 = 0$, $x_2 = 4$, $x_3 = 5/4$, $x_4 = 1$, $x_5 = 0$. Let $a = 0$ $b = 1$. Then

$$x_1 = 1/10,\ x_3 = -1/5,\ x_3 = -11/20,\ x_4 = 0,\ x_5 = 1.$$

Thus,

$$S_1 = (0,4,5/4,1,0),$$

$$S_2 = (1/10,\ -1/5,\ -11/20,\ 0,1) .$$

Hence, the solution set of the system (3) is

$$(-7/10,\ 12/5,\ 62/20,\ 0,0) + c_1 (0,4,5/4,1,0)$$
$$+ c_2 (1/10,\ -1/5,\ -11/20,0,1).$$

● **PROBLEM** 8-34

Show that the system of linear equations over the rational number field Q

$$2x_1 + 6x_2 - x_3 + x_4 = 2$$
$$x_1 + 3x_2 + x_3 \qquad = 5 \qquad\qquad (1)$$
$$-x_1 - 3x_2 - x_3 \qquad = 0$$

has no solution.

Solution: Let Q represent the rational number field. Let T be the linear transformation of the rational vector space Q^4 into the rational vector space Q^3 (denoted $V_4(Q)$ and $V_3(Q)$ respectively),

$$T(q_1,\ldots,q_4) = \begin{pmatrix} p_1 \\ p_2 \\ p_3 \end{pmatrix} \text{ where } \begin{array}{l} q_1,\ldots,q_4 \in Q \\ \\ p_1,\ldots,p_3 \in Q \end{array}$$

given by

$$T(q_1,\ldots,q_4) = \begin{array}{l} 2q_1 + 6q_2 - q_3 + q_4 \\ q_1 + 3q_2 - q_3 \\ -x_1 - 3x_2 - x_3 \end{array} .$$

Note that in $V_n(Q)$, vectors of n-tuples of rationals and scalars are rational numbers. Also recall that $V_n(Q)$ is an n-dimensional vector space, and the standard basis is $\{e_i\}_{i=1,\ldots,n}$, where $e_i = (0,\ldots,0,1,0,\ldots,0)$
Then the above system of linear equations has a solution if and only if

$$b = (b_1,\ldots,b_m) \in \text{Range } T$$

where $b_1,\ldots,b_m$ are the constants of the nonhomogeneous system of equations. That is, there exists a 4-tuple $(q_1,\ldots,q_4)$ such that $T(q_1,\ldots,q_4) = \begin{pmatrix} 2 \\ 5 \\ 0 \end{pmatrix}$. Thus, the system (1) has a solution only if the

vector $b = (2,5,0)$ is in the range of A. We know any vector $x \in V_4(Q)$ is of the form,

$$x = \sum_{i=1}^{4} a_i e_i \ ; \ a_i \in Q \ ,$$

and, hence,

$$Tx = T[\sum_{i=1}^{4} a_i e_i] = [\sum_{i=1}^{4} a_i T(e_i)] \text{ since } T \text{ is a linear}$$

transformation. Denote the vectors Te_i by v_i, $i=1,\ldots,n$. Then , if $x = \sum_{i=1}^{4} a_i e_i$ is any vector in $V_4(Q)$, we have shown that

$$Tx = \sum_{i=1}^{4} a_i v_i .$$

In other words, v_1,v_2,v_3,v_4 span range T. Now we can compute from (1) that

$$v_1 = Te_1 = (2,1,-1)$$
$$v_2 = Te_2 = (6,3,-3)$$
$$v_3 = Te_3 = (-1,1,-1)$$
$$v_4 = Te_4 = (1,0,0) .$$

Note that v_1 and v_4 are linearly independent, but that $v_2 = 3v_1$ and $v_3 = v_1 - 3v_4$. Hence, a basis for the range of T is (v_1,v_4).
A solution of the system exists only if $(2,5,0)$ is in the range of T. The question of the existence of a solution to (1) is, therefore, equivalent to the question of the existence of rational numbers c_1 and

c_2 for which

$$(2,5,0) = c_1(2,1,-1) + c_2(1,0,0) = (2c_1 + c_2, c_1, - c_1) .$$

Clearly, the two conditions $c_1 = 5$ and $-c_1 = 0$ are mutually incompatible. Therefore, system (1) has no solution.

● **PROBLEM** 8-35

Show, by example, how the method of Gauss elimination is related to row echelon matrices.

Solution: Gauss elimination is used to solve systems of linear equations. Consider the following set of equations:

$$\begin{aligned}
2x_1 - 3x_2 + 2x_3 + 5x_4 &= 3 \\
x_1 - x_2 + x_3 + 2x_4 &= 1 \\
3x_1 + 2x_2 + 2x_3 + x_4 &= 0 \\
x_1 + x_2 - 3x_3 - x_4 &= 0 .
\end{aligned} \qquad (1)$$

According to the method of Gauss elimination, use one of the equations to eliminate one of the unknowns from the remaining three equations. In (1) above, choose the first equation for use and x_1 for elimination.

By suitably multiplying this equation, x_1 can be successively eliminated from the other equations by subtraction. For example, multiplying the first equation by $\cdot 5$ and subtracting from the second eliminates x_1 from the second. In this manner we obtain

$$\begin{aligned}
2x_1 - 3x_2 + 2x_3 + 5x_4 &= 3 \\
.5x_2 \quad\quad - .5x_4 &= -.5 \\
6.5x_2 - x_3 - 6.5x_4 &= -4.5 \\
2.5x_2 - 4x_3 - 3.5x_4 &= -1.5 .
\end{aligned} \qquad (2)$$

Now use another equation from (2) and eliminate another unknown from the remaining two equations. Here, choose the second equation and x_2 for elimination. The result is

$$\begin{aligned}
2x_1 - 3x_2 + 2x_3 + 5x_4 &= 3 \\
.5x_2 + 0x_3 - .5x_4 &= -.5 \\
x_3 + 0x_4 &= -2 \\
4x_3 + x_4 &= -1 .
\end{aligned} \qquad (3)$$

Now use the third equation to eliminate x_3 from the last equation. Thus,

$$\begin{aligned}
2x_1 - 3x_2 + 2x_3 + 5x_4 &= 3 \\
.5x_2 + 0x_3 - .5x_4 &= -.5 \\
x_3 + 0x_4 &= -2 \\
- x_4 &= -7 .
\end{aligned} \qquad (4)$$

System (4) is in triangular form and is amenable to back-substitution. From the last equation, $x_4 = 7$; substituting this into the equation above, $x_3 = -2$. Proceeding to the next equation, $x_2 = 6$ and finally $x_1 = -5$. It is more convenient to work with the coefficient matrix of system (1). Form the augmented matrix

$$\begin{bmatrix} 2 & -3 & 2 & 5 & \vdots & 3 \\ 1 & -1 & 1 & 2 & \vdots & 1 \\ 3 & 2 & 2 & 1 & \vdots & 0 \\ 1 & 1 & -3 & -1 & \vdots & 0 \end{bmatrix} \qquad (5)$$

Using elementary row operations, we obtain the equivalent row echelon matrix

$$\begin{bmatrix} 1 & -1.5 & 1 & 2.5 & \vdots & 1.5 \\ 0 & 1 & 0 & -1 & \vdots & -1 \\ 0 & 0 & 1 & 0 & \vdots & -2 \\ 0 & 0 & 0 & 1 & \vdots & 7 \end{bmatrix} . \qquad (6)$$

Comparing (4) and (6), note that they differ only in that (6) has the leading coefficients of the unknowns equal to one. We conclude that the method of Gauss elimination is equivalent to using elementary row operations to reduce a matrix to row echelon form. In fact, the steps of a Gauss elimination process, in which we eliminate coefficients from the equations, correspond to the row operations used to reduce the matrix of coefficients or the augmented matrix.

● **PROBLEM** 8-36

If the method of Gauss elimination corresponds in its final form to an echelon matrix, what is the matrix analogue of the Gauss-Jordan method for solving linear systems of equations? Explain by example.

Solution: Consider the equations:

$$x_1 + x_2 + 2x_3 = 9$$
$$2x_1 + 4x_2 - 3x_3 = 1 \qquad (1)$$
$$3x_1 + 6x_2 - 5x_3 = 0 .$$

The augmented coefficient matrix of (1) is:

$$\begin{bmatrix} 1 & 1 & -2 & \vdots & 9 \\ 2 & 4 & -3 & \vdots & 1 \\ 3 & 6 & -5 & \vdots & 0 \end{bmatrix} \qquad (2)$$

Recall that in the method of Gauss elimination we successively eliminate unknowns from successive equations. Thus, for example, we can use the first equation of system (1) to eliminate x_1 from the remaining two equations. This amounts to applying elementary row operations on (2) to reduce the second and third elements of the first column to zero. This gives

$$\begin{bmatrix} 1 & 1 & 2 & \vdots & 9 \\ 0 & 2 & -7 & \vdots & -17 \\ 0 & 3 & -11 & \vdots & -27 \end{bmatrix} . \qquad (3)$$

In equation form this is:

$$x_1 + x_2 + 2x_3 = 9$$
$$2x_2 - 7x_3 = -17 \qquad (4)$$
$$3x_2 - 11x_3 = -27 .$$

According to the method of Gauss, we would now proceed to use the second equation of (4) to eliminate x_2 from the third equation. But, by Gauss–Jordan, we use the second equation of (4) to eliminate x_2 from the other two equations. Multiply the second row of (3) by .5 to obtain

$$\left[\begin{array}{ccc|c} 1 & 1 & 2 & 9 \\ 0 & 1 & -3.5 & -8.5 \\ 0 & 3 & -11 & -27 \end{array}\right] . \tag{5}$$

Now use elementary row operations on (5) to reduce the first and third elements of the second column to zero. This results in

$$\left[\begin{array}{ccc|c} 1 & 0 & 1.5 & 17.5 \\ 0 & 1 & -3.5 & -8.5 \\ 0 & 0 & -.5 & -1.5 \end{array}\right] . \tag{6}$$

In equation form this is:

$$\begin{aligned} x_1 + 1.5x_3 &= 17.5 \\ x_2 - 3.5x_3 &= -8.5 \\ - .5x_3 &= -1.5 \end{aligned} \quad . \tag{7}$$

The final step of the Gauss–Jordan method is elimination of x_3 from both the first and second equations of (7). We obtain:

$$\begin{aligned} x_1 &= 1 \\ x_2 &= 2 \\ x_3 &= 3 \end{aligned}$$

The coefficient matrix of this system is:

$$\left[\begin{array}{ccc|c} 1 & 0 & 0 & 1 \\ 0 & 1 & 0 & 2 \\ 0 & 0 & 1 & 3 \end{array}\right] .$$

Thus, it can be seen that the Gauss–Jordan method corresponds to applying elementary row operations to reduce a matrix to row reduced echelon form.

CHAPTER 9

POLYNOMIAL ALGEBRA

DEFINITION OF POLYNOMIALS

Define polynomial forms.

Solution: Let F be a field, $x^0 = 1 \in F$ and $x^{n+1} = x^n \cdot x$, where $n \geq 0$. The product $x^n \cdot x$ is understood to obey the same algebraic rules as if x were a member of a ring with identity. If $F[x]$ represents the set of all finite linear combinations of powers of x over F, then $F[x]$ is known as the set of all polynomial forms over F. Every member, $f(x)$, of $F[x]$ is known as a polynomial in one indeterminate, x over F. Each $f(x)$ has the form

$$f(x) = a_0 + a_1 x + a_2 x^2 + \ldots + a_{n-1} x^{n-1} + a_n x^n$$

where n is called the degree of f ($\deg.(f) = n$). a_n is known as the leading coefficient of f (if $a_n \neq 0$). A polynomial is monic if its leading coefficient is 1.

Let $f(x) = 2x^4 + x^3 + 4x^2 + 3x + 1$ and $g(x) = x^2 + 1$. Find the polynomials $q(x)$ and $r(x)$ such that $f(x) = q(x)g(x) + r(x)$.

Solution: We define the division algorithm or remainder theorem. This result is merely a precise statement of the familiar long division method. Division Algorithm: If $f(x)$ and $g(x)$ are polynomials with $g(x) \neq 0$, then there exist unique polynomials $q(x)$ and $r(x)$ such that

$$f(x) = q(x) \, g(x) + r(x) \qquad (1)$$

where either $r(x) = 0$ or $\deg r < \deg g$. Suppose $\deg f \geq \deg g$.
Let $f(x) = a_0 + a_1 x + \ldots + a_n x^n$, and $g(x) = b_0 + b_1 x + \ldots + b_m x^m$
where $a_n, b_m \neq 0$ and $n \geq m$. Now, form the polynomial:

$$f_1(x) = f(x) - \frac{a_n}{b_m} x^{n-m} g(x) . \qquad (2)$$

Then, $\deg f_1 < \deg f$. By induction, there exist polynomials $q_1(x)$ and $r(x)$ such that

$$f_1(x) = q_1(x) \, g(x) + r(x) .$$

Where either $r(x) = 0$ or deg $r <$ deg g. Substituting this into (2), and solving for $f(x)$ produces

$$f(x) = \left(q_1(x) + \frac{a_n}{b_m} x^{n-m}\right) g(x) + r(x)$$

or,

$$f(x) = q(x)\,g(x) + r(x)$$

where $q(x) = q_1(x) + \frac{a_n}{b_m} x^{n-m}$. When, for two given polynomials $f(x)$ and $g(x)$, $q(x)$ and $r(x)$ satisfy (1), we say that $f(x)$ has been divided by $g(x)$, that the quotient is $q(x)$ and that the remainder is $r(x)$. The given functions are:

$$f(x) = 2x^4 + x^3 + 4x^2 + 3x + 1 \quad \text{and} \quad g(x) = x^2 + 1.$$

Now,

$$f_1(x) = f(x) - \frac{a_n}{b_m} x^{n-m}\, g(x) = f(x) - 2x^2 g(x)$$

or,

$$f(x) = 2x^2 g(x) + f_1(x)\ . \tag{3}$$

But,

$$\begin{aligned} f_1(x) &= (2x^4 + x^3 + 4x^2 + 3x + 1) - 2x^2(x^2 + 1)\\ &= x^3 + 2x^2 + 3x + 1\ . \end{aligned}$$

Then,

$$f_2(x) = f_1(x) - xg(x)$$

or,

$$f_1(x) = xg(x) + f_2(x). \tag{4}$$

Also,

$$f_2(x) = x^3 + 2x^2 + 3x + 1 - x(x^2 + 1) = 2x^2 + 2x + 1.$$

Then,

$$f_3(x) = 2x^2 + 2x + 1 - 2(x^2 + 1) = 2x - 1.$$

Also,

$$f_2(x) = f_3(x) + 2g(x) = (2x - 1) + 2g(x)\ .$$

Substituting this into (4),

$$f_1(x) = xg(x) + (2x-1) + 2g(x).$$

Next, substituting this into (3) yields

$$f(x) = 2x^2 g(x) + xg(x) + (2x - 1) + 2g(x)$$

or,

$$f(x) = (2x^2 + x + 2)\,g(x) + (2x - 1)$$

Thus,

$$q(x) = 2x^2 + x + 2, \text{ and } r(x) = 2x - 1\ .$$

Another sometimes useful way to compute $q(x)$ and $r(x)$ is the following: Write down $g(x)\,\overline{f(x)}$, including the coefficients of each x^k where k is less than or equal to the degree of the polynomial. Then, proceed as in regular rational division.

Synthetic division:

$$2x^2 + x + 2$$

$$x^2 + 0x + 1 \overline{) 2x^4 + x^3 + 4x^2 + 3x + 1}$$

$$-(2x^4 + 0x^3 + 2x^2)\downarrow \quad \downarrow$$

$$0 + x^3 + 2x^2 + 3x \downarrow$$

$$-(x^3 + 0x^2 + x) \downarrow$$

$$0 + 2x^2 + 2x + 1$$

$$-(2x^2 + 0x + 2)$$

Remainder: $0 + 2x - 1$

Multiply the divisor by $2x^2$ and substract. Bring down the next term, $3x$. Now, multiply by x and subtract. When there are no more terms to bring down we are finished.

So, $q(x) = 2x^2 + x + 2$ and $r(x) = 2x - 1$. Thus, we have the same answer as found in the previous method.

$$f(x) = (2x^2 + x + 2) \, g(x) + (2x - 1).$$

CHARACTERISTIC AND MINIMUM POLYNOMIALS

● PROBLEM 9-3

(A)   Define the characteristic polynomial of the matrix A.

(B)   Let

$$A = \begin{bmatrix} 1 & 2 & -1 \\ 1 & 0 & 1 \\ 4 & -4 & 5 \end{bmatrix}.$$

Find the characteristic polynomial of A.

Solution: Let A be a $n \times n$ matrix with elements a_{ij} , for $i, j = 1, 2, \dots, n$. The matrix

$$\lambda I_n - A = \begin{bmatrix} \lambda - a_{11} & -a_{12} & \cdots & -a_{1n} \\ -a_{21} & \lambda - a_{22} & \cdots & -a_{2n} \\ \vdots & \vdots & & \vdots \\ -a_{n1} & -a_{n2} & \cdots & \lambda - a_{nn} \end{bmatrix}$$

where λ is an indeterminate and I_n is the n by n identity matrix is called the characteristic matrix of A. Its determinant

$$f(\lambda) = |\lambda I_n - A| \tag{1}$$

is called the characteristic polynomial of A.
   In order to find the highest term of this polynomial, use the fact that the value of the determinant is the sum of products of its elements, one taken from each row and each column. Thus, expanding $f(\lambda) = |\lambda I_n - A|$

yields a polynomial of degree n. The expression involving λ^n in the characteristic polynomial of A comes from the product $(\lambda-a_{11})(\lambda-a_{22})\ldots$ $(\lambda-a_{nn})$, the diagonal elements. All remaining products of the determinant have degree not higher than n-2, since, if one of the factors of such a product is $-a_{ij}$, where $i \neq j$, then this product does not contain the factors $\lambda-a_{ii}$, $\lambda-a_{jj}$. Thus, $f(\lambda) = (\lambda-a_{11})(\lambda-a_{22})\ldots(\lambda-a_{nn})$ + terms of degree not higher than n-2, or,

$$f(\lambda) = \lambda^n - (a_{11} + \ldots + a_{nn})\lambda^{n-1} + \ldots$$

Setting $\lambda = 0$ in (1) yields

$$f(0) = |-A| = (-1)^n |A| .$$

But, $f(0)$ is the constant term of the characteristic polynomial. Therefore the constant term of the characteristic polynomial of a matrix A is the determinant of this matrix multiplied by $(-1)^n$ where n is the order of A. A polynomial of degree n with real coefficients has n roots, some of which may be complex numbers.

(B)   From A, as given, compute the matrix

$$\lambda I - A \quad = \quad \begin{bmatrix} \lambda-1 & -2 & 1 \\ -1 & \lambda & -1 \\ -4 & 4 & \lambda-5 \end{bmatrix} .$$

Then, the characteristic polynomial of A is:

$$f(\lambda) = \det[\lambda I - A] = \begin{vmatrix} \lambda-1 & -2 & 1 \\ -1 & \lambda & -1 \\ -4 & 4 & \lambda-5 \end{vmatrix}$$

$$= (\lambda-1) \begin{vmatrix} \lambda & -1 \\ 4 & \lambda-5 \end{vmatrix} \quad -(-2) \begin{vmatrix} -1 & -1 \\ -4 & \lambda-5 \end{vmatrix} \quad +1 \begin{vmatrix} -1 & \lambda \\ -4 & 4 \end{vmatrix}$$

$$= (\lambda-1)[\lambda(\lambda-5) + 2[-1(\lambda-5)-4] + 1[-4+4\lambda]$$

$$f(\lambda) = (\lambda-1)[\lambda^2-5\lambda+4] - 2(\lambda-1) + 4(\lambda-1) = (\lambda-1)[\lambda^2-5\lambda+4-2+4]$$

$$= (\lambda-1)(\lambda-3)(\lambda-2)$$

or

$$f(\lambda) = \lambda^3 - 6\lambda^2 + 11\lambda - 6 .$$

It is worthwhile to note that $f(\lambda) = 0$ has roots 1, 3 and 2 which are called the characteristic roots or eigenvalues of A.

● PROBLEM 9-4

[A] Find the minimum polynomial $m(\lambda)$ of the matrix

$$A = \begin{bmatrix} 2 & 1 & 0 & 0 \\ 0 & 2 & 0 & 0 \\ 0 & 0 & 2 & 0 \\ 0 & 0 & 0 & 5 \end{bmatrix}$$

[B] Let A be a 3 by 3 matrix over the real field R. Show that A cannot be a zero of the polynomial, $\Phi(\lambda) = \lambda^2 + 1$.

Solution: Let A be an n-square matrix over a field R. The matrix $[\lambda I_n - A]$, where I_n is the n-square identity matrix and λ is an indeterminant, is called the characteristic matrix of A. Its determinant $f(\lambda) = \det(\lambda I_n - A)$ which is a polynomial in λ, is called the characteristic polynomial of A. The equation,

$$f(\lambda) = \det(\lambda I_n - A) = 0$$

is known as the characteristic equation of A.

The Cayley-Hamilton Theorem states that every matrix is a zero of its characteristic polynomial. Thus, if $f(\lambda) = \det(\lambda I_n - A)$, then $f(A) = 0$.

Consider all nonzero polynomials $\Phi(\lambda)$ which have the matrix A as a zero. Such a set is not empty since it contains at least, the characteristic polynomial of the matrix A. The nonzero polynomial of least degree with leading coefficient 1 is a member of the set $\varphi(\lambda)$ and is called the minimum polynomial of the matrix A. Every matrix A has a unique minimum polynomial. Also, the minimum polynomial of a matrix is a divisor of the characteristic polynomial of the matrix.

The characteristic polynomial of the given matrix A is $f(\lambda) = \det[\lambda I - A]$, or,

$$f(\lambda) = \det \begin{bmatrix} \lambda-2 & -1 & 0 & 0 \\ 0 & \lambda-2 & 0 & 0 \\ 0 & 0 & \lambda-2 & 0 \\ 0 & 0 & 0 & \lambda-5 \end{bmatrix}$$

The determinant of an upper triangular matrix is the product of the diagonal entries. Therefore,

$$f(\lambda) = (\lambda-2)(\lambda-2)(\lambda-2)(\lambda-5) .$$

The characteristic and minimum polynomials of a matrix A have the same irreducible factors. Thus, both $(\lambda-2)$ and $(\lambda-5)$ must be factors of $m(\lambda)$. Also, $m(\lambda)$ must divide $f(\lambda)$, hence, $m(\lambda)$ must be one of the following three polynomials:

$$m_1(\lambda) = (\lambda-2)(\lambda-5), \quad m_2(\lambda) = (\lambda-2)^2(\lambda-5),$$
$$m_3(\lambda) = (\lambda-2)^3(\lambda-5).$$

According to the Cayley-Hamilton Theorem, $m_3(\lambda) = f(\lambda)$ has the matrix A as a root. That is, $m_3(A) = f(A) = 0$.

Substituting A in the polynomial $m_1(\lambda)$ yields $m_1(A) = (A-2I)(A-5I)$

$$= \begin{bmatrix} 0 & 1 & 0 & 0 \\ 0 & 0 & 0 & 0 \\ 0 & 0 & 0 & 0 \\ 0 & 0 & 0 & 3 \end{bmatrix} \begin{bmatrix} -3 & 1 & 0 & 0 \\ 0 & -3 & 0 & 0 \\ 0 & 0 & -3 & 0 \\ 0 & 0 & 0 & 0 \end{bmatrix}$$

$$\neq 0 .$$

Therefore, $m_1(\lambda)$ cannot be the minimum polynomial of A since, by definition, A must be a root of its minimum polynomial:

$$m_2(\lambda) = m_2(A) = (A - 2I)^2(A - 5I)$$

$$m_2(A) = \begin{bmatrix} 0 & 0 & 0 & 0 \\ 0 & 0 & 0 & 0 \\ 0 & 0 & 0 & 0 \\ 0 & 0 & 0 & 9 \end{bmatrix} \begin{bmatrix} -3 & 1 & 0 & 0 \\ 0 & -3 & 0 & 0 \\ 0 & 0 & -3 & 0 \\ 0 & 0 & 0 & 0 \end{bmatrix}$$

$$= 0 .$$

Since $m_2(A) = 0$ and the degree of $m_2(\lambda)$ is less than the degree of $m_3(\lambda) = f(\lambda)$ we have

$$m_2(\lambda) = (\lambda-2)^2(\lambda-5) ,$$

the minimum polynomial of A.

[B] By the Cayley-Hamilton Theorem, the matrix A is a zero of its characteristic polynomial $f(\lambda)$. $f(\lambda)$ is of degree 3 (since A is a 3×3 matrix); hence, it has at least one real root. Let $\varphi(\lambda) = \lambda^2 + 1$. Suppose A is a zero of $\varphi(\lambda)$. Since $\varphi(\lambda)$ is irreducible, $\varphi(\lambda)$ must be the minimal polynomial of A. As we know, the equations $m_i(\lambda) = 0$ ($m_i(\lambda)$ are polynomials as described in part A) of this problem) and $f(\lambda) = 0$ have the same roots (but not necessarily with the same multiplicaties). Therefore, $\varphi(\lambda)$ and $f(\lambda)$ have the same roots. But the matrix A over the real field must have real roots while $\varphi(\lambda)$ has no real roots (since $\varphi(\lambda) = \lambda^2 + 1 = 0$ implies $\lambda = i$). This contradicts the fact that the characteristic and minimal polynomial have the same roots. Therefore, the 3×3 matrix A cannot be a zero of the polynomial $\varphi(\lambda) = \lambda^2 + 1$.

● **PROBLEM** 9-5

Find the characteristic polynomials and the eigenvalues of the matrices.

(i) $A = \begin{bmatrix} 2 & 3 \\ 1 & 4 \end{bmatrix}$; (ii) $B = \begin{bmatrix} \cos\alpha & \sin\alpha \\ -\sin\alpha & \cos\alpha \end{bmatrix}$;

(iii) $C = \begin{bmatrix} 1 & 2 & 3 \\ 2 & 1 & 3 \\ 3 & 3 & 6 \end{bmatrix}$

Solution: The characteristic polynomial of A is

$$f(\lambda) = \det(\lambda I - A) = \det \begin{bmatrix} \lambda-2 & -3 \\ -1 & \lambda-4 \end{bmatrix}$$

Therefore, $f(\lambda) = (\lambda-2)(\lambda-4) - 3 = \lambda^2 - 6\lambda + 5 = (\lambda-1)(\lambda-5)$.

The characteristic equation is $f(\lambda) = (\lambda-1)(\lambda-5) = 0$. Then the characteristic values are $\lambda = 1$ and $\lambda = 5$.

The zeros of the characteristic polynomial of a matrix are also called characteristic numbers or, proper values.

(ii) The characteristic polynomial of B is:

$$f(\lambda) = \det(\lambda I - B) = \det \begin{bmatrix} \lambda - \cos\alpha & -\sin\alpha \\ \sin\alpha & \lambda-\cos\alpha \end{bmatrix}$$

$$f(\lambda) = (\lambda - \cos\alpha)(\lambda - \cos\alpha) + \sin^2\alpha$$

$$= \lambda^2 - 2\lambda \cos\alpha + \cos^2\alpha + \sin^2\alpha .$$

But, $\sin^2\alpha + \cos^2\alpha = 1$. Therefore, $f(\lambda) = \lambda^2 - 2\lambda \cos \alpha + 1$. The characteristic equation is

$$f(\lambda) = \lambda^2 - 2 \cos \alpha \, \lambda + 1 = 0 .$$

We know the root of an equation, $ax^2 + bx + c = 0$ is

$$x = \frac{-b \pm \sqrt{b^2 - 4ac}}{2a} .$$

Thus,

$$\lambda = \frac{2\cos \alpha \pm \sqrt{4\cos^2\alpha - 4}}{2}$$

or,

$$\lambda = \frac{2\cos \alpha \pm 2\sqrt{\cos^2\alpha - 1}}{2}$$

$$= \cos \alpha \pm \sqrt{\cos^2\alpha - 1}$$

But, $\cos^2\alpha - 1 = -\sin^2\alpha$; therefore, $\lambda = \cos \alpha \pm \sqrt{-\sin^2\alpha}$ or,

$$\lambda = \cos \alpha \pm i \sin \alpha .$$

(iii) The characteristic polynomial of C is:

$$f(\lambda) = \det(\lambda I - C) = \det \begin{bmatrix} \lambda-1 & -2 & -3 \\ -2 & \lambda-1 & -3 \\ -3 & -3 & \lambda-6 \end{bmatrix}$$

$$= (\lambda-1) \begin{vmatrix} \lambda-1 & -3 \\ -3 & \lambda-6 \end{vmatrix} -(-2) \begin{vmatrix} -2 & -3 \\ -3 & \lambda-6 \end{vmatrix} + (-3) \begin{vmatrix} -2 & \lambda-1 \\ -3 & -3 \end{vmatrix}$$

$$f(\lambda) = (\lambda-1)[(\lambda-1)(\lambda-6) -9] + 2[-2(\lambda-6)-9] - 3[6 + 3(\lambda-1)]$$

$$= (\lambda-1)[\lambda^2 - 7\lambda - 3] + 2[-2\lambda + 3] - 3[3 + 3\lambda]$$

$$= \lambda^3 - 8\lambda^2 + 4\lambda + 3 - 4\lambda + 6 - 9 - 9\lambda$$

$$= \lambda^3 - 8\lambda^2 - 9\lambda = \lambda(\lambda-9)(\lambda+1) .$$

The characteristic equation is $f(\lambda) = \lambda(\lambda-9)(\lambda+1) = 0$. Then, the characteristic values are $\lambda = 0$, $\lambda = -1$ and $\lambda = 9$.

● PROBLEM 9-6

Find the minimal polynomials of the following matrices:

(i) $\begin{bmatrix} 3 & 1 \\ 0 & 3 \end{bmatrix}$; (ii) $\begin{bmatrix} 3 & 1 & 0 \\ 0 & 3 & 0 \\ 0 & 0 & 3 \end{bmatrix}$; (iii) $\begin{bmatrix} 2 & 0 & 0 \\ 0 & 3 & 1 \\ 0 & 0 & 3 \end{bmatrix}$;

(iv) $\begin{bmatrix} 2 & 0 & 0 & 0 \\ 0 & 2 & 0 & 0 \\ 0 & 0 & 3 & 0 \\ 0 & 0 & 0 & 3 \end{bmatrix}$

Solution: (i) The characteristic polynomial of the matrix A is

$$f(\lambda) = \det(\lambda I - A) = \det \begin{bmatrix} \lambda-3 & -1 \\ 0 & \lambda-3 \end{bmatrix}$$

$$= (\lambda-3)(\lambda-3) \ .$$

The minimum polynomial $m(\lambda)$ must divide $f(\lambda)$. Thus, $m(\lambda)$ is exactly one of the following:

$$m_1(\lambda) = (\lambda-3) \quad m_2(\lambda) = (\lambda-3)(\lambda-3) \ .$$

Now, $\qquad m_1(A) = (A-3I) \neq 0$.

$\qquad m_2(A) = f(A) = 0$ by the Cayley-Hamilton Theorem.

Therefore, the characteristic polynomial, $(\lambda-3)(\lambda-3)$, is the minimal polynomial of A.

(ii)

$$A = \begin{bmatrix} 3 & 1 & 0 \\ 0 & 3 & 0 \\ 0 & 0 & 3 \end{bmatrix}$$

The characteristic polynomial of A is

$$f(\lambda) = \det \begin{bmatrix} \lambda-3 & -1 & 0 \\ 0 & \lambda-3 & 0 \\ 0 & 0 & \lambda-3 \end{bmatrix}$$

$$= (\lambda-3)(\lambda-3)(\lambda-3),$$

since the determinant of an upper or lower triangular matrix is the product of the main diagonal elements. Now, $(\lambda-3)$ must be a factor of $m(\lambda)$, the minimal polynomial. Also, $m(\lambda)$ must divide $f(\lambda)$; hence, $m(\lambda)$ must be one of the following three polynomials:

$$m_1(\lambda) = (\lambda-3), \ m_2(\lambda) = (\lambda-3)(\lambda-3) \quad \text{or,}$$

$$m_3(\lambda) = (\lambda-3)(\lambda-3)(\lambda-3).$$

We see that $m_1(A) = (A-3I) \neq 0$, since

$$A-3I = \begin{bmatrix} 0 & 1 & 0 \\ 0 & 0 & 0 \\ 0 & 0 & 0 \end{bmatrix}$$

Therefore, $m_1(\lambda)$ is not the minimal polynomial. But,

$$m_2(A) = (A-3I)^2 = 0 \ .$$

Thus, $(\lambda-3)(\lambda-3)$ is a minimal polynomial of A.

(iii)

$$A = \begin{bmatrix} 2 & 0 & 0 \\ 0 & 3 & 1 \\ 0 & 0 & 3 \end{bmatrix}$$

The characteristic polynomial of A is

$$f(\lambda) = \det(\lambda I - A) = \det \begin{bmatrix} \lambda-2 & 0 & 0 \\ 0 & \lambda-3 & -1 \\ 0 & 0 & \lambda-3 \end{bmatrix}$$

$$= (\lambda-2)(\lambda-3)(\lambda-3) \ .$$

As we know, the minimum polynomial $m(\lambda)$ must be one of the following two polynomials:

$$m_1(\lambda) = (\lambda-2)(\lambda-3), \quad m_2(\lambda) = (\lambda-2)(\lambda-3)^2$$

$$m_1(A) = (A-2I)(A-3I)$$

$$= \begin{bmatrix} 0 & 0 & 0 \\ 0 & 1 & 1 \\ 0 & 0 & 1 \end{bmatrix} \begin{bmatrix} -1 & 0 & 0 \\ 0 & 0 & 1 \\ 0 & 0 & 0 \end{bmatrix} \neq 0 .$$

But, $m_2(A) = f(A) = 0$ (by the Cayley-Hamilton Theorem). Thus, $m_2(\lambda) = (\lambda-2)(\lambda-3)^2$ is the minimum polymial of A since $m_2(\lambda)$ is of least degree among the $m_i(\lambda)$'s and since $m_2(A) = 0$.

(iv)
$$A = \begin{bmatrix} 2 & 0 & 0 & 0 \\ 0 & 2 & 0 & 0 \\ 0 & 0 & 3 & 0 \\ 0 & 0 & 0 & 3 \end{bmatrix}$$

The determinant of an upper triangular matrix is the product of the diagonal entries. Therefore,

$$f(\lambda) = \det(\lambda I - A) = (\lambda-2)(\lambda-2)(\lambda-3)(\lambda-3).$$

Let $m_1(\lambda) = (\lambda-2)(\lambda-3)$. Then,

$$m_1(A) = (A-2I)(A-3I) = 0 .$$

Therefore, the minimum polymial of A is $(\lambda-2)(\lambda-3)$.

● **PROBLEM** 9-7

Find the characteristic and minimum polynomials of each of the following matrices

(a) $\begin{bmatrix} 3 & -1 \\ -1 & 3 \end{bmatrix}$ (b) $\begin{bmatrix} 1 & 1 \\ 0 & 2 \end{bmatrix}$ (c) $\begin{bmatrix} 1 & -2 \\ 1 & -1 \end{bmatrix}$ (d) $\begin{bmatrix} 1 & 1 \\ 0 & 1 \end{bmatrix}$

(e) $\begin{bmatrix} 0 & 1 & 0 & 0 \\ 0 & 0 & 0 & 0 \\ 0 & 0 & 1 & -2 \\ 0 & 0 & 1 & -1 \end{bmatrix}$ (f) $\begin{bmatrix} 3 & 1 & 0 & 0 \\ 0 & 3 & 0 & 0 \\ 0 & 0 & 2 & 1 \\ 0 & 0 & 1 & 2 \end{bmatrix}$

Solution: (a) The characteristic polynomial is

$$f(\lambda) = \det(\lambda I - A)$$

$$f(\lambda) = \det \begin{bmatrix} \lambda-3 & +1 \\ 1 & \lambda-3 \end{bmatrix}$$

$$= (\lambda-3)(\lambda-3) - 1$$

$$= \lambda^2 - 6\lambda + 8 = (\lambda-4)(\lambda-2) . \tag{1}$$

Then the characteristic equation is

$$f(\lambda) = \det(\lambda I - A) = 0 .$$

Thus,
$$(\lambda-4)(\lambda-2) = 0$$

and the characteristic values are $\lambda = 4$, and $\lambda = 2$.

Let $m(\lambda)$ be the minimum polynomial of A. Then, λ is a character-

istic value of A if and only if $m(\lambda) = 0$. Now, for the characteristic values, $\lambda = 4$ and $\lambda = 2$, we have

$$(\lambda-4)(\lambda-2) = 0 .$$

Therefore, the minimum polynomial $m(\lambda) = (\lambda-4)(\lambda-2)$.

(b) The characteristic polynomial is

$$f(\lambda) = \det \begin{bmatrix} \lambda-1 & -1 \\ 0 & \lambda-2 \end{bmatrix} = (\lambda-1)(\lambda-2)$$

and the characteristic values are $\lambda = 1$ and $\lambda = 2$. For these values,
$(\lambda-4)(\lambda-2)$ is equal to zero. Therefore, the minimum polynomial is
$(\lambda-4)(\lambda-2)$.

(c) $$f(\lambda) = \det \begin{bmatrix} \lambda-1 & 2 \\ -1 & \lambda+1 \end{bmatrix} = (\lambda-1)(\lambda+1) + 2 = \lambda^2 + 1 .$$

$f(\lambda) = \lambda^2 + 1$ has no real roots (roots of $\lambda^2 + 1$ are $\pm i$). Therefore, it is irreducible over R (that is, it cannot be factored into
any lesser degree polynomials in R[X], the set of polynomials with coefficients in the real numbers). The matrix A, however is a root of
$\lambda^2 + 1$ by the Cayley-Hamilton Theorem. Therefore, $f(\lambda) = \lambda^2 + 1$ is
the polynomial of least degree over the reals having A as a root. So,
$m(\lambda) = \lambda^2 + 1$.

(d) $$A = \begin{bmatrix} 1 & 1 \\ 0 & 1 \end{bmatrix}$$

Then,

$$f(\lambda) = \det \begin{bmatrix} \lambda-1 & -1 \\ 0 & \lambda-1 \end{bmatrix} = (\lambda-1)^2 .$$

The minimum polynomial must divide $f(\lambda)$ and each irreducible factor
of $f(\lambda)$ must be a factor of $m(\lambda)$. Therefore, the minimal polynomial
is either $m_1(\lambda) = (\lambda-1)$ or $m_2(\lambda) = (\lambda-1)^2$. But, $m_1(A) = (A - 1I) \neq 0$.
$m_2(A) = (A - 1I)^2 = 0$ (by the Cayley-Hamilton Theorem). Therefore,
$m(\lambda) = (\lambda-1)^2$ is the minimum polynomial.

(e)
$$f(\lambda) = \det \begin{bmatrix} \lambda & -1 & 0 & 0 \\ 0 & \lambda & 0 & 0 \\ 0 & 0 & \lambda-1 & 2 \\ 0 & 0 & -1 & \lambda+1 \end{bmatrix} = \lambda \begin{vmatrix} \lambda & 0 & 0 \\ 0 & \lambda-1 & 2 \\ 0 & -1 & \lambda+1 \end{vmatrix}$$

$$= \lambda[\lambda\{(\lambda-1)(\lambda+1) + 2\}]$$

$$= \lambda^2(\lambda^2 + 1) .$$

Then, $m(\lambda)$ is exactly one of the following:

$$m_1(\lambda) = \lambda(\lambda^2 + 1) , \quad m_2(\lambda) = \lambda^2(\lambda^2 + 1).$$

We have:

$$m_1(A) = A(A^2 + 1I)$$

$$= \begin{bmatrix} 0 & 1 & 0 & 0 \\ 0 & 0 & 0 & 0 \\ 0 & 0 & 1 & -2 \\ 0 & 0 & 1 & -1 \end{bmatrix} \begin{bmatrix} 1 & 0 & 0 & 0 \\ 0 & 1 & 0 & 0 \\ 0 & 0 & 0 & 0 \\ 0 & 0 & 0 & 0 \end{bmatrix}$$

$$= \begin{bmatrix} 0 & 1 & 0 & 0 \\ 0 & 0 & 0 & 0 \\ 0 & 0 & 0 & 0 \\ 0 & 0 & 0 & 0 \end{bmatrix} \neq 0$$

By the Cayley-Hamilton Theorem, $m_2(A) = f(A) = 0$. Therefore, $\lambda^2(\lambda^2 + 1)$ is a minimal polynomial of A.

(f)
$$f(\lambda) = \det \begin{bmatrix} \lambda-3 & -1 & 0 & 0 \\ 0 & \lambda-3 & 0 & 0 \\ 0 & 0 & \lambda-2 & -1 \\ 0 & 0 & -1 & \lambda-2 \end{bmatrix}$$

$$= \lambda-3 \begin{vmatrix} \lambda-3 & 0 & 0 \\ 0 & \lambda-2 & -1 \\ 0 & -1 & \lambda-2 \end{vmatrix}$$

$$= (\lambda-3)[(\lambda-3)\{(\lambda-2)^2 - 1\}]$$

$$= (\lambda-3)[(\lambda-3)(\lambda^2-4\lambda+3)]$$

$$= (\lambda-3)^3(\lambda-1) .$$

Therefore, $m(\lambda)$ must be one of the following:

$$m_1(\lambda) = (\lambda-3)(\lambda-1) , \quad m_2(\lambda) = (\lambda-3)^2(\lambda-1)$$

and

$$m_3(\lambda) = (\lambda-3)^3(\lambda-1) .$$

$m_1(A) = (A - 3I)(A - I)$

$$= \begin{bmatrix} 0 & 1 & 0 & 0 \\ 0 & 0 & 0 & 0 \\ 0 & 0 & -1 & 1 \\ 0 & 0 & 1 & -1 \end{bmatrix} \begin{bmatrix} 2 & 1 & 0 & 0 \\ 0 & 2 & 0 & 0 \\ 0 & 0 & 1 & 1 \\ 0 & 0 & 1 & 1 \end{bmatrix} \neq 0 .$$

$m_2(A) = (A - 3I)^2(A - I)$

$$= \begin{bmatrix} 0 & 1 & 0 & 0 \\ 0 & 0 & 0 & 0 \\ 0 & 0 & -1 & 1 \\ 0 & 0 & 1 & -1 \end{bmatrix}^2 \begin{bmatrix} 2 & 1 & 0 & 0 \\ 0 & 2 & 0 & 0 \\ 0 & 0 & 1 & 1 \\ 0 & 0 & 1 & 1 \end{bmatrix}$$

$$= \begin{bmatrix} 0 & 0 & 0 & 0 \\ 0 & 0 & 0 & 0 \\ 0 & 0 & 2 & -2 \\ 0 & 0 & -2 & 2 \end{bmatrix} \begin{bmatrix} 2 & 1 & 0 & 0 \\ 0 & 2 & 0 & 0 \\ 0 & 0 & 1 & 1 \\ 0 & 0 & 1 & 1 \end{bmatrix} = 0$$

We know that $m_3(A) = f(A) = 0$ by the Cayley-Hamilton Theorem. But, the degree of $m_2(\lambda)$ is less than the degree of $m_3(\lambda)$. Therefore, $m_2(\lambda) = (\lambda-3)^2(\lambda-1)$ is the minimum polynomial of A.

Let

$$A = \begin{bmatrix} 2 & 0 & 0 \\ 0 & 3 & 0 \\ 0 & 0 & 3 \end{bmatrix} \quad ; \quad B = \begin{bmatrix} 2 & 0 & 0 \\ 0 & 2 & 0 \\ 0 & 0 & 3 \end{bmatrix}$$

Show that A and B are not similar, but that they both have the same minimum polynomial.

Solution: Recall that two matrices, A and B, are similar if there exists a nonsingular matrix P such that $B = P^{-1}AP$. It is known that if matrices A and B are similar, they have the same characteristic equation. So, first determine the characteristic equations for A and B and see if they are unequal.

The characteristic polynomial of A is

$$f_A(\lambda) = \det \begin{bmatrix} \lambda-2 & 0 & 0 \\ 0 & \lambda-3 & 0 \\ 0 & 0 & \lambda-3 \end{bmatrix}$$

$$f_A(\lambda) = (\lambda-2)(\lambda-3)(\lambda-3) .$$

Now, the characteristic polynomial of B is

$$f_B(\lambda) = \det \begin{bmatrix} \lambda-2 & 0 & 0 \\ 0 & \lambda-2 & 0 \\ 0 & 0 & \lambda-3 \end{bmatrix}$$

$$= (\lambda-2)(\lambda-2)(\lambda-3) .$$

Since polynomials $f_A(\lambda)$ and $f_B(\lambda)$ are different, A and B are not similar. The minimal polynomial of A must coincide with one of the following polynomials:

$$m_1(\lambda) = (\lambda-2)(\lambda-3), \quad m_2(\lambda) = (\lambda-2)^2(\lambda-3).$$

Substituting A for λ, it is seen that $m_1(A) = (A - 2I)(A - 3I) = 0$ and by the Cayley-Hamilton Theorem, $m_2(A) = (\lambda-2)^2(\lambda-3) = 0$. But, the degree of $(\lambda-2)(\lambda-3)$ (i.e., 2) is less than the degree of $(\lambda-2)^2(\lambda-3)$ (i.e., 3). Therefore, $(\lambda-2)(\lambda-3)$ is the minimum polynomial of A.

In exactly the same way we find that the minimum polynomial of the matrix B is also the polynomial $(\lambda-2)(\lambda-3)$. Thus, the minimum polynomials of the matrices A and B are equal, but A and B are not similar. In general, nonsimilar matrices may or may not have the same minimum polynomial.

Let T be the linear operator on R^3 which is represented in the standard ordered basis by the matrix

$$A = \begin{bmatrix} 5 & -6 & -6 \\ -1 & 4 & 2 \\ 3 & -6 & -4 \end{bmatrix} .$$

Find the minimal polynomial of T.

Solution: The minimal polynomial of T is the monic polynomial of lowest degree $m(x)$, such that $m(T) = 0$. As a start towards finding $m(T)$, note that $|\lambda I - A|$, the characteristic polynomial of A (and, thus, T) also equals zero. Hence, the minimal polynomial either equals the characteristic polynomial or divides it.

The characteristic polynomial of A is

$$\begin{vmatrix} \lambda-5 & 6 & 6 \\ 1 & \lambda-4 & -2 \\ -3 & 6 & \lambda+4 \end{vmatrix} = 0$$

or, $(\lambda-5)[(\lambda-4)(\lambda+4)+12]-6[\lambda+4-6] + 6[6+3\lambda-12] = 0$

or, $(\lambda-1)(\lambda-2)^2 = 0$.

It may be verified that T is diagonalizable with $\lambda_1 = 1$; $\lambda_2 = 2$; $\lambda_3 = 2$ the entries along the diagonal. This fact can be used to unearth the minimal polynomial.

Let T be a diagonalizable linear operator and let $c_1, \ldots, c_k$ be the distinct eigenvalues of T. Then the minimal polynomial must be

$$m(x) = (x - c_1) \ldots (x - c_k) .$$

Hence, the minimal polynomial of A must be

$$m(\lambda) = (\lambda-1)(\lambda-2).$$

To review, if T is diagonalizable, the characteristic polynomial for T will be of the form

$$f(\lambda) = (\lambda - c_1)^{k_1} \ldots (\lambda - c_n)^{k_n}$$

where the c_i represent the eigenvalues and the k_i represents the multiplicities. The minimal polynomial will then be

$$m(\lambda) = (\lambda - c_1) \ldots (\lambda - c_n) .$$

● **PROBLEM** 9-10

Let $T: R^n \to R^n$ be defined by

$$Te_j = \begin{cases} e_{j+1} & 0 \le j \le n-1 \\ 0 & j = n \end{cases}$$

where e_i are the standard basis elements of R^n,

$$e_1 = (1,0,\ldots,0), \ e_2 = (0,1,\ldots,0) \ldots$$
$$\ldots e_j = (0,0,\ldots,1,0,\ldots,0) \ldots e_n = (0,0,\ldots,1).$$

Find the minimal polynomial of T.

Solution: First, explore the linear operator T. We have

$$Te_1 = e_2 \ ; \ Te_2 = e_3 \ ; \ \ldots \ Te_{n-1} = e_n \ ; \ Te_n = 0 .$$

Notice $T(T(e_1)) = T(e_2) = e_3$, so $T^2 e_1 = e_3$. e_j can also be expressed as the functions of the transformation. That is,

$$e_1 = Ie_1 = T^0 e_1 \ ; \ e_2 = T^1 e_1 \ , \ e_3 = T^2 e_1 \ldots e_n = T^{n-1} e_1 \ , \ 0 = T^n e_1 .$$

Having seen that $e_j = T^{j-1} e_1$ $(j = 1, \ldots, n)$, apply T^n to both sides of the equation. The result is $T^n e_j = T^n(T^{j-1} e_1) = T^{j-1}(T^n e_1)$.

Since $T^n e_1 = 0$, $T^n e_j = T^{j-1}(0) = 0$ for all $j = 1,...,n$. Therefore, $T^n \equiv 0$ and T satisfies a polynomial function of degree n. T cannot satisfy a polynomial of lower degree than n since $k < n$ implies

$$(T^k + a_{k-1}T^{k-1} + \ldots + a_1 T + a_0 I)e_1 = e_{k+1} + a_{k-1}e_k + \ldots + a_1 e_2 + a_0 e_1 \neq 0$$

(since the e_j are linearly independent). This shows that λ^n is the minimal function of T.

● **PROBLEM** 9-11

Show that the following theorem is true: If two matrices are similar, then they have the same characteristic polynomial. Then show, by means of a counter-example, that the converse is false.

Solution: It is true that the characteristic polynomials of similar matrices are equal. To show this, let A and B be similar matrices, i.e., there exists a nonsingular matrix X such that

$$A = X^{-1}B X$$

Then, the characteristic polynomial of A is

$$|\lambda I - A| = |\lambda I - X^{-1}BX| = |X^{-1}\lambda IX - X^{-1} BX|$$

$$= |X^{-1}(\lambda I - B)X| = |X^{-1}| \; |\lambda I - B| \; |X|$$

(by the properties of the determinant function, $|PQ| = |P||Q|$). But the product of the determinants, $|X^{-1}||X|$, is 1. Hence,

$$|\lambda I - A| = |\lambda I - B| .$$

The theorem we are asked to disprove has an inference from the equality of characteristic polynomials to the similarity of matrices. Thus, to show it is false, it is necessary to exhibit only two non-similar matrices that have the same characteristic equation.

Let

$$E = \begin{bmatrix} 1 & 0 \\ 0 & 1 \end{bmatrix} \quad \text{and} \quad A = \begin{bmatrix} 1 & 1 \\ 0 & 1 \end{bmatrix} .$$

The characteristic polynomials of E and A are, respectively,

$$|\lambda I - E| = (\lambda - 1)^2 \quad \text{and} \quad |\lambda I - A| = (\lambda - 1)^2 .$$

Thus, E and A have the same characteristic polynomial. However, they are not similar. The definition of similarity states that A is similar to B if there exists an invertible matrix P such that

$$A = P^{-1}BP .$$

But, for any invertible matrix X (since E is the 2×2 identity matrix,)

$$X^{-1}EX = X^{-1}X = E \neq A .$$

POLYNOMIAL OPERATIONS

If $F(X) = \begin{bmatrix} 1 & 0 & 1 \\ 2 & 1 & 1 \\ 1 & 1 & 1 \end{bmatrix} - \begin{bmatrix} 2 & 1 & 0 \\ -1 & 1 & 1 \\ 0 & 1 & 0 \end{bmatrix} X + \begin{bmatrix} 1 & 1 & 1 \\ 1 & 0 & 1 \\ 0 & 1 & 0 \end{bmatrix} X^2 ,$

and $B = \begin{bmatrix} 1 & 1 & 1 \\ 1 & 0 & 1 \\ 0 & 1 & 1 \end{bmatrix}$, find $F_L(B)$ and $F_R(B)$.

Solution: A matrix polynomial of order n is defined as:

$$F(X) = A_0 + A_1 X + A_2 X^2 + \ldots + A_m X^m \tag{1}$$

where A_0, A_1, ..., A_m are $n \times n$ matrices. The degree of the matrix polynomial (1) is the largest integer K for which $A_K \neq 0$. The co-erficient A_K of a polynomial $F(X)$ of degree K is called the leading coefficient of $F(X)$. Two matrix polynomials of order n,

$$F(X) = A_0 + A_1 X + A_2 X^2 + \ldots + A_m X^m ,$$

$$G(X) = B_0 + B_1 X + B_2 X^2 + \ldots + B_m X^m ,$$

are said to be equal if $A_i = B_i$; $i = 0, 1, 2, \ldots, m$.

If $F(X)$ is the matrix polynomial of order n given by (1), and B is an $n \times n$ matrix, then $F_L(B)$ and $F_R(B)$ are defined as follows:

$$F_L(B) = A_0 + B A_1 + B^2 A_2 + \ldots + B^m A_m$$

$$F_R(B) = A_0 + A_1 B + A_2 B^2 + \ldots + A_m B^m .$$

Now, from the given example, we have

$$A_0 = \begin{bmatrix} 1 & 0 & 1 \\ 2 & 1 & 1 \\ 1 & 1 & 1 \end{bmatrix}, \quad A_1 = -\begin{bmatrix} 2 & 1 & 0 \\ -1 & 1 & 1 \\ 0 & 1 & 0 \end{bmatrix}$$

and $A_2 = \begin{bmatrix} 1 & 1 & 1 \\ 1 & 0 & 1 \\ 0 & 1 & 0 \end{bmatrix}$.

Then,

$$F_L(B) = \begin{bmatrix} 1 & 0 & 1 \\ 2 & 1 & 1 \\ 1 & 1 & 1 \end{bmatrix} - \begin{bmatrix} 1 & 1 & 1 \\ 1 & 0 & 1 \\ 0 & 1 & 1 \end{bmatrix} \begin{bmatrix} 2 & 1 & 0 \\ -1 & 1 & 1 \\ 0 & 1 & 0 \end{bmatrix}$$

$$+ \begin{bmatrix} 1 & 1 & 1 \\ 1 & 0 & 1 \\ 0 & 1 & 1 \end{bmatrix}^2 \begin{bmatrix} 1 & 1 & 1 \\ 1 & 0 & 1 \\ 0 & 1 & 0 \end{bmatrix} .$$

267

Performing the matrix multiplication yields

$$F_L(B) = \begin{bmatrix} 1 & 0 & 1 \\ 2 & 1 & 1 \\ 1 & 1 & 1 \end{bmatrix} - \begin{bmatrix} 1 & 3 & 1 \\ 2 & 2 & 0 \\ -1 & 2 & 1 \end{bmatrix} + \begin{bmatrix} 4 & 5 & 4 \\ 3 & 3 & 3 \\ 2 & 3 & 2 \end{bmatrix}$$

$$= \begin{bmatrix} 4 & 2 & 4 \\ 3 & 2 & 4 \\ 4 & 2 & 2 \end{bmatrix} .$$

Then,

$$F_R(B) = \begin{bmatrix} 1 & 0 & 1 \\ 2 & 1 & 1 \\ 1 & 1 & 1 \end{bmatrix} - \begin{bmatrix} 2 & 1 & 0 \\ -1 & 1 & 1 \\ 0 & 1 & 0 \end{bmatrix} \begin{bmatrix} 1 & 1 & 1 \\ 1 & 0 & 1 \\ 0 & 1 & 1 \end{bmatrix}$$

$$+ \begin{bmatrix} 1 & 1 & 1 \\ 1 & 0 & 1 \\ 0 & 1 & 0 \end{bmatrix} \begin{bmatrix} 1 & 1 & 1 \\ 1 & 0 & 1 \\ 0 & 1 & 1 \end{bmatrix}^2$$

$$= \begin{bmatrix} 1 & 0 & 1 \\ 2 & 1 & 1 \\ 1 & 1 & 1 \end{bmatrix} - \begin{bmatrix} 3 & 2 & 3 \\ 0 & 0 & 1 \\ 1 & 0 & 1 \end{bmatrix} + \begin{bmatrix} 4 & 5 & 7 \\ 3 & 3 & 5 \\ 1 & 2 & 2 \end{bmatrix}$$

$$= \begin{bmatrix} 2 & 3 & 5 \\ 5 & 4 & 5 \\ 1 & 3 & 2 \end{bmatrix} .$$

● **PROBLEM** 9-13

Let
$$\varphi(\lambda) = -2 - 5\lambda + 3\lambda^2 , \quad A = \begin{bmatrix} 1 & 2 \\ 3 & 1 \end{bmatrix} . \quad \text{Show that}$$

$$\varphi(A) = \begin{bmatrix} 14 & 2 \\ 3 & 14 \end{bmatrix} .$$

Solution: Consider a polynomial in λ,

$$\Phi(\lambda) = a_0 + a_1\lambda + \dots + a_n\lambda^n ,$$

where $a_0, a_1, \dots, a_n$ are elements from a field R. The expression

$$a_0 I + a_1 A + \dots + a_n A^n ,$$

where A is an arbitrary square matrix with elements from R and I is an identity matrix, is called a polynomial in A and is denoted by $\varphi(A)$. The expression $\varphi(A)$ is also called the value of $\varphi(\lambda)$ for $\lambda = A$. For example, consider the equality

$$\lambda^2 - 1 = (\lambda-1)(\lambda+1) .$$

Evaluate the right and left sides for $\lambda = A$. Then,

$$A^2 - I = (A - I)(A + I) .$$

Similarly, for $\lambda^3 + 1 = (\lambda+1)(\lambda^2-\lambda+1)$ we have

$$A^3 + I = (A+I)(A^2-A+I) .$$

In general, from such relations between polynomials in λ, one obtains a matrix identity. Given

$$\varphi(\lambda) = -2-5\lambda + 3\lambda^2 ,$$

then

$$\varphi(\lambda) = 3\lambda^2 - 5\lambda - 2 = (\lambda-2)(3\lambda+1) .$$

Therefore,

$$\varphi(A) = (A - 2I)(3A + I) \qquad (1)$$

$$A = \begin{bmatrix} 1 & 2 \\ 3 & 1 \end{bmatrix}$$

$$A - 2I = \begin{bmatrix} 1 & 2 \\ 3 & 1 \end{bmatrix} -2 \begin{bmatrix} 1 & 0 \\ 0 & 1 \end{bmatrix} = \begin{bmatrix} -1 & 2 \\ 3 & -1 \end{bmatrix}$$

and

$$3A + I = 3 \begin{bmatrix} 1 & 2 \\ 3 & 1 \end{bmatrix} + \begin{bmatrix} 1 & 0 \\ 0 & 1 \end{bmatrix} = \begin{bmatrix} 4 & 6 \\ 9 & 4 \end{bmatrix}$$

Substitute this into (1). Then,

$$\varphi(A) = \begin{bmatrix} -1 & 2 \\ 3 & -1 \end{bmatrix} \begin{bmatrix} 4 & 6 \\ 9 & 4 \end{bmatrix} = \begin{bmatrix} -4+18 & -6+8 \\ 12-9 & 18-4 \end{bmatrix}$$

$$= \begin{bmatrix} 14 & 2 \\ 3 & 14 \end{bmatrix} .$$

● **PROBLEM 9-14**

Find the minimal polynomial of

$$A = \begin{bmatrix} 9 & -2 & 2 \\ -8 & 3 & -2 \\ -48 & 12 & -11 \end{bmatrix}$$

Solution: The method that is normally used for finding the minimal polynomial of a given matrix A is to find the characteristic polynomial $f(\lambda)$, to factor $f(\lambda)$ into linear factors, and then to find $m(\lambda)$, the monic divisor of $f(\lambda)$ of least degree for which $m(\lambda) = m(A) = 0$. This method does not always work, however, because sometimes it is difficult to carry out the needed factorization of $f(\lambda)$. Now, the following theorem describes another method for finding $m(\lambda)$.

Theorem: For each $n \times n$ matrix A, $m(\lambda) = f(\lambda)/d(\lambda)$, where $d(\lambda)$ is the monic greatest common divisor of the polynomial entries of $adj(\lambda I - A)$.
Given

$$A = \begin{bmatrix} 9 & -2 & 2 \\ -8 & 3 & -2 \\ -48 & 12 & -11 \end{bmatrix}$$

The characteristic polynomial $f(\lambda)$ is

$$f(\lambda) = \det(\lambda I - A) = \det \begin{bmatrix} \lambda-9 & 2 & -2 \\ 8 & \lambda-3 & 2 \\ 48 & -12 & \lambda+11 \end{bmatrix}$$

$$
\begin{aligned}
f(\lambda) &= (\lambda-9)[(\lambda-3)(\lambda+11) + 24] - 2[8(\lambda+11) - 96] - 2[-96 - 48(\lambda-3)] \\
&= (\lambda-9)[(\lambda+9)(\lambda-1)] - 16[(\lambda-1)] + 96[(\lambda-1)] \\
&= (\lambda-1)[(\lambda-9)(\lambda+9) - 16 + 96] \\
&= (\lambda-1)[\lambda^2 - 81 + 80] = (\lambda-1)(\lambda^2-1) \ .
\end{aligned}
$$

Now, the transpose of the matrix of cofactors of A is called the adjoint of A.

$$\lambda I - A = \begin{bmatrix} \lambda-9 & 2 & -2 \\ 8 & \lambda-3 & 2 \\ 48 & -12 & \lambda+11 \end{bmatrix} \ .$$

Then,

$$\text{adj}(\lambda I - A) = \begin{bmatrix} \lambda^2+8\lambda-9 & -2\lambda+2 & 2\lambda-2 \\ -8\lambda+8 & \lambda^2+2\lambda-3 & 2\lambda-2 \\ -48\lambda+48 & 12\lambda-12 & \lambda^2-12\lambda+11 \end{bmatrix}$$

or,

$$\text{adj}(\lambda I - A) = \begin{bmatrix} (\lambda+9)(\lambda-1) & -2(\lambda-1) & 2(\lambda-1) \\ -8(\lambda-1) & (\lambda+3)(\lambda-1) & 2(\lambda-1) \\ -48(\lambda-1) & 12(\lambda-1) & (\lambda-11)(\lambda-1) \end{bmatrix}$$

The greatest common divisor of the entries of $\text{adj}(\lambda I - A)$ is $d(\lambda) = \lambda-1$. Therefore,

$$m(\lambda) = f(\lambda)/d(\lambda) = (\lambda-1)(\lambda^2-1)/(\lambda-1)$$
$$= (\lambda^2-1) \ .$$

The above method for finding minimal polynomial $m(\lambda)$ always works, but it requires long computations. Thus, it would be used only when needed, as a general method, or when the characteristic polynomial $f(\lambda)$ is unfactorable.

● **PROBLEM** 9-15

Find the companion matrix of the following polynomials.
(i) $(\lambda-1)(\lambda-2)$, (ii) $(\lambda-1)^2$, (iii) $(\lambda-1)(\lambda-2)(\lambda-3)$.

Solution: Let
$$f(\lambda) = a_0 + a_1\lambda + \ldots + a_{n-1}\lambda^{n-1} + \lambda^n$$
be an arbitrary nonconstant polynomial with leading coefficient 1. Let the coefficients of this polynomial lie in a field R. The matrix

$$C = \begin{bmatrix} 0 & 0 & 0 & \ldots & 0 & -a_0 \\ 1 & 0 & 0 & \ldots & 0 & -a_1 \\ 0 & 1 & 0 & \ldots & 0 & -a_2 \\ \cdot & \cdot & \cdot & \cdot & \cdot & \cdot \\ 0 & 0 & 0 & \ldots & 0 & -a_{n-2} \\ 0 & 0 & 0 & \ldots & 1 & -a_{n-1} \end{bmatrix}$$

is called the companion matrix of the polynomial $f(\lambda)$.

(i) $f(\lambda) = (\lambda-1)(\lambda-2) = \lambda^2 - 3\lambda + 2$. Here, $a_0 = 2$, $a_1 = -3$ and $n = 2$. Thus, we have the 2×2 matrix,

$$C = \begin{bmatrix} 0 & -2 \\ 1 & 3 \end{bmatrix}$$

This is the companion matrix.

(ii) $f(\lambda) = (\lambda-1)^2 = \lambda^2 - 2\lambda + 1$. Here, $a_0 = 1$, $a_1 = -2$. Then, the companion matrix is

$$C = \begin{bmatrix} 0 & -1 \\ 1 & 2 \end{bmatrix} .$$

(iii) $f(\lambda) = (\lambda-1)(\lambda-2)(\lambda-3) = \lambda^3 - 6\lambda^2 + 11\lambda - 6$. Here, $a_0 = -6$, $a_1 = 11$, $a_2 = -6$. The companion matrix is

$$C = \begin{bmatrix} 0 & 0 & 6 \\ 1 & 0 & -11 \\ 0 & 1 & 6 \end{bmatrix}$$

● **PROBLEM** 9-16

Let $A = \begin{bmatrix} 9 & 5 & -4 \\ -8 & -4 & 4 \\ 2 & 2 & 0 \end{bmatrix}$. Find the square root of A.

Solution: Let A be an nxn matrix which is similar to a diagonal matrix. Suppose that the characteristic values of A are all positive real numbers. Then, there is one and only one nxn matrix B such that $B^2 = A$. Also, there is a polynomial $f(x)$ with real coefficients such that $B = f(A)$. In other words, one can find the square root B of any matrix A which satisfies this hypotheses.

Now, the characteristic polynomial of A is

$$f(\lambda) = \det(\lambda I - A) = \det \begin{bmatrix} \lambda-9 & -5 & 4 \\ 8 & \lambda+4 & -4 \\ -2 & -2 & \lambda \end{bmatrix}$$

$$= (\lambda-9) \begin{vmatrix} \lambda+4 & -4 \\ -2 & \lambda \end{vmatrix} - (-5) \begin{vmatrix} 8 & -4 \\ -2 & \lambda \end{vmatrix} + 4 \begin{vmatrix} 8 & \lambda+4 \\ -2 & -2 \end{vmatrix}$$

$$= (\lambda-9)[\lambda^2 + 4\lambda - 8] + 5[8\lambda - 8] + 4[2\lambda - 8]$$
$$= \lambda^3 - 5\lambda^2 + 4\lambda = \lambda(\lambda-1)(\lambda-4) .$$

Then, the characteristic equation of A is $\lambda(\lambda-1)(\lambda-4) = 0$. Hence, the characteristic values are 0,1, and 4. Since A has three distinct characteristic values, A is similar to a diagonal matrix. Then, from the given hypotheses, $B^2 = A$ and $B = f(A)$.

Let $f(x)$ be a scalar polynomial for which $f(\lambda_i) = \sqrt{\lambda_i}$, $i = 1,2,\ldots,n$. Thus, $f(0) = 0$, $f(1) = 1$, and $f(4) = 2$. If $f(x) = ax^2 + bx + c$, then

$$f(0) = c = 0$$
$$f(1) = a + b + c = 1$$

and

$$f(4) = 16a + 4b + c = 2 .$$

Solving the above equations yields

$$a = - 1/6, \quad b = 7/6 , \quad \text{and} \quad c = 0 .$$

Thus, $\qquad f(x) = - 1/6x^2 + 7/16x .$

Then,

$$B = f(A) = - 1/6 \ A^2 + 7/6 \ A$$

$$= - \frac{1}{6} \begin{bmatrix} 9 & 5 & -4 \\ -8 & -4 & 4 \\ 2 & 2 & 0 \end{bmatrix}^2 + \frac{7}{6} \begin{bmatrix} 9 & 5 & -4 \\ -8 & -4 & 4 \\ 2 & 2 & 0 \end{bmatrix}$$

$$= - \frac{1}{6} \begin{bmatrix} 33 & 17 & -16 \\ -32 & -16 & 16 \\ 2 & 2 & 0 \end{bmatrix} + \frac{7}{6} \begin{bmatrix} 9 & 5 & -4 \\ -8 & -4 & 4 \\ 2 & 2 & 0 \end{bmatrix}$$

$$= + \frac{1}{6} \begin{bmatrix} -33 & -17 & 16 \\ 32 & 16 & -16 \\ -2 & -2 & 0 \end{bmatrix} + \frac{1}{6} \begin{bmatrix} 63 & 35 & -28 \\ -56 & -28 & 28 \\ 14 & 14 & 0 \end{bmatrix}$$

$$B = f(A) = \frac{1}{6} \begin{bmatrix} 30 & 18 & -12 \\ -24 & -12 & 12 \\ 12 & 12 & 0 \end{bmatrix} = \begin{bmatrix} 5 & 3 & -2 \\ -4 & -2 & 2 \\ 2 & 2 & 0 \end{bmatrix}$$

Hence,

$$B = f(A) = \begin{bmatrix} 5 & 3 & -2 \\ -4 & -2 & 2 \\ 2 & 2 & 0 \end{bmatrix} \quad \text{is a square root of } A.$$

● **PROBLEM** 9-17

Let V be the vector space R^2 and let T be the operator defined by
$$T(xy) = (2x-y, \ x+y).$$
Let $f(x) = 2 + 3x$ and $g(x) = x + x^2$. Find $f(T)$ and $g(T)$.

Solution: Consider a polynomial $f(x)$ over a field R:
$$f(x) = a_0 + a_1 x + a_2 x^2 + \ldots + a_n x^n .$$
Now, suppose $T: V \to V$ is a linear operator on a vector space V over R . Then, define
$$f(T) = a_0 I + a_1 T + a_2 T^2 + \ldots + a_n T^n$$
where I is the identity mapping. Now, $f(x) = 2 + 3x$. Then, $f(T) = 2I + 3T$ or,
$$f(T)(x,y) = 2(x,y) + 3T(x,y).$$
We know that $\alpha T(x,y) = T(\alpha x, \alpha y)$ where α is any real scalar, and
$$(S+T)(x,y) = S(x,y) + T(x,y)$$

since T is a linear operator. Therefore,

$$f(T)(x,y) = 2(x,y) + (6x - 3y, 3x + 3y)$$
$$= (8x - 3y, 3x + 5y) .$$

Now,

$$g(x) = x + x^2 , \quad g(T) = T + T^2 .$$
$$T^2(x,y) = T(T(x,y)) = T(2x - y, x + y)$$
$$= (2(2x - y)-x-y, (2x - y) + (x + y)),$$

so

$$T^2(x,y) = (3x - 3y, 3x) .$$

Then,

$$g(T)(x,y) = T(x,y) + T^2(x,y)$$
$$= (2x - y, x + y) + (3x - 3y, 3x)$$
$$= (5x - 4y, 4x + y).$$

CHAPTER 10

EIGENVALUE PROBLEMS

FINDING THE EIGENVALUES OF A MATRIX

● PROBLEM 10-1

(1) Define an eigenvalue.
(2) Show that if u and v are eigenvectors of a linear operator f
which belong to λ and if a is a real number, then (a) u + v and (b)
au are also eigenvectors of f which belong to λ .

Solution: (1) A real number, λ, is an eigenvalue of f if and only if
there exists a non-zero vector u in V such that f(u) = λu .
    Thus, an eigenvalue is a number which acts as a scalar multiple of
some non-zero vector to give its f-image. If λ is an eigenvalue and
f(u) = λu , then u is called an eigenvector of f belonging to λ .
The set of eigenvectors of f which belong to a given eigenvalue, λ,
constitutes a subspace of R^n which is called the eigenspace of λ .
    Let P be the graph point of an eigenvector u which belongs to an
eigenvalue λ of an operator f on R^2 . If Q is the graph point of
λu, then $\overline{OP}$ and $\overline{OQ}$ are collinear. The eigenvalue, λ, signifies an
extension, contraction or reversal in direction of its eigenspace accord-
ing to whether its value is greater than 1, between 0 and 1 or less
than 0. (See Figure 1).

Fig. 1

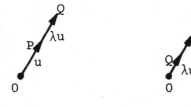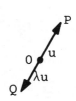

Extension	Contraction	Reversal of direction
$\lambda > 1$	$0 < \lambda < 1$	$\lambda < 0$

(2a) f(u+v) = f(u) + f(v) = λu + λv = λ(u+v) . Thus, u+v are also
eigenvectors of f which belong to λ .

(b) f(au) = af(u) = a(λu) = λ(au). Thus, each scalar multiple au is
an eigenvector.

Define the following types of symmetric matrices:
   (a)   Positive - definite.
   (b)   Positive - semi-definite.
   (c)   Negative-definite.
   (d)   Negative-semi-definite.
   (e)   Indefinite.

Solution: A matrix A is symmetric if its respective row and column vectors are equal. Symbolically, $A = A^T$.
   A symmetric matrix A is:

(a) Positive-definite if and only if all the eigenvalues are positive.
(b) Positive - semi-definite if and only if all the eigenvalues are non-negative.
(c) Negative - definite if and only if all the eigenvalues are negative.
(d) Negative - semi-definite if and only if all the eigenvalues are non-positive.
(e) Indefinite if and only if it has a positive and a negative eigen-value. For example, let

$$A = \begin{bmatrix} 1 & 2 & 0 \\ 2 & 1 & 0 \\ 0 & 0 & 3 \end{bmatrix} .$$

The matrix A has eigenvalues -1 and 3 and is, therefore, indefinite.
Let

$$A = \begin{bmatrix} 4 & 0 \\ 0 & 3 \end{bmatrix} .$$

The matrix A has eigenvalues 3 and 4 and is, therefore, positive-definite.

Find the real eigenvalues of the matrix,

$$A = \begin{bmatrix} -2 & -1 \\ 5 & 2 \end{bmatrix}$$

Solution: Let T be a linear transformation and A its matrix with respect to a given basis. Then λ is an eigenvalue if

$$AX = \lambda X , \tag{1}$$

where X is a non-zero vector. Choosing R^n as the underlying vector space, T: $R^n \rightarrow R^n$, (1) becomes

$$\begin{bmatrix} a_{11} & a_{12} & \cdots & a_{1n} \\ a_{21} & & & \cdot \\ \cdot & & & \cdot \\ \cdot & & & \cdot \\ \cdot & & & \cdot \\ a_{n1} & a_{n2} & \cdots & a_{nn} \end{bmatrix} \begin{bmatrix} x_1 \\ x_2 \\ \cdot \\ \cdot \\ \cdot \\ x_n \end{bmatrix} = \lambda \begin{bmatrix} x_1 \\ x_2 \\ \cdot \\ \cdot \\ \cdot \\ x_n \end{bmatrix} . \tag{2}$$

Expanding (2),

$$a_{11}x_1 + a_{12}x_2 + \ldots + a_{1n}x_n = \lambda x_1 \qquad (3)$$

$$a_{21}x_1 + a_{22}x_2 + \ldots + a_{2n}x_n = \lambda x_2$$

$$\vdots$$

$$a_{n1}x_1 + a_{n2}x_2 + \ldots + a_{nn}x_n = \lambda x_n$$

Rewriting (3),

$$(\lambda - a_{11})x_1 - a_{12}x_2 - \ldots - a_{1n}x_n = 0 \qquad (4)$$

$$-a_{21}x_1 + (\lambda - a_{22})x_2 - \ldots - a_{2n}x_n = 0$$

$$\vdots$$

$$-a_{n1}x_1 - a_{n2}x_2 - \ldots + (\lambda - a_{nn})x_n = 0 .$$

The set of linear homogeneous equations (4) can be expressed in matrix form as:

$$
\begin{bmatrix} \lambda & 0 & \ldots & 0 \\ 0 & \lambda & \ldots & 0 \\ \cdot & & & \\ \cdot & & & \\ \cdot & & & \\ 0 & 0 & \ldots & \lambda \end{bmatrix}
\begin{bmatrix} x_1 \\ x_2 \\ \cdot \\ \cdot \\ \cdot \\ x_n \end{bmatrix}
-
\begin{bmatrix} a_{11} & a_{12} & \ldots & a_{1n} \\ a_{21} & & & \\ \cdot & & & \\ \cdot & & & \\ a_{n1} & \ldots & & a_{nn} \end{bmatrix}
\begin{bmatrix} x_1 \\ x_2 \\ \cdot \\ \cdot \\ \cdot \\ x_n \end{bmatrix}
=
\begin{bmatrix} 0 \\ 0 \\ \cdot \\ \cdot \\ \cdot \\ 0 \end{bmatrix}
$$

or,

$$[\lambda I - A][X] = [0] . \qquad (5)$$

Recall now that a set of n linear homogeneous equations in n unknowns can have a non-trivial solution only if $\det[\lambda I - A] = 0$. The equation $\det[\lambda I - A] = 0$ is an nth degree polynomial in λ and its roots provide the eigenvalues of A and, thus, of T. By the Fundamental Theorem of Algebra there are n such roots in the complex field.

Form the matrix

$$\lambda I - A = \lambda \begin{bmatrix} 1 & 0 \\ 0 & 1 \end{bmatrix} - \begin{bmatrix} -2 & -1 \\ 5 & 2 \end{bmatrix}$$

$$= \begin{bmatrix} \lambda + 2 & 1 \\ -5 & \lambda - 2 \end{bmatrix}$$

Take its determinant to obtain the characteristic polynomial of A:

$$f(\lambda) = \det (\lambda I - A)$$

$$= \det \begin{vmatrix} \lambda + 2 & 1 \\ -5 & \lambda - 2 \end{vmatrix}$$

$$= (\lambda + 2)(\lambda - 2) + 5$$

$$= \lambda^2 + 1$$

The eigenvalues of A must, therefore, satisfy the quadratic equation $\lambda^2 + 1 = 0$. Since the only solutions to this equation are the imaginary numbers $\lambda = i$ and $\lambda = -i$.

A has no real eigenvalues.

Show that the characteristic values of an upper (or lower) triangular matrix are the diagonal entries of the matrix.

Solution: Let

$$A = \begin{bmatrix} 2 & 0 & 1 & 2 \\ 0 & 2 & -1 & 3 \\ 0 & 0 & -3 & 1 \\ 0 & 0 & 0 & 4 \end{bmatrix}.$$

The characteristic values of A are the roots of the characteristic polynomial $f(\lambda) = \det(\lambda I - A)$. Then,

$$f(\lambda) = \det(\lambda I - A) = \det \begin{vmatrix} \lambda-2 & 0 & -1 & -2 \\ 0 & \lambda-2 & +1 & -3 \\ 0 & 0 & \lambda+3 & -1 \\ 0 & 0 & 0 & \lambda-4 \end{vmatrix}$$

The determinant of an upper triangular matrix is the product of the diagonal entries. Therefore,

$$f(\lambda) = (\lambda-2)(\lambda-2)(\lambda+3)(\lambda-4) .$$

Hence, the characteristic values are 2,2,-3, and 4. In general, the characteristic values of an upper (or lower) triangular matrix are the diagonal entries of the matrix.

FINDING THE EIGENVECTORS OF A MATRIX

● **PROBLEM** 10-5

Find the real eigenvalues of A and their associated eigenvectors when

$$A = \begin{bmatrix} 1 & 1 \\ -2 & 4 \end{bmatrix}.$$

Solution: We wish to find all real numbers λ and all non-zero vectors $X = \begin{bmatrix} x_1 \\ x_2 \end{bmatrix}$ such that $AX = \lambda X$:

$$\begin{bmatrix} 1 & 1 \\ -2 & 4 \end{bmatrix} \begin{bmatrix} x_1 \\ x_2 \end{bmatrix} = \lambda \begin{bmatrix} x_1 \\ x_2 \end{bmatrix}$$

The above matrix equation is equivalent to the homogeneous system,

$$x_1 + x_2 = \lambda x_1$$

$$-2x_1 + 4x_2 = \lambda x_2$$

or,

$$(\lambda-1)x_1 - x_2 = 0 \tag{1}$$

$$+2x_1 + (\lambda-4)x_2 = 0 \quad .$$

Recall that a homogeneous system has a non-zero solution if and only if the determinant of the matrix of coefficients is zero. Thus,

$$\begin{vmatrix} \lambda-1 & -1 \\ 2 & \lambda-4 \end{vmatrix} = 0$$

or,

$$(\lambda-1)(\lambda-4) + 2 = 0 \quad .$$

Therefore,

$$\lambda^2 - 5\lambda + 6 = 0$$

or,

$$(\lambda-3)(\lambda-2) = 0 \quad .$$

Hence, $\lambda_1 = 2$ and $\lambda_2 = 3$ are the eigenvalues of A. To find an eigenvector of A associated with $\lambda_1 = 2$, form the linear system:

$$AX = 2X$$

or,

$$\begin{bmatrix} 1 & 1 \\ -2 & 4 \end{bmatrix} \begin{bmatrix} x_1 \\ x_2 \end{bmatrix} = 2 \begin{bmatrix} x_1 \\ x_2 \end{bmatrix}$$

This gives

$$\begin{array}{ll} x_1 + x_2 = 2x_1 & \quad (2-1)x_1 - x_2 = 0 \\ \quad\quad\quad\quad\quad\text{or} & \\ -2x_1 + 4x_2 = 2x_2 & \quad 2x_1 + (2-4)x_2 = 0 \end{array}$$

or,

$$\begin{array}{l} x_1 - x_2 = 0 \\ 2x_1 - 2x_2 = 0 \end{array} \quad \text{or, simply, } x_1 - x_2 = 0 \quad .$$

Observe that we could have obtained this last linear system by substituting $\lambda = 2$ in (1). It can be seen that any vector in R^2 of the form

$$x = k \begin{bmatrix} 1 \\ 1 \end{bmatrix}, \quad k \text{ a scalar, is an eigenvector of A}$$

associated with $\lambda_1 = 2$. Thus,

$$x_1 = \begin{bmatrix} 1 \\ 1 \end{bmatrix} \text{ is an eigenvector of A associated with}$$

$\lambda_1 = 2$. Similarly, for $\lambda_2 = 3$, we obtain from (1):

$$\begin{array}{ll} (3-1)x_1 - x_2 = 0 & \quad 2x_1 = x_2 = 0 \\ \quad\quad\quad\quad\quad\text{or,} & \\ 2x_1 + (3-4)x_2 = 0 & \quad 2x_1 - x_2 = 0 \quad . \end{array}$$

Thus, $x_2 = \begin{bmatrix} 1 \\ 2 \end{bmatrix}$ is an eigenvector of A associated with $\lambda_2 = 3$.

● **PROBLEM 10-6**

Find the eigenvalues and the corresponding eigenvectors of A where

$$A = \begin{bmatrix} 0 & \frac{1}{2} \\ \frac{1}{2} & 0 \end{bmatrix}$$

Solution: An eigenvalue of A is a scalar λ such that $AX = \lambda X$ for some non-zero vectors X. This may be converted to $(\lambda I - A)X = 0$ which implies $\det(\lambda I - A) = 0$, the characteristic equation. The roots of this equation yield the required eigenvalues.

$$\lambda I - A = \begin{bmatrix} \lambda & 0 \\ 0 & \lambda \end{bmatrix} - \begin{bmatrix} 0 & \frac{1}{2} \\ \frac{1}{2} & 0 \end{bmatrix} = \begin{bmatrix} \lambda & -\frac{1}{2} \\ -\frac{1}{2} & \lambda \end{bmatrix}$$

$\det(\lambda I - A) = \lambda^2 - \frac{1}{4}$.

Then, the characteristic equation is $\lambda^2 - \frac{1}{4} = 0$ and the eigenvalues are $\lambda_1 = \frac{1}{2}$ and $\lambda_2 = -\frac{1}{2}$. Substitute $\lambda = \frac{1}{2}$ in the equation $(\lambda I - A)x = 0$ to obtain the corresponding eigenvectors. $(\frac{1}{2}I - A)x = 0$

$$\begin{bmatrix} \frac{1}{2} & -\frac{1}{2} \\ -\frac{1}{2} & \frac{1}{2} \end{bmatrix} \begin{bmatrix} x_1 \\ x_2 \end{bmatrix} = \begin{bmatrix} 0 \\ 0 \end{bmatrix}$$

or,

$$\begin{array}{l} \frac{1}{2}x_1 - \frac{1}{2}x_2 = 0 \\ -\frac{1}{2}x_1 + \frac{1}{2}x_2 = 0 \end{array} \qquad \text{or, } x_1 - x_2 = 0 \ .$$

Thus,

$$x_1 = \begin{bmatrix} 1 \\ 1 \end{bmatrix} \quad \text{is an eigenvector of } A \text{ associated with the eigen-}$$

value $\lambda_1 = \frac{1}{2}$. Now, let $\lambda = -\frac{1}{2}$. Then,

$$\begin{bmatrix} -\frac{1}{2} & -\frac{1}{2} \\ -\frac{1}{2} & -\frac{1}{2} \end{bmatrix} \begin{bmatrix} x_1 \\ x_2 \end{bmatrix} = \begin{bmatrix} 0 \\ 0 \end{bmatrix}$$

therefore,

$$\begin{array}{l} -\frac{1}{2}x_1 - \frac{1}{2}x_2 = 0 \ ; \\ \\ -\frac{1}{2}x_1 - \frac{1}{2}x_2 = 0 \end{array} \qquad \text{or, } x_1 + x_2 = 0$$

Then,

$$x_2 = \begin{bmatrix} 1 \\ -1 \end{bmatrix} \quad \text{is an eigenvector of } A \text{ associated with the eigen-}$$

value $\lambda_2 = -\frac{1}{2}$. If we let $L: R^2 \to R^2$ be defined by

$$L(X) = AX = \begin{bmatrix} 0 & \frac{1}{2} \\ \frac{1}{2} & 0 \end{bmatrix} \begin{bmatrix} x_1 \\ x_2 \end{bmatrix}$$

then Figure 1 shows that X_1 and $L(X_1)$ are parallel and that X_2 and $L(X_2)$ are parallel also. This illustrates the fact that if X is an eigenvector of A, then X and AX are parallel.

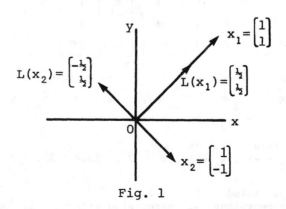

Fig. 1

Find the eigenvalues and the eigenvectors of the matrix A, where

$$A = \begin{bmatrix} 1 & 2 & -1 \\ 1 & 0 & 1 \\ 4 & -4 & 5 \end{bmatrix}.$$

Solution: First form the matrix $[\lambda I - A]$.

$$[\lambda I - A] = \begin{bmatrix} \lambda & 0 & 0 \\ 0 & \lambda & 0 \\ 0 & 0 & \lambda \end{bmatrix} - \begin{bmatrix} 1 & 2 & -1 \\ 1 & 0 & 1 \\ 4 & -4 & 5 \end{bmatrix}$$

$$= \begin{bmatrix} \lambda-1 & -2 & 1 \\ -1 & \lambda & -1 \\ -4 & 4 & \lambda-5 \end{bmatrix}.$$

Then,

$$\det(\lambda I - A) = \lambda-1 \begin{vmatrix} \lambda & -1 \\ 4 & \lambda-5 \end{vmatrix} -(-2) \begin{vmatrix} -1 & -1 \\ -4 & \lambda-5 \end{vmatrix} +1 \begin{vmatrix} -1 & \lambda \\ -4 & 4 \end{vmatrix}$$

$$\det(\lambda I - A) = (\lambda-1)[\lambda(\lambda-5) + 4] + 2[-\lambda+5 - 4] + 1[-4 + 4\lambda]$$

$$= (\lambda-1)[\lambda^2-5\lambda + 4] - 2\lambda + 2 - 4 + 4\lambda$$

$$= (\lambda-1)[\lambda^2-5\lambda + 4] + 2\lambda - 2$$

$$= (\lambda-1)[\lambda^2-5\lambda + 4 + 2]$$

$$= (\lambda-1)(\lambda-3)(\lambda-2).$$

The characteristic equation of A is $(\lambda-1)(\lambda-3)(\lambda-2) = 0$. The eigen-values of A are $\lambda = 1, \lambda = 2, \lambda = 3$. To find an eigenvector x_1 associated with $\lambda_1 = 1$, form the system:

$$(\lambda_1 I - A)X = 0, \text{ with } \lambda = 1:$$

$$\begin{bmatrix} 1-1 & -2 & 1 \\ -1 & 1 & -1 \\ -4 & 4 & 1-5 \end{bmatrix} \begin{bmatrix} x_1 \\ x_2 \\ x_3 \end{bmatrix} = \begin{bmatrix} 0 \\ 0 \\ 0 \end{bmatrix}$$

or,

$$\begin{bmatrix} 0 & -2 & 1 \\ -1 & 1 & -1 \\ -4 & 4 & -4 \end{bmatrix} \begin{bmatrix} x_1 \\ x_2 \\ x_3 \end{bmatrix} = \begin{bmatrix} 0 \\ 0 \\ 0 \end{bmatrix}$$

or,

$$\begin{array}{ll} -2x_2 + x_3 = 0 & 2x_2 - x_3 = 0 \\ -x_1 + x_2 - x_3 = 0 \qquad \text{or,} & x_1 - x_2 + x_3 = 0. \\ -4x_1 + 4x_2 - 4x_3 = 0 \end{array}$$

One solution to this system is
$$x_1 = -1, x_2 = 1, x_3 = 2. \text{ Thus, } X = \begin{bmatrix} -1 \\ 1 \\ 2 \end{bmatrix} \text{ is an eigen-}$$

vector of A associated with $\lambda = 1$.
  To find an eigenvector X_2 associated with $\lambda = 2$, form the system:

$$(\lambda I - A)X = 0 \text{, with } \lambda = 2.$$

$$\begin{bmatrix} 2\text{-}1 & -2 & 1 \\ -1 & 2 & -1 \\ -4 & 4 & 2\text{-}5 \end{bmatrix} \begin{bmatrix} x_1 \\ x_2 \\ x_3 \end{bmatrix} = \begin{bmatrix} 0 \\ 0 \\ 0 \end{bmatrix}$$

or,

$$x_1 - 2x_2 + x_3 = 0 \qquad\qquad x_1 - 2x_2 + x_3 = 0$$

$$-x_1 + 2x_2 - x_3 = 0 \qquad \text{or,} \quad 4x_1 - 4x_2 + 3x_3 = 0 .$$

$$-4x_1 + 4x_2 - 3x_3 = 0$$

Solving this system gives $x_1 = -2$, $x_2 = 1$, and $x_3 = 4$ as one solution.

Thus,

$$x_2 = \begin{bmatrix} -2 \\ 1 \\ 4 \end{bmatrix} \text{ is an eigenvector of } A \text{ associated}$$

with $\lambda = 2$.

To find an eigenvector, X for $\lambda = 3$, solve the system $(\lambda I - A)X = 0$ for X with $\lambda = 3$.
$\lambda = 3$

$$[3I - A]X = 0$$

or,

$$\begin{bmatrix} 3\text{-}1 & -2 & 1 \\ -1 & 3 & -1 \\ -4 & 4 & 3\text{-}5 \end{bmatrix} \begin{bmatrix} x_1 \\ x_2 \\ x_3 \end{bmatrix} = \begin{bmatrix} 0 \\ 0 \\ 0 \end{bmatrix} .$$

This is equivalent to

$$2x_1 - 2x_2 + x_3 = 0 \qquad\qquad 2x_1 - 2x_2 + x_3 = 0$$

$$-x_1 + 3x_2 - x_3 = 0 \qquad \text{or,} \quad x_1 - 3x_2 + x_3 = 0 .$$

$$-4x_1 + 4x_2 - 2x_3 = 0$$

Solving this system yields $x_1 = -1$, $x_2 = 1$, $x_3 = 4$. Thus, $X_3 = \begin{bmatrix} -1 \\ 1 \\ 4 \end{bmatrix}$

is an eigenvector of A associated with $\lambda = 3$.

● **PROBLEM** 10-8

The characteristic values of the matrix
$$A = \begin{bmatrix} 8 & 2 & -2 \\ 3 & 3 & -1 \\ 24 & 8 & -6 \end{bmatrix}$$
are 2 and 1. Find the characteristic vectors.

Solution: Let W_1 denote the set of characteristic vectors that belong to 2. A vector $V = (x_1, x_2, x_3)$ is in W_1 if and only if $VA = 2V$; that is, $V[A - 2I] = 0$ or,

$$[x_1,x_2,x_3] \begin{bmatrix} 6 & 2 & -2 \\ 3 & 1 & -1 \\ 24 & 8 & -8 \end{bmatrix} = [0\ 0\ 0]$$

or,

$$6x_1 + 3x_2 + 24x_3 = 0$$
$$2x_1 + x_2 + 8x_3 = 0 \qquad\qquad (1)$$
$$-2x_1 - x_2 - 8x_3 = 0 .$$

Thus, W_1 is the solution space of (1). The system (1) is clearly equivalent to $2x_1 + x_2 + 8x_3 = 0$. Since $x_2 + 8x_3 = -2x_1$ and $2x_1 + x_2 = -8x_3$, it follows that $v_1 = (1,-2,0)$ and $v_2 = (0,-8,1)$; $[v_1 v_2]$ is a basis of W_1 . The vectors v_1 and v_2 are characteristic vectors belonging to 2, and every characteristic vector belonging to 2 is a linear combination of v_1 and v_2 .

Let W_2 denote the set of characteristic vectors that belong to 1. A vector $V = (x_1,x_2,x_3)$ belongs to W_2 if and only if $VA = V$; that is,

$$v[A - 1I] = 0 ,$$

or,

$$[x_1 x_2 x_3] \begin{bmatrix} 7 & 2 & -2 \\ 3 & 2 & -1 \\ 24 & 8 & -7 \end{bmatrix} = [0\ 0\ 0]$$

or,

$$7x_1 + 3x_2 + 24x_3 = 0$$
$$2x_1 + 2x_2 + 8x_3 = 0$$
$$-2x_1 - x_2 - 7x_3 = 0$$

Solving this system gives $x_1 = 3$, $x_2 = 1$, $x_3 = -1$ as one uncomplicated solution. Thus, $v_3 = (3,1,-1)$ is a basis of w_2 . The vector v_3 is a characteristic vector belonging to 1, and every characteristic vector belonging to 1 is a scalar multiple of v_3 .

● **PROBLEM 10-9**

Given

$$A = \begin{bmatrix} 1 & 0 & -1 & 0 \\ 0 & 1 & 0 & -1 \\ 1 & 0 & -1 & 0 \\ 0 & 1 & 0 & -1 \end{bmatrix} \quad \text{and} \quad B = \begin{bmatrix} 4 & -1 & -1 & 0 \\ -1 & 4 & 0 & -1 \\ 1 & 0 & 2 & -1 \\ 0 & 1 & -1 & 2 \end{bmatrix},$$

Find a characteristic vector common to A and B.

Solution: If AB = BA, then A and B have a characteristic vector in common.

Let λ_1 be a characteristic root of A and let $x_1,\ldots,x_k$ be a basis for the null space of $\lambda_1 I_n - A$. Then, any non-zero linear combination of $x_1,\ldots,x_k$ is a characteristic vector corresponding to λ_1 . Conversely, any characteristic vector of A corresponding to λ_1 can be expressed as a linear combination of $(x_1, x_2, \ldots, x_k)$. Suppose that $Bx_i \neq 0$, $i = 1, \ldots, k$, then

$$\begin{aligned} A(Bx_i) &= BAx_i \\ &= B(\lambda_1 x_i) \\ &= \lambda_1 (Bx_i) \; ; \end{aligned}$$

thus,

$$Bx_i \in Sp(x_1, \ldots, x_k) \;, \quad i = 1, \ldots, k \;.$$

Let X be the matrix whose jth column is x_j, $j = 1, \ldots, k$. Therefore, the system of equations,

$$Xy_i = Bx_i \;, \quad i = 1, \ldots, k \;, \tag{1}$$

is the matrix whose ith column is y_i, $i = 1, \ldots, k$. The relation (1) can be written compactly as

$$XY = BX \;. \tag{2}$$

Let μ_1 be any characteristic root of Y and $z = (\alpha_1, \ldots, \alpha_k)$, a characteristic vector of Y corresponding to μ_1. Then, by (2),

and since
$$\begin{aligned} B(xz) &= XYz \\ Yz &= \mu_1 z \;, \end{aligned}$$

$$B(xz) = \mu_1 (Xz) \;.$$

Now,

$$Xz = \sum_{t=1}^{k} \alpha_t x_t \neq 0$$

because $z = (\alpha_1, \ldots, \alpha_k) \neq 0$ and $x_1, \ldots, x_k$ are linearly independent. Thus, Xz is a characteristic vector of B. But,

$$Xz = \sum_{t=1}^{k} \alpha_t x_t \in (x_1, \ldots, x_k)$$

and, therefore, Xz is a characteristic vector of A. Now, find the product of AB and BA.

$$AB = \begin{bmatrix} 3 & -1 & -3 & 1 \\ -1 & 3 & 1 & -3 \\ 3 & -1 & -3 & 1 \\ -1 & 3 & 1 & -3 \end{bmatrix} = BA \;.$$

Thus, $AB = BA$. Therefore, A and B have a common characteristic vector. The matrix A is clearly singular and, therefore, $\det(I\lambda - A) = 0$ implies $\lambda = 0$ (since $\det A = 0$). Since the rank of $\lambda_1 I_4 - A$ is 2, the dimension of its null space is $4-2 = 2$. The vector $x_1 = (1,0,1,0)$ and $x_2 = (0,1,0,1)$ form a basis for this null space

Thus,
$$Bx_1 = (3,-1,1,3); \; Bx_2 = (-1,3,-1,3)$$

$$Bx_1 = 3x_1 - x_2 \quad \text{and} \quad Bx_2 = -x_1 + 3x_2 \;.$$

Let

$$X = \begin{bmatrix} 1 & 0 \\ 0 & 1 \\ 1 & 0 \\ 0 & 1 \end{bmatrix} \quad \text{and} \quad Y = \begin{bmatrix} 3 & -1 \\ -1 & 3 \end{bmatrix} \;.$$

Then, as in (2),
$$XY = BX .$$

To find the characteristic root of Y, solve the equation, $\det(\lambda I - Y) = 0$. Then,

$$\det \begin{vmatrix} \lambda-3 & 1 \\ 1 & \lambda-3 \end{vmatrix} = 0$$

or,

$$(\lambda-3)^2 - 1 = 0$$

$$(\lambda-4)(\lambda-2) = 0 .$$

Therefore, $\mu_1 = 4$ is the characteristic root of Y, and $z = (1,-1)$ is the corresponding characteristic vector. Then,

$$Xz = \begin{bmatrix} 1 & 0 \\ 0 & 1 \\ 1 & 0 \\ 0 & 1 \end{bmatrix} \begin{bmatrix} 1 \\ -1 \end{bmatrix} = \begin{bmatrix} 1 \\ -1 \\ 1 \\ -1 \end{bmatrix}$$

Thus, $Xz = (1,-1,1,-1)$ is a characteristic vector common to A and B.

● **PROBLEM** 10-10

Find the eigenvalues of m'm:

(1) $m = \begin{bmatrix} 1 & 0 \\ 0 & 0 \end{bmatrix}$

(2) $m = \begin{bmatrix} 0 & 1 \\ -1 & 0 \end{bmatrix}$

(3) $m = \begin{bmatrix} 1 & 1 \\ 0 & 1 \end{bmatrix}.$

Solution: The map $X \rightarrow mX = Y$ maps the unit sphere in x-space onto an ellipsoid in y space, and the squares of the semi-axes of the image are equal to the eigenvalues of m'm. If we have the map $X \rightarrow mX$ and put $mX = Y$, then the x's which are mapped into $Y'Y = 1$ satisfy $X'm'mX = 1$. They lie on a quadric surface in the original space.

When the eigenvalues are all positive, the locus is an ellipsoid.

(1) $m = \begin{bmatrix} 1 & 0 \\ 0 & 0 \end{bmatrix}$

The locus is $X'm'mX = 1$ (in this case, the pair of lines $x_1 = \pm 1$). These map onto the two points $(\pm 1, 0)$. m'm has eigenvalues 1 and 0.

(2) $m = \begin{bmatrix} 0 & 1 \\ -1 & 0 \end{bmatrix}$

This is a rotation. The unit circle is rotated through a right angle. m'm = I. The unit circle is invariant.

(3) $\quad m = \begin{bmatrix} 1 & 1 \\ 0 & 1 \end{bmatrix}$

This is a shear. The unit circle becomes the ellipse

$$x^2 - 2x_1x_2 + 2x_2^2 = 1.$$

The major axis is in the direction $(2, -1 + \sqrt{5})$ and is of length $1 + \sqrt{5}$. The minor axis is in the direction $(2, -1 - \sqrt{5})$ and is of length $-1 + \sqrt{5}$.

$mm' = \begin{bmatrix} 1 & 1 \\ 1 & 2 \end{bmatrix}$. The eigenvalues are $\frac{1}{2}(3 \pm \sqrt{5})$. The ellipse

$x_1^2 + 2x_1x_2 + 2x_2^2 = 1$ is mapped onto the unit circle.

● **PROBLEM** 10-11

Show for the following matrix, A, that any column eigenvector corresponding to a particular eigenvalue is orthogonal to all row eigenvectors corresponding to other eigenvalues and vice versa.

$$A = \begin{bmatrix} -1 & 2 & 2 \\ -8 & 7 & 4 \\ -13 & 5 & 8 \end{bmatrix}$$

Solution: We know that if λ is an eigenvalue of a square matrix A and X is a corresponding eigenvector (column eigenvector), they satisfy the relation
$$AX = \lambda X .$$
The eigenvalues can be obtained as the roots of the characteristic equation $\det(A - \lambda I) = 0$.
   The eigenvectors corresponding to the various eigenvalues can be found by solving the equations
$$(A - \lambda I)X = 0$$
after substituting the appropriate values for λ . A similar course of action can be carried out to find the row eigenvectors of the matrix.
   A row eigenvector is a non-zero vector Y which satisfies the relation:
$$YA = \lambda Y .$$
Thus, if $Y = (y_1, y_2, y_3)$, the row eigenvector satisfies

$$[y_1, y_2, y_3] \begin{bmatrix} -1 & 2 & 2 \\ -8 & 7 & 4 \\ -13 & 5 & 8 \end{bmatrix} = \lambda [y_1, y_2, y_3] .$$

Hence,
$$(-y_1 - 8y_2 - 13y_3, \ 2y_1 + 7y_2 + 5y_3, \ 2y_1 + 4y_2 + 8y_3) = (\lambda y_1, \ \lambda y_2, \ \lambda y_3)$$

or

$$-y_1 - 8y_2 - 13y_3 = \lambda y_1$$
$$+2y_1 + 7y_2 + 5y_3 = \lambda y_2$$
$$2y_1 + 4y_2 + 8y_3 = \lambda y_3 \ .$$

Rearranging,

$$(-1-\lambda)y_1 - 8y_2 - 13y_3 = 0$$
$$2y_1 + (7-\lambda)y_2 + 5y_3 = 0 \qquad\qquad (1)$$
$$2y_1 + 4y_2 + (8-\lambda)y_3 = 0 \ .$$

It is known that a set of homogeneous linear equations has a solution in which the unknowns are not all zero if and only if the determinant of the coefficients is zero. We have

$$\begin{vmatrix} -1-\lambda & -8 & -13 \\ 2 & 7-\lambda & 5 \\ 2 & 4 & 8-\lambda \end{vmatrix} = 0$$

or,

$$(-1-\lambda)\begin{vmatrix} 7-\lambda & 5 \\ 4 & 8-\lambda \end{vmatrix} - (-8)\begin{vmatrix} 2 & 5 \\ 2 & 8-\lambda \end{vmatrix} -13\begin{vmatrix} 2 & 7-\lambda \\ 2 & 4 \end{vmatrix}$$

$$=(-1-\lambda)[(7-\lambda)(8-\lambda) - 20] + 8[2(8-\lambda) - 10] - 13[8-2(7-\lambda)]$$

or,

$$\lambda^3 - 14\lambda^2 + 63\lambda - 90 = 0$$
$$\lambda^2(\lambda-3) - 11\lambda(\lambda-3) + 30(\lambda-3) = 0$$
$$(\lambda-3)(\lambda-6)(\lambda-5) = 0 \ .$$

Therefore, the eigenvalues are $\lambda = 3$, $\lambda = 5$ and $\lambda = 6$.
When $\lambda = 3$, equations (1) become

$$-4y_1 - 8y_2 - 13y_3 = 0$$
$$-2y_1 + 4y_2 + 5y_3 = 0$$
$$2y_1 + 4y_2 + 5y_3 = 0 \ .$$

These have solutions

$$y_1 = -2, \ y_2 = 1, \ y_3 = 0 \ .$$

Thus, $Y_1 = (-2,1,0)$ is the row eigenvector associated with eigenvalue $\lambda = 3$. Putting $\lambda = 5$ in equations (1) yields

$$Y_2 = (7,-2,-2)$$

and, for $\lambda = 6$, $y_3 = (-3,1,1)$. The column eigenvectors are obtained by solving the equation $(A - \lambda I)X = 0$ for X.

For $\lambda = 3$,

$$(A - 3I)X = 0$$

$$\begin{bmatrix} -4 & 2 & 2 \\ -8 & 4 & 4 \\ -13 & 5 & 5 \end{bmatrix} \begin{bmatrix} x_1 \\ x_2 \\ x_3 \end{bmatrix} = \begin{bmatrix} 0 \\ 0 \\ 0 \end{bmatrix}$$

Thus, $X_1 = \begin{bmatrix} 0 \\ 1 \\ -1 \end{bmatrix}$ is an eigenvector corresponding to $\lambda = 3$.

Similarly, $X_2 = \begin{bmatrix} 1 \\ 2 \\ 1 \end{bmatrix}$ with eigenvalue $\lambda = 5$, $X_3 = \begin{bmatrix} 2 \\ 4 \\ 3 \end{bmatrix}$ with eigen-

value $\lambda = 6$. Now,

$$Y_1 X_2 = [-2,1,0] \begin{bmatrix} 1 \\ 2 \\ 1 \end{bmatrix} = 0$$

$$Y_1 X_3 = [-2,1,0] \begin{bmatrix} 2 \\ 4 \\ 3 \end{bmatrix} = 0$$

and, likewise, $Y_r X_s = 0$ where $r \neq s$. Thus, the column eigenvector corresponding to a particular eigenvalue is orthogonal to all row eigenvectors corresponding to other eigenvalues.

THE BASIS FOR AN EIGENSPACE

● **PROBLEM** 10-12

Find the eigenvalues of A and a basis for each eigenspace, where

$$A = \begin{bmatrix} 2 & 2 \\ -1 & 5 \end{bmatrix} .$$

Solution: Each eigenvalue of A has associated with it a set of eigenvectors, i.e., vectors X such that

$$AX = \lambda X \tag{1}$$

where λ is the eigenvalue. The set of eigenvectors forms a subspace of R^n called the eigenspace. From (1) we obtain the characteristic equation as follows:

$$(\lambda I - A) = \lambda \begin{bmatrix} 1 & 0 \\ 0 & 1 \end{bmatrix} - \begin{bmatrix} 2 & 2 \\ -1 & 5 \end{bmatrix} = \begin{bmatrix} \lambda-2 & -2 \\ 1 & \lambda-5 \end{bmatrix} .$$

$$\det(\lambda I - A) = \det \begin{vmatrix} \lambda-2 & -2 \\ 1 & \lambda-5 \end{vmatrix} = [(\lambda-2)(\lambda-5) + 2]$$

$$= \lambda^2 - 7\lambda + 12$$

$$= (\lambda-4)(\lambda-3) .$$

The characteristic equation of A is: $(\lambda-4)(\lambda-3) = 0$. Then, eigenvalues of A are $\lambda = 4$ and $\lambda = 3$. By definition, the eigenspace of A corresponding to λ is $(\lambda I - A)X = 0$. If $\lambda = 4$ then

$$(4I - A)X = \begin{bmatrix} 4-2 & -2 \\ 1 & 4-5 \end{bmatrix} \begin{bmatrix} x_1 \\ x_2 \end{bmatrix} = \begin{bmatrix} 0 \\ 0 \end{bmatrix}$$

or,

$$\begin{bmatrix} 2 & -2 \\ 1 & -1 \end{bmatrix} \begin{bmatrix} x_1 \\ x_2 \end{bmatrix} = \begin{bmatrix} 0 \\ 0 \end{bmatrix}$$

or, $2x_1 - 2x_2 = 0$

$x_1 - x_2 = 0$ or, $x_1 - x_2 = 0$.

The system has only one independent solution, i.e., $x_1 = 1$, $x_2 = 1$.
Thus, the eigenvectors corresponding to $\lambda = 4$ are the non-zero vectors
of the form

$$X = \alpha \begin{bmatrix} 1 \\ 1 \end{bmatrix} ,$$

where α is a scalar, so that $\begin{bmatrix} 1 \\ 1 \end{bmatrix}$ is a basis for the eigenspace corresponding to $\lambda = 4$. If $\lambda = 3$,

$$\begin{bmatrix} 3-2 & -2 \\ 1 & 3-5 \end{bmatrix} \begin{bmatrix} x_1 \\ x_2 \end{bmatrix} = \begin{bmatrix} 0 \\ 0 \end{bmatrix}$$

or,

$$\begin{bmatrix} 1 & -2 \\ 1 & -2 \end{bmatrix} \begin{bmatrix} x_1 \\ x_2 \end{bmatrix} = \begin{bmatrix} 0 \\ 0 \end{bmatrix} .$$

This gives

$x_1 - 2x_2 = 0$

$x_1 - 2x_2 = 0$ or, $x_1 - 2x_2 = 0$.

Thus, $X = \begin{bmatrix} 2 \\ 1 \end{bmatrix}$ is an eigenvector which generates and forms a basis of
the eigenspace of 3.

● **PROBLEM** 10-13

Find a basis for the eigenspace of

$$A = \begin{bmatrix} 3 & -2 & 0 \\ -2 & 3 & 0 \\ 0 & 0 & 5 \end{bmatrix} .$$

Solution: If λ is an eigenvalue of A, then the solution space for
the system of equations $(\lambda I - A)X = 0$ is called the eigenspace of A
corresponding to λ, and the non-zero vectors in the eigenspace are
called the eigenvectors of A corresponding to λ .

Form the matrix

$$\lambda I - A = \lambda \begin{bmatrix} 1 & 0 & 0 \\ 0 & 1 & 0 \\ 0 & 0 & 1 \end{bmatrix} - \begin{bmatrix} 3 & -2 & 0 \\ -2 & 3 & 0 \\ 0 & 0 & 5 \end{bmatrix}$$

$$= \begin{bmatrix} \lambda-3 & 2 & 0 \\ 2 & \lambda-3 & 0 \\ 0 & 0 & \lambda-5 \end{bmatrix} .$$

$$\det(\lambda I - A) = \det \begin{vmatrix} \lambda-3 & 2 & 0 \\ 2 & \lambda-3 & 0 \\ 0 & 0 & \lambda-5 \end{vmatrix}$$

$$= \lambda - 5 \begin{vmatrix} \lambda - 3 & 2 \\ 2 & \lambda - 3 \end{vmatrix}$$

$$= \lambda - 5 [(\lambda - 3)^2 - 4]$$

$$= \lambda - 5 [\lambda^2 - 6\lambda + 9 - 4]$$

$$= (\lambda - 5)[\lambda^2 - 6\lambda + 5]$$

$$= (\lambda - 5)(\lambda - 5)(\lambda - 1)$$

$$= (\lambda - 5)^2 (\lambda - 1) .$$

The characteristic equation of A is $(\lambda - 5)^2(\lambda - 1) = 0$, so that the eigenvalues of A are $\lambda = 1$ and $\lambda = 5$.

By definition,

$$X = \begin{bmatrix} x_1 \\ x_2 \\ x_3 \end{bmatrix}$$

is an eigenvector of A corresponding to λ if and only if x is a non-trivial solution of $(\lambda I - A)X = 0$. Thus,

$$\begin{bmatrix} \lambda - 3 & 2 & 0 \\ 2 & \lambda - 3 & 0 \\ 0 & 0 & \lambda - 5 \end{bmatrix} \begin{bmatrix} x_1 \\ x_2 \\ x_3 \end{bmatrix} = \begin{bmatrix} 0 \\ 0 \\ 0 \end{bmatrix} . \qquad (1)$$

If $\lambda = 5$, then equation (1) becomes

$$\begin{bmatrix} 2 & 2 & 0 \\ 2 & 2 & 0 \\ 0 & 0 & 0 \end{bmatrix} \begin{bmatrix} x_1 \\ x_2 \\ x_3 \end{bmatrix} = \begin{bmatrix} 0 \\ 0 \\ 0 \end{bmatrix} .$$

Solving this system yields $x_1 = -s$, $x_2 = s$, $x_3 = t$, where s and t are any scalars. Thus, the eigenvectors of A corresponding to $\lambda = 5$ are the non-zero vectors of the form

$$X = \begin{bmatrix} -s \\ s \\ t \end{bmatrix} = \begin{bmatrix} -s \\ s \\ 0 \end{bmatrix} + \begin{bmatrix} 0 \\ 0 \\ t \end{bmatrix} = s \begin{bmatrix} -1 \\ 1 \\ 0 \end{bmatrix} + t \begin{bmatrix} 0 \\ 0 \\ 1 \end{bmatrix}$$

Since $\begin{bmatrix} -1 \\ 1 \\ 0 \end{bmatrix}$ and $\begin{bmatrix} 0 \\ 0 \\ 1 \end{bmatrix}$ are linearly independent, they form a basis for the eigenspace corresponding to $\lambda = 5$. If $\lambda = 1$, then equation (1) becomes

$$\begin{bmatrix} -2 & 2 & 0 \\ 2 & 2 & 0 \\ 0 & 0 & -4 \end{bmatrix} \begin{bmatrix} x_1 \\ x_2 \\ x_3 \end{bmatrix} = \begin{bmatrix} 0 \\ 0 \\ 0 \end{bmatrix} .$$

Solving this system yields $x_1 = t$, $x_2 = t$, $x_3 = 0$; t is any scalar. Thus, the eigenvectors corresponding to $\lambda = 1$ are non-zero vectors of the form:

$$X = \begin{bmatrix} t \\ t \\ 0 \end{bmatrix} = t \begin{bmatrix} 1 \\ 1 \\ 0 \end{bmatrix} \text{ so that } \begin{bmatrix} 1 \\ 1 \\ 0 \end{bmatrix}$$

is a basis for the eigenspace corresponding to $\lambda = 1$.

Find the eigenvalues and an orthonormal basis for the eigenspace of A
where,

$$A = \begin{bmatrix} 1 & 2 & 0 \\ 2 & 1 & 0 \\ 0 & 0 & 3 \end{bmatrix} .$$

Solution: A is a symmetric matrix. An important result concerning
symmetric matrices is that all eigenvalues of a symmetric matrix are
real. Form the matrix

$$(\lambda I - A) = \begin{bmatrix} \lambda & 0 & 0 \\ 0 & \lambda & 0 \\ 0 & 0 & \lambda \end{bmatrix} - \begin{bmatrix} 1 & 2 & 0 \\ 2 & 1 & 0 \\ 0 & 0 & 3 \end{bmatrix}$$

$$= \begin{bmatrix} \lambda-1 & -2 & 0 \\ -2 & \lambda-1 & 0 \\ 0 & 0 & \lambda-3 \end{bmatrix} .$$

Now, expanding along the third column yields

$$\det(\lambda I - A) = (\lambda-3) \begin{vmatrix} \lambda-1 & -2 \\ -2 & \lambda-1 \end{vmatrix}$$

$$= (\lambda-3)(\lambda^2 - 2\lambda + 1 - 4)$$

$$= (\lambda-3)(\lambda-3)(\lambda+1) .$$

The characteristic equation of A is $(\lambda-3)^2(\lambda+1) = 0$ and, therefore,
the eigenvalues of A are $\lambda = 3$ and $\lambda = -1$.

Now find eigenvectors corresponding to $\lambda_1 = 3$. To do this, solve
$(\lambda I - A)X = 0$ for X with $\lambda = 3$.

$(3I - A)X = 0$

or,

$$\begin{bmatrix} 2 & -2 & 0 \\ -2 & 2 & 0 \\ 0 & 0 & 0 \end{bmatrix} \begin{bmatrix} x_1 \\ x_2 \\ x_3 \end{bmatrix} = \begin{bmatrix} 0 \\ 0 \\ 0 \end{bmatrix} .$$

This is equivalent to

$$\begin{matrix} 2x_1 - 2x_2 = 0 \\ -2x_1 + 2x_2 = 0 \end{matrix} \qquad \text{or,} \quad x_1 - x_2 = 0 .$$

Solving this system gives $x_1 = s$, $x_2 = s$, $x_3 = t$. Therefore,

$$X = \begin{bmatrix} s \\ s \\ t \end{bmatrix} = \begin{bmatrix} s \\ s \\ 0 \end{bmatrix} + \begin{bmatrix} 0 \\ 0 \\ t \end{bmatrix} = s \begin{bmatrix} 1 \\ 1 \\ 0 \end{bmatrix} + \begin{bmatrix} 0 \\ 0 \\ 1 \end{bmatrix}$$

or,

$$x_1 = \begin{bmatrix} 1 \\ 1 \\ 0 \end{bmatrix} \quad x_2 = \begin{bmatrix} 0 \\ 0 \\ 1 \end{bmatrix} .$$

Note that x_1 and x_2 are orthogonal to each other since $x_1 \cdot x_2 = 0$.
Next, normalize x_1 and x_2 to obtain the unit orthogonal solutions
by replacing x_i with

$$\frac{x_i}{|x_i|} \; .$$

Since $|x_1| = \sqrt{2}$ and $|x_2| = 1$,

$$u_1 = \begin{bmatrix} 1/\sqrt{2} \\ 1/\sqrt{2} \\ 0 \end{bmatrix}; \quad u_2 = \begin{bmatrix} 0 \\ 0 \\ 1 \end{bmatrix}$$

and they form a basis for the eigenspace corresponding to $\lambda = 3$. To
find the eigenvectors corresponding to $\lambda = -1$, solve $(\lambda I - A)X = 0$
for X with $\lambda = -1$.

$(-1I - A)X = 0$ or,

$$\begin{bmatrix} -2 & -2 & 0 \\ -2 & -2 & 0 \\ 0 & 0 & -4 \end{bmatrix} \begin{bmatrix} x_1 \\ x_2 \\ x_3 \end{bmatrix} = 0 \; .$$

Carrying out the indicated matrix multiplication,

$$\begin{matrix} -2x_1 - 2x_2 = 0 \\ -2x_1 - 2x_2 = 0 \\ -4x_3 = 0 \end{matrix} \quad \text{or,} \quad \begin{matrix} x_1 + x_2 = 0 \\ x_3 = 0 \end{matrix} \; .$$

Solving this system gives

$$x_3 = \begin{bmatrix} 1 \\ -1 \\ 0 \end{bmatrix} \; .$$

Now, normalize x_3 to obtain the unit orthogonal solution. Thus,

$$u_3 = \begin{bmatrix} 1/\sqrt{2} \\ -1\sqrt{2} \\ 0 \end{bmatrix}$$

forms a basis for the eigenspace corresponding to $\lambda = -1$. Since
$u_1 \cdot u_3 = 0$ and $u_2 \cdot u_3 = 0$, $\{u_1, u_2, u_3\}$ is an orthonormal basis of R^3.
In general, if A is a symmetric $n \times n$ matrix, then the eigenvectors of
A contain an orthonormal basis of R^n.

APPLICATIONS OF THE SPECTRAL THEOREM

● **PROBLEM** 10-15

Find an orthogonal matrix P such that $P^{-1}AP$ is a diagonal matrix B
where,

$$A = \begin{bmatrix} 3 & 1 \\ 1 & 3 \end{bmatrix} \; .$$

Solution: Recall that the transpose A^t of A is the matrix obtained from A by interchanging the rows and columns of A. We say that A is symmetric if $A^t = A$.

For symmetric matrices there is the following theorem:

If A is symmetric, there is an invertible matrix P such that $B = P^{-1}AP$ is a diagonal matrix.

This theorem is usually called the Spectral Theorem; it tells us, in particular, that if a matrix is symmetric, its characteristic polynomial must have only real roots. If A is symmetric, one may actually find an orthogonal matrix P such that $P^{-1}AP$ is diagonal. Recall that an orthogonal matrix is a matrix whose columns are orthonormal.

Now, consider the given matrix

$$A = \begin{bmatrix} 3 & 1 \\ 1 & 3 \end{bmatrix} .$$

A is symmetric. Form the matrix

$$(\lambda I - A) = \begin{bmatrix} \lambda & 0 \\ 0 & \lambda \end{bmatrix} - \begin{bmatrix} 3 & 1 \\ 1 & 3 \end{bmatrix}$$

$$= \begin{bmatrix} \lambda-3 & -1 \\ -1 & \lambda-3 \end{bmatrix} .$$

Then,

$$\det(\lambda I - A) = \det \begin{vmatrix} \lambda-3 & -1 \\ -1 & \lambda-3 \end{vmatrix}$$

$$= (\lambda-3)^2 - 1$$

$$= \lambda^2 - 6\lambda + 9 - 1$$

$$= \lambda^2 - 6\lambda + 8$$

$$= (\lambda-4)(\lambda-2) .$$

The characteristic equation of A is $(\lambda-4)(\lambda-2) = 0$ so that the characteristic values are $\lambda = 4$ and $\lambda = 2$. If $\lambda = 4$, then

$$4I - A = \begin{bmatrix} 4 & 0 \\ 0 & 4 \end{bmatrix} - \begin{bmatrix} 3 & 1 \\ 1 & 3 \end{bmatrix} = \begin{bmatrix} 1 & -1 \\ -1 & 1 \end{bmatrix} .$$

Now find the characteristic vectors (or eigenvectors).

$(4I - A)X = 0$

$$\begin{bmatrix} 1 & -1 \\ -1 & 1 \end{bmatrix} \begin{bmatrix} x_1 \\ x_2 \end{bmatrix} = 0$$

or, $x_1 - x_2 = 0$
$-x_1 + x_2 = 0$

or, $x_1 - x_2 = 0$. Thus, $x_1 = \begin{bmatrix} t \\ t \end{bmatrix}$ where $t \in R$ are

the eigenvectors. Clearly, $\begin{bmatrix} 1 \\ 1 \end{bmatrix}$ is a basis for the space. Since the

norm of $[1,1]$ is $([1,1] \cdot [1,1])^{\frac{1}{2}} = \sqrt{2}$, normalize $[1,1]$ to obtain

$$x_1 = \begin{bmatrix} 1/\sqrt{2} \\ 1/\sqrt{2} \end{bmatrix} .$$

This is an orthonormal basis for the null space of $4I - A$. Now let $\lambda = 2$. Then,

$$2I - A = \begin{bmatrix} 2 & 0 \\ 0 & 2 \end{bmatrix} - \begin{bmatrix} 3 & 1 \\ 1 & 3 \end{bmatrix} = \begin{bmatrix} -1 & -1 \\ -1 & -1 \end{bmatrix} .$$

Thus, $(4I - A)X = 0$,

$$\begin{bmatrix} -1 & -1 \\ -1 & -1 \end{bmatrix} \begin{bmatrix} x_1 \\ x_2 \end{bmatrix} = \begin{bmatrix} 0 \\ 0 \end{bmatrix}$$

or,

$$-x_1 - x_2 = 0$$
$$\qquad \text{or, } x_1 + x_2 = 0 .$$
$$-x_1 - x_2 = 0$$

Clearly, $\begin{bmatrix} -1 \\ 1 \end{bmatrix}$ is a basis for the space. If we normalize, $u_2 = \begin{bmatrix} -1/\sqrt{2} \\ 1/\sqrt{2} \end{bmatrix}$

is an orthonormal basis for the null space of $2I - A$.

Observe that $u_1 \cdot u_2 = 0$ so that u_1 and u_2 are an orthonormal basis for R^2. Thus, in general, if A is symmetric and λ_1 and λ_2 are distinct characteristic values of A, the corresponding characteristic vectors x_1, x_2 must be orthogonal. Now construct an orthogonal matrix P whose columns are orthonormal. Thus,

$$P = \begin{bmatrix} 1/\sqrt{2} & -1/\sqrt{2} \\ 1/\sqrt{2} & 1/\sqrt{2} \end{bmatrix} .$$

Since P is an orthogonal matrix, $P^{-1} = P^t$. Therefore,

$$P^{-1} = \begin{bmatrix} 1/\sqrt{2} & 1/\sqrt{2} \\ -1/\sqrt{2} & 1/\sqrt{2} \end{bmatrix} .$$

Then, $B = P^{-1}AP$ is a diagonal matrix.

$$B = \begin{bmatrix} 1/\sqrt{2} & 1/\sqrt{2} \\ -1/\sqrt{2} & 1/\sqrt{2} \end{bmatrix} \begin{bmatrix} 3 & 1 \\ 1 & 3 \end{bmatrix} \begin{bmatrix} 1/\sqrt{2} & -1/\sqrt{2} \\ 1/\sqrt{2} & 1/\sqrt{2} \end{bmatrix}$$

$$= \begin{bmatrix} 1/\sqrt{2} & 1/\sqrt{2} \\ -1/\sqrt{2} & 1/\sqrt{2} \end{bmatrix} \begin{bmatrix} 4/\sqrt{2} & -2/\sqrt{2} \\ 4/\sqrt{2} & 2/\sqrt{2} \end{bmatrix}$$

$$= \begin{bmatrix} 2+2 & -1+1 \\ -2+2 & 1+1 \end{bmatrix}$$

$$= \begin{bmatrix} 4 & 0 \\ 0 & 2 \end{bmatrix} .$$

● **PROBLEM** 10-16

Find an orthogonal matrix P such that $P^{-1}AP$ is a diagonal matrix B where,

$$A = \begin{bmatrix} 1 & 1 & 0 \\ 1 & 1 & 0 \\ 0 & 0 & 2 \end{bmatrix} .$$

Solution: A is a symmetric matrix.

$$\lambda I - A = \begin{bmatrix} \lambda & 0 & 0 \\ 0 & \lambda & 0 \\ 0 & 0 & \lambda \end{bmatrix} - \begin{bmatrix} 1 & 1 & 0 \\ 1 & 1 & 0 \\ 0 & 0 & 2 \end{bmatrix}$$

$$= \begin{bmatrix} \lambda-1 & -1 & 0 \\ -1 & \lambda-1 & 0 \\ 0 & 0 & \lambda-2 \end{bmatrix} .$$

To find $\det(\lambda I - A)$, expand along the third column.

$$\det(\lambda I - A) = (\lambda-2) \begin{vmatrix} \lambda-1 & -1 \\ -1 & \lambda-1 \end{vmatrix}$$

$$= (\lambda-2)[(\lambda-1)^2 - 1]$$

$$= (\lambda-2)[\lambda^2 - 2\lambda + 1 - 1]$$

$$= (\lambda-2)(\lambda^2 - 2\lambda)$$

$$= \lambda(\lambda-2)^2 .$$

Thus, the characteristic equation of A is $\lambda(\lambda-2)^2 = 0$ so that characteristic values are $\lambda = 0$ and $\lambda = 2$. If $\lambda = 0$, then

$$0I - A = \begin{bmatrix} -1 & -1 & 0 \\ -1 & -1 & 0 \\ 0 & 0 & -2 \end{bmatrix} ;$$

$(0I - A)X = 0$

$$\begin{bmatrix} -1 & -1 & 0 \\ -1 & -1 & 0 \\ 0 & 0 & -2 \end{bmatrix} \begin{bmatrix} x_1 \\ x_2 \\ x_3 \end{bmatrix} = \begin{bmatrix} 0 \\ 0 \\ 0 \end{bmatrix}$$

or,

$$\begin{aligned} -x_1 - x_2 &= 0 \\ -x_1 - x_2 &= 0 \\ -2x_3 &= 0 \end{aligned} \quad \text{or,} \quad \begin{aligned} x_1 + x_2 &= 0 \\ x_3 &= 0 . \end{aligned}$$

Therefore, the null space consists of vectors of the form

$$\alpha \begin{bmatrix} -1 \\ 1 \\ 0 \end{bmatrix} , \ \alpha \text{ a scalar, so } x_1 = \begin{bmatrix} 1/\sqrt{2} \\ -1/\sqrt{2} \\ 0 \end{bmatrix}$$

is an orthonormal basis for the null space of $0I - A$. Now let $\lambda = 2$. Then,

$$2I - A = \begin{bmatrix} 2 & 0 & 0 \\ 0 & 2 & 0 \\ 0 & 0 & 2 \end{bmatrix} - \begin{bmatrix} 1 & 1 & 0 \\ 1 & 1 & 0 \\ 0 & 0 & 2 \end{bmatrix}$$

$$= \begin{bmatrix} 1 & -1 & 0 \\ -1 & 1 & 0 \\ 0 & 0 & 0 \end{bmatrix} .$$

Then, $(2I - A)X = 0$;

$$\begin{bmatrix} 1 & -1 & 0 \\ -1 & 1 & 0 \\ 0 & 0 & 0 \end{bmatrix} \begin{bmatrix} x_1 \\ x_2 \\ x_3 \end{bmatrix} = \begin{bmatrix} 0 \\ 0 \\ 0 \end{bmatrix}$$

Solving this system yields $x_1 = s$, $x_2 = s$, $x_3 = t$. Thus, the null space of $2I - A$ consists of vectors of the form:

$$\begin{bmatrix} s \\ s \\ t \end{bmatrix} = s \begin{bmatrix} s \\ s \\ 0 \end{bmatrix} + \begin{bmatrix} 0 \\ 0 \\ t \end{bmatrix} = s \begin{bmatrix} 1 \\ 1 \\ 0 \end{bmatrix} + t \begin{bmatrix} 0 \\ 0 \\ 1 \end{bmatrix} .$$

Then,

$$\begin{bmatrix} 1 \\ 1 \\ 0 \end{bmatrix} \text{ and } \begin{bmatrix} 0 \\ 0 \\ 1 \end{bmatrix}$$

are an orthogonal basis for this null space. Normalizing shows that

$$x_2 = \begin{bmatrix} 1/\sqrt{2} \\ 1/\sqrt{2} \\ 0 \end{bmatrix} \text{ and } x_3 = \begin{bmatrix} 0 \\ 0 \\ 1 \end{bmatrix}$$

are an orthonormal basis for the null space of $2I - A$. Using the inner product function, $x_1 \cdot x_2 = 0$, $x_1 \cdot x_3 = 0$. Therefore, $\{x_1, x_2, x_3\}$ is a orthonormal basis consisting of characteristic vectors for A. Now, we construct an orthogonal matrix P whose columns are orthonormal.

$$P = \begin{bmatrix} 1/\sqrt{2} & 1/\sqrt{2} & 0 \\ -1/\sqrt{2} & 1/\sqrt{2} & 0 \\ 0 & 0 & 1 \end{bmatrix}$$

Then, since this is an orthogonal matrix,

$$P^{-1} = P^t = \begin{bmatrix} 1/\sqrt{2} & -1/\sqrt{2} & 0 \\ 1/\sqrt{2} & 1/\sqrt{2} & 0 \\ 0 & 0 & 1 \end{bmatrix} .$$

Recall that $B = P^{-1}AP$ is diagonal, and its diagonal entries are the characteristic values of A corresponding to the columns of P. Thus,

$$B = \begin{bmatrix} 0 & 0 & 0 \\ 0 & 2 & 0 \\ 0 & 0 & 2 \end{bmatrix}$$

which can be checked by calculating $P^{-1}AP$ directly.

● **PROBLEM** 10-17

Reduce the following quadratic equation to standard form:

$$ax^2 + bxy + cy^2 = d .$$

Solution: $ax^2 + bxy + cy^2 = d$.   (1)
Put

$$A = \begin{bmatrix} a & b/2 \\ b/2 & c \end{bmatrix} \quad \text{and} \quad \bar{u} = \begin{bmatrix} x \\ y \end{bmatrix} .$$

A is symmetric and

$$Au \cdot u = \begin{bmatrix} ax + b/2 \ y \\ b/2 \ x + cy \end{bmatrix} \cdot \begin{bmatrix} x \\ y \end{bmatrix}$$

$$= ax^2 + bxy + cy^2 ,$$

so we can rewrite equation (1) as

$$Au \cdot u = d .$$ (2)

Since A is symmetric, apply the spectral theorem to obtain an ortho-gonal matrix P and diagonal matrix B such that $B = P^{-1}AP$. Solve this for A to obtain $A = PBP^{-1}$. Substituting this in equation (2) gives

$$PBP^{-1}u \cdot u = d .$$ (3)

For any matrix A and any u and v, we have $Au \cdot v = u \cdot A^t v$.
Using this to shift P across the dot product yields

$$PBP^{-1}u \cdot u = BP^{-1}u \cdot P^t u .$$

Since P is orthogonal, we know that $P^t = P^{-1}$. Hence, write equation (3) as

$$BP^{-1}u \cdot P^{-1}u = d .$$ (4)

We can also write

$$B = \begin{bmatrix} b_1 & 0 \\ 0 & b_2 \end{bmatrix} \quad \text{and} \quad P^{-1}u = \begin{bmatrix} x_1 \\ y_1 \end{bmatrix}$$

Therefore, equation (3) can now be expressed as

$$\begin{bmatrix} b_1 & 0 \\ 0 & b_2 \end{bmatrix} \begin{bmatrix} x_1 \\ y_1 \end{bmatrix} \cdot \begin{bmatrix} x_1 \\ y_1 \end{bmatrix} = d$$

which, after carrying out these products, gives

$$b_1 x_2^2 + b_2 y_1^2 = d .$$ (5)

Thus, relative to the $x_1 y_1$ coordinate system, rewrite equation (1) in the form (4), in which no xy term appears. Equation (4) can easily be graphed in the $x_1 y_1$ system and, thereby, results in the graph of equa-tion (1).

This example illustrates the application of the spectral theorem to a geometric problem.

● **PROBLEM** 10-18

Eliminate the xy term by using the Spectrum Theorem and use this in-formation to draw a graph of $3x^2 + 2xy + 3y^2 = 1$.

Solution: The Spectral Theorem states that for a symmetric matrix A we can find an orthogonal matrix P as a function of the eigenvalues of A .

$$3x^2 + 2xy + 3y^2 = 1 .$$

Put

$$A = \begin{bmatrix} 3 & 1 \\ 1 & 3 \end{bmatrix} \quad ; \quad u = \begin{bmatrix} x \\ y \end{bmatrix} .$$

A is symmetric, therefore, apply the spectral theorem to obtain an orthogonal matrix P and a diagonal matrix B such that $B = P^{-1}AP$. Now, the characteristic polynomial of A is $f(\lambda) = \det(\lambda I - A)$

$$\begin{aligned} f(\lambda) &= \det(\lambda I - A) \\ &= \det \begin{vmatrix} \lambda-3 & -1 \\ -1 & \lambda-3 \end{vmatrix} \\ &= (\lambda-3)(\lambda-3) - 1 \\ &= \lambda^2 - 6\lambda + 9 - 1 \\ &= (\lambda-4)(\lambda-2) \quad . \end{aligned}$$

Therefore, the characteristic equation is $(\lambda-2)(\lambda-4) = 0$. Then the eigenvalues (or characteristic values) are $\lambda = 2$ and $\lambda = 4$. Substitute $\lambda = 2$ in $(\lambda I - A)X = 0$ to obtain the corresponding eigenvectors.

$(2I - A)X = 0$

or,

$$\begin{bmatrix} -1 & -1 \\ -1 & -1 \end{bmatrix} \begin{bmatrix} x_1 \\ x_2 \end{bmatrix} = \begin{bmatrix} 0 \\ 0 \end{bmatrix}$$

or, $\begin{matrix} -x_1 - x_2 = 0 \\ -x_1 - x_2 = 0 \end{matrix}$ or, $x_1 + x_2 = 0$.

Therefore, $x_1 = \begin{bmatrix} 1 \\ -1 \end{bmatrix}$ is an eigenvector of $\lambda = 2$. Next, normalize x_1 to obtain the unit orthogonal solution,

$$u_1 = \begin{bmatrix} 1/\sqrt{2} \\ -1/\sqrt{2} \end{bmatrix}.$$

Substitute $\lambda = 4$ in the equation $(\lambda I - A)X = 0$ to obtain the corresponding eigenvectors.

$(4I - A)X = 0$

or,

$$\begin{bmatrix} 1 & -1 \\ -1 & 1 \end{bmatrix} \begin{bmatrix} x_1 \\ x_2 \end{bmatrix} = \begin{bmatrix} 0 \\ 0 \end{bmatrix}$$

$\begin{matrix} x_1 - x_2 = 0 \\ -x_1 + x_2 = 0 \end{matrix}$ or, $x_1 - x_2 = 0$.

Thus, $x_2 = \begin{bmatrix} 1 \\ 1 \end{bmatrix}$ is an eigenvector of $\lambda = 4$. Normalize x_2 to obtain the unit orthogonal solution.

$$u_2 = \begin{bmatrix} 1/\sqrt{2} \\ 1/\sqrt{2} \end{bmatrix} . \quad \text{Note that } u_1 \cdot u_2 = 0 .$$

Construct the orthogonal matrix P whose columns are orthonormal. Thus,

$$P = \begin{bmatrix} 1/\sqrt{2} & 1/\sqrt{2} \\ -1/\sqrt{2} & 1/\sqrt{2} \end{bmatrix} .$$

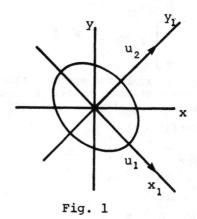

Fig. 1

Then, since P is an orthogonal matrix,

$$P^{-1} = P^t = \begin{bmatrix} 1/\sqrt{2} & -1/\sqrt{2} \\ 1/\sqrt{2} & 1/\sqrt{2} \end{bmatrix} .$$

Recall that $B = P^{-1}AP$ is diagonal and its diagonal entries are the characteristic values of A. Thus,

$$B = \begin{bmatrix} 2 & 0 \\ 0 & 4 \end{bmatrix} .$$

We can also write

$$BP^{-1}u \cdot P^{-1}u = d, \text{ (see previous example equation (4)).}$$

or,

$$\begin{bmatrix} 2 & 0 \\ 0 & 4 \end{bmatrix} \begin{bmatrix} x_1 \\ y_1 \end{bmatrix} \cdot \begin{bmatrix} x_1 \\ y_1 \end{bmatrix} = 1 .$$

Thus, $2x_1^2 + 4y_1^2 = 1 .$ (1)

To graph this equation in the (x_1, y_1) coordinate system, recall that x_1 and y_1 are the coordinates of (x,y) relative to the basis u_1 and u_2. Since the graph of equation (1) is an ellipse, the graph shown in Figure 1 is obtained.

EIGENVALUES AND EIGENVECTORS OF LINEAR OPERATORS

● **PROBLEM** 10-19

The matrix of the transformation, $(x,y)T = (y,x)$, is

$$A = \begin{bmatrix} 0 & 1 \\ 1 & 0 \end{bmatrix} .$$

Find the eigenvalues, the eigenvectors, and also the diagonal matrix of T.

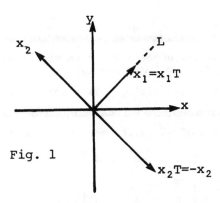

Fig. 1

Solution: To find eigenvalues, set $AX = \lambda X$. Then,

$$[A - \lambda I] = \begin{bmatrix} 0 & 1 \\ 1 & 0 \end{bmatrix} - \begin{bmatrix} \lambda & 0 \\ 0 & \lambda \end{bmatrix}$$

$$= \begin{bmatrix} -\lambda & 1 \\ 1 & -\lambda \end{bmatrix} .$$

Therefore, the characteristic equation $\det(A - \lambda I) = 0$ is $\lambda^2 - 1 = 0$, and the eigenvalues are $\lambda = 1$ and $\lambda = -1$. To find an eigenvector corresponding to $\lambda = 1$, solve $X(A - 1I) = 0$ for X. Thus,

$$\begin{bmatrix} x_1 \\ x_2 \end{bmatrix} \begin{bmatrix} -1 & 1 \\ 1 & -1 \end{bmatrix} = \begin{bmatrix} 0 \\ 0 \end{bmatrix}$$

or,

$$\begin{array}{c} -x_1 + x_2 = 0 \\ x_1 - x_2 = 0 \end{array} \quad \text{or,} \quad x_1 - x_2 = 0 .$$

Thus,

$$X_1 = \begin{bmatrix} 1 \\ 1 \end{bmatrix} \text{ is an eigenvector corresponding to } \lambda = 1.$$

To find an eigenvector corresponding to $\lambda = -1$, solve $X[A - (-1)I] = 0$ for X. Hence,

$$\begin{bmatrix} x_1 \\ x_2 \end{bmatrix} \begin{bmatrix} 1 & 1 \\ 1 & 1 \end{bmatrix} = \begin{bmatrix} 0 \\ 0 \end{bmatrix}$$

or,

$$\begin{array}{c} x_1 + x_2 = 0 \\ x_1 + x_2 = 0 \end{array} \quad \text{or,} \quad x_1 + x_2 = 0 .$$

So, in general, $x_1 = -x_2$. One specific solution is: $x_1 = -1$, $x_2 = +1$. Therefore,

$$X_2 = \begin{bmatrix} -1 \\ 1 \end{bmatrix} \text{ is an eigenvector corresponding to } \lambda = -1.$$

Since we have two distinct eigenvalues and a two-dimensional vector space, the eigenvectors form a basis for R^2 . Relative to this basis, T has the diagonal matrix,

$$B = \begin{bmatrix} 1 & 0 \\ 0 & -1 \end{bmatrix} ,$$

whose diagonal entries are the eigenvalues. T is now seen to be the reflection across the line $L = \{a(1,1): a \in R\}$ (Figure 1).

Find the eigenvalues and a basis for each of the eigenspaces of the linear operator $T: P_2 \rightarrow P_2$ defined by:

$$T(a + bx + cx^2) = (3a - 2b) + (-2a + 3b)x + (5c)x^2 .$$

Solution: The matrix of T with respect to the standard basis $B = \{1, x, x^2\}$ is

$$A = \begin{bmatrix} 3 & -2 & 0 \\ -2 & 3 & 0 \\ 0 & 0 & 5 \end{bmatrix} .$$

The eigenvalues of T are the eigenvalues of A. We form the matrix

$$\lambda I - A = \begin{bmatrix} \lambda & 0 & 0 \\ 0 & \lambda & 0 \\ 0 & 0 & \lambda \end{bmatrix} - \begin{bmatrix} 3 & -2 & 0 \\ -2 & 3 & 0 \\ 0 & 0 & 5 \end{bmatrix}$$

$$= \begin{bmatrix} \lambda-3 & 2 & 0 \\ 2 & \lambda-3 & 0 \\ 0 & 0 & \lambda-5 \end{bmatrix} .$$

Expanding along the third column yields

$$\det(\lambda I - A) = (\lambda-5) \begin{vmatrix} \lambda-3 & 2 \\ 2 & \lambda-3 \end{vmatrix}$$

$$= (\lambda-5) [(\lambda-3)^2 - 4]$$

$$= (\lambda-5)[\lambda^2 - 6\lambda + 5]$$

$$= (\lambda-5)(\lambda-5)(\lambda-1)$$

$$= (\lambda-5)^2(\lambda-1) .$$

The characteristic equation of A is $(\lambda-5)^2(\lambda-1) = 0$ so that the eigenvalues of A are $\lambda = 1$ and $\lambda = 5, 5$. If $\lambda = 1$, then

$$(1I - A)u = 0$$

or,

$$\begin{bmatrix} -2 & 2 & 0 \\ 2 & 2 & 0 \\ 0 & 0 & -4 \end{bmatrix} \begin{bmatrix} u_1 \\ u_2 \\ u_3 \end{bmatrix} = \begin{bmatrix} 0 \\ 0 \\ 0 \end{bmatrix} .$$

Solving this system yields $u_1 = t$, $u_2 = t$, $u_3 = 0$. Thus, the eigenvectors corresponding to $\lambda = 1$ are the non-zero vectors of the form

$$u = \begin{bmatrix} t \\ t \\ 0 \end{bmatrix} = t \begin{bmatrix} 1 \\ 1 \\ 0 \end{bmatrix} , \text{ so that } \begin{bmatrix} 1 \\ 1 \\ 0 \end{bmatrix} \text{ is a basis for the}$$

eigenspace corresponding to $\lambda = 1$. If $\lambda = 5$, then

$$(\lambda I - A)u = 0$$

or,

$$\begin{bmatrix} 2 & 2 & 0 \\ 2 & 2 & 0 \\ 0 & 0 & 0 \end{bmatrix} \begin{bmatrix} u_1 \\ u_2 \\ u_3 \end{bmatrix} = \begin{bmatrix} 0 \\ 0 \\ 0 \end{bmatrix} .$$

Solving this system gives $u_1 = -s$, $u_2 = s$, $u_3 = t$. Thus, the eigenvectors of A corresponding to $\lambda = 5$ are the non-zero vectors of the
form:

$$u = \begin{bmatrix} -s \\ s \\ t \end{bmatrix} = \begin{bmatrix} -s \\ s \\ 0 \end{bmatrix} + \begin{bmatrix} 0 \\ 0 \\ t \end{bmatrix} = s \begin{bmatrix} -1 \\ 1 \\ 0 \end{bmatrix} + t \begin{bmatrix} 0 \\ 0 \\ 1 \end{bmatrix}$$

Hence, the eigenspace of A corresponding to $\lambda = 5$ has the basis
$\{u_1, u_2\}$ and that corresponding to $\lambda = 1$ has the basis $\{u_3\}$ where

$$u_1 = \begin{bmatrix} -1 \\ 1 \\ 0 \end{bmatrix} \quad u_2 = \begin{bmatrix} 0 \\ 0 \\ 1 \end{bmatrix} \quad u_3 = \begin{bmatrix} 1 \\ 1 \\ 0 \end{bmatrix} .$$

These matrices are the coordinate matrices with respect to B of

$$P_1 = -1 + x , \quad P_2 = x^2 , \quad P_3 = 1 + x .$$

Thus, $\{-1 + x, x^2\}$ is a basis for the eigenspace of T corresponding
to $\lambda = 5$, and $\{1 + x\}$ is a basis for the eigenspace corresponding to
$\lambda = 1$.

● **PROBLEM** 10-21

T is given by the rule $(x\ y)T = (-4x + y, -5x + 2y)$. Find the eigenvalues and eigenvectors for T.

Solution: The usual matrix A for T is

$$A = \begin{bmatrix} -4 & -5 \\ 1 & 2 \end{bmatrix} .$$

Then,

$$\lambda I - A = \begin{bmatrix} \lambda & 0 \\ 0 & \lambda \end{bmatrix} - \begin{bmatrix} -4 & -5 \\ 1 & 2 \end{bmatrix} = \begin{bmatrix} \lambda+4 & 5 \\ -1 & \lambda-2 \end{bmatrix}$$

$$\det(\lambda I - A) = (\lambda+4)(\lambda-2) + 5$$
$$= \lambda^2 + 2\lambda - 8 + 5 = \lambda^2 + 2\lambda - 3$$
$$= (\lambda+3)(\lambda-1) .$$

Therefore, the characteristic equation of A is $(\lambda+3)(\lambda-1) = 0$. Also,
the eigenvalues are $\lambda = -3$ and $\lambda = 1$.

Now, find an eigenvector corresponding to $\lambda_1 = -3$. To do this,
solve $X(A - \lambda I) = 0$ for X with $\lambda = -3$.

$$A - \lambda I = \begin{bmatrix} -4-\lambda & -5 \\ 1 & 2-\lambda \end{bmatrix}$$

or,

$$X \begin{bmatrix} -4-\lambda & -5 \\ 1 & 2-\lambda \end{bmatrix} = 0 ;$$

that is,

$$X \begin{bmatrix} -1 & -5 \\ 1 & 5 \end{bmatrix} = 0$$

301

or,

$$(x_1 \; x_2) \begin{bmatrix} -1 & -5 \\ 1 & 5 \end{bmatrix} = (0 \; 0) .$$

This is the system of equations,

$$-1x_1 + 1x_2 = 0$$
$$-5x_1 + 5x_2 = 0 ,$$

whose matrix is

$$\begin{bmatrix} -1 & 1 \\ -5 & 5 \end{bmatrix}$$ which gives $X_1 = X_2$.

Therefore, $(1,1)$ is an eigenvector corresponding to $\lambda_1 = -3$. Similarly, to find an eigenvector corresponding to $\lambda_2 = 1$, solve $X(A - 1I) = 0$ for X. Thus,

$$X \begin{bmatrix} -5 & -5 \\ 1 & 1 \end{bmatrix} = 0 .$$

This gives a system of equations with matrix $\begin{bmatrix} -5 & 1 \\ -5 & 1 \end{bmatrix}$. That is,

$-5x_1 + x_2 = 0$; $-5x_1 + x_2 = 0$ which shows that every eigenvector of $\lambda_2 = 1$ is of the form $(k,5k) = k(1,5)$ with k a scalar. Thus, $(1,5)$ is an eigenvector corresponding to $\lambda_2 = 1$. Therefore, $x_1 = (1,1)$, $x_2 = (1,5)$. Since we have two distinct eigenvalues and a two-dimensional vector space, the eigenvectors form a basis for R^2; relative to this basis, T has the diagonal matrix

$$B = \begin{bmatrix} -3 & 0 \\ 0 & 1 \end{bmatrix}$$

whose diagonal entries are the eigenvalues. In general, a linear operator $T : V \rightarrow V$ can be represented by a diagonal matrix B if and only if V has a basis consisting of eigenvectors of T. In this case, the diagonal elements of B are the corresponding eigenvalues.

● **PROBLEM** 10-22

Reduce the matrix A of the linear transformation to a diagonal form where

$$A = \begin{bmatrix} 1 & 3 & 1 & 2 \\ 0 & -1 & 1 & 3 \\ 0 & 0 & 2 & 5 \\ 0 & 0 & 0 & -2 \end{bmatrix}$$

Solution: If the characteristic polynomial of a linear transformation of an n-dimensional linear space has n distinct real roots, then the matrix of the transformation reduces to diagonal form in an appropriate coordinate system. The characteristic values of A are the diagonal entries since A is an upper triangular matrix. Thus,

$$\lambda_1 = 1 , \quad \lambda_2 = -1 , \quad \lambda_3 = 2 , \quad \lambda_4 = -2 .$$

To find an eigenvector corresponding to $\lambda = 1$, solve the system $X(A - 1I) = 0$ for X . Thus,

302

$$[x_1 \ x_2 \ x_3 \ x_4] \begin{bmatrix} 0 & 3 & 1 & 2 \\ 0 & -2 & 1 & 3 \\ 0 & 0 & 1 & 5 \\ 0 & 0 & 0 & -3 \end{bmatrix} = 0$$

or,

$$3x_1 - 2x_2 = 0$$
$$x_1 + x_2 + x_3 = 0$$
$$2x_1 + 3x_2 + 5x_3 - 3x_4 = 0 \ .$$

Solving these equations yields: $x_1 = 2$, $x_2 = 3$, $x_3 = -5$, $x_4 = -4$.

Thus,

$X_1 = (2,3,-5,-4)$ is an eigenvector associated with $\lambda = 1$.

Similarly,

$$X_2 = (0,-3,1,-4)$$
$$X_3 = (0,0,4,5)$$
$$X_4 = (0,0,0,1)$$

are eigenvectors associated with $\lambda_2 = -1$; $\lambda_3 = 2$; $\lambda_4 = -2$. Since we have four linearly independent eigenvalues and a four-dimensional vector space, the eigenvectors form a basis for R^4 . The matrix of the transformation A reduces to diagonal form,

$$B = \begin{bmatrix} 1 & 0 & 0 & 0 \\ 0 & -1 & 0 & 0 \\ 0 & 0 & 2 & 0 \\ 0 & 0 & 0 & -2 \end{bmatrix}$$

whose diagonal entries are characteristic values of A.

● **PROBLEM** 10-23

Which of the following operators on R^2 is diagonalizable?

(1) $T_1(x,y) = (x - 2y, \ x - y)$.

(2) $T_2(x,y) = (x + y, \ y)$.

Solution: T is diagonalizable if and only if there is a basis for V consisting of characteristic vectors for T.

(a)   The matrix of T_1 is

$$A_1 = \begin{bmatrix} 1 & -2 \\ 1 & -1 \end{bmatrix}$$

Form a matrix,

$$(\lambda I - A) = \begin{bmatrix} \lambda-1 & 2 \\ -1 & \lambda+1 \end{bmatrix}$$

$$\det(\lambda I - A) = (\lambda-1)(\lambda+1) + 2$$
$$= \lambda^2 + 1 \ .$$

The characteristic equation is $\lambda^2 + 1 = 0$, and this equation has no real roots. Thus, T has no real characteristic values, so T cannot be diagonalizable.

303

(b)   The matrix of T_2 is
$$A = \begin{bmatrix} 1 & 1 \\ 0 & 1 \end{bmatrix}.$$

Then,
$$(\lambda I - A) = \begin{bmatrix} \lambda-1 & -1 \\ 0 & \lambda-1 \end{bmatrix}$$
$$\det(\lambda I - A) = (\lambda-1)^2.$$

The characteristic equation is $(\lambda-1)^2 = 0$. Thus, $\lambda = 1$ is the only real characteristic value of T. Therefore, if T is diagonalizable, then we must have at least two independent solutions to $T_2(x,y) = (x,y)$.

To obtain the eigenvectors, solve $(\lambda I - A)X = 0$ for X with $\lambda = 1$.
$(1I - A)X = 0$
$$\begin{bmatrix} 0 & -1 \\ 0 & 0 \end{bmatrix} \begin{bmatrix} x_1 \\ x_2 \end{bmatrix} = \begin{bmatrix} 0 \\ 0 \end{bmatrix}.$$

Solving this system gives $x_2 = 0$, $x_1 = t$, or,
$$X = \begin{bmatrix} t \\ 0 \end{bmatrix} = t \begin{bmatrix} 1 \\ 0 \end{bmatrix}$$

since T_2 has only one linearly independent eigenvector, T is not diagonalizable. It seems reasonable to ask whether the characteristic equation helps in determining diagonalizability. If $T = I$, then
$$\det(\lambda I - T) = \det \begin{vmatrix} \lambda-1 & 0 \\ 0 & \lambda-1 \end{vmatrix} = (\lambda-1)^2$$

The identity is clearly diagonalizable. Note from this that I and the operator of (b) have the same characteristic equation, yet I is diagonalizable and the operator of (b) is not. It can be concluded that the characteristic equation, even if it has real roots, does not give enough information to determine if T is diagonalizable. In spite of this, there is one situation in which the characteristic equation does give enough information:

If $\det(\lambda I - T) = 0$ has n distinct real roots, $d_1, d_2, \ldots, d_n$ (i.e., $d_i \neq d_j$ if $i \neq j$ and $\det(d_i I - T) = 0$ for $i = 1, 2, \ldots, n$) and $n = \dim V$, then T is diagonalizable.

● **PROBLEM** 10-24

(a) Let T_1 be the linear operator on R^2 defined by $T_1(xy) = (x'y')$ if and only if
$$\begin{bmatrix} 1 & 1 \\ 0 & 2 \end{bmatrix} \begin{bmatrix} x \\ y \end{bmatrix} = \begin{bmatrix} x' \\ y' \end{bmatrix}$$
Is T_1 diagonalizable?

(b) Let T be the linear operator on R^2 such that:
$T(xy) = (x, y')$ if and only if $\begin{bmatrix} 3 & -1 \\ -1 & 3 \end{bmatrix} \begin{bmatrix} x \\ y \end{bmatrix} = \begin{bmatrix} x' \\ y' \end{bmatrix}$.
Show that TT_1 is not diagonalizable.

(c) Let T_2 be the linear operator on R^2 defined by
$T_2(xy) = (x',y')$ if and only if $\begin{bmatrix} 1 & 100 \\ 0 & 2 \end{bmatrix} \begin{bmatrix} x \\ y \end{bmatrix} = \begin{bmatrix} x' \\ y' \end{bmatrix}$.
Show that $T + T_2$ is not diagonalizable.

Solution: (a) The matrix of T_1 is $A = \begin{bmatrix} 1 & 1 \\ 0 & 2 \end{bmatrix}$.

$$\det[\lambda I - A] = \det \begin{vmatrix} \lambda-1 & -1 \\ 0 & \lambda-2 \end{vmatrix}$$

$$= (\lambda-1)(\lambda-2) .$$

Therefore, the characteristic values for T_1 are $\lambda = 1$ and $\lambda = 2$. Thus, T_1 has two distinct real roots; T_1 is diagonalizable. The characteristic vectors for T_1 are obtained by solving the equations $[1I - A]X = 0$ and $[2I - A]X = 0$, or,

$$\begin{bmatrix} 0 & -1 \\ 0 & -1 \end{bmatrix} \begin{bmatrix} x_1 \\ x_2 \end{bmatrix} = \begin{bmatrix} 0 \\ 0 \end{bmatrix} \text{ and } \begin{bmatrix} 1 & -1 \\ 0 & 0 \end{bmatrix} \begin{bmatrix} x_1 \\ x_2 \end{bmatrix} = \begin{bmatrix} 0 \\ 0 \end{bmatrix}$$

Thus,

$$X_1 = \begin{bmatrix} 1 \\ 0 \end{bmatrix} \text{ and } X_2 = \begin{bmatrix} 1 \\ 1 \end{bmatrix}.$$

(b) The matrix of TT_1 with respect to the standard basis is

$$\begin{bmatrix} 3 & -1 \\ -1 & 3 \end{bmatrix} \begin{bmatrix} 1 & 1 \\ 0 & 2 \end{bmatrix} = \begin{bmatrix} 3 & 1 \\ -1 & 5 \end{bmatrix}$$

then,

$$\det(\lambda I - TT_1) = \det \begin{vmatrix} \lambda-3 & -1 \\ 1 & \lambda-5 \end{vmatrix}$$

$$= (\lambda-3)(\lambda-5) + 1$$

$$= (\lambda-4)^2 .$$

Therefore, $\lambda = 4$ is the only characteristic value of TT_1. Since nullity $(4I - TT_1) = $ nullity $\begin{bmatrix} 1 & -1 \\ 1 & -1 \end{bmatrix} = 1$, conclude that TT_1 is not diagonalizable.

(c) The matrix of $T + T_2$ with respect to the standard basis is

$$\begin{bmatrix} 3 & -1 \\ -1 & 3 \end{bmatrix} + \begin{bmatrix} 1 & 100 \\ 0 & 2 \end{bmatrix} = \begin{bmatrix} 4 & 99 \\ -1 & 5 \end{bmatrix}.$$

$$\det[\lambda I - (T + T_2)] = \begin{vmatrix} \lambda-4 & -99 \\ 1 & \lambda-5 \end{vmatrix}$$

$$= (\lambda-4)(\lambda-5) + 99$$

$$= \lambda^2 - 9\lambda + 119 .$$

Using the quadratic formula, conclude that $T + T_2$ has no real characteristic values. Therefore, $T + T_2$ is not diagonalizable.

● **PROBLEM** 10-25

Show that geometric methods can be used to find the characteristic vectors and characteristic values for rotations and projections.

Solution: Make use of the fact that a non-zero vector V is a characteristic vector for T if and only if TV is a multiple of V.

(a) Suppose T is a counterclockwise rotation in R^2 through the angle

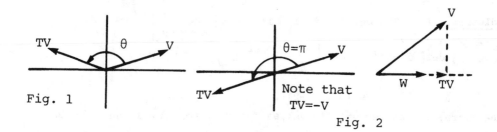

Fig. 1

Note that
TV=-V

Fig. 2

θ and that θ is not a multiple of π. If $V \neq 0$, TV cannot be a multiple of V, as shown in Figure 1. This fact establishes that:

A rotation through θ has no characteristic vectors if θ is not a multiple of π.

Suppose $\theta = \pi$. Then, as Figure 2 indicates, for every non-zero vector V, we have TV = -V. Since this is also true if θ is any odd multiple of π, we know that:

If θ is an odd multiple of π, then every non-zero vector is a characteristic vector belonging to the characteristic value $\lambda = -1$.

If θ is an even multiple of π, then T is just the identity operator. Thus, for every non-zero vector V, we have TV = V. In other words:

If θ is an even multiple of π, then every non-zero vector is a characteristic vector belonging to the characteristic value $\lambda = 1$.

(b) Suppose T is the projection in R^2 onto the non-zero vector W. A vector parallel to W is left fixed by T; that is, if V is a multiple of W, TV = V. This shows that every non-zero multiple of W is a characteristic vector belonging to the characteristic value $\lambda = 1$. If V is orthogonal to W, then TV = 0. Since 0 = 0V, every non-zero vector V which is orthogonal to W is a characteristic vector belonging to the characteristic value $\lambda = 0$. If V is neither parallel nor orthogonal to W, then (as Figure 3 indicates) TV is not parallel or perpendicular to V. Such a vector V cannot be a characteristic vector. To summarize, it has been shown that:

The projection T onto W has the characteristic values $\lambda = 0$ and $\lambda = 1$. The non-zero multiples of W are the characteristic vectors belonging to $\lambda = 1$, and the non-zero vectors, orthogonal to W, are the characteristic vectors belonging to $\lambda = 0$.

Note that TV is not parallel to V and $TV \neq \lambda V$ for any scalar λ.

● **PROBLEM** 10-26

Suppose that T is the $90°$ counterclockwise rotation of R^2 around the origin. Then T has

$$\begin{bmatrix} 0 & 1 \\ -1 & 0 \end{bmatrix}$$

as its usual matrix. Find the eigenvalues of T.

Solution: Note from the figure that the effect of T is to send the point (x,y) to the point (y,-x). To find eigenvalues, solve:

$$\det \begin{vmatrix} \lambda & -1 \\ +1 & \lambda \end{vmatrix} = 0$$

Thus,

$$\lambda^2 + 1 = 0,$$

which has no real roots. Hence, T has no eigenvalues (and, therefore, no eigenvectors). This fact can be seen geometrically. T does not stretch any vector, but rather T rotates them. Therefore, to conclude, T has no diagonal matrix.

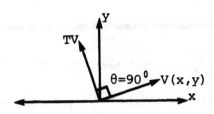

CHAPTER 11

DIAGONALIZATION OF A MATRIX

CHANGE OF BASIS

● PROBLEM 11-1

Let (1) $E = [e_1, e_2]$, the usual basis for R^2. (2) $T \overset{E}{\Leftrightarrow} A = \begin{bmatrix} 2 & 5 \\ 3 & 7 \end{bmatrix}$

is a linear transformation. (3) $F = [f_1, f_2]$ is another basis for R^2, where $f_1 = (1,1) = e_1 + e_2$; $f_2 = (-1,1) = -e_1 + e_2$.
Find the matrix of T relative to F.

<u>Solution</u>: Suppose T is a linear transformation relative to the usual basis. A is then the matrix of T with respect to a given basis. If B is the matrix of T with respect to another basis then $B = P^{-1}AP$. In other words, B is similar to A. The matrix P is the conversion matrix from the old to the new basis, or simply the conversion matrix. Thus, the matrix of T relative to the new basis = (conversion matrix)$^{-1}$ · (matrix of T relative to old basis) · (conversion matrix).

$F = PE$ and $T \overset{E}{\Leftrightarrow} A$ implies that the matrix of T under the basis f is given by $P^{-1}AP$. We must find P and P^{-1}.

$$F = [f_1, f_2] \quad f_1 = 1e_1 + 1e_2$$
$$f_2 = -1e_1 + 1e_2 .$$

Then,

$$P = \begin{bmatrix} 1 & -1 \\ 1 & 1 \end{bmatrix}$$

The adjoint method can be used to find P^{-1}. Thus,

$$P^{-1} = \frac{1}{\det P} \text{ adj } P = \begin{bmatrix} \frac{1}{2} & \frac{1}{2} \\ -\frac{1}{2} & \frac{1}{2} \end{bmatrix} = \frac{1}{2}\begin{bmatrix} 1 & 1 \\ -1 & 1 \end{bmatrix} .$$

Therefore,

$$T \overset{F}{\Leftrightarrow} P^{-1}AP = \frac{1}{2}\begin{bmatrix} 1 & 1 \\ -1 & 1 \end{bmatrix}\begin{bmatrix} 2 & 5 \\ 3 & 7 \end{bmatrix}\begin{bmatrix} 1 & -1 \\ 1 & 1 \end{bmatrix}$$

$$= \frac{1}{2}\begin{bmatrix} 1 & 1 \\ -1 & 1 \end{bmatrix}\begin{bmatrix} 7 & 3 \\ 10 & 4 \end{bmatrix}$$

$$= \frac{1}{2}\begin{bmatrix} 17 & 7 \\ 3 & 1 \end{bmatrix} = \begin{bmatrix} 17/2 & 7/2 \\ 3/2 & 1/2 \end{bmatrix} .$$

This is the matrix of T relative to the basis $\{f_1, f_2\}$.

Let T be the linear operator on R^2 such that
$$T(x,y) = (x',y') \quad \text{if and only if} \quad \begin{bmatrix} 3 & -1 \\ -1 & 3 \end{bmatrix}\begin{bmatrix} x \\ y \end{bmatrix} = \begin{bmatrix} x' \\ y' \end{bmatrix}.$$
Compute the matrix of T with respect to the basis $\{(1,1),(1,-1)\}$.

Solution: Now
$$T\begin{bmatrix} 1 \\ 1 \end{bmatrix} = \begin{bmatrix} 3 & -1 \\ -1 & 3 \end{bmatrix}\begin{bmatrix} 1 \\ 1 \end{bmatrix} = \begin{bmatrix} 2 \\ 2 \end{bmatrix} = 2\begin{bmatrix} 1 \\ 1 \end{bmatrix}$$
and
$$T\begin{bmatrix} 1 \\ -1 \end{bmatrix} = \begin{bmatrix} 3 & -1 \\ -1 & 3 \end{bmatrix}\begin{bmatrix} 1 \\ -1 \end{bmatrix} = \begin{bmatrix} 4 \\ -4 \end{bmatrix} = 4\begin{bmatrix} 1 \\ -1 \end{bmatrix}.$$
Thus,
$$T(1,1) = 2(1,1) \quad \text{and} \quad T(1,-1) = 4(1,-1).$$
The matrix of T with respect to $\{(1,1),(1,-1)\}$ is the matrix $\begin{bmatrix} 2 & 0 \\ 0 & 4 \end{bmatrix}$.

This is a diagonal matrix. Generalize this result, as follows: let V be a finite dimensional vector space, and let T be a linear operator on V. We say that T is diagonalizable if there is a basis $\{x_1, x_2, \ldots, x_n\}$ for V such that the matrix of T with respect to $\{x_1, x_2, \ldots, x_n\}$ is a diagonal matrix. Thus, the operator T in the above example is a diagonalizable operator on R^2. Let P be the diagonal matrix of T with respect to $\{x_1, x_2, \ldots, x_n\}$. Then

$$P = \begin{bmatrix} d_1 & 0 & 0 & \ldots & 0 \\ 0 & d_2 & 0 & \ldots & 0 \\ \cdot & & & & \\ \cdot & & & & \\ 0 & 0 & 0 & \ldots & d_n \end{bmatrix}.$$

We have $\det(P) = d_1 d_2 \ldots d_n$. Thus, T is invertible if and only if $d_i \neq 0$ for $i = 1, 2, \ldots, n$. T^{-1} is easily described in terms of $\{x_1, x_2, \ldots, x_n\}$ since the matrix of T^{-1} with respect to this basis is P^{-1}, and

$$P^{-1} = \begin{bmatrix} 1/d_1 & 0 & \ldots & 0 \\ 0 & 1/d_2 & \ldots & 0 \\ \cdot & & & \\ \cdot & & & \\ 0 & 0 & \ldots & 1/d_n \end{bmatrix} \quad \text{if } d_i \neq 0, \, i = 1, 2, \ldots, n.$$

If $e_1 = (1,0)$, $e_2 = (0,1)$, $f_1 = (1,3)$, $f_2 = (2,5)$, then $\{e_1,e_2\}$ and $\{f_1,f_2\}$ are bases of R^2 .

(a) Find the transition matrix P from $\{e_i\}$ to $\{f_i\}$.

(b) Find the transition matrix Q from $\{f_i\}$ to $\{e_i\}$ and verify that $Q = P^{-1}$.

(c) Show that $[v_f] = P^{-1}[v]_e$ for any vector $v \in R^2$.

(d) Show that $[T]_f = P^{-1}[T]_e P$ for the operator T defined on R^2 by $T(x,y) = (2y, 3x-y)$.

Solution: Suppose that $\{u_1, u_2, \ldots, u_n\}$ and $\{v_1, v_2, \ldots, v_n\}$ are two bases of the vector space V. Each linear transformation $T: V \to V$ is associated with two matrices: a matrix $A = (a_{ij})$ which represents T with respect to the basis $\{u_1, u_2, \ldots, u_n\}$, and a matrix $B = (b_{ij})$ which represents T with respect to the basis $\{v_1, v_2, \ldots, v_n\}$. Suppose:

$$v_1 = P_{11}u_1 + P_{12}u_2 + \cdots + P_{1n}u_n$$
$$v_2 = P_{21}u_1 + P_{22}u_2 + \cdots + P_{2n}u_n \tag{1}$$
$$\cdots \cdots \cdots \cdots$$
$$v_n = P_{n1}u_1 + P_{n2}u_2 + \cdots + P_{nn}u_n .$$

Then the matrix P, termed the transition matrix from one basis to the other, is

$$\begin{bmatrix} P_{11} & P_{21} & \cdots & P_{n1} \\ P_{12} & P_{22} & \cdots & P_{n2} \\ \cdots & \cdots & \cdots & \cdots \\ P_{1n} & P_{2n} & \cdots & P_{nn} \end{bmatrix} .$$

The matrix Q is determined by the relation

$$u_1 = q_{11}v_1 + q_{12}v_2 + \cdots + q_{1n}v_n$$
$$u_2 = q_{21}v_1 + q_{22}v_2 + \cdots + q_{2n}v_n \tag{2}$$
$$\cdots \cdots \cdots \cdots \cdots \cdots \cdots$$
$$u_n = q_{n1}v_1 + q_{n2}v_2 + \cdots + q_{nn}v_n$$

To show that $P^{-1} = Q$, write the system of equations (1) in matrix form:

$$P \begin{bmatrix} u_1 \\ u_2 \\ \cdot \\ \cdot \\ \cdot \\ u_n \end{bmatrix} = \begin{bmatrix} v_1 \\ v_2 \\ \cdot \\ \cdot \\ \cdot \\ v_n \end{bmatrix}$$

Assuming that P is nonsingular (that is, invertible) yields:

$$P^{-1}P \begin{bmatrix} u_1 \\ \cdot \\ \cdot \\ \cdot \\ u_n \end{bmatrix} = P^{-1} \begin{bmatrix} v_1 \\ \cdot \\ \cdot \\ \cdot \\ v_n \end{bmatrix} \quad . \quad \text{However,} \quad P^{-1}P = I \, ,$$

therefore,

$$\begin{bmatrix} u_1 \\ \cdot \\ \cdot \\ \cdot \\ u_n \end{bmatrix} = P^{-1} \begin{bmatrix} v_1 \\ \cdot \\ \cdot \\ \cdot \\ v_n \end{bmatrix} \quad .$$

From the system of equations (2) it is known

$$\begin{bmatrix} u_1 \\ \cdot \\ \cdot \\ \cdot \\ u_n \end{bmatrix} = Q \begin{bmatrix} v_1 \\ \cdot \\ \cdot \\ \cdot \\ v_n \end{bmatrix} \quad .$$

So,

$$P^{-1} \begin{bmatrix} v_1 \\ \cdot \\ \cdot \\ \cdot \\ v_n \end{bmatrix} = Q \begin{bmatrix} v_1 \\ \cdot \\ \cdot \\ \cdot \\ v_n \end{bmatrix} \quad .$$

Therefore, $P^{-1} = Q$ or $PP^{-1} = PQ$. This gives $I = PQ$.

(a) Here e_i and f_i are two bases of R^2 . Now $f_1 = (1,3) = (1,0)+3(0,1)$ $= e_1+3e_2$, $f_2 = (2,5) = 2(1,0) + 5(0,1) = 2e_1 + 5e_2$. Thus,

$$P = \begin{bmatrix} 1 & 2 \\ 3 & 5 \end{bmatrix} \quad .$$

(b) First find the coordinates of an arbitrary vector $(a,b) \in R^2$ with respect to the basis $\{f_i\}$. We have

$$(a,b) = x(1,3) + y(2,5) = (x+2y, 3x+5y),$$

or

$$x + 2y = a \quad \text{and} \quad 3x + 5y = b.$$

These equations must be solved for x and y in terms of a and b . Multiplying the first equation by -3 and adding to the second equation yields $-y = -3a+b$. Therefore, $y = 3a-b$. Substituting this value of y in the first equation results in $x + 2(3a-b) = a$, which gives $x = 2b-5$. Thus, $(a,b) = (2b-5a)f_1 + (3a-b)f_2$. Now, e_i can be found in terms of f_i:

$$e_1 = (1,0) = -5f_1 + 3f_2$$
$$e_2 = (0,1) = 2f_1 - f_2 \quad .$$

Thus,

$$Q = \begin{bmatrix} -5 & 2 \\ 3 & -1 \end{bmatrix} \quad .$$

Then,

$$PQ = \begin{bmatrix} 1 & 2 \\ 3 & 5 \end{bmatrix} \begin{bmatrix} -5 & 2 \\ 3 & -1 \end{bmatrix} = \begin{bmatrix} 1 & 0 \\ 0 & 1 \end{bmatrix} = I \quad .$$

Thus, $Q = P^{-1}$.

(c) Let $v = ae_1 + be_2$. Then the coordinate vector of v relative to $\{e_i\}$ is $[v]_e = \begin{bmatrix} a \\ b \end{bmatrix}$. Similarly, $[v]_f = \begin{bmatrix} 2b - 5a \\ 3a - b \end{bmatrix}$. Hence,

$$P^{-1}[v]_e = \begin{bmatrix} -5 & 2 \\ 3 & -1 \end{bmatrix} \begin{bmatrix} a \\ b \end{bmatrix} = \begin{bmatrix} -5a + 2b \\ 3a - b \end{bmatrix} = [v]_f .$$

(d) $T(x,y) = (2y, 3x-y)$. $[T]_e$ and $[T]_f$ are called the matrix representa-
tions of T, relative to the basis $\{e_i\}$ and $\{f_i\}$, respectively.
If $(a,b) \in R^2$, then $(a,b) = ae_1 + be_2$, $T(e_1) = T(1,0) = (0,3) = 0e_1 + 3e_2$,
$T(e_2) = T(0,1) = (2,-1) = 2e_1 - e_2$. Thus,

$$[T]_e = \begin{bmatrix} 0 & 2 \\ 3 & -1 \end{bmatrix} .$$

If $(a,b) \in R^2$ with respect to the basis $\{f_i\}$ then $(a,b) = (-5a + 2b)f_1 + (3a-b)f_2$, $T(f_1) = T(1,3) = (6,0) = -30f_1 + 18f_2$, $T(f_2) = T(2,5) = (10,1) = -48f_1 + 29f_2$. Hence,

$$[T]_f = \begin{bmatrix} -30 & -48 \\ 18 & 29 \end{bmatrix} .$$

Then

$$P^{-1}[T]_e P = \begin{bmatrix} -5 & 2 \\ 3 & -1 \end{bmatrix} \begin{bmatrix} 0 & 2 \\ 3 & -1 \end{bmatrix} \begin{bmatrix} 1 & 2 \\ 3 & 5 \end{bmatrix}$$

$$= \begin{bmatrix} -5 & 2 \\ 3 & -1 \end{bmatrix} \begin{bmatrix} 6 & 10 \\ 0 & 1 \end{bmatrix}$$

$$= \begin{bmatrix} -30 & -48 \\ 18 & 29 \end{bmatrix} = [T]_f .$$

● PROBLEM 11-4

Consider the following bases of R^3: $\{e_1 = (1,0,0),\ e_2 = (0,1,0),$
$e_3 = (0,0,1)\}$ and $\{f_1 = (1,1,1),\ f_2 = (1,1,0),\ f_3 = (1,0,0)\}$.
(a) Find the transition matrix P from $\{e_i\}$ to $\{f_i\}$.
(b) Find the transition matrix Q from $\{f_i\}$ to $\{e_i\}$. Verify that
$Q = P^{-1}$.
(c) Show that $[v]_f = P^{-1}[v]_e$ for any vector $v \in R^3$.
(d) Show that $[T]_f = P^{-1}[T]_e P$ for the T defined by $T(x,y,z) = (2y+z,\ x-4y,\ 3x)$.

<u>Solution:</u>   (a) The f_i must be expressed in terms of the e_i .
$$f_1 = (1,1,1) = (1,0,0) + (0,1,0), + (0,0,1)$$
$$= e_1 + e_2 + e_3 ,$$
$$f_2 = (1,1,0) = 1(1,0,0) + 1(0,1,0) + 0(0,0,1)$$
$$= e_1 + e_2 + 0e_3 ,$$

$$f_3 = (1,0,0) = 1(1,0,0) + 0(0,1,0) + 0(0,0,0)$$
$$= e_1 + 0e_2 + 0e_3 \ .$$

Hence,

$$P = \begin{bmatrix} 1 & 1 & 1 \\ 1 & 1 & 0 \\ 1 & 0 & 0 \end{bmatrix} \ .$$

(b) First find the coordinates of an arbitrary vector $(a,b,c) \in R^3$ with respect to the basis $\{f_1, f_2, f_3\}$.

$$(a,b,c) = x(1,1,1) + y(1,1,0) + z(1,0,0) = (x+y+z, \ x+y, \ x).$$

Hence,

$$x+y+z = a, \ x+y = b, \ x = c,$$

Solving for a, b and c yields

$$x = c, \ y = b-c, \ z = a-b.$$

Thus,

$$(a,b,c) = cf_1 + (b-c)f_2 + (a-b)f_3 \ .$$

Now,

$$e_1 = (1,0,0) = 0f_1 + 0f_2 + 1f_3$$

$$e_2 = (0,1,0) = 0f_1 + 1f_2 - 1f_3$$

$$e_3 = (0,0,1) = 1f_1 - 1f_2 + 0f_3 \ .$$

Hence,

$$Q = \begin{bmatrix} 0 & 0 & 1 \\ 0 & 1 & -1 \\ 1 & -1 & 0 \end{bmatrix}$$

$$PQ = \begin{bmatrix} 1 & 1 & 1 \\ 1 & 1 & 0 \\ 1 & 0 & 0 \end{bmatrix} \begin{bmatrix} 0 & 0 & 1 \\ 0 & 1 & -1 \\ 1 & -1 & 0 \end{bmatrix} = \begin{bmatrix} 1 & 0 & 0 \\ 0 & 1 & 0 \\ 0 & 0 & 1 \end{bmatrix} = I.$$

Therefore, $Q = P^{-1}$.

(c) If $v = (a,b,c)$, then $[v]_e = \begin{bmatrix} a \\ b \\ c \end{bmatrix}$ and $[v]_f = \begin{bmatrix} c \\ b-c \\ a-b \end{bmatrix}$. Then,

$$P^{-1}[v]_e = \begin{bmatrix} 0 & 0 & 1 \\ 0 & 1 & -1 \\ 1 & -1 & 0 \end{bmatrix} \begin{bmatrix} a \\ b \\ c \end{bmatrix} = \begin{bmatrix} c \\ b-c \\ a-b \end{bmatrix} = [v]_f$$

(d) $T(x,y,z) = (2y+z, \ x-4y, \ 3x)$. $T(e_1) = T(1,0,0) = (0,1,3) = 0e_1 + 1e_2 + 3e_3$. $T(e_2) = T(0,1,0) = (2,-4,0) = 2e_1 - 4e_2 + 0e_3$.
$T(e_3) = T(0,0,1) = (1,0,0) = 1e_1 + 0e_2 + 0e_3$. Hence,

$$[T]_e = \begin{bmatrix} 0 & 2 & 1 \\ 1 & -4 & 0 \\ 3 & 0 & 0 \end{bmatrix} \quad \text{is the matrix associated with } T$$

under the basis $\{e_i\}$.

If $(a,b,c) \in R^3$ with respect to the basis $\{f_i\}$ then $(a,b,c) = cf_1 + (b-c)f_2 + (a-b)f_3$.

$T(x,y,z) = (2y+z, \ x-4y, \ 3x)$. Then,

$$T(f_1) = T(1,1,1) = (3,-3,3) = 3f_1 - 6f_2 + 6f_3 \ .$$
$$T(f_2) = T(1,1,0) = (2,-3,3) = 3f_1 - 6f_2 + 5f_3 \ .$$
$$T(f_3) = T(1,0,0) = (0,1,3) = 3f_1 - 2f_2 - f_3 \ .$$

As a result,

$$[T]_f = \begin{bmatrix} 3 & 3 & 3 \\ -6 & -6 & -2 \\ 6 & 5 & -1 \end{bmatrix}$$

Thus,

$$P^{-1}[T]_e P = \begin{bmatrix} 0 & 0 & 1 \\ 0 & 1 & -1 \\ 1 & -1 & 0 \end{bmatrix} \begin{bmatrix} 0 & 2 & 1 \\ 1 & -4 & 0 \\ 3 & 0 & 0 \end{bmatrix} \begin{bmatrix} 1 & 1 & 1 \\ 1 & 1 & 0 \\ 1 & 0 & 0 \end{bmatrix}$$

$$= \begin{bmatrix} 0 & 0 & 1 \\ 0 & 1 & -1 \\ 1 & -1 & 0 \end{bmatrix} \begin{bmatrix} 3 & 2 & 0 \\ -3 & -3 & 1 \\ 3 & 3 & 3 \end{bmatrix} = \begin{bmatrix} 3 & 3 & 3 \\ -6 & -6 & -2 \\ 6 & 5 & -1 \end{bmatrix}$$

$$= [T]_f \ .$$

● **PROBLEM** 11-5

Let $M = \begin{bmatrix} 1 & 0 \\ 1 & 1 \end{bmatrix}$ and define the linear transformation $T: R^2 \rightarrow R^2$ by

$T(x) = \begin{bmatrix} 1 & 0 \\ 1 & 1 \end{bmatrix} x$. Let $B = \left\{ \begin{bmatrix} 1 \\ 1 \end{bmatrix}, \begin{bmatrix} 1 \\ -1 \end{bmatrix} \right\}$ be an ordered basis for

R^2 . Find another matrix M' such that M and M' are similar.

Solution: First express B in terms of the usual basis:

$$B = \left\{ \begin{bmatrix} 1 \\ 1 \end{bmatrix}, \begin{bmatrix} 1 \\ -1 \end{bmatrix} \right\} ;$$

then

$$i(e_1) = \begin{bmatrix} 1 \\ 0 \end{bmatrix} = \tfrac{1}{2}\begin{bmatrix} 1 \\ 1 \end{bmatrix} + \tfrac{1}{2}\begin{bmatrix} 1 \\ -1 \end{bmatrix}$$

and

$$i(e_2) = \begin{bmatrix} 0 \\ 1 \end{bmatrix} = \tfrac{1}{2}\begin{bmatrix} 1 \\ 1 \end{bmatrix} - \tfrac{1}{2}\begin{bmatrix} 1 \\ -1 \end{bmatrix} .$$

Therefore,

$$[B,i,N] = \begin{bmatrix} \tfrac{1}{2} & \tfrac{1}{2} \\ \tfrac{1}{2} & -\tfrac{1}{2} \end{bmatrix}$$

is the transition matrix from the given basis B to the normal basis N (e_1,e_2,e_3). Now,

$$T\begin{bmatrix} 1 \\ 1 \end{bmatrix} = \begin{bmatrix} 1 & 0 \\ 1 & 1 \end{bmatrix}\begin{bmatrix} 1 \\ 1 \end{bmatrix} = \begin{bmatrix} 1 \\ 2 \end{bmatrix} = \tfrac{3}{2}\begin{bmatrix} 1 \\ 1 \end{bmatrix} + -\tfrac{1}{2}\begin{bmatrix} 1 \\ -1 \end{bmatrix}$$

and

$$T\begin{bmatrix} 1 \\ -1 \end{bmatrix} = \begin{bmatrix} 1 & 0 \\ 1 & 1 \end{bmatrix}\begin{bmatrix} 1 \\ -1 \end{bmatrix} = \begin{bmatrix} 1 \\ 0 \end{bmatrix} = \tfrac{1}{2}\begin{bmatrix} 1 \\ 1 \end{bmatrix} + \tfrac{1}{2}\begin{bmatrix} 1 \\ -1 \end{bmatrix} .$$

Therefore,

$$(B,T,B) = \begin{bmatrix} 3/2 & \tfrac{1}{2} \\ -\tfrac{1}{2} & \tfrac{1}{2} \end{bmatrix} = M'$$

is the representation of the operator T relative to the basis

$$B = \left\{ \begin{bmatrix} 1 \\ 1 \end{bmatrix}, \begin{bmatrix} 1 \\ -1 \end{bmatrix} \right\}$$

The nxn matrices M and M' are similar, provided there exists an invertible nxn matrix A such that $A^{-1}(M'A) = M$. Let

$$A = \begin{bmatrix} \frac{1}{2} & \frac{1}{2} \\ \frac{1}{2} & -\frac{1}{2} \end{bmatrix} .$$

Then

$$\begin{bmatrix} \frac{1}{2} & \frac{1}{2} \\ \frac{1}{2} & -\frac{1}{2} \end{bmatrix}^{-1} \begin{bmatrix} 3/2 & \frac{1}{2} \\ -\frac{1}{2} & \frac{1}{2} \end{bmatrix} \begin{bmatrix} \frac{1}{2} & \frac{1}{2} \\ \frac{1}{2} & -\frac{1}{2} \end{bmatrix}$$

$$= \begin{bmatrix} 1 & +1 \\ 1 & -1 \end{bmatrix} \begin{bmatrix} 1 & \frac{1}{2} \\ 0 & -\frac{1}{2} \end{bmatrix} = \begin{bmatrix} 1 & 0 \\ 1 & 1 \end{bmatrix} = M .$$

Therefore, $M = (N,T,N) = \begin{bmatrix} 1 & 0 \\ 1 & 1 \end{bmatrix}$ and $M' = (B,T,B) = \begin{bmatrix} 3/2 & \frac{1}{2} \\ -\frac{1}{2} & \frac{1}{2} \end{bmatrix}$

are similar.

● **PROBLEM** 11-6

Show that the matrices M' and M are similar, where $M = \begin{bmatrix} 1 & 1 \\ 0 & 1 \end{bmatrix}$ and

$M' = \begin{bmatrix} 1 & 0 \\ 1 & 1 \end{bmatrix} .$

<u>Solution</u>: Two nxn matrices M and M' are similar if $A^{-1}(M'A) = M$,
where A is an invertible nxn matrix.
    To see that M and M' are similar, let

$$A = \begin{bmatrix} 0 & 1 \\ 1 & 0 \end{bmatrix} .$$

Then

$$A^{-1} = \begin{bmatrix} 0 & 1 \\ 1 & 0 \end{bmatrix} \quad \text{since } A^{-1} = \frac{1}{\det A} \text{ Adj } A , \text{ and}$$

$$A^{-1}(m'A) = \begin{bmatrix} 0 & 1 \\ 1 & 0 \end{bmatrix} \begin{bmatrix} 1 & 0 \\ 1 & 1 \end{bmatrix} \begin{bmatrix} 0 & 1 \\ 1 & 0 \end{bmatrix}$$

$$= \begin{bmatrix} 0 & 1 \\ 1 & 0 \end{bmatrix} \begin{bmatrix} 0 & 1 \\ 1 & 1 \end{bmatrix}$$

$$= \begin{bmatrix} 1 & 1 \\ 0 & 1 \end{bmatrix} = M .$$

FINDING A DIAGONAL MATRIX TO
REPRESENT A LINEAR TRANSFORMATION

● **PROBLEM** 11-7

Prove that an nxn matrix A is diagonalizable if and only if it has n linearly independent eigenvectors. In this case A is similar to a matrix D whose diagonal elements are the eigenvalues of A.

Solution: Suppose that A is similar to D. If this is the case, there exists a matrix P such that

$$P^{-1}AP = D$$

and

$$PP^{-1}AP = PD.$$

This implies

$$AP = PD . \tag{1}$$

Let

$$D = \begin{bmatrix} \lambda_1 & 0 & . & . & . & 0 \\ . & & & & & \\ . & & & & & . \\ . & & & & & . \\ 0 & & \lambda_2 & & & 0 \\ . & & & & . & \\ . & & & & . & 0 \\ . & & & & . & \\ 0 & . & . & . & 0 & \lambda_n \end{bmatrix} ,$$

and let x_j ; $j = 1,2,\ldots,n$ be the jth column of P. Note that the jth column of the matrix AP is Ax_j and the jth column of PD is $\lambda_j x_j$. Thus, from (1) we have

$$Ax_j = \lambda_j x_j \tag{2}$$

Since P is a nonsingular matrix, its columns are linearly independent and are therefore, all non-zero. Hence, λ_j is an eigenvalue of A, and x_j is a corresponding eigenvector. Moreover, since P is nonsingular, its column vectors are linearly independent.

Conversely, suppose that $\lambda_1, \lambda_2, \ldots, \lambda_n$ are n eigenvalues of A and that the corresponding eigenvectors $x_1, x_2, \ldots, x_n$ are linearly independent. Let P be the matrix whose jth column is x_j. P is nonsingular. From (2) we obtain (1) which implies that A is diagonalizable.

● PROBLEM 11-8

Given that

$$A = \begin{bmatrix} 1 & 1 \\ -2 & 4 \end{bmatrix} ,$$

find an invertible matrix P such that $P^{-1}AP$ is a diagonal matrix D.

Solution: The characteristic equation of the matrix A is $\det(\lambda I - A) = 0$. Thus,

$$\det \begin{vmatrix} \lambda-1 & -1 \\ 2 & \lambda-4 \end{vmatrix} = 0 ,$$

or

$$(\lambda-1)(\lambda-4) + 2 = 0 ,$$

$$\lambda^2 - 5\lambda + 6 = 0$$

or

$$(\lambda-3)(\lambda-2) = 0 .$$

The eigenvalues are therefore, $\lambda_1 = 2$ and $\lambda_2 = 3$. To obtain the eigenvector corresponding to the eigenvalue $\lambda_1 = 2$, solve the equation $(2I - A)X = 0$ for x:

$$\begin{bmatrix} 1 & -1 \\ 2 & -2 \end{bmatrix} \begin{bmatrix} x_1 \\ x_2 \end{bmatrix} = \begin{bmatrix} 0 \\ 0 \end{bmatrix},$$

or

$$x_1 - x_2 = 0$$
$$2x_1 - 2x_2 = 0$$

which gives $x_1 - x_2 = 0$ or $x_1 = x_2$.

Thus, $x_1 = \begin{bmatrix} 1 \\ 1 \end{bmatrix}$ is an eigenvector of 2. Similarly, an eigenvector x_2 corresponding to the eigenvalue $\lambda = 3$ is $\begin{bmatrix} 1 \\ 2 \end{bmatrix}$. Since the eigenvectors x_1 and x_2 are linearly independent A is diagonalizable. It is possible to see that x_1 and x_2 are independent because $a_1 x_1 + a_2 x_2 = 0$ has only the trivial solution, $a_1 = a_2 = 0$. Let P be the matrix whose columns are x_1 and x_2:

$$P = \begin{bmatrix} 1 & 1 \\ 1 & 2 \end{bmatrix}$$

and

$$P^{-1} = \begin{bmatrix} 2 & -1 \\ -1 & 1 \end{bmatrix}$$

(using the adjoint method). Thus,

$$P^{-1}AP = \begin{bmatrix} 2 & -1 \\ -1 & 1 \end{bmatrix} \begin{bmatrix} 1 & 1 \\ -2 & 4 \end{bmatrix} \begin{bmatrix} 1 & 1 \\ 1 & 2 \end{bmatrix} = \begin{bmatrix} 2 & 0 \\ 0 & 3 \end{bmatrix}$$

On the other hand, if we let $\lambda_1 = 3$ and $\lambda_2 = 2$, then $x_1 = \begin{bmatrix} 1 \\ 2 \end{bmatrix}$ and $x_2 = \begin{bmatrix} 1 \\ 1 \end{bmatrix}$. In that case,

$$P = \begin{bmatrix} 1 & 1 \\ 2 & 1 \end{bmatrix} \quad \text{and} \quad P^{-1} = \begin{bmatrix} -1 & 1 \\ 2 & -1 \end{bmatrix},$$

and

$$P^{-1}AP = \begin{bmatrix} -1 & 1 \\ 2 & -1 \end{bmatrix} \begin{bmatrix} 1 & 1 \\ -2 & 4 \end{bmatrix} \begin{bmatrix} 1 & 1 \\ 2 & 1 \end{bmatrix} = \begin{bmatrix} 3 & 0 \\ 0 & 2 \end{bmatrix}.$$

● **PROBLEM** 11-9

Find a nonsingular matrix P such that $P^{-1}AP$ is diagonal, given that

$$A = \begin{bmatrix} 1 & 1 \\ 3 & -1 \end{bmatrix}.$$

Solution: A matrix A of order n, with n distinct characteristic values $\lambda_1, \lambda_2, \ldots, \lambda_n$ is similar to $\text{diag}[\lambda_1, \lambda_2, \ldots, \lambda_n]$, a diagonal matrix. Consider T to be the linear operator represented by the matrix A. If

$v_1, v_2, \ldots, v_n$ are characteristic vectors of T belonging to characteristic values $\lambda_1, \lambda_2, \ldots, \lambda_n$, then $v_i T = \lambda_i v_i$; $i = 1, 2, \ldots, n$. Hence, if $[v_1, v_2, \ldots, v_n]$ is a basis of v_1, some vector space, then $\text{diag}(\lambda_1, \lambda_2, \ldots, \lambda_n)$ represents T with respect to $[v_1, v_2, \ldots, v_n]$. On the other hand, if $\text{diag}(\lambda_1, \lambda_2, \ldots, \lambda_n)$ represents T with respect to a basis $\{v_1, v_2, \ldots, v_n\}$ of v_1, then $v_i T = \lambda_i v_i$ for $i = 1, 2, \ldots, n$. This shows that $v_1, v_2, \ldots, v_n$ are characteristic vectors of v.

Let $\{v_1, v_2, \ldots, v_n\}$ be a basis of R^n consisting of characteristic vectors of A, such that v_i belongs to λ_i, $i = 1, 2, \ldots, n$. Set $P = \lambda(v_1, v_2, \ldots, v_n)$. Then $v_i A = \lambda_i v_i$, $e_i P = v_i$, $v_i P^{-1} = e_i$, $i = 1, 2, \ldots, n$. If $B = PAP^{-1}$, then B is similar to A, and $e_i B = e_i PAP^{-1} = v_i AP^{-1} = \lambda_i v_i P^{-1} = \lambda_i e_i$, $i = 1, 2, \ldots, n$. We have

$$B = \lambda(e_1 B, e_2 B, \ldots, e_n B) = \text{diag}(r_1, r_2, \ldots, r_n) .$$

Now, suppose that A is similar to $\text{diag}(r_1, r_2, \ldots, r_n)$. If $T: K^n \to K^n$ is the linear transformation represented by A with respect to the standard basis $\{e_1, e_2, \ldots, e_n\}$, then T and A have the same characteristic vectors, and $\text{diag}(r_1 r_2, \ldots, r_n)$ represents T with respect to a basis $\{v_1, v_2, \ldots, v_n\}$ of K^n. To find a nonsingular matrix P such that $P^{-1}AP$ is diagonal, an appropriate basis of characteristic vectors must be obtained.

Now, the characteristic equation of A is $\det(\lambda I - A) = 0$. Thus,

$$\det \begin{vmatrix} \lambda - 1 & -1 \\ -3 & \lambda + 1 \end{vmatrix} = 0 ,$$

or

$$(\lambda - 1)(\lambda + 1) - 3 = \lambda^2 - 4 = (\lambda + 2)(\lambda - 2) = 0 .$$

The characteristic values of A are $\lambda_1 = 2$, $\lambda_2 = -2$. Next, solve the equation $(\lambda I - A)x = 0$ to obtain characteristic vectors.

For $\lambda = 2$:

$$\begin{bmatrix} 1 & -1 \\ -3 & 3 \end{bmatrix} \begin{bmatrix} x_1 \\ x_2 \end{bmatrix} = \begin{bmatrix} 0 \\ 0 \end{bmatrix} ,$$

or $x_1 = x_2$. Thus, $X_1 = \begin{bmatrix} 1 \\ 1 \end{bmatrix}$ forms a basis for the characteristic vectors associated with $\lambda = 2$.

For $\lambda = -2$:

$$\begin{bmatrix} -3 & -1 \\ -3 & -1 \end{bmatrix} \begin{bmatrix} x_1 \\ x_2 \end{bmatrix} = \begin{bmatrix} 0 \\ 0 \end{bmatrix} ,$$

or $-3x_1 - x_2 = 0$. Thus, $X_2 = \begin{bmatrix} 1 \\ -3 \end{bmatrix}$ forms a basis for the eigenvector associated with $\lambda = -2$.

Then, the matrix P is given by

$$P = \begin{bmatrix} 1 & 1 \\ 1 & -3 \end{bmatrix} ,$$

$P^{-1} = \dfrac{1}{\det P} \text{adj } P$. Sometimes the following is a simpler method for finding P^{-1} . Write down the matrix formed by putting P next to I

(the identity matrix):

$$\begin{bmatrix} 1 & 1 & 1 & 0 \\ 1 & -3 & 0 & 1 \end{bmatrix}.$$

Using row operations, make the left 2×2 matrix (P) into the identity matrix. The right 2×2 matrix will then be P^{-1}. Replace the 2nd row with the 1st row minus the second:

$$\begin{bmatrix} 1 & 1 & 1 & 0 \\ 0 & 4 & 1 & -1 \end{bmatrix}.$$

Divide the second row by 4:

$$\begin{bmatrix} 1 & 1 & 1 & 0 \\ 0 & 1 & \frac{1}{4} & -\frac{1}{4} \end{bmatrix}.$$

Replace the first row with the first row minus the second:

$$\begin{bmatrix} 1 & 0 & 3/4 & \frac{1}{4} \\ 0 & 1 & \frac{1}{4} & -\frac{1}{4} \end{bmatrix}.$$

Now the left 2×2 matrix is the identity, and, therefore, the right 2×2 matrix is P^{-1}.

$$P^{-1} = \begin{bmatrix} 3/4 & \frac{1}{4} \\ \frac{1}{4} & -\frac{1}{4} \end{bmatrix}$$

Now,

$$P^{-1}AP = \begin{bmatrix} 3/4 & \frac{1}{4} \\ \frac{1}{4} & -\frac{1}{4} \end{bmatrix} \begin{bmatrix} 1 & 1 \\ 3 & -1 \end{bmatrix} \begin{bmatrix} 1 & 1 \\ 1 & -3 \end{bmatrix}$$

$$= \begin{bmatrix} 3/4 & \frac{1}{4} \\ \frac{1}{4} & -\frac{1}{4} \end{bmatrix} \begin{bmatrix} 2 & -2 \\ 2 & 6 \end{bmatrix} = \begin{bmatrix} 2 & 0 \\ 0 & -2 \end{bmatrix}.$$

Notice that the diagonal entries of $P^{-1}AP$ are the eigenvalues previously found.

● **PROBLEM** 11-10

Find a matrix P that diagonalizes

$$A = \begin{bmatrix} 3 & -2 & 0 \\ -2 & 3 & 0 \\ 0 & 0 & 5 \end{bmatrix}.$$

Solution: The following is the procedure for diagonalizing a diagonalizable n×n matrix A.
1. Find n linearly independent eigenvectors of A, $x_1, x_2, \ldots, x_n$.
2. Form the matrix P having $x_1, x_2, \ldots, x_n$ as its column vectors.
3. The matrix $P^{-1}AP$ will then be diagonal with $\lambda_1, \lambda_2, \ldots, \lambda_n$ as its successive diagonal entries where λ_i is the eigenvalue corresponding to x_i, $i = 1, 2, \ldots, n$. The characteristic equation of A is

$$\det \begin{vmatrix} \lambda-3 & 2 & 0 \\ 2 & \lambda-3 & 0 \\ 0 & 0 & \lambda-5 \end{vmatrix} = 0 .$$

Expanding along the third row,

$$(\lambda-5) \begin{vmatrix} \lambda-3 & 2 \\ 2 & \lambda-3 \end{vmatrix} = 0 ,$$

or

$$(\lambda-5)[(\lambda-3)^2 - 4] = 0 ,$$

or

$$(\lambda-5)(\lambda-1)(\lambda-5) = 0 .$$

The eigenvalues of A are $\lambda_1 = 1$ and $\lambda_2 = 5$.

Solve the equation $(5I - A)y = 0$ to obtain the eigenvectors corresponding to $\lambda = 5$:

$$\begin{bmatrix} 2 & 2 & 0 \\ 2 & 2 & 0 \\ 0 & 0 & 0 \end{bmatrix} \begin{bmatrix} y_1 \\ y_2 \\ y_3 \end{bmatrix} = \begin{bmatrix} 0 \\ 0 \\ 0 \end{bmatrix} .$$

Solving the above system yields:

$$y_1 = -s , \; y_2 = s , \; y_3 = t ,$$

or

$$y = \begin{bmatrix} -s \\ s \\ t \end{bmatrix} = \begin{bmatrix} -s \\ s \\ 0 \end{bmatrix} + \begin{bmatrix} 0 \\ 0 \\ t \end{bmatrix} = s \begin{bmatrix} -1 \\ 1 \\ 0 \end{bmatrix} + t \begin{bmatrix} 0 \\ 0 \\ 1 \end{bmatrix} .$$

Thus,

$$x_1 = \begin{bmatrix} -1 \\ 1 \\ 0 \end{bmatrix} \text{ and } x_2 = \begin{bmatrix} 0 \\ 0 \\ 1 \end{bmatrix}$$ are two linearly independent eigenvectors corresponding to $\lambda = 5$.

Similarly, we find that $x_3 = \begin{bmatrix} 1 \\ 1 \\ 0 \end{bmatrix}$ is an eigenvector associated with $\lambda = 1$. Thus, x_1, x_2 and x_3 are linearly independent vectors such that

$$P = \begin{bmatrix} -1 & 0 & 1 \\ 1 & 0 & 1 \\ 0 & 1 & 0 \end{bmatrix} .$$

To find P^{-1}, use the adjoint method:

$$P^{-1} = \frac{1}{\det P} \text{ adj. } P .$$

We find

$$P^{-1} = \begin{bmatrix} -\tfrac{1}{2} & \tfrac{1}{2} & 0 \\ 0 & 0 & 1 \\ \tfrac{1}{2} & \tfrac{1}{2} & 0 \end{bmatrix} .$$

It follows that $P^{-1}AP$ is the required diagonal matrix.

$$P^{-1}AP = \begin{bmatrix} -1/2 & 1/2 & 0 \\ 0 & 0 & 1 \\ 1/2 & 1/2 & 0 \end{bmatrix} \begin{bmatrix} 3 & -2 & 0 \\ -2 & 3 & 0 \\ 0 & 0 & 5 \end{bmatrix} \begin{bmatrix} -1 & 0 & 1 \\ 1 & 0 & 1 \\ 0 & 1 & 0 \end{bmatrix}$$

$$= \begin{bmatrix} 5 & 0 & 0 \\ 0 & 5 & 0 \\ 0 & 0 & 1 \end{bmatrix}$$

There is no preferred order for the columns of P. Since the ith diagonal entry of $P^{-1}AP$ is an eigenvalue for the ith column vector of P, changing the order of the columns of P merely changes the order of the eigenvalues on the diagonal of $P^{-1}AP$.

Thus, had we written

$$P = \begin{bmatrix} -1 & 1 & 0 \\ 1 & 1 & 0 \\ 0 & 0 & 1 \end{bmatrix}$$

in the last example, we would have obtained

$$P^{-1}AP = \begin{bmatrix} 5 & 0 & 0 \\ 0 & 1 & 0 \\ 0 & 0 & 5 \end{bmatrix} .$$

● **PROBLEM 11-11**

Diagonalize the matrix $A = \begin{bmatrix} 1 & 2 \\ -1 & 4 \end{bmatrix}$ and find a diagonalizer for it.

<u>Solution</u>: If A is diagonalizable, there exists a matrix P such that PAP^{-1} is a diagonal matrix. P is called the diagonalizer of A. The diagonal elements of PAP^{-1} are the eigenvalues of A. The characteristic equation of A is $\det(\lambda I - A) = 0$. Then

$$\det \begin{vmatrix} \lambda-1 & -2 \\ 1 & \lambda-4 \end{vmatrix} = 0 ,$$

or

$$(\lambda-1)(\lambda-4) + 2 = 0 ,$$

or

$$\lambda^2 - 5\lambda + 6 = (\lambda-3)(\lambda-2) = 0 .$$

It follows that the eigenvalues of A are $\lambda_1 = 2$ and $\lambda_2 = 3$. To find eigenvectors for A, form the vector $\begin{bmatrix} x_1 \\ x_2 \end{bmatrix}$ and set $(\lambda I - A)\begin{bmatrix} x_1 \\ x_2 \end{bmatrix} = 0$ for $\lambda = 2,3$. For $\lambda = 2$,

$$\begin{bmatrix} 1 & -2 \\ 1 & -2 \end{bmatrix} \begin{bmatrix} x_1 \\ x_2 \end{bmatrix} = \begin{bmatrix} 0 \\ 0 \end{bmatrix} ,$$

or $x_1 - 2x_2 = 0$. Thus, $x_1 = \begin{bmatrix} 2 \\ 1 \end{bmatrix}$ is an eigenvector corresponding to $\lambda = 2$. Similarly, for $\lambda = 3$, we find $\begin{bmatrix} 1 \\ 1 \end{bmatrix}$ is an eigenvector. Therefore,

$$P = \begin{bmatrix} 2 & 1 \\ 1 & 1 \end{bmatrix} \quad \text{diagonalizes} \quad A = \begin{bmatrix} 1 & 2 \\ -1 & 4 \end{bmatrix} ,$$

yielding the diagonal form

$$D = \begin{bmatrix} 2 & 0 \\ 0 & 3 \end{bmatrix} .$$

Find an invertible matrix P such that $P^{-1}AP$ is a diagonal matrix B where

$$A = \begin{bmatrix} 1 & 0 & -2 \\ 0 & 0 & 0 \\ -2 & 0 & 4 \end{bmatrix} .$$

Solution: The matrix P has as its columns linearly independent eigenvectors each one of which belongs to an eigenvalue of A. Thus, we form the matrix $[\lambda I - A]$:

$$[\lambda I - A] = \begin{bmatrix} \lambda & 0 & 0 \\ 0 & \lambda & 0 \\ 0 & 0 & \lambda \end{bmatrix} - \begin{bmatrix} 1 & 0 & -2 \\ 0 & 0 & 0 \\ -2 & 0 & 4 \end{bmatrix} = \begin{bmatrix} \lambda-1 & 0 & 2 \\ 0 & \lambda & 0 \\ 2 & 0 & \lambda-4 \end{bmatrix}$$

Since the characteristic equation of A is $\det(\lambda I - A) = 0$, the result is:

$$\det \begin{vmatrix} \lambda-1 & 0 & 2 \\ 0 & \lambda & 0 \\ 2 & 0 & \lambda-4 \end{vmatrix} = (\lambda-1)[\lambda(\lambda-4)] + 2[0 - 2\lambda] = 0 ,$$

or

$$\lambda[\lambda^2 - 5\lambda + 4 - 4] = 0 ,$$

$$\lambda^2 (\lambda-5) = 0 ,$$

and the characteristic values of A are $\lambda = 0$, $\lambda = 0$ and $\lambda = 5$.

The characteristic vector associated with $\lambda = 0$ is obtained by solving the equation $(0I - A)x = 0$. Thus,

$$\begin{bmatrix} -1 & 0 & 2 \\ 0 & 0 & 0 \\ 2 & 0 & -4 \end{bmatrix} \begin{bmatrix} x_1 \\ x_2 \\ x_3 \end{bmatrix} = \begin{bmatrix} 0 \\ 0 \\ 0 \end{bmatrix} ,$$

or

$$-x_1 + 2x_3 = 0$$

$$2x_1 - 4x_3 = 0 ,$$

or

$$X = \begin{bmatrix} 2a \\ b \\ a \end{bmatrix} \text{ with } a \text{ and } b \text{ arbitrary scalars (not both zero).}$$

Two such vectors which are also linearly independent are $\begin{bmatrix} 2 \\ 0 \\ 1 \end{bmatrix}$ and $\begin{bmatrix} 0 \\ 1 \\ 0 \end{bmatrix}$.

To find the eigenvector corresponding to $\lambda = 5$, solve the equation $(5I - A)x = 0$:

$$\begin{bmatrix} 4 & 0 & 2 \\ 0 & 5 & 0 \\ 2 & 0 & 1 \end{bmatrix} \begin{bmatrix} x_1 \\ x_2 \\ x_3 \end{bmatrix} = \begin{bmatrix} 0 \\ 0 \\ 0 \end{bmatrix} ,$$

and, therefore,

(1) $4x_1 + 2x_3 = 0$

(2) $5x_2 = 0$

(3) $2x_1 + x_3 = 0$.

Equation (2) yields $x_2 = 0$. Equations (1) and (3) are dependent, so, taking equation (3) yields $x_3 = -2x_1$. One vector which satisfies these conditions is $x = \begin{bmatrix} 1 \\ 0 \\ -2 \end{bmatrix}$. We have obtained

$$\begin{bmatrix} 2 \\ 0 \\ 1 \end{bmatrix}, \begin{bmatrix} 0 \\ 1 \\ 0 \end{bmatrix}, \text{ and } \begin{bmatrix} 1 \\ 0 \\ -2 \end{bmatrix}$$

as a basis for R^3 that consists of characteristic vectors of A. If we let P be the matrix with these basis vectors as columns, P is invertible and $P^{-1}AP$ is a diagonal matrix whose diagonal entries are the characteristic values of A. Thus,

$$P = \begin{bmatrix} 2 & 0 & 1 \\ 0 & 1 & 0 \\ 1 & 0 & -2 \end{bmatrix} , \quad P^{-1} = \frac{1}{\det P}(\text{adj } P) =$$

$$-\frac{1}{5} \cdot \begin{bmatrix} -2 & 3 & -1 \\ 0 & -5 & 0 \\ -1 & 0 & 2 \end{bmatrix} \text{ and}$$

$$P^{-1}AP = \begin{bmatrix} 0 & 0 & 0 \\ 0 & 0 & 0 \\ 0 & 0 & 5 \end{bmatrix} .$$

We know this to be true, but in order to check do the matrix arithmatic:

$$P^{-1}AP = -\frac{1}{5} \begin{bmatrix} -2 & 3 & -1 \\ 0 & -5 & 0 \\ -1 & 0 & 2 \end{bmatrix} \begin{bmatrix} 1 & 0 & -2 \\ 0 & 0 & 0 \\ -2 & 0 & 4 \end{bmatrix} \begin{bmatrix} 2 & 0 & 1 \\ 0 & 1 & 0 \\ 1 & 0 & -2 \end{bmatrix}$$

$$= -\frac{1}{5} \begin{bmatrix} 0 & 0 & 0 \\ 0 & 0 & 0 \\ -5 & 0 & 10 \end{bmatrix} \begin{bmatrix} 2 & 0 & 1 \\ 0 & 1 & 0 \\ 1 & 0 & -2 \end{bmatrix}$$

$$= -\frac{1}{5} \begin{bmatrix} 0 & 0 & 0 \\ 0 & 0 & 0 \\ 0 & 0 & -25 \end{bmatrix} = \begin{bmatrix} 0 & 0 & 0 \\ 0 & 0 & 0 \\ 0 & 0 & 5 \end{bmatrix} .$$

● **PROBLEM** 11-13

Show that the matrix A is diagonalizable where

$$A = \begin{bmatrix} 0 & 0 & 0 \\ 0 & 1 & 0 \\ 1 & 0 & 1 \end{bmatrix} .$$

323

Solution: The characteristic equation of A is $\det(\lambda I - A) = 0$. Thus,

$$\det \begin{vmatrix} \lambda & 0 & 0 \\ 0 & \lambda-1 & 0 \\ -1 & 0 & \lambda-1 \end{vmatrix} = 0,$$

or $\lambda(\lambda-1)(\lambda-1) = 0$. Then the eigenvalues are $\lambda_1 = 0$, $\lambda_2 = 1$, $\lambda_3 = 1$. Now, to find the eigenvectors corresponding to $\lambda = 1$, solve the equation $(1I - A)x = 0$ for x:

$$\begin{bmatrix} 1 & 0 & 0 \\ 0 & 0 & 0 \\ -1 & 0 & 0 \end{bmatrix} \begin{bmatrix} x_1 \\ x_2 \\ x_3 \end{bmatrix} = \begin{bmatrix} 0 \\ 0 \\ 0 \end{bmatrix}.$$

Solving the above system yields $x_1 = 0$, $x_2 = r$, $x_3 = s$, where r and s are any real numbers. Thus,

$$x = \begin{bmatrix} 0 \\ r \\ s \end{bmatrix} = \begin{bmatrix} 0 \\ r \\ 0 \end{bmatrix} + \begin{bmatrix} 0 \\ 0 \\ s \end{bmatrix} = r\begin{bmatrix} 0 \\ 1 \\ 0 \end{bmatrix} + s\begin{bmatrix} 0 \\ 0 \\ 1 \end{bmatrix}$$

Thus, v_2 and v_3 are $\begin{bmatrix} 0 \\ 1 \\ 0 \end{bmatrix}$ and $\begin{bmatrix} 0 \\ 0 \\ 1 \end{bmatrix}$, two linearly independent eigenvectors associated with $\lambda = 1$. Now look for an eigenvector associated with $\lambda_1 = 0$. We have to solve $(0I - A)x = 0$. Because $0I$ is the zero matrix we have only $-AX = 0$, or

$$\begin{bmatrix} 0 & 0 & 0 \\ 0 & -1 & 0 \\ -1 & 0 & -1 \end{bmatrix} \begin{bmatrix} x_1 \\ x_2 \\ x_3 \end{bmatrix} = \begin{bmatrix} 0 \\ 0 \\ 0 \end{bmatrix}.$$

A solution is any vector of the form $\begin{bmatrix} t \\ 0 \\ -t \end{bmatrix}$ for any real number t. Thus, $x_1 = \begin{bmatrix} 1 \\ 0 \\ -1 \end{bmatrix}$ is an eigenvector associated with $\lambda_1 = 0$. Since x_1, x_2, and x_3 are linearly independent, A can be diagonalized. Note that an $n \times n$ matrix may fail to be diagonalizable either because all the roots of its characteristic polynomial are not real numbers or because it does not have n linearly independent eigenvectors.

● **PROBLEM 11-14**

If $T: R^3 \to R^3$ is given by $[x_1, x_2, x_3]T = [-17x_1 - 17x_2 - 23x_3, 10x_1 + 11x_2 + 12x_3, 8x_1 + 7x_2 + 12x_3]$. Find a diagonal matrix which represents T with respect to a basis of R^3.

Solution: Let V be a vector space over R and let $T: V \to V$ be a linear transformation on V. A scalar $\lambda \in R$ is called a characteristic value of T if there is a non-zero vector $v \in V$ such that $vT = \lambda v$.

A given linear transformation $T: V \to V$ can be represented by a diagonal matrix if and only if there is a basis $\{v_1, v_2, \ldots, v_n\} \in V$ consisting of characteristic vectors of T. Then the matrix representing T with respect to the basis $\{v_1, v_2, \ldots, v_n\}$ is $\text{diag}(\lambda_1, \lambda_2, \ldots, \lambda_n)$ where λ_i is the characteristic value of T to which v_i belongs, $(i = 1, 2, \ldots, n)$.

The matrix of the given transformation, with respect to the normal basis, is

$$A = \begin{bmatrix} -17 & -17 & -23 \\ 10 & 11 & 12 \\ 8 & 7 & 12 \end{bmatrix} \qquad (1)$$

The characteristic roots of (1) are: $\lambda_1 = 1$, $\lambda_2 = 2$ and $\lambda_3 = 3$. Characteristic vectors associated with these values are

$$V_1 = [13, -7, -5], \quad V_2 = [-3, 2, 1] \quad \text{and} \quad V_3 = [-2, 1, 1], \text{ respectively.}$$

That is,

$$\begin{array}{rcl} [13, -7, -5]T & = & 1[13, -7, -5] \\ [-3, 2, 1]T & = & 2[-3, 2, 1] \\ [-2, 1, 1]T & = & 3[-2, 1, 1] . \end{array}$$

The three characteristic vectors are linearly independent since for any transformation, non-zero eigenvectors belonging to distinct $(\lambda_i \neq \lambda_j$ if $i \neq j)$ eigenvalues are linearly independent. It takes only 3 linearly independent vectors to span R^3. Hence, v_1, v_2, v_3 form a basis for R^3. Conclude, therefore, that the diagonal matrix

$$\begin{bmatrix} 1 & 0 & 0 \\ 0 & 2 & 0 \\ 0 & 0 & 3 \end{bmatrix}$$

represents T with respect to $\{v_1, v_2, v_3\}$.

● **PROBLEM 11-15**

Show that the matrix A is not diagonalizable where

$$A = \begin{bmatrix} -3 & 2 \\ -2 & 1 \end{bmatrix} .$$

<u>Solution</u>: The characteristic equation of A is $\det(\lambda I - A) = 0$; therefore,

$$\begin{vmatrix} \lambda+3 & -2 \\ +2 & \lambda-1 \end{vmatrix} = 0 ,$$

or

$$(\lambda+3)(\lambda-1) + 4 = 0$$

$$(\lambda+1)^2 = 0 .$$

Thus, $\lambda = -1$ is the only eigenvalue of A; the eigenvectors corresponding to $\lambda = -1$ are the solutions of $(-I-A)x = 0$. Thus,

$$\begin{bmatrix} 2 & -2 \\ 2 & -2 \end{bmatrix} \begin{bmatrix} x_1 \\ x_2 \end{bmatrix} = \begin{bmatrix} 0 \\ 0 \end{bmatrix} ,$$

or

$$2x_1 - 2x_2 = 0$$
$$2x_1 - 2x_2 = 0 \ .$$

The solutions of this system are $x_1 = t$, $x_2 = t$. Hence, the eigenspace consists of all vectors of the form

$$\begin{bmatrix} t \\ t \end{bmatrix} = t \begin{bmatrix} 1 \\ 1 \end{bmatrix} \ .$$

Since this space is 1-dimensional, A does not have two linearly independent eigenvectors and, therefore is not diagonalizable.

PROPERTIES OF DIAGONAL MATRICES

Verify the following rules by giving examples:
(a) If A is an nxn diagonal matrix and B is an nxn matrix, each row of AB is then just the product of the diagonal entry of A times the corresponding row of B.
(b) If B is a diagonal matrix, each column of AB is just the product of the corresponding column of A with the corresponding diagonal entry of B.

Solution: (a) Let

$$A = \begin{bmatrix} 2 & 0 & 0 \\ 0 & -1 & 0 \\ 0 & 0 & 3 \end{bmatrix} \quad \text{and} \quad B = \begin{bmatrix} 4 & 2 & 1 \\ -1 & 0 & 6 \\ 2 & 1 & -3 \end{bmatrix} \ .$$

Then

$$AB = \begin{bmatrix} 2 & 0 & 0 \\ 0 & -1 & 0 \\ 0 & 0 & 3 \end{bmatrix} \begin{bmatrix} 4 & 2 & 1 \\ -1 & 0 & 6 \\ 2 & 1 & -3 \end{bmatrix}$$

$$= \begin{bmatrix} 8+0+0 & 4+0+0 & 2+0+0 \\ 0+1+0 & 0+0+0 & 0-6+0 \\ 0+0+6 & 0+0+3 & 0+0-9 \end{bmatrix} = \begin{bmatrix} 8 & 4 & 2 \\ 1 & 0 & -6 \\ 6 & 3 & -9 \end{bmatrix} \ .$$

This shows that each row of AB is the product of the diagonal element of A and the corresponding row of B.
Let

$$AB = \begin{bmatrix} 4 & 0 & 0 & 0 \\ 0 & 0 & 0 & 0 \\ 0 & 0 & 3 & 0 \\ 0 & 0 & 0 & -2 \end{bmatrix} \quad \text{and} \quad B = \begin{bmatrix} 3 & 0 & -1 & 2 \\ -1 & 1 & 0 & 1 \\ 4 & -1 & -2 & 1 \\ 0 & 1 & 3 & -4 \end{bmatrix}$$

Then

$$AB = \begin{bmatrix} 12 & 0 & -4 & 8 \\ 0 & 0 & 0 & 0 \\ 12 & -3 & -6 & 3 \\ 0 & -2 & -6 & 8 \end{bmatrix}$$

Now use this rule to take the powers of a diagonal matrix. Find A^2 where
A is the first matrix above.

$$\begin{bmatrix} 2 & 0 & 0 \\ 0 & -1 & 0 \\ 0 & 0 & 3 \end{bmatrix}^2 = \begin{bmatrix} 2 & 0 & 0 \\ 0 & -1 & 0 \\ 0 & 0 & 3 \end{bmatrix}\begin{bmatrix} 2 & 0 & 0 \\ 0 & -1 & 0 \\ 0 & 0 & 3 \end{bmatrix}$$

$$= \begin{bmatrix} 4 & 0 & 0 \\ 0 & 1 & 0 \\ 0 & 0 & 9 \end{bmatrix}$$

First find A^3:

$$A^3 = \begin{bmatrix} 2 & 0 & 0 \\ 0 & -1 & 0 \\ 0 & 0 & 3 \end{bmatrix}^3 = \begin{bmatrix} 2 & 0 & 0 \\ 0 & -1 & 0 \\ 0 & 0 & 3 \end{bmatrix}\begin{bmatrix} 2 & 0 & 0 \\ 0 & -1 & 0 \\ 0 & 0 & 3 \end{bmatrix}^2$$

$$= \begin{bmatrix} 2 & 0 & 0 \\ 0 & -1 & 0 \\ 0 & 0 & 3 \end{bmatrix}\begin{bmatrix} 4 & 0 & 0 \\ 0 & 1 & 0 \\ 0 & 0 & 9 \end{bmatrix} = \begin{bmatrix} 8 & 0 & 0 \\ 0 & -1 & 0 \\ 0 & 0 & 2 \end{bmatrix}$$

Observe how easy it is to take the powers of a diagonal matrix. Now continue with the further powers.

$$\begin{bmatrix} 2 & 0 & 0 \\ 0 & -1 & 0 \\ 0 & 0 & 3 \end{bmatrix}^{10} = \begin{bmatrix} 2^{10} & 0 & 0 \\ 0 & 1 & 0 \\ 0 & 0 & 3^{10} \end{bmatrix} = \begin{bmatrix} 1024 & 0 & 0 \\ 0 & 1 & 0 \\ 0 & 0 & 59049 \end{bmatrix}$$

(b)

$$BA = \begin{bmatrix} 4 & 2 & 1 \\ -1 & 0 & 6 \\ 2 & 1 & 3 \end{bmatrix}\begin{bmatrix} 2 & 0 & 0 \\ 0 & -1 & 0 \\ 0 & 0 & 3 \end{bmatrix} = \begin{bmatrix} 8 & -2 & 3 \\ -2 & 0 & 18 \\ 4 & -1 & -9 \end{bmatrix}$$

$$BA = \begin{bmatrix} 3 & 0 & -1 & 2 \\ -1 & 1 & 0 & 1 \\ 4 & -1 & -2 & 1 \\ 0 & 1 & 3 & -4 \end{bmatrix}\begin{bmatrix} 4 & 0 & 0 & 0 \\ 0 & 0 & 0 & 0 \\ 0 & 0 & 3 & 0 \\ 0 & 0 & 0 & -2 \end{bmatrix}$$

$$= \begin{bmatrix} 12 & 0 & -3 & -4 \\ -4 & 0 & 0 & -2 \\ 16 & 0 & -6 & -2 \\ 0 & 0 & 9 & 8 \end{bmatrix}.$$

Thus, each column of BA is the product of a column of B and the corresponding diagonal element of A.

● PROBLEM 11-17

Given:
$$A = \begin{bmatrix} 1 & 1 \\ 0 & 1 \end{bmatrix} \quad \text{and} \quad P = \begin{bmatrix} 1 & 1 \\ 1 & -1 \end{bmatrix}.$$

(a) Find P^{-1} .

(b) Find $P^{-1}AP$.

(c) Verify that, if B is similar to A, then A is similar to B.

(d) Show that $B^k = P^{-1}A^kP$ if $B = P^{-1}AP$ where k is any positive integer.

<u>Solution:</u> (a) It is known that $P^{-1} = \dfrac{1}{\det P}$ adj P.

$$\det P = \begin{vmatrix} 1 & 1 \\ 1 & -1 \end{vmatrix} = -1-1 = -2 .$$

$$\text{adj } P = \begin{bmatrix} -1 & -1 \\ -1 & 1 \end{bmatrix} .$$

Therefore,

$$P^{-1} = \begin{bmatrix} \tfrac{1}{2} & \tfrac{1}{2} \\ \tfrac{1}{2} & -\tfrac{1}{2} \end{bmatrix} .$$

(b)

$$P^{-1}AP = \begin{bmatrix} \tfrac{1}{2} & \tfrac{1}{2} \\ \tfrac{1}{2} & -\tfrac{1}{2} \end{bmatrix} \begin{bmatrix} 1 & 1 \\ 0 & 1 \end{bmatrix} \begin{bmatrix} 1 & 1 \\ 1 & -1 \end{bmatrix}$$

$$= \begin{bmatrix} \tfrac{1}{2} & \tfrac{1}{2} \\ \tfrac{1}{2} & -\tfrac{1}{2} \end{bmatrix} \begin{bmatrix} 2 & 0 \\ 1 & -1 \end{bmatrix}$$

$$= \begin{bmatrix} 3/2 & -\tfrac{1}{2} \\ \tfrac{1}{2} & \tfrac{1}{2} \end{bmatrix} .$$

We say that the matrix B is similar to the matrix A if there is an invertible matrix P such that

$$B = P^{-1}AP .$$

Therefore, let

$$B = \begin{bmatrix} 3/2 & -\tfrac{1}{2} \\ \tfrac{1}{2} & \tfrac{1}{2} \end{bmatrix}$$

and then B is similar to A.

(c) If P is invertible and $B = P^{-1}AP$, then

$$PBP^{-1} = P(P^{-1}AP)P^{-1}$$
$$= (PP^{-1})A(PP^{-1})$$
$$= A \quad \text{since} \quad PP^{-1} = I .$$

Let $Q = P^{-1}$ so that $Q^{-1} = P$. Then $Q^{-1}BQ = A$.

$$Q = P^{-1} = \begin{bmatrix} \tfrac{1}{2} & \tfrac{1}{2} \\ \tfrac{1}{2} & -\tfrac{1}{2} \end{bmatrix} .$$

Thus,

$$Q^{-1} = \dfrac{1}{\det Q} \text{ adj } Q$$

$$Q^{-1} = \dfrac{1}{-\tfrac{1}{2}} \begin{bmatrix} -\tfrac{1}{2} & -\tfrac{1}{2} \\ -\tfrac{1}{2} & \tfrac{1}{2} \end{bmatrix} = \begin{bmatrix} 1 & 1 \\ 1 & -1 \end{bmatrix}$$

$$Q^{-1}BQ = \begin{bmatrix} 1 & 1 \\ 1 & -1 \end{bmatrix} \begin{bmatrix} 3/2 & -\frac{1}{2} \\ \frac{1}{2} & \frac{1}{2} \end{bmatrix} \begin{bmatrix} \frac{1}{2} & \frac{1}{2} \\ \frac{1}{2} & -\frac{1}{2} \end{bmatrix}$$

$$= \begin{bmatrix} 1 & 1 \\ 1 & -1 \end{bmatrix} \begin{bmatrix} \frac{1}{2} & 1 \\ \frac{1}{2} & 0 \end{bmatrix}$$

$$= \begin{bmatrix} 1 & 1 \\ 0 & 1 \end{bmatrix}$$

$$= A .$$

Thus, if B is similar to A, then A is similar to B.

(d) Let K = 2; check that $P^{-1}A^2P = B^2$. Then

$$B^2 = \begin{bmatrix} 3/2 & -\frac{1}{2} \\ \frac{1}{2} & \frac{1}{2} \end{bmatrix} \begin{bmatrix} 3/2 & -\frac{1}{2} \\ \frac{1}{2} & \frac{1}{2} \end{bmatrix}$$

$$= \begin{bmatrix} 2 & -1 \\ 1 & 0 \end{bmatrix} .$$

$$A^2 = \begin{bmatrix} 1 & 1 \\ 0 & 1 \end{bmatrix} \begin{bmatrix} 1 & 1 \\ 0 & 1 \end{bmatrix} = \begin{bmatrix} 1 & 2 \\ 0 & 1 \end{bmatrix} .$$

Then,

$$P^{-1}A^2P = \begin{bmatrix} \frac{1}{2} & \frac{1}{2} \\ \frac{1}{2} & -\frac{1}{2} \end{bmatrix} \begin{bmatrix} 1 & 2 \\ 0 & 1 \end{bmatrix} \begin{bmatrix} 1 & 1 \\ 1 & -1 \end{bmatrix}$$

$$= \begin{bmatrix} \frac{1}{2} & \frac{1}{2} \\ \frac{1}{2} & -\frac{1}{2} \end{bmatrix} \begin{bmatrix} 3 & -1 \\ 1 & -1 \end{bmatrix}$$

$$= \begin{bmatrix} 2 & -1 \\ 1 & 0 \end{bmatrix}$$

$$= B^2 .$$

Suppose K = 3:

$$A^3 = A^2A = \begin{bmatrix} 1 & 2 \\ 0 & 1 \end{bmatrix} \begin{bmatrix} 1 & 1 \\ 0 & 1 \end{bmatrix} = \begin{bmatrix} 1 & 3 \\ 0 & 1 \end{bmatrix} ,$$

$$B^3 = B^2B = \begin{bmatrix} 2 & -1 \\ 1 & 0 \end{bmatrix} \begin{bmatrix} 3/2 & -\frac{1}{2} \\ \frac{1}{2} & \frac{1}{2} \end{bmatrix} = \begin{bmatrix} 5/2 & -3/2 \\ 3/2 & -\frac{1}{2} \end{bmatrix} .$$

Then,

$$P^{-1}A^3P = \begin{bmatrix} \frac{1}{2} & \frac{1}{2} \\ \frac{1}{2} & -\frac{1}{2} \end{bmatrix} \begin{bmatrix} 1 & 3 \\ 0 & 1 \end{bmatrix} \begin{bmatrix} 1 & 1 \\ 1 & -1 \end{bmatrix}$$

$$= \begin{bmatrix} \frac{1}{2} & \frac{1}{2} \\ \frac{1}{2} & -\frac{1}{2} \end{bmatrix} \begin{bmatrix} 4 & -2 \\ 1 & -1 \end{bmatrix}$$

$$= \begin{bmatrix} 5/2 & -3/2 \\ 3/2 & -\frac{1}{2} \end{bmatrix}$$

$$= B^3 .$$

In general for any positive integer k, $B^k = P^{-1}A^kP$ if $B = P^{-1}AP$. To prove this rigorously for any matrices A and B and an invertible matrix P, use an inductive argument. Given $P^{-1}AP = B$, show $P^{-1}A^kP = B^k$. Take $n = 1$; $P^{-1}AP = B$. When $n = 2$; $P^{-1}AP = B$ gives $B^2 = P^{-1}APP^{-1}AP$ so $B^2 = P^{-1}A^2P$ since $PP^{-1} = I$.

Assume $P^{-1}A^kP = B^k$ is true for $k = n$; show that it is true for $k = n+1$. $P^{-1}A^nP = B^n$ so, since $B^{n+1} = B^n \cdot B = P^{-1}A^nPB = P^{-1}A^nPP^{-1}AP = P^{-1}A^{n+1}P$, $B^{n+1} = P^{-1}A^{n+1}P$.

From this it follows that if B is similar to A, then B^k is similar to A^k. Observe that the powers of A are easy to find. Direct calculation gives:

$$A^3 = \begin{bmatrix} 1 & 3 \\ 0 & 1 \end{bmatrix} ; \quad A^4 = \begin{bmatrix} 1 & 4 \\ 0 & 1 \end{bmatrix} .$$

In general, we obtain the formula

$$A^k = \begin{bmatrix} 1 & k \\ 0 & 1 \end{bmatrix} .$$

Again, to be rigorous, one would need to use an inductive argument. To find B^k, use the formula $B = P^{-1}A^kB$. Thus,

$$B^k = P^{-1} \begin{bmatrix} 1 & k \\ 0 & 1 \end{bmatrix} \begin{bmatrix} 1 & 1 \\ 1 & -1 \end{bmatrix}$$

$$= P^{-1} \begin{bmatrix} 1+k & 1-k \\ 1 & -1 \end{bmatrix}$$

$$= \begin{bmatrix} \frac{1}{2} & \frac{1}{2} \\ \frac{1}{2} & -\frac{1}{2} \end{bmatrix} \begin{bmatrix} 1+k & 1-k \\ 1 & -1 \end{bmatrix}$$

$$= \begin{bmatrix} 1+k/2 & -k/2 \\ k/2 & 1-k/2 \end{bmatrix} .$$

ORTHOGONAL DIAGONALIZATION

● **PROBLEM** 11-18

Define a symmetric matrix. Is every symmetric matrix similar to a diagonal matrix?

<u>Solution</u>: The transpose A^t of A is the matrix obtained from A by interchanging the rows and columns of A. A matrix A is symmetric if

$A^t = A$. For example

$$A = \begin{bmatrix} 2 & 1 \\ 1 & 3 \end{bmatrix} \quad \text{is symmetric since } A^t = \begin{bmatrix} 2 & 1 \\ 1 & 3 \end{bmatrix},$$

while

$$A = \begin{bmatrix} 2 & 1 \\ 2 & 3 \end{bmatrix} \quad \text{is not symmetric since } A^t = \begin{bmatrix} 2 & 2 \\ 1 & 3 \end{bmatrix} \neq A.$$

If a matrix is symmetric, then it is similar to a diagonal matrix. The characteristic polynomial of a symmetric matrix has only real roots and for each root of multiplicity k, one can find k independent characteristic vectors.

If A is symmetric we can actually find an orthogonal matrix P such that $P^{-1}AP$ is diagonal. The orthogonal matrix is a matrix whose columns are orthonormal. Thus, a symmetric matrix is always similar to a diagonal matrix.

● **PROBLEM** 11-19

Find an orthogonal matrix P that diagonalizes

$$A = \begin{bmatrix} 4 & 2 & 2 \\ 2 & 4 & 2 \\ 2 & 2 & 4 \end{bmatrix}.$$

Solution: The matrix A is symmetric. Construct a matrix P whose column vectors form an orthonormal set of eigenvectors of A. This can be done as follows:
1) Find a basis for each eigenspace of A.
2) Apply the Gram-Schmidt process to each of these bases to obtain an orthonormal basis for each eigenspace.
3) Form the matrix P whose columns are the basis vectors constructed in Step 2; this matrix orthogonally diagonalizes A. The characteristic equation of A is

$$\det(\lambda I - A) = \det \begin{vmatrix} \lambda-4 & -2 & -2 \\ -2 & \lambda-4 & -2 \\ -2 & -2 & \lambda-4 \end{vmatrix} = 0,$$

or

$$(\lambda-4)[(\lambda-4)^2 - 4] - (-2)[-2(\lambda-4) - 4] + (-2)[4 + 2(\lambda-4)] = 0.$$

$$(\lambda-4)(\lambda-6)(\lambda-2) - 4(\lambda-2) - 4(\lambda-2) = 0.$$

$$(\lambda-2)[\lambda^2 -10\lambda + 24 - 4 - 4] = (\lambda-2)^2(\lambda-8) = 0.$$

Thus, the eigenvalues of A are $\lambda = 2$ and $\lambda = 8$. To find the eigenvectors, solve the equation $(\lambda I - A)x = 0$ for x. First, with $\lambda = 2$, $(2I - A)x = 0$, or

$$\begin{bmatrix} -2 & -2 & -2 \\ -2 & -2 & -2 \\ -2 & -2 & -2 \end{bmatrix} \begin{bmatrix} x_1 \\ x_2 \\ x_3 \end{bmatrix} = \begin{bmatrix} 0 \\ 0 \\ 0 \end{bmatrix}.$$

Solving this system gives $x_1 + x_2 + x_3 = 0$. So, $X_1 = \begin{bmatrix} -1 \\ 1 \\ 0 \end{bmatrix}$ and $X_2 = \begin{bmatrix} -1 \\ 0 \\ 1 \end{bmatrix}$ are two linearly independent vectors of this form. X_1 and

X_2 form a basis for the eigenspace corresponding to $\lambda = 2$. Before proceeding, define the Gram-Schmidt process which enables us to obtain an orthonormal basis from any given basis. Let $\{u_1, u_2, \ldots, u_n\}$ be a basis of R^n. First, select any one of the original vectors u_1, for example, then set $v_1 = \dfrac{1}{|u_1|} u_1$, so v_1 has unit length. Thus, $v_1 \cdot v_1 = 1$. Set $w_2 = u_2 - (u_2 \cdot v_1)v_1$. Now $w_2 \neq 0$. The second member v_2 of the desired set of orthogonal unit vectors is obtained by dividing w_2 by its length. Thus, $v_2 = \dfrac{1}{|w_2|} \cdot w_2$: then $v_2 \cdot v_2 = 1$, $v_1 \cdot v_2 = 0$. In the

third step write $w_3 = u_3 - (u_3 \cdot v_1)v_3 - (u_3 \cdot v_2)v_2$ $(w_3 \neq 0)$. The third required vector v_3 is then given by $v_3 = \dfrac{1}{|w_3|} w_3$; then $v_3 \cdot v_3 = 1$,

$v_1 \cdot v_3 = v_2 \cdot v_3 = 0$. A continuation of this process finally determines the nth member of the required set in the form

$$v_n = \frac{1}{|w_n|} w_n$$

where $w_n = u_n - \sum_{k=1}^{n-1} (u_n \cdot v_k)v_k$. Applying the Gram-Schmidt process to $[X_1 ; X_2]$ yields the orthonormal eigenvectors. First, recall $|X_1|$ is defined to be $\sqrt{X_1 \cdot X_1} = \sqrt{x_a^2 + x_b^2 + x_c^2}$ where x_a, x_b, x_c are the components of X_1. So, in this case $|X| = \sqrt{(-1)^2 + (1)^2 + 0^2} = \sqrt{2}$.
Therefore,

$$v_1 = \frac{X_1}{|X_1|} = \frac{1}{\sqrt{2}} \begin{bmatrix} -1 \\ 1 \\ 0 \end{bmatrix} = \begin{bmatrix} -1/\sqrt{2} \\ 1/\sqrt{2} \\ 0 \end{bmatrix},$$

and,

$$w_2 = X_2 - (X_2 \cdot v_1)v_1 .$$

Therefore,

$$w_2 = \begin{bmatrix} -1 \\ 0 \\ 1 \end{bmatrix} - 1/\sqrt{2} \begin{bmatrix} -1/\sqrt{2} \\ 1/\sqrt{2} \\ 0 \end{bmatrix}, \qquad w_2 = \begin{bmatrix} -\frac{1}{2} \\ -\frac{1}{2} \\ 1 \end{bmatrix},$$

and, hence,

$$v_2 = \frac{w_2}{|w_2|} = \frac{1}{\sqrt{6}} \begin{bmatrix} -1 \\ -1 \\ 2 \end{bmatrix} = \begin{bmatrix} -1/\sqrt{6} \\ -1/\sqrt{6} \\ 2/\sqrt{6} \end{bmatrix} .$$

Now let $\lambda = 8$. Then $(8I - A)x = 0$, or

$$\begin{bmatrix} 4 & -2 & -2 \\ -2 & 4 & -2 \\ -2 & -2 & 4 \end{bmatrix} \begin{bmatrix} x_1 \\ x_2 \\ x_3 \end{bmatrix} = \begin{bmatrix} 0 \\ 0 \\ 0 \end{bmatrix} .$$

Thus, $X_3 = \begin{bmatrix} 1 \\ 1 \\ 1 \end{bmatrix}$ forms a basis for the eigenspace corresponding to $\lambda = 8$.

Applying the Gram-Schmidt process to X_3 yields

$$v_3 = \begin{bmatrix} 1/\sqrt{3} \\ 1/\sqrt{3} \\ 1/\sqrt{3} \end{bmatrix} .$$

By construction $<v_1, v_2> = 0$; further, $<v_1 \cdot v_3> = <v_2 \cdot v_3> = 0$ so that $\{v_1, v_2, v_3\}$ is an orthonormal set of eigenvectors. Thus,

$$P = \begin{bmatrix} -1/\sqrt{2} & -1/\sqrt{6} & 1/\sqrt{3} \\ 1/\sqrt{2} & -1/\sqrt{6} & 1/\sqrt{3} \\ 0 & 2/\sqrt{6} & 1/\sqrt{3} \end{bmatrix}$$

orthogonally diagonalizes A. Thus, P is an orthonormal set of eigen-vectors and $P^{-1}AP$ is a diagonal matrix.

● **PROBLEM** 11-20

Let $A = \begin{bmatrix} 2 & 1 & 1 \\ 1 & 2 & 1 \\ 1 & 1 & 2 \end{bmatrix}$. Find a (real) orthogonal matrix P such that $P^t AP$ is diagonal.

Solution: A is a symmetric matrix. Hence, it has real eigenvalues. First, find the characteristic polynomial of A:

$$f(\lambda) = \det(\lambda I - A) = \begin{vmatrix} \lambda-2 & -1 & -1 \\ -1 & \lambda-2 & -1 \\ -1 & -1 & \lambda-2 \end{vmatrix}$$

$$= (\lambda-2)\begin{vmatrix} \lambda-2 & -1 \\ -1 & \lambda-2 \end{vmatrix} - (-1)\begin{vmatrix} -1 & -1 \\ -1 & \lambda-2 \end{vmatrix} + (-1)\begin{vmatrix} -1 & \lambda-2 \\ -1 & -1 \end{vmatrix}$$

$$= (\lambda-2)[(\lambda-2)(\lambda-2) - 1] + [-(\lambda-2) - 1] - [1 + (\lambda-2)]$$

$$= (\lambda-2)(\lambda-3)(\lambda-1) - (\lambda-1) - (\lambda-1)$$

$$= (\lambda-1)[\lambda^2 - 5\lambda + 6 - 1 - 1]$$

$$= (\lambda-1)(\lambda-1)(\lambda-4).$$

Therefore, the characteristic equation of A is $(\lambda-1)^2 (\lambda-2) = 0$, and the eigenvalues are $\lambda = 1$ and $\lambda = 4$. To obtain eigenvectors corres-ponding to $\lambda = 1$, solve the equation $(1I - A)x = 0$:

$$\begin{bmatrix} -1 & -1 & -1 \\ -1 & -1 & -1 \\ -1 & -1 & -1 \end{bmatrix} \begin{bmatrix} x_1 \\ x_2 \\ x_3 \end{bmatrix} = \begin{bmatrix} 0 \\ 0 \\ 0 \end{bmatrix},$$

or $x_1 + x_2 + x_3 = 0$. Thus,

$$X_1 = \begin{bmatrix} 1 \\ -1 \\ 0 \end{bmatrix} \text{ and } X_2 = \begin{bmatrix} 1 \\ 1 \\ -2 \end{bmatrix}$$

are the eigenvectors corresponding to the eigenvalue $\lambda = 1$. For $\lambda = 4$, $(4I - A)x = 0$, or

$$\begin{bmatrix} +2 & -1 & -1 \\ -1 & 2 & -1 \\ -1 & -1 & 2 \end{bmatrix} \begin{bmatrix} x_1 \\ x_2 \\ x_3 \end{bmatrix} = \begin{bmatrix} 0 \\ 0 \\ 0 \end{bmatrix},$$

or

$$2x_1 - x_2 - x_3 = 0$$
$$-x_1 + 2x_2 - x_3 = 0$$
$$-x_1 - x_2 + 2x_3 = 0 .$$

Thus,

$$X_3 = \begin{bmatrix} 1 \\ 1 \\ 1 \end{bmatrix} \text{ is an eigenvector associated with } \lambda = 4.$$

Next, normalize X_1, X_2, X_3 to obtain the unit orthogonal solutions,

$$u_1 = \frac{X_1}{|X_1|} , \quad u_2 = \frac{X_2}{|X_2|} , \quad u_3 = \frac{X_3}{|X_3|} .$$

Thus,

$$u_1 = \begin{bmatrix} 1/\sqrt{2} \\ -1/\sqrt{2} \\ 0 \end{bmatrix} \quad u_2 = \begin{bmatrix} 1/\sqrt{6} \\ 1/\sqrt{6} \\ -2/\sqrt{6} \end{bmatrix} \quad u_3 = \begin{bmatrix} 1/\sqrt{3} \\ 1/\sqrt{3} \\ 1/\sqrt{3} \end{bmatrix} .$$

If P is the matrix whose columns are the u_i respectively,

$$P = \begin{bmatrix} 1/\sqrt{2} & 1/\sqrt{6} & 1/\sqrt{3} \\ -1/\sqrt{2} & 1/\sqrt{6} & 1/\sqrt{3} \\ 0 & -2/\sqrt{6} & 1/\sqrt{3} \end{bmatrix}$$

and

$$P^t AP = \begin{bmatrix} 1 & 0 & 0 \\ 0 & 1 & 0 \\ 0 & 0 & 4 \end{bmatrix} .$$

● **PROBLEM 11-21**

Let $A = \begin{bmatrix} 1 & 5 & 7 \\ 0 & 4 & 3 \\ 0 & 0 & 1 \end{bmatrix}$ and $B = \begin{bmatrix} 1 & 5 & 5 \\ 0 & 4 & 3 \\ 0 & 0 & 1 \end{bmatrix} .$

(a) Test A and B for diagonalizability.
(b) If diagonalizable, find their respective diagonal forms, diagonalizers and spectral decompositions.
(c) Express the projections of each decomposition as polynomials in the given matrix.

Solution: (a) If the matrix $A(T \overset{E}{\Leftrightarrow} A)$ has the minimal polynomial

$$m_\lambda(A) = \prod_{j=1}^{t} (\lambda - \lambda_j)^{k_j} ,$$

then

$$\prod_{j=1}^{t} (T - \lambda_j I)^{k_j} = \vec{0} .$$

The matrix A is diagonalizable if its minimal polynomial has no multiple (repeated) roots. The characteristic polynomial for A is

$$c(\lambda) = \det \begin{vmatrix} \lambda-1 & -5 & -7 \\ 0 & \lambda-4 & -3 \\ 0 & 0 & \lambda-1 \end{vmatrix} = (\lambda-1)^2 (\lambda-4) .$$

Similarly, the characteristic polynomial for B is $c(\lambda) = (\lambda-1)^2(\lambda-4)$.
Let $f(\lambda)$ be any polynomial satisfied by the linear operator
$T(T \overset{E}{\ominus} A)$.
For both A and B, $f(\lambda) = (\lambda-1)(\lambda-4)$. The characteristic values of A and B are $\lambda = 1$ and $\lambda = 4$. Then,

$$(1I - A)(4I - A) = \begin{bmatrix} 0 & -5 & -7 \\ 0 & -3 & -3 \\ 0 & 0 & 0 \end{bmatrix} \begin{bmatrix} 3 & -5 & -7 \\ 0 & 0 & -3 \\ 0 & 0 & 3 \end{bmatrix}$$

$$= \begin{bmatrix} 0 & 0 & -6 \\ 0 & 0 & 0 \\ 0 & 0 & 0 \end{bmatrix} \neq \vec{0} .$$

Therefore, A is not diagonalizable and $m(\lambda) = c(\lambda) = (\lambda-1)^2(\lambda-4)$.
For matrix B,

$$(1I - B)(4I - B) = \begin{bmatrix} 0 & -5 & -5 \\ 0 & -3 & -3 \\ 0 & 0 & 0 \end{bmatrix} \begin{bmatrix} 3 & -5 & -5 \\ 0 & 0 & -3 \\ 0 & 0 & 3 \end{bmatrix}$$

$$= \begin{bmatrix} 0 & 0 & 0 \\ 0 & 0 & 0 \\ 0 & 0 & 0 \end{bmatrix} = \vec{0} .$$

Therefore, B is diagonalizable with $m(\lambda) = f(\lambda) = (\lambda-1)(\lambda-4)$.

(b) Since B is diagonalizable, the diagonal form is

$$\begin{bmatrix} 1 & 0 & 0 \\ 0 & 1 & 0 \\ 0 & 0 & 4 \end{bmatrix} .$$

To find a diagonalizer for B, find three linearly independent eigenvectors for B.
For $\lambda = 1$, $(1I - B)x = 0$.
Therefore,

$$\begin{bmatrix} 0 & -5 & -5 \\ 0 & -3 & -3 \\ 0 & 0 & 0 \end{bmatrix} \begin{bmatrix} x_1 \\ x_2 \\ x_3 \end{bmatrix} = \begin{bmatrix} 0 \\ 0 \\ 0 \end{bmatrix}.$$

Solving this system yields $x_1 = a$, $x_2 = b$, $x_3 = -b$ where a and b are arbitrary. Thus,

$$X_1 = \begin{bmatrix} 1 \\ 1 \\ -1 \end{bmatrix} \text{ and } X_2 = \begin{bmatrix} 2 \\ 1 \\ -1 \end{bmatrix}$$

are two linearly independent eigenvectors corresponding to $\lambda = 1$.
For $\lambda = 4$,
            $(4I - B)X = 0$.
Then,

$$\begin{bmatrix} 3 & -5 & -5 \\ 0 & 0 & -3 \\ 0 & 0 & 3 \end{bmatrix} \begin{bmatrix} x_1 \\ x_2 \\ x_3 \end{bmatrix} = 0 ,$$

or $3x_1 - 5x_2 - 5x_3 = 0$

$ -3x_3 = 0 .$

Thus, $X_3 = \begin{bmatrix} 5 \\ 3 \\ 0 \end{bmatrix}$ is an eigenvector corresponding to $\lambda = 4$. Note that

X_1, X_2, X_3 are linearly independent since they belong to distinct eigen-values. Then a diagonalizer matrix P is the matrix whose columns are the three linearly independent eigenvectors of B. Thus,

$$P = \begin{bmatrix} 1 & 2 & 5 \\ 1 & 1 & 3 \\ -1 & -1 & 0 \end{bmatrix} ,$$

or

$$E_1 \overset{x}{\Leftrightarrow} \begin{bmatrix} 1 & 0 & 0 \\ 0 & 1 & 0 \\ 0 & 0 & 0 \end{bmatrix} ; \quad E_2 \overset{x}{\Leftrightarrow} \begin{bmatrix} 0 & 0 & 0 \\ 0 & 0 & 0 \\ 0 & 0 & 1 \end{bmatrix} .$$

Then the spectral decomposition of B is given by:

$$I = E_1 + E_2$$

$$B = \lambda_1 E_1 + \lambda_2 E_2 = 1E_1 + 4E_2 .$$

(c)

$$E_1^2 = \begin{bmatrix} 1 & 0 & 0 \\ 0 & 1 & 0 \\ 0 & 0 & 0 \end{bmatrix} = E_1 ,$$

$$E_2^2 = \begin{bmatrix} 0 & 0 & 0 \\ 0 & 0 & 0 \\ 0 & 0 & 1 \end{bmatrix} = E_2 ,$$

and $E_1 E_2 = E_2 E_1 = \vec{0}$ since $E_j = f_i(B) (j = 1,2)$ where f_j is the Lagrange polynomial satisfying $f_j(\lambda_k) = \delta_{jk}$ we state the spectral theorem:

Let $T: C^n \to C^n$ have the spectrum (set of all its distinct eigen-values $\{\lambda_1, \ldots, \lambda_n\}$. T is diagonalizable then linear operators $E_1, \ldots, E_n$ defined on C^n are such that

(1) $I = \sum_{j=1}^{n} E_j$

(2) $T = \sum_{j=1}^{n} \lambda_j E_j$

(3) $E_j E_k = E_k E_j = \vec{0}$ $(j \neq k ; j,k = 1, \ldots, n)$

(4) $E_j^2 = E_j$ $(j = 1, \ldots, n)$

(5)   The operators E_j $(j = 1, \ldots, n)$ are projection operators, not necessarily orthogonal, on C^n . The image of E_j is the space of eigen-vectors of T associated with the eigenvalue λ_j $(j = 1, \ldots, n)$ of T.

(6) $TE_j = E_j T = \lambda_j E_j$ and $(T - \lambda_j I)E_j = \vec{0}$ for all $j = 1, \ldots, t$.

Since the decomposition of T depends on the spectrum or set of all eigenvalues of T, it is called the spectral decomposition of T. We have

$$X_1 = \begin{bmatrix} 1 \\ 1 \\ -1 \end{bmatrix}, \qquad X_2 = \begin{bmatrix} 2 \\ 1 \\ -1 \end{bmatrix}, \qquad X_3 = \begin{bmatrix} 5 \\ 3 \\ 0 \end{bmatrix},$$

$X = \{X_1 \ X_2 \ X_3\}$. Then

$$E_1 x_1 = x_1 \qquad\qquad E_2 x_1 = 0$$

$$E_1 x_2 = x_2 \qquad\qquad E_2 x_2 = 0$$

$$E_1 x_3 = 0 \qquad\qquad E_2 x_3 = x_3$$

For E_1 we have $f_1(1) = 1$, $f_1(4) = 0$, and for E_2 we have $f_2(1) = 0$, $f_2(4) = 1$. The next step is to express the projections of each decomposition as polynomials in the given matrix. Let x_1 and x_2 be one pair of distinct members of the field F , y_1 and y_2 in F . Then the unique polynomial $f(x)$ of degree one satisfying $f(x_1) = y_1$, $f(x_2) = y_2$ is

$$f(x) = \frac{y_1(x - x_2)}{x_1 - x_2} + \frac{y_2(x - x_1)}{x_2 - x_1} .$$

Thus,

$$E_1 = f_1(B) = -\tfrac{1}{3}(B - 4I)$$

$$E_2 = f_2(B) = \tfrac{1}{3}(B - I) .$$

CHAPTER 12

THE JORDAN CANONICAL FORM

INVARIANT SUBSPACES

[A]   Suppose T: V → V is linear. Show that each of the following is invariant under T:
(i)   {0} , (ii) V . (iii)   kernel of T, (iv)   image of T .

[B]   Find all invariant subspace of

$$A = \begin{bmatrix} 2 & -5 \\ 1 & -2 \end{bmatrix}$$

viewed as an operator on R^2 .

Solution: Let T: V → V be linear. A subspace W of V is said to be invariant under T or T-invariant if $T(u) \in W$ for every $u \in W$. (T maps W into W.) In particular, a one-dimensional space <u> , $u \neq 0$, is invariant under T if $T(u) = \lambda u$ where λ is a scalar.

(i) For any linear operator T , $T(0) = 0 \in \{0\}$. Hence, {0} is invariant under T.
(ii) For every $u \in V$, $T(u) \in V$. Thus, V is invariant under T.
(iii) Let $u \in \ker T$. Then $T(u) = 0$, and $0 \in \ker T$ since $T(0) = 0$ for any linear operator T. Now the kernel of T is a subspace of V. Thus, ker T is invariant under T.
(iv) Since $T(u) \in \text{Im } T$ for every $u \in V$, $T(u)$ is certainly in Im T if $u \in \text{Im } T$. The image of T is also a subspace of V . Hence, the image of T is invariant under T.

[B] First of all, we know that R^2 and {0} are invariant under A. If A has any other invariant subspace W, then W must be 1-dimensional. Since A: $R^2 \to R^2$, no subspace of dimension 2 can be invariant. Thus, $T(u) = \lambda u$. If $u \in W$. But this implies u is an eigenvector of T.

Now, the characteristic polynomial of A is

$$F(\lambda) = \det(\lambda I - A) = \begin{bmatrix} \lambda-2 & 5 \\ -1 & \lambda+2 \end{bmatrix} = \lambda^2 + 1 ,$$

and $\lambda^2 + 1$ has no real roots. Hence, A has no eigenvalues (in R) and so A has no eigenvectors. Therefore, R^2 and {0} are the only subspaces invariant under A.

State the primary decomposition theorem. Verify that it is true by adducing an example.

Solution: Let $T: V \rightarrow V$ be a linear operator. The primary decomposition theorem uses the notions of minimal polynomial, T-invariant spaces and direct sum in showing that the space V can be broken down into subspaces. We can then study the effects of T on these subspaces. Precisely stated, let $T: V \rightarrow V$ be a linear operator with minimal polynomial

$$m(x) = (x-c_1)^{n_1}(x-c_2)^{n_2} \ldots (x-c_r)^{n_r}$$

where $(x-c_i)^{n_i}$ are distinct monic irreducible polynomials. Then V is the direct sum of T-invariant subspaces $W_1, \ldots, W_r$ where W_i is the kernel of $(T-c_i I)^{n_i}$.

As an application of this theorem, let $T: V_4 \rightarrow V_4$ be the operator whose matrix representation with respect to the standard basis is

$$A = \begin{bmatrix} 2 & 1 & 0 & 0 \\ 0 & 2 & 0 & 0 \\ 0 & 0 & 1 & 1 \\ 0 & 0 & -2 & 4 \end{bmatrix} .$$

The characteristic polynomial of A is $\det(A-tI) = (t-3)(t-2)^3$. The minimal polynomial must divide the characteristic polynomial and is the polynomial of lowest degree such that $m(A) = 0$. Thus, $m(t) = (t-3)(t-2)$ or $(t-3)(t-2)^2$ or $(t-3)(t-3)^3$. By trial and error, we find

$$m(A) = (A-3I)(A-2I)^2 = 0 .$$

Hence, $$m(x) = (x-3)(x-2)^2 = 0 .$$

The kernel of $(A-3I)$ is the solution set to

$$\begin{bmatrix} -1 & 1 & 0 & 0 \\ 0 & -1 & 0 & 0 \\ 0 & 0 & -2 & 1 \\ 0 & 0 & -2 & 1 \end{bmatrix} \begin{bmatrix} x_1 \\ x_2 \\ x_3 \\ x_4 \end{bmatrix} = \begin{bmatrix} 0 \\ 0 \\ 0 \\ 0 \end{bmatrix}$$

or vectors of the form $\begin{bmatrix} 0 \\ 0 \\ a \\ 2a \end{bmatrix}$ where a is any scalar. This is a subspace of R^4 of dimension 1. The kernel of $(A-2I)^3$ is the solution set to

$$\begin{bmatrix} 0 & 1 & 0 & 0 \\ 0 & 0 & 0 & 0 \\ 0 & 0 & -2 & 2 \\ 0 & 0 & -2 & 2 \end{bmatrix}^3 \begin{bmatrix} x_1 \\ x_2 \\ x_3 \\ x_4 \end{bmatrix} = \begin{bmatrix} 0 \\ 0 \\ 0 \\ 0 \end{bmatrix}$$

or, multiplying $[(A-2I)$ by $(A-2I)]$ by $(A-2I)$,

$$\begin{bmatrix} 0 & 0 & 0 & 0 \\ 0 & 0 & 0 & 0 \\ 0 & 0 & -1 & 1 \\ 0 & 0 & -2 & 2 \end{bmatrix} \begin{bmatrix} x_1 \\ x_2 \\ x_3 \\ x_4 \end{bmatrix} = \begin{bmatrix} 0 \\ 0 \\ 0 \\ 0 \end{bmatrix}.$$

Hence, the null space (kernel) of $(T-2I)^3$ consists of all vectors of the form

$$\begin{bmatrix} t_1 \\ t_2 \\ t_3 \\ t_3 \end{bmatrix} = t_1 \begin{bmatrix} 1 \\ 0 \\ 0 \\ 0 \end{bmatrix} + t_2 \begin{bmatrix} 0 \\ 1 \\ 0 \\ 0 \end{bmatrix} + t_3 \begin{bmatrix} 0 \\ 0 \\ 1 \\ 1 \end{bmatrix}$$

The kernel of $(T-2I)^3$ is a subspace of R^4 of dimension 3. Further-more, the two kernels are disjoint sets, i.e., $R^4 = \ker(T-3I) \oplus \ker(T-2I)^3$, as the theorem states.

Finally, note that the two kernels are T-invariant subspaces, i.e., when T is applied to a vector in $\ker(T-3I)$ or $\ker(T-2I)^3$, the result is a vector again in $\ker(T-3I)$ or $\ker(T-2I)^3$, respectively.

JORDAN BLOCK MATRICES

● PROBLEM 12-3

Define Jordan block and Jordan form matrix.

<u>Solution</u>: Let $M_n(F)$ be the set of $n \times n$ matrices with entries from a field F. A matrix $A \in M_n(F)$ of the form

$$A = \begin{bmatrix} \lambda & 1 & 0 & . & . & . & . & 0 \\ 0 & \lambda & 1 & . & . & . & . & 0 \\ 0 & 0 & \lambda & . & . & . & . & 0 \\ . & . & . & & & & & . \\ . & . & . & & & & & . \\ . & . & . & . & \lambda & 1 \\ 0 & 0 & & . & . & 0 & \lambda \end{bmatrix}$$

is called a Jordan block. For example, the following matrices are Jordan blocks.

$$\begin{bmatrix} 3i & 1 \\ 0 & 3i \end{bmatrix} \qquad \begin{bmatrix} 3 & 1 & 0 & 0 \\ 0 & 3 & 1 & 0 \\ 0 & 0 & 3 & 1 \\ 0 & 0 & 0 & 3 \end{bmatrix}$$

$$\begin{bmatrix} 5-2i & 1 & 0 \\ 0 & 5-2i & 1 \\ 0 & 0 & 5-2i \end{bmatrix} .$$

A Jordan form matrix is an $n \times n$ matrix made up of Jordan blocks strung along its main diagonal. All entries not in these blocks are zeros. For example,

$$\begin{bmatrix} i & 1 & 0 & 0 & 0 \\ 0 & i & 0 & 0 & 0 \\ 0 & 0 & 4 & 1 & 0 \\ 0 & 0 & 0 & 4 & 1 \\ 0 & 0 & 0 & 0 & 4 \end{bmatrix} \qquad \begin{bmatrix} 4 & 0 & 0 & 0 & 0 & 0 \\ 0 & 3 & 1 & 0 & 0 & 0 \\ 0 & 0 & 3 & 1 & 0 & 0 \\ 0 & 0 & 0 & 3 & 0 & 0 \\ 0 & 0 & 0 & 0 & 2-i & 1 \\ 0 & 0 & 0 & 0 & 0 & 2-i \end{bmatrix}$$

are Jordan form matrices. In general, if A is an $n \times n$ matrix with complex entries i.e., A represents a linear transformation of the complex vector space $\mathbb{C}^n$, then A is similar to a Jordan form matrix B. Furthermore, the diagonal entries of B are the complex eigenvalues of A, and they are repeated in B as often as they occur as roots of the characteristic equation.

● **PROBLEM** 12-4

Determine all possible Jordan canonical forms for a linear operator
T: V → V whose characteristic polynomial is $f(\lambda) = (\lambda-2)^3 (\lambda-5)^2$.

<u>Solution</u>: Let T: V → V be a linear operator whose characteristic and minimum polynomials are, respectively,

$$f(\lambda) = (\lambda - a_1)^{n_1} \ldots (\lambda - a_r)^{n_r} \quad \text{and} \quad m(\lambda) = (\lambda - a_1)^{m_1} \ldots (\lambda - a_r)^{m_r}$$

where a_i (i = 1,2,...,r) are distinct scalars. Then T has a block diagonal matrix representation J whose diagonal entries are blocks of the form

$$J_{ij} = \begin{bmatrix} a_i & 1 & 0 & \ldots & 0 & 0 \\ 0 & a_i & 1 & \ldots & 0 & 0 \\ \ldots & \ldots & \ldots & \ldots & \ldots & \ldots \\ 0 & 0 & 0 & \ldots & a_i & 1 \\ 0 & 0 & 0 & \ldots & 0 & a_i \end{bmatrix}$$

For each a_i , the corresponding blocks J_{ij} have the following properties:
(i) There is at least one J_{ij} of order m_i ; all other J_{ij} are of order $\leq m_i$.
(ii) The sum of the orders of the J_{ij} is n_i .
(iii) The number of J_{ij} equals the geometric multiplicity of a_i . The geometric multiplicity of a_i is the dimension of the eigenspace of a_i .
(iv) The number of J_{ij} of each possible order is uniquely determined by T.

The matrix J is called the Jordan canonical form of the operator T. A diagonal block J_{ij} is called a Jordan block belonging to the eigenvalue λ_i . We are given that the characteristic equation, ($f(\lambda)$ = $(\lambda-2)^3 (\lambda-5)^2$. Since $(\lambda-2)$ has exponent 3 in $f(\lambda)$, 2 must appear three times on the main diagonal. Similarly, 5 must appear twice. This follows from the fact that the number of J_{ij} equals the geometric multi-

341

plicity of λ_i . Thus, the possible Jordan canonical forms are

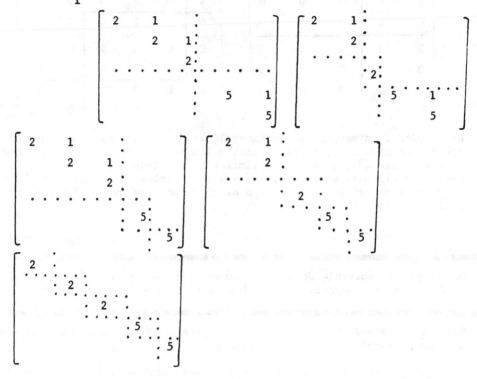

● PROBLEM 12-5

Determine all possible Jordan canonical forms J for a matrix of order 5 whose minimal polynomial is $m(\lambda) = (\lambda-2)^2$.

Solution: We can learn from the minimal polynomial what the order of the Jordan blocks along the main diagonal of the Jordan form will be. Now, $m(\lambda) = (\lambda-2)^2$; since the minimal polynomial divides the characteristic polynomial and has as factors every factor that the characteristic polynomial has, we see that the characteristic polynomial is $(\lambda-2)^5$. Furthermore, in the Jordan form, the largest block matrix must be of the same order as the highest power to which a factor in the minimal polynomial has been raised. In the present problem, this power is two. J must have one Jordan block of order 2 and the others must be of order 2 or 1. Thus,

$$J = \begin{bmatrix} 2 & 1 & & & \\ & 2 & & & \\ & & 2 & 1 & \\ & & & 2 & \\ & & & & 2 \end{bmatrix}$$

or,

$$J = \begin{bmatrix} 2 & 1 & & & \\ & 2 & & & \\ & & 2 & & \\ & & & 2 & \\ & & & & 2 \end{bmatrix}$$

342

Let

$$A = \begin{bmatrix} 0 & 1 & 1 & 0 & 1 \\ 0 & 0 & 1 & 1 & 1 \\ 0 & 0 & 0 & 0 & 0 \\ 0 & 0 & 0 & 0 & 0 \\ 0 & 0 & 0 & 0 & 0 \end{bmatrix}$$

Find the nilpotent matrix M in canonical form which is similar to A .

Solution: A matrix A is called nilpotent if some power of it is zero. If $A^k = 0$, k - 1 is called the index of nilpotentcy. If an nxn matrix is nilpotent, then its canonical form is

$$\begin{bmatrix} 0 & 1 & & & & & 0 \\ 0 & 0 & 1 & & & & \\ 0 & 0 & 0 & 1 & & & \\ & & & & \ddots & & \\ & & & & & & 1 \\ 0 & & & & & & 0 \end{bmatrix}$$

i.e., it has one's along the diagonal just above the main diagonal and zeros everywhere else.

$$A^2 = \begin{bmatrix} 0 & 0 & 1 & 1 & 1 \\ 0 & 0 & 0 & 0 & 0 \\ 0 & 0 & 0 & 0 & 0 \\ 0 & 0 & 0 & 0 & 0 \\ 0 & 0 & 0 & 0 & 0 \end{bmatrix} \quad , \text{ then } A^3 = 0 .$$

Hence, A is nilpotent of index 2. The rank of the matrix A is two. Therefore, the nullity of A = 5 - 2 = 3. Thus, M contains 3 diagonal blocks. Since A is nilpotent of index 2, M contains a diagonal block of order 2 and none greater than 2. Accordingly, M must contain 2 diagonal blocks of order 2 and 1 of order 1. Thus,

$$M = \begin{bmatrix} 0 & 1 & 0 & 0 & 0 \\ 0 & 0 & 0 & 0 & 0 \\ 0 & 0 & 0 & 1 & 0 \\ 0 & 0 & 0 & 0 & 0 \\ 0 & 0 & 0 & 0 & 0 \end{bmatrix}$$

Find the Jordan matrix of

$$A = \begin{bmatrix} 3 & 1 & -3 \\ -7 & -2 & 9 \\ -2 & -1 & 4 \end{bmatrix} .$$

Solution: First, construct the characteristic matrix $\lambda I - A$.

$$\lambda I - A = \begin{bmatrix} \lambda-2 & -1 & 3 \\ 7 & \lambda+2 & -9 \\ 2 & 1 & \lambda-4 \end{bmatrix} .$$

The characteristic polynomial is $f(\lambda) = \det[\lambda I - A]$

$$f(\lambda) = (\lambda-3) \begin{vmatrix} \lambda+2 & -9 \\ 1 & \lambda-4 \end{vmatrix} - (-1) \begin{vmatrix} 7 & -9 \\ 2 & \lambda-4 \end{vmatrix} + 3 \begin{vmatrix} 7 & \lambda+2 \\ 2 & 1 \end{vmatrix}$$

$$= (\lambda-3)[\lambda^2-2\lambda+1] + [7\lambda-10] + 3[3-2\lambda] = (\lambda-3)(\lambda-1)^2 + \lambda-1$$

$$= (\lambda-1)[\lambda^2-4\lambda+4] = (\lambda-1)(\lambda-2)^2 .$$

Next, find the minimum polynomial, $m(\lambda)$. This is the polynomial of lowest degree such that $m(A) = 0$. The minimum polynomial divides the characteristic polynomial. This determines $m(\lambda)$ in the given problem, and we find the minimum polynomial is $m(\lambda) = (\lambda-1)(\lambda-2)^2$. The highest power to which a factor is raised in the minimum polynomial is the order of the largest Jordan block in the Jordan matrix. Therefore, the Jordan matrix must have one Jordan block of order 2, and the other must be of order 1. Thus, the Jordan matrix has the form

$$\begin{bmatrix} 2 & 1 & 0 \\ 0 & 2 & 0 \\ \hline 0 & 0 & 1 \end{bmatrix} .$$

THE JORDAN CANONICAL FORM OF MATRICES & OPERATORS

● **PROBLEM** 12-8

Define an elementary Jordan matrix and give examples.

Solution: Let $T: V \to V$ be represented in some basis by A. Then, the characteristic equation of A is $\det(A - \lambda I) = 0$. The roots of this polynomial in λ are the eigenvalues associated with A. Each eigenvalue gives rise to an elementary Jordan matrix.

Definition: An elementary Jordan matrix is a matrix of type

$$\begin{bmatrix} \lambda & \cdot & \cdot & \cdot & \cdot & \cdot & 0 \\ 1 & \lambda & \cdot & \cdot & \cdot & 0 \\ \cdot & 1 & \cdot & & & \cdot \\ \cdot & & \cdot & \cdot & & \cdot \\ \cdot & & & \cdot & \cdot & \cdot \\ 0 & \cdot & \cdot & \cdot & 1 & \lambda \end{bmatrix} ;$$

that is, the eigenvalue λ is repeated on the diagonal with one's just below the diagonal and zeros everywhere else. The following are elementary Jordan matrices:

$$\begin{bmatrix} 0 & 0 & 0 \\ 1 & 0 & 0 \\ 0 & 1 & 0 \end{bmatrix} \begin{bmatrix} 0 & 0 & 0 & 0 \\ 1 & 0 & 0 & 0 \\ 0 & 1 & 0 & 0 \\ 0 & 0 & 1 & 0 \end{bmatrix} \begin{bmatrix} 5 & 0 & 0 & 0 \\ 1 & 5 & 0 & 0 \\ 0 & 1 & 5 & 0 \\ 0 & 0 & 1 & 5 \end{bmatrix}$$

The first two have eigenvalues zero, the last has eigenvalue 5.

● **PROBLEM 12-9**

A) Find the Jordan canonical form of the following matrices:

(i)
$$A = \begin{bmatrix} 1 & 5 & 7 \\ 0 & 4 & 3 \\ 0 & 0 & 1 \end{bmatrix}$$

(ii)
$$A = \begin{bmatrix} 0 & 0 & 1 \\ 1 & 0 & -3 \\ 0 & 1 & 3 \end{bmatrix}.$$

B) Let $m(\lambda) = (\lambda - \lambda_1)^4 (\lambda - \lambda_2)$, $f(\lambda) = (\lambda - \lambda_1)^5 (\lambda - \lambda_2)^3$ for the linear transformation T.
Find the Jordan Canonical form of T.

Solution: Let $T: c^n \to c^n$. The characteristic polynomial associated with T is

$$f(\lambda) = \prod_{j=1}^{t} (\lambda - \lambda_j)^{d_j} ;$$

the minimum polynomial being $m(\lambda) = \prod_{j=1}^{t} (\lambda - \lambda_j)^{k_j}$ where λ_j $(j = 1, \dots, t)$ are distinct scalars. Then there exists a basis z for c^n relative to which T has a matrix of the form

$$T \overset{z}{\longleftrightarrow} A = \begin{bmatrix} A_1 & & & \bigcirc \\ & A_2 & & \\ & & \ddots & \\ \bigcirc & & & \ddots \\ & & & & A_t \end{bmatrix} \qquad (1)$$

where A is the direct sum of the submatrices A_j of order d_j $(j = 1, \dots, t)$. Further, each submatrix A_j $(j = 1, \dots, t)$ is itself a direct sum of elementary Jordan matrices with eigenvalue λ_j and a scalar matrix $\lambda_j I$. Thus,

$$
A_j = \begin{bmatrix}
J_{11}(\lambda_j) & 0 & \cdots & \cdots & \cdots & \cdots & 0 \\
0 & & & & & & \cdot \\
\cdot & & J_{m1}(\lambda_j) & & & & \cdot \\
\cdot & & & J_{1\ell}(\lambda_j) & & & \cdot \\
\cdot & & & & J_{2\ell}(\lambda_j) & & \cdot \\
\cdot & & & & & J_{mk_j-1}(\lambda_j)\ 0 & \\
\cdot & & & & & & \\
0 & \cdots & \cdots & \cdots & \cdots & \cdots & \lambda_j I
\end{bmatrix} \qquad (2)
$$

$J_{p\ell}(\lambda_j)$ is the elementary Jordan matrix with eigenbalue λ_j having order $k_j - \ell + 1$ ($p = 1,\ldots,m_\ell$; $\ell = 1,\ldots,k_j-1$), and the scalar matrix $\lambda_j I$ is of order m_{kj} . The numbers m_ℓ ($\ell = 1,\ldots,k_j$) are associated with the linear operator $T - \lambda_j I = T_j - \lambda_j I$, nilpotent of degree k_j on

the space $W_j = K[(T - \lambda_j I)^{k_j}]$ whose dimension is d_j .

$A_j = N_j + \lambda_j I$ where N_j is the matrix of $T_j - \lambda_j I$ relative to a properly chosen basis on W_j . Therefore, the elementary Jordan matrices in A_j , the matrix of T_j relative to this basis, all have eigenvalue λ_j , and the numbers m_ℓ ($\ell = 1,\ldots,k_j$) determine these elementary

Jordan matrices as follows: The first m_1 elementary Jordan matrices in A_j are each of order k_j (the degree of nilpotence of $T_j-\lambda_j I$). The next m_2 are each of order $k_j - 1$, followed by m_3 , each of order $k_j - 2,\ldots,m_\ell$, each of order $k_j - \ell + 1$ ($\ell = 1,\ldots,k_j-1$) and a final scalar matrix $\lambda_j I$, of order m_{k_j} . Suppose $m_\ell = 0$ for any

$\ell = 2,\ldots,k_j-1$; A_j has no elementary Jordan matrices of the form $J_{p\ell}(\lambda_j)$ in the direct sum notations. If $m_{k_j} = 0$, A's subdiagonal

ends with a 1 rather than with a zero, and its final diagonal block is an elementary Jordan matrix with eigenvalue λ_j having order at least two rather than a scalar matrix $\lambda_j I$.

We call the form (1) subject to formula (2) and the above conditions the Jordan canonical form of T on c^n . It is unique up to rearrangement of the submatrices A_j (or equivalently the eigenvalues λ_j) ($j = 1,\ldots,t$).

Thus, a knowledge of the numbers m_ℓ ($\ell = 1,\ldots,k_j$) enables us to write down the matrix A_j of (2) without first finding the basis in W_j relative to which A_j has the above Jordan Canonical form.

346

[A]   We are given that

$$A = \begin{bmatrix} 1 & 5 & 7 \\ 0 & 4 & 3 \\ 0 & 0 & 1 \end{bmatrix} .$$

The characteristic polynomial of A is

$$f(\lambda) = \det[\lambda I - A] = \begin{vmatrix} \lambda-1 & -5 & -7 \\ 0 & \lambda-4 & -3 \\ 0 & 0 & \lambda-1 \end{vmatrix}$$

$$= (\lambda-1)(\lambda-4)(\lambda-1) = (\lambda-1)^2(\lambda-4) .$$

Then the eigenvalues are $\lambda = 1$ and $\lambda = 4$. The minimum polynomial of A is also $m(\lambda) = (\lambda-1)^2(\lambda-4)$.

Therefore, A is the direct sum $\begin{bmatrix} A_1 & 0 \\ 0 & A_2 \end{bmatrix}$ where A_1 is the

2×2 (since $d_2 = 2$) Jordan form of the operator $A_1 - 1I$, nilpotent of degree 2 (since $k_1 = 2$) on $W_1 = K[(A-2I)^2]$, and A_2 is a 1×1 matrix.

To find A_1 , we must first know the nullity of $(A_1 - 1I)$.

$$N(A_1 - 1I) = N(A - 1I) = n - rank(A - 1I)$$

$$A - 1I = \begin{bmatrix} 0 & 5 & 7 \\ 0 & 3 & 3 \\ 0 & 0 & 0 \end{bmatrix} .$$

Thus, $N(A - 1I) = 3 - 2 = 1$.

If T is nilpotent of degree 2 on c^n ($n \geq 2$), then the Jordan canonical form of T is determined, $N(T)$.

$$m_1 = n - N(T) = r(T)$$

and

$$m_2 = N(T) - r(T) .$$

Thus, for A_1 ,

$$m_1 = r(A_1 - 1I) = 2 - N(A_1 - 1I) = 2 - 1 = 1,$$

and

$$m_2 = N(A_1 - 1I) - r(A_1 - 1I) = 1 - 1 = 0 .$$

Thus, the elementary Jordan matrix in A_1 is

$$\begin{bmatrix} 1 & 0 \\ 1 & 1 \end{bmatrix} ,$$

and we know that A_2 is a 1×1 matrix with eigenvalue $\lambda_2 = 4$. Thus, the Jordan canonical form of A is

$$\left[\begin{array}{cc|c} 1 & 0 & 0 \\ 1 & 1 & 0 \\ \hline 0 & 0 & 4 \end{array} \right]$$

(ii)

$$A = \begin{bmatrix} 0 & 0 & 1 \\ 1 & 0 & -3 \\ 0 & 1 & 3 \end{bmatrix} .$$

The characteristic polynomial of A is

$$f(\lambda) = \det[\lambda I - A] = \begin{vmatrix} \lambda & 0 & -1 \\ -1 & \lambda & 3 \\ 0 & -1 & \lambda-3 \end{vmatrix}$$

$$= \lambda[\lambda(\lambda-3) + 3] - 1 = \lambda^3 - 3\lambda^2 + 3\lambda - 1$$

$$= (\lambda-1)^3 .$$

The eigenvalues are $\lambda = 1,1,1$. We also find $m(\lambda) = (\lambda-1)^3$. Therefore, the Jordan canonical form has one elementary Jordan matrix of order 3. Thus, the canonical form of A is

$$\begin{bmatrix} 1 & 0 & 0 \\ 1 & 1 & 0 \\ 0 & 1 & 1 \end{bmatrix} .$$

[B] We are given that

$$m(\lambda) = (\lambda-\lambda_1)^4(\lambda-\lambda_2) .$$

$$f(\lambda) = (\lambda-\lambda_1)^5(\lambda-\lambda_2)^3 .$$

Therefore, the matrix of the transformation A is the direct sum

$$\begin{bmatrix} A_1 & 0 \\ 0 & A_2 \end{bmatrix} .$$

Here, A_1 is the 5x5 Jordan form of the operator $(A-\lambda_1 I) = (A_1-\lambda_1 I)$, nilpotent of degree 4 on $W_1 = K[A - \lambda_1 I]^5$ of dimension 5; A_2 is the 3×3 Jordan form of the operator $(A_2 - \lambda_2 I)$, nilpotent of degree 1 .

The first elementary Jordan matrix in A_1 is of order 4 (since the degree of nilpotence $A_1 - \lambda_1 I$ is 4), and, therefore, the next elementary Jordan matrix in A_1 is 1×1.

The Jordan canonical form of T is then

$$\begin{bmatrix} \lambda_1 & 0 & 0 & 0 & 0 & 0 & 0 & 0 \\ 1 & \lambda_1 & 0 & 0 & 0 & 0 & 0 & 0 \\ 0 & 1 & \lambda_1 & 0 & 0 & 0 & 0 & 0 \\ 0 & 0 & 1 & \lambda_1 & 0 & 0 & 0 & 0 \\ 0 & 0 & 0 & 0 & \lambda_1 & 0 & 0 & 0 \\ 0 & 0 & 0 & 0 & 0 & \lambda_2 & 0 & 0 \\ 0 & 0 & 0 & 0 & 0 & 0 & \lambda_2 & 0 \\ 0 & 0 & 0 & 0 & 0 & 0 & 0 & \lambda_2 \end{bmatrix}$$

Observe that in the submatrix A_2, the subdiagonal entry of every row is zero since $k_1 = 1$ (the degree of nilpotence of $A_2 - A_2 I$).

● PROBLEM 12-10

Find the Jordan canonical form of

$$A = \begin{bmatrix} 2 & 0 & 0 & 1 & 0 \\ 0 & 2 & 0 & 0 & 1 \\ 0 & 0 & 2 & 0 & 0 \\ 0 & 0 & 0 & 2 & 0 \\ 0 & 0 & 0 & 0 & 1 \end{bmatrix}$$

<u>Solution:</u> First, find the characteristic polynomial and minimal polynomial of A. The characteristic polynomial of A is

$$f(\lambda) = \det[\lambda I - A] = \det \begin{bmatrix} \lambda-2 & 0 & 0 & -1 & 0 \\ 0 & \lambda-2 & 0 & 0 & -1 \\ 0 & 0 & \lambda-2 & 0 & 0 \\ 0 & 0 & 0 & \lambda-2 & 0 \\ 0 & 0 & 0 & 0 & \lambda-1 \end{bmatrix}.$$

The determinant of an uppertriangular matrix is the product of the diagonal entries. Thus, $f(\lambda) = (\lambda-2)^4(\lambda-1)$.

The minimal polynomial must be one of the following four polynomials:

$$m_1(\lambda) = (\lambda-2)(\lambda-1), \quad m_2(\lambda) = (\lambda-2)^2(\lambda-1),$$
$$m_3(\lambda) = (\lambda-2)^3(\lambda-1), \quad m_4(\lambda) = (\lambda-2)^4(\lambda-1).$$

Now, working up by degrees,

$$m_1(A) = (A-2I)(A-1I) \neq 0,$$

but

$$m_2(A) = (A-2I)^2(A-1I) = 0.$$

Therefore, $(\lambda-2)^2(\lambda-1)$ is the minimal polynomial of A.

The eigenvalues are $\lambda = 2$ and $\lambda = 1$. Therefore, A is the direct sum

$$\begin{bmatrix} A_1 & 0 \\ 0 & A_2 \end{bmatrix}$$

where A_1 is the 4×4 Jordan form of the operator $A - 2I = A_1 - 2I$, nilpotent of degree 2 (since $k_1 = 2$) on $W_1 = K[(A-2I)^4]$ of dimension 4 (since $d_1 = 4$). The matrix A_2 is the 1×1 matrix (since $d_2 = 1$). Now, to find A_1, we must know the nullity of $[A_1 -2I]$.

$$N[A_1 - 2I] = N(A - 2I) = n - r(A_1 - 2I) = 5 - 2 = 3.$$

We know

$$m_1 = n - N(T) = r(T) = n - N(A_1 - 2I) = 4 - 3 = 1,$$

and

$$m_2 = N(T) - r(T) = N(A_1 - 2I) - r(A_1 - 2I) = 3 - 1 = 2.$$

Thus, the first (m_1) elementary Jordan matrices in A_1 are of order 2 (since $k_1 = 2$), and the next m_2 are each of order 1 (since $k_j - 1 = 2 - 1 = 1$).

A_2 is the matrix of 1×1 with eigenvalue $\lambda_2 = 1$. Therefore, the Jordan Canonical form of A is

$$\begin{bmatrix} 2 & 0 & 0 & 0 & 0 \\ 1 & 2 & 0 & 0 & 0 \\ 0 & 0 & 2 & 0 & 0 \\ 0 & 0 & 0 & 2 & 0 \\ 0 & 0 & 0 & 0 & 1 \end{bmatrix} .$$

● PROBLEM 12-11

Let the linear operator T be nilpotent of degree 4 on C^6, $T \neq \vec{0}$. Find the Jordan canonical form of T.

Solution: Let T be nilpotent of degree K on C^n, $(1 < K \leq n)$. The rank of $T = d(TC^n) = m$. Let $K = (n-2)$, $T \neq \vec{0}$; then the m's associated with T (and, therefore, the Jordan Canonical form of T) are either uniquely determined or indeterminate according to the following formula: Let $K = n - 2$.

(i)   if $n = 4$, then $m_1 = 1$, $m_2 = 2$ or $m_1 = 2$.

(ii) if $n = 5+$, then $m_1 = m_{n-3} = 1$ or $m_1 = 1$, $m_{n-2} = 2$

      ($n = 5+$ means $n \geq 5$).

Now, T is nilpotent of degree 4 on C^6. Thus, $K = 4$ and $n = 6$ or $K = 6 - 2 = n - 2$. Therefore, we have $m_1 = m_{n-3} = 1$ or, $m_1 = m_3 = 1$. Also, $m_1 = 1$, $m_{n-2} = 2$ or, $m_1 = 1$, $m_4 = 2$. We know, too, that the nullity of $T = N(T) = \sum_{\ell=1}^{K} m_\ell$. Therefore, $N(T) = m_1 + m_3 = 1 + 1 = 2$ or $N(T) = m_1 + m_4 = 1 + 2 = 3$. We know the first m_1 elementary Jordan matrices in A are each of order K (the degree of nilpotence of T); the next m_2 are each of order $K - 1 \ldots m_\ell$, each of order $K - \ell + 1$ ($\ell = 1, \ldots, K-1$) and a final zero matrix of order m_k. Thus, $m_1 = 1$ and is of order 4 (since $K = 4$). $m_2 = m_4 = 0$, $m_3 = 1$ and is of order 2. Hence, the Jordan canonical form of T is

$$\begin{bmatrix} 0 & 0 & 0 & 0 & 0 & 0 \\ 1 & 0 & 0 & 0 & 0 & 0 \\ 0 & 1 & 0 & 0 & 0 & 0 \\ 0 & 0 & 1 & 0 & 0 & 0 \\ 0 & 0 & 0 & 0 & 0 & 0 \\ 0 & 0 & 0 & 0 & 1 & 0 \end{bmatrix}$$

When $m_1 = 1$ (order of $K = 4$), $m_2 = m_3 = 0$, $m_4 = 2$ (i.e., zero matrix of order two), the result is

$$\begin{bmatrix} 0 & 0 & 0 & 0 & | & 0 & 0 \\ 1 & 0 & 0 & 0 & | & 0 & 0 \\ 0 & 1 & 0 & 0 & | & 0 & 0 \\ 0 & 0 & 1 & 0 & | & 0 & 0 \\ \hline 0 & 0 & 0 & 0 & | & 0 & 0 \\ 0 & 0 & 0 & 0 & | & 0 & 0 \end{bmatrix}$$

● **PROBLEM** 12-12

Let $T: C^8 \to C^8$. The matrix of T with respect to the usual basis E is

$$T \overset{E}{\longleftrightarrow} A = \begin{bmatrix} 0 & 0 & 0 & 0 & 0 & 0 & 0 & 0 \\ 0 & 0 & 0 & 0 & 0 & 0 & 0 & 0 \\ 1 & 0 & 0 & 0 & 0 & 0 & 0 & 0 \\ 0 & 1 & 0 & 0 & 0 & 0 & 0 & 0 \\ 0 & 0 & 1 & 0 & 0 & 0 & 0 & 0 \\ 0 & 0 & 0 & 1 & 0 & 0 & 0 & 0 \\ 0 & 0 & 0 & 0 & 1 & 0 & 0 & 0 \\ 0 & 0 & 0 & 0 & 0 & 1 & 0 & 0 \end{bmatrix}$$

Find the Jordan Canonical form of T.

Solution: First find the degree of nilpotency of T. The rank of T is 6. Then

$$r(T^2) = 4$$
$$r(T^3) = 2$$
$$r(T^4) = 0 .$$

Therefore, T is nilpotent of degree 4 on C^8 (i.e., K = 4). State the relationship between the numbers m_ℓ and the rank of the various powers of T.

$$r(T^{K-\ell}) - r(T^{K-\ell+1}) = \sum_{p=1}^{\ell} m_p \quad (\ell = 1, \ldots, K) .$$

Using the above equation, we find

$$m_1 = r(T^3) - r(T^4) = 2 - 0 = 2 . \tag{i}$$
$$m_1 + m_2 = r(T^2) - r(T^3) = 4 - 2 = 2. \tag{ii}$$
$$m_1 + m_2 + m_3 = r(T) - r(T^2) = 6 - 4 = 2 . \tag{iii}$$

Now, nullity of $T = N(T) = \sum_{\ell=1}^{K} m_\ell$. But $N(T) = n - r(T) = 8 - 6 = 2$. Thus,

$$m_1 + m_2 + m_3 + m_4 = 2 . \tag{iv}$$

From the equations (i) to (iv), we have $m_1 = 2$, $m_2 = 0$, $m_3 = 0$, $m_4 = 0$. Thus, the first m_1 ($m_1 = 2$) elementary Jordan matrices are each of order K (K = 4). Therefore, the Jordan Canonical form of T is

$$
\left[
\begin{array}{cccc|cccc}
0 & 0 & 0 & 0 & 0 & 0 & 0 & 0 \\
1 & 0 & 0 & 0 & 0 & 0 & 0 & 0 \\
0 & 1 & 0 & 0 & 0 & 0 & 0 & 0 \\
0 & 0 & 1 & 0 & 0 & 0 & 0 & 0 \\
\hline
0 & 0 & 0 & 0 & 0 & 0 & 0 & 0 \\
0 & 0 & 0 & 0 & 1 & 0 & 0 & 0 \\
0 & 0 & 0 & 0 & 0 & 1 & 0 & 0 \\
0 & 0 & 0 & 0 & 0 & 0 & 1 & 0
\end{array}
\right]
$$

● **PROBLEM** 12-13

Let T be a linear operator on C^2 . Show that every 2×2 matrix over the field of complex numbers is similar to a matrix of the type

$$
\begin{bmatrix} c_1 & 0 \\ 0 & c_2 \end{bmatrix} \quad \text{or} \quad \begin{bmatrix} c & 0 \\ 1 & c \end{bmatrix} .
$$

Solution: If T is a linear operator for which the characteristic polynomial factors completely over the scalar field, then there is an ordered basis for V in which T is represented by a matrix which is in Jordan form. The characteristic polynomial for the given operator is $(x-c_1)(x-c_2)$ if c_1 and c_2 are distinct complex numbers. If $c_1 = c_2$ then it is $(x-c)^2$ for some c in C . In either case, we see that the polynomial factors into linear products. If the two eigenvalues are distinct, then the eigenvectors associated with these eigenvalues form a basis for c^2 . With respect to this basis, T is diagonalizable and is represented by

$$
\begin{bmatrix} c_1 & 0 \\ 0 & c_2 \end{bmatrix} . \tag{1}
$$

If the characteristic equation is $(x-c)^2$, then we must look to the minimal polynomial to help us in our search for the simplest matrix representation of T. Now, $m(x)$ may be $(x-c)$, in which case $T = cI$

$$
= \begin{bmatrix} c & 0 \\ 0 & c \end{bmatrix}
$$

or it may be $(x-c)^2$. In the latter case, the kernel of $(T-cI)^2$ is a T-invariant space, and T can be represented as the sum of a nilpotent matrix and a diagonal matrix. This is equivalent to asserting that T is represented in some ordered basis by the matrix

$$
\begin{bmatrix} c & 0 \\ 1 & c \end{bmatrix} . \tag{2}
$$

Thus, every 2×2 complex matrix is similar to a matrix of type (1) or (2), possibly with $c_1 = c_2$ in (1), i.e., if the single eigenvalue $c_1 = c_2$ has associated with it two linearly independent eigenvectors.

Find the Jordan Canonical form of

$$A = \begin{bmatrix} 5 & 0 & 0 & 0 & 0 & 0 & 0 & 0 & 0 & 0 & 0 & 0 \\ 0 & 5 & 0 & 0 & 0 & 0 & 0 & 0 & 0 & 0 & 0 & 0 \\ 1 & 0 & 5 & 0 & 0 & 0 & 0 & 0 & 0 & 0 & 0 & 0 \\ 0 & 1 & 0 & 5 & 0 & 0 & 0 & 0 & 0 & 0 & 0 & 0 \\ 0 & 0 & 1 & 0 & 5 & 0 & 0 & 0 & 0 & 0 & 0 & 0 \\ 0 & 0 & 0 & 1 & 0 & 5 & 0 & 0 & 0 & 0 & 0 & 0 \\ 0 & 0 & 0 & 0 & 1 & 0 & 5 & 0 & 0 & 0 & 0 & 0 \\ 0 & 0 & 0 & 0 & 0 & 1 & 0 & 5 & 0 & 0 & 0 & 0 \\ 0 & 0 & 0 & 0 & 0 & 0 & 1 & 0 & 2 & 0 & 0 & 0 \\ 0 & 0 & 0 & 0 & 0 & 0 & 0 & 1 & 0 & 2 & 0 & 0 \\ 0 & 0 & 0 & 0 & 0 & 0 & 0 & 0 & 1 & 0 & 2 & 0 \\ 0 & 0 & 0 & 0 & 0 & 0 & 0 & 0 & 0 & 1 & 0 & 2 \end{bmatrix}$$

Solution: First of all, we need the characteristic polynomial and min-imal polynomial of A .

$$f(\lambda) = \det(A - \lambda I) .$$

The matrix $[A - \lambda I]$ is a lower triangular matrix, therefore, the de-terminant of this matrix is the product of the diagonal entries. Thus, $f(\lambda) = (\lambda - 5)^8 (\lambda - 2)^4$. The minimum polynomial $m(\lambda)$ must divide $f(\lambda)$. Also, each irreducible factor of $f(\lambda)$, i.e., $(\lambda - 5)$ and $(\lambda - 2)$, must be a factor of $m(\lambda)$. We find $m(\lambda) = (\lambda - 5)^4 (\lambda - 2)^2$ for which $m(A) = 0$. Let A be a matrix of a linear operator T relative to E . Then A is the direct sum

$$\begin{bmatrix} A_1 & 0 \\ 0 & A_2 \end{bmatrix}$$

where A_1 is the 8×8 Jordan form of the operator $A - 5I = A_1 - 5I$, nilpotent of degree 4 of dimension 8 . As we know, the first ele-mentary Jordan matrices in A_1 are of order 4 [the degree of nil-potence of $(A_1 - 5I)$].

To find A_1 , we will need to know the respective ranks $r[(A-5I)]^j$ where $j = 1,2,3$. Then,

$$r[A - 5I] = 10 ,$$
$$r[A - 5I]^2 = 8 ,$$
$$r[A - 5I]^3 = 6 .$$

To find $m_1,\ldots,m_\ell$, we have the following formula:

$$\sum_{p=1}^{\ell} m_p = \begin{cases} r[(T - \lambda_j I)^{k_j - 1}] - (n - d_j) , & \ell = 1 \\ r[(T - \lambda_j I)^{k_j - \ell}] - r[(T - \lambda_j I)^{k_j - \ell + 1}], \\ \qquad (\ell = 2,\ldots,k_j) \end{cases}$$

Therefore, $m_1 = r[A - 5I]^3 - (n - d_1) = 6 - (12 - 8) = 2$. Then

$m_1 + m_2 = r[A - 5I]^2 - r[A - 5I]^3 = 8 - 6 = 2$.

$m_1 + m_2 + m_3 = r(A - 5I) - r[A - 5I]^2 = 10 - 8 = 2$.

$m_1 + m_2 + m_3 + m_4 = N(A - 5I) = n - r(A - 5I) = 12 - 10 = 2$.

We have $m_1 = 2$, $m_2 = 0$, $m_3 = 0$, $m_4 = 0$. Thus, A_1 has two elementary Jordan matrices.

To find A_2 , it is necessary to know only the following: $r(A - 2I)$. Now $r(A - 2I) = 10$. Then,

$$m_1 = r(A_2 - 2I) = 10 - (12 - 4) = 2$$
$$m_2 = N(A_2 - 2I) - r(A_2 - 2I) = 2 - 2 = 0 .$$

Thus, B_2 has two elementary Jordan matrices, each having dimension 2. Therefore, the Jordan canonical form of A is

$$
\begin{bmatrix}
5 & 0 & 0 & 0 & 0 & 0 & 0 & 0 & 0 & 0 & 0 & 0 \\
1 & 5 & 0 & 0 & 0 & 0 & 0 & 0 & 0 & 0 & 0 & 0 \\
0 & 1 & 5 & 0 & 0 & 0 & 0 & 0 & 0 & 0 & 0 & 0 \\
0 & 0 & 1 & 5 & 0 & 0 & 0 & 0 & 0 & 0 & 0 & 0 \\
0 & 0 & 0 & 0 & 5 & 0 & 0 & 0 & 0 & 0 & 0 & 0 \\
0 & 0 & 0 & 0 & 1 & 5 & 0 & 0 & 0 & 0 & 0 & 0 \\
0 & 0 & 0 & 0 & 0 & 1 & 5 & 0 & 0 & 0 & 0 & 0 \\
0 & 0 & 0 & 0 & 0 & 0 & 1 & 5 & 0 & 0 & 0 & 0 \\
0 & 0 & 0 & 0 & 0 & 0 & 0 & 0 & 2 & 0 & 0 & 0 \\
0 & 0 & 0 & 0 & 0 & 0 & 0 & 0 & 1 & 2 & 0 & 0 \\
0 & 0 & 0 & 0 & 0 & 0 & 0 & 0 & 0 & 0 & 2 & 0 \\
0 & 0 & 0 & 0 & 0 & 0 & 0 & 0 & 0 & 0 & 1 & 2 \\
\end{bmatrix}
$$

● **PROBLEM** 12-15

What is the Jordan form for the differentiation operator on V?

Solution: Let $a_0, \ldots, a_{n-1}$ be complex numbers, and let V be the space of all n times differentiable functions f on an interval of the real line which satisfy the differential equation

$$\frac{d^n f}{dx^n} + a_{n-1} \frac{d^{n-1} f}{dx^{n-1}} + \ldots + a_1 \frac{df}{dx} + a_0 f = 0 .$$

Let D be the differentiation operator. Then, V is invariant under D because V is the null space of $p(D)$ where $p = x^n + \ldots a_1 x + a_0$. Let $c_1, \ldots, c_k$ be the distinct complex roots of p:

$$p = (x - c_1)^{r_1} \ldots (x - c_k)^{r_k} .$$

Let V_i be the null space of $(D - c_i I)^{r_i}$, that is, the set of solutions to the differential equation

354

$$(D - c_i I)^{r_i} f = 0 \; .$$

The primary decomposition theorem tells us that

$$V = V_1 \oplus \ldots \oplus V_k \; .$$

Let N_i be the restriction of $D - c_i I$ to V_i . The Jordan form for
the operator D (on V) is then determined by the rational forms for
the nilpotent operators $N_1, \ldots, N_k$ on the spaces $V_1, \ldots, V_k$. So, we
must know the rational form for the operator $N = (D - cI)$ on the space
V_c . This consists of the solutions of the equation

$$(D - cI)^r f = 0 \; .$$

Next, we must know how many elementary nilpotent blocks will be in the
rational form for N . The number will be the nullity of N , i.e.,
the dimension of the characteristic space associated with the characteristic value c. That dimension is 1 because any function which satisfies the differential equation $Df = cf$ is a scalar multiple of the
expotential function, $h(x) = e^{cx}$. Therefore, the operator N (on the
space V_c) has a cyclic vector. A good choice for a cyclic vector is

$$g = x^{r-1} h:$$
$$g(x) = x^{r-1} e^{cx} \; .$$

This gives

$$Ng = (r-1) x^{r-2} h$$

$$\cdot$$
$$\cdot$$
$$\cdot$$

$$N^{r-1} g = (r-1)! h$$

Thus, we conclude that the Jordan form for D (on the space V) is
the direct sum of k elementary Jordan matrices, one for each root c_i.

CHAPTER 13

MATRIX FUNCTIONS

FUNCTIONS OF MATRICES

● PROBLEM 13-1

Find $f(A)$ where $A = \begin{pmatrix} 1 & -2 \\ 4 & 5 \end{pmatrix}$

and $f(t) = t^2 - 3t + 7$.

Solution: The general polynomial of a scalar variable t is denoted by

$$f(t) = a_n t^n + a_{n-1} t^{n-1} + \ldots + a_1 t + a_0 .$$

The general polynomial of a square matrix A of order n is

$$f(A) = a_n A^n + a_{n-1} A^{n-1} + \ldots + a_1 A + a_0 I ,$$

where I is an nxn identity matrix.

Given that

$$f(t) = t^2 - 3t + 7$$

then,

$$f(A) = A^2 - 3A + 7I .$$

$$A^2 = \begin{pmatrix} 1 & -2 \\ 4 & 5 \end{pmatrix} \begin{pmatrix} 1 & -2 \\ 4 & 5 \end{pmatrix} = \begin{pmatrix} 1 - 8 & -2 - 10 \\ 4 + 20 & -8 + 25 \end{pmatrix} = \begin{pmatrix} -7 & -12 \\ 24 & 17 \end{pmatrix}$$

Hence

$$f(A) = \begin{pmatrix} -7 & -12 \\ 24 & 17 \end{pmatrix} - 3 \begin{pmatrix} 1 & -2 \\ 4 & 5 \end{pmatrix} + 7 \begin{pmatrix} 1 & 0 \\ 0 & 1 \end{pmatrix}$$

356

$$= \begin{bmatrix} -3 & -6 \\ 12 & 9 \end{bmatrix} .$$

If $A = \begin{bmatrix} 0 & -1 \\ 4 & 4 \end{bmatrix}$, evaluate $f(A) = \sin^{-1}(\frac{1}{4})A$.

Solution: Let $m(\lambda)$ be the minimum polynomial of the matrix A. Suppose we can find a polynomial $r(\lambda)$ which has degree less than the degree of the minimum polynomial satisfying

$$r(\lambda) = f(\lambda) \text{ and } r'(\lambda) = f'(\lambda)$$

for each eigenvalue λ of A. Then, $r(A) = f(A)$, where $f(A)$ is a matrix polynomial. Now, to evaluate $\sin^{-1}(1/4)A$, we first find the minimum polynomial of A.

The characteristic polynomial of A is

$$C(\lambda) = \det(\lambda I - A) = \det \begin{bmatrix} \lambda & 1 \\ -4 & \lambda-4 \end{bmatrix}$$

$$= \lambda(\lambda - 4) + 4 = \lambda^2 - 4\lambda + 4 = (\lambda - 2)^2 ;$$

and the minimum polynomial is $(\lambda - 2)^2$. The eigenvalues are $\lambda_1 = 2$ and $\lambda_2 = 2$. The polynomial $r(\lambda)$ must be of degree less than the degree of the minimum polynomial, i.e.,
$r(\lambda) = \alpha + \beta\lambda$ satisfying $r(\lambda) = f(\lambda)$
or $r(2) = \sin^{-1} 1/4(2) = \sin^{-1} 1/2$, and
$r'(\lambda) = f'(\lambda) = r'(2) = \frac{d}{d\lambda} \sin^{-1} 1/4 \lambda|_{\lambda = 2}$

$$= 1/4 \left[1 - (2/4)^2 \right]^{-1/2} = \frac{\sqrt{3}}{6} ;$$

that is,

$$\alpha + 2\beta = \sin^{-1} 1/2 = \pi/6$$

and $\qquad \beta = \sqrt{3}/6$.

Solving for α yields
$\qquad \alpha = \pi/6 - 2\sqrt{3}/6 = 1/6[\pi - 2\sqrt{3}]$.
Now substitute these values of α and β into the equation
$\qquad r(\lambda) = \alpha + \beta\lambda$.
Thus,
$$r(\lambda) = \frac{1}{6}[\pi - 2\sqrt{3}] + \frac{\sqrt{3}}{6} \lambda$$

$$= \frac{1}{6} \left[(\pi - 2\sqrt{3}) + \sqrt{3}\lambda \right].$$

However, it is known that $r(A) = f(A) = \sin^{-1}(\frac{1}{4})A$.
Therefore,

$$\sin^{-1}(1/4)A = 1/6 \left[(\pi - 2\sqrt{3})I + \sqrt{3}\ A \right]$$

$$= \frac{1}{6} \left[(\pi - 2\sqrt{3}) \begin{bmatrix} 1 & 0 \\ 0 & 1 \end{bmatrix} + \sqrt{3} \begin{bmatrix} 0 & -1 \\ 4 & 4 \end{bmatrix} \right]$$

$$\sin^{-1}(1/4)A = 1/6 \begin{bmatrix} \pi - 2\sqrt{3} & -\sqrt{3} \\ \sqrt{3}(4) & \pi - 2\sqrt{3} + 4\sqrt{3} \end{bmatrix}$$

$$= 1/6 \begin{bmatrix} \pi - 2\sqrt{3} & -\sqrt{3} \\ 4\sqrt{3} & \pi + 2\sqrt{3} \end{bmatrix}$$

● **PROBLEM** 13-3

---

a) If $S_k = \begin{bmatrix} 1/k & 1 - 1/k^2 \\ 2 & 1 + 1/k \end{bmatrix}$ $k = 1, 2, \ldots$

find $\lim\limits_{k \to \infty} S_k$.

b) If $A_k = \begin{bmatrix} 1/k! & 0 \\ 1/2^k & 0 \end{bmatrix}$, $k = 0, 1, \ldots$

find $\sum\limits_{k=0}^{\infty} A_k$.

Solution: Both problems above involve infinite sequences
of matrices. Thus we must formulate an adequate definition
of convergence for sequences of matrices.
      Let $S_0, S_1, \ldots, S_k$ be a sequence of mxn matrices.
Then the sequence is said to converge to an mxn matrix
$C = [c_{ij}]$ if the sequence of numbers
$$S_{ij}^{(0)}, S_{ij}^{(1)}, \ldots, S_{ij}^{(k)}, \ldots$$
converges to c_{ij} for all i and j, that is, if

$$\lim_{k\to\infty} \left[s_{ij}^{(k)} \right] = \left[\lim_{k\to\infty} s_{ij}^{(k)} \right] . \qquad (1)$$

a) Using (1), it can be seen that

$$\lim_{k\to\infty} S_k = \begin{bmatrix} \lim_{k\to\infty} 1/k & \lim_{k\to\infty} 1 - 1/k^2 \\ \lim_{k\to\infty} 2 & \lim_{k\to\infty} 1 + 1/k \end{bmatrix}$$

The limits of $\frac{1}{k}$, $\frac{1}{k^2}$ as $k \to \infty$ are zero. Hence,

$$\lim_{k\to\infty} S_k = \begin{bmatrix} 0 & 1 \\ 2 & 1 \end{bmatrix} .$$

b) Consider an infinite series

$$A_0 + A_1 \ldots + A_k + \ldots$$

in which each of the terms A_0, A_1, ..., A_k .... is an mxn matrix. The partial sums are:

$$S_0 = A_0$$
$$S_1 = A_0 + A_1$$
$$S_2 = A_0 + A_1 + A_2$$
$$\vdots$$
$$S_k = A + A_1 + A_2 + \ldots + A_k .$$
$$\vdots$$

However, the S_k above represent elements of a sequence. If these partial sums converge to a matrix C, then the series converges to C, i.e.:

$$C = A_0 + A_1 + \ldots + A_k .$$

Let $A_k = [a_{ij}^{(k)}]$, $k = 0, 1, \ldots$ and $C = [c_{ij}]$. Then

$$c_{ij} = a_{ij}^{(0)} + a_{ij}^{(1)} + \ldots + a_{ij}^{(k)} + \ldots$$
$$i = 1, 2, \ldots, m; j = 1, 2, \ldots, n .$$

Thus the matrix to which the series $\sum_{k=0}^{\infty} A_k$ converges is found by computing the limits of the series of each element of A_k considered separately. For the given matrix,

$$\sum_{k=0}^{\infty} \frac{1}{k!} = \frac{1}{0!} + \frac{1}{1!} + \frac{1}{2!} + \frac{1}{3!} + \ldots + \frac{1}{\ell!} + \ldots = e$$

$$\sum_{k=0}^{\infty} 0 = 0 + 0 + \dots + 0 + \dots \qquad = 0$$

$$\sum_{k=0}^{\infty} \frac{1}{2^k} = \frac{1}{2^0} + \frac{1}{2^1} + \frac{1}{2^2} + \dots + \frac{1}{2^\ell} + \dots \qquad (1)$$

The series (1) is a geometric series with common ratio 1/2 . Its limit is

$$\frac{1}{1 - \frac{1}{2}} = 2 .$$

Thus, $\sum_{k=0}^{\infty} A_k = \begin{bmatrix} e & 0 \\ 2 & 0 \end{bmatrix}$

● **PROBLEM** 13-4

Given $A(t) = \begin{bmatrix} t^2 & \cos t \\ e^t & \sin t \end{bmatrix}$,

Find $\dfrac{dA}{dt}$

Solution: The elements of a matrix A are often functions of a dummy variable, say t. When it is necessary to exhibit the functional dependence of A on t, the matrix is written A(t) and the entry in the ith row in the jth column is $a_{ij}(t)$.

Now A(t) is said to be differentiable at t_o if each of the entries $a_{ij}(t)$ is differentiable at t_o . The derivative of A with respect to t is then

$$\frac{dA}{dt} = A'(t_o) = \begin{bmatrix} a'_{ij}(t_o) \end{bmatrix} .$$

Now we find the derivative of the given matrix A(t):

$$a_{11} = t^2 \quad \text{then,} \quad a'_{11} = \frac{da_{11}}{dt} = 2t$$

$$a_{12} = \cos t , \quad a'_{12} = -\sin t$$

$$a_{21} = e^t \qquad a'_{21} = e^t.$$

$$a_{22} = \sin t \qquad a'_{22} = \cos t .$$

Hence

$$\frac{dA}{dt} = A'(t) = \begin{bmatrix} 2t & -\sin t \\ e^t & \cos t \end{bmatrix}$$

Given $A(t) = \begin{bmatrix} \sin t & \cos t & t \\ \dfrac{\sin t}{t} & e^t & t^2 \\ 1 & 0 & t^3 \end{bmatrix}$ $(t \neq 0)$

Find $\dfrac{dA}{dt}$.

Solution: The elements of the given matrix are functions of a variable t which are differentiable. Therefore,

$$dA/dt = A'(t) = [a'_{ij}(t)] .$$

We find the derivative of each element $a_{ij}(t)$ with respect to t. Thus,

$a_{11} = \sin t$ $a_{12} = \cos t$

$\dfrac{da_{11}}{dt} = a'_{11}(t) = \cos t$ $\dfrac{da_{12}}{dt} = a'_{12}(t) = -\sin t$.

$a_{13} = t$, $\dfrac{da_{13}}{dt} = a'_{13} = 1$

$a_{21} = \dfrac{\sin t}{t}$.   We must find $\dfrac{da_{21}}{dt}$. It is known that the derivative of $\dfrac{u}{v}$ is given by

$$\frac{d}{dt}\left(\frac{u}{v}\right) = \frac{v\,\frac{du}{dt} - u\,\frac{dv}{dt}}{v^2} .$$

Let $v = t$, and $u = \sin t$.

Then,

$$a'_{21} = \frac{d}{dt}\left(\frac{\sin t}{t}\right) = \frac{t\cos t - \sin t}{t^2}$$

$$a_{22}' = \frac{d}{dt}(e^t) = e^t .$$

$$a_{23}' = \frac{d}{dt}(t^2) = 2t .$$

$$a_{31}' = \frac{d}{dt}(1) = 0$$

$$a_{32}' = \frac{d}{dt}(0) = 0$$

$$a_{33}' = \frac{d}{dt}(t^3) = 3t^2 .$$

Therefore,

$$\frac{dA}{dt} = \begin{bmatrix} \cos t & -\sin t & 1 \\ \dfrac{t \cos t - \sin t}{t^2} & e^t & 2t \\ 0 & 0 & 3t^2 \end{bmatrix}$$

● **PROBLEM** 13-6

Given

$$A = \begin{bmatrix} 1 & e^t \\ t^2 & t \end{bmatrix} .$$

find $\displaystyle\int_0^1 A(t) \, dt.$

Solution: When the elements of the matrix A, $[a_{ij}]$ are functions of a scalar variable t, the integral of A with respect to t taken between the limits a and b is defined as that matrix which has for its (i, j)th element

$$\int_a^b a_{ij}(t)\,dt. \quad \text{Therefore,}$$

$$\int_a^b A(t)\,dt = [c_{ij}]$$

362

where $[c_{ij}] = \int_a^b a_{ij}(t)\,dt.$

In defining matrix integration, we have assumed that the elements $[a_{ij}]$ of the matrix A are continuous real-valued functions of a real variable t.

Now consider the given matrix

$$A = \begin{bmatrix} 1 & e^t \\ t^2 & t \end{bmatrix}.$$

Then,

$$\int_0^1 A(t)\,dt = c_{ij}, \quad c_{ij} = \int_0^1 a_{ij}(t)\,dt.$$

Therefore,

$$c_{11} = \int_0^1 a_{11}(t)\,dt = \int_0^1 1\,dt.$$

$$c_{12} = \int_0^1 a_{12}(t)\,dt = \int_0^1 e^t\,dt$$

$$c_{21} = \int_0^1 a_{21}(t)\,dt = \int_0^1 t^2\,dt$$

$$c_{22} = \int_0^1 a_{22}(t)\,dt = \int_0^1 t\,dt.$$

Thus, the integral over (0, 1) of A(t) is

$$\int_0^1 A(t)\,dt = \begin{bmatrix} \int_0^1 1\,dt & \int_0^1 e^t\,dt \\ \int_0^1 t^2\,dt & \int_0^1 t\,dt \end{bmatrix}$$

Upon integration of each element of the above matrix we get

363

$$\int_0^1 A(t)\,dt = \begin{pmatrix} t\Big|_0^1 & & e^t\Big|_0^1 \\ & & \\ \dfrac{t^3}{3}\Big|_0^1 & & \dfrac{t^2}{2}\Big|_0^1 \end{pmatrix}$$

$$\int_0^1 A(t)\,dt = \begin{pmatrix} 1 & & e-1 \\ & & \\ 1/3 & & 1/2 \end{pmatrix}$$

CALCULATING THE EXPONENTIAL OF A MATRIX

• **PROBLEM** 13-7

Find $e^A = f(A)$, where

$$A = \begin{pmatrix} 3 & -3 & 3 \\ -1 & 5 & -2 \\ -1 & 3 & 0 \end{pmatrix}$$

Solution: We know that e^x can be defined by the following power series:

$$e^x = 1 + \frac{x}{1!} + \frac{x^2}{2!} + \frac{x^3}{3!} + \cdots + \frac{x^n}{n!} + \cdots .$$

Now the exponential function of a square matrix A is defined by the same power series as the exponential function of a scalar. Thus

$$e^A = I + \frac{A}{1!} + \frac{A^2}{2!} + \frac{A^3}{3!} + \cdots + \frac{A^n}{n!} + \cdots \qquad (1)$$

To compute e^A , the expansion (1), is inconvenient. It is possible to reduce (1) to the polynomial form. Thus, if A is a square matrix of order n, then

$$e^A = \alpha_{n-1}A^{n-1} + \alpha_{n-2}A^{n-2} + \cdots + \alpha_1 A + \alpha_0 I \qquad (2)$$

where α_{n-1} , α_{n-2} , $\cdots$ α_0 are constants.

To find α_0 , α_1 , $\cdots$, α_{n-1} we first compute the eigenvalues of A, λ_1 , λ_2 , $\cdots$ λ_n . For each eigenvalue we have

$$e^{\lambda_i} = r(\lambda_i)$$

where $r(\lambda) = \alpha_{n-1}\lambda^{n-1} + \ldots + \alpha_2\lambda^2 + \alpha_1\lambda + \alpha_o$. (3)

The characteristic equation of the given matrix A is det. $[\lambda I - A] = 0$. Therefore,

$$\det \begin{bmatrix} \lambda - 3 & 3 & -3 \\ 1 & \lambda - 5 & 2 \\ 1 & -3 & \lambda \end{bmatrix} = 0$$

or $\lambda - 3 \begin{vmatrix} \lambda - 5 & 2 \\ -3 & \lambda \end{vmatrix} - 3 \begin{vmatrix} 1 & 2 \\ 1 & \lambda \end{vmatrix} - 3 \begin{vmatrix} 1 & \lambda - 5 \\ 1 & -3 \end{vmatrix} = 0$

$\lambda - 3 \ [\lambda(\lambda-5) + 6] \ -3 \ [\lambda-2] \ -3 \ [-3 - (\lambda-5)] = 0$

$(\lambda-2) \ [(\lambda-3)^2 - 3 + 3] = 0$

or $(\lambda-2)(\lambda-3)^2 = 0$.

Then, the eigenvalues are $\lambda = 2$, $\lambda = 3$, and $\lambda = 3$.

Substitute these values into (3) to obtain

$$r(2) = \alpha_2 4 + \alpha_1 2 + \alpha_o \ .$$

Since $e^{\lambda_i} = r(\lambda_i)$, $e^2 = 4\alpha_2 + 2\alpha_1 + \alpha_o$. (i)

Similarly,

$$r(3) = 9\alpha_2 + 3\alpha_1 + \alpha_o$$ (ii)

or $e^3 = 9\alpha_2 + 3\alpha_1 + \alpha_o$.

Note that there are only two equations and three unknowns.

The third equation is given by

$$e^3 = \frac{d}{d\lambda} \ (r(\lambda))|_{\lambda=3}$$

$$e^3 = \frac{d}{d\lambda} \ (\alpha_2\lambda^2 + \lambda \alpha_1 + \alpha_o \ |_{\lambda=3}$$

$$= 2\alpha_2\lambda + \alpha_1 \ |_{\lambda=3}$$

$$= 6\alpha_2 + \alpha_1 \ .$$ (iii)

365

We have

$$e^2 = 4\alpha_2 + 2\alpha_1 + \alpha_o \tag{i}$$

$$e^3 = 9\alpha_2 + 3\alpha_1 + \alpha_o \tag{ii}$$

$$e^3 = 6\alpha_2 + \alpha_1 \tag{iii}$$

or $E = A\alpha$ where $E = \begin{pmatrix} e^2 \\ e^3 \\ e^3 \end{pmatrix}$ and

$$A = \begin{bmatrix} 4 & 2 & 1 \\ 9 & 3 & 1 \\ 6 & 1 & 0 \end{bmatrix} \qquad \text{and} \qquad \alpha = \begin{pmatrix} \alpha_2 \\ \alpha_1 \\ \alpha_o \end{pmatrix}$$

Use Cramer's rule to solve this system yields

$$\alpha_2 = \frac{\begin{vmatrix} e^2 & 2 & 1 \\ e^3 & 3 & 1 \\ e^3 & 1 & 0 \end{vmatrix}}{\begin{vmatrix} 4 & 2 & 1 \\ 9 & 3 & 1 \\ 6 & 1 & 0 \end{vmatrix}} \tag{4}$$

Now

$$\begin{vmatrix} 4 & 2 & 1 \\ 9 & 3 & 1 \\ 6 & 1 & 0 \end{vmatrix} = -1 \qquad \text{and}$$

$$\begin{vmatrix} e^2 & 2 & 1 \\ e^3 & 3 & 1 \\ e^3 & 1 & 0 \end{vmatrix} = \begin{vmatrix} e^3 & 3 \\ e^3 & 1 \end{vmatrix} - \begin{vmatrix} e^2 & 2 \\ e^3 & 1 \end{vmatrix} = -e^2$$

So (4) implies that $\alpha_2 = \dfrac{-e^2}{-1} = e^2$.

Similarly,

366

$$\alpha_1 = \frac{\begin{vmatrix} 4 & e^2 & 1 \\ 9 & e^3 & 1 \\ 6 & e^3 & 0 \end{vmatrix}}{\begin{vmatrix} 4 & 2 & 1 \\ 9 & 3 & 1 \\ 6 & 1 & 0 \end{vmatrix}} = \frac{-e^3 + 6e^2}{-1}$$

and

$$\alpha_0 = \frac{\begin{vmatrix} 4 & 2 & e^2 \\ 9 & 3 & e^3 \\ 6 & 1 & e^3 \end{vmatrix}}{\begin{vmatrix} 4 & 2 & 1 \\ 9 & 3 & 1 \\ 6 & 1 & 0 \end{vmatrix}} = \frac{-9e^2 + 2e^3}{-1}$$

Thus solving equations (i), (ii), and (iii), yields

$$\alpha_2 = e^2 \, , \ \alpha_1 = e^3 - 6e^2 \, , \ \alpha_0 = 9e^2 - 2e^3 \, .$$

In the given problem, since the matrix is of order 3 x 3, n = 3. From (2) we have

$$e^A = \alpha_2 A^2 + \alpha_1 A + \alpha_0 I \, .$$

$$A^2 = \begin{bmatrix} 3 & -3 & 3 \\ -1 & 5 & -2 \\ -1 & 3 & 0 \end{bmatrix} \begin{bmatrix} 3 & -3 & 3 \\ -1 & 5 & -2 \\ -1 & 3 & 0 \end{bmatrix}$$

$$= \begin{bmatrix} 9 + 3 - 3 & -9 - 15 + 9 & 9 + 6 + 0 \\ -3 - 5 + 2 & 3 + 25 - 6 & -3 - 10 + 0 \\ -3 - 3 + 0 & 3 + 15 + 0 & -3 - 6 + 0 \end{bmatrix}$$

$$= \begin{bmatrix} 9 & -15 & 15 \\ -6 & 22 & -13 \\ -6 & 18 & -9 \end{bmatrix} \, .$$

Thus,

$$e^A = \alpha_2 \begin{bmatrix} 9 & -15 & 15 \\ -6 & 22 & -13 \\ -6 & 18 & -9 \end{bmatrix} + \alpha_1 \begin{bmatrix} 3 & -3 & 3 \\ -1 & 5 & -2 \\ -1 & 3 & 0 \end{bmatrix}$$

$$+ \alpha_0 \begin{bmatrix} 1 & 0 & 0 \\ 0 & 1 & 0 \\ 0 & 0 & 1 \end{bmatrix}$$

$$e^A = \begin{bmatrix} 9\alpha_2 + 3\alpha_1 + \alpha_0 & -15\alpha_2 - 3\alpha_1 & 15\alpha_2 + 3\alpha_0 \\ 6\alpha_2 - \alpha_1 & 22\alpha_2 + 5\alpha_1 + \alpha_0 & -13\alpha_2 - 2\alpha_1 \\ -6\alpha_2 - \alpha_1 & 18\alpha_2 + 3\alpha_1 & -9\alpha_2 + \alpha_0 \end{bmatrix}$$

Now substitute the values of α_2 , α_1 and α_0 into the above matrix. This yields

$$e^A = \begin{bmatrix} 9e^2+3e^3-18e^2+9e^2-2e^3 & -15e^2-3e^3+18e^2 & 15e^2+3e^3-18e^2 \\ -6e^2-e^3+6e^2 & 22e^2+5e^3-30e^2+9e^2-2e^3 & -13e^2-2e^3+12e^2 \\ -6e^2-e^3+6e^2 & 18e^2+3e^3-18e^2 & -9e^2+9e^2-2e^3 \end{bmatrix}$$

or,

$$e^A = \begin{bmatrix} e^3 & -3e^3 + 3e^2 & 3e^3 - 3e^2 \\ -e^3 & 3e^3 + e^2 & -2e^3 - e^2 \\ -e^3 & 3e^3 & -2e^3 \end{bmatrix}$$

● **PROBLEM 13-8**

Given the matrix

$$A = \begin{bmatrix} 0 & 1 \\ 8 & -2 \end{bmatrix} \quad,$$

find e^{At} .

Solution: Just as e^x is defined as

$$1 + x + \frac{x^2}{2!} + \frac{x^3}{3!} + \ldots = \sum_{n=0}^{\infty} \frac{x^n}{n!} \quad,$$

so e^{At} is defined as

$$I + At + \frac{A^2 t^2}{2!} + \ldots = \sum_{n=0}^{\infty} \frac{1}{n!} A^n t^n \quad. \tag{a}$$

The infinite series (a) may be reduced to a polynomial. Thus,

$$e^{At} = \alpha_{n-1} A^{n-1} t^{n-1} + \alpha_{n-2} A^{n-2} t^{n-2} + \ldots + \alpha_1 At + \alpha_0 I , \qquad \text{(b)}$$

where $\alpha_0, \alpha_1, \ldots, \alpha_{n-1}$ are unknown functions of t. To find $\alpha_0, \alpha_1, \ldots, \alpha_{n-1}$ we first compute the eigenvalues of At, $\lambda_1, \lambda_2, \lambda_3, \ldots, \lambda_n$. Then, for each eigenvalue, $e^{\lambda} = r(\lambda)$ where $r(\lambda)$ is defined to be:

$$r(\lambda) = \alpha_{n-1} \lambda^{n-1} + \alpha_{n-2} \lambda^{n-2} + \ldots + \alpha_1 \lambda + \alpha_0 . \qquad \text{(c)}$$

The resulting set of linear equations

$$e^{\lambda i} = r(\lambda_i) , \quad i = 1, \ldots, n$$

can then be solved for the unknown $\alpha_0, \alpha_1, \ldots, \alpha_{n-1}$. Then these solutions are substituted into (b) to give e^{At}. Thus, in the given problem,

$$At = \begin{bmatrix} 0 & 1 \\ 8 & -2 \end{bmatrix} t = \begin{bmatrix} 0 & t \\ 8t & -2t \end{bmatrix} \qquad \text{(d)}$$

and $e^{At} =$

$$\alpha_1 \begin{bmatrix} 0 & t \\ 8t & -2t \end{bmatrix} + \alpha_0 \begin{bmatrix} 1 & 0 \\ 0 & 1 \end{bmatrix}$$

$$= \begin{bmatrix} \alpha_0 & \alpha_1 t \\ 8\alpha_1 t & -2\alpha_1 t + \alpha_0 \end{bmatrix} . \qquad \text{(e)}$$

We wish to find α_0 and α_1 in (e). From (d), the eigenvalues of At are found by solving

$$\det |At - \lambda I| = \det \begin{vmatrix} -\lambda & t \\ 8t & -2t-\lambda \end{vmatrix} = 0 . \qquad \text{(f)}$$

From (f), $\lambda^2 + 2t\lambda - 8t^2 = 0$ and hence $\lambda_1 = 2t$, $\lambda_2 = -4t$. Using (c),

$$e^{2t} = 2t\alpha_1 + \alpha_0$$
$$e^{-4t} = -4t\alpha_1 + \alpha_0 . \qquad \text{(g)}$$

Solving for α_0, α_1 in the system (g),

$$\alpha_1 = \frac{1}{6t}(e^{2t} - e^{-4t}); \quad \alpha_0 = \frac{1}{3}(2e^{2t} + e^{-4t}) .$$

Substituting these values into (e) and simplifying,

$$e^{At} = \frac{1}{6} \begin{bmatrix} 4e^{2t} + 2e^{-4t} & e^{2t} - e^{-4t} \\ 8e^{2t} - 8e^{-4t} & 2e^{2t} + 4e^{-4t} \end{bmatrix}$$

● **PROBLEM** 13-9

Given the matrix

$$A = \begin{bmatrix} 0 & 0 & 0 \\ 1 & 0 & 0 \\ 1 & 0 & 1 \end{bmatrix} ,$$

find e^{At} .

Solution: The infinite series:

$$e^{At} = I + At + \frac{A^2 t^2}{2!} + \ldots = \sum_{n=0}^{\infty} \frac{A^n t^n}{n!}$$

can be reduced to a polynomial of degree $n-1$ (where n is the order of the square matrix A). That is,

$$e^{At} = \alpha_{n-1} A^{n-1} t^{n-1} + \alpha_{n-2} A^{n-2} t^{n-2} + \ldots + \alpha_2 A^2 t^2 + \alpha_1 At + \alpha_0 I , \quad (a)$$

where $\alpha_0, \alpha_1, \ldots, \alpha_{n-1}$ are functions of t to be determined from A. To find $\alpha_0, \alpha_1, \ldots, \alpha_{n-1}$ we find the eigenvalues of At (if A is of order nxn, there will be n eigenvalues). Then, for each eigenvalue λ we set

$$e^{\lambda} = r(\lambda) = \lambda^{n-1} \alpha_{n-1} + \lambda^{n-2} \alpha_{n-2} + \ldots + \lambda \alpha_1 + \alpha_0 .$$

The n eigenvalues give rise to n equations in the n unknowns, $\alpha_0, \alpha_1, \ldots, \alpha_{n-1}$. Therefore, we can find unique solutions to $\alpha_0, \alpha_1, \ldots, \alpha_{n-1}$. When these are substituted into (a), we obtain the polynomial representation of e^{At} .

In the given problem,

$$e^{At} = \alpha_2 A^2 t^2 + \alpha_1 At + \alpha_0 I$$

$$= \alpha_2 \begin{bmatrix} 0 & 0 & 0 \\ 0 & 0 & 0 \\ 1 & 0 & 1 \end{bmatrix} t^2 + \alpha_1 \begin{bmatrix} 0 & 0 & 0 \\ 1 & 0 & 0 \\ 1 & 0 & 1 \end{bmatrix} t + \alpha_0 \begin{bmatrix} 1 & 0 & 0 \\ 0 & 1 & 0 \\ 0 & 0 & 1 \end{bmatrix}$$

$$= \begin{bmatrix} \alpha_0 & 0 & 0 \\ \alpha_1 t & \alpha_0 & 0 \\ \alpha_2 t^2 + \alpha_1 t & 0 & \alpha_2 t^2 + \alpha_1 t + \alpha_0 \end{bmatrix} . \quad (b)$$

We now proceed to find the eigenvalues of At.

$$A = \begin{bmatrix} 0 & 0 & 0 \\ t & 0 & 0 \\ t & 0 & t \end{bmatrix} .$$

$$At - \lambda I = \begin{bmatrix} -\lambda & 0 & 0 \\ t & -\lambda & 0 \\ t & 0 & t-\lambda \end{bmatrix} . \quad (c)$$

The eigenvalues of At are the roots of the characteristic equation, obtained by expanding the determinant of (c) and setting it equal to zero. Thus,

$$\begin{vmatrix} -\lambda & 0 & 0 \\ t & -\lambda & 0 \\ t & 0 & t-\lambda \end{vmatrix} = 0 . \qquad \text{(d)}$$

Evaluating (d) along the first row

$$-\lambda(-\lambda)(t-\lambda) = 0 . \qquad \text{(e)}$$

From (e), $\lambda_1 = 0$, $\lambda_2 = 0$, $\lambda_3 = t$. The case where an eigenvalue had multiplicity greater than one was not covered in the previous discussion. In this problem, for instance, we have only two equations,

$$e^0 = r(0) = 0^2 \alpha_2 + 0\alpha_1 + \alpha_0 \qquad \text{(f)}$$

$$e^t = r(\lambda) = t^2 \alpha_2 + t\alpha_1 + \alpha_0 , \qquad \text{(g)}$$

in the three unknowns, α_0, α_1, α_2. We need another equation. This is obtained as follows:

$$e^\lambda = \frac{d^n r(\lambda)}{d^n \lambda} \Big|_{\lambda=\lambda_i} . \qquad \text{(h)}$$

where n is the order of multiplicity. Using (h), where $\lambda = 0$,

$$e^0 = \frac{d}{d\lambda} (\alpha_2 \lambda^2 + \alpha_1 \lambda + \alpha_0) \Big|_{\lambda=0}$$

$$= 2\alpha_2 \lambda + \alpha_1 \big|_{\lambda=0}$$

$$= \alpha_1 . \qquad \text{(i)}$$

Combining (f), (g) and (i),

$$1 = \alpha_0$$
$$e^t = t^2 \alpha_2 + t\alpha_1 + \alpha_0$$
$$1 = \alpha_1 . \qquad \text{(j)}$$

The system (j) can be solved for α_0, α_1, α_2 to give:

$$\alpha_0 = 1; \ \alpha_1 = 1; \ \alpha_2 = \frac{e^t - t - 1}{t^2} .$$

These values are not substituted into (b) to give e^{At}. Thus,

$$e^{At} = \begin{bmatrix} 1 & 0 & 0 \\ t & 1 & 0 \\ e^t - 1 & 0 & e^t \end{bmatrix} .$$

● **PROBLEM** 13-10

Let $A = \begin{bmatrix} 0 & 1 \\ 0 & 0 \end{bmatrix}$, $B = \begin{bmatrix} 0 & 0 \\ -1 & 0 \end{bmatrix}$.

Find $e^{At} e^{Bt}$ and $e^{(A+B)t}$. Are they equal?

<u>Solution:</u> We first find $e^{At}e^{Bt}$ and then $e^{(A+B)t}$.

The polynomial representation of e^{At} is

$$e^{At} = \alpha_1 At + \alpha_0 I$$

$$= \begin{bmatrix} \alpha_0 & \alpha_1 t \\ 0 & \alpha_0 \end{bmatrix} . \tag{a}$$

To find α_0, α_1, we first find the eigenvalues of the matrix,

$$At = \begin{bmatrix} 0 & t \\ 0 & 0 \end{bmatrix} . \tag{b}$$

The eigenvalues of (b) are $\lambda_1 = \lambda_2 = 0$. Then, using the relation,

$$e^{\lambda} = r(\lambda) = \alpha_1 \lambda + \alpha_0 ,$$

we have:

$$e^0 = r(0) = \alpha_0 . \tag{c}$$

To obtain another equation in the unknowns, α_0, α_1, we use the relation

$$e^{\lambda} = \frac{d}{d\lambda} r(\lambda) \big|_{\lambda=0} .$$

Thus, since $r(\lambda) = \alpha_1 \lambda + \alpha_0$,

$$\frac{d}{d\lambda} r(\lambda) \big|_{\lambda=0} = \alpha_1 .$$

That is, $e^0 = \alpha_1$. $\tag{d}$

Combining (c) and (d),

$$\alpha_0 = 1 \; ; \; \alpha_1 = 1 .$$

Substituting into (a),

$$e^{At} = \begin{bmatrix} 1 & t \\ 0 & 1 \end{bmatrix} . \tag{e}$$

Now consider the matrix B:

$$e^{Bt} = \alpha_1 Bt + \alpha_0 I$$

$$= \begin{bmatrix} \alpha_0 & 0 \\ -\alpha_1 t & \alpha_0 \end{bmatrix} . \tag{f}$$

The eigenvalues of Bt are $\lambda_1 = \lambda_2 = 0$. The previous analysis may be repeated to give $\alpha_0 = 1$, $\alpha_2 = 1$. Then, substituting these values into (f),

$$e^{Bt} \begin{bmatrix} 1 & 0 \\ -t & 1 \end{bmatrix} \tag{g}$$

Multiplying the matrices of (e) and (g),

$$e^{At} \cdot e^{Bt} = \begin{bmatrix} 1 & t \\ 0 & 1 \end{bmatrix} \begin{bmatrix} 1 & 0 \\ -t & 1 \end{bmatrix}$$

$$= \begin{bmatrix} 1-t^2 & t \\ -t & 1 \end{bmatrix} \, . \qquad \text{(h)}$$

We now find $e^{(A+B)t}$. The polynomial representation of $e^{(A+B)t}$ is

$$e^{(A+B)t} = \alpha_1 (A+B)t + \alpha_0 I$$

$$= \begin{bmatrix} \alpha_0 & \alpha_1 t \\ -\alpha_1 t & \alpha_0 \end{bmatrix} , \qquad \text{(i)}$$

since

$$A+B = \begin{bmatrix} 0 & 1 \\ 0 & 0 \end{bmatrix} + \begin{bmatrix} 0 & 0 \\ -1 & 0 \end{bmatrix} = \begin{bmatrix} 0 & 1 \\ -1 & 0 \end{bmatrix}$$

and

$$(A+B)t - \lambda I = \begin{bmatrix} -\lambda & t \\ -t & -\lambda \end{bmatrix} = \lambda^2 + t^2 = 0 \, .$$

The eigenvalues of $(A+B)t$ are $\lambda = \pm it$. These are roots of multiplicity one. Thus we have two equations of the form

$$e^{it} = r(it) = \alpha_1 it + \alpha_0 \qquad \text{(j)}$$

$$e^{-it} = r(-it) = -\alpha_1 it + \alpha_0 \, . \qquad \text{(k)}$$

Subtracting (k) from (j)

$$e^{it} - e^{-it} = 2\alpha_1 it$$

or,

$$\alpha_1 = \frac{e^{it} - e^{-it}}{2it} \, . \qquad \text{(l)}$$

Since $\sin t = \dfrac{e^{it} - e^{-it}}{2i}$ (l) may be rewritten as

$$\alpha_1 = \frac{\sin t}{t} \, . \qquad \text{(m)}$$

Then, upon substitution into (j)

$$e^{it} = i \sin t + \alpha_0 \, . \qquad \text{(n)}$$

Euler's formula is:

$$e^{it} = \cos t + i \sin t \, .$$

We see that $\alpha_0 = \cos t$. Substituting these values of α_0, α_1 into (i),

$$e^{(A+B)t} = \begin{bmatrix} \cos t & \sin t \\ -\sin t & \cos t \end{bmatrix} \, . \qquad \text{(o)}$$

Comparing (h) and (o) we see that $e^{At} \cdot e^{Bt} \neq e^{(A+B)t}$. This is in contrast to the case where At and Bt are numbers, not matrices.

● **PROBLEM** 13-11

Show that $e^{At} e^{Bt} = e^{(A+B)t}$ if, and only if, the matrices A and B commute, i.e., $AB = BA$.

<u>Solution</u>: We use the infinite polynomial definition of e^{At} to solve this problem.

373

$$e^{At} = I + At + \frac{A^2 t^2}{2!} + \ldots = \sum_{n=0}^{\infty} \frac{A^n t^n}{n!} . \tag{a}$$

We also use the fact that, for matrices A and B,

$$(A+B)^2 = (A+B)(A+B) = A^2 + AB + BA + B^2 = A^2 + 2AB + B^2 \tag{b}$$

if, and only if, AB = BA, i.e., the matrices commute.

Equation (b) may be written as

$$\sum_{k=0}^{2} \binom{2}{k} A^{2-k} B^k$$

where,

$$\binom{2}{k} = \frac{2!}{(2-k)! k!} .$$

We may generalize to

$$(A+B)^n = \sum_{k=0}^{n} \binom{n}{k} A^{n-k} B^k . \tag{c}$$

Equation (c) is the binomial theorem. It is true for matrices only if AB = BA.

Now consider $e^{At} e^{Bt}$. By definition (a):

$$e^{At} \cdot e^{Bt} = \left(\sum_{n=0}^{\infty} \frac{1}{n!} A^n t^n \right) \left(\sum_{n=0}^{\infty} \frac{1}{n!} B^n t^n \right) . \tag{d}$$

The product of two infinite series is called the Cauchy product. Forming the Cauchy product, (d) may be written as

$$\sum_{n=0}^{\infty} \sum_{k=0}^{n} \frac{A^{n-k} t^{n-k}}{(n-k)!} \frac{B^k t^k}{k!} . \tag{e}$$

We now perform a series of operations on (e)

$$\sum_{n=0}^{\infty} \sum_{k=0}^{n} \frac{A^{n-k} t^{n-k}}{(n-k)!} \frac{B^k t^k}{k!} = \sum_{n=0}^{\infty} \left[\sum_{k=0}^{n} \frac{A^{n-k}}{(n-k)!} \frac{B^k}{k!} \right] t^n$$

$$= \sum_{n=0}^{\infty} \left[\sum_{k=0}^{n} \binom{n}{k} a^{n-k} B^k \right] \frac{t^n}{n!} . \tag{f}$$

We now turn to $e^{(A+B)t}$. By (a)

$$e^{(A+B)t} = \sum_{n=0}^{\infty} \frac{(A+B)^n t^n}{n!} = \sum_{n=0}^{\infty} (A+B)^n \frac{t^n}{n!} . \tag{g}$$

Comparing equations (c) and (f) we see that only if A and B commute can (f) and (g) be equal. That is,

$$e^{At} \cdot e^{Bt} = e^{(A+B)t} \quad \text{if, and only if,} \quad AB = BA .$$

● PROBLEM 13-12

Prove that for a given matrix A, and variables t and s,

$$e^{At} \cdot e^{-As} = e^{A(t-s)} . \tag{a}$$

Solution: We know that

$$e^{At} \cdot e^{Bt} = e^{(A+B)t} \qquad\qquad (b)$$

only if A and B commute, i.e., AB = BA . Letting t = 1 in (b),

$$e^A \cdot e^B = e^{(A+B)} .$$

We note that At and -As commute since,

$$(At)(-As) = AA(-ts) = AA(-st) = (-As)(At).$$

Therefore, $e^{At} \cdot e^{-As} = e^{(At-As)} = e^{A(t-s)}$ and the proof is complete.

● **PROBLEM** 13-13

Find $e^A = f(A)$ where

$$A = \begin{bmatrix} 3 & -3 & 3 \\ -1 & 5 & -2 \\ -1 & 3 & 0 \end{bmatrix}$$

and $f(z) = e^z$ for z , a scalar from the field K .

Solution: One method of finding e^A is to construct a polynomial p(z) that agrees with f(z) on the spectrum of A. The notion of the spectrum of a matrix is defined as follows:
    let A be an nxn matrix with minimal polynomial

$$m_A(z) = (z - r_1)^m (z - r_2)^{m_2} \cdots (z - r_d)^{m_d}$$

where $\sqrt{r}_1, \sqrt{r}_2, \ldots, \sqrt{r}_d$ are the eigenvalues of A.
    The set

$$\{ \underbrace{r_1, r_1, \ldots, r_1}_{m_1}, \underbrace{r_2, \ldots, r_2}_{m_2}, \underbrace{r_3, \ldots, r_3}_{m_3}, \ldots,$$

$$\underbrace{r_d, \ldots, r_d}_{m_d} \}$$

is called the spectrum of A. A function f is defined on the spectrum of A if

$$f(r_1), f'(r_1), f''(r_2) \ldots f^{(m_1-1)}_{(m_2-1)}(r_1);$$
$$f(r_2), f'(r_2), f''(r_2) \ldots f^{2}(r_2);$$

$$\cdots \cdots \cdots \cdots \cdots \cdots \cdots$$

$$f(r_d), f'(r_d), f''(r_d) \ldots f^{(m_d-1)}(r_d)$$

all exist. If p(z) is another function defined on the spectrum of A such that

$$p(\, r_1) = f(\, r_1); \; p'(\, r_1) = f'(\, r_1) \ldots . \; p^{(m_1-1)}(\, r_1) = f^{(m_1-1)}(\, r_1)$$

$$\vdots$$

$$p(\, r_d) = f(\, r_d); \; ;'(\, r_d) = f'(\, r_d) \ldots . \; p^{(m_d-1)}(\, r_d) = f^{(m_d-1)}(\, r_d)$$

then p(z) and f(z) agree on the spectrum of A. This agreement is denoted by p = Af. Now, if p = Af, then p(A) = f(A). Thus, to find f(A), we need only find p(z), where p(z) is any polynomial that agrees with f(z) on the spectrum of A.

The minimal polynomial of the given matrix A is found to be $(x - 2)(x - 3)^2$. Hence the spectrum of A is $(2,3)$.

$$\text{For } f(x) = e^z \, , \quad f(2) = e^2 \, ;$$
$$f(3) = e^3 \, , \quad f'(3) = e^3 \, .$$

Thus, any polynomial p(z) such that p(2) = f(2), p(3) = f(3) and $p'(3) = f'(3) = e^3$ agrees with f(z) on the spectrum of A. To find p(z) we can use the following fact: If f is defined on the spectrum of A, then there is only one polynomial r(z) which agrees with f on the spectrum of A and has degree less than the degree of the minimal polynomial of A.

Since deg $m_A(z) = 3$, let

$$p(z) = az^2 + bz + c. \quad \text{Then,}$$
$$p(2) = 4a + 2b + c = e^2$$
$$p(3) = 9a + 3b + c = e^3$$
$$p'(3) = 6a + b \quad\quad = e^3 \, .$$

Solving this system of linear equations for the unknowns a, b, c yields:

$$a = e^2, \; b = e^3 - 6e^2, \; c = 9e^2 - 2e^3.$$

Hence, $e^A = f(A) = p(A) = (e^2A + e^3 - 6e^2)A$

$$+ (9e^2 - 2e^3)I$$

$$= \begin{bmatrix} e^3 & -3e^3 + 3e^2 & 3e^3 - 3e^2 \\ -e^3 & 3e^3 + e^2 & -2e^3 - e^2 \\ -e^3 & 3e^3 & -2e^3 \end{bmatrix} .$$

CHAPTER 14

INNER PRODUCT SPACES

EXAMPLES OF INNER PRODUCT SPACES

● **PROBLEM** 14-1

Show geometrically that for each angle θ, the transformation $T_\theta : R^2 \rightarrow R^2$, defined by $(x,y)T_\theta = (x \cos \theta - y \sin \theta, x \sin \theta + y \cos \theta)$, is an orthogonal transformation.

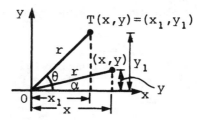

Solution: Let V be an inner product space. A linear transformation $T: V \rightarrow V$ is said to be orthogonal if $|vT| = |v|$ for each vector v in V. Thus, orthogonal transformations are those linear transformations which preserve distances.

The matrix of the transformation with respect to the standard basis is

$$A_\theta = \begin{bmatrix} \cos \theta & \sin \theta \\ -\sin \theta & \cos \theta \end{bmatrix}.$$

From the figure, it is seen that $x = r \cos \alpha$, $y = r \sin \alpha$. Then, $(x_1,y_1) = T(x,y)$ or,

$$x_1 = r \cos(\alpha + \theta) = r \cos \alpha \cos \theta - r \sin \alpha \sin \theta$$
$$= x \cos \theta - y \sin \theta$$
$$y_1 = r \sin(\alpha + \theta) = r \cos \alpha \sin \theta + r \sin \alpha \cos \theta$$
$$= x \sin \theta + y \cos \theta .$$

Thus,

$$(x_1,y_1) = (x,y) \begin{bmatrix} \cos \theta & \sin \theta \\ -\sin \theta & \cos \theta \end{bmatrix} = (x,y)T_\theta .$$

The figure describes a rotation through an angle θ, and from the figure we derive the given transformation. Hence, the given transformation is a rotation. A rotation preserves magnitudes; however, we can check explicitly that T_θ is orthogonal. Let $v = (x,y)$; $|v| = \sqrt{x^2 + y^2}$. Now,

$$vT_\theta = (x \cos \theta - y \sin \theta, x \sin \theta + y \cos \theta), \text{ so}$$
$$|vT_\theta| = \sqrt{x \cos \theta - y \sin \theta)^2 + (x \sin \theta + y \cos \theta)^2}$$

$$= \sqrt{\frac{x^2 \cos^2 \theta + y^2 \sin^2 \theta - 2xy \cos \theta \sin \theta + x^2 \sin^2 \theta + y^2 \cos^2 +}{2xy \cos \theta \sin \theta}}$$

$$= \sqrt{x^2 (\cos^2 \theta + \sin^2 \theta) + y^2 (\sin^2 \theta + \cos^2 \theta)}$$

$$= \sqrt{x^2 + y^2} \quad (\text{since} \quad \sin^2 \theta + \cos^2 \theta = 1)$$

$$= |v| .$$

● **PROBLEM** 14-2

Let P_n be the space of polynomials of degree less than n in a real variable. Define an inner product on P_n and find the inner product of the two functions $p(x) = x^2$ and $q(x) = 1 - x$.

Solution: P_n is a vector space, but the axioms for a vector space make no mention of the length of a vector or the distance between vectors. Inner products are used to define these concepts. Let u,v be elements of a vector space, V. Then, an inner product on V is a scalar-valued function of u and v such that

i) $<u,v>$ is a real number (where $<u,v>$ means the inner product of u and v).
ii) $<u,v> = <v,u>$.
iii) $<u,(v+w)> = <u,v> + <u,w>$ for $w \in V$.
iv) $<u,(av)> = a<u,v>$ for any real number, a.
v) $<u,u> > 0$ if $u \neq 0$; $<u,u> = 0$ if $u = 0$.

An inner product for P_n is

$$<f(x),g(x)> = \int_0^1 f(x)g(x)dx. \tag{1}$$

Each of the requirements for an inner product must be checked. First, realize that $f(x)$ and $g(x)$ are polynomials (therefore, integrable) which implies that the definite integral in (1) is a real number. Next, ii) is satisfied since

$$<f(x),g(x)> = \int_0^1 f(x)g(x)dx = \int_0^1 g(x)f(x)dx = <g(x),f(x)> .$$

Now, $<f(x),g(x)+w(x)> = \int_0^1 f(x)[g(x) + w(x)]dx = \int_0^1 f(x)g(x) + f(x)w(x)dx$

$$= \int_0^1 f(x)g(x)dx + \int_0^1 f(x)w(x)dx$$

$$= <f(x),g(x)> + <f(x),w(x)> ; \text{ therefore,}$$

iii) is satisfied. Next, $<f(x),\alpha g(x)>$ (α is a real number)

$$= \int_0^1 f(x)\alpha g(x)dx = \alpha \int_0^1 f(x)g(x)dx = \alpha<f(x),g(x)>.$$

Lastly investigate v). For any polynomial function, $f(x)$, we know $(f(x))^2 \geq 0$. If $f(x)$ is not constantly 0 on the interval $[0,1]$, $\int_0^1 (f(x))^2 dx \neq 0$. To see this, recall that this integral is the area

under the curve $[f(x)]^2$ from $x = 0$ to 1. So, $f(x)$, not always 0 on $[0,1]$, implies $<f(x),f(x)> > 0$. If $f(x) = 0$, $(f(x))^2 = 0$ and thus,

378

$$\int_0^1 (f(x))^2 \, dx = \int_0^1 0 \, dx = 0 \quad . \quad \text{Thus, (1) defines an inner product.}$$

It is often possible to construct new inner products from a given inner product. The method is as follows: Let V and W be vector spaces over F and let $<u,v>$ be an inner product on W. If T is a non-singular linear transformation from V into W, then the equation:

$$p_T(u,v) = <Tu \cdot Tv> \tag{1}$$

defines an inner product p_T on V. Give examples to illustrate this method.

Solution: Let V be a finite-dimensional vector space and let $B = \{\alpha_1, \ldots, \alpha_n\}$ be an ordered basis for V. Use equation (1) to show that there exists an inner product, $<\cdot,\cdot>$, on V such that for

$$X = X_1\alpha_1 + \ldots + X_n\alpha_n = \sum_j X_j\alpha_j \quad \text{and} \quad Y = Y_1\alpha_1 + \ldots + Y_n\alpha_n = \sum_k Y_k\alpha_k \ ,$$

$X,Y \in V$, $<X,Y> = X_1\bar{Y}_1 + \ldots + X_n\bar{Y}_n = \sum_{\ell=1}^{n} X_\ell\bar{Y}_\ell$ ($\bar{y}$ is the complex conjugate

of y. Notice that if y is a real number then $\bar{y} = y$). We show this making no assumptions about V or the basis for V. Therefore, it is true for any finite dimensional vector space with any valid basis. In order to do this let $e_1, \ldots, e_n$ be the standard basis vectors in F^n, and let T be the linear transformation from V into F^n defined by $T\alpha_j = e_j$, $j = 1, \ldots, n$. Taking the standard inner product on F^n ,

$$p_t(X,Y) = <T(X),T(Y)> = <T(\sum_j X_j\alpha_j),T(\sum_k Y_k\alpha_k)>$$

$$= <\sum_j X_j T(\alpha_j), \sum_k Y_k T(\alpha_k)>$$

(by properties of a linear transformation)

$$= <\sum_j X_j e_j, \sum_k Y_k e_k> \tag{2}$$

(since $T(\alpha_j) = e_j$, $j = 1, \ldots, n$) , $\sum_j X_j e_j$ and $\sum_k Y_k e_k$ are vectors in F^n

so let us use the standard inner product on F^n. Equation (2) becomes:

$$\sum_{j=1}^{n} X_j\bar{Y}_j \quad .$$

We have used p_t and the standard inner product on F^n to show that there exists an inner product in V , $p_T(X,Y) = \sum_{j=1}^{n} X_j\bar{Y}_j$. We can see that

$p_T(\alpha_j,\alpha_j) = <e_j,e_j> = 1$ and $p_T(\alpha_k,\alpha_j) = <e_j,e_k> = 0$, $j \neq k$. Thus, for any basis for V there is an inner product on V with the property

$$<\alpha_j,\alpha_k> = \delta_{jk}, \delta_{jk} = \begin{cases} 0 & j \neq k \\ 1 & j = k \end{cases} .$$

As another illustration, let $C[0,1]$ be the vector space of all continuous real-valued functions on the unit interval, $0 \le t \le 1$. Then an inner product on $C[0,1]$ is given by $<f,g> = \int_0^1 f(t)g(t)dt$. Let T be the inner operator that multiplies elements of $C[0,1]$ by t. Thus, $Tf(t) = tf(t)$. Since $T(f+g)t = tf(t) + tg(t) = t(f(t) + g(t))$ and $t(af)t = taf(t) = atf(t)$, T is a linear operator. Set

$$P_T<f,g> = \int_0^1 (Tf)(t)(Tg)(t)dt$$

$$= \int_0^1 f(t)g(t)t^2 dt . \qquad (1)$$

The integral (1) is a new inner product on $C[0,1]$.

● **PROBLEM** 14-4

What is the angle between a diagonal of a cube and one of its edges?

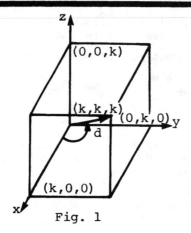

Fig. 1

Solution: Assume the cube is centered at the origin of a three dimensional coordinate system and let k be the length of an edge.

Now, find the length of the diagonal. It can be seen from Fig. 1 that the vector d (diagonal) has coordinates (k,k,k). But (k,k,k) can be written as a linear combination of the standard basis vectors for R^3 :

$$d = (k,k,k) = k(1,0,0) + k(0,1,0) + k(0,0,1).$$

$$= ku_1 + ku_2 + ku_3 .$$

Choose one of the edges for example, ku_1. The angle between the diagonal and this edge can be found by the angle formulation of the dot product. That is, if v_1, v_2 are two vectors in R^n, then

$$\cos \theta = \frac{v_1 \cdot v_2}{\|v_1\| \|v_2\|} \qquad (1)$$

where $\|v_1\|$ and $\|v_2\|$ denote the lengths of v_1 and v_2, respectively.

Thus,

$$\cos \theta = \frac{ku_1 \cdot d}{\|ku_1\| \|d\|} . \qquad (2)$$

In n-dimensional Euclidean space, the length of a vector $(x_1, x_2, \ldots, x_n)$ is given by

$$\sqrt{x_1^2 + x_2^2 + \ldots + x_n^2} = \left(\sum_{i=1}^n x_i^2 \right)^{\frac{1}{2}} .$$

Hence, $\|ku_1\| = k\|u_1\| = k(1 + 0 + 0)^{\frac{1}{2}} = k$; $\|d\| = \sqrt{k^2 + k^2 + k^2} = \sqrt{3k^2} = \sqrt{3} \, k$.

The dot product of two vectors $v, w \in R^n$ is defined as

$$v \cdot w = (v_1, v_2, \ldots, v_n) \cdot (w_1, w_2, \ldots, w_n)$$

$$= v_1 w_1 + v_2 w_2 + \ldots + v_n w_n = \sum_{i=1}^{n} v_i w_i \quad .$$

Hence, $k u_1 \cdot d = (k,0,0) \cdot (k,k,k) = k^2$.

Substitution of the above results into (2) yields

$$\cos \theta = \frac{k^2}{\sqrt{3} \, k^2} = \frac{1}{\sqrt{3}} \quad .$$

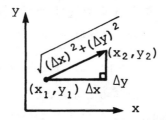

Fig. 2

The angle whose cosine is $1/\sqrt{3}$ is given by $\cos^{-1}\left(\frac{1}{\sqrt{3}}\right)$ and

$$\theta = \cos^{-1}\left(\frac{1}{\sqrt{3}}\right) = 54° \ 41' \quad .$$

● **PROBLEM** 14-5

Find the distance between the vectors u and v where i) $u = (1,7)$, $v = (6,-5)$; ii) $u = (3,-5,4)$, $v = (6,2,-1)$; iii) $u = (5,3,-2,-4,1)$. $v = (2,-1,0,-7,2)$.

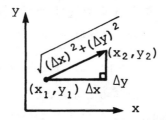

Solution: The Pythagorean theorem provides the foundation for the concept of distance in Euclidean space.
In R^2, $d(x,y) = \sqrt{(\Delta x)^2 + (\Delta y)^2}$ where $\Delta x = x_2 - x_1$ and $\Delta y = y_2 - y_1$.
Generalizing to R^n, let $x = (x_1, x_2, \ldots, x_n)$, $y = (y_1, y_2, \ldots, y_n)$ be two vectors in R^n . Then $d(x,y) = ((\Delta x_1)^2 + (\Delta x_2)^2 + \ldots + (\Delta x_n)^2)^{\frac{1}{2}}$ where
$\Delta x_1 = y_1 - x_1, \Delta x_2 = y_2 - x_2, \ldots, \Delta x_n = y_n - x_n$.
i) $d(u,v) = \sqrt{(1-6)^2 + (7+5)^2} = \sqrt{169} = 13.$
ii) $d(u,v) = \sqrt{9 + 49 + 24} = \sqrt{83}.$
iii) $d(u,v) = \sqrt{9 + 16 + 4 + 9 + 1} = \sqrt{39}.$
Note that $d(u,v) = d(v,u)$, i.e., the distance between two points is a symmetrical function of the points.

APPLICATIONS OF DOT PRODUCT

● **PROBLEM** 14-6

Compute $u \cdot v$ where i) $u = (2,-3,6)$, $v = (8,2,-3)$; ii) $u = (1,-8,0,5)$, $v = (3,6,4)$; iii) $u = (3,-5,2,1)$, $v = (4,1,-2,5)$.

Solution: In the vector space R^n, the dot product is defined as follows: for $X = (x_1, x_2, \ldots, x_n)$, $Y = (y_1, y_2, \ldots, y_n)$:

$$X \cdot Y = (x_1 y_1 + x_2 y_2 + \ldots + x_n y_n). \qquad (1)$$

Thus, to compute the dot product of two vectors from R^n, multiply corresponding components and add. The result of taking the dot product is a scalar.

i) $u \cdot v = (2, -3, 6) \cdot (8, 2, -3) = 2(8) - 3(2) + 6(-3) = -8$.

ii) $u \cdot v = (1 - 8, 0, 5) \cdot (3, 6, 4)$. But, here the dot product is not defined since $u \in R^4$ while $v \in R^3$.

iii) $u \cdot v = (3, -5, 2, 1) \cdot (4, 1, -2, 5) = 3(4) - 5(1) + 2(-2) + 1(5) = 8$.

● **PROBLEM 14-7**

Let C^n denote the vector space of complex n-tuples. Define a suitable inner product for C^n and find $u \cdot v$ and $\|u\|$ where $u = (2+3i, 4-i, 2i)$ and $v = (3-2i, 5, 4-6i)$.

Solution: In the vector space R^n, the inner or dot product is defined for $u = (u_1, u_2, \ldots, u_n)$ and $v = (v_1, v_2, \ldots, v_n)$ as

$$u \cdot v = u_1 v_1 + u_2 v_2 + \ldots + u_n v_n = \sum_{i=1}^{n} u_i v_i.$$

This inner product function satisfies the following properties:

i) $u \cdot v = v \cdot u$
ii) $(au + bv) \cdot w = a(u \cdot w) + b(v \cdot w)$ for $w \in R^n$ and a, b scalars from the field.
iii) $(u \cdot u) \geq 0$ and $(u \cdot u) = 0 \Leftrightarrow u = 0$.

Since the dot product is used for defining lengths and angles, an inner product on C^n should, if it is to be useful, enjoy the same properties.

But, here we run into difficulties. For example, let $ix \in C$. Then,

$$\|ix\| = (ix \cdot ix)^{\frac{1}{2}} = [(ix)(ix)]^{\frac{1}{2}}$$
$$= (i^2 x^2)^{\frac{1}{2}} = (-x^2)^{\frac{1}{2}} = -\|x\| \qquad (i^2 = -1).$$

Hence, for $x \neq 0$, $\|ix\| > 0$ implies $\|x\| < 0$ and $\|ix\| < 0$ implies $\|x\| > 0$. But, by property iii) above, length is nonnegative. Hence, we redefine the dot product for complex vector spaces. Let $u, v \in C$. Then,

$$(u \cdot v) = u\bar{v}$$

where the bar denotes the complex conjugate, i.e., $\overline{a+ib} = a - ib$. With this definition of the dot product,

$$\|ix\| = [(ix \cdot ix)]^{\frac{1}{2}} = [(ix)(-ix)]^{\frac{1}{2}} = (-i^2 x^2)^{\frac{1}{2}}$$
$$= \|x\| \geq 0.$$

But now the property of symmetry, (i), is lost since $(u \cdot v) = \overline{(v \cdot u)} \neq (v \cdot u)$. This new inner product function satisfies the following conditions:

i) $(u \cdot v) = \overline{(v \cdot u)}$
ii) $(au+bv) \cdot w = a(u \cdot w) + b(v \cdot w)$
iii) $(u \cdot u) \geq 0$; $(u \cdot u) = 0 \Leftrightarrow u = 0$.

In C^n, i.e., the set of all n-tuples of complex numbers, if $u = (z_1, z_2, \ldots, z_n)$ and $v = (w_1, w_2, \ldots, w_n)$, then

$$(u \cdot v) = z_1 \bar{w}_1 + z_2 \bar{w}_2 + \ldots + z_n \bar{w}_n = \sum_{i=1}^{n} z_i \bar{w}_i \ , \quad \text{and}$$

$$\|u\| = (u \cdot u)^{\frac{1}{2}} = \sqrt{z_1 \bar{z}_1 + z_2 \bar{z}_2 + \ldots + z_n \bar{z}_n} = \left(\sum_{i=1}^{n} |z_i|^2 \right)^{\frac{1}{2}} \ .$$

Turning to the given problem,

$$u = (2+3i, \ 4-i, 2i) \quad \text{and} \quad \bar{v} = (3+2i, 5, 4+6i).$$

$$(u \cdot v) = (2+3i)(3+2i) + (4-i)(5) + 2i(4+6i).$$

The product of two complex numbers $(a+ib)(c+id)$ is defined as $ac - bd + i(ad + bc)$. Thus,

$$(u \cdot v) = (6-6) + i(13) + 20 - i5 + (i8 - 12).$$

Addition of complex numbers $(a+ib) + (c+id)$ is defined as $(a+c) + i(b+d)$. Thus, $(u \cdot v) = 8 + i16$.

Next, find the norm of u.

$$\|u\| = \sqrt{u \cdot u} = [(2+3i)(2-3i) + (4-i)(4+i) + 2i(-2i)]^{\frac{1}{2}}$$
$$= [13 + i(0) + 17 + 4]^{\frac{1}{2}} = \sqrt{34} \doteq 5.83.$$

● **PROBLEM** 14-8

Show that the dot product can be derived from the theorem of Pythagoras and the law of cosines.

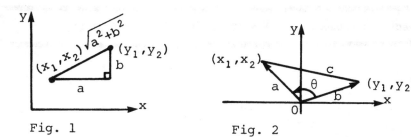

Fig. 1   Fig. 2

Solution: The Pythagorean theorem is used to derive the notion of distance between two points in the plane while the law of cosines enables angle measurements to be made.

Definition: $d(\alpha, \beta) = \sqrt{\alpha^2 + \beta^2}$, the distance between the origin and the point (α, β).

Notice that $a = |x_1 - y_1|$ and $b = |x_2 - y_2|$.

From Fig. 1, the distance between points (x_1, x_2) and (y_1, y_2) is

$$((x_1 - y_1)^2 + (x_2 - y_2)^2)^{\frac{1}{2}} = d((x_1 - y_1), \ (x_2 - y_2)).$$

So, any distance in R^2 can be expressed in terms of the function d. In Fig. 2, the distance of (x_1, x_2) from the origin is given by $a = d(x_1, x_2)$ while the distance of (y_1, y_2) from 0 is given by $b = d(y_1, y_2)$. The distance from (x_1, x_2) to (y_1, y_2) is given by c. Then, the law of cosines states that

$$\cos \theta = \frac{a^2 + b^2 - c^2}{2ab} \tag{1}$$

383

Since $a^2 + b^2 - c^2 = x_1^2 + x_2^2 + y_1^2 + y_2^2 - (x_1-y_1)^2 - (x_2-y_2)^2$

$$= 2(x_1y_1 + x_2y_2) \text{ , (1) may be rewritten as}$$

$$\cos \theta = \frac{x_1y_1 + x_2y_2}{d(x_1,x_2)d(y_1,y_2)} \qquad (2)$$

The dot product of two vectors $u = (x_1,x_2)$ and $v = (y_1,y_2)$ is defined as; $u \cdot v = x_1y_1 + x_2y_2$. Thus,

$$\cos \theta = \frac{u \cdot v}{d(x_1,x_2)d(y_1,y_2)}$$

Now notice that $d(x_1,x_2) = \sqrt{x_1^2 + x_2^2} = \sqrt{(x_1,x_2) \cdot (x_1,x_2)} = \sqrt{u \cdot u} = (u \cdot u)^{\frac{1}{2}}$.
Similarly, $d(y_1,y_2) = (v \cdot v)^{\frac{1}{2}}$. Hence,

$$\cos \theta = \frac{u \cdot v}{(u \cdot u)^{\frac{1}{2}}(v \cdot v)^{\frac{1}{2}}} \qquad (3)$$

From (2), the dot product can be expressed in terms of the angle between two vectors and the distance between them:

$$u \cdot v = x_1y_1 + x_2y_2 = (\cos \theta)(d(x_1,x_2)d(y_1,y_2)) \text{ .}$$

From (3), the length or norm of a vector can be expressed through the dot product.

● **PROBLEM** 14-9

Find a vector orthogonal to $A = (2,1,-1)$ and $B = (1,2,1)$.

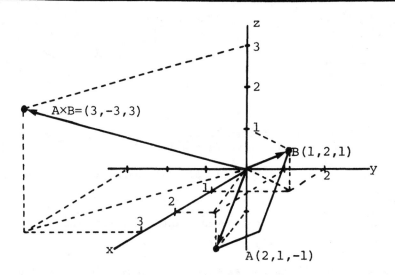

Solution: A vector $V = (v_1,v_2,v_3)$ is said to be orthogonal to A and B if $A \cdot V = 0$ and $B \cdot V = 0$. In other words,

$$(v_1,v_2,v_3) \cdot (2,1,-1) = 2v_1 + v_2 - v_3 = 0 \qquad (1)$$

and $\qquad (v_1,v_2,v_3) \cdot (1,2,1) = v_1 + 2v_2 + v_3 = 0 \qquad (2)$

Adding (1) and (2)

$$3v_1 + 3v_2 = 0 \text{ .}$$

Therefore $v_1 = -v_2$. Let $v_2 = -3$. From $v_1 = -v_2$ and $v_3 = -(v_1 + 2v_2)$, the result is $v_1 = 3$ and $v_3 = -(3-6) = 3$. Therefore, the vector $(3,-3,3)$ is orthogonal to $A = (2,1,-1)$ and $B = (1,2,1)$.

● **PROBLEM** 14-10

Let $(x_1, x_2, \ldots, x_n)$, $(y_1, y_2, \ldots, y_n)$, $(z_1, z_2, \ldots, z_n)$ be three vectors in R^n. Verify the following properties of the dot product using these vectors:

a) $X \cdot X \geq 0$; $X \cdot X = 0$ if and only if $X = 0$

b) $X \cdot Y = Y \cdot X$.

c) $(X+Y) \cdot Z = X \cdot Y + X \cdot Z$.

d) $(cX) \cdot Y = X \cdot (cY) = c(X \cdot Y)$

Solution: First define the dot product. If $X = (x_1, x_2, \ldots, x_n)$ and $Y = (y_1, y_2, \ldots, y_n)$ are vectors in R^n, then their dot product is defined by: $X \cdot Y = x_1 y_1 + x_2 y_2 + \ldots + x_n y_n$. The dot product is also known as the inner product since it does satisfy properties a) – d), the properties required of an inner product.

a) $X \cdot X = (x_1, x_2, \ldots, x_n) \cdot (x_1, x_2, \ldots, x_n) = x_1^2 + x_2^2 + \ldots + x_n^2$

$$= \sum_{i=1}^{n} x_i^2 \geq 0 \ .$$

If $X = 0$ $(X = (0,0,0,\ldots,0))$ then $X \cdot X = 0$, and if $X \neq 0$, $\exists x_i \neq 0$. This implies $X \cdot X = \sum_{i=1}^{n} x_i^2 \neq 0$.

b) $X \cdot Y = (x_1, x_2, \ldots, x_n) \cdot (y_1, y_2, \ldots, y_n) = x_1 y_1 + x_2 y_2 + \ldots + x_n y_n$

$= y_1 x_1 + y_2 x_2 + \ldots + y_n x_n = (y_1, y_2, \ldots, y_n) \cdot (x_1 x_2, \ldots, x_n) = Y \cdot X$.

c) $(X+Y) \cdot Z = [(x_1, x_2, \ldots, x_n) + (y_1, y_2, \ldots, y_n)] \cdot (z_1, z_2, \ldots, z_n)$

$= (x_1 + y_1, \ x_2 + y_2, \ldots, x_n + y_n) \cdot (z_1, z_2, \ldots, z_n)$

$= [(x_1 + y_1) z_1 + (x_2 + y_2) z_2 + \ldots + (x_n + y_n) z_n]$

$= [(x_1 z_1 + y_1 z_1) + (x_2 z_2 + y_2 z_2) + \ldots + (x_n z_n + y_n z_n)]$

(by the associative property of real numbers)

$= (x_1 z_1 + x_2 z_2 + \ldots + x_n z_n + y_1 z_1 + y_2 z_2 + \ldots + y_n z_n)$

$= (x_1 z_1 + x_2 z_2 + \ldots + x_n z_n) + (y_1 z_1 + y_2 z_2 + \ldots + y_n z_n)$

$= X \cdot Z + Y \cdot Z$.

d) Here, c is a scalar from the field over which R^n is defined.

$(cX) \cdot Y = [c(x_1, x_2, \ldots, x_n)] \cdot (y_1, y_2, \ldots, y_n)$

$= (cx_1, cx_2, \ldots, cx_n) \cdot (y_1, y_2, \ldots, y_n)$

$= cx_1 y_1 + cx_2 y_2 + \ldots + cx_n y_n.$ (1)

From (1), first obtain:

$$(1) = x_1 cy_1 + x_2 cy_2 + \ldots + x_n cy_n = X \cdot (cY) .$$

Returning to (1):

$$(1) = cx_1 y_1 + cx_2 y_2 + \ldots + cx_n y_n = c(X \cdot Y) .$$

The dot product is used to define the notions of length and distance in R^n. Thus, $\|X\| = (X \cdot X)^{\frac{1}{2}}$ and $\|X-Y\| = [(X-Y) \cdot (X-Y)]^{\frac{1}{2}}$.

The dot product is also useful for defining the angle between two vectors:

$$\cos \theta = \frac{X \cdot Y}{\|X\| \ \|Y\|} = \frac{(X \cdot Y)}{(X \cdot X)^{\frac{1}{2}} (Y \cdot Y)^{\frac{1}{2}}} .$$

● **PROBLEM** 14-11

Let $u = (4,0,1,2,0)$, $v = (2,1,-1,1,1)$. Find:
a) $u \cdot v$; b) $\|u\|, \|v\|$; c) the projection of u onto v and the projection of u orthogonal to v.

Solution: a) The dot product of two vectors $u = (u_1, \ldots, u_n)$, $v = (v_1, \ldots, v_n)$ in R^n is defined to be

$$u \cdot v = u_1 v_1 + u_2 v_2 + \ldots + u_n v_n = \sum_{i=1}^{n} u_i v_i .$$

Hence, for the given vectors u and v,

$$u \cdot v = (4,0,1,2,0) \cdot (2,1,-1,1,1)$$
$$= 4(2) + 0(1) + 1(-1) + 2(1) + 0(1) = 9 .$$

b) The dot product can be used to find the length of a vector in R^n. By a generalization of the Pythagorean theorem,

$$\|u\| = (u_1^2 + u_2^2 + \ldots + u_n^2)^{\frac{1}{2}} = \left[\sum_{i=1}^{n} u_i^2 \right]^{\frac{1}{2}} .$$

But, $(u \cdot u)^{\frac{1}{2}} = [(u_1, u_2, \ldots, u_n) \cdot (u_1, u_2, \ldots, u_n)]^{\frac{1}{2}}$

$$= (u_1^2 + u_2^2 + \ldots + u_n^2)^{\frac{1}{2}} = \left[\sum_{i=1}^{n} u_i^2 \right]^{\frac{1}{2}} .$$

Hence, $\|u\| = (u \cdot u)^{\frac{1}{2}}$. In this case,

$$\|u\| = [(4,0,1,2,0) \cdot (4,0,1,2,0)]^{\frac{1}{2}} = \sqrt{16 + 1 + 4} = \sqrt{21}$$

$$\|v\| = (v \cdot v)^{\frac{1}{2}} = [(2,1,-1,1,1) \cdot (2,1,-1,1,1)]^{\frac{1}{2}} = \sqrt{4 + 1 + 1 + 1 + 1} =$$
$$= \sqrt{8} .$$

c) The dot product can be used to decompose a vector into a sum of two perpendicular vectors. Two vectors are orthogonal (or perpendicular) if their dot product equals zero. Given two vectors u and v in R^n, use the dot product to find w_1 and w_2 such that

$$u = w_1 + w_2 \tag{1}$$

where w_1 is a scalar multiple of v and w_2 is orthogonal to v. w_1 is the projection of u on v. To show (1), let $w_1 = kv$. Then,

$$u = kv + w_2 .$$

Since $w_2 \cdot v = 0$, $u \cdot v = (kv + w_2) \cdot v = kv \cdot v + w_2 \cdot v = k \|v\|^2$. Hence, $k = \dfrac{u \cdot v}{\|v\|^2}$

and $w_1 = \dfrac{u \cdot v}{\|v\|^2} \, v$. Then, $w_2 = u - \dfrac{u \cdot v}{\|v\|^2} \, v$. The vector w_2 is the pro-

jection of u orthogonal to v .

For the given vectors,

$$w_1 = \dfrac{u \cdot v}{\|v\|^2} \, v = \dfrac{9}{8}(2,1,-1,1,1) = (\tfrac{18}{8}, \tfrac{9}{8}, \tfrac{-9}{8}, \tfrac{9}{8}, \tfrac{9}{8}) \ .$$

$$v_2 = u - w_1 = (4,0,1,2,0) - (\tfrac{18}{8}, \tfrac{9}{8}, \tfrac{-9}{8}, \tfrac{9}{8}, \tfrac{9}{8})$$

$$= (\tfrac{14}{8}, \tfrac{-9}{8}, \tfrac{17}{8}, \tfrac{7}{8}, \tfrac{-9}{8}) \ .$$

Since w_1 and w_2 are orthogonal, $w_1 \cdot w_2 = 0$. To check this result explicitly,

$$w_1 \cdot w_2 = (\tfrac{18}{8}, \tfrac{9}{8}, \tfrac{-9}{8}, \tfrac{9}{8}, \tfrac{9}{8}) \cdot (\tfrac{14}{8}, \tfrac{-9}{8}, \tfrac{17}{8}, \tfrac{7}{8}, \tfrac{-9}{8})$$

$$= \dfrac{1}{64}(18(14) - 81 - 9(17) + 9(7) - 81) = 0 \ .$$

● **PROBLEM** 14-12

Let $u = (2,-1,3)$ and $v = (4,-1,2)$ be vectors in R^3 . Find the orthogonal projection of u on v and the component of u orthogonal to v .

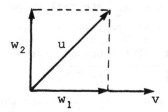

Solution: Two vectors in R^n are perpendicular if $u \cdot v = 0$. Using the dot product, it is possible to decompose a vector into the sum of two perpendicular vectors. That is, given u and v, u can be written as

$$u = w_1 + w_2$$

where w_1 is a scalar multiple of v and w_2 is perpendicular to v.

To obtain w_1 and w_2 , reason as follows: Since w_1 is a scalar multiple of v, write it in the form $w_1 = kv$. Thus,

$$u = w_1 + w_2 = kv + w_2 \ .$$

Since w_2 is perpendicular to v, we have $w_2 \cdot v = 0$. Using this and the properties of the dot product (an inner product), we can find k.

$$u \cdot v = (kv + w_2) \cdot v = (kv \cdot v) + (w_2 \cdot v) = (kv \cdot v) + 0 =$$

$$= k\|v\|^2 \ .$$

This implies $k = \dfrac{u \cdot v}{\|v\|^2}$, and, hence, $w_1 = \dfrac{u \cdot v}{\|v\|^2} \, v$. Now, $w_2 = u - \dfrac{u \cdot v}{\|v\|^2} \, v$.

The vector w_1 is the orthogonal projection of u on v, while the vector w_2 is the component of u orthogonal to v .

For the given vectors,

$$u \cdot v = (2)(4) + (-1)(-1) + 3(2) = 15.$$

$$\|v\|^2 = 4^2 + (-1)^2 + 2^2 = 21 \ .$$

Hence, the orthogonal projection of u on v is $w_1 = \frac{u \cdot v}{\|v\|^2} v = \frac{15}{21}(4,-1,2)$
$= (\frac{20}{7}, \frac{-5}{7}, \frac{10}{7})$. The component of u orthogonal to v is

$$w_2 = u - w_1 = (2,-1,3) - (\frac{20}{7}, \frac{-5}{7}, \frac{10}{7}) = (\frac{-6}{7}, \frac{-2}{7}, \frac{11}{7}) .$$

NORMS OF VECTORS

● **PROBLEM** 14-13

Distinguish between n-dimensional Euclidean space and the vector space of n-tuples.

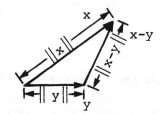

Solution: Let R^n denote the vector space of n-tuples of the form $x = (x_1, x_2, \ldots, x_n)$ where the x_i are real numbers. The axioms for a vector space do not provide any definition of distance. If we think of the length of a vector x in R^n as a function of its distance from the zero vector 0, then, by the Pythagorean theorem,

$$d(x,0) = \sqrt{x_1^2 + x_2^2 + \ldots + x_n^2} . \qquad (1)$$

Instead of the origin, or zero vector, consider the distance between any two vectors in R^n .

From the figure and using the Pythagorean theorem,

$$d(x,y) = \sqrt{(y_1-x_1)^2 + (y_2-x_2)^2 + \ldots + (y_n-x_n)^2}$$

where $x = (x_1,x_2,\ldots,x_n)$, $y = (y_1,y_2,\ldots,y_n)$. The distance function in combination with the vector space R^n results in Euclidean n-dimensional space.

The distance function above is also known as the Euclidean norm. A norm is a nonnegative valued function of a vector $x \in V$, where V is a vector space, denoted by $\|x\|$ such that

a) $\|x\| > 0$, $x \neq 0$ and $\|x\| = 0 \Leftrightarrow x = 0$.

b) $\|kx\| = |k| \|x\|$ for any scalar k ,

c) $\|x+y\| \leq \|x\| + \|y\|$ for $y \in V$.

Note that we can define other norms on R^n . The resulting structure will then no longer be Euclidean space. For example, $\|x\| = \max_i |x_i|$, where $|x_i|$ denotes the absolute value of x_i, is also a norm on R^n.

● **PROBLEM** 14-14

Consider the vector space $C[0,1]$ of all continuous functions defined on $[0,1]$. If $f \in C[0,1]$, show that $(\int_0^1 f^2(x)dx)^{\frac{1}{2}}$ defines a norm on all elements of this vector space.

Solution: Since f is continuous, f^2 is continuous. If a function is continuous, it is integrable; thus, $(\int_0^1 f^2(x)dx)^{\frac{1}{2}} = \|f\|$ is well-defined.

Let V be a vector space. A norm on V is a nonnegative number associated with every $v \in V$, denoted $\|v\|$, such that

a) $\|v\| \geq 0$ and $\|v\| = 0 \Leftrightarrow v = 0$,

b) $\|kv\| = |k| \|v\|$ for k, a scalar,

c) $\|v+w\| \leq \|v\| + \|w\|$, for $w \in V$.

It is necessary to show that $(\int_0^1 f^2(x)dx)^{\frac{1}{2}}$ satisfies conditions

a) - c) above.

a) If $f(x) \in C[0,1]$, then $f^2(x) \geq 0$ on $[0,1]$. But this implies that $(\int_0^1 f^2(x)dx)^{\frac{1}{2}} \geq 0$. If $\|f\| = (\int_0^1 f^2(x)dx)^{\frac{1}{2}} = 0$, then $f^2(x) = 0$.

This suggests that $f(x) = 0$ on $[0,1]$ since a nonnegative function continuous over the interval $[0,1]$ can have zero integral over $[0,1]$ only if that function is identically zero on $[0,1]$. So, $\|f\| = 0$ if and only if $f \equiv 0$.

b) Let k be a scalar from the field over which $C[0,1]$ is defined. Then,

$$\|kf\| = \int_0^1 [[kf(x)]^2 \ dx]^{\frac{1}{2}} = |k| [\int_0^1 f^2(x)dx]^{\frac{1}{2}} = |k| \|f\|.$$

c) First, show that $\|f+g\|$, $f,g \in C[0,1]$ is well-defined. Now,

$$\|f+g\| = \int_0^1 ([f(x) + g(x)]^2 \ dx)^{\frac{1}{2}}$$

exists since sums and squares of continuous functions in $C[0,1]$ are also in $C[0,1]$ and, therefore, integrable over $[0,1]$. Note also that

$fg = \frac{1}{2}[(f+g)^2 - f^2 - g^2]$ is also in $C[0,1]$ and integrable over $[0,1]$. It is now possible to show $\|f+g\| \leq \|f\| + \|g\|$.

$$\|f+g\| \leq \|f\| + \|g\| \Leftrightarrow \|f+g\|^2 \leq (\|f\| + \|g\|)^2 \ . \tag{1}$$

But,
$$\|f+g\|^2 = \int_0^1 [f(x) + g(x)]^2 \ dx$$

$$= \int_0^1 f^2(x)dx + 2\int_0^1 f(x)g(x)dx + \int_0^1 g^2(x)dx \tag{2}$$

and
$$(\|f\| + \|g\|)^2 = \|f\|^2 + 2\|f\| \|g\| + \|g\|^2$$

$$= \int_0^1 f^2(x)dx + 2(\int_0^1 f^2(x)dx)^{\frac{1}{2}}(\int_0^1 g^2(x)dx)^{\frac{1}{2}}$$

$$+ \int_0^1 g^2(x)dx \ . \tag{3}$$

Comparing (2) and (3), we see that (1) can hold if and only if

$$\int_0^1 f(x)g(x)dx \leq (\int_0^1 f^2(x)dx)^{\frac{1}{2}} (\int_0^1 g^2(x)dx)^{\frac{1}{2}} \tag{4}$$

By the properties of the absolute value functions,

$$\int_0^1 f(x)g(x)dx \leq \int_0^1 |f(x)| \cdot |g(x)| dx \ .$$

It can be proved that $\int_0^1 |f(x)| \cdot |g(x)| dx \le \|f\|\|g\|$.

Let λ be a real variable and form $\int_0^1 [|f(x)| + \lambda|g(x)|]^2 dx \ge 0$. But,

$$\int_0^1 [|f(x)| + \lambda|g(x)|]^2 dx = \lambda^2 \int_0^1 g^2(x) dx + 2\lambda \int_0^1 |f(x)||g(x)| dx + \int_0^1 f^2(x) dx$$

is a nonnegative quadratic polynomial in λ . Hence, it has no real roots and its discriminant is nonpositive; i.e.,

$$4[\int_0^1 |f(x)| \ |g(x)| dx]^2 - 4\int_0^1 g^2(x) dx \int_0^1 f^2(x) dx \le 0$$

which implies

$$\int_0^1 |f(x)||g(x)| dx \le (\int_0^1 f^2(x) dx)^{\frac{1}{2}} (\int_0^1 g^2(x) dx)^{\frac{1}{2}} ,$$

as was to be shown.

Thus, $(\int_0^1 f^2(x) dx)^{\frac{1}{2}}$ defines a norm on $C[0,1]$ known as the Euclidean norm.

● **PROBLEM** 14-15

Let $C[0,1]$ be the vector space of all real-valued continuous functions defined on $[0,1]$. Consider the polynomials $2x$ and $1 - 2x^2$. Show that they are orthogonal with respect to the Euclidean norm. Then find their lengths.

Solution: The inner product $<f,g> = \int_0^1 f(x) g(x) dx$ makes the space $C[0,1]$ a Euclidean space. Using this inner product, it is possible to find the length or norm of a vector: $\|f\| = [\int_0^1 [f(x)]^2 dx]^{\frac{1}{2}}$. This norm is analagous to the norm of a vector $u = (u_1, u_2, \ldots, u_n)$ in R^n ; $\|u\| = (\Sigma u_i^2)^{\frac{1}{2}}$. Two vectors are orthogonal if their inner product equals zero, i.e., $f(x)$ and $g(x)$ are orthogonal if $\int_0^1 f(x) g(x) dx = 0$. The inner product of $f(x) = 2x$ and $g(x) = 1 - 2x^2$

$$= \int_0^1 2x(1 - 2x^2) dx = -4\int_0^1 x^3 dx + 2\int_0^1 x \, dx$$

$$= -x^4 \Big|_0^1 + x^2 \Big|_0^1 = -1 + 1 = 0 .$$

Thus, the two given functions are orthogonal. The length of $2x$ is given by

$$\|2x\| = <2x,2x>^{\frac{1}{2}} = [\int_0^1 4x^2 \, dx]^{\frac{1}{2}} = [\frac{4}{3} x^3 \Big|_0^1]^{\frac{1}{2}} = \frac{2}{\sqrt{3}} .$$

Similarly,

$$\|1 - 2x^2\| = [\int_0^1 (4x^4 - 4x^2 + 1) dx]^{\frac{1}{2}} = [\frac{4}{5} x^5 - \frac{4}{3} x^3 + x \Big|_0^1]^{\frac{1}{2}} = \sqrt{7/15} .$$

THE CROSS-PRODUCT

● **PROBLEM** 14-16

1.)   Find $A \times B$ where $A = (1,2,-2)$ and $B = (3,0,1)$.

2.) Verify directly that $A \cdot (A \times B) = 0$ and $B \cdot (A \times B) = 0$ where
$A = (1,2,-2)$ and $B = (3,0,1)$.
3.) Show that $A \cdot (A \times B) = 0$ and $B \cdot (A \times B) = 0$ where A, B are any
vectors in R^3.

Solution: If $A = (a_1, a_2, a_3) \neq 0$ and $B = (b_1, b_2, b_3) \neq 0$ are vectors,
the vector $A \times B$, where
$$A \times B = (a_2 b_3 - a_3 b_2, a_3 b_1 - a_1 b_3, a_1 b_2 - a_2 b_1),$$
is called the cross product or vector product of A and B.
If $A = (a_1, a_2, a_3)$ and $B = (b_1, b_2, b_3)$ are vectors in R^3, then the
dot product is defined by: $A \cdot B = a_1 b_1 + a_2 b_2 + a_3 b_3$.

1.) $A = (1,2,-2)$ and $B = (3,0,1)$. By the definition of the cross product above, $A \times B = (1,2,-2) \times (3,0,1) = (2(1) - (-2)(0), (-2)(3) - (1)(1),$
$(1)(0) - 2(3)) = (2,-7,-6)$.

2.) In 1) we found $A \times B = (2,-7,-6)$. Therefore, the dot product $A \cdot (A \times B) =$
$(1,2,-2) \cdot (2,-7,-6) = 2 - 14 + 12 = 0$, and the dot product $B \cdot (A \times B) =$
$(3,0,1) \cdot (2,-7,-6) = 6 - 0 - 6 = 0$.

3.) By the definition of cross product, $A \times B = (a_2 b_3 - a_3 b_2, a_3 b_1 - a_1 b_3, a_1 b_2 - a_2 b_1)$
if $A = (a_1, a_2, a_3)$ and $B = (b_1, b_2, b_3)$. By the definition of dot product,

$A \cdot (A \times B) = (a_1, a_2, a_3) \cdot (a_2 b_3 - a_3 b_2, a_3 b_1 - a_1 b_3, a_1 b_2 - a_2 b_1) =$

$\qquad = a_1(a_2 b_3 - a_3 b_2) + a_2(a_3 b_1 - a_1 b_3) + a_3(a_1 b_2 - a_2 b_1) = 0$
$B \cdot (A \times B) = (b_1, b_2, b_3) \cdot (a_2 b_3 - a_3 b_2, a_3 b_1 - a_1 b_3, a_1 b_2 - a_2 b_1)$
$\qquad = b_1(a_2 b_3 - a_3 b_2) + b_2(a_3 b_1 - a_1 b_3) + b_3(a_1 b_2 - a_2 b_1) = 0$.

● **PROBLEM** 14-17

Let $X = 2i + j + 2k$ and $Y = 3i - j - 3k$. Find $X \times Y$.

Solution: Let $U = u_1 i + u_2 j + u_3 k$ and $V = v_1 i + v_2 j + v_3 k$ where i, j
and k are the unit vectors $(1,0,0)$, $(0,1,0)$ and $(0,0,1)$, respectively.
Then, their cross-product is the vector $U \times V$ defined by $U \times V =$
$(u_2 v_3 - u_3 v_2)i + (u_3 v_1 - u_1 v_3)j + (u_1 v_2 - u_2 v_1)k$.
For ease of computation, it is convenient to view the cross-product
as analogous to the determinant

$$U \times V = \begin{vmatrix} i & j & k \\ u_1 & u_2 & u_3 \\ v_1 & v_2 & v_3 \end{vmatrix} . \qquad (1)$$

We can find $X \times Y$ using (1). Thus,

$$X \times Y = \begin{vmatrix} i & j & k \\ 2 & 1 & 2 \\ 3 & -1 & -3 \end{vmatrix} = \begin{array}{l} i(-3+2) - j(-6-6) \\ + k(-2-3) \end{array}$$

$= -i + 12j - 5k$.

Find the area of the triangle determined by the points $P_1(2,2,0)$, $P_2(-1,0,1)$ and $P_3(0,4,3)$ by using the cross-product.

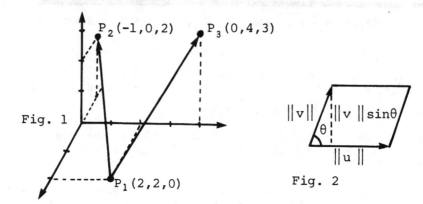

Fig. 1

Fig. 2

Solution: Fig. 1. The triangle whose area is to be found is given in the figure above.

To relate the area of a triangle to the notion of a cross-product, use the following argument:

Lagrange's Identity states that

$$\|u \times v\|^2 = \|u\|^2 \|v\|^2 - (u \cdot v)^2 \qquad (1)$$

where $\|w\|$ denotes the length of a vector in R^3. Thus, if $w = (w_1, w_2, w_3)$, $\|w\| = [\sum_{i=1}^{3} w_i^2]^{\frac{1}{2}}$. $(u \cdot v)$ is the dot product, $(u \cdot v) = \|u\| \|v\| \cos \theta$ where θ is the angle between u and v. Equation (1) may be rewritten as

$$\|u \times v\|^2 = \|u\|^2 \|v\|^2 - \|u\|^2 \|v\|^2 \cos^2 \theta$$

$$= \|u^2\| \|v^2\| (1 - \cos^2 \theta)$$

$$= \|u^2\| \|v^2\| (\sin^2 \theta)$$

and, hence,

$$\|u \times v\| = \|u\| \|v\| \sin \theta \qquad (2)$$

From Fig.2, observe that $\|v\| \sin \theta$ is the altitude of the parallelogram determined by u and v. The base of the parallelogram is u. Since

$$\text{Area of parallelogram} = (\text{base})(\text{altitude}),$$

it can be seen from (2) that the norm of $u \times v$ equals the area of the parallelogram determined by u and v.

Utilize this result to solve the given problem. From Fig. 1, the area of the triangle is $\frac{1}{2}$ the area of the parallelogram formed by the vectors $\overrightarrow{P_1 P_2}$ and $\overrightarrow{P_1 P_3}$. $\overrightarrow{P_1 P_2}$ is given by $P_2(-1,0,2) - P_1(2,2,0) = (-3,-2,2)$, and $\overrightarrow{P_1 P_3} = P_3(0,4,3) - P_1(2,2,0) = (-2,2,3)$. In R^3, the cross-product is defined as follows:

$$X \times Y = \begin{vmatrix} e_1 & e_2 & e_3 \\ x_1 & x_2 & x_3 \\ y_1 & y_2 & y_3 \end{vmatrix} \qquad \text{where}$$

$X = (x_1, x_2, x_3)$, $Y = (y_1, y_2, y_3)$ and $\{e_1, e_2, e_3\}$ is the standard basis

for R^3. Now, $\overrightarrow{P_1 P_2} \times \overrightarrow{P_1 P_3} = (-3, -2, 2) \times (-2, 2, 3) =$

$$
\begin{vmatrix} e_1 & e_2 & e_3 \\ -3 & -2 & 2 \\ -2 & 2 & 3 \end{vmatrix} = e_1 \begin{vmatrix} -2 & 2 \\ 2 & 3 \end{vmatrix} - e_2 \begin{vmatrix} -3 & 2 \\ -2 & 3 \end{vmatrix} +
$$

$$
e_3 \begin{vmatrix} -3 & -2 \\ -2 & 2 \end{vmatrix}
$$

$= -10e_1 + 5e_2 - 10e_3$. This is $(-10, 5, -10)$ when written as a point

in R^3. Since $\|(-10, 5, -10)\| = \sqrt{100 + 25 + 100} = 15$

$$
\text{Area (Triangle)} = \tfrac{1}{2} \|\overrightarrow{P_1 P_2} \times \overrightarrow{P_1 P_3}\|
$$

$$
= \tfrac{1}{2}(15) = 7.5 \ .
$$

GRAM-SCHMIDT ORTHOGONALIZATION PROCESS

● **PROBLEM** 14-19

Let $S = \text{Sp}\{(1, 0, 1), (0, 2, 1)\}$. Then S is a subspace of R^3. Find the orthogonal complement of S.

Solution: Let V be a vector space and W a subspace of V. A vector $v \in V$ is orthogonal to W if v is orthogonal to each vector in W. Then, the orthogonal complement of W in V is the set of all vectors of V that are orthogonal to W.

To find the orthogonal complement, first find a set that spans the subspace W. Then, a vector v is orthogonal to the subspace W if v is orthogonal to each vector in the spanning set.

Two vectors that span S are the given vectors, namely, $w_1 = (1, 0, 1)$ and $w_2 = (0, 2, 1)$. Let $V = (x_1, x_2, x_3) \in R^3$. Then, the condition that v be orthogonal to w_1 and w_2 is given by the solution set to the two homogeneous equations:

$$
v \cdot w_1 = (x_1, x_2, x_3) \cdot (1, 0, 1) = 0
$$

$$
v \cdot w_2 = (x_1, x_2, x_3) \cdot (0, 2, 1) = 0 \tag{1}
$$

or,

$$
x_1 + x_3 = 0
$$

$$
2x_2 + x_3 = 0
$$

or,

$$
x_1 = -x_3
$$

$$
2x_2 = -x_3 \ .
$$

Let $x_3 = -2$. Then, $x_2 = 1$ and $x_1 = 2$. Thus, the orthogonal complement of S in R^3 is the set

$$
S^\perp = \text{Sp}\{(2, -1, -2)\}
$$ since the system (1) has only one linearly independent solution. Note: $\dim S^\perp = 1$, and, since w_1 and w_2 are

independent, dim S = 2. Consequently, $\dim(S^1 \oplus S) = 3$. The set $\{(1,0,1),(0,2,1),(2,-1,-2)\}$ is a basis for R^3.

● **PROBLEM** 14-20

Show that the functions 1, cos πx, cos 2πx,...,cos nπx, form an orthogonal set over [0,1]. Then normalize them to obtain an orthogonal set.

Solution: A collection of vectors is an orthogonal set if the vectors are mutually perpendicular. If, in addition, the vectors are all of unit length, the set is orthonormal.

Consider the vector space V = C[0,1] of all real-valued continuous functions defined on the interval [0,1]. The functions 1, cos πx, cos 2πx, ...,cos nπx belong to this space. In order to define perpendicularity between elements of this space, it is necessary to define an inner product on C[0,1]. Let f,g ∈ C[0,1]. Then, the inner product, denoted by $\langle f,g \rangle$, is a real-valued function defined on V x V such that

a) $\langle kf,g \rangle = k\langle f,g \rangle$ for k, a scalar .

b) $\langle f+g,h \rangle = \langle f,g \rangle + \langle g,h \rangle$ for h∈V

c) $\langle f,g \rangle = \langle g,f \rangle$.

d) $\langle f,f \rangle > 0$, f ≠ 0; $\langle f,f \rangle = 0$ if and only if f = 0 .

A suitable inner product on C[0,1] is

$$\langle f,g \rangle = \int_0^1 f(x)g(x)dx .$$

To show that the given functions are pairwise orthogonal, it must be shown:

$$\langle 1, \cos m\pi x \rangle = 0;$$

$$\langle \cos m\pi x, \cos n\pi x \rangle = 0, \quad m \neq n ,$$

m an integer, n an integer. Now,

$$\langle 1, \cos m\pi x \rangle = \int_0^1 \cos m\pi t \, dt$$

$$= \frac{1}{m\pi} \sin m\pi t \Big|_0^1 = 0 .$$

Since $\langle f,g \rangle = 0$ implies f and g are orthogonal, the constant function 1 is orthogonal to all the other members in the set.

$$\langle \cos m\pi x, \cos n\pi x \rangle = \int_0^1 \cos m\pi t \cos n\pi t \, dt . \qquad (1)$$

But, $\cos m\pi x \cos n\pi x = \frac{1}{2}[\cos(m+n)\pi x + \cos(m-n)\pi x]$. Hence, (1) becomes

$$\frac{1}{2}\int_0^1 \cos(m+n)\pi t \, dt + \frac{1}{2}\int_0^1 \cos(m-n)\pi t \, dt$$

$$= \frac{\sin(m+n)\pi t}{2(m+n)\pi} \Big|_0^1 + \frac{\sin(m-n)\pi t}{2(m-n)\pi} \Big|_0^1 = 0 .$$

Thus, the functions 1, cos πx, cos 2πx,...,cos nπx ... form an orthogonal set. Next, to normalize this set, find the length of each of the functions. The norm of a function f in C[0,1] is given by

$$\|f\| = \left(\int_0^1 f^2(x)dx\right)^{\frac{1}{2}} = \langle f,f \rangle^{\frac{1}{2}} .$$

$$\|1\| = \langle 1,1 \rangle^{\frac{1}{2}} = \left(\int_0^1 dt\right)^{\frac{1}{2}} = 1 .$$

The function $f(x) = 1$ has norm 1.

$$\|\cos m\pi x\| = <\cos m\pi x, \cos m\pi x>^{\frac{1}{2}} = (\int_0^1 \cos^2 m\pi t \, dt)^{\frac{1}{2}} . \qquad (2)$$

By the half-angle formula, $\cos^2 m\pi x = \frac{1}{2} + \frac{1}{2} \cos 2m\pi x$. Hence, (2) becomes

$$(\frac{1}{2}\int_0^1 \cos 2m\pi t + 1 \, dt)^{\frac{1}{2}} = [\frac{1}{2}(\frac{\sin 2m\pi t}{2m\pi} + t) \Big|_0^1]^{\frac{1}{2}} = \frac{1}{\sqrt{2}} .$$

Thus, the norm of $\cos m\pi x$ is $1/\sqrt{2}$. Since this is true for arbitrary integral values of _m, the vectors $\sqrt{2} \cos m\pi x$ have unit norm. Thus, the functions $1, \sqrt{2} \cos \pi x, \sqrt{2} \cos 2\pi x, \ldots, \sqrt{2} \cos n\pi x$ form an orthonormal basis.

● **PROBLEM** 14-21

a) The vectors $\{(1,0),(0,1)\}$ form an orthonormal basis for R^2. Find another orthonormal basis for R^2.
b) The vectors $u_1 = (1,1,1,1)$, $u_2 = (1,-1,1,-1)$ and $u_3 = (1,2,-1,-2)$ are orthogonal. Orthonormalize them.

<u>Solution</u>: A set of vectors forms an orthogonal set if the vectors are mutually perpendicular. If, in addition, all the vectors in the set are of unit length, then the set is orthonormal.

In R^n, two vectors are orthogonal if their dot product equals zero. This can be seen from the relationship

$$\cos \theta = \frac{u \cdot v}{\|u\| \, \|v\|}$$

where $u, v \in R^n$. $\cos \theta = 0$ when $\theta = 90°$, i.e., $u \cdot v = 0$.

Let $u = (u_1, u_2, \ldots, u_n)$, $v = (v_1, v_2, \ldots, v_n) \in E^n$.
Then, $u \cdot v = (u_1, u_2, \ldots, u_n) \cdot (v_1, v_2, \ldots, v_n)$

$$= u_1 v_1 + u_2 v_2 + \ldots + u_n v_n .$$

The length of vectors in R^n is given by the formula $\|u\| = (u \cdot u)^{\frac{1}{2}}$.

a) The problem is to find two perpendicular vectors (thus, linearly independent vectors) in R^2 that are of unit length.

Let (u_1, u_2), $(v_1, v_2) \in R^2$. Then, $U \cdot V = 0$ gives $u_1 v_1 + u_2 v_2 = 0$ which implies $u_1 v_1 = -u_2 v_2$. Thus, all vectors u, v such that $u_1 = -u_2$, $v_1 = v_2$ will be orthogonal. Choose the simplest vectors $(1,-1)$ and $(1,1)$. Their lengths are $\|(1,-1)\|$ and $\|(1,1)\|$ or,

$$\|(1,-1)\| = [(1,-1) \cdot (1,-1)]^{\frac{1}{2}} = \sqrt{2} ;$$

$$\|(1,1)\| = [(1,1) \cdot (1,1)]^{\frac{1}{2}} = \sqrt{2} .$$

Therefore, we divide each vector by its magnitude to get a new vector pointing in the same direction but with magnitude one. An orthonormal basis for R^2 is, therefore,

$$\{(1/\sqrt{2}, -1/\sqrt{2}), (1/\sqrt{2}, 1/\sqrt{2})\} .$$

This means that any vector $(x,y) \in R^2$ can be expressed as a linear combination of the above basis.

b) First check that the three vectors in R^4 are mutually perpendicular.

$(1,1,1,1) \cdot (1,-1,1,-1) = ((1) - 1(1) + 1(1) - 1(1) = 0$

$(1,1,1,1) \circ (1,2,-1,-2) = 1(1) + 1(2) + 1(-1) + 1(-2) = 0$

$(1,-1,1,-1) \cdot (1,2,-1,-2) = 1(1) - 1(2) + 1(-1) - (1)(-2) = 0$.

The length of the vectors is given by the norm of U, $(U \cdot U)^{\frac{1}{2}}$:

$\|(1,1,1,1)\| = [(1,1,1,1) \cdot (1,1,1,1)]^{\frac{1}{2}} = \sqrt{4} = 2$

$\|(1,-1,1,-1)\| = [(1,-1,1,-1) \cdot (1,-1,1,-1)]^{\frac{1}{2}} = \sqrt{4} = 2$

$\|(1,2,-1,-2)\| = [(1,2,-1,-2)(1,2,-1,2)]^{\frac{1}{2}} = \sqrt{10}$.

Dividing each vector by its norm results in normalized vectors. Thus,

$\{(\frac{1}{2},\frac{1}{2},\frac{1}{2},\frac{1}{2}), (\frac{1}{2},-\frac{1}{2},\frac{1}{2},-\frac{1}{2}), (1/\sqrt{10},\ 2/\sqrt{10},\ -1/\sqrt{10},\ -2/\sqrt{10})\}$ is an orthonormal set.

● **PROBLEM 14-22**

Let $F = \{(1,0,0,1,0),(0,1,1,-1,0),(1,1,1,1,1)\}$ be a set of linearly independent vectors in E^5 . Construct an orthogonal set G from F.

Solution: First show that the vectors in F are not orthogonal. Two vectors, $u = (u_1 u_2,\ldots,u_n)$, $v = (v_1,v_2,\ldots,v_n)$, in R^n are orthogonal if the dot product

$$(u \cdot v) = u_1 v_1 + u_2 v_2 + \ldots + u_n v_n = 0 .$$

But, $(1,0,0,1,0) \circ (0,1,1-1,0) = -1 \neq 0$

$(1,0,0,1,0) \cdot (1,1,1,1,1) = 2 \neq 0$

$(0,1,1,-1,0) \cdot (1,1,1,1,1) = 1 \neq 0$.

Given a set of linearly independent vectors, the Gram-Schmidt procedure tells us how to obtain an orthonormal set. In general, let $B_1 = \{v_1,v_2,\ldots,v_m\}$ be a set of m linearly independent vectors. From this set construct an orthonormal set $B_2 = \{u_1,u_2,\ldots,u_m\}$ that has the same dimension as B_1 . Set

$$u_1 = \frac{v_1}{\|v_1\|}$$

where $\|v_1\| = (v_1 \cdot v_1)^{\frac{1}{2}}$ and denotes the length of v_1 . Then, set $w_2 = -(v_2 \cdot u_1)u_1 + v_2$. Now, $w_2 \neq 0$, for $w_2 = 0$ would imply that v_2 depends on u_1 and, hence, on v_1 . If $u_2 = \frac{1}{\|w_2\|} w_2$, then $u_2 \cdot u_2 = 1$, $u_1 \cdot u_2 = 0$. Set $w_3 = -(v_3 \cdot u_1)u_1 - (v_3 \cdot u_2)u_2 + v_3$. Similarly, $w_3 \neq 0$, for $w_3 = 0$ would imply that v_3 depends on u_1 and u_2 and, hence, on v_1 and v_2 . If $u_3 = \frac{1}{\|w_3\|} w_3$, then $u_3 \cdot u_3 = 1$, $u_1 \cdot u_3 = u_2 \cdot u_3 = 0$ and $\{u_1 u_2, u_3\}$ forms an orthonormal set. Continuing in this way after m steps results in an orthonormal set $\{u_1,u_2,\ldots,u_m\}$.

Turning to the given problem, let

$$g_1 = f_1 = (1,0,0,1,0) .$$

Let $g_2 = f_2 - \dfrac{(f_2 \cdot f_1)}{(f_1 \cdot f_1)} f_1 = (0,1,1,-1,0) + \frac{1}{2}(1,0,0,1,0) = (\frac{1}{2},1,1,-\frac{1}{2},0)$.

Set
$$g_3 = f_3 - \frac{(f_3 \cdot g_2)}{(g_2 \cdot g_2)} g_2 - \frac{(f_3 \cdot g_1)}{(g_1 \cdot g_1)} g_1$$
$$= (1,1,1,1,1) - \frac{2}{5/2} (\tfrac{1}{2},1,1,-\tfrac{1}{2},0) - 2/2(1,0,0,1,0)$$
$$= (-2/5, 1/5, 1/5, 2/5, 1).$$

Therefore, $G = \{(1,0,0,1,0), (\tfrac{1}{2},1,1,-\tfrac{1}{2},0), (-2/5,1/5,1/5,2/5,1)\}$.

The set G is orthogonal.

● **PROBLEM** 14-23

Find an orthonormalizing sequence v_1, v_2, v_3 for the following set of vectors in E^4 :
$$u_1 = (1,-1,1,-1); \quad u_2 = (5,1,1,1); \quad u_3 = (2,3,4,-1) .$$

<u>Solution</u>: First verify that the set $\{u_1, u_2, u_3\}$ is linearly independent. The equation $c_1 u_1 + c_2 u_2 + c_3 u_3 = 0$ determines a system of equations which has as its only solution $c_1 = c_2 = c_3 = 0$. Thus, u_1, u_2, u_3 form a linearly independent set. Now, from a set of m linearly independent vectors and using the Gram-Schmidt orthogonalization procedure, construct a set of m orthonormalized vectors. Suppose we wish to convert $S_1 = \{u_1, u_2, \ldots, u_m\}$ into a set of orthonormal vectors, $S_2 = \{v_1, v_2, \ldots, v_m\}$. Select any member of the first set, say $w_1 = u_1$, and divide it by its length. Thus, the first member of the desired set S_2 is:
$$v_1 = \frac{u_1}{\|u_1\|} .$$

Now, choose a second member from S_1, say u_2, and set $w_2 = u_2 - cv_1$. If w_2 is to be orthogonal to v_1, their dot product, $\langle v_1, w_2 \rangle$, must be zero, i.e.,
$$\langle v_1, (u_2 - cv_1) \rangle = 0 . \tag{1}$$
By the rules of the dot product, (1) is
$$\langle v_1, u_2 \rangle - c \langle v_1, v_1 \rangle = 0 .$$
But, since v_1 is a unit vector, $\langle v_1, v_1 \rangle = 1$. Hence, $c = \langle v_1, u_2 \rangle$ and $w_2 = u_2 - \langle v_1, u_2 \rangle v_1$. Thus,
$$v_2 = \frac{w_2}{\|w_2\|} .$$
For the third step, let $w_3 = u_3 - c_1 v_1 - c_2 v_2$. If w_3 is to be simultaneously orthogonal to v_1 and v_2, the values of c_1 and c_2 are accordingly determined and
$$w_3 = u_3 - \langle v_1, u_3 \rangle v_1 - \langle v_2, u_3 \rangle v_2 .$$
Then,
$$v_3 = w_3 / \|w_3\| .$$
The process may be continued until obtaining the mth member, $v_m = \frac{w_m}{\|w_m\|}$,
where
$$w_m = u_m - \sum_{k=1}^{m-1} \langle v_k, u_m \rangle v_k . \tag{2}$$

Applying this procedure in the given problem,

$$v_1 = \frac{u_1}{\|u_1\|} .$$

The norm of a vector $U = (x_1, x_2, \ldots, x_n)$ in E^n is given by

$$\|U\| = (\sum_{i=1}^{n} x_i^2)^{\frac{1}{2}} .$$

Hence,

$$v_1 = \frac{(1,-1,1,-1)}{\sqrt{4}} = \tfrac{1}{2}(1,-1,1,-1) .$$

Next, $v_2 = \dfrac{w_2}{\|w_2\|}$ where $w_2 = u_2 - \langle v_1 u_2 \rangle v_1 = (5,1,1,1) - 2(\tfrac{1}{2},-\tfrac{1}{2},\tfrac{1}{2},-\tfrac{1}{2})$

$$= (4,2,0,2).$$

Since $\|w_2\| = \sqrt{24}$, $v_2 = \dfrac{(4,2,0,2)}{\sqrt{4 \times 6}} = \dfrac{1}{\sqrt{6}} (2,1,0,1) .$

Finally, using (2) with $m = 3$, $v_3 = w_3 / \|w_3\|$ where

$$w_3 = u_3 - \langle v_1, u_3 \rangle v_1 - \langle v_2, u_3 \rangle v_2$$
$$= (2,3,4,-1) - 2(\tfrac{1}{2},-\tfrac{1}{2},\tfrac{1}{2},-\tfrac{1}{2}) - \sqrt{6}/\sqrt{6}(2,1,0,1) = (-1,3,3,-1).$$

Since $\|w_3\| = \sqrt{20}$, $v_3 = 1/\sqrt{20}(-1,3,3,-1)$.

The set $\{v_1, v_2, v_3\} = \{ (\tfrac{1}{2},-\tfrac{1}{2},\tfrac{1}{2},-\tfrac{1}{2}), (\dfrac{2}{\sqrt{6}}, \dfrac{1}{\sqrt{6}}, 0, \dfrac{1}{\sqrt{6}}), (\dfrac{-1}{\sqrt{20}}, \dfrac{3}{\sqrt{20}}, \dfrac{3}{\sqrt{20}}, \dfrac{-1}{\sqrt{20}}) \}$

is a set of orthonormal vectors constructed from $\{u_1, u_2, u_3\}$.

● **PROBLEM 14-24**

Show that the vectors $f_1 = (\tfrac{1}{2}, \sqrt{3}/2)$, $f_2 = (\sqrt{3}/2, -\tfrac{1}{2})$ form an ortho-normal basis for E^2 . Then find the coordinates of an arbitrary vector $(x_1, x_2) \in E^2$.

<u>Solution</u>: First show that the set $\{f_1, f_2\}$ is a basis for E^2 . Since E^2 has dimension two, any set containing two linearly independent vectors forms a basis for this space. Now, any set of orthonormal vectors is linearly independent. To see this, let $G = \{g_1, g_2, \ldots, g_n\}$ be ortho-normal. Suppose $a_1 g_1 + a_2 g_2 + \ldots + a_n g_n = \sum_{j=1}^{n} a_j g_j = 0.$ (1)

Take the inner product with respect to g_1 on both sides of equation (1).

$$\langle g_1, \sum_{j=1}^{n} a_j g_j \rangle = \langle g_1, 0 \rangle ,$$ (2)

but $\langle v, 0 \rangle = v \cdot 0 = 0$ for any vector v. So, (2) becomes

$$\langle g_1, \sum_{j=1}^{n} a_j g_j \rangle = 0 .$$ (3)

However, by the properties of an inner product

$$(i.e., \langle v, b_1 u_1 + b_2 u_2 \rangle = b_1 \langle v, u_1 \rangle + b_2 \langle v, u_2 \rangle) ,$$

we obtain from (3),

$$\langle g_1, \sum_{j=1}^{n} a_j g_j \rangle = \sum_{j=1}^{n} a_j \langle g_1, g_j \rangle = 0 .$$ (4)

G is an orthogonal set, so $\sum_{s \neq t} \langle g_s, g_t \rangle = 0$ and G is normalized. Thus, $\langle g_1, g_1 \rangle^{\frac{1}{2}} = \|g\| = 1$. This implies $\langle g_1, g_1 \rangle = 1$. Therefore, (4) can be written as

$$a_1 \langle g_1, g_1 \rangle + a_2 \langle g_1, g_2 \rangle + \dots + a_n \langle g_1, g_n \rangle = 0 .$$

This yields

$$a_1 (1) + a_2 (0) + \dots + a_n (0) = 0$$
$$a_1 = 0 .$$

Similarly, it is possible to find $a_i = 0$, $i = 2, n$. Beginning with equation (1), $\sum a_j \, g_j = 0$, we obtain

$$\langle g_i, \sum a_j \, g_j \rangle = \langle g_i, 0 \rangle = 0 .$$

Then, $\sum_{j=1}^{n} a_j \langle g_i, g_j \rangle = a_1 \langle g_i, g_1 \rangle + \dots + a_i \langle g_i, g_i \rangle + \dots + a_n \langle g_i, g_n \rangle$

$$= a_1 (0) + \dots + a_i (1) + \dots + a_n (0) = 0$$

which implies that $a_i = 0$. So, $\sum_{i=n}^{n} a_i \, g_i = 0 \Rightarrow a_i = 0$, $i = 1, \dots, n$ implies G is a set of linearly independent vectors.

Hence, if it is shown that f_1 and f_2 are orthogonal to each other and are normalized, $\{f_1, f_2\}$ will be a basis for E^2.

$$\langle f_1, f_2 \rangle = \langle (\tfrac{1}{2}, \sqrt{3}/2), (\sqrt{3}/2, -\tfrac{1}{2}) \rangle = \sqrt{3}/4 - \sqrt{3}/4 = 0 .$$

Thus, f_1 and f_2 are mutually orthogonal. The norm of a vector $u = (u_1, u_2, \dots, u_n)$ in E^n is given by $\|u\| = \sum_{i=1}^{n} u_i^2$.

$$\|f_1\| = \|(\tfrac{1}{2}, \sqrt{3}/2)\| = \sqrt{\tfrac{1}{4} + 3/4} = 1 .$$

$$\|f_2\| = \|(\sqrt{3}/2, -\tfrac{1}{2})\| = \sqrt{3/4 + \tfrac{1}{4}} = 1 .$$

Thus, f_1, f_2 are both of unit length, and $\{f_1, f_2\}$ forms a basis of E^2. Let $(x_1, x_2) \in E^2$. Then,

$$(x_1, x_2) = c_1 f_1 + c_2 f_2$$
$$x_1 = c_1/2 + \sqrt{3} c_2/2 \; ; \; x_2 = \sqrt{3} c_1/2 - c_2/2 .$$

Solving this system for c_1, c_2,

$$c_1 = \frac{x_1 + \sqrt{3} x_2}{2} \quad , \quad c_2 = \frac{\sqrt{3} x_1 - x_2}{2}$$

and $(x_1, x_2) = \dfrac{(x_1 + \sqrt{3} x_2) f_1}{2} + \dfrac{(x_1 \sqrt{3} - x_2) f_2}{2}$

● PROBLEM 14-25

Make the following two matrices orthogonal:

a) $\begin{bmatrix} 3/5 & a \\ 4/5 & b \end{bmatrix}$; b) $\begin{bmatrix} 2/3 & 0 & 6 \\ 2/3 & a & c \\ 1/3 & 2/\sqrt{5} & d \end{bmatrix}$.

Solution: A matrix A is said to be orthogonal if $A^{-1} = A^T$. It can be shown that when a matrix is orthogonal its column vectors are orthonormal. We utilize this clue to solve a) and b).

a) The condition that the column vectors form an orthonormal set gives

$$1 = \|(a,b)\| = \sqrt{a^2 + b^2} \qquad (1)$$

since the length of a vector $u = (u_1, u_2, \ldots, u_n)$ in E^n is given by $\|u\| = \sqrt{u_1^2 + u_2^2 + \ldots + u_n^2}$. Also, the dot product of the two column vectors is equal to zero:

$$\langle (3/4, 4/5), (a,b) \rangle = 1/5(3a + 4b) = 0. \qquad (2)$$

Solving the system (1) and (2),

$$a^2 + b^2 = 1$$

$$3/5\ a + 4/5\ b = 0$$

$a = -4/3\ b$; $16/9\ b^2 + b^2 = 1$. Hence, $b^2 = 9/25$ and $b = \pm 3/5$. Corresponding values of a are $a = \mp 4/5$. Thus, either of the two column vectors, $(4/5, -3/5)$, $(-4/5, 3/5)$, will make the given matrix orthogonal. As a check, let

$$A_1 = \begin{bmatrix} 3/5 & 4/5 \\ 4/5 & -3/5 \end{bmatrix}.$$

Then, $A^{-1} = \dfrac{1}{\det A} \mathrm{adj.}\ A$

$$= -\begin{bmatrix} -3/5 & -4/5 \\ -4/5 & 3/5 \end{bmatrix} = \begin{bmatrix} 3/5 & 4/5 \\ 4/5 & -3/5 \end{bmatrix} = A^T.$$

So $A^{-1} = A^T$, as expected.

b) Again employing the fact that an orthogonal matrix must have orthogonal columns, it can be seen that the dot product of the first two columns must be zero:

$$\langle (2/3, 2/3, 1/3), (0, a, 2/\sqrt{5}) \rangle = 0,$$

$$2/3\ a + 2/3\sqrt{5} = 0.$$

Hence, $a = -\sqrt{1/5}$. Now, recall that the cross-product of two orthogonal vectors is a vector that is orthogonal to the other two. Since the third column is orthogonal to each of the first two columns.

$$(b,c,d) = \pm\ (2/3, 2/3, 1/3) \times (0, -1/\sqrt{5}, 2/\sqrt{5}).$$

Compute the cross-product by imagining a matrix formed with the above two vectors and the unit vectors (i, j, k). The cross-product is then the following determinant:

$$\pm \begin{vmatrix} i & j & k \\ 2/3 & 2/3 & 1/3 \\ 0 & -1/\sqrt{5} & 2/\sqrt{5} \end{vmatrix} = \pm \frac{1}{3\sqrt{5}}\ (5, -4, -2).$$

The plus or minus sign shows that we can take either $(2/3, 2/3, 1/4) \times (0, -1/\sqrt{5}, 2/\sqrt{5}$ or, vice-versa, $(A \times B = -B \times A)$. Thus, $(b,c,d) = \pm \dfrac{1}{3\sqrt{5}}\ (5, -4, -2)$. Hence, either choice for (b,c,d) would give the desired orthogonal matrix.

Use the Gram-Schmidt process to transform $[(1,0,1),(1,2,-2),(2,-1,1)]$ into an orthogonal basis for R^3. Assume the standard inner product.

Solution: The standard inner product for R^n is the dot product defined as follows: Let $u = (u_1,u_2,...,u_n)$, $v = (v_1 v_2,...,v_n)$ be two vectors in R^n. Then, the inner or dot product of u and v, denoted $<u,v>$, is given by $<u,v> = <(u_1,u_2,...,u_n),(v_1,v_2,...,v_n)> = u_1 v_1 + u_2 v_2 +...+ u_n v_n = \sum_{i=1}^{n} u_i v_i$. The dot product is useful for defining both orthogonalization and the length of a vector. The norm of u, denoted $\|u\|$, is given by $\|u\| = <u,u>^{\frac{1}{2}}$. Two vectors u,v are orthogonal if $<u,v> = 0$.

The Gram-Schmidt process constructs a set of orthogonal vectors from a given set of linearly independent vectors. Since a basis for R^3 must contain three linearly independent vectors, we check that the given vectors are linearly independent. Set $c_1(1,0,1) + c_2(1,2,-2) + c_3(2,-1,1) = 0$.

Thus,
$$c_1 + c_2 + 2c_3 = 0$$
$$2c_2 - c_3 = 0$$
$$c_1 - 2c_2 + c_3 = 0.$$

The only solution to this system is $c_1 = c_2 = c_3 = 0$, i.e., the three vectors are linearly independent and form a basis for R^3.

Label the three vectors u_1,u_2,u_3 respectively. Choose u_1 and relabel it v_1. Thus, $v_1 = (1,0,1)$. The set now becomes $\{v_1,u_2,u_3\}$. We know u_1 and u_2 are independent, and the vector space they span is also spanned by $\{v_1,u_2\}$. We wish to construct a vector v_2 in this two-dimensional subspace of R^3 which is perpendicular to v_1. Choose v_2 as the candidate for this transfiguration. $v_2 \in Sp\{u_1,u_2\}$ implies $v_2 \in Sp\{v_1,u_2\}$ which, in turn, implies $v_2 = c_1 v_1 + c_2 u_2$ for some constants c_1 and c_2. The required orthogonality of v_1 and v_2 implies $<v_1,v_2> = 0$. Hence,

$$<v_1, c_1 v_1 + c_2 u_2> = <v_1, c_1 v_1> + <v_1, c_2 u_2>$$
$$= c_1<v_1,v_1> + c_2<v_1,u_2> = 0.$$

Assume that $c_2 = 1$. Then,

$$c_1 = \frac{-<v_1,u_2>}{<v_1,v_1>} = \frac{-<v_1,u_2>}{\|v_1\|^2}$$

and hence,

$$v_2 = c_2 u_2 + c_1 u_1 = u_2 - \frac{<v_1,u_2>}{\|v_1\|^2} v_1$$

Thus, $v_2 = (1,2,-2) \dfrac{-<(1,0,1),(1,2,-2)>}{\|(1,0,1)\|^2} (1,0,1) = (1,2,-2) - (-\frac{1}{2})(1,0,1)$

$$= (3/2,2,-3/2).$$

The set $\{v_1,v_2\}$ is orthogonal and forms a basis for the subspace of R^3.

Now seek a third vector v_3 that is orthogonal to both v_1 and v_2. Again, $Sp\{u_1, u_2, u_3\} = Sp\{v_1, v_2, v_3\}$, so $v_3 \in Sp\{u_1, u_2, u_3\}$ implies v_3 can be written as $v_3 = b_1 v_1 + b_2 v_2 + b_3 u_3$, subject to $\langle v_3, v_1 \rangle = \langle v_3, v_2 \rangle = 0$. That is,

$$\langle b_1 v_1 + b_2 v_2 + b_3 u_3, v_1 \rangle = 0$$
$$\langle b_1 v_1 + b_2 v_2 + b_3 u_3, v_2 \rangle = 0$$

By the rules of the dot product,

$$b_1 \langle v_1, v_1 \rangle + b_2 \langle v_2, v_1 \rangle + b_3 \langle u_3, u_1 \rangle = 0$$
$$b_1 \langle v_1, v_2 \rangle + b_2 \langle v_2, v_2 \rangle + b_3 \langle u_3, v_2 \rangle = 0$$

Since $\langle v_2, v_1 \rangle = \langle v_1, v_2 \rangle = 0$, the above two equations reduce to

$$b_1 \langle v_1, v_1 \rangle + b_3 \langle u_3, v_1 \rangle = 0$$
$$b_2 \langle v_2, v_2 \rangle + b_3 \langle u_3, v_2 \rangle = 0 \ .$$

Assume $b_3 = 1$:

$$b_1 = \frac{-\langle u_3, v_1 \rangle}{\|v_1\|^2} \quad \text{and} \quad b_2 = \frac{-\langle u_3, u_2 \rangle}{\|v_2\|^2} \ .$$

Thus,

$$v_3 = b_3 u_3 + b_2 v_2 + b_1 v_1 = u_3 - \frac{\langle u_3, v_2 \rangle v_2}{\|v_2\|^2} - \frac{\langle u_3, v_1 \rangle v_1}{\|v_1\|^2}$$

$$= (2,-1,1) - \frac{\langle (2,-1,1),(3/2,2,-3/2) \rangle}{34/4} (3/2,2,-3/2)$$

$$- \frac{\langle (2,-1,1),(1,0,1) \rangle}{2} (1,0,1)$$

$$= (2,-1,1) - \frac{(-\frac{1}{2})}{34/4} (3/2,2,-3/2) - 3/2(1,0,1)$$

$$= (2,-1,1) + (3/34, 2/17, -3/34) + (-3/2, 0, -3/2)$$

$$= (10/17, -15/17, -10/17) \ .$$

The required orthogonal basis is

$$\{(1,0,1),(3/2,2,-3/2),(10/17, -15/17, -10/17)\} \ .$$

Any scalar multiple of these basis vectors will have the same properties of orthogonality, and so, if we wish, we may take $\{(1,0,1),(3,4,-3),(2,-3,-2)\}$ as an orthogonal basis of R^3.

• PROBLEM 14-27

Let $T: R^3 \to R^3$ have the matrix representation

$$A = \begin{bmatrix} 2 & -1 & 1 \\ 0 & 1 & 1 \\ -1 & 1 & 1 \end{bmatrix}$$

with respect to the standard basis. Find an orthogonal matrix F such that FAF^T is a lower triangular matrix.

Solution: An orthogonal matrix is a nonsingular real matrix M for which $M^{-1} = M^T$, i.e., the inverse of M is just the transpose of M. A lower triangular matrix is a matrix of all whose elements above the main diagonal are zero.

We use the fact that if A is a real $n×n$ matrix whose character-istic values are real, then A is orthogonally similar to a lower-tri-angular matrix. Two real matrices A and B are orthogonally similar if there exists an orthogonal matrix C such that $B = CAC^T$.

The characteristic equation of A is given by $\det(I\lambda - A) = 0$ and is found to be $(\lambda-1)^2 (\lambda-2) = 0$. Hence, the eigenvalues $\lambda_1 = 1$ and $\lambda_2 = 2$ are real. To find the orthogonal matrix F, first construct an orthonormal basis of R^3 using the eigenvalues of A. The vector $v_1 = \sqrt{3}/3 (1,-1,1)$ is a characteristic vector of A belonging to 1 and $|v_1| = 1$. Let $v_2 = \sqrt{2}/2 (1,1,0)$ and $v_3 = \sqrt{6}/6 (1,-1,-2)$ be charac-teristic vectors associated with 1 and 2. Then, $\{v_1,v_2,v_3\}$ is an orthonormal basis of R^3 . The matrix

$$G = r(v_1,v_2,v_3) = \begin{bmatrix} \sqrt{3}/3 & -\sqrt{3}/3 & \sqrt{3}/3 \\ \sqrt{2}/2 & \sqrt{2}/2 & 0 \\ \sqrt{6}/6 & -\sqrt{6}/6 & -\sqrt{6}/3 \end{bmatrix}$$

is orthogonal and

$$B = GAG^T = \begin{bmatrix} 1 & 0 & 0 \\ 2\sqrt{6}/3 & 1 & -\sqrt{3}/3 \\ \sqrt{2} & 0 & 2 \end{bmatrix} = \begin{bmatrix} 1 & 0 & 0 \\ 2\sqrt{6}/3 & & \\ \sqrt{2} & & B_1 \end{bmatrix}$$

where

$$B_1 = \begin{bmatrix} 1 & -\sqrt{3}/3 \\ 0 & 2 \end{bmatrix} .$$

The characteristic polynomial of B_1 is $(x-1)(x-2)$. An orthonormal vector belonging to 1 is $w_1 = \frac{1}{2}(\sqrt{3},1)$ and $w_2 = \frac{1}{2}(1,-\sqrt{3})$ is an ortho-normal vector belonging to the eigenvalue 2. $\{w_1,w_2\}$ is an orthonormal basis of R^2 and $H_1 = r(w_1,w_2)$ is an orthogonal matrix. The matrix

$$C_1 = H_1B_1H_1^T = \begin{bmatrix} 1 & 0 \\ \sqrt{3}/3 & 2 \end{bmatrix}$$

is lower triangular. If

$$H = \begin{bmatrix} 1 & 0 & 0 \\ 0 & & H_1 \\ 0 & & \end{bmatrix} = \begin{bmatrix} 1 & 0 & 0 \\ 0 & \sqrt{3}/2 & \frac{1}{2} \\ 0 & \frac{1}{2} & -\sqrt{3}/2 \end{bmatrix}$$

then

$$C = HBH^T = \begin{bmatrix} 1 & 0 & 0 \\ 3\sqrt{2}/2 & 1 & 0 \\ -\sqrt{6}/6 & -\sqrt{3}/3 & 2 \end{bmatrix}$$

Hence, if

$$F = HG = \begin{bmatrix} \sqrt{3}/3 & -\sqrt{3}/3 & \sqrt{3}/3 \\ \sqrt{6}/3 & \sqrt{6}/6 & -\sqrt{6}/6 \\ 0 & \sqrt{2}/2 & \sqrt{2}/2 \end{bmatrix}$$

then F is orthogonal and

$$FAF^T = \begin{bmatrix} 1 & 0 & 0 \\ 3\sqrt{2}/2 & 1 & 0 \\ -\sqrt{6}/6 & -\sqrt{3}/3 & 2 \end{bmatrix}$$

is a lower triangular matrix.

CHAPTER 15

BILINEAR FORMS

BILINEAR FUNCTIONS

● PROBLEM 15-1

Show that the set of all bilinear forms on R^2 can be identified with the set of all 2 × 2 matrices with entries in R.

<u>Solution</u>: A bilinear form on a vector space V is a function

$$f: \quad V \times V \to R$$

such that $f(v_1, v_2)$ is linear in v_1 if v_2 is held fixed.
Also, f is linear in v_2 if v_1 is held fixed where $v_1 \in V$
and $v_2 \in V$. Suppose f is a bilinear form defined on R^2.
If $v_1 = (x_1, x_2)$ and $v_2 = (y_1, y_2)$ are elements of R^2, then,
using the standard basis $B = \{e_1 e_2\}$,

$$f(v_1, v_2) = f(x_1 e_1 + x_2 e_2, v_2)$$

$$= x_1 f(e_1, v_2) + x_2 f(e_2, v_2)$$

$$= x_1 f(e_1, y_1 e_1 + y_2 e_2) + x_2 f(e_2, y_1 e_1 + y_2 e_2)$$

$$= x_1 y_1 f(e_1, e_1) + x_1 y_2 f(e_1, e_2) + x_2 y_1 f(e_2, e_1)$$

$$+ x_2 y_2 f(e_2, e_2).$$

The above derivation follows from the definition of a bilinear form. Letting $a_{ij} = f(e_i, e_j)$,

$$f(v_1, v_2) = a_{11} x_1 y_1 + a_{12} x_1 y_2 + a_{21} x_2 y_1 + x_2 y_2 a_{22}$$

$$= \sum_{i=1}^{2} \sum_{j=1}^{2} a_{ij} x_i y_j. \tag{1}$$

If X and Y are the coordinate matrices of v_1 and v_2, and if
$A = (a_{ij})$, then (1) is equivalent to $f(v_1, v_2) = X^T A Y$. Thus,

any bilinear form f over R corresponds to a unique 2 × 2 matrix A = (a_{ij}) where a_{ij} = $f(e_i, e_j)$. On the other hand, any 2 × 2 matrix B = (b_{ij}) defines a bilinear form f on R^2 by the formula $f(v_1, v_2)$ = $X^T BY$ where X and Y are the coordinate matrices of v_1 and v_2, respectively.

Thus, the set of all bilinear forms defined on R^2 is equivalent to the set of all real 2 × 2 matrices.

● **PROBLEM** 15-2

Let f: $R^2 \times R^2 \to R$ be given by $f((x_1, x_2), (y_1, y_2))$ =
$2x_1 y_1 - 3x_1 y_2 + x_2 y_2$.

a) Find the matrix A of f in the basis
$\{u_1 = (1,0), u_2 = (1,1)\}$;

b) Find the matrix B of f in the basis
$\{v_1 = (2,1), v_2 = (1,-1)\}$;

c) Find the transition matrix P from the basis $\{u_i\}$ to the basis $\{v_i\}$ and verify that B = $P^T AP$.

Solution: The function f is an example of a bilinear form. A bilinear form on a vector space V is a scalar valued function that is linear in either of its arguments while the other is held fixed. That is, f: V x V → K satisfies:

i) $f(a\vec{w}_1 + b\vec{w}_2, \vec{z})$ = $af(\vec{w}_1, \vec{z}) + bf(\vec{w}_2, \vec{z})$

ii) $f(\vec{w}, a\vec{z}_1 + b\vec{z}_2)$ = $af(\vec{w}, \vec{z}_1) + bf(\vec{w}, \vec{z}_2)$.

Like linear transformations, bilinear forms can be identified by their effects on the basis vectors of V. For instance, if V = R^2 and $\{p_i\}$ is a basis of V, then the four values of $f(e_i, e_j)$ for i,j = 1,2 completely determine f. It is useful to form the matrix A = (a_{ij}) where a_{ij} = $f(e_i, e_j)$, and identify the bilinear form with this matrix.

a) Let A = (a_{ij}) where a_{ij} = $f(u_i, u_j)$.

a_{11} = $f(u_1, u_1)$ = $f((1,0), (1,0))$ = 2 - 0 + 0 = 2

a_{12} = $f(u_1, u_2)$ = $f((1,0), (1,1))$ = 2 - 3 + 0 = -1

a_{21} = $f(u_2, u_1)$ = $f((1,1), (1,0))$ = 2 - 0 + 0 = 2

a_{22} = $f(u_2, u_2)$ = $f((1,1), (1,1))$ = 2 - 3 + 1 = 0

Hence,

$$A = \begin{bmatrix} 2 & -1 \\ 2 & 0 \end{bmatrix}$$

is the matrix of f in the basis $\{u_1, u_2\}$.

      b) Let $B = (b_{ij})$ where $b_{ij} = f(v_i, v_j)$.

$b_{11} = f(v_1, v_1) = f((2,1), (2,1)) = 8 - 6 + 1 = 3$

$b_{12} = f(v_1, v_2) = f((2,1), (1,-1)) = 4 + 6 - 1 = 9$

$b_{21} = f(v_2, v_1) = f((1,-1), (2,1)) = 4 - 3 - 1 = 0$

$b_{22} = f(v_2, v_2) = f((1,-1), (1,-1) = 2 + 3 + 1 = 6.$

Thus,

$$B = \begin{bmatrix} 3 & 9 \\ 0 & 6 \end{bmatrix}$$

is the matrix of f in the basis $\{v_1, v_2\}$.

      By definition, the transition matrix is the matrix P such that

$$\begin{pmatrix} v_1 \\ v_2 \end{pmatrix} = P(u_1, u_2).$$

To find the transition matrix P, we must express each of the basis vectors v_i in terms of the basis $\{u_i\}$.

$$v_1 = (2,1) = c_1 u_1 + c_2 u_2 = c_1(1,0) + c_2(1,1).$$

$$2 = c_1 + c_2$$

$$1 = c_2.$$

Therefore, $c_1 = 1$. Hence, $v_1 = u_1 + u_2$. Similarly,

$$v_2 = (1,-1) = c_1 u_1 + c_2 u_2 = c_1(1,0) + c_2(1,1).$$

Therefore,

$$c_1 + c_2 = 1$$

$$c_2 = -1$$

$$c_1 = 1 - (-1) = 2,$$

and we find $c_1 = 2$, $c_2 = -1$. Thus,

$$v_2 = 2u_1 - u_2.$$

The columns of P are the coordinates of the v_i under the basis $\{u_1, u_2\}$.

$$P = \begin{bmatrix} 1 & 2 \\ 1 & -1 \end{bmatrix}.$$

Then
$$P^T = \begin{bmatrix} 1 & 1 \\ 2 & -1 \end{bmatrix}$$

and, using part a),

$$P^T A P = \begin{bmatrix} 1 & 1 \\ 2 & -1 \end{bmatrix}\begin{bmatrix} 2 & -1 \\ 2 & 0 \end{bmatrix}\begin{bmatrix} 1 & 2 \\ 1 & -1 \end{bmatrix} = \begin{bmatrix} 1 & 1 \\ 2 & -1 \end{bmatrix}\begin{bmatrix} 1 & 5 \\ 2 & 4 \end{bmatrix} = \begin{bmatrix} 3 & 9 \\ 0 & 6 \end{bmatrix} = B.$$

Thus, P is indeed the transition matrix.

QUADRATIC FORMS

• PROBLEM 15-3

What is a quadratic form?

Solution: A function of n variables $f(x_1, \ldots, x_n)$ is call-ed homogeneous of the second degree if:

$$f(tx_1, tx_2, \ldots, tx_n) = t^2 f(x_1, x_2, \ldots, x_n). \qquad (1)$$

A quadratic form is a homogeneous function of degree two that is of the form:

$$A = \sum_{i=1}^{n} \sum_{j=1}^{n} a_{ij} x_i x_j$$

where $a_{ij} = a_{ji}$ for $i = 1, \ldots, n$.

$$A = a_{11}x_1^2 + a_{12}x_1x_2 + a_{13}x_1x_3 + \ldots + a_{1n}x_1x_n$$

$$+ a_{21}x_2x_1 + a_{22}x_2^2 + a_{23}x_2x_3 + \ldots + a_{2n}x_2x_n + \ldots$$

$$+ a_{n1}x_nx_1 + a_{n2}x_nx_2 + a_{n3}x_nx_3 + \ldots + a_{nn}x_n^2. \qquad (2)$$

Equation (2) may be rewritten as:

$$A = a_{11}x_1^2 + a_{22}x_2^2 + \ldots + a_{nn}x_n^2 + 2a_{12}x_1x_2$$

$$+ 2a_{13}x_1x_3 + \ldots + 2a_{n-1,n}x_{n-1}x_n, \qquad (3)$$

where we have used the symmetry of $a_{ij} = a_{ji}$. Note that A can be given a matrix form; i.e.,

$$A = [x_1 x_2 \ldots x_n] \begin{bmatrix} a_{11} & a_{12} & \cdots & a_{1n} \\ a_{21} & a_{22} & \cdots & a_{2n} \\ \vdots & \vdots & & \vdots \\ a_{n1} & a_{n2} & \cdots & a_{nn} \end{bmatrix} \begin{bmatrix} x_1 \\ x_2 \\ \vdots \\ x_n \end{bmatrix} = X^T \Phi X \qquad (4)$$

where X and Φ are defined in the obvious way. It is always possible to construct a quadratic form from a given real symmetric matrix. For example, let

$$A = \begin{bmatrix} 1 & 2 \\ 2 & 3 \end{bmatrix}, \quad B = \begin{bmatrix} 1 & 3 & 5 \\ 3 & 2 & -1 \\ 5 & -1 & 4 \end{bmatrix}.$$

We have

$$Q_A \langle x_1, x_2 \rangle = [x_1 x_2] \begin{bmatrix} 1 & 2 \\ 2 & 3 \end{bmatrix} \begin{bmatrix} x_1 \\ x_2 \end{bmatrix}.$$

Upon multiplying the above, the result is the quadratic form:

$$x_1^2 + 2x_1x_2 + 2x_1x_2 + 3x_2^2 = x_1^2 + 4x_1x_2 + 3x_2^2.$$

Similarly,

$$Q_B \langle x_1, x_2, x_3 \rangle = [x_1 x_2 x_3] \begin{bmatrix} 1 & 3 & 5 \\ 3 & 2 & -1 \\ 5 & -1 & 4 \end{bmatrix} \begin{bmatrix} x_1 \\ x_2 \\ x_3 \end{bmatrix}$$

$$= (x_1 + 3x_2 + 5x_3, 3x_1 + 2x_2 - x_3, 5x_1 - x_2 + 4x_3) \begin{bmatrix} x_1 \\ x_2 \\ x_3 \end{bmatrix}$$

$$= x_1^2 + 3x_1x_2 + 5x_1x_3 + 3x_1x_2 + 2x_2^2 - x_3x_2$$

$$+ 5x_1x_3 - x_2x_3 + 4x_3^2$$

$$= x_1^2 + 2x_2^2 + 4x_3^2 + 6x_1x_2 + 10x_1x_3 - 2x_2x_3.$$

When a quadratic form is represented by a diagonal matrix, the terms x_ix_j, $i \neq j$ vanish, and the form is said to be in canonical form. Every real quadratic form can be put into canonical form using a suitable basis for the matrix.

● **PROBLEM** 15-4

Let $F(x_1, x_2, x_3) = x_1^2 - x_3^2 + 3x_1x_2 - 6x_2x_3$. What is the polar form of F?

Solution: First define what is meant by the polar form of a quadratic form. Then apply the definition to solve the given problem. A quadratic form in n-indeterminates is a function $f(x_1, \ldots, x_n)$ of the form $\sum_{i=1}^{n} \sum_{0=1}^{n} a_{ij}x_ix_j$ where $a_{ij} = a_{ji}$.

The set of quadratic forms in n-indeterminates can be identified with the set of all symmetric bilinear forms over R^n. A bilinear form is a scalar-valued function defined on a vector space over a field K that satisfies

i) $f(au_1 + bu_2, v) = af(u_1, v) + bf(u_2, v)$

ii) $f(u, av_1 + bv_2) = af(u, v_1) + bf(u, v_2)$

where $a, b \in K$ and $u_i, v_i \in V$. A bilinear form is said to be symmetric if $f(u, v) = f(v, u)$. If q: $V \rightarrow K$ is a quadratic form, then $q(v) = f(v, v)$ for some symmetric bilinear form f on V. The polar form of $q(v)$ is just the corresponding bilinear from, $f(u, v)$. It is obtained from $q(v)$ by the following formula:

$$f(u, v) = \frac{1}{2}(q(u + v) - q(u) - q(v)).$$

Examining the given function, we first see that it can be rewritten as a quadratic form, i.e.,

$$F(x_1, x_2, x_3) = x_1^2 + \frac{3}{2}x_1x_2 + \frac{3}{2}x_2x_1 - 3x_2x_3 - 3x_3x_2 - x_3^2. \quad (1)$$

By substituting in the polar form formula, we obtain:

$$f(x,y) = \frac{1}{2}\Big(F(x_1+y_1,\ x_2+y_2,\ x_3+y_3) - F(x_1,x_2,x_3)$$

$$- F(y_1,y_2,y_3)\Big).$$

Therefore,

$$f(x,y) = \frac{1}{2}\Big((x_1+y_1)^2 + \frac{3}{2}(x_1+y_1)(x_2+y_2) + \frac{3}{2}(x_2+y_2)(x_1+y_1)$$

$$- 3(x_2+y_2)(x_3+y_3) - 3(x_3+y_3)(x_2+y_2) - (x_3+y_3)^2$$

$$- (x_1^2 + \frac{3}{2}x_1x_2 + \frac{3}{2}x_2x_1 - 3x_2x_3 - 3x_3x_2 - x_3^2)$$

$$- (y_1^2 + \frac{3}{2}y_1y_2 + \frac{3}{2}y_2y_1 - 3y_2y_3 - 3y_3y_2 - y_3^2)\Big)$$

After simplification,

$$F(x,y) = x_1y_1 - x_3y_3 + \frac{3}{2}x_1y_2 + \frac{3}{2}x_2y_1 - 3x_2y_3 - 3x_3y_2. \qquad (2)$$

The function given by (2) is the required polar form. It corresponds to replacing the quadratic form in one determinate by a symmetric bilinear form in two unknowns.

● **PROBLEM** 15-5

A general quadric surface in three-dimenaional space with center at the origin has for its equation:

$$a_{11}x_1^2 + a_{22}x_2^2 + a_{33}x_3^2 + a_{12}x_1x_2 + a_{13}x_1x_3 + a_{23}x_2x_3 = 1.$$

Use a change of variables to reduce the quadratic form $Q(x_1,x_2,x_3) = x_1^2 - 2x_1x_2 + 4x_2x_3 - 2x_2^2 + 4x_3^2$ to a sum or difference of squares.

Solution: A quadric surface in R^3 is given by an equation $q(v) = 1$, where $q(v)$ is a quadratic form in $v = (x_1,x_2,x_3)$.

We can make each term of the quadratic form $q(V)$ a square and eliminate cross product terms by the familiar process of completing the square. This gives us an equation of the form $\pm z_1^2 \pm z_2^2 \pm z_3^2 = 1$, where the z_i represents a relabeling of the terms contained in parenthesis in the completed square (corresponding geometrically to a change of the corrdinate system).

We have developed a way of identifying quadric surfaces. By eliminating cross product terms and making the subsequent change of coordinates, we can easily identify the quadric form.

411

In the given problem, we complete the square by adding and subtracting the required terms. First, to $Q(x_1,x_2,x_3)$, add and subtract x_2^2.

$$Q(x_1,x_2,x_3) = x_1^2 - 2x_1x_2 + x_2^2 - 3x_2^2 + 4x_2x_3 + 4x_3^2$$

$$= (x_1 - x_2)^2 - 3x_2^2 + 4x_2x_3 + 4x_3^2. \qquad (1)$$

Now add and subtract x_2^2 to (1):

$$Q(x_1,x_2,x_3) = -4x_2^2 + x_2^2 + 4x_2x_3 + 4x_3^2 + (x_1 - x_2)^2$$

$$= -4x_2^2 + (x_2 + 2x_3)^2 + (x_1 - x_2)^2$$

$$= (x_1 - x_2)^2 - 4x_2^2 + (x_2 + 2x_3)^2. \qquad (2)$$

Equations (2) contains only sums and differences of squares. Thus, we have an equation of the form $z_1^2 - 4z_2^2 + z_3^2 = 1$ representing the original quadric surface. Finally, if we let $(z_2')^2 = 4z_2^2$, we have $z_1^2 \quad z_2'^2 + z_3^2 = 1$.

● **PROBLEM** 15-6

Reduce the quadratic form

$$Q(x_1,x_2,x_3) = x_1^2 + x_3^2 - 2x_1x_2 - 2x_1x_3 + 10x_2x_3$$

to the simplest form. What is the matrix of the transformation?

Solution: A quadratic form is in simplest form when all cross-product terms are eliminated. This is accomplished by completing the square and relabeling terms within parentheses by a change of variables. Geometrically, this relabeling corresponds to a coordinate transformation.
We have,
$$Q(x_1,x_2,x_3) = x_1^2 + x_3^2 - 2x_1x_2 - 2x_1x_3 + 10x_2x_3.$$

Now add and subtract x_2^2 to Q.

$$Q = (x_1^2 + x_2^2 + x_3^2 - 2x_1x_2 - 2x_1x_3 + 2x_2x_3) + 8x_2x_3 - x_2^2.$$

Note that

$$(x_1 - x_2 - x_3)^2 = x_1^2 + x_2^2 + x_3^2 - 2x_1x_2 - 2x_1x_3 + 2x_2x_3.$$

Therefore,

$$Q = (x_1 - x_2 - x_3)^2 + 8x_2x_3 - x_2^2. \tag{1}$$

Next, introduce new variables

$$y_1 = (x_1 - x_2 - x_3); \quad y_2 = x_2; \quad y_3 = x_3.$$

Then (1) becomes

$$Q' = y_1^2 - y_2^2 + 8y_2y_3.$$

Subtract and add $16y_3^2$ to Q': So,

$$Q' = y_1^2 - y_2^2 + 8y_2y_3 - 16y_3^2 + 16y_3^2$$

$$= y_1^2 - (y_2^2 - 8y_2y_3 + 16y_3^2) + 16y_3^2$$

$$= y_1^2 - (y_2 - 4y_3)^2 + 16y_3^2.$$

Introduce new variables once more:

$$z_1 = y_1; \quad z_2 = y_2 - 4y_3; \quad z_3 = 4y_3.$$

Then Q' is transformed into

$$Q'' = z_1^2 - z_2^2 + z_3^2.$$

The above transformations were all linear, i.e., they have matrices associated with them. Since $y_1 = x_1 - x_2 - x_3$, $y_2 = x_2$ and $y_3 = x_3$, the transformation from (x_1, x_2, x_3) to (y_1, y_2, y_3) is given by the matrix equation.

$$[x_1, x_2, x_3] \begin{bmatrix} 1 & 0 & 0 \\ -1 & 1 & 0 \\ -1 & 0 & 1 \end{bmatrix} = \begin{bmatrix} y_1 \\ y_2 \\ y_3 \end{bmatrix}$$

Similarly, the transformation from (y_1, y_2, y_3) to (z_1, z_2, z_3) is $T(y_1, y_2, y_3) = (y_1, y_2 - 4y_3, 4y_3)$. The matrix equation of the linear transformation is, therefore,

$$[y_1, y_2, y_3] \begin{bmatrix} 1 & 0 & 0 \\ 0 & 1 & 0 \\ 0 & -4 & 4 \end{bmatrix} = \begin{bmatrix} z_1 \\ z_2 \\ z_3 \end{bmatrix}.$$

413

The change of coordinates from (x_1, x_2, x_3) to (z_1, z_2, z_3) is, consequently, given by the product of the two matrices:

$$\begin{bmatrix} 1 & 0 & 0 \\ -1 & 1 & 0 \\ -1 & 0 & 1 \end{bmatrix} \begin{bmatrix} 1 & 0 & 0 \\ 0 & 1 & 0 \\ 0 & -4 & 4 \end{bmatrix} = \begin{bmatrix} 1 & 0 & 0 \\ -1 & 1 & 0 \\ -1 & -4 & 4 \end{bmatrix}.$$

● **PROBLEM** 15-7

Reduce the quadratic form:

$$F(x_1, x_2, x_3, x_4, x_5) = x_1^2 + 4x_2^2 + 8x_3^2 - x_4^2 - 4x_1 x_2 + 6x_1 x_3 - 12x_2 x_3$$

$$+ 2x_3 x_4 + x_2 x_5 - x_4 x_5 \qquad (1)$$

to diagonal form.

Solution: A quadratic form $\displaystyle\sum_{i=1}^{n} \sum_{j=1}^{n} a_{ij} x_i x_j$ is in diagonal form when the cross-product terms, i.e., $x_i x_j$ $(i \neq j)$, have been eliminated. One method of diagonalizing (1) is to re-write it as a symmetric matrix and find the eigenvalues. These eigenvalues then form a diagonal matrix, and multiplication by the x-vector (row and column-wise) yields the diagonal or canonical quadratic form.

Another method, which is used here, is to simplify the quadritic form by substitution and completing the square.

Consider $x_1^2 + 4x_2^2 + 8x_3^2$ in (1). We have

$$(x_1 - 2x_2 + 3x_3)^2 = x_1^2 + 4x_2^2 + 9x_3^2 + \text{(cross-product terms)}.$$

$$\qquad (2)$$

Since $(a + b + x)^2 = a^2 + b^2 + c^2 + 2(ab + bc + ca)$, (2) is equal to

$$x_1^2 + 4x_2^2 + 8x_3^2 - 4x_1 x_2 - 12x_2 x_3 + 6x_1 x_3 + x_3^2. \qquad (3)$$

Using (2) and (3), (1) can be written as

$$F = (x_1 - 2x_2 + 3x_3)^2 - x_3^2 - x_4^2 + 2x_3 x_4 + x_2 x_5 - x_4 x_5.$$

Let $y_1 = x_1 - 2x_2 + 3x_3$; $y_i = x_i$ $(i > 1)$. Then,

$$F_1(y_1, y_2, y_3, y_4, y_5) = y_1^2 - y_3^2 + 2y_3 y_4 - y_4^2 + y_2 y_5 + y_4 y_5$$

$$= y_1^2 - (y_3 - y_4)^2 + y_2 y_5 - y_4 y_5. \qquad (4)$$

414

Now, let $z_3 = y_3 - y_4$; $z_i = y_i$ $(i \neq 3)$. Then, (4) becomes

$$F_2 = z_1^2 - z_3^2 + z_2 z_5 - z_4 z_5.$$

Let $w_4 = z_2 - z_4 - z_5$; $w_i = z_i$ $(i \neq 4)$.

$$F_3 = w_1^2 - w_3^2 + (w_4 + w_5)w_5$$

$$= w_1^2 - w_3^2 + (w_5 + \tfrac{1}{2}w_4)^2 - \tfrac{1}{4}w_4^2.$$

Finally, let $v_2 = w_5 + \tfrac{1}{2}w_4$, $v_5 = w_2$, $v_i = w_i$ $(i = 1,3,4)$.

$$F_4 = v_1^2 + v_2^2 - v_3^2 - \tfrac{1}{4}v_4^2.$$

● **PROBLEM** 15-8

Let $5x_1^2 + 4x_1x_2 + 8x_2^2 = 9$ be a conic. Use the coordinate transformation equations

$$x_1 = y_1 + y_2$$

$$x_2 = -y_1 + y_2/2$$

to express the conic in canonical form.

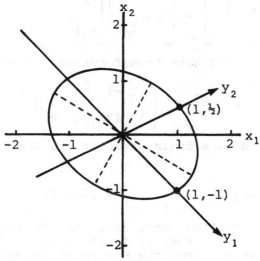

Solution: A quadratic form

$$A = \sum_{i=1}^{n} \sum_{j=1}^{n} a_{ij}x_{ij} = a_{11}x_1^2 + a_{12}x_1x_2 + \ldots + a_{nn}x_n^2,$$

is said to be in canonical form when the corss-product terms have been eliminated, i.e., when

$$A = b_{11}y_1^2 + b_{22}y_2^2 + \ldots + b_{nn}y_n^2,$$

where the y_i are the coordinates of the form in a new coordinate system.

Since a quadratic form can also be expressed in matrix form as:

$$A = [x_1 x_2 \ldots x_n] \begin{bmatrix} a_{11} & a_{12} & \cdots & a_{nn} \\ a_{21} & & & \\ \vdots & & & \\ a_{n1} & \cdots\cdots & & a_{nn} \end{bmatrix} \begin{bmatrix} x_1 \\ x_2 \\ \vdots \\ x_n \end{bmatrix},$$

the problem of conversion to canonical form can be regarded as the problem of choosing a suitable basis so that the matrix of the quadratic form is diagonal. Now, a given symmetric matrix can be diagonalized if there exists a matrix P such that $P^T AP$ is diagonal. The matrix P represents the transition matrix from the old to the new coordinate system.

$$P = \begin{bmatrix} 1 & 1 \\ -1 & 1/2 \end{bmatrix}$$

since we are given that

$$\begin{bmatrix} x_1 \\ x_2 \end{bmatrix} = \begin{bmatrix} 1 & 1 \\ -1 & 1/2 \end{bmatrix} \begin{bmatrix} y_1 \\ y_2 \end{bmatrix},$$

which is the coordinate transformation system. Hence,

$$P^T = \begin{bmatrix} 1 & -1 \\ 1 & 1/2 \end{bmatrix} \text{ and } P^T AP = \begin{bmatrix} 1 & -1 \\ 1 & 1/2 \end{bmatrix}\begin{bmatrix} 5 & 2 \\ 2 & 8 \end{bmatrix}\begin{bmatrix} 1 & 1 \\ -1 & 1/2 \end{bmatrix} = \begin{bmatrix} 9 & 0 \\ 0 & 9 \end{bmatrix}.$$

The equation of the conic, relative to the new axes, is, therefore,

$$9y_1^2 + 9y_2^2 = 9 \quad \text{or} \quad y_1^2 + y_2^2 = 1.$$

Although the equation looks like that of a circle, it can be seen from its graph that it is actually an ellipse. The discrepancy arises from the nature of the new basis vectors.

The matrix P showed that the unit points $(1,0)$ and $(0,1)$ are mapped onto $(1,-1)$ and $(1,1/2)$, respectively. These new basis vectors are neither orthogonal $((1,-1) \cdot (1,1/2) = 1/2)$ nor of unit length $(\sqrt{2}, \sqrt{5}/2)$. Hence, when transforming from a coordinate system with orthogonal axes, some distortions were produced.

Reduce the equation $5x_1^2 + 4x_1x_2 + 8x_2^2 = 9$ to canonical form and use the new equation to sketch its graph.

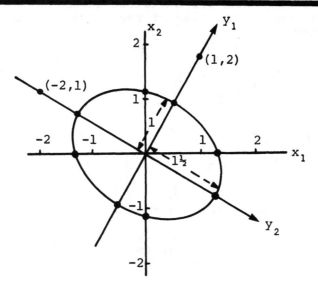

Solution: A quadratic form $f(x_1, x_2, \ldots, x_n)$

$$A = \sum_{i=1}^{n} \sum_{j=1}^{n} a_{ij} x_i{}^j{}_j$$

is said to be in canonical form if the cross-product terms have been eliminated, i.e.,

$$A = \sum_{i=j=1}^{n} a_{ij} x_i x_j.$$

The given equation represents a quadric surface since the expression on the left is the quadratic form

$$q(x_1, x_2) = 5x_1^2 + 2x_1x_2 + 2x_2x_1 + 8x_2^2.$$

This quadratic form can be expressed in matrix form as:

$$\begin{bmatrix} x_1 & x_2 \end{bmatrix} \begin{bmatrix} 5 & 2 \\ 2 & 8 \end{bmatrix} \begin{bmatrix} x_1 \\ x_2 \end{bmatrix} \qquad (1)$$

We would like to accomplish a transformation of the coordinates (x_1, x_2) to new coordinates (y_1, y_2) in which the quadratic form is reduced to canonical form. This corresponds to a change of basis where the matrix representing the quadratic form is reduced to a diagonal matrix. In other words, the problem of finding the canonical form of q is equivalent

417

to that of finding a diagonal matrix that is similar to the
symmetric matrix in (1). Any symmetric matrix is diagonal-
izeable; i.e., is similar to a diagonal matrix. This is the
Spectral Theorem of linear algebra. For real symmetric
matrices, we have the result that all the eigenvalues are
real and form the diagonal elements of the diagonalized matrix. To show this for the given quadratic form, proceed as
follows: The eigenvalues of the matrix in (1) are scalars
λ such that

$$\begin{bmatrix} 5 & 2 \\ 2 & 8 \end{bmatrix} \begin{bmatrix} x_1 \\ x_2 \end{bmatrix} = \lambda \begin{bmatrix} x_1 \\ x_2 \end{bmatrix} \qquad (2)$$

Therefore,

$$\begin{bmatrix} 5-\lambda & 2 \\ 2 & 8-\lambda \end{bmatrix} \begin{bmatrix} x_1 \\ x_2 \end{bmatrix} = 0.$$

To find the characteristic equation, take the determinant of

$\begin{bmatrix} 5-\lambda & 2 \\ 2 & 8-\lambda \end{bmatrix}$ and set it equal to 0. As a result,

the characteristic equation

$$(5 - \lambda)(8 - \lambda) - 4 = 0 \quad \text{or} \quad (\lambda - 9)(\lambda - 4) = 0.$$

Hence, the eigenvalues are $\lambda_1 = 9$ and $\lambda_2 = 4$. Associated
with $\lambda_1 = 9$, find an eigenvector by solving the system:

$$\begin{bmatrix} -4 & 2 \\ 2 & -1 \end{bmatrix} \begin{bmatrix} x_1 \\ x_2 \end{bmatrix} = \begin{bmatrix} 0 \\ 0 \end{bmatrix}.$$

The vector $[x_1, x_2] = [1,2]$ satisfies the system.
    Similarly, for $\lambda_2 = 4$, the system becomes:

$$\begin{bmatrix} 1 & 2 \\ 2 & 4 \end{bmatrix} \begin{bmatrix} x_1 \\ x_2 \end{bmatrix} = \begin{bmatrix} 0 \\ 0 \end{bmatrix}.$$

and $[-2,1]$ is an eigenvector associated with $\lambda_2 = 4$.
Furthermore, note that $[1,2] \cdot [-2,1] = 0$, i.e., the two
eigenvectors are orthogonal. Next, we normalize the two
vectors: that is, make them unit vectors by dividing them
by their respective lengths. Since the length of a vector
in R^n is given by $\|u\| = (u \cdot u)^{1/2}$,

$$([1,2] \cdot [1,2])^{1/2} = \sqrt{5}; \quad ([-2,1] \cdot [-2,1])^{1/2} = \sqrt{5}.$$

Now form a matrix P whose columns are composed of the components of the orthonormal eigenvectors. This orthogonal matrix P has the property that $P^{-1}AP$ is a diagonal matrix. Thus,

$$P = \begin{bmatrix} 1/\sqrt{5} & -2/\sqrt{5} \\ 2/\sqrt{5} & 1/\sqrt{5} \end{bmatrix}$$

and, since P is orthogonal, $P^{-1} = P^{T}$. Therefore, P

$$P^{-1}AP = \begin{bmatrix} 1/\sqrt{5} & 2/\sqrt{5} \\ -2/\sqrt{5} & 1/\sqrt{5} \end{bmatrix} \begin{bmatrix} 5 & 2 \\ 2 & 8 \end{bmatrix} \begin{bmatrix} 1/\sqrt{5} & -2/\sqrt{5} \\ 2/\sqrt{5} & 1/\sqrt{5} \end{bmatrix} = \begin{bmatrix} 9 & 0 \\ 0 & 4 \end{bmatrix}.$$

Thus, using the columns of P as the basis vectors of the new coordinate system, the given quadratic form can be written as:

$$9y_1^2 + 4y_2^2 = 9 \quad \text{or} \quad y_1^2 + \frac{4}{9}y_2^2 = 1.$$

Note that P is the transition matrix from the old coordinates (x_1, x_2), for which we had the usual basis $\{(1,0), (0,1)\}$, to the new coordinates (y_1, y_2).

The graph is that of an ellipse. Note that the transformation was orthogonal, i.e., the new axes are perpendicular to each other. Thus, the original figure has merely been rotated without suffering any distortion. In particular, we can now identify the major and minor axes of the ellipse.

● **PROBLEM** 15-10

The equation of a quadric surface is given as:

$$Q(x) = 3x_1^2 + 4x_2^2 + 2\sqrt{3}x_2x_3 + 6x_3^2 = 21. \tag{1}$$

Reduce the quadratic form to diagonal form and sketch the resulting surface.

Solution: A quadratic form can be written as

$$Q(x_1, \ldots, x_n) = [x_1 x_2 \cdots x_n] \begin{bmatrix} a_{11} & a_{12} & \cdots & a_{1n} \\ a_{21} & a_{22} & \cdots & a_{2n} \\ \vdots & \vdots & & \vdots \\ a_{n1} & a_{n2} & & a_{nn} \end{bmatrix} \begin{bmatrix} x_1 \\ x_2 \\ \vdots \\ x_n \end{bmatrix} \tag{2}$$

where the matrix is symmetric. The given equation (1) may be put into the form (2). Thus,

$$[x_1 x_2 x_3] \begin{bmatrix} 3 & 0 & 0 \\ 0 & 4 & \sqrt{3} \\ 0 & \sqrt{3} & 6 \end{bmatrix} \begin{bmatrix} x_1 \\ x_2 \\ x_3 \end{bmatrix} = 21. \tag{3}$$

The problem is to diagonalize the symmetric matrix in (3). We use the facts that the eigenvalues of a real n x n symmetric matrix are real, and that the eigenvectors associated with the eigenvalues form an orthogonal basis for R^n. This enables us to construct an orthogonal matrix P such that $P^{-1}AP$ is diagonal, where A is the matrix of the quadratic form. First, set $Ax = \lambda x$ or $(A - \lambda I)x = 0$ where

$$(a - \lambda\ J) = \begin{bmatrix} 3-\lambda & 0 & 0 \\ 0 & 4-\lambda & \sqrt{3} \\ 0 & \sqrt{3} & 6-\lambda \end{bmatrix}$$

and λ is a scalar. To find the characteristic equation, set $\det(A - \lambda I) = 0$, so that

$$\det(A - \lambda I) = (3 - \lambda)(4 - \lambda)(6 - \lambda) - \sqrt{3}\sqrt{3}(3 - \lambda)$$

$$= (3 - \lambda)(\lambda - 3)(\lambda - 7) = 0.$$

Therefore, our characteristic equation is

$$(3 - \lambda)(\lambda - 7)(\lambda - 3) = 0.$$

Hence, the three eigenvalues are 3, 3 and 7. Substituting $\lambda = 3$,

$$(A - 3I)x = 0 = \begin{bmatrix} 0 & 0 & 0 \\ 0 & 1 & \sqrt{3} \\ 0 & \sqrt{3} & 3 \end{bmatrix} \begin{bmatrix} x_1 \\ x_2 \\ x_3 \end{bmatrix}. \tag{4}$$

From (4), $\sqrt{3}x_2 + 3x_3 = 0$ or $-\sqrt{3}x_3 = x_2$. Since x_1 is arbitrary, eigenvectors for $\lambda = 3$ are of the form $(a, -\sqrt{3}b, b)$ where a and b are arbitrary real numbers. Let $a = 1$, $b = 0$; then $(1,0,0)$ is one eigenvector. Let $a = 0$, $b = 1$; then $(0,-\sqrt{3},1)$ is another eigenvector. Furthermore, $(1,0,0) \cdot (0,-\sqrt{3},1) = 0$, i.e., the two eigenvectors are orthogonal.

Next, let $\lambda = 7$ to obtain

$$(A - 7I)x = 0 = \begin{bmatrix} -4 & 0 & 0 \\ 0 & -3 & \sqrt{3} \\ 0 & \sqrt{3} & -1 \end{bmatrix} \begin{bmatrix} x_1 \\ x_2 \\ x_3 \end{bmatrix}$$

or,
$$-4x_1 = 0$$

$$-3x_2 + \sqrt{3}x_3 = 0$$

$$\sqrt{3}x_2 - x_3 = 0.$$

Thus, $x_1 = 0$, $x_2 = a$ and $x_3 = \sqrt{3}a$ where "a" is arbitrary. A simple eigenvector is $(0,\sqrt{3},3)$, and it is orthogonal to $(1,0,0)$ and $(0,\sqrt{3},-1)$.

Now, we must normalize the above three eigenvectors. Since $(u \cdot u)^{1/2} = \|u\|$ is the length of a vector in R^n, $\frac{u}{\|u\|}$ is a normalized vector.

$$[(1,0,0) \cdot (1,0,0)]^{1/2} = 1;$$

$$[(0,-\sqrt{3},1) \cdot (0,-\sqrt{3},1)]^{1/2} = 2;$$

$$[(0,\sqrt{3},3) \cdot (0,\sqrt{3},3)]^{1/2} = \sqrt{12} = 2\sqrt{3}.$$

Now form the matrix P whose columns are the components of the three orthonormalized vectors. Thus,

$$P = \begin{bmatrix} 1 & 0 & 0 \\ 0 & \sqrt{3}/2 & 1/2 \\ 0 & -1/2 & \sqrt{3}/2 \end{bmatrix}$$

since P is an orthogonal matrix, $P^{-1} = P^T$. Hence,

$$P^{-1}AP = \begin{bmatrix} 1 & 0 & 0 \\ 0 & \sqrt{3}/2 & -1/2 \\ 0 & 1/2 & \sqrt{3}/2 \end{bmatrix} \begin{bmatrix} 3 & 0 & 0 \\ 0 & 4 & \sqrt{3} \\ 0 & \sqrt{3} & 6 \end{bmatrix} \begin{bmatrix} 1 & 0 & 0 \\ 0 & \sqrt{3}/2 & 1/2 \\ 0 & -1/2 & \sqrt{3}/2 \end{bmatrix}$$

$$= \begin{bmatrix} 3 & 0 & 0 \\ 0 & 3 & 0 \\ 0 & 0 & 7 \end{bmatrix}.$$

The columns of the matrix P are the unit vectors of the new coordinate system within which the quadric surface has the equation

$$Q'(y_1,y_2,y_3) = 3y_1^2 + 3y_2^2 + 7y_3^2 = 21$$

or $\quad y_1^2/7 + y_2^2/7 + y_3^2/3 = 1.$ \hfill (5)

It can be seen from the figure that the surface is an ellipsoid with circular cross section in planes parallel to the $y_1 y_2$ plane.

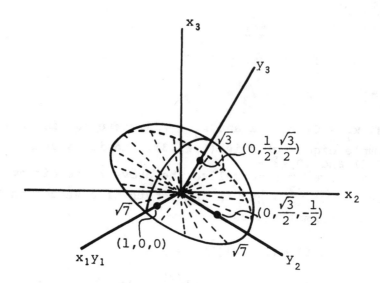

CONIC SECTIONS & QUADRATIC FORMS

● **PROBLEM** 15-11

Find the axes of symmetry of the following ellipsoid:

$$6x_1^2 + 8x_1x_2 - 4x_1x_3 + 12x_2^2 - 8x_2x_3 + 13x_3^2 = k. \qquad (1)$$

Solution: The principal axes or axes of symmetry of a quadric surface are mutually orthogonal. When the axes of symmetry are taken to be the coordinate axes, in terms of these coordinates, the equation representing the surface is in canonical form. This suggests converting the geometric problem of finding the axes of symmetry into the problem of finding the basis that diagonalizes the quadratic form. In fact, finding the basis of eigenvectors corresponding to the diagonalized quadratic form is equivalent to finding the axes of symmetry. Equation (1) can be written in matrix for as:

$$[x_1 x_2 x_3] \begin{bmatrix} 6 & 4 & -2 \\ 4 & 12 & -4 \\ -2 & -4 & 13 \end{bmatrix} \begin{bmatrix} x_1 \\ x_2 \\ x_3 \end{bmatrix}. \qquad (2)$$

The matrix A in (2) is symmetric, i.e., $A = A^T$. Thus, there is a basis consisting of eigenvectors of A. The eigenvectors of A are the axes of symmetry of the surface.

Let

$$Ax = \lambda x \quad \text{or} \quad (A - \lambda I)x = 0 \qquad (3)$$

We wish to find scalars λ and non-zero vectors x such that (3) is true with A given by (2). We obtain:

422

$$\begin{bmatrix} 6-\lambda & 4 & -2 \\ 4 & 12-\lambda & -4 \\ -2 & -4 & 13-\lambda \end{bmatrix} \begin{bmatrix} x_1 \\ x_2 \\ x_3 \end{bmatrix} = \begin{bmatrix} 0 \\ 0 \\ 0 \end{bmatrix}. \tag{4}$$

Equation (4) can have non-zero solutions for $[x_1, x_2, x_3]$ only if $\det[A - I\lambda] = 0$. Thus, we must find the roots of the characteristic equation:

$$\lambda^3 - 31\lambda^2 + 270\lambda - 648 = 0. \tag{5}$$

Equation (5) may be rewritten as:

$$\lambda(\lambda^2 - 31\lambda + 270) = 648. \tag{6}$$

648 can be decomposed into $+2 \times +2 \times +2 \times +3 \times +3 \times +3 \times +3$. $\lambda = 648$ is not a solution of (6) so the solutions must be multiplicative combinations of the prime factors above. By trial and error, we find $\lambda_1 = 4$, $\lambda_2 = 9$ and $\lambda_3 = 18$. When $\lambda = 4$, equation (4) becomes

$$2x_1 + 4x_2 - 2x_3 = 0$$

$$4x_1 + 8x_2 - 4x_3 = 0$$

$$-2x_1 - 4x_2 + 9x_3 = 0.$$

One solution to the above system is $[x_1, x_2, x_3] = [2, -1, 0]$.
Letting $\lambda = 9$ and $\lambda = 18$ in (4), we obtain, in a similar manner, the eigenvectors $[2, 4, 5]$ and $[1, 2, -2]$, respectively.
These three eigenvectors are mutually orthogonal, and they give the directions of the principal axes.

● **PROBLEM** 15-12

Construct an orthogonal matrix from the eigenvectors associated with the symmetric matrix:

$$A = \begin{bmatrix} 5 & 1 & 1 \\ 1 & 5 & -1 \\ 1 & -1 & 5 \end{bmatrix}. \tag{1}$$

How does the transformation $x = Hy$ affect the related quadric surface, where H is the orthogonal matrix?

<u>Solution</u>: An orthogonal matrix is a matrix H such that $H \cdot H^T = I$, where I is the identity matrix. If H is a matrix whose column vectors are mutually orthogonal, then H is an orthogonal matrix.
We know that any symmetric real $n \times n$ matrix is diagonalizable, and the eigenvectors of such a matrix form an or-

thogonal basis of R^n. Thus, from the given matrix A which is symmetric, we construct the associated orthogonal matrix H. This is done by finding the eigenvectors of A, normalizing them, and making them the column vectors of H.

Recall that these eigenvectors also form the basis of a coordinate system in which any quadric surface whose quadric form corresponds to A is in canonical form. That is, the trasformation x = Hy will give us a coordinate transformation that diagonalizes the quadratic form.

The quadratic form associated with the symmetric matrix (1) is:

$$[x_1 x_2 x_3] \begin{bmatrix} 5 & 1 & 1 \\ 1 & 5 & -1 \\ 1 & -1 & 5 \end{bmatrix} \begin{bmatrix} x_1 \\ x_2 \\ x_3 \end{bmatrix}$$

$$= 5x_1^2 + 2x_1 x_2 + 5x_2^2 + 5x_3^2 + 2x_1 x_3 - 2x_2 x_3.$$

The eigenvalues associated with (1) are scalars λ that satisfy the matrix equation

$$Ax = \lambda x. \tag{2}$$

From (2), we obtain

$$[A - I\lambda]x = 0$$

which implies $\det[A - \lambda I] = 0$. Thus, the result is the characteristic equation:

$$\det \begin{bmatrix} 5-\lambda & 1 & 1 \\ 1 & 5-\lambda & -1 \\ 1 & -1 & 5-\lambda \end{bmatrix} = 0. \tag{3}$$

Expanding (3) along the first row,

$$(5 - \lambda)[(5 - \lambda)^2 - 1] - 1[(5 - \lambda) + 1] + 1[-1 - (5 - \lambda)] = 0$$

$$\lambda^3 - 15\lambda^2 + 72\lambda - 108 = 0. \tag{4}$$

This can be factored as

$$(\lambda - x)(\lambda - y)(\lambda - z) = 0.$$

Thus, $\quad xyz = 108$

$$x + y + z = 15.$$

By trial and error, we find x = 3 and y = z = 6; hence, (4) is $(\lambda - 3)(\lambda - 6)(\lambda - 6) = 0$.

Thus, the eigenvalues are $\lambda_1 = 3$; $\lambda_2 = 6$; $\lambda_3 = 6$. An eigenvector associated with $\lambda_1 = 3$ is found by solving the system:

$$\begin{bmatrix} 2 & 1 & 1 \\ 1 & 2 & -1 \\ 1 & -1 & 2 \end{bmatrix} \begin{bmatrix} x_1 \\ x_2 \\ x_3 \end{bmatrix} = \begin{bmatrix} 0 \\ 0 \\ 0 \end{bmatrix}.$$

One solution is $[-1,1,1]$. Next, find mutually orthogonal eigenvectors corresponding to the double root $\lambda = 6$. Form the system

$$\begin{bmatrix} -1 & 1 & 1 \\ 1 & -1 & -1 \\ 1 & -1 & -1 \end{bmatrix} \begin{bmatrix} x_1 \\ x_2 \\ x_3 \end{bmatrix} = \begin{bmatrix} 0 \\ 0 \\ 0 \end{bmatrix}.$$

Two solutions to this system are $[1,1,0]$ and $[1,-1,2]$. Further, since $[1,1,0] \cdot [1,-1,2] = 0$, they are orthogonal.

The next stage is to form an orthogonal matrix whose columns are the components of the normalized eigenvectors. The norms of $[-1,1,1]$, $[1,1,0]$ and $[1,-1,2]$ are $\sqrt{3}$, $\sqrt{2}$ and $\sqrt{6}$, respectively. Hence, the required orthogonal matrix is

$$H = \begin{bmatrix} -1/\sqrt{3} & 1/\sqrt{2} & 1/\sqrt{6} \\ 1/\sqrt{3} & 1/\sqrt{2} & -1/\sqrt{6} \\ 1/\sqrt{3} & 0 & 2/\sqrt{6} \end{bmatrix}.$$

Since H is orthogonal, $H^{-1} = H^T$, and $H^{-1}AH = H^TAH$ is a 3×3 diagonal matrix with diagonal entries 3, 6, 6. The transformation

$$\begin{bmatrix} x_1 \\ x_2 \\ x_3 \end{bmatrix} = H \begin{bmatrix} y_1 \\ y_2 \\ y_3 \end{bmatrix}$$

transforms the equation

$$5x_1^2 + 2x_1x_2 + 2x_1x_3 + 5x_2^2 - 2x_2x_3 + 5x_3^2 = k$$

in the (x_1, x_2, x_3) coordinate system to the equation

$$3y_1^2 + 6y_2^2 + 6y_3^2 = k$$

in the (y_1, y_2, y_3) system.

Reduce the quadric surface

$$3x^2 + 5y^2 + 2z^2 + 4xy\sqrt{2} + 5x - 2y + 7z + 10 = 0 \qquad (1)$$

to canonical form.

Solution: A quadratic form

$$Q = \sum_{i=1}^{n} \sum_{j=1}^{n} a_{ij}x_ix_j$$

is said to be in canonical form when the cross-product terms x_ix_j, $i \neq j$ have been eliminated. Equation (1) above is not a quadratic form. However, the first few terms,

$$3x^2 + 5y^2 + 2z^2 + 4\sqrt{2}xy, \qquad (2)$$

are a quadratic form.

The procedure for reducing (2) to canonical form is as follows. First, find the matrix associated with the quadratic form (2):

$$[x \; y \; z] \begin{bmatrix} 3 & 2\sqrt{2} & 0 \\ 2\sqrt{2} & 5 & 0 \\ 0 & 0 & 2 \end{bmatrix} \begin{bmatrix} x \\ y \\ z \end{bmatrix}$$

since the matrix is symmetric, it has real eigenvalues. Form the system

$$\begin{bmatrix} 3-\lambda & 2\sqrt{2} & 0 \\ 2\sqrt{2} & 5-\lambda & 0 \\ 0 & 0 & 2-\lambda \end{bmatrix} \begin{bmatrix} x \\ y \\ z \end{bmatrix} = \begin{bmatrix} 0 \\ 0 \\ 0 \end{bmatrix}.$$

This system has a non-zero solution in $[x,y,z]$ only if the determinant of the coefficient matrix is zero, thus obtaining the characteristic equation by expansion along the third row:

$$(2 - \lambda)[(3 - \lambda)(5 - \lambda) - 8] = 0$$

or, $(2 - \lambda)[\lambda^2 - 8\lambda + 7] = (2 - \lambda)(\lambda - 7)(\lambda - 1) = 0. \qquad (3)$

From (3), $\lambda_1 = 2$, $\lambda_2 = 1$ and $\lambda_3 = 7$ are the eigenvalues. The eigenvalue $\lambda = 7$ yields

$$-4x_1 + 2\sqrt{2}x_2 = 0$$

$$2\sqrt{2}x_1 - 2x_2 = 0$$

$$-5x_3 = 0$$

and an eigenvector $(1,\sqrt{2},0)$. The norm of this eigenvector is $[(1,\sqrt{2},0) \cdot (1,\sqrt{2},0)]^{1/2} = \sqrt{3}$, and, hence, the normalized eigenvector has coordinates $(1/\sqrt{3},\sqrt{2}/\sqrt{3},0)$.

The eigenvalue $\lambda = 2$ yields the eigenvector $(0,0,1)$ which is already of unit length. The eigenvalue $\lambda = 1$ has associated with it the eigenvector $(\sqrt{2},-1,0)$. The norm of this eigenvector is $[(\sqrt{2},-1,0) \cdot (\sqrt{2},-1,0)]^{1/2} = \sqrt{3}$, and,

thus, the normalized eigenvector is $(\sqrt{2}/3,-1/\sqrt{3},0)$. Now form an orthogonal matrix whose columns are the components of the normalized eigenvectors:

$$H = \begin{bmatrix} 1/\sqrt{3} & 0 & \sqrt{2}/\sqrt{3} \\ \sqrt{2}/\sqrt{3} & 0 & -1/\sqrt{3} \\ 0 & 1 & 0 \end{bmatrix}.$$

The matrix product $H^{T}AH$ will now be the diagonal matrix,

$$\begin{bmatrix} 7 & 0 & 0 \\ 0 & 2 & 0 \\ 0 & 0 & 1 \end{bmatrix}$$

thus reducing the quadratic form to canonical form.

But what about the terms $5x - 2y + 7z$? Recall that the matrix H is actually a transition matrix from the old coordinates x, y, z to new coordinates in which the quadratic form part of (1) is diagonalized.

Form the coefficient vector $[5,-2,7] = B$. Applying H to B, we will have expressed the remaining part of (1) in terms of our new coordinates, and we can then express the entire equation in these terms. Thus,

$$BH = [5,-2,7] \begin{bmatrix} 1/\sqrt{3} & 0 & 2/\sqrt{3} \\ \sqrt{2}/\sqrt{3} & 0 & -1/\sqrt{3} \\ 0 & 1 & 0 \end{bmatrix} = \left[\frac{5-2\sqrt{2}}{\sqrt{3}}, 7, \frac{5\sqrt{2}+2}{\sqrt{3}} \right].$$

The equation of the surface becomes

$$7x_1^2 + 2y_1^2 + z_1^2 + \frac{(5-2\sqrt{2})}{\sqrt{3}}x_1 + 7y_1 + \frac{(5\sqrt{2}+2)}{\sqrt{3}}z_1 + 10 = 0.$$

Now this can be simplified even further by completing the square.

$$7x_1^2 + \frac{(5-2\sqrt{2})}{\sqrt{3}}x_1 + 2y_1^2 + 7y_1 + z_1^2 + \frac{(5\sqrt{2}+2)}{\sqrt{3}}z_1 + 10$$

$$= 7\left(x_1 + \frac{5-2\sqrt{2}}{14\sqrt{3}}\right)^2 + 2\left(y_1 + \frac{7}{4}\right)^2 + \left(z_1 + \frac{5\sqrt{2}+2}{2\sqrt{3}}\right)^2 - K = 0,$$

where $K = -\dfrac{(57 + 80\sqrt{2})}{56}$. . Let

$$x_1 + \frac{5-2\sqrt{2}}{14\sqrt{3}} = x_2$$

$$y_1 + \frac{7}{4} = y_2$$

$$z_1 + \frac{5\sqrt{2}+2}{2\sqrt{3}} = z_2.$$

Then (1) is reduced to the canonical form,

$$7x_2^2 + 2y_2^2 + z_2^2 - K = 0.$$

● **PROBLEM** 15-14

Reduce the quadric surface

$$x^2 + 4y^2 + 2z^2 + 4xy + 5x - 2y + 7z + w + 10 = 0 \qquad (1)$$

to canonical form.

Solution: Break up the equation of the surface into two parts. Rewrite (1) as

$$(x^2 + 4y^2 + 2z^2 + 4xy) + (5x - 2y + 7z + w + 10) = 0.$$

The expression within the first pair of parentheses is a quadratic form. It is equivalent to

$$\begin{bmatrix} x, & y, & z, & w \end{bmatrix} \begin{bmatrix} 1 & 2 & 0 & 0 \\ 2 & 4 & 0 & 0 \\ 0 & 0 & 2 & 0 \\ 0 & 0 & 0 & 0 \end{bmatrix} \begin{bmatrix} x \\ y \\ z \\ w \end{bmatrix}$$

which is of the form x'Ax. The matrix A is symmetric. Hence, it has real eigenvalues, and, associated with these eigenvalues, mutually orthogonal eigenvectors. To obtain the eigenvalues, find scalars λ such that

$$Ax = \lambda x$$

for non-zero vectors $x \varepsilon R^4$. This leads to the system of equations

$$[A - \lambda I]x = 0 \qquad (2)$$

which implies, for non-trivial solutions,

$$\det[A - \lambda I] = 0,$$

where I is the 4 × 4 identity matrix. Thus, we must evaluate

$$\begin{vmatrix} 1-\lambda & 2 & 0 & 0 \\ 2 & 4-\lambda & 0 & 0 \\ 0 & 0 & 2-\lambda & 0 \\ 0 & 0 & 0 & -\lambda \end{vmatrix} = 0.$$

Expanding along the first column and then the third column of the 3 × 3 determinant,

$$-\lambda\{(2 - \lambda)[(4 - \lambda)(1 - \lambda) - 4]\}$$

$$= \lambda^4 - 7\lambda^3 + 10\lambda^2$$

$$= \lambda^2[\lambda^2 - 7\lambda + 10] = 0.$$

Hence, $\lambda_1 = 5$; $\lambda_2 = 2$; $\lambda_3 = 0$; $\lambda_4 = 0$. Eigenvectors associated with these eigenvalues are found by successively substituting λ_i (i = 1,...,4) into (2) and finding the resulting non-zero solutions. Representative eigenvectors for each eigenvalue are:

$$[1,2,0,0], \ [0,0,1,0], \ [2,-1,0,0] \ \text{and} \ [0,0,0,1].$$

When normalized, the components of these eigenvectors become

$$[1/\sqrt{5}, \ 2/\sqrt{5}, \ 0,0], \ [0,0,1,0], \ [2/\sqrt{5}, \ -1/\sqrt{5},0,0] \ \text{and} \ [0,0,0,1].$$

The matrix whose columns are the normalized eigenvectors above is the required orthogonal matrix. It represents the coordinate transformation that will reduce the quadric surface:

$$H = \begin{bmatrix} 1/\sqrt{5} & 0 & 2/\sqrt{5} & 0 \\ 2/\sqrt{5} & 0 & -1/\sqrt{5} & 0 \\ 0 & 1 & 0 & 0 \\ 0 & 0 & 0 & 1 \end{bmatrix}.$$

Using H, the quadratic form $x^2 + 4y^2 + 2z^2 + 4xy$ is reduced to $5x_1^2 + 2y_1^2$ in the new coordinate system.

Next, simplify B = 5x - 2y + 7z + w.

$$B = [5 \ -2 \ 7 \ 1]\begin{bmatrix} x \\ y \\ z \\ w \end{bmatrix} = bX,$$

and $bH = [1/\sqrt{5}, \ 7, \ 12/\sqrt{5}, \ 1]$.

Thus, the equation of the surface is now

$$5x_1^2 + 2y_1^2 + x_1/\sqrt{5} + 7y_1 + (12/\sqrt{5})z_1 + w_1 + 10 = 0. \qquad (3)$$

Equation (3) is capable of further simplification.

$$5x_1^2 + x_1/\sqrt{5} + 2y_1^2 + 7y_1 + 12/\sqrt{5}z_1 + w_1 + 10$$

$$= 5\left[x_1 + \frac{1}{10\sqrt{5}}\right]^2 + 2\left[y_1 + \frac{7}{4}\right]^2 + \frac{12}{\sqrt{5}}z_1 + w_1 + 3.874 = 0.$$

Let $x_2 = x_1 + \frac{1}{10\sqrt{5}}$, $y_2 = y_1 + \frac{7}{4}$, $z_2 = \frac{12}{\sqrt{5}}z_1 + 3.874$, and $w_2 = w_1$. Then,

$$5x_2^2 + 2y_2^2 + z_2 + w_2 = 0.$$

Set $x = x_2$, $y = y_2$ and

$$\begin{bmatrix} z \\ w \end{bmatrix} = \begin{bmatrix} 1 & 1 \\ 0 & 1 \end{bmatrix} \begin{bmatrix} z_2 \\ w_2 \end{bmatrix}$$

Then, the original equation of the quadric surface reduces to the canonical form

$$5x^2 + 2y^2 + z = 0.$$

● **PROBLEM** 15-15

Identify the following quadratic forms by finding their eigenvalues:

1) $Q(x,y) = 2x^2 - 4xy - y^2$

2) $Q(x,y) = 9x^2 + 6xy + y^2$.

Solution: Every quadratic form has associated with it a symmetric matrix S. The coordinate change that reduces the quadric surface to a recognizable conic is exactly the change of basis which diagonalizes the matrix S. It is the diagonalized matrix whose elements are eigenvalues of S that represents the reduced quadratic form. The type of conic surface represented is determined by the nature of the eigenvalues. Thus, it makes sense to say that we can "identify" a quadratic form by its eigenvalues.

1) $Q(x,y) = 2x^2 - 4xy - y^2 = [x \quad y] \begin{bmatrix} 2 & -2 \\ -2 & -1 \end{bmatrix} \begin{bmatrix} x \\ y \end{bmatrix}.$ \qquad (1)

To find the eigenvalues of the matrix in (1), set

$$\begin{bmatrix} 2 & -2 \\ -2 & -1 \end{bmatrix} \begin{bmatrix} x \\ y \end{bmatrix} = \lambda \begin{bmatrix} x \\ y \end{bmatrix}$$

or,
$$\begin{bmatrix} 2-\lambda & 2 \\ -2 & -1-\lambda \end{bmatrix} \begin{bmatrix} x \\ y \end{bmatrix} = \begin{bmatrix} 0 \\ 0 \end{bmatrix}. \tag{2}$$

System (2) can have non-trivial solutions only if

$$\det \begin{bmatrix} 2-\lambda & -2 \\ -2 & -1-\lambda \end{bmatrix} = 0$$

or, $(\lambda - 3)(\lambda + 2) = 0.$

Hence, the eigenvalues are $\lambda = 3$ and $\lambda = -2$, and there exists, with respect to some basis, a diagonal matrix

$$D = \begin{bmatrix} 3 & 0 \\ 0 & -2 \end{bmatrix} \tag{30}$$

which represents the quadratic form. To show this, first find two orthogonal eigenvectors associated with the eigenvalues. If $\lambda = 3$, (2) becomes the system

$$\begin{bmatrix} -1 & -2 \\ -2 & -4 \end{bmatrix} \begin{bmatrix} x \\ y \end{bmatrix} = \begin{bmatrix} 0 \\ 0 \end{bmatrix}.$$

By solving the system, we find that an eigenvector associated with $\lambda = 3$ is $[x,y] = [2,-1]$. If $\lambda = -2$, by substitution (2) becomes

$$\begin{bmatrix} 4 & -2 \\ -2 & 1 \end{bmatrix} \begin{bmatrix} x \\ y \end{bmatrix} = \begin{bmatrix} 0 \\ 0 \end{bmatrix},$$

and solving, we find an eigenvector $[x,y] = [1,2]$. These two eigenvectors are linearly independent. In fact, they are orthogonal since $[1,2] \cdot [2,-1] = 0$. If normalized, they will form an orthonormal basis. Since both vectors are of norm $\sqrt{5}$,

$$B = \{1/\sqrt{5}[2,-1], \ 1/\sqrt{5}[1,2]\}$$

is an orthonormal basis. With respect to this basis, the quadratic form has the diagonal matrix

$$\begin{bmatrix} 3 & 0 \\ 0 & -2 \end{bmatrix}.$$

so, $\begin{bmatrix} 3 & 0 \\ 0 & -2 \end{bmatrix}\begin{bmatrix} x^2 \\ y^2 \end{bmatrix} = 3x^2 - 2y^2$ which is the equation for a hyperbola.

2). Note that in 1), the diagonal matrix had the eigenvalues of the matrix in (1) as its diagonal entries:

$$Q(x,y) = 9x^2 + 6xy + y^2 = [x \ \ y]\begin{bmatrix} 9 & 3 \\ 3 & 1 \end{bmatrix}\begin{bmatrix} x \\ y \end{bmatrix}. \qquad (4)$$

The eigenvalues of the matrix in (4) are found by solving

$$\begin{bmatrix} 9 & 3 \\ 3 & 1 \end{bmatrix}\begin{bmatrix} x \\ y \end{bmatrix} = \lambda \begin{bmatrix} x \\ y \end{bmatrix}$$

which yields

$$\begin{bmatrix} 9-\lambda & 3 \\ 3 & 1-\lambda \end{bmatrix}\begin{bmatrix} x \\ y \end{bmatrix} = \begin{bmatrix} 0 \\ 0 \end{bmatrix}.$$

Hence, $\lambda(\lambda - 10) = 0$ or, $\lambda = 0$ and $\lambda = 10$. Thus, the diagonal matrix representing the quadratic form is

$$D = \begin{bmatrix} 10 & 0 \\ 0 & 0 \end{bmatrix},$$

and the equation of the conic is $10x^2$. This is the equation of a parabola.

We could also say that if there are two eigenvalues which are both strictly positive, the surface is an ellipse.

● **PROBLEM** 15-16

Identify the following quadric curves and surfaces

a) $\dfrac{x^2}{2} + \dfrac{y^2}{3} + \dfrac{z^2}{7} = 1$ c) $x^2 + y^2 - \dfrac{z^2}{9} = 1$

b) $\dfrac{x^2}{3} + \dfrac{y^2}{2} = 1$ d) $xy = 1.$

<u>Solution</u>: The functions given above are either curves in R^2 or surfaces in R^3. In the Cartesian plane, the formulae for the ellipse, parabola and hyperbola are given by

$$\frac{(x - x_0)^2}{a^2} + \frac{(y - y_0)^2}{b^2} = 1; \quad (y - y_0)^2 = 4a(x - x_0)$$

or $\quad (x - x_0)^2 = 4a(y - y_0)$, and $\dfrac{(x - x_0)^2_+}{a^2} - \dfrac{(y - y_0)^2}{b^2} = 1$,

respectively. The corresponding surfaces in R^3 are called ellipsoids, paraboloids and hyperboloids.

   a) This equation is an ellipsoid in R^3.

   b) This equation represents an ellipse in R^2, but not an ellipsoid.

   c) This equation represents a hyperboloaid in R^3.

   d) In R^2, $y = 1/x$ is the equation of a hyperbola, and, in R^3, $xy = 1$ is a hyperbolic cylinder, but not a hyperboloid.

● PROBLEM 15-17

Suppose A is a symmetric matrix whose eigenvalues are all strictly positive.

Consider the locus x'Ax = k. Where on the locus is x'x a maximum? Then consider the locus x'x = 1. Given a symmetric matrix A as above, where on x'x = 1 is x'Ax a maximum?

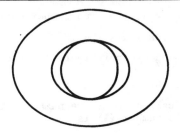

Solution: When the eigenvalues are all positive, the locus is an ellipsoid. Thus, the question asks which point on the locus is the furthest from the origin. Now, for an ellipse, such a point is provided by either end of the major axis. This axis is oriented in the direction of the eigenvector of A corresponding to the minimum eigenvalue.

   To answer the second question, use the figure: The locus x'x = 1 is a circle in two dimensions or a sphere in n-dimensional space. The locus x'Ax = k gives, for varying values of k, a family of ellipsoids. From the figure, for sufficiently large k, the ellipsoid is completely outside the sphere; however, by reducing k continuously, an ellipsoid that is tangent to the inscribed sphere at two points can be obtained. Furthermore, it touches at the end of its shortest axis of symmetry. Thus, x'Ax is a maximum at the two diametrically opposite points on the sphere located by the eigenvector corresponding to the largest eigenvalue of A.

CHAPTER 16

NUMERICAL METHODS OF LINEAR ALGEBRA

DEFINITIONS AND EXAMPLES

● PROBLEM 16-1

Give examples of the following concepts that arise frequently in the methods of numerical computations:
1) Approximate numbers
2) Range number
3) Significant numbers
4) Absolute and relative error

What are some important theorems on relative error?

Solution: 1) The approximate value of a number is some number that differs from the true value by some amount. If x represents the true number and x' the approximate number, then the error ε is given by

$$\varepsilon = \varepsilon(x) = x - x' = \Delta x = dx .$$

We do not know ε but we can specify that its absolute value be less than η . Thus $|\varepsilon| \leq \eta$.
   If x' = 112 and $\eta = 4$, the true number satisfies the relationship

$$108 \leq x \leq 116 \qquad (1)$$

2) Another representation of approximate numbers is given by $\begin{bmatrix} x_H \\ x_L \end{bmatrix}$ where

x_H is the highest possible value of x and x_L is the lowest possible value of x. This is called the range number. The range number representation of the approximate number (1) is

$$\begin{bmatrix} 116 \\ 108 \end{bmatrix} .$$

3) Quite frequently, we wish to operate with only a single number. In such cases the approximate number is recorded as a significant number. This significant number is correct to the last significant digit, that is, the error is at most one-half unit in the last recorded digit. For example, the five-figure approximation to π is 3.1416. This may be successively rounded off to a smaller number of significant figures, i.e., 3.142, 3.14, 3.1, 3.

4) The error defined above is the absolute error. In many situations it is not so much the error as the ratio of the error to the number that is important. The relative error of x, which is defined as

$$\varepsilon_r(x) = \frac{\varepsilon}{x} = \frac{x - x'}{x} = \frac{\Delta x}{x} = 1 - \frac{x'}{x}$$

may be used to measure this ratio.

Some important theorems on relative error are:
1) The absolute value of the relative error of a product (or quotient) is at most equal to the sum of the greatest absolute values of the relative errors of the numbers from which it is formed.
2) The absolute value of the relative error of a power (or root) is at most equal to the absolute value of the power (or the reciprocal of the root) times the greatest relative error of the number.

● **PROBLEM** 16-2

Apart from mistakes in entering or reading data, the two main sources of error in numerical analysis are truncation error and rounding error. The cumulative effect of such errors can make the final result suspect. Give illustrations of truncation and rounding errors.

Solution: Truncation error occurs when an approximate formula is used instead of an exact one. For example, the trigonometric functions (sin x, cos x and their relations) are defined in terms of infinite series. But all tables for these functions chop off an infinite series after a finite number of terms.

Differentiation gives us another example where a true formula may be replaced by an approximate one. Let $y = f(x)$. The value of dy/dx at the point $x = a$ is given by the formula

$$\frac{dy}{dx} = \lim_{h \to 0} \frac{f(a+h) - f(a)}{h} \ . \tag{1}$$

Assuming h is small, dy/dx is given approximately by the formula

$$\frac{f(a+h) - f(a)}{h} \ ,$$

the accuracy increasing as h is decreased. Let $y = e^x/x$. Then, by the rules of differentiation,

$$\frac{dy}{dx} = \frac{xe^x - e^x}{x^2}$$

and so, when x = 2, dy/dx = 1.8473. But now, using (1) we can find approximate values for dy/dx when a = 2 and h is, in turn, 0.2, 0.1, 0.05, 0.01. We find dy/dx $\approx$ 2.038, 1.941, 1.892, 1.85. Clearly, as h decreases,

$$\frac{f(a+h) - f(a)}{h}$$

approaches dy/dx.

Rounding error occurs when numbers involving an infinite sequence of decimals come up in calculations. For example, 1/3, 6/7, etc. must be rounded off when in their decimal form. Let x = 1.50, y = 1.48 and z = 0.02. Let us find 1/x, 1/z, 1/x+y and 1/x-y . We see that

$$\frac{1}{x} = \frac{1}{1.50} \cong 0.667 \tag{1}$$

$$\frac{1}{z} = \frac{1}{0.02} \cong 50 \tag{2}$$

$$\frac{1}{x+y} = \frac{1}{2.98} \cong 0.3356 \tag{3}$$

$$\frac{1}{x-y} = \frac{1}{0.02} \cong 50 \ . \tag{4}$$

Now, assume that those figures are all subject to a maximum error of 0.005 either way and find the greatest and least values of 1/x, 1/z, 1/x+y, and 1/x-y .

435

For (1): The greatest value will be obtained when $x = 1.495$ and the least value when $x = 1.505$. $1/x$ will then be

$$\frac{1}{1.495} \simeq 0.6689 \quad \text{and} \quad \frac{1}{1.505} \simeq 0.6646 \quad \text{respectively.}$$

For (2): The greatest and least values for $1/z$ will obtain when $z = 0.015$ and 0.025 respectively. $1/z$ will be $1/0.015 \simeq 66.7$ and $1/0.025 = 40$.

For (3): When $x = 1.495$ and $y = 1.475$, $1/x+y \simeq 0.3367$. When $x = 1.505$ and $y = 1.485$, $1/x+y \simeq 0.3344$.

For (4): When $x = 1.495$ and $y = 1.485$,

$$\frac{1}{x-y} = \frac{1}{0.01} = 100 . \tag{5}$$

when $x = 1.505$ and $y = 1.475$,

$$\frac{1}{x-y} = \frac{1}{0.03} \simeq 33.3 . \tag{6}$$

Comparing (4), (5) and (6) we see that the number $1/x-y$ can fluctuate wildly when x and y are subject to rounding error.

● PROBLEM 16-3

Reduce the matrix

$$A = \begin{bmatrix} 1 & 4 & 1 & 3 \\ 0 & -1 & 3 & -1 \\ 3 & 1 & 0 & 2 \\ 1 & -2 & 5 & 1 \end{bmatrix} \tag{1}$$

to triangular form.

Solution: A matrix having only zero elements either to the right or to the left of its principal diagonal is known as a triangular matrix.

To convert an $n \times n$ matrix into a triangular matrix we proceed as follows:

1) Choose any non-zero element of the top row, say the ith.

2) Reduce to zero all the other elements of the top row by subtracting suitable multiples of the ith column from the other columns in turn, and leave the ith column unchanged.

3) Next, choosing any convenient non-zero element of the second row of the new matrix, say the jth ($j \neq i$), and leaving both the ith and jth columns unchanged, we can in the same way annul all the elements of the second row other than the ith and the jth.

4) Repeat until all rows have been converted. The resulting matrix is triangular.

Note that we may obtain different triangular matrices corresponding to the same matrix. This occurs because the selection of elements in the rows is arbitrary. The operations 1) - 4) correspond to a sequence of postmultiplications on the given matrix.

We consecutively reduce the rows of (1):

$$\begin{bmatrix} ① & 4 & 1 & 3 \\ 0 & -1 & 3 & -1 \\ 3 & 1 & 0 & 2 \\ 1 & -2 & 5 & 1 \end{bmatrix} \begin{bmatrix} 1 & -4 & -1 & -3 \\ 0 & 1 & 0 & 0 \\ 0 & 0 & 1 & 0 \\ 0 & 0 & 0 & 1 \end{bmatrix} = \begin{bmatrix} 1 & 0 & 0 & 0 \\ 0 & -1 & 3 & -1 \\ 3 & -11 & -3 & -7 \\ 1 & -6 & 4 & -2 \end{bmatrix}$$

$$\begin{bmatrix} 1 & 0 & 0 & 0 \\ 0 & \boxed{-1} & 3 & -1 \\ 3 & -11 & -3 & -7 \\ 1 & -6 & 4 & -2 \end{bmatrix} \begin{bmatrix} 1 & 0 & 0 & 0 \\ 0 & 1 & 3 & -1 \\ 0 & 0 & 1 & 0 \\ 0 & 0 & 0 & 1 \end{bmatrix} = \begin{bmatrix} 1 & 0 & 0 & 0 \\ 0 & -1 & 0 & 0 \\ 3 & -11 & -36 & 4 \\ 1 & -6 & -14 & 4 \end{bmatrix}$$

$$\begin{bmatrix} 1 & 0 & 0 & 0 \\ 0 & -1 & 0 & 0 \\ 3 & -11 & \boxed{-36} & 4 \\ 1 & -6 & -14 & 4 \end{bmatrix} \begin{bmatrix} 1 & 0 & 0 & 0 \\ 0 & 1 & 0 & 0 \\ 0 & 0 & 1 & 1/9 \\ 0 & 0 & 0 & 1 \end{bmatrix} = \begin{bmatrix} 1 & 0 & 0 & 0 \\ 0 & -1 & 0 & 0 \\ 3 & -11 & -36 & 0 \\ 1 & -6 & -14 & 22/9 \end{bmatrix}$$

This last matrix is lower-triangular as required.

GAUSSIAN ELIMINATION WITH PARTIAL PIVOTING

● PROBLEM 16-4

Use a partial pivoting strategy to solve the system
$$(0.100)10^{-3} x_1 + (0.100)10^1 x_2 = (0.200)10^1 \qquad (1)$$
$$(0.100)10^1 \ x_1 + (0.100)10^1 x_2 = (0.300)10^1 \qquad (2)$$
Is the answer markedly superior to merely using Gauss elimination?

Solution: To illustrate partial pivoting consider the system

$$ax + by = p$$
$$cx + dy = q .$$

The pivot (i.e., the coefficient of one of the unknowns that is used to eliminate that unknown in the other equations) is   a.   Solving the system we obtain

$$y = \frac{(q - \frac{c}{a} p)}{(d - \frac{c}{a} b)} \quad , \quad x = \frac{p}{a} - \frac{b}{a} y .$$

Suppose that the coefficients   a,b,c,d,p   and   q   are varied by small amounts. Then, if $|a|$ is small compared to the magnitudes of the other coefficients, the corresponding variations in   x   and   y   could be very large. Since computers frequently make round-off errors on coefficients the method of Gauss elimination may lead to bad solutions if the initial pivoting element is small compared to the other coefficients in the system.
    In general it is advisable when practicing Gaussian elimination to keep the ratios

$$\left| a_{ik}^{(k-1)} \Big/ a_{kk}^{(k-1)} \right| \quad (i,k = 1,\ldots,n)$$

less than   1.   The method of partial pivoting ensures that this is so by interchanging equations   i   and ℓ after determining that $a_{i\ell}^{(\ell-1)}$ is

the element of greatest magnitude in column ℓ .
    Turning to the given system, note that

$$|a_{21}| = (0.100)10^1 \quad \text{is greater than}$$
$$|a_{11}| = (0.100)10^{-3} .$$

Therefore it is necessary to interchange equations (1) and (2) so that the new pivot is $(0.100)10^1$, the element of greatest magnitude in column 1. Eliminating x_1 from the second equation of the permuted system we obtain

$$(0.100)10^1 x_1 + (0.100)10^1 x_2 = (0.300)10^1$$

$$(0.100)10^1 x_2 = (0.200)10^1 \ ,$$

whence

$$x_2 = \frac{(0.200)10^1}{(0.100)10^1} = (0.200)10^1$$

$$x_1 = \frac{[(0.300)10^1 - (0.200)10^1]}{(0.100)10^1} = (0.100)10^1 \ .$$

Thus $x_1 = 1.00$, $x_2 = 2.00$ correct to two decimal places.

As a comparison, suppose we use $(0.100)10^{-3}$ as a pivot; we eliminate x_1 from (2) and obtain

$$(0.100)10^{-3} x_1 + (0.100)10^1 x_2 = (0.200)10^1$$

$$(-0.100)10^5 x_2 = (-0.200)10^5 \ .$$

Hence, by back substitution,

$$x_2 = \frac{(-0.200)}{(-0.100)} = (0.200)10^1$$

$$x_1 = \frac{[(0.200)10^1 - (0.200)10^1]}{(0.100)10^{-3}} = (0.000)10^{-2} \ .$$

Thus we obtain $x_1 = 0.00$, $x_2 = 2.00$ correct to two decimal places. Although the estimate of x_2 is satisfactory, that of x_1 is completely wrong.

● **PROBLEM** 16-5

Solve the linear system
$$14x_1 + 2x_2 + 4x_3 = -10$$
$$16x_1 + 40x_2 - 4x_3 = 55$$
$$-2x_1 + 4x_2 - 16x_3 = -38$$
by Gaussian elimination with partial pivoting.

Solution: Form the augmented matrix

$$\begin{bmatrix} 14 & 2 & 4 & \vdots & -10 \\ 16 & 40 & -4 & \vdots & 55 \\ -2 & 4 & -16 & \vdots & -38 \end{bmatrix} \ .$$

According to the method of partial pivoting the first element in the first row should be the largest in the column. Thus, we interchange the first and second rows, to obtain:

$$\begin{bmatrix} 16 & 40 & -4 & \vdots & 55 \\ 14 & 2 & 4 & \vdots & -10 \\ -2 & 4 & -16 & \vdots & -38 \end{bmatrix} \ .$$

Now, divide the first row by the pivot, 16:

$$\begin{bmatrix} 1 & 2.5 & -0.25 & \vdots & 3.438 \\ 14 & 2 & 4 & \vdots & -10 \\ -2 & 4 & -16 & \vdots & -38 \end{bmatrix} .$$

Next, eliminate the coefficient of the first variable from the second and third rows:

$$\begin{bmatrix} 1 & 2.5 & -.025 & \vdots & 3.438 \\ 0 & -33 & 7.5 & \vdots & -58.132 \\ 0 & 9 & -16.5 & \vdots & -31.124 \end{bmatrix}$$

-14 times the first row was added to the second row; 2 times the first row was added to its third row. The pivot is now -33, the element of largest modulus in the second column. Thus the second row is divided by -33. We obtain:

$$\begin{bmatrix} 1 & 2.5 & -0.25 & \vdots & 3.438 \\ 0 & 1 & -0.227 & \vdots & 1.762 \\ 0 & 9 & -16.5 & \vdots & -31.124 \end{bmatrix} .$$

Eliminate the coefficient of x_2 in all equations occurring below it, i.e., in the third row. -9 times the second row is added to the third row to obtain:

$$\begin{bmatrix} 1 & 2.5 & -0.25 & \vdots & 3.438 \\ 0 & 1 & -0.227 & \vdots & 1.762 \\ 0 & 0 & -14.457 & \vdots & -46.982 \end{bmatrix} .$$

Finally, pivot on -14.457 to obtain:

$$\begin{bmatrix} 1 & 2.5 & -0.25 & \vdots & 3.438 \\ 0 & 1 & -0.227 & \vdots & 1.762 \\ 0 & 0 & 1 & \vdots & 3.254 \end{bmatrix}$$

The solution is now obtained by back-substituting. The system is:

$$x_1 + 2.5x_2 - 0.25x_3 = 3.438$$
$$x_2 - 0.227x_3 = 1.762$$
$$x_3 = 3.254$$

from which $x_3 = 3.25$. Therefore:

$$x_2 = 1.762 + (0.227)(3.25) = 2.500$$
$$x_1 = 3.438 + (0.25)(3.25) - (2.5)(2.500)$$
$$= -1.999 .$$

The exact solution is

$$x_1 = -2, \quad x_2 = 2.5 \quad \text{and} \quad x_3 = 3.25 .$$

● **PROBLEM** 16-6

Find the inverse of the matrix

$$A = \begin{bmatrix} 0.866 & -0.500 & 0.000 \\ 0.500 & 0.866 & 0.000 \\ 0.000 & 0.000 & 1.000 \end{bmatrix}$$

using Gauss elimination with partial pivoting.

<u>Solution</u>: To show the general method of solution, let X be an n ✕ n
matrix with elements x_{ij} and let

$$AX = I \qquad\qquad (1)$$

where I is the unit matrix of order n . Then (1) may be written:

$$\sum_{k=1}^{n} a_{ik}x_{kj} = \delta_{ij} \qquad (i,j = 1,\ldots,n) \qquad\qquad (2)$$

where

$$\delta_{ij} = \begin{cases} 1 & i = j \\ 0 & i \neq j \end{cases} .$$

Let $x_j = [x_{1j},\ldots,x_{nj}]^T$, $e_j = [0,\ldots,0, 1^{(j)}, 0,\ldots,0]^T$

so that x_j is a column vector the elements of which constitute column j
of X and e_j is a column vector with elements the values of which are
all zero save that of the jth which is one. Then the n equations in
(2) are equivalent to the set of n linear systems:

$$Ax_j = e_j \qquad (j = 1,\ldots,n). \qquad\qquad (3)$$

We can solve these n linear systems for the n columns of X and hence
obtain X. But from (1) we see that $X = A^{-1}$. Hence we have a method of
estimating the elements of A^{-1} by solving the n linear systems (3).
    Turning to the given matrix

$$e_1 = [1,0,0]^T,\ e_2 = [0,1,0]^T,\ e_3 = [0,0,1]^T ,$$

and we must solve the three systems

$$Ax_j = e_j \qquad (j = 1,2,3) .$$

Since each system has the same matrix A we solve all three simultaneously,
writing three right-hand sides corresponding to j = 1,2,3. We have:

$$0.866x_{1j}\ -0.500x_{2j} + 0.000x_{3j} = 1\ 0\ 0$$
$$0.500x_{1j} + 0.866x_{2j} + 0.000x_{3j} = 0\ 1\ 0$$
$$0.000x_{1j} + 0.000x_{2j} + 1.000x_{3j} = 0\ 0\ 1 .$$

There is no need to interchange rows since the coefficient of x_{1j} is the
largest in its column. Eliminating x_{1j} from the second and third rows
yields:

$$0.866x_{1j} - 0.500x_{2j} + 0.000x_{3j} = \qquad 1 \qquad 0\ 0$$
$$1.5468x_{2j} + 0.000x_{3j} = -.57737\ 1\ 0$$
$$1.000x_{3j} = \qquad 0 \quad 0\ 1$$

By back substitution we obtain

$$x_{31} = 0.000 \qquad x_{32} = 0.000 \qquad x_{33} = 1.000$$
$$x_{21} = -0.500 \qquad x_{22} = 0.866 \qquad x_{23} = 0.000$$
$$x_{11} = 0.866 \qquad x_{12} = 0.500 \qquad x_{13} = 0.000$$

whence

$$A^{-1} = \begin{bmatrix} 0.866 & 0.500 & 0.000 \\ -0.500 & 0.866 & 0.000 \\ 0.000 & 0.000 & 1.000 \end{bmatrix}$$

correct to three decimal places.
    Note that the above method requires less computation than the ana-
lytic expression for A^{-1} , i.e.,

$$A^{-1} = \frac{1}{|A|} \text{ Adj } A.$$

CROUT METHOD

Solve numerically for x_1, x_2 and x_3 when

$$554.11x_1 - 281.91x_2 - 34.240x_3 = 273.02$$
$$-281.91x_1 + 226.81x_2 + 38.100x_3 = -63.965 \qquad (1)$$
$$- 34.240x_1 + 38.100x_2 + 80.221x_3 = 34.717 \; .$$

Solution: We can use the Crout method for solving sets of linear algebraic equations. Let

$$a_{11}x_1 + a_{12}x_2 + \ldots + a_{1n}x_n = C_1$$
$$a_{21}x_1 + a_{22}x_2 + \ldots + a_{2n}x_n = C_2$$
$$\vdots \qquad\qquad\qquad\qquad\qquad\qquad \vdots \qquad (2)$$
$$a_{n1}x_1 + a_{n2}x_2 + \ldots + a_{nn}x_n = C_n \; .$$

The augmented matrix of (2) is

$$M = \begin{bmatrix} a_{11} & a_{12} & \cdots\cdots\cdots & a_{1n} & \vdots & C_1 \\ a_{21} & a_{22} & \cdots\cdots\cdots & a_{2n} & \vdots & C_2 \\ \vdots & & & & \vdots & \vdots \\ a_{n1} & a_{n2} & \cdots\cdots\cdots & a_{nn} & \vdots & C_n \end{bmatrix} = [A \; \vdots \; C] \; . \qquad (3)$$

From (3) we obtain the auxiliary matrix

$$M' = \begin{bmatrix} a'_{11} & a'_{12} & \cdots\cdots\cdots & a'_{1n} & \vdots & C'_1 \\ a'_{21} & a'_{22} & \cdots\cdots\cdots & a'_{2n} & \vdots & C'_2 \\ \vdots & & & & \vdots & \vdots \\ a'_{n1} & a'_{n2} & \cdots\cdots\cdots & a'_{nn} & \vdots & C'_n \end{bmatrix} \equiv [A' \; \vdots \; C'] \qquad (4)$$

from which the solution vector is found. The steps for proceeding from (3) to (4) are as follows:

1. The elements of M' are determined in the following order: elements of the first column, then elements of the first row to the right of the first column; elements of the second column, then elements of the second row to the right of the first column, and so on until all elements are determined.

2. The first column of M' is identical to the first column of M. Each element of the first row of M' except the first is obtained by

dividing the corresponding element of M by the leading element a_{11}.

3. Each element a'_{ij} on or below the main diagonal of M' is obtained by subtracting from the corresponding element a_{ij} of M the sum of the products of elements in the ith row and corresponding elements in the jth column of M', all uncalculated elements being imagined to be zero. In symbols:

$$a'_{ij} = a_{ij} - \sum_{k=1}^{j-1} a'_{ik} a'_{kj} \qquad (i \geq j).$$

4. Each element a'_{ij} to the right of the principal diagonal is calculated by using the procedure in 3 and then dividing by the diagonal element a'_{ii} in M'. Thus

$$a'_{ij} = \frac{a_{ij} - \sum a'_{ik} a'_{kj}}{a'_{ii}} \qquad (i < j).$$

Next, the final solution vector x is obtained from M' using the following three rules:

1. The elements of x are determined in the reverse order $x_n, x_{n-1}, \ldots, x_2, x_1$.

2. The last element x_n is identical to the last element c'_n of c'.

3. Each remaining element x_i of x is obtained by subtracting from the corresponding element c'_i of c' the sum of the products of elements in the ith row of A' and the corresponding element of the column x. All uncalculated elements of x are imagined to be zero. Thus

$$x_i = c'_i - \sum_{k=i+1}^{n} a'_{ik} x_k .$$

Finally, checks on the calculations are available according to the following two rules:

1. In the auxiliary matrix, any element of the check column should exceed by unity the sum of the other elements in its row which lie to the right of the principal diagonal.

2. Each element of the check column associated with the solution vector should exceed by unity the corresponding element of the solution vector.

Turning to the given problem, the augmented matrix is:

$$M = \begin{bmatrix} 554.11 & -281.91 & -34.24 & \vdots & 273.02 \\ -281.91 & 226.81 & 38.100 & \vdots & -63.965 \\ -34.240 & 38.100 & 80.221 & \vdots & 34.717 \end{bmatrix} \begin{matrix} \text{Check} \\ 510.98 \\ -80.965 \\ 118.80 \end{matrix}$$

The auxiliary matrix and associated check column are:

$$M' = \begin{bmatrix} 554.11 & -0.50876 & -0.061793 & \vdots & 0.49272 \\ -281.91 & 83.385 & 0.24801 & \vdots & 0.89870 \\ -34.24 & 20.680 & 72.976 & \vdots & 0.45224 \end{bmatrix} \begin{matrix} \text{Check} \\ .92216 \\ 2.14668 \\ 1.45228 \end{matrix}$$

When all the elements in the first row except the first element are divided by 554.11 the results are as above. For example,

$$\frac{-281.91}{554.11} = -0.50876 .$$

To check, observe that $-.50876 - .061793 + .49272 = -.077833$ and $.922167 \approx -.077833 + 1.000$. The second element in the second column

(83.385) was obtained as follows:

$$a'_{22} = 226.81 - (-281.91)(-.50876) = 83.385.$$

Similarly,

$$a'_{32} = 38.100 - (-34.240)(-.50876) = 20.680.$$

The other entries are found using 3) and 4) as above.
The last element of c' in $M' = [A' \vdots c']$ is .45224. Hence $x_3 = $

.45224. Then $x_2 = .8987 - (.24801)(.45224) = .78654$ and

$$x_1 = .49272 - [(-.50876)(.78654) + (-.061793)(.45224)]$$

$$= .92083.$$

Thus, $x_1 = 0.92083$, $x_2 = 0.78654$ and $x_3 = 0.45224$. Note that since the coefficient matrix is symmetric, a'_{23} and a'_{32} need not be calculated independently.

● **PROBLEM** 16-8

Use the Crout method to solve the system

$$
\begin{aligned}
2x_1 - x_2 &= 6 \\
- x_1 + 3x_2 - 2x_3 &= 1 \\
-2x_2 + 4x_3 - 3x_4 &= -2 \qquad (1) \\
- 3x_3 + 5x_4 &= 1 \quad .
\end{aligned}
$$

Solution: A system of equations is said to be in tridiagonal form if the only non-zero elements are located either on the diagonal or adjacent to it. The system (1) is a tridiagonal system.
The general tridiagonal system is of the form

$$
\begin{aligned}
d_1 x_1 + f_1 x_2 &= c_1 \\
e_2 x_1 + d_2 x_2 + f_2 x_3 &= c_2 \\
e_3 x_2 + d_3 x_3 + f_3 x_4 &= c_3 \\
\cdots \cdots \cdots \cdots \cdots \cdots \\
e_n x_{n-1} + d_n x_n &= c_n \quad .
\end{aligned}
$$

According to the Crout method for solving sets of linear equations, one obtains from the original matrix A another matrix M and then from M the solution vector x. When the system is tridiagonal the auxiliary matrix M is also of tridiagonal form. The relationship between the original and auxiliary matrices are given by the forms

$$e'_i = e_i \qquad (i = 2,3,\ldots,n) \qquad\qquad (1)$$

$$d'_1 = d_1 \; , \quad d'_i = d_i - e'_i f'_{i-1} \quad (i = 2,3,\ldots,n) \quad (2)$$

$$f'_i = \frac{f_i}{d'_i} \qquad (i = 1,2,\ldots,n-1) \qquad\qquad (3)$$

and

$$c'_1 = \frac{c_1}{d'_1} \; , \quad c'_i = \frac{c_i - e'_i c'_{i-1}}{d'_i} \quad (i = 2,3,\ldots,n). \quad (4)$$

The solution vector then takes the form

$$x_n = c'_n \, , \quad x_i = c'_i - f'_i x_{i+1} \qquad (i = n-1, n-2, \ldots, 2, 1). \qquad (5)$$

The above facts can be expressed concisely as follows: Let the coefficients and constants of the system be given by the augmented matrix

$$P = \begin{bmatrix}
 & d_1 & f_1 & \vdots & c_1 \\
 & e_2 & d_2 & f_2 & \vdots & c_2 \\
 & \cdot & \cdot & \circ & \vdots & \cdot \\
P = & \cdot & \cdot & \cdot & \vdots & \cdot \\
 & \cdot & \cdot & \cdot & \vdots & \circ \\
 & e_{n-1} & d_{n-1} & f_{n-1} & \vdots & c_{n-1} \\
 & e_n & d_n & & \vdots & c_n
\end{bmatrix}$$

Here the diagonal elements are in the second column, the subdiagonal elements in the first column and the superdiagonal elements in the third column.

The array for the given problem is

$$P = \begin{bmatrix}
 & 2 & -1 & \vdots & 6 \\
-1 & 3 & -2 & \vdots & 1 \\
-2 & 4 & -3 & \vdots & -2 \\
-3 & 5 & & \vdots & 1
\end{bmatrix}$$

According to the Crout method, the first column remains the same while the first row is divided by the leading element (here 2). The remaining elements are found using the forms above. The auxiliary matrix is

$$P' = \begin{bmatrix}
 & d'_1 & f'_1 & \vdots & c'_1 \\
 & e'_2 & d'_2 & f'_2 & \vdots & c'_2 \\
 & e'_3 & d'_3 & f'_3 & \vdots & c'_3 \\
 & e'_4 & d'_4 & & \vdots & c'_4
\end{bmatrix} \qquad (6)$$

Using (1), (2), (3), and (4) the elements of P' can be obtained as follows:

$$e'_2 = e_2 = -1, \quad e'_3 = e_3 = -2, \quad e'_4 = e_4 = -3, \quad d'_1 = d_1 = 2.$$

Then,

$$f'_1 = \frac{f_1}{d'_1} = \frac{-1}{2} = -\frac{1}{2} \, , \quad c'_1 = \frac{c_1}{d'_1} = \frac{6}{2} = 3 \, .$$

Now,

$$d'_2 = d_2 - e'_2 f'_1 = 3 - (-1)(-\tfrac{1}{2}) = \frac{5}{2}$$

then,

$$f'_2 = \frac{f_2}{d'_2} = \frac{-2}{5/2} = -\frac{4}{5}$$

and

$$c'_2 = \frac{c_2 - e'_2 c'_1}{d'_2} = \frac{1 - (-1)(3)}{5/2} = \frac{8}{5} \, .$$

Similarly,

$$d'_3 = 12/5 \, , \quad f'_3 = -5/4 \, , \quad c'_3 = \tfrac{1}{2} \, , \quad d'_4 = 5/4, \quad \text{and} \quad c'_4 = 2 \, .$$

Therefore, (6) becomes:

$$P' = \begin{bmatrix} & 2 & -1/2 & \vdots & 3 \\ -1 & 5/2 & -4/5 & \vdots & 8/5 \\ -2 & 12/5 & -5/4 & \vdots & 1/2 \\ -3 & 5/4 & & \vdots & 2 \end{bmatrix}$$

Now, solution vector x is given by

$$x = \begin{bmatrix} x_1 \\ x_2 \\ x_3 \\ x_4 \end{bmatrix}$$

Using (5), obtain x_4, x_3, x_2 and x_1:

$$x_4 = c_4' = 2$$

$$x_3 = c_3' - f_3' x_4$$

$$= \tfrac{1}{2} - (-5/4)2 = 3$$

$$x_2 = c_2' - f_2' x_3 = 8/5 - (-4/5)(3) = 4$$

$$x_1 = c_1' - f_1' x_2 = 3 - (-\tfrac{1}{2})(4) = 5 .$$

Hence, the solution vector x is

$$x = \begin{bmatrix} 5 \\ 4 \\ 3 \\ 2 \end{bmatrix} .$$

CHOLESKI FACTORIZATION PROCESS

● PROBLEM 16-9

Use the Choleski factorization process to solve the system

$$\begin{aligned} 2x_1 \quad\quad + x_3 &= 4 \\ -3x_1 + 4x_2 - 2x_3 &= -3 \\ x_1 + 7x_2 - 5x_3 &= 6 . \end{aligned} \tag{1}$$

Solution: An important result in the theory of matrices is that any square matrix can be written as the product of a lower triangular and an upper triangular matrix. Moreover, this decomposition is not unique; in fact, it is possible to arbitrarily choose all the elements on the leading diagonal of either L or U. We can use this result to solve simultaneous linear equations. Let

$$AX = B \tag{2}$$

be a system of n linear equations in n unknowns.
When A is resolved into factors LU then (2) can be rewritten as

$$LUX = B . \tag{3}$$

Let $UX = W$, so that $LW = B$. Now W can be found and after that X follows immediately. To illustrate the procedure, consider (1). It may be written as:

$$\begin{bmatrix} 2 & 0 & 1 \\ -3 & 4 & -2 \\ 1 & 7 & -5 \end{bmatrix} \begin{bmatrix} x_1 \\ x_2 \\ x_3 \end{bmatrix} = \begin{bmatrix} 4 \\ -3 \\ 6 \end{bmatrix} . \qquad (4)$$

To factorize the coefficient matrix in (4) as the product of a lower triangular and an upper triangular matrix, write

$$\begin{bmatrix} 2 & 0 & 1 \\ -3 & 4 & -2 \\ 1 & 7 & -5 \end{bmatrix} = \begin{bmatrix} 1 & 0 & 0 \\ x_1 & 1 & 0 \\ x_2 & x_3 & 1 \end{bmatrix} \begin{bmatrix} y_1 & y_2 & y_3 \\ 0 & y_4 & y_5 \\ 0 & 0 & y_6 \end{bmatrix}$$

$$(5)$$

From (5), $2 = y_1$; $0 = y_2$; $1 = y_3$. Then

$$\begin{bmatrix} 1 & 0 & 0 \\ x_1 & 1 & 0 \\ x_2 & x_3 & 1 \end{bmatrix} \begin{bmatrix} 2 & 0 & 1 \\ 0 & y_4 & y_5 \\ 0 & 0 & y_6 \end{bmatrix} = \begin{bmatrix} 2 & 0 & 1 \\ -3 & 4 & -2 \\ 1 & 7 & -5 \end{bmatrix} \qquad (6)$$

From (6), $2x_1 = -3$ or $x_1 = -3/2$; $2x_2 = 1$ or $x_2 = 1/2$, and $x_3 y_4 = 7$. But $y_4 = 4$. Hence $x_3 = 7/4$. Proceeding, we find that

$$\begin{bmatrix} 1 & 0 & 0 \\ -3/2 & 1 & 0 \\ 1/2 & 7/4 & 1 \end{bmatrix} \begin{bmatrix} 2 & 0 & 1 \\ 0 & 4 & -1/2 \\ 0 & 0 & -37/8 \end{bmatrix} = \begin{bmatrix} 2 & 0 & 1 \\ -3 & 4 & -2 \\ 1 & 7 & -5 \end{bmatrix} \qquad (7)$$

Express (4) in the form (3):

$$\begin{bmatrix} 1 & 0 & 0 \\ -1.5 & 1 & 0 \\ 0.5 & 1.75 & 1 \end{bmatrix} \begin{bmatrix} 2 & 0 & 1 \\ 0 & 4 & -0.5 \\ 0 & 0 & -4.625 \end{bmatrix} \begin{bmatrix} x_1 \\ x_2 \\ x_3 \end{bmatrix} = \begin{bmatrix} 4 \\ -3 \\ 6 \end{bmatrix} .$$

If now $W = \{w_1, w_2, w_3\}$, using $LW = B$ we obtain:

$$\begin{bmatrix} 1 & 0 & 0 \\ -1.5 & 1 & 0 \\ 0.5 & 1.75 & 1 \end{bmatrix} \begin{bmatrix} w_1 \\ w_2 \\ w_3 \end{bmatrix} = \begin{bmatrix} 4 \\ -3 \\ 6 \end{bmatrix} .$$

The components of W can be found by forward substitution. Thus, $w_1 = 4$, $w_2 = 3$ and $w_3 = -1.25$. Since $UX = W$,

$$\begin{bmatrix} 2 & 0 & 1 \\ 0 & 4 & -0.5 \\ 0 & 0 & -4.625 \end{bmatrix} \begin{bmatrix} x_1 \\ x_2 \\ x_3 \end{bmatrix} = \begin{bmatrix} 4 \\ 3 \\ -1.25 \end{bmatrix} .$$

The elements of X are found by back-substitution. Thus, $x_3 = 0.270$, $x_2 = 0.784$ and $x_1 = 1.865$. The solution to the system is therefore

$$x_1 = 1.865 , \quad x_2 = 0.784 \quad \text{and} \quad x_3 = 0.270.$$

Use the Cholesky factorization process to find the inverse of the matrix

$$A = \begin{bmatrix} 0.7 & -5.4 & 1.0 \\ 3.5 & 2.2 & 0.8 \\ 1.0 & -1.5 & 4.3 \end{bmatrix}. \tag{1}$$

<u>Solution:</u>   According to the Cholesky factorization process, any $n \times n$ matrix can be written as the product of a lower triangular and an upper triangular matrix. Moreover, this decomposition is not unique and the elements along the main diagonal of either matrix can be chosen arbitrarily.

Finding the inverse of (1) means finding a matrix X such that

$$AX = B \tag{2}$$

where B is the 3×3 identity matrix. But since A is square, it can be expressed as

$$A = LU \tag{3}$$

where L and U are a lower and upper triangular matrix respectively. Substituting (3) into (2) yields:

$$LUX = B$$

or

$$LY = B \qquad (UX = Y). \tag{4}$$

Since B is the identity matrix, Y is the only unknown in (4). We can solve for Y and then for X, the required inverse.

Applying the above process to the given matrix A yields:

$$\begin{bmatrix} 0.7 & -5.4 & 1.0 \\ 3.5 & 2.2 & 0.8 \\ 1.0 & -1.5 & 4.3 \end{bmatrix} = \begin{bmatrix} 1 & 0 & 0 \\ \ell_1 & 1 & 0 \\ \ell_2 & \ell_3 & 1 \end{bmatrix} \begin{bmatrix} u_1 & u_2 & u_3 \\ 0 & u_4 & u_5 \\ 0 & 0 & u_6 \end{bmatrix},$$

where the simplest form of the main diagonal for L was chosen arbitrarily as $(1,1,1)$. Thus the matrix factors as

$$\begin{bmatrix} 0.7 & -5.4 & 1.0 \\ 3.5 & 2.2 & 0.8 \\ 1.0 & -1.5 & 4.3 \end{bmatrix} = \begin{bmatrix} 1 & 0 & 0 \\ 5 & 1 & 0 \\ 1.43 & 0.21 & 1 \end{bmatrix} \begin{bmatrix} 0.7 & -5.4 & 1.0 \\ 0 & 29.2 & -4.2 \\ 0 & 0 & 3.75 \end{bmatrix}.$$

Equation (4) becomes

$$\begin{bmatrix} 1 & 0 & 0 \\ 5 & 1 & 0 \\ 1.43 & 0.21 & 1 \end{bmatrix} \begin{bmatrix} y_{11} & y_{12} & y_{13} \\ y_{21} & y_{22} & y_{23} \\ y_{31} & y_{32} & y_{33} \end{bmatrix} = \begin{bmatrix} 1 & 0 & 0 \\ 0 & 1 & 0 \\ 0 & 0 & 1 \end{bmatrix} \tag{5}$$

From (5), $y_{11} = 1$, $y_{12} = y_{13} = 0$; $5y_{11} + y_{21} = 0$, hence $y_{21} = -5 \times 1 = -5$; $y_{22} = 1$, $y_{23} = 0$; $1.43y_{11} + 0.21y_{21} + y_{31} = 0$, hence $y_{31} = -0.38$; $0.21y_{22} + y_{32} = 0$, hence $y_{32} = -0.21$ and $y_{33} = 1$. The complete matrix Y is

$$\begin{bmatrix} 1 & 0 & 0 \\ -5 & 1 & 0 \\ -0.38 & -0.21 & 1 \end{bmatrix}.$$

The equation $UX = Y$ now becomes

$$\begin{bmatrix} 0.7 & -5.4 & 1.0 \\ 0 & 29.2 & -4.2 \\ 0 & 0 & 3.75 \end{bmatrix} \begin{bmatrix} x_{11} & x_{12} & x_{13} \\ x_{21} & x_{22} & x_{23} \\ x_{31} & x_{32} & x_{33} \end{bmatrix} = \begin{bmatrix} 1 & 0 & 0 \\ -5 & 1 & 0 \\ -0.38 & -0.21 & 1 \end{bmatrix}$$

Then,

$$\begin{bmatrix} 0.7x_{11} - 5.4x_{21} + x_{31} & .7x_{12} - 5.4x_{22} + x_{32} & .7x_{13} - 5.4x_{23} + x_{33} \\ 29.2x_{21} - 4.2x_{31} & 29.2x_{22} - 4.2x_{32} & 29.2x_{23} - 4.2x_{33} \\ 3.75x_{31} & 3.75x_{32} & 3.75x_{33} \end{bmatrix}$$

$$= \begin{bmatrix} 1 & 0 & 0 \\ -5 & 1 & 0 \\ -0.38 & -.21 & 1 \end{bmatrix}$$

Equating the two matrices, we obtain the following:

$0.7x_{11} - 5.4x_{21} + x_{31} = 1$, $.7x_{12} - 5.4x_{22} + x_{32} = 0$, $.7x_{13} - 5.4x_{23} + x_{33} = 0$

$29.2x_{21} - 4.2x_{31} = -5$, $29.2x_{22} - 4.2x_{32} = 1$, $29.2x_{23} - 4.2x_{33} = 0$

$3.75x_{31} = -.38$ $3.75x_{32} = -.21$, $3.75x_{33} = 1$

Solving these equations yields:

$x_{11} = 0.11$, $x_{12} = 0.32$, $x_{13} = -0.08$

$x_{21} = -0.19$, $x_{22} = 0.03$, $x_{23} = .04$

$x_{31} = -0.10$, $x_{32} = -0.06$, $x_{33} = 0.27$.

Hence, the matrix X is

$$X = \begin{bmatrix} 0.11 & 0.32 & -0.08 \\ -0.19 & 0.03 & 0.04 \\ -0.10 & -0.06 & 0.27 \end{bmatrix}$$

The product AX

$$= \begin{bmatrix} 1.003 & 0.002 & -0.020 \\ -0.113 & 1.138 & 0.024 \\ -0.035 & 0.017 & 1.021 \end{bmatrix}$$

is an approximation to the identity matrix. The inaccuracies are due to the fact that the calculations were done to 2 decimal places.

Lest the reader suspect that the above method is like using a cannon to kill a fly, note that many practical problems of matrix inversion (the input-output matrix of industrial economics, for example) involve matrices of order 100 or higher. In such cases analytic methods are too complex and numerical methods are preferred.

ITERATIVE METHODS FOR SOLVING SYSTEMS OF LINEAR EQUATIONS

● PROBLEM 16-11

> Give examples of ill-conditioned matrices.

Solution: Consider the linear system

$$100x_1 + 99x_2 = 398$$
$$99x_1 + 98x_2 = 394 . \qquad (1)$$

The exact solution of (1) is $x_1 = 2$, $x_2 = 2$. Now consider

$$100x_1 + 99x_2 = 398 \qquad (2)$$
$$99x_1 + 98x_2 = 393.98 .$$

The exact solution of (2) is

$$x_1 = +.02, \quad x_2 = 4 .$$

Now consider the two linear systems

$$x_1 + x_2 = 4 \qquad (3)$$
$$1.01x_1 + x_2 = 3.02$$

which have the exact solution $x_1 = 2$, $x_2 = 2$ and

$$x_1 + x_2 = 4 \qquad (4)$$
$$1.06125x_1 + x_2 = 3.02$$

which has the exact solution $x_1 = -16$, $x_2 = +20$.

If we regard (2) as having been obtained from (1) by a small variation of the right-hand side, then we see that the numerical solution of
(1) is extremely sensitive to such small variations.

Similarly a small variation in the coefficient matrix of (3) produces
a very large variation in the numerical solution of the system. Systems
(1) and (3) are examples of ill-conditioning. A system is said to be ill-
conditioned when small relative variations in the elements of A or B
produce large relative variations in the solution vector x.

Since physical measurements and computing machines always have small
errors, if a system of linear equations representing a physical or social
system is ill-conditioned, the solution vector x can be very much in
error.

● PROBLEM 16-12

> Show that the system of linear equations
>
> $$5x + 7y + 6z + 5u = 23$$
> $$7x + 10y + 8z + 7u = 32$$
> $$6x + 8y + 10z + 9u = 33 \qquad (1)$$
> $$5x + 7y + 9z + 10u = 31$$
>
> is ill-conditioned. How can ill-conditioning be measured?

Solution: A set of simultaneous linear equations is said to be ill-conditioned if small changes in the coefficients of the unknowns or the constants on the right-hand side result in large changes in the solution. By examination, $x = y = z = u = 1$ is an exact solution to (1). However, if we use the method of relaxation we find the approximate solution to be $x = 2.36$, $y = .18$, $z = .65$, $u = 1.21$ with the corresponding residuals $R_1 = .01$, $R_2 = -.01$, $R_3 = -.01$, $R_4 = .01$. Thus the equations are ill-conditioned since even with a small error in approximation, the iterated solution is far away from the actual solution.

To obtain measures of ill-conditioning note that Cramer's rule gives for the solution of the system

$$\sum_{j=1}^{n} a_{ij}x_j - c_i = 0 \quad (i = 1,2,\ldots,n),$$

$$x_j = \frac{D_j}{D} \tag{2}$$

(where D_j represents the determinant obtained by replacing the jth column of the coefficient matrix by c_i).

If $D = \det(a_{ij})$ is small in relation to D_j, (2) will fluctuate badly as D_j changes. This is one measure of ill-conditioning.

Another method, which avoids the computation of determinants is based on the following geometrical argument.

Let

$$R_1 = a_{11}x + a_{12}y - c_1$$

$$R_2 = a_{21}x + a_{22}y - c_2 . \tag{3}$$

Here the R_i's represent residues when approximate solutions for x and y are tried. When $R_1 = R_2 = 0$ we have the exact solution below which is the point of intersection of two straight lines. If we ask that the residuals only be small, then $|R_1| < \epsilon_1$, $|R_2| < \epsilon_2$ where ϵ_1 and ϵ_2 are small positive numbers. Then the solutions we are prepared to accept lie in the shaded parallelogram. Fig. 1 is an example of a well conditioned system since all acceptable solutions are close to the actual solution. But consider Fig. 2. The shaded parallelogram would be very long and narrow. Hence the system would be ill-conditioned since the set of acceptable solutions will contain members not at all close to the actual solutions.

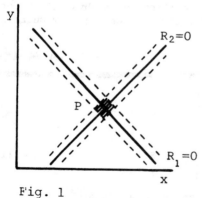

Fig. 1

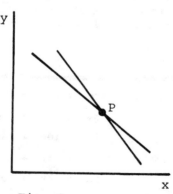

Fig. 2

450

If n = 3, the equations

$$\sum_{j=1}^{n} a_{ij}x_j - c_i = 0 \quad (i = 1,2,3)$$

represent three planes. The angle between any two of them, say one and two, is

$$\cos \theta_{12} = \frac{a_{11}a_{21} + a_{12}a_{22} + a_{13}a_{23}}{\sqrt{\sum a_{ij}^2} \sqrt{\sum a_{2j}^2}}$$

Generalizing this formula to n dimensions, n linear equations represent n hyperplanes and the $\binom{n}{2}$ angles between them are given by

$$\cos \theta_{ij} = \frac{\sum_{k=1}^{n} a_{ik}a_{jk}}{\sqrt{\sum a_{ik}^2} \sqrt{\sum a_{jk}^2}} \tag{4}$$

From (2) we see that if the angle between lines is very small the system they represent is ill-conditioned. Thus we conclude that if the cosines of θ_{ij} in (4) are nearly 1, a system is ill-conditioned. Applying this measure to the system (1) we obtain:

$$\cos^2\theta_{12} = \frac{188^2}{(135)(262)} = \frac{35344}{35370} = .999$$

$$\cos^2\theta_{13} = \frac{191^2}{(135)(281)} = \frac{36481}{37935} = .962$$

$$\cos^2\theta_{14} = \frac{178^2}{(135)(255)} = \frac{31684}{34425} = .920$$

$$\cos^2\theta_{23} = \frac{265^2}{(262)(281)} = \frac{70225}{73622} = .954$$

$$\cos^2\theta_{24} = \frac{247^2}{(262)(255)} = \frac{61009}{66810} = .913$$

$$\cos^2\theta_{34} = \frac{266^2}{(281)(255)} = \frac{70756}{71655} = .987$$

If we accept .95 as the critical value of $\cos^2\theta$, the system (1) is clearly ill-conditioned.

● **PROBLEM** 16-13

Solve the linear system

$$14x_1 + 2x_2 + 4x_3 = -10$$
$$16x_1 + 40x_2 - 4x_3 = 55$$
$$-2x_1 + 4x_2 - 16x_3 = -38$$

using the Jacobi iteration method.

Solution: In an iterative method, we begin with an initial approximation to the solution which we successively try to improve. If the successive approximations tend to approach the solution, the method converges. Otherwise it diverges.

451

The first step in the Jacobi method is to express x_i in the ith equation in terms of the remaining variables. Thus:

$$x_1 = -\frac{10}{14} - \frac{2}{14} x_2 - \frac{4}{14} x_3$$

$$x_2 = \frac{55}{40} - \frac{16}{40} x_1 + \frac{4}{40} x_3$$

$$x_3 = \frac{38}{16} - \frac{2}{16} x_1 + \frac{4}{16} x_2$$

or

$$x_1 = -0.714 - 0.143x_2 - 0.286x_3$$

$$x_2 = 1.375 - 0.400x_1 + 0.100x_3 \qquad (1)$$

$$x_3 = 2.375 - 0.125x_1 + 0.250x_2 \ .$$

Next, choose an initial approximation $x_1^{(0)}$, $x_2^{(0)}$,...,$x_n^{(0)}$ to the solution. In the absence of other information, let $x_1^{(0)} = x_2^{(0)} = \ldots = x_n^{(0)} = 0$. In the given problem

$$x_1^{(0)} = x_2^{(0)} = x_3^{(0)} = 0 \ .$$

For the third step, substitute the values of the variables calculated in the previous iteration [(k-1)th iteration] into the right side of (1) to obtain a new approximation. In the given problem

$$x_1^{(1)} = -0.714 - 0.143(0) - 0.286(0) = -0.714$$

$$x_2^{(1)} = 1.375 - 0.400(0) + 0.100(0) = 1.375$$

$$x_3^{(1)} = 2.375 - 0.125(0) + 0.250(0) = 2.375 \ .$$

The second approximation for x_1 is

$$x_1^{(2)} = -0.714 - 0.143(1.375) - 0.286(2.375) = -1.590$$

correct to three decimal places.
   The second approximations for $x_2^{(2)}$ and $x_3^{(2)}$ are

$$x_2^{(2)} = 1.899 \ , \quad x_3^{(2)} = 2.808.$$

Iterating in the above manner we obtain the first nine approximations (k = 8) in the table below:

Iteration	$x_1^{(k)}$	$x_2^{(k)}$	$x_3^{(k)}$
0	0	0	0
1	-0.714	1.375	2.375
2	-1.590	1.899	2.808
3	-1.789	2.292	3.049
4	-1.914	2.396	3.172
5	-1.964	2.458	3.213
6	-1.984	2.482	3.236
7	-1.994	2.493	3.244
8	-1.998	2.497	3.247

Our best approximation to the solution is

$$x_1 = -1.998 \ , \ x_2 = 2.497 \ , \ x_3 = 3.247.$$

452

Moreover, studying the data, we see that the differences between the kth and $(k-1)^{th}$ approximations are successively decreasing, suggesting convergence. A sufficient condition for convergence of the Jacobi algorithm is that

$$\sum_{\substack{j=1 \\ j \neq k}}^{n} |a_{kj}| < |a_{kk}|, \qquad (k = 1,\ldots,n),$$

i.e., the diagonal element in each row dominates the sum of the non-diagonal elements in that row.

For the given system

$$|14| > |2| + |4| = 6$$

$$|40| > |16| + |-4| = 20$$

$$|-16| > |-2| + |4| = 6.$$

Thus we know that as $k \to \infty$, the approximations will converge to the exact solution, i.e.,

$$\lim_{k \to \infty} |x^* - x_k| < \epsilon$$

where x^* is the exact solution vector and $\epsilon > 0$ is a challenge number. Using analytic methods the exact solution to the given problem is found to be

$$x_1 = -2, \; x_2 = 2.5, \; x_3 = 3.25.$$

● **PROBLEM** 16-14

Solve the linear system

$$\begin{aligned}
10x_1 + x_2 + x_3 &= 15 \\
x_1 + 10x_2 + x_3 &= 24 \\
x_1 + x_2 + 10x_3 &= 33
\end{aligned}$$

using the Jacobi method.

Solution: Consider the linear system

$$AX = B \tag{1}$$

where A is an $n \times n$ matrix and X and B are $n \times 1$ column vectors. We wish to solve (1) for the unknown vector X. Expanding yields:

$$\begin{aligned}
a_{11}x_1 + a_{12}x_2 + \ldots + a_{1n}x_n &= b_1 \\
a_{21}x_1 + a_{22}x_2 + \ldots + a_{2n}x_n &= b_2 \\
&\;\;\vdots \\
a_{n1}x_1 + a_{n2}x_2 + \ldots + a_{nn}x_n &= b_n.
\end{aligned} \tag{2}$$

Let $X^* = [x_1^*, x_2^*, \ldots, x_n^*]$ be the solution vector. Then the system (2) can be expressed in the form

$$x_k^* = \frac{1}{a_{kk}} \left(b_k - \sum_{\substack{j=1 \\ j \neq k}}^{n} a_{kj} x_j^* \right) \quad (k = 1,\ldots,n) \tag{3}$$

$a_{kk} \neq 0$.

Let $x^{(0)}$ be an estimate of x^*. Then we can generate a sequence $\{x^{(i)}\}$ of vectors from

$$x_k^{(i+1)} = \frac{1}{a_{kk}}\left[b_k - \sum_{\substack{j=1 \\ j \neq k}}^{n} a_{kj}\, x_j^{(i)} \right] \tag{4}$$

$(k = 1,\ldots,n)$, $(i = 0,1,2,\ldots)$. Write $\|x^* - x^{(i)}\| = \max_{1 \leq k \leq n}\{|x_k^* - x_k^{(i)}|\}$,

$i = (0,1,2,\ldots)$. This function defines a norm and thus we can measure the distance between the optimal solution and the ith approximation. Let

$$\alpha = \max_{1 \leq k \leq n}\left\{ \sum_{\substack{j=1 \\ j \neq k}}^{n} \left|\frac{a_{kj}}{a_{kk}}\right| \right\} . \tag{5}$$

Then, subtracting (4) from (3) and using the triangle inequality we obtain:

$$|x_k^* - x_k^{(i+1)}| \leq \sum_{\substack{j=1 \\ j \neq k}}^{k} \left|\frac{a_{kj}}{a_{kk}}\right| |x_j^* - x_j^{(1)}| \quad (k = 1,\ldots,n)$$

$$\leq \|x^* - x^{(i)}\| \sum_{\substack{j=1 \\ j \neq k}}^{n} \left|\frac{a_{kj}}{a_{kk}}\right| \quad (k = 1,\ldots,n) .$$

Hence $\|x^* - x^{(i+1)}\| \leq \alpha \| x^* - x^{(i)}\|$, $(i = 0,1,2,\ldots)$. By repeated use of this inequality we obtain:

$$\|x^* - x^{(i+1)}\| \leq \alpha^{i+1} \|x^* - x^{(0)}\| .$$

If $0 \leq \alpha < 1$, $\|x^* - x^{(i+1)}\| \to 0$ as $i \to \infty$. Hence $|x^* - x_k^{(i+1)}| \to 0$ as $i \to \infty$. From (5), if $\alpha < 1$:

$$\sum_{\substack{j=1 \\ j \neq i}}^{n} |a_{kj}| < |a_{kk}| \quad (k = 1,\ldots,n) . \tag{6}$$

A matrix whose elements satisfy (6) is called strictly diagonally dominant. Now consider the given system:

$$|a_{12}| + |a_{13}| = 2 < |a_{11}| = 10$$

$$|a_{21}| + |a_{23}| = 2 < |a_{22}| = 10$$

$$|a_{31}| + |a_{32}| = 2 < |a_{33}| = 10 .$$

Thus the matrix A of the system is strictly diagonally dominant. We take as the initial estimate

$$x_k^{(0)} = \frac{b_k}{a_{kk}} \quad (k = 1,\ldots,n).$$

Thus, $x_1^{(0)} = 1.5$, $x_2^{(0)} = 2.4$, $x_3^{(0)} = 3.3$

$$x_1^{(i+1)} = \frac{1}{a_{11}}\left[b_1 - a_{12}x_2^{(i)} - a_{13}x_3^{(i)} \right]$$

$$= \frac{1}{10}\left[15 - x_2^{(i)} - x_3^{(i)} \right]$$

$$x_2^{(i+1)} = \frac{1}{a_{22}} \left[b_2 - a_{21}x_1^{(i)} - a_{23}x_3^{(i)} \right]$$

$$= \frac{1}{10} \left[24 - x_1^{(i)} - x_3^{(i)} \right]$$

$$x_3^{(i+1)} = \frac{1}{a_{33}} \left[b_3 - a_{31}x_1^{(i)} - a_{32}x_2^{(i)} \right]$$

$$= \frac{1}{10} \left[33 - x_1^{(i)} - x_2^{(i)} \right]$$

Using these recurrence relations we obtain the results in the table below:

i	$x_1^{(i)}$	$x_2^{(i)}$	$x_3^{(i)}$
0	1.500000	2.400000	3.300000
1	0.930000	1.920000	2.910000
2	1.017000	2.016000	3.015000
3	0.996900	1.996800	2.996700
4	1.000650	2.000640	3.000630
5	0.999873	1.999872	2.999871
6	1.000026	2.000026	3.000026
7	0.999995	1.999995	2.999995

Note that the exact answer is

$$x_1 = 1, \ x_2 = 2, \ x_3 = 3.$$

The Jacobi method for the numerical solution of (1) may be expressed as an algorithm as follows:

1. Compute $x_k^{(1)}$ $(k = 1,\ldots,n)$ from (3) with $i = 0$, and go to 2.

2. If $\left| x_k^{(1)} - x_k^{(0)} \right| \le \epsilon$ $(k = 1,\ldots,n)$, $\epsilon > 0$, set $x_k^* = x_k^{(1)}$ $(k = 1,\ldots,n)$ and stop; otherwise go to (3).

3. Set $x_k^{(0)} = x_k^{(1)}$ $(k = 1,\ldots,n)$ and go to 1.

● **PROBLEM** 16-15

Use the Gauss-Seidel method to solve the following linear system:

$$\begin{aligned}
10x_1 + x_2 + x_3 &= 15 \\
x_1 + 10x_2 + x_3 &= 24 \\
x_1 + x_2 + 10x_3 &= 33
\end{aligned} \qquad (1)$$

Solution: The Gauss-Seidel method is a modification of the Jacobi method for solving sets of linear equations.
    Let
$$AX = b \qquad (2)$$
be a system of n linear equations in n unknowns. Here $A = \{a_{ij}\}$,

$X = \{x_j\}$ and $b = \{b_i\}$ are $n \times n$, $n \times 1$ and $n \times 1$ matrices respectively, $(i, j = 1, 2, .., n)$. Equation (2) can be expressed in the form

$$x_k^* = \frac{1}{a_{kk}} \left[b_k - \sum_{\substack{j=1 \\ j \neq k}}^{n} a_{kj} x_j^* \right] \quad (k = 1, \ldots, n)$$

$$a_{kk} \neq 0,$$

where x^* is the solution vector. Given an estimate $x^{(0)}$ of x^* we can generate a sequence $\{x^{(i)}\}$ of vectors from

$$x_k^{(i+1)} = \frac{1}{a_{kk}} [b_k - \sum_{\substack{j=1 \\ j \neq k}}^{n} a_{kj} x_j^{(i)}] \tag{3}$$

$$(k = 1, \ldots, n)(i = 0, 1, 2, \ldots).$$

Using (3), the Jacobi method may be expressed as an algorithm. But, examining (3) we note that at the stage when $x_k^{(i+1)}$ is to be computed, the values of $x_j^{(i+1)}$ $(j = 1, \ldots, k-1)$ have already been obtained. Thus we can use the value of $x_j^{(i+1)}$ rather than that of $x_j^{(i)}$ $(j = 1, \ldots, k-1)$ in computing $x_k^{(i+1)}$. Doing this we obtain, in place of (3):

$$x_1^{(i+1)} = \frac{1}{a_{11}} \left[b_1 - \sum_{j=2}^{n} a_{1j} x_j^{(i)} \right]$$

$$x_k^{(i+1)} = \frac{1}{a_{kk}} \left[b_k - \sum_{j=1}^{k-1} a_{kj} x_j^{(i+1)} - \sum_{j=k+1}^{n} a_{kj} x_j^{(i)} \right] \tag{4}$$

$$(k = 2, \ldots, n-1),$$

$$x_n^{(i+1)} = \frac{1}{a_{nn}} \left[b_n - \sum_{j=1}^{n-1} a_{nj} x_j^{(i+1)} \right].$$

The sequence of vectors $\{x^{(i)}\}$ converges to the solution x^* of (2). The following theorem applies:

If A is strictly diagonally dominant, then the sequence of vectors $\{x^{(i)}\}$ generated from (4) with $x^{(0)}$ (the initial approximation) chosen arbitrarily converges to the solution x^* of (2).

For the linear system (1), the equations (4) become:

$$x_1^{(i+1)} = \frac{1}{a_{11}} \left[b_1 - a_{12} x_2^{(i)} - a_{13} x_3^{(i)} \right]$$

$$= \frac{1}{10} \left[15 - x_2^{(i)} - x_3^{(i)} \right]$$

$$x_2^{(i+1)} = \frac{1}{a_{22}} \left[b_2 - a_{21} x_1^{(i+1)} - a_{23} x_3^{(i)} \right] \tag{5}$$

$$= \frac{1}{10} \left[24 - x_1^{(i+1)} - x_3^{(i)} \right].$$

$$x_3^{(i+1)} = \frac{1}{a_{33}} \left[b_3 - a_{31} x_1^{(i+1)} - a_{32} x_2^{(i+1)} \right]$$

$$= \frac{1}{10} \left[33 - x_1^{(i+1)} - x_2^{(i+1)} \right].$$

Let the initial approximation be

$$x_k^{(0)} = \frac{b_k}{a_{kk}} \quad (k = 1,2,3).$$

(This choice ensures quicker convergence in most cases than the general approximation $x^{(0)} = 0$). Using (5) repeatedly we obtain the following results:

i	$x_1^{(i)}$	$x_2^{(i)}$	$x_3^{(i)}$
0	1.500000	2.400000	3.300000
1	0.930000	1.977000	3.009300
2	1.001370	1.998933	2.999970
3	1.000110	1.999992	2.999990
4	1.000002	2.000001	3.000000

The use of the Jacobi method yielded after the 7th iteration:

$$x_1^{(7)} = 0.999995 , \quad x_2^{(7)} = 1.000005 , \quad x_3^{(7)} = 2.999995.$$

Needless to say, the exact answer, computed analytically, is

$$x_1 = 1.000000 , \quad x_2 = 2.000000 , \quad x_3 = 3.000000 .$$

● **PROBLEM 16-16**

Solve the linear system

$$14x_1 + 2x_2 + 4x_3 = -10$$
$$16x_1 + 40x_2 - 4x_3 = 55 \qquad (1)$$
$$-2x_1 + 4x_2 - 16x_3 = -38$$

using the Gauss-Seidel iteration method.

Solution: The Gauss-Seidel method is similar to the Jacobi iteration method. Let

$$a_{11}x_1 + a_{12}x_2 + \ldots + a_{1n}x_n = b_1$$
$$a_{21}x_1 + a_{22}x_2 + \ldots + a_{2n}x_n = b_2 \qquad (2)$$
$$\begin{matrix} \cdot & \cdot & \cdot & \cdot \\ \cdot & \cdot & \cdot & \cdot \\ \cdot & \cdot & \cdot & \cdot \end{matrix}$$
$$a_{n1}x_1 + a_{n2}x_2 + \ldots + a_{nn}x_n = b_n$$

be a system of n linear equations in n unknowns. Recall that the Jacobi method proceeds as follows:

Step 1: Write x_i in the ith equation in terms of the remaining variables in that equation. Rewriting (2) yields:

$$x_1 = \frac{1}{a_{11}}\left[b_1 - \sum_{j=2}^{n} a_{1j}x_j \right]$$

$$x_2 = \frac{1}{a_{22}} \left[b_2 - \sum_{\substack{j=1 \\ j \neq 2}}^{n} a_{2j} x_j \right] \tag{3}$$

$$\vdots$$

$$x_\ell = \frac{1}{a_{\ell\ell}} \left[b_\ell - \sum_{\substack{j=1 \\ j \neq \ell}}^{n} a_{\ell j} x_j \right]$$

$$\vdots$$

$$x_n = \frac{1}{a_{nn}} \left[b_n - \sum_{j=1}^{n-1} a_{nj} x_j \right] .$$

Step 2: Choose an initial approximation. $x_1^{(0)}, x_2^{(0)}, \ldots, x_n^{(0)}$ to the solution. A suitable a priori initial approximation is

$$x_1^{(0)} = x_2^{(0)} = \ldots x_n^{(0)} = 0 .$$

Step 3: Substitute the values of the variables calculated in the previous iteration into the right side of (3) to obtain a new approximation.

$$x_1^{(k)} = \frac{1}{a_{11}} \left[b_1 - \sum_{j=2}^{n} a_{1j} x_j^{(k-1)} \right]$$

$$x_2^{(k)} = \frac{1}{a_{22}} \left[b_2 - \sum_{\substack{j=1 \\ j \neq 2}}^{n} a_{2j} x_j^{(k-1)} \right] \tag{4}$$

$$\vdots$$

$$x_n = \frac{1}{a_{nn}} \left[b_n - \sum_{j=1}^{n-1} a_{nj} x_j^{(k-1)} \right]$$

where $k = 1, 2, \ldots$ are the successive iterations.

The Gauss-Seidel method modifies Step 3 in the following manner.

Step 3': Substitute the most recently calculated values of the variables, in the current iteration, into the right side of (3) to obtain a new approximation $x_1^{(k)}, x_2^{(k)}, \ldots, x_n^{(k)}$.

To see how this workds in practice, we first write (1) in the form (3):

$$x_1 = -0.714 - 0.143x_2 - 0.286x_3$$

$$x_2 = 1.375 - 0.400x_1 + 0.100x_3$$

$$x_3 = 2.375 - 0.125x_1 + 0.250x_2 .$$

Now, by the Jacobi method, the first approximation is

$$x_1^{(1)} = -0.714 - 0.143(0) - 0.286(0) = -0.714$$

$$x_2^{(1)} = 1.375 - 0.400(0) + (0.100)(0) = 1.375$$

$$x_3^{(1)} = 2.375 - 0.125(0) + (0.250)(0) = 2.375.$$

By the Gauss-Seidel method the first approximation is

$$x_1^{(1)} = -0.714 - 0.143(0) - 0.286(0) = -0.714$$

$$x_2^{(1)} = 1.375 - 0.400(-0.714) + 0.100)(0) = 1.661$$

$$x_3^{(1)} = 2.375 - 0.125(-0.714) + (0.250)(1.661) = 2.879 .$$

458

By the Jacobi method the second approximation is
$$x_1^{(2)} = -1.590, \quad x_2^{(2)} = 1.899, \quad x_3^{(2)} = 2.808,$$
while the Gauss-Seidel method yields:
$$x_1^{(2)} = -1.775, \quad x_2^{(2)} = 2.373, \quad x_3^{(2)} = 3.190.$$

The first six approximations to the solutions using the Gauss-Seidel method are shown below:

Iteration	$x_1^{(k)}$	$x_2^{(k)}$	$x_3^{(k)}$
0	0	0	0
1	-0.714	1.661	2.879
2	-1.775	2.373	3.190
3	-1.965	2.480	3.241
4	-1.996	2.497	3.249
5	-2.000	2.500	3.250

Thus the approximation to the solution is $x_1 \simeq -2.000$, $x_2 \simeq 2.500$, $x_3 \simeq 3.250$. As a comparison, the Jacobi method, after nine iterations, yielded: $x_1 \simeq -1.998$, $x_2 \simeq 2.497$, $x_3 \simeq 3.247$ while the exact analytic solution was found to be
$$x_1 = -2, \quad x_2 = 2.5 \quad \text{and} \quad x_3 = 3.25.$$

● **PROBLEM** 16-17

Let $AX = b$ be a system of equations in the unknown $n \times 1$ vector X. If iterative processes are applied to this system, under what conditions will the processes converge to the correct solution?

Solution: The two fundamental iterative methods used in solving sets of linear equations are Jacobi iteration and Gauss-Seidel iteration. This problem will discuss the convergence criteria of these two techniques.

Let the system $AX = b$ be written as

$$\begin{bmatrix} a_{11} & a_{12} & \cdots & a_{1n} \\ a_{21} & a_{22} & \cdots & a_{2n} \\ \vdots & \vdots & & \vdots \\ a_{n1} & a_{n2} & \cdots & a_{nn} \end{bmatrix} \begin{bmatrix} x_1 \\ x_2 \\ \vdots \\ x_n \end{bmatrix} = \begin{bmatrix} b_1 \\ b_2 \\ \vdots \\ b_n \end{bmatrix}.$$

Since every square matrix can be written as the sum of a lower triangular, upper triangular and diagonal matrix we obtain:
$$A = L + D + U .$$

Thus $(L + D + U)X = b$ whence $DX = -LX - UX + b$

or
$$X = -D^{-1}LX - D^{-1}UX + D^{-1}b , \qquad (1)$$

assuming D is invertible. The addition of superscripts, $(r+1)$ on the left and (r) on the right, leads to the recurrence relation

$$X^{(r+1)} = -D^{-1}(L + U)X^{(r)} + D^{-1}b . \qquad (2)$$

When this matrix equation is written out in full, it becomes the rearrangement of the original system of equations appropriate for Jacobi iteration,
in which all terms except those on the diagonal are taken over to the right
hand sides, and each resulting equation is then divided by the coefficient
of the diagonal term which remains on the left.

Alternatively, if we allocate different superscripts to the two terms
in X on the right-hand side of (2) we obtain:

$$X^{(r+1)} = -D^{-1}LX^{(r+1)} - D^{-1}UX^{(r)} + D^{-1}b \tag{3}$$

which is the Gauss-Seidel rearrangement of the original system. Recall
that in Gauss-Seidel iteration, when the value of $x_i^{(r+1)}$ is being calculated, the values of $x_1^{(r+1)}$, $x_2^{(r+1)}$, $\ldots$, $x_{i-1}^{(r+1)}$ are available. The coefficients of these components are the non-zero elements of L.

For analyzing convergence it is convenient to have (3) in a form that
gives $X^{(r+1)}$ explicitly in terms of $X^{(r)}$. Rewriting (3) and repeating
(2) we obtain:

$$\text{Jacobi:} \qquad X^{(r+1)} = -D^{-1}(L+U)X^{(r)} + D^{-1}b \tag{4}$$

$$\text{Gauss-Seidel:} \quad X^{(r+1)} = -(D+L)^{-1}UX^{(r)} + (D+L)^{-1}b . \tag{5}$$

Both (4) and (5) are of the form

$$X^{(r+1)} = MX^{(r)} + c \tag{5}$$

where M is an $n \times n$ matrix and c is an $n \times 1$ column vector. For
convergence of (5) we require that

$$\lim_{r \to \infty} X^{(r+1)} = \lim_{r \to \infty} X^{(r)} = X = A^{-1}b .$$

Hence the fixed point X of the process must satisfy

$$X = MX + c . \tag{6}$$

Let the error in the rth iterated solution be given by

$$e^{(r)} = X^{(r)} - X .$$

Subtracting (6) from (5) gives $e^{(r+1)} = Me^{(r)}$ whence

$$e^{(r)} = M^r e^{(0)} , \tag{7}$$

where $e^{(0)}$ is the error in the initial approximation. If (7) is to equal
zero as $r \to \infty$:

$$\lim_{r \to \infty} M^r = 0 \quad \text{(the null matrix)}. \tag{8}$$

A necessary and sufficient condition for (8) is that all the eigenvalues
of M be of modulus less than unity.

Denote $\max_i |\lambda_i|$, the spectral radius of M, by $\rho(M)$. Then an it-
erative process

$$X^{(r+1)} = MX^{(r)} + c$$

is convergent if and only if $\rho(M) < 1$.

● **PROBLEM** 16-18

Solve the following linear system using the method of relaxation:

$$3x + 9y - 2z = 11$$
$$4x + 2y + 13z = 24$$
$$11x - 4y + 3z = -8 \qquad (1)$$

Solution: The methods of Gauss elimination and Gauss elimination with partial pivoting obtain solutions for the unknowns one after another. The method of relaxation, however, obtains a solution for all unknown simultaneously.

The basic principle of the method is as follows. Let the equations be

$$\sum_{j=1}^{n} a_{ij}x_j - c_i = 0, \quad (i = 1,\ldots,n) . \qquad (2)$$

Set

$$R = \sum_{j=1}^{n} a_{ij}x_j - c_i .$$

Then the solution of the n equations is a set of numbers $x_1, x_2, \ldots, x_n$ which has the property that it makes all the R_i equal to zero. The relaxation method employs an iterative procedure which makes the R_i smaller and smaller at each step. The quantities R_i are known as the residuals of the n equations. To apply the method to (1) first rewrite it in the form (2):

$$3x + 9y - 2z - 11 = 0$$
$$4x + 2y + 13z - 24 = 0 \qquad (3)$$
$$11x - 4y + 3z + 8 = 0 .$$

Now construct an operations tableau:

Δx	Δy	Δz	ΔR_1	ΔR_2	ΔR_3
1	0	0	3	4	11
0	1	0	9	2	-4
0	0	1	-2	13	3

Entries in this tableau are interpreted, for example, as "An increment of 1 unit of x produces an increment of 3 units in R_1, 4 units in R_2 and 11 units in R_3."

Let $x = y = z = 0$ be the initial approximation. Then $R_1 = -11$, $R_2 = -24$ and $R_3 = 8$. If we choose to make $R_1 = 0$, the following tableau is obtained:

$x = 0$	$y = 0$	$z = 0$	$R_1 = -11$	$R_2 = -24$	$R_3 = 8$
0	0	2	-15	2	14
-1	0	0	-18	-2	3
0	2	0	0	2	-5
$x = -1$	$y = 2$	$z = 2$	0	2	-5

To understand this table observe from the operations tableau that a unit change in z produces a change of -2 in R_1. Thus when z goes from

0 to 2, R_1 decreases from -11 to -15. But a unit change in z produces a change of +13 in R_2. Hence when z changes from 0 to 2,
R_2 increases from -24 to +2. Next, when x changes from 0 to -1,
R_1 changes by -3. But because of the change in z, R_1 is now -15.
Hence a change of -1 in x makes the cumulative R_1 = -18. Finally,
when y changes by 2, R_1 changes by +18 and the cumulative total is
zero. The other columns are found in a similar manner.

In the next tableau, to avoid introducing decimal fractions of an increment, we multiply through by 10 and continue.

-10	20	20	0	20	-50
5	0	0	15	40	5
0	0	-3	21	1	-4
0	-2	0	3	-3	4
x = -5	y = 18	z = 17	3	-3	4

The third trial solution is $10x = -5$, $10y = 18$, $10z = 17$. To obtain the
second decimal place in the solution, multiply again by 10 and relax.

-50	180	170	30	-30	40
-4	0	0	-18	-46	-4
0	0	4	-10	6	8
0	1	0	- 1	8	4
0	0	-1	1	-5	1
-54	181	173	1	-5	1

Check by substituting $100x = -54$, $100y = 181$, $100z = 173$. We obtain the
residuals 21, -5, and 1, indicating an error. To continue the procedure, simply start with the correct residuals. Completing the solution
to five decimal places gives:

-550	1790	1740	-20	0	10
0	2	0	- 2	4	2
-5500	17920	17400	-20	40	20
0	2	0	- 2	44	12
0	0	-3	4	5	3
-55000	179220	173970	40	50	30
0	-4	0	4	42	46
-4	0	0	-8	26	2
0	0	-2	-4	0	-4
-55004	179216	173968	-4	0	-4

We conclude that $10^5 x = -55004$, $10^5 y = 179216$, $10^5 z = 173968$ and
$R_1 = -.00004$, $R_2 = 0$, $R_3 = -.00004$.

METHODS FOR FINDING EIGENVALUES

Find the smallest eigenvalue of the matrix

$$A = \begin{pmatrix} 4 & 1 & -1 \\ 2 & 3 & -1 \\ -2 & 1 & 5 \end{pmatrix} \ . \tag{1}$$

Solution: Recall that the eigenvalues of an $n \times n$ matrix A are scalars λ such that

$$AX = \lambda X \tag{2}$$

for $X \neq 0$. Rewriting (2) yields

$$[A - \lambda I]X = 0$$

which can have a non-zero solution if

$$\det[A - \lambda I] = 0 \ . \tag{3}$$

Equation (3) is an nth degree polynomial in λ and the roots may be found by some numerical process such as the Newton-Raphson method. The smallest root is therefore the smallest eigenvalue.

The above method is an indirect method for finding the eigenvalues of A. There is also a direct method as follows.

Assume that all the eigenvalues are positive. If λ is an eigenvalue and X is the associated eigenvector they satisfy (2). Subtracting pX from both sides of (2) gives

$$AX - pX = \lambda X - pX$$

or,

$$[A - pI]X = (\lambda - p)X \ . \tag{4}$$

From (4) we see that $\lambda - p$ is an eigenvalue of A - pI, and that X, the associated eigenvector, remains unchanged. If p is taken equal to the dominant eigenvalue of A, the dominant eigenvalue of A - pI will be p less than the numerically smallest eigenvalue of A. Thus, to find the smallest eigenvalue of A we find the largest eigenvalue of A - pI and add the largest eigenvalue of A to it. The largest eigenvalue of A is found by repeatedly postmultiplying A by an initially arbitrary non-zero vector. Let $X^{(0)} = [1,0,0]$. Then,

$$AX^{(0)} = \begin{bmatrix} 4 & 1 & -1 \\ 2 & 3 & -1 \\ -2 & 1 & 5 \end{bmatrix} \begin{bmatrix} 1 \\ 0 \\ 0 \end{bmatrix} = \begin{bmatrix} 4 \\ 2 \\ -2 \end{bmatrix} = 4 \begin{bmatrix} 1 \\ 0.5 \\ -0.5 \end{bmatrix} \ ;$$

$$AX^{(1)} = \begin{bmatrix} 4 & 1 & -1 \\ 2 & 3 & -1 \\ -2 & 1 & 5 \end{bmatrix} \begin{bmatrix} 1 \\ 0.5 \\ -0.5 \end{bmatrix} = \begin{bmatrix} 5 \\ 4 \\ -4 \end{bmatrix} = 5 \begin{bmatrix} 1 \\ 0.8 \\ 0.8 \end{bmatrix} \ ;$$

$$AX^{(2)} = 5.6 \begin{bmatrix} 1 \\ 0.93 \\ 0.93 \end{bmatrix} \ ; \qquad AX^{(3)} = 5.86 \begin{bmatrix} 1 \\ 0.98 \\ -0.98 \end{bmatrix} \ ;$$

463

$$AX^{(4)} = 5.96 \begin{bmatrix} 1 \\ 0.99 \\ -0.99 \end{bmatrix} \; ; \qquad AX^{(5)} = 5.98 \begin{bmatrix} 1 \\ 1 \\ -1 \end{bmatrix}$$

$$AX^{(6)} = 6 \begin{bmatrix} 1 \\ 1 \\ -1 \end{bmatrix} \; ; \qquad AX^{(7)} = 6 \begin{bmatrix} 1 \\ 1 \\ -1 \end{bmatrix} \; .$$

The largest eigenvalue is 6 and $[1,1,-1]$ is an eigenvector associated with it. Next, by a similar process, find the largest eigenvalue of $A - 6I$.

$$A - 6I = \begin{bmatrix} -2 & 1 & -1 \\ 2 & -3 & -1 \\ -2 & 1 & -1 \end{bmatrix} . \qquad (5)$$

Let the initial eigenvector be $[1,0,0]$.

$$AX^{(0)} = \begin{bmatrix} -2 & 1 & -1 \\ 2 & -3 & -1 \\ -2 & 1 & -1 \end{bmatrix} \begin{bmatrix} 1 \\ 0 \\ 0 \end{bmatrix} = \begin{bmatrix} -2 \\ 2 \\ -2 \end{bmatrix} = -2 \begin{bmatrix} 1 \\ -1 \\ 1 \end{bmatrix}$$

$$AX^{(1)} = \begin{bmatrix} -2 & 1 & -1 \\ 2 & -3 & -1 \\ -2 & 1 & -1 \end{bmatrix} \begin{bmatrix} 1 \\ -1 \\ 1 \end{bmatrix} = \begin{bmatrix} -4 \\ 4 \\ -4 \end{bmatrix} = -4 \begin{bmatrix} 1 \\ -1 \\ 1 \end{bmatrix}$$

The dominant eigenvalue of the matrix (5) is -4 and the corresponding eigenvector is $[1,-1,1]$. Thus the smallest eigenvalue of the original matrix is $-4 + 6 = 2$ and the corresponding eigenvector is $\lfloor 1,-1,1 \rfloor$.

● **PROBLEM** 16-20

Find the largest eigenvalue of the matrix

$$\begin{bmatrix} 5 & 2 \\ 2 & 8 \end{bmatrix} . \qquad (1)$$

Explain why the method works.

Solution: By ordinary analytic methods the characteristic equation of (1) is $\lambda^2 - 13\lambda + 36 = 0$, from which the eigenvalues are $\lambda = 9,4$ and the eigenvectors are $[1,2]$ for $\lambda = 9$ and $[2,-1]$ for $\lambda = 4$.

A direct method that constructs the dominant eigenvalue is the following iterative process. Let $X^{(0)} = [1 \; 0]$ be an approximation for the eigenvector associated with the largest eigenvalue. Then

$$AX^{(0)} = \begin{bmatrix} 5 & 2 \\ 2 & 8 \end{bmatrix} \begin{bmatrix} 1 \\ 0 \end{bmatrix} = \begin{bmatrix} 5 \\ 2 \end{bmatrix} = 5 \begin{bmatrix} 1 \\ .4 \end{bmatrix} .$$

Let $X^{(1)} = \begin{bmatrix} 1 \\ .4 \end{bmatrix}$ be the next approximation.

$$AX^{(1)} = \begin{bmatrix} 5 & 2 \\ 2 & 8 \end{bmatrix} \begin{bmatrix} 1 \\ .4 \end{bmatrix} = \begin{bmatrix} 5.8 \\ 5.2 \end{bmatrix} = 5.8 \begin{bmatrix} 1 \\ 0.90 \end{bmatrix}$$

$$AX^{(2)} = \begin{bmatrix} 5 & 2 \\ 2 & 8 \end{bmatrix} \begin{bmatrix} 1 \\ 0.90 \end{bmatrix} = \begin{bmatrix} 6.8 \\ 9.2 \end{bmatrix} = 9.2 \begin{bmatrix} 0.74 \\ 1 \end{bmatrix}$$

$$AX^{(3)} = \begin{bmatrix} 5.7 \\ 9.48 \end{bmatrix} = 9.48 \begin{bmatrix} 0.60 \\ 1 \end{bmatrix}$$

$$AX^{(4)} = 9.2 \begin{bmatrix} 0.54 \\ 1 \end{bmatrix} \qquad AX^{(5)} = 9.08 \begin{bmatrix} 0.52 \\ 1 \end{bmatrix}$$

$$AX^{(6)} = 9.04 \begin{bmatrix} 0.51 \\ 1 \end{bmatrix} \qquad AX^{(7)} = 9.020 \begin{bmatrix} 0.504 \\ 1 \end{bmatrix}$$

$$AX^{(8)} = 9.008 \begin{bmatrix} 0.502 \\ 1 \end{bmatrix}$$

The scalar multiple is tending towards 9 and the vector towards [0.5 1] or [1 2], i.e., the largest eigenvalue and the associated eigenvector.

Note that at each stage, the largest element in the eigenvector approximation was taken as unity.

To show why the method works, let A be an $n \times n$ matrix and $v^{(0)}$ an arbitrary non-zero $n \times 1$ column vector. Now A has n eigenvalues and there will be n corresponding eigenvectors. We wish to find λ_1, the dominant eigenvalue.

First, express $v^{(0)}$ as a linear combination of the n eigenvectors $(v_1, v_2, \ldots, v_n)$. Thus

$$v^{(0)} = k_1 v_1 + k_2 v_2 + \ldots + k_n v_n. \tag{2}$$

Premultiply (2) by A to obtain

$$Av^{(0)} = k_1 Av_1 + k_2 Av_2 + \ldots + k_n Av_n. \tag{3}$$

Assuming that v_1 corresponds to the largest eigenvalue, v_2 to the next largest eigenvalue, and so on we obtain:

$$Av_1 = \lambda_1 v_1 \; ; \; Av_2 = \lambda_2 v_2; \; \ldots \; Av_n = \lambda_n v_n,$$

and (3) becomes

$$Av^{(0)} = k_1 \lambda_1 v_1 + k_2 \lambda_2 v_2 + \ldots + k_n \lambda_n v_n. \tag{4}$$

Divide (4) by the largest element in $Av^{(0)}$, say b_0. Let $v^{(1)}$ be the next approximation. Then,

$$v^{(1)} = \frac{1}{b_0} \left(k_1 \lambda_1 v_1 + k_2 \lambda_2 v_2 + \ldots + k_n \lambda_n v_n \right)$$

which can be written as

$$v^{(1)} = \frac{\lambda_1}{b_0} \left(k_1 v_1 + k_2 \frac{\lambda_2}{\lambda_1} v_2 + \ldots + k_n \frac{\lambda_n}{\lambda_1} v_n \right). \tag{5}$$

Premultiply (5) by A:

$$Av^{(1)} = \frac{\lambda_1}{b_0} A \left(k_1 v_1 + k_2 \frac{\lambda_2}{\lambda_1} v_2 + \ldots + k_n \frac{\lambda_n}{\lambda_1} v_n \right). \tag{6}$$

But $Av_1 = \lambda_1 v_1$; $Av_2 = \lambda_2 v_2$; $\ldots$; $Av_n = \lambda_n v_n$ and (6) becomes

$$Av^{(1)} = \frac{\lambda_1}{b_0} \left(k_1 \lambda_1 v_1 + k_2 \frac{\lambda_2^2}{\lambda_1} v_2 + \ldots + k_n \frac{\lambda_n^2}{\lambda_1} v_n \right).$$

Once again, the largest element on the right-hand side is made unity, by say b_1. Then the next approximation is

$$v^{(2)} = \frac{\lambda_1}{b_0 b_1} \left[k_1 \lambda_1 v_1 + k_2 \frac{\lambda_2^2}{\lambda_1} v_2 + \dots + k_n \frac{\lambda_n^2}{\lambda_1} v_n \right]$$

$$= \frac{\lambda_1^2}{b_0 b_1} \left[k_1 v_1 + k_2 \left(\frac{\lambda_2}{\lambda_1}\right)^2 v_2 + \dots + k_n \left(\frac{\lambda_n}{\lambda_1}\right)^2 v_n \right].$$

Continuing this process we obtain the mth approximation:

$$v^{(m)} = \frac{\lambda_1^m}{\prod\limits_{i=0}^{m-1} b_i} \left[k_1 v_1 + k_2 \left(\frac{\lambda_2}{\lambda_1}\right)^m v_2 + \dots + k_n \left(\frac{\lambda_n}{\lambda_1}\right)^m v_n \right].$$

As $m \to \infty$ $\lambda_i / \lambda_1 \to 0$ $(i = 2,3,\dots,n)$ since λ_1 is the largest eigenvalue. Thus

$$\lim_{m \to \infty} v^{(m)} = \frac{\lambda_1^m}{\prod\limits_{i=0}^{m-1} b_i} k_1 v_1$$

and

$$\lim_{m \to \infty} A v^{(m)} = \frac{\lambda_1^m}{\prod\limits_{i=1}^{m-1} b_i} k_1 A v_1 = \lim_{m \to \infty} \lambda_1 \frac{\lambda_1^m}{\prod\limits_{i=1}^{m-1} b_i} k_1 v_1 .$$

Hence, $\lim\limits_{m \to \infty} A v^{(m)} = \lambda_1 v^{(m)}$. But this states that λ_1 is an eigenvalue of A and $v^{(m)}$ its associated eigenvector. Thus, we see that the initial approximation is arbitrary. Furthermore, the rate of convergence depends on the ratios λ_i / λ $(i = 2,3,\dots,n)$.

● **PROBLEM** 16-21

Find the largest characteristic value of λ for the system

$$x_1 + x_2 + x_3 = \lambda x_1$$

$$x_1 + 2x_2 + 2x_3 = \lambda x_2$$

$$x_1 + 2x_2 + 3x_3 = \lambda x_3 .$$

Solution: Consider the general problem of determining the dominant characteristic number for the system

$$AX = \lambda X , \tag{1}$$

where λ is a scalar, A is an $n \times n$ matrix and X is an $n \times 1$ column vector. To solve (1) we must determine the roots of the characteristic equation $|A - \lambda I| = 0$. The root with largest magnitude is the dominant root.

To start the procedure, choose an initial non-zero approximation to the corresponding characteristic vector, say $v^{(1)}$. A good choice is $\{1,1,\dots,1\}$. Put $v^{(1)}$ into the left-hand side of (1). Then

$$A v^{(1)} \approx \lambda v^{(1)}$$

or
$$y^{(1)} \approx \lambda v^{(1)} \qquad (2)$$

where $y^{(1)} = Av^{(1)}$. If the respective components of $v^{(1)}$ and $y^{(1)}$ are nearly in a constant ratio, the approximation is good and the ratio is an approximation to the true value of λ. We now obtain a second approximation $v^{(2)}$ by multiplying $y^{(1)}$ by a constant. This process is repeated until successive approximations differ by less than some preassigned ϵ.

Letting $v^{(r)}$ denote the rth approximation, the iteration can be specified by the relations

$$y^{(r)} = Av^{(r)}, \quad v^{(r+1)} = \alpha_r y^{(r)} \qquad (r = 1, 2, \ldots)$$

where α_r is a conveniently chosen multiplicative constant. In general,

$$y^{(r)} \sim \lambda_n v^{(r)}, \quad v^{(r)} \to c e_n \qquad (r \to \infty)$$

where '$\sim$' means "is asymptotically equivalent to". Here λ_n is the dominant root of A.

Applying the above analysis to the given problem, let $v^{(1)} = (1,1,1)$. Then,

$$y^{(1)} = \begin{bmatrix} 1 & 1 & 1 \\ 1 & 2 & 2 \\ 1 & 2 & 3 \end{bmatrix} \begin{bmatrix} 1 \\ 1 \\ 1 \end{bmatrix} = \begin{bmatrix} 3 \\ 5 \\ 6 \end{bmatrix} = 6 \begin{bmatrix} 1/2 \\ 5/6 \\ 1 \end{bmatrix} \qquad (3)$$

On the other hand $\lambda \begin{bmatrix} 1 \\ 1 \\ 1 \end{bmatrix} = \begin{bmatrix} \lambda \\ \lambda \\ \lambda \end{bmatrix}$. $\qquad (4)$

We wish to determine the largest λ such that the x_i components of (3) and (4) are equal. Thus $\lambda^{(1)} = 6$, and

$$x^{(2)} = \frac{1}{6} y^{(1)} = \begin{bmatrix} 1/2 \\ 5/6 \\ 1 \end{bmatrix} .$$

Then,

$$y^{(2)} = \begin{bmatrix} 1 & 1 & 1 \\ 1 & 2 & 2 \\ 1 & 2 & 3 \end{bmatrix} \begin{bmatrix} 1/2 \\ 5/6 \\ 1 \end{bmatrix} = \begin{bmatrix} 7/3 \\ 25/6 \\ 31/6 \end{bmatrix} = \frac{31}{6} \begin{bmatrix} 14/31 \\ 25/31 \\ 1 \end{bmatrix}$$

Thus,

$$\begin{bmatrix} 7/3 \\ 25/6 \\ 31/6 \end{bmatrix} \approx \lambda \begin{bmatrix} 1/2 \\ 5/6 \\ 1 \end{bmatrix} .$$

The third components will be equal when $\lambda = 31/6 \approx 5.17$. This is the second approximation to the dominant characteristic value. The third iteration then gives

$$y^{(3)} = \begin{bmatrix} 1 & 1 & 1 \\ 1 & 2 & 2 \\ 1 & 2 & 3 \end{bmatrix} \begin{bmatrix} 14/31 \\ 25/31 \\ 1 \end{bmatrix} = \begin{bmatrix} 70/31 \\ 126/31 \\ 157/31 \end{bmatrix} = \frac{157}{31} \begin{bmatrix} 70/157 \\ 126/157 \\ 1 \end{bmatrix}$$

and $\lambda^{(3)} = 157/31 \approx 5.06$.

The ratios $v_1 : v_2 : v_3$ according to the four approximations are $(1:1:1)$, $(0.5:0.833:1)$, and $(0.466:0.803:1)$. The next cycle leads to the value $\lambda^{(4)} \approx 5.05$ and to the ratios $0.445 : 0.802 : 1$ which hence are accurate to three significant digits.

Find the smallest eigenvalue of the matrix system

$$
\begin{bmatrix}
4 & 1 & -1 \\
2 & 3 & -1 \\
-2 & 1 & 5
\end{bmatrix}
\qquad (1)
$$

by first finding its inverse.

Solution: Let $AX = \lambda X$. The eigenvalues of A are the scalars λ such that the preceding relationship holds for some non-zero vectors X.
    Solving for X yields:

$$X = A^{-1}\lambda X = \lambda A^{-1}X$$

and $\lambda^{-1}X = A^{-1}X$, $(\lambda \neq 0)$. This shows that $1/\lambda$ is an eigenvalue of A^{-1}. Hence, by finding the largest eigenvalue of A^{-1} we can simultaneously find the smallest eigenvalue of A. The inverse of (1) can be readily found by the factorization or Doolittle process. Since any $n \times n$ matrix can be written as the product of a lower triangular and an upper triangular matrix with arbitrary elements along the principal diagonal of either matrix,

$$
\begin{bmatrix}
4 & 1 & -1 \\
2 & 3 & -1 \\
-2 & 1 & 5
\end{bmatrix}
=
\begin{bmatrix}
1 & 0 & 0 \\
\ell_1 & 1 & 0 \\
\ell_2 & \ell_3 & 1
\end{bmatrix}
\begin{bmatrix}
u_1 & u_2 & u_3 \\
0 & u_4 & u_5 \\
0 & 0 & u_6
\end{bmatrix}. \qquad (2)
$$

By the rules of matrix multiplication we obtain the set of equations

$$
\begin{array}{lll}
u_1 = 4 & \ell_1 u_1 = 2 & \ell_2 u_1 = -2 \\
u_2 = 1 & \ell_1 u_2 + u_4 = 3 & \ell_2 u_2 + \ell_3 u_4 = 1 \\
u_3 = -1 & \ell_1 u_3 + u_5 = -1 & \ell_2 u_3 + \ell_3 u_5 + u_6 = 5 .
\end{array}
$$

Thus,

$$
\begin{bmatrix}
1 & 0 & 0 \\
.5 & 1 & 0 \\
-0.5 & .6 & 1
\end{bmatrix}
\begin{bmatrix}
4 & 1 & -1 \\
0 & 2.5 & -.5 \\
0 & 0 & 4.8
\end{bmatrix}
=
\begin{bmatrix}
4 & 1 & -1 \\
2 & 3 & -1 \\
-2 & 1 & 5
\end{bmatrix}. \qquad (3)
$$

The inverse of a matrix A is a matrix B such that $AB = I$. Thus,

$$LUB = I, \qquad (4)$$

where L is a lower triangular matrix and U is an upper triangular matrix. Let $UB = Y$ in (4). Then

$$LY = I . \qquad (5)$$

Using (3) in (5) we obtain:

$$
\begin{bmatrix}
1 & 0 & 0 \\
0.5 & 1 & 0 \\
-0.5 & .6 & 1
\end{bmatrix}
\begin{bmatrix}
y_{11} & y_{12} & y_{13} \\
y_{21} & y_{22} & y_{23} \\
y_{31} & y_{32} & y_{33}
\end{bmatrix}
=
\begin{bmatrix}
1 & 0 & 0 \\
0 & 1 & 0 \\
0 & 0 & 1
\end{bmatrix}.
$$

We can find each y_{ij} by working along each row in turn from top to bottom:

$$Y = \begin{bmatrix} 1 & 0 & 0 \\ -.5 & 1 & 0 \\ 3.5 & -.6 & 1 \end{bmatrix}.$$

But $Y = UB$ or,

$$\begin{bmatrix} 1 & 0 & 0 \\ 0.5 & 1 & 0 \\ 3.5 & -.6 & 1 \end{bmatrix} = \begin{bmatrix} 4 & 1 & -1 \\ 0 & 2.5 & -.5 \\ 0 & 0 & 4.8 \end{bmatrix} \begin{bmatrix} b_{11} & b_{12} & b_{13} \\ b_{21} & b_{22} & b_{23} \\ b_{31} & b_{32} & b_{33} \end{bmatrix}.$$

We can find each b_{ij} by working along each row in turn from bottom to top. Thus,

$$\begin{bmatrix} 0.333 & -0.125 & 0.042 \\ -0.167 & 0.375 & 0.042 \\ 0.167 & -0.125 & 0.208 \end{bmatrix}. \tag{6}$$

We can find the largest eigenvalue of (6) using the power method. It is found to be 0.5 and has associated with it the eigenvector $[1,-1,1]$.

The smallest eigenvalue of the given matrix is thus 2 and the associated eigenvector is $[1,-1,1]$.

● **PROBLEM 16-23**

The largest eigenvalue of the matrix

$$A = \begin{bmatrix} 0 & 5 & -6 \\ -4 & 12 & -12 \\ -2 & -2 & 10 \end{bmatrix}$$

was found, using the power method, to be $\lambda_1 = 16$. An associated eigenvector was $v_1 = [0.5, 1.0, -0.5]$.

Find the remaining eigenvalues and eigenvectors of A by a deflation process.

Solution: The method of deflation proceeds in the following manner. First use the largest eigenvalue to reduce the order of the original n × n matrix. Then find the largest eigenvalue of this deflated (n-1) × (n-1) matrix by the power method. Use this eigenvalue to deflate the matrix to n-2 order. Continue in this way until all the eigenvalues are found.

Let A be an n × n matrix with largest eigenvalue λ_1 and associated eigenvector v_1. If v_1 does not have 1 as its element of largest modulus, multiply v_1 by a permutation matrix P which interchanges the largest element and the first element (note that $P = P^{-1}$). Suppose $Pv_1 = \omega_1$. We must find an elementary matrix R such that $R\omega_1 = e_1$, the elementary vector with first element 1 and all other elements 0.

Let B be the matrix

$$B = RPAP^{-1}R^{-1} = RPAPR^{-1}.$$

Then

$$Be_1 = RPAPR^{-1}e_1$$

$$= RPAP\omega_1$$

$$= RPAv_1$$

$$= \lambda_1 RPv_1 = \lambda_1 e_1 .$$

Thus, B has e_1 as an eigenvector associated with the eigenvalue λ_1. This means that B must be of the form

$$B = \begin{bmatrix} \lambda_1 & \vdots & x^1 \\ \hline 0 & \vdots & \\ 0 & \vdots & \\ \cdot & \vdots & B_1 \\ \cdot & \vdots & \\ \cdot & \vdots & \\ 0 & \vdots & \end{bmatrix} .$$

Here B_1 is an $(n-1)\times(n-1)$ matrix. Since B is similar to A it has the same eigenvalues as A. The eigenvalues $\lambda_2, \lambda_3, \ldots, \lambda_n$ are the eigenvalues of the deflated matrix B_1 which will be used for the next iterations by the power method.

Note that RPA may be computed by applying elementary row operations to the matrix A. $B = RPAP^{-1}R^{-1}$ can then be computed by applying the operations corresponding to P^{-1} and R^{-1} to the columns of RPA. Apply the above method to the given matrix. From $v_1 = [0.5, 1.0, -0.5]$, the appropriate deflation process is:

i) interchange rows 1 and 2 of A (this corresponds to bringing 1.0 to the first element in v_1).

ii) subtract 1/2 row 1 from row 2; add 1/2 row 1 to row 3 (now v_1 has been transformed to $e_1 = (1,0,0)$.

iii) interchange columns 1 and 2.

The resulting matrices are:

$$\begin{bmatrix} 0 & 5 & -6 \\ -4 & 12 & -12 \\ -2 & -2 & 10 \end{bmatrix} \xrightarrow{\text{i)}} \begin{bmatrix} -4 & 12 & -12 \\ 0 & 5 & -6 \\ -2 & -2 & 10 \end{bmatrix} \xrightarrow{\text{ii)}} \begin{bmatrix} -4 & 12 & -12 \\ 2 & -1 & 0 \\ -4 & 4 & 4 \end{bmatrix}$$

$$\xrightarrow{\text{iii)}} \begin{bmatrix} 12 & -4 & -12 \\ -1 & 2 & 0 \\ 4 & -4 & 4 \end{bmatrix}$$

The deflated matrix is

$$B_1 = \begin{bmatrix} 2 & 0 \\ -4 & 4 \end{bmatrix} .$$

Using the power method and iterating with matrix B_1 produces the eigenvalue $\lambda_2 = 4$ and associated eigenvector $v_2 = [0,1]$ of B_1. Applying the deflation process once again gives the matrix

$$c = \begin{bmatrix} 4 & -4 \\ 0 & 2 \end{bmatrix} .$$

The deflated $|x|$ matrix is thus $c_1 = [2]$ and $\lambda_3 = 2$ associated with $v_3 = [1]$. Next we must reconstruct from the eigenvectors of the deflated matrices the corresponding eigenvectors of A. The eigenvalue λ_2 and corresponding eigenvector y_2 of B_1 can be used to find the corresponding eigenvector of A. It is of the form $z = \begin{bmatrix} z \\ y_2 \end{bmatrix}$. The element z is determined by the defining condition $Bz = \lambda_2 z$. Hence $(\lambda_2 - \lambda_1)z = x'y_2$. Now B_1 has eigenvector $y_2 = (0,1)$ associated with $\lambda_2 = 4$. The corresponding eigenvector of B is then

$$Z = \begin{bmatrix} z \\ 0 \\ 1 \end{bmatrix}$$

where $-12z = (-4,-12)(0,1)$. Thus $z = 1$, and the eigenvector of B is $[1,0,1]$. The corresponding eigenvector for A is given by applying to this vector the operation's inverse to ii) and i), that is:

ii)'   add 1/2 row 1 to row 2; subtract 1/2 row 1 from row 3.

i)'   interchange rows 1 and 2.

The results are

$$\begin{bmatrix} 1 \\ 0 \\ 1 \end{bmatrix} \; \text{ii)'} \; \rightarrow \; \begin{bmatrix} 1 \\ 1/2 \\ 1/2 \end{bmatrix} \; \text{i)'} \; \rightarrow \; \begin{bmatrix} 1/2 \\ 1 \\ 1/2 \end{bmatrix} = v_2 \; .$$

Finally, performing similar calculations with the eigenvector $[1]$ results in $v_3 = [2,2,1]$.

● PROBLEM 16-24

Let

$$A = \begin{bmatrix} 4 & 1 & 2 \\ 2 & 4 & -3 \\ 3 & 1 & 3 \end{bmatrix} \qquad (1)$$

Find the dominant root of A from the limiting form of a high power of the matrix.

Solution: The limiting form of a high power of a matrix depends upon the number and the nature of the dominant characteristid values.

When there is a single dominant real root, and A^m with m large, the ratio of elements of A, a_{m+1}/a_m tends to λ_1, the characteristic root of largest magnitude.

When the dominant root is real and repeated, the typical equation for the root is

$$a_m \lambda^2 - 2a_{m+1} \lambda + a_{m+2} = 0 \; . \qquad (2)$$

Raising the given matrix (1) to high powers yields:

$$A^8 = \begin{bmatrix} 952149 & 625000 & 63476 \\ -463868 & -234375 & -161132 \\ 952148 & 625000 & 63477 \end{bmatrix}$$

$$A^9 = \begin{bmatrix} 5249024 & 3515625 & 219726 \\ -2807618 & -1562500 & -708007 \\ 5249023 & 3515625 & 219727 \end{bmatrix}$$

$$A^{10} = \begin{bmatrix} 28686524 & 19531250 & 610351 \\ -16479493 & -9765625 & -3051757 \\ 28686523 & 19531250 & 610352 \end{bmatrix} \quad .$$

Since there are two nearly equal ratios (the first row is nearly equal to the third row), we suspect there are two dominant roots associated with A.

Let m = 8 in (2). Then, taking the leading diagonal elements in A^8, A^9 and A^{10} we obtain:

$$952149 \lambda^2 - 2(5249024) \lambda + 28686524 = 0 \quad .$$

Solving this quadratic equation yields:

$$\lambda = 5.000008 \quad \text{or} \quad 6.025628.$$

Similarly, the last diagonal elements yield:

$$\lambda = 4.999966 \quad \text{or} \quad 1.923078 \quad .$$

Hence $\lambda = 5$.

As a check, by solving the characteristic equation of A we obtain the three characteristic roots $\lambda_1 = 5$, $\lambda_2 = 5$ and $\lambda_3 = 1$. Thus $\lambda = 5$ is the dominant root.

● **PROBLEM** 16-25

Eigensystems arise in the physical sciences when we study vibrations. In such systems the eigenvector corresponding to the smallest eigenvalue will have elements that are all of the same sign. Using this information, estimate the smallest eigenvalue of the following matrix by means of the Rayleigh quotient:

$$A = \begin{bmatrix} 1.7 & -1 & 0 \\ -1 & 2 & -1 \\ 0 & -1 & 2 \end{bmatrix}$$

Solution: Rayleigh's quotient arises in the study of optimizing quadratic forms subject to constraints. That is, optimize

$$(x, Ax)$$

subject to

$$(x, x) = 1$$

where A is an n × n Hermitian matrix (a matrix with complex elements whose complex conjugate is equal to its transpose), and (z, z) denotes the inner product.

The Rayleigh quotient corresponding to a Hermitian matrix A is the expression

$$\rho = \rho(x) = \frac{(x, Ax)}{(x, x)} \quad . \tag{1}$$

Furthermore, if A is Hermitian with eigenvalues $\lambda_1 \le \lambda_2 \le \dots \le \lambda_n$ and associated orthonormalized eigenvectors $x_1, \dots, x_n$, then

$$\lambda_1 \le \rho(x) \le \lambda_n$$

and

$$\lambda_1 = \min \rho(x) = \rho(x_1) \quad . \tag{2}$$

We can use (1) to compute approximate eigenvalues. Let x_j be an eigenvector associated with the (real) eigenvalue λ_i of a Hermitian matrix A. Define

$$x = x_i + \epsilon z \ . \tag{3}$$

Substituting (3) into (1) yields:

$$\rho(x) = \lambda_i + [\rho(z) - \lambda_i] \frac{(z,z)}{(x,x)} |\epsilon|^2 \ . \tag{4}$$

Since (3) is an approximation x to the eigenvector x_i, (4) tells us that $\rho(x)$ will give an approximation to λ_i that is accurate to second order in ϵ ($|\epsilon| > 0$ is a small number).

Turning to the given problem the Rayleigh quotient is

$$\rho(x) = \cfrac{[x_1,x_2,x_3] \begin{bmatrix} 1.7 & -1 & 0 \\ -1 & 2 & -1 \\ 0 & -1 & 2 \end{bmatrix} \begin{bmatrix} x_1 \\ x_2 \\ x_3 \end{bmatrix}}{[x_1,x_2,x_3] \begin{bmatrix} x_1 \\ x_2 \\ x_3 \end{bmatrix}}$$

$$= \frac{1.7x_1^2 + 2x_2^2 + 2x_3^2 - 2x_1x_2 - 2x_2x_3}{x_1^2 + x_2^2 + x_3^2} \ .$$

We know that the associated eigenvectors of the lowest eigenvalue will have elements of the same sign. Let $v_1 \approx [1,1,1]$ and then $v_1 \approx [1,2,1]$. We find $\rho(v_1) = 0.57$ and 0.62 respectively. These estimates are upper bounds on λ_1, so the best estimate here is 0.57 corresponding to a trial eigenvector $[1,1,1]$. By analytic methods we find that the exact eigenvalue is 0.5 corresponding to an eigenvector $[1,1.2,0.8]$.

● **PROBLEM** 16-26

Describe a matrix method for finding
a)   the cube root of 2 .
b)   the fourth root of 5 .

Solution: The method described does not converge very quickly.
To find the cube root of 2 we set up the matrix:

$$\begin{bmatrix} y & 2 & 2x \\ x & y & 2 \\ 1 & x & y \end{bmatrix} \ . \tag{1}$$

This matrix is constructed by building up diagonals in the following manner.   In the bottom left-hand corner a 1 is entered. Along the diagonal above, x's are entered where x is a guessed approximation to $3\sqrt{2}$. Along the main diagonal y's are entered, where y is a guessed approximation to $3\sqrt{2^2}$. The next diagonal contains 2's and the upper right-hand diagonal is 2x, which is 2 times the square root approximation of 2.

If we repeatedly multiply (1) by an arbitrary non-zero vector (i.e., containing no zeros) it will be found that the terms of the arbitrary column vector tend to the ratio

$$2^{2/3} : 2^{1/3} : 1 \ .$$

First, we must choose values for x and y. Let $x = 1$ and $y = 2$. Then A becomes

$$A = \begin{bmatrix} 2 & 2 & 2 \\ 1 & 2 & 2 \\ 1 & 1 & 2 \end{bmatrix} \tag{2}$$

Multiplying A by $v_1 = [1,1,1]^T$ we obtain:

$$Av_1 = \begin{bmatrix} 2 & 2 & 2 \\ 1 & 2 & 2 \\ 1 & 1 & 2 \end{bmatrix} \begin{bmatrix} 1 \\ 1 \\ 1 \end{bmatrix} = \begin{bmatrix} 6 \\ 5 \\ 4 \end{bmatrix} = v_2$$

$$Av_2 = \begin{bmatrix} 2 & 2 & 2 \\ 1 & 2 & 2 \\ 1 & 1 & 2 \end{bmatrix} \begin{bmatrix} 6 \\ 5 \\ 4 \end{bmatrix} = \begin{bmatrix} 30 \\ 24 \\ 19 \end{bmatrix} = v_3$$

$$Av_3 = \begin{bmatrix} 146 \\ 116 \\ 92 \end{bmatrix} \qquad Av_4 = \begin{bmatrix} 708 \\ 562 \\ 446 \end{bmatrix} \ .$$

Replacing 446 with 1, the ratios $\dfrac{562}{446}$, $\dfrac{708}{446}$ represent $2^{1/3}$ and $2^{2/3}$ respectively. Thus

$$2^{1/3} \approx 1.26 \quad \text{and} \quad 2^{2/3} \approx 1.587.$$

The error in these approximations is less than .0009 and .0004 respectively.

b) The matrix for computing the $4\sqrt{5}$ is

$$B = \begin{bmatrix} 4 & 5 & 5 & 10 \\ 2 & 4 & 5 & 5 \\ 1 & 2 & 4 & 5 \\ 1 & 1 & 2 & 4 \end{bmatrix} \ .$$

The 1's in the second diagonal are an approximation to $5^{1/4}$; the 2's approximate $5^{2/4}$; the 4's approximate $5^{3/4}$. The next diagonal contains 5's as an approximation to $5^{5/4}$. The final term is again 5 times the approximate square root of 5.

Again, letting $v_1 = \begin{bmatrix} 1 \\ 1 \\ 1 \\ 1 \end{bmatrix}$ we obtain the following sequence of v_i

where

$$v_{i+1} = Bv_1 : v_1 = \begin{bmatrix} 1 \\ 1 \\ 1 \\ 1 \end{bmatrix}, \ v_2 = \begin{bmatrix} 24 \\ 16 \\ 12 \\ 8 \end{bmatrix}, \ v_3 = \begin{bmatrix} 316 \\ 212 \\ 144 \\ 96 \end{bmatrix}, \ v_4 = \begin{bmatrix} 4004 \\ 2680 \\ 1796 \\ 1200 \end{bmatrix} \ .$$

The ratios of elements with respect to each other are $1 : 5^{1/4} : 5^{2/4} : 5^{3/4}$.
Replacing 1200 with 1, $\frac{1796}{1200} \approx 5^{1/4} = 1.4966$. If we compute $(1.4966)^4$
we obtain 5.0176 in place of 5. Hence the approximation has a relative error of 0.352%.

USE OF FLOW CHARTS IN NUMERICAL METHODS

● **PROBLEM** 16-27

Construct a flow chart for the simplex method of linear programming.

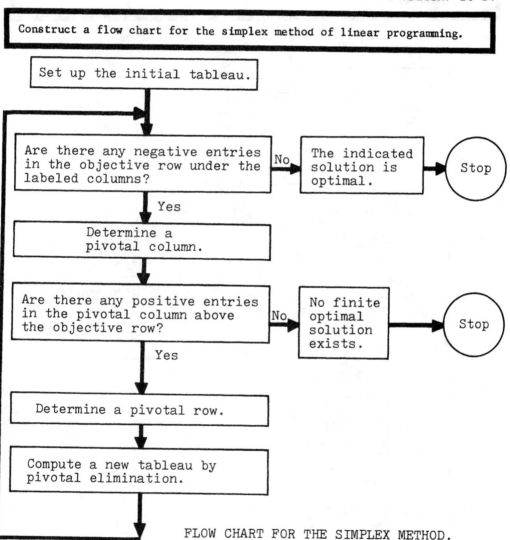

FLOW CHART FOR THE SIMPLEX METHOD.

Set up the initial tableau.

Are there any negative entries in the objective row under the labeled columns? — No → The indicated solution is optimal. → Stop

Yes

Determine a pivotal column.

Are there any positive entries in the pivotal column above the objective row? — No → No finite optimal solution exists. → Stop

Yes

Determine a pivotal row.

Compute a new tableau by pivotal elimination.

Solution: To start the simplex method we require an initial basic feasible solution. Then the simplex algorithm proceeds according to the following rules:
1) Find the largest negative entry in the objective row and choose that labeled column.
2) Compute the ratios of the solution values to the corresponding elements in the chosen column. Choose the row with minimum ratio and

475

pivot on that member for the rest of the column.
3) Repeat step 1) on the resulting tableau.
The algorithm stops according to the following rules:

a) If there are no negative entries in the objective row then the indicated solution is optimal.
b) If all the entries in the pivotal column are negative, then no finite optimal solution exists and the algorithm stops.
Thus we construct the following flow chart for the simplex method:

● **PROBLEM** 16-28

Construct a flow-chart for deriving the characteristic equation of an n x n matrix.

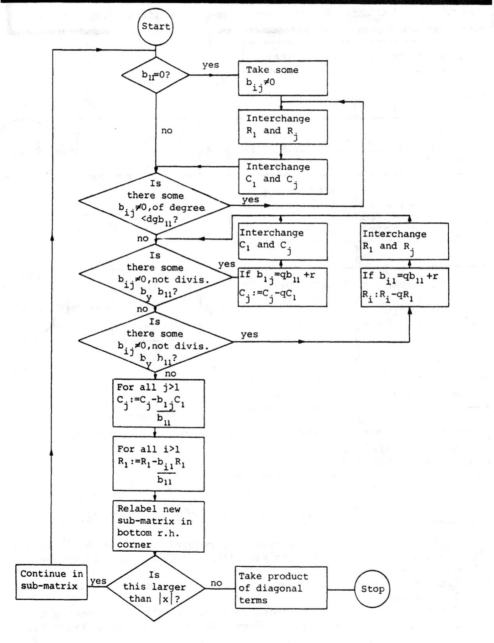

Solution: Let A be an n × n matrix. The eigenvalues of A are scalars λ such that Ax = λx where X is an n × 1 non-zero column vector. To find λ we rewrite the above equation as

$$(A - I\lambda)x = 0 .$$ (1)

Equation (1) can have non-trivial solutions only if

$$|A - \lambda I| = 0 .$$ (2)

Equation (2), when expanded is an nth degree polynomial in λ . By the Fundamental Theorem of Algebra it has n roots. These n roots are the eigenvalues associated with A. Equation (2) is the characteristic equation of the matrix A.

The flow-chart for deriving the characteristic equation is given below. Note that $B = A - \lambda I = \{b_{ij}\}$ (i,j = 1,2,...,n) is the input.

● **PROBLEM** 16-29

Construct a flow diagram for the solution of n simultaneous linear equations in n unknowns $x_1, x_2, \ldots, x_n$ using the Gauss-Seidel method.

Solution: The equations are taken as being in the form

$$\begin{bmatrix} a_{11} & a_{12} & \cdots & a_{1n} \\ a_{21} & a_{22} & \cdots & a_{2n} \\ \vdots & \vdots & & \vdots \\ a_{n1} & a_{n2} & \cdots & a_{nn} \end{bmatrix} \begin{bmatrix} x_1 \\ x_2 \\ \vdots \\ x_n \end{bmatrix} = \begin{bmatrix} b_1 \\ b_2 \\ \vdots \\ b_n \end{bmatrix}$$ (1)

where $|a_{ii}| > \Sigma |a_{ij}|$ (i = 1,2,...,n), i.e., the coefficient matrix is diagonally dominant. For obtaining X_i, the new value for x_i, the ith equation gives $X_i = (b_i - \Sigma_i)/a_{ii}$ where $\Sigma_i = \Sigma a_{ij} x_j$ with j taking all values from 1 to n except i.

Also, let ε > 0 be the chosen maximum possible difference between successive iterations. When $|X_i - x_i| < \epsilon$, the program stops.

The computer algorithm for the Gauss-Seidel iteration procedure j 'en outlined by the following flow-chart.

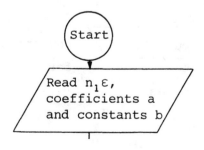

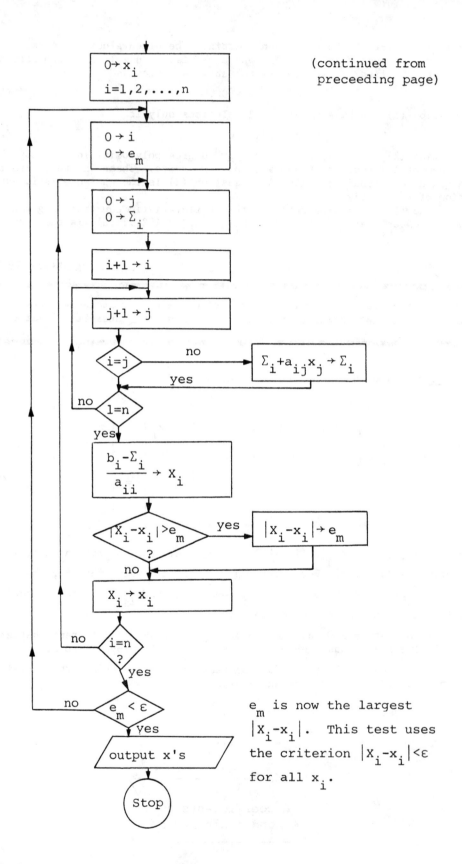

(continued from
preceeding page)

e_m is now the largest
$\left|X_i - x_i\right|$. This test uses
the criterion $\left|X_i - x_i\right| < \varepsilon$
for all x_i.

Construct a flow-chart for solving the set of simultaneous linear equations $AX = b$ by finding the inverse of the coefficient matrix.

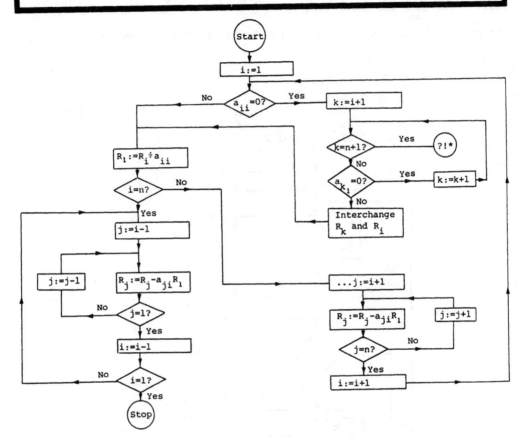

Solution: The system may be written in the form

$$
\begin{bmatrix}
a_{11} & a_{12} & \cdots \cdots & a_{1n} \\
a_{21} & a_{22} & \cdots \cdots & a_{2n} \\
\vdots & \vdots & & \vdots \\
a_{n1} & a_{n2} & \cdots \cdots & a_{nn}
\end{bmatrix}
\begin{bmatrix}
x_1 \\
\vdots \\
\vdots \\
x_n
\end{bmatrix}
=
\begin{bmatrix}
b_1 \\
b_2 \\
\vdots \\
b_n
\end{bmatrix}
$$

We can solve the set of simultaneous linear equations, $AX = b$ by multiplying each side by A^{-1}. Thus, $X = A^{-1}b$. X is the $n \times 1$ matrix of unknowns.

Construct a flow-chart for inversion of a matrix using the Gauss-Jordan method.

Solution: Let AX = B be a system of n linear equations in n unknowns. The inverse of the coefficient matrix A is an n x n matrix A^{-1} such that $AA^{-1} = I$, the n x n identity matrix. Form the partitioned matrix [A : I]. Then the Gauss-Jordan method uses elementary row operations on this partitioned matrix to convert it to [I : A^{-1}]. The following flow-chart catalogues the endomorphic structure for this procedure.

The matrix B is now [I : A^{-1}] so the ouput is the elements of A^{-1}.

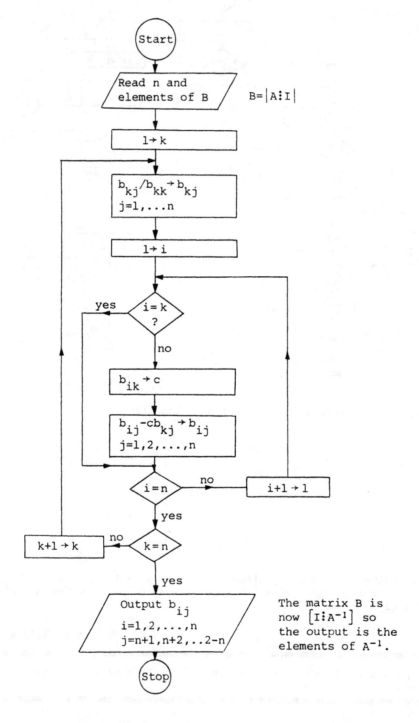

Start

Read n and elements of B $B = |A \vdots I|$

$1 \rightarrow k$

$b_{kj}/b_{kk} \rightarrow b_{kj}$
$j = 1, \ldots n$

$1 \rightarrow i$

yes   i = k ?

no

$b_{ik} \rightarrow c$

$b_{ij} - c b_{kj} \rightarrow b_{ij}$
$j = 1, 2, \ldots, n$

i = n   no $i + 1 \rightarrow 1$

yes

$k + 1 \rightarrow k$ no   k = n

yes

Output b_{ij}
$i = 1, 2, \ldots, n$
$j = n+1, n+2, \ldots 2-n$

The matrix B is now $[I \vdots A^{-1}]$ so the output is the elements of A^{-1}.

Stop

480

CHAPTER 17

LINEAR PROGRAMMING AND GAME THEORY

EXAMPLES, GEOMETRIC SOLUTIONS & THE DUAL

● PROBLEM 17-1

Give an example of a problem that is amenable to linear programming methods.

Solution: An appropriate description of linear programming would be the maximization of a linear function of n non-negative variables, subject to m linear constraints. The key word is linear. For maximizing a non-linear function subject to constraints, the method of Lagrange multipliers is available.

One of the first problems to be posed in linear programming is known as the diet problem. The essential idea is that a nutritionist wishes to find the cheapest diet that meets fixed nutritional requirements.

Suppose only two goods, A and B, are available. Let each ounce of food A contain 2 units of protein, 1 unit of iron and 1 unit of thiamine. Each ounce of food B contains 1 unit of protein, 1 unit of iron and three units of thiamine. Suppose that each ounce of A costs 30 cents while an ounce of B costs 40 cents. The nutritionist wants the meal to provide at least 12 units of protein, at least 9 units of iron and at least 15 units of thiamine. How many ounces of each of the foods should be used to minimize the cost of the meal?

We can cast the above problem into mathematical form. Notice that the two foods A and B have constant marginal cost.

Let x and y denote the number of ounces of foods A and B respectively. Then the number of units of protein supplied by the meal is
$$2x + y .$$
Thus, the first constraint is
$$2x + y \geq 12 \tag{1}$$
The inequality in (1) indicates that consumption of protein can exceed 12 units without destroying the solution of the problem.

Similarly we obtain the iron and thiamine constraints:
$$x + y \geq 9$$
$$x + 3y \geq 15.$$
Finally, we require that the amounts of food A and food B bought be non-negative, i.e.,
$$x \geq 0 , y \geq 0 .$$
We wish to minimize the cost of the meal which is
$$z = 30x + 40y \tag{2}$$
The equation (2) is the objective that we wish to minimize. It is called the objective function.

The linear programming problem is therefore: Find values of x and y that minimize
$$z = 30x + 40y$$

subject to the restrictions

$$2x + y \geq 12$$
$$x + y \geq 9$$
$$x + 3y \geq 15$$

$x \geq 0$, $y \geq 0$.

We could solve the above problem graphically but if the number of unknowns were higher, a numerical method of solution known as the simplex algorithm would be used.

● **PROBLEM** 17-2

Solve the following linear programming problem:

$$\text{Maximize} \quad 6L_1 + 11L_2 \quad \quad (1)$$

subject to:

$$2L_1 + L_2 \leq 104$$
$$L_1 + 2L_2 \leq 76 \quad \quad (2)$$

and $L_1 \geq 0$, $L_2 \geq 0$.

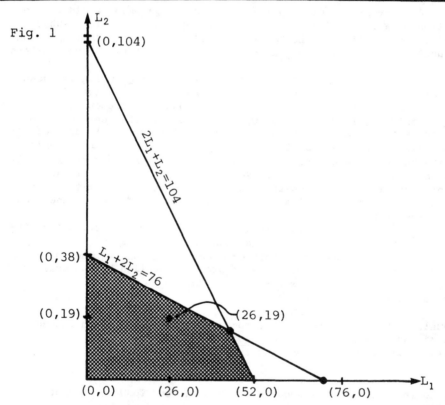

Fig. 1

Solution: Since there are only two variables $(L_1$ and $L_2)$, we can portray the above problem geometrically.

The two constraints $L_1 \geq 0$, $L_2 \geq 0$ generate the first quadrant of the Euclidean plane. The lines $L_1 = 0$, $L_2 = 0$ form the boundaries of this region.

Instead of considering the inequalities directly, we first find the boundaries of the constraint region. This is equivalent to graphing the two lines $2L_1 + L_2 = 104$ and $L_1 + 2L_2 = 76$.

482

The feasible region is the area that simultaneously satisfies all four constraints. In Fig. 1, it is the darkened region. The feasible region represents possible solutions to the maximization problem. For example, (26,19) is a feasible solution, and when substituted into the objective function yields a value of

$$6(26) + 11(19) = 365.$$

Yet a better solution is the point (0,38) on the boundary of the feasible region which yields a value of

$$6(0) + 11(38) = 418.$$

What we require is a method that will reduce the infinite set of feasible points to a finite number of points. Any point not on the boundary of the feasible region is non-optimal, since we can increase (1) by increasing one variable while keeping the other constant and still satisfy the constraints.

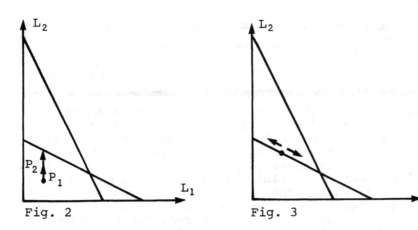

Fig. 2   Fig. 3

We can continue to do this until we meet a constraint line as in Fig. 2. Thus, the set of optimal points consists of points on the boundary of the feasible region. Now consider a point on the boundary. If movement along the boundary in one direction increases (1), then movement in the opposite direction decreases (1). Now we can keep on moving in the direction of increasing (1) until we meet another constraint. Thus the points at which (1) is locally maximized occur at the intersection of two or more constraints. Since the number of constraints is finite the set of such points is finite and every point in this set yields a greater value for (1) than any other point in the feasible region.

We now have a method for solving the given problem. First we find the intersection points of pairs of constraints. Then we check which points are feasible. Finally, from among these basic feasible points we find that point which maximizes the objective function.

The four constraints are:

$$L_1 = 0 \tag{3}$$
$$L_2 = 0 \tag{4}$$
$$2L_1 + L_2 = 104 \tag{5}$$
$$L_1 + 2L_2 = 76 \tag{6}$$

Taking these four equations two at a time we have $\binom{4}{2}$ or 6 points. For example, the solution to (3) and (4) is the point (0,0). The solution to (4) and (6) is (76,0). The point (0,0) is feasible but the point (76,0) is not. In this way we obtain the following table:

System of equations	(L_1, L_2)	Feasible Point
(3) and (4)	(0,0)	yes
(3) and (5)	(0,104)	no
(3) and (6)	(0,38)	yes
(4) and (5)	(52,0)	yes
(4) and (6)	(76,0)	no
(5) and (6)	(44,16)	yes

Next we compute the value of (1) for each of the above feasible points.

Feasible point	$f(L_1, L_2) = 6L_1 + 11L_2$
(0,0)	0
(0,38)	418
(52,0)	312
(44,16)	440

Thus, the objective function is maximized when $L_1 = 44$ and $L_2 = 16$. The value of the objective function is then 440.

● **PROBLEM** 17-3

A company makes desk organizers. The standard model requires 2 hours of the cutter's and one hour of the finisher's time. The deluxe model requires 1 hour of the cutter's time and 2 hours of the finisher's time. The cutter has 104 hours of time available for this work per month , while the finisher has 76 hours of time available for work. The standard model brings a profit of $6 per unit, while the deluxe one brings a profit of $11 per unit. The company, of course, wishes to make the most profit. Assuming they can sell whatever is made, how much of each model should be made in each month?

Solution: The company wishes to make the most profit within the given constraints. We graph the constraints and within the defined region we pick the point with the most profit. The profit is found by the formula:

$$\text{Profit} = \$6 \ X + \$11 \ Y$$

where X stands for the number of standard desk organizers and Y stands for the number of deluxe ones.

The constraints for this problem are:

(1) $X \geq 0$; we cannot have a negative number of standard units.

(2) $Y \geq 0$; we cannot have a negative number of deluxe units.

(3) The finisher has only 76 hours of time available. Since a standard model takes one hour of the finisher's time, and a deluxe model takes 2 hours of the finisher's time, we get the constraint

$$X + 2 \ Y \leq 76.$$

484

(4) The cutter has only 104 hours of time available.
A standard unit takes two hours of the cutter's time, and
a deluxe unit takes one hour of the cutter's time, thus,
we get the constraint

$$2 X + Y \leq 104.$$

We can now graph these constraints to get the region in
which we can choose our point of maximum profit.

Maximum profit

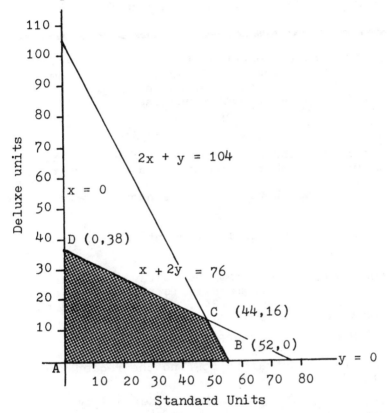

The shaded area of the graph is the area which conforms
to the constraints. Within this region we must pick the
point with the maximum profit. By a theorem of linear
programming we know that the point of maximum profit
occurs at a corner of the region. Thus, we need only
check the corners and take the point with the most profit.

A)	(0,0)	Profit = $6 (0)	+	$11 (0)	= $0
B)	(52,0)	Profit = $6 (52)	+	$11 (0)	= $312
C)	(44,16)	Profit = $6 (44)	+	$11 (16)	= $440
D)	(0,38)	Profit = $6 (0)	+	$11 (38)	= $418

By observation, we note that the point with the
largest profit is (44,16) (Point C). Thus, for the
company to make the maximum profit of $440, they must
produce 44 standard units and 16 deluxe ones.

A marketing manager wishes to maximize the number of people exposed to the company's advertising. He may choose television commercials, which reach 20 million people per commercial, or magazine advertising, which reaches 10 million people per advertisement. Magazine advertisements cost $40,000 each while a television advertisement costs $75,000. The manager has a budget of $2,000,000 and must buy at least 20 magazine advertisements. How many units of each type of advertising should be purchased?

Solution: We find the constraints of the problem, and graph them to find the region defined by them. From this region we will pick the point which maximizes the number of people exposed to the advertisements.

The constraints are:

Let T stand for the number of television commercials and M stand for the number of magazine advertisements.

(1) $T \geq 0$ We cannot have a negative number of television commercials

(2) $M \geq 20$ We must have at least twenty magazine advertisements.

(3) This constraint comes from the costs. In thousands the cost of a television commercial is $75 and the cost of a magazine advertisement is $40. He is budgeted to $2,000,000 so we get the constraint

$$40M + 75T \leq 2,000.$$

We now graph these constraints to find the region which is defined by them.

The shaded area is the region which is defined by the constraints. To find the point which yields the highest number of people exposed to the advertisement can be found by a theorem of linear programming which states that the point must be one of the corners of the region.

A) (0,20) number of people = 20 million × T + 10 million × M = 20 million × 0 + 10 million × 20 = 200 million.

B) (16,20) number of people = 20 million × 16 + 10 million × 20 = 520 million.

C) (0,50) number of people = 20 million × 0 + 10 million × 50

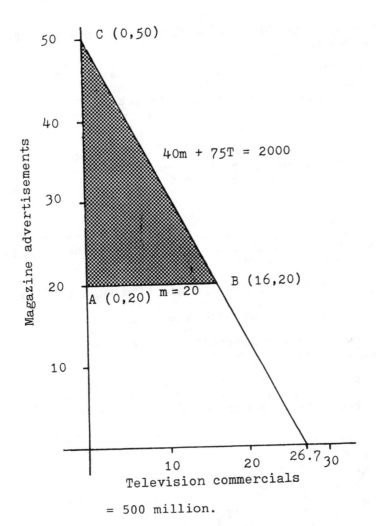

C (0,50)

$40m + 75T = 2000$

B (16,20)

A (0,20)   m = 20

Magazine advertisements

Television commercials

= 500 million.

Thus, the best thing for the manager to do is have 16 television commercials and 20 magazine advertisements.

● **PROBLEM** 17-5

A businessman needs 5 cabinets, 12 desks, and 18 shelves cleaned out. He has two part time employees Sue and Janet. Sue can clean one cabinet, three desks and three shelves in one day, while Janet can clean one cabinet, two desks and 6 shelves in one day. Sue is paid $25 a day, and Janet is paid $22 a day. In order to minimize the cost how many days should Sue and Janet be employed?

Solution: The businessman wishes to minimize the cost of cleaning out the office. To do this we must graph the constraints, and take the point which gives the minimum cost. To find the cost we use the formula

Cost = $25 X + $22 Y,

where X is the number of days Sue is employed and Y is

487

the number of days Janet is employed. We now will find
the constraints.

(1) Since Sue can do a cabinet in one day, and Janet can
do a cabinet in one day, and we must have at least 5
cabinets cleaned, we get the constraint

$$X + Y \geq 5.$$

(2) Since Sue can do 3 desks in one day, and Janet can
do 2 desks in one day, and we require 12 desks to be
cleaned, we have the constraint

$$3X + 2Y \geq 12.$$

(3) Similarly the constraint $3X + 6Y \geq 18$ comes from
Sue being able to clean 3 shelves and Janet being able
to clean 6 shelves in one day.

(4) $X \geq 0$ Sue cannot work a negative number of days

(5) $Y \geq 0$ Janet cannot work a negative number of days.

We now graph the constraints to find the region
described by them.

The shaded area is the region described by the
constraints. Note that this is an infinite region,
because they can work more days than is needed. To

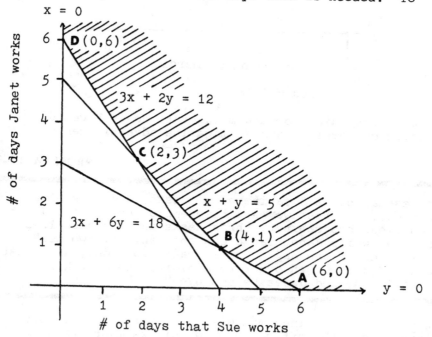

find the minimum point, we refer to a theorem of linear
programming which states that a minimum cost must occur
in one of the corners. Thus, we need only check which
of the four corners has the smallest cost, and we will
have the answer.

488

A)	(6,0)	$25 (6)	+	$22 (0) = $150
B)	(4,1)	$25 (4)	+	$22 (1) = $122
C)	(2,3)	$25 (2)	+	$22 (3) = $116
D)	(0,6)	$25 (0)	+	$22 (6) = $132

We can now see that the minimum cost is, with Janet working 3 days and Sue working 2 days, equal to $116. (Point C on the graph.)

● PROBLEM 17-6

Convert the following standard linear programming problem to canonical form:

Maximize
$$8x_1 + 15x_2 + 6x_3 + 20x_4$$

subject to:
$$x_1 + 3x_2 + x_3 + 2x_4 \leq 9$$
$$2x_1 + 2x_2 + 2x_3 + 3x_4 \leq 12$$
$$3x_1 + 3x_2 + 2x_3 + 5x_4 \leq 16$$

$x_1, x_2, x_3, x_4 \geq 0$.

Solution: The general standard linear programming problem takes the form

$$\text{Max} \quad f(x_1, \ldots, x_n) = b_1 x_1 + b_2 x_2 + \ldots + b_n x_n \quad (1)$$

subject to:
$$a_{11} x_1 + a_{12} x_2 + \ldots + a_{1n} x_n \leq c_1$$
$$a_{21} x_1 + a_{22} x_2 + \ldots + a_{2n} x_n \leq c_2 \quad (2)$$
$$\cdot \qquad \cdot \qquad \qquad \cdot \qquad \cdot$$
$$a_{m1} x_1 + a_{m2} x_2 + \ldots + a_{mn} x_n \leq c_n$$

and $x_i \geq 0$ $(i = 1, 2, \ldots, n)$. $\quad (3)$

The function $f(x_1, \ldots, x_n)$ is the objective function. It is a real-valued function whose argument comes from the set of all n-tuples that satisfy (2) and (3).

We can rewrite the above program in matrix form. Thus,
$$\text{Max} \quad f(X) = (B \cdot X)$$

subject to: $\quad AX \leq C$, $X \geq 0$,

where $B = (b_1, \ldots, b_n)$, $X = (x_1, x_2, \ldots, x_n)$, $C = (c_1, \ldots, c_n)$ and

$$A = \begin{bmatrix} a_{11} & a_{12} & \cdots & a_{1n} \\ a_{21} & & & \\ \cdot & & & \cdot \\ \cdot & & & \cdot \\ \cdot & & & \\ a_{m1} & a_{m2} & \cdots & a_{mn} \end{bmatrix} \cdot$$

The solution of the problem is facilitated by converting the inequalities

489

to equalities. We do this by introducing m new variables, one for each of the inequalities. Thus, let

$$x_{n+1} = c_1 - (a_{11}x_1 + a_{12}x_2 + \ldots + a_{1n}x_n)$$

$$\vdots$$

$$x_{n+m} = c_m - (a_{m1}x_1 + a_{m2}x_2 + \ldots + a_{mn}x_n)$$

Then the system of inequalities (2) becomes the system of equalities

$$a_{11}x_1 + a_{12}x_2 + \ldots + a_{1n}x_n + x_{n+1} = c_1$$

$$a_{21}x_1 + a_{22}x_2 + \ldots + a_{2n}x_n + x_{n+2} = c_2$$

$$\vdots$$

$$a_{m1}x_1 + a_{m2}x_2 + \ldots + a_{mn}x_n + x_{n+m} = c_m$$

By increasing the number of unknowns from n to n+m, we obtain a system of equalities: the analysis of such systems is much more developed than the study of systems of inequalities. When the inequalities in the constraint inequations are converted to equalities, the program is in canonical form.

The given problem in canonical form is:

Maximize

$$8x_1 + 15x_2 + 6x_3 + 20x_4$$

subject to

$$x_1 + 3x_2 + x_3 + 2x_4 + x_5 = 9$$

$$2x_1 + 2x_2 + 2x_3 + 3x_4 + x_6 = 12$$

$$3x_1 + 3x_2 + 2x_3 + 5x_4 + x_7 = 16$$

$$x_i \geq 0 \quad (i = 1,2,\ldots,7).$$

● **PROBLEM 17-7**

In order to produce 1000 tons of non-oxidizing steel for engine valves, at least the following units of manganese, chromium and molybdenum, will be needed weekly: 10 units of manganese, 12 units of chromium, and 14 units of molybdenum (1 unit is 10 pounds). These metals are obtainable from dealers in non-ferrous metals, who, to attract markets make them available in cases of three sizes, S, M and L. One S case costs $9 and contains 2 units of manganese, 2 units of chromium and 1 unit of molybdenum. One M case costs $12 and contains 2 units of manganese, 3 units of chromium, and 1 unit of molybdenum. One L case costs $15 and contains 1 unit of manganese, 1 unit of chromium and 5 units of molybdenum.

How many cases of each kind should be purchased weekly so that the needed amounts of manganese, chromium and molybdenum are obtained at the smallest possible cost? What is the smallest possible cost?

Solution: The general linear programming problem has the following form:

Minimize $z = c_1x_1 + c_2x_2 + \ldots + c_nx_n$ (1)

Subject to:

$$a_{11}x_1 + a_{12}x_2 + \ldots + a_{1n}x_n \geq b_1$$
$$\vdots \qquad \vdots \qquad \qquad \vdots \qquad \vdots \qquad\qquad (2)$$
$$a_{m1}x_1 + a_{m2}x_2 + \ldots + a_{mn}x_n \geq b_m$$

$x_i \geq 0 \quad (i = 1, 2, \ldots, n)$.

The linear function (1) is called the objective function while the inequalities (2) represent the constraints.

To cast the above problem into this form, we first find the objective function. Since we are buying cases we want to minimize

$$z = 9S + 12M + 15L . \qquad\qquad (3)$$

To obtain the constraints, we see that at least 10 units of manganese are required. Each small case contains 2 units, each medium case contains 2 units and each large case contains 1 unit. Thus, the manganese constraint is

$$2S + 2M + L \geq 10 .$$

Similarly we obtain the constraints on chromium and molybdenum:

$$2S + 3M + L \geq 12$$
$$S + M + 5L \geq 14 .$$

Thus the problem is:

Minimize $z = 9S + 12M + 15L$ $\qquad\qquad (3)$

Subject to:

$$2S + 2M + L \geq 10 \qquad\qquad (4)$$
$$2S + 3M + L \geq 12 \qquad\qquad (5)$$
$$S + M + 5L \geq 14 \qquad\qquad (6)$$

$S \geq 0 , M \geq 0 , L \geq 0$. $\qquad\qquad (7)$

The region common to the constraints (4)-(7) is called the feasible region. From the theory of linear programming we know that a point at which the objective function is optimized must be on the boundary. Hence we can convert the inequalities in (4)-(7) to equations. A fundamental theorem of linear programming is that if a solution exists it must be at a point of intersection of three equations. We have 6 constraints and three unknowns; hence the number of possible solutions is $\binom{6}{3}$ or 20. For example, the point (0,0,0), corresponding to the solution set of $S = M = L = 0$, is a possible solution. But since it fails to satisfy the constraints (4)-(6) it is not a feasible solution. On the other hand, the solution to $S = 0$, $2S + 2M + L = 10$ and $2S + 3M + L = 12$ is (0,2,6) which is a feasible point. Substituting into the objective function (3):

$$z = 9(0) + 12(2) + 15(6) = \$114.$$

Proceeding in this way for every triplet of equations we find that the solution to (4)-(6) is (2,2,2). This point satisfies all the constraints, i.e., it is feasible. Substituting $S = 2$, $M = 2$ and $L = 2$ in (3):

$$z = 9(2) + 12(2) + 15(2) = \$72.$$

This is the minimum cost and is the solution to the problem. We should buy 2 small cases, 2 medium cases and 2 large cases to obtain this minimum cost.

● PROBLEM 17-8

A manufacturer of electronic instruments produces two types of timer: a standard and a precision model with net profits of $10 and $15, respectively. His work force cannot produce more than 50 instruments per day. Moreover, the four main components used in production are in short supply so that the following stock constraints hold:

Component	Stock	Number used per timer	
		Standard	Precision
a	220	4	2
b	160	2	4
c	370	2	10
d	300	5	6

Graphically determine the point of optimum profit. If profits on the standard timer were to change, by how much could they change without altering the original solution?

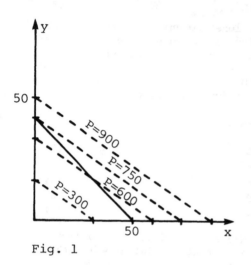

Fig. 1

Fig. 2

Solution: Let x and y denote the number of standard and precision timers produced respectively. Then the manufacturer's problem is to

$$\text{Maximize} \quad 10x + 15y$$

subject to
$$x + y \leq 50$$
$$4x + 2y \leq 220$$
$$2x + 4y \leq 160$$
$$2x + 10y \leq 370$$
$$5x + 6y \leq 300$$

$x, y \geq 0$.

Begin the analysis by considering only the first constraint and the profit line as in Fig. 1. The feasible region is that portion of the xy plane that satisfies $x \geq 0$, $y \geq 0$ and $x + y \leq 50$. Now consider $P = 10x + 15y$. $P(x,y)$ is a function of two variables and we would require a three dimensional graph to plot it. But if we let P be a constant P_0, we obtain the linear equation
$$P_0 = 10x + 15y , \qquad (1)$$

or,
$$y = \frac{P_0}{15} - \frac{2}{3} x .$$

Lines of the form (1) are isoprofit lines, i.e., at all points on this line, the profit is the same. By substituting various values for P_0 we obtain a series of parallel lines cutting the x and y axis at $P_0/10$ and $P_0/15$ respectively, and having a slope of $-10/15 = -2/3$. Note that as we move further away from the origin, P increases. If P = 300, all combinations of x and y are within the feasible region. This suggests increasing P until it touches the feasible region at as few points as possible. When P = 750, the profit line is tangent to the feasible region at y = 50. Hence profit is maximized when 50 precision timers and 0 standard timers are produced. Now, add in the other constraints to obtain Fig. 2.

The feasible region now consists of the polygon ABCDE. We would obtain the optimum profit point by again drawing in the isoprofit lines. Maximum profit occurs at C.

Now, suppose the profit on the standard timer is changed. Examining Fig. 2 we can see that the profit line will always pass through C provided its slope lies between those of the lines DC and CB. Suppose it becomes the same as that of DC($-\frac{1}{2}$). The profit line is then P = 7.5x + 15y and the optimum occurs anywhere along DC so that x = 20, y = 30 and x = 10, y = 35 both give the same maximum value for P of 600. Suppose the slope is varied again by reducing the profit on the standard model to 7, giving P = 7x + 15y (slope - 7/15); we find the optimum has now moved to the point D(x = 10, y = 35) with a corresponding maximum profit of $595.

● **PROBLEM** 17-9

The solution to the problem:

Minimize   45I + 50L + 60A   (1)

subject to

$$I + 2L + 4A \geq 6$$
$$3I + 3L + 2A \geq 3$$ (2)
$$I + 3L + A \geq 2$$

I ≥ 0, L ≥ 0, A ≥ 0, is

I = 0, L = 1/5, A = 7/5 .

This yields a value of 94 when substituted into the objective function (1). The solution to the program

Maximize   6B + 3E + 2M   (3)

subject to

$$B + 3E + M \leq 45$$
$$2B + 3E + 3M \leq 50$$ (4)
$$4B + 2B + B \leq 60$$

B ≥ 0, E ≥ 0, M ≥ 0, is

B = 13, E = 0 and M = 8.

Here again the value of the function is 94.
    What is the relationship between the two programs?

Solution: The minimization problem (1), (2) and the maximization problem (3),(4) are called mathematical duals of each other. To see how they are related, we rewrite the problems in matrix form. Thus,

Minimize $\begin{bmatrix} 45 & 50 & 60 \end{bmatrix} \begin{bmatrix} I \\ L \\ A \end{bmatrix}$ (1')

subject to:

$$\begin{pmatrix} 1 & 2 & 4 \\ 3 & 3 & 2 \\ 1 & 3 & 1 \end{pmatrix} \begin{pmatrix} I \\ L \\ A \end{pmatrix} \geq \begin{pmatrix} 6 \\ 3 \\ 2 \end{pmatrix}$$ (2')

I ≥ 0, L ≥ 0, A ≥ 0. The dual is:

Maximize $\begin{bmatrix} 6 & 3 & 2 \end{bmatrix} \begin{pmatrix} B \\ E \\ M \end{pmatrix}$ (3')

subject to:

493

$$\begin{pmatrix} 1 & 3 & 1 \\ 2 & 3 & 3 \\ 4 & 2 & 1 \end{pmatrix} \begin{pmatrix} B \\ E \\ M \end{pmatrix} \leq \begin{pmatrix} 45 \\ 50 \\ 60 \end{pmatrix}. \qquad (4')$$

The coefficients of the objective function, (1') in the minimization problem have become the constraint values for the maximization problem, in (4'). Conversely, the coefficients of the objective function of the maximization problem, (3') are the constraint values for the minimization problem. Notice, also, that the matrix of coefficients (4') is just the transpose of the matrix of coefficients in (2').

From the above example, we see that every linear program has a dual. In general, the dual of

$$\text{Maximize} \qquad b_1 x_1 + b_2 x_2 + \ldots + b_n x_n$$

subject to:

$$\begin{pmatrix} a_{11}x_1 + a_{12}x_2 + \ldots + a_{1n}x_n \\ a_{21}x_1 + a_{22}x_2 + \ldots + a_{2n}x_n \\ \vdots \\ a_{m1}x_1 + a_{m2}x_2 + \ldots + a_{mn}x_n \end{pmatrix} \leq \begin{pmatrix} c_1 \\ c_{.2} \\ \vdots \\ c_m \end{pmatrix}$$

$x_i \geq 0 \quad (i = 1,2,\ldots,n)$, is

$$\text{Minimize} \quad c_1 y_1 + c_2 y_2 + \ldots + c_m y_m$$

subject to:

$$\begin{pmatrix} a_{11}y_1 + a_{21}y_2 + \ldots + a_{m1}y_1 \\ \vdots \\ a_{1n}y_1 + a_{2n}y_2 + \ldots + a_{mn}y_m \end{pmatrix} \geq \begin{pmatrix} b_1 \\ \vdots \\ b_n \end{pmatrix}$$

Since a linear program and its dual are so intimately related, we can expect their solutions to at least be acquaintances. According to the Fundamental Theorem of linear programming, if a linear program and its dual both have a feasible point, then they both have the same optimal (= best feasible) point. If one has no feasible point, the other has no optimal point, i.e., the dual cannot be solved.

● PROBLEM 17-10

According to the Fundamental Theorem of linear programming, if either a linear program or its dual has no feasible point, then the other one has no solution. Illustrate this assertion with an example.

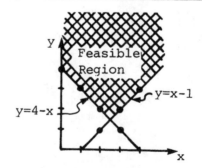

Consider the linear program

$$\text{Maximize} \quad x + y \tag{1}$$

subject to

$$x - y \le 1 \tag{2}$$

$$x + y \ge 4 \tag{3}$$

$$x \ge 0 , \; y \ge 0 . \tag{4}$$

We first convert the program into standard form. The second inequality, (3), needs to be reversed. We then obtain

$$-x - y \le -4 . \tag{5}$$

The dual to the program (1), (2), (4) and (5) is

$$\text{Minimize} \quad u - 4v \tag{6}$$

subject to

$$u - v \ge 1$$

$$-u - v \ge 1 \tag{7}$$

$$u \ge 0 , \; v \ge 0 .$$

But for non-negative u and v, the constraint $-u - v \ge 1$ can never be satisfied. Thus the minimization problem has no feasible solution. Hence the maximization problem has no best feasible solution.

We can see that the last statement is true without appealing to the Fundamental Theorem. Graphing the constraints to the maximization problem: (See fig.)

The feasible region is unbounded from above. This means that we can keep on increasing the value of the objective function while satisfying the constraints. Thus no maximum exists, i.e., there is no best feasible solution.

THE SIMPLEX ALGORITHM & SENSITIVITY ANALYSIS

● PROBLEM 17-11

Find a basic feasible solution to the problem:

$$\text{Maximize} \quad x_1 + 2x_2 + 3x_3 + 4x_4 \tag{1}$$

while satisfying the conditions

$$x_1 + 2x_2 + x_3 + x_4 = 3$$

$$x_1 - x_2 + 2x_3 + x_4 = 4 \tag{2}$$

$$x_1 + x_2 - x_3 - x_4 = -1 .$$

$$x_1, \; x_2, \; x_3, \; x_4 \ge 0 .$$

Solution: We can solve this problem using the simplex method. Note that the number of equations, 3, is less than the number of unknowns, 4. The system of linear equations can be written in vector form as:

$$x_1 \begin{pmatrix} 1 \\ 1 \\ 1 \end{pmatrix} + x_2 \begin{pmatrix} 2 \\ -1 \\ 1 \end{pmatrix} + x_3 \begin{pmatrix} 1 \\ 2 \\ -1 \end{pmatrix} + x_4 \begin{pmatrix} 1 \\ 1 \\ -1 \end{pmatrix} = \begin{pmatrix} 3 \\ 4 \\ -1 \end{pmatrix} \tag{3}$$

Any three of the vectors on the left can be used as a basis for R^3. Thus, by setting either $x_1 = 0$, or $x_2 = 0$ or $x_4 = 0$ we can still find $x_i, \; x_j, \; x_k$ (i,j,k = 1 or 2 or 3 or 4) that satisfy (3). Such solutions are called basic solutions, since they depend on the particular basis chosen when one of the x_i's is set equal to zero. A basic feasible solution (non-negative x_i), when substituted into the objective function will yield some value, not necessarily optimal. Now, the number of basic solutions is the number of ways of selecting a basis for R^3 from a set of four vectors, i.e.,

$$\binom{4}{3} = \frac{4!}{3! \; 1!} = 4 \; .$$

According to the theory of linear programming, if an optimal solution to (1) and (2) exists, then an optimal basic solution exists. Thus the optimal solution can be found by using one of the four bases of R^3 .
Let $v_1 = [1,1,1]$, $v_2 = [2,-1,1]$, $v_3 = [1,2,-1]$ and $v_4 = [1,1,-1]$.
Let $b = [3,4,-1]$. Then the system (3) becomes

$$x_1 v_1 + x_2 v_2 + x_3 v_3 + x_4 v_4 = b.$$

Letting $x_2 = 0$, we see that

$$x_1 = 1, \; x_2 = 0, \; x_3 = 1, \; x_4 = 1$$

is a basic feasible solution which depends on $\{v_1, v_3, v_4\}$. Corresponding to this solution we can construct a simplex tableau. The general form of the simplex tableau is as follows:

	v_1	v_2	$\cdots$	v_q	$\cdots$	v_n	b
v_{k1}	s_{11}	s_{12}	$\cdots s_{1q}$		$\cdots$	s_{1n}	x_{k1}
v_{k2}	s_{21}	s_{22}	$\cdots s_{2q}$		$\cdots$	s_{2n}	x_{k2}
$\vdots$	$\vdots$						$\vdots$
v_{kr}	s_{r1}	s_{r2}	$\cdots s_{rq}$		$\cdots$	s_{rn}	x_{kr}
	d_1	d_2	$\cdots d_q$		$\cdots$	d_n	D

The column vectors $v_1,\ldots,v_n$ can all be expressed as linear combinations of the basic vectors $v_{k1}, v_{k2},\ldots,v_{kr}$. The coordinates (s_{ij}), $i = 1,\ldots,r$, $j = 1,\ldots,n$ form the main body of the table. The numbers $d_1, d_2,\ldots,d_n$, D in the last row of the table are defined as follows:

$$d_j = (c_{k1} s_{1j} + c_{k2} s_{2j} + \ldots + c_{kr} s_{rj}) - c_j \; , \; j = 1,\ldots,n.$$

$$D = c_{k1} x_{k1} + c_{k2} x_{k2} + \ldots + c_{kr} x_{kr}$$

$$= c_1 x_1 + c_2 x_2 + \ldots + c_n x_n \; .$$

The c_i $(i = 1,2,\ldots,n)$ represent the coefficients of the objective function, $z = c_1 x_1 + c_2 x_2 + \ldots + c_n x_n$. The c_{kr} are the coefficients of the x_{kr} which are the coefficients of the basis $\{v_{k1},\ldots,v_{kr}\}$ chosen from $v_1, v_2,\ldots,v_n$. D is the quantity to be maximized. The d_j indicate when the solution is optimal or how to proceed to a more nearly optimal solution. Returning to the given problem,

$$b = v_1 + v_3 + v_4; \; v_1 = 1v_1 + 0v_3 + 0v_4;$$
$$v_2 = 3/2 \; v_1 - 3v_3 + 7/2 \; v_4; \; v_3 = 0v_1 + 1v_3 + 0v_4;$$
$$v_4 = 0v_1 + 0v_3 + 1v_4 \; .$$

$$d_1 = c_1 1 + c_3 (0) + c_4 (0) - c_1 = 0$$

$$d_2 = c_1 (3/2) + c_3 (-3) + c_4 (7/2) - c_2 = 9/2$$

$$d_3 = c_1(0) + c_3(1) + c_4(0) - c_3 = 0$$
$$d_4 = c_1(0) + c_3(0) + c_4(1) - c_4 = 0$$
$$D = c_1 x_1 + c_2 x_2 + c_3 x_3 + c_4 x_4 = 8$$

The simplex tableau for this solution is:

	v_1	v_2	v_3	v_4	b
v_1	1	3/2	0	0	1
v_3	0	-3	1	0	1
v_4	0	7/2	0	1	1
	0	9/2	0	0	8

Next, we would choose another basis for the basic solution and construct another tableau. If the new value of D were greater than before we would proceed yet again using another new basis. This is the simplex method. We see that rules for choosing a new basis and for deciding when a solution is optimal are needed. It is important to note the following criteria:
(1) If the numbers $d_1, d_2, \ldots, d_n$ are all non-negative, then the given solution is optimal.
(2) If, for some index q, d_q is negative and $s_{1q}, s_{2q}, \ldots, s_{rq}$ are all non-positive, then the given problem does not have an optimal solution.

● PROBLEM 17-12

Find a solution to the following problem by solving its dual:

$$\text{Minimize} \quad 9x_1 + 12x_2 + 15x_3 \tag{1}$$

subject to

$$2x_1 + 2x_2 + x_3 \geq 10$$
$$2x_1 + 3x_2 + x_3 \geq 12 \tag{2}$$
$$x_1 + x_2 + 5x_3 \geq 14$$

$$x_1 \geq 0, \; x_2 \geq 0, \; x_3 \geq 0. \tag{3}$$

Solution: The dual to (1), (2) and (3) is

$$\text{Maximize} \quad 10y_1 + 12y_2 + 14y_3 \tag{4}$$

subject to

$$2y_1 + 2y_2 + y_3 \leq 9$$
$$2y_1 + 3y_2 + y_3 \leq 12 \tag{5}$$
$$y_1 + y_2 + 5y_3 \leq 15$$

$$y_1 \geq 0, \; y_2 \geq 0, \; y_3 \geq 0. \tag{6}$$

We solve this problem using the simplex method. Converting the inequalities to equalities by adding slack variables

$$2y_1 + 2y_2 + y_3 + y_4 = 9$$
$$2y_1 + 3y_2 + y_3 + y_5 = 12$$
$$y_1 + y_2 + 5y_3 + y_6 = 15.$$

This may be rewritten as:

$$y_1 \begin{pmatrix} 2 \\ 2 \\ 1 \end{pmatrix} + y_2 \begin{pmatrix} 2 \\ 3 \\ 1 \end{pmatrix} + y_3 \begin{pmatrix} 1 \\ 1 \\ 5 \end{pmatrix} + y_4 \begin{pmatrix} 1 \\ 0 \\ 0 \end{pmatrix} + y_5 \begin{pmatrix} 0 \\ 1 \\ 0 \end{pmatrix}$$

$$+ y_6 \begin{pmatrix} 0 \\ 0 \\ 1 \end{pmatrix} = \begin{pmatrix} 9 \\ 12 \\ 15 \end{pmatrix}. \tag{7}$$

Since we have three equations in six unknowns, we can obtain a basic feasible solution by setting any three of the y_i's to zero. Let

$$v_1 = \begin{pmatrix} 2 \\ 2 \\ 1 \end{pmatrix}, \quad v_2 = \begin{pmatrix} 2 \\ 3 \\ 1 \end{pmatrix}, \quad v_3 = \begin{pmatrix} 1 \\ 1 \\ 5 \end{pmatrix}, \quad v_4 = \begin{pmatrix} 1 \\ 0 \\ 0 \end{pmatrix}, \quad v_5 = \begin{pmatrix} 0 \\ 1 \\ 0 \end{pmatrix}, \quad v_6 = \begin{pmatrix} 0 \\ 0 \\ 1 \end{pmatrix}$$

and

$b = \begin{pmatrix} 9 \\ 12 \\ 15 \end{pmatrix}$. Let $d_1 = 10$, $d_2 = 12$, $d_3 = 14$. Setting $y_1 = y_2 = y_3 = 0$, a basic feasible solution is $y_4 = 9$, $y_5 = 12$, $y_6 = 15$. We start the simplex algorithm with v_4, v_5 and v_6 in the basis.

	v_1	v_2	v_3	b	v_4	v_5	v_6
v_4	2	2	1	9	1	0	0
v_5	2	3	1	12	0	1	0
v_6	1	1	5	15	0	0	1
d	-10	-12	-14	0	0	0	0

$\uparrow$
D

To increase the value of D, we must choose another vector for the basis. We choose the column which has the most negative value in the row labelled 'd'. Here v_3 is the chosen column. Next we must decide which vector in the basis to discard. The rule here is to choose that row for which the ratio of b to v_3 is the smallest. Here the ratios are 9/1, 12/1 and 15/5. Hence we replace v_6 by v_3, by pivoting on 5. The process of conversion is carried out by first converting the 5 to a 1 and then using elementary row operations to reduce every other element under v_3 to zero. Thus we obtain the new tableau

	v_1	v_2	v_3	b	v_4	v_5	v_6
v_4	9/5	9/5	0	6	1	0	-1/5
v_5	9/5	14/5	0	9	0	1	-1/5
v_3	1/5	1/5	1	3	0	0	1/5
d	-36/5	-46/5	0	42	0	0	14/5

The value 42 was obtained by using $b = c_1 y_1 + c_2 y_2 + c_3 y_3 + c_4 y_4 + c_5 y_5 + c_6 y_6$ where the y_i are the coefficients of the vectors in the basis. Here the basis vectors are v_4, v_5, v_6. The coefficients are, from (4), $c_1 = 10$, $c_2 = 12$, $c_3 = 14$, $c_4 = c_5 = c_6 = 0$ and the y_i are obtained from the column labelled b. Hence

$$D = c_4(6) + c_5(9) + c_3(3) = 0(6) + 0(9) + 14(3) = 42.$$

Since the row d still contains negative entries we repeat the above procedure by pivoting on 14/5. We obtain

	v_1	v_2	v_3	b	v_4	v_5	v_6
v_4	9/14	0	0	3/14	1	-9/14	-1/14
v_2	9/14	1	0	45/14	0	5/14	-1/14
v_3	1/14	0	1	33/14	0	-1/14	3/14
d	-9/7	0	0	501/7	0	23/7	15/7

Next, we pivot on the first element in the first column, i.e., 9/14. The final simplex tableau is

	v_1	v_2	v_3	b	v_4	v_5	v_6
v_1	1	0	0	1/3	14/9	-1	-1/9
v_2	0	1	0	3	-1	1	0
v_3	0	0	1	7/3	-1/9	0	2/9
d	0	0	0	72	2	2	2

Thus the solution to the maximization problem (the dual of the given problem) is

$$y_1 = 1/3, \ y_2 = 3, \ y_3 = 7/3$$

with value 72. We know from duality theory that if a program has an optimal feasible point then so does its dual and both programs have the same value. The dual of (4), (5), (6) is (1), (2), (3). Thus the solution to the minimization problem has value 72. But what are the values of x_1, x_2, x_3 at this optimum point? From the Complementary Slackness Theorem of duality theory, these values are the values of the slack variables in the final simplex tableau of the dual problem. Thus $x_1 = 2$, $x_2 = 2$, $x_3 = 2$ is the required solution.

● **PROBLEM** 17-13

Find nonnegative numbers x_1, x_2, x_3, x_4 which maximize

$$4x_1 + 5x_2 + 3x_3 + 6x_4 \qquad (1)$$

and satisfy the inequalities

$$x_1 + 3x_2 + x_3 + 2x_4 \leq 2$$
$$3x_1 + 3x_2 + 2x_3 + 2x_4 \leq 4 \qquad (2)$$
$$3x_1 + 2x_2 + 4x_3 + 5x_4 \leq 6 \ .$$

Solution: We cannot use the simplex method directly to solve (1) subject to (2) because we need equalities in the constraint equations. Hence, introduce three additional variables, x_5, x_6, x_7 to convert each inequality to an equality. The system (2) then becomes

$$x_1 + 3x_2 + x_3 + 2x_4 + x_5 = 2$$
$$3x_1 + 3x_2 + 2x_3 + 2x_4 + x_6 = 4 \qquad (3)$$
$$3x_1 + 2x_2 + 4x_3 + 5x_4 + x_7 = 6 .$$

These new variables are called slack variables. We now have 3 equations in 7 unknowns. The system (3) can be written as

$$x_1\begin{pmatrix}1\\3\\3\end{pmatrix} + x_2\begin{pmatrix}3\\3\\2\end{pmatrix} + x_3\begin{pmatrix}1\\2\\4\end{pmatrix} + x_4\begin{pmatrix}2\\2\\5\end{pmatrix} + x_5\begin{pmatrix}1\\0\\0\end{pmatrix} + x_6\begin{pmatrix}0\\1\\0\end{pmatrix} + x_7\begin{pmatrix}0\\0\\1\end{pmatrix}$$

$$= \begin{pmatrix}2\\4\\6\end{pmatrix} , \quad \text{or}$$

$$x_1 v_1 + x_2 v_2 + x_3 v_3 + x_4 v_4 + x_5 v_5 + x_6 v_6 + x_7 v_7 = [2,4,6]. \qquad (4)$$

By setting any four of the unknowns in (4) to zero, we obtain a system of three equations in three unknowns with a unique solution. Say we let $x_{\ell_1} = x_{\ell_2} = x_{\ell_3} = x_{\ell_4} = 0$. Then $v_{\ell_5}, v_{\ell_6}, v_{\ell_7}$ form a basis for R^3.
The corresponding solution of x's is called a basic feasible solution. We can obtain

$$\binom{7}{3} = \frac{7!}{3! \; 4!} = 35$$

basis feasible solutions. According to the theory of linear programming if an optimal solution exists, one of these solutions is an optimal solution. The simplex method is a systematic way of changing the basis vectors and computing solutions. To get started on the simplex method we need a basic feasible solution. Let $x_1 = x_2 = x_3 = x_4 = 0$. Then, from (4),

$$x_5\begin{pmatrix}1\\0\\0\end{pmatrix} + x_6\begin{pmatrix}0\\1\\0\end{pmatrix} + x_7\begin{pmatrix}0\\0\\1\end{pmatrix} = \begin{pmatrix}2\\4\\6\end{pmatrix} .$$

Thus $x_1 = x_2 = x_3 = x_4 = 0$, $x_5 = 2$, $x_6 = 4$, $x_7 = 6$ is a basic feasible solution. We form the simplex tableau using this solution.

	v_1	v_2	v_3	v_4	b	v_5	v_6	v_7
v_5	1	3	1	2	2	1	0	0
v_6	3	3	2	2	4	0	1	0
v_7	3	2	4	5	6	0	0	1
	-4	-5	-3	-6	0	0	0	0

The column under v_1 was derived as follows: $v_1 = [1,3,3] = 1v_5 + 3v_6 + 3v_7$. The last element was computed using the formula $d_1 = c_1(1) + c_2(3) + c_3(3) - c_1$, where c_5, c_6, c_7 are the coefficients of the slack variables in the objective function

$$c_1 x_1 + c_2 x_2 + \ldots + c_3 x_3 + c_4 x_4 + c_5 x_5 + c_6 x_6 + c_7 x_7 .$$

But $c_1 = 4$, $c_2 = 5$, $c_3 = 3$, $c_4 = 6$, $c_5 = c_6 = c_7 = 0$. Thus, $d_1 = -4$.
The other columns of the simplex tableau were found in a similar manner. The value of the program, D, is zero. To find another basic feasible solution we use the following two rules to help us select another basis

for R^3.

1) Choose the column that has the most negative value of d_j (here, v_4).

2) Choose the row which has the smallest ratio of the jth element of b to the corresponding element of the column chosen in 1). (Here v_5 is the row since $2/2$ is smaller than $4/2$, $6/5$).

3) Remove the chosen basis vector and in its place put the chosen non-basic vector to obtain a new basis.

(Here, we remove v_5 and put v_4 in its place).

We must now find a new tableau associated with this new basis. To do this we use the following rule:

4) Divide the chosen row by the element in the chosen column. Then subtract multiples of the chosen row from each of the other rows so that the chosen column will have zeros everywhere except for a one in the chosen row.

Carrying out these steps on the tableau we obtain the new tableau

	v_1	v_2	v_3	v_4	b	v_5	v_6	v_7
v_4	1/2	3/2	1/2	1	1	1/2	0	0
v_6	2	0	1	0	2	-1	1	0
v_7	1/2	-11/2	3/2	0	1	-5/2	0	1
	-1	4	0	0	6	3	0	0

Now, according to 1), the first column has the most negative value and is chosen. Of the ratios $1\frac{1}{2}$ $2/2$ and $1\frac{1}{2}$, v_6 is the minimum ratio. Thus we remove v_6 and put v_1 in its place. The new tableau is

	v_1	v_2	v_3	v_4	b	v_5	v_6	v_7
v_4	0	3/2	1/4	1	1/2	3/4	-1/4	0
v_1	1	0	1/2	0	1	-1/2	1/2	0
v_7	0	-11/2	5/4	0	1/2	-9/4	-1/4	1
	0	4	1/2	0	7	5/2	1/2	0

Since all $d_j \geq 0$, the solution $x_1 = 1$, $x_2 = x_3 = 0$, $x_4 = \frac{1}{2}$, $x_5 = x_6 = 0$, $x_7 = \frac{1}{2}$ is an optimal solution. The value of the problem is 7.

● **PROBLEM** 17-14

Find nonnegative numbers x_1, x_2, x_3, x_4 which maximize

$$3x_1 + x_2 + 9x_3 - 9x_4$$

and satisfy the conditions

$$x_1 + x_2 + x_3 - 5x_4 = 4$$
$$x_1 - x_2 + 3x_3 + x_4 = 0 .$$

Solution: Use the simplex algorithm to solve the problem. The calculations are conveniently set forth in the form of a table.

c_j	solution variables	solution values	3 x_1	1 x_2	9 x_3	-9 x_4
3	x_1	4	1	1	1	-5
1	x_2	0	1	-1	3	1
	z_j					
	$c_j - z_j$					

Under each column (x_1, x_2, x_3, x_4) are written the coefficients from the constraint equations of the variables found in the heading. For example, under x_3 is written $(1,3)$. Under the column headed solution values, the constants of the constraints are written. The first row in the heading of the table contains the c_j's or the profit per unit (the coefficients of the variables in the profit equation).

Before filling in the rest of the table, we must identify an initial solution. A basic feasible solution is

$$x_1 = x_2 = 2, \ x_3 = x_4 = 0 \ .$$

The value associated with this solution is $3(2) + (1)2 + 9(0) - 9(0) = 8$. The terms x_1, x_2 are entered in the simplex table under the solution variables column, and their per-unit profits are entered in the first column under the c_j heading.

Finally, consider the computation of the z_j's and $c_j - z_j$'s. The z_j total of a column is the amount of profit which is given up by replacing some of the present solution mix with one unit of the item heading the column. It is found by multiplying the c_j of the row by the number in the row and jth column (the substitution coefficient) and adding.

c_j	solution variables	solution values	3 x_1	1 x_2	9 x_3	-9 x_4
3	x_1	4	1	1	1	-5
1	x_2	0	1	-1	3	1
	z_j	8	4	2	6	-14
	$c_j - z_j$		-1	-1	3	5

The $c_j - z_j$ row represents the net profit that is added by one unit of the product. x_3 and x_4 are the only positive profits. That means we want to replace some of x_1 or x_2 with one or more units of x_3. The next step is to determine which row (x_1 or x_2) is to be replaced by x_3. Divide each amount in the "Solution values" column by the amount in the comparable row of the x_3 column:

$$\text{for } x_1 \text{ row: } \frac{-4}{5}$$

$$\text{for } x_2 \text{ row: } \frac{0}{1} \ .$$

Since negative ratios don't count, we choose x_2 for elimination, i.e., pivot on 1. To obtain a new table, convert all other elements in the x_4 column to zero.

502

c_j	solution variables	solution values	3 x_1	1 x_2	9 x_3	-9 x_4
3	x_1	4	6	-4	16	0
9	x_3	0	1	-1	3	1
	z_j	12	27	-21	75	9
	$c_j - z_j$		-24	22	-66	-18

The z_j and $c_j - z_j$ are calculated as before. If all the $c_j - z_j$ were negative or zero, the solution would be optimal. But the column headed x_2 has a positive amount. On the other hand, both elements in this column are negative indicating that no pivoting is possible.

By the rules of the simplex algorithm the problem has no optimal solution.

● **PROBLEM** 17-15

Use the simplex algorithm to solve the following linear programming problem:

Maximize
$$z = 4x_1 + 8x_2 + 5x_3 \qquad (1)$$
subject to
$$x_1 + 2x_2 + 3x_3 \le 18$$
$$x_1 + 4x_2 + x_3 \le 6 \qquad (2)$$
$$2x_1 + 6x_2 + 4x_3 \le 15$$
$$x_1 \ge 0, \ x_2 \ge 0, \ x_3 \ge 0 . \qquad (3)$$

Solution: We first convert the inequalities to equalities by defining slack variables. Let
$$x_4 = 18 - x_1 - 2x_2 - 3x_3$$
$$x_5 = 6 - x_1 - 4x_2 - x_3$$
$$x_6 = 15 - 2x_1 - 6x_2 - 4x_3 .$$
Thus, the resource constraints (2) become:
$$x_1 + 2x_2 + 3x_3 + x_4 = 18$$
$$x_1 + 4x_2 + x_3 + x_5 = 6 \qquad (4)$$
$$2x_1 + 6x_2 + 4x_3 + x_6 = 15 .$$
Now use the tableau method to solve (1), (4) and (3). Treating (4) as a matrix, form the coefficient tableau

$$
\begin{array}{ccc|ccc|c}
1 & 2 & 3 & 1 & 0 & 0 & 18 \\
1 & 4 & 1 & 0 & 1 & 0 & 6 \\
2 & 6 & 4 & 0 & 0 & 1 & 15 \\
\hline
-4 & -8 & -5 & 0 & 0 & 0 & 0
\end{array} \qquad (5)
$$

The bottom row in (5) is the coefficient vector of the objective function. Now convert the matrix

$$\begin{bmatrix} 1 & 2 & 3 \\ 1 & 4 & 1 \\ 2 & 6 & 4 \end{bmatrix} \quad \text{to a matrix of 0's and 1's .}$$

Simultaneously we will obtain the maximum possible value of the objective function subject to the constraints.

In the conversion process use the following rules:

1) Locate the most negative number in the extra row and select the column in which this number occurs. Here, -8 is the most negative number which occurs under column 2.

2) Now choose an element in this column as the pivot element. Recall that the pivot element is the element that is converted to 1 and is then used to eliminate other elements in the column (by elementary row operations). The choice of the pivot is a distinguishing feature of the simplex method. Form the ratios of the constraint constants to the positive elements of the chosen column and pivot on that column element which is the denominator of the smallest ratio. Here the ratios are 18/2, 6/4 and 15/6 . Thus we choose the second element as the pivot.

3) Convert the remaining entries in the chosen column to zero.

4) Repeat 1)-3) for another column or until every column has a single 1 and the rest of its entries zero. The tableau (5) becomes

1/2	0	5/2	1	-1/2	0	15	
1/4	1	1/4	0	1/4	0	3/2	(6)
1/2	0	5/2	0	-3/2	1	6	
-2	0	-3	0	2	0	12	

The most negative entry in the last row is -3. The ratios are

$$\frac{15}{(5/2)}, \quad \frac{3/2}{(1/4)} \quad \text{and} \quad \frac{6}{(5/2)} .$$

The lowest ratio is $6/(5/2)$. We choose this element as pivot and convert all other entries in this column to zero to obtain

0	0	0	1	1	-1	9
1/5	1	0	0	2/5	-1/10	9/10
1/5	0	1	0	-3/5	2/5	12/5
-7/5	0	0	0	1/5	6/5	96/5

The lowest ratio in the first column is $\frac{9}{10}/\frac{1}{5}$. Pivoting on this element

0	0	0	1	1	-1	9
1	5	0	0	2	-1/2	9/2
0	-1	1	0	-1	1/2	3/2
0	7	0	0	3	1/2	51/2

Hence an optimal solution is
$$x_1 = 9/2, \quad x_2 = 0 \quad \text{and} \quad x_3 = 3/2 .$$

The slack variables are
$$x_4 = 9, \quad x_5 = 0 \quad \text{and} \quad x_6 = 0 .$$

The optimal value of z is 51/2 .

Show through the simplex method and then graphically that the following linear program has no solution.

$$\text{Maximize} \quad 2x + y$$

subject to

$$-x + y \leq 1$$
$$x - 2y \leq 2$$

$x, y \geq 0$.

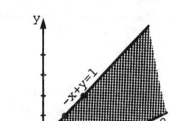

(0,1)

(0,0)

(2,0)

Solution: First convert the inequalities to equalities by the addition of slack variables. Thus

$$-x + y + s_1 = 1$$
$$x - 2y + s_2 = 2$$

where $s_1 = 1 - (-x+y)$ and $s_2 = 2 - (x-2y)$. This may be rewritten as

$$x \begin{bmatrix} -1 \\ 1 \end{bmatrix} + y \begin{bmatrix} 1 \\ -2 \end{bmatrix} + s_1 \begin{bmatrix} 1 \\ 0 \end{bmatrix} + s_2 \begin{bmatrix} 0 \\ 1 \end{bmatrix} = \begin{bmatrix} 1 \\ 2 \end{bmatrix} \quad (1)$$

There are 2 equations in four unknowns. Letting two of the unknowns equal zero yields two equations in two unknowns which will have a unique solution only if the two vectors with non-zero coefficients form a basis for R^2 . The corresponding coefficients then are said to be a basic feasible solution. For example, suppose $x = 0$, $s_1 = 0$ in (1). Then

$$y \begin{bmatrix} 1 \\ -2 \end{bmatrix} + s_2 \begin{bmatrix} 0 \\ 1 \end{bmatrix} = \begin{bmatrix} 1 \\ 2 \end{bmatrix}$$

$y = 1$, $s_2 = 4$ is the unique solution. Thus $\begin{bmatrix} 1 \\ -2 \end{bmatrix}$ and $\begin{bmatrix} 0 \\ 1 \end{bmatrix}$ form a basis

for R^2 and $y = 1$, $s_2 = 4$ is a basic feasible solution. Now form a simplex tableau using this solution. First find v_1, v_2, v_3 and v_4 in terms of the basis $\{v_2, v_4\}$. $[-1,1] = c_1[1,-2] + c_2[0,1]$ and $c_1 = -1$, $c_2 = -1$.

$$[1,-2] = 1[1,-2] + 0[0,1]$$
$$[1,0] = c_1[1,-2] + c_2[0,1]$$
$$\text{and} \quad c_1 = 1, \ c_2 = 2$$
$$[0,1] = 0[1,-2] + 1[0,1] \quad .$$

Let $b = [1,2] = c_1 v_2 + c_2 v_4 = c_1[1,-2] + c_2[0,1]$. Then $c_1 = 1$, $c_2 = 4$. The simplex tableau summarizes the above information:

	v_1	v_2	v_3	v_4	b
v_2	-1	1	1	0	1
v_4	-1	0	2	1	4
d	-2	-1	0	0	0

According to the rules, choose v_1 as the column for pivoting. But the minimum ratio rule cannot be applied to this column because both elements are nonpositive. Hence there is no solution to the maximum linear program (and, therefore, by the duality theorem, no solution to the minimum dual). Since there are only two variables we can graphically demonstrate the above problem. (See fig.)

The feasible region extends upwards indefinitely. Thus, whatever combination of x and y chosen to substitute into the objective function, we can always find another pair that will yield a higher value while satisfying the constraints.

● **PROBLEM** 17-17

The following problem is an illustration of degeneracy.

Maximize $P = 4x_1 + 3x_2$ (1)

subject to

$$4x_1 + 2x_2 \leq 10.0$$
$$2x_1 + 8/3x_2 \leq 8.0$$ (2)
$$x_1 \geq 0, \ x_2 \geq 1.8 \ .$$

What are the signs of degeneracy
a)   in the simplex tableau
b)   graphically?

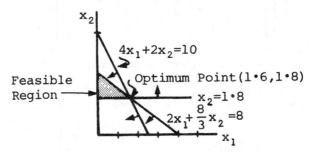

Solution: To apply the simplex method convert the inequalities in (2) to equalities by the addition of artificial and slack variables. In maximization problems, artificial variables are introduced so as to facilitate the simplex method of solution. One of the requirements of the method is that every equation contain a variable whose coefficient is 1 in that equation and zero in every other equation. Thus, the system (2) becomes

$$4x_1 + 2x_2 + x_3 \qquad = 10$$
$$2x_1 + 8/3 \ x_2 + x_4 = 8$$
$$x_2 + x_5 - x_6 = 1.8 \ .$$

Here x_3, x_4 and x_6 are slack variables, while x_5 is an artificial variable. To ensure that it will not appear in the final solution, assign it a profit factor coefficient of -M, where M is a very large number.

The initial simplex tableau with $x_1 = x_2 = x_6 = 0$ is

c_j	Solution variables	Solution values	4 x_1	3 x_2	0 x_3	0 x_4	-M x_5	0 x_6
0	x_3	10	4	2	1	0	0	0
0	x_4	8	2	8/3	0	1	0	0
-M	→ x_5	1.8	0	①	0	0	1	-1
	z_j	-1.8M	0	-M	0	0	-M	M
	$c_j - z_j$		4	M+3 ↑	0	0	0	-M

Replace x_5 by x_2 ; i.e., pivot on 1.

c_j	Solution variables	Solution values	4 x_1	3 x_2	x_3	x_4	x_5	x_6
0	x_3	6.4	4	0	1	0	-2	2
0	x_4	3.2	2	0	0	1	-8/3	8/3
3	x_2	1.8	0	1	0	0	1	-1
	z_j		0	3	0	0	3	-3
	$c_j - z_j$		4 ↑	0	0	0	-M-3	3

Here, both row x_3 and row x_4 are minimum ratios: $\frac{6.4}{4} = \frac{3.2}{2} = 1.6$. This
is the signal that degeneracy exists. By the rules of the simplex algori-
thm, we can arbitrarily replace either row. If the chosen row eventually
leads to no solution (the simplex tables begin to repeat themselves), then
choose the other row at the point where the degeneracy was discovered.

In the present problem, replacing row x_4 with x_1 leads immediately
to the final simplex tableau and optimal solution $x_1 = 1.6$, $x_2 = 1.8$ and
P = 11.8.

Replacing row x_3 with x_1 yields the same result but after an additional iteration.

b) The degeneracy situation can also be identified by examining the graph
of the problem. (See fig.)

The optimum point occurs at the intersection of the three constraint
equations. Since all the constraints are satisfied exactly, there is no
slack in any constraint. In a nondegenerate case, at least one of these
slack variables would be non-zero.

● **PROBLEM** 17-18

The optimum solution to the problem

$$\text{Maximize} \qquad P = 12x_1 + 9x_2 \qquad (1)$$

subject to

$$3x_1 + 2x_2 \le 7$$
$$3x_1 + x_2 \le 4 \qquad (2)$$
$$x_1 \ge 0, \ x_2 \ge 0$$

is $P = 9(7/2) = 31\frac{1}{2}$. The solution to the dual is $y_1 = 4\frac{1}{2}$, $y_2 = 0$.

Now assume the first constraint of (2) is changed from 7 to 8, i.e.,
$$3x_1 + 2x_2 \le 8 .$$
Find the increase in P. What is the dual for this new problem?

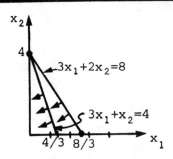
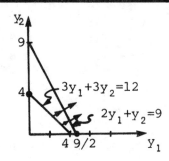

Solution: The new problem is:

$$\text{Maximize} \quad P = 12x_1 + 9x_2$$

subject to

$$3x_1 + 2x_2 \le 8$$
$$3x_1 + x_2 \le 4$$
$$x_1 \ge 0, \; x_2 \ge 0 .$$

The dual to this problem is

$$\text{Minimize} \quad C = 8y_1 + 4y_2$$

subject to

$$3y_1 + 3y_2 \ge 12$$
$$2y_1 + y_2 \ge 9$$
$$y_1, y_2 \ge 0 .$$

We can graph the new program and its dual: (See fig.)

If a program has an optimal solution it has an optimal solution at the intersection of two constraints. From the graph of the primal, P is maximized when $x_1 = 0$ and $x_2 = 4$. The value of P is

$$P = 9(4) = 36.$$

Note that P has increased by $4\frac{1}{2}$ units as predicted by the dual value $y_1 = 4\frac{1}{2}$. But, examining the graph of the new dual there is no unique optimal solution, since any values for y_1 and y_2 on the line going from $y_1 = 4\frac{1}{2}$ to $y_2 = 9$ will satisfy the constraints.

Suppose that another unit is added to the first constraint in the primal which now becomes:

$$\text{Maximize} \quad P = 12x_1 + 9x_2$$

subject to

$$3x_1 + 2x_2 \le 9$$
$$3x_1 + x_2 \le 4$$
$$x_1 \ge 0, \; x_2 \ge 0 .$$

The optimum value of P is still 36. The value does not change, since the other unchanged restraint acts to prevent an improvement unless both restraints are changed. The solution is degenerate; of the two primal ordinary variables (x_1 and x_2) and the two primal slack variables, only one of these four variables (x_2) is positive even though there are two constraints.

In a manufacturing process, the final product has a requirement that it must weigh exactly 150 pounds. The two raw materials used are A, with a cost of $4 per unit and B, with a cost of $8 per unit. At least 14 units of B and no more than 20 units of A must be used. Each unit of A weighs 5 pounds; each unit of B weighs 10 pounds.

How much of each type of raw material should be used for each unit of final product if we wish to minimize cost?

Fig. 1

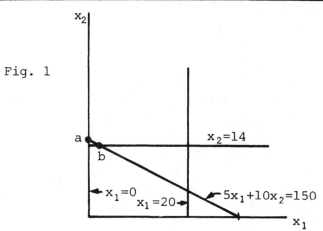

Solution: The objective function is

$$C = 4x_1 + 8x_2 .$$ (1)

The constraints are

$$5x_1 + 10x_2 = 150$$
$$x_1 \leq 20$$
$$x_2 \geq 14$$ (2)
$$x_1 \geq 0$$

We will take the graphical approach to this linear programming problem. The constraints, (2) are graphed in the figure.

Since the pertinent region lies within $0 \leq x_1 \leq 20$, $x_2 \geq 14$, and on $5x_1 + 10x_2 = 150$, we can immediately find 2 solutions, points a and b where a = (0,15) and b = (2,14).

	Solution 1	Solution 2
Raw material A, (x_1)	0	2
Raw material B, (x_2)	15	14
Total cost, $4x_1 + 8x_2$	120	120

This is an example of a problem having multiple solutions. In such problems, two or more corner points have the same optimum value.

● **PROBLEM** 17-20

Assume that two products x_1 and x_2 are manufactured on two machines 1 and 2. Product x_1 requires three hours on machine 1 and one-half hour on machine 2. Product x_2 requires two hours on machine 1 and 1 hour on machine 2. There are six hours of available capacity on machine 1 and four hours on machine 2. Finally, each unit of x_1 produces a net increase in profit of \$12.00 and each unit of x_2 an incremental profit of \$4.00. 1)Maximize the profit. 2)Obtain a solution to this problem using the simplex method. 3)Apply sensitivity analysis to the final tableau.

Solution: In linear programming form the problem is:

$$\text{Maximize} \quad P = 12x_1 + 4x_2 \qquad (1)$$

subject to

$$3x_1 + 2x_2 \leq 6$$
$$\tfrac{1}{2} x_1 + x_2 \leq 4 \qquad (2)$$
$$x_1 \geq 0, \; x_2 \geq 0 \; .$$

Introduce the slack variables x_3, x_4:

$$3x_1 + 2x_2 + x_3 = 6$$
$$\tfrac{1}{2} x_1 + x_2 + x_4 = 4 .$$

Let $v_1 = \begin{bmatrix} 3 \\ \tfrac{1}{2} \end{bmatrix}$, $v_2 = \begin{bmatrix} 2 \\ 1 \end{bmatrix}$, $v_3 = \begin{bmatrix} 1 \\ 0 \end{bmatrix}$, $v_4 = \begin{bmatrix} 0 \\ 1 \end{bmatrix}$. Let $x_1 = x_2 = 0$.

Then an initial feasible solution is $x_1 = x_2 = 0$, $x_3 = 6$, $x_4 = 4$. The initial simplex tableau is

	v_1	v_2	v_3	v_4	b	
v_3	③	2	1	0	6	
v_4	$\tfrac{1}{2}$	1	0	1	4	
d	-12	-4	0	0	0	D

Pivoting on 3, the next simplex tableau is

	v_1	v_2	v_3	v_4	b	
v_1	1	2/3	1/3	0	2	(3)
v_4	0	2/3	-1/6	1	3	
d	0	4	4	0	24	

Since all $d_j \geq 0$, the solution is $x_1 = 2$, $x_4 = 3$ and $P = 24$.

In order to carry out the sensitivity analysis we also need the dual of (1), (2). This is:

$$\text{Minimize} \quad U = 6y_1 + 4y_2$$

subject to

$$3y_1 + \tfrac{1}{2} y_2 \geq 12$$
$$2y_1 + y_2 \geq 4$$
$$y_1 \geq 0, \; y_2 \geq 0 \; .$$

The optimum solution to the dual is $y_1 = 4$, $y_4 = 4$; $U = 24$. The dual variables are known as shadow prices. y_1 has a value of \$4 which means an hour of machine 1 has a value of \$4.00 . But y_1 is also the slack

510

variable x_3 in the primal, i.e., unused hours on machine 1. Now if we could buy additional hours of machine 1 time for less than \$4.00, could we increase P indefinitely by adding more hours?

To answer this question, examine the v_3 column in (3). The meaning of the coefficients is that if we were to add one unit of x_3 it would replace one-third unit of x_1 and $-1/6$ unit of x_4. But adding a unit of x_3 is equivalent to reducing hours of production by machine 1 on x_1. How far can production be reduced before a change in the solution mix occurs?

	Solution values	x_3	Solution values x_3
x_1	2	1/3	6
x_4	3	-1/6	-18

We see that six units of x_3 can be introduced before variable x_1 goes out of the solution. This means we can cut back a maximum of six hours of those available on machine 1 before production of x_1 stops.

On the other hand, adding negative x_3 means that we can make additional hours available on machine 1. Here we can add 18 additional hours of time on machine 1 before we run out of slack time on machine 2 (i.e., x_4 goes to zero and out of the solution).

Make the same kind of analysis relative to the amount of time available on machine 2. Here the slack variable is x_4 with a value of 3. Thus we can reduce x_4 only by 3 before there arises a shortage of machine 2 time. Also, since there already is slack, the additional time can be added indefinitely without altering the solution mix.

The above sensitivity analysis is summarized below:

	Machine-hrs. available	Shadow price	Shadow lower	Range for valid price upper
Machine 1	6	\$4	0	24
Machine 2	4	0	1	no limit

● **PROBLEM** 17-21

Consider the following problem:

$$\text{Maximize} \quad P = 5x_1 + 8x_2 \tag{1}$$

subject to

$$2x_1 + x_2 \leq 14$$
$$x_1 + 3x_2 \leq 12 \tag{2}$$
$$x_2 \leq 3$$
$$x_1 \geq 0, \; x_2 \geq 0 \;.$$

Suppose that an additional constraint on x_1 and x_2 is imposed:

$$x_1 + x_2 \leq K \tag{3}$$

where K is some unspecified amount. How does the solution of (1), (2) and (3) change as K varies from zero to very large values?

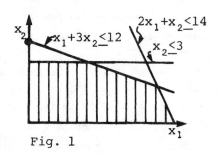

Fig. 1

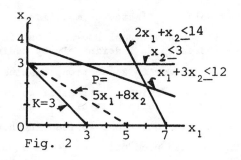

Fig. 2

Solution: We first graph the problem (1), (2), as in Fig. 1.

Assume the first two constraints represent time used on machine 1 and machine 2, respectively, to produce units of products x_1 and x_2. The third constraint indicates that not more than three units of product x_2 can be sold. Now consider the additional constraint $x_1 + x_2 \le K$.

This states that the total amount of working capital used must be less than an unspecified amount, K. If $K = 0$ the only solution is that of no production and $x_1 = x_2 = 0$. As K increases to one, the first dollar of working capital is used to produce one unit of x_2 since x_2 is the more profitable product. Since each unit of K spent on x_2 produces \$8 of profit as K increases, only x_2 is produced until the situation in Fig. 2 is reached.

When $K > 3$, the market constraint $(x_2 \le 3)$ becomes binding, and additional units of x_2 cannot be produced. Hence, production of x_1 now begins. Each dollar of working capital has an incremental value of \$5, the profit associated with selling one unit of x_1. Production of x_1 continues until the situation in Fig. 3 is reached.

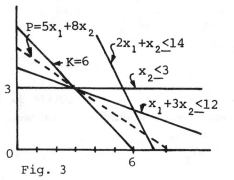

Fig. 3

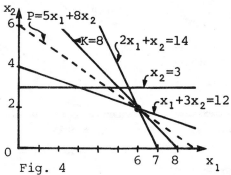

Fig. 4

In Fig. 3, the constraint on time available on machine 2 also becomes binding, at the point at which six units of K are available. Now, three units of x_1 and 9 units of x_2 are produced for a total profit of \$87. Note that P can still be increased by moving down the $x_1 + 3x_2 = 12$ constraint. But now each unit of working capital contributes only \$3.50 profit. To see this, observe that one unit of x_2 can be substituted for three units of x_1 yielding an additional profit of $\$5(3) - \$8(1) = \$7$. But this substitution requires three units for the new x_1 less one unit for reduced x_2. Hence the marginal value of K

is $7/2 = 3.50. Finally, the situation portrayed in Fig. 4 is reached. Here the constraints on both machine times are binding. Increasing the working capital, K, beyond this point would have no effect on the solution. This sort of analysis is useful to a business man who is trying to decide how much working capital to invest in this production operation. He would not invest more than $8. However, he might invest even less if he had profitable alternative uses for his funds. For example, if he could use the working capital elsewhere to return him $4 per unit, then he would not invest more than six units. If he could get a return of $6 per unit, he would not use more than 3 units.

● PROBLEM 17-22

Assume there are three factories $(F_1, F_2$ and $F_3)$ supplying goods to three warehouses (W_1, W_2, W_3). The amounts available in each factory, the amounts needed in each factory and the costs of shipping from factory i to warehouse j are given in the table below:

Source Destination	F_1	F_2	F_3	Units demanded
W_1	$.90	$1.00	$1.00	5
W_2	$1.00	$1.40	$.80	20
W_3	$1.30	$1.00	$.80	20
units available	20	15	10	45

Find the minimum cost of satisfying warehouse demands given that any factory may supply to any warehouse.

Solution: This is a transportation problem. It may be solved by iteration, i.e., we start with a solution and then use it to find more nearly optimal solutions.

An initial solution may be found by finding the box that has the lowest value in both its row and column. Place in that box the lower of demand or supply requirements. Next find the next lower value and repeat the placing of units shipped according to demand and supply requirements. We thus obtain

Source Destination	F_1	F_2	F_3	units demanded
W_1	.90 / 5	$1.00 / 0	$1.00 / 0	5
W_2	1.00 / 10	1.40 / 0	.80 / 10	20
W_3	1.30 / 5	1.00 / 15	.80 / 0	20
units available	20	15	10	45

The total cost of this program is

$$(.90)(5) + (1.00)(10) + (1.30)(5) + (1.00)(15) + (.80)(10) = $44.00.$$

Now check whether this solution is optimal. Pick a box with zero entry, say W_1F_2. This means that F_2 supplies no goods to W_1. The cost

513

of F_2 directly supplying W_1 is $1.00 per unit. But F_2 is also in-directly supplying W_1 since by supplying W_3 it allows F_1 to supply W_1. What is this indirect cost?

The cost of shipping one unit from F_2 to W_1 by this indirect route is:

        + $1.00   charge for shipping from F_2 to W_3

        - $1.30   every unit F_2 sends to W_3 saves the cost of supplying W_3 from F_1

        + $0.90   charge for shipping from F_1 to W_1
        ‾‾‾‾‾‾‾‾‾
        + $0.60

The indirect cost, i.e., the cost currently incurred is $0.60 whereas the direct cost is $1.00. Thus the current solution is cheaper.

The other zero boxes may be evaluated in a comparable manner. For example, the indirect shipment from F_2 to W_2 is the charge from F_2 to W_3 ($1.00) less the W_3F_1 charge ($1.30) plus the W_2F_1 charge ($1.00)= 70¢. Again, this is less than the cost of direct shipment ($1.40), so the current indirect route should be continued. In this way we obtain the following table:

Unused route	Cost of direct route	Cost of indirect route
W_1F_2	$1.00	$0.60
W_1F_3	$1.00	$0.70
W_2F_2	$1.40	$0.70
W_3F_3	$0.80	$1.10

By using W_3F_3 a saving of $.30 per unit can be obtained. But how many units can be shipped? The answer is the minimum number in any of the connections of the indirect route which must supply units for the transfer. This is five units, from box W_3F_1. Thus we ship 5 units by the direct route W_3F_3; since F_3 produces only 10 units, this imposes a reduction in the W_2F_3 box to 5. The new pattern is shown in the table below:

Source / Destination	F_1	F_2	F_3	Units demanded
W_1	.90 / 5	1.00 / 0	1.00 / 0	5
W_2	1.00 / 15	1.40 / 0	.80 / 5	20
W_3	1.30 / 0	1.00 / 15	.80 / 5	20
units available	20	15	10	45

Once again we must compare the cost of using the direct route to the cost of using the indirect route.

Unused route	Cost of using direct route	Cost of using indirect route
W_3F_1	$1.30	$1.00
W_1F_2	$1.00	$0.90
W_2F_2	$1.40	$1.00
W_1F_3	$1.00	$0.70

In every case the cost of using the indirect route is less than the cost of the direct route, indicating that we are minimizing the shipment costs. The total cost of shipment from factories to warehouses is:

$$(5)(\$0.90) + (15)(\$1.00) + (15)(\$1.00) + 5(\$0.80) + 5(\$0.80) = \$42.50.$$

INTEGER PROGRAMMING

• PROBLEM 17-23

The Brown Company has two warehouses and three retail outlets. Warehouse number one (which will be denoted by W_1) has a capacity of 12 units; warehouse number two (W_2) holds 8 units. These warehouses must ship the product to the three outlets, denoted by O_1, O_2, and O_3. O_1 requires 8 units. O_2 requires 7 units, and O_3 requires 5 units. Thus, there is a total storage capacity of 20 units, and also a demand for 20 units. The question is, which warehouse should ship how many units to which outlet? (The objective being, of course, to accomplish this at the least possible cost.)

Costs of shipping from either warehouse to any of the outlets are known and are summarized in the following table, which also sets forth the warehouse capacities and the needs of the retail outlets:

	O_1	O_2	O_3	Capacity
W_1	$3.00	$5.00	$3.00	12
W_2	2.00	7.00	1.00	8
Needs (units)	8	7	5	

Solution: This seems to be a linear programming problem with three variables. However, the third variable can be computed from the previous two. Let X be the number of units sent from warehouse 1 to outlet 1. Since outlet 1 requires only 8 units, we have the constraint $X \leq 8$. Obviously $X \geq 0$ means you either ship one or you don't. Let y be the number of units shipped from warehouse 1 to outlet 2. Similarly,

$$0 \leq y \leq 7$$

is another constraint. Because there are 12 units in warehouse 1, the number of units sent to outlet 3 is

$$12 - X - y.$$

Obviously, this must be larger or equal to zero. Thus,

$$12 - X - y \geq 0 \quad \text{or} \quad X + y \leq 12$$

is a constraint.

The amount of units sent from warehouse 2 to outlet 1 is the original eight less the X that was sent from warehouse 1, or 8 - X. Similarly we find all of the others.

	O_1	O_2	O_3
W_1	x	y	12 - x - y
W_2	8 - x	7 - y	x + y - 7

Note that the quantities shipped from both warehouses to Outlet 3 have been determined by simply subtracting the quantities shipped to Outlets 1 and 2 from the total capacities of the warehouses.

So we have the following constraints:

(1) $X \geq 0$
(2) $y \geq 0$
(3) $X \leq 8$
(4) $y \leq 7$
(5) $X+y \leq 12$
(6) $X+y \geq 7$ (from $O_3 - W_2 \geq 0$) .

We wish to minimize the cost function. It can be found by multiplying the number of units sent by their costs. We get

$$\text{Cost} = 3X + 5y + 3(12-X-y) + 2(8-X) + 7(7-y) +$$
$$1(X+y-7) = 94 - X - 4y.$$

To minimize this we plot the constraints to find the region that is defined by them.
The shaded region is the area that is defined by the constraints. By linear programming we know the minimum cost will appear at a corner of this region. We need only check these corners to pick the best point.

A   (8,0) $94 - 8 = \$86$
B   (7,0) $94 - 7 = \$87$
C   (0,7) $94 - 0 - (4 \times 7) = \$66$
D   (5,7) $94 - 5 - (4 \times 7) = \$61$
E   (8,4) $94 - 8 - (4 \times 4) = \$70$

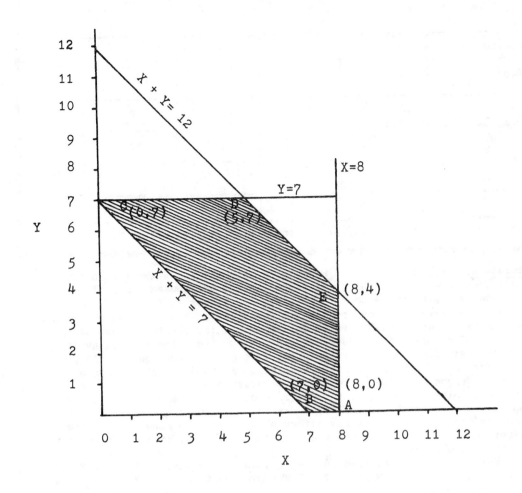

Thus the point with the least cost is point D. So for the lowest cost the shipping schedule should be

		O_1	O_2	O_3	
W_1	....	5	7	0	12
W_2	....	3	0	5	8
		8	7	5	

● PROBLEM 17-24

Suppose that in the course of solving a transportation problem, the following trial solution is computed:

Source \ Destination	F_1	F_2	F_3	Units demanded	
W_1	.90 / 0	1.00 / 5	1.00 / 0	5	
W_2	1.00 / 20	1.40 / 0	.80 / 0	20	(1)
W_3	1.30 / 0	1.00 / 10	.80 / 10	20	
	20	15	10	45	

How would you progress further to find the optimal solution?

Solution: When the number of boxes used in obtaining a trial solution is less than $F + W - 1$, (number of factories and warehouses minus 1) the problem of degeneracy appears. This problem arises as follows: To improve on a trial solution, consider alternative routes. An indirect route is the path a unit would have to follow from a factory to a given warehouse, using only established channels, (i.e., the shipment must avoid zero boxes; otherwise, we are shipping from a box which has no units, or introducing two new boxes into the solution instead of one). When a solution is degenerate, there are too many unused boxes.

To resolve the degeneracy case, record some very small amount, say d, in one of the zero boxes. This d is interpreted as a quantity of goods. The box with the d entry may either ship or receive goods but in the final solution the d is assigned a value of zero if it is still present in the calculations.

For the given trial solution, only 4 boxes are used while $W + F - 1 = 5$. Put d units in box $W_1 F_1$. Total shipping cost is $43. The zero boxes may now be evaluated in the standard manner.

Source \ Destination	F_1	F_2	F_3	Units demanded
W_1	.90 / d	1.00 / 5	1.00 / 0	5
W_2	1.00 / 20	1.40 / 0	.80 / 0	20
W_3	1.30 / 0	1.00 / 10	.80 / 10	20
Supply	20	15	10	45

Consider $W_3 F_1$. The cost of the direct route is $1.30. The indirect route is from $F_1 W_1$, which permits a reduction of shipment from F_2 to W_1; but this in turn requires an increase in the shipment from F_2 to W_3 . The costs are $W_1 F_1 + W_3 F_2 - W_1 F_2$ or $.90 . Since the indirect route is cheaper than the direct route it should continue to be used. The other zero boxes may be evaluated similarly with the result below:

518

Source / Destination	F_1	F_2	F_3	
W_1			1.00 / .80	
W_2		1.40 / 1.10	.80 / .90	
W_3	1.30 / .90			

Shipping from F_3 to W_3 has an indirect cost in excess of the direct cost. The maximum amount which can be shifted is five units, since this is the minimum amount in a box which must be reduced (box W_1F_2). The new trial solution is optimal. In this example degeneracy disappeared in one step. It is possible for degeneracy to remain through several iterations; in fact, the optimal solution may be degenerate.

● **PROBLEM** 17-25

Solve the following problem in integer programming:
   Find non-negative integers X_{ij} which will

$$\text{Minimize} \quad 200 X_{11} + 300 X_{12} + 250 X_{21} + 100 X_{22} + 250 X_{31}$$
$$+ 250 X_{32}$$

subject to

$$X_{11} + X_{21} + X_{31} \geq 30$$
$$X_{12} + X_{22} + X_{32} \geq 20$$
$$-X_{11} - X_{12} \geq -20$$
$$-X_{21} - X_{22} \geq -20$$
$$-X_{31} - X_{32} \geq -20$$

Solution: Since some of the constraints are negative we cannot apply the simplex algorithm to the given problem. However, the dual, which is a maximization problem, is receptive to the simplex technique. The dual is

$$\text{Maximize} \quad 30y_1 + 20y_2 - 20y_3 - 20y_4 - 20y_5$$

subject to:

$$y_1 - y_3 \leq 200$$
$$y_2 - y_3 \leq 300$$
$$y_1 - y_4 \leq 250$$
$$y_2 - y_4 \leq 100$$
$$y_1 - y_5 \leq 250$$
$$y_2 - y_5 \leq 250$$
$$y_1, y_2, \ldots, y_5 \geq 0$$

519

We can convert the inequalities to equalities by adding six slack variables
$$y_6, \; y_7, \; y_8, \; y_9, \; y_{10}, \; y_{11} \; .$$

An initial basic feasible solution is obtained by setting $y_1 = y_2 = y_3 = y_4 = y_5 = 0$. The simplex tableau is then

	v_1	v_2	v_3	v_4	v_5	v_6	v_7	v_8	v_9	v_{10}	v_{11}	b
v_6	1	0	-1	0	0	1	0	0	0	0	0	200
v_7	0	1	-1	0	0	0	1	0	0	0	0	300
v_8	1	0	0	-1	0	0	0	1	0	0	0	250
v_9	0	1	0	-1	0	0	0	0	1	0	0	100
v_{10}	1	0	0	0	-1	0	0	0	0	1	0	250
v_{11}	0	1	0	0	-1	0	0	0	0	0	1	250
d	-30	-20	20	20	20	0	0	0	0	0	0	0

Since -30 is the largest negative entry and $200/1$ is the minimum positive ratio we replace v_6 by v_1 in the basis on the left. Thus we pivot on the first element in the first column to obtain

	v_1	v_2	v_3	v_4	v_5	v_6	v_7	v_8	v_9	v_{10}	v_{11}	b
v_1	1	0	-1	0	0	1	0	0	0	0	0	200
v_7	0	1	-1	0	0	0	1	0	0	0	0	300
v_8	0	0	1	-1	0	-1	0	1	0	0	0	50
v_9	0	(1)	0	-1	0	0	0	0	1	0	0	100
v_{10}	0	0	1	0	-1	-1	0	0	0	1	0	50
v_{11}	0	1	0	0	-1	0	0	0	0	0	0	250
d	0	-20	-10	20	20	30	0	0	0	0	0	6,000

Now we replace v_9 by v_2 to obtain a new basic feasible solution. Pivoting on the encircled element:

	v_1	v_2	v_3	v_4	v_5	v_6	v_7	v_8	v_9	v_{10}	v_{11}	b
v_1	1	0	-1	0	0	1	0	0	0	0	0	200
v_7	0	0	-1	1	0	0	1	0	-1	0	0	200
v_8	0	0	1	-1	0	-1	0	1	0	0	0	50
v_2	0	1	0	-1	0	0	0	0	1	0	0	100
v_{10}	0	0	(1)	0	-1	-1	0	0	0	1	0	50
v_{11}	0	0	0	1	-1	0	0	0	-1	0	1	50
d	0	0	-10	0	20	30	0	0	20	0	0	8,000

520

The only negative entry is under v_3. Notice that the ratios of b to v_8 and of b to v_{10} are both the same, i.e., 50/1. By the rules of the simplex algorithm we can choose either v_8 or v_{10} for liquidation. We choose to replace v_{10} by v_3 and hence pivot on the encircled element. The next tableau is

	v_1	v_2	v_3	v_4	v_5	v_6	v_7	v_8	v_9	v_{10}	v_{11}	b
v_1	1	0	0	0	-1	0	0	0	0	1	0	250
v_7	0	0	0	1	-1	-1	1	0	1	1	0	250
v_8	0	0	0	-1	1	0	0	1	0	-1	0	0
v_2	0	1	0	-1	0	0	0	0	1	0	0	100
v_3	0	0	1	0	-1	-1	0	0	0	1	0	50
v_{11}	0	0	0	1	-1	0	0	0	-1	0	1	50
d	0	0	0	0	10	20	0	0	20	10	0	8,500

Since there are no more negative entries in the last row the algorithm has converged to a solution. The maximum feasible solution is 8,500. Therefore the minimum feasible solution is also 8,500. At this value, the values of the X_{ij} are read off from the slack variable values in the last row. Thus, $X_{11} = 20$, $X_{12} = 0$, $X_{21} = 0$, $X_{22} = 20$, $X_{31} = 10$, $X_{32} = 0$. Note that the main body of the simplex tableau consisted only of ones and zeros. This is a characteristic feature of integer programs. It makes pivoting easier and also ensures that the solution values will be integers.

● **PROBLEM** 17-26

Consider the following integer programming problem:

$$\text{Maximize} \quad P = 6x_1 + 3x_2 + x_3 + 2x_4 \qquad (1)$$

subject to

$$x_1 + x_2 + x_3 + x_4 \leq 8$$
$$2x_1 + x_2 + 3x_3 \leq 12$$
$$5x_2 + x_3 + 3x_4 \leq 6 \qquad (2)$$
$$x_1 \leq 1$$
$$x_2 \leq 1$$
$$x_3 \leq 4$$
$$x_4 \leq 2$$

x_1, x_2, x_3, x_4 all non-negative integers. Use the branch and bound algorithm to solve this problem.

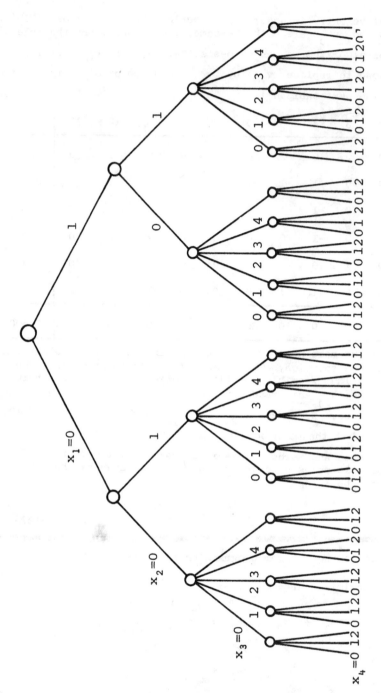

Solution: When a problem is required to have an integer solution, this means that there are a finite number of possible solution points. It is theoretically possible to enumerate and evaluate every possible solution to find the optimum.

We can list all possible solutions to the given problem by means of a tree diagram. The constraints $x_1 \le 1$, $x_2 \le 1$ imply that x_1 and x_2 can take on only the values zero or one. x_3 can take on five values while x_4 can take on 3 values. Thus, there are $2 \times 2 \times 5 \times 3 = 60$.

possible solutions.

Some of the possible solutions will not satisfy the remaining constraints. For example, the solution $x_2 = 1$, $x_3 = 3$ and $x_4 = 2$ fails to satisfy the constraint $5x_2 + x_3 + 3x_4 \le 6$. The branch and bound approach reduces the search by eliminating whole branches of the above tree. The principle used is: A branch can be eliminated if it can be shown to contain no feasible solution better than one already obtained.

Solving the problem as a linear programming problem (L.P.) yields an upper limit or bound on the possible integer solution. If the simplex method is used to solve the given problem we obtain the solution:

$$x_1 = 1; \ x_2 = 0; \ x_3 = 3.33; \ x_4 = 0.89$$

and $P = 11.11$. The integer solution must therefore be less than or equal to $P = 11$. An initial feasible integer solution is $x_1 = x_2 = x_3 = x_4 = 0$ and $P = 0$.

Now select an arbitrary variable and construct branches. Select x_4 for branching. Since the L.P. solution was $x_4 = .89$, we can let $x_4 = 0$ or $x_4 \ge 1$.

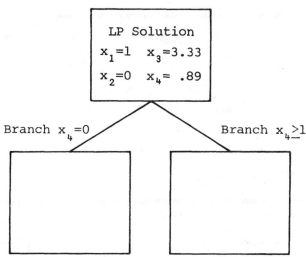

LP Solution
$x_1 = 1$ $x_3 = 3.33$
$x_2 = 0$ $x_4 = .89$

Branch $x_4 = 0$

Branch $x_4 \ge 1$

Consider the branch $x_4 = 0$. Replace the last constraint in the original problem $(x_4 \le 2)$ by the constraint $x_4 = 0$. Considered as a linear programming problem, the solution is $x_1 = 1$, $x_2 = .57$, $x_3 = 3.14$, $x_4 = 0$ and $P = 10.85$.

Now select x_2 as the branch variable. The two possible branches are $x_2 = 1$ and $x_2 = 0$. Let $x_2 = 1$ in the original problem (1),(2).

The solution to the L.P. problem is $x_1 = 1$, $x_2 = 1$, $x_3 = 1$, $x_4 = 0$ and $P = 10$. This is better than the original feasible solution $P = 0$.

Letting $x_2 = 0$ we obtain $P = 9.33$. Since the bound on profit in this branch is less than 10, we eliminate this branch.

Now move back to the branch $x_4 \ge 1$. The solution to the L.P. is

$$x_1 = 1, \ x_2 = 0, \ x_3 = 3, \ x_4 = 1 \quad \text{and} \quad P = 11.$$

This is the optimal solution since all branches have been searched.

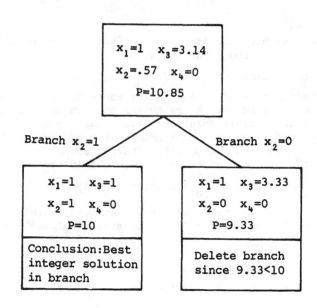

● **PROBLEM** 17-27

Use the branch and bound method to solve the integer programming problem

$$\text{Maximize} \quad P = 2x_1 + 3x_2 + x_3 + 2x_4$$

subject to

$$5x_1 + 2x_2 + x_3 + x_4 \leq 15$$
$$2x_1 + 6x_2 + 10x_3 + 8x_4 \leq 60$$
$$x_1 + x_2 + x_3 + x_4 \leq 8$$
$$2x_1 + 2x_2 + 3x_3 + 3x_4 \leq 16$$

$$x_1 \leq 3, \ x_2 \leq 7, \ x_3 \leq 5, \ x_4 \leq 5.$$

Solution: x_1 can take on any of the four values 0,1,2 or 3. Similarly, there are eight possibilities for x_2, six for x_3 and six for x_4. By the Fundamental Principle of Counting there are $4 \times 8 \times 6 \times 6 = 1,152$ possible solutions. The branch and bound procedure eliminates non-optimal solutions and thus reduces the amount of calculation. The steps in the procedure are as follows:

1) Find an initial feasible integer solution.
2) Branch: Select a variable and divide the possible solutions into two groups. Select one branch for investigation.
3) Find an upper bound or maximum value for the problem defined by the branch selected. This bound can be found by considering the problem as a linear programming problem.
4) Compare: Compare the bound obtained for the branch being considered with the best solution so far for the previous branches examined. If the bound is less, delete the whole new branch. If the bound is greater and an integer it becomes the new best solution so far. If the bound is greater but not an integer, continue in this same branch by branching further (Step 2).
5) Completion: When all branches have been examined, the best solution so far is the optimal solution.
   The general approach is illustrated for the given problem in the figure.

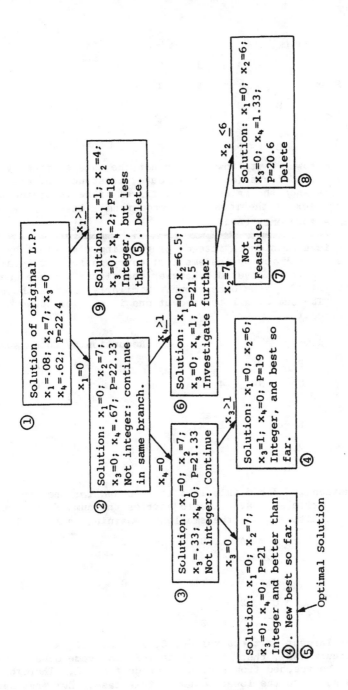

Thus the optimal solution to the integer program is $x_1 = 0$, $x_2 = 7$, $x_3 = 0$, $x_4 = 0$ and $P = 21$. When first selecting a variable for branching, a good rule to follow is to choose a variable whose linear programming solution is non-integer.

● **PROBLEM** 17-28

A travelling salesman must visit four towns. The distance between towns is given in the table below:

	To	A	B	C	D
From	A		1	4	5
	B	3		1	2
	C	2	4		3
	D	5	2	6	

The distance from town x to town y is not the same as from y to x because of necessary detours (one-way streets, construction, etc.). What is the minimum distance the salesman must travel if he is to touch every town and finish back at the town he started from? Assume he can touch each intermediate town only once.

Solution: This is a problem in integer programming. We start by reducing the entries in each row by the smallest entry in it. At least these distances will have to be covered, whatever the circuit, since each town will have to be left for some other town. Similarly, reduce the entries in each column. The sum of all the reductions is a lower bound on the length of any circuit.

Reduction by Rows

		A	B	C	d
(-1)	A		0	3	4
(-1)	B	2		0	1
(-2)	C	0	2		1
(-2)	D	3	0	4	

Reduction by Columns

	A	B	C	D
A		0	3	3
B	2		0	0
C	0	2		0
D	3	0	4	
				(-1)

The total reduction is 7, and this is a lower bound for the length of a circuit. Now we write against all zero entries the sum of the smallest remaining entry in its row and the smallest remaining entry in its column.

	A	B	C	D
A		3		
B			3	0
C	2			0
D		3		

Take one of the largest of these values, 3, in AB. If AB is not a link in the circuit ($\overline{AB}$), then A will be left for some other node than B, and B will be reached from some other node than A. Therefore, another 3 can be added to the lower bound in this case. But what happens if AB is used? We exclude the A row and B column and also BA because this would bring us back to A, without having been to all towns.

	A	C	D
B		0	0
C	0		0
D	3	4	

Reduce the last row by 3 and add this to the lower bound, if we use AB as a link.

526

	A	C	D
B		0	0
C	0		0
D	0	1	

which leads to

	A	C	D
B		1	0
C	0		0
D	1		

for the sum of the lowest remaining entries in row and column. If we do not take BC as a link, then the lower bound is increased by another 1. It is now 11. If we take BC, we omit row B, column C and CA (because this would close the short circuit ABCA).

	A	D
C		0
D	0	

The circuit is closed by CD, DA, total length 10. If we do not take BC, then we know already that the length of any circuit is at least 11, longer than 10. Going one step further back, we see that a circuit not including AB has a length of at least 10.

Thus, the salesman must cover at least 10 miles if he is to complete a circuit touching every town. One such circuit is ABCDA.

TWO PERSON ZERO-SUM GAMES

● PROBLEM 17-29

Two players, A and B, each call out one of the numbers 1 and 2 simultaneously. If they both call 1, no payment is made. If they both call 2, B pays A \$3.00. If A calls 1 and B calls 2, B pays A \$1.00. If A calls 2 and B calls 1, A pays B \$1.00.

What is the payoff matrix for this game? Is the game fair to both players?

Solution: This is a two person zero-sum game. It is zero-sum since whatever one player wins, the other must lose. Thus cooperation is impossible.

We construct the payoff matrix by listing A's strategies as rows and B's strategies as columns. The convention is to tabulate the payoffs to the row player.

A \ B	call 1	call 2
call 1	0	1
call 2	-1	3

The game is said to be fair if the value of the game is the same to both players To find the value of the game to A we assume he behaves rationally and see what conclusion he is forced to. A reasons that if he calls 1 his worst payoff is 0. If he calls 2, his worst payoff is \$-1. Thus, he will always call 1, so as not to lose anything.

B thinks that: If I call 1 I can at worst draw with A, if I call 2 the worst payoff is \$3.00. Thus I should call 1. Since the value of the game is the same to both A and B it is a fair game.

A new soda company, Super-Cola, recently entered the market. This company has three choices of advertising campaigns. Their major competitor, Cola-Cola, also has three counter campaigns of advertising to choose from in order to minimize the number of people switching from their soda to the new one. It has been found that their choices of campaign results in the following pay-off matrix:

Number, in 10,000's, of people switching from Cola-Cola to Super-Cola.			
	Cola-Cola		
Super-Cola	Counter-Compaign 1	Counter-Compaign 2	Counter-Compaign 3
Campaign 1	2	3	7
Campaign 2	1	4	6
Campaign 3	9	5	8

Find the best strategies for Super-Cola and Cola-Cola.

Solution: Each company wishes the strategy that is best for them. Super-Cola wishes to get the maximum amount of people from Cola-Cola, and Cola-Cola wishes to minimize their losses to Super-Cola. To do this, we use the minimax procedure. Super-Cola realizes that Cola-Cola will always look for the minimum losses, thus Super-Cola considers the minimum of what will happen for each choice of campaign. Super-Cola notices that the minimum gain for campaign 1 is 20,000 people, the minimum gain for campaign 2 is 10,000 people, but the minimum gain for campaign 3 is 50,000 people. Thus, Super-Cola will choose campaign 3 - the maximum of the minimums. Similarly Cola-Cola realizes that Super-Cola will want to choose the maximum for each of Cola-Cola's counter-campaigns. Thus, Cola-Cola only looks at the maximums. For counter-campaign 1 the maximum loss is 90,000 people, for counter-campaign 2 the maximum loss is 50,000 people, and for counter-campaign 3 the maximum loss is 80,000 people. Thus, Cola-Cola will choose counter-campaign 2 which is the minimum of the maximums. In this way they will minimize the losses. The point on the payoff matrix which they both choose is called the "saddle point". At this point neither company will change it's strategy for they are doing the best that they can. This type of a "game" is called a two-player zero-sum game, because whatever one player wins, the other player loses. Thus, the algebraic sum of the two is zero. Another way of looking at this problem is "pure-strategy". Super-Cola will look at the matrix and note that campaign 3 contains the largest numbers in each column. Thus, campaign 3 is the best choice regardless of which counter-campaign

Cola-Cola chooses. Cola-Cola will notice this also.
They will choose counter-campaign 2, for that one
contains the minimum of all their choices, given Super-Cola
will choose campaign 3.

● **PROBLEM** 17-31

Players A and B simultaneously call out either of the numbers 1 and
2. If their sum is even, B pays A that number of dollars, if odd, A
pays B. What kind of strategy should both players adopt?

Solution: Games may be classified according to the following criteria:
1) Number of players
2) Number of moves
3) Whether or not they are zero-sum
4) Whether or not they are of full information

Here the number of players is two. There are two moves to the game,
A's move and B's move. The game is zero-sum since A's gain is B's loss
and vice-versa. Finally, a game is said to be of full-information if, at
each stage of the game, all previous moves are known to both players. The
given game is not of full-information since both players play simultaneously.

The analysis of a game is simplified by constructing its payoff matrix.
For the given game

B:	1	2	Row Min.
A:			
1	2	-3	-3
2	-3	4	-3
col. max	2	4	

The optimal strategy for A is the maximum value of the game, i.e., the
maximum of the row minima. By calling out either 1 or 2 his maximum
value is -3. For B's optimum strategy we have the minimax value of the
game +2 (from calling 1). Thus A expects to lose 3 while B expects
to lose 2 (recall that positive entries in the payoff matrix are payments
by B to A). But since this is a zero-sum game, the above conclusion
cannot be true. The paradox arises because the game has no saddle point.
When there is one element that will clearly be chosen by both players, it
is called a saddle point. Thus, there is no predictable solution to a
single game. Over many games, both players play their strategies in a
random manner, the frequencies being chosen to give them their best payoffs over many games. We say that the players have moved from pure
strategies to mixed strategies. Note that if either player chose a pure
strategy he would be bound to lose over many games. Thus if B persists
in calling 1, A will continue to call 1.

● **PROBLEM** 17-32

Solve the following game

B:	B_1	B_2	B_3
A			
A_1	1	2	3
A_2	0	3	-1
A_3	-1	-2	4

Solution: From A's point of view, none of the strategies dominate each other. Similarly from B's point each strategy has some reward that cannot be exceeded by the other strategies. Thus, this is an irreducible 3 $\times$ 3 payoff matrix.

A knows that if he adopts strategy A_1 the worst that can happen is that he will receive 1 (if B plays B_1). Similarly the minimum returns from A_2 and A_3 are -1 and -2 respectively. Since A wants to maximize his payoff, he chooses the strategy that will yield him the maximum payoff amongst the minimum returns. Thus, he will play strategy A, where maximin = 1.

B, on the other hand wants to minimize the amount that he must pay A. If he plays B_1 the most he must pay A is 1 (if A plays A_1). Similarly, the maximum payments from playing B_2 and B_3 are 3 and 4 respectively (negative entries represent payments by A to B). B chooses the strategy that will yield him the minimum penalty amongst the maximum payments. Thus, he will play B_1 since minimax = 1.

The maximin and the minimax represent the values of the game to A and B, respectively. When they are equal, as in this case, the common value is the value of the game and is called a saddlepoint.

Every two-person zero-sum game with full information (i.e., each player knows the strategies of his opponent) has a saddlepoint. But if maximin $\neq$ minimax, the value of a single game is uncertain. By introducing the expected value of a game (i.e., the average value of the game when many games are played) even games without full information can be shown to have a unique value.

● **PROBLEM** 17-33

Consider the following payoff matrix:

	C_1	C_2	C_3	C_4
R_1	2	3	-3	2
R_2	1	3	5	2
R_3	9	5	8	10

Find the value of this game.

Solution: The game is in matrix form. Since there are two players, (R the row player and C the column player) and four choices for C with three choices for R, the matrix is of order 3 $\times$ 4. Each entry is considered as a payment by C to R if C pursues action C_j while R pursues action R_i. The value of the game is the choice that R and C make, provided this choice is common.

Assuming R and C behave rationally, we would observe the following behaviour. R, the row player reasons as follows: If I pick row 1, the worst that can happen is that I collect -$3 (this means R pays C, the column player, $3), if I choose row 2 the minimum I receive is $1.00 and if I select row 3 my minimum payoff will be $5.00. Thus, considering my three possible choices, selecting row 3 guarantees that I will receive $5.00 from C. This is the maximum of the minimum payoffs.

C, on the other hand, will reason as follows: If I pick column 1, it is possible that I may have to pay R $9.00. If I choose column 2, I may have to pay R $5.00, while choosing column 3 means the maximum I have to pay R is $8.00. Finally, if I decide to play column 4, my maximum payoff to R will be $10.00. Thus, to minimize the maximum payoff to R, I must choose column 2.

Hence R chooses row 3, C chooses column 2 and R receives a $5.00 payoff from C. The value of the game is 5.

● **PROBLEM** 17-34

Find the general solution to a 2 × 2 game using geometric methods.

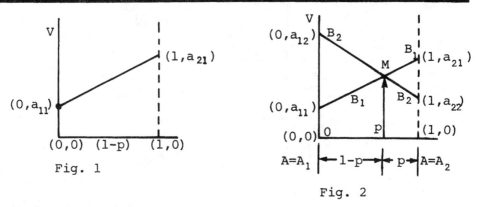

Fig. 1

Fig. 2

Solution: The general 2 × 2 payoff matrix is:

	B:	B_1	B_2
A:			
	A_1:	a_{11}	a_{12}
	A_2:	a_{21}	a_{22}

The payoff matrix may or may not have a saddlepoint. Assume it does not have a saddlepoint. If B just plays B_1 and A plays A_1 and A_2 in proportions p and 1-p (not necessarily optimal), then A will receive, on average,

$$V = a_{11}p + a_{21}(1-p) \tag{1}$$

We can graph (1) treating the a's as fixed, but the p's as unknowns. When (1-p) = 0, $V = a_{11}$. When (1-p) = 1, $V = a_{21}$. All other points in Fig. 1 are mixed strategies. Let

$$S_A^* = \begin{Bmatrix} A_1 & A_2 \\ p & (1-p) \end{Bmatrix} \qquad S_B^* = \begin{Bmatrix} B_1 & B_2 \\ q & (1-q) \end{Bmatrix}$$

be the optimal strategies for A and B. Thus, according to S_A^*, A plays A_1 with proportion p and A_2 the remaining proportion of the time. An important result in game theory is that if a player adopts an optimal mixed strategy, the value of the game to him does not depend on what strategy the other player chooses. Using this result we can solve for p and V from S_A^*:

$$V = a_{11}p + a_{21}(1-p) \qquad \text{(B plays } B_1 \text{ always)}$$

531

$$V = a_{12}p + a_{22}(1-p) \qquad \text{(B plays } B_2 \text{ always)}$$

We plot both of these lines (Fig. 2). The solution set is the point of intersection, p = PI, 1-p = OP, V = PM .

Suppose A were to choose strategy A_2 with probability less than 1-p . Then B, by choosing strategy B_1B_1 will lose less than MP.

Similarly if A chooses 1-p greater than OP, B can lessen V(A) by playing on B_2B_2. Hence A will choose (1-p) = OP since this will guarantee him most. A similar line of argument will give us B's optimum strategy from Fig. 3 (note the relabelling of the axes).

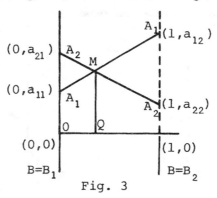

Fig. 3

● **PROBLEM 17-35**

Simplify the following payoff matrix

A's strategies	B's strategies B_1	B_2	B_3	B_4
A_1	0	-1	2	-4
A_2	1	3	3	6
A_3	2	-4	5	1

Solution: It is usual to try to reduce the size of a game in order to solve it more easily. There are two basic ways in which this may be accomplished. We eliminate:
1) Duplicate strategies
2) Dominated strategies

We can use 2) in the given problem to reduce the size of the payoff matrix. The strategy A_i is said to be dominated by A_j if the payoffs to A in A_i are less than or equal to the payoffs to A in A_j . It is strictly dominated if the inequality is a strict inequality. We see that if A_j dominates A_i, then A will never play A_j, assuming rational behaviour.

Examining the given payoff matrix, A_1 is strictly dominated by A_2 since every element in A_1 is less than the corresponding element in A_2. Hence A_1 may be eliminated from contention.

Similarly, from B the column player's point of view, a strategy B_i

is dominated by B_j if the payoffs to B in B_i are less than or equal to the payoffs to B in B_j. In the given payoff matrix B_2 dominates B_3 since the payoffs to B from playing B_2 are always greater or equal to the payoffs to B from playing B_3 (recall that a negative entry represents a payment by A to B and that positive elements are payments by B to A; hence from B's point of view, smaller numbers are preferable to larger numbers). Thus we obtain the reduced payoff matrix.'

	B:	B_1	B_2	B_4
A:				
A_2		1	3	6
A_3		2	-4	1

But now, since A_1 was eliminated, B_2 strictly dominates B_4 and further reduction is possible:

	B:	B_1	B_2
A:			
A_2		1	3
A_3		2	-4

Thus the matrix has been reduced from one of order 3×4 to a payoff matrix of order 2×2.

● **PROBLEM** 17-36

Consider the general 2×2 matrix game:

	B:	B_1	B_3
A:			
A_1		a_{11}	a_{12}
A_2		a_{21}	a_{22}

Show that if this game has a saddle point, it must have a dominance.

<u>Solution:</u> The game has a saddle point if the maximum of the minimum payoffs for A equals the minimum of the maximum payoffs for B. Let a_{11} be the saddle point for the game above. Thus a_{11} is less than or equal to $\max(a_{12}, a_{22})$. Further, a_{11} is the row minimum of A_1, giving

$$a_{11} \leq a_{12} . \tag{1}$$

Since a_{11} must be the column maximum of B_1,

$$a_{11} \geq a_{21} . \tag{2}$$

Also, $a_{11} \leq \max(a_{12}, a_{22})$, and $a_{11} \geq \min(a_{21}, a_{22})$. Now, either $a_{12} \geq a_{22}$ or $a_{12} \leq a_{22}$. If $a_{12} \geq a_{22}$, since $a_{11} \geq a_{21}$, A_1 dominates A_2. If $a_{12} \leq a_{22}$, since $a_{11} \leq a_{22}$ but $a_{11} \geq a_{21}$ we see that $a_{21} \leq a_{22}$. But from (1), $a_{11} \leq a_{12}$, hence B_1 dominates B_2. Thus the original payoff matrix reduces to the single saddle point a_{11}. For higher order games it is generally not true that a game with a saddle point must have a dominance.

NON-ZERO SUM GAMES & MIXED STRATEGIES

● **PROBLEM 17-37**

Give an example of a two-person non-zero-sum game.

Solution: The fundamental theorem of two-person games is that every zero-sum game with mixed strategies has a unique and identical value to both opponents. However, when the game is non-zero sum, some solutions will yield more joint satisfaction to the participants than others. Thus cooperation becomes possible and the solutions are non-unique.

A typical two person non-zero sum game is the "prisoner's dilemma" a parable designed to illustrate that rational behaviour is not always the most satisfying.

Assume two people, A and B, are being accused of jointly engineering a crime. They are kept in separate cells and separately called for an interview with the Grand Inquisitor who presents them with the following alternatives: If you both admit responsibility for the crime you will both be sentenced to 5 years incarceration. If one of you denies complicity while the other admits guilt, then the one who claims to be innocent will receive 10 years while the confesee will be set free. Finally, if both of you deny having committed the crime, you will both receive 2 years each. Are you innocent or guilty?

This is a non-zero sum game without full information and has the payoff matrix given below:

	B:	Admit (B_1)	Deny (B_2)
A:		-5 \ -5	-10 \ 0
	Admit (A_1)		
	Deny (A_2)	0 \ -10	-2 \ -2

From A's point of view, the strategy A_1 dominates A_2. Hence A will choose A_1, i.e., he will admit that he and B committed the crime. From B's point of view, the strategy B_1 dominates B_2 and hence B will also admit his guilt. Thus, by acting rationally, A and B will receive 5 years each. But from the payoff table, this is non-optimal since they both could have obtained 2 years each.

● **PROBLEM 17-38**

Show how a game with the payoff matrix below can be converted to a linear programming problem.

	B's strategies	
A's strategies	2	4
	6	1

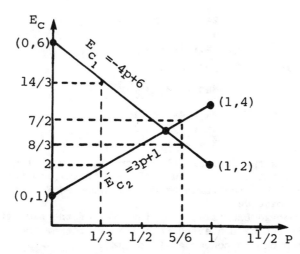

Solution: Since A's maximin strategy is to play row 1 while B's minimax strategy is to play column 2, we see that maximin $\neq$ minimax, i.e., no saddle point exists. However, if we let A choose row 1 with probability p and row 2 with probability $1-p$ then we can compute the expected value of the game to A. Now A has an expected value E_{c_1} against the column player playing column 1 of

$$E_{c_1} = 2p + 6(1-p) = -4p + 6.$$

Similarly, $E_{c_2} = 4p + (1)(1-p) = 3p + 1$. We can graph E_{c_i} versus p:

(See fig.)

For any choice of p, $0 \leq p \leq 1$, $E_{c_1}, E_{c_2} > 0$. Let M denote the minimum expected value to the row player corresponding to his choosing row 1 with probability p and row 2 with probability $1-p$. For example, if $p = 1/3$, $M = E_{c_2}$ since $E_{c_2} = 2$ is smaller than $E_{c_1} = 14/3$. Conversely, if $p = 5/6$, $M = E_{c_1}$ since $E_{c_1} = 10/3$ is smaller than $E_{c_2} = 7/2$. Define s and t as follows:

$$s = p/M \qquad t = \frac{1-p}{M} . \tag{1}$$

Then, $s+t = \frac{p}{M} + \frac{1-p}{M} = 1/M$. Thus, the maximization of M is equivalent to the minimization of $s+t$ where s and t are both non-negative. Further restrictions on s and t are obtained by noting that for any p, E_{c_1} and $E_{c_2} \geq M$, i.e.,

$$2p + 6(1-p) \geq M \tag{2}$$
$$4p + (1)(1-p) \geq M .$$

But, from (1), $p = sM$ and $1-p = tM$. Hence, the inequalities (2) become

$$2sM + 6tM \geq M$$
$$4sM + tM \geq M .$$

Dividing through by M,

$$2s + 6t \geq 1$$
$$4s + t \geq 1 .$$

Thus, A's problem is:

$$\text{Minimize } s + t$$

subject to

535

$$2s + 6t \geq 1$$
$$4s + t \geq 1$$
$$s \geq 0,\ t \geq 0 \ . \tag{3}$$

By analogous logic, B's problem is:

$$\text{Maximize } x + y$$

subject to

$$2x + 4y \leq 1 \tag{4}$$
$$6x + y \leq 1$$
$$x \geq 0,\ y \geq 0.$$

Note that (4) is the dual of (3). From the theory of duality we know that if they exist, the optimal values of (3) and (4) are the same. From game theory we know that the value of a game computed from either player's point of view must coincide.

Finally, George Dantzig, (the inventor of the simplex method), has shown that any game can be converted to a linear program and, conversely, any linear program can be converted to a game.

● **PROBLEM 17-39**

Consider a game with the following payoff matrix:

		B:	B_1	B_2	min
A:	A_1:		2	-3	-3
	A_2:		-3	4	-3
	max		2	4	

Find the value of the game when both players use mixed strategies.

Solution: The game has no saddle point since maximin = 2 ≠ minimax = -3. Thus for a single game, there is no predictable value. Now a fundamental result in the theory of games is that by using mixed strategies every finite two-person zero-sum game has a solution, this solution being at the same time the best for both players. To find the common value of the game we reason as follows: Let

$$S_A^* = \begin{pmatrix} A_1 & A_2 \\ p & 1-p \end{pmatrix} \qquad S_B^* = \begin{pmatrix} B_1 & B_2 \\ q & 1-q \end{pmatrix} \tag{1}$$

S_A^* in (1) is to be interpreted as saying that the strategy of A is to play A_1 with probability p and A_2 the remaining proportion of the time. S_B^* is the strategy of B. If S_A^* is optimal than $V(A)$ does not depend on the frequencies with which B_1, B_2 are used, where $V(A)$ denotes the value of the game to A. Hence,

$$V = p(2) + (1-p)(-3) \qquad \text{(if B uses } B_1 \text{ only)}$$
$$V = p(-3) + (1-p)(4) \qquad \text{(if B uses } B_2 \text{ only)}.$$

Equating these we get

$$p = 7/12,\ (1-p) = 5/12 \text{ and } V = -1/12 \ .$$

Similarly, if B keeps to S_B^*, then

$$V = 2(2) + (1-q)(-3) \qquad \text{(A uses } A_1\text{)}$$

$$V = q(-3) + (1-q)(4) \quad \text{(A uses } A_2\text{)}$$

giving $q = 7/12$, $(1-q) = 5/12$ and $V = -1/12$. Hence

$$S_A^* \left\{ \begin{array}{cc} A_1 & A_2 \\ 7/12 & 5/12 \end{array} \right\} \quad S_B^* = \left\{ \begin{array}{cc} B_1 & B_2 \\ 7/12 & 5/12 \end{array} \right\}$$

and the value of the game is $-1/12$ to A, i.e., B will win $1/12$.

● **PROBLEM** 17-40

How would you solve a game with the following payoff matrix:

B:		B_1	B_2	B_3
A:				
	A_1:	-1	0	1
	A_2:	3	2	-1
	A_3:	-3	1	0

Solution: Since the maximin of $A = 1 \not= $ minimax of $B = 1$, the game has no saddle point. Furthermore, it has no dominance. Hence, it is an irreducible 3 × 3 game and cannot be solved by graphical methods. To find a method of solution, first generalize to the general solution of an m × n zero-sum game. Consider the general m × n zero-sum game: A has moves $A_1,\ldots,A_m$; B has moves $B_1,\ldots,B_n$. The payoff matrix to A is then

B:		B_1	B_2	. . . B_j	. . . B_n
A:	A_1	a_{11}	a_{12}	. . . a_{1j}	. . . a_{1n}
	A_2	a_{21}	a_{22}	. . . a_{2j}	. . . a_{2n}
	.	.			.
	.	.			.
	.	.			.
	A_i	a_{i1}	a_{i2}	. . . a_{ij}	. . . a_{in}
	.	.			.
	.	.			.
	.	.			.
	A_m	a_{m1}	a_{m2}	. . . a_{mj}	. . . a_{mn}

Let S_A^* be A's optimum mixed strategy where the moves $A_1, A_2, \ldots, A_m$ are played with probability $p_1, p_2, \ldots, p_m$. Thus

$$S_A^* = \left\{ \begin{array}{cccc} A_1 & A_2 & \ldots & A_m \\ p_1 & p_2 & \ldots & p_m \end{array} \right\}$$

where $p_1 + p_2 + \ldots + p_m = 1$; $p_i \geq 0$ $(i = 1,2,\ldots,m)$. Similarly,

$$S_B^* = \left\{ \begin{array}{cccc} B_1 & B_2 & \ldots & B_n \\ q_1 & q_2 & \ldots & q_n \end{array} \right\}$$

where
$$q_1 + q_2 + \ldots + q_n = 1,$$
$$q_j \geq 0 \quad (j = 1,2,\ldots,n) .$$

Assume that the value of the game to A is positive (this can be ensured by adding some constant to every element in the payoff matrix so that every element is positive. Let $V(A)$ be the value of the game to A. Assume B plays only the pure strategy B_j. Then the average payoff to A is:

$$p_1 a_{1j} + p_2 a_{2j} + \ldots + p_m a_{mj} .$$

This is not less than V since B_j need not necessarily be s_B^*, the optimal strategy for B. Thus,

$$p_1 a_{1j} + p_2 a_{2j} + \ldots + p_m a_{mj} \geq V. \tag{1}$$

The relationship (1) must be true for all j $(j = 1,\ldots,n)$. Put $x_1 = p_1/V$, $x_2 = p_2/V, \ldots, x_m = p_m/V$. Then (1) becomes

$$a_{1j} x_1 + a_{2j} x_2 + \ldots + a_{mj} x_j \geq 1 \quad (j = 1,2,\ldots,n).$$

The condition $\sum\limits_{i=1}^{m} p_i = 1$ becomes

$$x_1 + x_2 + \ldots + x_m = 1/V , \quad x_i \geq 0, \quad (i = 1,2,\ldots,m).$$

A wishes to maximize $V(A)$, i.e., minimize $1/V$. Thus the game problem becomes the standard linear programming problem.

$$\text{Minimize} \quad (1/V) = x_1 + x_2 + \ldots + x_m \tag{2}$$

subject to

$$a_{11} x_1 + a_{21} x_2 + \ldots + a_{m1} x_m \geq 1$$
$$a_{12} x_1 + a_{22} x_2 + \ldots + a_{m2} x_m \geq 1 \tag{3}$$
$$\vdots \qquad\qquad \vdots \qquad\qquad \vdots$$
$$a_{1n} x_1 + a_{2n} x_n + \ldots + a_{mn} x_m \geq 1$$

$x_1 \geq 0, x_2 \geq 0, \ldots, x_m \geq 0$. Looking at the problem from B's point of view, the value of the game is $-V$ and, as before,

$$q_1 b_{1j} + q_2 b_{2j} + \ldots + q_n b_{nj} \geq -V$$

where the b's are the elements of B's payoff matrix $(b_{ij} = -a_{ji})$, and

$$q_1 + q_2 + \ldots + q_n = 1.$$

Putting $y_1 = q_1/V$, $y_2 = q_2/V, \ldots, y_m = q_m/V$ we obtain

$$y_1 b_{1j} + y_2 b_{2j} + \ldots + y_n b_{nj} \geq -1, \quad (j = 1,\ldots,m), \tag{4}$$

subject to $y_1 + y_2 + \ldots + y_n = 1/V$, $y_j \geq 0$ $(j = 1,\ldots,m)$. Now express the b_{ij} in terms of the payoff matrix elements a_{ji}. Then (4) becomes $-y_1 a_{j1} - y_2 a_{j2} - \ldots - y_n a_{jn} \geq -1$ or

$$a_{j1} y_1 + a_{j2} y_2 + \ldots + a_{jn} y_n \leq 1 \quad (j = 1,\ldots,m).$$

B wishes to minimize V, i.e., maximize $1/V$. Hence his problem is:

$$\text{Maximize} \quad (1/V) = y_1 + y_2 + \cdots + y_n \tag{5}$$

subject to

$$a_{11}y_1 + a_{12}y_2 + \cdots + a_{1n}y_n \leq 1$$

$$a_{21}y_1 + a_{22}y_2 + \cdots + a_{2n}y_n \leq 1 \tag{6}$$

$$\cdot$$
$$\cdot$$
$$\cdot$$

$$a_{m1}y_1 + a_{m2}y_2 + \cdots + a_{mn}y_n \leq 1$$

$y_1 \geq 0$, $y_2 \geq 0, \ldots, y_n \geq 0$. Observe that the two programs (2), (3) and (5), (6) are duals of each other. Thus the optimal solutions to these two linear programming problems must be such that they both give the same value for $1/V$. Turning to the given problem, we first ensure that V is positive by adding 4 to all terms in the original payoff matrix. We thus obtain the positive matrix:

	B:	B_1	B_2	B_3
A =	A_1:	3	4	5
	A_2:	7	6	3
	A_3:	1	5	4

Then the solution to the game is the solution to the linear programming problem

$$\text{Minimize} \quad 1/V = x_1 + x_2 + x_3$$

subject to

$$3x_1 + 7x_2 + x_3 \geq 1$$

$$4x_1 + 6x_2 + 5x_3 \geq 1$$

$$5x_1 + 3x_2 + 4x_3 \geq 1$$

$x_1, x_2, x_3 \geq 0$.

CHAPTER 18

APPLICATIONS TO DIFFERENTIAL EQUATIONS

THE DIFFERENTIAL OPERATOR

● **PROBLEM** 18-1

For the following linear differential equation, find the solution that satisfies the initial condition $y(-1) = -2$.

$$y' + 3y = 0 . \tag{1}$$

Solution: The first order linear differential equation may be expressed in general form as $y' - ay = 0$ where a is a constant. In operator form, this equation may be written as
$$(D - aI)y = 0$$
or $Ly = 0$ where $L = D - aI$. The general solution to $Ly = 0$ is then $y = ce^{ax}$. The given differential equation (1) is of the first order. We can write equation (1) in operator form as $(D + 3I)y = 0$, or $Ly = 0$, where $L = D + 3I$. The general solution is therefore
$$y = ce^{-3x} . \tag{2}$$
As verification, substitute (2) into (1). Now we find the solution to equation (1) that satisfies the condition $y(-1) = -2$. Substitute $x = -1$ in equation (2):
$$y(-1) = -2 = ce^{-3(-1)} = ce^3$$
or $ce^3 = -2$. Hence,
$$c = -2e^{-3} .$$
Substitute the value of c in equation (2):
$$y = -2e^{-3}e^{-3x}$$
$$= -2e^{-3(x+1)} .$$
Hence,
$$y = -2e^{-3(x+1)}$$
is the solution to equation (1) that satisfies the condition $y(-1) = -2$.

● **PROBLEM** 18-2

Find the general solution to each of the following differential equations:

[1] $y' - y = 0$.

[2] $y'' + 4y' - y = 0$.

[3] $y'' + 8y' + 16y = 0$.

Solution: Consider the general linear differential equation with constant coefficients:

$$y'' + by' + cy = 0 \ .$$

Let D be the differentiation operator and I the identity operator; then the above equation can be rewritten as follows:

$$(D^2 + bD + cI)y = 0 \ .$$

Let $L = D^2 + bD + cI$. Then,

$$Ly = 0$$

where L is known as the associated differential operator.

Consider the polynomial

$$p(\lambda) = \lambda^2 + b\lambda + c \ .$$

This polynomial is usually called the characteristic polynomial of L. Its roots are called the characteristic roots of L. If the roots are unequal and real then the general solution to $Ly = 0$ is given by

$$y = c_1 e^{\alpha x} + c_2 e^{\beta x} \ ,$$

where α and β are real roots of the polynomial $p(\lambda)$.

Now we will find the general solution of the given differential equations.

[1] $y'' - y = 0$.

This equation can be written as

$$(D^2 - I)y = 0$$

or $Ly = 0$ where $L = D^2 - I$ is an associated differential operator. The characteristic polynomial of L is

$$p(\lambda) = \lambda^2 - 1 = (\lambda+1)(\lambda-1) \ .$$

The roots of the above polynomial are $\alpha_1 = 1$ and $\beta = -1$. Since the roots are real and unequal, the general solution to $Ly = 0$ is

$$y = c_1 e^{\alpha x} + c_2 e^{\beta x} \ .$$

Upon substituting the values of α and β , we obtain the general solution to $y'' - y = 0$ as follows:

$$y = c_1 e^{x} + c_2 e^{-x} \ .$$

[2] $y'' + 4y' - y = 0$. Then,

$$(D^2 + 4D - I)y = 0$$

or $Ly = 0$ where $L = D^2 + 4D - I$. The characteristic polynomial of L is $p(\lambda) = \lambda^2 + 4\lambda - 1.$

The quadratic formula gives the roots

$$\alpha = \frac{-4 + \sqrt{20}}{2} \quad , \quad \beta = \frac{-4 - \sqrt{20}}{2}$$

$$= -2 + \sqrt{5} \qquad \beta = -2 - \sqrt{5} \ .$$

Since these roots are real and unequal, the general solution is

$$y = c_1 e^{(-2+\sqrt{5})x} + c_2 e^{(-2-\sqrt{5})x} \ .$$

[3] $y'' + 8y' + 16y = 0$.

The above equation is the same as $Ly = 0$, where $L = D^2 + 8D + 16I$. The characteristic polynomial of L is $p(\lambda) = \lambda^2 + 8\lambda + 16 = (\lambda+4)^2$ then the roots of $p(\lambda)$ are $\alpha = -4$, and $\beta = -4$. Here the roots are real and equal. In this case, the general solution to $Ly = 0$ is given by

$$y = c_1 e^{\alpha x} + c_2 x \ e^{\alpha x} \ .$$

Upon substituting the values of α , and β in the above equation, we obtain the general solution to $y'' + 8y' + 16y = 0$:

$$y = c_1 e^{-4x} + c_2 x \, e^{-4x} \ .$$

In general,

(1)   If the roots α, and β are real and unequal, then the general solution to $Ly = 0$ is

$$y = c_1 e^{\alpha x} + c_2 e^{\beta x} \ .$$

(2)   If the roots α and β are real and equal, then the general solution to $Ly = 0$ is

$$y = c_1 e^{\alpha x} + c_2 \, x \, e^{\alpha x} \ .$$

(3)   If the roots of the associated linear operator are either complex numbers or pure imaginary numbers the general solution takes on a form involving the trigonometric functions.

● **PROBLEM** 18-3

Solve the following initial value problem: (1)

$$y'' + 2y' + y \ = \ 0$$
$$y(1) \ = \ 0$$
$$y'(1) \ = \ 2 \ .$$

Solution: The equation (1) is a second order linear differential equation.   In operator form, equation (1) can be written as
$$(D^2 + 2D + I)y = 0 \ ,$$
or $Ly = 0$, where $L = D^2 + 2D + I$.   The characteristic polynomial of L is

$$\rho(\lambda) = \lambda^2 + 2\lambda + 1 = (\lambda + 1)^2 \ .$$

The roots are $\alpha = -1$, and $\beta = -1$.   Since the roots are real and equal, the general solution is

$$y = c_1 e^{-x} + c_2 x \, e^{-x} \ . \tag{2}$$

Now we find the value of c_1 and c_2 by using the given initial conditions, i.e., $y(1) = 0$, $y'(1) = 2$.   Differentiate the equation (2) with respect to x:

$$dy/dx = y' = -c_1 e^{-x} + c_2 e^{-x} - c_2 x e^{-x}$$
$$= -c_1 e^{-x} + c_2 (e^{-x} - x e^{-x}) \ . \tag{3}$$

Substitute $x = 1$ in equation (2) and (3):

$$y(1) = 0 = c_1 e^{-1} + c_2 1 e^{-1} \tag{4}$$

$$y'(1) = 2 = -c_1 e^{-1} + c_2 (e^{-1} - 1 e^{-1}) \tag{5}$$

or

$$c_1 e^{-1} + c_2 e^{-1} = 0$$

$$c_1 e^{-1} = -2 \ . \tag{6}$$

Solving the above system (6) yields:

$$c_1 = -2e^1 \ , \ c_2 = 2e^1 \ .$$

Substituting these values in equation (2) we get:

$$y = -2e^1 e^{-x} + 2e^1 x \, e^{-x}$$
$$= -2e^{-x+1} + 2xe^{-x+1}$$

which is the solution to the given initial value problem. The solution
may be verified by substitution into the differential equation.

● **PROBLEM** 18-4

Solve the following initial value problem:

$$y^{(5)} + y^{(4)} - y^{(3)} - y^{(2)} = 0 .$$

$y(0) = 1, \ y'(0) = -1, \ y''(0) = 0, \ y'''(0) = 1, \ \text{and} \ y''''(0) = 0 .$

<u>Solution:</u> The nature of the general solution to higher-order linear
operators depends upon the factorization of the characteristic polynomial of the operator L .
Let

$$y^{(n)} + a_{n-1}y^{(n-1)} + \ldots + a_1 y' + a_0 y = 0 \tag{1}$$

be a higher order linear differential equation with constant coefficients.
In operator form we can write the above equation as follows:

$$(D^n + a_{n-1}D^{(n-1)} + \ldots + a_1 D + a_0 I)y = 0$$

or $Ly = 0$ where $L = D^n + a_{n-1}D^{(n-1)} + \ldots + a_1 D + a_0 I.$ The characteristic polynomial of L is

$$f(\lambda) = \lambda^{(n)} + a_{n-1}\lambda^{(n-1)} + \ldots + a_1 \lambda + a_0 .$$

Using the theory of factorization, we can write

$$p(\lambda) = p_1(\lambda)^{n_1} p_2(\lambda)^{n_2} \ldots p_k(\lambda)^{n_k} . \tag{2}$$

Since each side of equation (2) must have the same degree, we have

$$n = n_1 \ \text{degree} \ p_1 + n_2 \ \text{degree} \ p_2 + \ldots + n_k \ \text{degree} \ p_k .$$

Suppose

$$p_i(\lambda)^{n_i} = \lambda - \alpha_i \ \text{and} \ n_i = 1 .$$

This factor contributes to the general solution of equation (1) the term
$e^{\alpha_i x}$. If $n_i = 2$, then we have the two solutions $e^{\alpha_i x}$ and $xe^{\alpha_i x}$.

In general, for $p_i(\lambda)^{n_i} = (\lambda - \alpha_i)^{n_i}$, we have the n_i independent solutions

$$e^{\alpha_i x}, \ xe^{\alpha_i x}, \ \ldots, x^{n_i - 1} e^{\alpha_i x} .$$

This process for obtaining the solutions for each factor $p_i(\lambda)^{n_i}$ gives
us n solutions to equation (1) (from equation (2)). Now consider the
given equation

$$y^{(5)} + y^{(4)} - y^{(3)} - y^{(2)} = 0 .$$

In operator form it can be written as

$$(D^5 + D^4 - D^3 - D^2)y = 0$$

or $Ly = 0$, where $L = D^5 + D^4 - D^3 - D^2$. The characteristic polynomial
of L is then

$$p(\lambda) = \lambda^5 + \lambda^4 - \lambda^3 - \lambda^2 = \lambda^2(\lambda^3 + \lambda^2 - \lambda - 1)$$
$$= \lambda^2(\lambda+1)^2(\lambda-1) .$$

Thus,

$$p_1(\lambda) = \lambda^2 \ \text{and} \ n_1 = 2 ;$$

therefore, we have two solutions, 1 and x.
$$p_2(\lambda) = (\lambda+1)^2 , n_2 = 2;$$
therefore, the factor $(\lambda+1)^2$ gives the two solutions e^{-x}, xe^{-x}.
$$p_3(\lambda) = (\lambda-1) , n_3 = 1;$$
therefore we have one solution, e^x. The general solution is therefore
$$y = c_1 + c_2x + c_3e^{-x} + c_4xe^{-x} + c_5e^x .$$
Using the given initial conditions we can find the values of the constants c_1, c_2, c_3, c_4 and c_5. We have
$$y = c_1 + c_2x + c_3e^{-x} + c_4xe^{-x} + c_5e^x . \tag{3}$$

Then,
$$y' = c_2 - c_3e^{-x} + c_4(e^{-x} - xe^{-x}) + c_5e^x$$
$$y'' = c_3e^{-x} + c_4(-2e^{-x} + xe^{-x}) + c_5e^x$$
$$y''' = -c_3e^{-x} + c_4(3e^{-x} - xe^{-x}) + c_5e^x$$
$$y'''' = c_3e^{-x} + c_4(-4e^{-x} + xe^{-x}) + c_5e^x$$

Setting $x = 0$ in each of these and using the initial conditions yields the following five equations:
$$1 = c_1 + c_3 + c_5$$
$$-1 = c_2 - c_3 + c_4 + c_5$$
$$0 = c_3 - 2c_4 + c_5$$
$$1 = -c_3 + 3c_4 + c_5$$
$$0 = c_3 - 4c_4 + c_5$$

Solving these equations yields the unique solution $c_1 = 1$, $c_2 = -2$, $c_3 = -1/2$, $c_4 = 0$ and $c_5 = 1/2$.

Upon substituting these values into equation (3) we obtain
$$y = 1 - 2x - \frac{1}{2}e^{-x} + \frac{1}{2}e^x$$
as the solution to the given initial value problem.

● **PROBLEM** 18-5

Solve the following:
$$y'' - 9y = 0 \tag{1}$$

Solution: Since equation (1) has constant coefficients, we will seek a solution of the form $y = e^{\lambda x}$. Substituting into (1) yields,
$$\lambda^2 e^{\lambda x} - 9e^{\lambda x} = (\lambda^2 - 9)e^{\lambda x} = 0 .$$
Now $e^{\lambda x}$ does not vanish for any finite x, let alone for all x in a given interval. Therefore, we set the $\lambda^2 - 9$ factor equal to zero. Thus $\lambda^2 - 9 = 0$ or $\lambda = \pm 3$. We have found two solutions, e^{3x} and e^{-3x}. We know that if $u_1, \ldots, u_n$ are solutions of an nth-order linear homogeneous differential equation $Ly = 0$, and if their Wronskian eval-

uated at $x = x_0$, $W(x_0) \neq 0$, then $u_1, \ldots, u_n$ are linearly independent. Therefore, the general solution of $Ly = 0$ is given by

$$y = c_1 u_1(x) + \ldots + c_n u_n(x) .$$

Now find the Wronskian of the two solutions, $u_1 = e^{3x}$ and $u_2 = e^{-3x}$ at the point $x = x_0$. Thus,

$$W(u_1, u_2) = \begin{vmatrix} u_1 & u_2 \\ u_1' & u_2' \end{vmatrix}$$

$$W(e^{3x_0}, e^{-3x_0}) = \begin{vmatrix} e^{3x_0} & e^{-3x_0} \\ 3e^{3x_0} & -3e^{-3x_0} \end{vmatrix}$$

$$= -e^{3x_0} \cdot 3e^{-3x_0} - 3e^{3x_0} e^{-3x_0}$$

$$= -6 \neq 0 .$$

Since $W(u_1, u_2) \neq 0$, u_1 and u_2 are linearly independent. Hence, the general solution of (1) is

$$y = c_1 e^{3x} + c_2 e^{-3x} .$$

SOLVING SYSTEMS BY FINDING EIGENVALUES

● PROBLEM 18-6

Find the solution of the following system of linear homogeneous differential equations:

$$x_1'(t) = x_1(t) - x_2(t) + 4x_3(t)$$

$$x_2'(t) = 3x_1(t) + 2x_2(t) - x_3(t) \tag{1}$$

$$x_3'(t) = 2x_1(t) + x_2(t) - x_3(t) .$$

Solution: Consider the system of first order linear homogeneous differential equations:

$$x_1'(t) = a_{11}x_1(t) + a_{12}x_2(t) + \ldots + a_{1n}x_n(t)$$

$$x_2'(t) = a_{21}x_1(t) + a_{22}x_2(t) + \ldots + a_{2n}x_n(t)$$

$$\cdots \cdots \cdots \cdots \cdots \cdots \cdots \cdots \cdots \tag{2}$$

$$x_n'(t) = a_{n1}x_1(t) + a_{n2}x_2(t) + \ldots + a_{nn}x_n(t) .$$

The system (2) can be rewritten in matrix notation as:

$$\begin{bmatrix} x_1'(t) \\ x_2'(t) \\ \cdot \\ \cdot \\ \cdot \\ x_n'(t) \end{bmatrix} = \begin{bmatrix} a_{11} & a_{12} & \cdots & a_{1n} \\ a_{21} & a_{22} & \cdots & a_{2n} \\ \cdots & \cdots & \cdots & \cdots \\ a_{n1} & a_{n2} & \cdots & a_{nn} \end{bmatrix} \begin{bmatrix} x_1(t) \\ x_2(t) \\ \cdot \\ \cdot \\ \cdot \\ x_n(t) \end{bmatrix}$$

or

$$X' = AX \qquad\qquad (3)$$

where

$$X' = \begin{bmatrix} x'_1(t) \\ x'_2(t) \\ . \\ . \\ . \\ x'_n(t) \end{bmatrix}, \qquad A = \begin{bmatrix} a_{11} & a_{12} & \cdots & a_{1n} \\ a_{2n} & a_{22} & \cdots & a_{2n} \\ \cdots & \cdots & \cdots & \cdots \\ a_{n1} & a_{n2} & \cdots & a_{nn} \end{bmatrix}$$

and

$$X = \begin{bmatrix} x_1(t) \\ x_2(t) \\ . \\ . \\ . \\ x_n(t) \end{bmatrix} .$$

We wish to find n linearly independent solutions. Recall that both the first order and second order linear homogeneous equations have exponential functions as solutions. Generalizing, let $x(t) = e^{\lambda t} v$ be a solution of equation (3), where v is a constant vector. Now observe that

$$\frac{d}{dt} e^{\lambda t} v = \lambda e^{\lambda t} v \quad \text{and} \quad A(e^{\lambda t} v) = e^{\lambda t} A v .$$

Therefore, $x(t) = e^{\lambda t} v$ is a solution of equation (3) if and only if $\lambda e^{\lambda t} v = e^{\lambda t} A v$. Dividing both sides of this equation by $e^{\lambda t}$ gives ($e^{\lambda t} \neq 0$ for any λt)

$$Av = \lambda v . \qquad\qquad (4)$$

Recall that a non-zero vector v satisfying equation (4) is called an eigenvector of A with eigenvalue λ.

Now for each eigenvector v_i of A with eigenvalue λ_i, we have $x_i(t) = e^{\lambda_i(t)} v_i$ as a solution of equation (3). If A has n linearly independent eigenvectors $v_1, \ldots, v_n$ with eigenvalues $\lambda_1, \ldots, \lambda_n$ respectively, then $x_i(t) = e^{\lambda_i t} v_i$, $i = 1, \ldots, n$ are n linearly independent solutions of equation (3). The general solution of (3) is then given by

$$x(t) = c_1 e^{\lambda_1 t} v_1 + c_2 e^{\lambda_2 t} v_2 + \ldots + c_n e^{\lambda_n t} v_n .$$

Turning to the given system, we first write (1) in matrix notation. Thus,

$$\begin{bmatrix} x'_1(t) \\ x'_2(t) \\ x'_3(t) \end{bmatrix} = \begin{bmatrix} 1 & -1 & 4 \\ 3 & 2 & -1 \\ 2 & 1 & -1 \end{bmatrix} \begin{bmatrix} x_1(t) \\ x_2(t) \\ x_3(t) \end{bmatrix}$$

or X' = AX where

$$A = \begin{pmatrix} 1 & -1 & 4 \\ 3 & 2 & -1 \\ 2 & 1 & -1 \end{pmatrix}.$$

We wish to find eigenvalues of A and eigenvectors corresponding to each eigenvalue.

The characteristic polynomial of the matrix A is

$$f(\lambda) = \det(\lambda I - A) = \det \begin{pmatrix} \lambda-1 & 1 & -4 \\ -3 & \lambda-2 & 1 \\ -2 & -1 & \lambda+1 \end{pmatrix}$$

$$f(\lambda) = \lambda-1 \begin{vmatrix} \lambda-2 & 1 \\ -1 & \lambda+1 \end{vmatrix} -1 \begin{vmatrix} -3 & 1 \\ -2 & \lambda+1 \end{vmatrix} -4 \begin{vmatrix} -3 & \lambda-2 \\ -2 & -1 \end{vmatrix}$$

$$= (\lambda-1)[(\lambda-2)(\lambda+1)+1] -1[-3(\lambda+1)+2] -4[3 + 2(\lambda-2)]$$

$$= (\lambda-1)[\lambda^2 - \lambda - 1] - 1[-3\lambda - 1] - 4[2\lambda - 1]$$

$$= (\lambda-1)[\lambda^2 - \lambda - 1] - 5(\lambda-1)$$

$$= (\lambda-1)[\lambda^2 - \lambda - 6] = (\lambda-1)(\lambda-3)(\lambda+2).$$

Then, the eigenvalues of A are

$$\lambda_1 = 1, \ \lambda_2 = 3, \ \lambda_3 = -2.$$

For $\lambda_1 = 1$, we seek a non-zero vector v such that

$$(1I - A)v = \begin{pmatrix} 0 & 1 & -4 \\ -3 & -1 & 1 \\ -2 & -1 & 2 \end{pmatrix} \begin{pmatrix} v_1 \\ v_2 \\ v_3 \end{pmatrix} = \begin{pmatrix} 0 \\ 0 \\ 0 \end{pmatrix}$$

or

$$v_2 - 4v_3 = 0$$

$$-3v_1 - v_2 + v_3 = 0$$

$$-2v_1 - v_2 + 2v_3 = 0.$$

Solving the above systems yields $v_1 = -v_3$ and $v_2 = 4v_3$. Hence, each

vector $v = a \begin{bmatrix} -1 \\ 4 \\ 1 \end{bmatrix}$ is an eigenvector of A associated with the eigenvalue

$\lambda_1 = 1$. Therefore, $ae^t \begin{bmatrix} -1 \\ 4 \\ 1 \end{bmatrix}$ is a solution of the differential equation

for any constant a. For simplicity, we take

$$x_1(t) = e^t \begin{bmatrix} -1 \\ 4 \\ 1 \end{bmatrix}.$$

For $\lambda_2 = 3$, we seek a non-zero vector v such that

$$(3I - A)v = \begin{pmatrix} 2 & 1 & -4 \\ -3 & 1 & 1 \\ -2 & -1 & 4 \end{pmatrix} \begin{pmatrix} v_1 \\ v_2 \\ v_3 \end{pmatrix} = \begin{pmatrix} 0 \\ 0 \\ 0 \end{pmatrix} ,$$

or

$$2v_1 + v_2 - 4v_3 = 0$$

$$-3v_1 + v_2 + v_3 = 0$$

$$-2v_1 - v_2 + 4v_3 = 0 .$$

Solving the above system yields: $v_1 = v_3$, $v_2 = 2v_3$. Thus each vector

$$v = a \begin{pmatrix} 1 \\ 2 \\ 1 \end{pmatrix}$$

is an eigenvector of A corresponding to $\lambda = 3$. Therefore,

$$x_2(t) = e^{\lambda_2 t} v_2 = e^{3t} \begin{pmatrix} 1 \\ 2 \\ 1 \end{pmatrix}$$

is a second solution of the differential equation. For $\lambda = -2$, we seek a non-zero vector v such that

$$(-2I - A)v = \begin{pmatrix} -3 & 1 & -4 \\ -3 & -4 & 1 \\ -2 & -1 & -1 \end{pmatrix} \begin{pmatrix} v_1 \\ v_2 \\ v_3 \end{pmatrix} = \begin{pmatrix} 0 \\ 0 \\ 0 \end{pmatrix}$$

or

$$-3v_1 + v_2 - 4v_3 = 0$$

$$-3v_1 - 4v_2 + v_3 = 0$$

$$-2v_1 - v_2 - v_3 = 0 .$$

Solving the above system yields: $v_1 = -v_3$, $v_2 = v_3$. Hence each vector $v = a \begin{pmatrix} -1 \\ 1 \\ 1 \end{pmatrix}$ is an eigenvector of A with eigenvalue -2. Therefore,

$$x_3(t) = e^{\lambda_3(t)} v_3 = e^{-2t} \begin{pmatrix} -1 \\ 1 \\ 1 \end{pmatrix}$$ is a third solution of the differential

equation. Since A has distinct eigenvalues, these solutions must be linearly independent. Therefore every solution $x(t)$ must be of the form

$$x(t) = c_1 e^t \begin{pmatrix} -1 \\ 4 \\ 1 \end{pmatrix} + c_2 e^{3t} \begin{pmatrix} 1 \\ 2 \\ 1 \end{pmatrix} + c_3 e^{-2t} \begin{pmatrix} -1 \\ 1 \\ 1 \end{pmatrix}$$

$$= \begin{pmatrix} -c_1 e^t + c_2 e^{3t} - c_3 e^{-2t} \\ 4c_1 e^t + 2c_2 e^{3t} + c_3 e^{-2t} \\ c_1 e^t + c_2 e^{3t} + c_3 e^{-2t} \end{pmatrix} .$$

● PROBLEM 18-7

Find the general solution to the system

$$x_1' = x_1$$
$$x_2' = x_1 + 2x_2 \tag{1}$$
$$x_3' = x_1 - x_3 .$$

<u>Solution:</u> The system (1) can be rewritten as

$$\begin{pmatrix} x_1' \\ x_2' \\ x_3' \end{pmatrix} = \begin{pmatrix} 1 & 0 & 0 \\ 1 & 2 & 0 \\ 1 & 0 & -1 \end{pmatrix} \begin{pmatrix} x_1 \\ x_2 \\ x_3 \end{pmatrix}$$

or $\qquad\qquad\qquad X' = AX \tag{2}$

where

$$A = \begin{pmatrix} 1 & 0 & 0 \\ 1 & 2 & 0 \\ 1 & 0 & -1 \end{pmatrix}$$

We need eigenvalues of the matrix A to obtain the eigenvectors. Note that the matrix A is lower triangular.

The eigenvalues of a lower triangular matrix are the diagonal entries of the matrix. Therefore, the eigenvalues are $\lambda_1 = 1$, $\lambda_2 = 2$, and $\lambda_3 = -1$.

Now, for $\lambda_1 = 1$, we must solve the vector equation $(\lambda I - A)v = 0$ to obtain the eigenvector associated with $\lambda_1 = 1$. Thus we obtain:

$$\begin{pmatrix} 0 & 0 & 0 \\ -1 & -1 & 0 \\ -1 & 0 & 2 \end{pmatrix} \begin{pmatrix} v_1 \\ v_2 \\ v_3 \end{pmatrix} = \begin{pmatrix} 0 \\ 0 \\ 0 \end{pmatrix}$$

or

$$-v_1 - v_2 = 0$$
$$-v_1 + 2v_3 = 0 .$$

Solving the above system yields: $v_1 = -v_2$, $v_1 = 2v_3$. Hence each vector $v = a \begin{pmatrix} 2 \\ -2 \\ 1 \end{pmatrix}$ is an eigenvector of A with eigenvalue 1. Consequently,

$$x_1(t) = e^{\lambda_1 t} v = e^t \begin{pmatrix} 2 \\ -2 \\ 1 \end{pmatrix}$$

is a solution of differential equation (2). For $\lambda_2 = 2$, we seek a non-zero vector v such that

$$[2I - A]v = \begin{pmatrix} 1 & 0 & 0 \\ -1 & 0 & 0 \\ -1 & 0 & 3 \end{pmatrix} \begin{pmatrix} v_1 \\ v_2 \\ v_3 \end{pmatrix} = \begin{pmatrix} 0 \\ 0 \\ 0 \end{pmatrix}$$

or

$$v_1 = 0$$
$$-v_1 + v_3 = 0 .$$

Thus $v_3 = 0$. Hence each vector $v = a \begin{pmatrix} 0 \\ 1 \\ 0 \end{pmatrix}$ is an eigenvector of A corresponding to $\lambda = 2$. Hence

$$x_2(t) = e^{\lambda_2(t)} v = e^{2t} \begin{pmatrix} 0 \\ 1 \\ 0 \end{pmatrix}$$

is a second solution of the differential equation.
    Similarly

$$v = \begin{pmatrix} 0 \\ 0 \\ 1 \end{pmatrix}$$

is an eigenvector corresponding to eigenvalue $\lambda_3 = -1$. Therefore,

$$x_3(t) = e^{\lambda_3(t)} v_3 = e^{-t} \begin{pmatrix} 0 \\ 0 \\ 1 \end{pmatrix}$$

is a third solution of the differential equation.
    Since A has distinct eigenvalues, these solutions must be linearly independent. Therefore, every solution $x(t)$ must be of the form:

$$x(t) = c_1 e^t \begin{pmatrix} 2 \\ -2 \\ 1 \end{pmatrix} + c_2 e^{2t} \begin{pmatrix} 0 \\ 1 \\ 0 \end{pmatrix} + c_3 e^t \begin{pmatrix} 0 \\ 0 \\ 1 \end{pmatrix}$$

$$= \begin{pmatrix} 2c_1 e^t \\ -2c_1 e^t + c_2 e^{2t} \\ c_1 e^t + c_3 e^{-t} \end{pmatrix} .$$

● **PROBLEM** 18-8

Solve the following:

$$\begin{aligned} x' &= -x + 2y \\ y' &= -2x - y \end{aligned} \tag{1}$$

Solution: The system (1) can be written as:

$$\begin{pmatrix} x' \\ y' \end{pmatrix} = \begin{pmatrix} -1 & 2 \\ -2 & -1 \end{pmatrix} \begin{pmatrix} x \\ y \end{pmatrix} . \tag{2}$$

Let

$$X = \begin{pmatrix} x \\ y \end{pmatrix} \quad \text{and} \quad A = \begin{pmatrix} -1 & 2 \\ -2 & -1 \end{pmatrix} .$$

Then,

$$\frac{dX}{dt} = AX . \tag{3}$$

To obtain the general solution of equation (3), we need its eigenvalues and corresponding eigenvectors.
    The characteristic polynomial of A is

$$f(\lambda) = \det(\lambda I - A) = \det \begin{pmatrix} \lambda+1 & -2 \\ 2 & \lambda+1 \end{pmatrix}$$

$$= (\lambda+1)^2 + 4$$

$$= \lambda^2 + 2\lambda + 5 .$$

The characteristic equation is $\lambda^2 + 2\lambda + 5 = 0$. Then the roots of this equations are:

$$\lambda_1 = \frac{-2+\sqrt{4-20}}{2} = -1 + 2i$$

$$\lambda_2 = \frac{-2 - \sqrt{4-20}}{2} = -1 - 2i \ .$$

The eigenvectors corresponding to $\lambda_1 = -1 + 2i$ and $\lambda_2 = -1 - 2i$ are

$\begin{bmatrix} 1 \\ i \end{bmatrix}$ and $\begin{bmatrix} 1 \\ -i \end{bmatrix}$ respectively. Then, the general solution of equation

(3) is

$$X = c_1 e^{(-1+2i)t} \begin{bmatrix} 1 \\ i \end{bmatrix} + c_2 e^{(-1-2i)t} \begin{bmatrix} 1 \\ -i \end{bmatrix}$$

(where c_1 and c_2 are arbitrary constants),

$$X = \begin{pmatrix} c_1 e^{(-1+2i)t} + c_2 e^{(-1-2i)t} \\ \\ ic_1 e^{(-1+2i)t} - ic_2 e^{(-1-2i)t} \end{pmatrix}$$

$$= \begin{pmatrix} c_1 e^{-t} e^{i2t} + c_2 e^{-t} e^{-i2t} \\ \\ ic_1 e^{-t} e^{2it} - ic_2 e^{-t} e^{-2it} \end{pmatrix}$$

We know that $e^{i\theta} = \cos \theta + i \sin \theta$; $e^{-i\theta} = \cos \theta - i \sin \theta$. Therefore,

$$X = \begin{pmatrix} c_1 e^{-t}[\cos 2t + \sin 2t] + c_2 e^{-t}[\cos 2t - i \sin 2t] \\ \\ ic_1 e^{-t}[\cos 2t + i \sin 2t] - ic_2 e^{-t}[\cos 2t - i \sin 2t] \end{pmatrix}$$

$$= \begin{pmatrix} e^{-t} \cos 2t[c_1 + c_2] + e^{-t} \sin 2t \ i[c_1 - c_2] \\ \\ e^{-t} \cos 2ti[c_1 - c_2] - e^{-t} \sin 2t[c_1 + c_2] \end{pmatrix}$$

Let $c_1 + c_2 = A$ and $i[c_1 - c_2] = B$. We know $X = \begin{bmatrix} x \\ y \end{bmatrix}$. Then,

$$\begin{pmatrix} x \\ y \end{pmatrix} = \begin{pmatrix} Ae^{-t} \cos 2t + Be^{-t} \sin 2t \\ \\ -Ae^{-t} \sin 2t + Be^{-t} \cos 2t \end{pmatrix}$$

Therefore the general solution of the system (1) is

$$x = e^{-t}[A \cos 2t + B \sin 2t] \ .$$
$$y = e^{-t}[-A \sin 2t + B \cos 2t] \ .$$

● **PROBLEM** 18-9

Solve the initial value problem

$$x_1' = x_1 + 12x_2$$
$$x_2' = 3x_1 + x_2 \ .$$ (1)

The initial conditions are $x_1(0) = 0$

$$x_2(0) = 1 \ .$$

Solution: In matrix notation the system (1) becomes

$$\begin{pmatrix} x_1' \\ x_2' \end{pmatrix} = \begin{pmatrix} 1 & 12 \\ 3 & 1 \end{pmatrix} \begin{pmatrix} x_1 \\ x_2 \end{pmatrix}.$$

Also,

$$X' = AX \tag{2}$$

where

$$A = \begin{pmatrix} 1 & 12 \\ 3 & 1 \end{pmatrix} \quad , \quad X = \begin{pmatrix} x_1 \\ x_2 \end{pmatrix}.$$

The characteristic polynomial of the matrix A is

$$f(\lambda) = \det(\lambda I - A) = \det \begin{pmatrix} \lambda-1 & -12 \\ -3 & \lambda-1 \end{pmatrix}$$

$$= (\lambda-1)^2 - 36$$

$$= \lambda^2 - 2\lambda - 35$$

$$= (\lambda-7)(\lambda+5) .$$

Hence the eigenvalues are $\lambda_1 = 7$ and $\lambda_2 = -5$.

Now, the eigenvector associated with $\lambda_1 = 7$ is obtained by solving the equation $(\lambda I - A)V = 0$ for $\lambda = 7$. Thus,

$$(7I - A)V = \begin{pmatrix} 6 & -12 \\ -3 & 6 \end{pmatrix} \begin{pmatrix} v_1 \\ v_2 \end{pmatrix} = \begin{pmatrix} 0 \\ 0 \end{pmatrix} , \quad \text{or}$$

$$6v_1 - 12v_2 = 0$$

$$-3v_1 + 6v_2 = 0 .$$

This implies that $v_1 = 2v_2$. Therefore the everyvector $v = a \begin{bmatrix} 2 \\ 1 \end{bmatrix}$ is an eigenvector of A with eigenvalue 7. Hence, $x_1(t) = e^{\lambda t}v_1 = e^{7t}\begin{bmatrix} 2 \\ 1 \end{bmatrix}$ is a solution of equation (2).

Similarly, the eigenvector corresponding to $\lambda = -5$ can be found by solving the equation $[-5I - A]V = 0$:

$$\begin{pmatrix} -6 & -12 \\ -3 & -6 \end{pmatrix} \begin{pmatrix} v_1 \\ v_2 \end{pmatrix} = \begin{pmatrix} 0 \\ 0 \end{pmatrix}$$

or

$$-6v_1 - 12v_2 = 0$$

$$-3v_1 - 6v_2 = 0 .$$

This implies that $v_1 = -2v_2$. Therefore every vector $v = a \begin{bmatrix} -2 \\ 1 \end{bmatrix}$ is an eigenvector of A with eigenvalue -5, and

$$x_2(t) = e^{\lambda_2 t}v_2 = e^{-5t}\begin{bmatrix} -2 \\ 1 \end{bmatrix}$$

is a second solution of the equation (2). Since A has distinct eigenvalues, these solutions are linearly independent. Hence the general solution of equation (2) is given by

$$X(t) = c_1 e^{7t} \begin{bmatrix} 2 \\ 1 \end{bmatrix} + c_2 e^{-5t} \begin{bmatrix} -2 \\ 1 \end{bmatrix}. \tag{3}$$

The constants c_1 and c_2 are determined from the initial conditions. The initial conditions, in matrix notation, take the form

$$x(0) = \begin{bmatrix} x_1(0) \\ x_2(0) \end{bmatrix} = \begin{bmatrix} 0 \\ 1 \end{bmatrix}.$$

Therefore,

$$X(0) = \begin{bmatrix} 0 \\ 1 \end{bmatrix} = c_1 \begin{bmatrix} 2 \\ 1 \end{bmatrix} + c_2 \begin{bmatrix} -2 \\ 1 \end{bmatrix}$$

$$\begin{bmatrix} 0 \\ 1 \end{bmatrix} = \begin{bmatrix} 2c_1 - 2c_2 \\ c_1 + c_2 \end{bmatrix}.$$

Hence,

$$2c_1 - 2c_2 = 0$$

$$c_1 + c_2 = 1.$$

Solving these equations yields: $c_1 = 1/2$ and $c_2 = 1/2$. Upon substitution of the values of c_1 and c_2 into equation (3) we obtain the unique solution of the system (1). Thus,

$$X(t) = 1/2\ e^{7t} \begin{bmatrix} 2 \\ 1 \end{bmatrix} + 1/2\ e^{-5t} \begin{bmatrix} -2 \\ 1 \end{bmatrix}$$

$$= 1/2 \begin{bmatrix} 2e^{7t} - 2e^{-5t} \\ e^{7t} + e^{-5t} \end{bmatrix}.$$

Then,

$$X(t) = \begin{bmatrix} x_1 \\ x_2 \end{bmatrix} = \begin{bmatrix} e^{7t} - e^{-5t} \\ (1/2)\ e^{7t} + (1/2)\ e^{-5t} \end{bmatrix}$$

Hence,

$$x_1 = e^{7t} - e^{-5t}$$

$$x_2 = (1/2)\ e^{7t} + (1/2)\ e^{-5t}.$$

● **PROBLEM** 18-10

Find the general solution of the following linear system of differential equations:

$$\dot{x}_1 = x_1$$

$$\dot{x}_2 = 3x_2 - 2x_3 \tag{1}$$

$$\dot{x}_3 = -2x_2 + 3x_3.$$

Solution: In matrix notation the system (1) becomes:

$$\begin{bmatrix} \dot{x}_1 \\ \dot{x}_2 \\ \dot{x}_3 \end{bmatrix} = \begin{bmatrix} 1 & 0 & 0 \\ 0 & 3 & -2 \\ 0 & -2 & 3 \end{bmatrix} \begin{bmatrix} x_1 \\ x_2 \\ x_3 \end{bmatrix}$$

or $\dot{X} = AX$ where

$$A = \begin{bmatrix} 1 & 0 & 0 \\ 0 & 3 & -2 \\ 0 & -2 & 3 \end{bmatrix} .$$

The characteristic polynomial of A is

$$f(\lambda) = \det(\lambda I - A) = \det \begin{bmatrix} \lambda-1 & 0 & 0 \\ 0 & \lambda-3 & 2 \\ 0 & 2 & \lambda-3 \end{bmatrix}$$

$$= \lambda-1 \begin{vmatrix} \lambda-3 & 2 \\ 2 & \lambda-3 \end{vmatrix}$$

$$= (\lambda-1)[(\lambda-3)^2 - 4]$$

$$= (\lambda-1)[\lambda^2 - 6\lambda + 5]$$

$$= (\lambda-1)(\lambda-1)(\lambda-5)$$

$$= (\lambda-1)^2 (\lambda-5) .$$

Hence the eigenvalues of A are $\lambda_1 = \lambda_2 = 1$ and $\lambda_3 = 5$. We seek three linearly independent eigenvectors associated with eigenvalues λ_1, λ_2 and λ_3 . However, $\lambda = 1$ is an eigenvalue of multiplicity 2. Can we find two linearly independent eigenvectors corresponding to $\lambda = 1$? As we know, if λ is an eigenvalue of A of multiplicity k, and the rank of the matrix $(\lambda I_n - A)$ is equal to n - k, then we can find k linearly independent eigenvectors of A associated with the eigenvalue λ . Now the rank of the matrix

$$(1\, I_3 - A) = \begin{bmatrix} 0 & 0 & 0 \\ 0 & -2 & 2 \\ 0 & 2 & -2 \end{bmatrix} \text{ is } 1 ,$$

and n - k = 3 - 2 = 1. Therefore we can find two linearly independent eigenvectors associated with $\lambda = 1$ with the multiplicity of 2. The eigenvector

$$V = \begin{bmatrix} v_1 \\ v_2 \\ v_3 \end{bmatrix}$$

corresponding to the eigenvalue $\lambda = 1$ must satisfy the equation

$$(1\, I - A)V = \begin{bmatrix} 0 & 0 & 0 \\ 0 & -2 & 2 \\ 0 & 2 & -2 \end{bmatrix} \begin{bmatrix} v_1 \\ v_2 \\ v_3 \end{bmatrix} = \begin{bmatrix} 0 \\ 0 \\ 0 \end{bmatrix}$$

Hence

$$-2v_2 + 2v_3 = 0$$
$$2v_2 - 2v_3 = 0 .$$

This implies that $v_2 = v_3$ and that v_1 is arbitrary. Therefore $\begin{bmatrix} 1 \\ 0 \\ 0 \end{bmatrix}$

and

$\begin{pmatrix} 0 \\ 1 \\ 1 \end{pmatrix}$ are two linearly independent eigenvectors associated with $\lambda = 1$. The eigenvector

$$V = \begin{pmatrix} v_1 \\ v_2 \\ v_3 \end{pmatrix} \qquad \text{corresponding to the eigenvalue}$$

$\lambda = 5$ must satisfy the equation

$$(5I - A)V = \begin{pmatrix} 4 & 0 & 0 \\ 0 & 2 & 2 \\ 0 & 2 & 2 \end{pmatrix} \begin{pmatrix} v_1 \\ v_2 \\ v_3 \end{pmatrix} = \begin{pmatrix} 0 \\ 0 \\ 0 \end{pmatrix}$$

Hence,

$$4v_1 = 0$$
$$2v_2 + 2v_3 = 0 .$$

This implies that $v_1 = 0$ and $v_3 = -v_2$. Therefore $V = \begin{pmatrix} 0 \\ 1 \\ -1 \end{pmatrix}$ is an

eigenvector associated with $\lambda = 5$. Now we have three linearly independent eigenvectors. Therefore, the general solution to system (1) is

$$X(t) = c_1 e^{\lambda_1 t} v_1 + c_2 e^{\lambda_2 t} v_2 + c_3 e^{\lambda_3 t} v_3$$

$$= c_1 e^t \begin{pmatrix} 1 \\ 0 \\ 0 \end{pmatrix} + c_2 e^t \begin{pmatrix} 0 \\ 1 \\ 1 \end{pmatrix} + c_3 e^{5t} \begin{pmatrix} 0 \\ 1 \\ -1 \end{pmatrix}$$

$$X(t) = \begin{pmatrix} c_1 e^t \\ c_2 e^t + c_3 e^{5t} \\ c_2 e^t - c_3 e^{5t} \end{pmatrix}$$

or

$$x_1(t) = c_1 e^t$$
$$x_2(t) = c_2 e^t + c_3 e^{5t}$$
$$x_3(t) = c_2 e^t - c_3 e^{5t} .$$

● **PROBLEM** 18-11

Solve the following initial value problem:

$$x_1' = 2x_1 + x_2 + 3x_3$$
$$x_2' = 2x_2 - x_3 \qquad\qquad (1)$$
$$x_3' = 2x_3 .$$

Initial conditions:
$$x_1(0) = 1$$
$$x_2(0) = 2$$
$$x_3(0) = 1 .$$

Solution: In matrix notation, system (1) can be written as:

$$\begin{pmatrix} x_1' \\ x_2' \\ x_3' \end{pmatrix} = \begin{pmatrix} 2 & 1 & 3 \\ 0 & 2 & -1 \\ 0 & 0 & 2 \end{pmatrix} \begin{pmatrix} x_1 \\ x_2 \\ x_3 \end{pmatrix}$$

or $X' = AX$ (2)

where

$$A = \begin{pmatrix} 2 & 1 & 3 \\ 0 & 2 & -1 \\ 0 & 0 & 2 \end{pmatrix}.$$

The characteristic polynomial of A is $f(\lambda) = \det[\lambda I - A]$

$$= \det \begin{pmatrix} \lambda-2 & -1 & -3 \\ 0 & \lambda-2 & 1 \\ 0 & 0 & \lambda-2 \end{pmatrix}$$

$$= (\lambda-2) \begin{vmatrix} \lambda-2 & +1 \\ 0 & \lambda-2 \end{vmatrix}$$

$$= (\lambda-2)(\lambda-2)(\lambda-2) = (\lambda-2)^3.$$

Therefore $\lambda = 2$ is an eigenvalue of A with multiplicity 3. The eigenvector corresponding to $\lambda = 2$ can be obtained by solving equation $(\lambda I - A)V = 0$ for $\lambda = 2$. Thus,

$$(2I - A)V = \begin{pmatrix} 0 & -1 & -3 \\ 0 & 0 & 1 \\ 0 & 0 & 0 \end{pmatrix} \begin{pmatrix} v_1 \\ v_2 \\ v_3 \end{pmatrix} = \begin{pmatrix} 0 \\ 0 \\ 0 \end{pmatrix}$$

or

$$-v_2 - 3v_3 = 0$$

$$v_3 = 0.$$

This implies that $v_2 = v_3 = 0$ and v_1 is arbitrary. Hence

$$x_1(t) = e^{\lambda_1 t} v = e^{2t} \begin{pmatrix} 1 \\ 0 \\ 0 \end{pmatrix}$$

is one solution of equation (2). The matrix A is 3×3, hence $n = 3$. Therefore the differential equation (2) has 3 linearly independent solutions of the form $e^{\lambda t} v$. Since A has only one linearly independent eigenvector, we have only one linearly independent solution of the form $e^{\lambda t} v$. To find the additional two solutions of equation (2), we proceed as follows:

(1)   Pick an eigenvalue λ of A and find all vectors v for which $(A - \lambda I)^2 v = 0$, but $(A - \lambda I)v \neq 0$. For each such vector v,

$$e^{At} v = e^{\lambda t} e^{(A-\lambda I)t} = e^{\lambda t}[I + t(A - \lambda I)]v$$

is a second solution of $X' = AX$.

(2)   For the third solution, we find all vectors v for which $(A - \lambda I)^3 v = 0$, but $(A - \lambda I)^2 \neq 0$. For each such vector v,

$$e^{At} v = e^{\lambda t}[v + t(A - \lambda I)v + \frac{t^2}{2!}(A - \lambda I)^2 v]$$

is the third solution of equation (2). Thus, we look for all solutions of the equation $(A - \lambda I)^2 V = 0$, for $\lambda = 2$:

$$(A - \lambda I)^2 V = \begin{pmatrix} 0 & 1 & 3 \\ 0 & 0 & -1 \\ 0 & 0 & 0 \end{pmatrix} \begin{pmatrix} 0 & 1 & 3 \\ 0 & 0 & -1 \\ 0 & 0 & 0 \end{pmatrix} \begin{pmatrix} v_1 \\ v_2 \\ v_3 \end{pmatrix} = \begin{pmatrix} 0 \\ 0 \\ 0 \end{pmatrix}$$

$$\begin{pmatrix} 0 & 0 & -1 \\ 0 & 0 & 0 \\ 0 & 0 & 0 \end{pmatrix} \begin{pmatrix} v_1 \\ v_2 \\ v_3 \end{pmatrix} = \begin{pmatrix} 0 \\ 0 \\ 0 \end{pmatrix} .$$

This implies that $v_3 = 0$ and both v_1 and v_2 are arbitrary. Now we have to find the vector v such that $(A - \lambda I)V \neq 0$, but $(A - \lambda I)^2 V = 0$. The vector

$$V = \begin{pmatrix} 0 \\ 1 \\ 0 \end{pmatrix}$$

satisfies $(A - \lambda I)^2 V = 0$, but $(A - \lambda I)V \neq 0$. Hence,

$$x_2(t) = e^{At} \begin{pmatrix} 0 \\ 1 \\ 0 \end{pmatrix} = e^{2t} e^{(A - 2I)t} \begin{pmatrix} 0 \\ 1 \\ 0 \end{pmatrix}$$

$$= e^{2t} [I + t(A - 2I)] \begin{pmatrix} 0 \\ 1 \\ 0 \end{pmatrix}$$

$$= e^{2t} \left[I + t \begin{pmatrix} 0 & 1 & 3 \\ 0 & 0 & -1 \\ 0 & 0 & 0 \end{pmatrix} \right] \begin{pmatrix} 0 \\ 1 \\ 0 \end{pmatrix}$$

$$= e^{2t} \left[\begin{pmatrix} 0 \\ 1 \\ 0 \end{pmatrix} + t \begin{pmatrix} 1 \\ 0 \\ 0 \end{pmatrix} \right] .$$

$x_2(t) = e^{2t} \begin{pmatrix} t \\ 1 \\ 0 \end{pmatrix}$ is a second solution of $X' = AX$. Since the equation

$(A - 2I)^2 V = 0$ has only two linearly independent solutions, we look for all solutions of the equation

$$(A - 2I)^3 V = \begin{pmatrix} 0 & 1 & 3 \\ 0 & 0 & -1 \\ 0 & 0 & 0 \end{pmatrix}^3 \begin{pmatrix} v_1 \\ v_2 \\ v_3 \end{pmatrix} = \begin{pmatrix} 0 \\ 0 \\ 0 \end{pmatrix} .$$

As we know from the Cayley-Hamilton theorem, every matrix is a zero of its characteristic polynomial. Thus,

$$f(A) = (A - 2I)^3 = 0 .$$

Therefore,

$$(A - 2I)^3 V = \begin{pmatrix} 0 & 0 & 0 \\ 0 & 0 & 0 \\ 0 & 0 & 0 \end{pmatrix} \begin{pmatrix} v_1 \\ v_2 \\ v_3 \end{pmatrix} = \begin{pmatrix} 0 \\ 0 \\ 0 \end{pmatrix}$$

This implies that every vector v is a solution of this equation. The vector

$$v = \begin{pmatrix} 0 \\ 0 \\ 1 \end{pmatrix} \quad \text{satisfies} \quad (A - 2I)^3 v = 0 , \text{ but}$$

$(A - 2I)^2 \neq 0$. Hence,

$$x_3(t) = e^{At}\begin{pmatrix} 0 \\ 0 \\ 1 \end{pmatrix} = e^{2t}e^{(A - 2I)t}\begin{pmatrix} 0 \\ 0 \\ 1 \end{pmatrix}$$

$$= e^{2t}[I + t(A - 2I) + \frac{t^2}{2}(A - 2I)^2]\begin{pmatrix} 0 \\ 0 \\ 1 \end{pmatrix}$$

$$= e^{2t}\left[\begin{pmatrix} 0 \\ 0 \\ 1 \end{pmatrix} + t\begin{pmatrix} 3 \\ -1 \\ 0 \end{pmatrix} + \frac{t^2}{2}\begin{pmatrix} -1 \\ 0 \\ 1 \end{pmatrix}\right]$$

$$= e^{2t}\begin{pmatrix} 3t - t^2/2 \\ -t \\ 1 \end{pmatrix}$$

is a third linearly independent solution. Therefore, the general solution of equation (2) is:

$$X(t) = c_1 e^{2t}\begin{pmatrix} 1 \\ 0 \\ 1 \end{pmatrix} + c_2 e^{2t}\begin{pmatrix} t \\ 1 \\ 0 \end{pmatrix} + c_3 e^{2t}\begin{pmatrix} 3t - t^2/2 \\ -t \\ 1 \end{pmatrix}$$

$$= e^{2t}\left[c_1\begin{pmatrix} 1 \\ 0 \\ 0 \end{pmatrix} + c_2\begin{pmatrix} t \\ 1 \\ 0 \end{pmatrix} + c_3\begin{pmatrix} 3t - t^2/2 \\ -t \\ 1 \end{pmatrix}\right]. \qquad (3)$$

The constants c_1, c_2 and c_3 are determined from the initial conditions. In matrix notation, the given initial conditions can be written as:

$$X(0) = \begin{pmatrix} x_1(0) \\ x_2(0) \\ x_3(0) \end{pmatrix} = \begin{pmatrix} 1 \\ 2 \\ 1 \end{pmatrix}.$$

Upon substitution of the initial conditions into equation (3) we obtain:

$$\begin{pmatrix} 1 \\ 2 \\ 1 \end{pmatrix} = c_1\begin{pmatrix} 1 \\ 0 \\ 0 \end{pmatrix} + c_2\begin{pmatrix} 1 \\ 1 \\ 0 \end{pmatrix} + c_3\begin{pmatrix} 0 \\ 0 \\ 1 \end{pmatrix}$$

$$\begin{bmatrix} 1 \\ 2 \\ 1 \end{bmatrix} = \begin{pmatrix} c_1 \\ c_2 \\ c_3 \end{pmatrix}.$$

Therefore $c_1 = 1$, $c_2 = 2$ and $c_3 = 1$. Substitute, these values of c_1, c_2 and c_3 into equation (3). We have

$$X(t) = e^{2t}\left[1\begin{pmatrix} 1 \\ 0 \\ 0 \end{pmatrix} + 2\begin{pmatrix} t \\ 1 \\ 0 \end{pmatrix} + 1\begin{pmatrix} 3t - t^2/2 \\ -t \\ 1 \end{pmatrix}\right]$$

$$= e^{2t}\begin{bmatrix} 1 + 5t - t^2/2 \\ 2-t \\ 1 \end{bmatrix}$$

Therefore, the solution to the initial value problem (1) is:

$$x_1 = e^{2t}[1 + 5t - t^2/2]$$

$$x_2 = e^{2t}[2 - t]$$

$$x_3 = e^{2t}.$$

CONVERTING A SYSTEM TO DIAGONAL FORM

● **PROBLEM** 18-12

(a) Find the general solution of the following system:

$$y_1' = 3y_1$$
$$y_2' = -2y_2 \qquad\qquad (1)$$
$$y_3' = 5y_3$$

(b) Find a unique solution of the system (1) which satisfies the initial conditions $y_1(0) = 1$, $y_2(0) = 4$, and $y_3(0) = -2$.

Solution: In matrix notation system (1) can be written as:

$$\begin{pmatrix} y_1' \\ y_2' \\ y_3' \end{pmatrix} = \begin{pmatrix} 3 & 0 & 0 \\ 0 & -2 & 0 \\ 0 & 0 & 5 \end{pmatrix} \begin{pmatrix} y_1 \\ y_2 \\ y_3 \end{pmatrix}$$

or

$$Y' = AY \qquad\qquad (2)$$

where

$$A = \begin{pmatrix} 3 & 0 & 0 \\ 0 & -2 & 0 \\ 0 & 0 & 5 \end{pmatrix} .$$

We seek a solution of equation (2) in the form $Y = e^{\lambda t}V$ where v is the eigenvector of λ. Since the matrix A is 3×3, (n = 3), there are three linearly independent solutions of equation (2). Therefore, we need three linearly independent eigenvectors. To find the eigenvectors we first determine the eigenvalues of A.

Since the matrix A is a diagonal matrix, the elements on the main diagonal are the eigenvalues. Thus this system is uncoupled, i.e., the rates of change of x_i depend only upon x_i. Therefore the eigenvalues of A are $\lambda_1 = 3$, $\lambda_2 = -2$ and $\lambda = 5$. Now, the eigenvector associated with $\lambda_1 = 3$ is obtained by solving the equation $(\lambda I - A)V = 0$ for $\lambda = 3$. Hence,

$$(3I - A)V = \begin{pmatrix} 0 & 0 & 0 \\ 0 & 5 & 0 \\ 0 & 0 & -2 \end{pmatrix} \begin{pmatrix} v_1 \\ v_2 \\ v_3 \end{pmatrix} = \begin{pmatrix} 0 \\ 0 \\ 0 \end{pmatrix}$$

or

$$5v_2 = 0 .$$
$$-2v_3 = 0 .$$

This implies that $v_2 = v_3 = 0$ and v_1 is arbitrary. Therefore each vector

$$v = a\begin{pmatrix} 1 \\ 0 \\ 0 \end{pmatrix}$$ is an eigenvector corresponding to $\lambda_1 = 3$,

and $Y_1(t) = e^{\lambda_1 t}v_1 = e^{3t}\begin{pmatrix} 1 \\ 0 \\ 0 \end{pmatrix}$ is a solution of the differential equation

(2). Similarly, the eigenvector corresponding to $\lambda = -2$ is obtained by solving the equation $(\lambda I - A)V = 0$, for $\lambda = -2$. Hence,

$$[-2I - A]V = \begin{pmatrix} -5 & 0 & 0 \\ 0 & 0 & 0 \\ 0 & 0 & -7 \end{pmatrix} \begin{pmatrix} v_1 \\ v_2 \\ v_3 \end{pmatrix} = \begin{pmatrix} 0 \\ 0 \\ 0 \end{pmatrix}$$

or

$$-5v_1 = 0$$
$$-7v_3 = 0 .$$

This implies that $v_1 = v_3 = 0$ and v_2 is arbitrary. Hence, every vector $v = a\begin{pmatrix} 0 \\ 1 \\ 0 \end{pmatrix}$ is an eigenvector associated with $\lambda = -2$, and $Y_2(t) = e^{-2t}\begin{pmatrix} 0 \\ 1 \\ 0 \end{pmatrix}$

is a second solution of equation (2). Now, we solve the equation $(\lambda I - A)V = 0$ for $\lambda = 5$, to obtain the eigenvector associated with 5:

$$[5I - A]V = \begin{pmatrix} 2 & 0 & 0 \\ 0 & 7 & 0 \\ 0 & 0 & 0 \end{pmatrix} \begin{pmatrix} v_1 \\ v_2 \\ v_3 \end{pmatrix} = \begin{pmatrix} 0 \\ 0 \\ 0 \end{pmatrix}$$

This implies that $v_1 = v_2 = 0$ and v_3 is arbitrary. Hence $v = a\begin{pmatrix} 0 \\ 0 \\ 1 \end{pmatrix}$ is an eigenvector associated with $\lambda = 5$, and $Y_3(t) = e^{5t}\begin{pmatrix} 0 \\ 0 \\ 1 \end{pmatrix}$ is a third

solution of equation (2).

Since A has 3 distinct eigenvalues, the general solution of equation (2) is $Y = c_1Y_1 + c_2Y_2 + c_3Y_3$

$$= c_1 e^{3t}\begin{pmatrix} 1 \\ 0 \\ 0 \end{pmatrix} + c_2 e^{-2t}\begin{pmatrix} 0 \\ 1 \\ 0 \end{pmatrix} + c_3 e^{5t}\begin{pmatrix} 0 \\ 0 \\ 1 \end{pmatrix},$$

$$Y = \begin{pmatrix} y_1 \\ y_2 \\ y_3 \end{pmatrix} = \begin{pmatrix} c_1 e^{3t} \\ c_2 e^{-2t} \\ c_3 e^{5t} \end{pmatrix}$$

or

$$y_1 = c_1 e^{3t}$$
$$y_2 = c_2 e^{-2t} \qquad\qquad (3)$$
$$y_3 = c_3 e^{5t} .$$

In general, if the constant matrix A is diagonal, then the solution of the differential equation Y' = AY, is immediately obtainable. In this case, the system in matrix notation is:

$$\begin{pmatrix} y_1' \\ y_2' \\ \vdots \\ \vdots \\ y_n' \end{pmatrix} = \begin{pmatrix} a_{11} & 0 & \cdots & \cdots & 0 \\ 0 & \ddots & & & \vdots \\ \vdots & & a_{22} & & \vdots \\ \vdots & & & \ddots & \vdots \\ 0 & \cdots & \cdots & \cdots & a_{nn} \end{pmatrix} \begin{pmatrix} y_1 \\ y_2 \\ \vdots \\ \vdots \\ y_n \end{pmatrix}$$

and the eigenvalues of A are $\lambda_1 = a_{11}$, $\lambda_2 = a_{22}, \ldots, \lambda_n = a_{nn}$. We immediately obtain n linearly independent solution vectors:

$$Y_1 = \begin{pmatrix} e^{\lambda_1 t} \\ 0 \\ \cdot \\ \cdot \\ \cdot \\ 0 \end{pmatrix}, \quad Y_2 = \begin{pmatrix} 0 \\ e^{\lambda_2 t} \\ \cdot \\ \cdot \\ \cdot \\ 0 \end{pmatrix} \cdots, \quad Y_n = \begin{pmatrix} 0 \\ 0 \\ \cdot \\ \cdot \\ \cdot \\ e^{\lambda_n t} \end{pmatrix}$$

Hence the general solution is $Y = \sum_{i=1}^{n} c_i Y_i$ or

$$Y = \begin{pmatrix} y_1 \\ y_2 \\ \cdot \\ \cdot \\ \cdot \\ y_n \end{pmatrix} = \begin{pmatrix} c_1 e^{\lambda_1 t} \\ c_2 e^{\lambda_2 t} \\ \cdot \\ \cdot \\ \cdot \\ c_n e^{\lambda_n t} \end{pmatrix}$$

(b) To obtain the unique solution we substitute the given initial conditions into equation (3) and obtain:

$$y_1(0) = 1 = c_1 e^{3(0)} = c_1$$
$$y_2(0) = 4 = c_2 e^{-2(0)} = c_2$$
$$y_3(0) = -2 = c_3 e^{5(0)} = c_3 \; .$$

Therefore, $c_1 = 1$, $c_2 = 4$ and $c_3 = -2$. Hence, from (3) the solution satisfying the initial condition is

$$y_1 = e^{3t}$$
$$y_2 = 4e^{-2t}$$
$$y_3 = -2e^{5t} \; .$$

● **PROBLEM** 18-13

(a) Write the system
$$x_1' = 3x_1$$
$$x_2' = -2x_2 \qquad\qquad (1)$$
$$x_3' = 4x_3$$
in matrix notation.
(b) Solve the system.
(c) Find a solution of the system which satisfies the initial conditions $x_1(0) = 3$, $x_2(0) = 5$, and $x_3(0) = 7$.

Solution: (a) In matrix notation, the system (1) becomes

$$\begin{pmatrix} x_1' \\ x_2' \\ x_3' \end{pmatrix} = \begin{pmatrix} 3 & 0 & 0 \\ 0 & -2 & 0 \\ 0 & 0 & 4 \end{pmatrix} \begin{pmatrix} x_1 \\ x_2 \\ x_3 \end{pmatrix}$$

or
$$X' = AX \qquad\qquad (2)$$

where

$$X' = \begin{pmatrix} x_1' \\ x_2' \\ x_3' \end{pmatrix}, \qquad A = \begin{pmatrix} 3 & 0 & 0 \\ 0 & -2 & 0 \\ 0 & 0 & 4 \end{pmatrix}$$

$$X = \begin{pmatrix} x_1 \\ x_2 \\ x_3 \end{pmatrix}.$$

(b) Since the matrix A is a diagonal matrix, the elements on the main diagonal are the eigenvalues of A. Thus, the eigenvalues are $\lambda_1 = 3$, $\lambda_2 = -2$ and $\lambda_3 = 4$. Then the general solution of the system (1) is:

$$X = \begin{pmatrix} x_1 \\ x_2 \\ x_3 \end{pmatrix} = \begin{pmatrix} c_1 e^{\lambda_1 t} \\ c_2 e^{\lambda_2 t} \\ c_3 e^{\lambda_3 t} \end{pmatrix} = \begin{pmatrix} c_1 e^{3t} \\ c_2 e^{-2t} \\ c_3 e^{4t} \end{pmatrix}$$

or

$$\begin{aligned} x_1 &= c_1 e^{3t} \\ x_2 &= c_2 e^{-2t} \\ x_3 &= c_3 e^{4t} \end{aligned} \qquad (3)$$

(c) If we substitute the given initial conditions into (3) we obtain:

$$\begin{aligned} x_1(0) &= 3 = c_1 e^0 = c_1 \\ x_2(0) &= 5 = c_2 e^0 = c_2 \\ x_3(0) &= 7 = c_3 e^0 = c_3 . \end{aligned}$$

Hence, $c_1 = 3$, $c_2 = 5$ and $c_3 = 7$. Substitute these values into system (3). The solution satisfying the initial conditions is:

$$\begin{aligned} x_1 &= 3e^{3t} \\ x_2 &= 5e^{-2t} \\ x_3 &= 7e^{4t} . \end{aligned}$$

● PROBLEM 18-14

Find the general solution of the following system of linear differential equations:

$$\begin{aligned} x_1' &= 4x_1 + x_2 \\ x_2' &= 3x_1 + 2x_2 . \end{aligned} \qquad (1)$$

Solution: We write system (1) in matrix notation as

$$\begin{pmatrix} x_1' \\ x_2' \end{pmatrix} = \begin{pmatrix} 4 & 1 \\ 3 & 2 \end{pmatrix} \begin{pmatrix} x_1 \\ x_2 \end{pmatrix}$$

or

$$X' = AX \tag{2}$$

where

$$A = \begin{pmatrix} 4 & 1 \\ 3 & 2 \end{pmatrix} .$$

Here, the matrix A is not a diagonal matrix. It is often possible to transform the matrix A into another matrix D which is diagonal. That is, we shall try to make a substitution for x that will yield a new system with a diagonal coefficient matrix. Next, we solve this new simpler system and use this solution to determine the solution of the original system. Thus we uncouple the given coupled system, making it easier to solve. Let

$$X = PY \tag{3}$$

where P is an $n \times n$ constant matrix. Then, $X' = PY'$. If we make the substitutions $X = PY$ and $X' = PY'$ in the original system $X' = AX$ we get

$$PY' = APY.$$

If P is nonsingular, then it is invertible. Therefore,

$$P^{-1}PY' = P^{-1}APY$$

or

$$Y' = P^{-1}APY$$

or

$$Y' = DY \tag{4}$$

where

$$D = P^{-1}AP .$$

Thus, if we can choose the matrix P such that D is diagonal, we can easily solve system (4), and finally obtain the solution X of the original system $X' = AX$, where $X = PY$.

Now, turning to equation (2), we must find the matrix P that diagonalizes A. Recall that the matrix A will be diagonalized by any matrix P whose columns are linearly independent eigenvectors of A.

Now

$$A = \begin{pmatrix} 4 & 1 \\ 3 & 2 \end{pmatrix} .$$

Then the characteristic polynomial of A is $f(\lambda) = \det(\lambda I - A)$. Substitute the values of A into the characteristic polynomial:

$$f(\lambda) = \det \begin{pmatrix} \lambda-4 & -1 \\ -3 & \lambda-2 \end{pmatrix}$$

$$= (\lambda-4)(\lambda-2) - 3$$

$$= (\lambda-1)(\lambda-5) .$$

Therefore the eigenvalues are $\lambda_1 = 1$ and $\lambda_2 = 5$. The eigenvector associated with $\lambda_1 = 1$ is obtained by solving equation $(\lambda I - A)V = 0$ for $\lambda_1 = 1$, as follows:

$$(1I - A)V = \begin{pmatrix} -3 & -1 \\ -3 & -1 \end{pmatrix} \begin{pmatrix} v_1 \\ v_2 \end{pmatrix} = \begin{pmatrix} 0 \\ 0 \end{pmatrix}$$

or $-3v_1 - v_2 = 0$. Then the eigenvector corresponding to $\lambda_1 = 1$ is $\begin{pmatrix} 1 \\ -3 \end{pmatrix}$. We can find the eigenvector associated with $\lambda = 5$ using the same method. We obtain $\begin{bmatrix} 1 \\ 1 \end{bmatrix}$ as an eigenvector with $\lambda = 5$.

The matrix P is then

$$P = \begin{pmatrix} 1 & 1 \\ -3 & 1 \end{pmatrix} ,$$

$$P^{-1} = \frac{1}{\det P} \text{ adj } P = \frac{1}{4}\begin{pmatrix} 1 & -1 \\ 3 & 1 \end{pmatrix}.$$

Then

$$D = P^{-1}AP = \frac{1}{4}\begin{bmatrix} 1 & -1 \\ 3 & 1 \end{bmatrix}\begin{bmatrix} 4 & 1 \\ 3 & 2 \end{bmatrix}\begin{bmatrix} 1 & 1 \\ -3 & 1 \end{bmatrix} = \begin{bmatrix} 1 & 0 \\ 0 & 5 \end{bmatrix}.$$

Therefore the substitution $X = PY$ and $X' = PY'$ yields the new diagonal system

$$Y' = DY = \begin{bmatrix} 1 & 0 \\ 0 & 5 \end{bmatrix} Y. \tag{5}$$

Then the general solution of the system (5) is

$$Y = \begin{pmatrix} c_1 e^t \\ c_2 e^{5t} \end{pmatrix}.$$

$X = PY$ yields, as the solution for equation (2),

$$X = \begin{pmatrix} x_1 \\ x_2 \end{pmatrix} = \begin{pmatrix} 1 & 1 \\ -3 & 1 \end{pmatrix}\begin{pmatrix} c_1 e^t \\ c_2 e^{5t} \end{pmatrix} = \begin{pmatrix} c_1 e^t + c_2 e^{5t} \\ -3c_1 e^t + c_2 e^{5t} \end{pmatrix}.$$

Thus the general solution of system (1) is

$$x_1 = c_1 e^t + c_2 e^{5t}$$

$$x_2 = -3c_1 e^t + c_2 e^{5t}.$$

● **PROBLEM** 18-15

Solve the following initial value problem:

$$y_1' = y_1 + y_2$$

$$y_2' = 4y_1 - 2y_2. \tag{1}$$

Initial conditions: $y_1(0) = 1$, $y_2(0) = 6$.

Solution: In matrix notation, system (1) can be written

$$\begin{pmatrix} y_1' \\ y_2' \end{pmatrix} = \begin{pmatrix} 1 & 1 \\ 4 & -2 \end{pmatrix}\begin{pmatrix} y_1 \\ y_2 \end{pmatrix}$$

or

$$Y' = AY \tag{2}$$

where

$$A = \begin{pmatrix} 1 & 1 \\ 4 & -2 \end{pmatrix}, \quad Y' = \begin{pmatrix} y_1' \\ y_2' \end{pmatrix}, \quad Y = \begin{pmatrix} y_1 \\ y_2 \end{pmatrix}.$$

The matrix A can be diagonalized by any matrix P whose columns are linearly independent eigenvectors of A.

The characteristic polynomial of A is

$$f(\lambda) = \det(\lambda I - A) = \det\begin{pmatrix} \lambda-1 & -1 \\ -4 & \lambda+2 \end{pmatrix}$$

$$= (\lambda-1)(\lambda+2) - 4$$

$$= (\lambda-2)(\lambda+3).$$

Hence the eigenvalues are $\lambda_1 = 2$ and $\lambda_2 = -3$. $V = \begin{pmatrix} v_1 \\ v_2 \end{pmatrix}$ is an eigenvector of A corresponding to λ if and only if V is a nontrivial solution of $(\lambda I - A)V = 0$, that is, of

$$\begin{pmatrix} \lambda - 1 & -1 \\ -4 & \lambda + 2 \end{pmatrix} \begin{pmatrix} v_1 \\ v_2 \end{pmatrix} = \begin{pmatrix} 0 \\ 0 \end{pmatrix}. \tag{3}$$

If $\lambda = 2$, system (3) becomes

$$\begin{pmatrix} 1 & -1 \\ -4 & 4 \end{pmatrix} \begin{pmatrix} v_1 \\ v_2 \end{pmatrix} = \begin{pmatrix} 0 \\ 0 \end{pmatrix}$$

or

$$v_1 - v_2 = 0$$
$$-4v_1 + 4v_2 = 0$$

or

$$v_1 = v_2 .$$

Therefore

$$\begin{pmatrix} v_1 \\ v_2 \end{pmatrix} = a \begin{pmatrix} 1 \\ 1 \end{pmatrix} \text{ is an eigenvector associated with } \lambda = 2.$$

For $\lambda = -3$, the system (3) becomes

$$\begin{pmatrix} -4 & -1 \\ -4 & -1 \end{pmatrix} \begin{pmatrix} v_1 \\ v_2 \end{pmatrix} = \begin{pmatrix} 0 \\ 0 \end{pmatrix}$$

or

$$-4v_1 - v_2 = 0$$
$$-4v_1 - v_2 = 0 .$$

Therefore $v_2 = -4v_1$. Hence,

$$\begin{pmatrix} v_1 \\ v_2 \end{pmatrix} = a \begin{pmatrix} 1 \\ -4 \end{pmatrix} \text{ is an eigenvector associated with } \lambda = -3.$$

Thus $\begin{pmatrix} 1 \\ 1 \end{pmatrix}$ and $\begin{pmatrix} 1 \\ -4 \end{pmatrix}$ are two linearly independent eigenvectors of A.
Then, the matrix P is

$$P = \begin{pmatrix} 1 & 1 \\ 1 & -4 \end{pmatrix} , \text{ and it diagonalizes } A.$$

Now,

$$P^{-1} = \frac{1}{\det P} \text{adj } P = \frac{1}{-5} \begin{pmatrix} -4 & -1 \\ -1 & 1 \end{pmatrix}$$

$$D = P^{-1}AP = -\frac{1}{5} \begin{pmatrix} -4 & -1 \\ -1 & 1 \end{pmatrix} \begin{pmatrix} 1 & 1 \\ 4 & -2 \end{pmatrix} \begin{pmatrix} 1 & 1 \\ 1 & -4 \end{pmatrix}$$

$$= -\frac{1}{5} \begin{pmatrix} -4 & -1 \\ -1 & 1 \end{pmatrix} \begin{pmatrix} 2 & -3 \\ 2 & 12 \end{pmatrix}$$

$$= \begin{pmatrix} 2 & 0 \\ 0 & -3 \end{pmatrix} .$$

Therefore the substitution

$$Y = PU \text{ and } Y' = PU' \text{ in equation (2), yields the new}$$

diagonal system:

$$PU' = APU$$

$$U' = P^{-1}APU = DU \text{ where } D = P^{-1}AP$$

$$= \begin{pmatrix} 2 & 0 \\ 0 & -3 \end{pmatrix} \begin{pmatrix} u_1 \\ u_2 \end{pmatrix}. \tag{4}$$

Therefore the solution of system (4) is:

$$u_1 = c_1 e^{2t}$$

$$u_2 = c_2 e^{-3t}$$

or

$$U = \begin{pmatrix} u_1 \\ u_2 \end{pmatrix} = \begin{pmatrix} c_1 e^{2t} \\ c_2 e^{-3t} \end{pmatrix}.$$

Therefore equation $Y = PU$ yields:

$$Y = \begin{pmatrix} y_1 \\ y_2 \end{pmatrix} = \begin{pmatrix} 1 & 1 \\ 1 & -4 \end{pmatrix} \begin{pmatrix} c_1 e^{2t} \\ c_2 e^{-3t} \end{pmatrix}.$$

Thus

$$y_1 = c_1 e^{2t} + c_2 e^{-3t}$$

$$y_2 = c_1 e^{2t} - 4c_2 e^{-3t}. \tag{5}$$

Substituting the given initial conditions into (5) we obtain:

$$y_1(0) = 1 = c_1 e^0 + c_2 e^0$$

$$y_2(0) = 6 = c_1 e^0 - 4c_2 e^0.$$

Hence,

$$c_1 + c_2 = 1$$

$$c_1 - 4c_2 = 6.$$

Solving this system we obtain $c_1 = 2$ and $c_2 = -1$. Substituting these values into (5) yields the solution to system (1) which satisfies the given initial conditions. Thus

$$y_1 = 2e^{2t} - e^{-3t}$$

$$y_2 = 2e^{2t} + 4e^{-3t}$$

is the unique solution to the system (1).

● **PROBLEM 18-16**

Solve the following initial value problem:

$$x' = 4x + y$$

$$y' = 3x + 2y. \tag{1}$$

Initial conditions: $x(0) = -1$

$$y(0) = 7.$$

Solution: We first write system (1) in matrix form:

566

$$\begin{pmatrix} x' \\ y' \end{pmatrix} = \begin{pmatrix} 4 & 1 \\ 3 & 2 \end{pmatrix} \begin{pmatrix} x \\ y \end{pmatrix} . \qquad (2)$$

Let $X = \begin{pmatrix} x \\ y \end{pmatrix}$ and $A = \begin{pmatrix} 4 & 1 \\ 3 & 2 \end{pmatrix}$. Then matrix equation (2) can be written as

$$\frac{dx}{dt} = AX . \qquad (3)$$

We write the initial conditions in matrix form as:

$$\begin{pmatrix} x(0) \\ y(0) \end{pmatrix} = \begin{pmatrix} -1 \\ 7 \end{pmatrix} .$$

Let

$$X(0) = \begin{pmatrix} x(0) \\ y(0) \end{pmatrix} \text{ and } X_0 = \begin{pmatrix} -1 \\ 7 \end{pmatrix} .$$

Then

$$X(0) = X_0 . \qquad (4)$$

Now we state the following fundamental theorem. Let A be a $n \times n$ constant matrix. Then the general solution of the differential equation

$$\frac{dX(t)}{dt} = AX(t)$$

is given by $X(t) = e^{tA}c$ where c is an arbitrary constant vector. The unique solution of the differential equation which also satisfies the initial condition $X(t_0) = X_0$ is given by $X(t) = e^{tA}X_0$.

Now to compute e^{tA} we first find the eigenvalues of A. If the eigenvalues of A are distinct, then the matrix A is similar to a diagonal matrix. Then, the diagonal matrix D is given by $D = P^{-1}AP$ where P is the matrix with the eigenvectors $v_1, v_2, \ldots, v_n$ as its columns. D is a diagonal matrix whose diagonal entries are eigenvalues $\lambda_1, \lambda_2, \ldots, \lambda_n$. Solving for A, we have $A = PDP^{-1}$. Then $e^A = e^{PDP^{-1}}$.

As we know, the matrix function e^A is defined as

$$e^A = \lim_{m \to \infty} \sum_{k=0}^{m} \frac{1}{k!} A^k .$$

Thus

$$e^A = e^{PDP^{-1}} = \lim_{n \to \infty} \sum_{k=0}^{n} \frac{1}{k!} (PDP^{-1})^k$$

$$e^A = \lim_{n \to \infty} \sum_{k=0}^{n} \frac{1}{k!} PD^k P^{-1} = P \left(\lim_{n \to \infty} \sum_{k=0}^{n} \frac{1}{k!} D^k \right) P^{-1}$$

$$= Pe^D P^{-1} .$$

$$D = \begin{pmatrix} \lambda_1 & 0 \cdots 0 \\ 0 & \lambda_2 \cdots 0 \\ \vdots & \vdots \ddots \vdots \\ 0 & 0 \cdots \lambda_n \end{pmatrix} .$$

e^D is easily evaluated as follows:

$$e^D = \begin{pmatrix} e^{\lambda_1} & 0 & \cdots & 0 \\ 0 & e^{\lambda_2} & \cdots & 0 \\ \vdots & & & \vdots \\ 0 & 0 & \cdots & e^{\lambda_n} \end{pmatrix}.$$

Then,

$$e^{tA} = e^{tPDP^{-1}}.$$

Now to obtain the solution of equation (3), we find e^{tA} where $A = \begin{pmatrix} 4 & 1 \\ 3 & 2 \end{pmatrix}$.

The characteristic polynomial of A is

$$f(\lambda) = \det(\lambda I - A) = \begin{pmatrix} \lambda-4 & -1 \\ -3 & \lambda-2 \end{pmatrix}$$

$$= (\lambda-4)(\lambda-2) - 3$$

$$= (\lambda-1)(\lambda-5).$$

The eigenvalues of A are $\lambda_1 = 1$, and $\lambda_2 = 5$ with corresponding eigenvectors $v_1 = \begin{bmatrix} 1 \\ -3 \end{bmatrix}$, and $v_2 = \begin{bmatrix} 1 \\ 1 \end{bmatrix}$. Since the matrix A has distinct eigenvalues, it is similar to a diagonal matrix D. Then the matrix P is given by

$$P = \begin{pmatrix} 1 & 1 \\ -3 & 1 \end{pmatrix}$$

whose columns are the eigenvectors v_1 and v_2, and

$$D = P^{-1}AP = \begin{pmatrix} 1 & 0 \\ 0 & 5 \end{pmatrix}$$

Then,

$$e^{tA} = e^{tPDP^{-1}} = Pe^{tD}P^{-1}$$

$$e^{tD} = \begin{pmatrix} e^t & 0 \\ 0 & e^{5t} \end{pmatrix}, \quad P^{-1} = \frac{1}{\det P} \text{adj } P = \frac{1}{4}\begin{pmatrix} 1 & -1 \\ 3 & 1 \end{pmatrix}.$$

Therefore,

$$e^{tA} = \begin{pmatrix} 1 & 1 \\ -3 & 1 \end{pmatrix}\begin{pmatrix} e^t & 0 \\ 0 & e^{5t} \end{pmatrix}\begin{pmatrix} \frac{1}{4} & -\frac{1}{4} \\ \frac{3}{4} & \frac{1}{4} \end{pmatrix}$$

$$e^{tA} = \begin{pmatrix} 1 & 1 \\ -3 & 1 \end{pmatrix}\begin{pmatrix} \frac{1}{4}e^t & -\frac{1}{4}e^t \\ \frac{3}{4}e^{5t} & \frac{1}{4}e^{5t} \end{pmatrix}$$

$$= \begin{pmatrix} \frac{1}{4}e^t + \frac{3}{4}e^{5t} & -\frac{1}{4}e^t + \frac{1}{4}e^{5t} \\ -\frac{3}{4}e^t + \frac{3}{4}e^{5t} & \frac{3}{4}e^t + \frac{1}{4}e^{5t} \end{pmatrix}$$

$$= \frac{1}{4}\begin{pmatrix} e^t + 3e^{5t} & -e^t + e^{5t} \\ -3e^t + 3e^{5t} & 3e^t + e^{5t} \end{pmatrix}.$$

Therefore the unique solution of the given initial value problem is:

$$X = e^{tA}X_0, \quad \text{where} \quad X_0 = \begin{pmatrix} -1 \\ 7 \end{pmatrix}.$$

Then,

$$X = \frac{1}{4}\begin{bmatrix} e^t + 3e^{5t} & -e^t + e^{5t} \\ -3e^t + 3e^{5t} & 3e^t + e^{5t} \end{bmatrix}\begin{bmatrix} -1 \\ 7 \end{bmatrix}$$

$$X = \frac{1}{4}\begin{bmatrix} -e^t - 3e^{5t} - 7e^t + 7e^{5t} \\ 3e^t - 3e^{5t} + 21e^t + 7e^{5t} \end{bmatrix}$$

$$= \frac{1}{4}\begin{bmatrix} -8e^t + 4e^{5t} \\ 24e^t + 4e^{5t} \end{bmatrix}$$

$$= \begin{bmatrix} -2e^t + e^{5t} \\ 6e^t + e^{5t} \end{bmatrix} .$$

But $X = \begin{pmatrix} x \\ y \end{pmatrix}$. Therefore, the solution of the system (1) is:

$$x = -2e^t + e^{5t}$$
$$y = 6e^t + e^{5t} .$$

THE FUNDAMENTAL MATRIX

• PROBLEM 18-17

(a) Find a fundamental matrix solution of the system

$$Y' = AY, \text{ if } A = \begin{bmatrix} 3 & 5 \\ -5 & 3 \end{bmatrix} .$$

(b) Find e^{tA} .

Solution: If A is a matrix of constant coefficients of a linear system and $v_1, v_2, \ldots, v_n$ are n linearly independent eigenvectors corresponding respectively to the eigenvalues $\lambda_1, \lambda_2, \ldots, \lambda_n$, then

$$Y(t) = [\exp(\lambda_1 t)V_1, \exp(\lambda_2 t)V_2, \ldots, \exp(\lambda_n t)V_n]$$

is a fundamental matrix solution of the linear system with constant co-efficients $Y' = AY$. In particular this is the case if the eigenvalues $\lambda_1, \lambda_2, \ldots, \lambda_n$ are distinct.

We now find the eigenvalues of the given matrix $A = \begin{bmatrix} 3 & 5 \\ -5 & 3 \end{bmatrix}$.

The characteristic polynomial of A is

$$f(\lambda) = \det[\lambda I - A] = \det\begin{bmatrix} \lambda-3 & -5 \\ 5 & \lambda-3 \end{bmatrix}$$

$$= (\lambda-3)^2 + 25$$

$$= \lambda^2 - 6\lambda + 34 .$$

Then, the characteristic equation is $\lambda^2 - 6\lambda + 34 = 0$.
The roots of this equation are

$$\frac{+6 + \sqrt{36 - (4)(34)}}{2} \quad \text{and} \quad \frac{+6 - \sqrt{36 - (4)(34)}}{2} .$$

Therefore the eigenvalues of A are $\lambda_1 = 3 + 5i$ and $\lambda_2 = 3 - 5i$. The eigenvector $V = \begin{pmatrix} v_1 \\ v_2 \end{pmatrix}$ corresponding to the eigenvalue $\lambda_1 = 3 + 5i$ must satisfy the linear homogeneous algebraic system

$$(A - \lambda_1 I)V = \begin{pmatrix} -5i & 5 \\ -5 & -5i \end{pmatrix}\begin{pmatrix} v_1 \\ v_2 \end{pmatrix} = \begin{pmatrix} 0 \\ 0 \end{pmatrix}$$

then,

$$-5iv_1 + 5v_2 = 0$$

$$-5v_1 - 5iv_2 = 0$$

or

$$-iv_1 + v_2 = 0$$

$$-v_1 - iv_2 = 0 \ .$$

This implies that $v_2 = iv_1$ and $v_1 = -iv_2$. Therefore, $V_1 = a\begin{pmatrix} 1 \\ i \end{pmatrix}$ is an eigenvector for any constant a. Similarly, the eigenvector $V_2 = \begin{pmatrix} v_1 \\ v_2 \end{pmatrix}$ corresponding to the eigenvalue $\lambda_2 = 3 - 5i$ is found to be $V_2 = k\begin{pmatrix} i \\ 1 \end{pmatrix}$ for any constant k. Thus,

$$V_1 = \begin{pmatrix} 1 \\ i \end{pmatrix}, \text{ and } V_2 = \begin{pmatrix} i \\ 1 \end{pmatrix}$$

are linearly independent eigenvectors corresponding to λ_1, λ_2, respectively. Then, the fundamental matrix solution of the system $Y' = AY$ is

$$Y(t) = \begin{bmatrix} e^{\lambda_1 t} V_1 , & e^{\lambda_2 t} V_2 \end{bmatrix}$$

or

$$Y(t) = \begin{bmatrix} e^{(3+5i)t}\begin{pmatrix} 1 \\ i \end{pmatrix} , & e^{(3-5i)t}\begin{pmatrix} i \\ 1 \end{pmatrix} \end{bmatrix}$$

$$Y(t) = \begin{bmatrix} e^{(3+5i)t} & i\,e^{(3-5i)t} \\ i\,e^{(3+5i)t} & e^{(3-5i)t} \end{bmatrix} .$$

(b) Consider a linear system with constant coefficients

$$Y' = AY \tag{2}$$

where A is a constant matrix. According to a theorem in linear algebra, e^{At} is also the fundamental matrix solution of (2).

In part (a) we show that $Y(t)$ is also a fundamental matrix of $Y' = AY$. Since e^{tA} and $Y(t)$ are both fundamental matrices of $Y' = AY$, there exists a nonsingular matrix C such that

$$e^{tA} = Y(t)C \tag{3}$$

Setting $t = 0$ in (3), we obtain $C = Y^{-1}(0)$. Thus

$$e^{tA} = Y(t)Y^{-1}(0) \tag{4}$$

By using equation (4), we can find the e^{tA} where $A = \begin{bmatrix} 3 & 5 \\ -5 & 3 \end{bmatrix}$.

We have,

$$Y(t) = \begin{bmatrix} e^{(3+5i)t} & i\,e^{(3-5i)t} \\ i\,e^{(3+5i)t} & e^{(3-5i)t} \end{bmatrix}$$

Then,

$$Y^{-1}(0) = \begin{pmatrix} 1 & i \\ i & 1 \end{pmatrix}^{-1} = \tfrac{1}{2}\begin{pmatrix} 1 & -i \\ -i & 1 \end{pmatrix}$$

Thus,

$$e^{tA} = Y(t)\ Y^{-1}(0)$$

$$= \begin{pmatrix} e^{(3+5i)t} & i\ e^{(3-5i)t} \\ i\ e^{(3+5i)t} & e^{(3-5i)t} \end{pmatrix}\ \tfrac{1}{2}\begin{pmatrix} 1 & -i \\ -i & 1 \end{pmatrix}$$

$$= \tfrac{1}{2}\begin{bmatrix} e^{(3+5i)t} + e^{(3-5i)t} & -i\ e^{(3+5i)t} + i\ e^{(3-5i)t} \\ i\ e^{(3+5i)t} - i\ e^{(3-5i)t} & e^{(3+5i)t} + e^{(3-5i)t} \end{bmatrix}$$

$$= \tfrac{1}{2}\begin{bmatrix} e^{3t}(e^{i5t} + e^{-i5t}) & -i\ e^{3t}(e^{i5t} - e^{-i5t}) \\ i\ e^{3t}(e^{i5t} - e^{-i5t}) & e^{3t}(e^{i5t} + e^{-i5t}) \end{bmatrix}$$

We know that

$$\sin \theta = \frac{e^{i\theta} - e^{-i\theta}}{2i}$$

$$\cos \theta = \frac{e^{i\theta} + e^{-i\theta}}{2}\ .$$

Thus,

$$e^{i5t} + e^{-i5t} = 2 \cos 5t$$

$$e^{i5t} - e^{-i5t} = 2i \sin 5t\ .$$

Therefore,

$$e^{tA} = \tfrac{1}{2}\begin{bmatrix} e^{3t}\ 2 \cos 5t & -i\ e^{3t}\ 2i \sin 5t \\ i\ e^{3t}\ 2i \sin 5t & e^{3t}\ 2 \cos 5t \end{bmatrix}$$

$$e^{tA} = e^{3t}\begin{bmatrix} \cos 5t & \sin 5t \\ -\sin 5t & \cos 5t \end{bmatrix}\ .$$

● **PROBLEM** 18-18

Show that $\varphi(t) = \begin{pmatrix} e^t & te^t \\ 0 & e^t \end{pmatrix}$ is a fundamental matrix for the

system. $Y' = AY$, where

$$A = \begin{pmatrix} 1 & 1 \\ 0 & 1 \end{pmatrix},\quad Y = \begin{bmatrix} y_1 \\ y_2 \end{bmatrix}$$

Solution: Consider the system

$$Y' = AY\ . \tag{1}$$

Now a matrix of n rows whose columns are solutions of (1) is called a solution matrix. If we form an n × n matrix using n linearly independent solutions as columns, we have a solution matrix whose columns are linearly independent. This solution matrix is called a fundamental matrix.

A solution matrix $\varphi(t)$ of $Y' = AY$ is a fundamental matrix if and only if $\det \varphi(t) \neq 0$. Now, given that

$$\varphi(t) = \begin{pmatrix} e^t & te^t \\ 0 & e^t \end{pmatrix},$$

we first show that $\varphi(t)$ is a solution matrix. Let $\varphi_1(t)$ denote the first column of $\varphi(t)$. Then $\varphi_1(t) = \begin{pmatrix} e^t \\ 0 \end{pmatrix}$ and $\varphi_1'(t) = \begin{pmatrix} e^t \\ 0 \end{pmatrix}$. Also,

$$\begin{pmatrix} 1 & 1 \\ 0 & 1 \end{pmatrix} \begin{pmatrix} e^t \\ 0 \end{pmatrix} = \begin{pmatrix} e^t \\ 0 \end{pmatrix} = \varphi_1'(t) .$$

Therefore,

$$\varphi_1'(t) = \begin{pmatrix} 1 & 1 \\ 0 & 1 \end{pmatrix} \varphi_1(t) .$$

Similarly, if $\varphi_2(t)$ denotes the second column of $\varphi(t)$, we have

$$\varphi_2(t) = \begin{pmatrix} te^t \\ e^t \end{pmatrix} \text{ and } \varphi_2'(t) = \begin{pmatrix} (t+1) & e^t \\ e^t \end{pmatrix} = \begin{pmatrix} 1 & 1 \\ 0 & 1 \end{pmatrix} \begin{pmatrix} te^t \\ e^t \end{pmatrix} = \begin{pmatrix} 1 & 1 \\ 0 & 1 \end{pmatrix} \varphi_2(t) .$$

Therefore, $\varphi(t) = [\varphi_1(t), \varphi_2(t)]$ is a solution matrix.

Now if $\varphi(t)$ is a fundamental matrix then $\det \varphi(t) \neq 0$.

$$\det \varphi(t) = \det \begin{pmatrix} e^t & te^t \\ 0 & e^t \end{pmatrix} = e^t e^t - 0 = e^{2t} . \text{ Thus}$$

$\det \varphi(t) \neq 0$, and $\varphi(t)$ is a fundamental matrix for the given system.

● **PROBLEM** 18-19

Show that if a real homogeneous system of two first-order equations has a fundamental matrix

$$\begin{pmatrix} e^{it} & e^{-it} \\ ie^{it} & -ie^{-it} \end{pmatrix}$$

then

$$\begin{pmatrix} \cos t & \sin t \\ -\sin t & \cos t \end{pmatrix} \text{ is also a fundamental matrix.}$$

Solution: We know that if $\varphi(t)$ and $\psi(t)$ are two fundamental matrices of $Y' = AY$, then there exists a nonsingular constant matrix C such that $\psi(t) = \varphi(t)C$. Therefore, we look for a nonsingular matrix

$$C = \begin{pmatrix} c_{11} & c_{12} \\ c_{21} & c_{22} \end{pmatrix} \text{ such that}$$

$$\begin{pmatrix} e^{it} & e^{-it} \\ ie^{it} & -ie^{-it} \end{pmatrix} \begin{pmatrix} c_{11} & c_{12} \\ c_{21} & c_{22} \end{pmatrix} = \begin{pmatrix} \cos t & \sin t \\ -\sin t & \cos t \end{pmatrix} .$$

Upon matrix multiplication, we obtain

$$\begin{pmatrix} c_{11}e^{it} + c_{21}e^{-it} & c_{12}e^{it} + c_{22}e^{-it} \\ c_{11}ie^{it} - ic_{21}e^{-it} & ic_{12}e^{it} - ic_{22}e^{-it} \end{pmatrix} = \begin{pmatrix} \cos t & \sin t \\ -\sin t & \cos t \end{pmatrix}$$

Therefore,

$$c_{11}e^{it} + c_{21}e^{-it} = \cos t \tag{1}$$

$$c_{12}e^{it} + c_{22}e^{-it} = \sin t \tag{2}$$

$$i(c_{11}e^{it} - c_{21}e^{-it}) = -\sin t \tag{3}$$

$$i(c_{12}e^{it} - c_{22}e^{-it}) = \cos t . \tag{4}$$

Now we use the elementary identities

$$\cos t = \frac{e^{it} + e^{-it}}{2} , \quad \sin t = \frac{e^{it} - e^{-it}}{2i} .$$

Substitute these values in equations (1) to (4). Then,

$$c_{11}e^{it} + c_{21}e^{-it} = \frac{e^{it} + e^{-it}}{2}$$

or

$$e^{it}(2c_{11} - 1) + e^{-it}(2c_{21} - 1) = 0 .$$

This implies that

$$2c_{11} - 1 = 0 \qquad\qquad c_{11} = 1/2$$
$$\text{or}$$
$$2c_{21} - 1 = 0 \qquad\qquad c_{21} = 1/2 .$$

Also,

$$c_{12}e^{it} + c_{22}e^{-it} = \frac{e^{it} - e^{-it}}{2i}$$

or

$$e^{it}(2ic_{12} - 1) + e^{-it}(2ic_{22} + 1) = 0 .$$

This implies that

$$2ic_{12} - 1 = 0 \qquad\qquad c_{12} = 1/2i$$
$$\text{or}$$
$$2ic_{22} + 1 = 0 \qquad\qquad c_{22} = -1/2i .$$

Thus we have

$$c_{11} = 1/2 \qquad\qquad c_{12} = 1/2i$$

$$c_{21} = 1/2 \qquad\qquad c_{22} = -1/2i .$$

Then

$$C = 1/2 \begin{bmatrix} 1 & 1/i \\ 1 & -1/i \end{bmatrix} = 1/2 \begin{bmatrix} 1 & -i \\ 1 & i \end{bmatrix} .$$

Now $\det C = 1/2(i - (-i)) = 2i/2 = i \neq 0$. Therefore the matrix C is nonsingular and hence the matrix

$$\begin{bmatrix} \cos t & \sin t \\ -\sin t & \cos t \end{bmatrix}$$

is also a fundamental matrix.

● **PROBLEM** 18-20

(a) Find the general solution to the following system:

$$\dot{x} = y + t$$
$$\dot{y} = -2x - 3y - t^2 . \tag{1}$$

573

<u>Solution:</u> The system (1) consists of linear nonhomogeneous first-order equations. In general, consider the system of linear nonhomogeneous differential equations:

$$\dot{x}_1 = a_{11}x_1 + a_{12}x_2 + \ldots + a_{1n}x_n + f_1$$

$$\dot{x}_2 = a_{21}x_1 + a_{22}x_2 + \ldots + a_{2n}x_n + f_2 \qquad (2)$$

$$\cdots \cdots \cdots \cdots \cdots \cdots \cdots \cdots$$

$$\dot{x}_n = a_{n1}x_1 + a_{n2}x_2 + \ldots + a_{nn}x_n + f_n \ .$$

The coefficients a_{ij} and f_i , $i,j = 1,2,\ldots,n$, are assumed to be continuous functions of t.

In order to write the system (2) in matrix form, define an $n \times n$ matrix A and vectors $X(t)$ and $F(t)$ as follows:

$$A(t) = \begin{pmatrix} a_{11} & a_{12} & \cdots & a_{1n} \\ a_{21} & a_{22} & \cdots & a_{2n} \\ \cdots & \cdots & \cdots & \cdots \\ a_{n1} & a_{n2} & \cdots & a_{nn} \end{pmatrix} , \quad X = \begin{pmatrix} x_1(t) \\ x_2(t) \\ \\ x_n(t) \end{pmatrix} , \quad F(t) = \begin{pmatrix} f_1(f) \\ f_2(t) \\ \\ f_n(t) \end{pmatrix}$$

$$\dot{X} = \begin{pmatrix} \dot{x}_1 \\ \dot{x}_2 \\ \cdot \\ \cdot \\ \cdot \\ \dot{x}_n \end{pmatrix} .$$
Then the system (2) can be written in the matrix form

$$\begin{pmatrix} \dot{x}_1 \\ \dot{x}_2 \\ \cdot \\ \cdot \\ \cdot \\ \dot{x}_n \end{pmatrix} = \begin{pmatrix} a_{11} & a_{12} & \cdots & a_{1n} \\ a_{21} & a_{22} & \cdots & a_{2n} \\ \cdots & \cdots & \cdots & \cdots \\ a_{n1} & a_{n2} & \cdots & a_{nn} \end{pmatrix} \begin{pmatrix} x_1 \\ x_2 \\ \cdot \\ \cdot \\ \cdot \\ x_n \end{pmatrix} + \begin{pmatrix} f_1(t) \\ f_2(t) \\ \cdot \\ \cdot \\ \cdot \\ f_n(t) \end{pmatrix}$$

or

$$\dot{X} = A(t)X + F(t) \qquad (3)$$

When $F(t) = 0$, the system (3) is homogeneous. Let $X(t)$ be a general solution of the homogeneous differential equation

$$, \ \dot{X} = A(t)X \ . \qquad (4)$$

Let $X = X(t)U$. $\qquad (5)$

Substituting this expression into the differential equation $\dot{X} = AX + F(t)$ yields
$$X = AX(t)U + F(t) \ . \qquad (6)$$

Differentiating (5), we have

$$\dot{X} = \dot{X}(t)U + X(t)\dot{U} \ .$$

Substituting the value of $\dot{X}$ into equation (6) yields

$$\dot{X}(t)U + X(t)\dot{U} = AX(t)U + F(t) \ . \qquad (7)$$
The matrix $X(t)$ is a fundamental matrix solution of equation (4). Hence $\dot{X}(t) = AX(t)$. Therefore, equation (7) becomes
$$AX(t)U + X\dot{U} = AX(t)U + F(t)$$

or
$$\dot{X}U = F(t) . \tag{8}$$

Recall that the columns of $X(t)$ are linearly independent eigenvectors of R^n. Hence its inverse X^{-1} exists. Therefore equation (8) can be written as

$$\dot{U} = X^{-1}(t)F(t) .$$

Integrating this expression between t_0 and t gives

$$U(t) = C + \int_{t_0}^{t} X^{-1}(t) F(t) \, dt ,$$

where c is a constant vector. Therefore a solution of (3) may be written

$$x(t) = X(t)\left[c + \int_{t_0}^{t} X^{-1}(t) \, f(t) \, dt \right] .$$

For initial conditions at $t = 0$ we have

$$x(0) = X(0)C .$$

Now turn to system (1). In matrix notation it becomes

$$\begin{pmatrix} \dot{x} \\ \dot{y} \end{pmatrix} = \begin{pmatrix} 0 & 1 \\ -2 & -3 \end{pmatrix} \begin{pmatrix} x_1 \\ y_1 \end{pmatrix} + \begin{pmatrix} t \\ -t^2 \end{pmatrix} .$$

Let

$$\dot{X} = \begin{pmatrix} \dot{x} \\ \dot{y} \end{pmatrix} , \quad A = \begin{pmatrix} 0 & 1 \\ -2 & -3 \end{pmatrix} , \quad X = \begin{pmatrix} x_1 \\ y_1 \end{pmatrix} \text{ and } F(t) \begin{pmatrix} t \\ -t^2 \end{pmatrix} ,$$

or

$$\dot{X} = A(t)X + F(t) . \tag{9}$$

We first find the solution of the homogeneous system

$$\dot{X} = A(t)X \tag{10}$$

where

$$A = \begin{pmatrix} 0 & 1 \\ -2 & -3 \end{pmatrix} .$$

The characteristic polynomial of A is:

$$f(\lambda) = \det(\lambda I - A) = \det \begin{pmatrix} \lambda & -1 \\ 2 & \lambda+3 \end{pmatrix}$$

$$= \lambda(\lambda+3) + 2$$

$$= \lambda^2 + 3\lambda + 2$$

$$= (\lambda+2)(\lambda+1) .$$

Therefore the eigenvalues are $\lambda_1 = -1$ and $\lambda_2 = -2$. The eigenvector corresponding to $\lambda_1 = -1$ is obtained by solving the equation $(\lambda I - A)V = 0$ for $\lambda = -1$ as follows:

$$(-1I - A)V = \begin{pmatrix} -1 & -1 \\ 2 & 2 \end{pmatrix} \begin{pmatrix} v_1 \\ v_2 \end{pmatrix} = \begin{pmatrix} 0 \\ 0 \end{pmatrix} .$$

This implies that $v_1 = -v_2$. Hence every vector $v = a\begin{pmatrix} 1 \\ -1 \end{pmatrix}$ is an eigenvector with $\lambda = -1$. Solve the equation $(-3I - A)V = 0$ to obtain the eigenvector $\begin{pmatrix} 1 \\ -2 \end{pmatrix}$ corresponding to $\lambda = -3$. Now since A has distinct eigenvalues and the eigenvectors are linearly independent, the general solution of equation (9) is:

$$X(t) = \begin{pmatrix} e^{-t} & e^{-2t} \\ -e^{-t} & -2e^{-2t} \end{pmatrix} .$$

Let $X^{-1}(t) = \begin{pmatrix} \alpha & \beta \\ \gamma & \delta \end{pmatrix}$. Recall that

$$X(t)X^{-1}(t) = I .$$

Thus,

$$\begin{pmatrix} e^{-t} & e^{-2t} \\ -e^{-t} & -2e^{-2t} \end{pmatrix} \begin{pmatrix} \alpha & \beta \\ \gamma & \delta \end{pmatrix} = \begin{pmatrix} 1 & 0 \\ 0 & 1 \end{pmatrix}$$

$$\begin{pmatrix} \alpha e^{-t} + \gamma e^{-2t} & \beta e^{-t} + \delta e^{-2t} \\ -\alpha e^{-t} - 2\gamma e^{-2t} & -\beta e^{-t} - 2\delta e^{-2t} \end{pmatrix} = \begin{pmatrix} 1 & 0 \\ 0 & 1 \end{pmatrix}$$

or

$$\alpha e^{-t} + \gamma e^{-2t} = 1 , \qquad \beta e^{-t} + \delta e^{-2t} = 0$$

$$-\alpha e^{-t} - 2\gamma e^{-2t} = 0 , \qquad -\beta e^{-t} - 2\delta e^{-2t} = 1 .$$

Solving these equations we obtain

$$\alpha = 2e^{t} , \qquad \beta = e^{t}$$

$$\gamma = -e^{2t} \qquad \delta = -e^{2t} .$$

Thus,

$$X^{-1}(t) = \begin{pmatrix} 2e^{t} & e^{t} \\ -e^{2t} & -e^{2t} \end{pmatrix} .$$

Then, the general solution of equation (9) is:

$$X(t) = X(t)C + X(t) \int_0^t \begin{pmatrix} 2e^{t} & e^{t} \\ -e^{2t} & -e^{2t} \end{pmatrix} \cdot \begin{pmatrix} t \\ -t^2 \end{pmatrix} dt$$

$$= X(t)C + X(t) \int_0^t \begin{pmatrix} 2te^{t} - t^2 e^{t} \\ -te^{2t} + t^2 e^{2t} \end{pmatrix} dt$$

$$= X(t)C + \begin{pmatrix} e^{-t} & e^{-2t} \\ -e^{-t} & -2e^{-2t} \end{pmatrix} \begin{pmatrix} t^2 e^{t} + 4te^{t} - 4e^{t} + 4 \\ \tfrac{1}{2}t^2 e^{2t} - te^{2t} + \tfrac{1}{2}e^{2t} - \tfrac{1}{2} \end{pmatrix}$$

$$= X(t)C + \begin{pmatrix} -t^2/2 + 3t - 7/2 + 4e^{-t} - \tfrac{1}{2}e^{-2t} \\ -2t + 3 - 4e^{-t} + e^{-2t} \end{pmatrix}$$

$$= \begin{pmatrix} e^{-t} & e^{-2t} \\ -e^{-t} & -2e^{-2t} \end{pmatrix} C + \begin{pmatrix} -t^2/2 + 3t - 7/2 + 4e^{-t} - \tfrac{1}{2}e^{-2t} \\ -2t + 3 - 4e^{-t} + e^{-2t} \end{pmatrix}$$

This is the general solution to the system (1).

Find the general solution of the following system:

$$\dot{x}_1 = x_1$$

$$\dot{x}_2 = 2x_1 + x_2 - 2x_3 \qquad (1)$$

$$\dot{x}_3 = 3x_1 + 2x_2 + x_3 + e^t \cos 2t .$$

Solution: The system (1) can be written in matrix notation as:

$$\begin{pmatrix} \dot{x}_1 \\ \dot{x}_2 \\ \dot{x}_3 \end{pmatrix} = \begin{pmatrix} 1 & 0 & 0 \\ 2 & 1 & -2 \\ 3 & 2 & 1 \end{pmatrix} \begin{pmatrix} x_1 \\ x_2 \\ x_3 \end{pmatrix} + \begin{pmatrix} 0 \\ 0 \\ e^t \cos 2t \end{pmatrix}$$

$$\dot{X} = AX + F(t) , \qquad (2)$$

where

$$\dot{X} = \begin{pmatrix} \dot{x}_1 \\ \dot{x}_2 \\ \dot{x}_3 \end{pmatrix}, \quad A = \begin{pmatrix} 1 & 0 & 0 \\ 2 & 1 & -2 \\ 3 & 2 & 1 \end{pmatrix}, \quad X = \begin{pmatrix} x_1 \\ x_2 \\ x_3 \end{pmatrix}, \quad \text{and}$$

$$F(t) = \begin{pmatrix} 0 \\ 0 \\ e^t \cos 2t \end{pmatrix} .$$

As we know, the general solution of equation (2) is:

$$x(t) = X(t)C + X(t) \int_0^t X^{-1}(t) F(t) \, dt \qquad (3)$$

where $X(t)$ is the solution of the homogeneous system. If $X(t)$ is the fundamental matrix solution of $\dot{X} = AX$ and e^{At} is also the fundamental matrix solution of $\dot{X} = AX$, then we can write $X(t) = e^{At}$. Since e^{At} and $X(t)$ are both fundamental matrices of $\dot{X} = AX$,

$$e^{At} = X(t) X^{-1}(0) .$$

Then equation (3) becomes:

$$x(t) = e^{At}C + e^{At} \int_0^t e^{-At} F(t) \, dt . \qquad (4)$$

We first find e^{At} where

$$A = \begin{pmatrix} 1 & 0 & 0 \\ 2 & 1 & -2 \\ 3 & 2 & 1 \end{pmatrix} .$$

Now the characteristic polynomial of A is:

$$f(\lambda) = \det(\lambda I - A) = \begin{pmatrix} \lambda-1 & 0 & 0 \\ -2 & \lambda-1 & 2 \\ -3 & -2 & \lambda-1 \end{pmatrix}$$

$$= (\lambda-1) \begin{vmatrix} \lambda-1 & 2 \\ -2 & \lambda-1 \end{vmatrix}$$

$$f(\lambda) = (\lambda-1)[(\lambda-1)^2 + 4]$$
$$= (\lambda-1)[\lambda^2 - 2\lambda + 5] .$$

Thus, the eigenvalues of A are

$$\lambda_1 = 1, \quad \lambda_2 = \frac{2 + \sqrt{4-20}}{2} = 1 + 2i$$

and

$$\lambda_3 = \frac{2 - \sqrt{4-20}}{2} = 1 - 2i .$$

For $\lambda = 1$ we seek eigenvectors such that

$$(A - \lambda I)V = (A - 1I)V = \begin{pmatrix} 0 & 0 & 0 \\ 2 & 0 & -2 \\ 3 & 2 & 0 \end{pmatrix} \begin{pmatrix} v_1 \\ v_2 \\ v_3 \end{pmatrix} = \begin{pmatrix} 0 \\ 0 \\ 0 \end{pmatrix} .$$

Then, $\quad 2v_1 - 2v_3 = 0$
$$3v_1 + 2v_2 = 0 .$$

This implies that $v_1 = v_3$ and $v_2 = -3v_1/2$, hence $V_1 = \begin{pmatrix} 2 \\ -3 \\ 2 \end{pmatrix}$ is an eigen-

vector of A with 1. For $\lambda = 1 + 2i$, we seek non-zero vectors V such that

$$(A - \lambda I)V = [A - (1 + 2i)I]V = \begin{pmatrix} -2i & 0 & 0 \\ 2 & -2i & -2 \\ 3 & 2 & -2i \end{pmatrix} \begin{pmatrix} v_1 \\ v_2 \\ v_3 \end{pmatrix} = \begin{pmatrix} 0 \\ 0 \\ 0 \end{pmatrix}$$

Then $\qquad\qquad -2iv_1 = 0$
$$2v_1 - 2iv_2 - 2v_3 = 0$$
$$3v_1 + 2v_2 - 2iv_3 = 0 .$$

This implies that $v_1 = 0$, and $v_3 = -iv_2$. Hence, $V_2 = \begin{pmatrix} 0 \\ 1 \\ -1 \end{pmatrix}$ is an eigen-

vector of A with eigenvalue $1 + 2i$. Similarly, the eigenvector V_3 corresponding to the eigenvalue $\lambda_3 = 1 - 2i$ is found to be

$V_3 = \begin{pmatrix} 0 \\ -i \\ 1 \end{pmatrix}$. Since the eigenvectors V_1, V_2 and V_3 are linearly indepen-

dent, the fundamental matrix solution of the system $\dot{X} = AX$, is

$$X(t) = [e^{\lambda_1 t} V_1, e^{\lambda_2 t} V_2, e^{\lambda_3 t} V_3] \tag{5}$$

Now

$$e^{\lambda_1 t} V_1 = e^t \begin{pmatrix} 2 \\ -3 \\ 2 \end{pmatrix} = \begin{pmatrix} 2e^t \\ -3e^t \\ 2e^t \end{pmatrix} .$$

$$e^{\lambda_2 t} V_2 = e^{(1+2i)t} V_2 = e^t(\cos 2t + i \sin 2t)\left[\begin{pmatrix} 0 \\ 1 \\ 0 \end{pmatrix} -i \begin{pmatrix} 0 \\ 0 \\ 1 \end{pmatrix} \right]$$

$$= e^t \left[\cos 2t \begin{pmatrix} 0 \\ 1 \\ 0 \end{pmatrix} + \sin 2t \begin{pmatrix} 0 \\ 0 \\ 1 \end{pmatrix} \right]$$

$$+ ie^t \left[\sin 2t \begin{pmatrix} 0 \\ 1 \\ 0 \end{pmatrix} - \cos 2t \begin{pmatrix} 0 \\ 0 \\ 1 \end{pmatrix} \right] .$$

Consequently $e^{\lambda_2 t} V_2 = e^t \begin{pmatrix} 0 \\ \cos 2t \\ \sin 2t \end{pmatrix}$ is a real valued solution of $\dot{X} = AX$.

Similarly, we can find

$$e^{\lambda_3 t} V_3 = e^t \begin{pmatrix} 0 \\ \sin 2t \\ -\cos 2t \end{pmatrix} \quad . \quad \text{Substitute the values of}$$

$e^{\lambda_1 t} V_1$, $e^{\lambda_2 t} V_2$, $e^{\lambda_3 t} V_3$ in equation (5). Then

$$X(t) = \begin{pmatrix} 2e^t & 0 & 0 \\ -3e^t & e^t \cos 2t & e^t \sin 2t \\ 2e^t & e^t \sin 2t & -e^t \cos 2t \end{pmatrix}$$

is a fundamental matrix solution of $\dot{X} = AX$. Computing

$$X^{-1}(0) = \begin{pmatrix} 2 & 0 & 0 \\ -3 & 1 & 0 \\ 2 & 0 & -1 \end{pmatrix}^{-1} = \begin{pmatrix} \frac{1}{2} & 0 & 0 \\ 3/2 & 1 & 0 \\ 1 & 0 & -1 \end{pmatrix}$$

we see that

$$e^{At} = \begin{pmatrix} 2e^t & 0 & 0 \\ -3e^t & e^t \cos 2t & e^t \sin 2t \\ 2e^t & e^t \sin 2t & -e^t \cos 2t \end{pmatrix} \begin{pmatrix} \frac{1}{2} & 0 & 0 \\ 3/2 & 1 & 0 \\ 1 & 0 & -1 \end{pmatrix}$$

$$= e^t \begin{pmatrix} 1 & 0 & 0 \\ -3/2 + (3/2)\cos 2t + \sin 2t & \cos 2t & -\sin 2t \\ 1 + (3/2)\sin 2t - \cos 2t & \sin 2t & \cos 2t \end{pmatrix}$$

Now

$$\int_0^t e^{-At} F(t)dt = \int_0^t e^{-t} \begin{pmatrix} 1 & 0 & 0 \\ -3/2 + (3/2)\cos 2t - \sin 2t & \cos 2t & \sin 2t \\ 1 - (3/2)\sin 2t - \cos 2t & -\sin 2t & \cos 2t \end{pmatrix} \begin{pmatrix} 0 \\ 0 \\ e^t \cos 2t \end{pmatrix} dt$$

$$= \int_0^t e^{-t} \begin{pmatrix} 0 \\ e^t \sin 2t \cos 2t \\ e^t \cos^2 2t \end{pmatrix} dt$$

$$= \int_0^t \begin{pmatrix} 0 \\ \sin 2t + \cos 2t \\ \cos^2 2t \end{pmatrix} dt$$

The integration of this matrix is obtained by integrating each element with respect to t between the limits 0 and t. We know that

$$\int \sin ax \cos ax \, dx = \frac{\sin^2 ax}{2a}$$

$$\int \cos^2 ax \, dx = x/2 + \frac{\sin 2ax}{2a} \quad .$$

579

Therefore,

$$\int_0^t \sin 2t \cos 2t \, dt = \frac{\sin^2 2t}{4} = \frac{1 - \cos 4t}{8}$$

$$\int_0^t \cos^2 2t \, dt = \frac{t}{2} + \frac{\sin 4t}{8} \, .$$

Therefore,

$$\int_0^t \begin{bmatrix} 0 \\ \sin 2t \cos 2t \\ \cos^2 2t \end{bmatrix} dt = \begin{bmatrix} 0 \\ \dfrac{1 - \cos 4t}{8} \\ t/2 + \dfrac{\sin 4t}{8} \end{bmatrix}$$

Hence, we have

$$\int_0^t e^{-At} F(t) dt = \begin{bmatrix} 0 \\ \dfrac{1 - \cos 4t}{8} \\ t/2 + \dfrac{\sin 4t}{8} \end{bmatrix} .$$

Now substitute the values of e^{At} and $\int_0^t e^{-At} F(t) dt$ into equation (4).

Then

$$X(t) = e^t \begin{bmatrix} 1 & 0 & 0 \\ -3/2 + (3/2)\cos 2t + \sin 2t & \cos 2t & \sin 2t \\ 1 - (3/2)\sin 2t - \cos 2t & -\sin 2t & \cos 2t \end{bmatrix} C$$

$$+ e^t \begin{bmatrix} 1 & 0 & 0 \\ -3/2 + (3/2)\cos 2t + \sin 2t & \cos 2t & -\sin 2t \\ 1 + (3/2)\sin 2t - \cos 2t & \sin 2t & \cos 2t \end{bmatrix} \begin{bmatrix} 0 \\ \dfrac{1 - \cos 4t}{8} \\ t/2 + \dfrac{\sin 4t}{8} \end{bmatrix}$$

is the general solution to the system (1).

● **PROBLEM** 18-22

Find the solution of the following system.

$$\dot{x}_1 = x_2$$
$$\dot{x}_2 = -2x_1 - 3x_2$$ (1)

Solution: Let us first consider the following matrix equation

$$\dot{X}(t) = AX$$ (2)

where A is an n x n matrix.

Taking the Laplace transform of both sides of equation (2), we obtain

$$sX(s) - X(o) = AX(s)$$

where $\qquad X(s) = L[X].$ Hence

$$(sI - A)X(s) = X(o)$$

Premultiplying both sides of this last equation by $(sI-A)^{-1}$, we obtain,

$$X(s) = (sI-A)^{-1}X(o)$$

The inverse Laplace transform of $X(s)$ gives the solution $X(t)$.

Thus,

$$X(t) = L^{-1}[(sI-A)^{-1}]X(o) \qquad (3)$$

The inverse Laplace transform of a matrix is the matrix consisting of the inverse Laplace transforms of all elements. Note that

$$(sI-A)^{-1} = \frac{I}{s} + \frac{A}{s^2} + \frac{A^2}{s^3} + \ldots$$

Hence, the inverse Laplace transform of $(sI-A)^{-1}$ is given by

$$L^{-1}[(sI-A)^{-1}] = I + At + \frac{A^2t^2}{2!} + \frac{A^3t^3}{3!} + \ldots = e^{At} \qquad (4)$$

From equations (3) and (4), the solution of equation (2) is obtained as

$$X(t) = e^{At}X(o)$$

where $\qquad e^{At} = L^{-1}[(sI-A)^{-1}].$

Now consider the given system (1). This system can be expressed in matrix form as

$$\begin{bmatrix} \dot{x}_1 \\ \dot{x}_2 \end{bmatrix} = \begin{bmatrix} 0 & 1 \\ -2 & -3 \end{bmatrix} \begin{bmatrix} x_1 \\ x_2 \end{bmatrix}$$

or $\qquad \dot{X} = AX$

where $\qquad A = \begin{bmatrix} 0 & 1 \\ -2 & -3 \end{bmatrix}$

The matrix e^{At} is given by

$$e^{At} = L^{-1}[(sI-A)^{-1}].$$

Now

$$sI - A = \begin{bmatrix} s & 0 \\ 0 & s \end{bmatrix} - \begin{bmatrix} 0 & 1 \\ -2 & -3 \end{bmatrix} = \begin{bmatrix} s & -1 \\ 2 & s+3 \end{bmatrix}$$

The inverse of (sI-A) is given by

$$(sI-A)^{-1} = \frac{1}{\det[sI-A]} \text{ adj } [sI-A]$$

$$\det[sI-A] = \det \begin{bmatrix} s & -1 \\ 2 & s+3 \end{bmatrix} = s(s+3) + 2 = s^2 + 3s + 2$$

$$= (s+1)(s+2)$$

and

$$\text{adj}[sI-A] = \begin{bmatrix} s+3 & 1 \\ -2 & s \end{bmatrix}$$

Therefore,

$$(sI-A)^{-1} = \frac{1}{(s+1)(s+2)} \begin{bmatrix} s+3 & 1 \\ -2 & s \end{bmatrix}$$

$$= \begin{bmatrix} \dfrac{s+3}{(s+1)(s+2)} & \dfrac{1}{(s+1)(s+2)} \\ \dfrac{-2}{(s+1)(s+2)} & \dfrac{s}{(s+1)(s+2)} \end{bmatrix}$$

The inverse Laplace transform of a matrix $[sI-A]^{-1}$ is the matrix consisting of the inverse Laplace transforms of its elements.

Thus,

$$L^{-1}[(sI-A)^{-1}] = \begin{bmatrix} L^{-1}\left[\dfrac{(s+3)}{(s+1(s+2)}\right] & L^{-1}\left[\dfrac{1}{(s+1)(s+2)}\right] \\ L^{-1}\left[\dfrac{-2}{(s+1)(s+2)}\right] & L^{-1}\left[\dfrac{s}{(s+1)(s+2)}\right] \end{bmatrix}$$

We know that (referring to a table of Laplace transforms),

$$L^{-1}\left[\frac{s+3}{(s+1)(s+2)}\right] = 2e^{-t} - e^{-2t}$$

$$L^{-1}\left[\frac{1}{(s+1)(s+2)}\right] = e^{-t} - e^{-2t}$$

$$L^{-1}\left(\frac{-2}{(s+1)(s+2)}\right) = -2e^{-t} + 2e^{-t}$$

$$L^{-1}\left(\frac{s}{(s+1)(s+2)}\right) = -e^{-t} + 2e^{-2t} .$$

Therefore

$$e^{At} = L^{-1}[(sI-A)^{-1}] = \begin{bmatrix} 2e^{-t}-e^{-2t} & e^{-t}-e^{-2t} \\ -2e^{-t}+2e^{-2t} & -e^{-t}+2e^{-2t} \end{bmatrix}$$

Hence, the solution of the given system is given by

$$X(t) = e^{At}X(o)$$

$$= \begin{bmatrix} 2e^{-t}-e^{-2t} & e^{-t}-e^{-2t} \\ -2e^{-t}+2e^{-2t} & -e^{-t}+2e^{-2t} \end{bmatrix} X(o)$$

where $X(o)$ is a constant matrix of initial conditions and

$$X(t) = \begin{bmatrix} x_1(t) \\ x_2(t) \end{bmatrix} .$$

THE Nth ORDER LINEAR DIFFERENTIAL EQUATION AS A FIRST ORDER LINEAR SYSTEM

● **PROBLEM** 18-23

Consider the differential equation

$$\dddot{x} - 6\ddot{x} + 11\dot{x} - 6x = 0 \tag{1}$$

Find the general solution.

Solution: The given equation (1) is a third order linear differential equation with constant coefficients. Now we first show that every n-th order linear differential equation with constant coefficients is equivalent to a system of n first-order linear differential equations with constant coefficients that can be expressed in terms of matrices.

Consider the following n-th order homogeneous linear differential equation with constant coefficients:

$$\frac{d^n x}{dt^n} + a_1 \frac{d^{n-1} x}{dt^{n-1}} + \dots + a_{n-1} \frac{dx}{dt} + a_n x = 0 , \tag{2}$$

where $a_1, a_2, \dots, a_n$ are real numbers. Replace the single dependent variable x by n new variables defined as follows:

$$x_1 = x$$
$$x_2 = \frac{dx}{dt} = \frac{dx_1}{dt}$$

$$x_3 = \frac{d^2 x}{dt^2} = \frac{dx_2}{dt}$$

$$\cdot$$
$$\cdot$$
$$\cdot$$

$$x_n = \frac{d^{n-1} x}{dt^{n-1}} = \frac{dx_{n-1}}{dt} \quad .$$

These equations of definition provide differential equations for x_1; $x_2, \ldots, x_{n-1}$, and the original differential equation (2) itself provides the equation for $dx_n/dt = dx^n/dt^n$. Thus we obtain the system

$$\dot{x}_1 = \frac{dx_1}{dt} = x_2$$

$$\dot{x}_2 = \frac{dx_2}{dt} = x_3 \qquad\qquad (3)$$

$$\cdot$$
$$\cdot$$
$$\cdot$$

$$\dot{x}_{n-1} = \frac{dx_{n-1}}{dt} = x_n$$

$$\dot{x}_n = \frac{dx_n}{dt} = -a_n x_1 - a_{n-1} x_2 - \ldots - a_1 x_n \quad .$$

If we let

$$X = \begin{pmatrix} x_1 \\ x_2 \\ \cdot \\ \cdot \\ \cdot \\ x_n \end{pmatrix} \qquad A = \begin{bmatrix} 0 & 1 & 0 & \ldots \ldots 0 \\ 0 & 0 & 1 & 0 \ldots 0 \\ \cdot \\ \cdot \\ 0 & 0 & 0 & \ldots \ldots 1 \\ -a_n & -a_{n-1} & & \ldots \ldots -a_1 \end{bmatrix} \qquad \text{and}$$

$$\dot{X} = \begin{pmatrix} \dot{x}_1 \\ \dot{x}_2 \\ \cdot \\ \cdot \\ \cdot \\ \dot{x}_n \end{pmatrix} , \qquad \text{system (3) becomes}$$

$$\dot{X} = AX \quad . \qquad\qquad (4)$$

Thus system (4) is equivalent to equation (2). It is clear that if X is a solution of (4) then the first component, $x_1(t) = x(t)$, is a solution of (2) and vice versa.

Now equation (1) is a 3rd order differential equation. Here, $a_3 = -6$, $a_2 = 11$, $a_1 = -6$. Therefore,

$$\dot{x}_1 = \frac{dx_1}{dt} = x_2$$

$$\dot{x}_2 = \frac{dx_2}{dt} = x_3$$

$$\dot{x}_3 = \frac{dx_3}{dt} = 6x_1 - 11x_2 + 6x_3 \quad .$$

Writing this in matrix form, we get

584

$$\begin{pmatrix} \dot{x}_1 \\ \dot{x}_2 \\ \dot{x}_3 \end{pmatrix} = \begin{pmatrix} 0 & 1 & 0 \\ 0 & 0 & 1 \\ 6 & -11 & 6 \end{pmatrix} \begin{pmatrix} x_1 \\ x_2 \\ x_3 \end{pmatrix}$$

or

$$\dot{X} = AX. \qquad (5)$$

This is the linear system of first-order differential equations which can be easily solved by obtaining linearly independent eigenvectors of the matrix A. The characteristic polynomial of A is

$$f(\lambda) = \det(\lambda I - A) = \det \begin{pmatrix} \lambda & -1 & 0 \\ 0 & \lambda & -1 \\ -6 & +11 & \lambda-6 \end{pmatrix}$$

$$= \lambda \begin{vmatrix} \lambda & -1 \\ 11 & \lambda-6 \end{vmatrix} - (-1) \begin{vmatrix} 0 & -1 \\ -6 & \lambda-6 \end{vmatrix}$$

$$f(\lambda) = \lambda(\lambda^2 - 6\lambda + 11) + 1 (-6)$$

$$= \lambda^3 - 6\lambda^2 + 11\lambda - 6 = (\lambda-1)(\lambda-2)(\lambda-3) .$$

Therefore, the eigenvalues are $\lambda_1 = 1$, $\lambda_2 = 2$ and $\lambda_3 = 3$. The eigenvectors associated respectively with these eigenvalues are

$$\begin{pmatrix} 1 \\ 1 \\ 1 \end{pmatrix}, \quad \begin{pmatrix} 2 \\ 4 \\ 8 \end{pmatrix}, \quad \begin{pmatrix} 1 \\ 3 \\ 9 \end{pmatrix} .$$

Since, the eigenvalues of A are real and distinct, and the eigenvectors associated with these eigenvalues are linearly independent, the general solution to system (5) is then

$$X = c_1 \begin{pmatrix} 1 \\ 1 \\ 1 \end{pmatrix} e^t + c_2 \begin{pmatrix} 2 \\ 4 \\ 8 \end{pmatrix} e^{2t} + c_3 \begin{pmatrix} 1 \\ 3 \\ 9 \end{pmatrix} e^{3t}$$

or

$$X = \begin{pmatrix} x_1 \\ x_2 \\ x_3 \end{pmatrix} = \begin{pmatrix} c_1 e^t + 2c_2 e^{2t} + c_3 e^{3t} \\ c_1 e^t + 4c_2 e^{2t} + 3c_3 e^{3t} \\ c_1 e^t + 8c_2 e^{2t} + 9c_3 e^{3t} \end{pmatrix} .$$

Hence, the general solution of equation (1) is:

$$x = x_1 = c_1 e^t + 2c_2 e^{2t} + c_3 e^{3t} .$$

Observe that the characteristic polynomial associated with

$$\dddot{x} - 6\ddot{x} + 11\dot{x} - 6x$$

is $f(\lambda) = \lambda^3 - 6\lambda^2 + 11\lambda - 6.$

● **PROBLEM 18-24**

Solve the following initial value problem:
$$\ddot{x} + 2\dot{x} - 8x = e^t . \qquad (1)$$
The initial conditions are $x(0) = 1$, $\dot{x}(0) = -4$.

585

<u>Solution:</u> We first put the given initial value problem into matrix form. Rewrite equation (1) as
$$\ddot{x} = -2\dot{x} + 8x + e^t .$$

Now define $x_1(t) = x$ and $x_2(t) = \dot{x}$ (since the differential equation is second order, we need only two new variables). Then, we obtain
$$\dot{x}_1 = \dot{x} = x_2 .$$

Also,
$$\dot{x}_2 = \frac{d^2x}{dt^2} = -2\dot{x} + 8x + e^t .$$

But
$$\dot{x} = x_2 \quad \text{and} \quad x = x_1 .$$

Therefore,
$$\dot{x}_2 = -2x_2 + 8x_1 + e^t ,$$

thus,
$$\dot{x}_1 = 0x_1 + 1x_2 + 0 \tag{2}$$

$$\dot{x}_2 = 8x_1 - 2x_2 + e^t .$$

Now we can write system (2) in matrix notation as

$$\begin{pmatrix} \dot{x}_1 \\ \dot{x}_2 \end{pmatrix} = \begin{pmatrix} 0 & 1 \\ 8 & -2 \end{pmatrix} \begin{pmatrix} x_1 \\ x_2 \end{pmatrix} + \begin{pmatrix} 0 \\ e^t \end{pmatrix} . \tag{3}$$

Let
$$\dot{X} = \begin{pmatrix} \dot{x}_1 \\ \dot{x}_2 \end{pmatrix} , \quad A = \begin{pmatrix} 0 & 1 \\ 8 & -2 \end{pmatrix} , \quad X = \begin{pmatrix} x_1 \\ x_2 \end{pmatrix} , \quad \text{and}$$

$$F(t) = \begin{pmatrix} 0 \\ e^t \end{pmatrix} .$$

The initial conditions take the form
$$C = X(0) = \begin{pmatrix} x(0) \\ \dot{x}(0) \end{pmatrix} = \begin{pmatrix} 1 \\ -4 \end{pmatrix} .$$

Then, the given initial value problem in matrix notation becomes
$$\dot{X} = AX + F(t); \ X(0) = C. \tag{4}$$

This is a nonhomogeneous first-order system. The solution of system (4) is given by
$$X(t) = e^{At}C + e^{At} \int_{t_0}^{t} e^{-As} F(s)\ ds . \tag{5}$$

Now, we compute e^{At} , by using the formula $e^{At} = e^{tPDP^{-1}} = Pe^{tD}P^{-1}$,

where P is a matrix which diagonalizes A and D is a diagonal matrix. Recall that if the eigenvalues of A are distinct, then a diagonalizing matrix P always exists. The matrix P has eigenvectors as its columns. The characteristic polynomial of A is

$$f(\lambda) = \det(\lambda I - A) = \det \begin{pmatrix} \lambda & -1 \\ -8 & \lambda+2 \end{pmatrix}$$

$$= \lambda(\lambda+2) - 8$$

$$= \lambda^2 + 2\lambda - 8 = (\lambda+4)(\lambda-2) .$$

Hence, the eigenvalues are $\lambda_1 = -4$, and $\lambda_2 = 2$. For $\lambda_1 = -4$, the eigenvector V_1 must satisfy the equation $(-4I - A)V_1 = 0$. Therefore,

$$\begin{pmatrix} -4 & -1 \\ -8 & -2 \end{pmatrix} \begin{pmatrix} v_1 \\ v_2 \end{pmatrix} = \begin{pmatrix} 0 \\ 0 \end{pmatrix}$$

or

$$-4v_1 - v_2 = 0$$
$$-8v_1 - 2v_2 = 0 .$$

This implies that $v_2 = -4v_1$. Therefore $V_1 = \begin{pmatrix} 1 \\ -4 \end{pmatrix}$ is an eigenvector

associated with $\lambda_1 = -4$. Similarly, $V_2 = \begin{pmatrix} v_1 \\ v_2 \end{pmatrix}$ is an eigenvector associated

with $\lambda_2 = 2$ and it is found to be

$$V_2 = \begin{pmatrix} 1 \\ 2 \end{pmatrix} .$$

Then, the matrix P is given by

$$P = \begin{bmatrix} 1 & 1 \\ -4 & 2 \end{bmatrix} ,$$

$$P^{-1} = \frac{1}{\det} \text{ adj } P = \frac{1}{6} \begin{pmatrix} 2 & -1 \\ 4 & 1 \end{pmatrix} .$$

Then

$$D = P^{-1}AP = \frac{1}{6} \begin{pmatrix} 2 & -1 \\ 4 & 1 \end{pmatrix} \begin{pmatrix} 0 & 1 \\ 8 & -2 \end{pmatrix} \begin{pmatrix} 1 & 1 \\ -4 & 2 \end{pmatrix}$$

$$= \frac{1}{6} \begin{pmatrix} 2 & -1 \\ 4 & 1 \end{pmatrix} \begin{pmatrix} -4 & 2 \\ 16 & 4 \end{pmatrix}$$

$$= \begin{pmatrix} -4 & 0 \\ 0 & 2 \end{pmatrix} .$$

Therefore,

$$e^{tA} = p e^{tD} p^{-1}$$

$$= \frac{1}{6} \begin{pmatrix} 1 & 1 \\ -4 & 2 \end{pmatrix} \begin{pmatrix} e^{-4t} & 0 \\ 0 & e^{2t} \end{pmatrix} \begin{pmatrix} 2 & -1 \\ 4 & 1 \end{pmatrix}$$

$$= \frac{1}{6} \begin{pmatrix} 1 & 1 \\ -4 & 2 \end{pmatrix} \begin{pmatrix} 2e^{-4t} & -e^{-4t} \\ -4e^{2t} & e^{2t} \end{pmatrix}$$

$$= \frac{1}{6} \begin{bmatrix} 2e^{-4t} + 4e^{2t} & -e^{-4t} + e^{2t} \\ -8e^{-4t} + 8e^{2t} & 4e^{-4t} + 2e^{2t} \end{bmatrix} .$$

Now,

$$e^{tA}C = \frac{1}{6} \begin{bmatrix} 4e^{2t} + 2e^{-4t} & e^{2t} - e^{-4t} \\ 8e^{2t} - 8e^{-4t} & 2e^{2t} + 4e^{-4t} \end{bmatrix} \begin{bmatrix} 1 \\ -4 \end{bmatrix} .$$

$$e^{tA}C = \frac{1}{6} \begin{bmatrix} 4e^{2t} + 2e^{-4t} - 4e^{2t} + 4e^{-4t} \\ 8e^{2t} - 8e^{-4t} - 8e^{2t} - 16e^{-4t} \end{bmatrix}$$

$$= \frac{1}{6} \begin{bmatrix} 6e^{-4t} \\ -24e^{-4t} \end{bmatrix} = \begin{bmatrix} e^{-4t} \\ -4e^{-4t} \end{bmatrix} .$$

Now,

$$e^{-As}F(s) = \frac{1}{6} \begin{bmatrix} 4e^{-2s} + 2e^{4s} & e^{-2s} - e^{4s} \\ 8e^{-2s} - 8e^{4s} & 2e^{-2s} + 4e^{4s} \end{bmatrix} \begin{bmatrix} 0 \\ e^{s} \end{bmatrix}$$

$$= \frac{1}{6} \begin{bmatrix} e^s(e^{-2s} - e^{4s} \\ e^s(2e^{-2s} + 4e^{4s}) \end{bmatrix}$$

$$= \begin{bmatrix} (1/6)e^{-s} -(1/6)e^{5s} \\ (2/6)e^{-s} +(4/6)e^{5s} \end{bmatrix}$$

Therefore,

$$\int_0^t e^{-As}F(s)ds = \int_0^t \begin{bmatrix} (1/6)e^{-s} -(1/6)e^{5s} \\ (2/6)e^{-s} +(4/6)e^{5s} \end{bmatrix} ds$$

$$= \begin{Bmatrix} \int_0^t \left((1/6)e^{-s} -(1/6)e^{5s} \right) ds \\ \int_0^t \left((2/6)e^{-s} +(4/6)e^{5s} \right) ds \end{Bmatrix}$$

$$= \frac{1}{30} \begin{bmatrix} -5e^{-t} - e^{5t} + 6 \\ -10e^{-t} + 4e^{5t} + 6 \end{bmatrix} .$$

Then,

$$e^{At} \int_0^t e^{-As}F(s)ds$$

$$= \frac{1}{(6)} \frac{1}{(30)} \begin{bmatrix} 4e^{2t} + 2e^{-4t} & e^{2t} - e^{-4t} \\ 8e^{2t} - 8e^{-4t} & 2e^{2t} + 4e^{-4t} \end{bmatrix} \begin{pmatrix} -5e^{-t} - e^{5t} + 6 \\ -10e^{-t} + 4e^{5t} + 6 \end{pmatrix}$$

$$= \frac{1}{180} \begin{bmatrix} (4e^{2t} + 2e^{-4t})(-5e^{-t} - e^{5t} + 6) + (e^{2t} - e^{-4t})(-10e^{-t} + 4e^{5t} + 6) \\ (8e^{2t} - 8e^{-4t})(-5e^{-t} - e^{5t} + 6) + (2e^{2t} + 4e^{-4t})(-10e^{-t} + 4e^{5t} + 6) \end{bmatrix}$$

$$= \frac{1}{30} \begin{bmatrix} -6e^t + 5e^{2t} + e^{-4t} \\ -6e^t + 10e^{2t} - 4e^{-4t} \end{bmatrix} .$$

Substitute the value of $e^{tA}C$ and $e^{tA} \int_0^t e^{-As}F(s)ds$ into equation (5).

We obtain:

$$X(t) = \begin{bmatrix} e^{-4t} \\ -4e^{-4t} \end{bmatrix} + \frac{1}{30} \begin{pmatrix} -6e^t + 5e^{2t} + e^{-4t} \\ -6e^t + 10e^{2t} - 4e^{-4t} \end{pmatrix}$$

$$X(t) = \begin{pmatrix} x_1(t) \\ x_2(t) \end{pmatrix} = \begin{pmatrix} \frac{31}{30}e^{-4t} +(1/6)e^{2t} -(1/5)e^t \\ -\frac{62}{15}e^{-4t} +(1/3)e^{2t} -(1/5)e^t \end{pmatrix}$$

and $x(t) = x_1(t) = (31/30)e^{-4t} +(1/6)e^{2t} -(1/5)e^t$.

● PROBLEM 18-25

Solve the following initial value problem:

$$\ddot{x} + x = 3 ,$$ (1)

$$x(\pi) = 1, \quad \dot{x}(\pi) = 2 .$$

Solution: Equation (1) can be written as

$$\ddot{x} = -x + 3 .$$

Now we define two new variables:

$$x_1(t) = x(t) \quad \text{and} \quad x_2(t) = \dot{x} .$$

Then

$$\dot{x}_1 = \dot{x} = x_2 .$$

Therefore,

$$\dot{x}_2 = \frac{d^2 x}{dt^2} = -x + 3$$

$$= -x_1 + 3 .$$

Thus,

$$\dot{x}_1 = 0x_1 + 1x_2 + 0$$

$$\dot{x}_2 = -x_1 + 0x_2 + 3 .$$

These equations can be written in matrix notation as

$$\begin{pmatrix} \dot{x}_1 \\ \dot{x}_2 \end{pmatrix} = \begin{pmatrix} 0 & 1 \\ -1 & 0 \end{pmatrix} \begin{pmatrix} x_1 \\ x_2 \end{pmatrix} + \begin{pmatrix} 0 \\ 3 \end{pmatrix} ,$$

or

$$\dot{X} = AX(t) + F(t) \tag{2}$$

where

$$A = \begin{pmatrix} 0 & 1 \\ -1 & 0 \end{pmatrix} , \quad X(t) = \begin{pmatrix} x_1(t) \\ x_2(t) \end{pmatrix} , \quad F(t) = \begin{pmatrix} 0 \\ 3 \end{pmatrix} .$$

The initial conditions in matrix notation take the form

$$\begin{pmatrix} x(\pi) \\ \dot{x}(\pi) \end{pmatrix} = \begin{pmatrix} 1 \\ 2 \end{pmatrix} = C .$$

Therefore, the solution of system (2) which satisfies the initial conditions is given by

$$X(t) = e^{A(t-t_0)} C + e^{At} \int_{t_0}^{t} e^{-As} f(s) \, ds$$

or

$$X(t) = e^{A(t-t_0)} C + \int_{t_0}^{t} e^{A(t-s)} f(s) \, ds .$$

Here, $t_0 = \pi$. Then,

$$X(t) = e^{A(t-\pi)} C + \int_{\pi}^{t} e^{A(t-s)} f(s) \, ds . \tag{3}$$

Now we compute e^{tA} where $A = \begin{pmatrix} 0 & 1 \\ -1 & 0 \end{pmatrix}$. Here $n = 2$; hence,

$$e^{At} = \alpha_1 At + \alpha_0 I . \tag{4}$$

Set

$$e^{\lambda_i} = \alpha_0 + \alpha_1 \lambda_i .$$

The eigenvalues of At are $\lambda_1 = it$, and $\lambda_2 = -it$. Thus,

$$e^{it} = \alpha_1(it) + \alpha_0$$

$$e^{-it} = \alpha_1(-it) + \alpha_0 \ . \ .$$

Solving these equations for α_1 and α_0 yields:

$$\alpha_1 = \frac{1}{2it}(e^{it} - e^{-it})$$

$$\alpha_0 = \tfrac{1}{2}(e^{it} + e^{-it}) \ .$$

But,

$$\frac{e^{it} - e^{-it}}{2i} = \sin t \ , \quad \text{and}$$

$$\frac{e^{it} + e^{-it}}{2} = \cos t \ .$$

Thus, $\alpha_1 = \dfrac{\sin t}{t}$, $\alpha_0 = \cos t$. Substituting these values into (4), we obtain

$$e^{At} = \frac{\sin t}{t} At + \cos t \, I \ ,$$

$$= \frac{\sin t}{t} \begin{bmatrix} 0 & t \\ -t & 0 \end{bmatrix} + \cos t \begin{bmatrix} 1 & 0 \\ 0 & 1 \end{bmatrix}$$

$$= \begin{bmatrix} \cos t & \sin t \\ -\sin t & \cos t \end{bmatrix}$$

Then,

$$e^{A(t-\pi)}C = \begin{bmatrix} \cos(t-\pi) & \sin(t-\pi) \\ -\sin(t-\pi) & \cos(t-\pi) \end{bmatrix} \begin{bmatrix} 1 \\ 2 \end{bmatrix}$$

$$= \begin{bmatrix} \cos(t-\pi) + 2\sin(t-\pi) \\ -\sin(t-\pi) + 2\cos(t-\pi) \end{bmatrix}$$

$$e^{A(t-s)}f(s) = \begin{bmatrix} \cos(t-s) & \sin(t-s) \\ -\sin(t-s) & \cos(t-s) \end{bmatrix} \begin{bmatrix} 0 \\ 3 \end{bmatrix}$$

$$= \begin{bmatrix} 3\sin(t-s) \\ 3\cos(t-s) \end{bmatrix} \ .$$

Then,

$$\int_\pi^t e^{A(t-s)}f(s) = \int_\pi^t \begin{bmatrix} 3\sin(t-s) \\ 3\cos(t-s) \end{bmatrix} ds$$

$$= \begin{bmatrix} \int_\pi^t 3\sin(t-s)ds \\ \int_\pi^t 3\cos(t-s)ds \end{bmatrix}$$

$$= \begin{pmatrix} 3 \cos(t-s) \Big|_{s=\pi}^{s=t} \\ -3 \sin(t-s) \Big|_{s=\pi}^{s=t} \end{pmatrix}$$

$$= \begin{pmatrix} 3 - 3 \cos(t-\pi) \\ 3 \sin(t-\pi) \end{pmatrix} .$$

Substitute the values of $e^{A(t-\pi)}C$ and $\int_{\pi}^{t} e^{A(t-s)} f(s)ds$, into equation (3) to obtain:

$$X(t) = \begin{pmatrix} \cos(t-\pi) + 2 \sin(t-\pi) \\ -\sin(t-\pi) + 2 \cos(t-\pi) \end{pmatrix} + \begin{pmatrix} 3 - 3 \cos(t-\pi) \\ 3 \sin(t-\pi) \end{pmatrix}$$

$$= \begin{pmatrix} 3 - 2 \cos(t-\pi) + 2 \sin(t-\pi) \\ 2 \cos(t-\pi) + 2 \sin(t-\pi) \end{pmatrix} .$$

But

$$X(t) = \begin{pmatrix} x_1(t) \\ x_2(t) \end{pmatrix} .$$

Therefore,

$$\begin{pmatrix} x_1(t) \\ x_2(t) \end{pmatrix} = \begin{pmatrix} 3 - 2 \cos(t-\pi) + 2 \sin(t-\pi) \\ 2 \cos(t-\pi) + 2 \sin(t-\pi) \end{pmatrix}$$

Thus,

$$x_1(t) = 3 - 2 \cos(t-\pi) + 2 \sin(t-\pi) .$$

However, we have defined $x_1(t) = x(t)$. Therefore, the solution of the given initial value problem is

$$x(t) = 3 - 2 \cos(t-\pi) + 2 \sin(t-\pi) .$$

Recall that $\cos(t-\pi) = \cos t \cos \pi + \sin t \sin \pi$. However, $\cos \pi = -1$, and $\sin \pi = 0$; therefore $\cos(t-\pi) = -\cos t$. Similarly, $\sin(t-\pi) = -\sin t$. Thus the solution may be written as:

$$x(t) = 3 + 2 \cos t - 2 \sin t .$$

● PROBLEM 18-26

Consider the system of differential equations

$$y_1'' - 3y_1' + 2y_1 + y_2' - y_2 = 0 \qquad (1)$$

$$y_1' - 2y_1 + y_2' + y_2 = 0 . \qquad (2)$$

(a) Show that this system is equivalent to the system of first-order equations $U' = AU$, where

$$U = \begin{pmatrix} u_1 \\ u_2 \\ u_3 \end{pmatrix} = \begin{pmatrix} y_1 \\ y_1' \\ y_2 \end{pmatrix} , \quad A = \begin{pmatrix} 0 & 1 & 0 \\ -4 & 4 & 2 \\ 2 & -1 & -1 \end{pmatrix} .$$

591

(b)   Find a fundamental matrix for the system in part (a).

(c)   Find the solution of the original system satisfying the initial conditions

$$y_1 = 0, \ y_1'(0) = 1, \ y_2(0) = 0 \ .$$

Solution: First rewrite the given equations (1) and (2) as

$$y_1'' = 3y_1' - 2y_1 + y_2 - y_2' \tag{3}$$

$$y_2' = -y_1' + 2y_1 - y_2 \ . \tag{4}$$

Substituting (4) into (1) we obtain:

$$y_1'' = -4y_1 + 4y_1' + 2y_2 \tag{5}$$

$$y_2' = + 2y_1 - y_1' - y_2 \ . \tag{6}$$

Let $u_1 = y_1$, $u_2 = y_1'$ and $u_3 = y_2$; $u_1' = y_1'$, $u_2' = y_1''$, $u_3' = y_2'$.
Then equations (5) and (6) can be written as:

$$\begin{pmatrix} u_1' \\ u_2' \\ u_3' \end{pmatrix} = \begin{pmatrix} 0 & 1 & 0 \\ -4 & 4 & 2 \\ 2 & -1 & -1 \end{pmatrix} \begin{pmatrix} u_1 \\ u_2 \\ u_3 \end{pmatrix}$$

or
$$U' = AU \tag{7}$$
where

$$A = \begin{pmatrix} 0 & 1 & 0 \\ -4 & 4 & 2 \\ 2 & -1 & -1 \end{pmatrix}, \quad U = \begin{pmatrix} u_1 \\ u_2 \\ u_3 \end{pmatrix}, \quad U' = \begin{pmatrix} u_1' \\ u_2' \\ u_3' \end{pmatrix}.$$

Thus, we have shown that the given system is equivalent to the system $U' = AU$.

(b)   To obtain the fundamental matrix for system (7) we seek three linearly independent eigenvectors of the matrix A. Now, the characteristic polynomial of A is

$$f(\lambda) = \det(\lambda I - A) = \det \begin{pmatrix} \lambda & -1 & 0 \\ 4 & \lambda-4 & -2 \\ -2 & 1 & \lambda+1 \end{pmatrix}$$

$$= \lambda[(\lambda-4)(\lambda+1) + 2] + 1[4(\lambda+1) - 4]$$

$$= \lambda(\lambda^2 - 3\lambda - 2) + 4\lambda$$

$$= \lambda(\lambda-1)(\lambda-2) \ .$$

Hence A has three distinct eigenvalues: $\lambda_1 = 0$, $\lambda_2 = 2$, $\lambda_3 = 1$.   Let V_1, V_2, V_3 denote the eigenvectors associated with $\lambda_1, \lambda_2, \lambda_3$, respectively. Then, $V_1 = \begin{pmatrix} v_1 \\ v_2 \\ v_3 \end{pmatrix}$ must satisfy the equation $(A - \lambda_1 I)V_1 = 0$.

Therefore,

$$\begin{pmatrix} 0 & 1 & 0 \\ -4 & 4 & 2 \\ 2 & -1 & -1 \end{pmatrix} \begin{pmatrix} v_1 \\ v_2 \\ v_3 \end{pmatrix} = \begin{pmatrix} 0 \\ 0 \\ 0 \end{pmatrix}$$

or
$$v_2 = 0$$
$$-4v_1 - 4v_2 + 2v_3 = 0$$
$$2v_1 - v_2 - v_3 = 0 .$$

This implies that $v_2 = 0$ and $v_3 = 2v_1$. Hence, $V_1 = \alpha \begin{pmatrix} 1 \\ 0 \\ 2 \end{pmatrix}$, where α

is constant. For $\lambda_2 = 2$, V_2 must satisfy the equation $(A - \lambda_2 I)V_2 = 0$. Then,

$$\begin{pmatrix} -2 & 1 & 0 \\ -4 & 2 & 2 \\ 2 & -1 & -3 \end{pmatrix} \begin{pmatrix} v_1 \\ v_2 \\ v_3 \end{pmatrix} = \begin{pmatrix} 0 \\ 0 \\ 0 \end{pmatrix}$$

or
$$-2v_1 + v_2 = 0$$
$$-4v_1 + 2v_2 + v_3 = 0$$
$$2v_1 - v_2 - 3v_3 = 0 .$$

This implies that $v_2 = 2v_1$ and $v_3 = 0$, and thus $V_2 = \beta \begin{pmatrix} 1 \\ 2 \\ 0 \end{pmatrix}$ where β

is a constant. Finally, for $\lambda_3 = 1$, V_3 must satisfy the equation $(A - \lambda_3 I)V_3 = 0$. Then,

$$\begin{pmatrix} -1 & 1 & 0 \\ -4 & 3 & 2 \\ 2 & -1 & -2 \end{pmatrix} \begin{pmatrix} v_1 \\ v_2 \\ v_3 \end{pmatrix} = \begin{pmatrix} 0 \\ 0 \\ 0 \end{pmatrix}$$

or
$$-v_1 + v_2 = 0$$
$$-4v_1 + 3v_2 + 2v_3 = 0$$
$$2v_1 - v_2 - 2v_3 = 0 .$$

This implies that $v_1 = v_2$ and $v_1 = 2v_3$. Thus $V_3 = \gamma \begin{pmatrix} 2 \\ 2 \\ 1 \end{pmatrix}$ where γ

is constant. Since the eigenvectors V_1, V_2 and V_3 are linearly independent, the fundamental matrix for the system $U' = AU$ is given by

$$\varphi(t) = \begin{pmatrix} e^{\lambda_1 t} V_1 & , & e^{\lambda_2 t} V_2 & , & e^{\lambda_3 t} V_3 \end{pmatrix}$$

$$= \begin{pmatrix} 1 & e^{2t} & 2e^t \\ 0 & 2e^{2t} & 2e^t \\ 2 & 0 & e^t \end{pmatrix} .$$

(c) If $U(t)$ is the fundamental matrix solution e^{At}, then
$$U(t) = e^{tA} = \varphi(t) \ \varphi^{-1}(0) .$$

Then

$$\varphi^{-1}(0) = \begin{pmatrix} 1 & 1 & 2 \\ 0 & 2 & 2 \\ 2 & 0 & 1 \end{pmatrix}^{-1} = -\frac{1}{2} \begin{pmatrix} 2 & -1 & -2 \\ 4 & -3 & -2 \\ -4 & 2 & 2 \end{pmatrix}$$

Therefore,

$$e^{tA} = -\frac{1}{2}\begin{pmatrix} 1 & e^{2t} & 2e^{t} \\ 0 & 2e^{2t} & 2e^{t} \\ 2 & 0 & e^{t} \end{pmatrix}\begin{pmatrix} 2 & -1 & -2 \\ 4 & -3 & -2 \\ -4 & 2 & 2 \end{pmatrix}$$

We know that the solution of equation $U' = AU$ which satisfies the initial conditions is given by

$$U(t) = e^{tA}C .$$

Here $C = \begin{pmatrix} 0 \\ 1 \\ 0 \end{pmatrix}$. Then,

$$U(t) = -\frac{1}{2}\begin{pmatrix} 1 & e^{2t} & 2e^{t} \\ 0 & 2e^{2t} & 2e^{t} \\ 2 & 0 & e^{t} \end{pmatrix}\begin{pmatrix} 2 & -1 & -2 \\ 4 & -3 & -2 \\ -4 & 2 & 2 \end{pmatrix}\begin{pmatrix} 0 \\ 1 \\ 0 \end{pmatrix}$$

$$U(t) = -\frac{1}{2}\begin{pmatrix} 1 & e^{2t} & 2e^{t} \\ 0 & 2e^{2t} & 2e^{t} \\ 2 & 0 & e^{t} \end{pmatrix}\begin{pmatrix} -1 \\ -3 \\ 2 \end{pmatrix}$$

$$= -\frac{1}{2}\begin{pmatrix} -1 - 3e^{2t} + 4e^{t} \\ -6e^{2t} + 4e^{t} \\ -2 + 2e^{t} \end{pmatrix} .$$

But

$$U(t) = \begin{pmatrix} u_1 \\ u_2 \\ u_3 \end{pmatrix} = \begin{pmatrix} y_1 \\ y_1' \\ y_2 \end{pmatrix} .$$

Hence

$$\begin{pmatrix} y_1 \\ y_1' \\ y_2 \end{pmatrix} = -\frac{1}{2}\begin{pmatrix} -1 - 3e^{2t} + 4e^{t} \\ -6e^{2t} + 4e^{t} \\ -2 + 2e^{t} \end{pmatrix} .$$

Thus,

$$y_1 = -\frac{1}{2}(-1 - 3e^{2t} + 4e^{t})$$

$$y_2 = -\frac{1}{2}(-2 + 2e^{t}) .$$

● **PROBLEM** 18-27

Find the general solution of the following system of differential equations:

$$3\frac{dx}{dt} + \frac{dy}{dt} = \sin t \tag{1}$$

$$\frac{dx}{dt} + 3\frac{dy}{dt} = \cos t .$$

Solution: The given system of differential equations can be written in matrix notation as follows:

$$\begin{pmatrix} 3 & 1 \\ 1 & 3 \end{pmatrix}\begin{pmatrix} \frac{dx}{dt} \\ \frac{dy}{dt} \end{pmatrix} = \begin{pmatrix} \sin t \\ \cos t \end{pmatrix} \tag{2}$$

Let

$$A = \begin{pmatrix} 3 & 1 \\ 1 & 3 \end{pmatrix} \qquad U = \begin{pmatrix} \frac{dx}{dt} \\ \frac{dy}{dt} \end{pmatrix} \qquad V = \begin{pmatrix} \sin t \\ \cos t \end{pmatrix} .$$

Thus

$$AU = V . \tag{3}$$

Recall that a matrix is called symmetric if $A^t = A$ where A^t is the transpose matrix of A obtained by interchanging the rows and columns of A. Thus the matrix A of equation (3) is symmetric. Now according to the spectral theorem, if A is symmetric there is an invertible matrix P such that $B = P^{-1}AP$ is a diagonal matrix where P is an orthogonal matrix. Solving for A we have

$$A = PBP^{-1} .$$

Substituting this into equation (3) yields:

$$PBP^{-1}U = V .$$

Now multiply by P^{-1} to obtain

$$BP^{-1}U = P^{-1}V . \tag{4}$$

To find P, recall that an orthogonal matrix is a matrix whose columns are orthonormal. Now,

$$A = \begin{pmatrix} 3 & 1 \\ 1 & 3 \end{pmatrix} .$$

The characteristic polynomial of A is

$$\begin{aligned}
f(\lambda) &= \det(\lambda I - A) \\
&= \det \begin{pmatrix} \lambda-3 & -1 \\ -1 & \lambda-3 \end{pmatrix} \\
&= (\lambda-3)^2 - 1 = \lambda^2 - 6\lambda + 8 \\
&= (\lambda-2)(\lambda-4) .
\end{aligned}$$

Therefore, the characteristic roots are $\lambda = 2$ and $\lambda = 4$. The characteristic vector corresponding to characteristic value $\lambda = 2$ is $\begin{pmatrix} 1 \\ -1 \end{pmatrix}$, and the characteristic vector corresponding to characteristic value $\lambda = 4$ is $\begin{pmatrix} 1 \\ 1 \end{pmatrix}$. If we normalize the above characteristic vectors, we have the following orthonormal vectors:

$$\begin{pmatrix} 1/\sqrt{2} \\ -1/\sqrt{2} \end{pmatrix}, \quad \begin{pmatrix} 1/\sqrt{2} \\ 1/\sqrt{2} \end{pmatrix} .$$

Therefore, the matrix P is given by

$$\begin{pmatrix} 1/\sqrt{2} & 1/\sqrt{2} \\ -1/\sqrt{2} & 1/\sqrt{2} \end{pmatrix} .$$

Since P is an orthogonal matrix, $P^{-1} = P^T$. Thus,

$$P^{-1} = \begin{pmatrix} 1/\sqrt{2} & -1/\sqrt{2} \\ 1/\sqrt{2} & 1/\sqrt{2} \end{pmatrix}$$

and

$$B = P^{-1}AP$$

$$= \begin{bmatrix} 1/\sqrt{2} & -1/\sqrt{2} \\ \\ 1/\sqrt{2} & 1/\sqrt{2} \end{bmatrix} \begin{bmatrix} 3 & 1 \\ \\ 1 & 3 \end{bmatrix} \begin{bmatrix} 1/\sqrt{2} & 1/\sqrt{2} \\ \\ -1/\sqrt{2} & 1/\sqrt{2} \end{bmatrix}$$

$$= \begin{bmatrix} 2 & 0 \\ \\ 0 & 4 \end{bmatrix}.$$

Let $V = \begin{pmatrix} x \\ y \end{pmatrix}$. Equation (4) states: $BP^{-1}U = P^{-1}V$. Let $\begin{pmatrix} x_1 \\ y_1 \end{pmatrix} = P^{-1}V =$

$P^{-1}\begin{pmatrix} x \\ y \end{pmatrix}$. Also, $\begin{pmatrix} x_1 \\ y_1 \end{pmatrix} = BP^{-1}U$. Thus,

$$\begin{pmatrix} x_1 \\ y_1 \end{pmatrix} = \begin{bmatrix} 1/\sqrt{2} & -1/\sqrt{2} \\ \\ 1/\sqrt{2} & 1/\sqrt{2} \end{bmatrix} \begin{pmatrix} x \\ y \end{pmatrix}.$$

By matrix multiplication, we obtain,

$$x_1 = (1/\sqrt{2})(x-y) , \quad y_1 = (1/\sqrt{2})(x+y) ;$$

we therefore have

$$P^{-1}U = \begin{bmatrix} (1/\sqrt{2})(dx/dt - dy/dt) \\ \\ (1/\sqrt{2})(dx/dt + dy/dt) \end{bmatrix} = \begin{bmatrix} dx_1/dt \\ \\ dy_1/dt \end{bmatrix}$$

since

$$P^{-1} = \begin{bmatrix} 1/\sqrt{2} & -1/\sqrt{2} \\ \\ 1/\sqrt{2} & 1/\sqrt{2} \end{bmatrix}$$

and

$$U = \begin{pmatrix} dx/dt \\ dy/dt \end{pmatrix}.$$

Also,

$$P^{-1}V = \begin{bmatrix} (1/\sqrt{2})(\sin t - \cos t) \\ \\ (1/\sqrt{2})(\sin t + \cos t) \end{bmatrix}.$$

Substitute these values into equation (4): $[BP^{-1}U = P^{-1}V]$

$$\begin{bmatrix} 2 & 0 \\ \\ 0 & 4 \end{bmatrix} \begin{bmatrix} dx_1/dt \\ \\ dy_1/dt \end{bmatrix} = \begin{bmatrix} (1/\sqrt{2})(\sin t - \cos t) \\ \\ (1/\sqrt{2})(\sin t + \cos t) \end{bmatrix}$$

Therefore,

$$2 \frac{dx_1}{dt} = (1/\sqrt{2})(\sin t - \cos t)$$

$$4 \frac{dy_1}{dt} = (1/\sqrt{2})(\sin t + \cos t) .$$

This system is easy to solve by integration. Recalling the formulas

$$\int \cos u \, du = \sin u + c, \quad \int \sin u \, du = -\cos u + c$$

we have,

$$x_1 = (1/2\sqrt{2})(-\cos t - \sin t) + c_1$$

$$y_1 = \left(1/4\sqrt{2}\right)(-\cos t + \sin t) + c_2$$

where c_1 and c_2 are arbitrary constants. In matrix form, the above system can be written as

$$\begin{pmatrix} x_1 \\ y_1 \end{pmatrix} = \begin{pmatrix} \left(1/2\sqrt{2}\right)(-\cos t - \sin t) \\ \left(1/4\sqrt{2}\right)(-\cos t + \sin t) \end{pmatrix} + \begin{pmatrix} c_1 \\ c_2 \end{pmatrix}.$$

We have assumed that

$$\begin{pmatrix} x_1 \\ y_1 \end{pmatrix} = P^{-1} \begin{pmatrix} x \\ y \end{pmatrix}.$$

Therefore,

$$\begin{pmatrix} x \\ y \end{pmatrix} = P \begin{pmatrix} x_1 \\ y_1 \end{pmatrix}$$

or

$$\begin{pmatrix} x \\ y \end{pmatrix} = P \begin{pmatrix} \left(1/2\sqrt{2}\right)(-\cos t - \sin t) \\ \left(1/4\sqrt{2}\right)(-\cos t + \sin t) \end{pmatrix} + P \begin{pmatrix} c_1 \\ c_2 \end{pmatrix}$$

we put

$$P \begin{pmatrix} c_1 \\ c_2 \end{pmatrix} = \begin{pmatrix} d_1 \\ d_2 \end{pmatrix}$$

Substitute the matrix which represents P in the above formula:

$$\begin{pmatrix} x \\ y \end{pmatrix} = \begin{pmatrix} 1/\sqrt{2} & 1/\sqrt{2} \\ -1/\sqrt{2} & 1/\sqrt{2} \end{pmatrix} \begin{pmatrix} \left(1/2\sqrt{2}\right)(-\cos t - \sin t) \\ \left(1/4\sqrt{2}\right)(-\cos t + \sin t) \end{pmatrix} + \begin{pmatrix} d_1 \\ d_2 \end{pmatrix}$$

$$= \begin{pmatrix} -\left(1/4\right)(-\cos t - \sin t) + \left(1/8\right)(-\cos t + \sin t) \\ -\left(1/4\right)(-\cos t - \sin t) + \left(1/8\right)(-\cos t + \sin t) \end{pmatrix} + \begin{pmatrix} d_1 \\ d_2 \end{pmatrix}$$

$$= \begin{pmatrix} -\left(3/8\right)\cos t - \left(1/8\right)\sin t + d_1 \\ \left(P/8\right)\cos t + \left(3/8\right)\sin t + d_2 \end{pmatrix}.$$

Therefore the general solution to system (1) is

$$x = -\left(1/8\right)\sin t - \left(3/8\right)\cos t + d_1$$
$$y = \left(3/8\right)\sin t + \left(1/8\right)\cos t + d_2.$$

STABILITY OF NON-LINEAR SYSTEMS

● PROBLEM 18-28

Consider the system

$$\frac{dx}{dt} = F(x,y), \quad \frac{dy}{dt} = G(x,y) , \tag{1}$$

where in general $F(x,y)$ and $G(x,y)$ are non-linear functions. Under what conditions can (1) be replaced by the linear system

$$\frac{dx}{dt} = ax + by$$

$$\frac{dy}{dt} = cx + dy \quad ? \tag{2}$$

What can be said about the critical points of (2) and, by extension, (1)?

Solution: Eliminating the parameter t in (1) we obtain

$$\frac{dy}{dx} = \frac{G(x,y)}{F(x,y)} .$$

A critical point of the system (1) is a point where $G(x,y) = F(x,y) = 0$. If (x_0, y_0) is a critical point of (1), then by the existence and uniqueness theorem, $x(t) = x_0$, $y(t) = y_0$ is the constant solution for all time. Thus a critical point is also known as an equilibrium point. Note that it is always possible to transform the critical point (x_0, y_0) to the point $(0,0)$ by the linear change of variables $u = x - x_0$, $v = y - y_0$.

By determining the critical points of (1) we can learn a great deal about the behavior of its solutions as time goes by.

But finding the critical points for (1) may often be very difficult. Hence we restrict our attention to the study of almost linear systems, i.e., $F(x,y)$ and $G(x,y)$ are in the form

$$F(x,y) = ax + by + F_1(x,y)$$

$$G(x,y) = cx + dy + G_1(x,y) \tag{3}$$

$$\begin{vmatrix} a & b \\ c & d \end{vmatrix} \neq 0 \quad \text{and} \quad \frac{F_1(x,y)}{(x^2+y^2)^{\frac{1}{2}}} , \quad \frac{G_1(x,y)}{(x^2+y^2)^{\frac{1}{2}}} \rightarrow 0$$

as $(x^2+y^2)^{\frac{1}{2}} \rightarrow 0$. Substituting (3) into (1) near the critical point $(x,y) = (0,0)$ we obtain

$$\frac{dx}{dt} = ax + by$$

$$\frac{dy}{dt} = cx + dy , \tag{4}$$

which is system (2). Rewriting (4) in matrix form yields:

$$\begin{bmatrix} x'(t) \\ y'(t) \end{bmatrix} = \begin{bmatrix} a & b \\ c & d \end{bmatrix} \begin{bmatrix} x \\ y \end{bmatrix} .$$

Since the system has constant coefficients we seek solution of the form

$$x = Ae^{rt} , \quad y = Be^{rt} .$$

By substitution we obtain

$$\begin{bmatrix} a-r & b \\ c & d-r \end{bmatrix} \begin{bmatrix} A \\ B \end{bmatrix} = 0 .$$

598

Thus, r must be a root of the characteristic equation

$$r^2 - (a+d)r + (ad-bc) = 0 .$$ (5)

A number of different solutions are possible depending on whether the roots of (5) are both real and positive, negative or equal; it is also possible for the roots to be complex with real part or pure imaginary. The behavior of the solutions near the critical point (0,0) are also different.

● **PROBLEM** 18-29

The general solution of the system

$$\frac{dx}{dt} = ax + by$$ (1)

$$\frac{dy}{dt} = cx + dy$$

is

$$x = A_1 e^{r_1 t} + A_2 e^{r_2 t} , \quad y = B_1 e^{r_1 t} + B_2 e^{r_2 t} .$$ (2)

Discuss, with the aid of examples, the nature of the critical point (0,0) when:

a) r_1 and r_2 are real, and of the same sign, but $r_1 \neq r_2$;

b) r_1 and r_2 are real, and of opposite signs, and $r_1 \neq r_2$.

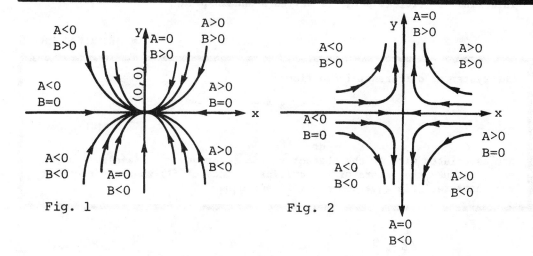

Fig. 1   Fig. 2

<u>Solution</u>: a) Suppose first that $r_1, r_2 < 0$. Then both x and y approach zero and the point (x,y) approaches the critical point (0,0) as $t \to \infty$ independent of the choice of the constants A_1, A_2, B_1 and B_2. Every solution of the system approaches the solution $x = 0$, $y = 0$ exponentially as $t \to \infty$.

To illustrate this case consider the system

$$\frac{dx}{dt} = -x , \quad \frac{dy}{dt} = -2y ,$$

which has the general solution

$$x = Ae^{-t} , \quad y = Be^{-2t} .$$ (3)

Fig. 1 shows a few trajectories of (2) for different values of A and B.

599

The arrows indicate the direction of the solutions as $t \to 0$. If $A \neq 0$, $B \neq 0$, then $y = (B/A^2)x^2$ (since $x^2 = A^2 e^{-2t}$) and we obtain the parabolas shown in the figure.

If the roots are positive, then the direction of the arrows is reversed, i.e., every solution tends to infinity as $t \to \infty$. This kind of critical point is called an improper node.

b) When $r_1 > 0$ and $r_2 < 0$, it is possible for the direction of motion to be toward the critical point on some trajectories and away from the critical point on other trajectories. As a typical example, consider

$$\frac{dx}{dt} = -x , \quad \frac{dy}{dt} = 2y$$

which has the general solution

$$x(t) = Ae^{-t} , \quad y(t) = Be^{2t} ,$$

where A and B are arbitrary. (See Fig. 2). For any initial point not on the x-axis, $x(t) \to 0$, $y(t) \to +\infty$ depending on the sign of B, as $t \to \infty$. But if $(x(0), y(0))$ is on the x-axis, then $x(t) \to 0$, $y(t) = 0$ as $t \to \infty$. Thus, only the trajectories on the x-axis approach the critical point; all others approach infinity when $A \neq 0$, $B \neq 0$. The points on the trajectories are given by

$$y = \frac{BA^2}{x^2} \quad \text{(hyperbolae)}.$$

This kind of critical point is called a saddle-point.

● **PROBLEM** 18-30

The system of differential equations

$$\frac{dx}{dt} = ax + by \qquad\qquad (1)$$

$$\frac{dy}{dt} = cx + dy$$

has associated with it the characteristic equation $r^2 - (a+d)r + (ad-bc) = 0$. Discuss the nature of the critical points of (1) when the roots of the characteristic equation are real and equal.

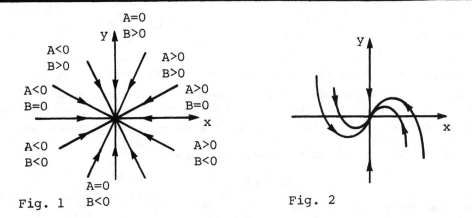

Fig. 1

Fig. 2

Solution: To see how we obtain the characteristic equation, rewrite the system (1) in matrix form:

$$\begin{pmatrix} x'(t) \\ y'(t) \end{pmatrix} = \begin{pmatrix} a & b \\ c & d \end{pmatrix} \begin{pmatrix} x \\ y \end{pmatrix} . \qquad (2)$$

Since the system has constant coefficients, let $x(t) = Ae^{rt}$, $y(t) = Be^{rt}$ be trial solutions. To find what conditions they must satisfy, differentiate $x(t)$, $y(t)$ and substitute into (2):

$$\begin{bmatrix} rAe^{rt} \\ rBe^{rt} \end{bmatrix} = \begin{bmatrix} a & b \\ c & d \end{bmatrix} \begin{bmatrix} Ae^{rt} \\ Be^{rt} \end{bmatrix}$$

or, since e^{rt} is never zero,

$$\begin{pmatrix} rA \\ rB \end{pmatrix} = \begin{pmatrix} a & b \\ c & d \end{pmatrix} \begin{pmatrix} A \\ B \end{pmatrix} \qquad \text{or,}$$

$$\begin{pmatrix} a-r & b \\ c & d-r \end{pmatrix} \begin{pmatrix} A \\ B \end{pmatrix} = \begin{pmatrix} 0 \\ 0 \end{pmatrix} . \qquad (3)$$

This system of homogeneous linear equations can have a non-trivial solution if and only if

$$\begin{vmatrix} a-r & b \\ c & d-r \end{vmatrix} = 0 .$$

Expanding the determinant we obtain the characteristic equation

$$r^2 - (a+d)r + (ad-bc) = 0 .$$

Since the system (1) has repeated eigenvalues the general solution is of the form

$$x(t) = (A_1 + A_2 t)e^{rt}, \quad y(t) = (B_1 + B_2 t)e^{rt} . \qquad (4)$$

If $r < 0$, then both $x(t)$ and $y(t)$ approach zero as $t \to \infty$ while if $r > 0$ the motion of all trajectories is away from the critical point. For simplicity, consider a system of the form (1) whose solution does not involve te^{rt}. For example, if

$$\frac{dx}{dt} = -x , \quad \frac{dy}{dt} = -y ,$$

the general solution is

$$x(t) = Ae^{-t} , \quad y(t) = Be^{-t} . \qquad (5)$$

Eliminating the parameter t in (5) we obtain:

$$y = Be^{-t} = B(x/A) = (B/A)x .$$

We see that the trajectories are straight lines through the origin with slope (B/A), as shown in Fig. 1. When $r > 0$, the direction of the arrows would be reversed. The critical point is called a proper node.

Now consider the case when the term te^{rt} is present in the solution of a system. A typical system is

$$\frac{dx}{dt} = -2x , \quad \frac{dy}{dt} = x - 2y$$

which has the general solution

$$x(t) = Ae^{-2t} , \quad y(t) = Be^{-2t} + Ate^{-2t} . \qquad (6)$$

We can eliminate the parameter t from (6) by separately considering $A > 0$, $A < 0$. Let $A > 0$. Then $e^{-2t} = x/A$ and hence

$$t = -\tfrac{1}{2} \ln(x/A) .$$

601

Then $y(x) = Bx/A - (x/2)\ln(x/A)$. For $A < 0$, $y(x) = -[Bx/A - (x/2)\ln(x/A]$.
The slope at any point can be obtained by recalling that

$$\frac{dy}{dx} = \frac{dy/dt}{dx/dt} = \frac{-2Be^{-2t} - 2Ate^{-2t} + Ae^{-2t}}{-2Ae^{-2t}}$$

$$= \frac{(-2B+A) - 2At}{-2A} .$$

For $t \to \infty$, $dy/dt \to \infty$; thus all of the trajectories enter the origin along the y-axis. (See Fig. 2). This type of critical point is called an improper node.

● **PROBLEM** 18-31

Let
$$\frac{dx}{dt} = ax + by$$

(1)

$$\frac{dy}{dt} = cx + dy .$$

Discuss the nature of the critical point of (1) when the roots of the associated characteristic equation are:
a)   complex;
b)   pure imaginary.

Solution: A complex number is an ordered pair of real numbers (a,b) that obeys the following multiplication rule:

$$(a,b)(c,d) = (ac - bd, ad + bc).$$

When $a = 0$, the number $(0,b)$ is said to be pure imaginary and

$$(0,b)(0,b) = -b^2 .$$

a)   In this case the general solution of (1) is

$$x = e^{\lambda t}[A_1 \cos \mu t + A_2 \sin \mu t] ,$$

$$y = e^{\lambda t}[B_1 \cos \mu t + B_2 \sin \mu t] .$$

To illustrate, consider the specific system

$$\frac{dx}{dt} = -x + 2y$$

$$\frac{dy}{dt} = -2x - y .$$

This can be written as

$$\begin{pmatrix} x' \\ y' \end{pmatrix} = \begin{pmatrix} -1 & 2 \\ -2 & -1 \end{pmatrix} \begin{pmatrix} x \\ y \end{pmatrix} .$$

Assuming that
$$\begin{bmatrix} x \\ y \end{bmatrix} = \begin{bmatrix} A \\ B \end{bmatrix} e^{rt} ,$$

we obtain the set of linear algebraic equations

$$\begin{pmatrix} -1-r & 2 \\ -2 & -1-r \end{pmatrix} \begin{pmatrix} A \\ B \end{pmatrix} = \begin{pmatrix} 0 \\ 0 \end{pmatrix}$$

(2)

which determine the eigenvalues and eigenvectors of the system.   The

602

characteristic equation is $r^2 + 2r + 5 = 0$ and the eigenvalues are $r_1 = -1 + 2i$, $r_2 = -1 - 2i$. From (2), the corresponding eigenvectors are

$$v_1 = \begin{bmatrix} -i \\ 1 \end{bmatrix}, \quad v_2 = \begin{bmatrix} 1 \\ -i \end{bmatrix}.$$

Hence a fundamental set of solutions is

$$\begin{pmatrix} x \\ y \end{pmatrix}^{(1)} = \begin{bmatrix} -i \\ 1 \end{bmatrix} e^{(-1+2i)t}, \quad \begin{pmatrix} x \\ y \end{pmatrix}^{(2)} = \begin{bmatrix} 1 \\ -i \end{bmatrix} e^{(-1-2i)t}$$

The general solution is therefore

$$x(t) = e^{-t}[A \cos 2t + B \sin 2t]$$
$$y(t) = e^{-t}[B \cos 2t - A \sin 2t] . \tag{3}$$

To sketch the corresponding trajectories, let
$$x = r \cos \theta , \quad y = r \sin \theta .$$

We must convert the integration constants into polar coordinates also. Let

$$R = (A^2 + B^2)^{\frac{1}{2}} , \quad R \cos \alpha = A, \quad R \sin \alpha = B.$$

After making these substitutions into (3) we obtain

$$r \cos \theta = Re^{-t} \cos (2t - \alpha)$$
$$r \sin \theta = -Re^{-t} \sin (2t - \alpha) ,$$

using the trigonometric identities $\cos(\beta - \gamma) = \cos \beta \cos \gamma + \sin \beta \sin \gamma$ and $\sin(\beta - \gamma) = \sin \beta \cos \gamma + \cos \alpha \sin \beta$. Thus,

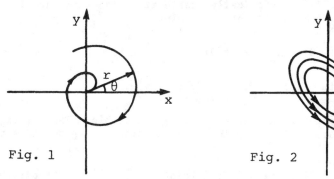

Fig. 1   Fig. 2

$r = Re^{-t}$; $\theta = -(2t - \alpha)$. Finally, eliminating t gives
$$r = Re^{(\theta - \alpha)/2} .$$

Since θ decreases with increasing t, the motion is clockwise. Since $\lambda = -1 < 0$, the direction of motion is toward the critical point $(0,0)$. The critical point shown in Fig. 1 is called a spiral point.

b) Here $\lambda = 0$ and the solutions are periodic in time and the trajectories are closed curves as shown in Fig. 2. The origin is called a center.

● **PROBLEM** 18-32

Let $\dfrac{dx}{dt} = F(x,y)$, $\dfrac{dy}{dt} = G(x,y)$ be an autonomous system. This system is said to be almost linear if

$$F(x,y) = ax + by + F_1(x,y) \qquad (1)$$
$$G(x,y) = cx + dy + G_1(x,y)$$

where $\dfrac{F_1(x,y)}{r} \to 0$, $\dfrac{G_1(x,y)}{r} \to 0$ as $r = (x^2 + y^2)^{\frac{1}{2}} \to 0$. The system
is linear if

$$F(x,y) = ax + by$$
$$G(x,y) = cx + dy. \qquad (2)$$

Discuss the relationships between the critical points of the almost
linear system (1) and the linear system (2).

Solution: A system of differential equations is said to be autonomous if
the parameter t does not appear on the right hand side.

Recall that for a linear system of the form (2), the nature of the
critical points depends upon the roots r_1, r_2 of the characteristic equation associated with (2). However, the nature of the critical point determines whether the system is stable, asymptotically stable, or unstable.
A critical point is stable if, given any $\epsilon > 0$, we can find a δ such
that every solution $x = \varphi(t)$, $y = \psi(t)$ of the autonomous system which at
$t = 0$ satisfies

$$\{ [\varphi(0) - x_0]^2 + [\psi(0) - y_0]^2 \}^{\frac{1}{2}} < \delta$$

exists and satisfies

$$\{ [\varphi(t) - x_0]^2 + [\psi(t) - y_0]^2 \}^{\frac{1}{2}} < \epsilon$$

for all $t \geq 0$.

A critical point is asymptotically stable if it is stable and if there
exists a δ_0, $0 < \delta_0 < \delta$ such that if a solution $x = \varphi(t)$, $y = \psi(t)$
satisfies

$$\{ [\varphi(0) - x_0]^2 + [\psi(0) - y_0]^2 \}^{\frac{1}{2}} < \delta_0$$

then

$$\lim_{t \to \infty} \varphi(t) = x_0, \quad \lim_{t \to \infty} \psi(t) = y_0.$$

Finally, a critical point is unstable if it is not stable. The relationships between the roots, critical point, and its stability properties are
as given below:

Roots	Type of Critical Point	Stability
1) $r_1, r_2 > 0$	Improper node	Unstable
2) $r_1, r_2 < 0$	Improper node	Asymptotically stable
3) $r_1 > 0, r_2 < 0$	Saddle-point	Unstable
4) $r_1 = r_2 > 0$	Node	Unstable
5) $r_1 = r_2 < 0$	Node	Asymptotically stable
6) $r_1, r_2 = \lambda \pm i\mu$	Spiral point	
$\lambda > 0$		Unstable
$\lambda < 0$		Asymptotically Unstable
7) $r_1 = i\mu, r_2 = -i\mu$	Center	Stable

604

Turning to the almost linear system, an advanced theorem in the theory of almost linear systems states that the types and stability of the critical point $(0,0)$ of the linear system (2) and the almost linear system (1) are identical except when the roots are equal or pure imaginary.

CHAPTER 19

ECONOMIC AND STATISTICAL APPLICATIONS

PROBLEMS IN PRODUCTION

Appalachian Creations has only three different daily production schedules. One given for the summer months of June, July and August; one for the winter months of December, January and February; and the third for the rest of the year:

Daily Production Table for June, July and August

Item	Factory		
	Danbury	Springfield	Rutland
Light sleeping bags	11	13	7
Heavy sleeping bags	3	4	0
Down Jackets	5	7	2
Two-person tents	8	15	0
Four-person tents	0	14	4

Daily Production Table for December, January and February

Item	Factory		
	Danbury	Springfield	Rutland
Light sleeping bags	2	5	0
Heavy sleeping bags	4	9	4
Down jackets	11	15	5
Two-person tents	12	18	0
Four-person tents	0	22	5

Daily Production Table for the other 6 months

Item	Factory		
	Danbury	Springfield	Rutland
Light sleeping bags	7	8	4
Heavy sleeping bags	4	6	3
Down Jackets	8	9	0
Two-person tents	10	16	0
Four-person tents	0	19	6

Construct a daily summer production matrix, a daily winter production matrix, and a daily off-season production matrix for the corresponding tables above.
2) What are the total summer production, total winter production, total off-season production and total yearly production matrices?

<u>Solution</u>: 1) Instead of having to always deal with a table in solving problems, matrices are usually helpful. A matrix is a rectangular array of numbers. The three daily production matrices are:

The daily summer production matrix:

$$\begin{bmatrix} 11 & 13 & 7 \\ 3 & 4 & 0 \\ 5 & 7 & 2 \\ 8 & 15 & 0 \\ 0 & 14 & 4 \end{bmatrix}$$

The daily winter production matrix:

$$\begin{bmatrix} 2 & 5 & 0 \\ 4 & 9 & 4 \\ 11 & 15 & 5 \\ 12 & 18 & 0 \\ 0 & 22 & 5 \end{bmatrix}$$

The daily off-season production matrix:

$$\begin{bmatrix} 7 & 8 & 4 \\ 4 & 6 & 3 \\ 8 & 9 & 0 \\ 10 & 16 & 0 \\ 0 & 19 & 6 \end{bmatrix}$$

2) Since each day's production is the same throughout the entire summer production period, we can use scalar multiplication to obtain the total summer production matrix. We simply multiply the daily summer matrix by 64, the total number of summer workdays in this particular year:

Total Summer Production Matrix:

$$64 \begin{bmatrix} 11 & 13 & 7 \\ 3 & 4 & 0 \\ 5 & 7 & 2 \\ 8 & 15 & 0 \\ 0 & 14 & 4 \end{bmatrix} = \begin{bmatrix} 704 & 832 & 448 \\ 192 & 256 & 0 \\ 320 & 448 & 128 \\ 512 & 960 & 0 \\ 0 & 896 & 256 \end{bmatrix}$$

Similarly, we obtain the total winter production matrix by multiplying by 63 and the total off-season production matrix by multiplying by 128:

Total Winter Production Matrix:

$$63 \begin{bmatrix} 2 & 5 & 0 \\ 7 & 9 & 4 \\ 11 & 15 & 5 \\ 12 & 18 & 0 \\ 0 & 22 & 5 \end{bmatrix} = \begin{bmatrix} 126 & 315 & 0 \\ 441 & 567 & 252 \\ 693 & 945 & 315 \\ 756 & 1134 & 0 \\ 0 & 1386 & 315 \end{bmatrix}$$

Total Off-Season Production Matrix:

$$128 \begin{bmatrix} 7 & 8 & 4 \\ 4 & 6 & 3 \\ 8 & 9 & 0 \\ 10 & 16 & 0 \\ 0 & 19 & 6 \end{bmatrix} = \begin{bmatrix} 896 & 1024 & 512 \\ 512 & 768 & 384 \\ 1024 & 1152 & 0 \\ 1280 & 2048 & 0 \\ 0 & 2432 & 768 \end{bmatrix}$$

To find the yearly production matrix we must add the three seasonal production matrices: Thus, the yearly production matrix is:

$$64 \begin{bmatrix} 11 & 13 & 7 \\ 3 & 4 & 0 \\ 5 & 7 & 2 \\ 8 & 15 & 0 \\ 0 & 14 & 4 \end{bmatrix} + 63 \begin{bmatrix} 2 & 5 & 0 \\ 7 & 9 & 4 \\ 11 & 15 & 5 \\ 12 & 18 & 0 \\ 0 & 22 & 5 \end{bmatrix} + 128 \begin{bmatrix} 7 & 8 & 4 \\ 4 & 6 & 3 \\ 8 & 9 & 0 \\ 10 & 16 & 0 \\ 0 & 19 & 6 \end{bmatrix}$$

$$= \begin{bmatrix} 704 & 832 & 448 \\ 192 & 256 & 0 \\ 320 & 448 & 128 \\ 512 & 960 & 0 \\ 0 & 896 & 256 \end{bmatrix} + \begin{bmatrix} 126 & 315 & 0 \\ 441 & 567 & 252 \\ 693 & 945 & 315 \\ 756 & 1134 & 0 \\ 0 & 1386 & 315 \end{bmatrix} + \begin{bmatrix} 896 & 1024 & 512 \\ 512 & 768 & 384 \\ 1024 & 1152 & 0 \\ 1280 & 2048 & 0 \\ 0 & 2432 & 768 \end{bmatrix}$$

$$= \begin{bmatrix} 704 + 126 + 896 & 832 + 315 + 1024 & 448 + 0 + 512 \\ 192 + 441 + 512 & 256 + 567 + 768 & 0 + 252 + 384 \\ 320 + 693 + 1024 & 448 + 945 + 1152 & 128 + 315 + 0 \\ 512 + 756 + 1280 & 960 + 1134 + 2048 & 0 + 0 + 0 \\ 0 + 0 + 0 & 896 + 1386 + 2432 & 256 + 315 + 768 \end{bmatrix}$$

$$= \begin{bmatrix} 1726 & 2171 & 960 \\ 1145 & 1591 & 636 \\ 2037 & 2545 & 443 \\ 2548 & 4142 & 0 \\ 0 & 4714 & 1339 \end{bmatrix}$$

● PROBLEM 19-2

The table below is the Springfield Plastics Company's production table:

Item	Factory			
	A	B	C	D
Wastebasket	2520	2000	1950	800
Hard Hat	3240	1780	1530	1360
Vacuum Cleaner Case	760	370	250	125

The table gives the price for each item as paid by each of 4 stores:

	Wastebasket	Hard Hat	Vacuum cleaner case
Store 1	$2	$1	$3
Store 2	4	3	2
Store 3	5	2	1
Store 4	3	2	4

Assuming each store buys one-fourth of each factory's output, use matrix operations to construct a table showing how much each store pays to each factory.

Solution: We can construct a production matrix corresponding to the production table:

$$\begin{bmatrix} 2520 & 2000 & 1950 & 800 \\ 3240 & 1780 & 1530 & 1360 \\ 760 & 370 & 250 & 125 \end{bmatrix} .$$

We can also construct a price matrix corresponding to the table of prices:

$$\begin{bmatrix} \$2 & \$1 & \$3 \\ 4 & 3 & 2 \\ 5 & 2 & 1 \\ 3 & 2 & 4 \end{bmatrix} .$$

Since each store buys one-fourth of each factory's output, then $\frac{1}{4} \times$ the production matrix, or

$$\frac{1}{4} \begin{bmatrix} 2520 & 2000 & 1950 & 800 \\ 3240 & 1780 & 1530 & 1360 \\ 760 & 370 & 250 & 125 \end{bmatrix}$$

is the amount each factory produces for each store. Therefore each store pays to each factory:

$$\begin{bmatrix} \$2 & \$1 & \$3 \\ 4 & 3 & 2 \\ 5 & 2 & 1 \\ 3 & 2 & 4 \end{bmatrix} \times \frac{1}{4} \begin{bmatrix} 2520 & 2000 & 1950 & 800 \\ 3240 & 1780 & 1530 & 1360 \\ 760 & 370 & 250 & 125 \end{bmatrix}$$

$$= \begin{bmatrix} \$2 & \$1 & \$3 \\ 4 & 3 & 2 \\ 5 & 2 & 1 \\ 3 & 2 & 4 \end{bmatrix} \begin{bmatrix} 630 & 500 & 487.5 & 200 \\ 810 & 445 & 382.5 & 340 \\ 190 & 92.5 & 62.5 & 31.25 \end{bmatrix} =$$

$$\begin{bmatrix} 1260 + 810 + 570 & 1000 + 445 + 277.5 & 975 + 382.5 + 187.5 & 400 + 340 + 93.75 \\ 2520 + 2430 + 380 & 2000 + 1335 + 185 & 1950 + 1147.5 + 125 & 800 + 1020 + 62.5 \\ 3150 + 1620 + 190 & 2500 + 890 + 92.5 & 2437.5 + 765 + 62.5 & 1000 + 680 + 31.25 \\ 1890 + 1620 + 760 & 1500 + 890 + 370 & 1462.5 + 765 + 250 & 600 + 680 + 125 \end{bmatrix}$$

$$= \begin{bmatrix} 2640 & 1722.5 & 1545 & 833.75 \\ 5330 & 3520 & 3222.5 & 1882.5 \\ 4960 & 3482.5 & 3265.0 & 1711.25 \\ 4270 & 2760 & 2477.5 & 1405 \end{bmatrix}$$

Therefore, a table showing how much each store pays to each factory is:

	Factory			
	A	B	C	D
Store 1	2640	1722.50	1545	833.75
Store 2	5330	3520	3222.5	1882.50
Store 3	4960	3482.50	3265.0	1711.25
Store 4	4270	2760	2477.5	1405

● **PROBLEM** 19-3

Suppose we are given the following model economy:

	Output		Final Demands	Gross outputs
	Agriculture	Manufacturing		
Agriculture	30	48	22	100
Manufacturing	60	24	36	120
Primary Imputs	10	48		

609

The table above is a model of a simple economy with only two industries.

Primary Inputs include land, labor, and any other inputs not currently produced by specific industries. Final Demands include those outputs that are not immediately plowed back into the production cycle, such as goods produced for households, government and export. The table says, for example, that there was $100 worth of total agricultural output and that to obtain this there was **$30 worth of agricultural investment, $60 worth of** manufacturing investment, and $10 worth of primary inputs. It also says that of the $100 worth of agricultural output, $30 is plowed back into agriculture, $48 worth is devoted to manufacturing, and $22 is available for households, government and export.

1) Using the above model economy describe an "open Leontief model".

2) Suppose we want to increase the agricultural sector of the demand vector from 22 to 24 while keeping the manufacturing sector of the final demands constant at 36. What must the gross outputs be if the Leontief matrix remains the same?

Solution: The Leontief matrix is that matrix obtained by dividing each element of the part of the above, enclosed in the inner rectangle by the total of the column in which the particular element lies. Thus the Leontief matrix for the model economy is

$$\begin{bmatrix} 30/100 & 48/120 \\ 60/100 & 24/120 \end{bmatrix} = \begin{bmatrix} .3 & .4 \\ .6 & .2 \end{bmatrix}$$

The action in the model economy can be written generally in matrix form:

$$\begin{bmatrix} \text{Leontief} \\ \text{Matrix} \end{bmatrix} \begin{bmatrix} \text{Gross} \\ \text{Outputs} \end{bmatrix} + \begin{bmatrix} \text{Final} \\ \text{Demands} \end{bmatrix} = \begin{bmatrix} \text{Gross} \\ \text{Outputs} \end{bmatrix} .$$

Specifically, the action in the model economy can be written as:

| Leontief Matrix | Gross Outputs | Final Demands | Gross Outputs |

$$\begin{bmatrix} .3 & .4 \\ .6 & .2 \end{bmatrix} \begin{bmatrix} 100 \\ 120 \end{bmatrix} + \begin{bmatrix} 22 \\ 36 \end{bmatrix} = \begin{bmatrix} 100 \\ 120 \end{bmatrix} .$$

This technological equation has matrix form $AX + D = X$ where A is the Leontief matrix, X is the gross outputs vector, and D is the final demands vector. The matrix A contains the technological coefficients a_{ij} which represent the amount of input from the ith industry going into the output of one unit of the jth industry. If all the elements of the matrix A in the equation $AX + D = X$ are all positive or zero and the sum of the elements in each column does not exceed 1 then A is called a Leontief matrix. The model economy is an open Leontief model because its technological equation includes final demands.

2) Since the agricultural sector of the demand vector increases from 22 to 24 and the manufacturing sector of the final demand vector remains the same at 36, the final demands vector D is now:

$$\begin{bmatrix} 24 \\ 36 \end{bmatrix}$$

We are looking for gross outputs, represented by the column vector X. Since our model economy is an open Leontief model it satisfies the technological equation $AX + D = X$ or $X - AX = D$ or $(I - A)X = D$. To solve for X multiply by $(I - A)^{-1}$ on both sides. Thus $X = (I - A)^{-1}D$.

610

In 1) we already found A to be
$$\begin{bmatrix} .3 & .4 \\ .6 & .2 \end{bmatrix}$$

Thus,

$$I-A = \begin{bmatrix} 1 & 0 \\ 0 & 1 \end{bmatrix} - \begin{bmatrix} .3 & .4 \\ .6 & .2 \end{bmatrix} = \begin{bmatrix} .7 & -.4 \\ -.6 & .8 \end{bmatrix}$$

Now, let us compute $(I - A)^{-1}$:

$$\left[\begin{array}{cc|cc} .7 & -.4 & 1 & 0 \\ -.6 & .8 & 0 & 1 \end{array}\right] \begin{array}{l} \text{Let } 10\text{x Row 1} \rightarrow \text{Row 1} \\ 10\text{x Row 2} \rightarrow \text{Row 2} \end{array}$$

$$\left[\begin{array}{cc|cc} 7 & -4 & 10 & 0 \\ -6 & 8 & 0 & 10 \end{array}\right] \text{Let Row 1 + Row 2} \rightarrow \text{Row 1}$$

$$\left[\begin{array}{cc|cc} 1 & 4 & 10 & 10 \\ -6 & 8 & 0 & 10 \end{array}\right] \text{Let Row 2 + 6 x Row 1} \rightarrow \text{Row 2}$$

$$\left[\begin{array}{cc|cc} 1 & 4 & 10 & 10 \\ 0 & 32 & 60 & 70 \end{array}\right] \text{Let } 1/32 \times \text{Row 2} \rightarrow \text{Row 2}$$

$$\left[\begin{array}{cc|cc} 1 & 4 & 10 & 10 \\ 0 & 1 & 15/8 & 35/16 \end{array}\right] \text{Let Row 1 - 4 x Row 2} \rightarrow \text{Row 1}$$

$$\left[\begin{array}{cc|cc} 1 & 0 & 5/2 & 5/4 \\ 0 & 1 & 15/8 & 35/16 \end{array}\right] \text{Thus,}$$

$$(I - A)^{-1} = \begin{bmatrix} 5/2 & 5/4 \\ 15/8 & 35/16 \end{bmatrix}$$

Therefore,
$$X = (I - A)^{-1} D = \begin{bmatrix} 5/2 & 5/4 \\ 15/8 & 35/16 \end{bmatrix} \begin{bmatrix} 24 \\ 36 \end{bmatrix} = \begin{bmatrix} 105 \\ 123.75 \end{bmatrix}$$

Thus, the gross outputs vector is now
$$\begin{bmatrix} 105 \\ 123.75 \end{bmatrix} .$$

The new technological equation is:
$$\begin{bmatrix} .3 & .4 \\ .6 & .2 \end{bmatrix} \begin{bmatrix} 105 \\ 123.75 \end{bmatrix} + \begin{bmatrix} 24 \\ 36 \end{bmatrix} = \begin{bmatrix} 105 \\ 123.75 \end{bmatrix} .$$

● **PROBLEM** 19-4

The table below describes the economy of an entire country. There are three
sectors: Industry, Government and Households.   The table below gives the
fractions of government budgets, industrial output and household budget
which is consumed by the government, industry and households.

		Fraction of			
		Government Budget	Industrial Output	Household Budget	
Consumed By	Government	.4	.2	.3	
	Industry	.3	.1	.1	
	Household	.3	.7	.6	

1) First, describe a closed Leontief model.
2) Supposing that the above economy is a closed Leontief model, find gross outputs.

Solution: A closed Leontief model is one in which no "final demands" vector and no "primary input" are included. What goes into the economy is the same as what goes out. The technological equation is then

$$\begin{pmatrix} \text{Leontied} \\ \text{Matrix} \end{pmatrix} \begin{pmatrix} \text{gross} \\ \text{outputs} \end{pmatrix} = \begin{pmatrix} \text{gross} \\ \text{outputs} \end{pmatrix}$$

or merely $AX = X$ where A is the Leontief matrix and X is gross outputs.

2) From the table above, we can see that the Leontief matrix, or

$$A = \begin{bmatrix} .4 & .2 & .3 \\ .3 & .1 & .1 \\ .3 & .7 & .6 \end{bmatrix}$$

Since we are considering a Leontief model, $AX = X$, $X = \begin{pmatrix} x \\ y \\ z \end{pmatrix}$ where

x = government's budget
y = value of industrial output
z = households' budget.

The equation $AX = X$ can be rewritten as:
$$AX - X = 0$$
then,
$$(A - I)X = 0 \tag{1}$$
Now,

$$A - I = \begin{bmatrix} .4 & .2 & .3 \\ .3 & .1 & .1 \\ .3 & .7 & .6 \end{bmatrix} - \begin{bmatrix} 1 & 0 & 0 \\ 0 & 1 & 0 \\ 0 & 0 & 1 \end{bmatrix} = \begin{bmatrix} -.6 & .2 & .3 \\ .3 & -.9 & .1 \\ .3 & .7 & -.4 \end{bmatrix}$$

Thus,

$$\begin{bmatrix} -.6 & .2 & .3 \\ .3 & -.9 & .1 \\ .3 & .7 & -.4 \end{bmatrix} \begin{pmatrix} x \\ y \\ z \end{pmatrix} = \begin{pmatrix} 0 \\ 0 \\ 0 \end{pmatrix}$$

Reduce
$$\begin{bmatrix} -.6 & .2 & .3 \\ .3 & -.9 & .1 \\ .3 & .7 & -.4 \end{bmatrix}$$
using elementary row operations:

$- \dfrac{10}{6} \times$ Row 1 $\rightarrow$ Row 1

$- 10 \times$ Row 2 $\rightarrow$ Row 2

$- 10 \times$ Row 3 $\rightarrow$ Row 3

$$\begin{bmatrix} 1 & -1/3 & -1/2 \\ -3 & 9 & -1 \\ -3 & -7 & 4 \end{bmatrix} \quad \begin{matrix} \text{Row } 2 + 3 \times \text{Row } 1 \rightarrow \text{Row } 2 \\ \text{Row } 3 + 3 \times \text{Row } 1 \rightarrow \text{Row } 3 \end{matrix}$$

$$\begin{bmatrix} 1 & -1/3 & -1/2 \\ 0 & 8 & -5/2 \\ 0 & -8 & 5/2 \end{bmatrix} \quad \text{Row } 3 + \text{Row } 2 \rightarrow \text{Row } 3$$

$$\begin{bmatrix} 1 & -1/3 & -1/2 \\ 0 & 8 & -5/2 \\ 0 & 0 & 0 \end{bmatrix} \quad 1/8 \times \text{Row } 2 \rightarrow \text{Row } 2$$

$$\begin{bmatrix} 1 & -1/3 & -1/2 \\ 0 & 1 & -5/16 \\ 0 & 0 & 0 \end{bmatrix} \quad \text{Row } 1 + 1/3 \times \text{Row } 2 \rightarrow \text{Row } 1$$

$$\begin{bmatrix} 1 & 0 & -29/48 \\ 0 & 1 & -5/16 \\ 0 & 0 & 0 \end{bmatrix}$$

From this last matrix, we have

$$\begin{bmatrix} 1 & 0 & -29/48 \\ 0 & 1 & -5/16 \\ 0 & 0 & 0 \end{bmatrix} \begin{bmatrix} x \\ y \\ z \end{bmatrix} = \begin{bmatrix} 0 \\ 0 \\ 0 \end{bmatrix}.$$

Thus, we conclude that $z = \alpha$, $x = 29/48\ \alpha$, $y = 5/16\ \alpha$. The economic interpretation is that an economy will satisfy the given equation if the government budget is $29/48$ times the total households' budget and the industrial budget is $5/16$ times the households' budget. Thus the gross output column vector is:

$$\begin{bmatrix} 29/48\ \alpha \\ 5/16\ \alpha \\ \alpha \end{bmatrix} \quad \text{for some} \quad \alpha > 0.$$

● **PROBLEM** 19-5

Suppose a building contractor has accepted orders for 5 ranch style houses, 7 Cape Cod houses and 12 colonial style houses. The table below gives the amount of each raw material going into each type of house expressed in convenient units:

	Steel	Wood	Glass	Paint	Labor
Ranch	5	20	16	7	17
Cape Cod	7	18	12	9	21
Colonial	6	25	8	5	13

1) Represent the builder's order by a row vector. Represent the table above in matrix form. Find how much steel, wood, glass, paint and labor the building needs.
2) Suppose that steel costs $15 per unit, wood cost $8 per unit, glass costs $5 per unit, paint costs $1 per unit and labor costs $10 per unit. Represent this as a column vector and find the total cost for the builder.

Solution: 1) The builder's order can be represented by means of a row vector: x = (5,7,12). Let

$$R = \begin{pmatrix} 5 & 20 & 16 & 7 & 17 \\ 7 & 18 & 12 & 9 & 21 \\ 6 & 25 & 8 & 5 & 13 \end{pmatrix}$$

The numbers in the matrix R give the amounts of each raw material going into each type of house. To find out how much steel, wood, glass, paint and labor the builder needs multiply:

$$xR = (5,7,12) \begin{pmatrix} 5 & 20 & 16 & 7 & 17 \\ 7 & 18 & 12 & 9 & 21 \\ 6 & 25 & 8 & 5 & 13 \end{pmatrix}$$

$$= (5 \cdot 5, + 7 \cdot 7 + 12 \cdot 6, 5 \cdot 20 + 7 \cdot 18 + 12 \cdot 25,$$

$$5 \cdot 16 + 7 \cdot 12 + 12 \cdot 8, 5 \cdot 7 + 7 \cdot 9 + 12 \cdot 5, 5 \cdot 17 + 7 \cdot 21 + 12 \cdot 13)$$

$$= (146, 526, 260, 158, 388) .$$

Thus, we see that the contractor should order 146 units of steel, 526 units of wood, 260 units of glass, 158 units of paint, and 388 units of labor.

2) From the steel, wood, glass, paint and labor costs, we can form a column vector

$$y = \begin{pmatrix} 15 \\ 8 \\ 5 \\ 1 \\ 10 \end{pmatrix}$$

To find the total cost, again multiply:

$$xRy = (5,7,12) \begin{pmatrix} 5 & 20 & 16 & 7 & 17 \\ 7 & 18 & 12 & 9 & 21 \\ 6 & 25 & 8 & 5 & 13 \end{pmatrix} \begin{pmatrix} 15 \\ 8 \\ 5 \\ 1 \\ 10 \end{pmatrix}$$

Therefore

$$xRy = (146, 526, 260, 158, 388) \begin{pmatrix} 15 \\ 8 \\ 5 \\ 1 \\ 10 \end{pmatrix}$$

$$= (146 \cdot 15 + 526 \cdot 8 + 260 \cdot 5 + 158 \cdot 1 + 388 \cdot 10)$$

$$= 11,736$$

Therefore, the total cost is $11,736.

● PROBLEM 19-6

Consider a firm with two production departments P_1 and P_2 and three service departments S_1, S_2 and S_3 as in the table below:

Department	Total Costs	Direct Costs, Dollars	Indirect Costs for Services from Department		
			S_1	S_2	S_3
S_1	x_1	600	$.25x_1$	$.15x_2$	$.15x_3$
S_2	x_2	1100	$.35x_1$	$.20x_2$	$.25x_3$
S_3	x_3	600	$.10x_1$	$.10x_2$	$.35x_3$
P_1	x_4	2100	$.15x_1$	$.25x_2$	$.15x_3$
P_2	x_5	1500	$.15x_1$	$.30x_2$	$.10x_3$
Totals			x_1	x_2	x_3

The total monthly costs of these departments are unknown and are denoted by x_1, x_2, x_3, x_4, x_5 in the table above. The direct monthly costs of the five departments are shown in the third column. The fourth, fifth and sixth columns show the allocation of charges for the services of S_1, S_2, S_3 to the various departments. Find the total costs for each department.

Solution: Since the total cost of a department is the direct plus indirect costs, the first three rows of the table yields the total costs for the three service departments. Thus,

$$
\begin{aligned}
x_1 &= 600 + .25x_1 + .15x_2 + .15x_3 \\
x_2 &= 1100 + .35x_1 + .20x_2 + .25x_3 \\
x_3 &= 600 + .10x_1 + .10x_2 + .35x_3
\end{aligned}
\tag{1}
$$

Let X, C and D denote the following matrices:

$$
X = \begin{pmatrix} x_1 \\ x_2 \\ x_3 \end{pmatrix} \quad
C = \begin{pmatrix} .25 & .15 & .15 \\ .35 & .20 & .25 \\ .10 & .10 & .35 \end{pmatrix} \quad
D = \begin{pmatrix} 600 \\ 1100 \\ 600 \end{pmatrix}
$$

Then the set of equations (1) can be written in matrix notation:

$$X = D + C \cdot X$$

which is equivalent to:

$$(I_3 - C) \cdot X = D .$$

The total costs of the three service departments can be obtained by solving the matrix equation for X. The solution is

$$X = (I_3 - C)^{-1} \cdot D .$$

Now,

$$
I_3 - C = \begin{pmatrix} 1 & 0 & 0 \\ 0 & 1 & 0 \\ 0 & 0 & 1 \end{pmatrix} - \begin{pmatrix} .25 & .15 & .15 \\ .35 & .20 & .25 \\ .10 & .10 & .35 \end{pmatrix}
$$

$$
= \begin{pmatrix} .75 & -.15 & -.15 \\ -.35 & .80 & -.25 \\ -.10 & -.10 & .65 \end{pmatrix}
$$

615

From the methods we have learned we can find

$$(I_3 - C)^{-1} = \begin{bmatrix} 1.57 & .36 & .50 \\ .79 & 1.49 & .76 \\ .36 & .28 & 1.73 \end{bmatrix}$$

Since $X = (I_3 - C)^{-1} \cdot D$ then

$$X = \begin{bmatrix} 1.57 & .36 & .50 \\ .79 & 1.49 & .76 \\ .36 & .28 & 1.73 \end{bmatrix} \cdot \begin{bmatrix} 600 \\ 1100 \\ 600 \end{bmatrix} = \begin{bmatrix} 1638.00 \\ 2569.00 \\ 1562.00 \end{bmatrix}$$

Thus $x_1 = \$1638.00$, $x_2 = \$2569.00$ and $x_3 = \$1562.00$.

Now, looking back at the table let us substitute what the actual values of x_1, x_2 and x_3 for the variables. Therefore our table now looks like:

Department	Total Costs	Direct Costs Dollars	Indirect Costs for Services from Departments, dollars		
			S_1	S_2	S_3
S_1	1638.00	600	409.50	385.35	234.30
S_2	2569.00	1100	573.30	513.80	390.50
S_3	1562.00	600	163.80	256.90	546.70
P_1	x_4	2100	245.70	642.25	234.30
P_2	x_5	1500	245.70	770.70	156.20

Now x_4 = total costs of P_1

= (direct costs + indirect costs) of P_1

= 2100 + 245.70 + 642.25 + 234.30

= 3222.25

and

x_5 = total costs of P_2

= (direct costs + indirect costs) of P_2

= 1500 + 245.70 + 770.70 + 156.20

= 2672.60 .

● PROBLEM 19-7

Take as our economy a chicken farm. Our goods are chickens and eggs. There are two possible processes for the chickens: laying eggs and hatching them. Let us assume that in a given month a chicken lays an average of 12 eggs if we use it for laying eggs. If used for hatching, it will hatch an average of four eggs per month.
1) Using this information construct two matrices - one for input and one for output.
2) If a farmer starts with three chickens and eight eggs ready for hatching, how many chickens will be used for laying eggs and how many will be used for hatching eggs? What will be the outputs?

616

Solution: 1) Let Goods represent columns in a matrix and Processes represent rows. Let Good 1 be a chicken and Good 2 be an egg. Process 1 is "laying eggs" and Process 2 is "hatching". For Process 1 only chickens are necessary. Therefore for 1 chicken and 0 eggs, the output will be 12 eggs and 1 (the original) chicken. For Process 2, for each chicken, 4 eggs is necessary. And the output will be 5 chickens (the original one and 4 hatched). Therefore the input matrix A and output matrix B are:

$$\begin{array}{cc} & \text{Chicken} \quad \text{Egg} \\ \begin{array}{l}\text{Laying eggs:} \\ \text{Hatching:}\end{array} A = & \begin{pmatrix} 1 & 0 \\ 1 & 4 \end{pmatrix} \end{array} , \quad B = \begin{array}{cc} \text{Chicken} \quad \text{Egg} \\ \begin{pmatrix} 1 & 12 \\ 5 & 0 \end{pmatrix}\end{array}$$

2) Let x_1 be the number of chickens which lay eggs and x_2 be the number of chickens which hatch. Therefore,

$$(x_1, x_2) \begin{pmatrix} 1 & 0 \\ 1 & 4 \end{pmatrix} = (x_1 + x_2, \ 4x_2)$$

where $x_1 + x_2$ are the number of chickens we are starting with and $4x_2$ are the number of eggs ready for hatching.

We are given that we are starting with 3 chickens and 8 eggs. Therefore,

$$(x_1, x_2) \begin{pmatrix} 1 & 0 \\ 1 & 4 \end{pmatrix} = (3 \quad 8)$$

$$(x_1, x_2) = (3 \quad 8) \begin{pmatrix} 1 & 0 \\ 1 & 4 \end{pmatrix}^{-1}$$

but

$$\begin{pmatrix} 1 & 0 \\ 1 & 4 \end{pmatrix}^{-1} = \begin{pmatrix} 1 & 0 \\ -1/4 & 1/4 \end{pmatrix} \quad \text{because}$$

$$\begin{pmatrix} 1 & 0 \\ 1 & 4 \end{pmatrix} \begin{pmatrix} 1 & 0 \\ -1/4 & 1/4 \end{pmatrix} = \begin{pmatrix} 1 & 0 \\ 0 & 1 \end{pmatrix} .$$

Therefore, $(x_1, x_2) = (3 \quad 8) \begin{pmatrix} 1 & 0 \\ -1/4 & 1/4 \end{pmatrix} = (1, 2)$.

1 chicken is used for laying eggs; 2 chickens are used for hatching.

For x_1, x_2 as above:

$$(x_1, x_2) B = (x_1, x_2) \begin{pmatrix} 1 & 12 \\ 5 & 0 \end{pmatrix} = (x_1 + 5x_2, \ 12x_1)$$

where $x_1 + 5x_2$ are the number of chickens that result and $12x_1$ are the number of eggs which result. Therefore,

$$(1, 2) \begin{pmatrix} 1 & 12 \\ 5 & 0 \end{pmatrix} = (11, \ 12)$$

The farmer now has 11 chickens and 12 eggs.

● **PROBLEM** 19-8

Consider a simple society consisting of a farmer who produces all the food, a tailor who makes all the clothes, and a builder who builds all the homes in the community. Suppose that during the year the portion of each commodity consumed by each of the individuals is given by the following table:

		Goods produced by		
		Farmer	Carpenter	Tailor
Goods	Farmer	7/16	1/2	3/16
consumed	Carpenter	5/16	1/6	5/16
by	Tailor	1/4	1/3	1/2

Suppose also that one unit of each commodity is produced during the year. Assume that every one pays the same price for a commodity. Our aim is to achieve a state of equilibrium: No one makes money or loses money. In order to achieve this state of equilibrium what must be the incomes of the farmer, builder and tailor?

Solution: Let p_1, p_2, and p_3 be the annual incomes received by the farmer, builder, and tailor, respectively. From our table we find the farmer's expenditures to be:

$$7/16 \ p_1 + 1/2 \ p_2 + 3/16 \ p_3 \ .$$

In order to achieve a state of equilibrium, the farmer's expenditures must equal his income. Therefore,

$$7/16 \ p_1 + 1/2 \ p_2 + 3/16 \ p_3 = p_1 \qquad (1)$$

Similarly for the carpenter:

$$5/16 \ p_1 + 1/6 \ p_2 + 5/16 \ p_3 = p_2 \qquad (2)$$

and the tailor:

$$1/4 \ p_1 + 1/3 \ p_2 + 1/2 \ p_3 = p_3 \ . \qquad (3)$$

The equations (1), (2) and (3) can be rewritten in matrix form as

$$AP = P \qquad (4)$$

where

$$A = \begin{bmatrix} 7/16 & 1/2 & 3/16 \\ 5/16 & 1/6 & 5/16 \\ 1/4 & 1/3 & 1/2 \end{bmatrix} \quad P = \begin{bmatrix} p_1 \\ p_2 \\ p_3 \end{bmatrix}$$

(4) can be rewritten as:

$$\begin{bmatrix} 7/16 & 1/2 & 3/16 \\ 5/16 & 1/6 & 5/16 \\ 1/4 & 1/3 & 1/2 \end{bmatrix} \begin{bmatrix} p_1 \\ p_2 \\ p_3 \end{bmatrix} - \begin{bmatrix} 1 & 0 & 0 \\ 0 & 1 & 0 \\ 0 & 0 & 1 \end{bmatrix} \begin{bmatrix} p_1 \\ p_2 \\ p_3 \end{bmatrix} = 0$$

Therefore,

$$\begin{bmatrix} (7/16)-1 & 1/2 & 3/16 \\ 5/16 & (1/6)-1 & 5/16 \\ 1/4 & 1/3 & (1/2)-1 \end{bmatrix} \begin{bmatrix} p_1 \\ p_2 \\ p_3 \end{bmatrix} = \begin{bmatrix} -9/16 & 1/2 & 3/16 \\ 5/16 & -5/6 & 5/16 \\ 1/4 & 1/3 & -1/2 \end{bmatrix} \begin{bmatrix} p_1 \\ p_2 \\ p_3 \end{bmatrix}$$

$$= 0$$

$$\qquad (5)$$

$$\begin{bmatrix} -9/16 & 1/2 & 3/16 \\ 5/16 & -5/6 & 5/16 \\ 1/4 & 1/3 & -1/2 \end{bmatrix} \text{ is equivalent to } \begin{bmatrix} 1 & 0 & -1 \\ 0 & 1 & -3/4 \\ 0 & 0 & 0 \end{bmatrix}$$

by elementary row operations. Therefore (5) is equivalent to:

$$\begin{bmatrix} 1 & 0 & -1 \\ 0 & 1 & -3/4 \\ 0 & 0 & 0 \end{bmatrix} \begin{bmatrix} p_1 \\ p_2 \\ p_3 \end{bmatrix} = 0$$

and $p_1 = + p_3$

$p_2 = 3/4 \, p_3$.

Let r be any real number then the solution to our original system of equations is:

$$P = r \begin{pmatrix} 1 \\ 3/4 \\ 1 \end{pmatrix}$$

Therefore this shows that the farmer and tailor have the same income. The carpenter's income is three-fourths of the farmer or tailor's income.

● **PROBLEM** 19-9

Suppose that n countries $C_1, C_2, \ldots, C_n$ are engaged in trading with each other and that a common currency is in use. Assume that prices are fixed and that C_j's income y_j comes entirely from selling its goods either internally or to other countries. A fraction of C_j's income that is spent on imports from C_i is a fixed number a_{ij} which does not depend upon C_j's income y_j .

1) First, find the income matrix (a matrix which includes the income of each country) for the n countries, whose international trade model is described above.

2) Consider the international trade model consisting of three countries C_1, C_2 and C_3 . Suppose that the fraction of C_1's income spent on imports from C_1 is 1/4, from C_2 is 1/2 and from C_3 is 1/4; that the fraction of C_2's income spent on imports from C_1 is 2/5, from C_2 is 1/5 and from C_3 is 2/5; the fraction of C_3's income spent on imports from C_1 is 1/2, from C_2 is 1/2 and from C_3 is 0. Find the income of each country.

Solution: 1) We were already given that a_{ij} is the fraction of C_j's income that is spent on imports from C_i . Therefore $a_{ij} \geq 0$ and

$$a_{1j} + a_{2j} + a_{3j} + \ldots + a_{nj} = 1 \tag{1}$$

for $j = 1, 2, 3, \ldots, n$.

Statement (1) must be true: Each a_{ij} represents a fraction of C_j's income. Therefore the sum of the fractions of C_j's income must add up to 1. Note that we treat domestic production of C_j as imports from C_j . Therefore we can now introduce our exchange matrix A:

$$A = \begin{pmatrix} a_{11} & a_{12} & a_{13} & \cdots & a_{1n} \\ a_{21} & a_{22} & a_{23} & \cdots & a_{2n} \\ a_{31} & a_{32} & a_{33} & \cdots & a_{3n} \\ \hline a_{n1} & a_{n2} & a_{n3} & \cdots & a_{nn} \end{pmatrix}$$

Now, we must determine the total income for each country C_i: Since a_{ij} is the fraction of C_j's income that is spent on imports from C_i then

$a_{ij}y_j$ is the value of C_i's exports to C_j. Therefore the total income of C_i is $a_{i1}y_1 + a_{i2}y_2 + \ldots + a_{in}y_n$. But, y_i is the total income of C_i. Therefore, $a_{i1}y_1 + a_{i2}y_2 + \ldots + a_{in}y_n = y_i$. Therefore the incomes of $C_1, C_2, \ldots, C_n$ are:

$$y_1 = a_{11}y_1 + a_{12}y_2 + \ldots + a_{1n}y_n$$

$$y_2 = a_{21}y_1 + a_{22}y_2 + \ldots + a_{2n}y_n \qquad (2)$$
$$- - - - - - - - - - - - - - - -$$
$$y_n = a_{n1}y_1 + a_{n2}y_2 + \ldots + a_{nn}y_n .$$

Then, we can write (2) in matrix form:

$$\begin{bmatrix} a_{11} & a_{12} & \cdots & a_{1n} \\ a_{21} & a_{22} & \cdots & a_{2n} \\ - & - & - & - \\ a_{n1} & a_{n2} & \cdots & a_{nn} \end{bmatrix} \begin{bmatrix} y_1 \\ y_2 \\ \cdot \\ y_n \end{bmatrix} = \begin{bmatrix} y_1 \\ y_2 \\ \cdot \\ y_n \end{bmatrix} \qquad (3)$$

Letting $\begin{bmatrix} y_1 \\ y_2 \\ \vdots \\ y_n \end{bmatrix} = P$, we can rewrite (3) as $AP = P$ or as $(A - I)P = 0$ where P is the income matrix.

2) From the definition of a_{ij} above we can readily see that we can say:

$$a_{11} = 1/4 \qquad a_{12} = 2/5 \qquad a_{13} = 1/2$$
$$a_{21} = 1/2 \qquad a_{22} = 1/5 \qquad a_{23} = 1/2$$
$$a_{31} = 1/4 \qquad a_{32} = 2/5 \qquad a_{33} = 0 .$$

Therefore our exchange matrix is:

$$A = \begin{bmatrix} 1/4 & 2/5 & 1/2 \\ 1/2 & 1/5 & 1/2 \\ 1/4 & 2/5 & 0 \end{bmatrix}$$

From our previous example we now know: $AP = P$. Therefore,

$$\begin{bmatrix} 1/4 & 2/5 & 1/2 \\ 1/2 & 1/5 & 1/2 \\ 1/4 & 2/5 & 0 \end{bmatrix} \begin{bmatrix} y_1 \\ y_2 \\ y_3 \end{bmatrix} = \begin{bmatrix} y_1 \\ y_2 \\ y_3 \end{bmatrix}$$

and therefore:

$$\begin{bmatrix} 1/4 -1 & 2/5 & 1/2 \\ 1/2 & 1/5 -1 & 1/2 \\ 1/4 & 2/5 & -1 \end{bmatrix} \begin{bmatrix} y_1 \\ y_2 \\ y_3 \end{bmatrix} = \begin{bmatrix} 0 \\ 0 \\ 0 \end{bmatrix} \qquad (3)$$

We will now use elementary row operations to reduce the matrix:

$$\begin{bmatrix} -3/4 & 2/5 & 1/2 \\ 1/2 & -4/5 & 1/2 \\ 1/4 & 2/5 & -1 \end{bmatrix} \text{ to echelon form.}$$

Multiply the first row by $-4/3$:

$$\begin{bmatrix} 1 & -8/15 & -2/3 \\ 1/2 & -4/5 & 1/2 \\ 1/4 & 2/5 & -1 \end{bmatrix}$$

Add $-1/2$ the first row to the second row:
Add $-1/4$ the first row to the third row:

$$\begin{bmatrix} 1 & -8/15 & -2/3 \\ 0 & -8/15 & 5/6 \\ 0 & 8/15 & -5/6 \end{bmatrix}$$

Add the second row to the third row:

$$\begin{bmatrix} 1 & -8/15 & -2/3 \\ 0 & -8/15 & 5/6 \\ 0 & 0 & 0 \end{bmatrix}$$

Therefore (3) now becomes

$$\begin{bmatrix} 1 & -8/15 & -2/3 \\ 0 & -8/15 & 5/6 \\ 0 & 0 & 0 \end{bmatrix} \begin{bmatrix} y_1 \\ y_2 \\ y_3 \end{bmatrix} = \begin{bmatrix} 0 \\ 0 \\ 0 \end{bmatrix} \qquad (4)$$

Let $y_3 = Y$. From the second row of (4) we get

$$-8/15 \; y_2 + 5/6 \; y_3 = -8/15 \; y_2 + 5/6 \; Y = 0 \; .$$

Therefore,
$$y_2 = \frac{5}{6} \cdot \frac{15}{8} \, Y = \frac{25}{16} \, Y \; .$$

From the first row of (4) we get:

$$y_1 - \frac{8}{15} \, y_2 - \frac{2}{3} \, y_3 = 0$$
$$y_1 = \frac{8}{15} \, y_2 + \frac{2}{3} \, y_3$$
$$= \frac{8}{15} \, (\frac{25}{16} \, Y) + \frac{2}{3} \, Y$$
$$= (\frac{5}{6} \, Y + \frac{2}{3} \, Y) = \frac{9}{6} \, Y = \frac{3}{2} \, Y \; .$$

Therefore our income matrix becomes:

$$P = \begin{bmatrix} y_1 \\ y_2 \\ y_3 \end{bmatrix} = Y \begin{bmatrix} 3/2 \\ 25/16 \\ 1 \end{bmatrix}$$

This shows that the income of C_1 is $3/2$ × income of C_3 and the income of C_2 is $25/16$ × income of C_3 .

PROBLEMS IN CONSUMPTION

Construct a simple model economy in which there are three industries: the crude oil industry, the refining industry that produces gasoline and the utility industry that supplies electricity. Suppose there are three types of consumers: the general public, the U.S. government and the export firms. The following chart gives the number of units that consumers need of crude oil, gasoline and electricity:

	crude oil	gasoline	electricity
crude oil industry	0	4	2
refining industry	8	0	6
utility industry	1	6	0
public	1	9	5
U.S. government	8	8	8
export firms	7	2	0

The price of crude oil is \$4 per unit, the price of gasoline is \$3 per unit and the price of electricity is \$2 per unit.
1) Find the demand vectors for the industries and consumers.
2) Find the total demand vector.
3) Find the price vector.
4) Find total costs for the industries.
5) Find the income vector.
6) Find the profits (or losses) of the three industries.

Solution: 1) Let the demand vector be α_{demand} where $\alpha_{demand} = (a,b,c)$;

a is the number of units needed of crude oil; b is the number of units needed of gasoline; c is the number of units needed of electricity. Therefore, from the chart we can find the demand vectors.

$\alpha_{crude\ oil} = (0,4,2)$

$\alpha_{refining} = (8,0,6)$

$\alpha_{utility} = (1,6,0)$

$\alpha_{public} = (1,9,5)$

$\alpha_{gov't} = (8,8,8)$

$\alpha_{export} = (7,2,0)$

2) The total demand is found by vector addition:

$\alpha_{total} = (0,4,2) + (8,0,6) + (1,6,0) + (1,9,5) + (8,8,8) + (7,2,0)$

$= (25,29,21)$.

3) Since crude oil's price is \$4 per unit, gasoline's price is \$3 per unit, electricity's price is \$2 per unit, the price vector can be expressed as $\beta = (4,3,2)$, where the entries in B correspond to the entries in α; i.e., the 1st entry corresponds to crude oil, the 2nd entry corresponds to gasoline, and the 3rd entry corresponds to electricity.
4) Note that (price of an item) x (quantity of an item) = cost incurred by that item. Therefore total cost = sum of costs incurred by each item. This is the same as the dot product of two vectors. Therefore, total cost = $\alpha_{demand} \cdot \beta$ total costs for crude oil industry: $(0,4,2) \cdot (4,3,2) =$

$0 + 12 + 4 = 16$ total costs for refining industry: $(8,0,6) \cdot (4,3,2) =$

$32 + 0 + 12 = 44$. Total costs for utility industry: $(1,6,0) \cdot (0,4,2) = 0 + 24 + 0 = 24$.

5) Income vector $= \alpha_{total} \cdot \beta$

$= (25,29,21) \cdot (4,3,2)$

$= (100,87,42)$

Therefore the income of the crude oil industry is 100, the income of the refining industry is 87 and the income of the utility industry is 42.

6) Since we found the total costs to be 16, 44, 24 for crude oil industry, refining industry and utility industry respectively, then the total cost vector can be written as: $(16,44,24)$. Therefore profits (or losses) = income - total cost

$= (100,87,42) - (16,44,24)$

$= (84,43,38)$.

Eash industry made a profit: crude oil industry's profit is $84, refining industry's profit is $43 and utility industry's profit is $38.

● **PROBLEM** 19-11

A certain fruit grower in Florida has a boxcar loaded with fruit ready to be shipped north. The load consists of 900 boxes of oranges, 700 boxes of grapefruit, and 400 boxes of tangerines. The market prices, per box, of the different types of fruit in various cities are given by the following chart.

	Oranges	Grapefruit	Tangerines
New York	$4 per box	$2 per box	$3 per box
Cleveland	$5 per box	$1 per box	$2 per box
St. Louis	$4 per box	$3 per box	$2 per box
Oklahoma City	$3 per box	$2 per box	$5 per box

To which city should the carload of fruit be sent in order for the grower to get maximum gross receipts for his fruit?

Solution: Consider the chart above as a price matrix:

$$\begin{bmatrix} 4 & 2 & 3 \\ 5 & 1 & 2 \\ 4 & 3 & 2 \\ 3 & 2 & 5 \end{bmatrix}$$

Now, form the quantity matrix:

$$\begin{bmatrix} 900 \text{ boxes} \\ 700 \text{ boxes} \\ 400 \text{ boxes} \end{bmatrix}.$$

Notice price of an item $\times$ quantity of an item gives you the income made on each item. Therefore, the income matrix can be obtained by matrix multiplication:

$$\begin{bmatrix} 4 & 2 & 3 \\ 5 & 1 & 2 \\ 4 & 3 & 2 \\ 3 & 2 & 5 \end{bmatrix} \begin{bmatrix} 900 \\ 700 \\ 400 \end{bmatrix} = \begin{bmatrix} 3600 + 1400 + 1200 \\ 4500 + 700 + 800 \\ 3600 + 2100 + 800 \\ 2700 + 1400 + 2000 \end{bmatrix} = \begin{bmatrix} 6200 \\ 6000 \\ 6500 \\ 6100 \end{bmatrix}$$

Note that in our income matrix the largest entry is 6500. The third entry represents St. Louis. Therefore the greatest income will come from St. Louis.

623

The I. N. Vestor Company loaned out a total of $30,000, part at 6% and the rest at 9%. The annual dividends from both investments was the same amount as that earned by the total loan if invested at 7%. Find the amount loaned out at each rate.

Solution: Let x be the amount loaned at 6% and y be the amount at 9%. Then from the given conditions, we are able to set up the system of equations:

$$x + y = 30,000$$
$$.06x + .09y = .07(30,000)$$

which reduces to the equivalent system of equations:

$$x + y = 30,000$$
$$2x + 3y = 70,000 \qquad (1)$$

Let $A = \begin{pmatrix} 1 & 1 \\ 2 & 3 \end{pmatrix}$, $X = \begin{pmatrix} x \\ y \end{pmatrix}$, and $C = \begin{pmatrix} 30,000 \\ 70,000 \end{pmatrix}$. Then (1) can be rewritten using matrices: $AX = C \qquad (2)$

Now, multiply each side of (2) by A^{-1}:

$$X = A^{-1}C .$$

Thus, we must find A^{-1}. First write the matrix $[A \mid I_n]$: $\begin{pmatrix} 1 & 1 & \vdots & 1 & 0 \\ 2 & 3 & \vdots & 0 & 1 \end{pmatrix}$

Replace the second row by the second row plus $-2x$ the first row:

$$\begin{pmatrix} 1 & 1 & \vdots & 1 & 0 \\ 0 & 1 & \vdots & -2 & 1 \end{pmatrix} .$$

Replace the first row by the first row plus $-1x$ the second row:

$$\begin{pmatrix} 1 & 0 & \vdots & 3 & -1 \\ 0 & 1 & \vdots & -2 & 1 \end{pmatrix}$$

Thus

$$A^{-1} = \begin{pmatrix} 3 & -1 \\ -2 & 1 \end{pmatrix} .$$

Therefore, $X = A^{-1}C$ becomes

$$\begin{pmatrix} x \\ y \end{pmatrix} = \begin{pmatrix} 3 & -1 \\ -2 & 1 \end{pmatrix} \begin{pmatrix} 30,000 \\ 70,000 \end{pmatrix} .$$

And by matrix multiplication we obtain:

$$\begin{pmatrix} x \\ y \end{pmatrix} = \begin{pmatrix} 90,000 - 70,000 \\ -60,000 + 70,000 \end{pmatrix} = \begin{pmatrix} 20,000 \\ 10,000 \end{pmatrix} .$$

The solution matrix,

$$\begin{pmatrix} 20,000 \\ 10,000 \end{pmatrix} ,$$

tells us that $x = \$20,000$ and $y = \$10,000$ are the desired amounts.

A rancher sold 25 hogs and 60 sheep to Mr. Kay for $3450. At the same prices, he sold 35 hogs and 50 sheep to Mr. Bea for $3300. Find the price of each hog and sheep.

<u>Solution</u>: Let y be the price of one sheep and x be the price of a hog. Since the rancher sold 25 hogs and 60 sheep for $3450 then 25x + 60y = 3450. Similarly, 35x + 50y = 3300. By Cramer's Rule the linear system,

$$a_1x + b_1y = c_1$$
$$a_2x + b_2y = c_2$$

is solved by using determinants:

$$x = \frac{\begin{vmatrix} c_1 & b_1 \\ c_2 & b_2 \end{vmatrix}}{\begin{vmatrix} a_1 & b_1 \\ a_2 & b_2 \end{vmatrix}} \quad \text{and} \quad y = \frac{\begin{vmatrix} a_1 & c_1 \\ a_2 & c_2 \end{vmatrix}}{\begin{vmatrix} a_1 & b_1 \\ a_2 & b_2 \end{vmatrix}}$$

Therefore our linear system:

$$25x + 60y = 3450$$
$$35x + 50y = 3300$$

can be easily solved by Cramer's Rule.

$$x = \frac{\begin{vmatrix} 3450 & 60 \\ 3300 & 50 \end{vmatrix}}{\begin{vmatrix} 25 & 60 \\ 35 & 50 \end{vmatrix}} \quad \text{and} \quad y = \frac{\begin{vmatrix} 25 & 3450 \\ 35 & 3300 \end{vmatrix}}{\begin{vmatrix} 25 & 60 \\ 35 & 50 \end{vmatrix}}$$

Thus,

$$x = \frac{3450 \cdot 50 - 3300 \cdot 60}{25 \cdot 50 - 35 \cdot 60}, \quad y = \frac{25 \cdot 3300 - 35 \cdot 3450}{25 \cdot 50 - 35 \cdot 60}$$

$$x = 30 \qquad\qquad y = 45$$

Therefore the rancher sells hogs for $30 and sheep for $45.

● **PROBLEM 19-14**

Ms. Wong has a total of $4200 invested in securities A,B, and C. The rates of annual dividends are 4% , 6% and 5% respectively, yielding total annual dividends of $214. If the sum of A and B is twice C, find the amount invested in each security.

<u>Solution</u>: From the given conditions, we can form the system of linear equations:

$$A + B + C = 4200$$
$$.04A + .06B + .05C = 214$$
$$A + B - 2C = 0 \ .$$

By multiplying through by 100, the second equation is equivalent to 4A + 6B + 5C = 21400. We can rewrite the equations above in matrix form:

$$\begin{pmatrix} 1 & 1 & 1 \\ 4 & 6 & 5 \\ 1 & 1 & -2 \end{pmatrix} \begin{pmatrix} A \\ B \\ C \end{pmatrix} = \begin{pmatrix} 4200 \\ 21400 \\ 0 \end{pmatrix}$$

We can solve for A,B and C by Cramer's Rule. Cramer's Rule states that for a system of equations

$$a_{11}x_1 + a_{12}x_2 + a_{13}x_3 = c_1$$
$$a_{21}x_1 + a_{22}x_2 + a_{23}x_3 = c_2$$
$$a_{31}x_1 + a_{32}x_2 + a_{33}x_3 = c_3 \quad,$$

$$x_1 = \frac{\begin{vmatrix} c_1 & a_{12} & a_{13} \\ c_2 & a_{22} & a_{23} \\ c_3 & a_{32} & a_{33} \end{vmatrix}}{\begin{vmatrix} a_{11} & a_{12} & a_{13} \\ a_{21} & a_{22} & a_{23} \\ a_{31} & a_{32} & a_{33} \end{vmatrix}} \quad ; \quad x_2 = \frac{\begin{vmatrix} a_{11} & c_1 & a_{13} \\ a_{21} & c_2 & a_{23} \\ a_{31} & c_3 & a_{33} \end{vmatrix}}{\begin{vmatrix} a_{11} & a_{12} & a_{13} \\ a_{21} & a_{22} & a_{23} \\ a_{31} & a_{32} & a_{33} \end{vmatrix}} \quad ;$$

$$x_3 = \frac{\begin{vmatrix} a_{11} & a_{12} & c_1 \\ a_{21} & a_{22} & c_2 \\ a_{31} & a_{23} & c_3 \end{vmatrix}}{\begin{vmatrix} a_{11} & a_{12} & a_{13} \\ a_{21} & a_{22} & a_{23} \\ a_{31} & a_{32} & a_{33} \end{vmatrix}} \quad ;$$

Thus,

$$A = \frac{\begin{vmatrix} 4200 & 1 & 1 \\ 21400 & 6 & 5 \\ 0 & 1 & -2 \end{vmatrix}}{\begin{vmatrix} 1 & 1 & 1 \\ 4 & 6 & 5 \\ 1 & 1 & -2 \end{vmatrix}}$$

$$= \frac{4200(6)(-2) + 1(5)(0) + 1(21400)(1) - 1(6)(0) - 1(21400)(-2) - 4200(5)(1)}{1(6)(-2) + 1(5)(1) + 1(4)(1) - 1(6)(1) - 1(4)(-2) - 1(5)(1)}$$

$$= \frac{-7200}{-6} = 1200 \quad.$$

$$B = \frac{\begin{vmatrix} 1 & 4200 & 1 \\ 4 & 21400 & 5 \\ 1 & 0 & 2 \end{vmatrix}}{\begin{vmatrix} 1 & 1 & 1 \\ 4 & 6 & 5 \\ 1 & 1 & -2 \end{vmatrix}}$$

$$= \frac{1(21400)(2) + 4200(5)(1) + 1(4)(0) - 1(21400)(1) - 4200(4)(2) - 1(5)(0)}{-6}$$

$$= \frac{-9600}{-6} = 1600$$

626

$$C = \frac{\begin{vmatrix} 1 & 1 & 4200 \\ 4 & 6 & 21400 \\ 1 & 1 & 0 \end{vmatrix}}{\begin{vmatrix} 1 & 1 & 1 \\ 4 & 6 & 5 \\ 1 & 1 & -2 \end{vmatrix}}$$

$$= \frac{1(6)(0) + 1(21400)(1) + 4200(4)(1) - 4200(6)(1) - 1(4)(0) - 1(21400)(1)}{-6}$$

$$= \frac{-8400}{-6}$$

$$= 1400 \ .$$

Thus, Ms. Wong has $1200 invested in security A, $1600 invested in security B, and $1400 invested in security C.

● **PROBLEM** 19-15

United States Oil is a producer of gasoline, oil and natural gas. To produce the gasoline, oil, and natural gas, some of these products are used by the company. Suppose that to produce 1 unit of gasoline, the company used 0 units of gasoline, 1 unit of oil, and 1 unit of natural gas. To produce 1 unit of oil, the company uses 0 unit of gasoline, 1/5 unit of oil and 2/5 unit of natural gas. For 1 unit of natural gas, the company uses 1/5 unit of gasoline, 2/5 unit of oil and 1/5 unit of natural gas. United States Oil has buyers for 100 units of gasoline, 100 units of oil, and 100 units of natural gas. How many units of each of these should be produced to meet the demand?

Solution: From the above data we can form a consumption matrix C.

$$C = \begin{matrix} & \text{Gasoline} & \text{Oil} & \text{Gas} \\ & \begin{pmatrix} 0 & 0 & 1/5 \\ 1 & 1/5 & 2/5 \\ 1 & 2/5 & 1/5 \end{pmatrix} & \begin{matrix} \text{Gasoline used} \\ \text{Oil used} \\ \text{Natural Gas used} \end{matrix} \end{matrix}$$

Let x,y, and z denote the daily production of units of gasoline, oil, and natural gas, respectively. The production matrix P is

$$P = \begin{pmatrix} x \\ y \\ z \end{pmatrix}$$ The product CP gives the

amount of gasoline, oil, and natural gas used by the company. Let D denote the demand. Since United States Oil has buyers for 100 units of gasoline, 100 units of oil, and 100 units of natural gas then:

$$D = \begin{pmatrix} 100 \\ 100 \\ 100 \end{pmatrix} \ .$$

To calculate the needed production, the amount used by the company for production must be subtracted from the amount produced. Therefore, P - CP = D or

$$\begin{pmatrix} 1 & 0 & 0 \\ 0 & 1 & 0 \\ 0 & 0 & 1 \end{pmatrix} \begin{pmatrix} x \\ y \\ z \end{pmatrix} - \begin{pmatrix} 0 & 0 & 1/5 \\ 1 & 1/5 & 2/5 \\ 1 & 2/5 & 1/5 \end{pmatrix} \begin{pmatrix} x \\ y \\ z \end{pmatrix} = \begin{pmatrix} 100 \\ 100 \\ 100 \end{pmatrix}$$

Thus,

$$\begin{pmatrix} 1 & 0 & -1/5 \\ -1 & 4/5 & -2/5 \\ -1 & -2/5 & 4/5 \end{pmatrix} \begin{pmatrix} x \\ y \\ z \end{pmatrix} = \begin{pmatrix} 100 \\ 100 \\ 100 \end{pmatrix}$$

Therefore:

$$\begin{pmatrix} x \\ y \\ z \end{pmatrix} = \begin{pmatrix} 1 & 0 & -1/5 \\ -1 & 4/5 & -2/5 \\ -1 & -2/5 & 4/5 \end{pmatrix}^{-1} \begin{pmatrix} 100 \\ 100 \\ 100 \end{pmatrix} \quad (1)$$

First let us find

$$\begin{pmatrix} 1 & 0 & -1/5 \\ -1 & 4/5 & -2/5 \\ -1 & -2/5 & 4/5 \end{pmatrix}^{-1} \quad (2)$$

To find the inverse of (2) first write

$$\begin{pmatrix} 1 & 0 & -1/5 & 1 & 0 & 0 \\ -1 & 4/5 & -2/5 & 0 & 1 & 0 \\ -1 & -2/5 & 4/5 & 0 & 0 & 1 \end{pmatrix} \quad (3)$$

Using elementary row operations reduce (3) to a form such as

$$\left(\begin{matrix} 1 & 0 & 0 \\ 0 & 1 & 0 & M \\ 0 & 0 & 1 \end{matrix} \right) \quad \text{where } M \text{ is a } 3 \times 3 \text{ matrix}$$

whose inverse is (2). From (3): Replace the second row by the second row plus the first row: Replace the third row by the third row plus the first row:

$$\begin{pmatrix} 1 & 0 & -1/5 & 1 & 0 & 0 \\ 0 & 4/5 & -3/5 & 1 & 1 & 0 \\ 0 & -2/5 & 3/5 & 1 & 0 & 1 \end{pmatrix}$$

Replace the second row by 5/4 × the second row:

$$\begin{pmatrix} 1 & 0 & -1/5 & 1 & 0 & 0 \\ 0 & 1 & -3/4 & 5/4 & 5/4 & 0 \\ 0 & -2/5 & 3/5 & 1 & 0 & 1 \end{pmatrix}$$

Replace the third row by (5/2 × the third row) + the second row:

$$\begin{pmatrix} 1 & 0 & -1/5 & 1 & 0 & 0 \\ 0 & 1 & -3/4 & 5/4 & 5/4 & 0 \\ 0 & 0 & 3/4 & 15/4 & 5/4 & 5/2 \end{pmatrix}$$

Replace the third row by 4/3 × the third row:

$$\begin{pmatrix} 1 & 0 & -1/5 & 1 & 0 & 0 \\ 0 & 1 & -3/4 & 5/4 & 5/4 & 0 \\ 0 & 0 & 1 & 5 & 5/3 & 10/3 \end{pmatrix}$$

Replace the first row by the first row plus 1/5 × the third row: Replace the second row by the second row plus 3/4 × the third row:

$$\begin{pmatrix} 1 & 0 & 0 & 2 & 1/3 & 2/3 \\ 0 & 1 & 0 & 5 & 5/2 & 5/2 \\ 0 & 0 & 1 & 5 & 5/3 & 10/3 \end{pmatrix}$$

Therefore:

$$
\begin{pmatrix} 1 & 0 & -1/5 \\ -1 & 4/5 & -2/5 \\ -1 & -2/5 & 4/5 \end{pmatrix}^{-1} = \begin{pmatrix} 2 & 1/3 & 2/3 \\ 5 & 5/2 & 5/2 \\ 5 & 5/3 & 10/3 \end{pmatrix}
$$

Substitute this result in (1):

$$
\begin{pmatrix} x \\ y \\ z \end{pmatrix} = \begin{pmatrix} 2 & 1/3 & 2/3 \\ 5 & 5/2 & 5/2 \\ 5 & 5/3 & 10/3 \end{pmatrix} \begin{pmatrix} 100 \\ 100 \\ 100 \end{pmatrix}
$$

By matrix multiplication we have:

$$
\begin{pmatrix} x \\ y \\ z \end{pmatrix} = \begin{pmatrix} 300 \\ 1000 \\ 1000 \end{pmatrix}
$$

The company must produce 300 units of gasoline, 1000 units of oil, and 1000 units of natural gas to meet the demand D.

● PROBLEM 19-16

Given the consumption matrix, $C = \begin{pmatrix} -1 & 3 \\ 1 & -1 \end{pmatrix}$ and given a demand vector $D = [d_1, d_2]$, can we obtain a production vector $X = [x_1, x_2]$ such that the demand D is met without any surplus?

Solution: Let us first define: comsumption matrix, production vector, demand vector and net production. Suppose that we have n goods $G_1, G_2, \ldots, G_n$ and n activities $M_1, M_2, \ldots, M_n$. Assume that each activity M_i produces only one good G_i and that G_i is only produced by M_i. Let c_{ij} be the amount of G_j that has to be consumed to produce one unit of G_i. The matrix $C = [c_{ij}]$ is called the consumption matrix. $X = [x_1, x_2, \ldots, x_n]$ is called the production vector if x_i is the number of units of G_i produced in a fixed period of time. $D = [d_1, d_2, \ldots, d_n]$ is called the demand vector where $d_i \geq 0$ are the demands for the goods G_i. Since C_i gives the amounts necessary to produce one unit of G_i, the amounts necessary to produce x_i units of G_i are given by $x_i C_i$. Thus, the total amount consumed by the model is $x_1 C_1 + x_2 C_2 + \ldots + x_n C_n = XC$. Net production is the total produced minus the total consumed; that is:

$$ X - XC = X(I_n - C) . $$

In order for the demand to be met without any surplus, net production must equal demand:

$$ X(I_n - C) = D . \tag{1} $$

Given the consumption matrix,

$$
C = \begin{pmatrix} -1 & 3 \\ 1 & -1 \end{pmatrix}, \quad I_2 - C = \begin{pmatrix} 1 & 0 \\ 0 & 1 \end{pmatrix} - \begin{pmatrix} -1 & 3 \\ 1 & -1 \end{pmatrix} = \begin{pmatrix} 2 & -3 \\ -1 & 2 \end{pmatrix} .
$$

Therefore equation (1) becomes:

$$[x_1, x_2] \begin{pmatrix} 2 & -3 \\ -1 & 2 \end{pmatrix} = [d_1, d_2] \quad \text{and}$$

$$[x_1, x_2] = [d_1, d_2] \begin{pmatrix} 2 & -3 \\ -1 & 2 \end{pmatrix}^{-1} \tag{2}$$

Now, note: $\begin{pmatrix} 2 & -3 \\ -1 & 2 \end{pmatrix}^{-1} = \begin{pmatrix} 2 & 3 \\ 1 & 2 \end{pmatrix}$ since $\begin{pmatrix} 2 & 3 \\ 1 & 2 \end{pmatrix}\begin{pmatrix} 2 & -3 \\ -1 & 2 \end{pmatrix} =$

$$\begin{pmatrix} 4-3 & -6+6 \\ 2-2 & -3+4 \end{pmatrix} = \begin{pmatrix} 1 & 0 \\ 0 & 1 \end{pmatrix} .$$

(2) becomes:

$$[x_1, x_2] = [d_1, d_2] \begin{pmatrix} 2 & 3 \\ 1 & 2 \end{pmatrix} .$$

Since d_1, d_2 are ≥ 0 and the matrix

$$\begin{pmatrix} 2 & 3 \\ 1 & 2 \end{pmatrix}$$

has all positive entries then $X = [x_1, x_2] \geq 0$. Thus we can obtain a production vector for any given demand vector.

● **PROBLEM** 19-17

Let $C = \begin{pmatrix} 0 & 2 \\ 2 & 0 \end{pmatrix}$ be a consumption matrix. Let $D = [d_1, d_2]$ where $d_1, d_2 \geq 0$ be a demand vector. For C and D above can we obtain a production vector $X = [x_1, x_2]$ such that the demand D is met without any surplus?

Solution: To have no surplus:

$$\text{demand} = \text{net production}$$
$$= \text{total produced} - \text{total consumed}.$$

Therefore,

$$D = X - XC = X(I_n - C)$$

and so,

$$X = D(I_n - C)^{-1} \tag{1}$$

$$I_2 - C = \begin{pmatrix} 1 & 0 \\ 0 & 1 \end{pmatrix} - \begin{pmatrix} 0 & 2 \\ 2 & 0 \end{pmatrix} = \begin{pmatrix} 1 & -2 \\ -2 & 1 \end{pmatrix} .$$

Note:

$$\begin{pmatrix} 1 & -2 \\ -2 & 1 \end{pmatrix}^{-1} = \begin{pmatrix} -1/3 & -2/3 \\ -2/3 & -1/3 \end{pmatrix} \quad \text{since}$$

$$\begin{pmatrix} -1/3 & -2/3 \\ -2/3 & -1/3 \end{pmatrix} \begin{pmatrix} 1 & -2 \\ -2 & 1 \end{pmatrix} = \begin{pmatrix} -1/3+4/3 & +2/3-2/3 \\ -2/3+2/3 & +4/3-1/3 \end{pmatrix} = \begin{pmatrix} 1 & 0 \\ 0 & 1 \end{pmatrix} .$$

Therefore, $(I_2 - C)^{-1} = \begin{pmatrix} -1/3 & -2/3 \\ -2/3 & -1/3 \end{pmatrix} .$

Thus, (1) becomes:

$$[x_1, x_2] = [d_1, d_2] \begin{pmatrix} -1/3 & -2/3 \\ -2/3 & -1/3 \end{pmatrix}$$

or,

$$X = D \begin{pmatrix} -1/3 & -2/3 \\ -2/3 & -1/3 \end{pmatrix} .$$

For $D \not= 0$, X is not a production vector, since all its components are negative. Thus, the problem has no solution. If $D = 0$, we do have a solution, namely $X = 0$, which means that if there is no demand, nothing is produced. In general, if $(I_n - C)^{-1} \geq 0$, then $X = D(I_n - C)^{-1} \geq 0$ is a production vector for any given demand vector.

● **PROBLEM** 19-18

R_1, R_2, R_3 and R_4 are 4 regions which trade in a certain nonrenewable commodity. The trade of this commodity is represented by the matrix

$$C = \begin{pmatrix} .1 & .3 & 0 & 0 \\ .7 & .6 & 0 & 0 \\ .1 & 0 & 1 & 0 \\ .1 & .1 & 0 & 1 \end{pmatrix}$$

where c_{ij} is the fraction of the commodity present at R_j which is shipped to R_i. If v_1, v_2, v_3 and v_4 are the amounts at the commodity present at R_1, R_2, R_3 and R_4, respectively, how much of the commodity is present at each location after n days?

Solution: $c_{ij} v_j$ is the amount of the commodity that will be shipped from R_j to R_i on the first day. Therefore, after one day the amount present at R_i will be $\sum_{j=1}^{4} c_{ij} v_j$. (The sum of the amounts shipped from each R_j to R_i). The matrix formed by placing in its ij^{th} entry, $\sum_{j=1}^{4} c_{ij} v_j$ is the matrix which represents the distribution of the commodity after one day. But the matrix formed with these entries is simply the matrix CV where V is the column vector of v_1, v_2, v_3 and v_4. To find the amount present after two days we again employ C to find the amounts of the commodity that will be shipped. So the amount present at R_i after two days will be $\sum_{j=1}^{4} c_{ij} (\sum_{j=1}^{4} c_{ij} v_j)$ or $\sum_{j=1}^{4} c_{ij} (CV)$ which is in fact the matrix product $C^2 V$ (a column vector of four entries where in the ith row is the amount of the commodity present at R_i). In general the amount of the commodity present at R_i after n days will be the amount in the ith row of the column vector $C^n V$. The problem then becomes computation of C^n. This task can be somewhat simplified if we notice that C can be rewritten in block matrix form.

$$C = \begin{pmatrix} X & 0 \\ Y & I \end{pmatrix}$$ where 0 and I are the 2 x 2 0 and identity matrices, respectively and

$$X = \begin{pmatrix} .1 & .3 \\ .7 & .6 \end{pmatrix} \quad \text{and} \quad Y = \begin{pmatrix} .1 & 0 \\ .1 & .1 \end{pmatrix} .$$

We can perform operations on a block matrix in the same way we do on an ordinary matrix.

$$C^2 = \begin{pmatrix} XX + 0Y & X0 + 0I \\ XY + IY & 0I + II \end{pmatrix} = \begin{pmatrix} X^2 & 0 \\ (X+I)Y & I \end{pmatrix}$$

Similarly,

$$C^n = \begin{pmatrix} X^n & 0 \\ \left(\sum_{j=0}^{n-1} X^j\right)Y & I \end{pmatrix} .$$

For example to find the amount of the commodity in each location after two days we proceed as follows.

$$X^2 = \begin{pmatrix} (.1)^2 + (.3)(.7) & .1(.3) + .3(.6) \\ .1(.7) + .6(.7) & .3(.7) + (.6)^2 \end{pmatrix}$$

$$= \begin{pmatrix} .22 & .21 \\ .49 & .57 \end{pmatrix}$$

$$(X+I)Y = \begin{pmatrix} 1.1 & .3 \\ .7 & 1.6 \end{pmatrix} \begin{pmatrix} .1 & 0 \\ .1 & .1 \end{pmatrix} = \begin{pmatrix} .14 & .03 \\ .23 & .16 \end{pmatrix} .$$

Thus

$$C^2 V = \begin{pmatrix} .22 & .21 & 0 & 0 \\ .49 & .57 & 0 & 0 \\ .14 & .03 & 1 & 0 \\ .23 & .16 & 0 & 1 \end{pmatrix} \begin{pmatrix} v_1 \\ v_2 \\ v_3 \\ v_4 \end{pmatrix}$$

So, in R_3 for example, there will be

$$.14v_1 + .03v_2 + v_3 \quad \text{units of the commodity after two days.}$$

● PROBLEM 19-19

There are two suppliers, firms R and C, of a new specialized type of tire that has 100,000 customers. Initially C has all 100,000 customers. Each company can advertise its product on TV or in the newspapers. A marketing firm determines that if both firms advertise on TV, then firm R gets 40,000 customers (and C gets 60,000). If they both use newspapers then each gets 50,000 customers. If R uses newspapers and C uses TV, then R gets 60,000 customers. If R uses TV and C uses newspapers, they each get 50,000 customers.
(1) Find the payoff matrix and discuss the game.
(2) Define saddle point, and strictly determined. Now, find the best course of action.

Solution: The payoff matrix is:

		T.V.	Newspapers
Firm R	T.V.	40,000	50,000
	Newspapers	60,000	50,000

The entries in the matrix indicate the number of customers secured by firm R. a_{ij} represents the number of customers given up by C to R if R chooses his i^{th} move and C chooses his j^{th} move. If player R plays his i^{th} move, he is assured of winning at least the smallest entry in the ith row of the matrix, no matter what C does. Thus R's best course of action is to choose that move which will maximize his assured winnings in spite of C's best countermove. Player R will get his largest payoff by maximizing his smallest gain. Player C's goals are in direct conflict with those of player R: he is trying to keep R's winnings to a minimum. If C plays his j^{th} move, he is assured of losing no more than the largest entry in the jth column of the matrix, no matter what R does. Thus, C's best course of action is to choose that move which will minimize his assured losses in spite of R's best countermove. Player C will do his best by minimizing his largest loss.

2) First, let us define saddle point and strictly determined:
 Definition:
 If the payoff matrix of a matrix game contains an entry a_{rs}, which is at the same time the minimum of row r and the maximum of column s, then a_{rs} is called a saddle point. Also, a_{rs} is called the value of the game, and if the value is zero, the game is said to be fair.

Definition: A matrix game is said to be strictly determined if its payoff matrix has a saddle point.
 Consider again the advertising game with payoff matrix:

		Firm C		
		T.V.	Newspapers	Row Minima
Firm R	T.V.	40,000	50,000	40,000
	Newspapers	60,000	50,000	50,000
Column maxima		60,000	50,000	

The entry $a_{22} = 50,000$ is a saddle point. 50,000 is the minimum of the second row and the maximum of the second column. Therefore the best course of action for both firms is to advertise in newspapers. The game is strictly determined with value 50,000.

GAMES & PROGRAMMING

● PROBLEM 19-20

1) Two stores, R and C, are planning to locate in one of two towns. Town 1 has 60 percent of the population while town 2 has 40 percent. If both stores locate in the same town they will split the total business of both towns equally, but if they locate in different towns each will get the business of that town. Where should each store locate?

2) Let us consider an extension of the above example. Stores R and C are trying to locate in one of three towns. The matrix game is:

```
                    Store  C  locates in

                      |  1        2        3
                      |
      Store R      1  |  50       50       80
      locates in   2  |  50       50       80
                   3  |  20       20       80
```

The entries in the matrix above represent the percentages of business that store R gets in each case. Where should each store locate?

Solution: 1) By the information given, we know that our payoff matrix is:

```
                    Store  C  locates in
                      |  1        2
                      |
      Store R      1  |  50       60
      locates in   2  |  40       50
```

The entries of the matrix represent the percentages of business that store R gets in each case.

Definition: A game defined by a matrix is said to be strictly determined if and only if there is an entry of the matrix that is the smallest element in its row and is also the largest element in its column. This entry is then called a saddle point and is the value of the game.

Therefore, let us find the maxima of the columns and the minima of the rows.

```
                 Store  C
                   |  1       2      Row Minima
                   |
   Store R   1     |  50      60        50
                   |
           2       |  40      50        40

   Column
   Maxima             50      60
```

Entry $a_{11} = 50$ is a saddle point. Hence it is the best strategy for both stores to locate in Town 1.

2) Let us examine the matrix game for column maxima and row minima.

```
                  Store  C  locates in
                    |  1     2     3     Row Minima:
                    |
   Store R     1    |  50    50    80       50
   locates     2    |  50    50    80       50          (1)
      in       3    |  20    20    50       20
   Column Maxima:   |  50    50    80
```

Note that each of the four 50 entries in the 2 × 2 matrix in the upper left-hand corner of (1) above is a saddle value of the matrix, since each is simultaneously the minimum of its row and maximum of its column. Note the 50 entry in the lower right-hand corner is not a saddle value. The game is strictly determined with optimal strategies:

For store R: "Locate in either town 1 or town 2"
For store C: "Locate in either town 1 or town 2".

634

Suppose that two costume companies each make clown, skeleton, and space costumes. They all sell for the same amount and use the same machinery and workmanship. Furthermore, the market for costumes is fixed; a certain given number of total costumes will be sold this Halloween. But each company has its own individual styles which affect how the costumes sell. On the basis of past experience, the following matrix has been inferred. It indicates, for example, that if both make clown outfits, then for every 20 that are sold, company I will lose 2 sales to company II. Similarly, if company I makes clown outfits and company II makes space suits, then for each 20 sold, company I will sell 4 more than company II.

		Company II		
		Clown	Skeleton	Space
Company	Clown	-2	0	4
I	Skeleton	0	2	1
	Space	-1	-4	0

How should each company plan its manufacturing?

<u>Solution:</u> Company I can quickly decide not to make space outfits, because it will do better with skeleton costumes, no matter what company II does. Similarly, company II should not make space outfits, because it can always be better off by making clown outfits, no matter what the choice of company I. Thus we have reduced the matrix as follows:

$$\begin{bmatrix} -2 & 0 & 4 \\ 0 & 2 & 1 \\ -1 & -1 & 0 \end{bmatrix}$$

With the remaining 2 by 2 matrix, it is clear that company I will profit most from playing the second row, and that company II should always play the first column. Thus this game also has a saddle point. It is $a_{21} = 0$.

The value of the game is 0; company I should always make skeleton outfits and company II should always make clown costumes.

Consider a buyer who at the beginning of each month decides whether to buy brand 1 or brand 2 that month. Each month the buyer will select either brand 1 or 2. The selection he makes at the beginning of any one month depends at most on the selection he made in the immediately preceding month and not on any other previous selections. Let p_{ij} denote the probability that at the beginning of any month the buyer selects brand j given that he bought brand i the preceding month. Suppose we know that $p_{11} = 3/4$, $p_{12} = 1/4$, $p_{21} = 1/2$ and $p_{22} = 1/2$.

1) First draw the transition diagram.
2) Assuming that the first month under consideration is January, determine the distribution of probabilities for April.

<u>Solution:</u> 1) The transition diagram is:

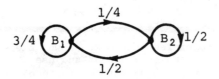

1/4

3/4 B_1 B_2 1/2

1/2

(Note: B_1 and B_2 represent brand 1 and brand 2, respectively.)

2) Since $P_{11} = 3/4$, $P_{12} = 1/4$, $P_{21} = 1/2$ and $P_{22} = 1/2$ then the transition matrix is:

$$P = \begin{pmatrix} 3/4 & 1/4 \\ 1/2 & 1/2 \end{pmatrix}$$

If $p^{(0)} = [p_1{}^{(0)} \quad p_2{}^{(0)}]$ denotes the probability vector for January, then we know the probability vector for February is $p^{(0)}P$, the probability vector for March is $p^{(0)}P^2$, etc. Since April is in the third step of the process, its probability vector is $p^{(0)}P^3$. We calculate that

$$P^3 = \begin{pmatrix} 3/4 & 1/4 \\ 1/2 & 1/2 \end{pmatrix} \begin{pmatrix} 3/4 & 1/4 \\ 1/2 & 1/2 \end{pmatrix} \begin{pmatrix} 3/4 & 1/4 \\ 1/2 & 1/2 \end{pmatrix}$$

$$= \begin{pmatrix} 43/64 & 21/64 \\ 21/32 & 11/32 \end{pmatrix} \quad ,$$

so

$$p^{(3)} = p^{(0)} \begin{pmatrix} 43/64 & 21/64 \\ 21/32 & 11/32 \end{pmatrix}$$

$$= [p_1{}^{(0)} \quad p_2{}^{(0)}] \begin{pmatrix} 43/64 & 21/64 \\ 21/32 & 11/32 \end{pmatrix}$$

$$= [43/64 \ p_1{}^{(0)} + 21/32 \ p_2{}^{(0)} \quad 21/64 \ p_1{}^{(0)} + 11/32 \ p_2{}^{(0)}]$$

● **PROBLEM** 19-23

Consider a market that is controlled by two brands B_1 and B_2. Suppose it is known that 10% of the buyers of each brand switch to the other brand during any given month. At the beginning of a given month 500 buyers are divided so that 300 purchase B_1 and 200 purchase B_2. What will be the number of buyers who will purchase B_1 and B_2 during the given month?

Solution: The movement of buyers in the market during the month can be described by the product of a buyer vector, B_1 and a transition matrix, T. Let $B = [b_1, b_2]$ where for $i = 1, 2$, b_i represents the number of buyers who purchase brand B_i at the beginning of the month. The transition matrix, T, on the other hand, is

$$T = [p_{ij}] = \begin{pmatrix} p_{11} & p_{12} \\ p_{21} & p_{22} \end{pmatrix} \quad ,$$

where for $i = 1, 2$ and $j = 1, 2$, p_{ij} is the probability that a current buyer of brand B_i will buy brand B_j during the month in question. In this example we then have: $B = [300 \quad 200]$ and

$$T = \begin{bmatrix} .9 & .1 \\ .1 & .9 \end{bmatrix}.$$

Therefore, $BT = [300 \quad 200] \begin{bmatrix} .9 & .1 \\ .1 & .9 \end{bmatrix}$

$$= [270+20 \quad 30+180]$$

$$= [290 \quad 210] .$$

The entries in the product matrix therefore represent the number of buyers who will purchase B_1 and B_2 in the given month. Therefore 290 will buy product B_1 and 210 will buy product B_2.

● **PROBLEM** 19-24

A petrol company recently conducted an advertising campaign in which a plastic reproduction of the head of a footballer was given away with each purchase. There were 16 heads in the set. How many purchases had to be made on the average in order to collect a complete set?

Solution: Let us assume that the various footballers were randomly distributed in the petrol stations, and that we can take no account of people making the task easier by swapping. There are seventeen states, which may be labelled 0 to 16, according to the number of footballers you already have. If you already have r footballers, the probability that you will get a duplicate next time and remain in the same state is $r/16$; the probability that you will get a new one and move into the next state is $(16-r)/16$. Therefore, the matrix of transition is a 17×17 matrix:

$$Q = \begin{bmatrix} 0 & 16 & 0 & 0 & . & . & . \\ 0 & 1 & 15 & 0 & . & . & . \\ 0 & 0 & 2 & 14 & . & . & . \\ 0 & 0 & 0 & 3 & . & . & . \\ . & . & . & . & . & . & . \\ . & . & . & . & . & 15 & 1 \\ . & . & . & . & . & 0 & 1 \end{bmatrix} \div 16$$

Here, we are now dealing with Absorbing Markhov chains. Let us cit some definitions and theorems and then we will proceed:

A state in a Markhov chain is in an absorbing state if it is impossible to leave it. A Markhov chain is absorbing if
(1)   It has at least one absorbing state and,
(2)   From every state it is possible to go to an absorbing state (not necessarily in one step).
Theorem: $N = (I - Q)^{-1}$ is the fundamental matrix for an absorbing chain. The entries of N give the mean number of times in each nonabsorbing state for each possible nonabsorbing state.
Lemma: If $N = (I - Q)^{-1}$ as in the theorem above, and we add the entries in a row, we shall have the mean number of times in any of the nonabsorbing states for a given starting time -- that is the mean time required before being absorbed. From the theorem above, $N = (I - Q)^{-1}$. By methods we have studied $N = (I - Q)^{-1}$ can be found to be:

$$16 \begin{bmatrix} 1/16 & 1/15 & 1/14 & \cdot & \cdot & 1/2 & 1 \\ 0 & 1/15 & 1/14 & \cdot & \cdot & 1/2 & 1 \\ 0 & 0 & 1/14 & \cdot & \cdot & 1/2 & 1 \\ \cdot & \cdot & \cdot & \cdot & \cdot & \cdot & \cdot \\ 0 & 0 & 0 & \cdot & \cdot & 0 & 1 \end{bmatrix}$$

From the Lemma above the average number of purchases required to go from state 0 to state 16 is therefore:

$$16(1 + \tfrac{1}{2} + 1/3 + \ldots + 1/16) .$$

A good approximation to the term in parenthesis can be obtained from the formula for Euler's limit:

$$1 + 1/2 + 1/3 + \ldots + 1/n\text{-}1 - \log n \to .577.$$

Therefore:

$1 + 1/2 + 1/3 + \ldots + 1/n\text{-}1$ is approximately $\ln n + .577$.

Thus:

$$16(1 + 1/2 + 1/3 + \ldots + 1/16) \approx 16 \cdot (\ln 17 + .577)$$

$$16(1 + 1/2 + 1/3 + \ldots + 1/16) \approx 16 \cdot (3.4)$$

$$16(1 + 1/2 + 1/3 + \ldots + 1/16) \approx 54$$

Therefore, the average number of purchases required is about 54.

● **PROBLEM** 19-25

A certain textile mill finishes cotton cloth obtained from weaving mills. The mill turns out two styles of cloth, a lightly printed style and a heavily printed one. The mill's output during a week is limited only by the capacity of its equipment for two of the finishing operations -- printing and bleaching -- and not by demand considerations. The maximum weekly output of the printing machinery is 800 thousand yards of cloth if the light pattern is printed exclusively, 400 thousand yards if the heavy pattern is printed exclusively, or any combination on the printing line $L + 2H = 800$ (where L represents light pattern and H heavy pattern). In a week, the maximum the bleaching equipment can handle is 500 thousand yards of the light-patterned cloth exclusively, 550 thousand yards of the heavy-patterned cloth exclusively, or any combination on the bleaching line, $1.1 L + H = 550$. The mill gained \$300 and \$290 per thousand yards of the light -- and heavy-patterned cloths, respectively.
1) Draw the graph of the two lines described above.
2) Solve the linear programming problem of maximizing the gain from a week's production.

__Solution:__ 1)

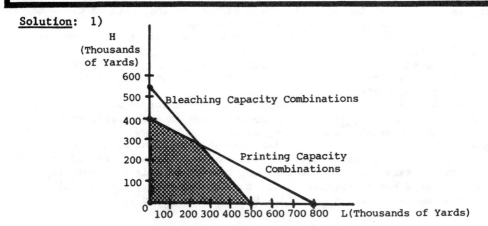

2)   We will use the graphical approach to solve this linear programming problem. We will utilize the graph above. Looking at the graph, we notice that any combination of L and H that falls in the area between the two axes and the printing line can be printed during a week. Again, any combination of L and H that falls in the area between the two axes and the bleaching line can be bleached in one week.

The rules for solving a linear programming problem through a graphical approach are:

i)   Graph all the constraining inequalities to obtain a picture of the feasible region.

ii)   Solve the corresponding equations to find the vertices, or corners of the feasible area.

iii) Find the vertex point which yields the maximum (or minimum) value of the function under consideration.

Thus, the vertices are:

Output (Thousands of Yards)

Light Pattern	Heavy Pattern
0	400
250	275
500	0
0	0

The vertices (0,400), (500,0), (0,0) are obtained easily from the graph. The fourth vertex (250, 275) is obtained from the simultaneous equations:

$$L + 2H = 800$$
$$1.1L + H = 550$$

From the first equation, $L = 800 - 2H$. Substituting this into the second equation yields

$$1.1(800 - 2H) + H = 550$$

or

$$880 - 2.2H + H = 550$$
$$- 1.2H = -330$$
$$1.2H = 330$$
$$H = 330/1.2$$
$$H = 275 .$$

Thus, $L + 2H = 800$ becomes $L + 2 \times 275 = 800$, $L = 250$.

The mill gained $300 and $290 per thousand yards of the light- and heavy-patterned cloths, respectively. Thus, for the four vertices, the gain from the week's production is:

at (0,400) : $300.0 + $290·400 = $116,000
at (250,275) : $300·250 + $290·275 = $155,000
at (500,0) : $300·500 + $290·0 = $150,000
at (0,0) : $300·0 + $290·0 = $ 0.

Thus, if the textile mill wants to obtain the highest possible gain from the week's operations, the combination 250 light pattern, 275 heavy pattern should be selected.

● **PROBLEM** 19-26

A mutual fund is deciding how to divide its investments among bonds, preferred stock, and speculative stock. It does not want to exceed a combined risk rate of 3 when the bonds have been assigned a risk rate of 1, the preferred stock, 3, and the speculative stock, 5. However, the fund does want a total annual yield of at least 10 percent. If the interest rate of the bonds is 8 percent, of the preferred stock is 12 percent, and of the speculative stock is 20 percent, how should the assets be distributed for the greatest annual yield?

Solution: The total amount to be invested does not matter. We let x,y, and z be the fractions of the whole that will be invested in bonds, pre-ferred stock, and speculative stock, respectively. Clearly these fractions must total the whole portfolio: $x + y + z = 1$. The other two constraints indicate the restrictions on the risk rate and total yield:

$$x + 3y + 5z \leq 3 \quad \text{(risk rate constraint)}$$

$$8x + 12y + 20z \geq 10 \quad \text{(annual yield constraint)} .$$

The annual yield constraint can be omitted, since whether it is there or not, we will make the left side of the inequality as large as possible. If it must be less than 10, there is no feasible solution to the problem as stated, and the fund must reconsider its goals. If it can equal or exceed 10 subject to the other constraints, that constraint is automatically satisfied. Thus we can summarize the problem:

$$\text{Maximize} \quad P = 8x + 12y + 20z$$

subject to $x + y + z = 1$ and $x + 3y + 5z \leq 3$. The non-negativity con-straints, as usual, are assumed. The first constraint needs no slack variable, but it does need an artificial variable. Recall that a slack variable s_i is used to convert the ith inequality to an equality. An artificial variable is introduced to make the number of unknowns in each row the same. This makes it possible to start with the basic feasible solution $x = y = z = 0$ while not violating the constraints.
    The second is given a slack variable as usual:

$$x + y + z + a = 1$$

$$x + 3y + 5z + s = 3 .$$

Step 1: The first project is to rid the problem of a, since it has no real-world interpretation. We do this by finding the minimum of $A = a$ (which minimum had better be zero), or, equivalently, we find the maximum of $-A = -a$. We might write:

	x	y	z	s	a	Solution
a	①	1	1	0	1	1
s	1	3	5	1	0	3
-A	0	0	0	0	-1	0
-A	1	1	1	0	0	1

The row above the dashed line is a summary of the equation $-A = -a$. There is a non-zero number in the last row of the a-column. This is not allowed, because a is a basic variable, and each basic variable must have 1 in its column and the rest of the numbers 0. To obtain a 0 in the last row in the a-column, we add the a-row to the last row and write the sum below; then we cross out the original row. Since all three posi-tive numbers in the last row are equal, any one of the first three columns may become the pivot column. We choose the first column to be the pivot column. Now, divide the positive elements of the inner rectangle in the pivot column into their corresponding elements in the "solutions" column. Since $1/1 < 3/1$, then the first row is the pivot row. Thus, the 1 in the upper left is the pivot. Now in the next table, replace the letter to the left of the pivot row by the letter above the pivot column. Thus x replaces a; a must leave. The pivot row is divided by the pivot. Add the second row of the inner rectangle to $-1x$ the first row of the inner rectangle. Replace the second row with this result.

	x	y	z	s	a	Solution
x	1	1	1	0	1	1
s	0	2	④	1	-1	2
-A	0	0	0	0	-1	0

Thus, since our present objective function, A is 0, we can proceed to Step 2 where the objective function in the original problem was

$$P = 8x + 12y + 20z .$$

Thus, our table is now:

	x	y	z	s	Solution
x	1	1	1	0	1
s	0	2	4	1	2
P	8	12	20	0	0

Since the largest entry in the last row is 20, the z-column is the pivot column. Since 2/4 < 1/1, the s-row is the pivot row and thus the 4 is the pivot. Now we must replace the letter to the left of the pivot row (s) by the letter above the pivot column (z). Divide the pivot row by the pivot, 4.

	x	y	z	s	Solution
x	1	1	1	0	1
z	0	1/2	1	1/4	1/2
P	8	12	20	0	0

Now, it is left for us to add the appropriate multiples of the pivot row to the other rows of numbers in such a way as to obtain zeros in the rest of the pivot column. First of all, add the P row to (-1)·x-row to obtain:

	x	y	z	s	Solution
x	1	1	1	0	1
z	0	1/2	1	1/4	1/2
P	0	4	12	0	-8

Now add the P-row to (-12)·z-row. Add the x-row to (-1)·z-row.

	x	y	z	s	Solution
x	1	1/2	0	-1/4	1/2
z	0	1/2	1	1/4	1/2
P	0	-2	0	-3	-14

Since there are no positive numbers in the bottom row, we have obtained the optimum solution. It is $P_{max} = 14$ for $x = \frac{1}{2}$, $y = 0$, and $z = \frac{1}{2}$.

Thus the maximum possible yield under these conditions is 14 percent, well above the 10 percent desired minimum.

The Red Tomato Company operates two plants for canning their tomatoes and has three warehouses for storing the finished products until they are purchased by retailers. The Company wants to arrange its shipments from the plants to the warehouses so that the requirements of the warehouses are met and so that shipping costs are kept at a minimum. The schedule below represents the per case shipping cost from plant to warehouse: (Table A)

Each week, plant I can produce up to 850 cases and plant II can produce up to 650 cases of tomatoes. Also, each week warehouse A requires 300 cases, warehouse B, 400 cases, and warehouse C, 500 cases. If we represent the number of cases shipped from plant I to warehouse A by x_1, from plant I to warehouse B by x_2, and so on, the above data can be represented by the table: (Table B)

TABLE A

	Warehouse		
	A	B	C
Plant I	$.25	$.17	$.18
Plant II	$.25	$.18	$.14

TABLE B

	Warehouse		
	A	B	C
Plant I	x_1	x_2	x_3
Plant II	x_4	x_5	x_6
Total Demand	300	400	500

Solution: The linear programming problem is stated as follows: Minimize the cost function
$$C = .25x_1 + .17x_2 + .18x_3 + .25x_4 + .18x_5 + .14x_6$$

subject to the conditions

$$x_1 + x_2 + x_3 \le 850 \qquad x_1 \ge 0, \ x_2 \ge 0$$
$$x_4 + x_5 + x_6 \le 650 \qquad x_3 \ge 0, \ x_4 \ge 0 \qquad (1)$$
$$x_1 + x_4 = 300 \qquad x_5 \ge 0, \ x_6 \ge 0$$
$$x_2 + x_5 = 400$$
$$x_3 + x_6 = 500 \ .$$

Before we begin to find a solution, notice that the linear objective function contains six variables. Also, the number of constraints is eleven. The simplex method requires that the constraints of a linear programming problem be given as linear equations, not as linear inequalities. Thus, before preceding further, we must change the inequalities of (1) to equalities. In order to do this, we introduce slack variables. For example, since $x_1 + x_2 + x_3 \le 850$, there is some nonnegative real number x_7, so that $x_1 + x_2 + x_3 + x_7 = 850$, $x_7 \ge 0$. Here x_7 is a slack variable. Similarly, there is a nonnegative integer x_8 so that $x_4 + x_5 + x_6 + x_8 = 650$, $x_8 \ge 0$.

Thus, by introducing slack variables, the linear programming problem can be restated as: Minimize the cost function.
$$C = .25x_1 + .17x_2 + .18x_3 + .25x_4 + .18x_5 + .14x_6$$
subject to the conditions

$$x_1 + x_2 + x_3 + x_7 = 850 \ , \ x_1 \geq 0 \ , \ x_2 \geq 0$$
$$x_4 + x_5 + x_6 + x_8 = 650 \ , \ x_3 \geq 0 \ , \ x_4 \geq 0 \qquad (2)$$
$$x_1 + x_4 = 300 \ , \qquad\qquad x_5 \geq 0 \ , \ x_6 \geq 0$$
$$x_2 + x_5 = 400 \ , \qquad\qquad x_7 \geq 0 \ , \ x_8 \geq 0$$
$$x_3 + x_6 = 500 \ .$$

Now the constraints have been expressed as linear equalities. Physically, the two slack variables x_7 and x_8 can be interpreted as cases of to-matoes produced at plant I and plant II, respectively, but not shipped to any warehouse. It is clear that the shipping cost of not shipping is zero so that the slack variables introduced cannot affect the objective (cost) function to be minimized. From the system of equations (2) we have

$$x_1 = 300 - x_4$$
$$x_2 = 400 - x_5$$
$$x_6 = 500 - x_3$$

Now, $x_1 = 850 - x_2 - x_3 - x_7$ but $x_1 = 300 - x_4$, $x_2 = 400 - x_5$. There-fore, $x_7 = 850 - 300 + x_4 - 400 + x_5 - x_3$ so, $x_7 = 150 - x_3 + x_4 + x_5$. Also, $x_8 = 650 - x_4 - x_5 - x_6$ but $x_6 = 500 - x_3$, thus,

$$x_8 = 650 - x_4 - x_5 - 500 + x_3$$

$$x_8 = 150 + x_3 - x_4 - x_5 \ .$$

The cost function C then becomes $C = .25x_1 + .17x_2 + .18x_3 + .25x_4 + .18x_5 + .14x_6$ but $x_1 = 300 - x_4$, $x_2 = 400 - x_5$, and $x_6 = 500 - x_3$. Therefore $C = .25(300 - x_4) + .17(400 - x_5) + .18x_3 + .25x_4 + .18x_5 + .14(500 - x_3)$.
$C = 75 - .25x_4 + 68 - .17x_5 + .18x_3 + .25x_4 + .18x_5 + 70 - .14x_3$.
$C = 213 + .04x_3 + .01x_5$. Therefore we have to minimize
$C = 213 + .04x_3 + .01x_5$ subject to $x_1 = 300 - x_4$

$$x_2 = 400 - x_5$$
$$x_6 = 500 - x_3$$
$$x_7 = 150 - x_3 + x_4 + x_5$$

$x_8 = 150 + x_3 - x_4 - x_5$ in which each constant term is ≥ 0 . The matrix representing the linear programming problem is:

	1	x_3	x_4	x_5
$-C$	-213	-.04	0	-.01
x_1	300	0	-1	0
x_2	400	0	0	-1
x_6	500	-1	0	0
x_7	150	-1	1	1
x_8	150	1	-1	-1

in which $-C$ is to be maximized. Since every entry in the $-C$ row is negative or zero, we need go no further. The maximum value for $-C$ is -213. The minimum cost C is then $213. The values $x_1, x_2, x_3, x_4, x_5, x_6$ giving the minimum cost of $213 are

643

$$x_1 = 300 \qquad x_2 = 400 \qquad x_3 = 0$$
$$x_4 = 0 \qquad x_5 = 0 \qquad x_6 = 500 \; .$$

Suppose that we have 2 factories and 3 warehouses. Factory I makes 40 widgets. Factory II makes 50 widgets. Warehouse A stores 15 widgets. Warehouse B stores 45 widgets. Warehouse C stores 30 widgets. It costs $80 to ship one widget from Factory I to warehouse A, $75 to ship one widget from Factory I to warehouse B, $60 to ship one widget from Factory I to warehouse C, $65 per widget to ship from Factory II to warehouse A, $70 per widget to ship from Factory II to warehouse B, and $75 per widget to ship from Factory II to warehouse C.
1) Set up the linear programming problem to find the shipping pattern which minimizes the total cost.
2) Find a feasible (but not necessarily optimal) solution to the problem of finding a shipping pattern using the Northwest Corner Algorithm.
3) Use the Minimum Cell Method to find a feasible solution to the shipping problem.

Solution: 1) Let x_{11} = number of widgets shipped from Factory I to warehouse A, x_{12} = number of widgets shipped from Factory I to warehouse B, x_{13} = number of widgets shipped from Factory I to warehouse C, x_{21} = number of widgets shipped from Factory II to warehouse A, x_{22} = number of widgets shipped from Factory II to warehouse B, x_{23} = number of widgets shipped from Factory II to warehouse C. Thus, the linear programming problem can be formulated as follows:

Minimize $C = 80x_{11} + 75x_{12} + 60x_{13} + 65x_{21} + 70x_{22} + 75x_{23}$

subject to

$$x_{11} + x_{12} + x_{13} \le 40 \; ,$$
$$x_{21} + x_{22} + x_{23} \le 50 \; , \quad x_{11} + x_{21} = 15,$$
$$x_{12} + x_{22} = 45 \quad \text{and} \quad x_{13} + x_{23} = 30 \; .$$

This linear programming problem may be solved using the simplex method.

2) The facts of the problem can be diagrammed in the following table, where the amounts the factories produce are written on the right, the amounts the warehouses can store are written on the bottom, and the numbers in the boxes are the costs of shipping from the factory on the left to the warehouse above.

Warehouse

		A	B	C	
Factory	I	80	75	60	40
	II	65	70	75	50
		15	45	30	

The Northwest corner algorithm first allocates as many widgets as possible to the upper left box (the northwest box). Next, proceed to the nearest box into which something can still be placed, and allocate as much as possible to that one. Then the process continues, each time moving either one box to the right, or one down, or one diagonally down, depending on

how the shipments can be made. Since the 15 at the bottom of the first
column is less than the 40 at the right of the first row, Factory I can
ship only 15 widgets to warehouse A, so we write a 15 in the upper left
box. Then nothing else can go to warehouse A; that is, nothing else will
be written in the boxes of the first column.

Warehouse

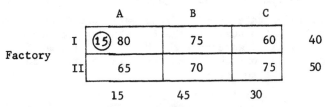

Thus, we move right from 15, making x_{12} = 40 - 15 = 25. Now the capacity of factory I has been exhausted, so there can be no more numbers
written in the first row. Moving down, we next set x_{22} = 45 - 25 = 20
to fill warehouse B. Now we must move right and set x_{23} = 50 - 20 = 30.
The results are as follows:

Warehouse

		A		B		C	
Factory	I	(15) 80	(25) 75			60	40
	II	65	(20) 70	(30) 75			50
		15	45	30			

This table tells us we can ship 15 widgets from Factory I to warehouse A,
25 widgets from Factory I to warehouse B, 20 widgets from Factory II to
warehouse B, and 30 widgets from Factory II to warehouse C. This is not
the optimum solution (that is, it is not the cheapest), but it is feasible.
The total cost is C = 15 · 80 + 25·75 + 20·70 + 30·75 = 6725 .

3) The Northwest Corner Algorthm ignores the costs. The Minimum Cell
Algorithm is another method of finding a feasible solution; unlike the
Northwest Corner Algorithm, it does take the cost into account. The Minimum Cell Algorithm finds the cheapest possible rate, and it sends as much
as possible at that rate. Again, the problem can be summarized in the
following table:

Warehouse

		A	B	C	
Factory	I	80	75	60	40
	II	65	70	75	50
		15	45	30	

Since 60 is the cheapest possible rate, we decide to send as much as possible at that rate. Since warehouse C needs 30 widgets and Factory I has
40 widgets, we can send at most 30 widgets from Factory I to warehouse C;
we do so and write it in the box. To show that we have accounted for all
30 items of warehouse C, we cross off the 30. To show that 30 of the 40
items in Factory I have been used, we cross off the 40 and write a 10
beside it to show there are 10 items left in Factory I.

Warehouse

		A	B		C		
Factory	I	80	75	(30)	60	4̶0̶ 10	
	II	65	70		75	50	
		15	45	3̶0̶			

The next cheapest shipping rate is 65. Since warehouse A needs only 15 widgets, we write a 15 in the box with the 65, showing that we will ship 15 widgets from Factory II to warehouse A. We cross out the 15 below the first column. And we cross out the 50 at the right of the second row and write a   50 - 15 = 35   next to it.

Warehouse

		A	B		C		
Factory	I	80	75	(30)	60	4̶0̶ 10	
	II	(15) 65	70		75	5̶0̶ 35	
		1̶5̶	45	3̶0̶			

The next cheapest shipping rate is 70. We have only 35 of the widgets produced by Factory II remaining to send to warehouse B, so we write a 35 in the box with the 70. Then we cross out the 35 at the right of the second row, and replace the 45 below the second column with a 45 - 35 = 10.

Warehouse

		A	B		C		
Factory	I	80	75	(30)	60	4̶0̶ 10	
	II	(15) 65	(35) 70		75	5̶0̶ 3̶5̶	
		1̶5̶	4̶5̶ 10	3̶0̶			

Since 75 is the lowest remaining rate, we ship the remaining 10 widgets from Factory I to warehouse B and indicate the feasible solution we obtain as follows:

Warehouse

		A	B		C		
Factory	I	80	(10) 75	(30)	60	4̶0̶ 1̶0̶	
	II	(15) 65	(35) 70		75	5̶0̶ 3̶5̶	
		1̶5̶	4̶5̶ 1̶0̶	3̶0̶			

This shipping pattern yields a cost of   15·65 + 35·70 + 10·75 + 30·60 = 5975.

● **PROBLEM** 19-29

A small-trailer manufacturer wishes to determine how many camper units and how many house trailers he should produce in order to make optimal use of his available resources. Suppose he has available 11 units of aluminum, 40 units of wood, and 52 person-weeks of work. (The preceding data are expressed in convenient units. We assume that all other needed resources are available and have no effect on his decision.) The table below gives the amount of each resource needed to manufactur each camper and each trailer.

	Aluminum	Wood	Person-weeks
Per camper	2	1	7
Per trailer	1	8	8

Suppose further that based on his previous year's sales record the manufacturer has decided to make no more than 5 campers. If the manufacturer
realized a profit of $300 on a camper and $400 on a trailer, what should
be his production in order to maximize his profit?

<u>Solution</u>: Letting x_1 represent the number of camper units, and x_2 the
number of house trailers, we consider first the constraints. From the
table we see that the manufacturer uses 2 units of aluminum per camper and
1 unit of aluminum per trailer. Thus he needs a total of $2x_1 + x_2$ units
of aluminum. This fact, along with the fact that he has available only
11 units of aluminum, gives us the inequality $2x_1 + x_2 \leq 11$. Similarly,
he needs a total of $x_1 + 8x_2$ units of wood. And since he has available
only 40 units of wood, we get $x_1 + 8x_2 \leq 40$. The total number of person-
weeks needed to build x_1 campers and x_2 trailers is $7x_1 + 8x_2$. And
since only 52 weeks are available, $7x_1 + 8x_2 \leq 52$. Since he wants to produce no more than 5 campers, we have $x_1 \leq 5$. And finally, there exists a
constraint that is unrelated to the numbers actually appearing in the statement of the problem. Certainly it is physically impossible for the manufacturer to produce a negative number of campers or trailers. Thus, we
need that $x_1, x_2 \geq 0$. We want, of course, to maximize the total profit
attained from x_1 campers and x_2 trailers, namely $300x_1 + 400x_2$. Thus,
we have reduced the given problem to the following:

Maximize $300x_1 + 400x_2$

subject to the conditions that

$$2x_1 + x_2 \leq 11$$
$$x_1 + 8x_2 \leq 40 \qquad\qquad (1)$$
$$7x_1 + 8x_2 \leq 52$$
$$x_1 \leq 5$$
$$x_1, x_2 \geq 0 .$$

The type of problem as above is called a maximum problem of linear programming. Now, we wish to determine the extreme points of the feasible solution
set. One way to make the process of finding the extreme points efficient
is to introduce slack variables. The purpose is to convert the inequalities
of (1) to equalities. Specifically, we let $x_3 = 11 - (2x_1 + x_2)$, $x_4 = 40 -$
$(x_1 + 8x_2)$, $x_5 = 52 - (7x_1 + 8x_2)$, and $x_6 = 5 - x_1$, and consider the system
of equations:

$$2x_1 + x_2 + x_3 \qquad\qquad\qquad = 11$$
$$x_1 + 8x_2 \qquad + x_4 \qquad\qquad = 40 \qquad (2)$$
$$7x_1 + 8x_2 \qquad\qquad + x_5 \qquad = 52$$
$$x_1 \qquad\qquad\qquad\qquad + x_6 = 5$$

647

and we still require that $x_1, x_2 \geq 0$. Moreover, the original inequality constraints will be satisfied if we require also that $x_3, x_4, x_5, x_6 \geq 0$. Observe, for example, that $x_3 = 11 - (2x_1 + x_2) \geq 0$ if and only if $2x_1 + x_3 \leq 11$. We first form the augmented matrix for (2). The function written below the matrix reminds us what we must maximize:

$$
\left(
\begin{array}{cccccc|c}
2 & 1 & 1 & 0 & 0 & 0 & 11 \\
1 & 8 & 0 & 1 & 0 & 0 & 40 \\
7 & 8 & 0 & 0 & 1 & 0 & 52 \\
1 & 0 & 0 & 0 & 0 & 1 & 5
\end{array}
\right)
$$

$$300x_1 + 400x_2 + 0x_3 + 0x_4 + 0x_5 + 0x_6$$

Thus, our starting tableau is:

	x_1	x_2	x_3	x_4	x_5	x_6		
x_3	2	1	1	0	0	0	11	11/1 = 11
x_4	1	(8)	0	1	0	0	40	40/8 = 5
x_5	7	8	0	0	1	0	52	52/8 = 6.5
x_6	1	0	0	0	0	1	5	
	-300	-400	0	0	0	0	0	

To determine the pivot element: The elements of the last row of the tableau are called indicators. We begin by finding the negative indicator having the largest absolute value. In the tableau above the indicator is clearly -400, which appears in the second column. We therefore call the second column the pivot column. We now consider the ratio of each element in the last column to the corresponding element in the pivot column, if the pivot column is positive. The row associated with the smallest of these ratios is called the pivot row. In our example the pivot column contains three positive elements: a 1 in the first row, an 8 in the second row, and an 8 in the third row. Thus the ratios we must compare are 1/1 = 11, 40/8 = 5 and 52/8 = 6.5. Since 5 is the smallest of the ratios, the second row is the pivot row. The pivot element is the element common to the pivot column and the pivot row, namely the 8 that is circled in the tableau above. Now, we must use elementary row operations to transform the tableau into one having a 1 in the place of the pivot element and 0's elsewhere in the pivot column. To accomplish this, we first multiply each element in the pivot row by the reciprocal of the pivot element to get:

	x_1	x_2	x_3	x_4	x_5	x_6	
x_3	2	1	1	0	0	0	11
x_4	1/8	1	0	1/8	0	0	5
x_5	7	8	0	0	1	0	52
x_6	1	0	0	0	0	1	5
	-300	-400	0	0	0	0	0

We then multiply the pivot row by -1 and add it to the first row, by -8 and add it to the third row, and by 400 and add it to the fifth row. The result is:

	x_1	x_2	x_3	x_4	x_5	x_6		
x_3	15/8	0	1	-1/8	0	0	6	$6/\frac{15}{8} = 16/5$
x_2	1/8	1	0	1/8	0	0	5	$5/\frac{1}{8} = 40$
x_5	6	0	0	-1	1	0	12	$12/6 = 2$
x_6	1	0	0	0	0	1	5	$5/1 = 5$
	-250	0	0	50	0	0	2000	

In the tableau above we replaced the x_4 in the notation column by x_2.
This replacement indicates that the x_2 variable was brought into the
solution and the x_4 variable was eliminated. Now, we must examine the
last row of the tableau above. Since -250 is the only negative indicator,
the first column is the pivot column. Comparing

$$6/\left(\frac{15}{8}\right) = \frac{16}{5}, \quad 5/\left(\frac{1}{8}\right) = 40, \quad \frac{12}{6} = 2, \quad \text{and} \quad 5/1 = 5,$$

we see that the third row is the pivot row. Thus the pivot element is 6,
which is circled in the tableau above. We first multiply the pivot row by
1/6 so a 1 appears in the pivot position. We then multiply this new
pivot row by -15/8, -1/8, -1, and 250, adding the results to the first,
second, fourth, and fifth rows, respectively, to get the following tableau:

	x_1	x_2	x_3	x_4	x_5	x_6	
x_3	0	0	1	3/16	-5/16	0	9/4
x_2	0	1	0	7/48	-1/48	0	19/4
x_1	1	0	0	-1/6	1/6	0	2
x_6	0	0	0	1/6	-1/6	1	3
	0	0	0	25/3	125/3	0	2500

Note in the tableau above: Since our pivot element was in the first column
and third row, we placed an x_1 in the third row of the notation column.

Since the tableau above include no negative indicators, we are done. Thus,
$x_1 = 2$, $x_2 = 19/4$ is the point at which the function assumes its maximum
value, namely, 2500.

STATISTICAL APPLICATIONS

● **PROBLEM** 19-30

1) Define sample space, random variable and expected value.
2) The probability that an American man in his early fifties will survive
for another year is about .99. How much should such a man pay for $20,000
worth of term insurance, exclusive of administrative costs and profit for
the company?
3) Suppose that the probability of finding 0,1,2, and 3 people in line
ahead of you at the Apex Supermarket is .2, .4, .25, and .15, respectively.
If the probability of finding 0,1,2, and 3 people in line at the B&Q Supermarket across the street is .3, .3, .2, and .2, respectively, at which
supermarket would you find shorter lines in the long run? (Assume that
both markets have the practice of opening up a new line as soon as one
exceeds 3 people in length.)

Solution: 1) A set of basic outcomes in a probability experiment is called a sample space of that experiment. If the sample space of a probability experiment consists of numbers, the variable x that results from performing the experiment is called a random variable. We can associate with any random variable a number called the expected value (or expectation) in such a way as to give the average value of the random variable in the long run -- that is, if the experiment is repeated many times. The expected value, E(x), of any random variable x that takes on a finite number of values can be computed by multiplying the probability vector by a column vector that lists the basic outcomes matching each probability. Thus, if a random variable x takes the values x_1, x_2 and x_3 with probabilities p_1, p_2, p_3, respectively,

$$E(x) = [p_1, p_2, p_3] \begin{pmatrix} x_1 \\ x_2 \\ x_3 \end{pmatrix} = p_1 x_1 + p_2 x_2 + p_3 x_3 .$$

2) There are two basic outcomes -- getting $0 by living through the year or getting $20,000 by dying in it. The probability vector is [.99, .01], so the expected value is

$$[.99, .01] \begin{pmatrix} 0 \\ 20,000 \end{pmatrix} = 0 + 200 = \$200 .$$

Thus, $200 is the amount to be paid for the insurance itself, exclusive of administrative costs and profits.

3) The probability vectors of the Apex Supermarket and B & Q Supermarket are:

[.2, .4, .25, .15], [.3, .3, .2, .2]

respectively. Thus, the expected values are:

Apex Super market   B & Q Supermarket

$$[.2, .4, 25, .15] \begin{pmatrix} 0 \\ 1 \\ 2 \\ 3 \end{pmatrix} = 1.35, \quad [.3, .3, .2, .2] \begin{pmatrix} 0 \\ 1 \\ 2 \\ 3 \end{pmatrix} = 1.3 .$$

Since the expected length of a line in Apex is 1.35 and that in B & Q is only 1.3, there is slightly less waiting on the average in B & Q than in Apex.

● **PROBLEM** 19-31

1) Show that the standard deviation of X and Y corresponds to vector addition (according to the parallelogram law).
2) Solve the following problems:
   (a) Suppose that 3 resistors are placed in series. The standard deviations of the given resistances are 20, 20 and 10 ohms, respectively. What is the standard deviation of the resistance of the series combination?
   (b) Gears A,B and C's widths have standard deviations of 0.001, 0.004 , and .002 in., respectively. If these are assembled side by side on a single shaft what is the standard deviation of this total width?

650

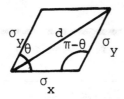

Solution: 1) Let variance $x = \sigma_x^2$, where σ_x is the standard deviation of x. Let variance $y = \sigma_y^2$ where σ_y is the standard deviation of y. From Statistics we know that variance $(x + y) = \sigma_x^2 + \sigma_y^2 + 2\sigma_{x,y}$ where $\sigma_{x,y}$ is called the covariance. In the diagram below let

$$\theta = \cos^{-1} \frac{\sigma_{x,y}}{\sigma_x \sigma_y} .$$

From the law of cosines we know that: $d^2 = \sigma_x^2 + \sigma_y^2 - 2\sigma_x \sigma_y \cos(\pi - \theta)$.
Use the following trigonometric identities:

$$\cos(A - B) = \cos A \cos B + \sin A \sin B$$

$$\cos(-\theta) = \cos \theta$$

Therefore,

$$\cos(\pi - \theta) = \cos \pi \cos(-\theta) + \sin \pi \sin \theta = -\cos \theta ,$$

and

$$d^2 = \sigma_x^2 + \sigma_y^2 - 2\sigma_x \sigma_y (-\cos \theta) .$$

Since

$$\theta = \cos^{-1} \frac{\sigma_{x,y}}{\sigma_x \sigma_y} , \quad \cos \theta = \frac{\sigma_{x,y}}{\sigma_x \sigma_y}$$

and

$$d^2 = \sigma_x^2 + \sigma_y^2 + 2\sigma_{x,y}$$

$$d^2 = \text{variance } (x + y)$$

$$d^2 = \sigma_{x+y}^2$$

$$d = \sigma_{x+y} .$$

Therefore, the diagonal equals the standard deviation of x and y. Standard deviation of x and y does indeed correspond to vector addition.

2a) The standard deviation of a sum of random variables corresponds to vector addition. If $\sigma_x = 20$, $\sigma_y = 20$ and $\sigma_z = 10$, then by vector analysis

$$\sigma_{x+y+z} = \sqrt{(20)^2 + (20)^2 + (10)^2}$$

$$\sigma_{x+y+z} = 30 \text{ ohms.}$$

2b) Let $\sigma_x = .001$, $\sigma_y = .004$ and $\sigma_z = .002$. Then by vector analysis,

$$\sigma_{x+y+z} = \sqrt{(.001)^2 + (.004)^2 + (.002)^2}$$

$$= \sqrt{.000001 + .000016 + .000004}$$

$$= \sqrt{.000021} = 10^{-3} \times \sqrt{21} .$$

651

A sales organization obtains the following data relating the number of salesmen to annual sales:

Number of salesmen	5	6	7	8	9	10
Annual sales (millions of dollars)	2.3	3.2	4.1	5.0	6.1	7.2

Let x denote the number of salesmen and let y denote the annual sales (in millions of dollars).
a) First describe the method of least squares in general and then;
b) Find the line of best fit relating x and y using the least square method.

Solution: a) Least Squares is a method which fits a line to a set of data. We begin by comparing the ordered pair (x_i, y_i) with $(x_i, f(x_i))$ where $f(x)$ is the line we are to fit with the data. In least squares we minimize

$$\sum_{i=1}^{n} |f(x_i) - y_i|^2$$

where $f(x) = Ax + B$ is a line with slope A and y-intercept, B and n represents the number of points in the data. Set

$$S(A,B) = \sum_{i=1}^{n} |Ax_i + B - y_i|^2 .$$

We wish to minimize $S(A,B)$. Therefore let us set the first partial derivatives equal to 0. Thus,

$$\partial S/\partial A = 2 \sum_{i=1}^{n} |Ax_i + B - y_i| x_i = 0 \qquad (1)$$

$$\partial S/\partial B = 2 \sum_{i=1}^{n} |Ax_i + B - y_i| 1 = 0 \qquad (2)$$

Now, we are left with a system of two linear equations:

From (1):

$$(Ax_1 + B - y_1)x_1 + (Ax_2 + B - y_2)x_2 + \ldots + (Ax_n + B - y_n)x_n$$

$$= A(x_1^2 + x_2^2 + \ldots + x_n^2) + B(x_1 + x_2 + \ldots + x_n) - (y_1 x_1 + y_2 x_2 + \ldots + y_n x_n)$$

$$= A \sum_{i=1}^{n} x_i^2 + B \sum_{i=1}^{n} x_i - \sum_{i=1}^{n} y_i x_i = 0 .$$

From (2):

$$(Ax_1 + B - y_1) + (Ax_2 + B - y_2) + \ldots + (Ax_n + B - y_n)$$

$$= A(x_1 + x_2 + \ldots + x_n) + (B + B + \ldots + B) - \sum_{i=1}^{n} y_i$$

$$= A \sum_{i=1}^{n} x_i + nB - \sum_{i=1}^{n} y_i = 0 .$$

Therefore we are left with the system:

$$A \sum_{i=1}^{n} x_i^2 + B \sum_{i=1}^{n} x_i = \sum_{i=1}^{n} y_i x_i$$

$$A \sum_{i=1}^{n} x_i + nB = \sum_{i=1}^{n} y_i$$

And it is left for us to solve for A and B. In matrix form:

$$\begin{pmatrix} \sum_{i=1}^{n} x_i^2 & \sum_{i=1}^{n} x_i \\ \sum_{i=1}^{n} x_i & n \end{pmatrix} \begin{pmatrix} A \\ B \end{pmatrix} = \begin{pmatrix} \sum_{i=1}^{n} y_i x_i \\ \sum_{i=1}^{n} y_i \end{pmatrix}$$

b) Using a) and the given table x_i are the number of salesmen, y_i are the annual sales and n is 6 for the number of points. Thus,

$$\sum_{i=1}^{n} x_i = 5 + 6 + 7 + 8 + 9 + 10 = 45$$

$$\sum_{i=1}^{n} x_i^2 = 5^2 + 6^2 + 7^2 + 8^2 + 9^2 + 10^2 = 355$$

$$\sum_{i=1}^{n} y_i = 2.3 + 3.2 + 4.1 + 5.0 + 6.1 + 7.2 = 27.9$$

$$\sum_{i=1}^{n} x_i y_i = 5(2.3) + 6(3.2) + 7(4.1) + 8(5.0) + 9(6.1) + 10(7.2) = 226.3$$

Therefore,

$$\begin{pmatrix} 355 & 45 \\ 45 & 6 \end{pmatrix} \begin{pmatrix} A \\ B \end{pmatrix} = \begin{pmatrix} 226.3 \\ 27.9 \end{pmatrix} \quad , \quad \text{or,}$$

$$355A + 45B = 226.3$$
$$45A + 6B = 27.9 .$$

To solve the above system, multiply the first row by 2 and the second row by -15:

$$710A + 90B = 452.6$$
$$-675A - 90B = -418.5 .$$

Now, adding together:

$$35A + 0B = 34.1$$
$$A = .974 .$$

But,
$$45A + 6B = 27.9 \quad \text{or}$$
$$6B = 27.9 - 45(.974)$$
$$6B = 27.9 - 43.83$$
$$6B = -15.93$$
$$B = -2.655 .$$

Therefore, the best-fitting line by Least Squares is:

$$f(x) = .974x - 2.655.$$

In an experiment designed to determine the extent of a person's natural orientation, a subject is put in a special room and kept there for a certain length of time. He is then asked to find his way out of a maze and a record is made of the time it takes him to accomplish this task. The following data are obtained:

Time in room (hours)	1	2	3	4	5	6
Time to find way out of maze (minutes)	0.8	2.1	2.6	2.0	3.1	3.3

Let x denote the number of hours in the room, and let y denote the number of minutes that it takes the subject to find his way out.
(a) Find the line of best fit relating x and y.
(b) Use the equation obtained in (a) to estimate the time it will take the subject to find his way out of the maze after 16 hours in the room.

Solution: a) We will use the method of least squares to find the line of best fit. By least squares the line of best fit is $f(x) = Ax + B$ where A and B are found from:

$$\begin{pmatrix} \sum_{i=1}^{n} x_i^2 & \sum_{i=1}^{n} x_i \\ \sum_{i=1}^{n} x_i & n \end{pmatrix} \begin{pmatrix} A \\ B \end{pmatrix} = \begin{pmatrix} \sum_{i=1}^{n} y_i x_i \\ \sum_{i=1}^{n} y_i \end{pmatrix} \quad (1)$$

The x_i's represent the time in the room; the y_i's represent the time to find the way out of the maze. $n = 6$ represents the number of points given in the data. Therefore,

$$\sum_{i=1}^{n} x_i = 1 + 2 + 3 + 4 + 5 + 6 = 21$$

$$\sum_{i=1}^{n} x_i^2 = 1 + 4 + 9 + 16 + 25 + 36 = 91$$

$$\sum_{i=1}^{n} y_i = .8 + 2.1 + 2.6 + 2.0 + 3.1 + 3.3 = 13.9$$

$$\sum_{i=1}^{n} x_i y_i = .8 + 2(2.1) + 3(2.6) + 4(2.0) + 5(3.1) + 6(3.3) = 56.1$$

From (1) we now have:

$$\begin{pmatrix} 91 & 21 \\ 21 & 6 \end{pmatrix} \begin{pmatrix} A \\ B \end{pmatrix} = \begin{pmatrix} 56.1 \\ 13.9 \end{pmatrix}$$

Thus,

$$91A + 21B = 56.1$$

$$21A + 6B = 13.9 .$$

Multiply the first row by 2; the second by -7 :

$$182A + 42B = 112.2$$

$$-147A - 42B = -97.3 .$$

Adding the two equations together yields:

$$35A - 0B = 14.9$$
$$A = .426 .$$

Substituting $A = .426$ in $21A + 6B = 13.9$ gives us:

$$21(.426) + 6B = 13.9$$
$$8.946 + 6B = 13.9$$
$$6B = 4.954$$
$$B = .826 .$$

Therefore our best fitting line is $f(x) = .426x + .826$.

(b) Since, $f(16) = .426(16) + .826$
$$= .6816 + .826$$
$$= 7.642$$

the subject will find his way out of the maze in 7.642 minutes, after 16 hours in the room.

● PROBLEM 19-34

Consider a random sample of pairs of heights and weights as in the following table:

Height	X	Pounds
5'3"	0	130
5'4"	1	145
5'5"	2	150
5'6"	3	165
5'7"	4	170

1) Fit a sample regression line by method of least squares. Let weight (y) be a function of height (x). (Note in the table above we assigned an x value to each entry of height).
2) For each of the entries under height, give the estimated expected value of the weight.

Solution: 1) By the method of least squares the sample regression line is given by: $f(x) = Ax + B$ where A and B are found from:

$$\begin{pmatrix} \sum_{i=1}^{n} x_i^2 & \sum_{i=1}^{n} x_i \\ \sum_{i=1}^{n} x_i & n \end{pmatrix} \begin{pmatrix} A \\ B \end{pmatrix} = \begin{pmatrix} \sum_{i=1}^{n} y_i x_i \\ \sum_{i=1}^{n} y_i \end{pmatrix} \quad (1)$$

Note the x's represent a number 0,1,2,3 or 4 assigned to an entry of height. The y' represent weight in pounds. We can solve for A and B in (1) by Cramer's Rule. By Cramer's Rule:

655

$$A = \frac{\begin{vmatrix} \sum\limits_{i=1}^{n} y_i x_i & \sum\limits_{i=1}^{n} x_i \\[6pt] \sum\limits_{i=1}^{n} y_i & n \end{vmatrix}}{\begin{vmatrix} \sum\limits_{i=1}^{n} x_i^2 & \sum\limits_{i=1}^{n} x_i \\[6pt] \sum\limits_{i=1}^{n} & n \end{vmatrix}}$$

and

$$B = \frac{\begin{vmatrix} \sum\limits_{i=1}^{n} x_i^2 & \sum\limits_{i=1}^{n} y_i x_i \\[6pt] \sum\limits_{i=1}^{n} x_i & \sum\limits_{i=1}^{n} y_i \end{vmatrix}}{\begin{vmatrix} \sum\limits_{i=1}^{n} x_i^2 & \sum\limits_{i=1}^{n} x_i \\[6pt] \sum\limits_{i=1}^{n} x_i & n \end{vmatrix}}$$

Therefore,

$$A = \frac{n \sum\limits_{i=1}^{n} y_i x_i - \left(\sum\limits_{i=1}^{n} x_i\right)\left(\sum\limits_{i=1}^{n} y_i\right)}{n \sum\limits_{i=1}^{n} x_i^2 - \left(\sum\limits_{i=1}^{n} x_i\right)^2}$$

and

$$B = \frac{\left(\sum\limits_{i=1}^{n} x_i^2\right)\left(\sum\limits_{i=1}^{n} y_i\right) - \left(\sum\limits_{i=1}^{n} x_i\right)\left(\sum\limits_{i=1}^{n} y_i x_i\right)}{n \sum\limits_{i=1}^{n} x_i^2 - \left(\sum\limits_{i=1}^{n} x_i\right)^2}$$

From our table of heights and weights we know that: $n = 5$ since there are 5 data points,

$$\sum_{i=1}^{n} x_i = 0 + 1 + 2 + 3 + 4 = 10 \ ,$$

$$\sum_{i=1}^{n} x_i^2 = 0 + 1 + 4 + 9 + 16 = 30 \ ,$$

$$\sum_{i=1}^{n} y_i = 130 + 145 + 150 + 165 + 170 = 760 \ ,$$

$$\sum_{i=1}^{n} x_i y_i = 0 \cdot 130 + 1 \cdot 145 + 2 \cdot 150 + 3 \cdot 165 + 4 \cdot 170 = 1620.$$

Thus,

$$A = \frac{5(1620) - 10(760)}{5(30) - (10)^2} = 10 \quad \text{and}$$

$$B = \frac{30(760) - 10(1620)}{5(30) - (10)^2} = 132 \ .$$

Therefore the sample regression line of y on x is $f(x) = 10x + 132$.

656

2) Now, we will find the estimated expected values of the weights. At $x = 0$ or when the height is 5'3" the estimated expected value of the weight is $f(0) = 132 + 10 \cdot 0 = 132$ lbs. At $x = 1$ or when the height if 5'4", the estimated expected value of the weight is $f(1) = 132 + 10 \cdot 1 = 142$ lbs. Likewise, at 5'5" the expected weight is $f(2) = 132 + 10(2) = 152$ lbs. The last two estimated expected values are similarly calculated.

The following table is a guide which the manufacturers supply with a certain gas cooker:

Weight of bird	Roasting time
5 - 10 lb	25 min per lb plus 25 min over
10 - 14 lb	20 min per lb plus 20 min over
14 - 18 lb	16 min per lb plus 16 min over
18 - 25 lb	14 min per lb plus 14 min over

If we drew the graph of cooking time against weight we get four line segments with rather surprising discontinuities. Therefore we want to find one formula which fits reasonably well over the whole range. Find the line of regression of y on x.

<u>Solution:</u> To find the line of regression of y on x we will use least squares. By the least squares method the line of regression of y on x is given by $f(x) = Ax + B$ where A and B are given by:

$$A = \frac{\begin{vmatrix} \sum\limits_{i=1}^{n} y_i x_i & \sum\limits_{i=1}^{n} x_i \\ \sum\limits_{i=1}^{n} y_i & n \end{vmatrix}}{\begin{vmatrix} \sum\limits_{i=1}^{n} x_i^2 & \sum\limits_{i=1}^{n} x_i \\ \sum\limits_{i=1}^{n} x_i & n \end{vmatrix}} \qquad (1)$$

and

$$B = \frac{\begin{vmatrix} \sum\limits_{i=1}^{n} x_i^2 & \sum\limits_{i=1}^{n} x_i y_i \\ \sum\limits_{i=1}^{n} x_i & \sum\limits_{i=1}^{n} y_i \end{vmatrix}}{\begin{vmatrix} \sum\limits_{i=1}^{n} x_i^2 & \sum\limits_{i=1}^{n} x_i \\ \sum\limits_{i=1}^{n} x_i & n \end{vmatrix}} \qquad (2)$$

We will let the x_i's represent the midpoints of the four ranges of the weights of the bird. Therefore the four midpoints of the ranges of weights are 7.5, 12, 16 and 21.5. According to the instructions for roasting times then:

Weight of Bird (x_i)	Roasting Time (y_i)
7.5	25 × 7.5 + 25 = 212.5
12	20 × 12 + 20 = 260
16	16 × 16 + 16 = 272
21.5	14 × 21.5 + 14 = 315

Thus, the y_i's represent the roasting time for each weight x_i. Therefore, we aim to fit the four points (7.5, 212.5), (12,260), (16,272) and (21.5, 315) by the least squares method. Let x_i represent weights of the bird and y_i represent roasting time. n is 4 since there are 4 weights considered, $\sum_{i=1}^{4} x_i = 7.5 + 12 + 16 + 21.5 = 57$

$$\sum_{i=1}^{4} y_i = 212.5 + 260 + 272 + 315 = 1059.5$$

$$\sum_{i=1}^{4} x_i y_i = 7.5(212.5) + 12(260) + 16(272) + 21.5(315)$$

$$= 15838$$

$$\sum_{i=1}^{4} x_i^2 = (7.5)^2 + (12)^2 + (16)^2 + (21.5)^2 = 918.5$$

Thus, from (1) we solve for A:

$$A = \frac{n \sum_{i=1}^{n} y_i x_i - \sum_{i=1}^{n} x_i \left(\sum_{i=1}^{n} y_i \right)}{n \sum_{i=1}^{n} x_i^2 - \left(\sum_{i=1}^{n} x_i \right)^2}$$

Substituting,

$$A = \frac{4(15838) - 57(1059.5)}{4(918.5) - (57)^2} \approx 7$$

Likewise, from (2) solve for B:

$$B = \frac{\left(\sum_{i=1}^{n} y_i \right)\left(\sum_{i=1}^{n} x_i^2 \right) - \left(\sum_{i=1}^{n} x_i \right)\left(\sum_{i=1}^{n} x_i y_i \right)}{\sum_{i=1}^{n} x_i^2 - \left(\sum_{i=1}^{n} x_i \right)^2}$$

Substituting,

$$B = \frac{(1059.5)(918.5) - (57)(15838)}{4(918.5) - (57)^2} \approx 166$$

Thus, the line of regression of y on x is f(x) = 7x + 166. This means that a roasting time of 166 min + 7 min per lb might be reasonable.

The technique used to estimate a relationship of the form $Y = \beta_0 + \beta X$ is known as simple regression. Generalize this technique to estimate a relationship of the form,

$$Y = \beta_0 + \beta_1 X_1 + \beta_2 X_2 + \ldots + \beta_K X_K ,$$

where K is some positive integer. This new technique is known as multiple regression.

Solution: To effect such a generalization, we reformulate the problem using some of the mathematical ideas found under the topic of linear algebra.

Our problem is the following: we have n observations of the variable Y, each observation has been taken in conjunction with observations of K other variables $X_1, X_2, \ldots X_K$. $(K < n)$.

We introduce the following notation, let Y_i be the ith observation of random variable Y. This observation corresponds with $X_{i1}, X_{i2}, X_{i3}, \ldots X_{iK}$, the ith observations of the K variables, $X_1, \ldots X_K$. The index i varies from 1 to n.

As in the case of simple regression we must make an assumption about the probability distribution of the Y_i observations. We will assume that the Y_i are independent and are distributed with mean

$$\beta_0 + \beta_1 X_1 + \ldots + \beta_K X_K \qquad \text{and variance } \sigma^2.$$

This can be written,

$$E(Y_1) = \beta_0 + \beta_1 X_{11} + \ldots + \beta_K X_{1K}$$

$$E(Y_2) = \beta_0 + \beta_1 X_{21} + \ldots + \beta_K X_{2K}$$

$$\vdots$$

$$E(Y_n) = \beta_0 + \beta_1 X_{n1} + \ldots + \beta_K X_{nK}$$

and $\text{Var } Y_i = \sigma^2$ $\qquad \text{Cov } (Y_i, Y_j) = 0$ for $i \neq j$.

Another way to express this assumption about the Y observations is to write each Y_i as the sum of a fixed and a random component. Let ε_i represent the random component in each equation.

Thus, $Y_i = \beta_0 + \beta_1 X_{i1} + \beta_2 X_{i2} + \ldots + \beta_K X_{iK} + \varepsilon_i$

for $i = 1, 2, \ldots n$. To maintain our previous assumptions, we require ε_i to be independent and distributed with mean zero and variance σ^2. These new assumptions give,

$$E(Y_i) = E\left[\beta_0 + \beta_1 X_{i1} + \ldots + \beta_K X_{iK} + \varepsilon_i\right]$$

$$= \beta_0 + \beta_1 X_{i1} + \ldots + \beta_K X_{iK} + E(\varepsilon_i)$$

$$= \beta_0 + \beta_1 X_{i1} + \ldots + \beta_K X_{ik} + 0$$

and

$$\text{Var } Y_i = \text{Var}\left[\beta_0 + \beta_1 X_{i1} + \ldots + \beta_K X_{iK} + \varepsilon_i\right]$$

$$= \text{Var } \varepsilon_i = \sigma^2 .$$

Thus we assume,

$$Y_1 = \beta_0 + \beta_1 X_{11} + \beta_2 X_{12} + \ldots + \beta_K X_{1K} + \varepsilon_1$$

$$Y_2 = \beta_0 + \beta_1 X_{21} + \beta_2 X_{22} + \ldots + \beta_K X_{2K} + \varepsilon_2$$

$$\vdots \qquad \vdots \qquad \vdots \qquad \qquad \vdots$$

$$Y_n = \beta_0 + \beta_1 X_{n1} + \beta_2 X_{n2} + \ldots + {}_K X_{nK} + \varepsilon_n .$$

This system of n linear equations may be rewritten with the help of vector and matrix notation. Let the column vectors be defined as follows,

$$\underline{Y} = \begin{pmatrix} Y_1 \\ Y_2 \\ \cdot \\ \cdot \\ \cdot \\ Y_n \end{pmatrix} \qquad \underline{\varepsilon} = \begin{pmatrix} \varepsilon_1 \\ \varepsilon_2 \\ \cdot \\ \cdot \\ \cdot \\ \varepsilon_n \end{pmatrix} , \qquad \underline{\beta} = \begin{pmatrix} \beta_0 \\ \beta_1 \\ \beta_2 \\ \cdot \\ \cdot \\ \cdot \\ \beta_K \end{pmatrix} .$$

Also define the nX K + 1 matrix X to be

$$X = \begin{pmatrix} 1 & X_{11} & X_{12} & \ldots & X_{1K} \\ 1 & X_{21} & X_{22} & \ldots & X_{2K} \\ \cdot & \cdot & \cdot & & \cdot \\ \cdot & \cdot & \cdot & & \cdot \\ \cdot & \cdot & \cdot & & \cdot \\ 1 & X_{n1} & X_{n2} & \ldots & X_{nK} \end{pmatrix} .$$

This matrix consists of an array of the X-observa-

tions and the column vector $\begin{pmatrix} 1 \\ 1 \\ \cdot \\ \cdot \\ \cdot \\ 1 \end{pmatrix}$. The matrix X is

often called the design matrix. This name reflects the fact that the X-observations are often chosen by the experimenter.

   With this new notation we express the system of n linear equations as

$$\underline{Y} = X\underline{\beta} + \underline{\varepsilon} \ .$$

   Our new problem is to choose a vector $\hat{\underline{\beta}}$ and produce a new vector $\underline{\hat{Y}} = X\hat{\underline{\beta}}$. We will choose $\hat{\underline{\beta}}$ such that it satisfies the least squares criterion,

$$\sum_{i=1}^{n} (Y_i - \hat{Y}_i)^2 \quad \text{is a minimum}$$

where Y_i are the observed Y values and $\hat{Y}_i$ are the values computed from the estimate $\hat{\underline{\beta}}$.

$\sum_{i=1}^{n} (Y_i - \hat{Y})^2$ can be written as the dot product

of the vector $\underline{Y} - \hat{\underline{Y}}$ with itself.

   Thus we wish to choose $\hat{\underline{\beta}}$ such that

$$\sum_{i=1}^{n} (Y_i - \hat{Y}_i)^2 \text{ is a minimum where } \hat{Y}_i = \sum_{j=0}^{K} \beta_0 X_{ij} \ .$$

   To minimize this expression, we take partial derivatives with respect to each component of $\hat{\underline{\beta}}$ and set each derivative equal to zero. Thus,

$$\frac{\partial}{\partial \beta_m} \left[\sum_{i=1}^{n} \left(Y_i - \sum_{j=0}^{K} \hat{\beta}_j X_{ij} \right)^2 \right] = 0$$

for m = 0, 1, ...K

   Let $X_{io} = 1$ for i = 1, ... n.

   Differentiating with respect to β_m, we see that,

$$\sum_{i=1}^{n} -2 X_{im} \left(Y_i - \sum_{j=0}^{n} \hat{\beta}_j X_{ij} \right) = 0$$

for m = 0, 1, 2, ..., K.

These K + 1 equations are called the normal equations and the solutions $\hat{\beta_0}, \hat{\beta_1}, \ldots, \hat{\beta_K}$, are the least squares estimates of the parameters β_0, β_1, ... β_K. The solution to this system are the values $\hat{\beta_0}$, $\hat{\beta_1}$, ... $\hat{\beta_K}$ that satisfy

$$\sum_{i=1}^{n} X_{im} \left(Y_i - \sum_{j=0}^{n} \hat{\beta}_j X_{ij} \right) = 0$$

for m = 0, 1, 2, ... K.

or

$$\sum_{i=1}^{n} Y_i X_{im} = \sum_{i=1}^{n} X_{im} \sum_{j=0}^{K} \hat{\beta}_j X_{ij}$$

for m = 0, 1, 2, ... K.

Writing this system in vector form we see that

$$\begin{pmatrix} \sum_{i=1}^{n} Y_i X_{i0} \\\\ \sum_{i=1}^{n} Y_i X_{i1} \\ \vdots \\\\ \sum_{i=1}^{n} Y_i X_{iK} \end{pmatrix} = \begin{pmatrix} \sum_{i=1}^{n} X_{i0} \sum_{j=0}^{K} \hat{\beta}_j X_{ij} \\\\ \sum_{i=1}^{n} X_{i1} \sum_{j=0}^{K} \hat{\beta}_j X_{ij} \\ \\ \sum_{i=1}^{n} X_{iK} \sum_{j=0}^{K} \hat{\beta}_j X_{ij} \end{pmatrix}.$$

This suggests that each component in these two vectors can be written as the dot product of two vectors.

Let $\underline{X}_j$ = the jth column vector of the design matrix X. Thus

$$X = (\underline{X}_0, \underline{X}_1, \ldots \underline{X}_K) \qquad \text{and each vector } \underline{X}_j$$

can be written $\underline{X}_j = \begin{pmatrix} X_{1j} \\ \cdot \\ \cdot \\ \cdot \\ X_{nj} \end{pmatrix}$ for j = 0, ..., K

Each of the components in the vector on the left side of the system of normal equations can be written as

$$\underline{X}'_j \cdot \underline{Y} = (X_{1j}, \ \ldots \ X_{nj}) \cdot \begin{pmatrix} Y_1 \\ \cdot \\ \cdot \\ \cdot \\ Y_n \end{pmatrix}$$

where $\underline{X}_j{}'$ is the transpose of the column vector $\underline{X}_j$.

$$\underline{X}'_j \cdot \underline{Y} = \sum_{i=1}^{n} Y_i X_{ij}$$

Thus

$$\begin{pmatrix} \underline{X}'_0 \cdot \underline{Y} \\ \cdot \\ \cdot \\ \underline{X}'_K \quad \underline{Y} \end{pmatrix} = \begin{pmatrix} \sum_{i=1}^{n} X_{i0} \sum_{j=0}^{K} \hat{\beta}_j X_{ij} \\ \sum_{i=1}^{n} X_{i1} \sum_{j=0}^{K} \hat{\beta}_j X_{ij} \\ \cdot \\ \cdot \\ \sum_{i=1}^{n} X_{iK} \sum_{j=0}^{K} \hat{\beta}_j X_{ij} \end{pmatrix}$$

or

$$\begin{pmatrix} \underline{X}'_0 \\ \cdot \\ \cdot \\ \cdot \\ \underline{X}'_K \end{pmatrix} \underline{Y} = \begin{pmatrix} \sum_{i=1}^{n} \sum_{j=0}^{K} X_{i0} X_{ij} \hat{\beta}_j \\ \cdot \\ \cdot \\ \cdot \\ \sum_{i=1}^{n} \sum_{j=0}^{K} X_{iK} X_{ij} \hat{\beta}_j \end{pmatrix} .$$

But $\underline{X}'_0, \ \ldots \underline{X}'_K$ are the rows of X', the transpose of the design matrix. Let $\underline{X}_i$ be the ith row vector of the matrix X, then

$$\sum_{i=1}^{n} X_{im} \left[\sum_{j=0}^{K} X_{ij} \hat{\beta}_j \right] = \sum_{i=1}^{n} X_{im} \underline{X}_i \underline{\hat{\beta}}$$

for $m = 0, \ \ldots \ K$. Combining these two results we notice that

$$X' \underline{Y} = \begin{pmatrix} \sum_{i=1}^{n} X_{i0} \underline{X}_i \underline{\hat{\beta}} \\ \cdot \\ \cdot \\ \cdot \\ \sum_{i=1}^{n} X_{iK} \underline{X}_i \underline{\hat{\beta}} \end{pmatrix} .$$

We also note that $\left(X_{im}, \ldots, X_{nm}\right)$ is the mth row vector of the transpose of X for m = 0, ... K. Thus,

$$
\begin{pmatrix}
\sum\limits_{i=1}^{n} X_{i0}\, \underline{X}_i\, \hat{\underline{\beta}} \\
\vdots \\
\sum\limits_{i=1}^{n} X_{iK}\underline{X}_i\, \beta
\end{pmatrix}
=
\begin{pmatrix}
\sum\limits_{i=1}^{n} X_{i0}\, \underline{X}_i \\
\vdots \\
\sum\limits_{i=1}^{n} X_{iK}\, \underline{X}_i
\end{pmatrix}
\qquad \hat{\underline{\beta}} = X'\, X\, \hat{\underline{\beta}}
$$

and the normal equations can be written as,

$$X'\, \underline{Y} = X'\, X\, \hat{\underline{\beta}}$$

If the design matrix X is invertible, then the product $X'\, X$ will be invertible and we solve for $\hat{\underline{\beta}}$,

$$\hat{\underline{\beta}} = [X'\, X]^{-1}\, X'\, \underline{Y} \ .$$

The error sum of squares is

$$\sum_{i=1}^{n} (Y_i - \hat{Y}_i)^2 \qquad\qquad \text{or}$$

$$(\underline{Y} - X\hat{\underline{\beta}})'\ (\underline{Y} - X\hat{\underline{\beta}})$$

in the sense of matrix multiplication.

We can also find the mean and covariance matrix of the vector $\hat{\underline{\beta}}$.

First define the mean vector and covariance matrix of a vector whose components are random variables.

$$
E(\hat{\underline{\beta}}) =
\begin{pmatrix}
E(\beta_0) \\
E(\beta_1) \\
\vdots \\
E(\beta_K)
\end{pmatrix}
\cdot
$$

That is, the expected value of a vector is a vector whose components are the expected values of the original components.
The covariance matrix of a random vector is the matrix with the i, jth component equal to $\text{Cov}(X_i, X_j)$ where X_i, X_j are the ith and jth components of the random vector. Thus the covariance matrix of the vector $\underline{Y}$ is

$$M_{\underline{Y}} = \begin{pmatrix} \text{Cov}(Y_1, Y_1) & \text{Cov}(Y_1, Y_2) & \cdots & \text{Cov}(Y_1, Y_n) \\ \text{Cov}(Y_2, Y_1) & & & \cdot \\ \cdot & & & \cdot \\ \cdot & & & \cdot \\ \cdot & & & \cdot \\ \text{Cov}(Y_n, Y_1) & & & \text{Cov}(Y_n, Y_n) \end{pmatrix}$$

but since all the Y_i are independent, all the components off the main diagonal vanish.

$$\text{Cov}(Y_i, Y_i) = \text{Var } Y_i = \sigma^2,$$

a constant, for $i = 1, \ldots N$. Therefore all elements on the main diagonal are σ^2 and $M_Y = \sigma^2 I$ where I is the identity matrix.

We now find the mean vector and covariance vector of $\hat{\underline{\beta}}$. $E(\hat{\underline{\beta}}) = E[(X'X)^{-1}X' \underline{Y}]$

$$= (X'X)^{-1}X' E(\underline{Y})$$

but $E(\underline{Y}) = X \underline{\beta}$ by assumption and

$$E(\hat{\underline{\beta}}) = (X'X)^{-1} X' (X \underline{\beta}) = (X'X)^{-1} (X'X) \underline{\beta} = \underline{\beta} .$$

This shows that $E(\hat{\beta}_0) = \beta_0$

$$E(\hat{\beta}_1) = \beta_1$$
$$\cdot$$
$$\cdot$$
$$\cdot$$
$$E(\hat{\beta}_K) = \beta_K .$$

Thus our vector of estimates $\hat{\underline{\beta}}$ is an unbiased estimator of $\underline{\beta}$.

The covariance matrix of $\hat{\underline{\beta}}$ can be found with the aid of the following result. If a vector $\underline{X} = A\underline{Y}$ where A is a matrix and $\underline{Y}$ a vector, the covariance matrix of $\underline{X}$ will be

$$M_{\underline{X}} = AM_{\underline{Y}} A' .$$

Thus $\hat{\underline{\beta}} = (X'X)^{-1} X'\underline{Y}$

implies $M_{\hat{\underline{\beta}}} = (X'X)^{-1} X' M_Y [(X'X)^{-1} X']^1$

$$= (X'X)^{-1} X' \sigma^2 I [X[(X'X)^{-1}]^1]$$
$$= \sigma^2 (X'X)^{-1} X'X (X'X)^{-1} = \sigma^2 (X'X)^{-1}.$$

A highly specialized industry builds one device each month. The total monthly demand is a random variable with the following distribution.

Demand	0	1	2	3
P(D)	1/9	6/9	1/9	1/9

When the inventory level reaches 3, production is stopped until the inventory drops to 2. Let the states of the system be the inventory level. The transition matrix is found to be

$$
P = \begin{array}{c} 0 \\ 1 \\ 2 \\ 3 \end{array}
\begin{pmatrix}
 & 0 & 1 & 2 & 3 \\
8/9 & 1/9 & 0 & 0 \\
2/9 & 6/9 & 1/9 & 0 \\
1/9 & 1/9 & 6/9 & 1/9 \\
1/9 & 1/9 & 6/9 & 1/9
\end{pmatrix} \tag{1}
$$

Assuming the industry starts with zero inventory find the transition matrix as $n \to \infty$.

Solution: This problem involves stochastic processes. Since both the random variable and the time intervals are discrete and there is no carry-over effect, we consider a Markov chain.

The transition matrix (1) gives the probabilities of changing from one state to another in one step. For example, the probability of the inventory level changing from 1 to 2 is 1/9. The relationship between the initial transition matrix and the transition matrix after n steps is given by the matrix equation

$$P(n) = P^n .$$

For example after two steps, the transition matrix is given by
$P(2) = P^2$

$$
= \begin{pmatrix}
8/9 & 1/9 & 0 & 0 \\
2/9 & 6/9 & 1/9 & 0 \\
1/9 & 1/9 & 6/9 & 1/9 \\
1/9 & 1/9 & 6/9 & 1/9
\end{pmatrix}
\begin{pmatrix}
8/9 & 1/9 & 0 & 0 \\
2/9 & 6/9 & 1/9 & 0 \\
1/9 & 1/9 & 6/9 & 1/9 \\
1/9 & 1/9 & 6/9 & 1/9
\end{pmatrix}
$$

$$
= \frac{1}{81} \begin{pmatrix}
66 & 14 & 1 & 0 \\
29 & 39 & 12 & 1 \\
17 & 14 & 43 & 7 \\
17 & 14 & 43 & 7
\end{pmatrix} . \tag{2}
$$

The elements of (2) give the probabilities of proceeding from the initial state to another in exactly 2 steps. Similarly, after 3 steps
$P(3) = P^3$

$$
= \begin{pmatrix}
8/9 & 1/9 & 0 & 0 \\
2/9 & 6/9 & 1/9 & 0 \\
1/9 & 1/9 & 6/9 & 1/9 \\
1/9 & 1/9 & 6/9 & 1/9
\end{pmatrix}^3
= \frac{1}{729} \begin{pmatrix}
557 & 151 & 20 & 1 \\
323 & 276 & 117 & 13 \\
214 & 151 & 314 & 50 \\
214 & 151 & 314 & 50
\end{pmatrix}
$$

We are interested in finding the value of $P(n)$ as $n \to \infty$. First, however, we must include the initial inventory level in our calculations. The probabilities of reaching the various states in n steps are given by $p'(n) = p'(0)P(n)$. Thus,

666

$$p'(1) = (1,0,0,0) \begin{pmatrix} 8/9 & 1/9 & 0 & 0 \\ 2/9 & 6/9 & 1/9 & 0 \\ 1/9 & 1/9 & 6/9 & 1/9 \\ 1/9 & 1/9 & 6/9 & 1/9 \end{pmatrix}$$

$$= (8/9, 1/9, 0, 0)$$
$$p'(2) = (66/81, 14/81, 1/81, 0/81)$$
$$p'(3) = (557/729, 151/729, 20/729, 1/729)$$

We note that the probability of zero inventory is decreasing. Furthermore, for $n > 3$, all the elements of P^n are greater than zero. For

$$S_D = \frac{\sum\limits_{i=1}^{n} (D_i - \bar{D})^2}{n-1} = \sqrt{\frac{\sum\limits_{i=1}^{n} [(Y_i - X_i) - (\bar{Y} - \bar{X})]^2}{n-1}}$$

To compute S_D note that

$$S_D^2 = \frac{\sum\limits_{i=1}^{n} [(Y_i - X_i) - (\bar{Y} - \bar{X})]^2}{n-1}$$

$$= \frac{\sum\limits_{i=1}^{n} (Y_i - \bar{Y})^2}{n-1} - \frac{2\sum\limits_{i=1}^{n} (Y_i - \bar{Y})(X_i - \bar{X})}{n-1} + \frac{\sum\limits_{i=1}^{n} (X_i - \bar{X})^2}{n-1}$$

$$= S_1^2 - 2 r_{12} S_1 S_2 + S_2^2$$

where S_1^2 and S_2^2 are the unbiased sample variances of the first and second groups of test scores and r_{12} is the sample correlation coefficient between the first and second sets of scores.

If $\mu_D = 0$, then this statistic is distributed with approximately a t-distribution with n-1 degrees of freedom. If this statistic is too large, that is, if

$$\frac{\sqrt{n}\,\bar{D}}{S_D} > t(0.95, n-1),$$

the 95th percentile of the t-distribution with n-1 degrees of freedom, then we reject H_0.

From the data given in the problem, we have $\bar{D} = 50 - 45 = 5$ and $S_D = \sqrt{36 + 25 - 2(0.6)(6)(5)}$
$$= \sqrt{25} = 5.$$

Therefore, the test statistic is

$$\frac{\sqrt{n}\,\bar{D}}{S_D} = \frac{8(5)}{5} = 8.$$

The 95th percentile of the t-distribution with 64 - 1 = 63 degrees of freedom is approximately 2. Therefore, since $8 > 2$, then we reject H_0 and conclude that there is a significant difference between the mean scores on the two exams.

Markov matrices having this property, P^{11} approaches a probability matrix A, where each row of A is the same probability vector α' (where α' denotes the transpose of the row vector α).

667

To find α, we solve the equations

$$\begin{pmatrix} 8/9 & 2/9 & 1/9 & 1/9 \\ 1/9 & 6/9 & 1/9 & 1/9 \\ 0 & 1/9 & 6/9 & 6/9 \\ 0 & 0 & 1/9 & 1/9 \end{pmatrix} \begin{pmatrix} \alpha_1 \\ \alpha_2 \\ \alpha_3 \\ \alpha_4 \end{pmatrix} = \begin{pmatrix} \alpha_1 \\ \alpha_2 \\ \alpha_3 \\ \alpha_4 \end{pmatrix} \qquad (3)$$

subject to the requirement $\alpha_i \geq 0$, $i = 1,4$ and $\sum_{i=1}^{n} \alpha_i = 1$. Using either Cramer's rule or Gauss-Jordan elimination, the solution to (3) is found to be

$$\alpha' = (45/72, 18/72, 8/82, 1/72) .$$

The limit matrix is then

$$A = \begin{pmatrix} 45/72 & 18/72 & 8/72 & 1/72 \\ 45/72 & 18/72 & 8/72 & 1/72 \\ 45/72 & 18/72 & 8/72 & 1/72 \\ 45/72 & 18/72 & 8/72 & 1/72 \end{pmatrix} .$$

Thus, if the initial inventory level is zero, the system approaches the state where the probability of zero inventory is 45/72.

● **PROBLEM** 19-38

1) Generalize the least squares method of curve-fitting to a quadratic function.
2) Fit the following data points to a quadratic function: (-5,2), (-4,7), (-3,9), (-2,12), (-1,13), (0,14), (1,14), (2,13), (3,10), (4,8), (5,4).

Solution: 1) The least squares method of curve-fitting is equally applicable to linear and non-linear functions. If a set of data points suggests that a parabola might make a good fit, then we are dealing with a quadratic function of the form:

$$f(x_i) = a + b_1 x_i + b_2 x_i^2 .$$

The least squares estimators a, b_1 and b_2 would be the values a, b_1 and b_2 resulting in the minimization of the sum of the squared deviations between the actual values and the predicted values:

$$\text{minimize } S = \sum_{i=1}^{N} [y_i - (a + b_1 x_i + b_2 x_i^2)]^2 ,$$

where the data consist of N pairs of values (x_i, y_i). To find the values of a, b_1 and b_2, take the partial derivatives of X with respect to a, b_1 and b_2, set these partial derivatives equal to zero, and solve the resulting equations. Thus,

$$\frac{\partial S}{\partial a} = 2 \sum_{i=1}^{N} [y_i - (a + b_1 x_i + b_2 x_i^2)] = 0$$

$$\frac{\partial S}{\partial b_1} = 2 \sum_{i=1}^{N} [y_i - (a + b_1 x_i + b_2 x_i^2)] x_i = 0$$

and

$$\frac{\partial S}{\partial b_2} = 2 \sum_{i=1}^{N} [y_i - (a + b_1 x_i + b_2 x_i^2)] x_i^2 = 0 .$$

From the three equations above, we can obtain:

$$\sum_{i=1}^{N} y_i - \left(\sum_{i=1}^{N} a + \sum_{i=1}^{N} b_1 x_i + \sum_{i=1}^{N} b_2 x_i^2 \right) = 0$$

$$\sum_{i=1}^{N} y_i x_i - \left(\sum_{i=1}^{N} a x_i + \sum_{i=1}^{N} b_1 x_i^2 + \sum_{i=1}^{N} b_2 x_i^3 \right) = 0$$

$$\sum_{i=1}^{N} y_i x_i^2 - \left(\sum_{i=1}^{N} a x_i^2 + \sum_{i=1}^{N} b_1 x_i^3 + \sum_{i=1}^{N} b_2 x_i^4 \right) = 0 .$$

Therefore, we have a system of equations in three unknowns:

$$a N + b_1 \sum_{i=1}^{N} x_i + b_2 \sum_{i=1}^{N} x_i^2 = \sum_{i=1}^{N} y_i$$

$$a \sum_{i=1}^{N} x_i + b_1 \sum_{i=1}^{N} x_i^2 + b_2 \sum_{i=1}^{N} x_i^3 = \sum_{i=1}^{N} x_i y_i$$

$$a \sum_{i=1}^{N} x_i^2 + b_1 \sum_{i=1}^{N} x_i^3 + b_2 \sum_{i=1}^{N} x_i^4 = \sum_{i=1}^{N} y_i x_i^2 .$$

In matrix form, these equations can be written as:

$$\begin{bmatrix} N & \sum_{i=1}^{N} x_i & \sum_{i=1}^{N} x_i^2 \\ \sum_{i=1}^{N} x_i & \sum_{i=1}^{N} x_i^2 & \sum_{i=1}^{N} x_i^3 \\ \sum_{i=1}^{N} x_i^2 & \sum_{i=1}^{N} x_i^3 & \sum_{i=1}^{N} x_i^4 \end{bmatrix} \begin{bmatrix} a \\ b_1 \\ b_2 \end{bmatrix} = \begin{bmatrix} \sum_{i=1}^{N} y_i \\ \sum_{i=1}^{N} x_i y_i \\ \sum_{i=1}^{N} y_i x_i^2 \end{bmatrix}$$

2) In order to fit the eleven data points to a quadratic function, we must calculate the sums from the data, insert these values in the equations we have obtained in 1), and then solve for a, b_1 and b_2. Therefore let us construct a table in order to make this task easier:

x_i	y_i	$x_i y_i$	x_i^2	x_i^3	x_i^4	$x_i^2 y_i$
-5	2	-10	25	-125	625	50
-4	7	-28	16	- 64	256	112
-3	9	-27	9	- 27	81	81
-2	12	-24	4	- 8	16	48
-1	13	-13	1	- 1	1	13
0	14	0	0	0	0	0
1	14	14	1	1	1	14
2	13	26	4	8	16	52
3	10	30	9	27	81	90
4	8	32	16	64	256	128
5	4	20	25	125	625	100
0	106	20	110	0	1958	688

Our N equals 11 since there are 11 data points.
Now, let us insert our values in the equations we have obtained in 1).

669

Thus our system of equations becomes:

$$11a + 0b_1 + 110b_2 = 106$$
$$0a + 110b_1 + 0b_2 = 20$$
$$110a + 0b_1 + 1958b_2 = 688 \ .$$

From our second equation above, we have $110b_1 = 20$, or $b_1 = 20/110$ or

$b_1 \approx .18$.

From our first and third equations above, we have $11a + 110b_2 = 106$

and $110a + 1958b_2 = 688$. Multiply $(11a + 110b_2 = 106)$ by -10 and add

to $(110a + 1958b_2 = 688)$. Thus,

$$-110a - 1100b_2 = -1060$$
$$+110a + 1958b_2 = 688$$
$$\overline{858b_2 = -372}$$

$$b_2 = \frac{-372}{858}$$

$$b_2 \approx -.43 \ .$$

Substituting $b_2 = -.43$ into $110a + 1958b_2 = 688$ yields

$$110a + 1958x -.43 = 688$$
$$110a \approx 688 + 842$$
$$110a \approx 1530$$
$$a \approx 13.9 \ .$$

Thus, $a = 13.9$, $b_1 = .18$ and $b_2 = -.43$.

The estimated regression curve is $f(x) = 13.9 + .18x - .43x^2$

● **PROBLEM** 19-39

1) Suppose we want to fit an exponential curve to a set of data. Explain how we could use the least squares method.
2) Fit an exponential curve to the following data: (1,1.00), (2,1.20), (3,1.80), (4,2.50), (5,3.60), (6,4.70), (7,6.60), (8,9.10).

Solution: 1) Suppose we want to fit an exponential curve to a set of data. An exponential curve would be of the form $f(x) = \alpha \beta^x$.

Now, take the natural logarithm of both sides:

$$\ln f(x) = \ln \alpha + x \ln \beta \ .$$

We can use a transformation of variables: $\alpha' = \ln \alpha$, $\beta' = \ln \beta$ and $V = \ln f$. Thus $\ln f(x) = \ln \alpha + x \ln \beta$ becomes $V(x) = \alpha' + \beta'x$, a linear function. We can now use the least squares method to estimate the parameters of the new equation. Thus, to find the best fitting exponential curve we must solve for α' and β' from:

$$\begin{pmatrix} \sum\limits_{i=1}^{N} x_i^2 & \sum\limits_{i=1}^{N} x_i \\ \sum\limits_{i=1}^{N} x_i & N \end{pmatrix} \begin{pmatrix} \beta' \\ \alpha' \end{pmatrix} = \begin{pmatrix} \sum\limits_{i=1}^{N} V_i x_i \\ \sum\limits_{i=1}^{N} V_i \end{pmatrix}$$

where the data consist of N pairs of values (x_i, y_i), $V_i = \ln y_i$, $\alpha' = \ln \alpha$ and $\beta' = \ln \beta$. The matrix representation of least squares above was developed in earlier problems.

2) Let us make a table of values:

x_i	y_i	$V_i = \ln y_i$	$x_i V_i$	x_i^2
1	1.00	0.00	0.00	1
2	1.20	0.18	0.36	4
3	1.80	0.59	1.77	9
4	2.50	0.92	3.68	16
5	3.60	1.28	6.40	25
6	4.70	1.55	9.30	36
7	6.60	1.89	13.23	49
8	9.10	2.21	17.68	64
36	30.50	8.62	52.42	204

From 1) we know:

$$\begin{pmatrix} \sum\limits_{i=1}^{N} x_i^2 & \sum\limits_{i=1}^{N} x_i \\ \sum\limits_{i=1}^{N} x_i & N \end{pmatrix} \begin{pmatrix} \beta' \\ \alpha' \end{pmatrix} = \begin{pmatrix} \sum\limits_{i=1}^{N} V_i x_i \\ \sum\limits_{i=1}^{N} V_i \end{pmatrix}$$

$V_i = \ln y_i$, $\alpha' = \ln \alpha$ and $\beta' = \ln \beta$. Thus, substituting in our values:

$$\begin{pmatrix} 204 & 36 \\ 36 & 8 \end{pmatrix} \begin{pmatrix} \beta' \\ \alpha' \end{pmatrix} = \begin{pmatrix} 52.42 \\ 8.62 \end{pmatrix}$$

By Cramer's Rule,

$$\beta' = \frac{\begin{vmatrix} 52.42 & 36 \\ 8.62 & 8 \end{vmatrix}}{\begin{vmatrix} 204 & 36 \\ 36 & 8 \end{vmatrix}} = \frac{8(52.42) - 36(8.62)}{8(204) - (36)^2}$$

Thus, $\beta' = .325$. Similarly,

$$\alpha' = \frac{\begin{vmatrix} 204 & 52.42 \\ 36 & 8.62 \end{vmatrix}}{\begin{vmatrix} 204 & 36 \\ 36 & 8 \end{vmatrix}} = \frac{204(8.62) - 36(52.42)}{204(8) - (36)^2}$$

Thus, $\alpha' = -.383$. But $\beta' = \ln \beta$ and $\alpha' = \ln \alpha$, and $f(x) = \alpha\beta^x$.
Thus, $\beta = e^{\beta'}$ and $\alpha = e^{\alpha'}$ or $\beta = e^{.325} = 1.39$ and $\alpha = e^{-.383} = .68$.
Using the transformation we arrive at the estimated exponential curve,
$f(x) = \alpha\beta^x$ or $f(x) = .68(1.39)^x$.

● **PROBLEM** 19-40

Suppose a safety expert is interested in the relationship between the number of licensed vehicles in a community and the number of accidents per year in that community. In particular, he wishes to use the number of licensed vechicles to predict the number of accidents per year. He takes a random sample of ten communities and obtains the results as in the Table below:

Community	x_i Licensed Vehicles (thousands)	y_i Number of accidents (hundreds)
1	4	1
2	10	4
3	15	5
4	12	4
5	8	3
6	16	4
7	5	2
8	7	1
9	9	4
10	10	2

1) First, generalize the least squares method for multiple regression.
2) Find the estimated regression line for the data above.

Solution: The linear model for multiply regression may be stated as follows:

$$y_i = b_1 + b_2 x_{2i} + b_3 x_{3i} + \ldots + b_K x_{Ki} + e_i \quad (i = 1, \ldots, N) \qquad (1)$$

where there are K variables denoted by y_i, x_{2i}, x_{3i}, ..., $x_{(K-1)i}$ and x_{Ki}. y_i is the dependent variable and the x_{ji} $(j = 2, 3, \ldots, K)$ are the $K-1$ independent variables and the e_i are simply random-error terms.

Because we will not in general be able to achieve perfect prediction, there will be a difference between the actual and predicted values of y_i. Given the values x_{2i}, x_{3i}, ..., x_{Ki}, the predicted value of y_i is $\hat{y}_i$ where $\hat{y}_i = b_1 + b_2 x_{2i} + b_3 x_{3i} + \ldots + b_K x_{Ki}$. The difference between the predicted value and the actual value is thus simply the random-error term, $e_i = y_i - \hat{y}_i$. We do not know the values of the K parameters, $b_1, b_2, \ldots, b_K$. The problem is to estimate these parameters. Applying the least squares criterion, we want to find $b_1, b_2, \ldots, b_K$ such as to minimize

$$S = \sum_{i=1}^{N} [y_i - (b_1 + b_2 x_{2i} + b_3 x_{3i} + \ldots + b_K x_{Ki})]^2 ,$$

or to minimize

$$S = \sum_{i=1}^{N} (y_i - \hat{y}_i)^2 ,$$

or to minimize

$$S = \sum_{i=1}^{N} e_i^2 .$$

Finding the least squares estimators involves taking K partial derivatives, setting them equal to zero, and solving the resulting set of K equations in K unknowns. This can be quite tedious, especially when K is large. We have already done this least squares method for $K = 2$ and $K = 3$. For $K = 2$ we had:

$$\sum_{i=1}^{N} y_i = N b_1 + b_2 \sum_{i=1}^{N} x_{2i}$$

$$\sum_{i=1}^{N} y_i x_{2i} = b_1 \sum_{i=1}^{N} x_{2i} + b_2 \sum_{i=1}^{N} x_{2i}^2$$

For $K = 3$ we had:

$$\sum_{i=1}^{N} y_i = Nb_1 + b_2 \sum_{i=1}^{N} x_{2i} + b_3 \sum_{i=1}^{N} x_{3i}$$

$$\sum_{i=1}^{N} x_{2i} y_i = b_1 \sum_{i=1}^{N} x_{2i} + b_2 \sum_{i=1}^{N} x_{2i}^2 + b_3 \sum_{i=1}^{N} x_{2i} x_{3i}$$

$$\sum_{i=1}^{N} x_{3i} y_i = b_1 \sum_{i=1}^{N} x_{3i} + b_2 \sum_{i=1}^{N} x_{2i} x_{3i} + b_3 \sum_{i=1}^{N} x_{3i}^2$$

By noticing the pattern to the equations above, we can write the set of K equations in K unknowns for the general multivariate linear model. The equations are:

$$\sum_{i=1}^{N} y_i = Nb_1 + b_2 \sum_{i=1}^{N} x_{2i} + b_3 \sum_{i=1}^{N} x_{3i} + \ldots + b_K \sum_{i=1}^{N} x_{Ki}$$

$$\sum_{i=1}^{N} x_{2i} y_i = b_1 \sum_{i=1}^{N} x_{2i} + b_2 \sum_{i=1}^{N} x_{2i}^2 + b_3 \sum_{i=1}^{N} x_{2i} x_{3i} + \ldots + b_K \sum_{i=1}^{N} x_{2i} x_{Ki}$$

$$\vdots$$

$$\sum_{i=1}^{N} x_{Ki} y_i = b_1 \sum_{i=1}^{N} x_{Ki} + b_2 \sum_{i=1}^{N} x_{Ki} x_{2i} + b_3 \sum_{i=1}^{N} x_{Ki} x_{3i} + \ldots + b_K \sum_{i=1}^{N} x_{Ki}^2$$

We can write the above equations in matrix form:

$$
\begin{pmatrix}
\sum_{i=1}^{N} y_i \\
\sum_{i=1}^{N} x_{2i} y_i \\
\vdots \\
\sum_{i=1}^{N} x_{Ki} y_i
\end{pmatrix}
=
\begin{pmatrix}
N & \sum_{i=1}^{N} x_{2i} & \sum_{i=1}^{N} x_{3i} & \cdots & \sum_{i=1}^{N} x_{Ki} \\
\sum_{i=1}^{N} x_{2i} & \sum_{i=1}^{N} x_{2i}^2 & \sum_{i=1}^{N} x_{2i} x_{3i} & \cdots & \sum_{i=1}^{N} x_{2i} x_{Ki} \\
\vdots & \vdots & \vdots & & \vdots \\
\sum_{i=1}^{N} x_{Ki} & \sum_{i=1}^{N} x_{Ki} x_{2i} & \sum_{i=1}^{N} x_{Ki} x_{3i} & \cdots & \sum_{i=1}^{N} x_{Ki}^2
\end{pmatrix}
\begin{pmatrix}
b_1 \\
b_2 \\
\vdots \\
b_K
\end{pmatrix}
\quad (2)
$$

Note:

$$
\begin{pmatrix}
1 & x_{21} & x_{31} & \cdots & x_{K1} \\
1 & x_{22} & x_{32} & \cdots & x_{K2} \\
1 & x_{23} & x_{33} & \cdots & x_{K3} \\
\vdots & \vdots & \vdots & & \vdots \\
1 & x_{2N} & x_{3N} & \cdots & x_{KN}
\end{pmatrix}^{t}
\begin{pmatrix}
y_1 \\
y_2 \\
y_3 \\
\vdots \\
\dot{y}_N
\end{pmatrix}
=
$$

$$
\begin{bmatrix}
1 & 1 & 1 \cdots & 1 \\
x_{21} & x_{22} & x_{23} & x_{2N} \\
x_{31} & x_{32} & x_{33} & x_{3N} \\
\cdot & \cdot & \cdot & \cdot \\
\cdot & \cdot & \cdot & \cdot \\
\hline
x_{K1} & x_{K2} & x_{K3} \cdots & x_{KN}
\end{bmatrix}
\begin{bmatrix}
y_1 \\ y_2 \\ y_3 \\ \cdot \\ \cdot \\ \cdot \\ y_N
\end{bmatrix}
=
\begin{bmatrix}
\sum\limits_{i=1}^{N} y_i \\[2ex]
\sum\limits_{i=1}^{N} x_{2i} y_i \\[2ex]
\sum\limits_{i=1}^{N} x_{Ki} y_i
\end{bmatrix}
\qquad (3)
$$

Also note,

$$
\begin{bmatrix}
1 & x_{21} & x_{31} \cdots & x_{K1} \\
1 & x_{22} & x_{32} \cdots & x_{K2} \\
1 & x_{23} & x_{33} \cdots & x_{K3} \\
\cdot & \cdot & \cdot & \cdot \\
\cdot & \cdot & \cdot & \cdot \\
\cdot & \cdot & \cdot & \cdot \\
1 & x_{2N} & x_{3N} \cdots & x_{KN}
\end{bmatrix}^{t}
\begin{bmatrix}
1 & x_{21} & x_{31} \cdots & x_{K1} \\
1 & x_{22} & x_{32} \cdots & x_{K2} \\
1 & x_{23} & x_{33} \cdots & x_{K3} \\
\cdot & \cdot & \cdot & \cdot \\
\cdot & \cdot & \cdot & \cdot \\
\cdot & \cdot & \cdot & \cdot \\
1 & x_{2N} & x_{3N} \cdots & x_{KN}
\end{bmatrix}
=
$$

$$
\begin{bmatrix}
1 & 1 & 1 \cdots & 1 \\
x_{21} & x_{22} & x_{23} \cdots & x_{2N} \\
x_{31} & x_{32} & x_{33} \cdots & x_{3N} \\
\cdot & \cdot & \cdot & \cdot \\
\cdot & \cdot & \cdot & \cdot \\
x_{K1} & x_{K2} & x_{K3} \cdots & x_{KN}
\end{bmatrix}
\begin{bmatrix}
1 & x_{21} & x_{31} \cdots & x_{K1} \\
1 & x_{22} & x_{32} \cdots & x_{K2} \\
1 & x_{23} & x_{33} \cdots & x_{K3} \\
\cdot & \cdot & \cdot & \cdot \\
\cdot & \cdot & \cdot & \cdot \\
1 & x_{2N} & x_{3N} \cdots & x_{KN}
\end{bmatrix}
\,,\quad =
$$

$$
\begin{bmatrix}
N & \sum\limits_{i=1}^{N} x_{2i} & \sum\limits_{i=1}^{N} x_{3i} \cdots & \sum\limits_{i=1}^{N} x_{Ki} \\[3ex]
\sum\limits_{i=1}^{N} x_{2i} & \sum\limits_{i=1}^{N} x_{2i}^{2} & \sum\limits_{i=1}^{N} x_{2i} x_{3i} \cdots & \sum\limits_{i=1}^{N} x_{2i} x_{Ki} \\[3ex]
\cdot & \cdot & \cdot & \cdot \\
\cdot & \cdot & \cdot & \cdot \\
\sum\limits_{i=1}^{N} x_{Ki} & \sum\limits_{i=1}^{N} x_{Ki} x_{2i} & \sum\limits_{i=1}^{N} x_{Ki} x_{3i} \cdots & \sum\limits_{i=1}^{N} x_{Ki}^{2}
\end{bmatrix}
\qquad (4)
$$

From (1) we have:

$$
y_i = b_1 + b_2 x_{2i} + b_3 x_{3i} + \cdots + b_K x_{Ki} + e_i \ .
$$

Suppose that we have a sample of size N, that is, a sample of N sets of values $(y_i, x_{2i}, x_{3i}, \ldots, x_{Ki})$, where $i = 1, 2, \ldots, N$. Then we have N equations of the form of (1), where the subscript i goes from 1 to N:

$$y_1 = b_1 + b_2 x_{21} + b_3 x_{31} + \ldots + b_K x_{K1} + e_1$$
$$y_2 = b_1 + b_2 x_{22} + b_3 x_{32} + \ldots + b_K x_{K2} + e_2$$
$$y_3 = b_1 + b_2 x_{23} + b_3 x_{33} + \ldots + b_K x_{K3} + e_3$$
$$\vdots$$
$$y_N = b_1 + b_2 x_{2N} + b_3 x_{3N} + \ldots + b_K x_{KN} + e_N$$

By using vectors and matrices, it is possible to express these N equations in much simpler form. Thus, let:

$$
Y = \begin{bmatrix} y_1 \\ y_2 \\ y_3 \\ \vdots \\ y_n \end{bmatrix}, \quad
X = \begin{bmatrix}
1 & x_{21} & x_{31} & \cdots & x_{K1} \\
1 & x_{22} & x_{32} & \cdots & x_{K2} \\
1 & x_{23} & x_{33} & \cdots & x_{K3} \\
\vdots & \vdots & \vdots & & \vdots \\
1 & x_{2N} & x_{3N} & \cdots & x_{KN}
\end{bmatrix},
$$

$$
B = \begin{bmatrix} b_1 \\ b_2 \\ b_3 \\ \vdots \\ b_n \end{bmatrix}, \quad
E = \begin{bmatrix} e_1 \\ e_2 \\ e_3 \\ \vdots \\ e_N \end{bmatrix}
$$

Thus, the set of N equations can be written as $Y = XB + E$. Now, we want to find our least squares estimates, B, in terms of Y and X. We can rewrite (3) using X and Y :

$$
X^t Y = \begin{bmatrix}
\sum\limits_{i=1}^{N} y_i \\[2ex]
\sum\limits_{i=1}^{N} x_{2i} y_i \\[2ex]
\vdots \\[1ex]
\sum\limits_{i=1}^{N} x_{Ki} y_i
\end{bmatrix}
\tag{3'}
$$

We can rewrite (4) using X and X^t :

$$X^tY = \begin{pmatrix} N & \sum\limits_{i=1}^{N} x_{2i} & \sum\limits_{i=1}^{N} x_{3i} & \cdots & \sum\limits_{i=1}^{N} x_{Ki} \\ \sum\limits_{i=1}^{N} x_{2i} & \sum\limits_{i=1}^{N} x_{2i}^2 & \sum\limits_{i=1}^{N} x_{2i}x_{3i} & \cdots & \sum\limits_{i=1}^{N} x_{2i}x_{Ki} \\ \vdots & \vdots & \vdots & & \vdots \\ \sum\limits_{i=1}^{N} x_{Ki} & \sum\limits_{i=1}^{N} x_{Ki}x_{2i} & \sum\limits_{i=1}^{N} x_{Ki}x_{3i} & \cdots & \sum\limits_{i=1}^{N} x_{Ki}^2 \end{pmatrix}$$ (4')

Substituting (3') and (4') into (2):

$$X^tY = (X^tX)B .$$

Thus, we want to solve for our least squares estimators, B:

$$B = (X^tX)^{-1} X^tY .$$ (5)

2) Looking at our table of data in the beginning of the problem, we note that $N = 10$ and that $K = 2$. Moreover, we can write down our two matrices Y and X:

$$Y = \begin{pmatrix} 1 \\ 4 \\ 5 \\ 4 \\ 3 \\ 4 \\ 2 \\ 1 \\ 4 \\ 2 \end{pmatrix} \quad , \quad X = \begin{pmatrix} 1 & 4 \\ 1 & 10 \\ 1 & 15 \\ 1 & 12 \\ 1 & 8 \\ 1 & 16 \\ 1 & 5 \\ 1 & 7 \\ 1 & 9 \\ 1 & 10 \end{pmatrix}$$

In order to apply equation (5), $B = (X^tX)^{-1} X^tY$, we must calculate $(X^tX)^{-1}$.
First,

$$X^tY = \begin{pmatrix} 1 & 1 & 1 & 1\ldots1 & 1 \\ 4 & 10 & 15 & 12\ldots9 & 10 \end{pmatrix} \begin{pmatrix} 1 & 4 \\ 1 & 10 \\ 1 & 15 \\ 1 & 12 \\ \vdots & \vdots \\ \vdots & \vdots \\ 1 & 9 \\ 1 & 10 \end{pmatrix} =$$

$\begin{pmatrix} 10 & 96 \\ 96 & 1060 \end{pmatrix}$ by matrix multiplication. To find the inverse of $\begin{pmatrix} 10 & 96 \\ 96 & 1060 \end{pmatrix}$:

$\left(\begin{array}{cc|cc} 10 & 96 & 1 & 0 \\ 96 & 1060 & 0 & 1 \end{array}\right)$ Divide the first row by 10:

$\left(\begin{array}{cc|cc} 1 & 9.6 & .1 & 0 \\ 96 & 1060 & 0 & 1 \end{array}\right)$ Replace the second row by the second row + (-96) × the first row:

$\left(\begin{array}{cc|cc} 1 & 9.6 & .1 & 0 \\ 0 & 138.4 & -9.6 & 1 \end{array}\right)$ Divide the second row by 138.4:

$$\begin{pmatrix} 1 & 9.6 & \vdots & .1 & 0 \\ 0 & 1 & \vdots & -96/1384 & 10/1384 \end{pmatrix}$$ Replace the first row by the first row plus $(-9.6) \times$ the second row:

$$\begin{pmatrix} 1 & 0 & \vdots & 1060/1384 & -96/1384 \\ 0 & 1 & \vdots & -96/1384 & 10/1384 \end{pmatrix}$$ Thus,

$$\begin{pmatrix} 10 & 96 \\ 96 & 1060 \end{pmatrix}^{-1} = \begin{pmatrix} 1060/1384 & -96/1384 \\ -96/1384 & 10/1384 \end{pmatrix} , \quad \text{or}$$

$$\begin{pmatrix} +.766 & -.069 \\ -.069 & .007 \end{pmatrix} .$$ We also need to determine $X^t Y$ by matrix multiplication:

$$X^t Y = \begin{pmatrix} 1 & 1 & 1 & \ldots & 1 \\ 4 & 10 & 15 & \ldots & 10 \end{pmatrix} \begin{pmatrix} 1 \\ 4 \\ 5 \\ \cdot \\ \cdot \\ \cdot \\ 4 \\ 2 \end{pmatrix} = \begin{pmatrix} 30 \\ 328 \end{pmatrix}$$

We are now ready to apply equation (5), $B = (X^t X)^{-1} X^t Y$. By substitution,

$$B = \begin{pmatrix} +.766 & -.069 \\ -.069 & .007 \end{pmatrix} \begin{pmatrix} 30 \\ 328 \end{pmatrix} = \begin{pmatrix} .22 \\ .29 \end{pmatrix}$$

Therefore, the estimated regression line is $y = .22 + .29x$.

● **PROBLEM** 19-41

Consider once again the accidents-vehicles example. Suppose that the safety expert decides to introduce another independent variable to the linear model: the number of men on the community's police force. He now wants to predict y_i , the number of accidents per year. x_{2i} represents the number of licensed vehicles in the community. x_{3i} represents the number of men on the police force. He is able to obtain data regarding x_{3i} for the ten communities for which he already had data on y_i and x_{2i} , all of the data is presented in the table below:

Community	y_i	x_{2i}	x_{3i}
1	1	4	20
2	4	10	6
3	5	15	2
4	4	12	8
5	3	8	9
6	4	16	8
7	2	5	12
8	1	7	15
9	4	9	10
10	2	10	10

Find the estimated regression line.

<u>Solution:</u> We will use the theory developed in the previous problem to find the estimated regression line. The regression coefficients are the entries of the column vector B, where $B = (X^t X)^{-1} X^t Y$. The matrices and vectors, X, Y, and B are defined below:

Y is a row vector with N elements y_1, y_2, y_3,...., and y_n:

$$Y = \begin{pmatrix} y_1 \\ y_2 \\ y_3 \\ \cdot \\ \cdot \\ \cdot \\ y_n \end{pmatrix} \quad .$$

B is a row vector with K elements b_1, b_2,....,b_K . Its entries represent the regression coefficients:

$$B = \begin{pmatrix} b_1 \\ b_2 \\ b_3 \\ \cdot \\ \cdot \\ \cdot \\ b_K \end{pmatrix} \quad .$$

Finally, X denotes an N × K matrix:

$$X = \begin{pmatrix} 1 & x_{21} & x_{31} & \cdot & \cdot & \cdot & x_{K1} \\ 1 & x_{22} & x_{32} & \cdot & \cdot & \cdot & x_{K2} \\ 1 & x_{23} & x_{33} & \cdot & \cdot & \cdot & x_{K3} \\ \cdot & \cdot & \cdot & & & & \cdot \\ \cdot & \cdot & \cdot & & & & \cdot \\ \cdot & \cdot & \cdot & & & & \cdot \\ 1 & x_{2N} & x_{3N} & \cdot & \cdot & \cdot & x_{KN} \end{pmatrix} \quad \cdot$$

Thus from the table, it is possible to determine the vector Y and the matrix X:

$$Y = \begin{pmatrix} 1 \\ 4 \\ 5 \\ 4 \\ 3 \\ 4 \\ 2 \\ 1 \\ 4 \\ 2 \end{pmatrix} \quad , \quad X = \begin{pmatrix} 1 & 4 & 20 \\ 1 & 10 & 6 \\ 1 & 15 & 2 \\ 1 & 12 & 8 \\ 1 & 8 & 9 \\ 1 & 16 & 8 \\ 1 & 5 & 12 \\ 1 & 7 & 15 \\ 1 & 9 & 10 \\ 1 & 10 & 10 \end{pmatrix}$$

Thus by matrix multiplication,

$$X^tX = \begin{bmatrix} 1 & 1 & 1 & \dots & 1 \\ 4 & 10 & 15 & \dots & 10 \\ 20 & 6 & 2 & \dots & 10 \end{bmatrix} \begin{bmatrix} 1 & 4 & 20 \\ 1 & 10 & 6 \\ 1 & 15 & 2 \\ 1 & 12 & 8 \\ \cdot & \cdot & \cdot \\ \cdot & \cdot & \cdot \\ \cdot & \cdot & \cdot \\ 1 & 10 & 10 \end{bmatrix} =$$

$$\begin{bmatrix} 10 & 96 & 100 \\ 96 & 1060 & 821 \\ 100 & 821 & 1218 \end{bmatrix} .$$

Thus, to find $(X^tX)^{-1}$: the cofactors of the nine elements of (X^tX) are:

$$(X^tX)_{11} = + \begin{vmatrix} 1060 & 821 \\ 821 & 1218 \end{vmatrix} = 617039$$

$$(X^tX)_{12} = - \begin{vmatrix} 96 & 821 \\ 100 & 1218 \end{vmatrix} = -34828$$

$$(X^tX)_{13} = + \begin{vmatrix} 96 & 1060 \\ 100 & 821 \end{vmatrix} = -27184$$

$$(X^tX)_{21} = - \begin{vmatrix} 96 & 100 \\ 821 & 1218 \end{vmatrix} = -34828$$

$$(X^tX)_{22} = + \begin{vmatrix} 10 & 100 \\ 100 & 1218 \end{vmatrix} = 2180$$

$$(X^tX)_{23} = - \begin{vmatrix} 10 & 96 \\ 100 & 821 \end{vmatrix} = +1390$$

$$(X^tX)_{31} = + \begin{vmatrix} 96 & 100 \\ 1060 & 821 \end{vmatrix} = -27194$$

$$(X^tX)_{32} = - \begin{vmatrix} 10 & 100 \\ 96 & 821 \end{vmatrix} = +1390$$

$$(X^tX)_{33} = + \begin{vmatrix} 10 & 96 \\ 96 & 1060 \end{vmatrix} = 1384$$

We form the transpose of the above matrix of cofactors to obtain the classical adjoint of (X^tX) :

$$adj(X^tX) = \begin{bmatrix} 617039 & -34828 & -27194 \\ -34828 & 2180 & 1390 \\ 027184 & 1390 & 1384 \end{bmatrix} .$$

We know that for any square matrix A, $A^{-1} = \frac{1}{|A|} (\text{adj } A)$. Thus,

$$|X^tX| = 10(1060)(1218) + 96(821)(100) + (100)(96)(821)$$
$$- 100(1060)(100) - 96(96)(1218) - 10(821)(821)$$
$$= 108502 \ .$$

Therefore,

$$(X^tX)^{-1} = \frac{1}{108502} \begin{pmatrix} 617039 & -34828 & -27194 \\ -34828 & 2180 & 1390 \\ -27184 & 1390 & 1384 \end{pmatrix}$$

$$= \begin{pmatrix} 5.68689 & -.32099 & -.25054 \\ -.32099 & .02009 & .01281 \\ -.25054 & .01281 & .01276 \end{pmatrix}$$

Also, $X^tY =$

$$\begin{pmatrix} 1 & 1 & 1 & \ldots & 1 \\ 4 & 10 & 15 & \ldots & 10 \\ 20 & 6 & 2 & \ldots & 10 \end{pmatrix} \begin{pmatrix} 1 \\ 4 \\ 5 \\ 4 \\ \cdot \\ \cdot \\ \cdot \\ 4 \\ 2 \end{pmatrix} = \begin{pmatrix} 30 \\ 328 \\ 244 \end{pmatrix}$$

To obtain the least squares estimate of the vector of regression coefficients, compute

$$B = (X^tX)^{-1}(X^tY)$$

$$= \begin{pmatrix} 5.68689 & -.32099 & -.25054 \\ -.32099 & .02009 & .01281 \\ -.25054 & .01281 & .01276 \end{pmatrix} \begin{pmatrix} 30 \\ 328 \\ 244 \end{pmatrix}$$

By matrix multiplication,

$$B = \begin{pmatrix} 4.190 \\ .085 \\ -.201 \end{pmatrix} \ .$$

The estimated regression line is therefore given by:

$$y = 4.190 + .085x_2 - .201x_3 \ .$$

CHAPTER 20

APPLICATIONS TO PHYSICAL SYSTEMS I (MECHANICS, CHEMICAL)

REPRESENTING FORCES BY VECTORS

● PROBLEM 20-1

Use vector addition in R^2 to compute the actual speed and direction of an airplane subject to wind conditions.
i) Suppose the plane is flying 200 mi/hr. due north with a tailwind which is 50 mi/hr. north.
ii) The plane is flying 120 mi/hr. north and the wind has velocity 50 mi/hr. east.

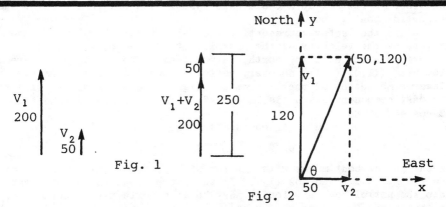

Fig. 1   Fig. 2

Solution: In general velocities can be represented by vectors in a vector-space and the sum of 2 velocities is the vector sum of the vectors representing them.
i)

Using the tail to head method of addition we see that the airplane subject to these conditions is flying north with a speed of 250 mi/hr.
ii)

v_1 and v_2 determine a parallelogram which specifies the desired sum $v_1 + v_2$. In the xy plane v_1 and v_2 have coordinates $(0,120)$ and $(50,0)$ respectively. The airplane will have velocity $v_1 + v_2 = (0,120) + (50,0) = (50,120)$. Thus the airplane is flying $\theta = $ arc tan 120/50 degrees north of east with speed

$$\|v_1 + v_2\| = \sqrt{(120)^2 + 50^2} = 130 .$$

Note there are three ways to find vector sums;
1) The tail to head method used in i);
2) The coordinate method used in ii); and
3) The parallelogram method, which we can see in fig. 2.

A ship is sighted 10 nautical miles north of a patrol boat; the ship is traveling N.E. at 12 knots. The patrol boat has a maximum speed of 20 knots; find i) the minimum time required for the patrol boat to intercept the ship and ii) the course the patrol boat must follow to do this.

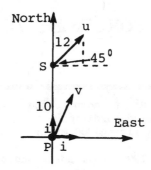

Solution: First set up a coordinate system with origin the point where the patrol boat is now.

u is the vector representing the velocity of the ship S and v represents the velocity of the patrol boat, P. Also, i and j are unit vectors pointing east and north, respectively. Now i and j are basis elements ((1,0),(0,1)) so every vector in the plane can be written as a linear combination of them. From the figure and the geometry of a triangle the east component of u is 12 sin 45° and the north component is 12 cos 45° . So

$$u = (12 \cos 45°)i + (12 \sin 45°)j$$
$$= 6\sqrt{2}\, i + 6\sqrt{2}\, j$$

If the patrol boat moves with maximum speed at an angle $\theta°$ north of east, $v = (20 \sin \theta)i + (20 \cos \theta)j$. If the patrol boat is to intercept the ship the patrol boat's velocity must be in a northerly direction relative to the ship. So the vector quantity $v - u$ must be in the j(north) direction. Therefore, the i component of $v - u$ must be 0.

$$v - u = (20 \sin \theta)i + (20 \cos \theta)j - 6\sqrt{2}\, i - 6\sqrt{2}\, j$$
$$= (20 \sin \theta - 6\sqrt{2})i + (20 \cos \theta - 6\sqrt{2})j .$$

So $20 \sin \theta - 6\sqrt{2} = 0$ which implies

$$\theta = \sin^{-1}\left(\frac{3\sqrt{2}}{10}\right) \approx 25°6' .$$

ii) Thus the course the patrol boat must follow is 25°6' north of east. We have

$$\sin \theta = \frac{3\sqrt{2}}{10} .$$

So we know

$$\cos \theta = \sqrt{1 - \sin^2 \theta} = \sqrt{1 - \frac{18}{100}} \approx .9055.$$

Therefore

$$v - u = (20 \cos \theta - 6\sqrt{2})j \approx 9.63\, j .$$

Since the actual northerly distance that the patrol boat must cover is 10 mi. (See fig) the interception will take place in $10/9.63 = 1.038$ hrs or approximately $62\frac{1}{2}$ minutes.

A 75-lb force acting at an angle of 20° above an inclined plane moves a 100-lb body up the plane at a constant speed. Find the force of friction if the plane is inclined at 30°.

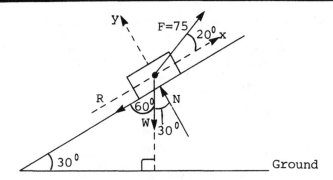

Solution: Represent each force by a vector in a free-body diagram. Choose the positive x-axis in the direction of motion.

In the figure, R is the retarding force of friction, W is the downward force of the weight of the body and N is the normal force exerted by the plane on the body (perpendicular to the plane).

Newton's Law of Motion states that the sum of all external forces acting on a body equals the product of the mass of the body and its acceleration (F = ma). Since the body moves with a constant speed its acceleration is zero. Therefore in any direction the sum of the forces must be zero. Since the frictional force is along the x-axis find the x-component of each force. F is 75-lb at an angle of 20° to the x-axis; therefore the x-component is 75 cos 20 ≐ 70.5. The normal force is perpendicular to the x-axis and thus gives no contribution in the x direction. The force of the weight of the body is at an angle of −(90 + 30)° = −120° to the x-axis. Therefore its component in the x direction is 100 cos (−120) = −50. The vector R lies along the x-axis in the negative direction and hence has x-component −R. The sum of the forces in the x-direction is 70.5 − 50 − R = 0 which implies that the force of friction, R = 20.5 lb.

● **PROBLEM** 20-4

i) Find the work done by moving an object along the vector 2i + 6j if the force acting on the particle is i + 2j.

ii) A uniform bar 6 ft. by 40 lbs rotates about a fixed horizontal axis at 0. Compute the work done by the weight of the bar as the bar rotates from the position vector 3i ft. to the position vector -1.8i - 2.4j ft. where the position vector of the bar is the vector from the origin to the center of mass of the bar.

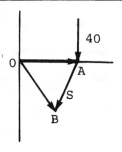

Solution: The work done by a constant force F through a displacement S
is defined to be the dot product, F·S .
i)   The force in this case is i + 2j and the displacement, S is 2i + 6j
where i and j are orthogonal unit vectors in the plane. This means
that a particle under the influence of the force i + 2j moves in the
plane on a straight line from the point (0,0) to the point (2,6). The
work done is (i + 2j)·(2i + 6j) = 2 + 12 = 14 units.
ii)
From the figure we see that the center of mass of the bar falls from
position vector OA to position vector OB under the 40 lb force of its
own weight. The position vector to A is given to be 3i and the position vector to B is given to be -1.81i - 2.4j. Therefore, the displacement is

$$S = (-1.81i - 2.4j) - (3i)$$

$$= -4.81i - 2.4j .$$

The work done is force · displacement which is, in this case,

$$(-40j)·(-4.81i - 2.4j)$$

$$= 0(-4.81) + (-40)(-2.4) = 96.0 \text{ ft.lb.}$$

● **PROBLEM** 20-5

Find the work done by a force (2,1,1) if its point of application moves
in a straight line from A(1,1,1) to B(2,1,3).

Solution: Define the physical notion of work. Physical forces and displacements can be represented by vectors. The work done by a constant
vector force F during a displacement d is defined to be F·d . If a
point moves from A(1,1,1) to B(2,1,3) it has experienced a displacement B - A = (2,1,3) - (1,1,1) = (1,0,2) . The work done, therefore, is
F·d = (2,1,1)·(1,0,2) = 2 + 2 = 4 units.

In physical systems work done is also the **change** in kinetic energy. Thus
the notion of dot products can also be used to find the **change** in kinetic
energy of a particle.

● **PROBLEM** 20-6

A particle is at the origin of the x,y and z axis in the standard coordinate system in R^3 . The particle is subject to the following forces:

   F_1: force of 5 units acting along Ox

   F_2: force of 3 units acting along zO

   F_3: force of 2 units acting along Oy

   F_4: force of $2\sqrt{2}$ units acting towards O

at an angle of $\pi/4$ to the x and y axes in the xy plane. Find the
resultant force on the particle.

Solution: Each of the forces can be represented by a vector in R^3 .
Using the standard basis vectors for R^3 , $e_1(1,0,0)$, $e_2 = (0,1,0)$ and
$e_3 = (0,0,1)$ the first three forces become:

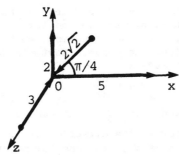

$$F_1 = 5e_1 + 0e_2 + 0e_3$$

$$F_2 = 0e_1 + 0e_2 - 3e_3 \quad \text{(minus 3 since the force is in the negative direction on the z-axis)}$$

$$F_3 = 0e_1 + 2e_2 + 0e_3 \ .$$

The fourth force must be resolved into its components. We can do this using these two properties of the dot product;

$$\|v\| = (v \cdot v)^{\frac{1}{2}} \quad \text{and} \quad u \cdot v = \|u\| \ \|v\| \cos \theta \ ,$$

where θ is the angle between u and v. Since F_4 is in the xy plane $F_4 = -(\alpha e_1 + \beta e_2 + 0e_3)$, with α, β positive real numbers.

$$\|F\| = 2\sqrt{2} = F \cdot F = (\alpha^2 + \beta^2)^{\frac{1}{2}} \tag{1}$$

F_4 makes an angle $\pi/4$ with the x-axis, so

$$F_4 \cdot e_2 = \|F_4\| \ \|e\| \cos \pi/4 \tag{2}$$

But

$$F_4 \cdot e_2 = -\alpha \cdot 0 + \beta \cdot 1 = -\beta \tag{3}$$

(2) and (3) imply $-\beta = -2\sqrt{2} \cdot 1 \cos \pi/4$ so $\beta = 2\sqrt{2} \cdot \sqrt{2}/2 = 2$
(since $\cos \pi/4 = \sqrt{2}/2$). Now from (1) $\sqrt{\alpha^2 + 2^2} = 2\sqrt{2}$ which implies $\alpha = \pm 2$ (α is positive so reject $\alpha = -2$). We then have

$$f_4 = -(2e_1 + 2e_2 + 0e_3) \ .$$

The resultant of a series of forces is the vector sum of the forces. Thus the resultant $F = F_1 + F_2 + F_3 + F_4 = (5e_1) - (3e_3) + (2e_2) - (2e_1 + 2e_2) = 3e_1 + 0e_2 - 3e_3$ which is a force of magnitude $3\sqrt{2}$ at an angle of 45° in the xz plane.

● **PROBLEM** 20-7

A particle P of constant mass m moves under the influence of a force which is always directed towards a fixed point O. Show that the angular momentum about O is constant. i) Deduce that the particle moves in a fixed plane through O and ii) that the component of its velocity perpendicular to OP is inversely proportional to OP.

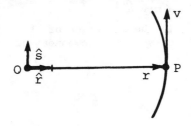

Solution: Let F be the force on P and let v be the velocity of P. F and v can be represented by vectors and let r be the vector OP (directed from O to P); then the situation can be seen in the figure.

Here P is moving along the curve. Newton's Second Law of motion for a particle of constant mass is

$$F = m \frac{dv}{dt} \qquad (1)$$

Since F is always directed towards O it can be written as $F = y(-\hat{r})$, where $\hat{r}$ is the unit vector in the direction of r and y is the magnitude of F. Thus $F = -y\, \hat{r}$. The angular momentum (call it h) of P about O is defined to be the cross product, $r \times mv$. To show h is constant we will show that the rate of change of h, (dh/dt) is O. Differentiating both sides of $h = r \times mv$ with respect to t yields

$$\frac{dh}{dt} = \frac{d}{dt} (r \times mv) .$$

Differentiating across the product we then obtain

$$\frac{dh}{dt} = \frac{dr}{dt} \times mv + r \times m \frac{dv}{dt} \qquad (2)$$

$$= v \times mv + r \times m \frac{dv}{dt}$$

since $\frac{dr}{dt} = v$. Now the cross product of two vectors is a vector perpendicular to the two vectors with magnitude the area of the parallelogram which the two vectors determine. Therefore, since v and mv do not determine a rectangle (they are collinear), $v \times mv = 0$. So,

$$\frac{dh}{dt} = r \times m \frac{dv}{dt} \qquad (3)$$

but $m \frac{dv}{dt}$ is just the force F which was determined to be $-y\, \hat{r}$. We have

$$\frac{dh}{dt} = r \times (-y\, \hat{r}) .$$

r and $-y\, \hat{r}$ are coincident so $r \times (-y\, \hat{r}) = 0$. Thus we have h is constant, $h \equiv h_0 = r_0 \times mv_0$ where h_0, r_0 and v_0 are the initial values for h, r and v respectively. At each time, t, $r \cdot h = 0$ since the cross product $r \times mv$ yields a vector perpendicular to both r and mv. But since h is constant $r \cdot h_0 = 0$ for each r. i) Therefore r always remains in the plane of O perpendicular to the constant h_0.

ii) First resolve v in the plane of motion. $v = v_1 \hat{r} + v_2 \hat{s}$, where $\hat{r}$ and $\hat{s}$ are perpendicular unit vectors in the plane of motion (see figure). Then equation (3) gives

$$h = mr \times v = mr \times (v_1 \hat{r} + v_2 \hat{s})$$

$$= (mr \times v_1 \hat{r}) + mr \times v_2 \hat{s}$$

$$= 0 + mr \times v_2 \hat{s}$$

$$= mrv_2 \hat{h}$$

where $\hat{h}$ is a unit vector in the direction of h. H is constant so r is inversely proportional to v_2, the component of the velocity perpendicular to OP.

VIBRATING SYSTEMS

Consider a mechanical system consisting of two masses connected by three springs as shown in the figure. Assume that the motion of the masses takes place in a straight line on a smooth frictionless plane. Determine the natural frequencies of the system:

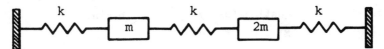

Solution: To formulate this mathematically, we shall use Newton's Second Law of motion and Hooke's law. Hooke's law states that the spring exerts a restoring force toward the (unstretched) equilibrium position; the magnitude of this force is proportional to the displacement of the spring from equilibrium and the constant of proportionality $k > 0$ is called the spring constant.

Let x_1 and x_2 denote the displacement of the masses from their equilibrium positions. Now suppose that at time t the system is in the position shown in the figure; then the only forces acting on the particle of mass m are the restoring force of the first spring $-kx_1$ (by Hooke's law); and the restoring force of the second spring. The net extension of the second spring is $x_2 - x_1$, because it is stretched x_2 units by the second weight and compressed x_1 units by the first weight. Thus, the restoring force of the second spring is $-k(x_2 - x_1)$. Now, Newton's Second Law of motion applied to the moving particle of mass m states that

$$\frac{d}{dt} (mv) = \text{sum of the forces acting on the particle} \qquad (1)$$

Here, v is the velocity vector and mv is the momentum. Since the motion of the present system is restricted to a line, the vectors involved are one dimensional and vector notation can be eliminated. Because the mass m is constant and in this case $v = x(t)$, we obtain, using (1):

$$m\ddot{x}_1 = -kx_1 - k(x_1 - x_2) . \qquad (2)$$

Similarly, the only forces acting on the particle of mass $2m$ are the restroring force $-k(x_2 - x_1)$ of the second spring whose net extension is $x_2 - x_1$ units, and the restoring force of the third spring $-kx_2$. Thus, Newton's law applied to the second particle of mass $2m$ yields

$$2m\ddot{x}_2 = -k(x_2 - x_1) - kx_2 . \qquad (3)$$

The equations (2) and (3) can be written as

$$m\ddot{x}_1 + 2kx_1 - kx_2 = 0$$
$$2m\ddot{x}_2 - kx_1 + 2kx_2 = 0 .$$

These are two simultaneous equations for the displacements x_1 and x_2 . Now these equations can be expressed in matrix notation as

$$\begin{bmatrix} m & 0 \\ 0 & 2m \end{bmatrix} \begin{bmatrix} \ddot{x}_1 \\ \ddot{x}_2 \end{bmatrix} + \begin{bmatrix} 2k & -k \\ -k & 2k \end{bmatrix} \begin{bmatrix} x_1 \\ x_2 \end{bmatrix} = \begin{bmatrix} 0 \\ 0 \end{bmatrix} \qquad (4)$$

or

$$M\ddot{X} + KX = 0 \tag{5}$$

where

$$M = \begin{bmatrix} m & 0 \\ 0 & 2m \end{bmatrix} = \text{mass matrix} \quad , \quad \ddot{X} = \begin{pmatrix} \ddot{x}_1 \\ \ddot{x}_2 \end{pmatrix}$$

$$K = \begin{bmatrix} 2k & -k \\ -k & 2k \end{bmatrix} = \text{stiffness matrix and} \quad x = \begin{pmatrix} x_1 \\ x_2 \end{pmatrix} .$$

If we premultiply the equation (5) by M^{-1} we obtain,

$$M^{-1}M\ddot{X} + M^{-1}KX = 0 .$$

But $M^{-1}M = I$ so

$$I\ddot{X} + AX = 0 \quad \text{where} \quad A = M^{-1}K \tag{6}$$

Because of the absence of damping and frictional forces, we expect the motion to be combinations of simple harmonic motions. We look for the simplest such motions, known as normal modes of vibration. These are motions where both masses move with the same frequency, called the natural frequency with possibly different amplitudes. Assuming harmonic motion,

$$\ddot{X} = -\lambda X \quad \text{where} \quad \lambda = \omega^2 , \quad (\omega = \text{natural frequency}) \text{ equation (6) becomes}$$

$$-I\lambda X + AX = 0$$

or

$$[A - I\lambda]X = 0 . \tag{7}$$

Then the characteristic equation of the system is $\det[A - \lambda I] = 0$. The roots λ_i of the characteristic equation are called eigenvalues and the natural frequencies of the system are determined from them by the relationship

$$\lambda_i = \omega_i^2 .$$

Now we compute A

$$A = M^{-1}k = \begin{bmatrix} m & 0 \\ 0 & 2m \end{bmatrix}^{-1} \begin{bmatrix} 2k & -k \\ -k & 2k \end{bmatrix}$$

$$A = \begin{bmatrix} \dfrac{1}{m} & 0 \\ 0 & \dfrac{1}{2m} \end{bmatrix} \begin{bmatrix} 2k & -k \\ -k & 2k \end{bmatrix}$$

$$= \begin{bmatrix} \dfrac{2k}{m} & \dfrac{-k}{m} \\ \dfrac{-k}{2m} & \dfrac{k}{m} \end{bmatrix} .$$

Then,

$$[A - \lambda I] = \begin{bmatrix} \dfrac{2k}{m} - \lambda & \dfrac{-k}{m} \\ -\dfrac{1}{2}\dfrac{k}{m} & \dfrac{k}{m} - \lambda \end{bmatrix} .$$

Substituting this value in equation (7), we have

688

$$\begin{bmatrix} \dfrac{2k}{m} - \lambda & \dfrac{-k}{m} \\[2ex] -\tfrac{1}{2}\dfrac{k}{m} & \dfrac{k}{m} - \lambda \end{bmatrix} \begin{bmatrix} x_1 \\[2ex] x_2 \end{bmatrix} = 0$$

Then the characteristic equation from the determinant of the above matrix is

$$\lambda^2 - \frac{3k}{m}\lambda + \frac{3}{2}\left(\frac{k}{m}\right)^2 = 0 \ .$$

Using the quadratic formula the eigenvalues are found to be

$$\lambda = \frac{-3k/m \overset{+}{-} \sqrt{9k^2/m^2 - 4(3/2\ k^2/m^2)}}{2} = k/m\left[\frac{3}{2} \overset{+}{-} \frac{\sqrt{3}}{2}\right] \ .$$

$\lambda_1 = .634\ k/m$ and $\lambda_2 = 2.366\ k/m$. But $\lambda = \omega^2$ so the natural frequencies of the system are

$$\omega_1 = \lambda_1^{\frac{1}{2}} = \sqrt{.634\ k/m} = .796\ \sqrt{k/m}$$

and

$$\omega_2 = \lambda_2^{\frac{1}{2}} = \sqrt{2.366\ k/m} = 1.538\ \sqrt{k/m}$$

● **PROBLEM 20-9**

Consider the symmetrical system with two degrees of freedom shown in figure 1, Compute the natural frequencies of the system.

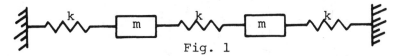
Fig. 1

Solution: In order to formulate this problem mathematically make the following assumptions:
i)   The spring has zero mass.
ii) The weight can be treated as though it were a point mass of mass m.
iii) The spring satisfies Hooke's Law. That is, the force exerted by each spring is proportional to the displacement of the spring from its position of equilibrium.
iv) Neither the table nor the surrounding medium offer any resistance to the motion of the system.
v)   The system obeys Newton's Second Law of motion. Force equals mass times acceleration ($\ddot{x}$).

Now derive the equations of motion by applying Newton's Second Law of motion to each moving weight. Let x_1 and x_2 denote the displacement of the masses from their equilibrium positions. A free-body diagram of the masses is shown in figure 2. The only forces acting on the particle of mass m are the restoring force of the first spring $-kx_1$ (by Hooke's law); and the restoring force $k(x_2 - x_1)$ of the second spring whose net extension is $x_2 - x_1$. Applying Newton's Second Law to the particle of mass m, we therefore obtain

$$m\ddot{x}_1 = -kx_1 + k(x_2 - x_1) \tag{1}$$

The only forces acting on the second particle of mass m are the restoring force $-k(x_2 - x_1)$ of the second spring whose net extension is $x_2 - x_1$

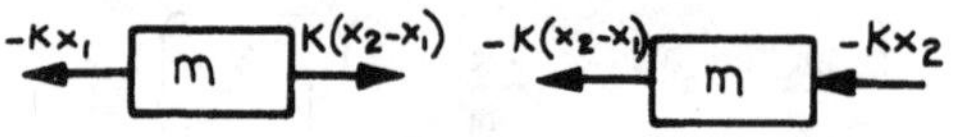

FIG. 2

and the restoring force of the third spring $-kx_2$. Thus Newton's Second Law applied to the second particle yields

$$m\ddot{x}_2 = -k(x_2 - x_1) - kx_2 \tag{2}$$

Now the equations (1) and (2) can be rewritten as

$$m\ddot{x}_1 + 2kx_1 - kx_2 = 0$$

$$m\ddot{x}_2 - kx_1 + 2kx_2 = 0 \ .$$

These equations can be expressed in matrix notation as

$$\begin{bmatrix} m & 0 \\ 0 & m \end{bmatrix} \begin{bmatrix} \ddot{x}_1 \\ \ddot{x}_2 \end{bmatrix} + \begin{bmatrix} 2k & -k \\ -k & 2k \end{bmatrix} \begin{bmatrix} x_1 \\ x_2 \end{bmatrix} = 0$$

or
$$M\ddot{X} + KX = 0 \tag{3}$$

where
$$M = \begin{bmatrix} m & 0 \\ 0 & m \end{bmatrix}, \quad K = \begin{bmatrix} 2k & -k \\ -k & 2k \end{bmatrix}, \quad \ddot{X} = \begin{bmatrix} \ddot{x}_1 \\ \ddot{x}_2 \end{bmatrix} \text{ and }$$

$$X = \begin{bmatrix} x_1 \\ x_2 \end{bmatrix} \ .$$

If we premultiply the equation (3) by M^{-1}, we obtain the following terms.

$$M^{-1}M = I \quad \text{(a unit matrix)}$$

$$M^{-1}k = A = \begin{bmatrix} m & 0 \\ 0 & m \end{bmatrix}^{-1} \begin{bmatrix} 2k & -k \\ -k & 2k \end{bmatrix}$$

$$= \begin{bmatrix} \dfrac{1}{m} & 0 \\ 0 & \dfrac{1}{m} \end{bmatrix} \begin{bmatrix} 2k & -k \\ -k & 2k \end{bmatrix}$$

$$= \begin{bmatrix} \dfrac{2k}{m} & \dfrac{-k}{m} \\ \dfrac{-k}{m} & \dfrac{2k}{m} \end{bmatrix}$$

and hence equation (3) becomes

$$I\ddot{X} + AX = 0 \ . \tag{4}$$

Assuming that the motion is harmonic, $\ddot{x} = -\lambda x_1$ where $\lambda = \omega^2$, equation (4) becomes

$$[A - \lambda I]x = 0 \ . \tag{5}$$

Then,

$$\begin{bmatrix} \dfrac{2k}{m} - \lambda & \dfrac{-k}{m} \\ \dfrac{-k}{m} & \dfrac{2k}{m} - \lambda \end{bmatrix} \begin{bmatrix} x_1 \\ x_2 \end{bmatrix} = 0 \ .$$

From equation (5), we form the determinant

$$\det[A - \lambda I] = 0 \ ,$$

which is the characteristic equation of the system. Thus,

$$\det \begin{bmatrix} \dfrac{2k}{m} - \lambda & \dfrac{-k}{m} \\[3mm] \dfrac{-k}{m} & \dfrac{2k}{m} - \lambda \end{bmatrix} = 0$$

or

$$\lambda^2 - \frac{4k}{m}\lambda + \frac{3k^2}{m^2} = 0 \ .$$

Factoring this yields

$$(\lambda - \frac{k}{m})(\lambda - \frac{3k}{m}) = 0 \ .$$

Then the eigenvalues are $\lambda_1 = k/m$ and $\lambda_2 = 3k/m$. But $\lambda = \omega^2$. Therefore natural frequencies of the system are

$$\omega_1 = \lambda_1^{\frac{1}{2}} = \sqrt{k/m}$$

and

$$\omega_2 = \lambda_2^{\frac{1}{2}} = \sqrt{3k/m} \ .$$

● **PROBLEM** 20-10

Given the mass-spring system in the figure, set up a system of equations that could be used to solve for the net force on each of the three masses if the displacement of each mass is known. Show how this system could be used to solve for the displacements of each mass if the force on each mass was known. Define the elasticity and stiffness matrices.

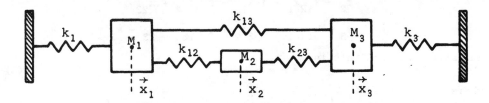

Solution: Let x_1, x_2 and x_3 be the displacements of M_1, M_2 and M_3 , respectively. The force exerted by a spring is proportional to the distance the spring has been stretched or compressed.

$$F = kx \ , \tag{1}$$

the constants for each spring are given in the figure. Spring k_1 is stretched a distance x_1 . Spring k_{12} is stretched a distance $x_2 - x_1$. For each spring we have these displacements:

$$k_1 : x_1$$
$$k_{12} : x_2 - x_1$$
$$k_{13} : x_3 - x_1$$
$$k_{23} : x_3 - x_2$$
$$k_3 : -x_3$$

691

Note that for k_3 the displacement is negative since the spring is being compressed, and x_3 is positive. From equation (1), since we know the displacement of spring k_1 and the constant associated with it (k_1) we can compute the force exerted by it. $F = k_1 x_1$. The forces exerted by each spring are:

$$k_1 \; : \; k_1 x_1$$
$$k_{12} \; : \; k_{12}(x_2 - x_1)$$
$$k_{13} \; : \; k_{13}(x_3 - x_1)$$
$$k_{23} \; : \; k_{23}(x_3 - x_2)$$
$$k_3 \; : \; -k_3 x_3 \; .$$

Let f_1, f_2 and f_3 be the net forces on each of the three masses. M_1 experiences a force from springs k_1, k_{12} and k_{13}. The force from spring k_1 is in the opposite direction of the forces exerted by k_{12} and k_{13}. Since we assume that to the left is the negative direction we have

$$f_1 = -k_1 x_1 + k_{12}(x_2 - x_1) + k_{13}(x_3 - x_1) \; .$$

Similarly, find f_2 and f_3. Thus,

$$f_1 = -k_1 x_1 + k_{12}(x_2 - x_1) + k_{13}(x_3 - x_1)$$
$$f_2 = -k_{12}(x_2 - x_1) + k_{23}(x_3 - x_2) \qquad\qquad (2)$$
$$f_3 = -k_{13}(x_3 - x_1) - k_{23}(x_3 - x_2) - k_3 x_3$$

Rewriting these equations after collecting terms yields:

$$f_1 = (-k_1 - k_{12} - k_{13})x_1 + k_{12}x_2 + k_{13}x_3$$
$$f_2 = k_{12}x_1 + (-k_{12} - k_{23})x_2 + k_{23}$$
$$f_3 = k_{13}x_1 + K_{23}x_2 + (-k_{13} - k_{23} - k_3)x_3$$

In matrix notation this is

$$F = KX \qquad\qquad (3)$$

where

$$F = \begin{bmatrix} f_1 \\ f_2 \\ f_3 \end{bmatrix} \;, \quad X = \begin{bmatrix} x_1 \\ x_2 \\ x_3 \end{bmatrix}$$

and

$$K = \begin{bmatrix} -(k_1 + k_{12} + k_{13}) & k_{12} & k_{13} \\ k_{12} & -(k_{12} + k_{23}) & k_{23} \\ k_{13} & k_{23} & -(k_{13} + k_{23} + k_3) \end{bmatrix}$$

Hence, given any values for x_1, x_2 and x_3 (and given the spring constants) the forces f_1, f_2 and f_3 can be found simply by performing the matrix multiplication KX.

A true spring always exerts a force when it is displaced. Therefore every spring constant is greater than 0. Now, to show K is non-singular compute $\det K$.

$$\det K = -(k_1 + k_{12} + k_{13})(k_{12} + k_{23})(k_{13} + k_{23} + k_3) + k_{12}k_{23}k_{13}$$

$$+ k_{13}k_{12}k_{23} - (-k_{13}(k_{12} + k_{23})k_{13} + k_{23}^2(k_1 + k_{12} + k_{13})$$

$$+ k_{12}^2(k_{13} + k_{23} + k_3)) \; .$$

Multiplying and gathering terms yields a non-zero expression in spring constants. Therefore, $\det K \neq 0$ and the matrix K^{-1} exists. So to find a system that can be used to solve for the displacements, if f_1, f_2 and f_3 are known, we solve equation (3) for X. $F = KX$ and K non-singular implies $X = K^{-1}F$.

The stiffness matrix of the system is a matrix in which the ij^{th} entry is the force applied to the ith mass as a result of a unit displacement of the jth mass. In the given problem suppose M_1 is displaced one unit $(x_2 = x_3 = 0, \; x_1 = 1)$; then from the first equation in system (2) the force on M_1 is $-(k_1 + k_{12} + k_{13})$ which is the 1,1 entry in K . In fact, K is the stiffness matrix of the system.

The elasticity matrix of a system is a matrix in which the ij^{th} entry is the displacement produced in the ith mass as a result of a unit force applied to the jth mass. The elasticity matrix is the inverse of the stiffness matrix. In the problem K^{-1} is the elasticity matrix.

● PROBLEM 20-11

Consider the system of three springs and two masses shown in the figure below. Find the natural frequencies of vibration of the system if, in suitable units,

$$m_1 = 1, \; m_2 = 4, \; k_1 = 1, \; k_2 = 4, \; k_3 = 4 \; .$$

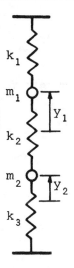

Solution: First derive the equations of motion by applying Newton's second law of motion to each mass. Let y_1 and y_2 denote the displacement of the masses from their equilibrium positions. Now the restoring force of the first spring is $-k_1 y_1$ (by Hooke's law) and the restoring force of the second spring is $-k_2(y_2 - y_1)$. Thus the forces acting on the mass m_1 are $-k_1 y_1$ and $-k_2(y_1 - y_2)$. Apply Newton's second law of motion to the mass m_1

$$m_1\ddot{y}_1 = \Sigma F = -k_1 y_1 - k_2(y_1 - y_2) . \tag{1}$$

Similarly, the restoring force of the second spring is $k_2(y_1 - y_2)$. Since its net extension is $y_1 - y_2$, and the restoring force of the third spring is $-k_3 y_2$. The forces acting on the mass m_2 , therefore are $k_2(y_1 - y_2)$ and $-k_3 y_2$. Thus, Newton's law applied to the second mass m_2 yields,

$$m_2\ddot{y}_2 = k_2(y_1 - y_2) - k_3 y_2 \tag{2}$$

Now substituting the given values of m_1, m_2, k_1, k_2 and k_3 , into equations (1) and (2), we get,

$$\ddot{y}_1 = - y_1 - 4(y_1 - y_2)$$

$$4\ddot{y}_2 = 4(y_1 - y_2) - 4y_2$$

or,

$$\ddot{y}_1 = -5y_1 + 4y_2 \tag{3}$$

$$\ddot{y}_2 = y_1 - 2y_2 \tag{4}$$

For harmonic motion, $\ddot{y}_1 = -\omega^2 y_1$ and $y_2 = -\omega^2 y_2$. Upon substitution of these values in equation (3) and (4), we have,

$$-\omega^2 y_1 = -5y_1 + 4y_2$$

$$-\omega^2 y_2 = y_1 - 2y_2 \tag{5}$$

Substituting

$$\omega^2 = \lambda \tag{6}$$

for simplicity, gives,

$$5y_1 - 4y_2 = \lambda y_1$$

$$-y_1 + 2y_2 = \lambda y_2$$

This system of linear equations, in matrix notation, can be written as,

$$\begin{bmatrix} 5 & -4 \\ -1 & 2 \end{bmatrix} \begin{pmatrix} y_1 \\ y_2 \end{pmatrix} = \lambda \begin{pmatrix} y_1 \\ y_2 \end{pmatrix}$$

or

$$AY = \lambda Y \tag{7}$$

where

$$A = \begin{bmatrix} 5 & -4 \\ -1 & 2 \end{bmatrix} .$$

Recall that a vector V is an eigenvector for A belonging to the eigenvalue λ , if $AV = \lambda V$, $(V \neq 0)$. Thus in equation (7), Y is a non-zero vector for each eigenvalue of matrix A. The eigenvalues of A are the roots of the characteristic equation of A. Now the characteristic polynomial of A is given by

$$f(\lambda) = \det(\lambda I - A) = \det \begin{bmatrix} \lambda-5 & 4 \\ 1 & \lambda-2 \end{bmatrix}$$

$$= (\lambda-5)(\lambda-2) - 4$$

$$= (\lambda-6)(\lambda-1)$$

Then the characteristic equation of A is $(\lambda-1)(\lambda-6) = 0$. Therefore, the eigenvalues of A are $\lambda_1 = 1$ and $\lambda_2 = 6$. Now from (6), we have $\omega^2 = \lambda$. Therefore the natural frequencies of the given system are

$$\omega_1 = (\lambda_1)^{\frac{1}{2}} = (1)^{\frac{1}{2}} = 1$$

and

$$\omega_2 = (\lambda_2)^{\frac{1}{2}} = (6)^{\frac{1}{2}} = 2.449 .$$

● **PROBLEM** 20-12

Consider the system of three springs (with moduli k_1, k_2, and k_3) and three masses m_1, m_2, and m_3 shown in figure 1. The two masses m_1 and m_2 are connected by a rigid "massless" rod. Mass m_3 is connected by a spring (k_3) to the mid-point of the rod. Suppose that in suitable units the masses and moduli have the values $m_1 = 1$, $m_2 = 1$, $m_3 = 4/3$, $k_1 = 1$, $k_2 = 1$, $k_3 = 4$. Find the natural frequencies of the system.

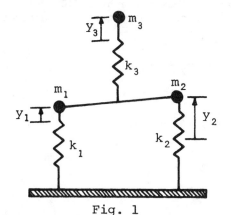

Fig. 1

Solution: The vertical upward displacements of the three masses from their equilibrium positions are y_1, y_2 and y_3 as illustrated. Hence the extensions, from the equilibrium configuration, of the springs of moduli k_1, k_2, and k_3 are respectively y_1, y_2 and $y_3 - \frac{1}{2}(y_1+y_2)$. By Hooke's Law the force exerted by a spring is proportional to the extension (or compression) of the spring. Since the extensions and the spring constants are known the forces exerted by each spring can be calculated. The forces acting on the mass m_1 are $-k_1 y_1$ and $\frac{1}{2}k_3[y_3 - \frac{1}{2}(y_1+y_2)]$. Applying Newton's Second Law to the mass m_1 gives

$$m_1 \ddot{y}_1 = -k_1 y_1 + \frac{1}{2}k_3[y_3 - \frac{1}{2}(y_1+y_2)] . \tag{1}$$

Now the forces acting on the second mass are $-k_2 y_2$, and $\frac{1}{2}k_3[y_3-\frac{1}{2}(y_1+y_2)]$. Thus Newton's Second Law applied to the mass m_2 yields,

$$m_2 \ddot{y}_2 = -k_2 y_2 + \frac{1}{2}k_3[y_3 - \frac{1}{2}(y_1 + y_2)] . \tag{2}$$

Now the only force acting on the mass m_3 is $-k_3[y_3 - \frac{1}{2}(y_1+y_2)]$. Applying Newton's Second Law to the mass m_3, we have

695

$$m_3 \ddot{y}_3 = -k_3 [y_3 - \tfrac{1}{2}(y_1 + y_2)] . \tag{3}$$

After rearrangement, the equations (1), (2), and (3) may be written

$$\ddot{y}_1 = - \frac{1}{m_1}(k_1 + \tfrac{1}{4}k_3)y_1 - \frac{1}{4m_1} k_3 y_2 + \frac{1}{2m_1} k_3 y_3$$

$$\ddot{y}_2 = - \frac{1}{4m_2} k_3 y_1 - \frac{1}{m_2}(k_2 + \tfrac{1}{4}k_3)y_2 + \frac{1}{2m_2} k_3 y_3$$

$$\ddot{y}_3 = \frac{1}{2m_3} k_3 y_1 + \frac{1}{2m_3}k_3 y_2 - \frac{1}{m_3} k_3 y_3 .$$

Now substitute the given values of m_1, m_2, m_3, k_1, k_2, and k_3 in the above equations. Thus,

$$\ddot{y}_1 = - 2y_1 - y_2 + 2y_3$$

$$\ddot{y}_2 = - y_1 - 2y_2 + 2y_3 \tag{4}$$

$$\ddot{y}_3 = (3/2)y_1 + (3/2)y_2 - 3y_3$$

For a system such as this, in which there is no damping, our experience indicates that the masses, once set in motion, may undergo undamped oscillation of a fixed frequency. Thus for a harmonic motion, $\ddot{y} = -\omega^2 y$ where ω = natural frequency. Therefore the system (4) can be written as

$$-\omega^2 y_1 = -2y_1 - y_2 + 2y_3$$

$$-\omega^2 y_2 = -y_1 - 2y_2 + 2y_3$$

$$-\omega^2 y_3 = (3/2)y_1 + (3/2)y_2 - 3y_3 .$$

For simplicity, putting $\omega^2 = \lambda$, gives

$$-\lambda y_1 = -2y_1 - y_2 + 2y_3$$

$$-\lambda y_2 = -y_1 - 2y_2 + 2y_3$$

$$-\lambda y_3 = (3/2)y_1 + (3/2)y_2 - 3y_3$$

or,

$$2y_1 + y_2 - 2y_3 = \lambda y_1$$

$$y_1 + 2y_2 - 2y_3 = \lambda y_3$$

$$(-3/2)y_1 - (3/2)y_2 + 3y_3 = \lambda y_3 .$$

The above system can be written in matrix form as

$$\begin{bmatrix} 2 & 1 & -2 \\ 1 & 2 & -2 \\ -3/2 & -3/2 & 3 \end{bmatrix} \begin{bmatrix} y_1 \\ y_2 \\ y_3 \end{bmatrix} = \lambda \begin{bmatrix} y_1 \\ y_2 \\ y_3 \end{bmatrix}$$

or,

$$AY = \lambda Y \tag{5}$$

where

$$A = \begin{bmatrix} 2 & 1 & -2 \\ 1 & 2 & -2 \\ -3/2 & -3/2 & 3 \end{bmatrix} .$$

Now, we are seeking a non-zero solution for Y. Recall that a vector V is an eigenvector for A belonging to the eigenvalue λ, if $AV = \lambda V$ ($V \neq 0$). Thus in equation (5), λ is an eigenvalue of A. Now the characteristic polynomial of A is

$$f(\lambda) = \det(\lambda I - A) = \det \begin{bmatrix} \lambda-2 & -1 & 2 \\ -1 & \lambda-2 & 2 \\ 3/2 & 3/2 & \lambda-3 \end{bmatrix}$$

$$= (\lambda-2) \begin{vmatrix} \lambda-2 & 2 \\ 3/2 & \lambda-3 \end{vmatrix} + 1 \begin{vmatrix} -1 & 2 \\ 3/2 & \lambda-3 \end{vmatrix} + 2 \begin{vmatrix} -1 & \lambda-2 \\ 3/2 & 3/2 \end{vmatrix}$$

$$= (\lambda-2)[(\lambda-2)(\lambda-3)-3] + 1[-1(\lambda-3)-3] + 2[-3/2-(3/2)(\lambda-2)]$$

$$= \lambda^3 - 7\lambda^2 + 9\lambda - 3$$

$$= \lambda^3 - \lambda^2 - 6\lambda^2 + 6\lambda + 3\lambda - 3$$

$$= (\lambda-1)(\lambda^2 - 6\lambda + 3) \ .$$

The characteristic equation of A is $(\lambda-1)(\lambda^2-6\lambda+3) = 0$. Then, the eigenvalues of A are, $\lambda_1 = 1$, $\lambda_2 = 3 + \sqrt{6}$, and $\lambda_3 = 3 - \sqrt{6}$. But $\lambda = \omega^2$. Therefore the natural frequencies of the given system are

$$\omega_1 = \sqrt{\lambda_1} = \sqrt{1} = 1$$

$$\omega_2 = \sqrt{\lambda_2} = \sqrt{3 + \sqrt{6}} \doteq 2.33$$

$$\omega_3 = \sqrt{\lambda_3} = \sqrt{3 - \sqrt{6}} \doteq .741 \ .$$

● **PROBLEM** 20-13

Consider the motion of a short train composed of three cars, each of mass m, coupled to each other by elastic couplings of stiffness k, as shown in figure below. Assume the train is able to roll along the track without any frictional forces acting upon it. Find the natural frequencies of the system.

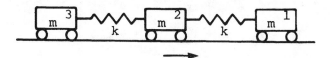

Solution: To formulate this problem mathematically, we shall apply Newton's Second Law of motion to each moving car. Let x_1, x_2, and x_3 denote the displacement of the cars from their equilibrium positions. The only force acting on the first car is the restoring force $-k(x_1 - x_2)$ of the first spring whose net extension is $x_1 - x_2$. Applying Newton's Second Law to the first car of mass m, we therefore obtain

$$m\ddot{x}_1 = -k(x_1 - x_2)$$

or,

$$m\ddot{x}_1 = -kx_1 + kx_2 \ . \tag{1}$$

Similarly, the forces acting on the second car are the restoring force $k(x_1 - x_2)$ of the first spring, and the restoring force $-k(x_2 - x_3)$ of the second spring whose net extension is $x_2 - x_3$ units. Thus, Newton's Second Law applied to the second car yields

$$m\ddot{x}_2 = k(x_1 - x_2) - k(x_2 - x_3)$$

or,

$$m\ddot{x}_2 = kx_1 - 2kx_2 + kx_3 \ . \tag{2}$$

Now the only force acting on the third car is the restoring force $k(x_2 - x_3)$ of the second spring. Thus Newton's Second Law applied to the third car yields

$$m\ddot{x}_3 = k(x_2 - x_3)$$

or,

$$m\ddot{x}_3 = kx_2 - kx_3 \ . \tag{3}$$

The equations (1), (2) and (3) can be written as

$$m\ddot{x}_1 + kx_1 - kx_2 = 0$$

$$m\ddot{x}_2 - kx_1 + 2kx_2 - kx_3 = 0 \tag{4}$$

$$m\ddot{x}_3 - kx_2 + kx_3 = 0 \ .$$

Now for harmonic motion $\ddot{x} = -\omega^2 x$, where ω = natural frequency. Therefore

$$\ddot{x}_1 = -\omega^2 x_1$$

$$\ddot{x}_2 = -\omega^2 x_2$$

$$\ddot{x}_3 = -\omega^2 x_3 \ .$$

Substituting these values in system (4), we have

$$-m\omega^2 x_1 + kx_1 - kx_2 = 0$$

$$-m\omega^2 x_2 - kx_1 + 2kx_2 - kx_3 = 0$$

$$-m\omega^2 x_3 - kx_2 + kx_3 = 0 \ .$$

Dividing each equation by k and putting $m\omega^2/k = \lambda$, for simplicity, gives

$$-\lambda x_1 + x_1 - x_2 = 0$$

$$-\lambda x_2 - x_1 + 2x_2 - x_3 = 0$$

$$-\lambda x_3 - x_2 + x_3 = 0 \ ,$$

or

$$x_1 - x_2 = \lambda x_1$$

$$-x_1 + 2x_2 - x_3 = \lambda x_2$$

$$- x_2 + x_3 = \lambda x_3 \ .$$

This set of equations can be written in matrix form as

$$\begin{bmatrix} 1 & -1 & 0 \\ -1 & 2 & -1 \\ 0 & -1 & 1 \end{bmatrix} \begin{bmatrix} x_1 \\ x_2 \\ x_3 \end{bmatrix} = \lambda \begin{bmatrix} x_1 \\ x_2 \\ x_3 \end{bmatrix}$$

or

$$AX = \lambda X \tag{5}$$

where

$$A = \begin{bmatrix} 1 & -1 & 0 \\ -1 & 2 & -1 \\ 0 & -1 & 1 \end{bmatrix} \quad .$$

Recall that a vector V (written as a column matrix) is an eigenvector for A belonging to the eigenvalue λ, if

$$AV = \lambda V \qquad (V \neq 0) \ .$$

Thus in equation (5), λ is an eigenvalue of A. Now the characteristic polynomial of A is

$$f(\lambda) = \det(\lambda I - A) = \det \begin{bmatrix} \lambda-1 & 1 & 0 \\ 1 & \lambda-2 & 1 \\ 0 & 1 & \lambda-1 \end{bmatrix}$$

$$= (\lambda-1)[(\lambda-2)(\lambda-1) - 1] - 1[\lambda-1]$$

$$= (\lambda-1)[\lambda^2 - 3\lambda + 2 - 1 - 1] = \lambda(\lambda-1)(\lambda-3) \ .$$

The characteristic equation of A is $\lambda(\lambda-1)(\lambda-3) = 0$. The eigenvalues of A are the roots of the characteristic equation of A. Thus, the eigenvalues of A are, $\lambda_1 = 0$, $\lambda_2 = 1$, and $\lambda_3 = 3$. But

$$\lambda = m\omega^2 / k \ .$$

Therefore, the natural frequencies of the system are

$$(m/k)\omega_1^2 = \lambda_1 = 0$$

$$(m/k)\omega_2^2 = \lambda_2 = 1$$

$$(m/k)\omega_3^2 = \lambda_3 = 3 \ .$$

Thus, $\omega_1 = 0$, $\omega_2 = \sqrt{k/m}$, and $\omega_3 = \sqrt{3k/m}$ are the natural frequencies of the system.

● **PROBLEM** 20-14

Consider the oscillations of three flywheels, each with moment of inertia I, fixed to a central shaft which is free to rotate on its bearings. Given that the torsional stiffness of the shaft is C find the natural frequencies for oscillations of the flywheels.

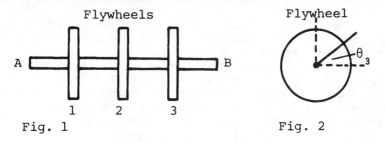

Fig. 1   Fig. 2

Solution: Let tne angular displacements for flywheels 1, 2, and 3 be θ_1, θ_2 and θ_3 respectively.

Figure 1 shows the positions of the flywheels 1, 2 and 3 on a central shaft. The shaft is free to rotate at points A and B.

Figure 2 shows an end view of the system from point B in figure 1.
In figure 2 we can see the angular displacement, θ_3, of flywheel 3.
Using Newton's law of force $f = m\ddot{s}$, we can derive the equations of motion
of the system.   In this case the displacement s is angular and for the
mass we use the moment of inertia of the flywheel.   The only force on flywheel 1 is the force of its rotation θ_1 and the rotation θ_2, of flywheel 2.   Also, the force necessary to produce an angle of rotation is
proportional to the torsional stiffness C.   We have for flywheel 1

$$I\ddot{\theta}_1 = C(\theta_2 - \theta_1) . \tag{1}$$

On the second flywheel the contribution from the angular displacement
both in flywheel 1 and in flywheel 3 must be considered.

$$I\ddot{\theta}_2 = C(\theta_1 - \theta_2) + C(\theta_3 - \theta_2) . \tag{2}$$

Similarly

$$I\ddot{\theta}_3 = C(\theta_2 - \theta_3) . \tag{3}$$

The displacement angle θ of each flywheel is undergoing oscillatory
motion of the form $\theta = k \sin \omega t$, where ω is the frequency and k is
the maximum possible angular displacement.   Now $\theta = k \sin \omega t$ implies
that $\ddot{\theta} = -\omega^2\theta$.   So substitute $\ddot{\theta}_i = -\omega^2\theta_i$, $i = 1,2,3$ in equations
(1), (2) and (3), and simplify to get

$$-I\omega^2\theta_1 = -C\theta_1 + C\theta_2$$
$$-I\omega^2\theta_2 = C\theta_1 - 2C\theta_2 + C\theta_3$$
$$-I\omega^2\theta_3 = C\theta_2 - C\theta_3 .$$

Let

$$\lambda = \frac{I\omega^2}{C} \tag{4}$$

We obtain

$$\theta_1 - \theta_2 = \lambda\theta_1$$
$$-\theta_1 + 2\theta_2 - \theta_3 = \lambda\theta_2$$
$$ - \theta_2 + \theta_3 = \lambda\theta_3 .$$

The system in matrix form is

$$A\theta = \lambda\theta \tag{5}$$

where

$$A = \begin{bmatrix} 1 & -1 & 0 \\ -1 & 2 & -1 \\ 0 & -1 & 1 \end{bmatrix} \text{ and } \theta = \begin{bmatrix} \theta_1 \\ \theta_2 \\ \theta_3 \end{bmatrix} .$$

For a matrix A the vectors, V such that $AV = \lambda V$ are called the eigenvectors of A and the possible values of λ (one for each class of eigenvectors are called the eigenvalues of A.
Thus the solution to this problem can be found by finding
the eigenvalues of A. $A\theta = \lambda\theta$ implies that $\lambda\theta - A\theta = 0$ or
$(\lambda I - A)\theta = 0$, where I is the 3 by 3 identity matrix.   This equation
has solutions when $\det(\lambda I - A) = 0$.   Hence, to find the eigenvalues of
A, find the roots of the characteristic equation of A, $\det(\lambda I - A) = 0$.

$$\det(\lambda I - A) = \begin{vmatrix} \lambda-1 & 1 & 0 \\ 1 & \lambda-2 & 1 \\ 0 & 1 & \lambda-1 \end{vmatrix}$$

$$= (\lambda-1)[(\lambda-2)(\lambda-1)-1] - (\lambda-1)$$

$$= (\lambda-1)[\lambda^2-3\lambda+1] - (\lambda-1)$$

$$= (\lambda-1)\lambda(\lambda-3) .$$

The eigenvalues of A are 1,0 and 3. From equation (4), $\lambda = I\omega^2/C$ so the possible natural frequencies are $\omega = \sqrt{C\lambda/I}$ with λ = 1,0, or 3. That is ω is $\sqrt{C/I}$ or 0 or $\sqrt{3C/I}$.

● **PROBLEM** 20-15

Three identical simple pendula are connected by two identical elastic springs as shown in the figure below. Find the natural frequencies of the system.

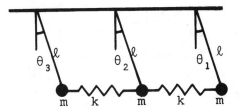

Solution: Let θ_1, θ_2 and θ_3 denote the angular displacements of the pendula. It will be assumed that the oscillations are small so that we may make the following approximations:

$$\sin \theta_i = \theta_i , \quad \cos \theta_i = 1 - \frac{\theta_i^2}{2} , \quad i = 1,2,3.$$

The horizontal component of gravity on the first pendulum is $-mg \sin \theta_1 \doteq$ $-mg\, \theta_1$, where g is the gravitational constant. The net angular displacement of the first pendulum is $\theta_1 - \theta_2$. From trigonometry we have $\sin(\theta_1 - \theta_2) = x/\ell$ where x is the linear displacement of the first pendulum with respect to the second. Hence the linear displacement (which is needed to compute the restoring force of the spring) is

$$x = \ell \sin(\theta_1 - \theta_2) \doteq \ell(\theta_1 - \theta_2) .$$

The forces acting on the first pendulum are $-mg\,\theta_1$ and $-k\ell(\theta_1 - \theta_2)$. Applying Newton's Second Law to the first pendulum we therefore obtain

$$m\ell\ddot{\theta}_1 = -mg\,\theta_1 - k\ell(\theta_1 - \theta_2) . \tag{1}$$

The forces acting on the second pendulum are the restoring force $k\ell(\theta_1 - \theta_2)$ of the first spring, the restoring force $-k\ell(\theta_2 - \theta_3)$ of the second spring, and the force $-mg\theta_2$. Thus, by applying Newton's Second Law, we obtain,

$$m\ell\ddot{\theta}_2 = -mg\,\theta_2 + k\ell(\theta_1 - \theta_2) - k\ell(\theta_2 - \theta_3) \tag{2}$$

Similarly, the forces acting on the third pendulum are $-mg\,\theta_3$ and $k\ell(\theta_2 - \theta_3)$. Thus Newton's Second Law applied to the third pendulum yields

$$m\ell\ddot{\theta}_3 = -mg\,\theta_3 + k\ell(\theta_2 - \theta_3) . \tag{3}$$

The equations (1), (2) and (3) can be rewritten as

$$m\ell\ddot{\theta}_1 + (mg + k\ell)\,\theta_1 - k\ell\theta_2 = 0$$

$$m\ell\ddot{\theta}_2 - k\ell\theta_1 + (mg + 2k\ell)\,\theta_2 - k\ell\theta_3 = 0$$

$$m\ell\ddot{\theta}_3 - k\ell\theta_2 + (mg + k\ell)\,\theta_3 = 0 .$$

These equations can be written in matrix form as

$$
\begin{bmatrix} m\ell & 0 & 0 \\ 0 & m\ell & 0 \\ 0 & 0 & m\ell \end{bmatrix} \begin{bmatrix} \ddot{\theta}_1 \\ \ddot{\theta}_2 \\ \ddot{\theta}_3 \end{bmatrix} + \begin{bmatrix} mg+k\ell & -k\ell & 0 \\ -k\ell & (mg+2k\ell) & -k\ell \\ 0 & -k\ell & (mg+k\ell) \end{bmatrix} \begin{bmatrix} \theta_1 \\ \theta_2 \\ \theta_3 \end{bmatrix}
$$

$$
= \begin{bmatrix} 0 \\ 0 \\ 0 \end{bmatrix}
$$

or

$$M\ddot{\theta} + K\theta = 0 \tag{4}$$

where

$$
M = \begin{bmatrix} m\ell & 0 & 0 \\ 0 & m\ell & 0 \\ 0 & 0 & m\ell \end{bmatrix}
$$

and

$$
K = \begin{bmatrix} mg+k\ell & -k\ell & 0 \\ -k\ell & mg+2k\ell & -k\ell \\ 0 & -k\ell & mg+k\ell \end{bmatrix}
$$

If we premultiply the equation (4) by M^{-1}, we obtain

$$M^{-1}M\ddot{\theta} + M^{-1}K\theta = 0 \tag{5}$$

Now $MM^{-1} = I$ (a unit matrix) and,

$$
M^{-1}K = A = \begin{bmatrix} m\ell & 0 & 0 \\ 0 & m\ell & 0 \\ 0 & 0 & m\ell \end{bmatrix}^{-1} \begin{bmatrix} mg+k\ell & -k\ell & 0 \\ -k\ell & mg+2k\ell & -k\ell \\ 0 & -k\ell & mg+k\ell \end{bmatrix}
$$

Since M is a diagonal matrix, M^{-1} is a diagonal matrix whose elements are the multiplicative inverses of the non-zero elements of M. Thus,

$M^{-1}K$

$$
= \begin{bmatrix} 1/m\ell & 0 & 0 \\ 0 & 1/m\ell & 0 \\ 0 & 0 & 1/m\ell \end{bmatrix} \begin{bmatrix} mg+k\ell & -k\ell & 0 \\ -k\ell & mg+2k\ell & -k\ell \\ 0 & -k\ell & mg+k\ell \end{bmatrix}
$$

$$
= \begin{bmatrix} \left(\dfrac{g}{\ell} + \dfrac{k}{m}\right) & \dfrac{-k}{m} & 0 \\ \dfrac{-k}{m} & \left(\dfrac{g}{\ell} + \dfrac{2k}{m}\right) & \dfrac{-k}{m} \\ 0 & \dfrac{-k}{m} & \left(\dfrac{g}{\ell} + \dfrac{k}{m}\right) \end{bmatrix}
$$

Let

$$a = g/\ell, \quad b = k/m . \tag{6}$$

Then, the matrix A can be written in the more compact form as

$$
A = \begin{bmatrix} (a+b) & -b & 0 \\ -b & (a+2b) & -b \\ 0 & -b & (a+b) \end{bmatrix}
$$

702

Recall that simple pendula experience undamped motion. Assuming harmonic motion, $\ddot{\theta} = -\lambda\theta$, where $\lambda = \omega^2$, and ω is the frequency of the oscillations, equation (5) becomes

$$[A - \lambda I]\theta = 0 .$$

Suppose $\theta = \theta_i$ is a non-zero solution to $(A - \lambda I)\theta = 0$, it follows that, θ_i is an eigenvector of A belonging to an eigenvalue λ , and as we know, the eigenvalues of A are the roots of the characteristic equation of A. Now, the characteristic polynomial of A is

$$f(\lambda) = \det[A - \lambda I] = \det \begin{pmatrix} (a+b)-\lambda & -b & 0 \\ -b & (a+2b)-\lambda & -b \\ 0 & -b & (a+b)-\lambda \end{pmatrix}$$

$$= [(a+b)-\lambda]\begin{vmatrix} (a+2b)-\lambda & -b \\ -b & (a+b)-\lambda \end{vmatrix} + b \begin{vmatrix} -b & -b \\ 0 & (a+b)-\lambda \end{vmatrix}$$

$$= [(a+b)-\lambda][a^2 + 3ab - 2\lambda a - 3\lambda b + \lambda^2 + b^2]$$
$$-b^2[(a+b)-\lambda]$$

$$= [(a+b)-\lambda][a^2 + 3ab - 2\lambda a - 3\lambda b + \lambda^2]$$

$$= [(a+b)-\lambda] \quad (a-\lambda)[(a+3b)-\lambda]$$

The eigenvalues of A are therefore, $\lambda = a$, $\lambda = a+b$, and $\lambda = a + 3b$. Now, from (6), we have
$$a = g/\ell , \quad b = k/m .$$
Thus,
$$\lambda_1 = g/\ell, \quad \lambda_2 = \frac{g}{\ell} + k/m \quad \text{and} \quad \lambda_3 = \frac{g}{\ell} + 3k/m .$$
Also, recall $\lambda = \omega^2$. Therefore, the natural angular frequencies of the given system are

$$\omega_1 = (\lambda_1)^{\frac{1}{2}} = (g/\ell)^{\frac{1}{2}}$$

$$\omega_2 = (\lambda_2)^{\frac{1}{2}} = \left(\frac{g}{\ell} + k/m\right)^{\frac{1}{2}}$$

$$\omega_3 = (\lambda_3)^{\frac{1}{2}} = \left(\frac{g}{\ell} + 3k/m\right)^{\frac{1}{2}} .$$

● **PROBLEM** 20-16

Three particles each of unit mass, are equally spaced at distances ℓ on a light horizontal string fastened rigidly at the ends and resting on a smooth horizontal table. What happens to the particles if
i)   they are subject to constant horizontal forces f_1, f_2, f_3 ;
ii) they are subject to periodic horizontal forces $p_1 \sin kt$, $p_2 \sin kt$, $p_3 \sin kt$;
iii) they are allowed to oscillate freely in the plane.

Fig. 1

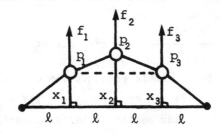

Solution: Let the particles p_1, p_2 and p_3 have displacements x_1, x_2 and x_3. For small displacements, tension in the string is not changed so let it be a constant, T. p_1 is subject to three forces -- f_1 and its two string tensions. The tension on p_1 to the right is

$$T \frac{(x_2 - x_1)}{\ell}$$

and to the left is $\frac{-Tx_1}{\ell}$. Thus the equation of motion on p_1 is

$$f_1 - \frac{Tx_1}{\ell} + \frac{T(x_2 - x_1)}{\ell} = \ddot{x}_1 .$$

Similarly, p_2 and p_3 have equations of motion

$$f_2 - \frac{T(x_2 - x_1)}{\ell} - \frac{T(x_2 - x_3)}{\ell} = \ddot{x}_2$$

$$f_3 + \frac{T(x_2 - x_3)}{\ell} - \frac{Tx_3}{\ell} = \ddot{x}_3 ,$$

where $\ddot{x} = \frac{\partial^2 x}{\partial t^2}$. To simplify the algebra let the physical constant $T/\ell = 1$. We have

$$f_1 - x_1 + x_2 - x_1 = \ddot{x}_1$$

$$f_2 - x_2 + x_1 - x_2 + x_3 = \ddot{x}_2$$

$$f_3 + x_2 - x_3 - x_3 = \ddot{x}_3$$

which yields

$$f_1 - (2x_1 - x_2) = \ddot{x}_1$$

$$f_2 - (-x_1 + 2x_2 - x_3) = \ddot{x}_2$$

$$f_3 - (-x_2 + 2x_3) = \ddot{x}_3 .$$

Writing this in matrix form we get $F - AX = \ddot{X}$ where F is the column vector f_1, f_2, f_3; X is the column vector x_1, x_2, x_3; $\ddot{X}$ is the column vector $\ddot{x}_1, \ddot{x}_2, \ddot{x}_3$ and

$$A = \begin{bmatrix} 2 & -1 & 0 \\ -1 & 2 & -1 \\ 0 & -1 & 2 \end{bmatrix}$$

the coefficient matrix of X_i.

i) If we have constant forces the system is at equilibrium and $\ddot{X} = 0$ so $F = AX$ is the relationship between the forces and their displacements.

ii) If $f_i = p_i \sin kt$ it is possible to find a solution $x_i = y_i \sin kt$ $(i = 1,2,3)$. This implies that

$$\ddot{x}_i = \frac{\partial^2 y_i \sin kt}{\partial t^2} = -k^2 y_i \sin kt .$$

So we have in matrix notation a simple presentation of this solution

$$P \sin kt - AY \sin kt = -K^2 Y \sin kt \quad \text{or} \quad (A - K^2 I)Y = P .$$

iii) We have a formula for regular oscillations; free oscillations occur when

$$(A - K^2 I)Y = 0 \tag{1}$$

(1) yields

$$AY = K^2 Y . \tag{2}$$

It is here that we seriously apply linear algebra. We see that if $k^2 = \lambda$, then, the values for k^2 are simply the eigenvalues of A. To find eigenvalues of A expand

$$\det(\lambda I - A) = (\lambda-2)^3 - [(\lambda-2) + (\lambda-2)]$$
$$= (\lambda-2)[(\lambda-2)^2 - 2]$$
$$= (\lambda-2)(\lambda^3 - 4\lambda + 2) .$$

The characteristic polynomial has roots $\lambda = 2$ and

$$\lambda = \frac{4 \pm \sqrt{16-8}}{2} = 2 \pm \sqrt{2} .$$

These are the possible values for K^2 in (2). The possible values for Y are the eigenvectors of A. The equations

$$(2I - A)Y_1 = 0$$
$$(\alpha I - A)Y_2 = 0 , \qquad \alpha = 2 - \sqrt{2}$$
$$(\beta I - A)Y_3 = 0 , \qquad \beta = 2 + \sqrt{2}$$

define the eigenvectors of A. $(2I - A)Y_1 = 0$ yields

$$\begin{bmatrix} 0 & 1 & 0 \\ 1 & 0 & 1 \\ 0 & 1 & 0 \end{bmatrix} \begin{bmatrix} c_1 \\ c_2 \\ c_3 \end{bmatrix} = 0 \qquad \text{if } Y_1 = (c_1, c_2, c_3) .$$

Thus $c_2 = 0$ and $c_1 = -c_3$. One such vector is $Y_1 = (1,0,-1)$. Similarly, we find $Y_2 = (1,\sqrt{2}, 1)$ and $Y_3 = (1,-\sqrt{2},1)$. This tells us that there are three different classes of free oscillations. Recalling that K^2 (the square of the frequency of the periodic motion) can be each of the eigenvalues and $x = y_i \sin kt$, $i = 1,2,3$ defines the displacement of a particle we find that the three normal modes are the following:

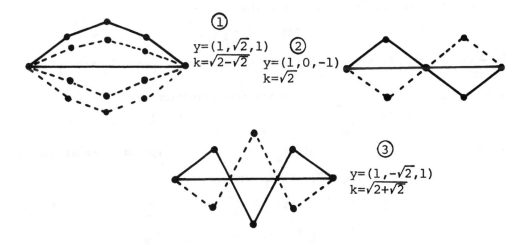

① $y=(1,\sqrt{2},1)$ ② $k=\sqrt{2-\sqrt{2}}$ $y=(1,0,-1)$ $k=\sqrt{2}$

③ $y=(1,-\sqrt{2},1)$ $k=\sqrt{2+\sqrt{2}}$

Three particles of mass m are spaced at distances l on a light vertical string which hangs freely from a point of support. All subsequent displacements are in a fixed vertical plane. How do the particles behave if:
i) they are displaced by constant horizontal forces of magnitudes f_1, f_2 and f_3 ; (ii) they are subject to horizontal forces with magnitudes $p_1 \sin kt$, $p_2 \sin kt$ and $p_3 \sin kt$; (iii) they oscillate freely?

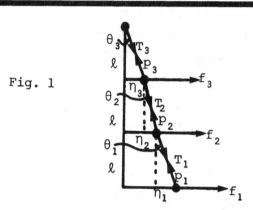

Fig. 1

Solution: In order to simplify the equations we will work only to a first order approximation which will adequately describe the motion of the particles for small vibrations of the system. With some tedious investigation it can be seen that the vertical accelerations of the particles are of second order (θ^2), so we ignore them and assume the net vertical forces are 0. Therefore, on p_1, the downward force of gravity cancels out the tension in the string T_1 , i.e., $T_1 = mg$. On p_2 gravity and the tension T_1 cancel out T_2 and $T_2 = T_1 + mg$. Since we already know $T_1 = mg$ we have $T_2 = 2mg$. Similarly, $T_3 = 3mg$. Next, resolve the horizontal forces on p_1 . p_1 experiences a force f_1 in one direction and the force $T_1 \sin \theta_1$ in the opposite direction. Using Newton's law $F = ma$, where F is the net force acting on an object, m, is the mass of the object and a is the acceleration (second derivative of the displacement with respect to time) notice that

$$f_1 - T_1 \sin \theta_1 = m \ddot{\eta}_1$$

$(\ddot{\eta} = \dfrac{d^2 \eta}{dt^2})$. p_2 is subject to horizontal forces f_2 and $T_1 \sin \theta_1$ in one direction and $T_2 \sin \theta_2$ in the opposite direction so

$$f_2 + T_1 \sin \theta_1 - T_2 \sin \theta_2 = m \ddot{\eta}_2 .$$

Similarly, $f_3 + T_2 \sin \theta_2 - T_3 \sin \theta_3 = m \ddot{\eta}_3$. In the system of equations

$$f_1 - T_1 \sin \theta_1 = m \ddot{\eta}_1$$

$$f_2 + T_1 \sin \theta_1 - T_2 \sin \theta_2 = m \ddot{\eta}_2 \qquad (1)$$

$$f_3 + T_2 \sin \theta_2 - T_3 \sin \theta_3 = m \ddot{\eta}_3$$

which were derived. Substitute the values found for T_1, T_2 and T_3 (i.e., mg, 2mg and 3mg, respectively). From Fig. 1 notice that $\sin \theta_3 = \eta_3/\ell$ and $\sin \theta_2 = (\eta_2 - \eta_3)/\ell$ and $\sin \theta_1 = (\eta_1 - \eta_2)/\ell$. These must also be substituted in the system (1). Thus system (1) becomes:

$$\ell f_1 - mg(\eta_1 - \eta_2) = m\ell\ddot{\eta}_1$$

$$\ell f_2 + mg(\eta_1 - \eta_2) - 2mg(\eta_2 - \eta_3) = m\ell\ddot{\eta}_2 \qquad (2)$$

$$\ell f_3 + 2mg(\eta_2 - \eta_3) - 3mg\,\eta_3 = m\ell\ddot{\eta}_3 \ .$$

The distance ℓ is a constant and f_i, $i = 1,2,3$ are applied forces so let f_1 be ℓf_1, $f_2 = \ell f_2$ and $f_3\ \ell f_3$. Choose physical units such that $mg = m\ell = 1$. Then the system (2) becomes

$$f_1 - \eta_1 + \eta_2 = \ddot{\eta}_1$$

$$f_2 + \eta_1 - 3\eta_2 + 2\eta_3 = \ddot{\eta}_2 \qquad (3)$$

$$f_3 + 2\eta_2 - 5\eta_3 = \ddot{\eta}_3 \ .$$

In matrix notation this is

$$F - A\eta = \ddot{\eta} \qquad (4)$$

where F, η and $\ddot{\eta}$ are the column vectors of f_1, f_2, f_3 and η_1, η_2, η_3 and $\ddot{\eta}_1, \ddot{\eta}_2, \ddot{\eta}_3$ respectively and

$$A = \begin{bmatrix} 1 & -1 & 0 \\ -1 & 3 & -2 \\ 0 & -2 & 5 \end{bmatrix}$$

If the system is in equilibrium under the action of the forces then $\ddot{\eta} = 0$.

(i) to find the behavior, when f_1, f_2 and f_3 are constant horizontal forces we need only solve the equations $f - A\eta = 0$ which is $f = A\eta$.

(ii) when the forces are $f_1 = p_1 \sin kt$, $f_2 = p_2 \sin kt$ and $p_3 \sin kt$ ($f = P \sin kt$, with P the column vector of p_1, p_2 and p_3) we can find a solution $\eta = Y \sin kt$ (Y is a column vector of constants y_1, y_2 and y_3). This would imply that

$$\ddot{\eta} = \frac{d^2(Y \sin kt)}{dt^2} = -k^2 Y \sin kt.$$

In this case equation (4) becomes

$$P \sin kt - AY \sin kt = -k^2 Y \sin kt \ .$$

When $\sin kt \neq 0$ divide by $\sin kt$ and obtain:

$$P - AY = -k^2 Y$$

which yields

$$(A - k^2 I)Y = P \qquad (5)$$

If the forces (p sin kt) are given, finding the displacement of the particles as a function of time is accomplished by inverting the matrix $A - k^2 I$; $Y = (A - k^2 I)^{-1}P$. So $\eta = Y \sin kt$, the equation of the displacements of the particles, is completely determined.

(iii) Free oscillations occur when $P = 0$ which implies that $(A - k^2 I)Y = 0$

or

$$AY = k^2Y \qquad (6)$$

Thus the problem of free oscillations amounts to finding the eigenvalues of A. Each eigenvalue is a possible value for k^2 and each eigenvector a possibility for Y. The eigenvalues of A are the roots of the character- istic equation of A, $\det(\lambda I - A) = 0$.

$$\det(\lambda I - A) = (\lambda-1)(\lambda-3)(\lambda-5) - [4(\lambda-1) + (\lambda-5)]$$

$$= \lambda^3 - 9\lambda^2 + 18\lambda - 6 . \qquad (7)$$

Using numerical methods solve (7), approximately, for its roots. Thus, the possible values for k^2 are .416, 2.29 and 6.29. To each value of k^2 there corresponds an eigenvector Y so that equation (6) can be solved. For example, if $k^2 = .416$ $(A - k^2I)Y = 0$ becomes $(A - .416I)Y = 0$. This can be solved numerically for Y. The column vector (.843 .493 .215) is a solution and, therefore, the eigenvector which corresponds to .416. Similarly (-.528 .683 .505) and (.102 -.540 .836) are eigenvectors corresponding to 2.29 and 6.29 respectively. Actually substituting values found for k and Y in the solution $\eta = Y \sin kt$ allows calcula- tions of the position of any one particle at any given time. For example, if $\eta = Y \sin \sqrt{2.29}\, t$,

$$Y = \begin{bmatrix} -.528 \\ .683 \\ .505 \end{bmatrix}$$

then a picture of the vibration at a specific time t might be:

In fact the three modes (one for each eigenvalue, and class of eigenvectors) are

Fig. 2

Assume the following facts about the two story frame in figure 1.

   A) The masses of the columns may be neglected with regard to the masses of the horizontal girders (m_1, m_2).

   B) The horizontal girders are very stiff (unbending) compared to the columns which have stiffnesses k_1 and k_2.

   C) All joints are rigid.

   D) The vibrations of the frame are undamped. (i.e., no friction). Consider vibrations of the frame in its plane (see figure 2). Note that the horizontal girders remain horizontal during vibration.

   1) Compute the natural frequencies of vibration in the plane.
   2) Depict the natural modes of vibration.

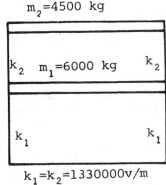

Fig. 1

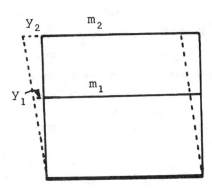

Fig. 2

Solution: 1) The columns of the steel frame obey Hooke's Law. Each column feels a restoring force towards its point of equilibrium which is proportional to its displacement from that point. (F = -ky), The constant of proportionality is called the stiffness of the column. Also recall Newton's law of motion which says that the force on an object is equal to the mass times the acceleration of the object. (F = mÿ). In figure 2 the upper beam is displaced a distance y_2 and therefore Newton's law implies that the force on that beam is $m_2\ddot{y}_2$. The upper part of the column exerts a restoring force proportional to the deflection of the upper part of the column. That deflection is $y_2 - y_1$. Equating the force on the upper girder by Newton's law with the force on the upper girder by Hooke's law we have

$$m_2\ddot{y}_2 = -k_2(y_2 - y_1) \ ,$$

which is

$$m_2\ddot{y}_2 + k_2(y_2 - y_1) = 0 \tag{1}$$

A similar calculation on the lower girder leads to the equation

$$m_1\ddot{y}_1 + k_1y_1 - k_2(y_2 - y_1) = 0 \tag{2}$$

(2) and (1) form a system of equations which can be written in matrix form as

$$M\ddot{Y} + KY = 0 \tag{3}$$

where

$$M = \begin{bmatrix} m_1 & 0 \\ 0 & m_2 \end{bmatrix} , \qquad \ddot{Y} = \begin{bmatrix} y_1 \\ y_2 \end{bmatrix}$$

$$K = \begin{bmatrix} k_1 + k_2 & -k_2 \\ -k_2 & k_2 \end{bmatrix} \qquad Y = \begin{bmatrix} y_1 \\ y_2 \end{bmatrix}$$

Since the motion is an undamped free (no outside force) vibration it will be simple harmonic motion. That is $\ddot{y} = -\omega^2 y$ where ω is the natural frequency of the motion. Substitute this in equation (3).

$$M(-\omega^2)Y + KY = 0 .$$

Multiply both sides of the equation by M^{-1} and obtain

$$M^{-1}M(-\omega^2)Y + M^{-1}KY = 0$$

which is

$$-\omega^2 Y + M^{-1}KY = 0 \tag{4}$$

Notice that M^{-1} must exist since M is non-singular, det $M = m_1 m_2 \neq 0$ since $m_1, m_2 > 0$. Multiply both sides of equation (4) by -1 and use the right distributive law to obtain

$$(\omega^2 I - M^{-1}K)Y = 0 . \tag{5}$$

Equation (5) is in the form $(\lambda I - A)Y = 0$, $\lambda = \omega^2$ and $A = M^{-1}K$ and therefore we can find the natural frequencies (values of ω) by computing the eigenvalues (λ) of A .

$$M^{-1} = \frac{1}{\det M} \text{ Adj } M = \frac{1}{m_1 m_2} \begin{bmatrix} m_2 & 0 \\ 0 & m_1 \end{bmatrix} = \begin{bmatrix} 1/m_1 & 0 \\ 0 & 1/m_2 \end{bmatrix}$$

$$A = M^{-1}K = \begin{bmatrix} 1/m_1 & 0 \\ 0 & 1/m_2 \end{bmatrix} \begin{bmatrix} k_1 + k_2 & -k_2 \\ -k_2 & k_2 \end{bmatrix}$$

$$= \begin{bmatrix} \dfrac{k_1 + k_2}{m_1} & \dfrac{-k_2}{m_1} \\ \dfrac{-k_2}{m_2} & \dfrac{k_2}{m_2} \end{bmatrix}$$

The eigenvalues of A are the roots of its characteristic equation:

$$\det(\lambda I - A) = 0$$

$$\det(\lambda I - A) = \begin{vmatrix} \lambda - \dfrac{k_1 + k_2}{m_1} & \dfrac{k_2}{m_1} \\ \dfrac{k_2}{m_2} & \lambda - \dfrac{k_2}{m_2} \end{vmatrix}$$

$$= \begin{vmatrix} \dfrac{m_1\lambda - (k_1+k_2)}{m_1} & \dfrac{k_2}{m_1} \\[2ex] \dfrac{k_2}{m_2} & \dfrac{\lambda m_2 - k_2}{m_2} \end{vmatrix}$$

$$= \frac{1}{m_1 m_2}[(m_1\lambda - (k_1+k_2))(\lambda m_2 - k_2) - k_2^2]$$

Equating this to zero and multiplying both sides by $m_1 m_2$ yields

$[m_1\lambda - (k_1+k_2)](\lambda m_2 - k_2) - k_2^2 = 0$. Multiply and gather terms to get

$m_1 m_2 \lambda^2 - (m_1 k_2 + m_2 k_1 + m_2 k_2)\lambda + k_2 k_1$. Using the constants from figure 1 this polynomial equation in λ is

$$6000(4500)\lambda^2 - [1330000(6000 + 4500 + 4500)]\lambda + (1330000)^2 = 0 .$$

This is $.27(10^8)\lambda^2 - 199.5(10^8)\lambda + 17689(10^8) = 0$. Divide by 10^8 to obtain:

$.27\lambda^2 - 199.5\lambda + 17689 = 0$. Using the quadratic formula we have

$$\lambda = 103 \quad \text{or} \quad \lambda = 635 .$$

Now $\omega^2 = x$ so the natural frequencies of the system are $\omega = \sqrt{103} = 10.15$ rad/sec. and $\omega = \sqrt{635} = 25.20$ rad/sec.

2) We can understand the natural modes of vibration by comparing the possible displacements, y_1 and y_2 . From equation (5) we see that the possible values of $Y = \begin{bmatrix} y_1 \\ y_2 \end{bmatrix}$ are the elements of the vector space of eigenvectors of A associated with any $\lambda = \omega^2$. First, substitute $\omega^2 = 103$ in equation (5) to obtain:

$$\begin{bmatrix} \dfrac{m_1(103) - (k_1+k_2)}{m_1} & \dfrac{k_2}{m_1} \\[3ex] \dfrac{k_2}{m_2} & \dfrac{103(m_2) - k_2}{m_2} \end{bmatrix} \begin{bmatrix} y_1 \\[3ex] y_2 \end{bmatrix} = \begin{bmatrix} 0 \\[3ex] 0 \end{bmatrix} .$$

Multiply the first row by m_1 and the second row by m_2 and substitute the values for $m_1, m_2, k_1 = k_2$. The result is:

$$\begin{bmatrix} -2042(10^3) & 1330(10^3) \\[2ex] 1330(10^3) & -866.5(10^3) \end{bmatrix} \begin{bmatrix} y_1 \\[2ex] y_2 \end{bmatrix} = \begin{bmatrix} 0 \\[2ex] 0 \end{bmatrix}$$

which is equivalent to the system of equations

$$-2042 y_1 + 1330 y_2 = 0 \qquad \text{(A)}$$

$$1330 y_1 - 866.5 y_2 = 0 \qquad \text{(B)}$$

Equation (A) yields
$$y_2 = \frac{2042 y_1}{1330} \doteq 1.535 y_1$$

Equation (B) yields

$$y_2 = \frac{1330 y_1}{866.5} \doteq 1.535 y_1 \quad .$$

The two equations are dependent and the eigenvector of A, with $\lambda = 103$ is the space of vectors y_i such that $y_2/y_1 = 1.535$. This means that the upper girder moves in the same direction as the lower girder at any instant since 1.535 is positive, but the upper girder moves 1.535 times as far as the lower girder. (See figure 3). If we substitute $\omega^2 = 635$ into equation (5) and proceed in a similar manner we obtain the matrix equation

$$\begin{bmatrix} 1150\,(10^3) & 1330\,(10^3) \\ 1330\,(10^3) & 1527.5\,(10^3) \end{bmatrix} \begin{bmatrix} y_1 \\ y_2 \end{bmatrix} = \begin{bmatrix} 0 \\ 0 \end{bmatrix}$$

or the system of equations

$$1150 y_1 + 1330 y_2 = 0$$

$$1330 y_1 + 1527.5 y_2 = 0 \quad .$$

From each of these equations we obtain approximately the same result, $y_2 = .87 y_1$. In this case the ratio y_2/y_1 is negative so the girders move in opposite directions and the top girder moves less than the lower one (about $1-.87y$ or $.13y$ less). This motion is depicted in figure 4.

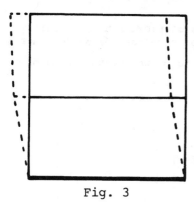

Fig. 3

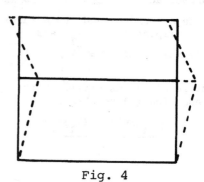

Fig. 4

● **PROBLEM** 20-19

The three masses shown in figure 1 are initially displaced so that

$$(x_1)_0 = 2, \ (x_2)_0 = -1, \ (x_3)_0 = 1 \quad .$$

From these positions they begin to move with initial velocities

$$(v_1)_0 = 1, \ (v_2)_0 = 2, \ (v_3)_0 = 0 \quad .$$

Assuming that there is no friction in the system, determine (i) the natural frequencies and normal modes of the system, (ii) subsequent motion of each mass.

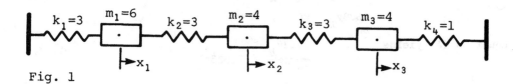

$k_1 = 3 \quad m_1 = 6 \quad k_2 = 3 \quad m_2 = 4 \quad k_3 = 3 \quad m_3 = 4 \quad k_4 = 1$

$x_1 \qquad x_2 \qquad x_3$

Fig. 1

Solution: Let x_1, x_2 and x_3 denote the displacement of the masses. Since friction is assumed to be negligible the only forces acting are those transmitted to the masses by the springs directly attached to them. Since the displacements of the masses are x_1, x_2 and x_3, the lengths of the springs have changed from their unstretched, initial lengths by the respective amounts

$$x_1 \;,\; x_2 - x_1 \;,\; x_3 - x_2 \;,\; -x_3 \;.$$

Hence, by Hooke's law, the forces instantaneously exerted by the springs are respectively,

$$k_1 x_1 \;,\; k_2(x_2 - x_1) \;,\; k_3(x_3 - x_2) \;,\; -k_4 x_3$$

plus signs indicating that the springs are in tension, minus signs that the springs are in compression. Now applying Newton's law to each of the masses in turn, yields the three differential equations:

$$
\begin{aligned}
m_1 \ddot{x}_1 &= -k_1 x_1 + k_2(x_2 - x_1) \\
m_2 \ddot{x}_2 &= -k_2(x_2 - x_1) + k_3(x_3 - x_2) \\
m_3 \ddot{x}_3 &= -k_3(x_3 - x_2) - k_4 x_3 \;.
\end{aligned}
\tag{1}
$$

Substituting the values of constants from figure 1, into the differential equations (1) yields:

$$
\begin{aligned}
6\ddot{x}_1 &= -3x_1 + 3(x_2 - x_1) \\
4\ddot{x}_2 &= -3(x_2 - x_1) + 3(x_3 - x_2) \\
4\ddot{x}_3 &= -3(x_3 - x_2) - x_3 \;.
\end{aligned}
$$

Dividing the first equation by 6; and the second and third equations by 4, yields:

$$
\begin{aligned}
\ddot{x}_1 &= -\tfrac{1}{2}x_1 + \tfrac{1}{2}(x_2 - x_1) \\
\ddot{x}_2 &= -(3/4)(x_2 - x_1) + (3/4)(x_3 - x_2) \\
\ddot{x}_3 &= -(3/4)(x_3 - x_2) - (1/4)(x_3)
\end{aligned}
$$

or, collecting terms in x_1, x_2, x_3,

$$
\begin{aligned}
\ddot{x}_1 &= -x_1 + (1/2)\left(x_2\right) \\
\ddot{x}_2 &= (3/4)\left(x_1\right) - (3/2)\left(x_2\right) + (3/4)\left(x_3\right) \\
\ddot{x}_3 &= + (3/4)\left(x_2\right) - x_3 \;.
\end{aligned}
\tag{2}
$$

Since there is no dissipation of energy through friction, it is clear that each mass must vibrate around its equilibrium position with constant amplitude. Hence, as a solution assume,

$$
X = Y \cos \omega t = \begin{pmatrix} y_1 \\ y_2 \\ y_3 \end{pmatrix} \cos \omega t
$$

that is,

$$x_1 = y_1 \cos \omega t, \; x_2 = y_2 \cos \omega t, \; x_3 = y_3 \cos \omega t$$

where ω is a frequency to be determined and y_1, y_2 and y_3 are the as yet unknown amplitudes through which the respective masses oscillate at this frequency. Also $X = -Y \, \omega^2 \cos \omega t$ substituting these into the differential equations (2), yields:

$$-y_1 \omega^2 \cos \omega t = -y_1 \cos \omega t + (1/2) y_2 \cos \omega t$$

$$-y_2 \omega^2 \cos \omega t = (-3/4) y_1 \cos \omega t - (3/2) y_2 \cos \omega t +$$

$$+(3/4) y_3 \cos \omega t$$

$$-y_3 \omega^2 \cos \omega t = (3/4) y_2 \cos \omega t - y_3 \cos \omega t \ .$$

Dividing out the common factor $\cos \omega t$, we obtain:

$$-y_1 \omega^2 = -y_1 + (1/2) y_2$$

$$-y_2 \omega^2 = (3/4) y_1 - (3/2) y_2 + (3/4) y_3 \qquad (3)$$

$$-y_3 \omega^2 = (3/4) y_2 - y_3 \ .$$

For simplicity, let $\omega^2 = \lambda$, and after rearrangement, we have

$$y_1 - (1/2) y_2 = \lambda y_1$$

$$(-3/4) y_1 + (3/2) y_2 - (3/4) y_3 = \lambda y_2$$

$$(-3/4) y_2 + y_3 = \lambda y_3 \ .$$

In matrix notation, this is

$$AY = \lambda Y \qquad (4)$$

where

$$A = \begin{bmatrix} 1 & -1/2 & 0 \\ -3/4 & 3/2 & -3/4 \\ 0 & -3/4 & 1 \end{bmatrix} \quad \text{and} \quad Y = \begin{bmatrix} y_1 \\ y_2 \\ y_3 \end{bmatrix}$$

Recall that a vector V is an eigenvector for A belonging to the eigenvalue λ if $AV = \lambda V$ $(V \neq 0)$. Thus in equation (3), Y is a non-zero vector for each eigenvalue of the matrix A. The characteristic equation of A is

$$\det(\lambda I - A) = 0 \ .$$

So,

$$\det[\lambda I - A] = \det \begin{bmatrix} \lambda-1 & 1/2 & 0 \\ 3/4 & \lambda-3/2 & 3/4 \\ 0 & 3/4 & \lambda-1 \end{bmatrix} = 0 \qquad (5)$$

or,

$$\lambda-1 \begin{vmatrix} \lambda-3/2 & 3/4 \\ 3/4 & \lambda-1 \end{vmatrix} - 1/2 \begin{vmatrix} 3/4 & 3/4 \\ 0 & \lambda-1 \end{vmatrix} = 0$$

or,

$$(\lambda-1)[(\lambda-3/2)(\lambda-1) - (3/4)(3/4)] - 1/2[3/4(\lambda-1) - 0] = 0 \ ,$$

$$(\lambda-1)[\lambda^2 - 5/2 \, \lambda + 15/16] - 3/8(\lambda-1) = 0 \ ,$$

$$(\lambda-1)[\lambda^2 - 5/2 \, \lambda + 9/16] = 0 \ ,$$

or,

$$(\lambda-1)(\lambda - 1/4)(\lambda - 9/4) = 0 \ .$$

714

Therefore, the eigenvalues of A are $\lambda_1 = 1/4$, $\lambda_2 = 1$, and $\lambda_3 = 9/4$.

Since,

$$\omega^2 = \lambda_1$$

then

$$\omega_1^2 = \lambda_1 = 1/4$$

$$\omega_2^2 = \lambda_2 = 1$$

$$\omega_3^2 = \lambda_3 = 9/4$$

or,

$$\omega_1 = 1/2$$

$$\omega_2 = 1$$

and

$$\omega_3 = 3/2 .$$

Now $Y = \begin{bmatrix} y_1 \\ y_2 \\ y_3 \end{bmatrix}$ is an eigenvector of A corresponding to λ, if and

only if Y is a nontrivial solution of $(\lambda I - A)Y = 0$.

For $\lambda = 1/4$, $(1/4\, I - A)Y = 0$, or,

$$\begin{bmatrix} -3/4 & 1/2 & 0 \\ 3/4 & -5/4 & 3/4 \\ 0 & 3/4 & -3/4 \end{bmatrix} \begin{bmatrix} y_1 \\ y_2 \\ y_3 \end{bmatrix} = \begin{bmatrix} 0 \\ 0 \\ 0 \end{bmatrix}$$

(This can be found quickly by looking at equation (5).) Then,

$$\left(-3/4\right) y_1 + \left(1/2\right) y_2 = 0$$

$$\left(3/4\right) y_1 - \left(5/4\right) y_2 + \left(3/4\right) y_3 = 0$$

$$\left(3/4\right) y_2 - \left(3/4\right) y_3 = 0 .$$

This implies that $y_1 = 2/3\, y_2$, and $y_2 = y_3$. Therefore,

$$Y_1 = \begin{bmatrix} 2 \\ 3 \\ 3 \end{bmatrix} \text{ is an eigenvector associated with}$$

$\lambda = 1/4$.

For $\lambda_2 = 1$,

$$\begin{bmatrix} 0 & 1/2 & 0 \\ 3/4 & -1 & 3/4 \\ 0 & 3/4 & 0 \end{bmatrix} \begin{bmatrix} y_1 \\ y_2 \\ y_3 \end{bmatrix} = \begin{bmatrix} 0 \\ 0 \\ 0 \end{bmatrix}$$

or,

$$\left(1/2\right) y_2 = 0$$

$$\left(-3/4\right) y_1 - y_2 + \left(3/4\right) y_3 = 0$$

$$\left(-3/4\right) y_2 = 0 .$$

This implies that $y_1 = -y_3$ and $y_2 = 0$. Therefore,

$$Y_2 = \begin{bmatrix} 1 \\ 0 \\ -1 \end{bmatrix} \quad \text{is an eigenvector corresponding to} \quad \lambda_2 = 1$$

For $\lambda_3 = 9/4$, then:

$$\begin{bmatrix} 5/4 & 1/2 & 0 \\ 3/4 & 3/4 & 3/4 \\ 0 & 3/4 & 5/4 \end{bmatrix} \begin{bmatrix} y_1 \\ y_2 \\ y_3 \end{bmatrix} = \begin{bmatrix} 0 \\ 0 \\ 0 \end{bmatrix}.$$

or

$$\left(5/4\right) y_1 + \left(1/2\right) y_2 = 0$$
$$\left(3/4\right) y_1 + \left(3/4\right) y_2 + \left(3/4\right) y_3 = 0$$
$$\left(3/4\right) y_2 + \left(5/4\right) y_3 = 0 .$$

This implies that $y_2 = \left(-5/2\right) y_1$ and $y_3 = \left(-3/4\right) y_2$. Thus,

$$Y_3 = \begin{bmatrix} 2 \\ -5 \\ 3 \end{bmatrix} \quad \text{is an eigenvector associated with} \quad \lambda_3 = 9/4.$$

We have already identified the three values $\omega = 1/2, 1, 3/2$ as the natural frequencies of the system, i.e., the only frequencies at which free vibrations of the system are possible. The vectors Y_1, Y_2 and Y_3 associated with the frequencies $\omega = 1/2$, $\omega = 1$, and $\omega = 3/2$ are called the normal modes of the system.

We have also found three particular solution vectors for the system (2), namely,

$$x_1 = y_1 \cos \omega_1 t = \begin{bmatrix} 2 \\ 3 \\ 3 \end{bmatrix} \cos \left(1/2\right) t$$

$$x_2 = y_2 \cos \omega_2 t = \begin{bmatrix} 1 \\ 0 \\ -1 \end{bmatrix} \cos t$$

$$x_3 = y_3 \cos \omega_3 t = \begin{bmatrix} 2 \\ -5 \\ 3 \end{bmatrix} \cos \left(3/2\right) t .$$

Clearly, if we had begun with the assumptions $x_1 = y_1 \sin \omega t$, $x_2 = y_2 \sin \omega t$, $x_3 = y_3 \sin \omega t$ we would also have obtained the algebraic equations (3) and hence the same three values of ω and the same three solution vectors. Therefore we have three more particular solutions:

$$x_4 = \begin{bmatrix} 2 \\ 3 \\ 3 \end{bmatrix} \sin t/2 , \quad x_5 = \begin{bmatrix} 1 \\ 0 \\ -1 \end{bmatrix} \sin t , \quad x_6 = \begin{bmatrix} 2 \\ -5 \\ 3 \end{bmatrix} \sin \left(3/2\right) t$$

and finally, the complete solution

$$x = c_1 x_1 + c_2 x_2 + c_3 x_3 + c_4 x_4 + c_5 x_5 + c_6 x_6$$

or,

$$x = c_1 \begin{bmatrix} 2 \\ 3 \\ 3 \end{bmatrix} \cos t/2 + c_2 \begin{bmatrix} 1 \\ 0 \\ -1 \end{bmatrix} \cos t + c_3 \begin{bmatrix} 2 \\ -5 \\ 3 \end{bmatrix} \cos \left(3/2\right) t$$

$$+ c_4 \begin{bmatrix} 2 \\ 3 \\ 3 \end{bmatrix} \sin t/2 + c_5 \begin{bmatrix} 1 \\ 0 \\ -1 \end{bmatrix} \sin t + c_6 \begin{bmatrix} 2 \\ -5 \\ 3 \end{bmatrix} \sin \left(3/2\right) t \qquad (6)$$

where the c's are arbitrary scalar coefficients. To determine the values
of the c's we must of course, use the given initial conditions. The
initial displacement vector $x(0)$ is given by

$$x(0) \; = \; \begin{pmatrix} 2 \\ -1 \\ 1 \end{pmatrix}$$

We set $t = 0$ in (6) and substitute the initial displacement vector for
$x(0)$, getting

$$\begin{pmatrix} 2 \\ -1 \\ 1 \end{pmatrix} \; = \; c_1 \begin{pmatrix} 2 \\ 3 \\ 3 \end{pmatrix} + \; c_2 \begin{pmatrix} 1 \\ 0 \\ -1 \end{pmatrix} + \; c_3 \begin{pmatrix} 2 \\ -5 \\ 3 \end{pmatrix}$$

or,

$$\begin{pmatrix} 2 & 1 & 2 \\ 3 & 0 & -5 \\ 3 & -1 & 3 \end{pmatrix} \begin{pmatrix} c_1 \\ c_2 \\ c_3 \end{pmatrix} \; = \; \begin{pmatrix} 2 \\ -1 \\ 1 \end{pmatrix} \quad .$$

Let

$$B = \begin{pmatrix} 2 & 1 & 2 \\ 3 & 0 & -5 \\ 3 & -1 & 3 \end{pmatrix} \quad C = \begin{pmatrix} c_1 \\ c_2 \\ c_3 \end{pmatrix} \quad \text{and} \quad K = \begin{pmatrix} 2 \\ -1 \\ 1 \end{pmatrix}$$

Then,

$$BC = K \; .$$

Multiplying by B^{-1}, (assuming it exists) we have

$$B^{-1}BC = B^{-1}K \; ,$$

since

$$B^{-1}B = I \; ,$$
$$C = B^{-1}K \; .$$

Now,

$$B^{-1} = \frac{1}{\det B} \; \text{adj } B$$

$$\det B = \det \begin{pmatrix} 2 & 1 & 2 \\ 3 & 0 & -5 \\ 3 & -1 & 3 \end{pmatrix}$$

Add the first row to the third row.

$$\det B = \det \begin{pmatrix} 2 & 1 & 2 \\ 3 & 0 & -5 \\ 5 & 0 & 5 \end{pmatrix}$$

Expand the determinant along the second column.

$$\det B = -1 \; \begin{vmatrix} 3 & -5 \\ 5 & 5 \end{vmatrix} \; = -1 \; [15 + 25] = -40 \; .$$

$$\text{adjB} = \begin{pmatrix} + \begin{vmatrix} 0 & -5 \\ -1 & 3 \end{vmatrix} & - \begin{vmatrix} 3 & -5 \\ 3 & 3 \end{vmatrix} & + \begin{vmatrix} 3 & 0 \\ 3 & -1 \end{vmatrix} \\[12pt] - \begin{vmatrix} 1 & 2 \\ -1 & 3 \end{vmatrix} & + \begin{vmatrix} 2 & 2 \\ 3 & 3 \end{vmatrix} & - \begin{vmatrix} 2 & 1 \\ 3 & -1 \end{vmatrix} \\[12pt] + \begin{vmatrix} 1 & 2 \\ 0 & -5 \end{vmatrix} & - \begin{vmatrix} 2 & 2 \\ 3 & -5 \end{vmatrix} & + \begin{vmatrix} 2 & 1 \\ 3 & 0 \end{vmatrix} \end{pmatrix}^T$$

$$= \begin{bmatrix} -5 & -24 & -3 \\ -5 & 0 & 5 \\ -5 & 16 & -3 \end{bmatrix}^T$$

$$\text{adj } B = \begin{bmatrix} -5 & -5 & -5 \\ -24 & 0 & 16 \\ -3 & 5 & -3 \end{bmatrix}$$

then,

$$B^{-1} = \frac{1}{-40} \begin{bmatrix} -5 & -5 & -5 \\ -24 & 0 & 16 \\ -3 & 5 & -3 \end{bmatrix}$$

Therefore,

$$C = B^{-1}K = \frac{1}{-40} \begin{bmatrix} -5 & -5 & -5 \\ -24 & 0 & 16 \\ -3 & 5 & -3 \end{bmatrix} \begin{bmatrix} 2 \\ -1 \\ 1 \end{bmatrix}$$

$$= \frac{1}{-40} \begin{bmatrix} -10 & + & 5 & - & 5 \\ -48 & + & 16 & & \\ -6 & - & 5 & - & 3 \end{bmatrix} = \frac{1}{-40} \begin{bmatrix} -10 \\ -32 \\ -14 \end{bmatrix}$$

or,

$$\begin{pmatrix} c_1 \\ c_2 \\ c_3 \end{pmatrix} = -\frac{1}{40} \begin{pmatrix} -10 \\ -32 \\ -14 \end{pmatrix}$$

Therefore, $c_1 = 1/4$, $c_2 = 4/5$ and $c_3 = 7/20$. To find c_4, c_5 and c_6, we first differentiate equation (6), getting

$$\frac{dx}{dt} = \left(-1/2\right) c_1 \begin{pmatrix} 2 \\ 3 \\ 3 \end{pmatrix} \sin t/2 - c_2 \begin{pmatrix} 1 \\ 0 \\ -1 \end{pmatrix} \sin t - \left(3/2\right) c_3 \begin{pmatrix} 2 \\ -5 \\ 3 \end{pmatrix} \sin\left(3/2\right) t$$

$$+ \left(1/2\right) c_4 \begin{pmatrix} 2 \\ 3 \\ 3 \end{pmatrix} \cos t/2 + c_5 \begin{pmatrix} 1 \\ 0 \\ -1 \end{pmatrix} \cos t + \left(3/2\right) c_6 \begin{pmatrix} 2 \\ -5 \\ 3 \end{pmatrix} \cos\left(3/2\right) t.$$

Then, setting $t = 0$ and substituting the initial velocity vector $\begin{pmatrix} 1 \\ 2 \\ 0 \end{pmatrix}$ for $\left.\frac{dx}{dt}\right|_{t=0}$, we have

$$\begin{pmatrix} 1 \\ 2 \\ 0 \end{pmatrix} = (1/2) c_4 \begin{pmatrix} 2 \\ 3 \\ 3 \end{pmatrix} + c_5 \begin{pmatrix} 1 \\ 0 \\ -1 \end{pmatrix} + (3/2) c_6 \begin{pmatrix} 2 \\ -5 \\ 3 \end{pmatrix}$$

or

$$c_4 + c_5 + 3c_6 = 1$$

$$3/2\ c_4 \qquad\quad - 15/2c_6 = 2$$

$$3/2\ c_4 - c_5 + 9/2c_6 = 0$$

Solving this system, we find $c_4 = 3/4$, $c_5 = 3/5$, and $c_6 = -7/60$.

With the c's determined, the solution is now complete, and

$$x = 1/4 \begin{pmatrix} 2 \\ 3 \\ 3 \end{pmatrix} \cos t/2 + 4/5 \begin{pmatrix} 1 \\ 0 \\ -1 \end{pmatrix} \cos t + 7/20 \begin{pmatrix} 2 \\ -5 \\ 3 \end{pmatrix} \cos (3/2) t$$

$$+ 3/4 \begin{pmatrix} 2 \\ 3 \\ 3 \end{pmatrix} \sin t/2 + 3/5 \begin{pmatrix} 1 \\ 0 \\ -1 \end{pmatrix} \sin t - 7/40 \begin{pmatrix} 2 \\ -5 \\ 3 \end{pmatrix} \sin (3/2) t$$

or, explicitly,

$$x_1 = (1/2) \cos t/2 + (4/5) \cos t + (7/10) \cos (3/2) t + (3/2) \sin t/2 + (3/5) \sin t \\ (- 7/20) \sin (3/2) t$$

$$x_2 = (3/4) \cos t/2 - (7/4) \cos (3/2) t + (9/4) \sin t/2 + (7/8) \sin (3/2) t$$

$$x_3 = (3/4) \cos t/2 - (4/5) \cos t + (21/20) \cos (3/2) t + (9/4) \sin t/2 \\ (- 3/5) \sin t - (21/40) \sin (3/2) t .$$

STRESS PROBLEMS

● **PROBLEM** 20-20

The two ropes **AB** and **BC** support the 500 lb weight as shown in the figure below. Determine the tensions in the ropes **AB** and **BC**.

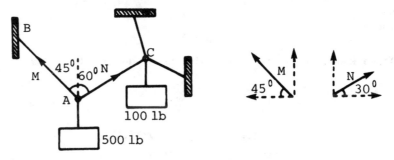

Solution: Let **M** and **N** represent tensions in the ropes **AB** and **AC** respectively. We shall first find the horizontal and vertical components of **M** and **N**.

The horizontal component N_h acts to the right and is equal to $N \cos 30°$. The vertical component N_v acts up and is equal to $N \sin 30°$.

Similarly, the horizontal component M_h acts to the left and is equal to

719

M cos 45° . The vertical component M_v acts up and is equal to M sin 45°.
Now for static equilibrium the following equations must be satisfied.

$$\Sigma\, F_h = 0$$

$$\Sigma\, F_v = 0$$

where,

$\Sigma\, F_h$ = algebraic sum of the horizontal

components of the forces on A ;

$\Sigma\, F_v$ = algebraic sum of the vertical

component of the forces on A.

Thus,

$$N \cos 30^{\circ} - M \cos 45^{\circ} = 0$$

$$N \sin 30^{\circ} + M \sin 45^{\circ} - 500 = 0 \ .$$

Substitute the values of the trigonometric functions into the above equations. Then,

$$\frac{\sqrt{3}\,N}{2} - \frac{\sqrt{2}\,M}{2} = 0$$

$$\tfrac{1}{2} N + \frac{\sqrt{2}\,M}{2} - 500 = 0$$

or

$$0.866N - 0.707M = 0 \tag{1}$$

$$0.5N \ + \ .707M = 500$$

The system (1) in matrix notation is:

$$\begin{bmatrix} .866 & -0.707 \\ .5 & .707 \end{bmatrix} \begin{bmatrix} N \\ M \end{bmatrix} = \begin{bmatrix} 0 \\ 500 \end{bmatrix}$$

or

$$AX = B \tag{2}$$

where

$$A = \begin{bmatrix} .866 & -0.707 \\ .5 & 0.707 \end{bmatrix} \quad X = \begin{bmatrix} N \\ M \end{bmatrix} \text{ and } B = \begin{bmatrix} 0 \\ 500 \end{bmatrix}$$

Recall that if A is invertible, we can solve equation (2) for X by multiplying by A^{-1} . We have

$$A^{-1}AX = A^{-1}B \ .$$

But

$$A^{-1}A = I \ ,$$

therefore,

$$IX = A^{-1}B \ . \tag{3}$$

As we know, if A is invertible, then det A $\neq$ 0 . Now

$$\det A = \det \begin{bmatrix} .866 & -0.707 \\ .5 & .707 \end{bmatrix}$$

$$= [(.866)(.707)] - [(-0.707)(.5)]$$

$$= 0.6123 + 0.3535 = 0.9658$$

since det A $\neq$ 0, the matrix A is invertible. It is known that

$$A^{-1} = \frac{1}{\det A} \text{ adj } A \ ,$$

where adj A is the transpose of the matrix of cofactors.

$$\text{adj } A = \begin{bmatrix} .707 & -0.5 \\ +0.707 & 0.866 \end{bmatrix}^{T} = \begin{bmatrix} 0.707 & 0.707 \\ -0.5 & 0.866 \end{bmatrix}$$

Then,

$$A^{-1} = \frac{1}{\det A} \begin{bmatrix} 0.707 & 0.707 \\ -0.5 & 0.866 \end{bmatrix}$$

$$= \frac{1}{0.9658} \begin{bmatrix} 0.707 & 0.707 \\ -0.5 & 0.866 \end{bmatrix}.$$

Upon substituticn of the value of A^{-1} and B into equation (3) yields:

$$X = \frac{1}{0.9658} \begin{bmatrix} 0.707 & 0.707 \\ -0.5 & 0.866 \end{bmatrix} \begin{bmatrix} 0 \\ 500 \end{bmatrix}$$

$$= \frac{1}{0.9658} \begin{bmatrix} 0.707 \times 500 \\ 0.866 \times 500 \end{bmatrix}$$

$$= \frac{1}{0.9658} \begin{bmatrix} 353.5 \\ 433.0 \end{bmatrix}$$

$$= \begin{bmatrix} 366 \\ 448 \end{bmatrix}$$

But

$$X = \begin{bmatrix} N \\ M \end{bmatrix}$$

therefore,

$$\begin{bmatrix} N \\ M \end{bmatrix} = \begin{bmatrix} 366 \\ 448 \end{bmatrix}$$

or

$$N = 366$$
$$M = 448$$

Therefore the tensions in AC and AB are 366 lb and 448 lb, respectively.

● **PROBLEM** 20-21

Consider the engineering mechanics, static equilibrium problem of figure 1. The two cables AB and BC support the 200 lb weight. Find the forces M and N.

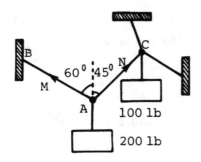

Solution: We shall first find the vertical and horizontal components of the forces M and N. The horizontal component M_h acts to the left and is equal to $M \cos 30°$. The vertical component M_v acts up and is equal to $M \sin 30°$. Similarly, the horizontal component N_h acts to the right and is equal to $N \cos 45°$. The vertical component N_v acts up and is equal to $N \sin 45°$. Now for static equilibrium. $\Sigma F_h = 0$ and $\Sigma F_v = 0$. Therefore,

$$N \cos 45° - M \cos 30° = 0$$

and

$$N \sin 45° + M \sin 30° - 200 = 0$$

Substitute the values of the trigonometric functions into the above equations and put the set of equations in their general form. Thus,

$$0.707N - .866M = 0 \tag{1}$$

$$0.707N + 0.5M = 200$$

The system (1) can be written in matrix notation as:

$$\begin{bmatrix} 0.707 & -.866 \\ 0.707 & 0.5 \end{bmatrix} \begin{bmatrix} N \\ M \end{bmatrix} = \begin{bmatrix} 0 \\ 200 \end{bmatrix}$$

or,

$$AX = B$$

where:

$$A = \begin{bmatrix} 0.707 & -.866 \\ 0.707 & 0.5 \end{bmatrix} \quad X = \begin{bmatrix} N \\ M \end{bmatrix}$$

and

$$B = \begin{bmatrix} 0 \\ 200 \end{bmatrix}.$$

To solve this system, use Cramer's rule. Thus,

$$N = \frac{\Delta_1}{\det A}$$

where Δ_1 is the determinant of the matrix A_1 obtained by replacing the first column of A by B. Therefore,

$$A_1 = \begin{bmatrix} 0 & -0.866 \\ 200 & 0.5 \end{bmatrix}$$

Now,

$$\det A = \det \begin{bmatrix} 0.707 & -0.866 \\ 0.707 & 0.5 \end{bmatrix}$$

$$= [(0.707) \times (0.5) - (-0.866) \times (0.707)]$$

$$= 0.3535 + .6123 = 0.9658$$

and

$$\det A_1 = \det \begin{bmatrix} 0 & -0.866 \\ 200 & 0.5 \end{bmatrix}$$

$$= 0 + .866 \times 200 = 173.2 .$$

Therefore,

$$N = \frac{\det A_1}{\det A} = \frac{173.2}{0.9658} = 179.3 \ .$$

Similarly,

$$M = \frac{\det A_2}{\det A}$$

where A_2 is the matrix obtained by replacing the second column of A by B. Thus,

$$A_2 = \begin{pmatrix} 0.707 & 0 \\ 0.707 & 200 \end{pmatrix}$$

then,

$$\det A_2 = \det \begin{pmatrix} 0.707 & 0 \\ 0.707 & 200 \end{pmatrix}$$

$$= 0.707 \times 200 - 0 = 141.4 \ .$$

Therefore,

$$M = \frac{\det A_2}{\det A} = \frac{141.4}{0.9658} = 146.4$$

Thus the forces M and N are

$$M = 146.4 \ \text{lb.}$$

$$N = 179.3 \ \text{lb.}$$

● **PROBLEM** 20-22

Calculate the stresses and strains in the plane mechanical framework shown in the figure below, subject to given external forces F_1 and F_2 acting at point A as shown. Each of the members of the framework labeled (1) to (5), makes a fixed angle θ_1 to θ_5, respectively, with the horizontal.

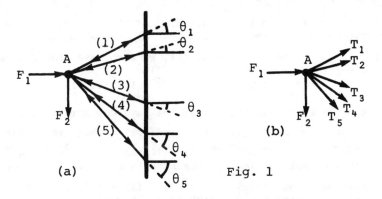

(a)   (b)   Fig. 1

<u>Solution:</u> In order to formulate this problem mathematically, make the following assumptions:
(i) The frame is pin-jointed; that is, its members are joined together at A and joined to the wall by pins, in such a way that the ends are free to rotate.
(ii) Forces at the joints are applied along the five members, and no bending moment is transmitted.
(iii) The weights of all members are negligible.

(iv) If the forces F_1 and F_2 are both zero, then there are no stresses in the members.

(v) Each member obeys Hooke's law: When a force is applied along its length, the extension e of the member is directly proportional to the force; that is, $e = kT$, where T is the force (tension) and k is a constant of proportionality called the flexibility.

(vi) The changes in the angles θ_i (i = 1,...,5) produced by application of the external forces are negligible.

Let T_i be the tension, e_i the extension, and k_i the flexibility of the ith member (i = 1,2,...,5) when the forces F_1 and F_2 are applied at A. The equations of force-equilibrium at the joint A are obtained by resolving all forces horizontally and vertically. The arrows in figure 1(a) indicate that the tension in a rod is positive when the rod is being extended, negative when the rod is being compressed. The force diagram at point A is therefore given by figure 1(b), which leads to the following equations of equilibrium

$$-T_1 \cos \theta_1 - T_2 \cos \theta_2 - \ldots - T_5 \cos \theta_5 = F_1 \qquad (1)$$

$$+ T_1 \sin \theta_1 + T_2 \sin \theta_2 + \ldots + T_5 \sin \theta_5 = F_2$$

These equations relate the internal and external forces in the framework. Here we are considering horizontal forces to the left as positive. In figure 1, the first two vertical components are positive since $\theta_1 > 0$, $\theta_2 > 0$, and the last three vertical components are negative since $\theta_3 < 0, \ \theta_4 < 0, \ \theta_5 < 0$.

We now wish to relate internal and external displacements in the framework. Suppose that when the forces F_1 and F_2 are applied, the point A moves d_1 units horizontally and d_2 units vertically, measured in the same direction as the respective forces F_1 and F_2 . The extensions e_i are given in terms of the d_i by the following formulae:

$$e_i = -d_1 \cos \theta_i + d_2 \sin \theta_i \quad (i = 1,2,\ldots,5) \qquad (2)$$

These equations simply state that the extension of the ith member is given by adding the components obtained by resolving d_1, d_2 along the rod, taking account of sign. From Hooke's law (v), we also have that for each member of the framework

$$e_i = k_i T_i \qquad (i = 1,2,\ldots,5) \qquad (3)$$

where the k_i are the known flexibilities. We have now derived the basic equations for the problem, namely (1), (2), and (3). These are twelve equations in the twelve unknowns quantities $T_1, T_2, T_3, T_4, T_5, \ e_1, e_2, e_3, e_4, e_5, \ d_1, d_2$. Thus the mathematical problem in this case is to solve the linear system of algebraic equations consisting of (1), (2) and (3).

So we have the following system which is to be solved.

$$-T_1 \cos \theta_1 - T_2 \cos \theta_2 - T_3 \cos \theta_3 - T_4 \cos \theta_4 - T_5 \cos \theta_5 = F_1$$

$$T_1 \sin \theta_1 + T_2 \sin \theta_2 + T_3 \sin \theta_3 + T_4 \sin \theta_4 + T_5 \sin \theta_5 = F_2$$

$$-d_1 \cos \theta_i + d_2 \sin \theta_i = e_i \quad (i = 1,\ldots,5) \qquad (4)$$

$$k_i T_i = e_i \quad (i = 1,\ldots,5)$$

Define,

$$T = \begin{pmatrix} T_1 \\ T_2 \\ T_3 \\ T_4 \\ T_5 \end{pmatrix} \qquad E = \begin{pmatrix} e_1 \\ e_2 \\ e_3 \\ e_4 \\ e_5 \end{pmatrix} \qquad F = \begin{pmatrix} F_1 \\ \\ \\ \\ F_2 \end{pmatrix}, \qquad D = \begin{pmatrix} d_1 \\ \\ \\ \\ d_2 \end{pmatrix}$$

$$A = \begin{pmatrix} -\cos\theta_1 & -\cos\theta_2 & -\cos\theta_3 & -\cos\theta_4 & -\cos\theta_5 \\ \sin\theta_1 & \sin\theta_2 & \sin\theta_3 & \sin\theta_4 & \sin\theta_5 \end{pmatrix}$$

$$K = \begin{pmatrix} k_1 & 0 & \cdot & \cdot & \cdot & 0 \\ 0 & k_2 & & & & 0 \\ \cdot & & k_3 & & & \cdot \\ \cdot & & & k_4 & & \\ 0 & & & & & 0 \\ 0 & \cdot & \cdot & \cdot & 0 & k_5 \end{pmatrix}$$

Then system (4) is conveniently written as follows. The first two equations are clearly expressible in the form

$$AT = F \tag{5}$$

To write the next five equations in matrix vector form, note that their co-efficient matrix is precisely A^T, the transpose of the matrix A. Then the five equations relating d_1, d_2 with $e_1, \ldots, e_5$ are

$$A^T D = E . \tag{6}$$

Finally, the last five equations are represented by the matrix equation

$$KT = E \tag{7}$$

Set up equations for the desired unknowns D. The E and F are also unknowns, so eliminate them as follows. For physical reasons the quanti-ties $k_i > 0$, $i = 1,2,\ldots,5$. Hence, K^{-1} exists and (7) yields

$$T = K^{-1}E$$

Hence (5) becomes

$$AK^{-1}E = F.$$

But from (6), $E = A^T D$ and therefore, D can be determined as the solution of the system

$$AK^{-1}A^T D = F . \tag{8}$$

Since $K \in R_{55}$ is diagonal matrix, so is $K^{-1} \in R_{55}$. Now, $A \in R_{25}$; hence $AK^{-1} \in R_{25}$ and $AK^{-1}A^T \in R_{22}$.

Now.

$$K^{-1} = \begin{pmatrix} k_1^{-1} & 0 & \cdot & \cdot & 0 \\ 0 & k_2^{-1} & & & 0 \\ \cdot & & k_3^{-3} & & \\ \cdot & & & k_4^{-4} & \cdot \\ \cdot & & & & \cdot \\ 0 & \cdot & \cdot & \cdot & 0 & k_5^{-1} \end{pmatrix}$$

Hence, by matrix multiplication

$$AK^{-1}A^T = \begin{pmatrix} \displaystyle\sum_{i=1}^{5}\left(\frac{\cos^2\theta_i}{k_i}\right) & \displaystyle -\sum_{i=1}^{5}\left(\frac{\cos\theta_i\sin\theta_i}{k_i}\right) \\ \displaystyle -\sum\left(\frac{\cos\theta_i\sin\theta_i}{k_i}\right) & \displaystyle \sum_{i=1}^{5}\left(\frac{\sin^2\theta_i}{k_i}\right) \end{pmatrix}$$

Let

$$C = AK^{-1}A^T .$$

Then (8) becomes

$$CD = F$$

or

$$D = C^{-1}F$$

which is all that is of interest physically. Having determined D we can find E from (6) as $E = A^T D$. (Note that this requires only matrix multiplication). Then we find T from (7) as

$$T = K^{-1}E = K^{-1}A^T D .$$

● **PROBLEM** 20-23

Figures 1, 2 and 3 represent three related pin-pointed frame-works.
i) x_B , y_B , x_C and y_C are given loads applied to joints B and C in figure 1. Find values for the unknown tensions, t_1, t_2, t_3 and t_4 so that the system will be in equilibrium.
ii) Strengthen the framework in figure 1 by the addition of the member CE (see figure 2). Again, find the equations of equilibrium of the system. Since this will yield four equations in five unknowns (no unique solution) assume that the members of the framework also obey Hooke's Law, LT = EAe where T is the tension, e the extension, L the length, A the cross-sectional area and E a numerical constant. Let E and A be the same for each member and call EA the constant k. Suppose that each member is fixed but free to rotate at the points A, D and E. Then find the unique solution for t_1, t_2, t_3, t_4 and t_5.
iii) In figure 3 AC and CD have been removed. Find a solution for t_1, t_2 and t_5 so that the framework will be in equilibrium.

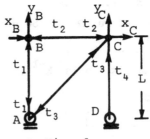

Fig. 1

Solution: i)
Since the framework is in equilibrium the resultant horizontal and vertical components of force on B and C are 0. First resolve horizontally on point B. Assuming the directions of the forces are as in the figure, $-t_2 + x_B$ must be 0. That is,

$$t_2 = x_B .$$ (1)

726

Resolving vertically at B yields $t_1 + y_B = 0$ or

$$t_1 = -y_B \, . \tag{2}$$

Now resolving horizontally on C we have three tensions to deal with. Since figure 2 is a square, angle ACD is 45° and the horizontal component of t_3 is

$$t_3 \cos 45° = \frac{\sqrt{2}}{2} t_3 \, .$$

Thus $t_2 + \frac{\sqrt{2}}{2} \ t_3 + x_C = 0$ or

$$t_2 + \frac{\sqrt{2}}{2} t_3 = -x_C \, . \tag{3}$$

Vertically on C we have $y_C + \frac{\sqrt{2}}{2} \ t_3 + t_4 = 0$ which yields

$$\frac{\sqrt{2}}{2} t_3 + t_4 = -y_C \, . \tag{4}$$

Equations (1), (2), (3) and (4) form a uniquely solvable system of four equations in four unknown.

$$
\begin{aligned}
t_2 && = x_B \\
t_1 && = -y_B \\
t_2 + \frac{\sqrt{2}}{2} t_3 && = -x_C \\
\frac{\sqrt{2}}{2} t_3 + t_4 && = -y_C
\end{aligned}
\tag{5}
$$

The system (5) can be expressed in matrix notation as

$$
\begin{pmatrix}
0 & 1 & 0 & 0 \\
1 & 0 & 0 & 0 \\
0 & 1 & \sqrt{2}/2 & 0 \\
0 & 0 & \sqrt{2}/2 & 1
\end{pmatrix}
\begin{pmatrix}
t_1 \\ t_2 \\ t_3 \\ t_4
\end{pmatrix}
=
\begin{pmatrix}
x_B \\ -y_B \\ -x_C \\ -y_C
\end{pmatrix}
$$

or

$$AT = B \, , \tag{6}$$

where

$$
\begin{pmatrix}
0 & 1 & 0 & 0 \\
1 & 0 & 0 & 0 \\
0 & 1 & \sqrt{2}/2 & 0 \\
0 & 0 & \sqrt{2}/2 & 1
\end{pmatrix}
, \quad
T =
\begin{pmatrix}
t_1 \\ t_2 \\ t_3 \\ t_4
\end{pmatrix}
, \quad \text{and} \quad
B =
\begin{pmatrix}
x_B \\ -y_B \\ -x_C \\ -y_C
\end{pmatrix}
$$

to solve system (6), we first form the augmented matrix [A:B], and then reduce it to echelon form.

The augmented matrix of the system (6) is

$$
\left(
\begin{array}{cccc|c}
0 & 1 & 0 & 0 & x_B \\
1 & 0 & 0 & 0 & -y_B \\
0 & 1 & \sqrt{2}/2 & 0 & -x_C \\
0 & 0 & \sqrt{2}/2 & 1 & -y_C
\end{array}
\right)
$$

Add -1 times the first row to the third row to obtain

$$\left[\begin{array}{cccc|c} 0 & 1 & 0 & 0 & x_B \\ 1 & 0 & 0 & 0 & -y_B \\ 0 & 0 & \sqrt{2}/2 & 0 & -x_C - x_B \\ 0 & 0 & \sqrt{2}/2 & 1 & -y_C \end{array}\right]$$

Add -1 times the third row to the fourth row to obtain

$$\left[\begin{array}{cccc|c} 0 & 1 & 0 & 0 & x_B \\ 1 & 0 & 0 & 0 & -y_B \\ 0 & 0 & \sqrt{2}/2 & 0 & -x_C - x_B \\ 0 & 0 & 0 & 1 & -y_C + x_C + x_B \end{array}\right]$$

which is the augmented matrix of the system

$$t_2 = x_B$$
$$t_1 = -y_B$$
$$\sqrt{2}/2\, t_3 = -x_C - x_B$$
$$t_4 = -y_C + x_C + x_B$$

Hence given x_B, y_B, x_C and y_C, the tensions required for equilibrium are:

$$t_1 = -y_B$$
$$t_2 = x_B$$
$$t_3 = -\sqrt{2}(x_B + x_C)$$

and

$$t_4 = x_B + x_C - y_C$$

ii)

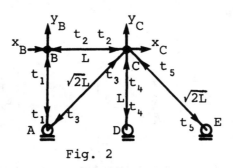

Fig. 2

In this case the system (A) is changed to

$$t_2 = x_B$$
$$t_1 = -y_B$$
$$t_2 + \frac{\sqrt{2}}{2}t_3 - \frac{\sqrt{2}}{2}t_5 = -x_C \qquad (A')$$
$$\frac{\sqrt{2}}{2}t_3 + t_4 + \frac{\sqrt{2}}{2}t_5 = -y_C$$

Since we have 4 equations in 5 unknowns we apply Hooke's Law. Let H_B and V_B be the horizontal and vertical components of the displacement of the point B. The corresponding displacements at C will be H_C and V_C. At the point B, applying Hooke's Law, $LT = ke$, we have vertically, $Lt_1 = kV_B$. Horizontally, the point B actually has displacement $H_C - H_B$ since the displacement at C will be in the opposite direction. Hence $Lt_2 = k(H_C - H_B)$. The equations for t_3, t_4 and t_5 are derived in a similar manner. Hooke's law and the framework in figure 2 have given rise to the following system of equations.

$$Lt_1 = kV_B$$
$$Lt_2 = k(H_C - H_B)$$
$$\sqrt{2}\, L\, t_3 = \frac{\sqrt{2}}{2}\, k(H_C + V_C) \qquad (B)$$
$$Lt_4 = kV_C$$
$$\sqrt{2}Lt_5 = \frac{\sqrt{2}}{2}\, k(V_C - H_C) .$$

Systems (A') and (B), together, are a system of 9 equations in 9 unknowns with a unique solution. Thus when assuming Hooke's law this structure is also determinate. To find the solution first solve each equation in (B) for t_i and then substitute those results in system (A').

$$t_1 = \frac{k}{L}\, V_B$$
$$t_2 = \frac{k}{L}\, (H_C - H_B)$$
$$t_3 = \frac{k}{2L}\, (H_C + V_C) \qquad (B')$$
$$t_4 = \frac{k}{L}\, V_C$$
$$t_5 = \frac{k}{2L}\, (V_C - H_C)$$

Substituting in (A') yields:

$$(k/L)\, V_B = -y_B$$
$$(k/L)(H_C - H_B) = x_B$$
$$(k/L)(H_C - H_B) + \left(\sqrt{2}/2\right)(k/2L)(H_C + V_C) - \left(\sqrt{2}/2\right)(k/2L)(V_C - H_C)$$
$$= -x_C$$
$$\left(\sqrt{2}/2\right)(k/2L)(H_C + V_C) + (k/L)\, V_C + \left(\sqrt{2}/2\right)(k/2L)(V_C - H_C) = -y_C .$$

With some algebraic manipulation this becomes:

$$V_B = (-L/k)\, y_B$$
$$H_C - H_B = (L/k)\, x_B \qquad (C)$$
$$\frac{2+\sqrt{2}}{2}\, H_C - H_B = (-L/k)\, x_C$$
$$\left(\frac{2+\sqrt{2}}{2}\right) V_C = (-L/k)\, y_C .$$

729

The second equation in (C) is, simply $H_C = (L/k)x_B + H_B$ which can be substituted in the third equation.

$$\frac{2+\sqrt{2}}{2}\Big((L/k)x_B + H_B\Big) - H_B = -(L/k)x_C \ .$$

This implies $H_B = (L/k)\Big[(1 + \sqrt{2})x_B - \sqrt{2}\,x_C\Big]$ and then

$$H_C = (L/k)\Big[(2 + \sqrt{2})x_B - \sqrt{2}\,x_C\Big] \ .$$

Now substitute the values for V_C, V_B, H_C and H_B in the system (B') to find t_i , $i = 1,\dots,5$. A little more algebra yields the following

unique solution $t_1 = -y_B$, $t_2 = x_B$, $t_3 = \frac{1}{2}\left[(2 + \sqrt{2})x_b - \sqrt{2}\,x_C + \dfrac{2y_C}{2+\sqrt{2}}\right]$,

$t_4 = \dfrac{-2y_C}{2+\sqrt{2}}$ and $t_5 = \frac{1}{2}\left[\dfrac{2y_C}{2+\sqrt{2}} - (2 + \sqrt{2})x_B - \sqrt{2}\,x_C\right]$. These are the necessary

tensions for equilibrium.

iii)

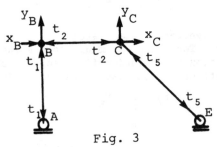

Fig. 3

In this case the equations of equilibrium for the joints B and C are

$$t_1 = -y_B$$

$$t_2 = x_B$$

$$t_2 - \frac{\sqrt{2}}{2}t_5 = -x_C$$

$$\frac{\sqrt{2}}{2}t_5 = -y_C \ .$$

Now we have 4 equations in 3 unknowns. In matrix notation, this system takes the form,

$$\begin{pmatrix} 1 & 0 & 0 \\ 0 & 1 & 0 \\ 0 & 1 & -\sqrt{2}/2 \\ 0 & 0 & \sqrt{2}/2 \end{pmatrix} \begin{pmatrix} t_1 \\ t_2 \\ t_5 \end{pmatrix} = \begin{pmatrix} -y_B \\ x_B \\ -x_C \\ -y_C \end{pmatrix}$$

or,

$$AX = B \tag{7}$$

where

$$\begin{pmatrix} 1 & 0 & 0 \\ 0 & 1 & 0 \\ 0 & 1 & -\sqrt{2}/2 \\ 0 & 0 & \sqrt{2}/2 \end{pmatrix}, \quad X = \begin{pmatrix} t_1 \\ t_2 \\ t_5 \end{pmatrix}, \quad B = \begin{pmatrix} -y_B \\ x_B \\ -x_C \\ -y_C \end{pmatrix}$$

To solve system (7), first write the augmented matrix $[A:B]$ of the system.
Then reduce it to echelon form. The augmented matrix of the system is

$$\begin{pmatrix} 1 & 0 & 0 & \vdots & -y_B \\ 0 & 1 & 0 & \vdots & x_B \\ 0 & 1 & -\sqrt{2}/2 & \vdots & -x_C \\ 0 & 0 & \sqrt{2}/2 & \vdots & -y_C \end{pmatrix}$$

Add the fourth row to the third row

$$\begin{pmatrix} 1 & 0 & 0 & \vdots & -y_B \\ 0 & 1 & 0 & \vdots & x_B \\ 0 & 1 & 0 & \vdots & -x_C - y_C \\ 0 & 0 & \sqrt{2}/2 & \vdots & -y_C \end{pmatrix}$$

Add -1 times the second row to the third row to obtain

$$\begin{pmatrix} 1 & 0 & 0 & \vdots & -y_B \\ 0 & 1 & 0 & \vdots & x_B \\ 0 & 0 & 0 & \vdots & -x_B - x_C - y_C \\ 0 & 0 & \sqrt{2}/2 & \vdots & -y_C \end{pmatrix}$$

which is the augmented matrix of the system

$$t_1 = -y_B$$

$$t_2 = x_B$$

$$0t_1 + 0t_2 + 0t_5 = x_B + x_C + y_C$$

$$\left(\sqrt{2}/2\right) t_5 = -y_C$$

This system only has a solution in the special case $x_B + y_C + x_C = 0$. This
reflects the physically obvious fact that the framework in figure 3 will
collapse, in general no matter what tensions are imposed unless the given
loads happen to be such that $x_B + y_C + x_C = 0$.

● **PROBLEM** 20-24

Find the reactions R_1, R_2 and R_3 for the truss shown in the figure
below. Assume that $P_1 = P_2 = P_3 = P_4 = P_5 = P_6 = 1 \cdot 0$

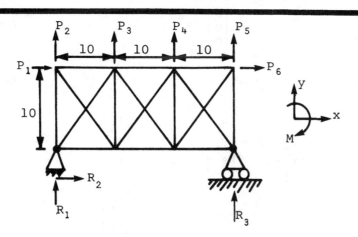

Solution: We first define truss. A truss is a structure made up of a number members fastened together at their ends in such a manner as to form a rigid body. A truss may be used to support a larger load or to span a greater distance than can be done effectively by a single beam or column. Now for external equilibrium, the sums of the forces of the system and sums of the moments of the forces with respect to a point (or to certain axes) must be equal to zero.

To formulate the problem mathematically, first write down the equilibrium equations. The external forces acting on the truss in the x direction are P_1, P_6, R_2 and in the y direction are P_2, P_3, P_4, P_5, R_1 and R_3. Now for external equilibrium the following equations must be satisfied.

$$\Sigma F_x = P_1 + P_6 + R_2 = 0$$

$$\Sigma F_y = P_2 + P_3 + P_4 + P_5 + R_1 + R_3 = 0 .$$

The moment of a force F with respect to a point A is defined as a vector with a magnitude equal to the product of the perpendicular distance from A to F and the magnitude of the force. The moment of a force about a point or axis is a measure of its tendency to turn or rotate a body about the point or axis.

For equilibrium the sum of the moments is equal to zero. When the moment is taken in the counterclockwise direction, it is positive according to the right-hand screw rule. Thus, taking moments about the left-hand support,

$$\Sigma M = 10P_1 - 10P_3 - 20P_4 - 30P_5 + 10P_6 - 30R_3 = 0 .$$

Thus for external equilibrium, the following equations are obtained:

$$R_2 + P_1 + P_6 = 0$$

$$R_1 + R_3 + P_2 + P_3 + P_4 + P_5 = 0 \qquad (1)$$

$$-30R_3 + 10P_1 - 10P_3 - 20P_4 - 30P_5 + 10P_6 = 0$$

This is a linear system of three equations. The system (1) can be written in matrix form as:

$$\begin{pmatrix} 0 & 1 & 0 \\ 1 & 0 & 1 \\ 0 & 0 & -30 \end{pmatrix} \begin{pmatrix} R_1 \\ R_2 \\ R_3 \end{pmatrix} + \begin{pmatrix} 1 & 0 & 0 & 0 & 0 & 1 \\ 0 & 1 & 1 & 1 & 1 & 0 \\ 10 & 0 & -10 & -20 & -30 & 10 \end{pmatrix} \begin{pmatrix} P_1 \\ P_2 \\ P_3 \\ P_4 \\ P_5 \\ P_6 \end{pmatrix}$$

$$= \begin{pmatrix} 0 \\ 0 \\ 0 \end{pmatrix} \qquad (2)$$

or

$$AR + BP = 0 \qquad (3)$$

where

$$A = \begin{pmatrix} 0 & 1 & 0 \\ 1 & 0 & 1 \\ 0 & 0 & -30 \end{pmatrix} \quad R = \begin{pmatrix} R_1 \\ R_2 \\ R_3 \end{pmatrix} , \quad P = \begin{pmatrix} P_1 \\ P_2 \\ P_3 \\ P_4 \\ P_5 \\ P_6 \end{pmatrix}$$

and

$$B = \begin{pmatrix} 1 & 0 & 0 & 0 & 0 & 1 \\ 0 & 1 & 1 & 1 & 1 & 0 \\ 10 & 0 & -10 & -20 & -30 & 10 \end{pmatrix} .$$

Equation (3) can be written as $AR = -BP$. Since the matrix A is non-singular, it has an inverse. Thus,

$$A^{-1}AR = -A^{-1}BP$$

But $A^{-1}A = I$ then

$$R = (-)A^{-1}BP .$$

Now,

$$A = \begin{pmatrix} 0 & 1 & 0 \\ 1 & 0 & 1 \\ 0 & 0 & -30 \end{pmatrix}$$

$$A^{-1} = \frac{1}{\det A} \text{ adj } A = \frac{1}{30} \begin{pmatrix} 0 & 30 & 1 \\ 30 & 0 & 0 \\ 0 & 0 & -1 \end{pmatrix}^{T}$$

$$= \begin{pmatrix} 0 & 1 & \frac{1}{30} \\ 1 & 0 & 0 \\ 0 & 0 & \frac{-1}{30} \end{pmatrix}$$

We are given that $P_1 = P_2 = P_3 = P_4 = P_5 = P_6 = 1 \cdot 0$; therefore,

$$R = (-) \begin{pmatrix} 0 & 1 & \frac{1}{30} \\ 1 & 0 & 0 \\ 0 & 0 & \frac{-1}{30} \end{pmatrix} \begin{pmatrix} 1 & 0 & 0 & 0 & 0 & 1 \\ 0 & 1 & 1 & 1 & 1 & 0 \\ 10 & 0 & -10 & -20 & -30 & 10 \end{pmatrix} \begin{pmatrix} 1 \\ 1 \\ 1 \\ 1 \\ 1 \\ 1 \end{pmatrix}$$

$$R = (-) \begin{pmatrix} 0 & 1 & \frac{1}{30} \\ 1 & 0 & 0 \\ 0 & 0 & \frac{-1}{30} \end{pmatrix} \begin{pmatrix} 2 \\ 4 \\ -40 \end{pmatrix} = (-) \begin{pmatrix} 8/3 \\ 2 \\ 4/3 \end{pmatrix}$$

But

$$R = \begin{pmatrix} R_1 \\ R_2 \\ R_3 \end{pmatrix} \qquad \text{thus,}$$

$$\begin{pmatrix} R_1 \\ R_2 \\ R_3 \end{pmatrix} = \begin{pmatrix} -8/3 \\ -2 \\ -4/3 \end{pmatrix}$$

Hence,

$$R_1 = -8/3, \; R_2 = -2, \; R_3 = -4/3 .$$

Consider a uniform strut of length ℓ under an axial force P at each end. Divide the strut into six equal intervals and let y_i denote the respective deflections.

Describe the bending of the strut by finding the possible deflections for y_i , $i = 1, 5$.

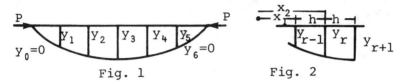

Fig. 1   Fig. 2

Solution: Assuming that there is symmetry about the midpoint one needs only to investigate half the strut.

Moving from left to right in figure 2, notice that the slope of the tangent to the displaced strut (d^2y/dx^2 , where x is the distance from the left to the point of deflection y) increases as y, the displacement, increases. This intuitively justifies the equation for the bending moment

$$\frac{d^2y}{dx^2} = Ky$$

where k is ome constant. In this case,

$$EI\ \frac{d^2y}{dx^2} = -Py \tag{1}$$

where t is the deflection at a distance x from one end, EI is the constant of rigidity of the strut and P is the axial force. A good approximation for d^2y/dx^2 is

$$\frac{y_{r+1} - 2y_r + y_{r-1}}{h^2} \tag{2}$$

(using the notation from figure 2).

Equation (2) can be applied to the given problem when considering $y = y_1$, y_2, y_3 and $h = \ell/6$. Since the strut is divided into six equal parts the results are:

$$\frac{d^2y_1}{dx_1^2} \approx \frac{y_2 - 2y_1 + y_0}{(\ell/6)^2} = \frac{y_2 - 2y}{\ell^2/36} \qquad (y_0 = 0)$$

$$\frac{d^2y_2}{dx_2^2} \approx \frac{y_3 - 2y_2 + y_1}{(\ell/6)^2} = \frac{y_3 - 2y_2 + y_1}{\ell^2/36}$$

$$\frac{d^2y_3}{dx_3^2} \approx \frac{y_4 - 2y_3 + y_2}{(\ell/6)^2} = \frac{2y_2 - 2y_2}{\ell^2/36} \qquad (y_4 = y_2)$$

Substitution of these approximations in equation (1) yields:

$$\frac{y_2 - 2y_1}{\ell^2/36} = \frac{-P}{EI}\ y_1$$

$$\frac{y_3 - 2y_2 + y_1}{\ell^2/36} = \frac{-P}{EI}\ y_2$$

$$\frac{2y_2 - 2y_3}{\ell^2/36} = \frac{-P}{EI} \, y_3$$

Substituting λ for the constant $P\ell^2/36EI$ these equations become

$$y_2 - 2y_1 = -\lambda y_1$$
$$y_3 - 2y_2 + y_1 = -\lambda y_2$$
$$2y_2 - 2y_3 = -\lambda y_3$$

or

$$2y_1 - y_2 \qquad\qquad = \lambda y_1$$
$$-y_1 + 2y_2 - y_3 \qquad = \lambda y_2$$
$$\qquad - 2y_2 + 2y_3 = \lambda y_3$$

In matrix notation this system is

$$\begin{bmatrix} 2 & -1 & 0 \\ -1 & 2 & -1 \\ 0 & -2 & 2 \end{bmatrix} \begin{pmatrix} y_1 \\ y_2 \\ y_3 \end{pmatrix} = \lambda \begin{pmatrix} y_1 \\ y_2 \\ y_3 \end{pmatrix}$$

which is of the form $AY = \lambda Y$. λ is an eigenvalue of A and the possible solutions for y are the eigenvectors of A. The eigenvalues of A are found by expanding $\det(\lambda I - A) =$

$$\det \begin{bmatrix} \lambda-2 & 1 & 0 \\ 1 & \lambda-2 & 1 \\ 0 & 2 & \lambda-2 \end{bmatrix} = (\lambda-2)^3 - [2(\lambda-2) + \lambda-2]$$

$$= (\lambda-2)(\lambda^2 - 4\lambda + 1).$$

The roots of this, the characteristic polynomial of A, are the eigenvalues of A. They are 2, $2-\sqrt{3}$ and $2 + \sqrt{3}$ (approximately 2, $.27$ and 3.73). To find the eigenvector of A associated with $\lambda = 2$ we must solve the equation $(2I - A)Y = 0$ which is

$$\begin{bmatrix} 0 & 1 & 0 \\ 1 & 0 & 1 \\ 0 & 2 & 0 \end{bmatrix} Y = 0 .$$

This yields the system

$$y_2 = 0$$
$$y_1 + y_3 = 0 .$$

A solution is $(1,0,-1)$. So if $(P\ell^2/36EI) = 2$, i.e., the axial pressure,

$$P = \frac{72EI}{\ell^2} ,$$

then the displacements y_1, y_2 and y_3 will be a multiple of $(1,0,-1)$. The strut might bend something like this:

Similarly with eigenvalues $2 - \sqrt{3}$, $2 + \sqrt{3}$ other solutions to the problem, can be determined.

CHEMICAL REACTIONS & MIXING PROBLEMS

A chemist needs 100 liters of a solution which is 25% hydrochloric acid (HCL). Suppose he has available a 10% HCL solution, a 20% HCL solution and a 40% HCL solution how many liters of each should be mixed to produce the required solution?

Solution: Let x_1, x_2 and x_3 be the amounts of 10, 20 and 40% solutions, respectively. Since 100 liters is required

$$x_1 + x_2 + x_3 = 100 . \tag{1}$$

Now the amount of HCL in 100 liters solution will be equal to $.25 \times 100 = 25$. Similarly, the amounts of HCL in 10%, 20% and 40% are $.1x_1$, $.2x_2$ and $.4x_3$, respectively. The material balance equation for HCL is

$$.1x_1 + .2x_2 + .4x_3 = 25 \tag{2}$$

Recall the following theorem. For the system of equations: $AX = B$ (A is the matrix of coefficients, X is the column vector of the variables and B is the column vector of constants) suppose

Rank A = Rank $[A:B] = r$, where $[A:B]$ (the augmented matrix) is formed by appending the column B to A. If m is the number of unknowns $m \leq r$ implies the system has a unique solution and $m > r$ implies there are infinitely many solutions. For the system

$$x_1 + x_2 + x_3 = 100$$

$$.1x_1 + .2x_2 + .4x_3 = 25$$

$$\text{Rank} \begin{pmatrix} 1 & 1 & 1 \\ .1 & .2 & .4 \end{pmatrix} = \text{Rank} \begin{pmatrix} 1 & 1 & 1 & \vdots & 100 \\ .1 & .2 & .4 & \vdots & 25 \end{pmatrix} = 2$$

and the number of unknowns is 3 so there are infinitely many solutions. Let $x_3 = a$ be any real constant. The system yields

$$x_1 + x_2 = 100 - a$$

$$x_1 + 2x_2 = 250 - 4a$$

which is equivalent to

$$x_1 = 2a - 50$$

$$x_2 = -3a + 150 .$$

Some possible solutions are the following:

liters 10% HCL	liters 20% HCL	liters 40% HCL(a)
10	60	30
20	45	35
25	37.5	37.5
26	36	38
46	6	48
50	0	50

Derive the conversion formula which can be used to change temperatures measured in degrees centigrade into degrees Fahrenheit. Then derive the reverse formula (i.e., change degrees Fahrenheit into degrees centigrade).

Solution: The Fahrenheit scale is simply the centigrade scale shifted and stretched along the axis of the scale. Hence the conversion function will be a linear function. Let the linear function be in the form

$$F^\circ = C^\circ x + b \qquad (1)$$

where F° is degrees Fahrenheit, C° is degrees centigrade, b is the amount of shift of the scale and x is the stretch or conversion factor. The freezing point of water, 0° centigrade, is 32° Fahrenheit and the boiling point of water, 100° centigrade is 212° Fahrenheit. Substituting these known values into equation (1)

$$0x + b = 32$$

$$100x + b = 212$$

or in matrix form

$$AX = B \quad (2) \quad \text{where}$$

$$A = \begin{pmatrix} 0 & 1 \\ 100 & 1 \end{pmatrix}, \quad X = \begin{pmatrix} x \\ b \end{pmatrix} \quad \text{and} \quad B = \begin{pmatrix} 32 \\ 212 \end{pmatrix}.$$

If A^{-1} exists, multiply equation (2) by A^{-1} on both sides to obtain $A^{-1}AX = A^{-1}B$ which is

$$X = A^{-1}B \qquad (3)$$

Now

$$A^{-1} = \frac{1}{\det A} \text{ adj } A,$$

$$A^{-1} = \frac{1}{0-100} \begin{bmatrix} 1 & -100 \\ -1 & 0 \end{bmatrix}^T$$

$$= \frac{-1}{100} \begin{bmatrix} 1 & -1 \\ -100 & 0 \end{bmatrix} = \begin{bmatrix} -1/100 & 1/100 \\ 1 & 0 \end{bmatrix}$$

Therefore equation (3) is

$$X = \begin{bmatrix} -1/100 & 1/100 \\ 1 & 0 \end{bmatrix} \begin{bmatrix} 32 \\ 212 \end{bmatrix} = \begin{bmatrix} -32/100 + 212/100 \\ 32 + 0 \end{bmatrix}$$

$$= \begin{bmatrix} 180/100 \\ 32 \end{bmatrix}$$

So,

$$X = \begin{bmatrix} x \\ b \end{bmatrix} = \begin{bmatrix} 9/5 \\ 32 \end{bmatrix},$$

or $x = 9/5$ and $b = 32$ which yields a conversion formula

$$F^\circ = C^\circ \frac{9}{5} + 32 \quad \text{(from equation (1))}.$$

To convert degrees Fahrenheit into degree centigrade, only solve this equation for C°.

$$F^\circ = C^\circ \frac{9}{5} + 32 \quad \text{implies} \quad C^\circ = (F^\circ - 32) \frac{5}{9}.$$

Determine the number of dimensionless groups in a problem of transient heat conduction in a rod with initial temperature T_i and end temperature T_0.

Solution: In a well-posed physical problem described by mathematical equations, the equations, and hence the solution may be written in terms of dimensionless groups, the number of dimensionless groups being considerably fewer than the number of variables and parameters in the system.

Let the physical quantities which have relevance in the problem be denoted by P_1, $P_2, \ldots, P_n$ where these may be entities like viscosity, thermal conductivity, surface tension, diameter, heat capacity, etc., and suppose the fundamental quantities are m_1, $m_2, \ldots, m_n$ where these are mass, length, time, temperature, etc. The expression $[P_j]$ will be the dimension of the physical quantity P_j so that

$$[P_j] = m_1^{\alpha_{ij}} m_2^{\alpha_{2j}} \ldots m_m^{\alpha_{mj}}$$

where α_{ij} might be the number of length units in P_j. The dimensional matrix may be written

	$[P_1]$	$[P_2]$	$[P_3]$	$[P_n]$
m_1	α_{11}	α_{12}	α_{13} . . .	α_{1n}
m_2	α_{21}	α_{22}	α_{23} . . .	α_{2n}
.				
.				
.				
m_m	α_{m1}	α_{m2}	α_{m3} . . .	α_{mn}

where α_{ij} is the number of fundamental quantities m_i involved in P_j (for example, if P_j were velocity and m_i were length, α_{ij} would be one). Considered as a vector space, the number of linear independent vectors is r and each of the remaining $n-r$ vectors can be expressed as a linear combination of the r independent ones, thus

$$[P_j] = \sum_{i=1}^{r} w_{ij}[P_i], \quad j = r+1, r+2, \ldots, n$$

where the w_{ij} are constants. The equation written in terms of the physical quantities themselves is:

$$P_j = \prod_{i=1}^{r} P_i^{w_{ij}}$$

or in other words

$$P_j^{-1} = \prod_{i=1}^{r} P_i^{w_{ij}}$$

is a dimensionless group. There cannot be more than $(n-r)$ independent ones, and, further, this is the minimum number of independent ones since one contains a quantity which does not appear in the other.

Consider the given problem, one is aware that in a problem of heat conduction in a rod in the transient state the length, initial temperature, heat capacity per unit of volume, thermal conductivity, time, end temperatures, and position in the rod are related, and we assume that the variables and parameters must appear in the final formula in dimensionless groups.

The pertinent variables with their dimensions are

$$
\begin{aligned}
L &= \text{length of rod} & [L] &= \ell \\
x &= \text{position of rod} & [x] &= \ell \\
\tau &= \text{time} & [\tau] &= t \\
c_p &= \text{heat capacity per unit} & & \\
& \quad \text{of volume} & [c_p] &= m/\ell t^2 \theta \\
T_0 - T_i &= \text{end temperatures} & [T_0 - T_1] &= \theta \\
T - T_i &= \text{temperature at position} & & \\
& \quad x \text{ at time } \tau, & [T - T_i] &= \theta \\
k &= \text{thermal conductivity} & [k] &= m\ell/t^3\theta
\end{aligned}
$$

where the fundamental units used are mass (m), length (ℓ), time (t), and temperature θ. Then the dimension matrix is

	L	x	τ	k	c_p	$T-T_i$	T_0-T_i
m	0	0	0	1	1	0	0
ℓ	1	1	0	1	-1	0	0
t	0	0	1	-3	-2	0	0
θ	0	0	0	-1	-1	1	1

This is a 4×8 matrix and can have at most 4 linearly independent rows.

The rank of this matrix is four, hence there must be three dimensionless groups. In this case the dimensionless groups may be determined by inspection of the table. A suitable set of three dimensionless groups is:

$$
\frac{T - T_i}{T_0 - T_i} \quad , \quad \frac{x}{L} \quad , \quad \frac{k\tau}{L^2 c_p} \quad .
$$

● **PROBLEM** 20-29

Consider a mixture of CO, H_2 and CH_4 fed into a furnace where it is burned with oxygen to form CO, CO_2, and H_2O. The following reactions are believed to occur:

$$
CO + \tfrac{1}{2}O_2 = CO_2
$$

$$
H_2 + \tfrac{1}{2}O_2 = H_2O
$$
(1)

$$
CH_4 + 2O_2 = CO_2 + 2H_2O
$$

$$
CH_4 + 3/2\,O_2 = CO + 2H_2O \ .
$$

Determine the minimum number of chemical reactions necessary to completely describe this system. Determine one set of independent reactions.

Solution: The given chemical reactions can be rewritten as

$$
CO + \tfrac{1}{2}O_2 - CO_2 = 0
$$

739

$$H_2 + \tfrac{1}{2} O_2 - H_2O \qquad = 0$$

$$CH_4 + 2O_2 - CO_2 - 2H_2O = 0 \qquad\qquad (2)$$

$$CH_4 + 3/2\, O_2 - CO - 2H_2O = 0 \ .$$

Now let us number the molecular species as follows:

$$A_1 = CO, \ A_2 = H_2 \ , \ A_3 = CH_4$$

$$A_4 = O_2 \ , \ A_5 = CO_2 \ , \ A_6 = H_2O$$

Then we may write the above reaction equations in the symbolic form as

$$A_1 \qquad\qquad + \tfrac{1}{2}A_4 - A_5 \ = \ 0$$

$$A_2 \qquad + \tfrac{1}{2}A_4 \quad - A_6 \ = \ 0$$

$$A_3 \quad + 2A_4 - A_5 - 2A_6 = 0$$

$$-A_1 + A_3 \qquad + 3/2\, A_4 - 2A_6 \quad = 0 \quad .$$

This is a linear system of four equations and six unknowns. The coefficient matrix of the above system is

$$A = \begin{pmatrix} 1 & 0 & 0 & \tfrac{1}{2} & -1 & 0 \\ 0 & 1 & 0 & \tfrac{1}{2} & 0 & -1 \\ 0 & 0 & 1 & 2 & -1 & -2 \\ -1 & 0 & 1 & 3/2 & 0 & -2 \end{pmatrix}$$

Recall that the non-zero rows of a matrix in echelon form are linearly
independent. It is also known that the number of non-zero rows in the
echelon form of a matrix is its rank. Thus the minimum number of primary
reactions is equal to the rank of the coefficient matrix A.

Now the matrix A may be reduced to echelon form using the elementary
row operations.

Add the first row to the fourth row.

$$\begin{pmatrix} 1 & 0 & 0 & \tfrac{1}{2} & -1 & 0 \\ 0 & 1 & 0 & \tfrac{1}{2} & 0 & -1 \\ 0 & 0 & 1 & 2 & -1 & -2 \\ 0 & 0 & 1 & 2 & -1 & -2 \end{pmatrix}$$

Add -1 times the third row to the fourth row.

$$\begin{pmatrix} 1 & 0 & 0 & \tfrac{1}{2} & -1 & 0 \\ 0 & 1 & 0 & \tfrac{1}{2} & 0 & -1 \\ 0 & 0 & 1 & 2 & -1 & -2 \\ 0 & 0 & 0 & 0 & 0 & 0 \end{pmatrix}$$

The above matrix is a row-reduced echelon matrix. Since the echelon matrix
has three non-zero rows, the rank of the matrix is three. Therefore there
are three independent reactions. Thus only three independent reactions
are needed to describe the reaction system, and these could be taken as
the first three equations in system (1). These reactions are

$$CO + \tfrac{1}{2} O_2 = CO_2$$

$$H_2 + \tfrac{1}{2} O_2 = H_2O$$

$$CH_4 + 2O_2 = CO_2 + 2H_2O \ .$$

During the catalytic oxidation of ammonia to nitric oxide the following reactions are believed to occur

$$4NH_3 + 5O_2 = 4NO + 6H_2O$$

$$4NH_3 + 3O_2 = 2N_2 + 6H_2O$$

$$4NH_3 + 6NO = 5N_2 + 6H_2O$$

$$2NO + O_2 = 2NO_2$$

$$2NO = N_2 + O_2$$

$$N_2 + 2O_2 = 2NO_2 \quad .$$

Determine the minimum number of chemical reactions necessary to completely
describe this system.

Solution: It is convenient to write the given equation in symbolic form as,

$$4NH_3 + 5O_2 - 4NO - 6H_2O = 0$$
$$4NH_3 + 3O_2 - 2N_2 - 6H_2O = 0$$
$$4NH_3 - 6NO - 5N_2 - 6H_2O = 0$$
$$2NO + O_2 - 2NO_2 = 0$$
$$2NO - N_2 - O_2 = 0$$
$$N_2 + 2O_2 - 2NO_2 = 0$$

Now, number the molecular species as follows: $A_1 = NH_3$, $A_2 = O_2$, $A_3 = NO$,
$A_4 = H_2O$, $A_5 = N_2$, and $A_6 = NO_2$. Then write the above reactions as the
linear equations

$$4A_1 + 5A_2 - 4A_3 - 6A_4 = 0$$
$$4A_1 + 3A_2 \quad\quad - 6A_4 - 2A_5 = 0$$
$$4A_1 \quad\quad + 6A_3 - 6A_4 - 5A_5 = 0$$
$$A_2 + 2A_3 \quad\quad\quad -2A_6 = 0$$
$$-A_2 + 2A_3 \quad -A_5 = 0$$
$$+2A_2 \quad\quad\quad +A_5 - 2A_6 = 0$$

This is a linear system of six equations and six unknowns. Now the coefficient matrix of the above equation is,

$$A = \begin{pmatrix} 4 & 5 & -4 & -6 & 0 & 0 \\ 4 & 3 & 0 & -6 & -2 & 0 \\ 4 & 0 & 6 & -6 & -5 & 0 \\ 0 & 1 & 2 & 0 & 0 & -2 \\ 0 & -1 & 2 & 0 & -1 & 0 \\ 0 & 2 & 0 & 0 & 1 & -2 \end{pmatrix}$$

Now if a set of reactions is linearly dependent (that is, each can be expressed as a linear combination of the others), some reactions are redundant and may be excluded from consideration. Always replace a system of equations by a system of independent equations, such as a system in echelon form. Then, the non-zero rows of a matrix in echelon form are linearly independent. Recall that the number of non-zero rows in the echelon form of a matrix is known as its rank. Thus the minimum number of primary reactions is equal to the rank of the coefficient matrix A.

Now the matrix A can be reduced to echelon form using the elementary row operations. Add -1 times the first row to the second and third rows to obtain:

$$\begin{pmatrix} 4 & 5 & -4 & -6 & 0 & 0 \\ 0 & -2 & 4 & 0 & -2 & 0 \\ 0 & -5 & 10 & 0 & -5 & 0 \\ 0 & 1 & 2 & 0 & 0 & -2 \\ 0 & -1 & 2 & 0 & -1 & 0 \\ 0 & 2 & 0 & 0 & 1 & -2 \end{pmatrix}$$

Divide the first row by 4.

$$\begin{pmatrix} 1 & 5/4 & -1 & -3/2 & 0 & 0 \\ 0 & -2 & 4 & 0 & -2 & 0 \\ 0 & -5 & 10 & 0 & -5 & 0 \\ 0 & 1 & 2 & 0 & 0 & -2 \\ 0 & -1 & 2 & 0 & -1 & 0 \\ 0 & 2 & 0 & 0 & 1 & -2 \end{pmatrix}$$

Divide the second row by -2

$$\begin{pmatrix} 1 & 5/4 & -1 & -3/2 & 0 & 0 \\ 0 & 1 & -2 & 0 & 1 & 0 \\ 0 & -5 & 10 & 0 & -5 & 0 \\ 0 & 1 & 2 & 0 & 0 & -2 \\ 0 & -1 & 2 & 0 & -1 & 0 \\ 0 & 2 & 0 & 0 & 1 & -2 \end{pmatrix}$$

Add 5 times the second row to the third row, -1 times the second row to the fourth row, 1 times the second row to the fifth row and -2 times the second row to the sixth row.

$$\begin{pmatrix} 1 & 5/4 & -1 & -3/2 & 0 & 0 \\ 0 & 1 & -2 & 0 & 1 & 0 \\ 0 & 0 & 0 & 0 & 0 & 0 \\ 0 & 0 & 4 & 0 & -1 & -2 \\ 0 & 0 & 0 & 0 & 0 & 0 \\ 0 & 0 & 4 & 0 & -1 & -2 \end{pmatrix}$$

Add -1 times the fourth row to the sixth row to obtain,

$$\begin{pmatrix} 1 & 5/4 & -1 & -3/2 & 0 & 0 \\ 0 & 1 & -2 & 0 & 1 & 0 \\ 0 & 0 & 0 & 0 & 0 & 0 \\ 0 & 0 & 4 & 0 & -1 & -2 \\ 0 & 0 & 0 & 0 & 0 & 0 \\ 0 & 0 & 0 & 0 & 0 & 0 \end{pmatrix}$$

Interchanging the third and fourth rows yields,

$$\begin{pmatrix} 1 & 5/4 & -1 & -3/2 & 0 & 0 \\ 0 & 0 & -2 & 0 & 1 & 0 \\ 0 & 0 & 4 & 0 & -1 & -2 \\ 0 & 0 & 0 & 0 & 0 & 0 \\ 0 & 0 & 0 & 0 & 0 & 0 \\ 0 & 0 & 0 & 0 & 0 & 0 \end{pmatrix}$$

Since the echelon form matrix has three non-zero rows, the rank of the matrix is three. Therefore, we need three primary chemical reactions to completely describe the given system. Note we would have to choose three independent reactions.

● **PROBLEM** 20-31

A chemical reaction is governed by the equations

$$A_1 + A_2 \xrightarrow{\;k_1\;} A_3$$

$$A_3 + A_2 \xrightarrow{\;k_2\;} A_4$$

$$A_5 + A_2 \xrightarrow{\;k_3\;} A_6$$

$$A_6 + A_2 \xrightarrow{\;k_4\;} A_7$$

where A_4 is the desired product and A_6 and A_7 are undesirable by-products. These reactions are assumed to proceed at constant volume and temperature. Find the number of independent kinetic differential equations which are required to describe the reaction.

Solution: The rate of change of the concentration of each substance in any given reaction is proportional to the product of the concentrations of the substances being used up in the reaction. Consider the reaction $A + B \rightarrow C + D$. Then, the change in the concentration of A with respect to time is given by

$$\frac{dx_A}{dt} = -kx_A x_B$$

where k is the rate constant or reaction rate constant. x_A and x_B are concentrations of A and B (in moles per unit volume) respectively. Returning to the given chemical reactions, let x_i be the concentration of A_i in moles per unit volume and k_i be the rate constant in liters per gram-mole second. Then, the differential equations for the chemical reactions are as follows:

$$\frac{dx_1}{dt} = -k_1 x_1 x_2$$

$$\frac{dx_2}{dt} = -k_1 x_1 x_2 - k_2 x_2 x_3 - k_3 x_2 x_5 - k_4 x_2 x_6$$

$$\frac{dx_3}{dt} = k_1 x_1 x_2 - k_2 x_2 x_3$$

$$\frac{dx_4}{dt} = k_2 x_2 x_3$$

$$\frac{dx_5}{dt} = -k_3 x_2 x_5$$

$$\frac{dx_6}{dt} = k_3 x_2 x_5 - k_4 x_2 x_6$$

$$\frac{dx_7}{dt} = k_4 x_2 x_6 \tag{1}$$

We arrange the system (1) in matrix form, symbolically, in the following way:

$$
\frac{d}{dt}
\begin{pmatrix} x_1 \\ x_2 \\ x_3 \\ x_4 \\ x_5 \\ x_6 \\ x_7 \end{pmatrix}
=
\begin{pmatrix}
-k_1 & 0 & 0 & 0 \\
-k_1 & -k_2 & -k_3 & -k_4 \\
k_1 & -k_2 & 0 & 0 \\
0 & k_2 & 0 & 0 \\
0 & 0 & -k_3 & 0 \\
0 & 0 & k_3 & -k_4 \\
0 & 0 & 0 & k_4
\end{pmatrix}
\begin{matrix} x_1 x_2 & x_2 x_3 & x_2 x_5 & x_2 x_6 \end{matrix}
\tag{2}
$$

(column labels: $x_1 x_2$, $x_2 x_3$, $x_2 x_5$, $x_2 x_6$)

We now wish to reduce the matrix on the right to diagonal form by elementary row operations. The element in the upper left corner is non-zero. The remaining non-zero elements in the first column can be eliminated in the following manner. Add -1 times the first row to the second row and add the first row to the third row. Thus (2) becomes

$$
\frac{d}{dt}
\begin{pmatrix} x_1 \\ x_2 - x_1 \\ x_3 + x_1 \\ x_4 \\ x_5 \\ x_6 \\ x_7 \end{pmatrix}
=
\begin{pmatrix}
-k_1 & 0 & 0 & 0 \\
0 & -k_2 & -k_3 & -k_4 \\
0 & -k_2 & 0 & 0 \\
0 & k_2 & 0 & 0 \\
0 & 0 & -k_3 & 0 \\
0 & 0 & k_3 & -k_4 \\
0 & 0 & 0 & k_4
\end{pmatrix}
\begin{matrix} x_1 x_2 & x_2 x_3 & x_2 x_5 & x_2 x_6 \end{matrix}
\tag{3}
$$

(column labels: $x_1 x_2$, $x_2 x_3$, $x_2 x_5$, $x_2 x_6$)

By inspection note that the simplest way to obtain a diagonal matrix is to interchange the second and third rows, then add the fifth row to the sixth row. This gives the new system,

$$
\frac{d}{dt}
\begin{pmatrix}
x_1 \\
x_3 + x_1 \\
x_2 - x_1 \\
x_4 \\
x_5 \\
x_5 + x_6 \\
x_7
\end{pmatrix}
=
\begin{array}{cccc}
x_1x_2 & x_2x_3 & x_2x_5 & x_2x_6 \\
\end{array}
\begin{pmatrix}
-k_1 & 0 & 0 & 0 \\
0 & -k_2 & 0 & 0 \\
0 & -k_2 & -k_3 & -k_4 \\
0 & k_2 & 0 & 0 \\
0 & 0 & -k_3 & 0 \\
0 & 0 & 0 & -k_4 \\
0 & 0 & 0 & k_4
\end{pmatrix}
$$

Now interchange rows three and five and rows four and six. Thus

$$
\frac{d}{dt}
\begin{pmatrix}
x_1 \\
x_3 + x_1 \\
x_5 \\
x_5 + x_6 \\
x_2 - x_1 \\
x_4 \\
x_7
\end{pmatrix}
=
\begin{array}{cccc}
x_1x_2 & x_2x_3 & x_2x_5 & x_2x_6 \\
\end{array}
\begin{pmatrix}
-k_1 & 0 & 0 & 0 \\
0 & -k_2 & 0 & 0 \\
0 & 0 & -k_3 & 0 \\
0 & 0 & 0 & -k_4 \\
0 & -k_2 & -k_3 & -k_4 \\
0 & k_2 & 0 & 0 \\
0 & 0 & 0 & k_4
\end{pmatrix}
\qquad (4)
$$

Now add −1 times the second row, −1 times the third row and −1 times the fourth row to the fifth row. Then add the second row to the sixth row and the fourth row to the seventh row. Thus system (4) becomes

$$
\frac{d}{dt}
\begin{pmatrix}
x_1 \\
x_1 + x_3 \\
x_5 \\
x_5 + x_6 \\
(-3x_1 + x_2 - x_3 \\ \quad -2x_5 - x_6) \\
x_1 + x_3 + x_4 \\
x_5 + x_6 + x_7
\end{pmatrix}
=
\begin{array}{cccc}
x_1x_2 & x_2x_3 & x_2x_5 & x_2x_6 \\
\end{array}
\begin{pmatrix}
-k_1 & 0 & 0 & 0 \\
0 & -k_2 & 0 & 0 \\
0 & 0 & -k_3 & 0 \\
0 & 0 & 0 & -k_4 \\
0 & 0 & 0 & 0 \\
0 & 0 & 0 & 0 \\
0 & 0 & 0 & 0
\end{pmatrix}
$$

Thus there are four independent equations and these equations are as follows:

$$\frac{dx_1}{dt} = - k_1 x_1 x_2$$

$$\frac{d(x_1 + x_3)}{dt} = - k_2 x_2 x_3$$

$$\frac{dx_5}{dt} = - k_3 x_2 x_5$$

$$\frac{d(x_5 + x_6)}{dt} = - k_4 x_2 x_6 \ .$$

● **PROBLEM** 20-32

In a given plant operation, four streams are mixed to form a single stream with the desired composition. The four inlet streams to the mixer and the single exit stream (final product) have compositions as shown in the figure. Determine the mass flow rate of the individual streams for making 2000 lb/hr. of final product.

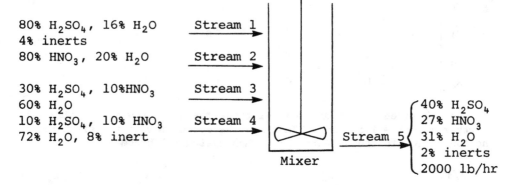

Mixer

Solution: In order to obtain algebraic equations, we shall make mass balances of each component. Mass balances are a direct application of the law of conservation of mass. This law states that matter is neither created nor destroyed. Now, let

$$
\begin{aligned}
w &= \text{stream 1 \ lb/hr.} \\
x &= \text{stream 2 \ lb/hr.} \\
y &= \text{stream 3 \ lb/hr.} \\
z &= \text{stream 4 \ lb/hr.}
\end{aligned}
$$

Now the mass flow rate of the exit stream is 2000 lb/hr. From the given composition of the exit stream, calculate the mass flow rate of each component. The exit stream has the following composition:

40% H_2SO_4 , 27% HNO_3 , 31% H_2O and 2% inert.

Therefore, the mass flow rate of H_2SO_4 is 40% of 2000, .4(2000) = 800 lb/hr.
Similarly, mass flow rate of HNO_3 = 2000 × .27 = 540 lb/hr.
mass flow rate of H_2O = 2000 × .31 = 620 lb/hr.

mass flow rate of inert = 2000 × .02 = 40   lb/hr.

Mass balances are based on the law of conservation of mass. For a given system,

(Rate of accumulation) = (rate of mass input)

- (rate of mass output)

For a steady state process, there can be no accumulation in the system, and the mass balance reduces to

input = output   (1)

We shall apply equation (1), to each component to obtain algebraic equations. The total mass flow rate of H_2SO_4 is 800 lb/hr. Stream 1 flows at w lb/hr and 80% of stream 1 is H_2SO_4 so stream 1 contributes 80% of w or .8w of the H_2SO_4 rate to the final exit rate of H_2SO_4 . Similarly, stream 2 contributes nothing to the flow of H_2SO_4 , stream 3 contributes 30% of y to the final flow rate and stream 4 contributes 10% of z to the final flow rate. Thus the H_2SO_4 mass balance is

$$.8w + 0x + 0.3y + 0.1z = 800 \qquad\qquad (2)$$

In the same manner we find the other rate contributions by balancing the masses. HNO_3 mass balance:

$$0w + .8x + 0.1y + 0.1z = 540 \qquad\qquad (3)$$

H_2O mass balance:

$$0.16w + 0.2x + 0.6y + 0.72z = 620 \qquad\qquad (4)$$

Inert mass balance:

$$0.04w + 0x + 0y + 0.08z = 40 \qquad\qquad (5)$$

We have obtained four linear equations in four unknowns. As we know, this system of linear equations can be solved by using the Gauss-Jordan method. The equations (1) to (5) can be expressed in matrix form as

$$\begin{bmatrix} .8 & 0 & 0.3 & 0.1 \\ 0 & 0.8 & 0.1 & 0.1 \\ .16 & .2 & 0.6 & .72 \\ .04 & 0 & 0 & .08 \end{bmatrix} \begin{pmatrix} w \\ x \\ y \\ z \end{pmatrix} = \begin{pmatrix} 800 \\ 540 \\ 620 \\ 40 \end{pmatrix}$$

or

$$AX = B \qquad\qquad (6)$$

where

$$A = \begin{bmatrix} .8 & 0 & 0.3 & 0.1 \\ 0 & 0.8 & .1 & 0.1 \\ .16 & .2 & .6 & .72 \\ .04 & 0 & 0 & .08 \end{bmatrix}$$

$$B = \begin{pmatrix} 800 \\ 540 \\ 620 \\ 40 \end{pmatrix}$$

The procedure for solving system (6), is as follows: Form the augmented matrix [A:B] of the system and proceed to reduce this matrix to a form

[I:B']. The augmented matrix of the system is:

$$\left[\begin{array}{cccc|c} .8 & 0 & .3 & .1 & 800 \\ 0 & .8 & .1 & .1 & 540 \\ .16 & .2 & .6 & .72 & 620 \\ .04 & 0 & 0 & .08 & 40 \end{array}\right]$$

Now perform the following row operations divide the first row by .8

$$\left[\begin{array}{cccc|c} 1 & 0 & .375 & .125 & 1000 \\ 0 & .8 & .1 & .1 & 540 \\ .16 & .2 & .6 & .72 & 620 \\ .04 & 0 & 0 & .08 & 40 \end{array}\right]$$

Add −.16 times the first row to the third row, and −.04 times the first row to the fourth row.

$$\left[\begin{array}{cccc|c} 1 & 0 & .375 & .125 & 1000 \\ 0 & .8 & .1 & .1 & 540 \\ 0 & .2 & .540 & .7 & 460 \\ 0 & 0 & -.015 & .075 & 0 \end{array}\right]$$

Divide the second row by .8

$$\left[\begin{array}{cccc|c} 1 & 0 & .375 & .125 & 1000 \\ 0 & 1 & .125 & .125 & 675 \\ 0 & .2 & .54 & .7 & 460 \\ 0 & 0 & -.015 & .075 & 0 \end{array}\right]$$

Add −0.2 times the second row to the third row.

$$\left[\begin{array}{cccc|c} 1 & 0 & 0.375 & .125 & 1000 \\ 0 & 1 & 0.125 & .125 & 675 \\ 0 & 0 & 0.515 & 0.675 & 325 \\ 0 & 0 & -0.015 & 0.075 & 0 \end{array}\right]$$

Divide the third row by 0.515.

$$\left[\begin{array}{cccc|c} 1 & 0 & 0.375 & .125 & 1000 \\ 0 & 1 & 0.125 & .125 & 675 \\ 0 & 0 & 1 & 1.310 & 631 \\ 0 & 0 & -0.015 & .075 & 0 \end{array}\right]$$

Add (i) −0.375 times the third row to the first row, (ii) −0.125 times the third row to the second row and (iii) .015 times the third row to the fourth row

$$\left[\begin{array}{cccc|c} 1 & 0 & 0 & -0.36625 & 763.375 \\ 0 & 1 & 0 & -0.03875 & 596.125 \\ 0 & 0 & 1 & 1.31000 & 631 \cdot 000 \\ 0 & 0 & 0 & 0.09465 & 9.465 \end{array}\right]$$

Divide the fourth row by 0.09465. Then add (i) −1.31 times the fourth row to the third row, (ii) +.03875 times the fourth row to the second row and (iii) 0.36625 times the fourth row to the first row.

$$
\begin{bmatrix}
1 & 0 & 0 & 0 & | & 800 \\
0 & 1 & 0 & 0 & | & 600 \\
0 & 0 & 1 & 0 & | & 500 \\
0 & 0 & 0 & 1 & | & 100
\end{bmatrix}
$$

We have reduced the matrix $[A:B]$ to a desired form $[I:B']$. The corresponding system of equations is therefore,

$$
\begin{aligned}
w &= 800 \\
x &= 600 \\
y &= 500 \\
z &= 100
\end{aligned}
$$

Thus, the mass flow rate of the individual streams for making 2000 lb/hr. of final product are as follows:

$$
\begin{aligned}
w &= \text{stream 1} = 800 \text{ lb/hr.} \\
x &= \text{stream 2} = 600 \text{ lb/hr.} \\
y &= \text{stream 3} = 500 \text{ lb/hr.} \\
z &= \text{stream 4} = 100 \text{ lb/hr.}
\end{aligned}
$$

● **PROBLEM** 20-33

Figure 1 shows two 50-gallon tanks connected by flow pipes with inlet and outlet all having the rates of flow as marked in gallons per minute (gal./m). The flow rates are arranged so that each tank is maintained at its capacity at all times. The left-hand tank receives salt solution at a concentration of 1 pound per gallon, and the right-hand tank receives pure water. How much salt is in each tank at any time $t > 0$?

Fig. 1

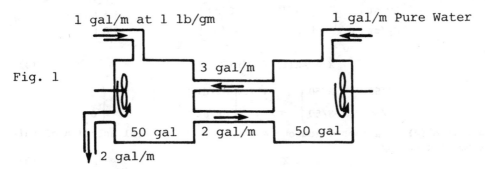

Solution: We assume that each tank is kept thoroughly mixed at all times, so that the concentration of salt is the same throughout the whole tank at any time. Let x and y denote the amount of salt in the left-hand tank and right-hand tank respectively, at time t. The rate of change of the amount of salt in the left tank is then dx/dt. If IN denotes the rate at which salt enters the tank and OUT denotes the rate at which it leaves, the basic equation is

$$\frac{dx}{dt} = \text{IN} - \text{OUT} . \qquad (1)$$

Since each tank is maintained at its capacity at all times, each tank contains 50 gal. of the salt solution at any time t. Now the left-hand 50 gal. tank contains x lb. of salt at time t, and so the concentration of salt at time t is

$$\frac{1}{50} x \text{ lb/gal.}$$

Similarly, the concentration of salt in the right-hand tank at any time t

749

is

$$\frac{1}{50} \, y \text{ lb/gal.}$$

The salt solution enters the left-hand tank from the right-hand tank, at the rate of 3 gal/min. and each gallon contains y/50 lb. of salt. Also the left-hand tank receives salt solution at the rate of 1 gal/min., and each gallon contains 1 lb. of salt. Thus,

$$\text{IN} = \left(3 \, \frac{\text{gal.}}{\text{min.}}\right)\left(\frac{y}{50} \, \frac{\text{lb}}{\text{gal}}\right) + \left(1 \, \frac{\text{gal.}}{\text{min.}}\right)\left(1 \, \frac{\text{lb}}{\text{gal.}}\right)$$

$$= \frac{3}{50} \, y + 1 \text{ lb/min.}$$

Since the salt solution flows out from the left-hand tank at the rate of 2 gal/min. and there are two out flows, hence,

$$\text{OUT} = \left(2 \, \frac{\text{gal.}}{\text{min.}}\right)\left(\frac{x}{50} \, \frac{\text{lb}}{\text{gal.}}\right) + \left(2 \, \frac{\text{gal.}}{\text{min.}}\right)\left(\frac{x}{50} \, \frac{\text{lb}}{\text{gal.}}\right)$$

$$= \frac{4}{50} \, x \, \frac{\text{lb}}{\text{min.}}$$

Then differential equation (1) becomes,

$$\frac{dx}{dt} = \frac{3}{50} \, y + 1 - \frac{4}{50} \, x$$

or

$$\frac{dx}{dt} = -\frac{4}{50} \, x + \frac{3}{50} \, y + 1 \tag{2}$$

In the same manner, obtain the following equation for the right-hand tank.

$$\frac{dy}{dt} = \frac{2}{50} \, x - \frac{3}{50} \, y \ . \tag{3}$$

Thus, a system of differential equations, can be written in matrix notation as

$$\begin{pmatrix} dx/dt \\ dy/dt \end{pmatrix} = \begin{bmatrix} -4/50 & 3/50 \\ 2/50 & -3/50 \end{bmatrix} \begin{pmatrix} x \\ y \end{pmatrix} + \begin{pmatrix} 1 \\ 0 \end{pmatrix}$$

or

$$\dot{X} = AX(t) + F(t) \tag{4}$$

where

$$A = \begin{bmatrix} -4/50 & 3/50 \\ 2/50 & -3/50 \end{bmatrix}, \quad \dot{X} = \begin{pmatrix} dx/dt \\ dy/dt \end{pmatrix}, \quad F(t) = \begin{pmatrix} 1 \\ 0 \end{pmatrix}$$

The system (4) is a nonhomogeneous system and the general solution of this system is given by

$$X(t) = X_h + X_p \ . \tag{5}$$

where X_h is a solution of the corresponding homogeneous system and X_p is the particular solution.

The corresponding homogeneous system is $\dot{X} = AX$, and the general solution is given by

$$X_h(t) = c_1 e^{\lambda_1 t} V_1 + c_2 e^{\lambda_2 t} V_2$$

where

(i) c_1 and c_2 are arbitrary constants.

(ii) λ_1 and λ_2 are distinct eigenvalues of the matrix A.

(iii) V_1 and V_2 are linearly independent eigenvectors corresponding to the eigenvalues λ_1 and λ_2 respectively.

First find the eigenvalues of A and then obtain eigenvectors associated with the eigenvalues. The characteristic polynomial of A is

$$f(\lambda) = \det[\lambda I - A] = \det \begin{pmatrix} \lambda + 4/50 & -3/50 \\ -2/50 & \lambda + 3/50 \end{pmatrix}$$

$$= (\lambda + 4/50)(\lambda + 3/50) - (3/50)(2/50)$$

$$= \lambda^2 + (7/50)\lambda + 12/(50)^2 - 6/(50)^2$$

$$= (\lambda + 1/50)(\lambda + 6/50)$$

The characteristic equation is then $(\lambda + 1/50)(\lambda + 6/50) = 0$. Therefore, the eigenvalues are $\lambda_1 = -1/50$, and $\lambda_2 = -6/50$.

$$V = \begin{pmatrix} v_1 \\ v_2 \end{pmatrix}$$ is an eigenvector of A corresponding to λ if and only if

V is a nontrivial solution of $(\lambda I - A)V = 0$, that is of

$$\begin{pmatrix} \lambda + 4/50 & -3/50 \\ -2/50 & \lambda + 3/50 \end{pmatrix} \begin{pmatrix} v_1 \\ v_2 \end{pmatrix} = \begin{pmatrix} 0 \\ 0 \end{pmatrix} \qquad (6)$$

For $\lambda = -1/50$, the system (6) becomes

$$\begin{pmatrix} 3/50 & -3/50 \\ -2/50 & 2/50 \end{pmatrix} \begin{pmatrix} v_1 \\ v_2 \end{pmatrix} = \begin{pmatrix} 0 \\ 0 \end{pmatrix}$$

or

$$\begin{array}{ll} (3/50)\,v_1 - (3/50)\,v_2 = 0 \\ (-2/50)\,v_1 + (2/50)\,v_2 = 0 \end{array} \Longrightarrow \begin{array}{ll} 3v_1 - 3v_2 = 0 \\ -2v_1 + 2v_2 = 0 \end{array}$$

This implies that $v_1 = v_2$. Therefore $V_1 = \begin{pmatrix} 1 \\ 1 \end{pmatrix}$ is an eigenvector as-

sociated with $\lambda_1 = -1/50$. For $\lambda_2 = -6/50$, the system (6) becomes,

$$\begin{pmatrix} -2/50 & -3/50 \\ -2/50 & -3/50 \end{pmatrix} \begin{pmatrix} v_1 \\ v_2 \end{pmatrix} = \begin{pmatrix} 0 \\ 0 \end{pmatrix}$$

or

$$-2/50\,v_1 - 3/50\,v_2 = 0$$

$$-2/50\,v_1 - 3/50\,v_2 = 0 \ .$$

This implies that $v_1 = -3/2\,v_2$. Hence

$$V_2 = \begin{pmatrix} -3/2 \\ 1 \end{pmatrix}$$ is an eigenvector associated with

$\lambda_2 = -6/50$. Thus,

$$X_h(t) = c_1 e^{(-1/50)t} \begin{pmatrix} 1 \\ 1 \end{pmatrix} + c_2 e^{(-6/50)t} \begin{pmatrix} -3/2 \\ 1 \end{pmatrix}$$

or

$$X_h(t) = c_1 \begin{pmatrix} e^{(-1/50)t} \\ e^{(-1/50)t} \end{pmatrix} + c_2 \begin{pmatrix} (-3/2)\,e^{(-6/50)t} \\ e^{(-6/50)t} \end{pmatrix}$$

751

To find a particular solution, we try

$$\begin{pmatrix} x_p(t) \\ y_p(t) \end{pmatrix} = \begin{pmatrix} a \\ b \end{pmatrix}$$

which we substitute into the given system:

$$0 = (-4/50)\, a + (3/50)\, b + 1$$
$$0 = (2/50)\, a - (3/50)\, b$$

solving this system yields:

$$a = 50/2 \ , \ b = 50/3 \ .$$

Then,

$$X_p(t) = \begin{pmatrix} 50/2 \\ 50/3 \end{pmatrix}$$

Substituting the values of $X_h(t)$ and $X_p(t)$ into equation (5), gives

$$X(t) = c_1 \begin{pmatrix} e^{-(1/50)\,t} \\ e^{-(1/50)\,t} \end{pmatrix} + c_2 \begin{pmatrix} -(3/2)\, e^{-(6/50)\,t} \\ e^{-(6/50)\,t} \end{pmatrix} + \begin{pmatrix} 50/2 \\ 50/3 \end{pmatrix}$$

or

$$\begin{pmatrix} x(t) \\ y(t) \end{pmatrix} = \begin{pmatrix} c_1 e^{-(1/50)\,t} - (3/2)\, c_2 e^{-(6/50)\,t} + 50/2 \\ c_1 e^{-(1/50)\,t} + c_2 e^{-(6/50)\,t} + 50/3 \end{pmatrix}$$

Thus, the general solution of system (4) is

$$x(t) = c_1 e^{(1/50)\,t} - 3/2\, c_2 e^{-(6/50)\,t} + 50/2$$

$$y(t) = c_1 e^{(1/50)\,t} + c_2 e^{-(6/50)\,t} + 50/3$$

From these equations we can obtain the salt concentration in pounds per gallon at any time > 0 . The constants c_1 and c_2 depend on the initial values $x(0)$ and $y(0)$. Thus, the equations

$$x(0) = c_1 - (3/2) c_2 + 50/2$$

$$y(0) = c_1 + c_2 + 50/3$$

determine c_1 and c_2 when $x(0)$ and $y(0)$ are the known amounts of salt at $t = 0$.

752

CHAPTER 21

APPLICATIONS TO PHYSICAL SYSTEMS II (ELECTRICAL)

KIRCHHOFF'S VOLTAGE LAW

● **PROBLEM** 21-1

Calculate the values of I_1 and I_2 for the circuit in figure 1, if R_1 = 4 ohms, and R_2 = R_3 = 3 ohms.

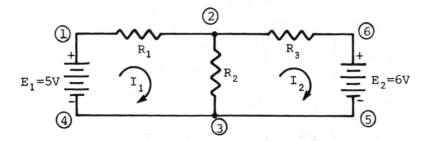

Solution: Kirchhoff's voltage law states that around any closed loop of a circuit the sum of the voltage drops must be 0. The loop ① ② ③ ④ ① has a battery which provides a voltage E_1. The equation V = IR (Ohm's law) says that when a current I passes through a resistance R the voltage drop is the product IR. Thus, when Kirchhoff's voltage law is applied to the loop ① ② ③ ④ ①, the following equation is obtained.

$$E_1 - I_1 R_1 - (I_1 - I_2) R_2 = 0$$

or

$$E_1 - I_1(R_1 + R_2) + I_2 R_2 = 0 .$$

Next, substitute the given values of E_1, R_1, and R_2 in this equation. Then,

$$5 - 7I_1 + 3I_2 = 0$$

or

$$7I_1 - 3I_2 = 5 . \tag{1}$$

Similarly, applying Kirchhoff's law to the loop ⑤ ③ ② ⑥ ⑤ results in equation (2)

$$3I_1 - 6I_2 = 6 . \tag{2}$$

Thus, we have obtained two linear equations in two unknown currents. In matrix notation, the equations (1) and (2) take the form

$$\begin{bmatrix} 7 & -3 \\ 3 & -6 \end{bmatrix} \begin{bmatrix} I_1 \\ I_2 \end{bmatrix} = \begin{bmatrix} 5 \\ 6 \end{bmatrix} \tag{3}$$

Let

$$A = \begin{bmatrix} 7 & -3 \\ 3 & -6 \end{bmatrix} \qquad X = \begin{bmatrix} I_1 \\ I_2 \end{bmatrix} \qquad \text{and} \qquad B = \begin{bmatrix} 5 \\ 6 \end{bmatrix}$$

then equation (3) becomes

$$AX = B . \tag{4}$$

We know that if A is invertible, we can solve the equation (4) for I by multiplying by A^{-1}. Thus,

$$A^{-1}AX = A^{-1}B ,$$

also, $A^{-1}A = I$, therefore,

$$IX = A^{-1}B . \tag{5}$$

Recall that if $\det A \neq 0$, then A is invertible. Now,

$$\det A = \det \begin{bmatrix} 7 & -3 \\ 3 & -6 \end{bmatrix}$$

$$= -42 + 9 = -33.$$

Since $\det A \neq 0$, A is invertible. Then,

$$A^{-1} = \frac{1}{\det A} \text{ adj } A .$$

Adj A is the transpose of the matrix of cofactors.

$$\text{Adj } A = \begin{bmatrix} -6 & -3 \\ 3 & 7 \end{bmatrix}^T = \begin{bmatrix} -6 & 3 \\ -3 & 7 \end{bmatrix}$$

So,

$$A^{-1} = \frac{1}{-33} \begin{bmatrix} -6 & 3 \\ -3 & 7 \end{bmatrix} .$$

Substitute the values of A^{-1} and B into (5). Then,

$$X = \frac{1}{-33} \begin{bmatrix} -6 & 3 \\ -3 & 7 \end{bmatrix} \begin{bmatrix} 5 \\ 6 \end{bmatrix}$$

$$= \frac{1}{-33} \begin{bmatrix} -30 + 18 \\ -15 + 42 \end{bmatrix}$$

$$X = \frac{1}{-33} \begin{bmatrix} -12 \\ 27 \end{bmatrix}$$

$$X = \begin{bmatrix} 4/11 \\ -9/11 \end{bmatrix}$$

or

$$\begin{bmatrix} I_1 \\ I_2 \end{bmatrix} = \begin{bmatrix} 4/11 \\ -9/11 \end{bmatrix}$$

Therefore, the values of the two currents I_1 and I_2 are

$$I_1 = 4/11 \text{ amps.}$$

and

$$I_2 = -9/11 \text{ amps.}$$

Consider the electric circuit of figure 1, given that $R_1 = 10$ ohms, $R_2 = 30$ ohms, $R_3 = 20$ ohms, $E_1 = 6V$ and $E_2 = 24V$, find the values of currents I_1 and I_2 .

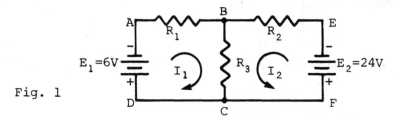

Fig. 1

Solution: Applying Kirchhoff's voltage law to the loops A B C D A and E B C F E , we have

$$E_1 - I_1R_1 - (I_1+I_2)R_3 = 0 \qquad (1)$$

and

$$E_2 - I_2R_2 - (I_2+I_1)R_3 = 0 \qquad (2)$$

Combining terms and rearranging, equations (1) and (2) become

$$E_1 - (R_1+R_3)I_1 - R_3I_2 = 0$$

$$E_2 - R_3I_1 - (R_2+R_3)I_2 = 0 .$$

Substitute the given values of R_1, R_2, R_3, E_1 and E_2 . Then,

$$6 - 30I_1 - 20I_2 = 0$$

$$24 - 20I_1 - 50I_2 = 0$$

or

$$30I_1 + 20I_2 = 6$$

$$20I_1 + 50I_2 = 24 \qquad (3)$$

Now divide each equation by 10 in order to keep the numbers small. Thus,

$$3I_1 + 2I_2 = .6$$

$$2I_1 + 5I_2 = 2.4$$

This system can be written in matrix notation as

$$\begin{bmatrix} 3 & 2 \\ 2 & 5 \end{bmatrix} \begin{bmatrix} I_1 \\ I_2 \end{bmatrix} = \begin{bmatrix} .6 \\ 2.4 \end{bmatrix}$$

or

$$AX = B \qquad (4)$$

where

$$A = \begin{bmatrix} 3 & 2 \\ 2 & 5 \end{bmatrix} \quad X = \begin{bmatrix} I_1 \\ I_2 \end{bmatrix} \quad \text{and} \quad B = \begin{bmatrix} .6 \\ 2.4 \end{bmatrix}$$

If A is invertible, we can solve equation (4) for X by multiplying by A^{-1} . Then, we have

$$A^{-1}AX = A^{-1}B$$

and as

$$A^{-1}A = I ,$$

$$X = A^{-1}B .$$ (5)

Now we compute det A.

$$\det A = \det \begin{bmatrix} 3 & 2 \\ 2 & 5 \end{bmatrix}$$

$$= [15 - 4] = 11$$

since det A ≠ 0, the matrix A is invertible. Then,

$$A^{-1} = \frac{1}{\det A} \text{ adj A} .$$

adj A is the transpose of the matrix of cofactors.

$$\text{adj A} = \begin{bmatrix} 5 & -2 \\ -2 & 3 \end{bmatrix}^T = \begin{bmatrix} 5 & -2 \\ -2 & 3 \end{bmatrix}$$

Therefore,

$$A^{-1} = 1/11 \begin{bmatrix} 5 & -2 \\ -2 & 3 \end{bmatrix} = \begin{bmatrix} 5/11 & -2/11 \\ -2/11 & 3/11 \end{bmatrix}$$

Substitute the values of A^{-1} and B into equation (5) .

$$X = \begin{bmatrix} 5/11 & -2/11 \\ -2/11 & 3/11 \end{bmatrix} \begin{bmatrix} .6 \\ 2.4 \end{bmatrix}$$

$$= \begin{bmatrix} .272 - 0.436 \\ -0.109 + 0.654 \end{bmatrix}$$

$$= \begin{bmatrix} -0.164 \\ 0.545 \end{bmatrix}$$

or

$$I_1 = \begin{bmatrix} -0.164 \\ 0.545 \end{bmatrix}$$
$$I_2$$

Therefore, the values of the currents I_1 and I_2 are

$$I_1 = -0.164 \text{ A}$$
$$I_2 = 0.545 \text{ A} .$$

● PROBLEM 21-3

Determine the current I_1 and I_2 for the circuit of figure 1, if $R_1 = 10$ ohms, $R_2 = 30$ ohms, and $R_3 = 20$ ohms.

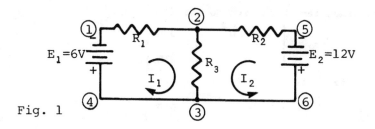

Fig. 1

Solution: Kirchhoff's voltage law is applied to the loops ① ② ③ ④ ①, and ⑤ ② ③ ⑥ ⑤ ; the following two equations in two unknown currents result:

$$E_1 - I_1 R_1 - (I_1 + I_2) R_3 = 0 \qquad (1)$$

$$E_2 - R_2 I_2 - R_3 (I_1 + I_2) = 0 . \qquad (2)$$

Equations (1) and (2) can be rewritten as

$$E_1 - (R_1 + R_3) I_1 - I_2 R_3 = 0$$

$$E_2 - R_3 I_1 - (R_2 + R_3) I_2 = 0 .$$

Substitution of the values of E_1, E_2, R_1 and R_2 yields

$$30 I_1 + 20 I_2 = 6$$

$$20 I_1 + 50 I_2 = 12.$$

Now divide each equation by 10. Then,

$$3 I_1 + 2 I_2 = .6$$

$$2 I_1 + 5 I_2 = 1.2 . \qquad (3)$$

The system (3) is a nonhomogeneous linear system. As we know, a procedure for solving a nonhomogeneous system is as follows:

First form the augmented matrix of the system. Then, reduce this matrix to an echelon matrix by performing row operations. Now the augmented matrix of system (3) is

$$\begin{bmatrix} 3 & 2 & | & .6 \\ 2 & 5 & | & 1.2 \end{bmatrix}$$

Now, multiply the first row by 2 and the second row by 3 to obtain

$$\begin{bmatrix} 6 & 4 & | & 1.2 \\ 6 & 15 & | & 3.6 \end{bmatrix} .$$

Add -1 times the first row to the second row

$$\begin{bmatrix} 6 & 4 & | & 1.2 \\ 0 & 11 & | & 2.4 \end{bmatrix}$$

Divide the second row by 11.

$$\begin{bmatrix} 6 & 4 & | & 1.2 \\ 0 & 1 & | & .218 \end{bmatrix}$$

Now add −4 times the second row to the first row.

$$\begin{bmatrix} 6 & 0 & | & .328 \\ 0 & 1 & | & .218 \end{bmatrix}$$

Dividing the first row by 6 yields

$$\begin{bmatrix} 1 & 0 & | & .055 \\ 0 & 1 & | & .218 \end{bmatrix}$$

This is the augmented matrix of

$$I_1 = .055$$

$$I_2 = .218 .$$

Therefore, the values of the current I_1 and I_2 are

$$I_1 = .055 \text{ A} = 55 \text{ mA}$$

$$I_2 = .218 \text{ A} = 218 \text{ mA} .$$

● **PROBLEM** 21-4

Obtain the values of I_1, I_2, I_3, and I_4 in the given circuit.

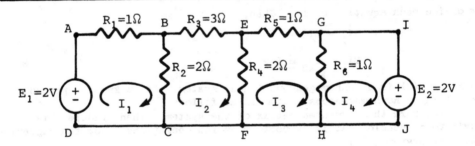

Solution: Kirchhoff's voltage law to the loop A B C D A; the following equation is obtained.

$$2 - 1I_1 - 2(I_1 - I_2) = 0 .$$

After rearrangement,

$$3I_1 - 2I_2 = 2 . \tag{1}$$

Similarly, for loop B E F C B,

$$3I_2 + 2(I_2 - I_3) + 2(I_2 - I_1) = 0$$

or

$$-2I_1 + 7I_2 - 2I_3 = 0 \tag{2}$$

For loop E G H F E,

$$1I_3 + 1(I_3 - I_4) + 2(I_3 - I_2) = 0$$

or

$$-2I_2 + 4I_3 - I_4 = 0 . \tag{3}$$

For loop G I J H G,

$$2 - 1(I_4 - I_3) = 0$$

758

or
$$-I_3 + I_4 = 2 \ . \tag{4}$$

So we have obtained four linear equations in four unknown currents. These equations are:

$$\begin{aligned}
3I_1 - 2I_2 &= 0 \\
-2I_1 + 7I_2 - 2I_3 &= 0 \\
- 2I_2 + 4I_3 - I_4 &= 0 \\
-I_3 + I_4 &= 2
\end{aligned} \tag{5}$$

Now write system (5) in matrix form,

$$\begin{bmatrix} 3 & -2 & 0 & 0 \\ -2 & 7 & -2 & 0 \\ 0 & -2 & 4 & -1 \\ 0 & 0 & -1 & 1 \end{bmatrix} \begin{bmatrix} I_1 \\ I_2 \\ I_3 \\ I_4 \end{bmatrix} = \begin{bmatrix} 2 \\ 0 \\ 0 \\ 2 \end{bmatrix} \tag{6}$$

Let

$$A = \begin{bmatrix} 3 & -2 & 0 & 0 \\ -2 & 7 & -2 & 0 \\ 0 & -2 & 4 & -1 \\ 0 & 0 & -1 & 1 \end{bmatrix}, \quad X = \begin{bmatrix} I_1 \\ I_2 \\ I_3 \\ I_4 \end{bmatrix}, \quad \text{and} \quad B = \begin{bmatrix} 2 \\ 0 \\ 0 \\ 2 \end{bmatrix}$$

Then system (6) becomes,
$$AX = B \ .$$

To solve this nonhomogeneous system, write the augmented matrix of the system and then reduce it to row-reduced echelon form. The augmented matrix of the system is

$$[A:B] = \begin{bmatrix} 3 & -2 & 0 & 0 & | & 2 \\ -2 & 7 & -2 & 0 & | & 0 \\ 0 & -2 & 4 & -1 & | & 0 \\ 0 & 0 & -1 & 1 & | & 2 \end{bmatrix}$$

Add the fourth row to the third row to obtain

$$\begin{bmatrix} 3 & -2 & 0 & 0 & | & 2 \\ -2 & 7 & -2 & 0 & | & 0 \\ 0 & -2 & 3 & 0 & | & 2 \\ 0 & 0 & -1 & 1 & | & 2 \end{bmatrix}$$

Adding 2/3 times the first row to the second row yields

$$\begin{bmatrix} 3 & -2 & 0 & 0 & | & 2 \\ 0 & 17/3 & -2 & 0 & | & 4/3 \\ 0 & -2 & 3 & 0 & | & 2 \\ 0 & 0 & -1 & 1 & | & 2 \end{bmatrix}$$

Add 2/3 times the third row to the second row,

$$\begin{bmatrix} 3 & -2 & 0 & 0 & | & 2 \\ 0 & 13/3 & 0 & 0 & | & 8/3 \\ 0 & -2 & 3 & 0 & | & 2 \\ 0 & 0 & -1 & 1 & | & 2 \end{bmatrix}$$

Multiply the second row by 3/13. Then add 2 times the resulting second row to the first row to obtain,

$$\begin{bmatrix} 3 & 0 & 0 & 0 & | & 42/13 \\ 0 & 1 & 0 & 0 & | & 24/39 \\ 0 & -2 & 3 & 0 & | & 2 \\ 0 & 0 & -1 & 1 & | & 2 \end{bmatrix}$$

Add 2 times the second row to the third row,

$$\begin{bmatrix} 3 & 0 & 0 & 0 & | & 42/13 \\ 0 & 1 & 0 & 0 & | & 24/39 \\ 0 & 0 & 3 & 0 & | & 126/39 \\ 0 & 0 & -1 & 1 & | & 2 \end{bmatrix}$$

Divide the first row and the third row by 3,

$$\begin{bmatrix} 1 & 0 & 0 & 0 & | & 42/39 \\ 0 & 1 & 0 & 0 & | & 24/39 \\ 0 & 0 & 1 & 0 & | & 42/39 \\ 0 & 0 & -1 & 1 & | & 2 \end{bmatrix}$$

Add the third row to the fourth row to obtain,

$$\begin{bmatrix} 1 & 0 & 0 & 0 & | & 42/39 \\ 0 & 1 & 0 & 0 & | & 24/39 \\ 0 & 0 & 1 & 0 & | & 42/39 \\ 0 & 0 & 0 & 1 & | & 120/39 \end{bmatrix}$$

This is the augmented matrix of the system

$$I_1 = 42/39$$
$$I_2 = 24/39$$
$$I_3 = 42/39$$
$$I_4 = 120/39$$

Therefore the values of the four currents are,

$$I_1 = 42/39$$
$$I_2 = 24/39$$
$$I_3 = 42/39$$
$$I_4 = 120/39 .$$

● PROBLEM 21-5

Find the power delivered by the 2-ampere source in the circuit shown in figure 1, and also obtain the values of currents I_2 and I_3 .

760

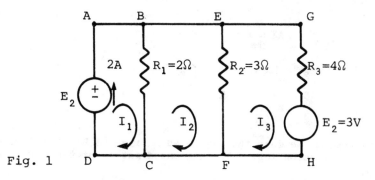

Fig. 1

Solution: The relationship between the current of 2 amperes from the current source and the loop current I_1 can only be satisfied by the equation

$$I_1 = 2 \text{ amperes}$$

since a current source has infinite resistance. Applying Kirchhoff's voltage law to the loop ABCDA yields

$$E_1 - R_1(I_1 - I_2) = 0 .$$

Since $R_1 = 2$ ohms and $I_1 = 2$ amperes, these values may be substituted to obtain

$$E_1 - 4 + 2I_2 = 0 . \tag{1}$$

Similarly, for loop BEFCB,

$$R_1(I_2 - I_1) + R_2(I_2 - I_3) = 0$$

or

$$-R_1 I_1 + I_2(R_1 + R_2) - R_2 I_3 = 0 .$$

Substituting $R_1 = 2$, $R_2 = 3$, and $I_1 = 2$ gives

$$-4 + 5I_2 - 3I_3 = 0 . \tag{2}$$

Similarly, for the loop EGHFE,

$$E_2 + R_3 I_3 + R_2(I_3 - I_2) = 0$$

or

$$E_2 - I_2 R_2 + I_3(R_3 + R_2) = 0 .$$

Substitute the given values of E_2, R_2, and R_3. The equation then becomes

$$3 - 3I_2 + 7I_3 = 0 . \tag{3}$$

The equations (1), (2), and (3) can be rewritten as

$$
\begin{aligned}
E_1 + 2I_2 \qquad &= 4 \\
5I_2 - 3I_3 &= 4 \\
3I_2 - 7I_3 &= 3
\end{aligned}
\tag{4}
$$

The system (4) can be written in matrix form as:

$$
\begin{bmatrix} 1 & 2 & 0 \\ 0 & 5 & -3 \\ 0 & 3 & -7 \end{bmatrix}
\begin{bmatrix} E_1 \\ I_2 \\ I_3 \end{bmatrix}
=
\begin{bmatrix} 4 \\ 4 \\ 3 \end{bmatrix}
$$

or

$$AX = B$$

where

$$A = \begin{bmatrix} 1 & 2 & 0 \\ 0 & 5 & -3 \\ 0 & 3 & -7 \end{bmatrix} \quad X = \begin{bmatrix} E_1 \\ I_2 \\ I_3 \end{bmatrix} \quad \text{and} \quad B = \begin{bmatrix} 4 \\ 4 \\ 3 \end{bmatrix}$$

To solve this system, we form the augmented matrix $[A:B]$, and then reduce it to row-reduced echelon form. The augmented matrix of the system is

$$\begin{bmatrix} 1 & 2 & 0 & \vline & 4 \\ 0 & 5 & -3 & \vline & 4 \\ 0 & 3 & -7 & \vline & 3 \end{bmatrix}$$

Add $-3/5$ times the second row to the third row. This yields

$$\begin{bmatrix} 1 & 2 & 0 & \vline & 4 \\ 0 & 5 & -3 & \vline & 4 \\ 0 & 0 & -26/5 & \vline & 3/5 \end{bmatrix}$$

Multiply the third row by $-5/26$, to obtain

$$\begin{bmatrix} 1 & 2 & 0 & \vline & 4 \\ 0 & 5 & -3 & \vline & 4 \\ 0 & 0 & 1 & \vline & -3/26 \end{bmatrix}$$

Add 3 times the third row to the second row. This yields

$$\begin{bmatrix} 1 & 2 & 0 & \vline & 4 \\ 0 & 5 & 0 & \vline & 95/26 \\ 0 & 0 & 1 & \vline & -3/26 \end{bmatrix}$$

Divide the second row by 5 then add -2 times the resulting second row to the first row to obtain

$$\begin{bmatrix} 1 & 0 & 0 & \vline & 66/26 \\ 0 & 1 & 0 & \vline & 19/26 \\ 0 & 0 & 1 & \vline & -3/26 \end{bmatrix}$$

This is the augmented matrix of the system

$$E_1 = 66/26$$

$$I_2 = 19/26$$

$$I_3 = -3/26$$

Therefore, the power delivered by the 2-ampere source is $33/13$ volts, and the values of the currents I_2 and I_3 are $19/26$ amp and $-3/26$ amp. respectively.

Use matrix multiplication to approximate the behavior of a leaky cable.
Use the fact that a leaky cable can be approximated by the following
circuit: where the current flows through the resistances r and leaks
away to the earth through the resistances R. Compute the output of this
circuit if the input is

$$w = \begin{bmatrix} v_0 \\ i_0 \end{bmatrix}$$

where i is a current, v is a voltage drop and $R = r = 1$.

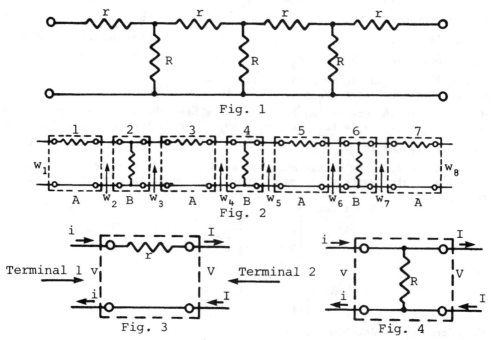

Fig. 1

Fig. 2

Fig. 3   Fig. 4

<u>Solution</u>: The circuit in figure 1 is more easily analyzed if it is
broken down into 7 separate systems, as in figure 2.
      Now, look at one section at a time.
      In fig. 3, i is the current flowing in and out of terminal 1 and
I is the current flowing in and out at the terminal 2. Let v be the
voltage difference between the input terminals and let V be the voltage
difference between the output terminals. A current, i, flowing through a
resistance r experiences a potential drop ri . (This is from Ohm's law
$i = v/r$). Thus $V = v - ri$. The current through the bottom (in figure 3)
has no resistance so $i = I$. If we measure input with the vector

$$w = \begin{bmatrix} v \\ i \end{bmatrix}$$

and output with the vector

$$W = \begin{bmatrix} V \\ I \end{bmatrix}$$

our equations, $V = v - ri$, become $I = i$, $W = Aw$ where

$$A = \begin{bmatrix} 1 & -r \\ 0 & 1 \end{bmatrix} .$$

The second link in the chain is shown in fig. 4.
      In fig. 4, v, V, i and I as before. Since the input and output term-
inals are directly connected there can be no change of voltage, i.e.,
$v = V$. Using Ohm's law we know a current v/R passes down the resistance

763

R, so there is that much less current to pass the output I. Thus,

$$I = i - (\frac{v}{R}) .$$

In this case our system of equations,

$$V = v$$

$$I = - (\frac{1}{R})v + i ,$$

in matrix form is $W = Bw$ where

$$B = \begin{bmatrix} 1 & 0 \\ -1/R & 1 \end{bmatrix} .$$

Let the inputs and outputs of each box in figure 2 be w_i, $i = 1,8$. That is (in figure 2) w_1 is the input to box 1, w_2 is the output of box 1 (therefore input to box 2), w_3 is the output of box 2 (therefore, the input to box 3) etc. In figure 2 boxes 1,3,5 and 7 are of the form which has output $W = Aw$. Thus $w_2 = Aw_1$, $w_4 = Aw_3$, $w_6 = Aw_5$ and $w_8 = Aw_7$.

Boxes 2,4 and 6 are of the form which has been shown to have output $W = Bw$ so $w_3 = Bw_2$, $w_5 = Bw_4$ and $w_7 = Bw_6$. It follows that the output of the whole system is $w_8 = A B A B A B Aw_1$. Thus the effect of a leaky cable can be found by matrix multiplication. For long cables (say 1000 boxes) we could easily find the output using computers to do the matrix multiplication.

Now, if $r = R = 1$ we have

$$A = \begin{bmatrix} 1 & -1 \\ 0 & 1 \end{bmatrix} \text{ and } B = \begin{bmatrix} 1 & 0 \\ -1 & 1 \end{bmatrix}$$

so

$$AB = \begin{bmatrix} 2 & -1 \\ -1 & 1 \end{bmatrix} .$$

Therefore, $w_8 = ABABABAw_1 =$

$$\begin{bmatrix} 2 & -1 \\ -1 & 1 \end{bmatrix} \begin{bmatrix} 2 & -1 \\ -1 & 1 \end{bmatrix} \begin{bmatrix} 2 & -1 \\ -1 & 1 \end{bmatrix} \begin{bmatrix} 1 & -1 \\ 0 & 1 \end{bmatrix} w_1$$

$$= \begin{bmatrix} 13 & -21 \\ -8 & 13 \end{bmatrix} \begin{bmatrix} v \\ i \end{bmatrix} .$$

The output will be a voltage $13v - 21i$ and a current $-8v + 13i$.

● PROBLEM 21-7

Consider the circuit in figure 1 with resistances and batteries as shown. Obtain the values of the currents i_1, i_2, and i_3.

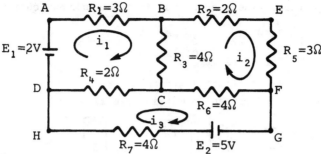

Fig. 1

Solution: Kirchhoff's voltage law states that the sum of the voltage drops around any closed loop is 0. To obtain algebraic equations, we shall apply Kirchhoff's voltage law to each closed loop. Consider loop A B C D A. Applying Kirchhoff's law yields,

$$E_1 - R_1 i_1 - R_3(i_1 - i_2) - R_4(i_1 - i_3) = 0 \qquad (1')$$

where E_1 is the voltage generated by an attached battery. $R_1 i_1$ is the voltage drop calculated using Ohm's Law (since $\Delta V = ri$) that is, when a current i_1 passes through a resistance r_1, the resulting voltage drop is $r_1 i_1$. Since i_1 and i_2 were chosen to be in opposite directions, the current passing through R_3 is $i_1 - i_2$ and hence the voltage drop through R_3 is $R_3(i_1 - i_2)$. Rearrangement of (1') results in

$$E_1 - i_1(R_1 + R_3 + R_4) + R_3 i_2 + R_4 i_3 = 0 .$$

Substituting the values of E_1, R_1, R_3, and R_4 gives,
$$2 - 9i_1 + 4i_2 + 2i_3 = 0$$
or

$$9i_1 - 4i_2 - 2i_3 = 2 . \qquad (1)$$

Similarly, for loop B E F C B,
$$R_2 i_2 + R_5 i_2 + R_6(i_2 - i_3) + R_3(i_2 - i_1) = 0$$
or

$$-R_3 i_1 + i_2(R_2 + R_3 + R_5 + R_6) - R_6 i_3 .$$

Substitute the given values of R_2, R_3, R_5, and R_6 into the equation to obtain

$$-4i_1 + 13i_2 - 4i_3 = 0 . \qquad (2)$$

Applying Kirchhoff's voltage law to loop D F G H D, yields

$$E_2 - R_4(i_3 - i_1) - R_6(i_3 - i_2) - R_7 i_3 = 0$$

or

$$E_2 + R_4 i_1 + R_6 i_2 - i_3(R_4 + R_6 + R_7) = 0 .$$

By substituting the values of E_2, R_4, R_6, and R_7 , the equation becomes
$$5 + 2i_1 + 4i_2 - 10i_3 = 0$$
or

$$-2i_1 - 4i_2 + 10i_3 = 5 . \qquad (3)$$

We have obtained three linear equations in three unknown currents. These equations are

$$9i - 4i_2 - 2i_3 = 2$$
$$-2i_1 - 4i_2 + 10i_3 = 5$$
$$-4i + 13i_2 - 4i_3 = 0 .$$

The above system can be written in matrix form as

$$\begin{bmatrix} 9 & -4 & -2 \\ -2 & -4 & 10 \\ -4 & 13 & -4 \end{bmatrix} \begin{bmatrix} i_1 \\ i_2 \\ i_3 \end{bmatrix} = \begin{bmatrix} 2 \\ 5 \\ 0 \end{bmatrix}$$

or

$$AI = B \tag{4}$$

where

$$A = \begin{bmatrix} 9 & -4 & -2 \\ -2 & -4 & 10 \\ -4 & 13 & -4 \end{bmatrix} , \quad I = \begin{bmatrix} i_1 \\ i_2 \\ i_3 \end{bmatrix} , \quad \text{and} \quad B = \begin{bmatrix} 2 \\ 5 \\ 0 \end{bmatrix} .$$

To solve this system, we first form the augmented matrix [A:B] and then reduce it to echelon form.

The augmented matrix of the system is

$$\left[\begin{array}{ccc|c} 9 & -4 & -2 & 2 \\ -2 & -4 & 10 & 5 \\ -4 & 13 & -4 & 0 \end{array} \right] .$$

Add 2/9 times the first row to the second row and 4/9 times the first row to the third row to obtain,

$$\left[\begin{array}{ccc|c} 9 & -4 & -2 & 2 \\ 0 & -4.8889 & -4.8889 & 5.4444 \\ 0 & 11.2222 & -4.8889 & 0.8889 \end{array} \right] .$$

Here we have to work with inexact arithmetic, rounding-off to the required number of decimal places. In such cases, try to arrange the rows so that the scale factors used in subtracting rows are less than 1 numerically. In this way each time a row is multiplied by a factor (less than one), the previous round-off error is multiplied by a number less than one and thus the round-off error decreases rather than increases. For example, if the number 2.52 is rounded-off to 2.5 the round-off error is .02. If 2.5 is then multiplied by a number r, the round-off error is $r \times .02$. So we try to make r as small as possible in order to minimize the compounding of error. In the case of a system of equations this can be done by partially pivoting. That is, at each stage we work with the largest coefficient of the variable to be eliminated (the largest coefficient is called the pivot).

Now, the largest coefficient of i_2 in the second and third equations is 11.2222, so we rearrange the last two rows as

$$\left[\begin{array}{ccc|c} 9 & -4 & -2 & 2 \\ 0 & 11.2222 & -4.8889 & 0.8889 \\ 0 & -4.8889 & 9.5556 & 5.4444 \end{array} \right]$$

Now add 4.8889/11.2222 times the second row to the third row to obtain

$$\begin{bmatrix} 9 & -4 & -2 & | & 2 \\ 0 & 11.2222 & -4.8889 & | & 0.8889 \\ 0 & 0 & 7.4258 & | & 5.8317 \end{bmatrix}$$

The above matrix is an augmented matrix of the system

$$9i_1 - 4i_2 - 2i_3 = 2$$

$$11.2222i_2 - 4.8889i_3 = 0.8889 \tag{5}$$

$$7.4258i_3 = 5.8317 .$$

The system (5) can be now easily solved. From the last equation

$$i_3 = \frac{5.8317}{7.4258} = 0.7853.$$

Substituting this value of i_3 into the second equation of system (5), yields

$$11.2222i_2 - 4.8889 \times (.7853) = 0.8889 .$$

Solving for i_2 , gives

$$i_2 = .4213 .$$

Now substitute the values of i_1 and i_2 into the first equation, thus,

$$9i_1 - 4 \times (.4213) - 2(.7853) = 2 .$$

Solving for i_1, yields,

$$i_1 = 0.5840 .$$

Therefore, the values of the three currents are

$$i_1 = 0.5840$$
$$i_2 = 0.4213$$
$$i_3 = 0.7853.$$

Observe that the scale factors used in this example were 2/9, 4/9, and 4.8889/11.2222, all of them less than 1 in magnitude, thus limiting the spreading of round-off error as far as possible.

VOLTAGE & CURRENT LAWS

● **PROBLEM** 21-8

Find the three currents I_a, I_b, and I_c in the circuit of figure 1.

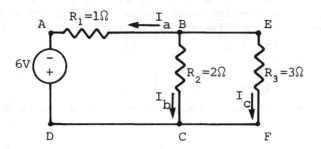

Fig. 1

Solution: Three rules are needed in order to write a system of equations which can be used to find the current in each loop of a resistance circuit. These three rules are:

1) Kirchhoff's Voltage Law, which states that the sum of the voltage drops around any closed loop is 0. In equation form, this would be

$$\Sigma V_i = 0 .$$

2) Ohm's Law, which states that when a current passes through a resistor, the resulting voltage drop is proportional to the rate of flow of the current and the constant of proportionality is the resistance of the resistor. Thus, $V = IR$.

3) Kirchhoff's Current Law, i.e., the net sum of the currents entering and leaving any given node is 0.

Using laws 1) and 2) on loop A B C D A, yields

$$6 - R_1 I_a + R_2 I_b = 0 .$$

Substituting the values of R_1 and R_2 , we get

$$6 - 1I_a + 2I_b = 0$$

or

$$I_a - 2I_b = 6 . \tag{1}$$

Similarly, applying Kirchhoff's voltage law to the loop A E F D A yields

$$I_a - 3I_c = 6 . \tag{2}$$

We have obtained two linear equations in three unknown currents. To obtain a third equation, we shall use Kirchhoff's current law. Applying this law at node B, the following equation is obtained

$$I_a + I_b + I_c = 0 . \tag{3}$$

The equations (1), (2), and (3) can be rewritten as

$$
\begin{aligned}
I_a - 2I_b \quad\quad &= 6 \\
I_a \quad\quad -3I_c &= 6 \\
I_a + I_b +I_c &= 0 .
\end{aligned}
\tag{4}
$$

The system (4) can be written in matrix form as

$$
\begin{bmatrix} 1 & -2 & 0 \\ 1 & 0 & -3 \\ 1 & 1 & 1 \end{bmatrix}
\begin{bmatrix} I_a \\ I_b \\ I_c \end{bmatrix}
=
\begin{bmatrix} 6 \\ 6 \\ 0 \end{bmatrix}
$$

or

$$AI = B$$

where

$$
A = \begin{bmatrix} 1 & -2 & 0 \\ 1 & 0 & -3 \\ 1 & 1 & 1 \end{bmatrix}, \quad
I = \begin{bmatrix} I_a \\ I_b \\ I_c \end{bmatrix}, \quad
B = \begin{bmatrix} 6 \\ 6 \\ 0 \end{bmatrix} .
$$

To solve this system form the augmented matrix $[A:B]$ and then reduce it to row-reduced echelon form. Therefore, the augmented matrix of the system is

$$
\begin{bmatrix} 1 & -2 & 0 & \vdots & 6 \\ 1 & 0 & -3 & \vdots & 6 \\ 1 & 1 & 1 & \vdots & 0 \end{bmatrix} .
$$

Add -1 times the first row to the second row and to the third row to obtain

$$\begin{bmatrix} 1 & -2 & 0 & | & 6 \\ 0 & 2 & -3 & | & 0 \\ 0 & 3 & 1 & | & -6 \end{bmatrix}$$

Add -3/2 times the second row to the third row. This yields

$$\begin{bmatrix} 1 & -2 & 0 & | & 6 \\ 0 & 2 & -3 & | & 0 \\ 0 & 0 & 11/2 & | & -6 \end{bmatrix}$$

Multiplying the third row by 2/11 gives,

$$\begin{bmatrix} 1 & -2 & 0 & | & 6 \\ 0 & 2 & -3 & | & 0 \\ 0 & 0 & 1 & | & -12/11 \end{bmatrix}$$

Add 3 times the third row to the second row to obtain

$$\begin{bmatrix} 1 & -2 & 0 & | & 6 \\ 0 & 2 & 0 & | & -36/11 \\ 0 & 0 & 1 & | & -12/11 \end{bmatrix}$$

Add the second row to the first row to obtain

$$\begin{bmatrix} 1 & 0 & 0 & | & 30/11 \\ 0 & 2 & 0 & | & -36/11 \\ 0 & 0 & 1 & | & -12/11 \end{bmatrix}$$

Dividing the second row by 2 yields

$$\begin{bmatrix} 1 & 0 & 0 & | & 30/11 \\ 0 & 1 & 0 & | & -18/11 \\ 0 & 0 & 1 & | & -12/11 \end{bmatrix}$$

This is the augmented matrix of the system

$$I_a = 30/11$$

$$I_b = -18/11$$

$$I_c = -12/11$$

Therefore, the values of the three currents are

$$I_a = 30/11 \text{ amperes}$$

$$I_b = -18/11 \text{ amperes}$$

and

$$I_c = -12/11 \text{ amperes .}$$

● **PROBLEM** 21-9

Determine the current flow through R_1 and R_2, and the source voltage E in the circuit of figure 1. $R_1 = 2$ ohms, $R_2 = 4$ ohms and $I = 9$ amp.

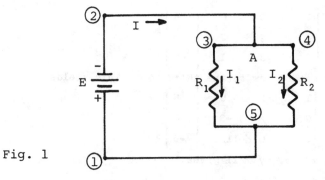

Fig. 1

Solution: We shall use Kirchhoff's current law and Kirchhoff's voltage law to derive the algebraic equations. Kirchhoff's current law states that the net flow of current into any junction must be zero. Applying this to junction A, gives $I - I_1 - I_2 = 0$.

Substitute the value of I into the equation. Rearrangement yields,

$$I_1 + I_2 = 9 . \tag{1}$$

Kirchhoff's voltage law states that the algebraic sum of the voltage drops around a closed circuit must equal zero. By applying Kirchhoff's voltage law to the loop ①②③⑤① one obtains,

$$E - I_1 R_1 = 0$$

and as

$$R_1 = 2 \text{ ohms},$$

$$E - 2I_1 = 0 . \tag{2}$$

Similarly, applying Kirchhoff's voltage law to the loop ①②④⑤① gives

$$E - I_2 R_2 = 0 ,$$

and as $R_2 = 4$ ohms,

$$E - 4I_2 = 0 . \tag{3}$$

Thus, we have obtained the following three linear equations:

$$
\begin{aligned}
E - 2I_1 \quad\quad &= 0 \\
E \quad\quad - 4I_2 &= 0 \\
I_1 + I_2 &= 9 .
\end{aligned} \tag{4}
$$

The system (4) is a nonhomogeneous system of three linear equations in three unknowns. The system (4) can be written in matrix form as

$$
\begin{bmatrix} 1 & -2 & 0 \\ 1 & 0 & -4 \\ 0 & 1 & 1 \end{bmatrix}
\begin{bmatrix} E \\ I_1 \\ I_2 \end{bmatrix}
=
\begin{bmatrix} 0 \\ 0 \\ 9 \end{bmatrix} . \tag{5}
$$

Let

$$
A = \begin{bmatrix} 1 & -2 & 0 \\ 1 & 0 & -4 \\ 0 & 1 & 1 \end{bmatrix} , \quad
X = \begin{bmatrix} E \\ I_1 \\ I_2 \end{bmatrix} , \quad
B = \begin{bmatrix} 0 \\ 0 \\ 9 \end{bmatrix} .
$$

Then (5) becomes,

$$AX = B . \tag{6}$$

To solve this system, we first form the augmented matrix and then reduce it to row-reduced echelon form. The augmented matrix of the system is

$$[A:B] = \begin{bmatrix} 1 & -2 & 0 & | & 0 \\ 1 & 0 & -4 & | & 0 \\ 0 & 1 & 1 & | & 9 \end{bmatrix}$$

Add -1 times the first row to the second row,

$$\begin{bmatrix} 1 & -2 & 0 & | & 0 \\ 0 & 2 & -4 & | & 0 \\ 0 & 1 & 1 & | & 9 \end{bmatrix}.$$

Add 4 times the third row to the second row,

$$\begin{bmatrix} 1 & -2 & 0 & | & 0 \\ 0 & 6 & 0 & | & 36 \\ 0 & 1 & 1 & | & 9 \end{bmatrix}.$$

Divide the second row by 6,

$$\begin{bmatrix} 1 & -2 & 0 & | & 0 \\ 0 & 1 & 0 & | & 6 \\ 0 & 1 & 1 & | & 9 \end{bmatrix}$$

Add 2 times the second row to the first row and -1 times the second row to the third row.

$$\begin{bmatrix} 1 & 0 & 0 & | & 12 \\ 0 & 1 & 0 & | & 6 \\ 0 & 0 & 1 & | & 3 \end{bmatrix}$$

This is an augmented matrix of the system

$$E = 12$$
$$I_1 = 6$$
$$I_2 = 3$$

Therefore the source voltage of the given circuit is $E = 12$ v. The currents I_1 and I_2 are

$$I_1 = 6 \text{ amp.}$$

and

$$I_2 = 3 \text{ amp.}$$

● **PROBLEM** 21-10

Consider the following circuit. Derive the system of equations that can be used to find the currents i_1 through i_5 in terms of the initial current input (i_0) and the resistances R_i, $i = 1,\ldots,5$. Solve the system to find the currents, in terms of the initial current, in each of the pieces of conducting wire if the resistances are each 1 ohm. That is, $R_1 = R_2 = R_3 = R_4 = R_5 = 1$.

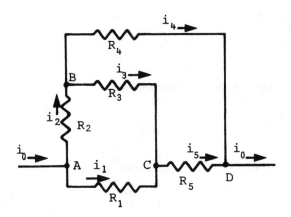

Solution: Kirchhoff's first law states that the net flow of current into any junction must be 0. At junction A the current i_0 flows in and currents i_1 and i_2 flow out. Therefore,

$$i_0 = i_1 + i_2 . \tag{1}$$

At B, i_2 flows in and i_3 and i_4 flow out so $i_3 + i_4 = i_2$ or

$$i_2 - i_3 - i_4 = 0 . \tag{2}$$

Similarly at junctions C and D Kirchhoff's law yields $i_1 + i_3 - i_5 = 0$ and $i_4 + i_5 = i_0$. Thus we have the system:

$$
\begin{aligned}
i_1 + i_2 &= i_0 \\
i_2 - i_3 - i_4 &= 0 \\
i_1 + i_3 - i_5 &= 0 \\
i_4 + i_5 &= i_0
\end{aligned}
\tag{3}
$$

Notice that the system (3) is not independent. The first equation is the sum of the other three. Therefore, an equivalent system is

$$
\begin{aligned}
i_2 - i_3 - i_4 &= 0 \\
i_1 + i_3 - i_5 &= 0 \\
i_4 + i_5 &= i_0
\end{aligned}
\tag{3'}
$$

This system is not sufficient to completely define i_n , $n = 1,\dots,5$ since it has many solutions. Apply Kirchhoff's second law which states that the sum of the products of the current with the resistance in any closed circuit, without batteries, is zero. In the circuit A B C A , i_2 and i_3 are in one direction while i_1 is in the opposite direction, in equation form this is

$$-R_1 i_1 + R_2 i_2 + R_3 i_3 = 0 . \tag{4}$$

In the circuit B C D B note that

$$-R_3 i_3 + R_4 i_4 - R_5 i_5 = 0 . \tag{5}$$

The system (3'), along with equations (4) and (5), completely determines i_n , $n = 1,\dots,5$.

$$i_2 - i_3 - i_4 \qquad\qquad = 0$$
$$i_1 \quad + i_3 \qquad\qquad i_5 \quad = 0$$
$$i_4 + i_5 \quad = i_0 \qquad (6)$$
$$-R_1 i_1 + R_2 i_2 + R_3 i_3 \qquad\qquad = 0$$
$$- R_3 i_3 + R_4 i_4 - R_5 i_5 = 0$$

In matrix form the system (6) is $AI = B$ where

$$A = \begin{bmatrix} 0 & 1 & -1 & -1 & 0 \\ 1 & 0 & 1 & 0 & -1 \\ 0 & 0 & 0 & 1 & 1 \\ -R_1 & R_2 & R_3 & 0 & 0 \\ 0 & 0 & -R_3 & R_4 & -R_5 \end{bmatrix}$$

$$B = \begin{bmatrix} 0 \\ 0 \\ i_0 \\ 0 \\ 0 \end{bmatrix} \qquad \text{and} \qquad I = \begin{bmatrix} i_1 \\ i_2 \\ i_3 \\ i_4 \\ i_5 \end{bmatrix}$$

In the special case $R_i = 1$, $i = 1,\ldots,5$, A becomes

$$\begin{bmatrix} 0 & 1 & -1 & -1 & 0 \\ 1 & 0 & 1 & 0 & -1 \\ 0 & 0 & 0 & 1 & 1 \\ -1 & 1 & 1 & 0 & 0 \\ 0 & 0 & -1 & 1 & -1 \end{bmatrix}$$

To solve this system write down the augmented matrix, $[A:I]$ and reduce it to echelon form. To simplify this process first reorder the rows of $[A:I]$

$$\left[\begin{array}{ccccc|c} -1 & 1 & 1 & 0 & 0 & 0 \\ 1 & 0 & 1 & 0 & -1 & 0 \\ 0 & 0 & -1 & 1 & -1 & 0 \\ 0 & 0 & 0 & 1 & 1 & i_0 \\ 0 & 1 & -1 & -1 & 0 & 0 \end{array}\right]$$

Add the first row to the second row to obtain,

$$\left[\begin{array}{ccccc|c} -1 & 1 & 1 & 0 & 0 & 0 \\ 0 & 1 & 2 & 0 & -1 & 0 \\ 0 & 0 & -1 & 1 & -1 & 0 \\ 0 & 0 & 0 & 1 & 1 & i_0 \\ 0 & 1 & -1 & -1 & 0 & 0 \end{array}\right]$$

Now add (-1) times the second row to the fifth row,

$$\begin{bmatrix} -1 & 1 & 1 & 0 & 0 & \vdots & 0 \\ 0 & 1 & 2 & 0 & -1 & \vdots & 0 \\ 0 & 0 & -1 & 1 & -1 & \vdots & 0 \\ 0 & 0 & 0 & 1 & 1 & \vdots & i_0 \\ 0 & 0 & -3 & -1 & 1 & \vdots & 0 \end{bmatrix}$$

Multiply the third row by -3 and add it to the fifth row,

$$\begin{bmatrix} -1 & 1 & 1 & 0 & 0 & \vdots & 0 \\ 0 & 1 & 2 & 0 & -1 & \vdots & 0 \\ 0 & 0 & -1 & 1 & -1 & \vdots & 0 \\ 0 & 0 & 0 & 1 & 1 & \vdots & i_0 \\ 0 & 0 & 0 & -4 & 4 & \vdots & 0 \end{bmatrix}$$

Multiply the fourth row by 4 and add it to the fifth row,

$$\begin{bmatrix} -1 & 1 & 1 & 0 & 0 & \vdots & 0 \\ 0 & 1 & 2 & 0 & -1 & \vdots & 0 \\ 0 & 0 & -1 & 1 & -1 & \vdots & 0 \\ 0 & 0 & 0 & 1 & 1 & \vdots & i_0 \\ 0 & 0 & 0 & 0 & 8 & \vdots & 4i_0 \end{bmatrix}$$

From the fifth row of this echelon matrix it can be seen that $8i_5 = 4i_0$, which means $i_5 = i_0/2$. Back-substitution yields $i_5 = i_0/2$, $i_4 = i_0/2$, $i_3 = 0$, $i_2 = i_0/2$ and $i_1 = i_0/2$. With the given resistances, the initial current splits in half at junction A into i_2 and i_3.

No current flows through BC and $i_0/2$ reunites with $i_0/2$ at D.

● **PROBLEM** 21-11

Determine E_1, E_2, and E_3 in the circuit of figure 1.

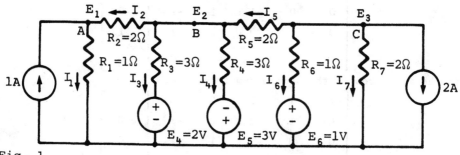

Fig. 1

Solution: We shall use Kirchhoff's current law and Ohm's law to obtain algebraic equations.

Node A: By Kirchhoff's current law,

current into node = current out of node

$$1 + I_2 = I_1 \tag{1}$$

Ohm's law is used next to express each current in terms of a component value in ohms and the voltage across it. This law states that the voltage drop E across a resistor is proportional to the current I flowing through the resistor. The constant of proportionality is called the resistance R. Thus, $E = RI$, or,

$$I = E/R . \tag{b}$$

Since the current I_2 passes through the resistance R_2, and the voltage drop across this resistor is $E_2 - E_1$, the value of I_2 is

then equal to $\dfrac{E_2 - E_1}{R_2}$. or, $I_2 = \dfrac{E_2 - E_1}{2}$.

Similarly,

$$I_1 = \frac{E_1}{R_1} = \frac{E_1}{1} .$$

Substituting the values of I_1 and I_2 into equation (1), gives,

$$1 + \frac{E_2 - E_1}{2} = \frac{E_1}{1} ,$$

or,

$$3E_1 - E_2 = 2 . \tag{2}$$

Node B: By Kirchhoff's current law,

$$\text{current into node = current out of node}$$

$$I_5 = I_2 + I_3 + I_4 . \tag{3}$$

Using equation (b) we have

$$I_5 = \frac{E_3 - E_2}{R_5} = \frac{E_3 - E_2}{2}$$

$$I_3 = \frac{E_2 - E_4}{R_3} = \frac{E_2 - 2}{3}$$

and

$$I_4 = \frac{E_2 + E_5}{R_4} = \frac{E_2 + 3}{3} .$$

Substitute these values into equation (3),

$$\frac{E_3 - E_2}{2} = \frac{E_2 - E_1}{2} + \frac{E_2 - 2}{3} + \frac{E_2 + 3}{3}$$

or,

$$3(E_3 - E_2) = 3(E_2 - E_1) + 2(E_2 - 2) + 2(E_2 + 3),$$

$$3E_1 - 10E_2 + 3E_3 = 2 . \tag{4}$$

For nodeC: Current into node = current out of node

$$0 = I_5 + I_6 + I_7 + 2 \tag{5}$$

Use of Ohm's law yields

$$I_6 = \frac{E_3 - E_6}{R_6} = \frac{E_3 - 1}{1}$$

$$I_7 = \frac{E_3}{R_7} = \frac{E_3}{2} .$$

Substituting the values of I_5, I_6 and I_7 into equation (5), yields

$$0 = \frac{E_3 - E_2}{2} + \frac{E_3 - 1}{1} + \frac{E_3}{2} + 2$$

or,

$$0 = E_3 - E_2 + 2E_3 - 2 + E_3 + 4$$

$$E_2 - 4E_3 = 2 . \qquad (6)$$

Thus, we have obtained three equations in three unknowns. These equations are:

$$3E_1 - E_2 \qquad\qquad = 2$$

$$3E_1 - 10E_2 + 3E_3 = 2 \qquad\qquad (7)$$

$$E_2 - 4E_3 = 2$$

The system (7) can be expressed in matrix notation as

$$\begin{bmatrix} 3 & -1 & 0 \\ 3 & -10 & 3 \\ 0 & 1 & -4 \end{bmatrix} \begin{bmatrix} E_1 \\ E_2 \\ E_3 \end{bmatrix} = \begin{bmatrix} 2 \\ 2 \\ 2 \end{bmatrix}$$

or,

$$AX = B \qquad\qquad (8)$$

where

$$A = \begin{bmatrix} 3 & -1 & 0 \\ 3 & -10 & 3 \\ 0 & 1 & -4 \end{bmatrix} , \quad X = \begin{bmatrix} E_1 \\ E_2 \\ E_3 \end{bmatrix} , \quad B = \begin{bmatrix} 2 \\ 2 \\ 2 \end{bmatrix} .$$

To solve system (8), Cramer's Rule can be used. According to this rule,

$$E_i = \frac{\Delta_i}{\Delta} , \qquad i = 1,2,3 .$$

where Δ_i = determinant of the matrix obtained by replacing the ith column of the matrix A by the column B.

Δ = determinant of matrix A.

First compute the determinant Δ of the matrix A

$$\Delta = \det \begin{bmatrix} 3 & -1 & 0 \\ 3 & -10 & 3 \\ 0 & 1 & -4 \end{bmatrix}$$

$$= 3 \begin{vmatrix} -10 & 3 \\ 1 & -4 \end{vmatrix} - (-1) \begin{vmatrix} 3 & 3 \\ 0 & -4 \end{vmatrix}$$

$$= 3 [40 - 3] + 1 [-12]$$

$$= 99 .$$

Now,

$$E_1 = \frac{\Delta_1}{\Delta} .$$

$$\Delta_1 = \det \begin{bmatrix} 2 & -1 & 0 \\ 2 & -10 & 3 \\ 2 & 1 & -4 \end{bmatrix}$$

$$= 2 \begin{vmatrix} -10 & 3 \\ 1 & -4 \end{vmatrix} - (-1) \begin{vmatrix} 2 & 3 \\ 2 & -4 \end{vmatrix}$$

$$= 2[40 - 3] + 1[-8 - 6]$$

$$= 60.$$

Thus,

$$E_1 = \frac{\Delta_1}{\Delta} = \frac{60}{99} = \frac{20}{33} .$$

Similarly,

$$E_2 = \frac{\Delta_2}{\Delta}$$

$$\Delta_2 = \det \begin{bmatrix} 3 & 2 & 0 \\ 3 & 2 & 3 \\ 0 & 2 & -4 \end{bmatrix}$$

$$= 3 \begin{vmatrix} 2 & 3 \\ 2 & -4 \end{vmatrix} - 2 \begin{vmatrix} 3 & 3 \\ 0 & -4 \end{vmatrix}$$

$$= 3[-8 - 6] - 2[-12]$$

$$= -18 .$$

Therefore,

$$E_2 = \frac{\Delta_2}{\Delta} = \frac{-18}{99} = -\frac{2}{11} .$$

Now,

$$E_3 = \frac{\Delta_3}{\Delta}$$

$$\Delta_3 = \det \begin{bmatrix} 3 & -1 & 2 \\ 3 & -10 & 2 \\ 0 & 1 & 2 \end{bmatrix}$$

Expand the determinant using the first column

$$\Delta_3 = 3 \begin{vmatrix} -10 & 2 \\ 1 & 2 \end{vmatrix} - 3 \begin{vmatrix} -1 & 2 \\ 1 & 2 \end{vmatrix}$$

$$= 3[-20 - 2] - 3[-2 - 2]$$

$$= -54.$$

Then,

$$E_3 = \frac{\Delta_3}{\Delta} = \frac{-54}{99} = -\frac{6}{11} .$$

Therefore, the values of E_1, E_2 and E_3 are:

$$E_1 = 20/33 \text{ volt.}$$
$$E_2 = -2/11 \text{ volt.}$$
$$E_3 = -6/11 \text{ volt.}$$

For the circuit shown in the figure below, find the values of the currents i_1, i_2, i_3, and i_4 .

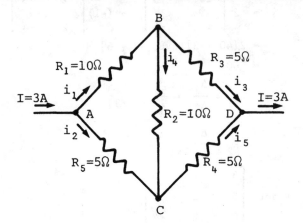

Solution: To derive algebraic equations, use Kirchhoff's voltage law and Kirchhoff's current law.

Kirchhoff's current law states that the net flow of current into any junction must be zero. At junction A, the current I flows in and currents i_1 and i_2 flow out. Therefore,

$$i_1 + i_2 = I \ .$$

As

$$I = 3,$$

$$i_1 + i_2 = 3 \ . \tag{1}$$

At B, i_1 flows in and i_3 and i_4 flow out. Therefore,

$$i_3 + i_4 = i_1$$

or

$$i_1 - i_3 - i_4 = 0 \ . \tag{2}$$

Similarly, applying Kirchhoff's current law to the junctions C and D in turn gives,

$$i_2 + i_4 - i_5 = 0 \tag{3}$$

$$i_3 + i_5 = 3 \ . \tag{4}$$

We have obtained four equations. Observe that the four equations are not independent. Rewrite these four equations as follows,

$$i_1 + i_2 \qquad\qquad\qquad = 3$$

$$i_3 \qquad + i_5 = 3$$

$$i_1 \qquad -i_3 - i_4 \qquad = 0$$

$$i_2 \qquad + i_4 - i_5 = 0 \ .$$

Add the second equation to the third equation. This yields,

$$i_1 + i_2 \qquad\qquad = 3$$

$$i_3 \qquad + i_5 = 3$$

$$i_1 \qquad\quad -i_4 + i_5 = 3$$

$$i_2 \qquad +i_4 - i_5 = 0$$

Now add the third equation to the fourth equation,

$$i_1 + i_2 \qquad\qquad = 3$$

$$i_3 \qquad + i_5 = 3$$

$$i_1 \qquad\quad -i_4 + i_5 = 3$$

$$i_1 + i_2 \qquad\qquad = 3$$

The first and fourth equations are the same. Therefore, there are three independent equations which are as follows,

$$i_1 + i_2 \qquad\qquad = 3$$

$$i_3 \qquad + i_5 = 3$$

$$i_1 \qquad\quad -i_4 + i_5 = 3 \ .$$

These equations are not sufficient to determine the currents $i_1, i_2, \ldots, i_5$.
To obtain a complete system of equations apply Kirchhoff's voltage law to the loops A B C A and B C D B.
For loop A B C A,

$$10i_1 + 10i_4 - 5i_2 = 0 \tag{5}$$

and for loop B C D B,

$$10i_4 + 5i_5 - 5i_3 = 0 \ . \tag{6}$$

We have obtained a system of five linear equations in five unknown currents.

$$i_1 + i_2 \qquad\qquad = 3$$

$$i_3 \qquad + i_5 = 3$$

$$i_1 \qquad\quad - i_4 + i_5 = 3$$

$$10i_1 - 5i_2 \quad +10i_4 \qquad = 0$$

$$-5i_3 +10i_4 +5i_5 = 0 \ .$$

In matrix notation this system takes the form

$$\begin{bmatrix} 1 & 1 & 0 & 0 & 0 \\ 0 & 0 & 1 & 0 & 1 \\ 1 & 0 & 0 & -1 & 1 \\ 10 & -5 & 0 & 10 & 0 \\ 0 & 0 & -5 & 10 & 5 \end{bmatrix} \begin{bmatrix} i_1 \\ i_2 \\ i_3 \\ i_4 \\ i_5 \end{bmatrix} = \begin{bmatrix} 3 \\ 3 \\ 3 \\ 0 \\ 0 \end{bmatrix}$$

or

$$AX = B \ ,$$

where

$$A = \begin{bmatrix} 1 & 1 & 0 & 0 & 0 \\ 0 & 0 & 1 & 0 & 1 \\ 1 & 0 & 0 & -1 & 1 \\ 10 & -5 & 0 & 10 & 0 \\ 0 & 0 & -5 & 10 & 5 \end{bmatrix}, \quad X = \begin{bmatrix} i_1 \\ i_2 \\ i_3 \\ i_4 \\ i_5 \end{bmatrix}, \quad B = \begin{bmatrix} 3 \\ 3 \\ 3 \\ 0 \\ 0 \end{bmatrix}.$$

To solve this system write the augmented matrix $[A\!:\!B]$, and then reduce it to echelon form. The augmented matrix of the system is

$$\begin{bmatrix} 1 & 1 & 0 & 0 & 0 & | & 3 \\ 0 & 0 & 1 & 0 & 1 & | & 3 \\ 1 & 0 & 0 & -1 & 1 & | & 3 \\ 10 & -5 & 0 & 10 & 0 & | & 0 \\ 0 & 0 & -5 & 10 & 5 & | & 0 \end{bmatrix}$$

Add -1 times the first row to the third row and -10 times the first row to the fourth row,

$$\begin{bmatrix} 1 & 1 & 0 & 0 & 0 & | & 3 \\ 0 & 0 & 1 & 0 & 1 & | & 3 \\ 0 & -1 & 0 & -1 & 1 & | & 0 \\ 0 & -15 & 0 & 10 & 0 & | & -30 \\ 0 & 0 & -5 & 10 & 5 & | & 0 \end{bmatrix}$$

Add -15 times the third row to the fourth row,

$$\begin{bmatrix} 1 & 1 & 0 & 0 & 0 & | & 3 \\ 0 & 0 & 1 & 0 & 1 & | & 3 \\ 0 & -1 & 0 & -1 & 1 & | & 0 \\ 0 & 0 & 0 & 25 & -15 & | & -30 \\ 0 & 0 & -5 & 10 & 5 & | & 0 \end{bmatrix}$$

Add 5 times the second row to the fifth row,

$$\begin{bmatrix} 1 & 1 & 0 & 0 & 0 & | & 3 \\ 0 & 0 & 1 & 0 & 1 & | & 3 \\ 0 & -1 & 0 & -1 & 1 & | & 0 \\ 0 & 0 & 0 & 25 & -15 & | & -30 \\ 0 & 0 & 0 & 10 & 10 & | & 15 \end{bmatrix}$$

Divide the fourth row by 25 and the fifth row by 10,

$$\begin{bmatrix} 1 & 1 & 0 & 0 & 0 & | & 3 \\ 0 & 0 & 1 & 0 & 1 & | & 3 \\ 0 & -1 & 0 & -1 & 1 & | & 0 \\ 0 & 0 & 0 & 1 & -3/5 & | & -6/5 \\ 0 & 0 & 0 & 1 & 1 & | & 3/2 \end{bmatrix}$$

Add -1 times the fourth row to the fifth row,

$$\begin{bmatrix} 1 & 1 & 0 & 0 & 0 & | & 3 \\ 0 & 0 & 1 & 0 & 1 & | & 3 \\ 0 & -1 & 0 & -1 & 1 & | & 0 \\ 0 & 0 & 0 & 1 & -3/5 & | & -6/5 \\ 0 & 0 & 0 & 0 & 8/5 & | & 27/10 \end{bmatrix}$$

Multiply the fifth row by 5/8,

$$\begin{bmatrix} 1 & 1 & 0 & 0 & 0 & | & 3 \\ 0 & 0 & 1 & 0 & 1 & | & 3 \\ 0 & -1 & 0 & -1 & 1 & | & 0 \\ 0 & 0 & 0 & 1 & -3/5 & | & -6/5 \\ 0 & 0 & 0 & 0 & 1 & | & 27/16 \end{bmatrix}$$

Add -1 times the fifth row to the second row and to the third row. Also add 3/5 times the fifth row to the fourth row,

$$\begin{bmatrix} 1 & 1 & 0 & 0 & 0 & | & 3 \\ 0 & 0 & 1 & 0 & 0 & | & 21/16 \\ 0 & -1 & 0 & -1 & 0 & | & -27/16 \\ 0 & 0 & 0 & 1 & 0 & | & -3/16 \\ 0 & 0 & 0 & 0 & 1 & | & 27/16 \end{bmatrix}$$

Add the fourth row to the third row,

$$\begin{bmatrix} 1 & 1 & 0 & 0 & 0 & | & 3 \\ 0 & 0 & 1 & 0 & 0 & | & 21/16 \\ 0 & -1 & 0 & 0 & 0 & | & -30/16 \\ 0 & 0 & 0 & 1 & 0 & | & -3/16 \\ 0 & 0 & 0 & 0 & 1 & | & 27/16 \end{bmatrix}$$

Add the third row to the first row and then interchange rows two and three to obtain,

$$\begin{bmatrix} 1 & 0 & 0 & 0 & 0 & | & 18/16 \\ 0 & -1 & 0 & 0 & 0 & | & -30/16 \\ 0 & 0 & 1 & 0 & 0 & | & 21/16 \\ 0 & 0 & 0 & 1 & 0 & | & -3/16 \\ 0 & 0 & 0 & 0 & 1 & | & 27/16 \end{bmatrix}$$

This is the augmented matrix of the system

$$
\begin{aligned}
i_1 &= 18/16 \\
-i_2 &= -30/16 \\
i_3 &= 21/16 \\
i_4 &= -3/16 \\
i_5 &= 27/16
\end{aligned}
$$

Therefore, the values of the currents are,

$$
\begin{aligned}
i_1 &= 18/16 \\
i_2 &= 30/16 \\
i_3 &= 21/16
\end{aligned}
$$

$$i_4 = -3/16$$
$$i_5 = 27/16 \ .$$

BRIDGE NETWORKS

Consider an unbalanced bridge network shown in the figure below. Assume that the voltage source v is a battery of 120 volts (i.e., $v = 120$), and given that $R_1 = 0.5$ ohm, $R_2 = R_4 = R_5 = R_6 = 2$ ohms and $R_3 = 1$ ohm. Find the values of currents I_1, I_2, and I_3 .

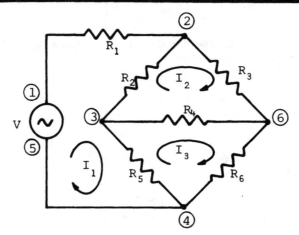

Solution: Kirchhoff's voltage law states that the algebraic sum of the voltages around a loop of an electric network is zero.

Applying Kirchhoff's voltage law to the loop ① ② ③ ④ ⑤ ①, yields

$$v - I_1R_1 - (I_1 - I_2)R_2 - (I_1 - I_3)R_5 = 0$$

or

$$v - (R_1 + R_2 + R_5)I_1 + R_2I_2 + R_5I_3 \ .$$

Substitute the values of v, R_1, R_2, and R_5 into the above equation.

Thus,

$$120 - 4.5I_1 + 2I_2 + 2I_3 = 0$$

or

$$4.5I_1 - 2I_2 - 2I_3 = 120 \ . \tag{1}$$

Now consider the loop ② ⑥ ③ ② . Applying Kirchhoff's voltage law to this loop yields

$$(I_2 - I_1)R_2 + (I_2 - I_3)R_4 + R_3I_2 = 0$$

or

$$-R_2I_1 + (R_2 + R_4 + R_3)I_2 - R_4I_3 = 0 \ .$$

Upon substitution of the values of R_2, R_4, and R_3, the equation becomes

$$-2I_1 + 5I_2 - 2I_3 = 0 \ . \tag{2}$$

Kirchhoff's voltage law applied to the loop ③ ⑥ ④ ③ yields

$$(I_3 - I_1)R_5 + I_3R_6 + (I_3 - I_2)R_4 = 0 \ .$$

Rearrangement yields

$$-R_5I_1 - R_4I_2 + (R_5 + R_4 + R_6)I_3 = 0 \ .$$

Substitution of the values of R_4, R_5, and R_6 into the above equation yields

$$-2I_1 - 2I_2 + 6I_3 = 0 . \tag{3}$$

Thus far we have obtained the following three equations in the three unknown currents:

$$4.5I_1 - 2I_2 - 2I_3 = 120$$
$$-2I_1 + 5I_2 - 2I_3 = 0 \tag{4}$$
$$-2I_1 - 2I_2 + 6I_3 = 0$$

System (4) may be solved by using Cramer's Rule. System (4) can be written in matrix form as:

$$\begin{bmatrix} 4.5 & -2 & -2 \\ -2 & 5 & -2 \\ -2 & -2 & 6 \end{bmatrix} \begin{bmatrix} I_1 \\ I_2 \\ I_3 \end{bmatrix} = \begin{bmatrix} 120 \\ 0 \\ 0 \end{bmatrix}$$

or,

$$AI = B$$

where

$$A = \begin{bmatrix} 4.5 & -2 & -2 \\ -2 & 5 & -2 \\ -2 & -2 & 6 \end{bmatrix}, \quad I = \begin{bmatrix} I_1 \\ I_2 \\ I_3 \end{bmatrix}, \quad \text{and} \quad B = \begin{bmatrix} 120 \\ 0 \\ 0 \end{bmatrix}$$

From Cramer's rule,

$$I_1 = \frac{\Delta_1}{\det A} \tag{5}$$

where Δ_1 is the determinant of the matrix obtained by replacing the first column A by the column B. Thus,

$$\Delta_1 = \det \begin{bmatrix} 120 & -2 & -2 \\ 0 & 5 & -2 \\ 0 & -2 & 6 \end{bmatrix}$$

Now if the determinant is expanded along the first column, all terms which contain 0 may be neglected. Therefore,

$$\Delta_1 = 120 \begin{vmatrix} 5 & -2 \\ -2 & 6 \end{vmatrix} = 120(30 - 4) = 3120 .$$

Now,

$$\det A = \det \begin{bmatrix} 4.5 & -2 & -2 \\ -2 & 5 & -2 \\ -2 & -2 & 6 \end{bmatrix}$$

Recall that if one row of A is added to a multiple of another row, the determinant is not changed.

Add -1 times the first row to the second row, and 3 times the first row to the third row. Then,

$$\det A = \det \begin{bmatrix} 4.5 & -2 & -2 \\ -6.5 & 7 & 0 \\ -11.5 & -8 & 0 \end{bmatrix}$$

Now expanding the determinant using the third column yields

$$\det A = -2 \begin{vmatrix} -6.5 & 7 \\ 11.5 & -8 \end{vmatrix}$$

$$= -2 \ [52 - 80.5] = 57 \ .$$

Substitute the values of Δ_1 and $\det A$ into equation (5)

$$I_1 = \frac{3120}{57} = 54.7 \ .$$

Similarly,

$$I_2 = \frac{\Delta_2}{\det A}$$

where Δ_2 is the determinant of the matrix obtained by replacing the second column of A by the column of B. Thus,

$$\Delta_2 = \det \begin{bmatrix} 4.5 & 120 & -2 \\ -2 & 0 & -2 \\ -2 & 0 & 6 \end{bmatrix}$$

Expand the determinant using the second column,

$$\Delta_2 = 120 \begin{matrix} -2 & -2 \\ -2 & 6 \end{matrix}$$

$$= -120(-12 - 4) = 1920 \ .$$

Then,

$$I_2 = \frac{\Delta_2}{\det A} = \frac{1920}{57} = 33.7 \ .$$

Now,

$$I_3 = \frac{\Delta_3}{\det A}$$

where Δ_3 is the determinant of the matrix obtained by replacing the third column A by the column B. Thus,

$$\Delta_3 = \det \begin{bmatrix} 4.5 & -2 & 120 \\ -2 & 5 & 0 \\ -2 & -2 & 0 \end{bmatrix}$$

Expanding the determinant by the third column, yields

$$\Delta_3 = 120 \begin{bmatrix} -2 & 5 \\ -2 & -2 \end{bmatrix}$$

$$= 120(4 + 10) = 1680.$$

Therefore,

$$I_3 = \frac{\Delta_3}{\det A} = \frac{1680}{57} = 29.5 \ .$$

The values of the three currents, I_1, I_2 and I_3 are

$$I_1 = 54.7$$

$$I_2 = 33.7$$

$$I_3 = 29.5$$

Find the values of the currents I_1, I_2, and I_3 in the circuit shown in the figure below.

Solution: Using Kirchhoff's voltage law on the loop E A B C D E, yields

$$100 - R_1(I_1 - I_2) - R_4(I_1 - I_3) = 0$$

or

$$100 - I_1(R_1 + R_4) + R_1 I_2 + R_4 I_3 = 0 .$$

Substituting the values of R_1 and R_4 yields

$$100 - 12I_1 + 4I_2 + 8I_3 = 0$$

or

$$12I_1 - 4I_2 - 8I_3 = 100 . \qquad (1)$$

Considering the loop A F B A,

$$R_2 I_2 + R_3(I_2 - I_3) + R_1(I_2 - I_1) = 0 .$$

Rearrangement yields

$$-R_1 I_1 + I_2(R_2 + R_3 + R_1) - R_3 I_3 = 0 .$$

Substituting the values of R_1, R_2, and R_3 yields

$$-4I_1 + 19I_2 - 5I_3 = 0 \qquad (2)$$

Now consider the third loop B F C B . Applying Kirchhoff's voltage law yields,

$$R_3(I_3 - I_2) + R_5 I_3 + R_4(I_3 - I_1) = 0 .$$

Collecting terms in I_1, I_2, and I_3,

$$-R_4 I_1 - R_3 I_2 + I_3(R_3 + R_5 + R_4) = 0 .$$

Substitute the values of R_3, R_4, and R_5 . Then,

$$-8I_1 - 5I_2 + 15I_3 = 0 . \qquad (3)$$

We have obtained three linear equations in three unknown currents. These equations are:

$$12I_1 - 4I_2 - 8I_3 = 100$$
$$-4I_1 + 19I_2 - 5I_3 = 0 \qquad (4)$$
$$-8I_1 - 5I_2 + 15I_3 = 0 .$$

The system (4) can be written in matrix form as

$$\begin{bmatrix} 12 & -4 & -8 \\ -4 & 19 & -5 \\ -8 & -5 & 15 \end{bmatrix} \begin{bmatrix} I_1 \\ I_2 \\ I_3 \end{bmatrix} = \begin{bmatrix} 100 \\ 0 \\ 0 \end{bmatrix} \qquad (5)$$

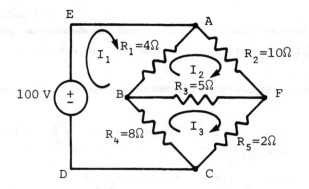

Let

$$A = \begin{bmatrix} 12 & -4 & -8 \\ -4 & 19 & -5 \\ -8 & -5 & 15 \end{bmatrix}, \quad I = \begin{bmatrix} I_1 \\ I_2 \\ I_3 \end{bmatrix}, \quad \text{and} \quad B = \begin{bmatrix} 100 \\ 0 \\ 0 \end{bmatrix}.$$

Then, equation (5) becomes

$$AI = B .$$

To solve this system Cramer's Rule can be used. According to this rule,

$$I_i = \frac{\Delta_i}{\Delta} , \quad i = 1,2,3,$$

where Δ_i = determinant of the matrix obtained by replacing the ith column of matrix A by the column of B ;

$$\Delta = \det A .$$

Now,

$$\det A = \det \begin{bmatrix} 12 & -4 & -8 \\ -4 & 19 & -5 \\ -8 & -5 & 15 \end{bmatrix} .$$

Recall that, if we add to one row of A a multiple of another row, the determinant, is not changed.

Add 3 times the second row to the first row and -2 times the second row to the third row. Then,

$$\det A = \det \begin{bmatrix} 0 & 53 & -23 \\ -4 & 19 & -5 \\ 0 & -43 & 25 \end{bmatrix} .$$

786

Expand the determinant using the first column.

$$\det A = -(-4) \begin{vmatrix} 53 & -23 \\ -43 & 25 \end{vmatrix}$$

$$= 4[1325 - 989]$$

$$= 1344.$$

Then,

$$\Delta_1 = \det \begin{bmatrix} 100 & -4 & -8 \\ 0 & 19 & -5 \\ 0 & -5 & 15 \end{bmatrix}$$

$$= 100 \begin{bmatrix} 19 & -5 \\ -5 & 15 \end{bmatrix}$$

$$= 100[285 - 25]$$

$$= 26000.$$

I_1 can now be calculated

$$I_1 = \frac{\Delta_1}{\Delta} = \frac{26000}{1344} = 19.35 \text{ amperes.}$$

Similarly,

$$I_2 = \frac{\Delta_2}{\Delta}$$

where Δ_2 = determinant of the matrix obtained by replacing the second column of A by a column of B. Thus,

$$\Delta_2 = \det \begin{bmatrix} 12 & 100 & -8 \\ -4 & 0 & -5 \\ -8 & 0 & 15 \end{bmatrix} .$$

Expanding the determinant using the second column yields

$$\Delta_2 = -100 \begin{vmatrix} -4 & -5 \\ -8 & 15 \end{vmatrix}$$

$$= -100 [-60-40]$$

$$= 10000 .$$

Therefore,

$$I_2 = \frac{10000}{1344} = 7.44 \text{ amperes.}$$

Now,

$$I_3 = \frac{\Delta_3}{\Delta}$$

where Δ_3 = determinant of the matrix obtained by replacing the third column of A by the column B. Thus,

$$\Delta_3 = \det \begin{bmatrix} 12 & -4 & 100 \\ -4 & 19 & 0 \\ -8 & -5 & 0 \end{bmatrix} .$$

Expand the determinant by the third column, neglecting terms containing zero:

$$\Delta_3 = 100 \begin{vmatrix} -4 & 19 \\ -8 & -5 \end{vmatrix}$$

$$= 100 \ [20 + 152]$$

$$= 17200$$

Then,

$$I_3 = \frac{\Delta_3}{\Delta} = \frac{17200}{1344} = 12 \cdot 80 \text{ amperes.}$$

Therefore the values of I_1, I_2, and I_3 are

$$I_1 = 19 \cdot 4 \text{ amperes}$$

$$I_2 = 7 \cdot 44 \text{ amperes}$$

$$I_3 = 12 \cdot 8 \text{ amperes.}$$

It is important to note that Cramer's rule is only applicable to a system with the same number of equations as unknowns, and that the solution can only be reached when $\Delta \neq 0$ (i.e., det A $\neq$ 0).

● **PROBLEM** 21-15

Find the three currents I_1, I_2 and I_3 for the bridged-tee network of figure 1.

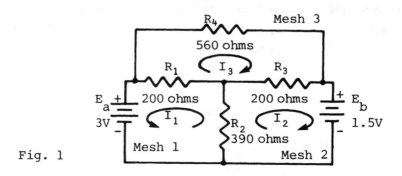

Fig. 1

Solution: In order to obtain equations in terms of the three currents, apply Kirchhoff's voltage law to each mesh. The law states: The algebraic sum of the voltages around a loop of an electric network is zero. Consider mesh 1. Remember that if two currents flow in opposite directions through a resistor, the net current is the difference between the two currents. We shall use Ohm's law to calculate the voltage drop across each resistor of mesh 1.

Now the voltage drops across the resistors R_1 and R_2 are $(I_1+I_3)R_1$ and $(I_1+I_2)R_2$ respectively. Applying Kirchhoff's voltage law to mesh 1, we therefore obtain

$$E_a - (I_1+I_3)R_1 - (I_1+I_2)R_2 = 0 \ .$$

We are given that $E_a = 3V$. Therefore,

$$(I_1+I_3)R_1 + (I_1+I_2)R_2 = 3$$

or

$$(R_1+R_2)I_1 + R_2I_2 + R_1I_3 = 3 \ .$$

Substitution of the given values of R_1, R_2 and R_3 yields

$$(200 + 390)I_1 + 390I_2 + 200I_3 = 3$$

or

$$590I_1 + 390I_2 + 200I_3 = 3 . \tag{1}$$

Now consider mesh 2. The voltage drops across the resistors R_2 and R_3 are $(I_1+I_2)R_2$ and $(I_2-I_3)R_3$ respectively. Applying Kirchhoff's voltage law to mesh 2, we therefore obtain

$$E_b - (I_1+I_2)R_2 - (I_2-I_3)R_3 = 0$$

or

$$(I_1+I_2)R_2 + (I_2-I_3)R_3 = E_b$$

$$R_2 I_1 + (R_2+R_3)I_2 - I_3 R_3 = E_b .$$

Substitution of the given values of R_2, R_3 and E_b yields

$$390I_1 + (390+200)I_2 - 200I_3 = 1.5$$

or

$$390I_1 + 590I_2 - 200I_3 = 1.5 . \tag{2}$$

The equation for mesh 3 is obtained in a like manner. The voltage drops across the resistors R_1, R_3, and R_4 are $(I_1+I_3)R_1$, $(I_3-I_2)R_3$, and $I_3 R_4$ respectively. Thus Kirchhoff's voltage law applied to this mesh yields

$$(I_1+I_3)R_1 + (I_3-I_2)R_3 + I_3 R_4 = 0$$

or

$$R_1 I_1 - R_3 I_2 + (R_1+R_3+R_4)I_3 = 0 . \tag{3}$$

Substitute the given values of R_1, R_3, and R_4 into equation (3),

$$200I_1 - 200I_2 + (200+200+560)I_3 = 0$$

or

$$200I_1 - 200I_2 + 960I_3 = 0 . \tag{4}$$

Thus, we have obtained three equations in three unknowns. Now each member of the three equations is divided by 100 to keep the numbers small. Thus,

$$5 \cdot 9 I_1 + 3 \cdot 9 I_2 + 2 I_3 = 0 \cdot 03$$

$$3 \cdot 9 I_1 + 5 \cdot 9 I_2 - 2 I_3 = 0.015 \tag{5}$$

$$2 I_1 - 2 I_2 + 9 \cdot 6 I_3 = 0 .$$

The system (5) can be written in matrix notation as:

$$\begin{bmatrix} 5 \cdot 9 & 3 \cdot 9 & 2 \\ 3 \cdot 9 & 5 \cdot 9 & -2 \\ 2 & -2 & 9 \cdot 6 \end{bmatrix} \begin{bmatrix} I_1 \\ I_2 \\ I_3 \end{bmatrix} = \begin{bmatrix} 0 \cdot 03 \\ 0 \cdot 015 \\ 0 \end{bmatrix}$$

or

$$AI = B \tag{6}$$

where

$$A = \begin{bmatrix} 5 \cdot 9 & 3 \cdot 9 & 2 \\ 3 \cdot 9 & 5 \cdot 9 & -2 \\ 2 & -2 & 9 \cdot 6 \end{bmatrix} , \quad B = \begin{bmatrix} 0 \cdot 03 \\ 0.015 \\ 0 \end{bmatrix} \quad \text{and}$$

$$I = \begin{bmatrix} I_1 \\ I_2 \\ I_3 \end{bmatrix}.$$

The system (5) can be solved by using Cramer's Rule. According to this rule, if $\det A \neq 0$, then the solution $I = [I_i]$ of the system of linear equations $AI = B$ is given by

$$I_i = \Delta_i / \Delta \quad (i = 1,2,3)$$

where $\Delta = \det A$ and $\Delta_i = \det A_i$, where A_i is the matrix obtained by replacing the ith column of A by B.

First compute $\det A$

$$\det A = \det \begin{bmatrix} 5 \cdot 9 & 3 \cdot 9 & 2 \\ 3 \cdot 9 & 5 \cdot 9 & -2 \\ 2 & -2 & 9 \cdot 6 \end{bmatrix}$$

$$= 5 \cdot 9 \begin{vmatrix} 5 \cdot 9 & -2 \\ -2 & 9 \cdot 6 \end{vmatrix} - 3 \cdot 9 \begin{vmatrix} 3 \cdot 9 & -2 \\ 2 & 9 \cdot 6 \end{vmatrix} + 2 \begin{vmatrix} 3 \cdot 9 & 5 \cdot 9 \\ 2 & -2 \end{vmatrix}$$

$$= 310.576 - 161.616 - 39.2$$

$$= 109.8 .$$

Then,

$$I_1 = \frac{\Delta_1}{\Delta} = \frac{\det A_1}{\det A} .$$

The matrix A_1 is obtained by replacing the first column of A by B. Thus,

$$A_1 = \begin{bmatrix} .03 & 3 \cdot 9 & 2 \\ .015 & 5 \cdot 9 & -2 \\ 0 & -2 & 9 \cdot 6 \end{bmatrix}$$

then,

$$\det A_1 = \begin{vmatrix} .03 & 3 \cdot 9 & 2 \\ .015 & 5 \cdot 9 & -2 \\ 0 & -2 & 9 \cdot 6 \end{vmatrix}$$

Recall that the value of a determinant is unchanged if a multiple of one row is added to another row. Adding -2 times the second row to the first row yields

$$\det A_1 = \det \begin{bmatrix} 0 & -7 \cdot 9 & 6 \\ .015 & 5 \cdot 9 & -2 \\ 0 & -2 & 9 \cdot 6 \end{bmatrix} .$$

Now it is easy to compute $\det A_1$ by expanding the determinant using the first column. Thus,

$$\det A_1 = -.015 \begin{vmatrix} -7 \cdot 9 & 6 \\ -2 & 9 \cdot 6 \end{vmatrix}$$

$$= -0 \cdot 015[(-75 \cdot 84) - (-12)]$$

$$= -0 \cdot 015(-63 \cdot 84)$$

$$= 0.958 .$$

I_1 can now be calculated,

$$I_1 = \frac{\det A_1}{\det A}$$

$$= \frac{0 \cdot 958}{109 \cdot 8} = .0087 \text{ amp} = 8 \cdot 7 \text{ ma}.$$

Similarly,

$$I_2 = \frac{\Delta_2}{\Delta} = \frac{\det A_2}{\det A} \, .$$

The matrix A_2 is obtained by replacing the second column of A by B, thus,

$$A_2 = \begin{bmatrix} 5 \cdot 9 & .03 & 2 \\ 3 \cdot 9 & .015 & -2 \\ 2 & 0 & 9 \cdot 6 \end{bmatrix}$$

Adding -2 times the second row to the first row gives

$$A_2 = \begin{bmatrix} -1 \cdot 9 & 0 & 6 \\ 3 \cdot 9 & .015 & -2 \\ 2 & 0 & 9 \cdot 6 \end{bmatrix} \, .$$

Then compute the $\det A_2$ by expanding the determinant using the second column. Thus,

$$\det A_2 = .015 \begin{vmatrix} -1 \cdot 9 & 6 \\ 2 & 9 \cdot 6 \end{vmatrix} = .015(-18 \cdot 24 - 12)$$

$$= -.4536$$

I_2 may now be calculated,

$$I_2 = \frac{\det A_2}{\det A} = \frac{-0.4536}{109 \cdot 8} = -0.0041 \text{ amp} \, .$$

or

$$I_2 = -4 \cdot 1 \text{ ma} \quad (\text{because } 1 \text{ amp} = 10^{-3} \text{ ma}).$$

Now,

$$I_3 = \frac{\Delta_3}{\Delta} = \frac{\det A_3}{\det A}$$

The matrix A_3 is obtained by replacing the third column of A by B, thus,

$$A_3 = \begin{bmatrix} 5 \cdot 9 & 3 \cdot 9 & .03 \\ 3 \cdot 9 & 5 \cdot 9 & 0 \cdot 015 \\ 2 & -2 & 0 \end{bmatrix}$$

Add -2 times the second row to the first row.

$$A_3 = \begin{bmatrix} -1 \cdot 9 & -7 \cdot 9 & 0 \\ 3 \cdot 9 & 5.9 & 0.015 \\ 2 & -2 & 0 \end{bmatrix}$$

$$\det A_3 = \begin{vmatrix} -1.9 & -7 \cdot 9 & 0 \\ 3 \cdot 9 & 5 \cdot 9 & 0.015 \\ 2 & -2 & 0 \end{vmatrix}$$

Expand the determinant by the third column.

$$\det A_3 = -0.015 \begin{vmatrix} -1.9 & -7.9 \\ 2 & -2 \end{vmatrix}$$

$$= -0.294.$$

Therefore,

$$I_3 = \frac{\det A_3}{\det A} = \frac{-0.294}{109.8} = -.0027 \text{ amp}$$

or

$$I_3 = -2.7 \text{ ma} \quad .$$

Thus, the values of three currents I_1, I_2, and I_3 are as follows:

$$I_1 = 8.7 \text{ ma}$$

$$I_2 = -4.1 \text{ ma}$$

$$I_3 = -2.7 \text{ ma}.$$

The currents I_2 and I_3 are negative because their directions are opposite to the directions chosen arbitrarily for them in figure 1.

● **PROBLEM** 21-16

Consider the electrical network shown in figure 1. Find the values of all the labeled currents, if $R_1 = R_2 = R_4 = R_5 = R_7 = 1$ ohm, and $R_3 = R_6 = 2$ ohms.

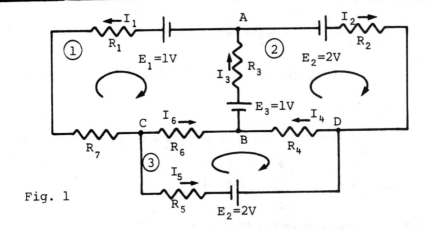

Fig. 1

Solution: We wish to determine six currents. Therefore, six independent equations will be needed to determine the six currents. Three equations can be obtained by applying Kirchhoff's voltage law to the loops ①, ②, and ③ .

For loop ① :

$$E_1 - I_1 R_1 - I_1 R_7 - I_6 R_6 + E_3 - I_3 R_3 = 0$$

or

$$I_1(R_1 + R_7) + I_3 R_3 + I_6 R_6 = E_1 + E_3 \quad .$$

Substituting the given values of R_1, R_3, R_6, R_7, E_1, and E_3 gives

$$2I_1 + 2I_3 + 2I_6 = 2 \quad . \tag{1}$$

Now, moving in a direction opposite to the assumed current flow, we obtain a voltage rise which is the negative of the voltage drop occurring in the direction of the current flow.

For loop ②:

$$-E_2 + I_2R_2 + I_4R_4 - E_3 + I_3R_3 = 0$$

or

$$I_2R_2 + I_3R_3 + I_4R_4 = E_3 + E_3 .$$

Substitution of the given values of resistances and voltages yields,

$$I_2 + 2I_3 + I_4 = 3 . \tag{2}$$

For loop ③:

$$E_5 - I_5R_5 + I_6R_6 - I_4R_4 = 0$$

or

$$I_4R_4 + I_5R_5 - I_6R_6 = E_5$$

or

$$I_4 + I_5 - 2I_6 = 2 . \tag{3}$$

The remaining three equations are found by means of Kirchhoff's current law, which states that at each node the net sum of the currents is 0. Note that a current entering the node must have the opposite sign of a current leaving the node.

For node A,

$$I_1 + I_2 - I_3 = 0 . \tag{4}$$

For node B,

$$I_3 - I_4 - I_6 = 0 . \tag{5}$$

For node C,

$$I_5 + I_6 - I_1 = 0 . \tag{6}$$

Equations (1) through (6) provide a set of six simultaneous linear equations to be solved for the currents. These equations can be rewritten as

$$
\begin{array}{rcrcrcrcrcrcl}
2I_1 & + & 2I_3 & & & & & & & + & 2I_6 & = & 2 \\
& & I_2 & + & 2I_3 & + & I_4 & & & & & = & 3 \\
& & & & & & I_4 & + & I_5 & - & 2I_6 & = & 2 \\
I_1 & + & I_2 & - & I_3 & & & & & & & = & 0 \\
& & & & I_3 & - & I_4 & & & - & I_6 & = & 0 \\
-I_1 & & & & & & & + & I_5 & + & I_6 & = & 0
\end{array} \tag{7}
$$

In matrix notation, system (7) becomes

$$
\begin{bmatrix}
2 & 0 & 2 & 0 & 0 & 2 \\
0 & 1 & 2 & 1 & 0 & 0 \\
0 & 0 & 0 & 1 & 1 & -2 \\
1 & 1 & -1 & 0 & 0 & 0 \\
0 & 0 & 1 & -1 & 0 & -1 \\
-1 & 0 & 0 & 0 & 1 & 1
\end{bmatrix}
\begin{bmatrix}
I_1 \\ I_2 \\ I_3 \\ I_4 \\ I_5 \\ I_6
\end{bmatrix}
=
\begin{bmatrix}
2 \\ 3 \\ 2 \\ 0 \\ 0 \\ 0
\end{bmatrix}
$$

or

$$AX = B \tag{8}$$

793

where

$$A = \begin{bmatrix} 2 & 0 & 2 & 0 & 0 & 2 \\ 0 & 1 & 2 & 1 & 0 & 0 \\ 0 & 0 & 0 & 1 & 1 & -2 \\ 1 & 1 & -1 & 0 & 0 & 0 \\ 0 & 0 & 1 & -1 & 0 & -1 \\ -1 & 0 & 0 & 0 & 1 & 1 \end{bmatrix}, \quad X = \begin{bmatrix} I_1 \\ I_2 \\ I_3 \\ I_4 \\ I_5 \\ I_6 \end{bmatrix}$$

and

$$B = \begin{bmatrix} 2 \\ 3 \\ 2 \\ 0 \\ 0 \\ 0 \end{bmatrix}.$$

To solve the system (8), form the augmented matrix $[A:B]$ and then reduce it to an echelon matrix. The augmented matrix of the system is

$$\begin{bmatrix} 2 & 0 & 2 & 0 & 0 & 2 & | & 2 \\ 0 & 1 & 2 & 1 & 0 & 0 & | & 3 \\ 0 & 0 & 0 & 1 & 1 & -2 & | & 2 \\ 1 & 1 & -1 & 0 & 0 & 0 & | & 0 \\ 0 & 0 & 1 & -1 & 0 & -1 & | & 0 \\ -1 & 0 & 0 & 0 & 1 & 1 & | & 0 \end{bmatrix}$$

Divide the first row by 2. Then add the first row to the sixth row. Multiply the first row by -1 and then add it to the fourth row, to obtain,

$$\begin{bmatrix} 1 & 0 & 1 & 0 & 0 & 1 & | & 1 \\ 0 & 1 & 2 & 1 & 0 & 0 & | & 3 \\ 0 & 0 & 0 & 1 & 1 & -2 & | & 2 \\ 0 & 1 & -2 & 0 & 0 & -1 & | & -1 \\ 0 & 0 & 1 & -1 & 0 & -1 & | & 0 \\ 0 & 0 & 1 & 0 & 1 & 2 & | & 1 \end{bmatrix}$$

Add -1 times the second row to the fourth row and -1 times the fifth row to the sixth row.

$$\begin{bmatrix} 1 & 0 & 1 & 0 & 0 & 1 & | & 1 \\ 0 & 1 & 2 & 1 & 0 & 0 & | & 3 \\ 0 & 0 & 0 & 1 & 1 & -2 & | & 2 \\ 0 & 0 & -4 & -1 & 0 & -1 & | & -4 \\ 0 & 0 & 1 & -1 & 0 & -1 & | & 0 \\ 0 & 0 & 0 & 1 & 1 & 3 & | & 1 \end{bmatrix}$$

Add -1 times the third row to the sixth row and 4 times the fifth row to the fourth row,

$$\left[\begin{array}{cccccc|c}
1 & 0 & 1 & 0 & 0 & 1 & 1 \\
0 & 1 & 2 & 1 & 0 & 0 & 3 \\
0 & 0 & 0 & 1 & 1 & -2 & 2 \\
0 & 0 & 0 & -5 & 0 & -5 & -4 \\
0 & 0 & 1 & -1 & 0 & -1 & 0 \\
0 & 0 & 0 & 0 & 0 & 5 & -1
\end{array}\right]$$

Divide the fourth row by -5 and the fifth row by 5,

$$\left[\begin{array}{cccccc|c}
1 & 0 & 1 & 0 & 0 & 1 & 1 \\
0 & 1 & 2 & 1 & 0 & 0 & 3 \\
0 & 0 & 0 & 1 & 1 & -2 & 2 \\
0 & 0 & 0 & 1 & 0 & 1 & 4/5 \\
0 & 0 & 1 & -1 & 0 & -1 & 0 \\
0 & 0 & 0 & 0 & 0 & 1 & -1/5
\end{array}\right]$$

Add

(i)   -1 times the sixth row to the first row
(ii)   2 times the sixth row to the third row
(iii) -1 times the sixth row to the fourth row
(iv)   the sixth row to the fifth row.

this yields,

$$\left[\begin{array}{cccccc|c}
1 & 0 & 1 & 0 & 0 & 0 & 6/5 \\
0 & 1 & 2 & 1 & 0 & 0 & 3 \\
0 & 0 & 0 & 1 & 1 & 0 & 8/5 \\
0 & 0 & 0 & 1 & 0 & 0 & 1 \\
0 & 0 & 1 & -1 & 0 & 0 & -1/5 \\
0 & 0 & 0 & 0 & 0 & 1 & -1/5
\end{array}\right]$$

Add

(i)   -1 times the fourth row to the second row and to the third row.

(ii) fourth row to the fifth row.

$$\left[\begin{array}{cccccc|c}
1 & 0 & 1 & 0 & 0 & 0 & 6/5 \\
0 & 1 & 2 & 0 & 0 & 0 & 2 \\
0 & 0 & 0 & 0 & 1 & 0 & 3/5 \\
0 & 0 & 0 & 1 & 0 & 0 & 1 \\
0 & 0 & 1 & 0 & 0 & 0 & 4/5 \\
0 & 0 & 0 & 0 & 0 & 1 & -1/5
\end{array}\right]$$

Add (i) -1 times the fifth row to the first row.

(ii) -2 times the fifth row to the second row, to obtain

$$\begin{bmatrix} 1 & 0 & 0 & 0 & 0 & 0 & \vline & 2/5 \\ 0 & 1 & 0 & 0 & 0 & 0 & \vline & 2/5 \\ 0 & 0 & 0 & 0 & 1 & 0 & \vline & 3/5 \\ 0 & 0 & 0 & 1 & 0 & 0 & \vline & 1 \\ 0 & 0 & 1 & 0 & 0 & 0 & \vline & 4/5 \\ 0 & 0 & 0 & 0 & 0 & 1 & \vline & -1/5 \end{bmatrix}$$

which is the augmented matrix of

$$I_1 = 2/5$$
$$I_2 = 2/5$$
$$I_5 = 3/5$$
$$I_4 = 1$$
$$I_3 = 4/5$$
$$I_6 = -1/5$$

Hence, the values of the currents are

$$I_1 = 2/5$$
$$I_2 = 2/5$$
$$I_3 = 4/5$$
$$I_4 = 1$$
$$I_5 = 3/5$$
$$I_6 = -1/5 \quad .$$

RLC NETWORKS

● PROBLEM 21-17

Consider the electrical circuit of figure 1. Find the natural frequencies of the system.

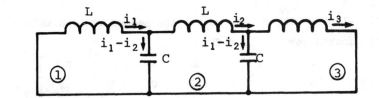

Fig. 1

Solution: Let i stand for current and q stand for charge. The voltage drop across an inductor is given by

$$v_L(t) = L \frac{di}{dt} \quad . \tag{1}$$

The voltage drop across a capacitor is given by

$$v_C = \frac{1}{C} q \quad . \tag{2}$$

796

Kirchhoff's voltage law states that the algebraic sum of the instantaneous voltage drops around a closed circuit in a specific direction is zero. Apply Kirchhoff's voltage law to the first loop. Using (1) and (2), the following equation can be derived,

$$L\frac{di_1}{dt} + \frac{1}{C}q_1 = 0 \tag{3}$$

where q_1 is the instantaneous charge on the capacitor. Since $i = dq/dt$, it can be written,

$$(i_1 - i_2) = \frac{dq_1}{dt} .$$

Now differentiate equation (3) with respect to t,

$$L\frac{d^2i_1}{dt^2} + \frac{1}{C}\frac{dq_1}{dt} = 0 .$$

Substituting the value of dq_1/dt yields,

$$L\frac{d^2i_1}{dt^2} + \frac{1}{C}(i_1 - i_2) = 0$$

or

$$L\frac{d^2i_1}{dt^2} = -\frac{i_1}{C} + \frac{i_2}{C} . \tag{4}$$

Similarly, applying Kirchhoff's voltage law to the second and third loops, the following equations are obtained.

$$L\frac{d^2i_2}{dt^2} = \frac{i_1}{C} - \frac{2i_2}{C} + \frac{i_3}{C} \tag{5}$$

$$L\frac{d^2i_3}{dt^2} = \frac{i_2}{C} - \frac{i_3}{C} \tag{6}$$

In an L-C-circuit without a battery, current as a function of time is a simple periodic function similar to the simple harmonic motion of a spring system. Therefore, $d^2i/dt^2 = -\omega^2 i$ where ω is the natural frequency of the system.
Let

$$\frac{d^2i_1}{dt^2} = -\omega^2 i_1, \quad \frac{d^2i_2}{dt^2} = -\omega^2 i_2, \quad \frac{d^2i_3}{dt^2} = -\omega^2 i_3 .$$

Then equations (4), (5) and (6) become

$$-L\omega^2 i_1 = -\frac{i_1}{C} + \frac{i_2}{C}$$

$$-L\omega^2 i_2 = \frac{i_1}{C} - \frac{2i_2}{C} + \frac{i_3}{C}$$

$$-L\omega^2 i_3 = \frac{i_2}{C} - \frac{i_3}{C} .$$

After rearranging terms this is:

$$i_1 - i_2 = CL\omega^2 i_1$$

$$-i_1 + 2i_2 - i_3 = CL\omega^2 i_2$$

$$-i_2 + i_3 = CL\omega^2 i_3 .$$

For simplicity substitute $\lambda = \omega^2 CL$ to obtain

797

$$i_1 - i_2 = \lambda i_1$$

$$-i_1 + 2i_2 - i_3 = \lambda i_2$$

$$- i_2 + i_3 = \lambda i_3$$

In matrix notation, the above system becomes,

$$\begin{bmatrix} 1 & -1 & 0 \\ -1 & 2 & -1 \\ 0 & -1 & 1 \end{bmatrix} \begin{bmatrix} i_1 \\ i_2 \\ i_3 \end{bmatrix} = \lambda \begin{bmatrix} i_1 \\ i_2 \\ i_3 \end{bmatrix}$$

or

$$AX = \lambda X ,$$

where

$$A = \begin{bmatrix} 1 & -1 & 0 \\ -1 & 2 & -1 \\ 0 & -1 & 1 \end{bmatrix} , \quad X = \begin{bmatrix} i_1 \\ i_2 \\ i_3 \end{bmatrix} .$$

We are seeking non-zero solutions for X. Thus X is an eigenvector for A belonging to the eigenvalue λ, if $AX = \lambda X (X \neq 0)$. Therefore, we shall obtain eigenvalues of A .   The characteristic polynomial of A is

$$f(\lambda) = \det (\lambda I - A) = \det \begin{bmatrix} \lambda-1 & 1 & 0 \\ 1 & \lambda-2 & 1 \\ 0 & 1 & \lambda-1 \end{bmatrix}$$

$$= (\lambda-1)[(\lambda-2)(\lambda-1) - 1] - 1[\lambda-1]$$

$$= (\lambda-1)[\lambda^2-3\lambda + 2 - 1 - 1] = \lambda(\lambda-1)(\lambda-3) .$$

The characteristic equation of A is $\lambda(\lambda-1)(\lambda-3) = 0$. Therefore, the eigenvalues of A are $\lambda_1 = 0$, $\lambda_2 = 1$, and $\lambda_3 = 3$. As $\lambda = \omega^2 CL$, then

$$\lambda_1 = \omega_1^2 CL = 0$$

$$\lambda_2 = \omega_2^2 CL = 1$$

$$\lambda_3 = \omega_3^2 CL = 3$$

or

$$\omega_1 = 0$$

$$\omega_2 = \sqrt{1/CL}$$

$$\omega_3 = \sqrt{3/CL} .$$

● **PROBLEM** 21-18

In the circuit shown in the figure below, find the current in each loop as a function of time, given that all charges and currents are zero when the switch is closed at t = 0 .

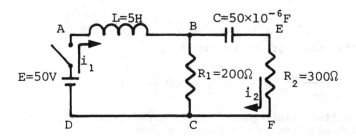

Solution: To solve this problem, the three laws concerning the voltage drops across resistors, inductors, and capacitors should first be noted:

1) The voltage drop across a resistor is given by

$$E_R = Ri \ ,$$

where R is a constant of proportionality called the resistance and i is the current through the resistor.

2) The voltage drop across an inductor is given by

$$E_L = L \frac{di}{dt} \ ,$$

where L is a constant of proportionality called the inductance, and i again denotes the current.

3) The voltage drop across a capacitor is given by

$$E_c = \frac{1}{c} q \ ,$$

where c is a constant of proportionality called the capacitance and q is the instantaneous charge on the capacitor. Since i = dq/dt, this is often written as

$$E_c = \frac{1}{c} \int i \ dt$$

Kirchhoff's voltage law states that the sum of the voltage drops around any closed circuit is zero. By applying Kirchhoff's voltage law to the loop A B C D A , the following equation can be derived,

$$E - E_L - E_{R_1} = 0 \qquad\qquad\qquad \text{(a)}$$

Using the laws 1) and 2) one can obtain,

$$E_{R_1} = R_1(i_1 - i_2)$$

and

$$E_L = L \frac{di_1}{dt}$$

Substituting these values of E_{R_1} and E_L , into equation (a) yields,

$$E - L \frac{di_1}{dt} - R_1(i_1 - i_2) = 0 \ .$$

Since E = 50 v , L = 0·5 H, and R_1 = 200 ohms, the above equation becomes,

$$50 - 0·5 \frac{di_1}{dt} - 200(i_1 - i_2) = 0 \qquad\qquad \text{(1)}$$

Consider the loop B E F C B . By applying Kirchhoff's voltage law and using the laws 1) and 3) for voltage drops, the following equation can be obtained,

$$\frac{1}{c} \int i_2 dt + i_2 R_2 + R_1(i_2 - i_1) = 0 \ ,$$

or

$$\frac{1}{c} \int i_2 dt - i_1 R_1 + i_2(R_1 + R_2) = 0 \ .$$

Letting $q = \int i_2 dt$, and $i_2 = dq/dt$,

$$\frac{1}{c}q - i_1 R_1 + \frac{dq}{dt}(R_1 + R_2) = 0 .$$

Substituting the values of c, R_1, and R_2,

$$\frac{1}{50 \times 10^{-6}}q - 200\, i_1 + 500\,\frac{dq}{dt} = 0 .$$

Dividing this differential equation by 500, and rearrange the terms to obtain the differential equation in the following form,

$$\frac{dq}{dt} - .4\, i_1 + 40\, q = 0 . \tag{2}$$

Since $i_2 = dq/dt$ the equation (1) can be rewritten as

$$50 - 0.5\,\frac{di_1}{dt} - 200 i_1 + 200\,\frac{dq}{dt} = 0 \tag{3}$$

From equation (2),

$$\frac{dq}{dt} = -40q + .4i_1 . \tag{4}$$

Substitution of this value of dq/dt into equation (3) yields,

$$50 - 0.5\,\frac{di_1}{dt} - 200 i_1 + 200(-40q + .4i_1) = 0$$

or

$$50 - 0.5\,\frac{di_1}{dt} - 120 i_1 - 8000q = 0 .$$

Divide the above equation by 0.5, to obtain

$$100 - \frac{di_1}{dt} - 240 i_1 - 16000q = 0 . \tag{5}$$

Thus equations (4) and (5) are two linear differential equations. These equations may be rewritten as

$$\frac{di_1}{dt} = -240 i_1 - 16000q + 100$$

$$\frac{dq}{dt} = .4 i_1 - 40q . \tag{6}$$

System (6) of linear differential equations can be expressed in matrix notation as

$$\begin{bmatrix} \dfrac{di_1}{dt} \\[2mm] \dfrac{dq}{dt} \end{bmatrix} = \begin{bmatrix} -240 & -16000 \\[2mm] .4 & -40 \end{bmatrix} \begin{bmatrix} i_1 \\[2mm] q \end{bmatrix} + \begin{bmatrix} 100 \\[2mm] 0 \end{bmatrix}$$

or

$$\dot{X} = AX + F(t) \tag{7}$$

where

$$A = \begin{bmatrix} -240 & -16000 \\[2mm] -0.4 & -40 \end{bmatrix}, \quad X = \begin{bmatrix} i_1 \\[2mm] q \end{bmatrix}, \quad \text{and} \quad F(t) = \begin{bmatrix} 100 \\[2mm] 0 \end{bmatrix}.$$

System (7) is a nonhomogeneous system. The corresponding homogeneous system is given by

$$\dot{X} = AX .$$

Now the general solution of the homogeneous system is

$$x_h(t) = c_1 e^{\lambda_1 t} V_1 + c_2 e^{\lambda_2 t} V_2$$

where:

(1) c_1 and c_2 are arbitrary constants.

(2) λ_1 and λ_2 are distinct eigenvalues of the matrix A.

(3) V_1 and V_2 are linearly independent eigenvectors corresponding to the eigenvalues λ_1 and λ_2 respectively.

To obtain the eigenvalues of A, write the characteristic equation of A. The characteristic equation of A is,

$$\det[\lambda I - A] = 0 .$$

Therefore,

$$\det \begin{bmatrix} \lambda + 240 & 16,000 \\ -.4 & \lambda + 40 \end{bmatrix} = 0$$

or

$$[(\lambda + 240)(\lambda + 40)] - [(16000)(-.4)] = 0$$

$$\lambda^2 + 280\lambda + 16000 = 0$$

or

$$(\lambda + 80)(\lambda + 200) = 0 .$$

Therefore, the eigenvalues of A are $\lambda_1 = -80$ and $\lambda_2 = -200$.

$V = \begin{bmatrix} v_1 \\ v_2 \end{bmatrix}$ is an eigenvector of A corresponding to λ if and only if

V is a nontrivial solution of $(\lambda I - A)V = 0$. So, for $\lambda = -80$,

$$(-80I - A)V = 0$$

or

$$\begin{bmatrix} 160 & 16,000 \\ -.4 & -40 \end{bmatrix} \begin{bmatrix} v_1 \\ v_2 \end{bmatrix} = \begin{bmatrix} 0 \\ 0 \end{bmatrix}$$

Then,

$$160v_1 + 16,000v_2 = 0$$

$$-.4v_1 - 40v_2 = 0 .$$

This implies that $v_1 = -100v_2$. Therefore, $V_1 = \begin{bmatrix} -100 \\ 1 \end{bmatrix}$ is an eigenvector associated with $\lambda = -80$. For $\lambda = -200$,

$$\begin{bmatrix} 40 & 16,000 \\ -.4 & -160 \end{bmatrix} \begin{bmatrix} v_1 \\ v_2 \end{bmatrix} = \begin{bmatrix} 0 \\ 0 \end{bmatrix}$$

or,

$$40v_1 + 16000v_2 = 0$$

$$-.4v_1 - 160v_2 = 0 .'$$

This implies that $v_1 = -400v_2$. Therefore, $V_2 = \begin{bmatrix} -400 \\ 1 \end{bmatrix}$ is an eigenvector corresponding to $\lambda_2 = -200$. Thus,

$$X_h(t) = c_1 e^{-80t} \begin{bmatrix} -100 \\ 1 \end{bmatrix} + c_2 e^{-200t} \begin{bmatrix} -400 \\ 1 \end{bmatrix}$$

or

$$X_h(t) = c_1 \begin{bmatrix} -100e^{-80t} \\ e^{-80t} \end{bmatrix} + c_2 \begin{bmatrix} -400e^{-200t} \\ e^{-200t} \end{bmatrix}$$

To find a particular solution of system (7) assume that

$$X_p = \begin{bmatrix} i_1 \\ q \end{bmatrix} , = \begin{bmatrix} a \\ b \end{bmatrix} .$$

Substitution of this into the nonhomogeneous system of differential equations (6) yields,

$$0 = -240a - 16000b + 100$$

$$0 = \cdot 4a - 40b .$$

Solving this system yields,

$$a = 1/4 , b = 1/400 ,$$

and therefore a complete solution of the system (7) is

$$X(t) = X_h + X_p .$$

Hence,

$$X(t) = c_1 \begin{bmatrix} -100e^{-80t} \\ e^{-80t} \end{bmatrix} + c_2 \begin{bmatrix} -400e^{-200t} \\ e^{-200t} \end{bmatrix} + \begin{bmatrix} \frac{1}{4} \\ \frac{1}{400} \end{bmatrix}$$

or,

$$\begin{bmatrix} i_1(t) \\ q(t) \end{bmatrix} = \begin{bmatrix} -100c_1 e^{-80t} - 400c_2 e^{-200t} + \frac{1}{4} \\ c_1 e^{-80t} + c_2 e^{-200t} + \frac{1}{400} \end{bmatrix}$$

Hence,

$$i_1(t) = -100c_1 e^{-80t} - 400c_2 e^{-200t} + \frac{1}{4} \qquad (8)$$

$$q(t) = c_1 e^{-80t} + c_2 e^{-200t} + \frac{1}{400}$$

The constants c_1 and c_2 depend on the initial values of $i_1(0)$ and $q(0)$. When $t = 0$, $i_1(0) = 0$, and $q(0) = 0$. Substituting $t = 0$ in system (8), yields,

$$i_1(0) = 0 = -100c_1 - 400c_2 + \frac{1}{4}$$

$$q(0) = 0 = c_1 + c_2 + \frac{1}{400} .$$

Solving these equations, one can find that $c_1 = -\frac{1}{240}$ and $c_2 = \frac{1}{600}$.

802

Therefore, system (8) can be rewritten as,

$$i_1(t) = \frac{5}{12} e^{-80t} - \frac{2}{3} e^{-200t} + \frac{1}{4}$$

$$q(t) = -\frac{1}{240} e^{-80t} + \frac{1}{600} e^{-200t} + \frac{1}{400} .$$

Knowing that $i_2 = dq/dt$, the value of current i_2 can be found by differentiating the 2nd equation of the above system with respect to t. Thus,

$$i_2 = \frac{dq}{dt} = \frac{1}{3} e^{-80t} - \frac{1}{3} e^{-200t} .$$

The required currents are therefore

$$i_1 = \frac{5}{12} e^{-80t} - \frac{2}{3} e^{-200t} + \frac{1}{4}$$

$$i_2 = \frac{1}{3} e^{-80t} - \frac{1}{3} e^{-200t} .$$

Evidently, $i_2 = 0$ when $t = 0$, as required.

● **PROBLEM** 21-19

Consider the electrical circuit shown in the figure below in which the known time-varying source current $i_s(t)$ is connected to nodes A and C. $v_1(t)$ is the variable voltage (unknown) across a capacitor of 5/3-farad. Let $i_1(t)$ be the variable current through the 3/5-henry inductor. The polarities are indicated in figure 1. When $v_1(t) > 0$, the potential of node A is larger than that of node C (measured with respect to a common reference). When $i_1(t) > 0$, the current flows from node A to node B . Suppose that at t = 0, we are given $v_1(0) = 0.6$ volts, $i_1(0) = 1.0$ ampere, $v_2(0) = 1.2$ volts.

Determine the voltage $v_1(t)$, $v_2(t)$, and the current $i_1(t)$ as functions of time (in terms of the given source current $i_s(t)$).

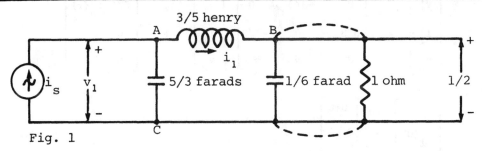

Fig. 1

Solution: In order to formulate this problem mathematically we shall use Ohm's Law. According to this law, v = iR, where v = voltage, i = current, and R = resistance. It is also known that i = Cv'(t). Here C is the capacitance, and v' = dv/dt . According to Kirchhoff's law of currents, the sum of the currents entering and leaving a given node is zero. Suppose that at time t the source current $i_s(t)$ leaves node C and enters node A as shown in figure 1. At the same time the current i_1 leaves node A (through the inductance of 3/5 henry). The current leaving node A through the capacitance of 5/3 farads is $5/3\ v_1'(t)$. (By using the formula $i = Cv'(t)$). Thus, when Kirchhoff's law is applied to node A,

$$i_s(t) - i_1(t) - 5/3\ v_1'(t) = 0 \ . \qquad (1)$$

Similarly, a current i_1 enters node B, and current $1/6\ v_2'(t)$ (because $i = Cv'(t) = 1/6\ v'(t)$) leaves node B through the capacitance of 1/6 farads. Also, the current $v_2(t)/1$ (because $i = v/R = v(t)/1$) leaves node B through the 1-ohm resistor. Thus, Kirchhoff's law applied to node B gives,

$$i_1(t) - 1/6\ v_2'(t) - v_2(t) = 0 \ . \qquad (2)$$

The voltage across the inductance of 3/5 henry is obtained by using the formula, $v(t) = Li_1'(t)$. Thus, $v(t) = 3/5\ i_1'(t)$. Then in the middle loop of the circuit shown in figure 1, since the sum of the voltage drops must also be zero (another one of Kirchhoff's laws), it can be seen that

$$v_1(t) - 3/5\ i_1'(t) - v_2(t) = 0 \ . \qquad (3)$$

Solving each equation (1), (2) and (3) for the quantities v_1', v_2', and i_1' respectively, yields

$$v_1' = -\ 3/5\ i_1 + 3/5\ i_s(t)$$

$$i_1' = 5/3\ v_1 - 5/3\ v_2 \qquad (4)$$

$$v_2' = 6i_1 - 6v_2 \ .$$

The system (4) is a system of first order linear differential equations. Note that the mathematical problem is to solve the following initial value problem for the unknown functions v_1, i_1, v_2, where $i_s(t)$ is a given source current.

$$v_1' = -\ 3/5\ i_1 + 5/3\ i_s(t), \quad v_1(0) = 0\cdot 6 \text{ volt.}$$

$$i_1' = 5/3\ v_1 - 5/3\ v_2 \ , \quad i_1(0) = 1 \text{ ampere.} \qquad (5)$$

$$v_2' = 6i_1 - 6v_2 \ , \quad v_2(0) = 1\cdot 2 \text{ volts.}$$

The system (5) can be written in matrix notation as

$$\begin{bmatrix} v_1' \\ i_1' \\ v_2' \end{bmatrix} = \begin{bmatrix} 0 & -3/5 & 0 \\ 5/3 & 0 & -5/3 \\ 0 & 6 & -6 \end{bmatrix} \begin{bmatrix} v_1 \\ i_1 \\ v_2 \end{bmatrix} + \begin{bmatrix} 5/3\ i_s(t) \\ 0 \\ 0 \end{bmatrix}$$

and

$$\begin{bmatrix} v_1(0) \\ i_1(0) \\ v_2(0) \end{bmatrix} = \begin{bmatrix} 0\cdot 6 \\ 1 \\ 1\cdot 2 \end{bmatrix} \qquad (6)$$

Let

$$Y = \begin{bmatrix} v_1 \\ i_1 \\ v_2 \end{bmatrix}, \quad A = \begin{bmatrix} 0 & -3/5 & 0 \\ 5/3 & 0 & -5/3 \\ 0 & 6 & -6 \end{bmatrix}, \quad F(t) = \begin{bmatrix} 5/3\ i_s(t) \\ 0 \\ 0 \end{bmatrix}$$

and

$$C = \begin{bmatrix} 0 \cdot 6 \\ 1 \\ 1 \cdot 2 \end{bmatrix} .$$

Then (6) has the form,

$$Y' = AY + F(t) , \quad C = \begin{bmatrix} 0 \cdot 6 \\ 1 \\ 1 \cdot 2 \end{bmatrix} . \tag{7}$$

It is known that the solution of system (7) is given by

$$Y(t) = e^{At} C + \int_0^t e^{A(t-s)} F(s) ds . \tag{8}$$

Now e^{At} is a fundamental matrix of the homogeneous system $Y' = AY$. To find the value of e^{At}, the eigenvalues of A must first be found.

The eigenvalues are the roots of the characteristic polynomial.

$$f(\lambda) = \det (\lambda I - A) = \begin{bmatrix} \lambda & 3/5 & 0 \\ -5/3 & \lambda & 5/3 \\ 0 & -6 & \lambda+6 \end{bmatrix} .$$

Therefore,

$$f(\lambda) = \lambda \begin{vmatrix} \lambda & 5/3 \\ -6 & \lambda+6 \end{vmatrix} - 3/5 \begin{vmatrix} -5/3 & 5/3 \\ 0 & \lambda+6 \end{vmatrix}$$

$$= \lambda [\lambda(\lambda+6) + 10] + (3/5)(5/3)(\lambda+6)$$

$$= \lambda^3 + 6\lambda^2 + 11\lambda + 6$$

$$= (\lambda+1)(\lambda+2)(\lambda+3) .$$

Hence the eigenvalues are $\lambda_1 = -1$, $\lambda_2 = -2$, and $\lambda_3 = -3$. Next, compute an eigenvector corresponding to each eigenvalue. Corresponding to the eigenvalue $\lambda_1 = -1$, consider the system

$$(-I - A)X = \begin{bmatrix} -1 & 3/5 & 0 \\ -5/3 & -1 & 5/3 \\ 0 & -6 & 5 \end{bmatrix} \begin{bmatrix} x_1 \\ x_2 \\ x_3 \end{bmatrix} = \begin{bmatrix} 0 \\ 0 \\ 0 \end{bmatrix}$$

or

$$-x_1 + 3/5 \, x_2 = 0$$

$$-(5/3)x_1 - x_2 + (5/3)x_3 = 0$$

$$-6x_2 + 5x_3 = 0 .$$

Solving this system yields

$$V_1 = \begin{bmatrix} 3/5 \\ 1 \\ 3/5 \end{bmatrix}$$

which is an eigenvector associated with $\lambda_1 = 1$. Similarly,

$$V_2 = \begin{bmatrix} 3/10 \\ 1 \\ 3/2 \end{bmatrix} , \quad V_3 = \begin{bmatrix} 1/10 \\ 1/2 \\ 1 \end{bmatrix}$$

are eigenvectors corresponding to the eigenvalues $\lambda_2 = -2$ and $\lambda_3 = -3$ respectively. Recall that

$$\varphi(t) = \left[e^{\lambda_1 t} v_1 \;,\; e^{\lambda_2 t} v_2 \;,\; e^{\lambda_3 t} v_3 \right]$$

is a fundamental matrix of the system $Y' = AY$, and also that

$$e^{At} = \varphi(t)\,\varphi^{-1}(0) \;.$$

Thus,

$$\varphi(t) = \begin{bmatrix} (3/5)e^{-t} & (3/10)e^{-2t} & (1/10)e^{-3t} \\ e^{-t} & e^{-2t} & (1/2)e^{-3t} \\ (6/5)e^{-t} & (3/2)e^{-2t} & e^{-3t} \end{bmatrix}$$

Then,

$$\varphi(0) = \begin{bmatrix} 3/5 & 3/10 & 1/10 \\ 1 & 1 & 1/2 \\ 6/5 & 3/2 & 1 \end{bmatrix}$$

and

$$\varphi^{-1}(0) = \frac{1}{\det \varphi(0)}\ \text{adj}\ \varphi(0) = \begin{bmatrix} 25/6 & -15/6 & 5/6 \\ -20/3 & 8 & -10/3 \\ 5 & -9 & 5 \end{bmatrix}$$

Hence,

$$e^{At} = \begin{bmatrix} (3/5)e^{-t} & (3/10)e^{-2t} & (1/10)e^{-3t} \\ e^{-t} & e^{-2t} & (1/2)e^{-3t} \\ (6/5)e^{-t} & (3/2)e^{-2t} & e^{-3t} \end{bmatrix} \begin{bmatrix} 25/6 & -2.5 & 5/6 \\ -20/3 & 8 & -10/3 \\ 5 & -9 & 5 \end{bmatrix}$$

$$= \begin{bmatrix} (5/2)e^{-t}-2e^{-2t}+(1/2)e^{-3t} & -(3/2)e^{-t}+(12/5)e^{-2t}-(9/10)e^{-3t} & (1/2)e^{-t}-e^{-2t}+(1/2)e^{-3t} \\ (25/6)e^{-t}-(20/3)e^{-2t}+(5/2)e^{-3t} & -2.5e^{-t}+8e^{-2t}-(9/2)e^{-3t} & (5/6)e^{-t}-(10/3)e^{-2t}+(5/2)e^{-3t} \\ 5e^{-t}-10e^{-2t}+5e^{-3t} & -3e^{-t}+12e^{-2t}-9e^{-3t} & e^{-t}-5e^{-2t}+5e^{-3t} \end{bmatrix}$$

Then, $e^{At}C =$

$$\begin{bmatrix} (5/2)e^{-t}-2e^{-2t}+(1/2)e^{-3t} & -(3/2)e^{-t}+(12/5)e^{-2t}-(9/10)e^{-3t} & (1/2)e^{-t}-e^{-2t}+(1/2)e^{-3t} \\ (25/6)e^{-t}-(20/3)e^{-2t}+(5/2)e^{-3t} & -2.5e^{-t}+8e^{-2t}-(9/2)e^{-3t} & (5/6)e^{-t}-(10/3)e^{-2t}+(5/2)e^{-3t} \\ 5e^{-t}-10e^{-2t}+5e^{-3t} & -3e^{-t}+12e^{-2t}-9e^{-3t} & e^{-t}-5e^{-2t}+5e^{-3t} \end{bmatrix}$$

$$\begin{bmatrix} 0.6 \\ 1 \\ 1.2 \end{bmatrix} =$$

806

$$
\begin{bmatrix}
1.5e^{-t} - 1.2e^{-2t} + .3e^{-3t} - 1.5e^{-t} + 2.4e^{-2t} - .9e^{-3t} + .6e^{-t} - 1.2e^{-2t} + .6e^{-3t} \\
2.5e^{-t} - 4e^{-2t} + 1.5e^{-3t} - 2.5e^{-t} + 8e^{-2t} - 4.5e^{-3t} + 1e^{-t} - 4e^{-2t} + 3e^{-3t} \\
3e^{-t} - 6e^{-2t} + 3e^{-3t} - 3e^{-t} + 12e^{-2t} - 9e^{-3t} + 1.2e^{-t} - 6e^{-2t} + 6e^{-3t}
\end{bmatrix}
$$

Therefore,

$$
e^{tA}C = \begin{bmatrix} 0.6\, e^{-t} \\ e^{-t} \\ 1.2\, e^{-t} \end{bmatrix}
$$

Now,

$$
e^{A(t-s)}F(s) = e^{A(t-s)} \begin{bmatrix} 5/3\, i_s(s) \\ 0 \\ 0 \end{bmatrix}
$$

$$
= 5/3 \begin{bmatrix}
(5/2)e^{-(t-s)} \quad -2e^{-2(t-s)} \quad +(1/2)e^{-3(t-s)} \\
(25/6)e^{-(t-s)} \quad -(20/3)e^{-2(t-s)} \quad +(5/2)e^{-3(t-s)} \\
5\,e^{-(t-s)} \quad -10e^{-2(t-s)} \quad + 5e^{-3(t-s)}
\end{bmatrix} i_s(s)\,ds
$$

Substituting the values of $e^{At}C$, and $e^{A(t-s)}F(s)$ into equation (8), yields

$$
Y(t) = \begin{bmatrix} v_1(t) \\ i_1(t) \\ v_2(t) \end{bmatrix} =
$$

$$
e^{-t}\begin{bmatrix} 0.6 \\ 1 \\ 1.2 \end{bmatrix} + 5/3 \int_0^t \begin{bmatrix}
(5/2)e^{-(t-s)} \quad - 2e^{-2(t-s)} \quad +(1/2)e^{-3(t-s)} \\
(25/6)e^{-(t-s)} \quad - (20/3)e^{-2(t-s)} \quad +(5/2)e^{-3(t-s)} \\
5e^{-(t-s)} \quad - 10e^{-2(t-s)} \quad + 5e^{-3(t-s)}
\end{bmatrix} i_s(s)\,ds \quad .
$$

$$(9)$$

This is a solution to the system (5). If the source current $i_s(t)$ is given, the solution (9) can be simplified further.

● **PROBLEM** 21-20

Consider an electrical circuit as shown in the figure below, consisting of a resistor, an inductor, and a capacitor connected in series. In this RLC circuit let $R = 3$ ohms, $L = 1$ henry, and $c = 0.5$ farad. Assume that there is no initial current and no initial charge (at $t = 0$) when the voltage is first applied. Find the subsequent current i_L in the system and the voltage $v_c(t)$ across the capacitor.

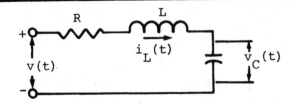

Solution: To set this problem up mathematically, first write the voltage-current relationship in the R, L, and C elements. The voltage-current relation in a resistor can be written using Ohm's law,

$$v_R(t) = Ri_L(t) \tag{1}$$

The current-voltage relationships for linear inductors can be written as

$$v_L(t) = L \frac{di_L(t)}{dt} \quad . \quad \text{(Faraday's law)} \tag{2}$$

According to Coulomb's law, the current-voltage relationship for a capacitor can be expressed as

$$i_C(t) = C \frac{dv_C(t)}{dt} \quad . \tag{3}$$

Now, Kirchhoff's voltage law governs the voltage relationships in a closed electrical circuit. The law states: at any given instant of time, the algebraic sum of the voltages in a closed electrical circuit is zero. Thus, Kirchhoff's voltage law applied to the given RLC circuit, yields,

$$v(t) - v_R(t) - v_L(t) - v_C(t) = 0 \quad . \tag{4}$$

From equations (1) and (2), substitute the values of $v_R(t)$ and $v_L(t)$ into equation (4). This yields,

$$v(t) - Ri_L(t) - L \frac{di_L(t)}{dt} - v_C(t) = 0$$

or,

$$L \frac{di_L(t)}{dt} = v(t) - Ri_L(t) - v_C(t) \quad . \tag{5}$$

Equations (5) and (3) can be rewritten as,

$$\frac{di_L(t)}{dt} = -\frac{R}{L} i_L(t) - \frac{1}{L} v_C(t) + \frac{1}{L} v(t)$$

$$\frac{dv_C(t)}{dt} = \frac{1}{C} i_L(t) \quad .$$

These two equations can be expressed in matrix form as,

$$
\begin{bmatrix} \dfrac{di_L(t)}{dt} \\[2ex] \dfrac{dv_C(t)}{dt} \end{bmatrix}
=
\begin{bmatrix} -R/L & -1/L \\[2ex] 1/C & 0 \end{bmatrix}
\begin{bmatrix} i_L(t) \\[2ex] v_C(t) \end{bmatrix}
+
\begin{bmatrix} 1/L \\[2ex] 0 \end{bmatrix}
v(t) \quad . \tag{6}
$$

Let

$$
\dot{X} = \begin{bmatrix} \dfrac{di_L(t)}{dt} \\[2ex] \dfrac{dv_C(t)}{dt} \end{bmatrix}
\quad , \quad
A = \begin{bmatrix} -R/L & -1/L \\[2ex] 1/C & 0 \end{bmatrix} \quad ,
$$

$$
X = \begin{bmatrix} i_L(t) \\[2ex] v_C(t) \end{bmatrix}
\quad , \quad
B = \begin{bmatrix} 1/L \\[2ex] 0 \end{bmatrix}
\quad , \quad r(t) = [v(t)] \quad .
$$

808

Then matrix equation (6) becomes,

$$\dot{X} = AX + Br(t) .$$

This is a nonhomogeneous equation and the solution of this equation for $t > 0$ is given by

$$X(t) = e^{At}X(0) + \int_0^t e^{A(t-\tau)} Br(\tau)d\tau . \tag{7}$$

Now before computing e^{At}, substitute the given values of R, L, and C into the matrix A and B. It is given that R = 3 ohms, L = 1h., C = .5f. Thus,

$$A = \begin{bmatrix} -3/1 & -1/1 \\ 1/5 & 0 \end{bmatrix} = \begin{bmatrix} -3 & -1 \\ 2 & 0 \end{bmatrix}$$

and

$$B = \begin{bmatrix} 1 \\ 0 \end{bmatrix} .$$

The matrix A is 2 X 2, therefore,

$$e^{At} = \alpha_1 At + \alpha_0 I$$

$$= \alpha_1 \begin{bmatrix} -3t & -t \\ 2t & 0 \end{bmatrix} + \alpha_0 \begin{bmatrix} 1 & 0 \\ 0 & 1 \end{bmatrix}$$

$$= \begin{bmatrix} -3\alpha_1 t & -\alpha_1 \\ 2\alpha_1 t & 0 \end{bmatrix} + \begin{bmatrix} \alpha_0 & 0 \\ 0 & \alpha_0 \end{bmatrix}$$

$$= \begin{bmatrix} -3\alpha_1 t + \alpha_0 & -\alpha_1 \\ 2\alpha_1 t & \alpha_0 \end{bmatrix} . \tag{8}$$

$r(\lambda)$ can be defined as $r(\lambda) = \alpha_1 \lambda + \alpha_0$, where λ is an eigenvalue of At. The characteristic polynomial of A is

$$f(\lambda) = \det[\lambda I - A] = \det \begin{bmatrix} \lambda+3 & 1 \\ -2 & \lambda \end{bmatrix} = \lambda^2 + 3\lambda + 2$$

$$= (\lambda+1)(\lambda+2) .$$

Thus the eigenvalues of A are $\lambda_1 = -1$ and $\lambda_2 = -2$. The eigenvalues of At are $\lambda_1 = -t$, and $\lambda_2 = -2t$. Substituting these values successively into the equation $e^{\lambda_1} = r(\lambda_1)$, these two equations are obtained,

$$e^{-t} = \alpha_1(-t) + \alpha_0 .$$

$$e^{-2t} = \alpha_1(-2t) + \alpha_0 .$$

809

Solving these equations for α_1 and α_0, it is found that

$$\alpha_0 = (2e^{-t} - e^{-2t}), \quad \alpha_1 = \frac{1}{t}(e^{-t} - e^{-2t}) .$$

Substituting these values into (8), and simplifying, yields

$$e^{At} = \begin{bmatrix} -e^{-t} + 2e^{-2t} & -e^{-t} + e^{-2t} \\ 2e^{-t} - 2e^{-2t} & 2e^{-t} - e^{-2t} \end{bmatrix} .$$

From the given initial conditions, ($i_L = 0$ and $v_C(0) = 0$), it can be seen that

$$X(0) = \begin{bmatrix} 0 \\ 0 \end{bmatrix} .$$

Thus,

$$e^{At}X(0) = \begin{bmatrix} -e^{-t} + 2e^{-2t} & -e^{-t} + e^{-2t} \\ 2e^{-t} - 2e^{-2t} & 2e^{-t} - e^{-2t} \end{bmatrix} \begin{bmatrix} 0 \\ 0 \end{bmatrix} = \begin{bmatrix} 0 \\ 0 \end{bmatrix} .$$

Now $v(t)$ is the input voltage of the circuit. If the input voltage $v(t)$ is a unit step function applied at $t = 0$, then,

$$r = [v(t)] = [1] .$$

Now,

$$\int_0^t e^{A(t-\tau)} Br(\tau) d\tau$$

$$= \int_0^t \begin{bmatrix} -e^{-(t-\tau)} + 2e^{-2(t-\tau)} & -e^{-(t-\tau)} + e^{-2(t-\tau)} \\ 2e^{-(t-\tau)} - 2e^{-2(t-\tau)} & 2e^{-(t-\tau)} - e^{-2(t-\tau)} \end{bmatrix} \begin{bmatrix} 1 \\ 0 \end{bmatrix} d\tau$$

$$= \int_0^t \begin{bmatrix} -e^{-(t-\tau)} + 2e^{-2(t-\tau)} \\ 2e^{-(t-\tau)} - 2e^{-2(t-\tau)} \end{bmatrix} d\tau$$

$$= \begin{bmatrix} e^{-t} - e^{-2t} \\ 1-2e^{-t} + e^{-2t} \end{bmatrix} .$$

Substitute the values of $e^{At}X(0)$, and $\int_0^t e^{A(t-\tau)} Br(\tau)d\tau$ into equation (7). This yields

$$X(t) = \begin{bmatrix} e^{-t} - e^{-2t} \\ 1-2e^{-t} + e^{-2t} \end{bmatrix} .$$

But

$$X(t) = \begin{bmatrix} i_L(t) \\ v_C(t) \end{bmatrix} .$$

810

Therefore,

$$i_L(t) = e^{-t} - e^{-2t}$$

and

$$v_C(t) = 1 - 2e^{-t} + e^{-2t} .$$

● **PROBLEM** 21-21

In the circuit in figure 1, suppose that $R_{13} = R_{12} = R_{23} = 50$ ohms and $C_1 = C_2 = C_3 = .02$ farads. Find, as a function of time, the currents i_1, i_2, and i_3 if at time $t = 0$, $i_1 = 0$, $i_2 = 3$, and $i_3 = -3$.

(Sum of the voltages around a loop must equal zero)

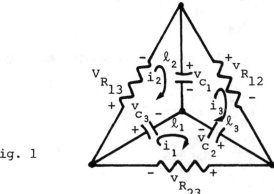

Fig. 1

Solution: We choose q_1, q_2, and q_3 to represent the charges associated with each of the currents i_1, i_2, and i_3. The currents in the circuit can be written in terms of the charge,

$$i_1 = \frac{dq_1}{dt}$$

$$i_2 = \frac{dq_2}{dt}$$

$$i_3 = \frac{dq_3}{dt} .$$

Writing the independent loop equations for the circuit yields (see fig. 1)

$$\ell_1: \quad -v_{C_2} + v_{C_3} + v_{R_{23}} = 0$$

$$\ell_2: \quad v_{C_1} - v_{C_3} + v_{R_{13}} = 0$$

$$\ell_3: \quad -v_{C_1} + v_{C_2} + v_{R_{12}} = 0$$

Since each capacitor voltage may be written as q_c/c, where q_c is the charge in the capacitor, then

811

$$v_{c_1} = \frac{q_2 - q_3}{c_1}$$

$$v_{c_2} = \frac{q_3 - q_1}{c_2}$$

$$v_{c_3} = \frac{q_1 - q_2}{c_3} \quad .$$

Also,

$$v_{R_{23}} = i_1 R_{23} = \frac{dq_1}{dt} R_{23}$$

$$v_{R_{13}} = i_2 R_{13} = \frac{dq_2}{dt} R_{13}$$

$$v_{R_{12}} = i_3 R_{12} = \frac{dq_3}{dt} R_{12} \quad .$$

Substituting the above voltage relations into the independent loop equations gives,

$$\frac{q_1 - q_3}{c_2} + \frac{q_1 - q_2}{c_3} + \frac{dq_1}{dt} R_{23} = 0$$

$$\frac{q_2 - q_3}{c_1} + \frac{q_2 - q_1}{c_3} + \frac{dq_2}{dt} R_{13} = 0$$

$$\frac{q_3 - q_2}{c_1} + \frac{q_3 - q_1}{c_2} + \frac{dq_3}{dt} R_{12} = 0 \quad .$$

If $R_{23} = R_{13} = R_{12} = R$ and $c_1 = c_2 = c_3 = c$, then the equations can be written as

$$-2q_1 + q_2 + q_3 = \frac{dq_1}{dt} Rc$$

$$q_1 - 2q_2 + q_3 = \frac{dq_2}{dt} Rc$$

$$q_1 + q_2 - 2q_3 = \frac{dq_3}{dt} Rc \quad .$$

In matrix form,

$$\begin{bmatrix} -2 & 1 & 1 \\ 1 & -2 & 1 \\ 1 & 1 & -2 \end{bmatrix} \begin{bmatrix} q_1 \\ q_2 \\ q_3 \end{bmatrix} = \begin{bmatrix} \dot{q}_1 \\ \dot{q}_2 \\ \dot{q}_3 \end{bmatrix} Rc \quad . \tag{1}$$

In this problem $R = 50$ and $c = .02$ are given. Hence, $Rc = 1$ and so system (1) becomes

$$AQ = \dot{Q}$$

where

$$A = \begin{bmatrix} -2 & 1 & 1 \\ 1 & -2 & 1 \\ 1 & 1 & -2 \end{bmatrix}, \quad Q = \begin{bmatrix} q_1 \\ q_2 \\ q_3 \end{bmatrix} \quad \text{and} \quad \dot{Q} = \begin{bmatrix} \dot{q}_1 \\ \dot{q}_2 \\ \dot{q}_3 \end{bmatrix} \quad .$$

A general solution to a differential equation of this form is

$$Q(t) = c_1 e^{\lambda_1 t} v_1 + c_2 e^{\lambda_2 t} v_2 + c_3 e^{\lambda_3 t} v_3 \qquad (2)$$

where λ_i, $(i = 1,2,3)$ are the eigenvalues of A with corresponding eigenvectors v_i. c_i $(i = 1,2,3)$ are arbitrary constants which are dependent on the initial conditions. The eigenvalues of A are the roots of the characteristic polynomial.

$$\det(\lambda I - A) = 0$$

$$\begin{vmatrix} \lambda+2 & -1 & -1 \\ -1 & \lambda+2 & -1 \\ -1 & -1 & \lambda+2 \end{vmatrix} = 0$$

$$(\lambda+2)^3 - 2 - [(\lambda+2) + (\lambda+2) + (\lambda+2)] = 0$$
$$(\lambda+2)^3 - 3\lambda - 8 = 0$$
$$\lambda^3 + 6\lambda^2 + 9\lambda = 0$$
$$\lambda(\lambda+3)^2 = 0 .$$

Therefore, the eigenvalues of A are 0, -3, -3. The eigenvectors associated with $\lambda = 0$ are the solutions of the equation $(\lambda I - A)V = 0$. Let

$$V = \begin{bmatrix} x_1 \\ x_2 \\ x_3 \end{bmatrix} ; \quad \text{then we have}$$

$$\begin{bmatrix} 0+2 & -1 & -1 \\ -1 & 0+2 & -1 \\ -1 & -1 & 0+2 \end{bmatrix} \begin{bmatrix} x_1 \\ x_2 \\ x_3 \end{bmatrix} = \begin{bmatrix} 0 \\ 0 \\ 0 \end{bmatrix} . \qquad (3)$$

Perform row operations on

$$\begin{bmatrix} 2 & -1 & -1 \\ -1 & 2 & -1 \\ -1 & -1 & 2 \end{bmatrix} \quad \text{to yield}$$

$$\begin{bmatrix} 2 & -1 & -1 \\ 0 & 3 & -3 \\ 0 & 0 & 0 \end{bmatrix} \quad \text{(i.e., 1) Replace the 2nd row with}$$

2 times the 2nd row plus the first. 2) Replace the 3rd row with 2 times the third row plus the first and; 3) Replace the 3rd row with the 2nd row plus the 3rd row.) So solutions to (3) are simply the solutions of the system

$$2x_1 - x_2 - x_3 = 0$$
$$3x_2 - 3x_3 = 0 .$$

This yields $x_1 = x_2 = x_3$. One such eigenvector is $v_1 = \begin{bmatrix} 1 \\ 1 \\ 1 \end{bmatrix}.$

For equation (2) to hold, two linearly independent eigenvectors associated with $\lambda = 3$ must be found. There is a theorem in linear algebra which states that if λ is an eigenvalue of A with multiplicity k and the rank of the matrix $(\lambda I - A)$ is $n-k$ (n is the number of rows, or columns, of A) then there exist k linearly independent eigenvectors associated with λ. In this case -3 has a multiplicity of 2 and the rank of

$$(-3I - A) = \begin{bmatrix} -1 & -1 & -1 \\ -1 & -1 & -1 \\ -1 & -1 & -1 \end{bmatrix} \quad \text{is 1.}$$

Also $n-k = 3 - 2 = 1$. Thus there are two linearly independent solutions to $(-3I - A)V = 0$. $(-3I - A)V = 0$ yields,

$$\begin{bmatrix} -1 & -1 & -1 \\ -1 & -1 & -1 \\ -1 & -1 & -1 \end{bmatrix} \begin{bmatrix} x_1 \\ x_2 \\ x_3 \end{bmatrix} = \begin{bmatrix} 0 \\ 0 \\ 0 \end{bmatrix}$$

which is equivalent to $-x_1 - x_2 - x_3 = 0$. Two vectors which satisfy this requirement are $v_2 = \begin{bmatrix} -2 \\ 1 \\ 1 \end{bmatrix}$ and $v_3 = \begin{bmatrix} -1 \\ 0 \\ 1 \end{bmatrix}$. For equation (2) to hold the

eigenvectors v_1, v_2, and v_3 must be linearly independent. Now
$c_1(1,1,1) + c_2(-2,1,1) + c_3(1,0,1) = 0$ implies

$$c_1 - c_2 + c_3 = 0$$
$$c_1 + c_2 = 0$$
$$-c_1 + c_2 + c_3 = 0$$

which has as its only solution $c_1 = c_2 = c_3 = 0$. v_1, v_2, and v_3 are linearly independent. As a function of time we have, from (2),

$$Q(t) = c_1 e^{0t} \begin{bmatrix} 1 \\ 1 \\ 1 \end{bmatrix} + c_2 e^{-3t} \begin{bmatrix} -2 \\ 1 \\ 1 \end{bmatrix} + c_3 e^{-3t} \begin{bmatrix} -1 \\ 0 \\ 1 \end{bmatrix}$$

To find the current, differentiate the charge with respect to time.

$$\frac{dQ(t)}{dt} = 0 - 3c_2 e^{-3t} \begin{bmatrix} -2 \\ 1 \\ 1 \end{bmatrix} - 3c_3 e^{-3t} \begin{bmatrix} -1 \\ 0 \\ 1 \end{bmatrix} \quad \text{or}$$

$$\begin{bmatrix} i_1 \\ i_2 \\ i_3 \end{bmatrix} = -3c_2 e^{-3t} \begin{bmatrix} -2 \\ 1 \\ 1 \end{bmatrix} - 3c_3 e^{-3t} \begin{bmatrix} -1 \\ 0 \\ 1 \end{bmatrix}.$$

At $t = 0$,

$$\begin{bmatrix} i_1 \\ i_2 \\ i_3 \end{bmatrix} = \begin{bmatrix} 0 \\ 3 \\ -3 \end{bmatrix}$$

therefore,

$$\begin{bmatrix} 0 \\ 3 \\ -3 \end{bmatrix} = -3c_2 e^{0} \begin{bmatrix} -2 \\ 1 \\ 1 \end{bmatrix} - 3c_3 e^{0} \begin{bmatrix} -1 \\ 0 \\ 1 \end{bmatrix}.$$

This yields

$$\begin{bmatrix} 0 \\ 3 \\ -3 \end{bmatrix} = -3c_2 \begin{bmatrix} -2 \\ 1 \\ 1 \end{bmatrix} -3c_3 \begin{bmatrix} -1 \\ 0 \\ 1 \end{bmatrix}$$

which is the system of equations

$$+6c_2 + 3c_3 = 0$$

$$-3c_2 = 3$$

$$-3c_2 - 3c_3 = -3 .$$

Equation two gives $c_2 = -1$ and thus equation one is $+6(-1) + 3c_3 = 0$ and $c_3 = 2$. Hence,

$$\begin{bmatrix} i_1 \\ i_2 \\ i_3 \end{bmatrix} = -3(-1)e^{-3t} \begin{bmatrix} -2 \\ 1 \\ 1 \end{bmatrix} -3(2)e^{-3t} \begin{bmatrix} -1 \\ 0 \\ 1 \end{bmatrix}$$

which is

$$\begin{bmatrix} i_1 \\ i_2 \\ i_3 \end{bmatrix} = 3e^{-3t} \begin{bmatrix} \begin{bmatrix} -2 \\ 1 \\ 1 \end{bmatrix} - 2 \begin{bmatrix} -1 \\ 0 \\ 1 \end{bmatrix} \end{bmatrix} .$$

CHAPTER 22

LINEAR FUNCTIONALS

EXAMPLES OF LINEAR FUNCTIONALS

● **PROBLEM** 22-1

Let V be a vector space over a field K. Give examples of
linear functionals ϕ: V → K.

Solution: A mapping ϕ: V → K is termed a linear functional
(or linear form) if, for every u,v ε V and every a,b ε K,

$$\phi(au + bv) = a\phi(u) + b\phi(v). \tag{1}$$

Thus, a linear functional on V is a linear mapping from V
into K.
Examples of linear functionals are the following:

a) Let π_i: K^n → K be the ith projection mapping i.e.
$\pi_i(a_1, a_2, \ldots, a_n) = a_i$. Then π_i is linear since

$$\pi_i(c_1(a_1, a_2, \ldots, a_n) + c_2(b_1, b_2, \ldots, b_n))$$

$$= \pi_i((c_1a_1, c_1a_2, \ldots, c_1a_n) + (c_2b_1, c_2b_2, \ldots, c_2b_n))$$

$$= \pi_i(c_1a_1 + c_2b_1, \ c_1a_2 + c_2b_2, \ldots, c_1a_n + c_2b_n)$$

$$= c_1a_i + c_2b_i$$

$$= c_1\pi_i(a_1, a_2, \ldots, a_n) + c_2\pi_i(b_1, b_2, \ldots, b_n).$$

b) Let V be the vector space of polynomials in t over R.
Let I: V → R be the integral operator defined by I(p(t)) =
$\int_0^1 p(t)dt$. By the rules of calculus,

$$I(c_1(p(t)) + c_2(q(t))) = \int_0^1 c_1p(t) + c_2q(t)dt$$

$$= c_1 \int_0^1 p(t)\,dt + c_2 \int_0^1 q(t)\,dt$$

for $p(t)$, $q(t) \in V$.

Thus $I: V \to R$ is a linear functional.

c) Let $V = R^4$, $K = R$. Then

$$\phi(x_1, x_2, x_3, x_4) = ax_1 + bx_2 + cx_3 + dx_4$$

where $a, b, c, d \in R$ is a linear functional since (letting $u = (x_1, x_2, x_3, x_4)$, $v = (y_1, y_2, y_3, y_4)$)

$$\phi(c, u + v) = \phi_1((c_1 x_1, c_1 x_2, c_1 x_3, c_1 x_4) + (y_1, y_2, y_3, y_4))$$

$$= \phi(c_1 x_1 + y_1,\ c_1 x_2 + y_2,\ c_1 x_3 + y_3,\ c_1 x_4 + y_4)$$

$$= a(c_1 x_1 + y_1) + b(c_1 x_2 + y_2) + c(c_1 x_3 + y_3)$$

$$+ d(c_1 x_4 + y_4)$$

$$= (ac_1 x_1 + bc_1 x_2 + cc_1 x_3 + dc_1 x_4)$$

$$+ (ay_1 + by_2 + cy_3 + dy_4)$$

$$= c_1(ax_1 + bx_2 + cx_3 + dx_4) + (ay_1 + by_2 + cy_3 + dy_4)$$

$$= c_1 \phi(u) + \phi(v).$$

● **PROBLEM 22-2**

Let A be an n × n matrix with entries in the field F. Show that the trace operator is a linear functional.

Solution: A function $\phi: V \to K$, i.e. with domain V, the vector space, and range K, the field over which V is defined, is called a linear functional if $\phi(au + bv) = a\phi(u) + b\phi(v)$, $a, b \in K$, $u, v \in V$. This condition may also be shortened to $\phi(au + v) = a\phi(u) + \phi(v)$ since if the function is linear in an arbitrary argument u it will be linear in v.   Let

$$A = \begin{bmatrix} a_{11} & a_{12} & \cdots & a_{1n} \\ a_{21} & a_{22} & \cdots & a_{2n} \\ \cdot & \cdot & & \cdot \\ \cdot & \cdot & & \cdot \\ \cdot & \cdot & & \cdot \\ a_{n1} & a_{n2} & & a_{nn} \end{bmatrix}$$

The trace of A is the scalar

$$\text{tr } A = a_{11} + a_{22} + \ldots + a_{nn} = \sum_{i=1}^{n} a_{ii}.$$

Let

$$B = \begin{bmatrix} b_{11} & b_{12} & \cdots & b_{1n} \\ b_{21} & b_{22} & \cdots & b_{2n} \\ \cdot & \cdot & & \cdot \\ \cdot & \cdot & & \cdot \\ \cdot & \cdot & & \cdot \\ b_{n1} & b_{n2} & \cdots & b_{nn} \end{bmatrix}$$

Then, for tr(CA + B) we obtain:

$$\text{tr}\left\{ C \begin{bmatrix} a_{11} & a_{12} & \cdots & a_{1n} \\ a_{21} & a_{22} & & a_{2n} \\ \cdot & & & \cdot \\ \cdot & & & \cdot \\ \cdot & & & \cdot \\ a_{n1} & a_{n2} & & a_{nn} \end{bmatrix} + \begin{bmatrix} b_{11} & b_{12} & \cdots & b_{1n} \\ b_{21} & b_{22} & & b_{2n} \\ \cdot & & & \cdot \\ \cdot & & & \cdot \\ \cdot & & & \cdot \\ b_{n1} & b_{n2} & & b_{nn} \end{bmatrix} \right\}$$

$$\text{tr}\left\{ \begin{bmatrix} ca_{11} & ca_{12} & \cdots & ca_{1n} \\ ca_{21} & ca_{22} & \cdots & ca_{2n} \\ \cdot & & & \cdot \\ \cdot & & & \cdot \\ \cdot & & & \cdot \\ ca_{n1} & ca_{n2} & & ca_{nn} \end{bmatrix} + \begin{bmatrix} b_{11} & b_{12} & \cdots & b_{1n} \\ b_{21} & b_{22} & \cdots & b_{2n} \\ \cdot & & & \cdot \\ \cdot & & & \cdot \\ \cdot & & & \cdot \\ b_{n1} & b_{n2} & \cdots & b_{nn} \end{bmatrix} \right\}$$

$$= \text{tr} \begin{bmatrix} ca_{11} + b_{11} & ca_{12} + b_{12} & \cdots & ca_{1n} + b_{1n} \\ ca_{21} + b_{21} & ca_{22} + b_{22} & \cdots & ca_{2n} + b_{2n} \\ \cdot & & & \\ \cdot & & & \\ \cdot & & & \\ ca_{n1} + b_{n1} & ca_{n2} + b_{n2} & \cdots & ca_{nn} + b_{nn} \end{bmatrix}$$

$$= ca_{11} + b_{11} + ca_{22} + b_{22} + \ldots + ca_{nn} + b_{nn}$$

$$= \sum_{i=1}^{n} ca_{ii} + b_{ii}$$

$$= c \sum_{i=1}^{n} a_{ii} + \sum_{i=1}^{n} b_{ii}$$

$$= (ca_{11} + ca_{22} + \ldots + ca_{nn}) + b_{11} + b_{22} + \ldots + b_{nn}$$

$$= c \operatorname{tr} A + \operatorname{tr} B.$$

Thus the trace function is a linear functional.

● PROBLEM 22-3

If A and B are n × n matrices over the field F, show that trace (AB) = trace (BA).

Solution: Let

$$A = \begin{bmatrix} a_{11} & a_{12} & \cdots & a_{1n} \\ a_{21} & a_{22} & \cdots & a_{2n} \\ \cdot & & & \\ \cdot & & & \\ \cdot & & & \\ a_{n1} & a_{n2} & \cdots & a_{nn} \end{bmatrix}$$

$$B = \begin{bmatrix} b_{11} & b_{12} & \cdots & b_{1n} \\ b_{21} & b_{22} & \cdots & b_{2n} \\ \cdot & & & \\ \cdot & & & \\ \cdot & & & \\ b_{n1} & b_{n2} & \cdots & b_{nn} \end{bmatrix}$$

where a_{ij}, b_{ij} are entries from the field F $(i,j = 1,2,\ldots,n)$.
Now,

$$AB = \begin{bmatrix} \sum_{j=1}^{n} a_{1j}b_{j1} & \sum_{j=1}^{n} a_{1j}b_{j2} & \cdots & \sum_{j=1}^{n} a_{1j}b_{jn} \\ \sum_{j=1}^{n} a_{2j}b_{j1} & \cdots\cdots\cdots\cdots\cdots & & \sum_{j=1}^{n} a_{2j}b_{jn} \\ \sum_{j=1}^{n} a_{nj}b_{j1} & \cdots\cdots\cdots\cdots\cdots & & \sum_{j=1}^{n} a_{nj}b_{jn} \end{bmatrix}$$

$$\text{and } BA = \begin{bmatrix} \sum_{j=1}^{n} b_{1j}a_{j1} & \sum_{j=1}^{n} b_{1j}a_{j2} & \cdots & \sum_{j=1}^{n} b_{1j}a_{jn} \\[2ex] \sum_{j=1}^{n} b_{2j}a_{j1} & \sum_{j=1}^{n} b_{2j}a_{j2} & \cdots & \sum_{j=1}^{n} b_{2j}a_{jn} \\[2ex] \vdots & & & \\[2ex] \sum_{j=1}^{n} b_{nj}a_{j1} & \sum_{j=1}^{n} b_{nj}a_{j2} & \cdots & \sum_{j=1}^{n} b_{nj}a_{jn} \end{bmatrix}$$

The trace of an $n \times n$ matrix is the sum of its diagonal entries. The trace is a linear functional from $V_{n\times n}$ to F. To show this,

$$\text{tr}(cA + B) = \sum ca_{ii} + b_{ii} = c\sum a_{ii} + \sum b_{ii} = c \text{ tr } A + \text{ tr } B.$$

Now,

$$\text{tr } AB = \sum_{j=1}^{n} a_{1j}b_{j1} + \sum_{j=1}^{n} a_{2j}b_{j2} + \cdots + \sum_{j=1}^{n} a_{nj}b_{jn}$$

$$= (a_{11}b_{11} + a_{12}b_{21} + \cdots + a_{1n}b_{n1})$$

$$+ (a_{21}b_{12} + a_{22}b_{22} + \cdots + a_{2n}b_{n2}) + \cdots$$

$$+ (a_{n1}b_{1n} + a_{n2}b_{2n} + \cdots + a_{nn}b_{nn})$$

$$= (a_{11}b_{11} + a_{21}b_{12} + \cdots + a_{n1}b_{1n})$$

$$+ (a_{12}b_{21} + a_{22}b_{22} + \cdots + a_{n2}b_{2n}) + \cdots$$

$$+ (a_{1n}b_{n1} + a_{2n}b_{n2} + \cdots + a_{nn}b_{nn})$$

$$= (b_{11}a_{11} + b_{12}a_{21} + \cdots + b_{1n}a_{n1})$$

$$+ (b_{21}a_{12} + b_{22}a_{22} + \cdots + b_{2n}a_{n2}) + \cdots$$

$$+ (b_{n1}a_{1n} + b_{n2}a_{2n} + \cdots + b_{nn}a_{nn})$$

$$= \sum_{j=1}^{n} b_{1j}a_{j1} + \sum_{j=1}^{n} b_{2j}a_{j2} + \cdots + \sum_{j=1}^{n} b_{nj}a_{jn}$$

$$= \text{tr } BA.$$

This shows that the trace function is a symmetric function of its arguments, i.e.

$$\phi(x,y) = \phi(y,x).$$

Let L be a linear functional defined on a finite-dimensional real inner product space. Prove that this functional must be the inner product.

Solution: Let R^n denote the vector space of real n-tuples, i.e. vectors of the form $(x_1, \ldots, x_n)$. Define an inner product as

$$(x_1, \ldots, x_n) \ (y_1, \ldots, y_n) = x_1 y_1 + x_2 y_2 + \ldots + x_n y_n) \quad (1)$$

where $(x_1, \ldots, x_n)$, $(y_1, \ldots, y_n) \in R^n$. Then R^n is a real inner product space. The inner product defined by (1) is a linear functional from R^n to R. Actually, the inner product is a bilinear functional (or form). That is, it obeys the conditions

$$\phi(ax_1 + bx_2, y) = a\phi(x_1, y) + b\phi(x_2, y)$$

$$\phi(x, ay_1 + by_2) = a\phi(x, y_1) + b\phi(x, y_2).$$

We say that the inner product is linear in one of its arguments when the other is held fixed. On the other hand, the inner product (x,y) for $\begin{Bmatrix} y \\ x \end{Bmatrix}$ fixed is a linear functional on V in the variable $\begin{Bmatrix} x \\ y \end{Bmatrix}$. Observe that

$$[c(x_1, \ldots, x_n) + (y_1, \ldots, y_n)] \cdot (z_1, z_2, \ldots, z_n)$$

$$= (cx_1 + y_1, cx_2 + y_2, \ldots, cx_n + y_n) \ (z_1, z_2, \ldots, z_n)$$

$$= (cx_1 + y_1)z_1 + (cx_2 + y_2)z_2 + \ldots + (cx_n + y_n)z_n$$

$$= cx_1 z_1 + y_1 z_1 + cx_2 z_2 + y_2 z_2 + \ldots + cx_n z_n + y_n z_n$$

$$= (cx_1 z_1 + cx_2 z_2 + \ldots + cx_n z_n) + (y_1 z_1 + y_2 z_2 + \ldots + y_n z_n)$$

$$= c(x_1 z_1 + \ldots + x_n z_n) + (y_1 z_1 + \ldots + y_n z_n)$$

$$= c(x_1, \ldots, x_n) \cdot (z_1, \ldots, z_n) + (y_1, \ldots, y_n) \cdot (z_1, \ldots, z_n).$$

Recall that a linear functional from a vector space to a field is a mapping of the form ϕ: $V \to K$ that satisfies $\phi(cu + v) = c\phi(u) + \phi(v)$ for $u, v \in V$ and $c \in K$. Thus the inner product is a linear functional.

To show that it is the only linear functional, proceed as follows: Let $E = \{e_1, e_2, \ldots, e_n\}$ be the usual basis for R^n, i.e. vectors of the form $(0, 0, \ldots, 1, 0, 0, \ldots 0)$. Then $x \in R^n$ implies that there exist unique scalars

$x_1, \ldots, x_n$ such that $x = \sum_{j=1}^{n} x_j e_j$. Since L is a linear functional

$$L(x) = L(x_1, \ldots, x_n) = L \sum x_j e_j = \sum x_j L(e_j).$$

Thus the values of L on a basis for V determine it uniquely on V. These values are scalars and hence are coordinates of some vector in V. Let

$$y = \sum L(e_j)e_j.$$

Then $L(x) = \langle x, y \rangle$ (the inner product of a variable x with a fixed y.) Thus the only linear functional defined on a finite dimensional real inner product space is the scalar product. Note that this conclusion can be extended to complex inner product spaces and to infinite dimensional spaces. In its full generality the above statement is known as the Riesz representation theorem of functional analysis.

● **PROBLEM** 22-5

Let $\phi: R^2 \to R$ and $\sigma: R^2 \to R$ be the linear functionals defined by $\phi(x,y) = x + 2y$ and $\sigma(x,y) = 3x - y$. Find
i) $\phi + \sigma$, ii) 4ϕ iii) $2\phi - 5\sigma$.

<u>Solution</u>: Since linear functionals are linear transformations whose range is the underlying field, we define addition and scalar multiplication as follows:

a) $(\phi_1 + \phi_2)v = \phi_1 v + \phi_2 v$.

b) $\phi(cv) = c\phi(v)$

for $\phi_1, \phi_2 \in (R^2)*$ (the dual space), $v \in R^2$ and $c \in R$.

i) Let $v = (x,y)$. Then,

$$(\phi + \sigma)(v) = (\phi + \sigma)(x,y)$$

$$= \phi(x,y) + \sigma(x,y)$$

$$= (x + 2y) + (3x - y)$$

$$= 4x + y.$$

ii) $4\phi(v) = 4\phi(x,y)$

$$= 4(x + 2y) = 4x + 8y.$$

iii) $(2\phi - 5\sigma)(v) = (2\phi - 5\sigma)(x,y)$

$$= 2\phi(x,y) - 5\sigma(x,y)$$

$$= \phi(2x, 2y) - \sigma(5x, 5y)$$

$$= 2x + 4y - (15x - 5y)$$

$$= -13x + 9y.$$

THE DUAL SPACE & DUAL BASIS

Let $\{(1,-1,3),\ (0,1,-1),\ (0,3,-2)\}$ be a basis of R^3. Find the dual basis $\{\phi_1, \phi_2, \phi_3\}$.

Solution: Let V be a vector space defined over a field K. The set of all linear functionals $\phi:\ V \to K$ is itself a vector space. To show this, define addition of linear functionals as

$$(\phi_1 + \phi_2)u = \phi_1(u) + \phi_2(u) \tag{1}$$

and scalar multiplication as

$$c\phi_1(u) = \phi(cu), \tag{2}$$

where ϕ_1, ϕ_2 are linear functionals and $c \in K$. Let V* denote the set of all linear functionals on V. V* is closed under addition and scalar multiplication. For, $(\phi_1 + \phi_2) = \phi_1(u) + \phi_2(u)$ is in K since $\phi_1(u)$, $\phi_2(u)$ are in K. Furthermore, letting $\phi_1(u) = k$, $c\phi_1(u) = ck \in K$ since a field is closed under scalar multiplication. Next, let 0 be the linear functional defined by $\phi(u) = 0$ for all $u \in V$. We know that the set of all functions, S, from an arbitrary set to a field is a vector space under the operations of function addition and scalar multiplication as in (1) and (2). Hence V* is a subspace of S and is therefore a vector space. V* is called the dual space.

Since V* is a vector space it has a basis. To obtain the dimension of V*, use the following theorem. If V and W are n and m-dimensional vector spaces respectively, L(V,W), the set of all linear transformations from V into W, is a vector space of dimension mn. Letting W be the one-dimensional vector space K, V* has dimension n. Next, use the given basis for R^3 to find the basis for V*. This requires, as justification, the following theorem. If $\{\alpha_1,\ \alpha_2,\ \dots,\ \alpha_n\}$ is an ordered basis for V and $\{\beta_1,\ \dots,\ \beta_n\}$ is any set of vectors in W, then there is one and only one linear transformation L: V $\to$ W such that

$$L\alpha_j = \beta_j \quad j = 1,\dots,n. \tag{1}$$

According to this theorem there is (for each i) a unique linear functional f_i on V such that

$$f_i(\alpha_j) = \delta_{ij} \quad (\delta_{ij} = 1,\ i = j;\ 0,\ i \neq j) \tag{2}$$

Comparing (1) and (2), note that

$$\beta_j = [0, 0, \ldots \underset{\substack{\uparrow \\ \text{jth} \\ \text{place}}}{1,} 0, \ldots 0].$$

In this way one can obtain from the basis of V a set of n distinct linear functionals. These functionals are also linearly independent, since, setting

$$f = \sum_i c_i f_i \qquad \text{yields:}$$

$$f(\alpha_j) = \sum_i c_i f_i(\alpha_j)$$

$$= \sum_i c_i \delta_{ij} = c_1(0) + \ldots + c_j(1) + \ldots + c_n(0)$$

$$= c_j. \quad (j = 1, \ldots, n).$$

Setting $f = 0$ yields $f(\alpha_j) = 0$ for each j. Hence the scalars c_j are all 0. Thus, f_i (i = 1,2, ..., n) form a basis for V* called the dual basis.

In the given problem linear functionals need to be found.

$$\phi_1(x,y,z) = a_1 x + a_2 y + a_3 y$$

$$\phi_2(x,y,z) = b_1 x + b_2 y + b_3 y$$

$$\phi_3(x,y,z) = c_1 x + c_2 y + c_3 y$$

such that $\quad \phi_1(v_1) = 1 \quad \phi_1(v_2) = 0 \quad \phi_1(v_3) = 0$

$$\phi_2(v_1) = 0 \quad \phi_2(v_2) = 1 \quad \phi_2(v_3) = 0$$

$$\phi_3(v_1) = 0 \quad \phi_3(v_2) = 0 \quad \phi_3(v_3) = 0$$

where $v_1 = (1,-1,3)$, $v_2 = (0,1,-1)$, $v_3 = (0,3,-2)$. One can find ϕ_1 as follows:

$$\phi_1(v_1) = \phi_1(1,-1,3) = a_1 - a_2 + 3a_3 = 1$$

$$\phi_1(v_2) = \phi_1(0,1,-1) = \qquad a_2 - a_3 = 0$$

$$\phi_1(v_3) = \phi_1(0,3,-2) = \qquad 3a_2 - 2a_3 = 0.$$

Solving this system of equations, yields $a_1 = 1$, $a_2 = 0$, $a_3 = 0$. Thus $\phi_1(x,y,z) = x$. Next, find ϕ_2:

$$\phi_2(v_1) = \phi_2(1,-1,3) = b_1 - b_2 + 3b_3 = 0$$

824

$$\phi_2(v_2) = \phi_2(0,1,-1) = \qquad b_2 - b_3 = 1$$

$$\phi_2(v_3) = \phi_2(0,3,-2) = \qquad 3b_2 - 2b_3 = 0.$$

Solving the system yields $b_1 = 7$, $b_2 = -2$, $b_3 = -3$. Hence $\phi_2(x,y,z) = 7x - 2y - 3z$. Finally, find ϕ_3:

$$\phi_3(v_1) = \phi_3(1,-1,3) = c_1 - c_2 + 3c_3 = 0$$

$$\phi_3(v_2) = \phi_3(0,1,-1) = \qquad c_2 - c_3 = 0$$

$$\phi_3(v_3) = \phi_3(0,3,-2) = \qquad 3c_2 - 2c_3 = 1.$$

Solving the system, yields $c_1 = -2$, $c_2 = 1$, $c_3 = 1$. Thus $\phi_3(x,y,z) = -2x + y + z$. The dual basis to $\{(1,-1,3), (0,1,-1), (0,3,-2)\}$ is therefore $\{\phi_1 = x, \phi_2 = 7x - 2y - 3z, \phi_3 = -2x + y + z\}$.

● **PROBLEM 22-7**

Give an example to illustrate the following theorem. Let V be a finite-dimensional vector space over the field F and let $B = \{\alpha_1, \ldots, \alpha_n\}$ be a basis for V. Then there is a unique dual basis $B^* = \{f_1, \ldots, f_n\}$ for V* such that $f_i(\alpha_j) = \delta_{ij}$. For each linear functional f on V we have

$$f = \sum_{i=1}^{n} f(\alpha_i) f_i \qquad (1)$$

and for each vector α in V we have

$$\alpha = \sum_{i=1}^{n} f_i(\alpha) \alpha_i.$$

Solution: Let $V = R^3$, $F = R$ and let $B = \{(1,-1,3), (0,1,-1), (0,3,-2)\}$ be an ordered basis for V. Then the dual basis B' consists of the three functionals

$$\{\phi_1(x_1,x_2,x_3), \phi_2(x_1,x_2,x_3), \phi_3(x_1,x_2,x_3)\}$$

where $\phi_1(x_1,x_2,x_3) = a_{11}x_1 + a_{12}x_2 + a_{13}x_3$. Since $\phi_i(\alpha_j) = \delta_{ij}$ ($\delta_{ij} = 1$, $i = j; 0$ $i \neq j$),

$$\phi_1(1,-1,3) = x_1 - x_2 + 3x_3 = 1$$

$$\phi_1(0,1,-1) = \qquad x_2 - x_3 = 0$$

$$\phi_1(0,3,-2) = \qquad 3x_2 - 2x_3 = 0.$$

825

Thus, $\phi_1(x_1,x_2,x_3) = x_1$. Similarly, it can be found that $\phi_2 = 7x_1 - 2x_2 - 3x_3$ and $\phi_3 = -2x_1 + x_2 + x_3$.

The functionals $\{\phi_1,\phi_2,\phi_3\}$ form a basis for V^*. More importantly, they are dual to the basis $B = \{\alpha_1,\alpha_2,\alpha_3\}$ for V. This means that any functional f in the dual space can be expressed as a linear combination of ϕ_1,ϕ_2 and ϕ_3, the coordinates being given by the effects of f on the basis vectors $\{\alpha_1,\alpha_2,\alpha_3\}$ for V. Conversely, it is obvious that any $\alpha \in V$ can be expressed as a linear combination of $\alpha_1, \alpha_2, \alpha_3$. Note, however, that the coordinates are given by the effects of ϕ_1, ϕ_2 and ϕ_3 on the vector α.

Let $\quad \alpha = (1,3,2)$

and let $f(x_1,x_2,x_3) = 4x_1 + x_2 - 3x_3$.

Then $\quad f(x_1,x_2,x_3) = \sum_{i=1}^{3} f(\alpha_i)\phi_i$

$$= f(1,-1,3)(x_1)$$

$$+ f(0,1,-1)(7x_1 - 2x_2 - 3x_3)$$

$$+ f(0,3,-2)(-2x_1 + x_2 + x_3)$$

$$= -6x_1 + 28x_1 - 8x_2 - 12x_3 - 18x_1 + 9x_2 + 9x_3$$

$$= 4x_1 + x_2 - 3x_3 = f(x_1,x_2,x_3).$$

Similarly,

$$\alpha = \sum_{i=1}^{3} \phi_i(\alpha)\alpha_i$$

$$= \phi_1(1,3,2)(1,-1,3) + \phi_2(1,3,2)(0,1,-1)$$

$$+ \phi_3(1,3,2)(0,3,-2)$$

$$= (1,-1,3) - 5(0,1,-1) + 3(0,3,-2)$$

$$= (1,3,2) = \alpha.$$

Note that if the standard basis for R^3 was chosen $B = \{(1,0,0), (0,1,0), (0,0,1)\}$, the corresponding dual would have been $\{\phi_1 = x_1, \phi_2 = x_2, \phi_3 = x_3\}$.

In R^3, let $\alpha_1 = (1,0,1)$, $\alpha_2 = (0,1,-2)$, $\alpha_3 = (-1,-1,0)$.

a) If f is a linear functional on R^3 such that $f(\alpha_1) = 1$, $f(\alpha_2) = -1$, $f(\alpha_3) = 3$, and if $\alpha = (a,b,c)$, find $f(\alpha)$.

b) Find a linear functional f on R^3 such that $f(\alpha_1) = f(\alpha_2) = 0$ but $f(\alpha_3) \neq 0$.

c) Let f be any linear functional such that $f(\alpha_1) = f(\alpha_2) = 0$ and $f(\alpha_3) \neq 0$.

If $\alpha = (2,3,-1)$, show that $f(\alpha) \neq 0$.

<u>Solution</u>: a) The vectors $\{\alpha_1, \alpha_2, \alpha_3\}$ form a basis for R^3. Thus

$$(a,b,c) = c_1\alpha_1 + c_2\alpha_2 + c_3\alpha_3$$

$$= c_1(1,0,1) + c_2(0,1,-2) + c_3(-1,-1,0)$$

$$= (c_1,0,c_1) + (0,c_2,-2c_2) + (-c_3 - c_3,0)$$

$$= (c_1 - c_3,\ c_2 - c_3,\ c_1 - 2c_2).$$

Solving for c_1, c_2 and c_3 yields:

$$c_1 = 2a - 2b - c$$

$$c_2 = a - b - c$$

$$c_3 = a - c - 2b.$$

Thus, $(a,b,c) = (2a - ab - c)(1,0,1) + (a - b - c)(0,1,-2)$

$$+ (a - 2b - c)(-1,-1,0).$$

Then,

$f(a,b,c) = (2a - 2b = c)f(1,0,1) + (a - b - c)f(0,1,-2)$

$$+ (a - 2b - c)f(-1,-1,0)$$

$$= (2a - 2b - c)(1) + (a - b - c)(-1)$$

$$+ (a - 2b - c)(3)$$

$$= 4a - 7b - 3c.$$

b) We have $f(1,0,1) = 0$, $f(0,1,-2) = 0$ and $f(-1,-1,0) =$

$c \neq 0$. Since we have a basis for R^3, we can find a basis for V^*, the dual space. This basis consists of unique linear functionals f_i such that

$$f_i(\alpha_j) = \delta_{ij} \qquad (1)$$

where $\delta_{ij} = 1$, $i = j$ and 0 otherwise.

For the given α_j (1) becomes:

$f_1(\alpha_1) = f_1(1,0,1) = 1 \qquad f_2(\alpha_1) = f_2(1,0,1) = 0$

$f_1(\alpha_2) = f_1(0,1,-2) = 0 \; ; \; f_2(\alpha_2) = f_2(0,1,-2) = 1 \; ;$

$f_1(\alpha_3) = f_1(-1,-1,0) = 0 \qquad f_2(\alpha_3) = f_2(-1,-1,0) = 0$

$f_3(\alpha_1) = f_3(1,0,1) = 0$

$f_3(\alpha_2) = f_3(0,1,-2) = 0 \qquad (2)$

$f_3(\alpha_3) = f_3(-1,-1,0) = 1$

Examining (2), note that $f_3(\alpha_1) = f_3(\alpha_2) = 0$ but $f_3(\alpha_3) = 1 \neq 0$. Now, since f_3 is a linear functional from R^3 to R, it is of the form $f_3(x_1,x_2,x_3) = c_1 x_1 + c_2 x_2 + c_3 x_3$ where $(x_1,x_2,x_3) \in R^3$ and $c_i \in R$.

Then, $\qquad f_3(1,0,1) \quad = c_1 \qquad + c_3 = 0$

$\qquad\qquad f_3(0,1,-2) = \qquad\quad c_2 - 2c_3 = 0$

$\qquad\qquad f_3(-1,-1,0) = -c_1 - c_2 \qquad = 1.$

Form the augmented matrix

$$\begin{bmatrix} 1 & 0 & 1 & | & 0 \\ 0 & 1 & -2 & | & 0 \\ -1 & -1 & 0 & | & 0 \end{bmatrix}$$

Applying elementary row operations we obtain the equivalent system:

$$c_1 + c_3 = 0$$

$$c_2 - 2c_3 = 0$$

$$-c_3 = 1.$$

828

By back-substitution $c_3 = -1$, $c_2 = -2$ and $c_1 = 1$. Thus,

$$f_3(x_1,x_2,x_3) = x_1 - 2x_2 - x_3, \tag{3}$$

and $f_3(\alpha_1) = f_3(\alpha_2) = 0$ but $f_3(\alpha_3) = 1 \neq 0$.

c) We must find the set of all linear functionals such that $f(\alpha_1) = f(\alpha_2) = 0$ but $f(\alpha_3) \neq 0$. That is,

$f(1,0,1) = 0$; $f(0,1,-2) = 0$; $f(-1,-1,0) = a$, $a \neq 0$.

From (3) we have: $f(x_1,x_2,x_3) = ax_1 - 2ax_2 - ax_3$. Let $\alpha = (2,3,-1)$. Then, $f(\alpha) = 2a - 2(3)a - a(-1) = -3a \neq 0$, since $a \neq 0$ by assumption.

● **PROBLEM** 22-9

Let V be the vector space of polynomials over R of degree ≤ 1, i.e. $V = \{a + bt: a,b \in R\}$. Let $\phi_1: V \to R$ and $\phi_2: V \to R$ be defined by

$$\phi_1(f(t)) = \int_0^1 f(t)dt \quad \text{and} \quad \phi_2(f(t)) = \int_0^2 f(t)dt.$$

Find the basis $\{v_1,v_2\}$ of V which is dual to $\{\phi_1,\phi_2\}$.

<u>Solution</u>: The vector space V has dimension 2 since $\{1,x\}$ forms a basis. The set of all linear functionals from V to R forms a vector space known as the dual space (denoted by V*). To find the dimension of V* we reason as follows.

Let V, W be vector spaces of dimensions n and m respectively. Then the set of all linear transformations from V to W forms a vector space of dimension nm. Since linear functionals have domain V and range F (a field is always a vector space over itself). The dimension of V* is n. In the given problem dim V* = 2. To show that ϕ_1 and ϕ_2 form a basis for V* it is sufficient to show that they are linearly independent. Let

$$c_1 \int_0^1 (a+bt)dt + c_2 \int_0^2 (a+bt)dt = 0.$$

Integrating and evaluating yields

$$c_1[a + b/2] + c_2[2a + 2b]$$

or $$(c_1 + 2c_2)a + (c_1/2 + 2c_2)b.$$

Assuming that $a,b \neq 0$

$$c_1 + 2c_2 = 0$$

$$\tfrac{1}{2}c_1 + 2c_2 = 0$$

which implies $c_1 = c_2 = 0$. Thus $\{\phi_1, \phi_2\}$ forms a basis for V*. We can use this basis to obtain the dual basis for V by applying the following theorem. Let $\{f_1, f_2, \ldots, f_n\}$ be a basis for V*. Then there is a unique dual basis for V, $B = \left(e_1, e_2, \ldots, e_n\right)$ such that

$$f_1(e_j) = \delta_{ij} = \begin{cases} 1, & \text{if } i = j \\ 0, & \text{if } i \neq j. \end{cases}$$

Thus, $f_1(e_1) = 1;; f_1(e_2) = 0 ; \ldots f_1(e_n) = 0$

$$f_2(e_1) = 0 ; f_2(e_2) = 1 ; \ldots f_2(e_n) = 0$$

$$\vdots$$

$$f_n(e_1) = 0 ; f_n(e_2) = 0 ; \ldots f_n(e_n) = 1$$

Turning to the given problem, let $e_1 = a_1 + b_1 t$, $e_2 = a_2 + b_2 t$. Then

$$f_1(e_1) = \int_0^1 a_1 + b_1 t \, dt = 1 \qquad f_2(e_1) = \int_0^2 a_1 + b_1 t \, dt = 0$$

$$f_1(e_2) = \int_0^1 a_2 + b_2 t \, dt = 0 \qquad f_2(e_2) = \int_0^2 a_2 + b_2 t \, dt = 1.$$

Evaluating the definite integrals yields

$$a_1 + \frac{b_1}{2} = 1 \qquad a_2 + \frac{b_2}{2} = 0$$

$$a_1 + b_1 = 0 \qquad a_2 + b_2 = 1,$$

and hence $a_1 = 2, b_1 = -2; a_2 = -1, b_2 = 2$. Thus the basis dual to $\left\{\phi_1, \phi_2\right\}$ is given by the vectors 2-2x. and -1 + 2x.

830

Let V be the vector space of all polynomial functions p from R into R which have degree 2 or less:

$$p(x) = c_0 + c_1 x + c_2 x^2.$$

Define three linear functionals on V by

$$f_1(p) = \int_0^1 p(x)dx, \quad f_2(p) = \int_0^2 p(x)dx, \quad f_3(p) = \int_0^{-1} p(x)dx.$$

Show that $\{f_1, f_2, f_3\}$ is a basis for V* by exhibiting the basis for V of which it is the dual.

<u>Solution</u>: Let $\{\alpha_1, \alpha_2, \alpha_3\}$ be a basis for V, i.e. the α_i are polynomials of degree two or less. If the given functionals form a basis for V*, the dual space, then there is a unique correspondence between $\{\alpha_1, \alpha_2, \alpha_3\}$ and $\{f_1, f_2, f_3\}$. We have

$$f_i(\alpha_j) = \delta_{ij} \quad (\delta_{ij} = 1, \ i = j; \ 0, i \neq j). \tag{1}$$

Let $\alpha_1 = a_0 + a_1 x + a_2 x^2$

$\alpha_2 = b_0 + b_1 x + b_2 x^2$

$\alpha_3 = d_0 + d_1 x + d_2 x^2.$

Then, from (1):

$$f_1(\alpha_1) = 1 \ ; \ f_1(\alpha_2) = 0 \ ; \ f_1(\alpha_3) = 0$$

$$f_2(\alpha_1) = 0 \ ; \ f_2(\alpha_2) = 1 \ ; \ f_2(\alpha_3) = 0$$

$$f_3(\alpha_1) = 0 \ ; \ f_3(\alpha_2) = 0 \ ; \ f_3(\alpha_3) = 1.$$

Now $f_1(p) = \int_0^1 p(x)dx = \int_0^1 c_0 + c_1 x + c_2 x^2 dx$

$$= c_0 x + \frac{c_1}{2}x^2 + \frac{c_2}{3}x^3 \Big|_0^1 = c_0 + \frac{c_1}{2} + \frac{c_2}{3}. \tag{2}$$

$$f_2(p) = \int_0^2 p(x)dx = \int_0^2 c_0 + c_1 x + c_2 x^2 dx$$

$$= c_0x + \frac{c_1}{2}x^2 + \frac{c_2}{3}x^3 \Big|_0^2 = 2c_0 + 2c_1 + \frac{8}{3}c_2. \qquad (3)$$

$$f_3(p) = \int_0^{-1} c_0 + c_1x + c_2x^2 dx$$

$$= c_0x + \frac{c_1}{2}x^2 + \frac{c_2}{3}x^3 \Big|_0^{-1} = -c_0 + \frac{c_1}{2} - \frac{c_2}{3}. \qquad (4)$$

Using (2), (3) and (4) we can evaluate $f_1(\alpha_1)$, $f_2(\alpha_1)$ and $f_3(\alpha_1)$. We obtain the following system in the unknowns a_0, a_1 and a_2:

$$a_0 + \frac{a_1}{2} + \frac{a_2}{3} = 1$$

$$2a_0 + a_1 + \frac{8}{3}a_2 = 0$$

$$-a_0 + \frac{a_1}{2} - \frac{a_2}{3} = 0.$$

Form the augmented matrix of coefficients:

$$\begin{bmatrix} 1 & 1/2 & 1/3 & \vdots & 1 \\ 2 & 1 & 8/3 & \vdots & 0 \\ -1 & 1/2 & -1/3 & \vdots & 0 \end{bmatrix} \qquad (5)$$

Reduce (5) to row echelon form:

$$\begin{bmatrix} 1 & 1/2 & 1/3 & \vdots & 1 \\ 0 & 0 & 2 & \vdots & -1 \\ 0 & 1 & 0 & \vdots & 1 \end{bmatrix} \longleftrightarrow \begin{bmatrix} 1 & 1/2 & 1/3 & \vdots & 1 \\ 0 & 1 & 0 & \vdots & 1 \\ 0 & 0 & 2 & \vdots & -1 \end{bmatrix}.$$

Thus, $a_2 = -1/2$; $a_1 = 1$ and $a_0 = 2/3$, and

$\alpha_1 = 2/3 + x - (1/2)x^2$. Next, find α_2 from the relations
$f_1(\alpha_2) = 0$; $f_2(\alpha_2) = 1$; $f_3(\alpha_2) = 0$, and using (2), (3) and
(4). The system is

$$b_0 + \frac{b_1}{2} + \frac{b_2}{3} = 0$$

$$2b_0 + b_1 + \frac{8}{3}b_2 = 1$$

$$-b_0 + \frac{b_1}{2} - \frac{b_2}{3} = 0.$$

The echelon matrix for this system is

$$\begin{bmatrix} 1 & 1/2 & 1/3 & \vline & 0 \\ 0 & 1 & 0 & \vline & 0 \\ 0 & 0 & 2 & \vline & 1 \end{bmatrix}.$$

Thus $b_2 = 1/2$, $b_1 = 0$ and $b_0 = -1/6$, and $\alpha_2 = -1/6 + (1/2)x^2$. Finally,

$$\alpha_3 = d_0 + d_1 x + d_2 x^2 = -\tfrac{1}{2} + x.$$

To verify that $\{\alpha_1, \alpha_2, \alpha_3\}$ forms a basis for V, recall that $\{1, x, x^2\}\}$ is a basis, and that an arbitrary vector in V has the form $a_0 + a_1 x + a_2 x^2$. Since V has dimension 3, if we show that $\{\alpha_1, \alpha_2, \alpha_3\}$ is a linearly independent set, then it is a basis. Alternatively, we may show that $a_0 + a_1 x + a_2 x^2$ can be written as a linear combination of α_1, α_2 and α_3, or, finally that $\{1, x, x^2\}$ can be written as a linear combination of α_1, α_2, α_3. Following the second alternative, set

$$a_0 + a_1 x + a_2 x^2 = c_1 \alpha_1 + c_2 \alpha_2 + c_3 \alpha_3.$$

Two polynomials are equal when the coefficients of like powers of x are equal. Thus,

$$a_0 = (2/3)c_1 - (1/6)c_2 - (1/2)c_3$$

$$a_1 = c_1 \qquad + \qquad c_3$$

$$a_2 = -(1/2)c_1 + (1/2)c_2.$$

Form the coefficient matrix and reduce to echelon form:

$$\begin{bmatrix} 2/3 & -1/6 & -1/2 \\ 1 & 0 & 1 \\ -1/2 & 1/2 & 0 \end{bmatrix} \longleftrightarrow \begin{bmatrix} 1 & -1/4 & -3/4 \\ 0 & 1 & 7 \\ 0 & 0 & -3 \end{bmatrix}.$$

Since the echelon matrix contains no row of zeros, the unknowns c_1, c_2, c_3 may be found in terms of the a_i. Thus $\{\alpha_1, \alpha_3, \alpha_3\}$ forms a basis for V and hence $\{f_1, f_2, f_3\}$ forms a basis for V*, the space of linear functionals from V to R.

Let V be the vector space of all polynomial functions from R into R with degree less than or equal to 2. Find a basis for V by using the following procedure:

  i) Find three linear functionals on V
 ii) Use these functionals as a basis for V*, the dual of V.
iii) Use the functionals to find a basis for V.

Solution: We use the fact that there is a relationship between a vector space and its dual in terms of the bases for each.

i) Three linear functionals on V are given as follows. Let t_1, t_2, t_3 be any three distinct numbers and let

$$L_i(p) = p(t_i)$$

where $p \, \varepsilon \, V$, i.e. polynomials of degree less than or equal to two.
    The $L_i(p)$ are linear functionals since

$$L_i(ap + q) = (ap + q)(t_i) = ap(t_i) + q(t_i) = aL_i(p) + L_i(q).$$

The $L_i(p)$ are linearly independent. To show this, set

$$L = c_1 L_1 + c_2 L_2 + c_3 L_3.$$

If $L = 0$, i.e. if $L(p) = 0$ for each $p \, \varepsilon \, V$, then applying L to the particular vectors, $1, x, x^2$ yields

$$L = c_1 L_1(1) + c_2 L_2(1) + c_3 L_3(1)$$

i.e. $c_1 + c_2 + c_3 = 0.$ (1)

$$L = c_1 L_1(x) + c_2 L_2(x) + c_3 L_3(x)$$

$$= c_1 t_1 + c_2 t_2 + c_3 t_3$$

i.e., $c_1 t_1 + c_2 t_2 + c_3 t_3 = 0.$ (2)

$$L = c_1 L_1(x^2) + c_2 L_2(x^2) + c_3 L_3(x^2)$$

$$= c_1 t_1^2 + c_2 t_2^2 + c_3 t_3^2$$

i.e., $c_1 t_1^2 + c_2 t_2^2 + c_3 t_3^2 = 0.$ (3)

The only values that satisfy (1) through (3) are
$c_1 = c_2 = c_3 = 0$. Thus the L_i are independent and since V
has dimension 3, these functionals form a basis for V^*.
(Recall that the dimension of the dual space is the same as
the dimension of the primal space). This follows from the
theorem that if V has dimension n and W has dimension m,
then the set of all linear transformations from V to W,
$Hom(V,W)$, has dimension mn. Here $\dim W = \dim K = 1$ and
thus, $\dim V^* = 1$.

Next, we must find the basis for V of which V^* is the
dual. Such a basis $\{p_1,p_2,p_3\}$ for V must satisfy

$$L_i(p_j) = \delta_{ij} \qquad\qquad (4)$$

or $$p_j(t_i) = \delta_{ij}.$$

Equation (4) arises from the following facts. Let $V(F)$ be
a finite-dimensional vector space over F and let
$\{\alpha_1,\dots,\alpha_n\}$ be a basis for V. Then there is a unique dual
basis $\left[f_1,f_2,\dots f_n\right]$ for V^* such that $f_i(\alpha_i) = \delta_{ij}$.

From (4) (since $\delta_{ij} = 1$, $i = j$ and 0 i $\neq$ j):

$$L_1(p_1) = 1;\ L_2(p_1) = 0;\ L_3(p_1) = 0$$

i.e. $$p_1(t_1) = 1;\ p_1(t_2) = 0;\ p_1(t_3) = 0.$$

A polynomial in x which satisfies these requirements is:

$$p_1(x) = \frac{(x - t_2)(x - t_3)}{(t_1 - t_2)(t_1 - t_3)} \qquad\qquad (5)$$

Similarly, $L_1(p_2) = 0;\ L_2(p_2) = 1;\ L_3(p_2) = 0$

i.e. $$p_2(t_1) = 0;\ p_2(t_2) = 1;\ p_2(t_3) = 0.$$

$$p_2(x) = \frac{(x - t_1)(x - t_3)}{(t_2 - t_1)(t_2 - t_3)} \qquad\qquad (6)$$

Finally, $$p_3(x) = \frac{(x - t_1)(x - t_2)}{(t_3 - t_1)(t_3 - t_2)}$$

The polynomials (5), (6) and (7) form a basis for V.

ANNIHILATORS, TRANSPOSES & ADJOINTS

Find the subspace of R^4 which the linear functionals below annihilate.

$$f_1(x_1, x_2, x_3, x_4) = x_1 + 2x_2 + 2x_3 + x_4$$

$$f_2(x_1, x_2, x_3, x_4) = 2x_2 + x_4$$

$$f_3(x_1, x_2, x_3, x_4) = -2x_1 - 4x_3 + 3x_4.$$

Solution: Let V be a vector space over F and of dimension n, and let W be a subspace of V. The annihilator of W is the set W^0 of linear functionals f on V such that $f(\alpha) = 0$ for every α in W. Note that the zero functional is automatically in W^0 since $0\alpha = 0$ for all $\alpha \, \varepsilon \, W$. Also, if $f \, \varepsilon \, W^0$, and $\alpha, \beta \, \varepsilon \, W$ and $c \, \varepsilon \, F$:

$$f(c\alpha + \beta) = cf(\alpha) + f(\beta) = 0 + 0 = 0, \text{ i.e.}$$

W^0 is closed under addition and scalar multiplication. Thus the set W^0 is a subspace of V^*, the dual space of V. To find the subspace W, consider the system of linear equations

$$a_{11}x_1 + \dots + a_{1n}x_n = 0$$

$$\vdots \qquad\qquad \vdots$$

$$a_{m1}x_1 + \dots + a_{mn}x_n = 0$$

for which we wish to find the solutions. That is, we wish to find the maximal subspace of R^n spanned by the solutions $(x_1, \dots, x_n)$. Let f_i $(i = 1, \dots, m)$ be the linear functionals on F^n defined by

$$f_i(x_1, \dots, x_n) = a_{i1}x_1 + \dots + a_{in}x_n.$$

We seek the subspace of F^n of all α such that $f_i(\alpha) = 0$ $i = 1, \dots, m$. That is, we are seeking the subspace annihilated by $f_1, \dots, f_m$. Row-reduction of the coefficient matrix provides us with a systematic method of finding this subspace. The n-tuple $(a_{i1}, \dots, a_{in})$ gives the coordinates of the linear functionals relative to the basis which is the dual to the standard basis for F^n. The row space of the coefficient matrix may thus be regarded as the space of linear functionals spanned by $f_1, \dots, f_m$. The solution space is the subspace annihilated by this space of functionals.

Turning to the given problem, the subspace which the linear functionals annihilate may be obtained by finding the row echelon form of the matrix

$$A = \begin{bmatrix} 1 & 2 & 2 & 1 \\ 0 & 2 & 0 & 1 \\ -2 & 0 & -4 & 3 \end{bmatrix}.$$

Multiply the first row by +2 and add the result to the third row. Then multiply the second row by 1/2. Then multiply -2 times the second row and add to the first row. Finally, multiply the third row by 1/5 to obtain the row-echelon matrix.

$$R = \begin{bmatrix} 1 & 0 & 2 & 0 \\ 0 & 1 & 0 & 0 \\ 0 & 0 & 0 & 1 \end{bmatrix}.$$

Therefore, the linear functionals

$$g_1(x_1, x_2, x_3, x_4) = x_1 + 2x_3$$

$$g_2(x_1, x_2, x_3, x_4) = x_2$$

$$g_3(x_1, x_2, x_3, x_4) = x_4$$

span the same subspace of $(R^4)^*$ and annihilate the same subspace of R^4 as do f_1, f_2, f_3. The subspace annihilated, since $g_1 = g_2 = g_3 = 0$, consists of the vectors with

$$x_1 = -2x_3; \quad x_2 = x_4 = 0.$$

● **PROBLEM 22-13**

Let W be the subspace of R^4 spanned by $v_1 = (1, 2, -3, 4)$ and $v_2 = (0, 1, 4, -1)$. Find a basis of the annihilator of W.

Solution: We first show that if $\phi \in V^*$ annihilates a subset S of V, then ϕ annihilates the linear span L(S) of S.

Suppose $v \in L(S)$. Then there exist $w_1, \dots, w_r \in S$ for which $v = a_1 w_1 + a_2 w_2 + \dots + a_r w_r$. Since ϕ is a linear functional,

$$\phi(v) = a_1 \phi(w_1) + a_2 \phi(w_2) + \dots + a_r \phi(w_r)$$

$$= a_1 0 + a_2 0 + \dots + a_r 0 = 0.$$

Since v was an arbitrary element of L(S), ϕ annihilates L(S), as claimed.

Now, to find a basis of the annihilator of W it is sufficient to find a basis of the set of linear functionals $\phi(x,y,z,w) = ax + by + cz + dw$ for which $\phi(v_1) = 0$ and $\phi(v_2) = 0$. This is so because $\{v_1, v_2\}$ is a subset of V and hence the above lemma is applicable. Thus,

$$\phi(1,2,-3,4) = a + 2b - 3c + 4d = 0$$

$$\phi(0,1,4,-1) = \qquad b + 4c - d = 0.$$

The system of equations in the unknowns a,b,c,d is in echelon form with free variables c and d. Set c = 1, d = 0 to obtain the solution a = 11, b = -4, c = 1, d = 0 and hence the linear functional $\phi_1(x,y,z,w) = 11x - 4y + z$.

Set c = 0, d = -1 to obtain the solution a = 6, b = -1, c = 0, d = -1 and hence the linear functional $\phi_2(x,y,z,w) = 6x - y - w$. The set of linear functionals $\{\phi_1, \phi_2\}$ is a basis of W°, the annihilator of W.

● **PROBLEM** 22-14

Let W be the subspace of R^5 which is spanned by the vectors

$\alpha_1 = (2,-2,3,4,-1)$, $\alpha_2 = (-1,1,2,5,2)$

$\alpha_3 = (0,0,-1,-2,3)$, $\alpha_4 = (1,-1,2,3,0)$.

Describe W°, the annihilator of W.

Solution: The annihilator of W is the set of all linear functionals f: $R^5 \to R$ such that f(α) = 0 for $\alpha \in$ W. The annihilator is a subspace of V*, the dual space. To describe it one need only find its dimension and a basis for it. Let f(α) = $c_1 x_1 + c_2 x_2 + c_3 x_3 + c_4 x_4 + c_5 x_5$. Then the system

$$f(\alpha_1) = 0$$

$$f(\alpha_2) = 0$$

$$f(\alpha_3) = 0$$

$$f(\alpha_4) = 0$$

has as its solution set the space spanned by the f.

We find the row space of the matrix formed by the α_i.

Let

$$A = \begin{bmatrix} 2 & -2 & 3 & 4 & -1 \\ -1 & 1 & 2 & 5 & 2 \\ 0 & 0 & -1 & -2 & 3 \\ 1 & -1 & 2 & 3 & 0 \end{bmatrix}.$$

Multiply the first row by 1/2. Then add it to the second row; multiply by -1 and add to the fourth row. Continuing the elementary row operations the row echelon matrix is obtained,

$$R = \begin{bmatrix} 1 & -1 & 0 & -1 & 0 \\ 0 & 0 & 1 & 2 & 0 \\ 0 & 0 & 0 & 0 & 1 \\ 0 & 0 & 0 & 0 & 0 \end{bmatrix}.$$

If f is a linear functional on R^5,

$$f(x_1,\ldots,x_5) = \sum_{j=1}^{5} c_j x_j$$

Therefore f is in W° if and only if $f(\alpha_i) = 0$ $i = 1,2,3,4$, i.e., if and only if

$$\sum_{j=1}^{5} a_{ij} c_j = 0 \qquad 1 \leq i \leq 4. \tag{1}$$

Note that $x_j = a_{1j}(\alpha_1) + a_{2j}(\alpha_2) + a_{3j}(\alpha_3) + a_{4j}(\alpha_4)$.

(1) is equivalent to

$$\sum_{j=1}^{5} R_{ij} c_j = 0 \qquad 1 \leq i \leq 3$$

where the R_{ij} are the entries in the row echelon matrix. This may be written as:

$$c_1 - c_2 - c_4 = 0$$

$$c_3 + 2c_4 = 0$$

$$c_5 = 0.$$

We obtain all such linear functionals f by assigning arbitrary values to c_2 and c_4, for example, $c_2 = a$ and $c_4 = b$. We then find the corresponding $c_1 = a + b$, $c_3 = -2b$,

$c_5 = 0$. So W^o consists of all linear functionals f of the form

$$f(x_1, x_2, x_3, x_4, x_5) = (a + b)x_1 + ax_2 - 2bx_3 + bx_4.$$

The dimension of W^o is 2 and a basis $\{f_1, f_2\}$ for W^o can be found by first substituting $a = 1$, $b = 0$ and then $a = 0$, $b = 1$:

$$f_1(x_1, \ldots, x_5) = x_1 + x_2$$

$$f_2(x_1, \ldots, x_5) = x_1 - 2x_3 + x_4.$$

The above general f in W^o is $f = af_1 + bf_2$.

● **PROBLEM** 22-15

Let ϕ be the linear functional on R^2 defined by $\phi(x,y) = x - 2y$. For each of the following linear operators T on R^2, find $(T^t(\phi))(x,y)$ if: i) $T(x,y) = (x,0)$; ii) $T(x,y) = (y, x+y)$; iii) $T(x,y) = (2x - 3y, 5x + 2y)$.

Solution: We first define the transpose of a linear mapping. Let $T: V \to U$ be a linear transformation where V and U are vector spaces over a field F. Let $\phi \in U^*$, the dual space of U. (That is, ϕ is a linear functional from U into F.)
The composition $\phi \circ T$ is a linear mapping from V into K so $\phi \circ T \in V^*$. Thus correspondence

$$\phi: \quad \phi \circ T$$

is a mapping from U^* to V^*. This mapping is denoted by T^t and is called the transpose of T. That is,

$$T^t: \quad U^* \to V^*$$

is defined by

$$T(\phi) = \phi \circ T$$

and $\quad (T^t(\phi)(v)) = \phi(T(v)) \quad$ for $v \in V$. $\qquad$ (1)

i) $(T^t(\phi)(x,y) = \phi(T(x,y)) = \phi(x,0) = x.$

ii) First note that the transpose mapping T^t as defined by (1) is linear. To see this, observe that for any scalars $a, b \in K$ and any linear functionals $\phi, \sigma \in U^*$,

$$T(a\phi + b\sigma) = (a\phi + b\sigma) \circ T$$

$$= a(\phi \circ T) + b(\sigma \circ T)$$

$$= aT^t(\phi) + bT^t(\sigma).$$

Now, $(T^t(\phi))(x,y) = \phi(T(x,y))$

$$= \phi(y,x+y) = y - 2(x+y) = -2x - y.$$

iii) $(T^t(\phi))(x,y) = \phi(T(x,y))$

$$= \phi(2x - 3y, 5x + 2y)$$

$$= (2x - 3y) - 2(5x + 2y) = -8x - 7y.$$

• **PROBLEM 22-16**

Why is the transpose mapping T^t so named?

Solution: Let T: V → W be a linear transformation. We
define T^t, the transpose of T, by the following procedure.
V* and W* are the dual spaces of V and W, i.e. the sets of
all linear functionals f: V → K and g: W → K respectively,
where K is the underlying field. Let $\phi \in$ W*. Then ϕ o T(v)
is a linear mapping from V into K, i.e. (ϕ o T) $\in$ V*. Thus
the mapping ϕ —> ϕ o T is a mapping from W* into V*. It is
denoted by T^t and is called the transpose of T. To justify
this name recall that the transpose of an m × n matrix

$$A = \begin{bmatrix} a_{11} & a_{12} & \cdots & a_{1n} \\ a_{21} & a_{22} & \cdots & a_{2n} \\ \cdot & & & \\ \cdot & & & \\ \cdot & & & \\ a_{m1} & a_{m2} & \cdots & a_{mn} \end{bmatrix}$$

is

$$A^t = \begin{bmatrix} a_{11} & a_{21} & \cdots & a_{m1} \\ a_{12} & & & \cdot \\ & & & \cdot \\ \cdot & & & \cdot \\ \cdot & & & \\ a_{1n} & a_{2n} & \cdots & a_{mn} \end{bmatrix}$$

The following theorem shows the relationship between the
transpose of a linear transformation and the transpose of
a matrix:
    Let T: V → W be linear, and let A be the matrix re-
presentation of T relative to bases $\{v_i\}$ of V and $\{w_i\}$ of
W. Then the transpose matrix A^t is the matrix representation of T^t: W* → V* relative to the bases dual to $\{u_i\}$ and
$\{w_i\}$.

Let V be an n-dimensional vector space over the field F and
let T be a linear operator on V. Suppose $B = \{\alpha_1, \ldots, \alpha_n\}$
and $B' = \{\alpha'_1, \ldots, \alpha'_n\}$ are two ordered bases for V. Show how
the transpose transformation T^t can be used to derive the
formula for changing bases, that is, to find T in the ordered basis B' given the matrix of T in B.

Solution: First define the transpose of a linear transformation. Let T: $V \to W$ be a linear transformation from an
n-dimensional vector space V to an m-dimensional vector
space W. Let g be a linear functional from W to F, the
underlying field. Thus $g \in W^*$. If $\alpha \in V$, define

$$f(\alpha) = g(T\alpha) \tag{1}$$

Examining (1), observe that $T\alpha \in W$ and thus $g(T\alpha) \in F$.
Thus, associated with T is a linear transformation,
T^t, such that for every linear functional $g \in W^*$ there exists a linear functional $f \in V^*$, i.e. T^t: $W^* \to V^*$ T^t is
called the transpose of T.
   Since T is an operator, $V = W$ and $W^* = V^*$. The matrix
of T in the ordered basis B is defined as the $n \times n$ matrix
A such that

$$T\alpha_j = \sum_{j=1}^{n} a_{ij}\alpha_i.$$

The dual basis for V^* corresponding to B also gives us the
coordinates of a vector $\alpha \in V$ under T. That is,

$$a_{ij} = f_i(T\alpha_j),$$

where $\{f_1, \ldots, f_n\}$, is the dual basis of B.
   We wish to change the basis. Suppose

$$B' = \{\alpha'_1, \ldots, \alpha'_n\}$$

is another ordered basis for V, with dual basis $\{f'_1, \ldots, f'_n\}$.
If B is the matrix of T in the ordered basis B', then

$$b_{ij} = f'_i(T\alpha'_j).$$

Let U be the invertible linear operator such that $U\alpha_j = \alpha'_j$.
Then the transpose of U is given by $U^t f'_i = f_i$. Since U is
invertible, so is U^t and $(U^t)^{-1} = (U^{-1})^t$. To see this,
note that $\det A^t = \det A$. Now,

$$(U^t)^{-1} = \frac{1}{\det A^t}(\text{Adj. } A^t). \tag{3}$$

On the other hand

$$(U^{-1})^t = \left(\frac{1}{\det A} \text{ Adj. } A\right)^t = \frac{1}{\det A}(\text{Adj. } A)^t. \tag{4}$$

But Adj. $A^t = (\text{Adj. } A)^t$. Thus (3) = (4) and $(U^t)^{-1} = (U^{-1})^t$.

Thus $f_t' = (U^{-1})^t f_i$ $i = 1, \ldots, n$.   Therefore,

$$b_{ij} = \left[(U^{-1})^t f_i\right](T\alpha_j')$$

$$= f_i(U^{-1}T\alpha_j')$$

$$= f_i(U^{-1}TU\alpha_j). \tag{2}$$

Now (2) gives the i, j entry of the matrix of $U^{-1}TU$ in the ordered basis B. This scalar is also the i, j entry of the matrix of T in the ordered basis B'. In other words

$$[T]_{B'} = [U^{-1}TU]_B$$

$$= [U^{-1}]_B[T]_B[U]_B$$

$$= [U]_B^{-1}[T]_B[U]_{B'}$$

which is the change-of-basis formula $P^{-1}AP$.

● **PROBLEM** 22-18

Give an example to illustrate the following theorem. Let V be a finite-dimensional inner product space, and let T be a linear operator on V. In any orthonormal basis for V the matrix of T* is the conjugate transpose of the matrix of T.

Solution: This theorem is dependent on the following theorem. Let V be a finite-dimensional inner product space and let $\beta = \{\alpha_1, \ldots, \alpha_n\}$ be an (ordered) orthonormal basis for V. Let T be a linear operator on V and let A be the matrix of T in the ordered basis β. Then $a_{ij} = (T\alpha_j/\alpha_k)$. If we assume that this is true, we can prove the given theorem.

Let $B = \{\alpha_1, \ldots, \alpha_n\}$ be an orthonormal basis for V, let $A = [T]_B$ and $B = [T^*]_B$. Then according to the previous theorem

$$a_{kj} = (T\alpha_j/\alpha_k)$$

$$b_{kj} = (T^*\alpha_j/\alpha_k).$$

By the definition of T* we then have

$$b_{kj} = (T^*\alpha_j/\alpha_k).$$

By the definition of T* we then have

$$b_{kj} = (T^*\alpha_j/\alpha_k)$$

$$= \overline{(\alpha_k/T^*\alpha_j)}$$

$$= \overline{a_{jk}}.$$

To illustrate the theorem let V be a finite-dimensional inner product space and E the orthogonal projection of V on a subspace W. Then for any vectors α and β in V,

$$(E\alpha/\beta) = (E\alpha/E\beta + (1 - E)\beta)$$

$$= (E\alpha/E\beta)$$

$$= E\alpha + (1 - E)\alpha/E\beta)$$

$$= (\alpha/E\beta).$$

From the uniqueness of the operator E* it follows that E* = E. Now consider the projection

$$E = \frac{-14}{154}(3,12,-1). \quad \text{Then}$$

$$A = \frac{1}{154} \begin{bmatrix} 9 & 36 & -3 \\ 36 & 144 & -12 \\ -3 & -12 & 1 \end{bmatrix}$$

is the matrix of E in the standard orthonormal basis. Since $E = E^*$, A is also the matrix of E*, and because A = A* this shows the truth of the theorem.

● **PROBLEM** 22-19

Let T: $R^2 \to R^2$ be a linear operator of the form $T(x,y) = (a_{11}x + a_{12}y, a_{21}x + a_{22}y)$. What is the adjoint of T? Assume the standard inner product.

Solution: The adjoint of a linear operator T in an inner product space V is a linear operator T* such that $\langle T\alpha, \beta \rangle = \langle \alpha, T^*\beta \rangle$ for all $\alpha, \beta \in V$.
   Let $\alpha = (\alpha_1, \alpha_2)$; $\beta = (\beta_1, \beta_2)$ and
$T^*(x,y) = (b_{11}x + b_{12}y, b_{21}x + b_{22}y)$. Now
$T\alpha = (a_{11}\alpha_1 + a_{12}\alpha_2, a_{21}\alpha_1 + a_{22}\alpha_2)$ and

$$\langle T\alpha, \beta \rangle = (a_{11}\alpha_1 + a_{12}\alpha_2)\beta_1 + (a_{21}\alpha_1 + a_{22}\alpha_2)\beta_2. \quad (1)$$

On the other hand,

$$T^*\beta = (b_{11}\beta_1 + b_{12}\beta_2, \ b_{21}\beta_1 + b_{22}\beta_2)$$

and $\langle \alpha, T^*\beta \rangle = \alpha_1(b_{11}\beta_1 + b_{12}\beta_2) + \alpha_2(b_{21}\beta_1 + b_{22}\beta_2). \quad (2)$

The condition that $\langle T\alpha, \beta \rangle = \langle \alpha, T^*\beta \rangle$ means that (1) = (2). That is,

$$a_{11}\alpha_1\beta_1 + a_{12}\alpha_2\beta_1 + a_{21}\alpha_1\beta_2 + a_{22}\alpha_2\beta_2 \quad (3)$$

$$= b_{11}\alpha_1\beta_1 + b_{12}\alpha_1\beta_2 + b_{21}\alpha_2\beta_1 + b_{22}\alpha_2\beta_2. \quad (4)$$

Since α, β and $T(x,y)$ are known we need to find $b_{11}, b_{12}, b_{21}, b_{22}$. Comparing (3) and (4) it is seen that

$$b_{11} = a_{11} \ ; \ b_{12} = a_{21} \ ; \ b_{21} = a_{12} ; ; \ b_{22} = a_{22}.$$

Thus $T^*(x,y) = (a_{11}x + a_{21}y, \ a_{12}x + a_{22}y).$

● **PROBLEM** 22-20

Let V be the space of polynomials over the field of complex numbers, with the inner product

$$(f/g) = \int_0^1 f(t)\overline{g(t)} \, dt,$$

(where the bar indicates complex conjugation). Consider the operator 'multiplication by f', that is, the linear operator M_f defined by $M_f(g) = fg$. What is the adjoint of this operator?

Solution: Let V be an inner product space and T a linear operator on V. The adjoint of T is a linear operator T* such that $\langle T\alpha, \beta \rangle = \langle \alpha, T^*\beta \rangle$ for all $\alpha, \beta \ \epsilon \ V$.

Since f is a polynomial $f = \sum a_k x^k$ and $\overline{f} = \sum \overline{a}_k x^k$. Here, if $a_k = \alpha_k + i\beta_k$, $\overline{a}_k = \alpha_k - i\beta_k$. Thus, $\overline{f}$ is the polynomial whose associated polynomial function is the complex conjugate of the polynomial function for f:

$$\overline{f}(t) = \overline{f(t)}, \quad t \text{ real}.$$

Consider the operator 'multiplication by f', that is, the linear operator M_f defined by $M_f(g) = fg$. Then this operator has an adjoint namely multiplication by $\overline{f}$. For

$$(M_f(g)/h) = (fg/h)$$

$$= \int_0^1 f(t)g(t)\overline{h(t)}\,dt$$

$$= \int_0^1 g(t)\overline{[\overline{f(t)}h(t)]}\,dt$$

$$= (g/\overline{f}h)$$

$$= (g/M_{\overline{f}}(h))$$

and so $(M_{\overline{f}})^* = M_f$.

Note that this vector space is infinite dimensional. It sometimes happens that a linear operator on an infinite dimensional space does not have an adjoint although this cannot happen in a finite dimensional space.

● **PROBLEM** 22-21

Let V be $K^{n \times n}$ with the inner product $(A/B) = \mathrm{tr}(B*A)$. Let M be a fixed n x n matrix over K (a field). What is the adjoint of left multiplication by M?

Solution: The adjoint of left multiplication by M is left multiplication by M*. But left multiplication by M* is the linear operator L_M defined by $L_M(A) = MA$.

$$(L_M(A)/B) = \mathrm{tr}(B*(MA))$$

$$= \mathrm{tr}(MAB*)$$

$$= \mathrm{tr}(AB*M)$$

$$= \mathrm{tr}(A(M*B)*)$$

$$= (A/L_M*(B)).$$

Thus $(L_M)^* = L_{M*}$.

Note that in the computation above we used the characteristic property of the trace function : $\mathrm{tr}(AB) = \mathrm{tr}(BA)$.

● **PROBLEM** 22-22

Let V be the space of polynomials over the field of complex numbers, with the inner product

$$(f/g) = \int_0^1 f(t)\overline{g(t)}\,dt .$$

Let D be the differentiation operator on C[x]. Show that D has no adjoint.

Solution: Integration by parts shows that
$\overline{(Df/g)}$ = f(1)g(1) - f(0)g(0) - (f/Dg). Let g be a fixed
function. Assume that D has an adjoint, i.e., that there is
a polynomial D*g such that (Df/g) = (f/D*g) for all f.   If
such a D*g exists, we shall have
(f/D*g) = f(1)g(1) - f(0)g(0) - (f/Dg) or
(f/(D*g + Dg)) = f(1)g(1) - f(0)g(0). With g fixed,
L(f) = f(1)g(1) - f(0)g(0) is a linear functional and cannot
be of the form L(f) = (f/h) unless L = 0. If D*g exists,
then with h = D*g + Dg we do have L(f) = (f/h), and so
g(0) = g(1) = 0. The existence of a suitable polynomial D*g
implies g(0) = g(1) = 0. Conversely, if g(0) = g(1) = 0,
the polynomial D*g = - Dg satisfies ((Df/g) = (f/D*G) for
all f. If any g is chosen for which g(0) ≠ 0 or g(1) ≠ 0,
D*g cannot suitably be defined, and so one can conclude
that D has no adjoint.

MULTILINEAR FUNCTIONALS

● **PROBLEM** 22-23

Find all 2-linear functions on 2 x 2 matrices over K where
K is a commutative ring with identity.

Solution: A commutative ring with identity satisfies all
the axioms for a field except the requirement that every
element have a multiplicative inverse. We use this structure when there is no need for division (of functions or
elements).
        Next, define n-linear functions from the domain of n x n
matrices to the range K. Let D be a function which assigns
to every n x n matrix A over K a scalar D(A) in K. We say
that D is n-linear if for each i, $1 \leq i \leq n$. D is a linear
function of the ith row when the other (n-1) rows are held
fixed.
        To understand this definition, let α_1, α_2, . . . , α_n
be the rows of A.   Then D(A) = D(α_1, α_2, ... , α_n).   The
statement that D is n-linear means that

$$D(\alpha_1, \alpha_2, \ldots, C\alpha_i + \alpha'_i, \ldots, \alpha_n)$$
$$= C(D(\alpha_1, \alpha_2, \ldots \alpha_i, \ldots, \alpha_n) + D(\alpha_1, \alpha_2, \ldots,$$
$$\alpha'_i, \ldots, \alpha_n). \tag{1}$$

        Turning to the given problem, let D be a 2-linear func-
tion. Denoting the rows of the 2 x 2 identity matrix by
ϵ_1, ϵ_2 we have D(A) = D($A_{11}\epsilon_1 + A_{12}\epsilon_2$, $A_{21}\epsilon_1 + A_{22}\epsilon_2$).   (2)

Applying (1) to (2) yields:
$$D(A) = A_{11}D(\epsilon_1, A_{21}\epsilon_1 + A_{22}\epsilon_2) + A_{12}D(\epsilon_2, A_{21}\epsilon_1 + A_{22}\epsilon_2)$$
$$= A_{11}A_{21} \, D(\epsilon_1, \epsilon_1) + A_{11}A_{22} \, D(\epsilon_1, \epsilon_2)$$

847

$$+ A_{12}A_{21} \, D(\varepsilon_2, \varepsilon_1) + A_{12}A_{22} \, D(\varepsilon_2, \varepsilon_2).$$

Thus D is completely determined by the four scalers

$$D(\varepsilon_1, \varepsilon_1) \quad D(\varepsilon_1, \varepsilon_2), \quad D(\varepsilon_2, \varepsilon_1) \quad \text{and} \quad D(\varepsilon_2, \varepsilon_2).$$

Let a,b,c,d, be any four scalars in K. Define

$$D(A) = A_{11}A_{21} \, a + A_{11}A_{22} \, b + A_{12}A_{21} \, c + A_{12}A_{22} \, d \; .$$

Then D is a 2-linear function on 2 x 2 matrices over K and

$$D(\varepsilon_1, \varepsilon_1) = a \qquad\qquad D(\varepsilon_1, \varepsilon_2) = b$$

$$D(\varepsilon_2, \varepsilon_1) = c \qquad\qquad D(\varepsilon_2, \varepsilon_2) = d \; .$$

● **PROBLEM 22-24**

Show that the determinant of a 2 x 2 matrix is a 2-linear function and belongs to a subspace of all functions from $V (= K^{n \times n})$ into K.

Solution: First solve the second part of the problem. Show that a linear combination of n-linear functions is n-linear. A function D from the set of n x n matrices defined over K to the set K is n-linear if for each i, $1 \le i \le n$, D is a linear function of the ith row when the other (n-1) rows are held fixed.

Let D and E be n-linear functions. If a and b belong to K, the linear combination aD + bE is defined by $(aD + bE)(A) = aD(A) = bE(A)$. Hence if we fix all rows except row i we obtain

$$(aD + bE)(c\alpha_i + \alpha'_i)$$

$$= aD(c\alpha_i + \alpha'_i) + bE(c\alpha_i + \alpha'_i)$$

$$= acD(\alpha_i) + aD(\alpha'_i) + bcE(\alpha_i) + bE(\alpha'_i)$$

$$= c(aD + bE)\,\alpha_i + (aD + bE)(\alpha'_i) \; .$$

Since a linear combination of two n-linear functions is n-linear it follows that a linear combination of n-linear functions is n-linear. Also, note that the zero function from $K^{n \times n}$ to K is n-linear. Thus the set of n-linear functions on V is a subspace of the space of all functions from V into K.

Next, let D be the function defined on 2 x 2 matrices over K by

$$D(A) = A_{11}A_{22} - A_{12}A_{21}$$

where
$$A = \begin{bmatrix} A_{11} & A_{12} \\ A_{21} & A_{22} \end{bmatrix} .$$

Now D is the sum of two 2-linear functions:
$$D = D_1 + D_2 . \tag{2}$$

To see this, let $k_1, \ldots . k_n$ be positive integers $1 \le k_i \le n$ and let a be an element of K. FOr each n x n matrix A over K define
$$D(A) = aA(1,k_1) \ldots . A(n,k_n) . \tag{1}$$

The function defined by (1) is n-linear since if we regard D as a function of the ith row of A, the others being fixed, we may write
$$D(\alpha_i) = A(i,k_i)b$$

where b is some fixed element of K. Let
$$\alpha'_i = (A'_{i1}, \ldots . , A'_{in}). \quad \text{Then}$$
$$D(c\alpha_i + \alpha'_i) = [cA(i,k_i) + A'(i,k_i)]b$$
$$= cD(\alpha_i) + D(\alpha'_i) .$$

Thus D is a linear function of each of the rows of A.

From (2):
$$D = D_1 + D_2$$

where $D_1(A) = A_{11}A_{22}$

and $D_2(A) = - A_{12}A_{21} .$

Thus D itself, being a linear combination of 2-linear functions, is a 2-linear function. But D is just the definition of a 2 x 2 determinant as was required.

● **PROBLEM** 22-25

Let F be a field and let D be any alternating 3-linear function on 3 x 3 matrices over the polynomial ring F[x]. Let
$$A = \begin{bmatrix} x & 0 & -x^2 \\ 0 & 1 & 0 \\ 1 & 0 & x^3 \end{bmatrix}$$
Show that $D(A) = (x^4 + x^2)D(\epsilon_1, \epsilon_2, \epsilon_3) .$

Solution: Let $K^{n \times n}$ be the set of $n \times n$ matrices defined over the field K and let D be an n-linear function from $K^{n \times n}$ into K. D is alternating if:

a) $D(A) = 0$ whenever two rows of A are equal (where A is an $n \times n$ matrix);

b) $D(A') = -D(A)$ where A' is a matrix obtained from A by interchanging two rows of A.

Let α_1, α_2, α_3 denote the rows of A. Then a 3-linear function is a function D such that

$$D(c\alpha_1 + \alpha_1' , \alpha_2, \alpha_3) = cD(\alpha_1, \alpha_2, \alpha_3)$$
$$+ D(\alpha_1', \alpha_2, \alpha_3) .$$

To find $D(A)$, let the rows of the 3 x 3 identity matrix

$$\begin{bmatrix} 1 & 0 & 0 \\ 0 & 1 & 0 \\ 0 & 0 & 1 \end{bmatrix}$$

be given by $\varepsilon_1 = (1,0,0)$ $\varepsilon_2 = (0,1,0)$ $\varepsilon_3 = (0,0,1)$.
Then

$$D(A) = D(X\varepsilon_1 - X^2\varepsilon_3, \varepsilon_2, \varepsilon_1 + X^3 \varepsilon_3) .$$

Since D is linear as a function of each row

$$D(A) = D(X\varepsilon_1, \varepsilon_2, \varepsilon_1 + X^3 \varepsilon_3) - X^2 D(\varepsilon_3, \varepsilon_2, \varepsilon_1 + X^3 \varepsilon_3)$$
$$= XD(\varepsilon_1, \varepsilon_2, \varepsilon_3) + X^4 D(\varepsilon_1, \varepsilon_2, \varepsilon_3)$$
$$- X^2 D(\varepsilon_3, \varepsilon_2, \varepsilon_1) - X^5 (\varepsilon_3, \varepsilon_2, \varepsilon_3) . \tag{1}$$

Since D is alternating, $D(\varepsilon_3, \varepsilon_2, \varepsilon_3) = D(\varepsilon_1, \varepsilon_2, \varepsilon_1) = 0$

and hence (1) becomes $D(A) = (X^2 + X^4)D(\varepsilon_1, \varepsilon_2, \varepsilon_3)$.

(Note that since $D(\varepsilon_3, \varepsilon_2, \varepsilon_1)$ means that rows 1 and 3 were interchanged,

$$- X^2 D(\varepsilon_3, \varepsilon_2, \varepsilon_1) = X^2 D(\varepsilon_1, \varepsilon_2, \varepsilon_3) .$$

● **PROBLEM** 22-26

Obtain an algebraic expression for the general r-linear form on the vector space V^n.

Solution: Let V be a vector space over a field F. If r is a positive integer, a function L from $V^r = V \times \ldots \times V$

into F is called multilinear if $L(\alpha_1, \alpha_2, \ldots, \alpha_r)$ is linear as a function of each α_i when the other α_j's are held fixed. That is, $L(\alpha_1, \alpha_2 \ldots \alpha_r)$ is linear if for each i $L(\alpha_1, \ldots, c\alpha_i + \beta_i, \ldots, \alpha_r) =$
$cL(\alpha_1, \ldots, \alpha_i, \ldots, \alpha_r) +$
$+ L(\alpha_1, \ldots, \beta_i, \ldots, \alpha_r)$. A multilinear function on V^r is also called an r-linear form on V.

If $\alpha_1, \ldots, \alpha_r$ are vectors in V and A is the r x n matrix with rows $\alpha_1, \ldots, \alpha_r$ then for any function L in $M^r(K^n)$,

$$L(\alpha_1, \ldots, \alpha_r) = L(\sum_{j=1}^{n} a_{1j}\epsilon_j, \alpha_2, \ldots, \alpha_r). \qquad (1)$$

To see this, let

$$A = \begin{bmatrix} a_{11} & \cdots & & & a_{1n} \\ a_{21} & \cdots & & & a_{2n} \\ \vdots & & & & \vdots \\ a_{r1} & \cdots & & & a_{rn} \end{bmatrix} .$$

Then $\alpha_1 = [a_{11} \cdots a_{1n}]$.

Let $\epsilon_1 = [1, 0, \ldots, 0]$, $\epsilon_2 = [0, 1, 0, \ldots 0] \ldots$

$\epsilon_n = [0, \ldots, 1]$.

Then $\alpha_1 = a_{11}\epsilon_1 + a_{12}\epsilon_2 + \ldots + a_{1n}\epsilon_n$

$$= \sum_{j=1}^{n} a_{1j}\epsilon_j .$$

Since L is a multilinear function (1) becomes

$L(\sum a_{1j}\epsilon_j, \alpha_2, \ldots, \alpha_r)$

$$= \sum a_{1j}L(\epsilon_j, \alpha_2, \ldots, \alpha_r) \qquad (2)$$

i.e., $a_{11}L(\epsilon_1, \alpha_2, \ldots, \alpha_r) + a_{12}L(\epsilon_2, \alpha_2, \ldots, \alpha_r)$

$+ \ldots + a_{1n}L(\epsilon_n, \alpha_2, \ldots, \alpha_r)$.

Let $\alpha_2 = \sum_{k=1}^{n} a_{2k}\epsilon_k$. Then (2) becomes

$$\sum_{j=1}^{n} a_{1j}L(\epsilon_j, \alpha_2, \ldots, \alpha_r) =$$

$$\sum_{j=1}^{n} a_{1j} L(\varepsilon_j, \sum_{k=1}^{n} A_{2k} \varepsilon_k, \ldots, \alpha_r)$$

$$= \sum_{j=1}^{n} \sum_{k=1}^{n} a_{1j} a_{2k} L(\varepsilon_j, \varepsilon_k, \ldots \alpha_3, \ldots, \alpha_r).$$

$$= \sum_{j,k=1}^{n} a_{1j} a_{2k} L(\varepsilon_j, \varepsilon_k, \alpha_3, \ldots, \alpha_r). \tag{3}$$

If we continue replacing $\alpha_3, \ldots, \alpha_r$ in turn by their expressions as linear combinations of the basis vectors, (3) becomes $L(\alpha_1, \ldots, \alpha_r)$

$$= \sum_{j_1, \ldots, j_r=1}^{n} a(1, j_1) \ldots a(r, j_r) L(\varepsilon_{j1}, \ldots, \varepsilon_{jr}) \tag{3}$$

where $a(ij_k) = a_{ij_k}$.

In (3) there is one term for each r-tuple $J = (j_1, \ldots, j_r)$ of positive integers between 1 and n. Since there are n choices for j_1, n for j_2, $\ldots$, n for j_r, the total number of r-tuples is n^r. Thus L is completely determined by (3) and the particular values

$$C_J = L(\varepsilon_{j_1}, \ldots, \varepsilon_{j_r})$$

assigned to the n^r elements $(\varepsilon_{j_1}, \ldots, \varepsilon_{j_r})$.

From this we see that if for each r-tuple J we choose an element C_J of K then

$$L(\alpha_1, \ldots, \alpha_r) = \sum_J a(1, j_1) \ldots a(r, j_r) C_J$$

defines an r-linear form on K^n.

● **PROBLEM 22-27**

Describe a general method for associating an alternating form with a multilinear form.

Solution: First, it is necessary to define multilinear and alternating forms. A multilinear form is a function from
$$\underbrace{V \times V \times \ldots \times V}_{r - times} \text{ into F that is linear in one argument}$$

while the remaining (r-1) arguments are held fixed. That is, L is a multilinear form if

$$L(\alpha_1, \alpha_2, \ldots, c\alpha_i + \alpha_i', \ldots, \alpha_r)$$

$$= cL(\alpha_1, \alpha_2, \ldots, \alpha_i, \ldots, \alpha_r)$$
$$+ L(\alpha_1, \alpha_2, \ldots \alpha_i', \ldots, \alpha_r) .$$

Next, L is called alternating if $L(\alpha_1, \ldots, \alpha_r) = 0$ whenever $\alpha_i = \alpha_j$, with $i \neq j$. This is equivalent to asserting that if L is an alternating multilinear function on V^r, then

$$L(\alpha_1, \ldots, \alpha_i, \ldots, \alpha_j, \ldots, \alpha_r) =$$
$$- L(\alpha_1, \ldots, \alpha_j, \ldots, \alpha_i, \ldots, \alpha_r) .$$

Let L be an r-linear form on V and let σ be a permutation of $\{1, \ldots, r\}$; there are n! such permutations. If L is alternating, the number of transpositions determines the sign of L, i.e.,

$$L_\sigma(\alpha_1, \ldots, \alpha_r) = L(\alpha_{\sigma 1}, \ldots \alpha_{\sigma r})$$
$$L_\sigma = (\text{sgn}\sigma)L .$$

Recall that the set of all multilinear forms forms a subspace $M^r(V)$. For each $L \in M^r(V)$ define $\Pi rL \in M^r(V)$ by

$$\Pi_\sigma L = \sum_\sigma (\text{sgn}\sigma)L_\sigma . \tag{1}$$

Rewriting (1) yields

$$(\Pi_r L)(\alpha_1, \ldots, \alpha_r) = \sum_\sigma (\text{sgn } \sigma)L(\alpha_{\sigma 1}, \ldots, \alpha_{\sigma r})$$

Now ΠrL is a linear transformation from $M^r(V)$ into $\Lambda^r(V)$, the set of all alternating r-forms on V. To see this, let τ be any permutation of $\{1, \ldots, r\}$. Then

$$(\Pi rL)(\alpha_\tau, \ldots, \alpha_{\tau r}) = \sum_\sigma (\text{sgn } \sigma)L(\alpha_{\tau\sigma 1}, \ldots \alpha_{\tau\sigma r})$$
$$= (\text{sgn } \tau) \sum_\sigma (\text{sgn } \sigma)L(\alpha_{\tau\sigma 1}, \ldots, \alpha_{\tau\sigma r}) .$$

Since $\tau\sigma(r)$ is a composition of permutations, as σ runs over all permutations of $\{1, \ldots, r\}$ so does $\tau\sigma$. Hence,

$$(\Pi rL)(\alpha_{r1}, \ldots, \alpha_{rr}) = (\text{sgn } \tau)(\Pi_r L)(\alpha_1, \ldots, \alpha_r)$$

and hence ΠrL is an alternating form. This shows that we can always associate an alternating form with a multilinear form.

CHAPTER 23

GEOMETRICAL APPLICATIONS

EQUATIONS FOR LINES & PLANES

● **PROBLEM 23-1**

Find the unique line passing through (2,4) and (3,-2) in the form

$$L = u + r(v-u).$$

Solution: There is a unique line passing through two points P and Q in the Euclidean plane. From the figure, or by successively setting $r = 0$ and then $r = 1$ it can be seen that the line

$$\overline{OP} + r(\overline{OQ} - \overline{OP})$$

contains both P and Q. Hence in R^2, the vectors u,v are contained in the set L,

$$L = u + r(v-u)$$

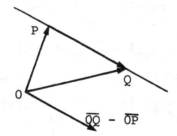

Turning to the given problem note that

$$L = [2,4] + r\{[3,-2] - [2,4]\} = [2,4] + r[1,-6].$$

● **PROBLEM 23-2**

Find the equation of the plane through the points $P_1(1,2,-1)$, $P_2(2,3,1)$, and $P_3(3,-1,2)$.

Solution: The general equation of the plane in R^3 is

$$ax + by + cz + d = 0. \qquad (1)$$

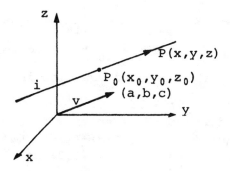

Since the three given points lie in the plane, their coordinates must satisfy (1),

$$a + 2b - c + d = 0$$
$$2a + 3b + c + d = 0 \qquad (2)$$
$$3a - b + 2c + d = 0$$

This system can be solved by the row echelon method. The coefficient matrix of system (2) is

$$A = \begin{bmatrix} 1 & 2 & -1 & 1 \\ 2 & 3 & 1 & 1 \\ 3 & -1 & 2 & 1 \end{bmatrix}$$

Add −2 times the first row to the second row and −3 times the first row to the third row to obtain:

$$\begin{bmatrix} 1 & 2 & -1 & 1 \\ 0 & -1 & 3 & -1 \\ 0 & -7 & 5 & -2 \end{bmatrix}$$

Add −7 times the second row to the third row.

$$\begin{bmatrix} 1 & 2 & -1 & 1 \\ 0 & -1 & 3 & -1 \\ 0 & 0 & -16 & 5 \end{bmatrix}$$

This is the coefficient matrix of the system

$$a + 2b - c + d = 0$$
$$- b + 3c - d = 0 \qquad (3)$$
$$-16c + 5d = 0$$

From (3), $-16c + 5d = 0$. Hence,

$$c = \frac{+5}{16} d.$$ Setting $d = t$ (a free variable), $c = \frac{5}{16} t.$

Back substituting yields

$$-b + \frac{15}{16}t - t = 0 \quad \text{or} \quad b = -\frac{t}{16} \; . \quad \text{Finally, we find}$$

$a - \frac{2t}{16} - \frac{5}{16}t + t = 0$ and $a = -\frac{9}{16}t$. A convenient value

of t is -16. Then, the required equation is:

$$9x + y - 5z - 16 = 0. \tag{4}$$

An alternative method of solving this problem is as
follows: Since $P_1(1,2,-1)$, $P_2(2,3,1)$, and $P_3(3,-1,2)$ lie
in the plane, the vectors $\overrightarrow{P_1P_2} = (1,1,2)$ and $\overrightarrow{P_1P_3} = (2,-3,3)$
are parallel to the plane. By the rules of the cross-
product,

$$\overrightarrow{P_1P_2} \times \overrightarrow{P_1P_3} = (1,1,2) \times (2,-3,3) = (9,1,-5)$$

is normal to the plane. Hence, by the point normal form,
the plane passing through $P_1(1,2,-1)$ and perpendicular to
the normal $(9,1,-5)$ is given by

$$9(x-1) + (y-2) - s(z+1) = 0$$

or, $\qquad 9x + y - 5z - 16 = 0.$ $\qquad\qquad$ (5)

● **PROBLEM** 23-3

Find the equation of the plane passing through the point
$P_o(x_o,y_o,z_o)$ and having the nonzero vector n = (a,b,c) as
a normal.

Solution: The normal to a plane of points is the vector
that is perpendicular to the plane. The required plane,
therefore, consists of all vectors that are orthogonal to
the normal vector n = (a,b,c). Letting $\overrightarrow{P_oP}$ represent all

vectors in space passing through $P_o(x_o,y_o,z_o)$ the following
equation can be derived,

$$\overrightarrow{P_oP} = (x-x_o, \; y-y_o, \; z-z_o).$$

The subset of these vectors that contains vectors orthogonal
to n is

$$\overrightarrow{P_oP} \cdot n = 0, \; (x-x_o, \; y-y_o, \; z-z_o) \cdot (a,b,c) = 0$$

or, $\qquad a(x-x_o) + b(y-y_o) + c(z-z_o) = 0$ $\qquad\qquad$ (1)

by the rules of the dot-product.

Equation (1) is the point normal form of the equation of a plane.

For example, let it be required to find the equation of the plane passing through the point (3,-1,7) and

856

perpendicular to the vector n = (4,2,-5).

Substituting the appropriate values in (1),

$$4(x-3) + 2(y+1) - 5(z-7) - 0,$$

is the desired equation.

● **PROBLEM 23-4**

List the different ways in which lines can be defined in R^3.

Solution: There are four ways in which a line in three-dimensional space can be depicted.   These are: (1) The geometrical approach; (2) Parametric equations; (3) Affine equations; (4) Solutions to systems of linear equations.

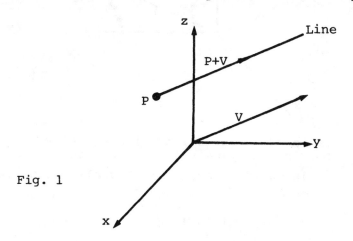

Fig. 1

Let P be a point in R^3.   Then "the line through P" is the set of all points of the form P + V where V is in a one-dimensional subspace W of R^3.   W is the direction space.

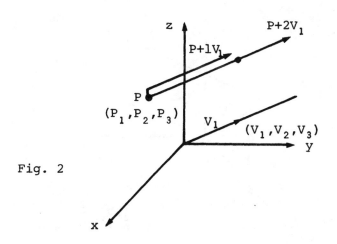

Fig. 2

857

Now the line through P is all points $P + tV_1$, V_1 fixed ($\neq 0$) $t \in R$ (t is called the parameter). In terms of coordinates, all points (x,y,z) with

$$x = P_1 + tv_1$$

$$y = P_2 + tv_2$$

$$z = P_3 + tv_3$$

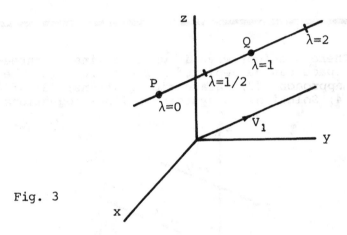

Fig. 3

The affine equations are all points $(1-\lambda)P + \lambda Q$ where $\lambda \in R$. The basis V_1 for the subspace W in the geometrical approach is $V_1 = Q - P$. In coordinates, a line is all points (x,y, z) with

$$x = (1-\lambda)P_1 + \lambda g_1$$

$$y = (1-\lambda)P_2 + \lambda g_2$$

$$z = (1-\lambda)P_3 + \lambda g_3$$

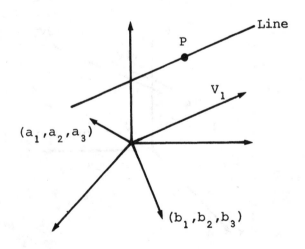

Fig. 4

Let X = [x,y,z]. Then the line through P is defined to be all solutions to

$$a_1x + a_2y + a_3z = c_1$$

$$b_1x + b_2y + b_3z = c_2$$

Here the basis V_1 for the one-dimensional subspace in the geometrical approach consists of a nonzero solution to the homogeneous system AX = 0. The point P is any particular solution, so AP = Y. Given a line in geometrical form, P + V, V in W, take a basis

$$V_1 = (v_1, v_2, v_3)$$

for W. Then the rows (a_1, a_2, a_3) and (b_1, b_2, b_3) of A are any independent solutions to

$$av_1 + bv_2 + cv_3 = 0$$

and Y = AP.

• **PROBLEM 23-5**

What are the different approaches to planes in R^3?

Solution: Planes in R^3 may be defined in the following four ways: (1) Geometrically; (2) Parametrically; (3) As affine equations; (4) As solutions to a system of linear equations.

   (1) Geometrically, a plane consists of all points of the form P + V, where P is a point in R^3 and V is in a two dimensional subspace W.

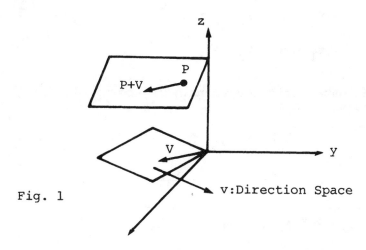

Fig. 1

   (2) Let t_1, t_2 ε R. Then a plane may be defined as all points (x,y,z) with

859

$$x = P_1 + t_1 v_1 + t_2 v_1' \; ;$$

$$y = P_2 + t_1 v_2 + t_2 v_2' \; ;$$

$$z = P_3 + t_1 v_3 + t_2 v_3' \; .$$

Thus a plane is the set of all points $P + t_1 V_1 + t_2 V_2$, where V_1, V_2 are linearly independent. Here $\{V_1, V_2\}$ is a basis for the two dimensional subspace in the geometrical approach.

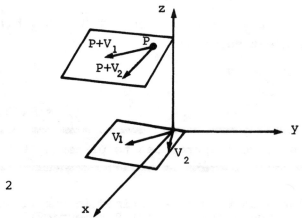

Fig. 2

(3) A plane may also be defined in terms of affine equations as all points $\lambda_1 P + \lambda_2 Q + \lambda_3 R$ where $\lambda_1 + \lambda_2 + \lambda_3 = 1$. Substituting into the affine equation

$$V_1 = Q - P, \quad V_2 = R - P$$

with

$$t_1 = \lambda_2, \quad t_2 = \lambda_3$$

yields the parametric equations. In coordinate form, the affine equations are

$$x = \lambda_1 P_1 + \lambda_2 q_1 + \lambda_3 r_1$$

$$y = \lambda_1 P_2 + \lambda_2 q_2 + \lambda_3 r_2$$

$$z = \lambda_1 P_3 + \lambda_2 q_3 + \lambda_3 r_3$$

where

$$\lambda_1 + \lambda_2 + \lambda_3 = 1.$$

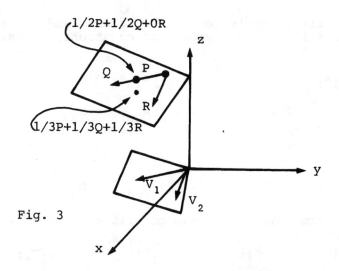

Fig. 3

(4)   A plane consists of all solutions to $AX = Y$, $X = (x,y,z)$, A of rank 1.   Thus, it is the set of all solutions to $ax + by + cz = d$. $A = (a,b,c)$ is a basis for the solution space of

$$av_1 + bv_2 + cv_3 = 0$$

$$av_1' + bv'_2 + cv_3' = 0$$

where $V_1 = (v_1,v_2,v_3)$ and $V_2 = (v_1', v_2', v_3')$ are, as above, a basis for the direction space and $d = AP$.

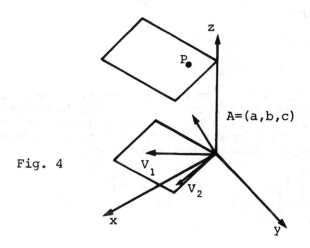

Fig. 4

● PROBLEM 23-6

Describe geometrically the solutions to

$$x_1 - 2x_2 + 4x_3 = 1,$$

$$3x_1 + x_2 - x_3 = 0.$$

<u>Solution:</u> Linear varieties (lines, planes and k-flats) arise from solutions to systems of equations. Consider a system of equations

$$a_{11}x_1 + \ldots \ldots \ldots + a_{1n}x_n = b_1$$

$$a_{21}x_1 + \ldots \ldots \ldots + a_{2n}x_n = b_2$$

$$\vdots \qquad\qquad\qquad \vdots \quad \vdots$$

$$a_{m1}x_1 + \ldots \ldots \ldots + a_{mn}x_n = b_m .$$

In matrix form this system can be written AX = B, where

$$A = \begin{bmatrix} a_{11} & \cdots & a_{1n} \\ a_{21} & \cdots & a_{2n} \\ \vdots & & \vdots \\ a_{m1} & \cdots & a_{mn} \end{bmatrix}, \quad X = \begin{bmatrix} x_1 \\ x_2 \\ \vdots \\ x_n \end{bmatrix}, \quad B = \begin{bmatrix} b_1 \\ b_2 \\ \vdots \\ b_m \end{bmatrix} .$$

This system has either i) no solutions, or ii) solution(s). In case ii) the set of all solutions constitutes a k-flat in R^n, where k = n-r, r the rank of A. The direction space is the kernel of T(X) = AX.

We must solve

$$x_1 - 2x_2 + 4x_3 = 1,$$

$$3x_1 + x_2 - x_3 = 0. \tag{1}$$

The general solution is the sum of any particular solution to (1) and the set of general solutions of the homogeneous system corresponding to (1). To find a particular solution, use elementary row operations to eliminate the coefficient of x_1 from the second equation. Thus,

$$\begin{bmatrix} 1 & -2 & 4 & \vdots & 1 \\ 3 & 1 & -1 & \vdots & 0 \end{bmatrix} \leftrightarrow \begin{bmatrix} 1 & -2 & 4 & \vdots & 1 \\ 0 & 7 & -13 & \vdots & -3 \end{bmatrix}$$

$$\leftrightarrow \begin{bmatrix} 1 & -2 & 4 & \vdots & 1 \\ 0 & 1 & -13/7 & \vdots & -3/7 \end{bmatrix} .$$

An equivalent equation system is therefore

$$x_1 = 2x_2 + 4x_3 = 1,$$

$$x_2 - 13/7x_3 = -3/7. \tag{2}$$

862

Let $x_3 = 0$; $x_2 = -3/7$ and $x_1 = 1/7$. Thus a particular solution to the nonhomogeneous system is $(1/7, -3/7, 0)$.

The general homogeneous solution follows if we set the right hand side of (2) equal to zero. Thus,

$$x_1 - 2x_2 + 4x_3 = 0$$

$$x_2 - 13/7x_3 = 0.$$

Let $x_3 = 1$; $x_2 = 13/7$ and $x_1 = -2/7$.

The solutions to (1) are of the form

$$(1/7, -3/7, 0) + c(-2/7, 13/7, 1). \qquad (3)$$

But (3) is in the form of a line $L = P + sp\{V_1\}$ where $V_1 \; \varepsilon$ W, a one-dimensional space. That is, the solutions lie on a line through $(1/7, -3/7, 0)$ with direction space spanned by $(-2/7, 13/7, 1)$.

● **PROBLEM 23-7**

Define and give examples of k-flats.

<u>Solution</u>: Let R^n be the set of all n-tuples. A k-flat F (also known as a linear variety or affine subspace) in R^n is the set of all n-tuples of the form $P + V$, where P is some point of R_n and V is from a k-dimensional subspace W of R^n. W is called the direction space of F and is written F = P + W.

A point is a 0-flat. For, let $\{p_1, p_2, \ldots, p_n\}$ be a point in R^n and W the subspace of R^n consisting of the zero vector alone. Then $P + V = P$. A line is a 1-flat. Here P is a point in R^n and V is from a one-dimensional subspace of R^n, for instance $V = (v_1, v_2, \ldots, v_n)$.

Similarly a plane is a 2-flat since it has $P = (p_1, p_2, \ldots, p_n)$ and V from a two dimensional subspace of R^n. That is to say, V can be expressed as the linear combination $c_1(x_1, x_2, \ldots x_n) + c_2(y_1, y_2, \ldots y_n)$ where $(x_1, x_2, \ldots x_n)$ and $(y_1, y_2, \ldots y_n)$ are linearly independent.

● **PROBLEM 23-8**

Let F be all points $P = P_o + sV + tW$ where $V = (s, -1, 0)$, $W = (2, 0, -1)$ and $P_o = (1, 1, 2)$. Find the equation describing F.

Solution: Here we have three points in R^3. In general, let $P_o, P_1, \ldots P_m$ be m column vectors in R^n. The minimal flat containing these m points can be described as the solution set for a system of linear equations $AX = Y$.

The direction space of this flat is spanned by $V_1 = P_1 - P_o, V_2 = P_2 - P_o, \ldots, V_m = P_m - P_o$. Furthermore, these direction space vectors satisfy the associated homogeneous system $AX = 0$.

The general method for finding a flat is as follows.

Let $V_1 = (v_{11}, \ldots, v_{1n}), \ldots, V_m = (v_{m1}, \ldots, v_{mn})$ and consider the system of homogeneous equations

$$z_1 v_{11} + \ldots + z_n v_{1n} = 0$$

$$\vdots \qquad \qquad \vdots \qquad (1)$$

$$z_1 v_{m1} + \ldots + z_n v_{mn} = 0.$$

The solutions to (1) form a subspace of R^n. Let $A_1 = (a_{11}, \ldots a_{1n}), \ldots, A_r = (a_{r1} \ldots a_{rn})$ be a basis for this subspace. Let

$$A = \begin{bmatrix} A_1 \\ \vdots \\ A_r \end{bmatrix} \qquad (\text{rank } A = r).$$

Then

$$AV_i = \begin{bmatrix} A_1 V_i \\ A_2 V_i \\ \vdots \\ A_r V_i \end{bmatrix} = \begin{bmatrix} 0 \\ 0 \\ \vdots \\ 0 \end{bmatrix} = 0.$$

Consider the solutions to $AX = Y$, where $Y = AP_o$. This system has solutions (e.g. $X = P_o$) and hence the set of all solutions forms a k-flat with $k = n - r$. This flat is of the form $P_o + K$, where K is the kernel of $T(X) = AX$. Now, since $AV_i = 0$ the V_i are in K and span K. Hence each $P_i = P_o + (P_i - P_o) = P_o + V_i$ is in F.

Turning to the given problem, to find the smallest k-flat containing P_o, V and W, let $V_1 = V$, $V_2 = W$. Then,

864

$$P_1 = V_1 + P_0 = (2,-1,0) + (1,1,2) = (3,0,2)$$
$$P_2 = V_2 + P_0 = (2,0,-1) + (1,1,2) = (3,1,1).$$

Consider the system of linear equations

$$2z_1 - z_2 = 0$$
$$2z_1 - z_3 = 0 \qquad \qquad (2)$$

whose solution set is spanned by $(1,2,2) = A = A_1$. Let $Y = AP_0 = (1,1,2)\begin{pmatrix} 1 \\ 2 \\ 2 \end{pmatrix} = 7$. Then the desired flat consists of all solutions X to the equation $AX = Y$. Thus $(1,2,2)X = 7$.

$$x + 2y + 2z = 7$$

is the equation describing F.

● **PROBLEM** 23-9

Show that the points $P = (0,2,0)$, $Q = (1,2,1)$ and $R = (2,2,1)$ are noncollinear. Then find the unique plane passing through these points.

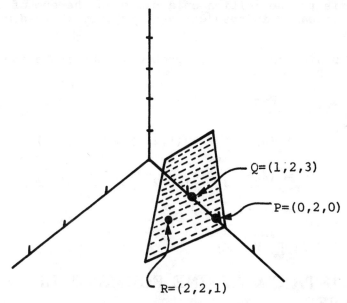

Solution: Three points, P, Q and R, are collinear if they lie on the same line in R^n. Another definition of "collinear" is the following: P, Q and R are collinear if and only if the vectors $Q - P$ and $R - P$ are not linearly independent.

Substituting into this definition the values given above for P, Q, and R yields:

$$Q-P = (1,2,1) - (0,2,0) = (1,0,1),$$

865

$$R-P = (2,2,1) - (0,2,0) = (2,0,1).$$

Since the equation

$$c_1(1,0,1) + c_2(2,0,1) = (0,0,0)$$

implies $c_1 = c_2 = 0$, Q-P and R-P are linearly independent. Hence, P, Q and R are noncollinear.

The unique plane passing through P, Q and R consists of all points whose coordinates are of the form

$$\lambda_1 P + \lambda_2 Q + \lambda_3 R$$

where $\lambda_1 + \lambda_2 + \lambda_3 = 1.$

To show this, let V = Q-P, W = R-P. Then, Q = P+V, and R = P+W.

Hence, $\lambda_1 P + \lambda_2 Q + \lambda_3 R = \lambda_1 P + \lambda_2(P+V) + \lambda_3(P+W)$

$$= (\lambda_1 + \lambda_2 + \lambda_3)P + \lambda_2 V + \lambda_3 W$$

$$= P + \lambda_2 V + \lambda_3 W. \tag{1}$$

But (1) is the definition of a plane as the sum of a point and two linearly independent vectors from a two-dimensional subspace.

Turning to the given problem, substitute the values for P, Q, R into

$$\lambda_1 P + \lambda_2 Q + \lambda_3 R$$

to obtain

$$\lambda_1(0,2,0) + \lambda_2(1,2,1) + \lambda_3(2,2,1)$$

$$= (\lambda_2 + 2\lambda_3, \ 2\lambda_1 + 2\lambda_2 + 2\lambda_3, \ \lambda_2 + \lambda_3)$$

$$= (\lambda_2 + 2\lambda_3, \ 2, \ \lambda_2 + \lambda_3).$$

This is the plane y = 2.

PARAMETRIC & AFFINE FORMS OF LINES & PLANES

● PROBLEM 23-10

Give an example to show that two non-parallel lines in R^3 need not intersect.

Solution: Consider the two lines given by the parametric equations

$$x_1 = -2+t \qquad\qquad y_1 = 7-3s$$

$$x_2 = 3-2t \qquad\qquad y_2 = 1+2s$$

$$x_3 = 1+5t \qquad\qquad y_3 = 4-s.$$

If the two lines are to intersect there must exist a point (a_1, a_2, a_3) common to both lines. That is,

$$a_1 = -2+t = 7-3s, \quad a_2 = 3-2t = 1+2s, \quad a_3 = 1+5t = 4-s,$$

or

$$-2+t = 7-3s$$

$$3-2t = 1+2s$$

$$1+5t = 4-s.$$

Since there are only two unknowns, t and s, select two equations and try to find t and s. Take, for example, the equations

$$-2+t = 7-3s \quad \text{and} \quad 3-2t = 1+2s.$$

Rearranging and combining like terms yields

$$t + 3s = 9 \quad \text{and} \quad 2t + 2s = 2.$$

Solving these two equations yields

$$s = 4, \quad \text{and} \quad t = -3.$$

But t = -3 gives the point

$$x_1 = -2+(-3) = -5$$

$$x_2 = 3-(-6) = 9$$

$$x_3 = 1+(-15) = -14$$

while s = 4 gives

$$y_1 = 7-3(4) = -5$$

$$y_2 = 1+2(4) = 9$$

$$y_3 = 4-4 = 0.$$

Thus the two lines cannot intersect. Note that the two lines are not parallel either since the direction vector d = (1,-2,5) of the first line is not parallel to the direction vector d'=(-3,2,-1) of the second.

● **PROBLEM 23-11**

Find the equation of the plane in R^3 which passes through (-1,2,1) and has the (orthogonal) direction vector d = (1,-3,2).

Solution: The usual parametric definition of a line requires the presence of a direction vector that is parallel
to the line through a point. Thus, let $a = (a_1,a_2,a_3)$ be
a point in R^3. The line in R^3 with parametric equations

$$x_1 = a_1 + d_1 t, \quad x_2 = a_2 + d_2 t, \quad x_3 = a_3 + d_3 t$$

can be described as the set of all points $x = (x_1,x_2,x_3)$ in
R^3 such that the vector

$$x-a = (x_1-a_1, \; x_2-a_2, \; x_3-a_3)$$

is parallel to the direction vector $d = (d_1,d_2,d_3)$ of the
line. Consider now the set of all points (x_1,x_2,x_3) in R^3
such that the vector x-a is orthogonal to a nonzero vector
$d = (d_1,d_2,d_3)$. By definition, these two vectors are orthogonal if and only if

$$d_1(x_1-a_1) + d_2(x_2-a_2) + d_3(x_3-a_3) = 0. \tag{1}$$

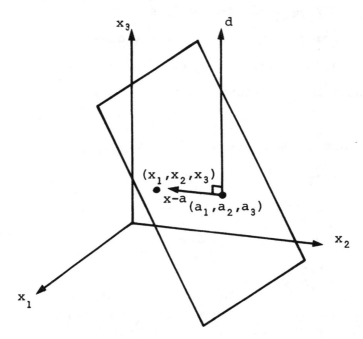

Regarding d and x-a as geometric vectors emanating
from the point (a_1,a_2,a_3), note from the figure that all
points (x_1,x_2,x_3) in R^3 satisfying Eq.(1) lie in a plane
through (a_1,a_2,a_3) which is perpendicular to d. Thus Eq.
(1) is the equation of a plane in R^3 passing through $(a_1,$
$a_2,a_3)$ and having the orthogonal direction vector d.

For the given problem, Eq.(1) becomes

$$1(x_1+1) + (-3)(x_2-2) + 2(x_3-1) = 0$$

868

or $\qquad$ $x_1 - 3x_2 + 2x_3 = -5.$

This is the required equation of the plane.

● **PROBLEM** 23-12

Find a parametric representation of the line passing through P and in the direction of u where (a) P = (2,5) and u = (-3,4); (b) P = (4,-2,3,1) and u = (2,5,-7,11).

Solution: Let $P = (a_1, a_2, \ldots, a_n)$ be a point in R^n. The line ℓ in R^n passing through the point $P = (a_1, a_2, \ldots, a_n)$ and in the direction of $u = (u_1, u_2, \ldots, u_n)$ consists of the points $X = P + tu$, $t\varepsilon R$, that is, consists of the points

$$X = (x_1, x_2, \ldots, x_n) \text{ obtained from}$$

$$x_1 = a_1 + u_1 t$$

$$x_2 = a_2 + u_2 t \qquad\qquad (1)$$

$$\vdots$$

$$x_n = a_n + u_n t$$

where t takes on all real values. The variable t is called a parameter and (1) is called a parametric representation of l.

(a) Substituting the given values into (1):

$$x = 2 - 3t$$

$$y = 5 + 4t.$$

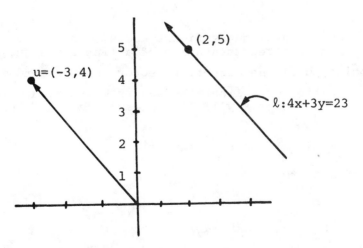

Eliminate t: From the first equation 3t = 2-x, or t = $\frac{2-x}{3}$. Substituting this into the second equation:

869

$$y = 5 + 4\left(\frac{2-x}{3}\right) = \frac{23}{3} - \frac{4}{3}x.$$

Thus, the line can also be written as $4x + 3y = 23$.

(b) Substituting into (1):

$$x_1 = 4 + 2t$$

$$x_2 = -2 + 5t$$

$$x_3 = 3 - 7t$$

$$x_4 = 1 + 11t.$$

Since this is a line in 4-space it cannot be graphed by humans.

● **PROBLEM** 23-13

Describe analytically the line segment between two points (a_1,a_2,a_3) and (b_1,b_2,b_3) in R^3.

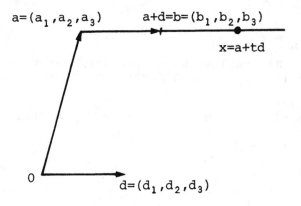

Solution: Let $b = a+d$ in the figure above. This figure shows the parametric representation of a line since (a_1,a_2, a_3) is a point, d is the direction vector and $x = a+td$ is the vector equation of a line. We see from the figure that if $0 \le t \le 1$, the point x will lie between a and b. That is,

$$x = a + t(b-a) \quad \text{where } 0 \le t \le 1$$

or, in coordinate form

$$x_1 = a_1 + t(b_1-a_1)$$
$$x_2 = a_2 + t(b_2-a_2)$$
$$x_3 = a_3 + t(b_3-a_3).$$

When $t = 0$, $x = a$ and when $t = 1$,

$$x = a + b - a = b.$$

The point $a + \frac{1}{2} (b-a)$

$$= (a_1, a_2, a_3) + \frac{1}{2} (b_1-a_1,\ b_2-a_2,\ b_3-a_3)$$

$$= \frac{a_1+b_1}{2},\ \frac{a_2+b_2}{2},\ \frac{a_3+b_3}{2}$$

is halfway from (a_1, a_2, a_3) to (b_1, b_2, b_3) along the line segment; that is, the point is the mid-point of the line segment.

As examples of line segments in R^3 consider the following:

1) The midpoint of the line segment joining $(-1,3,2)$ and $(3,1,-1)$ is $(1,2,1/2)$.

2) The point one-third of the way from $(-1,3,2)$ to $(3,1,-1)$ is obtained by putting $t = 1/3$. The desired point is $(1/3, 7/3, 1)$.

● **PROBLEM 23-14**

Write the parametric equations for the line in R^3 passing through $(-1,0,2)$ with direction vector $d = (2,-3,1)$.

Solution: We first find the general parametric representation of a line in R^3. Let (a_1, a_2, a_3) be a point of R^3 and let $d = (d_1, d_2, d_3)$ be a non-zero vector in R^3. We wish to describe the points on the line through (a_1, a_2, a_3) having direction given by d. From the figure it may be seen that a point x is on the line if, and only if, x=a+td for tεR.

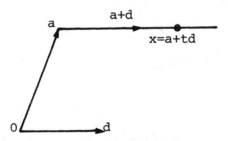

The line in R^3 through $a = (a_1, a_2, a_3)$ with non-zero direction vector $d = (d_1, d_2, d_3)$ is the set of all points $x = (x_1, x_2, x_3)$ such that

$$x = a + td \qquad t \ \varepsilon \ R.$$

The coordinate equations in parametric form are:

871

$$x_1 = a_1 + td_1$$

$$x_2 = a_2 + td_2$$

$$x_3 = a_3 + td_3 \ .$$

Thus, for the given point and direction vector, the parametric equations are:

$$x_1 = -1 + 2t$$

$$x_2 = \quad -3t$$

$$x_3 = \quad 2 + t.$$

For example, if $t = -1$, the point $(-3,3,1)$ is on the line.

To summarize, one way of obtaining a line is to specify two points. Another way is to specify a point and a direction vector. Note that in both cases, two pieces of information are required.

● **PROBLEM** 23-15

Find all the points of intersection in R^3 of the plane with the equation $3x + 5y - z = -2$ and the line with parametric equations $x = -3 + 2t$, $y = 4 + t$, $z = -1 - 3t$.

<u>Solution</u>: If the point (x_0, y_0, z_0) lies on the line we have $x_0 = -3+2t$, $y_0 = 4+t$ and $z_0 = -1-3t$. If x_0, y_0, z_0 also lie in the plane we must have $3x_0 + 5y_0 - z_0 = -2$.

Substituting the values for x_0, y_0, z_0 given above, we obtain:

$$3(-3+2t) + 5(4+t) - (-1-3t) = -2$$

Simplifying yields:

$$-9 + 6t + 20 + 5t + 1 + 3t = -2.$$

or $\qquad 14t = -14;$

hence, $\qquad t = -1.$

Thus, the only point of intersection is the point

$$(x_0, y_0, z_0) = (-3+(2)(-1),\ 4+(-1),\ -1-3(-1))$$

$$= (-5,3,2).$$

HYPERPLANES

● **PROBLEM** 23-16

Find an equation of the hyperplane H in the vector space

R^4 if: (i) H passes through P = [3,-2,1,-4] and is normal to u = [2,5,-6,-2];
ii) H passes through P = [1,-2,3,5] and is parallel to the hyperplane H' determined by 4x-5y+2z+w = 11.

Solution: The set H of elements in R^n which are solutions of a linear equation in n unknowns $x_1,\ldots,x_n$ of the form

$$c_1 x_1 + c_2 x_2 + \ldots + c_n x_n = b \tag{1}$$

with u = $(c_1,c_2,\ldots,c_n) \neq 0$ in R^n is called a hyperplane of R^n, and (1) is called an equation of H. The coefficient vector u is said to be normal to the hyperplane.

(i) It is given that u = [2,5,-6,-2]. Hence the equation of the hyperplane that is normal to u is H

$$= 2x + 5y - 6z - 2w = b. \tag{2}$$

To find b, reason as follows: Since H passes through P we can substitute P = [3,-2,1,-4] into (2) to obtain b. Thus,

$$2(3) - 5(2) - 6(1) - 2(-4) = -2,$$

and $2x + 5y - 6z - 2w = -2$

is an equation of H.

(ii) H and H' are parallel if, and only if, corresponding normal vectors are in the same or opposite directions. Hence an equation of H is of the form 4x - 5y + 2z + w = k. Substitution of P into this equation, yields k = 25. Thus an equation of H is 4x - 5y + 2z + w = 25.

● PROBLEM 23-17

Let $c_1 x_1 + c_2 x_2 + \ldots + c_n x_n = b$ (1)
be the equation of a hyperplane. Let set H be the elements in R^n which are solutions of a linear equation in n unknowns $x_1,x_2,\ldots,x_n$ of the form of (1) with u = $(c_1,c_2,\ldots,c_n) \neq 0$ in R^n. Show that the directed line segment $\overrightarrow{PQ}$ of any pair of points P,Q ϵ H is orthogonal to the coefficient vector u.

Solution: Suppose P = $(a_1,\ldots,a_n)$ and Q = $(b_1,\ldots,b_n)$. Then the a_i and the b_i are solutions of the given equation:

$$c_1 a_1 + c_2 a_2 + \ldots + c_n a_n = b$$

$$c_1 b_1 + c_2 b_2 + \ldots + c_n b_n = b.$$

873

Let $v = \overrightarrow{PQ} = Q - P = (b_1 - a_1, \ldots b_n - a_n)$.

Then $u \cdot v = (c_1, c_2, \ldots, c_n) \cdot (b_1 - a_1, \ldots b_n - a_n)$

$$= c_1(b_1 - a_1) + c_2(b_2 - a_2) + \ldots + c_n(b_n - a_n)$$

$$= c_1 b_1 - c_1 a_1 + c_2 b_2 - c_2 a_2 + \ldots + c_n b_n - c_n a_n$$

$$= (c_1 b_1 + c_2 b_2 + \ldots + c_n b_n) - (c_1 a_1 + \ldots + c_n a_n)$$

$$= b - b = 0.$$

Hence v, that is, $\overrightarrow{PQ}$ is orthogonal to u.

● **PROBLEM** 23-18

Show how the set of solutions to $2x + y = 4$ can be expressed as a one-dimensional subspace of R^2.

Solution: First it is necessary to define points and vectors in R^n. A point P in R^n is an n-tuple $(a_1, a_2, \ldots, a_n)$. A vector PQ from the point $P = (a_1, a_2, \ldots, a_n)$ to the point $Q = (b_1, b_2, \ldots, b_n)$ (or with tail at P and head at Q) is defined to be the n-tuple

$$PQ = (b_1 - a_1, \ldots, b_n - a_n).$$

Thus PQ = Q-P. In particular, the vector OP from the origin, O, to P has the same coordinates as the point P.

The general equation of a line is the set of all points (x,y) which satisfy $ax + by = c$, with a,b not both equal to zero. If $c=0$, $y = -\frac{a}{b} x$ is a one dimensional vector of R^2 passing through the origin. But if $c \neq 0$, the set of solutions L to $ax + by = c$ is a line which has been translated from the origin. In fact, if any point $P = (x_o, y_o)$ on L, is chosen, note that we may translate back by the vector OP and obtain a one-dimensional subspace W. That is, subtract OP from vector OQ, with Q on L to obtain the set

$$\{OQ - OP \mid Q \text{ on } L\} = \{PQ = Q-P \mid Q \text{ on } L\}.$$

This set is a one-dimensional subspace.

Turning to the given problem, let L be the set of solutions to $2x + y = 4$, and let $P = (1,2)$ be the chosen point. For any $Q = (x,y)$ on L consider $PQ = (x-1, y-2)$. Since the equation for L can also be written as

$$2(x-1) + y - 2 = 0,$$

it can be seen that

$$W = \{PQ \mid P \text{ fixed on } L, Q \text{ arbitrary on } L\}$$

is a one-dimensional subspace of R^2. W consists of the set of solutions of a non-trivial homogeneous linear equation; W is the kernel of a linear transformation of rank 1.

● **PROBLEM** 23-19

Find the equation of the hyperplane in R^4 through the point $X_o = \begin{bmatrix} 2 \\ 3 \\ -1 \\ 4 \end{bmatrix}$ and orthogonal to $A = \begin{bmatrix} 1 \\ 2 \\ 7 \\ 5 \end{bmatrix}$.

<u>Solution</u>: The equation of a hyperplane in R^n through a point $\beta = \begin{bmatrix} \beta_1 \\ \beta_2 \\ \cdot \\ \cdot \\ \cdot \\ \beta_n \end{bmatrix}$ and orthogonal to $\alpha = \begin{bmatrix} \alpha_1 \\ \alpha_2 \\ \cdot \\ \cdot \\ \cdot \\ \alpha_n \end{bmatrix}$ is $\alpha \cdot (X-B) = 0$

where $X = \begin{bmatrix} x_1 \\ x_2 \\ \cdot \\ \cdot \\ \cdot \\ x_n \end{bmatrix}$, an arbitrary point.

The vector α is called a normal to the hyperplane. If β is the zero vector, the hyperplane passes through the origin.

Let $X = \begin{bmatrix} x_1 \\ x_2 \\ x_3 \\ x_4 \end{bmatrix}$ be any point on the hyperplane. The equation of the hyperplane is then:

$$\begin{bmatrix} 1 \\ 2 \\ 7 \\ 5 \end{bmatrix} \cdot \left\{ \begin{bmatrix} x_1 \\ x_2 \\ x_3 \\ x_4 \end{bmatrix} - \begin{bmatrix} 2 \\ 3 \\ -1 \\ 4 \end{bmatrix} \right\} = 0,$$

$$\begin{pmatrix} 1 \\ 2 \\ 7 \\ 5 \end{pmatrix} \cdot \begin{pmatrix} x_1 - 2 \\ x_2 - 3 \\ x_3 + 1 \\ x_4 - 4 \end{pmatrix} = 0,$$

$$(x_1 - 2) + 2(x_2 - 3) + 7(x_3 + 1) + 5(x_4 - 4) = 0,$$

or, $\quad x_1 + 2x_2 + 7x_3 + 5x_5 = 21.$

● **PROBLEM** 23-20

What is the determinantal expression for a hyperplane in i) R^2 ii) R^3 iii) R^n?

Solution: Hyperplanes are $(n-1)$ flats in R^n. Recall that a k-flat F in R^n is the set of all n-tuples of the form $P+V$ where P is some point of R^n and V is from a k-dimensional subspace W of R^n. In R^3 a hyperplane is just a plane, while in R^2 a hyperplane is a line.

(i) Given two distinct points

$$P = (p_1, p_2) \quad Q = (q_1, q_2) \text{ in } R^2 ,$$

the equation for the line through P and Q is

$$\det \begin{bmatrix} p_1 & q_1 & x \\ p_2 & q_2 & y \\ 1 & 1 & 1 \end{bmatrix} = 0. \qquad (1)$$

Expanding (1) we obtain

$$p_1 q_2 + q_1 y + x p_2 - x q_2 - q_1 p_2 - p_1 y = 0,$$

or

$$y = \frac{p_2 - q_2}{p_1 - q_1} x + \frac{p_1 q_2 - q_1 p_2}{p_1 - q_1} .$$

This is the equation of a line in the form $y = mx + b$.

(ii) Given three noncollinear points

$P = (p_1, p_2, p_3), \quad Q = (q_1, q_2, q_3), \text{ and } R = (r_1, r_2, r_3)$

in R^3, the equation for the plane through these points is

$$\det \begin{bmatrix} p_1 & q_1 & r_1 & x \\ p_2 & q_2 & r_2 & y \\ p_3 & q_3 & r_3 & z \\ 1 & 1 & 1 & 1 \end{bmatrix} = 0. \qquad (2)$$

To show this, assume that

$$\det \begin{vmatrix} P & Q & R & X \\ 1 & 1 & 1 & 1 \end{vmatrix} = 0 \qquad (3)$$

is in fact the equation of a hyperplane in R^3. Substitution of P for X yields

$$\det \begin{vmatrix} P & Q & R & P \\ 1 & 1 & 1 & 1 \end{vmatrix} ,$$

which is 0 since two of its columns are the same. Similarly Q and R are on the plane; therefore, if (2) is the equation of a plane it is the desired plane. Expansion of (3) by cofactors using the last column yields one equation in three unknowns. Thus (3) is the equation of a plane if it is nontrivial and consistent, i.e. if at least one of the coefficients of x, y or z is nonzero. Note that

$$\text{rank} \begin{vmatrix} P & Q & R \\ 1 & 1 & 1 \end{vmatrix} = \text{rank} \begin{vmatrix} P & Q-P & R-P \\ 1 & 0 & 0 \end{vmatrix} ,$$

since the rank of a matrix is unchanged by elementary row operations. The rank of the matrix is 3 since P, Q and R are non-collinear. Thus we can add a vector (s,o) and still have linear independence (since $c_4=0$, $c_1 P + c_2 Q + c_3 R$ will still imply $c_1 = c_2 = c_3 = 0$ through elementary row operations on the coefficient matrix). Hence

$$\det \begin{vmatrix} P & Q & R & S \\ 1 & 1 & 1 & 0 \end{vmatrix} \neq 0.$$

Expanding the determinant by the last column shows that one of the coefficients of x,y or z in (2) is nonzero as required.

(iii) In general, given n+1 points $P_1, \dots, P_{n+1}$ in R^n which do not lie in an (n-1)-flat, the equation for the hyperplane through these points is

$$\det \begin{vmatrix} P_1 & P_2 & \cdots & P_{n+1} & X \\ 1 & 1 & & 1 & 1 \end{vmatrix} = 0$$

where $P_i = (p_1^{(i)}, \dots, p_n^{(i)})$ and $X = (x_1, \dots, x_n)$.

ORTHOGONAL RELATIONS

Classify the rigid motions in the plane.

Solution: Rigid motions in R^n are defined as functions that preserve distances. Thus if $f: R^n \to R^n$ is a rigid motion (not necessarily linear), then for P, $Q \varepsilon R^n$.

$$|f(P) - f(Q)| = |P - Q|.$$

The fundamental fact concerning rigid motions is the following: Any rigid motion of R^n is an orthogonal mapping followed by a translation. Thus, if F is a rigid motion of R^n, then for any P in R^n,

$$F(P) = P_0 + T(P)$$

where P_0 is a fixed point in R^n and $T(P)$ is an orthogonal linear transformation.

Now, relative to an orthonormal basis the matrix of any orthogonal mapping of R^2 has one of the two forms:

$$A = \begin{vmatrix} \cos\theta & -\sin\theta \\ \sin\theta & \cos\theta \end{vmatrix} \quad \text{or} \quad B = \begin{vmatrix} \cos\theta & \sin\theta \\ \sin\theta & -\cos\theta \end{vmatrix} \quad (1)$$

A represents rotation about the origin through an angle θ while B is reflection about a line through the origin and making an angle of $\theta/2$ with the x-axis.

Suppose we wish to rotate about some point $Q \neq$ the origin. This can be accomplished by first translating from Q to 0 (i.e. adding -Q), rotating about 0, and then translating back from 0 to Q (i.e. adding +Q). Thus rotation about Q through an angle θ is the mapping

$$F(P) = A(P-Q) + Q$$

where A is as in Eq. (1).

Next, reflection about a line

$$L = Q + W$$

can be done by translating by -Q, reflecting about W, and translating back by +Q. Thus reflection about L = Q+W is the transformation

$$F(P) = B(P-Q) + Q$$

where B, as in Eq. (1), is reflection about the line W through 0 and making an angle of $\theta/2$ with the positive x-axis.

878

Finally, glide reflection along the line $L = Q+W$ in an amount V_o is defined to be reflection in L followed by translation by V_o where V_o is in W. Thus,

$$G(P) = BP + (P_o + V_o)$$

where $BP_o = -P_o$ and $BV_o = V_o$.

It is now possible to classify rigid motions in the plane. Any rigid motion F in R^2 is one of the following four types (with A and B as in (1):

   i) $F(P) = P + P_o$ (translation by a vector) $P_o = F(0)$.

   ii) $F(P) = AP + P_o$, $\theta \neq 2k\pi$ (rotation through an angle θ about a point Q, where $P_o = Q - AQ$).

   iii) $F(P) = BP + P_o$ with $BP_o = -P_o$ (reflection about the line $L = 1/2 P_o + W$; if $P_o \neq 0$, then W is the set of all vectors perpendicular to P_o).

   iv) $F(P) = BP + P_o$, with $BP_o \neq -P_o$ (glide reflection along the line $L = \frac{1}{4}(P_o - BP_o) + W$, $W = $ span $\{P_o + BP_o\}$, in an amount $1/2(P_o + BP_o)$).

● **PROBLEM 23-22**

Classify the rigid motions in R^3.

<u>Solution</u>: Rigid motions in R^3 are of the form $F(P) = T(P) + F(0)$ where T is an orthogonal linear transformation and $F(0) = P_o$, a fixed point in R^3. To classify the linear rigid motions T, analyze the dimension of the subspace W of vectors left fixed by T:

$$v = \{V \text{ in } R^3 \mid T(V) = V\}$$

   i) If dim W = 3, every vector in R^3 is left fixed by T, i.e. T is the identity transformation.

   ii) Dimension W = 2. Let $\{V_1, V_2\}$ be an orthonormal basis for W and extend it to an orthonormal basis $\{V_1, V_2, U\}$ for R^3. This can be done by the Gram-Schmidt orthonormalization process. Then, since U is perpendicular to V_1 and V_2, T(u) will be perpendicular to $T(V_1) = V_1$ and to $T(V_2) = V_2$. Hence T(u) is in $W^\perp$ and thus $T(u) = ku$ (since $W^\perp$ is spanned by u). Since T preserves length $k = \pm 1$.

   But if k = 1, this leads to (i). Hence, T is reflection in a plane and the matrix of T is

879

$$B = \begin{bmatrix} 1 & 0 & 0 \\ 0 & 1 & 0 \\ 0 & 0 & -1 \end{bmatrix}$$

iii) Dimension W = 1. Let V span W with $|V|$ and extend to an orthonormal basis $\{V, W_1, W_2\}$ for R^3 where $W^\perp$ = span $\{W_1, W_2\}$. Then T maps $W^\perp$ onto $W^\perp$ and, restricted to W , T is an orthogonal mapping of this plane. It is thus either a rotation about 0 or a reflection in a line through 0. But if it were a reflection in a line L (spanned by V', say), then W would contain V' (orthogonal to V) and the dimension of W would be two. Hence T acts as a rotation in the plane $W^\perp$ and the matrix of T has the form

$$A = \begin{vmatrix} 1 & 0 & 0 \\ 0 & \cos\theta & -\sin\theta \\ 0 & \sin\theta & \cos\theta \end{vmatrix} \quad \cos\theta \neq 1$$

iv) Dimension W = 0. Here $T(V) \neq 0$ for all $V \neq 0$. Since the characteristic polynomial of T has degree three it has a real eigenvalue $\lambda = +1$ and since $\lambda \neq 1$, $\lambda = -1$. Let w = span $\{V\}$ where V is an eigenvector corresponding to this eigenvalue and extend to an orthonormal basis $\{V, W_1, W_2\}$ for R^3 so that $w^\perp$ = span $\{W_1, W_2\}$. As in the previous case, T cannot be a reflection on $w^\perp$ since then dimension W > 0. Hence, T is a rotation in the plane $w^\perp$ together with a reflection in this plane (a rotary reflection). The matrix of T is

$$C = \begin{vmatrix} -1 & 0 & 0 \\ 0 & \cos\theta & -\sin\theta \\ 0 & \sin\theta & \cos\theta \end{vmatrix} \quad \cos\theta \neq 1.$$

● **PROBLEM 23-23**

Give an example of a linear transformation using the quaternions.

Solution: Quaternions are vectors of the form

$$q = q_0 + a_1 i + a_2 j + a_3 k, \quad a_i \epsilon R.$$

The set of all quaternions forms a vector space over R with basis 1, i, j, k. Multiplication is defined so as to be distributive and satisfy

$$ij = -ji = k, \quad jk = -kj = i, \quad ki = -ik = j$$
$$i^2 = j^2 = k^2 = -1.$$

We can write

$$q = a_o + V \text{ where } V = a_1 i + a_2 j + a_3 k \text{ is a vector in } R^3.$$

Then $\bar{q} = a_o - V$ (the conjugate of q). The quaternion norm of q is then

$$|q| = \sqrt{q\bar{q}} .$$

Now, let V be a unit vector in R^3 and $q = a + bV$ a unit quaternion $(a^2 + b^2 = 1)$.

If X is a vector in R^3, then consider the mapping

$$X \to Y \text{ where } Y = qX\bar{q}$$

$$= (a + bV)X(a - bV). \qquad (1)$$

Since $\bar{Y} = q (-X)\bar{q} = -Y$, Y is also a vector in R^3. Thus, (1) is a linear transformation from R^3 into R^3.

● **PROBLEM** 23-24

Show that the linear transformation $T(V) = BV$, where

$$B = \begin{bmatrix} 1 & 0 \\ 0 & -1 \end{bmatrix}$$

is the matrix of T with respect to the usual basis, is an orthogonal mapping.

Solution: A linear transformation $T: V \to V$ is said to be orthogonal if

$$T(v) \cdot T(w) = v \cdot w$$

for v, w ϵ V. Orthogonal transformations preserve distances and angles.

To show that a given transformation is orthogonal use the following fact:

A linear transformation $T: R^n \to R^n$ is orthogonal if it maps orthonormal bases onto orthonormal bases.

A basis $\{V_1, V_2, \ldots, V_n\}$ for R^n is said to be orthonormal if

$$V_i \cdot V_j = \delta_{ij}$$

where

$$\delta_{ij} = \begin{cases} 1 & i = j \\ 0 & i \neq j . \end{cases}$$

The usual basis for R^2 is $\{e_1, e_2\} = \{(1,0),(0,1)\}$.

Now,
$$B[e_1] = \begin{bmatrix} 1 & 0 \\ 0 & -1 \end{bmatrix}\begin{bmatrix} 1 \\ 0 \end{bmatrix} = \begin{bmatrix} 1 \\ 0 \end{bmatrix}$$

$$B[e_2] = \begin{bmatrix} 1 & 0 \\ 0 & -1 \end{bmatrix}\begin{bmatrix} 0 \\ 1 \end{bmatrix} = \begin{bmatrix} 0 \\ -1 \end{bmatrix}$$

It is necessary to check that $\{(1,0),(0,-1)\}$ forms an orthonormal basis for R^2.

Let (x_1, x_2) be an arbitrary point in R^2. Then

$$(x_1, x_2) = c_1(1,0) + c_2(0,-1),$$

$$x_1 = c_1$$

$$x_2 = -c_2.$$

Thus $(x_1, x_2) = x_1(1,0) - x_2(0,-1)$,

and $\{(1,0), (0,-1)\}$ forms a basis for R^2.

Since $(1,0) \cdot (0,-1) = 1(0) + 0(-1) = 0$

the two basis vectors are orthogonal.

Since $\|(1,0)\| = \sqrt{1^2 + 0^2} = 1$

$$\|(0,-1)\| = \sqrt{0^2 + (-1)^2} = 1$$

the vectors $(1,0)$, $(0,-1)$ form an orthonormal basis for R^2.

The given linear transformation is orthogonal.

● **PROBLEM 23-25**

Determine the nearest point in $U = \text{span } \{V_1, V_2\}$ to Y, where $V_1 = (2,1,0)$ $V_2 = (-1,2,0)$ $Y = (1,2,3)$.

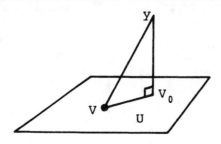

Solution: This problem requires the concept of orthogonality.   Two vectors V_1, V_2 in a Euclidean space are orthogonal only if $V_1 \cdot V_2 = 0$.   If U is any subspace of R^n, the
orthogonal complement U of U is the set of all vectors in
R^n which are perpendicular to every vector in U:

$$U^{\perp} = \{W \varepsilon R^n | W \cdot V = 0 \ \forall \ V \varepsilon U\}.$$

We can use the orthogonal complement to obtain the
following facts:

1)   Any vector Y in R^n can be wtitten in the form

$$Y = V + W \quad V \varepsilon U, \quad W \varepsilon U^{\perp}.$$

In particular, if $Y = V_o + W$, with V_o in U, W in $U^{\perp}$, V_o is
called the projection of Y on U (and W is called the perpendicular from U to Y).

2)   Given a subspace U and a vector Y in R^n, the projection
V_o of Y on U is the vector of U which is closest to Y in
that the distance from Y to vectors of V is at a minimum
at V_o.

Visually, (2) may be seen from the figure.

Apply the Pythagorean theorem to triangle VV_oY to
obtain:

$$|V-Y|^2 = |V-V_1|^2 + |V_o-Y|^2.$$

Since $|V-V_o|^2 \geq 0$,

$$|V-Y| \geq |V_o-Y| \text{ for any } V \text{ in } U \text{ and thus } Y \text{ is at a}$$
minimum distance from V_o.

To actually compute the distance for the given problem, first take an orthogonal basis for U, say, $\{(2,1,0)$
$(-1,2,0)\}$.   To extend this to an orthogonal basis for
R^3 add on a vector orthogonal to $\{V_1, V_2\}$, say, $W = (0,0,1)$.
W forms a basis for $U^{\perp}$.

Then $Y = aV_1 + bV_2 + cW$.   (1)

To find the coordinates a,b,c take the dot product of Y
with respect to V_1, V_2 and W as follows:

$$Y \cdot V_1 = (1,2,3) \cdot (2,1,0) = 4.$$ (2)

Since V_1 is orthogonal to V_2 and W, $V_1 \cdot V_2 = V_1 \cdot W = 0$.
Hence, from (1):

$$Y \cdot V_1 = aV_1 \cdot V_1 = a(2,1,0) \cdot (2,1,0) = 5a.$$ (3)

From (2) and (3), $5a = 4$ or $a = 4/5$.

Similarly, $Y \cdot V_2 = (1,2,3) \cdot (-1,2,0) = 3$

and $\qquad bV_2 \cdot V_2 = b(-1,2,0) \cdot (-1,2,0) = 5b$.

Since $\qquad Y \cdot V_2 = 3 = bV_2 \cdot V_2 = 5b, \quad b = 3/5$.

Finally, $\quad Y \cdot W = (1,2,3) \cdot (0,0,1) = 3$

and $\qquad cW \cdot W = c(0,0,1) \cdot (0,0,1) = c$.

Since $\qquad Y \cdot W = 3 = cW \cdot W = c, \quad c = 3$.

　　Substituting the values of a, b, and c into equation (1) yields,

$$Y = \frac{4}{5} V_1 + \frac{3}{5} V_2 + 3W$$

and the projection of Y on U is

$$V_o = aV_1 + bV_2$$

$$= \frac{4}{5}(2,1,0) + \frac{3}{5}(-1,2,0)$$

$$= (1,2,0)$$

while the perpendicular from Y to U is $3W = (0,0,3)$.

　　The distance from Y to U is

$$|Y - V_o| = |3W| = 3.$$

GEOMETRICAL PROBLEMS

● PROBLEM 23-26

Find the plane through the points $(1,0,3)$, $(2,-1,5)$, and $(3,4,6)$.

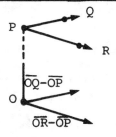

Solution: The plane containing the points P, Q, and R is

$$\overline{OP} + r(\overline{OQ} - \overline{OP}) + s(\overline{OR} - \overline{OP}). \tag{1}$$

This can be seen by the geometric argument in the figure.

We can also see that (1) is true by successively setting r = 0, s = 0; r = 1, s = 0; and r = 0, s = 1.

When r = 0, s = 0, (1) becomes $\overline{OP}$.

When r = 1, s = 0, (1) becomes $\overline{OQ}$.

When r = 0, s = 1, (1) becomes $\overline{OR}$.

Thus the noncollinear points P, Q, R are all in the plane,

$$\overline{OP} + r(\overline{OQ} - \overline{OP}) + s(\overline{OR} - \overline{OP}).$$

Thus, the plane through the given points, (1,0,3), (2,-1,5), (3,4,6) is given by the equation:

$$(1,0,3) + r((2,-1,5)-(1,0,3)) + s((3,4,6)-(1,0,3))$$

or by:

$$(1,0,3) + r(1,-1,2) + s(2,4,3).$$

● **PROBLEM 23-27**

Show that the points A(1,2,1), B(2,3,2), C(3,3,-2) are the vertices of a right triangle. Then find the angles of the triangle and its area.

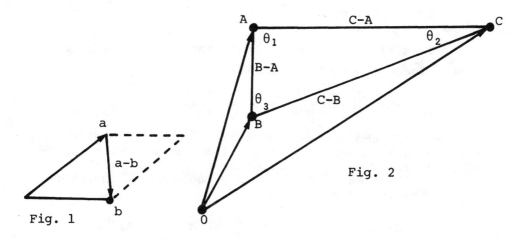

Fig. 1

Fig. 2

Solution: Recall that if a and b are two vectors then a-b is the diagonal of the parallelogram they determine. See Fig. 1.

Notice also that a-b has a tail at a and head at b. For b-a the situation is reversed. Fig. 2 illustrates the given problem, where A, B and C are the position vectors to the three given points. So to find the sides of the tri- angle we compute

$$C-A = (3,3,-2) - (1,2,1) = (2,1,-3)$$

$$C-B = (3,3,-2) - (2,3,2) = (1,0,-4)$$

$$B-A = (2,3,2) - (1,2,1) = (1,1,1).$$

$u \cdot v = \|u\| \; \|v\| \cos\theta$ where θ is the angle between u and v. Thus if $u \cdot v = 0$ and u and v are both non-zero vectors we have $\cos\theta = 0$ or $\theta = 90°$.

$$(C-A) \cdot (B-A) = (2,1,-3) \cdot (1,1,1) = 2(1)+1(1)+(-3)1 = 0$$

Therefore $\theta_1 = 90°$ and $\triangle ABC$ is a right triangle. Angle θ_2 is given by the equation

$$(C-A) \cdot (C-B) = \|C-A\| \; \|C-B\| \; \cos\theta_2$$

which yields

$$2(1) + -3(-4) = (\sqrt{2^2+1^2+(-3)^2})(\sqrt{1^2+(-4)^2}) \cos\theta_2.$$

So $\quad \cos\theta_2 = \dfrac{14}{\sqrt{14}(\sqrt{17})} = \dfrac{\sqrt{14}}{\sqrt{17}}$.

Thus $\quad \theta_2 = \cos^{-1} \dfrac{\sqrt{14}}{\sqrt{17}}$.

The third angle can be found similarly. It is easier, however, to say

$$\theta_3 = 180 - \theta_1 - \theta_2 , \quad \text{or}$$

$$\theta_3 = 90 - \theta_2 \quad \text{since the triangle is a right tri-}$$

angle. The area of the triangle is

$$\tfrac{1}{2} \|C-A\| \; \|B-A\| = \tfrac{1}{2}\sqrt{14}\sqrt{3} = \tfrac{1}{2}\sqrt{42} .$$

Geometrically the cross-product of two vectors yields a vector perpendicular to the two given vectors with magnitude the area of the parallelogram determined by the two vectors. The area of the triangle could also, therefore, be computed by the expression

$$\tfrac{1}{2} \| (C-A) \times (C-B) \| \quad \text{or} \quad \tfrac{1}{2} \| (B-A) \times (C-A) \| .$$

● **PROBLEM 23-28**

Prove that the lines joining the mid-points of the opposite edges of a tetrahedron bisect each other.

Solution: Let one vertex of the tetrahedron be at the origin, O, of a coordinate system in R^3, and let a, b and c be the position vectors of the other vertices, A, B and C respectively (see Fig. 1).

886

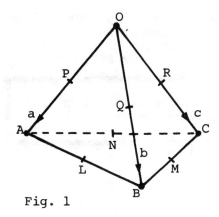

Fig. 1

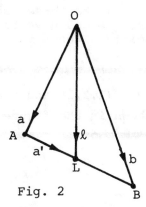

Fig. 2

Look in the plane of OAB to find the position vector to L, the mid-point of AB.

If vector a is added to vector a' (see Fig. 2) by the tail to head method of vector addition the result is ℓ, the position vector to the midpoint of AB. I.e. a+a' = ℓ. a', however, is on the vector b-a (by definition of vector subtraction) so $a + \frac{1}{2}(b-a) = \ell$. This yields

$$a + \frac{1}{2} b - \frac{1}{2} a = \ell, \quad \text{or} \quad \ell = \frac{1}{2}(a + b). \qquad (1)$$

The position vector of the mid-point of a segment is the average of the position vectors to the endpoints of the segment. Therefore the position vectors m and n to M and N the mid-points of BC and AC, respectively are as follows:

$$m = \frac{1}{2} (b + c), \qquad (2)$$

$$n = \frac{1}{2} (a + c). \qquad (3)$$

The position vector r to R the mid-point of OC is

$$r = \frac{1}{2} c. \qquad (4)$$

Now, the mid-point of the segment joining the mid-point of OC to the mid-point of AB (i.e. the mid-point of RL) has as position vector the average of the position vectors to R and L, which is $\frac{1}{2} (r+\ell)$. To evaluate this, substitute the values for r and ℓ obtained in equations (1) and (4) into $\frac{1}{2} (r+\ell)$:

$$\frac{1}{2} (r+\ell) = \frac{1}{2} (\frac{1}{2} c + \frac{1}{2} (a+b) = \frac{1}{4} (a+b+c).$$

Similarly the mid-point of NQ has position vector

$$\frac{1}{2} (n+b) = \frac{1}{4} (a+b+c)$$

and the mid-point of MP has position vector

$$\frac{1}{2} (m+p) = \frac{1}{4} (a+b+c).$$

Therefore the three segments joining the mid-points of the opposite edges have the same position vector for their midpoints. Hence they bisect each other.

● **PROBLEM** 23-29

Let P = (1,2,1) and v = sp{e_1,e_3} be the xz-plane in R^3, as shown in the figure. Find P + U.

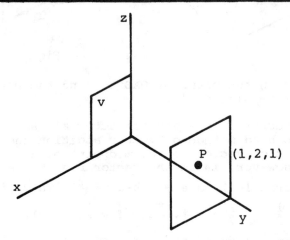

<u>Solution</u>: The subspace spanned by {e_1,e_3} consists of all vectors of the form $c_1(1,0,0) + c_3(0,0,0) = (c_1,0,c_3)$.

Now U is of dimension 2 and so $U^\perp$ is spanned by the single vector W = (w_1,w_2,w_3). Thus, all the vectors of U are of the form k(w_1,w_2,w_3). We know that if U is any subspace of R^3, then the orthogonal complement $U^\perp$ of U is the set of all vectors in R^3 which are perpendicular to every vector in U. This can be expressed as follows:

$$U^\perp = \{W \text{ in } R^3 | W \cdot V = 0 \text{ for all } V \text{ in } U\}.$$

Thus to find $U^\perp$:

$$W \cdot V = (kw_1, kw_2, kw_3) \cdot (c_1, 0, c_3) = (kc_1 w_1, \ 0, \ kc_3 w_3).$$

However, $W \cdot V = 0$; therefore, $kc_1 w_1 = 0$, $kc_3 w_3 = 0$. Thus, $w_1, w_3 = 0$.

$$U^\perp = sp\{(0,1,0)\}, \quad \text{or}, \quad sp\{e_2\}.$$

Now, P+v consists of all X in R^3 for which $W \cdot X = W \cdot P$ for all W in $U^\perp$. Thus, P+U = $\{X | e_2 \cdot X = e_2 \cdot P\}$, or $\{X = (x,y,z) | e_2 \cdot (x,y,z) = e_2 \cdot (1,2,1)\}$.

But, $e_2 \cdot (1,2,1) = (0,1,0) \cdot (1,2,1) = 0 + 2 + 0 = 2$

and $e_2 \cdot (x,y,z) = (0.1.0) \cdot (x,y,z) = 0 + y + 0 = y.$

888

Thus, $P + U = \{X \mid e_2 \cdot X = 2\}$,   or

$P + U = \{X = (x,y,z) \mid y = 2\}$.

● **PROBLEM** 23-30

Find the volume of the parallelepiped determined by the vectors $u = (2,3,5)$, $v = (-4,2,6)$ and $w = (1,0,3)$ in xyz-space.

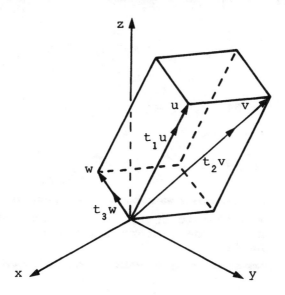

Solution: The parallelepiped can be represented as the set $\{t_1 u + t_2 v + t_3 w\}$ where $0 \leq t_i \leq 1$ for $i = 1,2,3$, where the parallelepiped is outlined by the three vectors u, v, w.

The formula for the volume of a parallelepiped is the absolute value of the 3x3 determinant of that matrix formed by using the three determining vectors as its rows. In this case

$$\text{Volume} = \text{absolute value of } \begin{vmatrix} 2 & 3 & 5 \\ -4 & 2 & 6 \\ 1 & 0 & 3 \end{vmatrix} . \qquad (1)$$

Recall that performing row operations on a matrix does not change the value of its determinant.

$$\begin{vmatrix} 2 & 3 & 5 \\ -4 & 2 & 6 \\ 1 & 0 & 3 \end{vmatrix} = \begin{vmatrix} 0 & 3 & -1 \\ 0 & 2 & 18 \\ 1 & 0 & 3 \end{vmatrix} . \text{ This equation is a}$$

result of the following two row operations:

889

1)   Multiply the third row of the matrix in equation (1) by 4 and add it to the second row.

2)   Multiply the third row by (-2) and add it to the first row.

Expanding along the first column yields:

$$\begin{vmatrix} 2 & 3 & 5 \\ -4 & 2 & 6 \\ 1 & 0 & 3 \end{vmatrix} = \begin{vmatrix} 0 & 3 & -1 \\ 0 & 2 & 18 \\ 1 & 0 & 3 \end{vmatrix}$$

$$= 0 + 0 + 1 \begin{vmatrix} 3 & -1 \\ 2 & 18 \end{vmatrix}$$

$$= 54 - (-2) = 56.$$

From equation (1) we obtain

Volume = 56 units.

● **PROBLEM 23-31**

Find the centroid or geometric center and the area of the parallelogram outlined by the vectors (1,1) and (-1,2) in the xy-plane (see the figure).

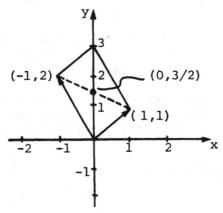

Solution: The parallelogram outlined or defined by the vectors u and v in R^2 can be described as the set of points $\{t_1 u + t_2 v\}$ where $0 \leq t_i \leq 1$, i = 1,2.   The given plane is, therefore,

$$t_1(1,1) + t_2(-1,2) \qquad 0 \leq t_i \leq 1. \tag{1}$$

The centroid or geometric center of a parallelogram is defined as the point where the diagonals of the parallelogram intersect.   The given representation (equation (1)) of a parallelogram is useful since the fact that t_i varies from

0 to 1 implies that the outer vertex of the parallelogram is the point where $t_1 = t_2 = 1$ and that the centroid is the point where $t_1 = t_2 = \frac{1}{2}$. Substitute $t_1 = t_2 = \frac{1}{2}$ into equation (1) to obtain the centroid:

$$\tfrac{1}{2}(1,1) + \tfrac{1}{2}(-1,2) = (0,\tfrac{3}{2}).$$

Note that the upper vertex was found by letting $t_1 = t_2 = 1$. That is, $(1,1) + (-1,2) = (0,3)$. The area of a parallelogram determined by two vectors u and v is the cross product u x v. Now u x v = $\|u\|\ \|v\|\ \sin\ \theta$, where θ is the angle between u and v. Therefore

$$(\text{area})^2 = \|u\|^2\ \|v\|^2\ \sin^2\theta$$

$$= \|u\|^2\ \|v\|^2\ (1-\cos^2\theta)$$

$$= \|u\|^2\ \|v\|^2 - \|u\|^2\ \|v\|^2\ \cos^2\theta$$

$$= \|u\|^2\ \|v\|^2 - (u \cdot v)^2$$

since the dot product can be defined as $u \cdot v = \|u\|\ \|v\|\ \cos\theta$ or $u \cdot v = u_1 v_1 + u_2 v_2$ where $u = (u_1,u_2)$ and $v = (v_1,v_2)$. Hence, if $u = (u_1,u_2)$ and $v = (v_1,v_2)$,

$$(\text{area})^2 = (u_1^2 + u_2^2)(v_1^2 + v_2^2) - (u_1 v_1 + u_2 v_2)^2$$

$$= u_1^2 v_1^2 + u_1^2 v_2^2 + u_2^2 v_1^2 + u_2^2 v_2^2$$

$$- u_1^2 v_1^2 - 2u_1 v_1 u_2 v_2 - u_2^2 v_2^2$$

$$= (u_1 v_2 - u_2 v_1)^2$$

and the area $= |u_1 v_2 - u_2 v_1|$. Notice that this is the absolute value of the determinant of the matrix with rows u and v.

$$\begin{vmatrix} u_1 & u_2 \\ v_1 & v_2 \end{vmatrix} = |u_1 v_2 - u_2 v_1| .$$

The area of the parallelogram in the figure is

$$\begin{vmatrix} 1 & 1 \\ -1 & 2 \end{vmatrix} = |2 - (-1)| = 3.$$

LOCAL LINEAR MAPPINGS

Find the local coordinates of the point with regular coordinates (5/2,-4) if the point (3,-5) is chosen as the
local origin.

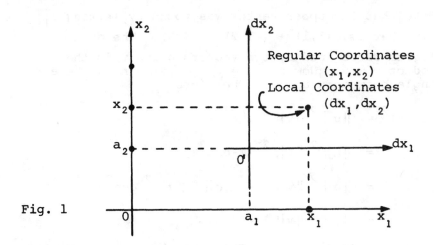

Fig. 1

Solution: Let $f: R^n \to R^n$ be a mapping. Let $a = (a_1, a_2, \ldots, a_n)$ be a point in the domain of f. Now restrict the domain of f to the vicinity of a and let a be the new local origin as shown in Fig. 1.

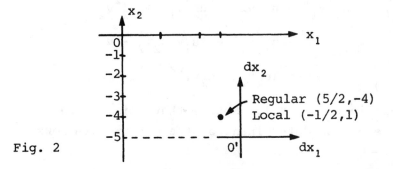

Fig. 2

The new local coordinate axes are parallel to the regular coordinate axes for R^n. As indicated in Fig. 1, a point $x = (x_1, x_2, \ldots, x_n)$ in R^n has the coordinate n-tuple $(x_1, \ldots x_n)$ and also a local coordinate n-tuple $(dx_1, \ldots, dx_n)$. The relations between the two coordinate systems are given by

$$x_i = a_i + dx_i$$

or

$$dx_i = x_i - a_i \qquad (i = 1, 2, \ldots, n).$$

892

Since the given point is in R^2, the local coordinates in terms of the regular coordinates are given by

$$dx_1 = x_1 - a_1$$

$$dx_2 = x_2 - a_2 .$$

(1)

We have $a_1 = 3$, $a_2 = -5$, $x_1 = 5/2$ and $x_2 = -4$. Thus, using (1):

$$dx_1 = 5/2 - 3 = -1/2$$

$$dx_2 = -4 - (-5) = 1$$

and the point in local coordinates is $(-1/2,1)$ as shown in Fig. 2.

● **PROBLEM 23-33**

Estimate $(2.05)^3$ by approximating $f(x) = x^3$ using a local linear map at $a = 2$.

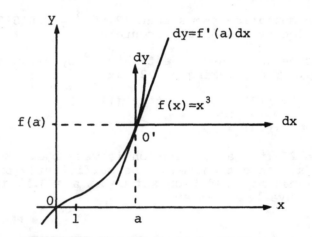

Solution: Let $f: R \rightarrow R$ be a mapping from a subset of R into R. Assume that $f'(a)$, the value of the derivative at a exists, where $a \in$ dom f. The differential df is defined to be $f'(a)dx$.

In the figure is shown the graph of f and local coordinate axes at the point $(a, f(a))$. Now, the tangent line to f at $(a, f(a))$ measures the slope of f. Since it passes through the origin of the local coordinate system, its equation in local coordinates is

$$dy = f'(a)dx.$$

(1)

The regular coordinate map for the tangent line at $(a, f(a))$ is

$$y - f(a) = f'(a)(x-a)$$

or, $y = f(a) + f'(a)(x-a)$.   (2)

Equation (2) is not in general a linear mapping, but equation (1) always is. Thus (1) is known as a local linear map. Use (2) to approximate f for x sufficiently close to a. That is,

$$f(x) \approx f(a) + f'(a)(x-a).$$ (3)

Take local coordinates so that

$x = a + dx$ and $x - a = dx$. Then (3) is

$$f(a+dx) \approx f(a) + f'(a)(dx),$$

or $f(a+dx) - f(a) \approx f'(a)dx$.   (4)

Turning to the given problem, if $f(x) = x^3$, $f'(x) = 3x^2$ and $f'(2) = 12$. Taking $a = 2$ and $dx = 0.05$ yields, from (4):

$$f(2.05) - f(2) = (2.05)^3 - 8 \approx f'(2)(0.05)$$

$$= 12(0.05) = 0.6$$

and hence, $(2.05)^3 \approx 8.6$.

A direct computation shows that $(2.05)^3 = 8.615125$, so the approximation is reasonably accurate.

Assume we wished to estimate 3^3 by using this same linear map. Then, taking $a = 2$, $dx = 1$ in 4:

$$f(3) - f(2) \approx f'(2)(1) = 12$$

or, $f(3) = 20$.

Since $3^3 = 27$, the approximation is very bad. But this is because $|dx|$ was not chosen to be sufficiently small. If a local linear map had been set up at $a = 3.1$, the approximation would have been better.

● **PROBLEM** 23-34

Let $f:R^2 \to R^3$ be defined by $(u,v,w) = f(x,y) = (3xy^2+1,$ $\sin xy, e^{xy}+y^4)$. Estimate, using a local linear map, $f(0.01, 0.98)$.

Solution: Let $f:R^n \to R^m$ be a mapping. For each x in the domain of f, the value f(x) is a point $y = (y_1, \ldots, y_m)$ in R^m. Each y_i is a function of $(x_1, \ldots, x_n)$. Thus there are mn partial derivatives, namely

$$\frac{\partial f_i}{\partial x_j} \qquad i = 1, \ldots, m \qquad j = 1, \ldots, n.$$

894

Arranging these derivatives in a matrix and evaluating at a yields the mxn Jacobian matrix of f at a:

$$f'(a) = \frac{\partial f_i}{\partial x_j}(a)$$

$$= \begin{bmatrix} \frac{\partial f_1}{\partial x_1}(a) & \frac{\partial f_1}{\partial x_2}(a) & \cdots & \frac{\partial f_1}{\partial x_n}(a) \\[2ex] \frac{\partial f_2}{\partial x_1}(a) & \frac{\partial f_2}{\partial x_2}(a) & \cdots & \frac{\partial f_2}{\partial x_n}(a) \\[2ex] \vdots & & & \\[2ex] \frac{\partial f_m}{\partial x_1}(a) & \frac{\partial f_m}{\partial x_2}(a) & \cdots & \frac{\partial f_m}{\partial x_n}(a) \end{bmatrix} \quad (1)$$

The matrix (1) corresponds to a linear map of R^n into R^m with respect to the standard ordered bases $\{e_1, \ldots, e_n\}$, $\{e_1, \ldots, e_m\}$. Choose local coordinates dx_j at a in R^n and local coordinates dy_i at f(a) in R^m. Let dx and dy be local coordinate column n-tuples and consider the local linear map given by

$$dy = f'(a)dx$$

or, written out,

$$\begin{bmatrix} dy_1 \\ dy_2 \\ \vdots \\ dy_m \end{bmatrix} = \begin{bmatrix} \frac{\partial y_1}{\partial x_1}(a) & \frac{\partial y_1}{\partial x_2}(a) & \cdots & \frac{\partial y_1}{\partial x_n}(a) \\[2ex] \vdots & & & \\[2ex] \frac{\partial y_m}{\partial x_1}(a) & \frac{\partial y_m}{\partial x_2}(a) & \cdots & \frac{\partial y_m}{\partial x_n}(a) \end{bmatrix} \begin{bmatrix} dx_1 \\ dx_2 \\ \vdots \\ dx_n \end{bmatrix}$$

This local linear map is the differential $df|_a$ of f at a and provides a good local linear approximation to the difference map $f(a+dx) - f(a)$ for $\|dx\|$ sufficiently small.

Applying the above machinery to the given problem, let dx = 0.01, dy = -0.02. Estimate, using the differential $df|_{(0,1)}$ which is the local linear map

$$\begin{bmatrix} du \\ dv \\ dw \end{bmatrix} = f'(0,1) \begin{bmatrix} dx \\ dy \end{bmatrix}$$

895

Computing partial derivatives of the coordinate functions yields

$$\frac{\partial f_1}{\partial x} = 3y^2 \qquad\qquad \frac{\partial f_1}{\partial y} = 6xy$$

$$\frac{\partial f_2}{\partial x} = y \cos xy \qquad\qquad \frac{\partial f_2}{\partial y} = x \cos xy$$

$$\frac{\partial f_3}{\partial x} = ye^{xy} \qquad\qquad \frac{\partial f}{\partial y} = xe^{xy} + 4y^3 .$$

Evaluating these partial derivatives at $(0,1)$, it is found that the 3x2 Jacobian matrix is

$$f'(0,1) \;=\; \begin{bmatrix} 3 & 0 \\ 1 & 0 \\ 1 & 4 \end{bmatrix}.$$

Thus,

$$\begin{bmatrix} du \\ dv \\ dw \end{bmatrix} = \begin{bmatrix} 3 & 0 \\ 1 & 0 \\ 1 & 4 \end{bmatrix} \begin{bmatrix} dx \\ dy \end{bmatrix}$$

$$= \begin{bmatrix} 3 & 0 \\ 1 & 0 \\ 1 & 4 \end{bmatrix} \begin{bmatrix} 0.01 \\ -0.02 \end{bmatrix} = \begin{bmatrix} 0.03 \\ 0.01 \\ -0.07 \end{bmatrix} .$$

Since $f(0,1) = (1,0,2)$ we obtain

$$f[(0,1) + (0.01,-0.02)]$$

$$= f(0.01,0.98) \approx \begin{bmatrix} 1 \\ 0 \\ 2 \end{bmatrix} + \begin{bmatrix} 0.03 \\ 0.01 \\ -0.07 \end{bmatrix} = \begin{bmatrix} 1.03 \\ 0.01 \\ 1.93 \end{bmatrix} .$$

● **PROBLEM 23-35**

Consider the map $f : R^2 \to R^3$ where

$$(x,y,z) = f(u,v) = (u,v,u^2+v^2).$$

Find the n-volume of a small region A about $(1,1)$ that is mapped by f into R^3.

Solution: Let $\phi : R^n \to R^n$ be a linear map. Then, if $G \subset R^n$ is a region of n-volume V, the image of G under ϕ has n-volume

896

$$|\det\ (M(\phi))\,|V. \tag{1}$$

The above result can be generalized to cover the case when $\phi:R^n \to R^m$, $m \geq n$. This time $M(\phi)$ is an mxn matrix. The n-volume of the image of G in R^m under ϕ is

$$(\sqrt{\det\ (M(\phi)^T \cdot M(\phi))}\ )\ V.$$

In the given problem, f is not linear but it is differentiable. Recall that the process of differentiation can be considered as a linear operator and that for small changes in x, the map f may be approximated around a by taking the linear map $df|_a$ of $dx = x-a$ and then translating the result by adding the vector f(a). These facts can be used to find the required n-volume.

Let f map a subset of R^n into R^m and let the coordinate functions $f_1, f_2, \ldots, f_m$ of f have continuous partial derivatives, so that f is a continuously differentiable map. The derivative of f at a point a in its domain is the mxn Jacobian matrix

$$f'(a) = \left| \frac{\partial f_i}{\partial x_i}\ (a) \right|.$$

The differential $df|_a$ of f at a is the linear map of R^n into R^m with matrix representation $f'(a)$ relative to the standard bases. Thus for a column vector dx in R^n we have

$$df|_a dx = f'(a) dx$$

and $$f(a+dx) \approx f(a) + f'(a) dx.$$

The linear map $df|_a$ with matrix $f'(a)$ carries a region of R^n into a region in R^m whose n-volume is $\sqrt{\det(f'(a))^T \cdot f'(a)}$ times as large. This n-volume is unchanged by adding the vector f(a) to each point in the region.

Thus $\sqrt{\det(f'(a)^T \cdot f'(a))}$ is the local volume change factor of f at a.

We have $$f:R^2 \to R^3.$$

The matrix of first order partial derivatives is

$$f'(u,v)\ =\ \begin{bmatrix} \partial x|\partial u & \partial x|\partial v \\ \partial y|\partial u & \partial y|\partial v \\ \partial z|\partial u & \partial z|\partial v \end{bmatrix} = \begin{bmatrix} 1 & 0 \\ 0 & 1 \\ 2u & 2v \end{bmatrix}.$$

The Jacobian matrix at (1,1) is

$$f'(1,1)\ =\ \begin{bmatrix} 1 & 0 \\ 0 & 1 \\ 2 & 2 \end{bmatrix}.$$

Then $(f'(1,1))^T f'(1,1) = \begin{bmatrix} 1 & 0 & 2 \\ 0 & 1 & 2 \end{bmatrix} \begin{bmatrix} 1 & 0 \\ 0 & 1 \\ 2 & 2 \end{bmatrix} = \begin{bmatrix} 5 & 4 \\ 4 & 5 \end{bmatrix}$

and hence

$$\sqrt{\det(f'(1,1)^T f'(1,1))} = \sqrt{25-16} = \sqrt{9} = 3.$$

Thus a small disk about $(1,1)$ of area A is mapped by f into a piece of surface of area approximately 3A.

● **PROBLEM** 23-36

(i) Find the length of one turn of the helix in R^3 which is described parametrically by $x = a \cos t$, $y = a \sin t$, $z = t$ (see Figure 1).

(ii) The ramp in Figure 2 is defined to be the image in R^3 of the rectangular region $[1,2] \times [0,2\pi]$ in R^2 under the map $f: R^2 \to R^3$ given by $f(x,y) = (x \cos y, x \sin y, y)$. Find the area of the ramp.

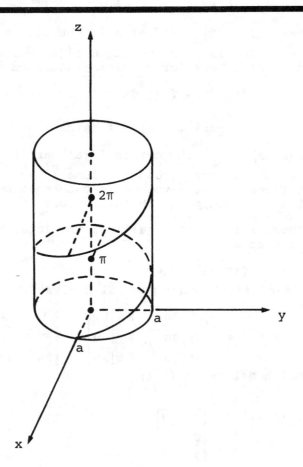

Fig. 1

898

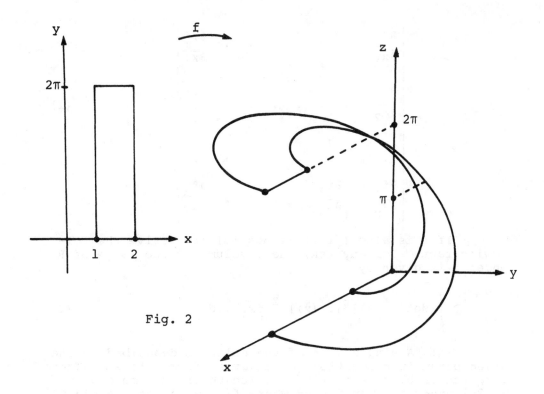

Fig. 2

Solution: From the theory of linear maps we have the following theorem: Let f be a continuously differentiable map
from R^n into R^m and let α be a point in the domain of f.

If $S = (\det[f'(\alpha)]^t[f'(\alpha)])^{\frac{1}{2}} \neq 0$ then, a sufficiently small
region R containing α, is mapped by f into a region of R^m
of n-volume approximately S times as great as the n-volume
of R.

That is,

n-volume of $f(R) = (\det[f'(\alpha)]^t[f'(\alpha)])^{\frac{1}{2}} \cdot$ n-volume of R.

Recall that the derivative of f at a point x in its domain
is defined to be the mxn Jacobian matrix,

$$f'(x) = \left[\frac{\partial f_i}{\partial x_i}\right].$$

That is, if $f(x) = (f_1(x),\ f_2(x),\ \ldots,\ f_m(x))$

with $x = (x_1, x_2, \ldots, x_n)$

then $f'(x) =$

899

$$
\begin{pmatrix}
\dfrac{\partial f_1}{\partial x_1} & \dfrac{\partial f_2}{\partial x_1} & \cdot & \cdot & \cdot & \cdot & \dfrac{\partial f_m}{\partial x_1} \\[2mm]
\dfrac{\partial f_1}{\partial x_2} & \cdot & & & & & \cdot \\[2mm]
& \cdot & \cdot & \cdot & \cdot & \cdot & \cdot \\[1mm]
\cdot & \cdot & & & & & \cdot \\[1mm]
\cdot & \cdot & & & & & \\[1mm]
\dfrac{\partial f_1}{\partial x_n} & \dfrac{\partial f_2}{\partial x_n} & \cdot & \cdot & \cdot & \cdot & \dfrac{\partial f_m}{\partial x_n}
\end{pmatrix}
\qquad (1)
$$

If f is also 1-1 it is possible to derive the following fact. For any $x \varepsilon R$, the m-volume of the image of R under f is given by

$$
\int_R (\det [f'(x)]^t [f'(x)])^{\frac{1}{2}} \, dx_1 \ldots dx_n \; . \qquad (2)
$$

(i) A single turn of the helix is described by the given parametric equations as t varies from 0 to 2π. Therefore the problem is to find the length of the image of $[0, 2\pi]$ under the map $f : R \rightarrow R^3$ where $f(t) = (a \cos t, a \sin t, t)$.

Now,

$$
[f'(t)]^T [f'(t)] = [-a \sin t, a \cos t, 1]
\begin{bmatrix}
-a \sin t \\
a \cos t \\
1
\end{bmatrix}
$$

$$
= a^2 \sin^2 t + a^2 \cos^2 t + 1
$$

$$
= a^2 (\sin^2 t + \cos^2 t) + 1 = a^2 + 1
$$

and hence

$$
(\det [f'(t)]^t [f'(t)])^{\frac{1}{2}} = (\det a^2 + 1)^{\frac{1}{2}} = \sqrt{a^2 + 1} \; .
$$

From equation (2) the arc length is then

$$
\int_0^{2\pi} (\det [f'(t)]^t [f'(t)])^{\frac{1}{2}} \, dt
$$

$$
= \int_0^{2\pi} \sqrt{a^2 + 1} \, dt = 2\pi \sqrt{a^2 + 1} \; .
$$

(ii) In the case of the ramp, again employ equation (1). First, equation (1) gives

900

$$f'(x,y) = \begin{bmatrix} \cos y & -x \sin y \\ \sin y & x \cos y \\ 0 & 1 \end{bmatrix}$$

so

$$[f'(x,y)]^T [f'(x,y)] = \begin{bmatrix} \cos y & \sin y & 0 \\ -x \sin y & x \cos y & 1 \end{bmatrix} \begin{bmatrix} \cos y & -x \sin y \\ \sin y & x \cos y \\ 0 & 1 \end{bmatrix}$$

$$= \begin{bmatrix} \cos^2 y + \sin^2 y + 0 & (\cos y)(-x \sin y) + (\sin y) x \cos y + 0 \\ -x \sin y \cos y + x \cos y \sin y + 0 & x^2 \sin^2 y + x^2 \cos^2 y + 1 \end{bmatrix}$$

$$= \begin{bmatrix} 1 & 0 \\ 0 & x^2 + 1 \end{bmatrix} .$$

Therefore,

$$(\det[f'(x,y)]^T [f'(x,y)])^{\frac{1}{2}} = (1 \cdot (x^2+1) - 0)^{\frac{1}{2}} = \sqrt{x^2+1}$$

and by equation (2) the area of the ramp is

$$\int_R \sqrt{x^2+1}\ dxdy = \int_0^{2\pi} \int_1^2 \sqrt{x^2+1}\ dxdy$$

$$= \int_0^{2\pi} \left. \frac{x\sqrt{x^2+1}}{2} + \frac{1}{2} \ln\ (x+\sqrt{x^2+1}) \right|_1^2 dy$$

$$= \int_0^{2\pi} (\sqrt{5} + \frac{1}{2} \ln(2+\sqrt{5}) - \frac{1}{2}\sqrt{2} - \frac{1}{2} \ln(1+\sqrt{2}))\ dy$$

$$= 2\pi (\sqrt{5} - \frac{1}{2}\sqrt{2} + \frac{1}{2} \ln \frac{2+\sqrt{5}}{1+\sqrt{2}})$$

$$= 11.373.$$

CHAPTER 24

MISCELLANEOUS APPLICATIONS

APPLICATIONS OF MATRIX ARITHMETIC

● **PROBLEM** 24-1

The following data matrix for the time interval 1940 – 1955 was based on the 1940 United States Census for a specific region.

Age	No. Females Alive in 1940	Females alive 15 years later (in 1955)	Daughters born in 15 year interval (1940 – 1955)
0 - 14	14, 459	16, 428	4, 651
15 - 29	15, 264	14, 258	10, 403
30 - 44	11, 346	14, 836	1, 374

Use matrix analysis on this data to obtain predictions of:

a) The number of females in the above categories in 1970

b) The number of females in the above categories in 1985.

Solution: To project population trends three sets of data are needed: (1) the number of people living in different age categories on a given date, (2) the number of people surviving in these age categories over a time interval from that date, (3) the number of people born in the time interval to people in a given age category. These data are given in the table. For example, the first entry in the second column, 16, 428 represents all the daughters born to the three age groups. The second entry is the number of survivors of the 0 - 14 group. Note that the number of survivors from the 30 - 44 group is not included in the table.

Now form the frequency matrix by the following procedure, i) Divide the number of daughters born in the 15 year interval by the number of females in each category to obtain rates of birth for the three groups. ii) Next divide survivors in a given age group by the original number of people in the previous age group (i.e., 15 years earlier). This gives the survival frequencies for the data. Thus, we obtain the following frequency matrix:

	0 - 14	15 - 29	30 - 44
Frequency for birth	$\dfrac{4,651}{14,459}$	$\dfrac{10,403}{15,264}$	$\dfrac{1,374}{11,346}$
Survival frequency 0 - 14 to 15 - 29	$\dfrac{14,258}{14,459}$	0	0
Survival frequency 15 - 29 to 30 - 44	0	$\dfrac{14,836}{15,264}$	0

The product of the frequency matrix F by the column matrix for females alive in 1940 gives the number of females alive in 1955. That is,

$$\begin{bmatrix} 0.32167 & 0.68154 & 0.12110 \\ 0.98610 & 0 & 0 \\ 0 & 0.97196 & 0 \end{bmatrix} \begin{bmatrix} 14,459 \\ 15,264 \\ 11,346 \end{bmatrix} = \begin{bmatrix} 16,428 \\ 14,258 \\ 11,028 \end{bmatrix}$$

To obtain a projection for 1970, which is 15 years later than 1955, square the frequency matrix and multiply by the column matrix for females in 1940.

$$\begin{bmatrix} 0.32167 & 0.68154 & 0.12110 \\ 0.98610 & 0 & 0 \\ 0 & 0.97196 & 0 \end{bmatrix}^2 \begin{bmatrix} 14,459 \\ 15,264 \\ 11,346 \end{bmatrix} = \begin{bmatrix} 16,799 \\ 16,200 \\ 13,858 \end{bmatrix}$$

Finally, to obtain a projection for 1985 raise the frequency matrix to the third power and multiply by the column matrix for females in 1940.

$$\begin{bmatrix} 0.32167 & 0.68154 & 0.12110 \\ 0.98610 & 0 & 0 \\ 0 & 0.97196 & 0 \end{bmatrix}^3 \begin{bmatrix} 14,459 \\ 15,264 \\ 11,346 \end{bmatrix} = \begin{bmatrix} 18,123 \\ 16,565 \\ 15,747 \end{bmatrix}$$

Thus in 1985 there will be 18, 123 females in the 0 - 14 group, 16,565 in the 15 - 29 group and 15, 747 in the 30 - 44 group.

● PROBLEM 24-2

Assume that rabbits do no reproduce during the first month of their lives but that beginning with the second month each pair of rabbits has one pair of offspring per month. Assuming that none of the rabbits die and beginning with one pair of newborn rabbits, how many pairs of rabbits are alive after n months?

Solution: Describe the growth pattern by means of the following figure

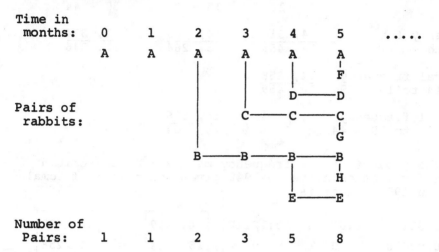

Time in
months: 0   1   2   3   4   5   .....

Pairs of rabbits:

Number of
Pairs: 1   1   2   3   5   8

When t = 0, A is the only pair of rabbits. When t = 1, there is still only A. At t = 2, we have two pairs, the original pair, A and their offspring B. When t = 3, there are three pairs, A and their offspring B and C. At time 4 we have five pairs A, B, C, the new pair, D produced by A and the pair E produced by B. When t = 5, we have A, B, C, D, E, the new offspring F of A, H of B and G of C. Thus the number of pairs after 5 months is 8.

After n months the number of pairs is given by the recursion formula

$$x_n = x_{n-1} + x_{n-2} \tag{1}$$

To see the reasoning behind (1), note that at t = n,

$$x_n = x_{n-1} + \text{# of pairs born in month n.}$$

Since one pair of rabbits is born to each pair alive 2 months previously, the number of newborn pairs is x_{n-2} whence (1) obtains.

Formula (1) requires knowledge of x_{n-1} and x_{n-2} to compute x_n. Using linear algebra we can compute x_n directly. Equation (1) can be written in matrix form as

$$(x_n, \ x_{n-1}) = (x_{n-1}, \ x_{n-2}) \begin{bmatrix} 1 & 1 \\ 1 & 0 \end{bmatrix} \tag{2}$$

Now list a few of the terms by the recurrence relation (2):

$$(x_2, \ x_1) = (x_1, \ x_0) \begin{bmatrix} 1 & 1 \\ 1 & 0 \end{bmatrix}$$

$$(x_3, x_2) = (x_2, x_1) \begin{bmatrix} 1 & 1 \\ 1 & 0 \end{bmatrix} = (x_1, x_0) \begin{bmatrix} 1 & 1 \\ 1 & 0 \end{bmatrix}^2$$

$$(x_4, x_3) = (x_3, x_2) \begin{bmatrix} 1 & 1 \\ 1 & 0 \end{bmatrix} = (x_1, x_0) \begin{bmatrix} 1 & 1 \\ 1 & 0 \end{bmatrix}^3 .$$

In general $(x_{n+1}, x_n) = (x_n, x_{n-1}) \begin{bmatrix} 1 & 1 \\ 1 & 0 \end{bmatrix}$

$$= (x_1, x_0) \begin{bmatrix} 1 & 1 \\ 1 & 0 \end{bmatrix}^n \qquad (3)$$

For the given problem, $x_1 = 1$, $x_0 = 1$ and (3) becomes

$$(x_{n+1}, x_n) = (1, 1) \begin{bmatrix} 1 & 1 \\ 1 & 0 \end{bmatrix}^n \qquad (4)$$

Thus, to obtain x_n we raise the matrix in (4) to the nth power, carry out the indicated multiplication and compare coordinates. But finding the nth power of a non-diagonal matrix can be time consuming. If we can find a diagonal matrix B that is similar to A (the matrix in (4)) our task is simplified. If B is similar to A, there exists a non-singular matrix P such that

$$A = P^{-1} BP.$$

Now, $A^n = P^{-1} B^n P$ $\qquad (5)$

To see that (5) is true:

$$A^n = (P^{-1} BP) (P^{-1} BP) \dots (P^{-1} BP). \qquad (6)$$

Since multiplication of matrices is associative, we can remove the brackets.

$$A^n = P^{-1} BPP^{-1} BPP^{-1} \dots PP^{-1} BP$$

$$= P^{-1} BIBI \dots IBP = P^{-1} B^n P.$$

The characteristic equation of $\begin{bmatrix} 1 & 1 \\ 1 & 0 \end{bmatrix}$ is

$$(1 - \lambda) (- \lambda) - 1 = \lambda^2 - \lambda - 1 = 0$$

905

with eigenvalues $\lambda = \dfrac{1 + \sqrt{5}}{2}, \dfrac{1 - \sqrt{5}}{2}$.

Since these eigenvalues are linearly independent,

$$B = \begin{bmatrix} \dfrac{1 + \sqrt{5}}{2} & 0 \\[2ex] 0 & \dfrac{1 - \sqrt{5}}{2} \end{bmatrix}$$

Since B is diagonal, B^n is obtained by raising the diagonal entries of B to the nth power. Hence,

$$B^n = \begin{bmatrix} \left(\dfrac{1 + \sqrt{5}}{2}\right)^n & 0 \\[3ex] 0 & \left(\dfrac{1 - \sqrt{5}}{2}\right)^n \end{bmatrix}$$

Next we must find the matrices P and P^{-1}. Corresponding to the two eigenvalues $\dfrac{1 + \sqrt{5}}{2}, \dfrac{1 - \sqrt{5}}{2}$, we have the systems of equations

$$\begin{bmatrix} 1 - \left(\dfrac{1 + \sqrt{5}}{2}\right) & 1 \\[2ex] 1 & -\left(\dfrac{1 + \sqrt{5}}{2}\right) \end{bmatrix} \begin{bmatrix} x_1 \\[2ex] x_2 \end{bmatrix} = \begin{bmatrix} 0 \\[2ex] 0 \end{bmatrix}$$

$$\begin{bmatrix} 1 - \left(\dfrac{1 - \sqrt{5}}{2}\right) & 1 \\[2ex] 1 & -\left(\dfrac{1 - \sqrt{5}}{2}\right) \end{bmatrix} \begin{bmatrix} x_1 \\[2ex] x_2 \end{bmatrix} = \begin{bmatrix} 0 \\[2ex] 0 \end{bmatrix} .$$

Thus two eigenvectors are

$(1 + \sqrt{5}, 2)$ and $(1 - \sqrt{5}, 2)$. Then

$$P = \begin{bmatrix} 1 + \sqrt{5} & 2 \\[2ex] 1 - \sqrt{5} & 2 \end{bmatrix}$$

and

$$P^{-1} = \begin{bmatrix} \dfrac{1}{2\sqrt{5}} & \dfrac{-1}{2\sqrt{5}} \\[3ex] \dfrac{-1 + \sqrt{5}}{4\sqrt{5}} & \dfrac{1 + \sqrt{5}}{4\sqrt{5}} \end{bmatrix}$$

Therefore, since $(x_{n+1}, x_n) = (1, 1)A^n$

$$= (1, 1)P^{-1}B^nP$$

it is necessary to only find $P^{-1}B^nP$ and multiply by $(1, 1)$ to obtain the formula for x_n

$$P^{-1}B^nP = \begin{bmatrix} \dfrac{1}{2\sqrt{5}} & \dfrac{-1}{2\sqrt{5}} \\[3mm] \dfrac{-1+\sqrt{5}}{4\sqrt{5}} & \dfrac{1+\sqrt{5}}{4\sqrt{5}} \end{bmatrix} \begin{bmatrix} \left(\dfrac{1+\sqrt{5}}{2}\right)^n & 0 \\[3mm] 0 & \left(\dfrac{1-\sqrt{5}}{2}\right)^n \end{bmatrix} \begin{bmatrix} 1+\sqrt{5} & 2 \\[3mm] 1-\sqrt{5} & 2 \end{bmatrix}$$

$$= \begin{bmatrix} \dfrac{1}{2\sqrt{5}}\left(\dfrac{1+\sqrt{5}}{2}\right)^n & \dfrac{-1}{2\sqrt{5}}\left(\dfrac{1-\sqrt{5}}{2}\right)^n \\[3mm] \dfrac{-1+\sqrt{5}}{4\sqrt{5}}\left(\dfrac{1+\sqrt{5}}{2}\right)^n & \dfrac{1+\sqrt{5}}{4\sqrt{5}}\left(\dfrac{1-\sqrt{5}}{2}\right)^n \end{bmatrix} \begin{bmatrix} 1+\sqrt{5} & 2 \\[3mm] 1-\sqrt{5} & 2 \end{bmatrix}$$

$$= \begin{bmatrix} \dfrac{1}{2\sqrt{5}}\left(\dfrac{1+\sqrt{5}}{2}\right)^n(1+\sqrt{5}) - \dfrac{1}{2\sqrt{5}}\left(\dfrac{1-\sqrt{5}}{2}\right)^n \\[3mm] (1+\sqrt{5})\left(\dfrac{-1+\sqrt{5}}{4\sqrt{5}}\right)\left(\dfrac{1+\sqrt{5}}{2}\right)^n + \left(\dfrac{1+\sqrt{5}}{4\sqrt{5}}\right)\left(\dfrac{1-\sqrt{5}}{2}\right)^n(1-\sqrt{5}) \end{bmatrix}$$

$$\begin{bmatrix} \dfrac{1}{\sqrt{5}}\left(\dfrac{1+\sqrt{5}}{2}\right)^n - \dfrac{1}{\sqrt{5}}\left(\dfrac{1-\sqrt{5}}{2}\right)^n \\[3mm] \dfrac{-1+\sqrt{5}}{2\sqrt{5}}\left(\dfrac{1+\sqrt{5}}{2}\right)^n + \left(\dfrac{1+\sqrt{5}}{2\sqrt{5}}\right)\left(\dfrac{1-\sqrt{5}}{2}\right)^n \end{bmatrix}$$

Since $(x_{n+1}\ x_n) = (1, 1)A^n = (1, 1)P^{-1}B^nP$ then from the above matrix $P^{-1}B^nP$ observe that,

$$x_n = \dfrac{1}{\sqrt{5}}\left(\dfrac{1+\sqrt{5}}{2}\right)^n - \dfrac{1}{\sqrt{5}}\left(\dfrac{1-\sqrt{5}}{2}\right)^n + \dfrac{(-1+\sqrt{5})}{2\sqrt{5}}\left(\dfrac{1+\sqrt{5}}{2}\right)^n$$

$$+ \dfrac{(1+\sqrt{5})}{2\sqrt{5}}\left(\dfrac{1-\sqrt{5}}{2}\right)^n$$

$$= \left(\dfrac{1+\sqrt{5}}{2}\right)^n\left(\dfrac{1}{\sqrt{5}} + \dfrac{(-1+\sqrt{5})}{2\sqrt{5}}\right) + \left(\dfrac{1-\sqrt{5}}{2}\right)^n\left(\dfrac{(1+\sqrt{5})}{2\sqrt{5}} - \dfrac{1}{\sqrt{5}}\right)$$

$$= \left(\frac{1 + \sqrt{5}}{2}\right)^{n}\left(\frac{2 - 1 + \sqrt{5}}{2\sqrt{5}}\right) + \left(\frac{1 - \sqrt{5}}{2}\right)^{n}\left(\frac{1 + \sqrt{5} - 2}{2\sqrt{5}}\right)$$

$$= \frac{1}{\sqrt{5}}\left(\frac{1 + \sqrt{5}}{2}\right)^{n+1} - \frac{1}{\sqrt{5}}\left(\frac{1 - \sqrt{5}}{2}\right)^{n+1}$$

For example, when n = 36 there are approximately 24 million pairs of rabbits and after 5 years (n = 60) about 2.5 trillion pairs of rabbits.

● **PROBLEM 24-3**

Show how matrices may be used to encode the message

"Return at noon."

Solution: First construct a one-to-one mapping from the set of letters in the alphabet to the set of real numbers. Let a → 1, b → 2, ..., z → 26 be this mapping. The message is therefore

18, 5, 20, 21, 18, 14, 1, 20, 14, 15, 15, 14.

Next, pick a 3x3 matrix A subject to the three restrictions:

1) The elements of A are integers

2) A is invertible

3) The elements of A^{-1}, the inverse of A are integers.

The matrix A is used to complicate the encoding process. Thus, it ensures that a person who does not know what A is will be unable to unravel the message.

$$\text{Let} \quad A = \begin{bmatrix} 1 & 2 & -1 \\ 2 & 5 & 2 \\ -1 & -2 & 2 \end{bmatrix}$$

$$\text{Then} \quad A^{-1} = \begin{bmatrix} 14 & -2 & 9 \\ -6 & 1 & -4 \\ -1 & 0 & 1 \end{bmatrix}$$

To use A, break up the message into three dimensional vectors. Let

W = [18, 5, 20]   X = [21, 18, 14]

Y = [1, 20, 14]   Z = [15, 15, 14]

908

Now form the four products WA, XA, YA and ZA.

$$WA = \begin{bmatrix} 18 & 5 & 20 \end{bmatrix} \begin{bmatrix} 1 & 2 & -1 \\ 2 & 5 & 2 \\ -1 & -2 & 2 \end{bmatrix} = \begin{bmatrix} 8 & 21 & 32 \end{bmatrix}$$

Similarly XA = [43 104 43]

YA = [27 74 67] and ZA = [31 77 43]

Thus the message sent is

8, 21, 32, 43, 104, 43, 27, 74, 67, 31, 77, 43

The receiver now breaks this string of numbers back into vectors with three entries each and multiplies each vector on the right by A^{-1}. Since $(WA)A^{-1} = W(AA^{-1}) = W$, etc., the 12 numbers sent, 8, 21, ..., 43 would be translated back to the original string of numbers 18, 5, ..., 14 and these would then be translated back to their letter equivalents r, e, t, u, ...

Break up the message:

[8, 21, 32] [43 104 43] [27 74 67] [31 77 43]

Now multiply on the right by A^{-1}:

$$\begin{bmatrix} 8 & 21 & 32 \end{bmatrix} \begin{bmatrix} 14 & -2 & 9 \\ -6 & 1 & -4 \\ 1 & 0 & 1 \end{bmatrix} = \begin{bmatrix} 18, & 5, & 20 \end{bmatrix}$$

The other vectors are decoded similarly.

● PROBLEM 24-4

a) Using the one-to-one correspondence

A ↔ 26, B ↔ 25, ... Y ↔ 2, Z ↔ 1

and the matrix

$$A = \begin{bmatrix} 2 & 3 \\ 1 & 2 \end{bmatrix}$$

encode the following message:

"Beware the Ides of March"

909

b) Use the correspondence

$$A \leftrightarrow 1, B \leftrightarrow 2, \ldots. Y \leftrightarrow 2, Z \leftrightarrow 1$$

and the matrix $\begin{bmatrix} 1 & 0 & 0 \\ 3 & 1 & 5 \\ -2 & 0 & 1 \end{bmatrix}$

to encode

"The end is near"

Solution: In these two applications of matrix analysis to cryptography we transform the message by multiplying each matrix by a vector of numbers representing bits of the message.

a) In order to be conformable for multiplication, the message must be broken up into bits of two letters each. Thus BE WA RE TH EI DE SO FM AR CH. Now write each pair as a 2x1 vector using the given correspondence:

$$\begin{bmatrix} B \\ E \end{bmatrix} = \begin{bmatrix} 25 \\ 22 \end{bmatrix} \quad \begin{bmatrix} W \\ A \end{bmatrix} = \begin{bmatrix} 4 \\ 26 \end{bmatrix} \quad \begin{bmatrix} R \\ E \end{bmatrix} = \begin{bmatrix} 9 \\ 22 \end{bmatrix} \quad \begin{bmatrix} T \\ H \end{bmatrix} = \begin{bmatrix} 7 \\ 19 \end{bmatrix}$$

$$\begin{bmatrix} E \\ I \end{bmatrix} = \begin{bmatrix} 22 \\ 18 \end{bmatrix} \quad \begin{bmatrix} D \\ E \end{bmatrix} = \begin{bmatrix} 23 \\ 22 \end{bmatrix} \quad \begin{bmatrix} S \\ O \end{bmatrix} = \begin{bmatrix} 8 \\ 12 \end{bmatrix} \quad \begin{bmatrix} F \\ M \end{bmatrix} = \begin{bmatrix} 21 \\ 14 \end{bmatrix}$$

$$\begin{bmatrix} A \\ R \end{bmatrix} = \begin{bmatrix} 26 \\ 9 \end{bmatrix} \quad \begin{bmatrix} C \\ H \end{bmatrix} = \begin{bmatrix} 24 \\ 19 \end{bmatrix}$$

Now multiply each of these vectors by the given matrix A.

$$A \cdot \begin{bmatrix} B \\ E \end{bmatrix} = \begin{bmatrix} 2 & 3 \\ 1 & 2 \end{bmatrix} \begin{bmatrix} 25 \\ 22 \end{bmatrix} = \begin{bmatrix} 116 \\ 69 \end{bmatrix}$$

$$A \cdot \begin{bmatrix} W \\ A \end{bmatrix} = \begin{bmatrix} 2 & 3 \\ 1 & 2 \end{bmatrix} \begin{bmatrix} 4 \\ 26 \end{bmatrix} = \begin{bmatrix} 86 \\ 56 \end{bmatrix}$$

$$A \cdot \begin{bmatrix} R \\ E \end{bmatrix} = \begin{bmatrix} 2 & 3 \\ 1 & 2 \end{bmatrix} \begin{bmatrix} 9 \\ 22 \end{bmatrix} = \begin{bmatrix} 84 \\ 53 \end{bmatrix}$$

$$A \cdot \begin{bmatrix} T \\ H \end{bmatrix} = \begin{bmatrix} 71 \\ 45 \end{bmatrix}, \quad A \cdot \begin{bmatrix} E \\ I \end{bmatrix} = \begin{bmatrix} 98 \\ 58 \end{bmatrix}, \quad A \cdot \begin{bmatrix} D \\ E \end{bmatrix} = \begin{bmatrix} 112 \\ 67 \end{bmatrix}$$

$$A \cdot \begin{bmatrix} S \\ O \end{bmatrix} = \begin{bmatrix} 52 \\ 32 \end{bmatrix} \quad A \cdot \begin{bmatrix} F \\ M \end{bmatrix} = \begin{bmatrix} 84 \\ 49 \end{bmatrix} \quad A \cdot \begin{bmatrix} A \\ R \end{bmatrix} = \begin{bmatrix} 79 \\ 44 \end{bmatrix}$$

$$A \cdot \begin{bmatrix} C \\ H \end{bmatrix} = \begin{bmatrix} 105 \\ 62 \end{bmatrix}$$

The coded message is

116 69 86 56 84 53 71 45 98 58 112 67 52 32 84

49 79 44 105 62.

To decode or unscramble the above message multiply each pair of numbers by A^{-1} on the left.

Since $A = \begin{bmatrix} 2 & 3 \\ 1 & 2 \end{bmatrix}$, $\quad A^{-1} = \begin{bmatrix} 2 & -3 \\ -1 & 2 \end{bmatrix}$

Then, $A^{-1} = \begin{bmatrix} 116 \\ 69 \end{bmatrix} = \begin{bmatrix} 2 & -3 \\ -1 & 2 \end{bmatrix} \begin{bmatrix} 116 \\ 69 \end{bmatrix} = \begin{bmatrix} 25 \\ 22 \end{bmatrix} = \begin{bmatrix} B \\ E \end{bmatrix}$

$A^{-1} = \begin{bmatrix} 86 \\ 56 \end{bmatrix} = \begin{bmatrix} 2 & -3 \\ -1 & 2 \end{bmatrix} \begin{bmatrix} 86 \\ 56 \end{bmatrix} = \begin{bmatrix} 4 \\ 26 \end{bmatrix} = \begin{bmatrix} W \\ A \end{bmatrix}$

and so on to obtain the original message.

b) Since we have a 3x3 matrix, the message is divided into triplets of letters.

THE END ISN EAR

Since $A = \begin{bmatrix} 1 & 0 & 0 \\ 3 & 1 & 5 \\ -2 & 0 & 1 \end{bmatrix}$, $\quad A^{-1} = \begin{bmatrix} 1 & 0 & 0 \\ -13 & 1 & -5 \\ 2 & 0 & 1 \end{bmatrix}$

Note that the matrix A is not only invertible but has an inverse consisting of integers. Thus the choice of the matrix to be used for encoding is restricted by these two requirements.

The encoded message is obtained by multiplying the matrix A times each column vector of the original message. Thus, we obtain

$$
A \cdot \begin{bmatrix} T \\ H \\ E \end{bmatrix} = \begin{bmatrix} 1 & 0 & 0 \\ 3 & 1 & 5 \\ -2 & 0 & 1 \end{bmatrix} \begin{bmatrix} 20 \\ 8 \\ 5 \end{bmatrix} = \begin{bmatrix} 20 \\ 93 \\ -35 \end{bmatrix}
$$

$$
A \cdot \begin{bmatrix} E \\ N \\ D \end{bmatrix} = \begin{bmatrix} 1 & 0 & 0 \\ 3 & 1 & 5 \\ -2 & 0 & 1 \end{bmatrix} \begin{bmatrix} 5 \\ 14 \\ 4 \end{bmatrix} = \begin{bmatrix} 5 \\ 49 \\ -6 \end{bmatrix}
$$

$$
A \cdot \begin{bmatrix} I \\ S \\ N \end{bmatrix} = \begin{bmatrix} 1 & 0 & 0 \\ 3 & 1 & 5 \\ -2 & 0 & 1 \end{bmatrix} \begin{bmatrix} 9 \\ 19 \\ 14 \end{bmatrix} = \begin{bmatrix} 9 \\ 116 \\ -4 \end{bmatrix}
$$

$$
A \cdot \begin{bmatrix} E \\ A \\ R \end{bmatrix} = \begin{bmatrix} 1 & 0 & 0 \\ 3 & 1 & 5 \\ -2 & 0 & 1 \end{bmatrix} \begin{bmatrix} 5 \\ 1 \\ 18 \end{bmatrix} = \begin{bmatrix} 5 \\ 106 \\ 8 \end{bmatrix}
$$

Thus the coded message is

20   93   -35   5   49   -6   9   116   -4

To decode the message form 3x1 column vectors of these numbers and multiply on the left by A^{-1}. For example,

$$
A^{-1} \begin{bmatrix} 20 \\ 93 \\ -35 \end{bmatrix} = \begin{bmatrix} 1 & 0 & 0 \\ -13 & 1 & -5 \\ 2 & 0 & 1 \end{bmatrix} \begin{bmatrix} 20 \\ 93 \\ -35 \end{bmatrix} = \begin{bmatrix} 20 \\ 8 \\ 5 \end{bmatrix} = \begin{bmatrix} T \\ H \\ E \end{bmatrix}
$$

Fit a quadratic polynomial to the data given below

x	y	x'	y'	x'y'	x'^2	x'^3	x'^4	x'^2y'
0	1.2	− 2	−1.8	3.6	4	−8	16	− 7.2
1	2.1	− 1	−0.9	0.9	1	−1	1	− 0.9
2	3.8	0	0.8	0	0	0	0	0
3	2.9	1	−0.1	−0.1	1	−8	1	− 0.1
4	5.0	2	2.0	4.0	4	−1	16	8.0
10	15.0	0	0	8.4	10	0	34	− 0.2

Here $x' = x - \bar{x}$ and $y' = y - \bar{y}$ where $\bar{x}, \bar{y} = \frac{\Sigma x}{n}, \frac{\Sigma y}{n}$ respectively.

<u>Solution</u>:   This is a problem in the theory of least-squares. The general least-square problem is:

Given a matrix A and a vector Y, find the vector X (if it exists) for which AX is as close to Y as possible, i.e., $|AX - Y|$ is as small as possible.

To solve this problem use the following facts. Given a point y and a subspace v of R^n with basis $\{V_1, \ldots, V_k\}$, let

$$A = (V_1, \ldots, V_k)$$

be the nxk matrix whose columns are this basis. Then the point of v nearest Y (that is, the projection of Y on v) has coordinates $V_0 = AX$, where X is the unique solution to

$$(A^TA) X = A^TY.$$

We wish to fit a quadratic polynomial $a + bx + cx^2$ to the data (x_i, y_i) $i = 1, 2, \ldots, n$ $(n > 3)$.   Then

$$A = \begin{bmatrix} 1 & x_1 & x_1^2 \\ \cdot & \cdot & \cdot \\ \cdot & \cdot & \cdot \\ \cdot & \cdot & \cdot \\ \cdot & \cdot & \cdot \\ \cdot & \cdot & \cdot \\ 1 & x_n & x_n^2 \end{bmatrix} \qquad Y = \begin{bmatrix} y_1 \\ \cdot \\ \cdot \\ \cdot \\ \cdot \\ \cdot \\ y_n \end{bmatrix} \qquad X = \begin{bmatrix} a \\ b \\ c \end{bmatrix}$$

913

$$A^TA = \begin{bmatrix} n & \Sigma x_i & \Sigma x_i{}^2 \\ \Sigma x_i & \Sigma x^2{}_i & \Sigma x_i{}^3 \\ \Sigma x_i{}^2 & \Sigma x_i{}^3 & \Sigma x_i{}^4 \end{bmatrix} \qquad A^TY = \begin{bmatrix} \Sigma y_i \\ \Sigma x_i y_i \\ \Sigma x_i{}^2 y_i \end{bmatrix}$$

Now use the data in the table (substituting x' for x) to obtain

$$A = \begin{bmatrix} 1 & -2 & 4 \\ 1 & -1 & 1 \\ 1 & 0 & 0 \\ 1 & -1 & 1 \\ 1 & -2 & 4 \end{bmatrix} \qquad Y = \begin{bmatrix} -1.8 \\ -0.9 \\ 0.8 \\ -0.1 \\ 2.0 \end{bmatrix} \qquad A^TY = \begin{bmatrix} 0 \\ 8.4 \\ -0.2 \end{bmatrix}$$

Set $(A^TA)X = A^TY$ to obtain

$$\begin{bmatrix} 5 & 0 & 10 \\ 0 & 10 & 0 \\ 10 & 0 & 34 \end{bmatrix} \begin{bmatrix} a \\ b \\ c \end{bmatrix} = \begin{bmatrix} 0 \\ 8.4 \\ -0.2 \end{bmatrix}$$

Solving this system of 3 equations in 3 unknowns yields

$$a = \frac{1}{35} \qquad b = 0.84 \qquad c = -\frac{1}{70}$$

Thus the quadratic polynomial which minimizes the squared deviations of the data from itself is

$$\frac{1}{35} + 0.84x - \frac{1}{70} x^2.$$

● **PROBLEM 24-6**

Apply Gershgorin's theorem to estimate the eigenvalues of the matrix

$$A = \begin{bmatrix} \frac{5}{2} & -\frac{1}{2} & 1 \\ \frac{3}{2} & \frac{1}{2} & -1 \\ 0 & 0 & 0 \end{bmatrix}$$

<u>Solution</u>: Let A be an nxn matrix:

$$A = \begin{bmatrix} a_{11} & a_{12} & \cdots\cdots & a_{1n} \\ & & & \cdot \\ a_{21} & & & \cdot \\ \cdot & & & \cdot \\ \cdot & & & \cdot \\ \cdot & & & \cdot \\ \cdot & & & \\ a_{n1} & a_{n2} & \cdots\cdots & a_{nn} \end{bmatrix}$$

In general, the eigenvalues of A will be complex numbers. Let

$$r_i = |a_{i1}| + \cdots + |a_{i,\ i-1}| + |a_{i,\ i+1}| + \cdots + |a_{i,n}|$$

denote the sum of the absolute values (or modulii) in the complex case of the elements of the ith row except for the ii element a_{ii}. Let

$$D_i = \{z: \ |z - a_{ii}| \le r_i\}$$

be the disk around a_{ii} with radius r_i. Then Gershgorin's theorem states that the eigenvalues of A are to be found in the union of the discs D_i i = 1, ..., n.

Applying this theorem to the given matrix (1):

$$D_1 = \{z: \ |z - 5/2| \le |-1/2| + |1|\}$$

$$= \{z: \ |z - 5/2| \le 3/2\}$$

$$D_2 = \{z: \ |z - 1/2| \le |3/2| + |-1|\}$$

$$= \{z: \ |z - 1/2| \le 5/2\}$$

$$D_3 = \{z: \ |z - 0| \le |0| + |0|\}$$

$$= \{z: \ |z - 0| \le 0\}.$$

Solving the three inequalities in D_1, D_2 and D_3 for z we obtain

$$1 \le z \le 4$$

$$-2 \le z \le 3$$

$$0 \le z \le 0 \quad \text{or } z = 0.$$

Setting up and solving the characteristic equation yields the actual roots 0, 1 and 2. Gershgorin's theorem is a method for obtaining a rough estimate of the eigenvalues of a given matrix. Its proof involves another theorem known as the Dominant Diagonal Theorem which is interesting in its own right. According to this theorem, a matrix A is invertible if each diagonal entry is larger in absolute value then the sum of the absolute values of the other entries in its row.

● **PROBLEM** 24-7

Let $p(x) = 6 + 2x^2 - 6x^4 + 4x^5 - 3x^6 + x^8$. Apply Cauchy's polynomial root theorem to find a circle of radius r within which all the roots of $p(x)$ lie.

Solution: This problem shows how eigenvalues can be applied to the study of roots of polynomials.

Let $p(x) = a_0 + a_1 x + \ldots + a_{n-1} x^{n-1} + x_n$.
The companion matrix P of $p(x)$ is the nxn matrix

$$P = \begin{bmatrix} 0 & 1 & 0 & \ldots & 0 \\ 0 & 0 & 1 & \ldots & 0 \\ \vdots & & & & \vdots \\ \vdots & & & & \vdots \\ \vdots & & & & \vdots \\ 0 & 0 & \ldots & \ldots & 1 \\ -a_0 & -a_1 & \ldots & -a_{n-2} & -a_{n-1} \end{bmatrix}$$

The companion matrix has the characteristic polynomial $\pm p(x)$ and thus its eigenvalues are the roots of $p(x)$. This fact is used to prove Cauchy's polynomial root theorem which states:

Let $p(x) = x^n + a_{n-1} x^{n-1} + \ldots + a_1 x + a_0$
be any polynomial with real or complex coefficients. Then all the roots of $p(x)$ lie within the circle about the origin in the complex plane of radius

$$r = \max \{|a_0|, 1 + |a_1|, 1 + |a_2|, \ldots, 1 + |a_{n-1}|\}$$

To prove the above theorem, simply apply Gershgorin's theorem to the companion matrix.

Turning to the given problem, the companion matrix is:

$$P = \begin{bmatrix} 0 & 1 & 0 & 0 & 0 & 0 & 0 & 0 \\ 0 & 0 & 1 & 0 & 0 & 0 & 0 & 0 \\ 0 & 0 & 0 & 1 & 0 & 0 & 0 & 0 \\ 0 & 0 & 0 & 0 & 1 & 0 & 0 & 0 \\ 0 & 0 & 0 & 0 & 0 & 1 & 0 & 0 \\ 0 & 0 & 0 & 0 & 0 & 0 & 1 & 0 \\ 0 & 0 & 0 & 0 & 0 & 0 & 0 & 1 \\ -6 & 0 & -2 & 0 & 6 & -4 & 3 & 0 \end{bmatrix}$$

To show that the characteristic polynomial of P is p(x), we must evaluate det (P - xI):

$$\begin{vmatrix} -x & 1 & 0 & 0 & 0 & & & \\ 0 & -x & 1 & 0 & 0 & & & \\ 0 & 0 & -x & 1 & 0 & & & \\ 0 & 0 & 0 & -x & 1 & & & \\ 0 & 0 & 0 & 0 & -x & 1 & & \\ 0 & 0 & 0 & 0 & 0 & -x & 1 & \\ 0 & 0 & 0 & 0 & 0 & 0 & -x & 1 \\ -6 & 0 & -2 & 0 & 6 & -4 & 3 & -x \end{vmatrix} \qquad (1)$$

Add x (column 2), x^2 (column 3) .... x^7 (column 8) to column 1. The value of the determinant is unchanged but the first column is [0, 0, ..., - p(x)]. Expanding (1) along this column by cofactors yields

det (P - xI) = - (- 1)^9 p(x) = p(x).

Now, r = max {|6|, 1, 1 + |2|, 1, 1 + |- 6|, 1 + |4|,

1, 1 + |- 3|, 1, 1}

= max {6, 1, 3, 7, 5, 4} = 7

Thus p(x) has all roots within a circle of radius 7.

● **PROBLEM** 24-8

Some societies are divided into clans in such a fashion that a man in clan i can only marry women in clan j while his children will belong to clan k. Show how matrix analysis can be used to indicate the sets of potential brides and grooms for different members of such societies.

917

<u>Solution</u>: Permutation matrices are helpful in analyzing kinship systems. A permutation matrix is an nxn matrix which obeys:

1)   every row has only one 1 and all other elements zero

2)   every column has only one 1 and all other elements zero.

The simplest nxn permutation matrix is the nxn identity matrix. Since there are n choices for placing a 1 in the first row, n-1 choices for the second row .... and 1 choice for the last row, the number of nxn permutation matrices is n!

Furthermore, the product of two nxn permutation matrices is a permutation matrix and the inverse of a permutation matrix is a permutation matrix.

Suppose there are n clans in a society. Let W be the matrix which has a 1 as its i, j entry if a man in clan i must marry a woman in clan j and has zeros otherwise in the ith row and in the jth column:

$$
W = \begin{bmatrix}
& & \vdots & & \\
& & \vdots & & \\
& & \vdots & & \\
\cdots & \cdots & 1 & \cdots & \cdots \\
& & \vdots & & \\
& & \vdots & & \\
& & \vdots & &
\end{bmatrix}
\begin{matrix} j \end{matrix}
$$

Similarly, let C be the matrix which has a 1 as its i, k entry if a man in clan i has children in clan k, and 0's otherwise in the ith row and jth column.

Children's
Clan

$$
C = \begin{bmatrix}
& & \vdots & & \\
& & \vdots & & \\
& & \vdots & & \\
\cdots & \cdots & 1 & \cdots & \cdots \\
& & \vdots & & \\
& & \vdots & & \\
& & \vdots & &
\end{bmatrix}
\begin{matrix} k \end{matrix}
$$

The rules of these societies imply that W and C are permutation matrices.

For example, amongst a tribe known as the Kariera, there are four clans. The matrices are

$$W: \text{Clan of Husband} \quad \begin{array}{c c} & \begin{array}{c c c c} 1 & 2 & 3 & 4 \end{array} \\ \begin{array}{c} 1 \\ 2 \\ 3 \\ 4 \end{array} & \left[\begin{array}{c c c c} 0 & 1 & 0 & 0 \\ 1 & 0 & 0 & 0 \\ 0 & 0 & 0 & 1 \\ 0 & 0 & 1 & 0 \end{array} \right] \end{array}$$

$$\text{Clan of Husband} \quad \begin{array}{c c} & \begin{array}{c} \text{Clan of Children} \\ \begin{array}{c c c c} 1 & 2 & 3 & 4 \end{array} \end{array} \\ \begin{array}{c} 1 \\ 2 \\ 3 \\ 4 \end{array} & \left[\begin{array}{c c c c} 0 & 0 & 1 & 0 \\ 0 & 0 & 0 & 1 \\ 1 & 0 & 0 & 0 \\ 0 & 1 & 0 & 0 \end{array} \right] \end{array}$$

These indicate, for example, that a man in clan 1 must marry a woman in clan 2, while his children will belong to clan 3.

● **PROBLEM** 24-9

In a certain community there are four clans, A, B, C and D. The rules of marriage and descent are summarized as follows:

An	A	man marries a	B	woman and has	A	children
	B		C		B	
	C		D		C	
	D		A		D	

Does this community allow:

a) matrilateral cross-cousin marriage

b) patrilateral cross-cousin marriage

c) After how many generations will a man's descendants belong to his clan again.

Solution: The first step is to set up the marriage matrix and the descendant matrix. Thus,

Wife's clan

A B C D

Husband's clan

	A	B	C	D
A	0	1	0	0
B	0	0	1	0
C	0	0	0	1
D	1	0	0	0

= W

Clan of children

A B C D

Clan of Father

	A	B	C	D
A	1	0	0	0
B	0	1	0	0
C	0	0	1	0
D	0	0	0	1

= C

The matrix product WC is

$$\begin{bmatrix} 0 & 1 & 0 & 0 \\ 0 & 0 & 1 & 0 \\ 0 & 0 & 0 & 1 \\ 1 & 0 & 0 & 0 \end{bmatrix} \begin{bmatrix} 1 & 0 & 0 & 0 \\ 0 & 1 & 0 & 0 \\ 0 & 0 & 1 & 0 \\ 0 & 0 & 0 & 1 \end{bmatrix} = \begin{bmatrix} 0 & 1 & 0 & 0 \\ 0 & 0 & 1 & 0 \\ 0 & 0 & 0 & 1 \\ 1 & 0 & 0 & 0 \end{bmatrix}$$

The matrix product CW is

$$\begin{bmatrix} 1 & 0 & 0 & 0 \\ 0 & 1 & 0 & 0 \\ 0 & 0 & 1 & 0 \\ 0 & 0 & 0 & 1 \end{bmatrix} \begin{bmatrix} 0 & 1 & 0 & 0 \\ 0 & 0 & 1 & 0 \\ 0 & 0 & 0 & 1 \\ 1 & 0 & 0 & 0 \end{bmatrix} = \begin{bmatrix} 0 & 1 & 0 & 0 \\ 0 & 0 & 1 & 0 \\ 0 & 0 & 0 & 1 \\ 1 & 0 & 0 & 0 \end{bmatrix}$$

Thus WC = CW (since C is the 4x4 identity matrix this follows immediately from the axioms for a vector space).

Next, find $W^T C$:

$$\begin{bmatrix} 0 & 0 & 0 & 1 \\ 1 & 0 & 0 & 0 \\ 0 & 1 & 0 & 0 \\ 0 & 0 & 1 & 0 \end{bmatrix} \begin{bmatrix} 1 & 0 & 0 & 0 \\ 0 & 1 & 0 & 0 \\ 0 & 0 & 1 & 0 \\ 0 & 0 & 0 & 1 \end{bmatrix} = \begin{bmatrix} 0 & 0 & 0 & 1 \\ 1 & 0 & 0 & 0 \\ 0 & 1 & 0 & 0 \\ 0 & 0 & 1 & 0 \end{bmatrix}$$

Here, $CW \neq W^T C$.

Consider the interpretations of WC, CW and $W^T C$.

If the i,j entry of WC is $\neq 0$ then a man of clan i marries a woman of clan k and a man of clan k has children in clan j. Thus WC lists the clan of the brother-in-law's children.

On the other hand, if the i,j entry of CW is $\neq 0$ a man of clan i has children in clan k and a man of clan k marries a woman of clan j. Thus CW lists the clan of the son's wife.

Finally, W^T lists the clan of the husband according to that of the wife.

a) Matrilateral cross-cousin marriage occurs when a man's son's daughter can marry his daughter's son. This is allowed if and only if W and C commute, i.e., WC = CW. We see that the given community allows matrilateral cross cousins to marry.

b) When a man's son's son can marry his daughter's daughter patrilateral cross marriage is allowed. This occurs if and only if $CW = W^T C$. Since $CW \neq W^T C$ here, patrilateral cross cousins are not allowed to marry.

c) A man's children are automatically of the same clan as he. His son's children will also be of the same clan. But his daughter's children will belong to clan D. The female children will have, after marriage, children of the C clan who will have children of the B clan who will then have children of the A clan. Thus, it will take at least five generations for a man's descendants to belong to his clan again.

● **PROBLEM** 24-10

In certain societies there are rules as to when marriages are permissible. A person is of a certain marriage type and can only marry someone of that type. Suppose that there are three marriage types t_1, t_2, t_3 in a given society.

Parents of a given type have children of different types as given below:

Type of both parents	Type of their son	Type of their daughter
t_1	t_2	t_3
t_2	t_3	t_1
t_3	t_1	t_2

a) Can a man marry his father's brother's daughter?
b) Can a man marry his mother's brother's daughter?

Use matrix analysis to answer the above two questions.

Solution: First draw family trees to show the relationships. Define the following symbols:

```
Δ  :  Male

O  :  Female

=  :  Marriage

|  :  Descendant

⌐¬  :  Sibling
```

A man's father's brother's daughter is his first cousin.
The family tree for this type of first-cousin relationship
is given below

(1)

Thus, the original couple are grandparents of the pro-
spective bride and groom. In the given problem, there are
three possible types for the original couple. Working out
the three cases:

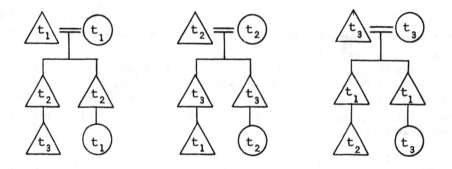

Observe that marriages between first cousins of this
kind are never allowed since the marriage types are different.

In general, society chooses a number, say n, of marriage
types, $t_1, t_2, \ldots, t_n$.

Consider the marriage type of sons. The parents must
be of the same marriage type. If his parents are of type
t_i, he will be of type t_j. Furthermore, if some other boy
has parents of a type different from t_i, then the boy will
be of type different from t_j. This defines a permutation

of the marriage types. The type of a son is obtained from
the type of his parents by a permutation specified by the
rule of the society.

For the given marriage rules, the permutation matrices
for the sons and daughters of parents of type 1, 2 or 3 are

$$S = \begin{bmatrix} 0 & 1 & 0 \\ 0 & 0 & 1 \\ 1 & 0 & 0 \end{bmatrix} \qquad D = \begin{bmatrix} 0 & 0 & 1 \\ 1 & 0 & 0 \\ 0 & 1 & 0 \end{bmatrix}$$

Let the t-vector be given by

$$t = [t_1, t_2, t_3].$$

Then from diagram 1, (the tree for a man to marry his
father's brother's daughter) we have

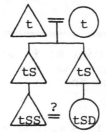

For a man to marry a woman the two people must be of
the same marriage type. Thus tSS would have to equal tSD,
or equivalently SS = SD.

$$\text{But, } SS = \begin{bmatrix} 0 & 1 & 0 \\ 0 & 0 & 1 \\ 1 & 0 & 0 \end{bmatrix} \begin{bmatrix} 0 & 1 & 0 \\ 0 & 0 & 1 \\ 1 & 0 & 0 \end{bmatrix} = \begin{bmatrix} 0 & 0 & 1 \\ 1 & 0 & 0 \\ 0 & 1 & 0 \end{bmatrix}$$

$$\text{and } SD = \begin{bmatrix} 0 & 1 & 0 \\ 0 & 0 & 1 \\ 1 & 0 & 0 \end{bmatrix} \begin{bmatrix} 0 & 0 & 1 \\ 1 & 0 & 0 \\ 0 & 1 & 0 \end{bmatrix} = \begin{bmatrix} 1 & 0 & 0 \\ 0 & 1 & 0 \\ 0 & 0 & 1 \end{bmatrix}$$

SS $\neq$ SD.   Therefore, we have verified that a man cannot
marry his father's brother's daughter.   In the second case,
the family tree for a man to marry his mother's brother's
daughter is:

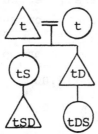

$$SD = \begin{bmatrix} 1 & 0 & 0 \\ 0 & 1 & 0 \\ 0 & 0 & 1 \end{bmatrix} \quad \text{and}$$

$$DS = \begin{bmatrix} 0 & 0 & 1 \\ 1 & 0 & 0 \\ 0 & 1 & 0 \end{bmatrix} \begin{bmatrix} 0 & 1 & 0 \\ 0 & 0 & 1 \\ 1 & 0 & 0 \end{bmatrix} = \begin{bmatrix} 1 & 0 & 0 \\ 0 & 1 & 0 \\ 0 & 0 & 1 \end{bmatrix}$$

Since SD = DS, then a man could marry his mother's brother's daughter in this society.

APPLICATIONS TO BIOLOGICAL SCIENCE

● **PROBLEM** 24-11

Three species of bacteria coexist in a test tube and feed on three resources. Suppose that a bacterium of the ith species consumes on average an amount c_{ij} of the jth resource per day.

Define $c_i = [c_{i1}, c_{i2}, c_{i3}]$ to be the consumption vector for the ith species. Suppose that $c_1 = [1, 1, 1]$, $c_2 = [1, 2, 3]$ and $c_3 = [1, 3, 5]$ and suppose that there are 15,000 units of the first resource supplied each day to the test tube, 30,000 units of the second resource and 45,000 units of the third resource.

Assuming that all the resources are consumed, what are the populations of the three species that can coexist in this environment?

Solution: Let x_1, x_2 and x_3 be the populations of the three species that can be supported by the resources. The x_1 individuals of the first species consume x_1 units of each resource, as given by the consumption vector $c_1 = [c_{11}, c_{12}, c_{13}]$. The x_2 bacteria of the second species consume x_2, $2x_2$ and $3x_2$ units of the first, second and third resources, respectively. The corresponding consumptions of the third species are x_3, $3x_3$ and $5x_3$ units. Equating the total consumption of each resource to the amount available, obtains

$$x_1 + x_2 + x_3 = 15,000$$
$$x_1 + 2x_2 + 3x_3 = 30,000 \qquad (1)$$
$$x_1 + 3x_2 + 5x_3 = 45,000.$$

Rewriting (1):

$$\begin{bmatrix} 1 & 1 & 1 \\ 1 & 2 & 3 \\ 1 & 3 & 5 \end{bmatrix} \begin{bmatrix} x_1 \\ x_2 \\ x_3 \end{bmatrix} = \begin{bmatrix} 15,000 \\ 30,000 \\ 45,000 \end{bmatrix} \qquad (2)$$

Apply elementary row operations on the augmented co-efficient matrix:

$$\left[\begin{array}{ccc:c} 1 & 1 & 1 & 15,000 \\ 1 & 2 & 3 & 30,000 \\ 1 & 3 & 5 & 45,000 \end{array}\right] \leftrightarrow \left[\begin{array}{ccc:c} 1 & 1 & 1 & 15,000 \\ 0 & 1 & 2 & 15,000 \\ 0 & 2 & 4 & 30,000 \end{array}\right]$$

$$\leftrightarrow \left[\begin{array}{ccc:c} 1 & 1 & 1 & 15,000 \\ 0 & 1 & 2 & 15,000 \\ 0 & 0 & 0 & 0 \end{array}\right]$$

Since the system does not have a unique solution, the reduced system of equations is:

$$x_1 + x_2 + x_3 = 15,000$$
$$x_2 + 2x_3 = 15,000. \qquad (2)$$

Examining (2), observe that

$$x_3 = \tfrac{1}{2} (15,000 - x_2)$$

from the second equation. Substituting into the first equation:

$$x_1 + x_2 + \tfrac{1}{2} (15,000 - x_2) = 15,000$$

$$x_1 + \tfrac{1}{2} x_2 + \frac{15,000}{2} = 15,000$$

$$x_1 = 15,000 - \frac{15,000}{2} - \tfrac{1}{2} x_2$$

$$= \tfrac{1}{2} (15,000 - x_2).$$

Thus $x_1 = x_3 = \tfrac{1}{2} (15,000 - x_2)$.

Since the number of bacteria must be nonnegative, $0 \leq x_2 \leq 15,000$, and $0 \leq x_1 = x_3 \leq 7500$. The total population that can coexist is 15,000 and the populations of the three species satisfy $x_1 = x_3$ and $x_2 = 15,000 - 2x_1$ if all resources are consumed.

● PROBLEM 24-12

Construct a matrix model to determine the quantities of pesticides that pass through crops to herbivorous animals, then to carnivorous animals and finally to humans.

Solution: Let there be m types of pesticides $(P_i \quad i = 1, \dots, m)$ sprayed on n types of plants, $(v_j \quad j = 1, \dots, n)$. Then the amount of pesticide i in the average plant of species j is given by the coefficient matrix:

$$A = \begin{bmatrix} a_{11} & a_{12} & \cdots\cdots & a_{1n} \\ a_{21} & a_{22} & \cdots\cdots & a_{2n} \\ \vdots & & & \\ a_{m1} & a_{m2} & \cdots\cdots & a_{mn} \end{bmatrix} \qquad (1)$$

Next, let there be p species of herbivores. Then

$$B = \begin{bmatrix} b_{11} & b_{12} & \cdots\cdots & b_{1p} \\ b_{21} & b_{22} & \cdots\cdots & b_{2p} \\ \vdots & & & \\ b_{n1} & b_{n2} & \cdots\cdots & b_{np} \end{bmatrix} \qquad (2)$$

where b_{ij} is the number of plants of species i that a herbivore of species j eats per year. If we multiply (1) and (2) then the i, jth entry of AB, $a_{i1} b_{1j} + a_{i2} b_{2j} + \cdots + a_{in} b_{nj}$ gives the total amount of pesticide of type i that a herbivore of species j consumes per year.

To analyze the next link in the food chain, consider q species of carnivores. Let

926

$$
C = \begin{bmatrix} c_{11} & c_{12} & \cdots\cdots & c_{1q} \\[2mm] c_{21} & c_{22} & \cdots\cdots & c_{2q} \\[1mm] \vdots & & & \\[1mm] c_{p1} & c_{p2} & \cdots\cdots & c_{pq} \end{bmatrix}
$$

where c_{ij} is the number of herbivores of species i that are eaten per year by a carnivore of species j. If the carnivores eat only the herbivores of species 1, 2, ..., p, then the i, j entry of the product ABC gives the total amount of pesticide i that a carnivore of species j consumes per year.

The final link in the chain is the number of carnivores of species i that an average human consumes per year. Let the column vector

$$
D = \begin{bmatrix} d_1 \\[2mm] d_2 \\[1mm] \vdots \\[1mm] d_q \end{bmatrix}
$$

denote this n-tuple of numbers.

The product ABCD

$$
= \begin{bmatrix} e_1 \\[2mm] e_2 \\[1mm] \vdots \\[1mm] e_m \end{bmatrix}
$$

gives, for the entry e_i, the amount of pesticide i that an average person consumes per year by virtue of eating carnivorous animals.

Since most animals eaten by humans are herbivores let us analyze the product ABD.

Let D be the column vector,

$$D = \begin{bmatrix} d_1 \\ d_2 \\ \vdots \\ dp \end{bmatrix}, \text{(where } d_i \text{ is the number of herbivores of species } j \text{ that an average human consumes per year)}$$

Thus, ABD gives the amount of pesticide i that an average person consumes per year by virtue of eating herbivorous animals.

● PROBLEM 24-13

Suppose that three persons have contracted a contagious disease. A second group of six persons is questioned to determine who has been in contact with the three infected persons. A third group of seven persons is then questioned to determine contacts with any of the six persons in the second group. Show how matrix analysis may be applied to find the number of people who may be infected.

Solution: The first order contact to a contagious disease is given by a 3x6 matrix of 1's and 0's. A 1 in the ith row and jth column indicates contact while a 0 indicates the reverse. For example, we could have

$$A = \begin{bmatrix} 0 & 0 & 1 & 0 & 1 & 0 \\ 1 & 0 & 0 & 1 & 0 & 0 \\ 0 & 0 & 1 & 1 & 0 & 1 \end{bmatrix} \qquad (1)$$

In this case, $a_{24} = 1$ which means that the fourth person in the second group came into contact with the second person in the group that originally contracted the disease.

To represent the contacts between people in the second and third groups a 6x7 matrix is required. In this matrix, $B = (b_{ij})$, $b_{ij} = 1$ if the jth person in the third group has had contact with the ith person in the second group and $b_{ij} = 0$ otherwise.

As an instance, let

928

$$B = \begin{bmatrix} 0 & 0 & 1 & 0 & 0 & 1 & 0 \\ 0 & 0 & 1 & 1 & 0 & 0 & 0 \\ 1 & 0 & 0 & 0 & 0 & 1 & 1 \\ 0 & 0 & 1 & 1 & 0 & 0 & 0 \\ 0 & 1 & 0 & 1 & 0 & 0 & 0 \\ 1 & 0 & 0 & 0 & 0 & 1 & 0 \end{bmatrix} \qquad (2)$$

The ith row of B represents the contacts or non-contacts between the ith person in the second group and the members of the third group. Here, $b_{33} = 0$ is interpreted to mean that the third person in the third group has not had contact with the third person in the second group.

Armed with (1) and (2) we can study the indirect or second order contacts between the seven persons in the third group and the three infected people. The matrix product C = AB gives the number of second order contacts between the jth person in the third group and the ith person in the infected group. Thus,

$$C = AB = \begin{bmatrix} 1 & 1 & 0 & 1 & 0 & 1 & 1 \\ 0 & 0 & 2 & 1 & 0 & 1 & 0 \\ 2 & 0 & 1 & 1 & 0 & 2 & 0 \end{bmatrix}$$

The component $c_{23} = 2$ implies that there are two second order contacts between the third person in the third group and the second contagious person.

Note that the sixth person in the third group has had $1 + 1 + 2 = 4$ indirect contacts with the infected group. Only the fifth person has had no contacts.

● **PROBLEM** 24-14

According to a psychological study, the personality traits emotional stability and sociability are related. Specifically, the study indicates that the relationship between these two traits can be represented by the matrix S = $[x_{ij}]$ where:

	High sociability	Average sociability	Low sociability
High emotional stability	0.5	0.3	0.2
Average emotional stability	0.4	0.3	0.3
Low emotional stability	0.1	0.4	0.5

= S,

and for i = 1, 2, 3 and j = 1, 2, 3, x_{ij} represents the pro-
bility that a person with rating i in emotional stability
will have rating j in sociability. In a group of 1000
randomly selected individuals, 300 are rated high in emotional stability, 500 are rated average in emotional stability,
and 200 are rated low in emotional stability. If the entries
in S are correct, how many of these individuals can be expected to fall into each of the sociability rating categories?

Solution: Let E = [300   500   200] represent
$$\begin{array}{cccc} & \text{High ES} & \text{Average ES} & \text{Low ES} \end{array}$$

the distribution of the 1000 individuals among the three
categories for emotional stability. To find the number of
individuals expected to be in each of the sociability rating
categories, simply multiply E times S. Thus, since

$$ES = [300 \quad 500 \quad 200] \begin{bmatrix} .5 & .3 & .2 \\ .4 & .3 & .3 \\ .1 & .4 & .5 \end{bmatrix} = [370 \quad 320 \quad 310],$$

Expect that out of the 1000 randomly selected individuals, 370 will have a high sociability rating, 320 will have
an average sociability rating, and 310 will have a low sociability rating.

● **PROBLEM** 24-15

Markov chains may be applied to genetics. According to the
Mendelian theory of genetics, many traits of an offspring
are determined by the genes of the parents.

Suppose the genotype of one parent is unknown and the genotype of the other parent is hybrid (heterozygous). Their
offspring is mated with a person whose genotype is hybrid.
This mating procedure is continued. In the long run, what
is the genotype of the offspring?

Solution: A given individual is either dominant, with genotype AA, hybrid with genotype Aa or recessive with genotype
aa.

An offspring inherits one gene from each parent in its
own genotype. For example if both parents are dominant, the
offspring will also be dominant

Parents   AA   AA
Offspring   A   A

We know one parent is hybrid, i.e., Aa. Hence the
various possibilities are

Aa   AA   Aa   Aa   Aa   aa

$$P(A) = \frac{1}{2} \quad\quad\quad P(A) = \frac{1}{2} \quad P(A) = \frac{1}{2} \quad\quad\quad P(A) = \frac{1}{2}$$

$$\quad\quad\quad\quad\quad A \quad\quad\quad\quad\quad\quad\quad\quad\quad\quad\quad\quad\quad\quad\quad\quad\quad a.$$

$$P(a) = \frac{1}{2} \quad\quad\quad P(a) = \frac{1}{2} \quad P(a) = \frac{1}{2} \quad\quad\quad P(a) = \frac{1}{2}$$

$$P(AA) = \frac{1}{2} \quad\quad\quad P(AA) = P(aa) = \frac{1}{4} \quad\quad\quad P(Aa) = \frac{1}{2}$$

$$P(Aa) = \frac{1}{2} \quad\quad\quad\quad\quad\quad P(Aa) = \frac{1}{2} \quad\quad\quad\quad\quad\quad P(aa) = \frac{1}{2}$$

Thus if the unknown parent is dominant, the probability of the offspring being dominant or hybrid is 1/2 respectively. If the unknown parent is recessive, the offspring will be either recessive or hybrid with probability 1/2. Finally, if the parent is hybrid the offspring will be dominant or recessive with probability 1/4 respectively or hybrid with probability 1/2. The transition matrix is therefore

$$
P = \quad
\begin{array}{c|ccc}
 & D & H & R \\
\hline
D & \frac{1}{2} & \frac{1}{2} & 0 \\
H & \frac{1}{4} & \frac{1}{2} & \frac{1}{4} \\
R & 0 & \frac{1}{2} & \frac{1}{2}
\end{array}
$$

Since $P^2 =$

$$
\begin{bmatrix}
\frac{3}{8} & \frac{1}{2} & \frac{1}{8} \\
\frac{1}{4} & \frac{1}{2} & \frac{1}{4} \\
\frac{1}{8} & \frac{1}{2} & \frac{3}{8}
\end{bmatrix}
$$

the transition matrix P is regular. Thus we know that there exists a fixed probability vector t which all the rows of P^n approach as n increases.

Thus
$$
[t_1 \ t_2 \ t_3]
\begin{bmatrix}
\frac{3}{8} & \frac{1}{2} & \frac{1}{8} \\
\frac{1}{4} & \frac{1}{2} & \frac{1}{4} \\
\frac{1}{8} & \frac{1}{2} & \frac{3}{8}
\end{bmatrix}
= [t_1 \ t_2 \ t_3]
$$

Solving this system yields

$$
t = \begin{bmatrix} \frac{1}{4} & \frac{1}{2} & \frac{1}{4} \end{bmatrix}.
$$

In the long run, no matter what the genotype of the unknown parent, the probabilities for the genotype of the offspring to be dominant is 1/4, to be hybrid is 1/2 and to be
recessive is 1/4.

• PROBLEM 24-16

In the brother-sister mating problem, two individuals are
mated, and from among their direct descendants two individuals of opposite sex are selected at random.   These are
mated, and the process continues indefinitely.

a)   What is the transition matrix for this experiment?

b)   What are the absorbing states?

Solution: The three possible genotypes for the parents are
AA, Aa and aa.   The offspring can then be of the types
E_1: AA x AA E_2: AA x Aa E_3: Aa x Aa E_4:   Aa x aa

E_5: aa x aa E_6: AA x aa. For example, E_4:   Aa x aa

indicates the mating of a hybrid (Aa) with a recessive (aa).
The transition matrix for this experiment is:

	E_1	E_2	E_3	E_4	E_5	E_6
E_1	1	0	0	0	0	0
E_2	$\frac{1}{4}$	$\frac{1}{2}$	$\frac{1}{4}$	0	0	0
E_3	$\frac{1}{16}$	$\frac{1}{4}$	$\frac{1}{4}$	$\frac{1}{4}$	$\frac{1}{16}$	$\frac{1}{8}$
E_4	0	0	$\frac{1}{4}$	$\frac{1}{2}$	$\frac{1}{4}$	0
E_5	0	0	0	0	1	0
E_6	0	0	1	0	0	0

The states E_1 and E_5 are absorbing states. E_6 is not
an absorbing state (there is always a transition from E_6 to
E_3).

To find the expected number of generations needed to
pass from a nonabsorbing state to either absorbing state,
first compute the fundamental matrix associated with the
transition matrix.

Rewrite the transition matrix so that the absorbing
states are in the first two rows.

932

	E_1	E_5	E_2	E_3	E_4	E_6
E_1	1	0	0	0	0	0
E_5	0	1	0	0	0	0
E_2	$\frac{1}{4}$	0	$\frac{1}{2}$	$\frac{1}{4}$	0	0
E_3	$\frac{1}{16}$	$\frac{1}{16}$	$\frac{1}{4}$	$\frac{1}{4}$	$\frac{1}{4}$	$\frac{1}{8}$
E_4	0	$\frac{1}{4}$	0	$\frac{1}{4}$	$\frac{1}{2}$	0
E_6	0	0	0	1	0	0

This matrix is now in the form

$$\begin{bmatrix} I_2 & 0 \\ S & Q \end{bmatrix}$$

and the fundamental matrix is

$$T = [I_4 - Q]^{-1}$$

		E_2	E_3	E_4	E_6
	E_2	$\frac{8}{3}$	$\frac{4}{3}$	$\frac{2}{3}$	$\frac{1}{6}$
=	E_3	$\frac{4}{3}$	$\frac{8}{3}$	$\frac{4}{3}$	$\frac{1}{3}$
	E_4	$\frac{2}{3}$	$\frac{4}{3}$	$\frac{8}{3}$	$\frac{1}{6}$
	E_6	$\frac{4}{3}$	$\frac{8}{3}$	$\frac{4}{3}$	$\frac{4}{3}$

The product of the fundamental matrix and S is:

933

$$T \cdot S = \begin{bmatrix} \frac{8}{3} & \frac{4}{3} & \frac{2}{3} & \frac{1}{6} \\ \frac{4}{3} & \frac{8}{3} & \frac{4}{3} & \frac{1}{3} \\ \frac{2}{3} & \frac{4}{3} & \frac{8}{3} & \frac{1}{6} \\ \frac{4}{3} & \frac{8}{3} & \frac{4}{3} & \frac{4}{3} \end{bmatrix} \begin{bmatrix} \frac{1}{4} & 0 \\ \frac{1}{16} & \frac{1}{16} \\ 0 & \frac{1}{4} \\ 0 & 0 \end{bmatrix} = \begin{matrix} & E_1 & E_5 \\ E_2 & \frac{3}{4} & \frac{1}{4} \\ E_3 & \frac{1}{2} & \frac{1}{2} \\ E_4 & \frac{1}{4} & \frac{3}{4} \\ E_6 & \frac{1}{2} & \frac{1}{2} \end{matrix}$$

Genetically, $T \cdot S$ can be interpreted to say that after a large number of inbred matings, a person is either in state E_1 or state E_5.

BUSINESS APPLICATIONS

● PROBLEM 24-17

There are three dairies in a community which supply all the milk consumed: Abbot's Dairy, Branch Dairy Products Company, and Calhoun's Milk Products, Inc. For simplicity, let's refer to them hereafter as A, B, and C. Each of the dairies knows that consumers switch from dairy-to-dairy over time because of advertising, dissatisfaction with service, and other reasons. To further simplify the mathematics necessary, let's assume that no new customers enter and no old customers leave the market during this period.

Consider now the following data on the flow of customers for each of the dairy companies as determined by their respective Operations Research Departments:

Flow of Customers

June 1		Gains			Losses			July 1
		From	From	From	To	To	To	
Dairy	Customers	A	B	C	A	B	C	Customers
A	200	0	35	25	0	20	20	220
B	500	20	0	20	35	0	15	490
C	300	20	15	0	25	20	0	290

Given the above data as well as all the assumptions made, determine

a) The one-step transition probabilities

b) Interpret the rows and the columns of the matrix of transition probabilities

Solution: This is an illustration of the use of Markov
chain methods in broad-switching analysis. The first step
would involve the computation of transition probabilities
for all three of our dairies which are nothing more than
the probabilities that a certain seller C, a dairy, in this
case, will retain, gain, and lose customers. In other words,
dairy B observes from the above table that it loses 50 customers this month; this is the same as saying that it has a
probability of .9 of retaining customers; similarly, dairy
A has a probability of .8 of retaining its customers; dairy
C has a probability of .85 of retaining its customers. These
transition probabilities for the retention of customers are
calculated in the following table.

Transition Probabilities for Retention of Customers

Dairy	June 1 customers	Number lost	Number retained	Probability of retention
A	200	40	160	160/200 = .8
B	500	50	450	450/500 = .9
C	300	45	255	255/300 = .85

In the matrix of transition probabilities in this problem, We will include for each dairy the retention probability
and the probability of its loss of customers to its two com-
petitors. The rows in this matrix show the retention of
customers and the loss of customers; the columns represent
the retention of customers and the gain of customers. These
probabilities have been calculated to three decimal places.

Retention and Loss
—————————————————————→

	A	B	C	
A	.800	.100	.100	Retention and Gain
B	.070	.900	.030	
C	.083	.067	.850	

Below is a matrix of the same dimensions as the one
above illustrating exactly how each probability was determined:

	A	B	C
A	160/200 = .800	20/200 = .100	20/200 = .100
B	35/500 = .070	450/500 = .900	15/500 = .030
C	25/300 = .083	20/300 = .067	255/300 = .850

The rows of the matrix of transition probabilities can
be read as follows:

935

Row 1 indicates that dairy A retains .8 of its customers (160), loses .1 of its customers (20) to dairy B, and loses .1 of its customers (20) to dairy C.

Row 2 indicates that dairy B retains .9 of its customers (450), loses .07 of its customers (35) to dairy A, and loses .03 of its customers (15) to dairy C.

Row 3 indicates that diary C retains .85 of its customers (255), loses .083 of its customers (25) to dairy A, and loses .067 of its customers (20) to dairy B.

Reading the columns yields the following interpretation:

Column 1 indicates that dairy A retains .8 of its customers (160), gains .07 of B's customers (35), and gains .083 of C's customers (25).

Column 2 indicates that dairy B retains .9 of its customers (450), gains .1 of A's customers (20), and gains .067 of C's customers (20).

Column 3 indicates that diary C retains .85 of its customers (255), gains .1 of A's customers (20), and gains .03 of B's customers (15).

● PROBLEM 24-18

Consider again the following original matrix of transition probabilities in the three-dairy problem:

$$
\begin{array}{c c c c}
 & A & B & C \\
A & .800 & .100 & .100 \\
B & .070 & .900 & .030 \\
C & .083 & .067 & .850 \\
\end{array}
$$

a) Find out what the final or equilibrium shares of the market will be

b) Prove that an equilibrium has actually been reached in part (a).

Solution: a) First of all, an equilibrium has been defined as a position that a Markov process reaches in the long run after which no further net change occurs. This condition can only be reached if no dairy takes action which alters the matrix of transition probabilities. From a marketing point of view, we would want the answer to the question: "What would the three final or equilibrium shares of the market be?"

To determine A's share of the market in the equilibrium period (labelling this unspecified future period the eq. period), we use the following relationship:

.800 times the share A had in the (eq. - 1) period (i.e.,
the period immediately preceding equilibrium)

+ .070 times the share B had in the (eq. - 1) period

+ .083 times the share C had in the (eq. - 1) period.

Writing this relationship as an equation, we have

$$A_{eq} = .800\ A_{eq-1} + .070\ B_{eq-1} + .083\ C_{eq-1}$$

The following two equations are B's and C's shares of
the equilibrium market.

$$B_{eq} = .100\ A_{eq-1} + .900\ B_{eq-1} + .067\ C_{eq-1}$$

$$C_{eq} = .100\ A_{eq-1} + .030\ B_{eq-1} + .850\ C_{eq-1}$$

Rewrite the three equations as follows:

$$A = .800\ A + .070\ B + .083\ C$$
$$B = .100\ A + .900\ B + .067\ C$$
$$C = .100\ A + .030\ B + .850\ C$$

Because the sum of the three market shares equals 1.0,
we can add another equation

$$1.0 = A + B + C.$$

Solving the equations simultaneously for the equilibrium
market shares, yields

$$A = .273$$
$$B = .454$$
$$C = .273$$

b) To prove that an equilibrium has been reached, multiply
the equilibrium market share (A, .273; B, .454; C, .273)
by the matrix of transition probabilities:

$$(.273 \quad .454 \quad .273) \times \begin{pmatrix} .800 & .100 & .100 \\ .070 & .900 & .030 \\ .083 & .067 & .850 \end{pmatrix}$$

$$= (.272 \quad .454 \quad .273)$$

Therefore, an equilibrium market condition has been reached.

To recapitulate on the equilibrium market shares calculated, it is important to bear in mind that they are based
upon the assumption that the matrix of transition probabilities remains fixed, and the propensities of all three dairies
to retain, gain, and lose customers do not change over time.

In many circumstances, those assumptions may be somewhat
invalid, but the general method remains the same. For the
period during which the transition probabilities are stable,
we can calculate the equilibrium which will result. However,
if there is good reason to believe that the transition probabilities are indeed changing because of some action by
management, use the new transition probabilities and calculate the equilibrium market shares which will result. In
that manner, we are essentially using Markov analysis as a
short-or-intermediate-run tool.

● **PROBLEM** 24-19

Refer to the three dairies of the previous problem and assume that the matrix of transition probabilities remains
fairly stable and that the July 1 market shares are these:
A = 22 percent, B = 49 percent, C = 29 percent. Calculate:

a)   the total market likely to be held by each of the
     dairies by August 1

b)   the probable market share by September 1

c)   the probable market shares which will occur after
     the third and sixth periods.

Solution: a) To calculate the probable shares of the total
market likely to be held by each of the dairies on August 1,
we would simply set up the July 1 market shares as a matrix
and multiply this matrix by the matrix of transition probabilities as follows:

July 1 market shares	Transition probabilities	Probable August 1 market shares

$$(.22 \quad .49 \quad .29) \times \begin{pmatrix} .800 & .100 & .100 \\ .070 & .900 & .030 \\ .083 & .067 & .850 \end{pmatrix} = (.234 \quad .483 \quad .283)$$

Total = 1.00   Total = 1.000

The matrix multiplication is explained in detail below.

Row 1 x column 1:

A's share of market x A's propensity to retain its customers
= .22 x .800 = .176

B's share of market x A's propensity to attract B's customers
= .49 x .070 = .034

C's share of market x A's propensity to attract C's customers
= .29 x .083 = .024

A's share of market on August 1 = .176 + .034 + .024 = .234

938

Row 1 x column 2:

A's share of market x B's propensity to attract A's customers
= .22 x .100 = .022

B's share of market x B's propensity to retain its customers
= .49 x .900 = .441

C's share of market x B's propensity to attract C's customers
= .29 x .067 = .020

B's share of market on August 1 = .022 + .441 + .020 = .483

Row 1 x column 3:

A's share of market x C's propensity to attract A's customers
= .22 x .100 = .022

B's share of market x C's propensity to attract B's customers
= .49 x .030 = .015

C's share of market x C's propensity to retain its customers
= .29 x .850 = .246

C's share of market on August 1 = .283

b)   The probable market share on September 1 can also be
calculated by squaring the matrix of transition probabilities
and multiplying the squared matrix by the July 1 market share:

$$\text{Method 1:} \quad (.22 \quad .49 \quad .29) \ \text{x} \begin{pmatrix} .800 & .100 & .100 \\ .070 & .900 & .030 \\ .083 & .067 & .850 \end{pmatrix}^2$$

= probable Sept. 1 market shares

or by multiplying the matrix of transition probabilities by
the market shares on August 1:

$$\text{Method 2:} \quad (.234 \quad .483 \quad .283) \ \text{x} \begin{pmatrix} .800 & .100 & .100 \\ .070 & .900 & .030 \\ .083 & .067 & .850 \end{pmatrix}$$

= probable Sept. 1 market shares

Method 1

The logic behind method 1 can be explained this way. By
squaring the original matrix of transition probabilities,
the probabilities of retention are calculated. Gain, and
loss can be multiplied by the original market shares (22,
49, and 29 percent) to yield the market shares which will
be obtained on September 1. As an example, to obtain the
column 1 - row 1 term X in the product, multiply row 1 by

column 1:

$$(.800 \quad .100 \quad .100) \times \begin{pmatrix} .800 \\ .070 \\ .083 \end{pmatrix} = (X)$$

Row 1 x column 1:

A's propensity to retain its own customers times A's propensity to retain its own customers equals that proportion of its original customers it holds for both periods

$$.8 \times .8 = .64$$

B's propensity to gain customers from A times A's propensity to gain customers from B equals A's regain of its customers from B

$$.1 \times .07 = .007$$

C's propensity to gain customers from A times A's propensity to gain customers from C equals A's regain of its own customers from C

$$.1 \times .083 = .0083$$

We get the x term in the product by adding together the results of the three calculations:

$$\begin{array}{r} .6400 \\ .0070 \\ \underline{.0083} \\ .6553 \end{array} = \text{portion of A's original customers A retains on September 1}$$

Similarly, the other eight terms in the square of the matrix can be explained and calculated. The resulting matrix for use in method 1 is

$$\begin{pmatrix} .6553 & .1767 & .1680 \\ .1215 & .8190 & .0595 \\ .1416 & .1256 & .7328 \end{pmatrix}$$

To complete method 1, we multiply the squared matrix by the July 1 market shares:

$$(.22 \quad .49 \quad .29) \times \begin{pmatrix} .6553 & .1767 & .1680 \\ .1215 & .8190 & .0595 \\ .1416 & .1256 & .7328 \end{pmatrix}$$

with the result

A   .245

B   .477   probable market shares on September 1

C   .278

Total 1.000

    For clarification the following computations and explanations are presented:

$$(.22 \quad .49 \quad .29) \times \begin{pmatrix} .6553 \\ .1215 \\ .1416 \end{pmatrix} = (.245)$$

A's original market share times A's propensity to retain its own customers after two periods equals A's share of its original customers on September 1

$$.22 \times .6553 = .144$$

B's original market share times A's propensity to gain B's original customers after two periods equals A's share of B's original customers on September 1

$$.49 \times .1215 = .060$$

C's original market share times A's propensity to gain C's original customers after two periods equals A's share of C's original customers on September 1

$$.29 \times .1416 = .041.$$

    Summing the three computations we have:

    $.144 + .060 + .041 = .245 =$ A's probable market share on September 1

Method 2

Multiplication of the original matrix of transition probabilities by the August 1 market share yields the same result as method 1. Reproducing the two matrices and explaining one of the multiplications, we have as follows:

$$(.234 \quad .483 \quad .283) \times \begin{pmatrix} .800 & .100 & .100 \\ .070 & .900 & .030 \\ .083 & .067 & .850 \end{pmatrix}$$

Row 1 x column 1:

A's share of the market at the end of the last period times A's propensity to retain its own customers equals A's retained share of its own customers it had at the end of the last period

$$.234 \times .800 = .187$$

B's share of the market at the end of the last period times A's propensity to gain customers from B equals A's gain of the customers B had at the end of the last period

$$.483 \times .070 = .034$$

C's share of the market at the end of the last period times A's propensity to gain customers from C equals A's gain of the customers C had at the end of the last period

$$.283 \times .083 = .024$$

$.187 + .034 + .024 = .245 = $ A's probable share of market on September 1.

c) Market shares after three periods:

July 1
market shares

Matrix of transition
probabilities cubed

$$(.22 \quad .49 \quad .29) \times \begin{pmatrix} .800 & .100 & .100 \\ .070 & .900 & .030 \\ .083 & .067 & .850 \end{pmatrix}^3$$

= probable market shares on October 1

and the market shares after six periods would be:

July 1
market shares

Matrix of transition probabilities to the sixth power

$$(.22 \quad .49 \quad .29) \times \begin{pmatrix} .800 & .100 & .100 \\ .070 & .900 & .030 \\ .083 & .067 & .850 \end{pmatrix}^6$$

= probable market shares on January 1

To summarize the uses of the two alternative methods of computing market shares for future periods, yields the following:

Method 1 would be employed if we simply wanted the market shares for the specified future period, while we would choose method 2 if we wanted to observe the changes which were occurring in the market shares during all the intervening periods.

● PROBLEM 24-20

There are three types of grocery stores in a given community. Within this community (with a fixed population) there always exists a shift of customers from one grocery store to another.

On January 1, 1/4 shopped at store I, 1/3 at store II and 5/12 at store III. Each month store I retains 90% of its customers and loses 10% of them to store II. Store II retains 5% of its customers and loses 85% of them the store I and 10% of them to store III. Store III retains 40% of its customers and loses 50% of them to store I and 10% to store II.

a) Find the transition matrix.

b) What proportion of customers will each store retain by Feb. 1 and Mar. 1?

c) Assuming the same pattern continues, what will be the long-run distribution of customers among the three stores?

Solution: a) The transition matrix will represent the probabilities of customers changing stores or remaining at the same store. Thus,

$$
\begin{array}{cccc}
 & I & II & III \\
I & .90 & .10 & 0 \\
P = \quad II & .85 & .05 & .10 \\
III & .50 & .10 & .40
\end{array}
$$

b) The initial distribution is $A^{(0)} = [.25, .33, .42]$. Thus, by Feb. 1, the percentage of customers shopping at the three stores will be

$$
A^{(0)}p = [.25 \quad .33 \quad .42]
\begin{bmatrix}
.90 & .10 & 0 \\
.85 & .05 & .10 \\
.50 & .10 & .40
\end{bmatrix}
$$

$$
= [.7166 \quad .0832 \quad .1999]
$$

To find the probability distribution after two months, note that $A^{(2)} = A^{(0)}p^2$

$$
p^2 =
\begin{bmatrix}
.90 & .10 & 0 \\
.85 & .05 & .10 \\
.50 & .10 & .40
\end{bmatrix}
\begin{bmatrix}
.90 & .10 & 0 \\
.85 & .05 & .10 \\
.50 & .10 & .40
\end{bmatrix}
$$

$$
=
\begin{bmatrix}
.895 & .095 & .010 \\
.857 & .098 & .045 \\
.735 & .095 & .170
\end{bmatrix}.
$$

Interpret p^2 as follows. The first element is composed

of the sum (.90)(.90) + (.10)(.85) + 0(.50).(.90)(.90) is
the probability of people who shop at store I continuing to
shop at shop I.   (.10)(.85) represents the probability of
people switching from store I to store II and then back to
store I.   (0)(.50) is the probability of people switching
from store I to store IIIand then back to store I.   The sum
is therefore the probability of people who were shopping at
store I winding up shopping there after two months.

$$A^{(2)} = A^{(0)}p^2 = [.25 \quad .33 \quad .42] \begin{bmatrix} .895 & .095 & .010 \\ .857 & .098 & .045 \\ .735 & .095 & .170 \end{bmatrix}$$

$$= [.8155 \quad .0956 \quad .0882].$$

c)   To determine the long-run shopping behaviour of this
community the following facts are needed:

A transition matrix, P, is regular if for some power
of p, all the entries are positive.   If P is a regular
matrix, then all the rows of P^n will be identically equal
to some probability vector t for all $m \geq n$.

In the given problem, p^2 contains positive elements
only.   Hence p is regular, i.e., there exists a fixed probability vector $[t_1 \quad t_2 \quad t_3]$ where $t_1 + t_2 + t_3 = 1$ such that

$$[t_1 \quad t_2 \quad t_3] \begin{bmatrix} .9 & .10 & .00 \\ .85 & .05 & .10 \\ .50 & .10 & .40 \end{bmatrix} = [t_1 \quad t_2 \quad t_3].$$

We obtain the set of linear equations

$$.9t_1 + .85t_2 + .50t_3 = t_1$$

$$.10t_1 + .05t_2 + .10t_3 = t_2$$

$$.00t_1 + .10t_2 + .40t_3 = t_3$$

The solution vector is

$$[t_1 \quad t_2 \quad t_3] = [.8888 \quad .0952 \quad .0158]$$

In the long run store I will have about 89% of all
customers, store II 9.5% and store III 1.5%.

● **PROBLEM 24-21**

Union Industries, a manufacturer of ladies' sleepwear classifies its sewing operators into four categories depending
upon their productivity during the preceding month; the

lowest category is 1 and the highest is 4. Historically, the sewing work force has been distributed across the four categories as follows:

$$1 = 30\%$$
$$2 = 35\%$$
$$3 = 25\%$$
$$4 = 10\%$$

A year ago, Union has introduced a new organizational system into its Idaho plant, one of its largest units, with 450 operators. The new system in effect groups the operators into voluntary work units which not only elect their own supervisors but also determine their own work schedules. Production records compiled since the new plan was adopted have enabled Mr. John Hayward, Plant Manager, to construct the following matrix of transition probabilities illustrating month-to-month changes in employee productivity:

		Lowest			Highest
		1	2	3	4
Lowest	1	.5	.3	.2	0
	2	.3	.4	.3	0
	3	.1	.2	.2	.5
Highest	4	.1	.1	.1	.7

It is also known that operators earn an average of \$700 a month and that productivity losses for the four categories of employee are 40 percent, 25 percent, 15 percent, and 5 percent for categories 1, 2, 3, and 4 respectively.

Suppose you were hired by the management of Union Industries to evaluate the effectiveness as well as efficiency of the new system, what will be your conclusion insofar as savings in productivity losses in its Idaho plant are concerned.

Solution: This problem is an application of Markov Analysis in employee productivity. We first determine the equilibrium probabilities of Mr. Hayward's matrix by solving the following equations simultaneously.

$$x_1 = .5x_1 + .3x_2 + .1x_3 + .1x_4$$

$$x_2 = .3x_1 + .4x_2 + .2x_3 + .1x_4$$

$$x_3 = .2x_1 + .3x_2 + .2x_3 + .1x_4$$

$$x_4 = \qquad\qquad\qquad .5x_3 + .7x_4$$

$$1 = x_1 + x_2 + x_3 + x_4$$

where $x_1 = 1$ (the highest)

$\quad\quad x_4 = 4$ (the lowest)

The equilibrium probabilities are as follows:

Highest	1	.247
	2	.241
	3	.192
Lowest	4	.320
		1.000

Then set up the cost comparison of the old and new organization system

	Employee catetory	Percent of employees		Productivity loss		
Old organiza- tional system	1	30	x	40%	=	12.00%
	2	35	x	25	=	8.75
	3	25	x	15	=	3.75
	4	10	x	5	=	.50
						25.00%

25% x $700/mo. x 450 employees = $78,750

New organiza- tional system	1	24.7	x	40%	=	9.88%
	2	24.1	x	25	=	6.03
	3	19.2	x	15	=	2.88
	4	32.0	x	5	=	1.60
						20.39%

20.39% x $700/mo. x 450 employees = $64,229

Therefore, in the above cost comparison of the old and new organzation system, it appears that the new organization system has the potential to save Union over $14,000 per month in productivity losses in its Idaho plant.

● **PROBLEM 24-22**

Randolph Raleigh is the Coca-Cola dealer in Pittsburgh, Pennsylvania. His warehouse manager inspects his soft drink crates (these are the wooden crates that hold 24 bottles) each week and classifies them as "just rebuilt this week," "in good working condition," "in fair condition," or "damaged beyond use." If a crate is damaged beyond use, it is sent to the repair area, where it is usually out of use for a week. Randolph's warehouse records indicate that this is the appropriate matrix of transition probabilities for his soft drink crates:

	Rebuilt	Good	Fair	Damaged
Rebuilt	0	.8	.2	0
Good	0	.6	.4	0
Fair	0	0	.5	.5
Damaged	1.0	0	0	0

Randolph's accountant informs him that it costs $2.50 to rebuild a crate, and the company incurs a loss of $1.85 in production efficiency each time a crate is found to be damaged beyond use. This efficiency is lost because broken crates slow down the truckloading process.

a) Given the above information calculate the expected weekly cost of both rebuilding and loss of production efficiency.

b) Assuming that Randolph wants to consider rebuilding crates whenever they are inspected and found to be in fair shape, determine the new matrix of transition probabilities and the average weekly cost of rebuilding and loss of production efficiency under these circumstances.

Solution: This is an illustration of Markov analysis to equipment repair. To calculate the expected weekly cost of both rebuilding and loss of production efficiency, we will need the equilibrium probabilities of Randolph's matrix. Here are 5 equations with 4 unknowns as shown below.

$$x_1 = 0x_1 + 0x_2 + 0x_3 + x_4$$

$$x_2 = .8x_1 + .6x_2 + 0x_3 + 0x_4$$

$$x_3 = .2x_1 + .4x_2 + .5x_3 + 0x_4$$

$$x_4 = 0x_1 + 0x_2 + .5x_3 + 0x_4$$

$$1.0 = x_1 + x_2 + x_3 + x_4$$

or

$$x_1 = \qquad\qquad\qquad x_4$$

$$x_2 = .8x_1 + .6x_2$$

$$x_3 = .2x_1 + .4x_2 + .5x_3$$

$$x_4 = \qquad\qquad .5x_3$$

$$1.0 = x_1 + x_2 + x_3 + x_4$$

where x_1 = Rebuilt

x_2 = Good

x_3 = Fair

x_4 = Damaged

Solving simultaneously the above equations, we arrive at the following transition probabilities

Rebuilt $\qquad \frac{1}{6}$ = .167

Good $\qquad \frac{1}{3}$ = .333

Fair $\qquad \frac{1}{3}$ = .333

Damaged $\qquad \frac{1}{6}$ = .167
$\qquad\qquad\qquad\qquad\qquad$ 1.000

The average weekly cost of rebuilding and loss of production efficiency is then:

Rebuilding cost + Damange (out of use) loss

$\frac{1}{6}$ x $2.50 $\qquad$ + $\qquad \frac{1}{6}$ x $1.85 = $ .725 per crate per week

b) Suppose now that Randolph wants to consider rebuilding crates whenever they are inspected and found to be in fair shape. This eliminates the possibility of a crate being damaged. In this instance, the new matrix of transition probabilities would be:

	Rebuilt	Good	Fair
Rebuilt	0	.8	.2
Good	0	.6	.4
Fair	1.0	0	0

The equilibrium probabilities for this matrix are found to be

x_1 = Rebuilt $\quad \frac{1}{4}$ = .25

x_2 = Good $\qquad \frac{1}{2}$ = .50

x_3 = Fair $\qquad \frac{1}{4}$ = .25
$\qquad\qquad\qquad\qquad\qquad$ 1.00

The average weekly cost of rebuilding and loss of production efficiency under these circumstances is:

$\qquad$ Rebuilding $\qquad$ Damage
$\qquad\qquad$ Cost $\qquad +\qquad$ (out of use)
$\qquad\qquad\qquad\qquad\qquad\qquad$ loss

$\frac{1}{4}$ x $2.50 $\quad$ + $\quad$ 0 $\quad$ = $.625 per crate per week

Therefore, rebuilding crates as soon as they are found to be in "fair" shape will save Randolph a little over 10 cents per week. Since Randolph owns over six thousand crates, this is a substantial saving for him.

● **PROBLEM** 24-23

Hamilton's Clothing Cupboard is a clothing-store catering to college students. Mr. Hamilton divides his accounts receivable into two classifications: 0-60 days old and 61-180 days old. He has currently $6,500 in accounts receivable and from analysis of his past records, he has been able to provide us with the following matrix of transition probabilities (the matrix can be thought of in terms of what happens to one dollar of accounts receivable):

	Paid	Bad debt	0-60 days	61-180 days
Paid	1	0	0	0
Bad debt	0	1	0	0
0-60 days	.5	0	.3	.2
61-180 days	.4	.3	.2	.1

a) Determine the probability that a dollar of 0-60 day or 61-180 day receivables would eventually find its way into either paid bills or bad debts

b) Forecast the future of Mr. Hamilton's $6,500 of accounts receivable given that $4,500 is in the 0-60 day category and $2,000 is in the 61-180 day category.

Solution: This is an application of the Markov chain analysis in the area of accounts receivable specifically to the estimation of that portion of the accounts receivable which will eventually become uncollectible (bad debts).

First determine the four probabilities of interest to Mr. Hamilton which is done in four steps:

Step 1 - First, we partition Hamilton's original matrix of transition probabilities into four matrices, each identified by a letter:

$$\begin{pmatrix} 1 & 0 & | & 0 & 0 \\ 0 & 1 & | & 0 & 0 \\ - & - & + & - & - \\ .5 & 0 & | & .3 & .2 \\ .4 & .3 & | & .2 & .1 \end{pmatrix}$$

$$I = \begin{pmatrix} 1 & 0 \\ 0 & 1 \end{pmatrix} \qquad O = \begin{pmatrix} 0 & 0 \\ 0 & 0 \end{pmatrix}$$

$$K = \begin{pmatrix} .5 & 0 \\ .4 & .3 \end{pmatrix} \qquad\qquad M = \begin{pmatrix} .3 & .2 \\ .2 & .1 \end{pmatrix}$$

Step 2 - Subtract matrix M from matrix I to get a new matrix which we have called R.

$$\underset{I}{\begin{pmatrix} 1 & 0 \\ 0 & 1 \end{pmatrix}} - \underset{M}{\begin{pmatrix} .3 & .2 \\ .2 & .1 \end{pmatrix}} = \begin{pmatrix} .7 & -.2 \\ -.2 & .9 \end{pmatrix}$$

Step 3 - Find the inverse of matrix R which is found to be

$$\begin{pmatrix} 1.5254 & .3390 \\ .3390 & 1.1864 \end{pmatrix}$$

Step 4 - Multiply this inverse by matrix K from Step 1. This multiplication is

$$\begin{pmatrix} 1.5254 & .3390 \\ .3390 & 1.1864 \end{pmatrix} \begin{pmatrix} .5 & 0 \\ .4 & .3 \end{pmatrix} = \begin{pmatrix} .8983 & .1017 \\ .6441 & .3559 \end{pmatrix}$$

Interpreting the answer for Mr. Hamilton, we have as follows:

The top row in the answer is the probability that $1 of his accounts receivable in the 0-60 day category will end up in the "paid" and "bad debt" categories. Specifically, there is a .8983 probability that $1 currently in the 0-60 day category will be paid and a .1017 probability that it will eventually become a bad debt. Consider the second row. These two entries represent the probability that $1 now in the 61-180 day category will end up in the "paid" and the "bad debt" categories.

He observes from this row that there is a .6441 probability that $1 currently in the 61-180 day category will be paid and a .3559 probability that it will eventually become a bad debt.

b) If Mr. Hamilton would want to know the future of his $6,500 of accounts receivable given that his accountant tells him that $4,500 is in the 0-60 day category and $2,000 is in the 61-180 day category, the following matrix multiplication has to be performed:

$$(\$4,500 \quad \$2,000) \begin{pmatrix} .8983 & .1017 \\ .6441 & .3559 \end{pmatrix}$$

$$(\$5,330.55 \quad \$1,169.45)$$

950

The above computation would indicate that $5,330.55 of his current accounts receivable is likely to wind up being paid and $1,169.45 is likely to become bad debts. Therefore, if he follows the standard practice of setting up a reserve for "doubtful" accounts, his accountant would set up $1,169.45 as the best estimate for this category.

● **PROBLEM** 24-24

A camera store stocks a particular model camera that can be ordered weekly. Let D_1, D_2,..., represent the demand for this camera during the first week, second week, ..., respectively. It is assumed that the D_i are independent and identically distributed random variables having a known probability distribution. Let X_0 represent the number of camers on hand at the outset, X_1 the number of cameras on hand at the end of week one, X_2 the number of cameras on hand at the end of week two, and so on. Assume that $X_0 = 3$. On Saturday, night, the store places an order that is delivered in time for the opening of the store on Monday. The store uses the following (s,S) ordering policy. If the number of cameras on hand at the end of the week is less than s = 1 (no cameras in stock), the store orders (up to) S = 3. Otherwise, the store does not order (if there are any cameras in stock, no order is placed). It is assumed that sales are lost when demand exceeds the inventory on hand. Assuming further that X_t is the number of cameras in stock at the end of the tth week (before an order is received, and that each D_t has a Poisson distribution with parameter $\lambda = 1$, determine

    a)   the one-step transition probabilities

    b)   Given that there are two cameras left in stock at the end of a week, what is the probability that there will be three cameras in stock two weeks and four weeks later.

    c)   Compute the expected time until the cameras are out of stock, assuming the process is started when there are three cameras available; i.e., the expected first passage time, μ_{30}, is to be obtained.

    d)   Determine the long-run steady-state probabilities.

Solution: This is an application of Markov Chains to the problem of having an optimal inventory policy. The principle of (s,S) ordering policy as given in the above problem refers to the periodic review policy that calls for ordering up to S units whenever the inventory level dips below s (S $\geq$ s). If the inventory level is s or greater, then no order is placed.

    (a)   to obtain the one-step transition probabilities,

Summation of Terms of the Poisson Distribution
1,000 P (Poisson with parameter $\lambda \le c$)

c \ λ	0.01	0.02	0.03	0.04	0.05	0.06	0.07	0.08	0.09
0	990	980	970	961	951	942	932	923	914
1	1000	1000	1000	999	999	998	998	997	996
2				1000	1000	1000	1000	1000	1000

c \ λ	0.10	0.15	0.20	0.25	0.30	0.35	0.40	0.45	0.50
0	905	861	819	779	741	705	670	638	607
1	995	990	982	974	963	951	938	925	910
2	1000	999	999	998	996	994	992	989	986
3		1000	1000	1000	1000	1000	999	999	998
4							1000	1000	1000

c \ λ	0.55	0.60	0.65	0.70	0.75	0.80	0.85	0.90	0.95	1.00
0	577	549	522	497	472	449	427	407	387	368
1	894	878	861	844	827	809	791	772	754	736
2	982	977	972	966	959	953	945	937	929	920
3	998	997	996	994	993	991	989	987	984	981
4	1000	1000	999	999	999	999	998	998	997	996
5			1000	1000	1000	1000	1000	1000	1000	999
6										1000

c \ λ	1.05	1.10	1.15	1.20	1.25	1.30	1.35	1.40	1.45	1.50
0	350	333	317	301	287	273	259	247	235	223
1	717	699	681	663	645	627	609	592	575	558
2	910	900	890	879	868	857	845	833	821	809
3	978	974	970	966	962	957	952	946	940	934
4	996	995	993	992	991	989	988	986	984	981
5	999	999	999	998	998	998	997	997	996	996
6	1000	1000	1000	1000	1000	1000	999	999	999	999
7							1000	1000	1000	1000

c \ λ	1.55	1.60	1.65	1.70	1.75	1.80	1.85	1.90	1.95	2.00
0	212	202	192	183	174	165	157	150	142	135
1	541	525	509	493	478	463	448	434	420	406
2	796	783	770	757	744	731	717	704	690	677
3	928	921	914	907	899	891	883	875	866	857
4	979	976	973	970	967	964	960	956	952	947
5	995	994	993	992	991	990	988	987	985	983
6	999	999	998	998	998	997	997	997	996	995
7	1000	1000	1000	1000	1000	999	999	999	999	999
8						1000	1000	1000	1000	1000

the following elements of the transition matrix would be utilized.

$$P = \begin{bmatrix} P_{00} & P_{01} & P_{02} & P_{03} \\ P_{10} & P_{11} & P_{12} & P_{13} \\ P_{20} & P_{21} & P_{22} & P_{23} \\ P_{30} & P_{31} & P_{32} & P_{33} \end{bmatrix}$$

To determine P_{00}, which is the conditional probability of having 0 cameras at the end of the week given that we have

O cameras on hand, it is necessary to evaluate $P\{X_t = 0 \mid X_{t-1} = 0\}$.

Therefore if $X_t = 0$, then the demand during the week has to be 3 or more since if we don't have cameras, we order up to 3. Hence $p_{00} = P\{D_t \geq 3\}$. This is just the probability that a Poisson random variable with parameter $\lambda = 1$ takes on a value of 3 or more, which is obtained from a table on the Summation of Terms of the Poisson Distribution, so that $p_{00} = 0.08$. $p_{10} = P\{X_t = 0 \mid X_t - 1 = 1\}$ can be obtained in a

similar way. To have $X_t = 0$, the demand during the week has to be 1 or more. Hence $p_{10} = P\{D_t \geq 1\} = 0.632$. To find $p_{21} = P\{X_t = 1 \mid X_{t-1} = 2\}$, then the demand during the week has to be exactly 1. Hence $p_{21} = P\{D_t = 1\} = 0.368$. The remaining entries are obtained in a similar manner, which yields the following (one step) transition matrix:

$$
P =
\begin{array}{c c}
 & \begin{array}{cccc} 0 & \quad 1 & \quad 2 & \quad 3 \end{array} \\
\begin{array}{c} 0 \\ 1 \\ 2 \\ 3 \end{array} &
\left[
\begin{array}{cccc}
0.080 & 0.184 & 0.368 & 0.368 \\
0.632 & 0.368 & 0 & 0 \\
0.264 & 0.368 & 0.368 & 0 \\
0.080 & 0.184 & 0.368 & 0.368
\end{array}
\right]
\end{array}
$$

(b)   The probabilities that there will be three cameras in stock two weeks and four weeks later given that there are two cameras left in stock at the end of a week are obtained by using the Chapman-Kolmogorov difference Equations which provide a method for computing the n-step transition probabilities, i.e., 2 and 4 weeks.

The two-step transition matrix is given by

$$P^{(2)} = P^2 =$$

$$
\begin{bmatrix}
0.080 & 0.184 & 0.368 & 0.368 \\
0.632 & 0.368 & 0 & 0 \\
0.264 & 0.368 & 0.368 & 0 \\
0.080 & 0.184 & 0.368 & 0.368
\end{bmatrix}
\begin{bmatrix}
0.080 & 0.184 & 0.368 & 0.368 \\
0.632 & 0.368 & 0 & 0 \\
0.264 & 0.368 & 0.368 & 0 \\
0.080 & 0.184 & 0.368 & 0.368
\end{bmatrix}
$$

$$
=
\begin{array}{c c}
 & \begin{array}{cccc} 0 & \quad 1 & \quad 2 & \quad 3 \end{array} \\
\begin{array}{c} 0 \\ 1 \\ 2 \\ 3 \end{array} &
\left[
\begin{array}{cccc}
0.249 & 0.286 & 0.300 & 0.165 \\
0.283 & 0.252 & 0.233 & 0.233 \\
0.351 & 0.319 & 0.233 & 0.097 \\
0.249 & 0.286 & 0.300 & 0.165
\end{array}
\right]
\end{array}
$$

Thus, given that there are two cameras left in stock at the end of a week, the probability is 0.097 that there will be three cameras in stock 2 weeks later; that is, $p_{23}^{(2)}$ = 0.097.

The four-step transition matrix can also be obtained as follows:

$$p^{(4)} = p^4 = p^{(2)} \cdot p^{(2)}$$

$$= \begin{bmatrix} 0.249 & 0.286 & 0.300 & 0.165 \\ 0.283 & 0.252 & 0.233 & 0.233 \\ 0.351 & 0.319 & 0.233 & 0.097 \\ 0.249 & 0.286 & 0.300 & 0.165 \end{bmatrix} \begin{bmatrix} 0.249 & 0.286 & 0.300 & 0.165 \\ 0.283 & .252 & 0.233 & 0.233 \\ 0.351 & .319 & 0.233 & 0.097 \\ 0.249 & .286 & 0.300 & 0.165 \end{bmatrix}$$

$$= \begin{bmatrix} 0.289 & 0.286 & 0.261 & 0.164 \\ 0.282 & 0.285 & 0.268 & 0.166 \\ 0.284 & 0.283 & 0.263 & 0.171 \\ 0.289 & 0.286 & 0.261 & 0.164 \end{bmatrix}$$

Therefore, given that there will be two cameras in stock at the end of a week, the probability is 0.171 that there will be three cameras in stock 4 weeks later; that is, $p_{23}^{(4)} = 0.171$.

(c) The first passage time refers to the length of time it will take the process in going from state i to state j for the first time. When j = 1, this first passage time is just the number of transitions until the process returns to the initial state i. In this case, the first passage time is called the recurrence time for state i which has to be assumed in this particular problem. Given the equation

$$\mu_{ij} = 1 + \sum_{k \neq j} p_{ik} \mu_{kj}, \text{ yields}$$

$$\mu_{30} = 1 + p_{31} \mu_{10} + p_{32} \mu_{20} + p_{33} \mu_{30},$$

$$\mu_{20} = 1 + p_{21} \mu_{10} + p_{22} \mu_{20} + p_{23} \mu_{30},$$

$$\mu_{10} = 1 + p_{11} \mu_{10} + p_{12} \mu_{20} + p_{13} \mu_{30},$$

or

$$\mu_{30} = 1 + 0.184 \mu_{10} + 0.368 \mu_{20} + 0.368 \mu_{30}$$

$$\mu_{20} = 1 + 0.368 \mu_{10} + 0.368 \mu_{20}$$

$$\mu_{10} = 1 + 0.368 \mu_{10}$$

The simultaneous solution to this system of equations is

$$\mu_{10} = 1.58 \text{ weeks,}$$

$$\mu_{20} = 2.51 \text{ weeks,}$$

$$\mu_{30} = 3.50 \text{ weeks}$$

so that the expected time until the cameras are out of stock is 3.50 weeks given that there are 3 stocks of cameras on hand.

(d) The long-run steady-state probabilities can be obtained by using the following steady-state equations as follows:

$$\pi_0 = \pi_0 P_{00} + \pi_1 P_{10} + \pi_2 P_{20} + \pi_3 P_{30}$$

$$\pi_1 = \pi_0 P_{01} + \pi_1 P_{11} + \pi_2 P_{21} + \pi_3 P_{31}$$

$$\pi_2 = \pi_0 P_{02} + \pi_1 P_{12} + \pi_2 P_{22} + \pi_3 P_{32}$$

$$\pi_3 = \pi_0 P_{03} + \pi_1 P_{13} + \pi_2 P_{23} + \pi_3 P_{33}$$

$$1 = \pi_0 + \pi_1 + \pi_2 + \pi_3$$

Substituting values for p_{ij} into these equations leads to the equations

$$\pi_0 = (0.080)\pi_0 + (0.632)\pi_1 + (0.264)\pi_2 + (0.080)\pi_3,$$

$$\pi_1 = (0.184)\pi_0 + (0.368)\pi_1 + (0.368)\pi_2 + (0.184)\pi_3,$$

$$\pi_2 = (0.368)\pi_0 \qquad\qquad + (0.368)\pi_2 + (0.368)\pi_3,$$

$$\pi_3 = (0.368)\pi_0 \qquad\qquad\qquad\qquad\qquad + (0.368)\pi_3,$$

$$1 = \pi_0 + \pi_1 + \pi_2 + \pi_3$$

Solving the last four equations provides the simultaneous solutions

$$\pi_0 = 0.285$$

$$\pi_1 = 0.285$$

$$\pi_2 = 0.264$$

$$\pi_3 = 0.166$$

Thus, after many weeks, the probability of finding zero, one, two, and three cameras in stock tends to 0.285, 0.285, 0.264, and 0.166 respectively. The corresponding expected recurrence times are

$$\mu_{00} = \frac{1}{\pi_0} = 3.51 \text{ weeks,}$$

955

$$\mu_{11} = \frac{1}{\pi_1} = 3.51 \text{ weeks},$$

$$\mu_{22} = \frac{1}{\pi_2} = 3.79 \text{ weeks},$$

$$\mu_{33} = \frac{1}{\pi_3} = 6.02 \text{ weeks}.$$

PROBABILITY

The following table shows the probabilities that a son may or may not follow in his father's footsteps in a given society:

	1	2	3	4	5	6	7	Class
1	.388	.146	.202	.062	.140	.047	.015	1. Professional and high administrative
2	.107	.267	.227	.120	.206	.053	.020	2. Managerial and Executive
3	.035	.101	.188	.191	.357	.067	.061	3. Higher grade supervisory & nonmanual
P = 4	.021	.039	.112	.212	.430	.124	.062	4. Lower grade supervisory & nonmanual
5	.009	.024	.075	.123	.473	.171	.125	5. Skilled manual & routine nonmanual
6	.000	.013	.041	.088	.391	.312	.155	6. Semi-skilled manual
7	.000	.008	.036	.083	.364	.235	.274	7. Unskilled manual

Interpret the above table. (Assume each father has only one son).

Solution: This problem is an application of Markov chains to social mobility. Examining the body of the table observe that the probability of the son of a manager becoming a skilled manual worker is .206. Note that the matrix of probabilities is a stochastic matrix, i.e., it is a square matrix with nonnegative entries whose rows each sum to one. The p_{ij} represent the probability that a father in class i has a son in class j.

Let $e_1 = [1, 0, 0, \ldots, 0]$, $e_2 = [0, 1, \ldots, 0]$ $\ldots, e_7 = [0, 0, \ldots, 0, 1]$.

Then $e_i P = [p_{i1}, \ldots, p_{i7}]$. Thus, given a father in class i, e_iP lists the probability that his son will be in class 1, class 2, ..., class 7. Similarly, e_iP^2 will list the probability that a father in class i will have grandsons in class 1, ..., class 7.

For another interpretation of the probability distribution, suppose p_1 of society is the percentage currently in class 1, ..., p_7 of society is the percentage in class 7, so that the probability vector $P = [p_1, ..., p_7]$ gives the current distribution of society among the various classes. Then pP will give the probability distribution a generation hence, pP^2 the probability distribution two generations hence, etc. Let the current distribution be

$$p = [.037, .043, .098, .148, .432, .131, .111]$$

Then the distribution in the next generation is

$$pP = [.029, .046, .094, .131, .409, .170, .121].$$

Also, with the help of the following facts, find the long-run distribution of the various classes.

If, for some j, P^j contains all positive entries the matrix P is said to be regular. If P is regular, there exists a probability vector $(t_1, t_2, ..., t_n)$ such that

$$[t_1, t_2, ..., t_n] \begin{bmatrix} p_{11} \cdots \cdots p_{1n} \\ \vdots \\ \vdots \\ p_{n1} \cdots \cdots p_{nn} \end{bmatrix}^m = [t_1, t_2, ..., t_n],$$

for all $k \geq m$. The probability vector $[t_1, t_2, ..., t_n]$ is a fixed vector toward which the distribution tends. In fact, all the rows of the transition matrix will be identically equal to this vector.

Since P^2 contains all positive entries, the given transition matrix is regular. The distribution approached is $\bar{p} = [.023, .042, .088, .127, .409, .182, .129]$. This was found by solving the set of simultaneous linear equations $[t_1 t_2 t_3 t_4 t_5 t_6 t_7]P = [t_1 t_2 t_3 \cdots t_7]$ subject to the requirement $\sum_{i=1}^{7} t_i = 1$.

● **PROBLEM** 24-26

Let S be the system consisting of two players A and B who begin with $2 apiece and match coins until one or the other of them has no more money. If the states of the system are

957

defined by the number of dollars in A's possession, i.e., the system is in the state S_{i+1} whenever A has i dollars (i = 0, 1, 2, 3, 4) find the matrix of one-step probabilities. Then, by raising this matrix to the second, third and fourth powers, find the matrices of corresponding step transition probabilities for S. What is the probability that A will be ruined in at most four turns? What is the probability that A will be ruined in exactly four turns?

Solution: Assume that whenever one of the players is bankrupt the game finishes. Also, at each turn of the game one player must lose, and the other win $1, i.e., no draws are possible.

Now examine A's possibilities on a given turn. If he has $0, he will continue to have $0. If he has $1, then on the next round he will have $0 with probability 1/2 $1 with probability 0, $2 with probability 1/2, $3 with probability 0 and $4 with probability 0. Similarly, if he has $2, then on the next round he will have $0 with probability 0, $1 with probability 1/2, $2 with probability 0, $3 with probability 1/2 and $4 with probability 0.

All the probabilities for each round are given by the transition matrix below:

$$P = \begin{bmatrix} 1 & 0 & 0 & 0 & 0 \\ \frac{1}{2} & 0 & \frac{1}{2} & 0 & 0 \\ 0 & \frac{1}{2} & 0 & \frac{1}{2} & 0 \\ 0 & 0 & \frac{1}{2} & 0 & \frac{1}{2} \\ 0 & 0 & 0 & 0 & 1 \end{bmatrix} \qquad (1)$$

The matrix (1) represents the matrix of 1-step transition probabilities. Next, to find the 2-step transition probabilities reason as follows. If A has $0 then after 2 rounds he will still have $0 with probability one. If A has $1, after one round he will have either $0 or $2 with probabilities 1/2. But from (1) a player with $2 will have $1 or $3 with probabilities 1/2. Thus a player with $1 will, after two rounds, have $0 with probability 1/2, $1 with probability 1/4, $2 with probability 0, $3 with probability 1/4 and $4 with probability 0.

If we multiply (1) with itself yields all the two-step transition probabilities. Thus,

$$P^2 = \begin{bmatrix} 1 & 0 & 0 & 0 & 0 \\ \frac{1}{2} & 0 & \frac{1}{2} & 0 & 0 \\ 0 & \frac{1}{2} & 0 & \frac{1}{2} & 0 \\ 0 & 0 & \frac{1}{2} & 0 & \frac{1}{2} \\ 0 & 0 & 0 & 0 & 1 \end{bmatrix} \begin{bmatrix} 1 & 0 & 0 & 0 & 0 \\ \frac{1}{2} & 0 & \frac{1}{2} & 0 & 0 \\ 0 & \frac{1}{2} & 0 & \frac{1}{2} & 0 \\ 0 & 0 & \frac{1}{2} & 0 & \frac{1}{2} \\ 0 & 0 & 0 & 0 & 1 \end{bmatrix}$$

$$= \begin{bmatrix} 1 & 0 & 0 & 0 & 0 \\ \frac{1}{2} & \frac{1}{4} & 0 & \frac{1}{4} & 0 \\ \frac{1}{4} & 0 & \frac{1}{2} & 0 & \frac{1}{4} \\ 0 & \frac{1}{4} & 0 & \frac{1}{4} & \frac{1}{2} \\ 0 & 0 & 0 & 0 & 1 \end{bmatrix}$$

Similarly the three step probabilities are given by P^3 and the four-step transition probabilities are given by P^4:

$$P^3 = \begin{bmatrix} 1 & 0 & 0 & 0 & 0 \\ \frac{5}{8} & 0 & \frac{1}{4} & 0 & \frac{1}{8} \\ \frac{1}{4} & \frac{1}{4} & 0 & \frac{1}{4} & \frac{1}{4} \\ \frac{1}{8} & 0 & \frac{1}{4} & 0 & \frac{5}{8} \\ 0 & 0 & 0 & 0 & 1 \end{bmatrix}; \quad P^4 = \begin{bmatrix} 1 & 0 & 0 & 0 & 0 \\ \frac{5}{8} & \frac{1}{8} & 0 & \frac{1}{8} & 0 \\ \frac{3}{8} & 0 & \frac{1}{4} & 0 & \frac{3}{8} \\ \frac{1}{8} & \frac{1}{8} & 0 & \frac{1}{8} & \frac{5}{8} \\ 0 & 0 & 0 & 0 & 1 \end{bmatrix}$$

The probability that A is ruined in at most four turns is simply the probability of a four-step transition from S_3 to S_1, namely, $p_{31}^{(4)} = \frac{3}{8}$, since among such transitions are included those in which the system reaches S_1 in less than four steps and then remains there. The probability that A is ruined in four turns and not before is the probability that S reaches S_1 in four steps but does not reach it in three steps or less, namely $p_{31}^{(4)} - p_{31}^{(3)} = \frac{3}{8} - \frac{1}{4} = \frac{1}{8}$.

Consider the following two-person game. Player 1 has $3.00
and player 2 has $2.00. They flip a fair coin; if it is a
head, player 1 pays player 2 $1.00 and if it is a tail,
player 2 pays player 1 $1.00. How long will it take for one
of the players to go broke or win all the money?

Solution: How much money a player has after any given flip
of the coin depends on how much he had after the previous
flip and not directly on how much he had in the preceding
stages of the game. Thus, this game can be represented
by a Markov chain.

The coin being flipped is fair so that a probability
of 1/2 is assigned to each event. The states at each step
are the amounts of money each player has at each stage of
the game. Thus a player can have $0 to $5 at any stage.
Each player can increase or decrease the amount of money he
has by only $1 at a time. The transition matrix is then of
the form

$$
P = \begin{array}{c c} & \begin{array}{cccccc} 0 & 1 & 2 & 3 & 4 & 5 \end{array} \\ \begin{array}{c} 0 \\ 1 \\ 2 \\ 3 \\ 4 \\ 5 \end{array} & \begin{pmatrix} 1 & 0 & 0 & 0 & 0 & 0 \\ \frac{1}{2} & 0 & \frac{1}{2} & 0 & 0 & 0 \\ 0 & \frac{1}{2} & 0 & \frac{1}{2} & 0 & 0 \\ 0 & 0 & \frac{1}{2} & 0 & \frac{1}{2} & 0 \\ 0 & 0 & 0 & \frac{1}{2} & 0 & \frac{1}{2} \\ 0 & 0 & 0 & 0 & 0 & 1 \end{pmatrix} \end{array}
$$

An absorbing state of a Markov chain occurs when the
probability of that state in the transition matrix is one.
The transition matrix here has two absorbing states, when
the player is broke or when he has all the money. Rearrange
the transition matrix so that the two absorbing states will
appear in the first two rows. Then

$$
P = \begin{array}{c c} & \begin{array}{cccccc} 0 & 5 & 1 & 2 & 3 & 4 \end{array} \\ \begin{array}{c} 0 \\ 5 \\ \\ 1 \\ 2 \\ 3 \\ 4 \end{array} & \left(\begin{array}{cc|cccc} 1 & 0 & 0 & 0 & 0 & 0 \\ 0 & 1 & 0 & 0 & 0 & 0 \\ \hline \frac{1}{2} & 0 & 0 & \frac{1}{2} & 0 & 0 \\ 0 & 0 & \frac{1}{2} & 0 & \frac{1}{2} & 0 \\ 0 & 0 & 0 & \frac{1}{2} & 0 & \frac{1}{2} \\ 0 & \frac{1}{2} & 0 & 0 & \frac{1}{2} & 0 \end{array}\right) \end{array} \qquad (1)
$$

The partitioned matrix is in the form

$$P = \begin{bmatrix} I_2 & | & O \\ -- & -|- & - \\ S & | & Q \end{bmatrix}$$

Let $T = (I_S - Q)^{-1}$.

The entries of T give the expected number of times the process is in each nonabsorbing state, provided the process began in a nonabsorbing state. The matrix T is called the fundamental matrix of an absorbing Markov chain.

For the given problem, the fundamental matrix T is

$$T = \left[\begin{bmatrix} 1 & 0 & 0 & 0 \\ 0 & 1 & 0 & 0 \\ 0 & 0 & 1 & 0 \\ 0 & 0 & 0 & 1 \end{bmatrix} - \begin{bmatrix} 0 & \frac{1}{2} & 0 & 0 \\ \frac{1}{2} & 0 & \frac{1}{2} & 0 \\ 0 & \frac{1}{2} & 0 & \frac{1}{2} \\ 0 & 0 & \frac{1}{2} & 0 \end{bmatrix} \right]^{-1}$$

	1	2	3	4
1	1.6	1.2	.8	.4
= 2	1.2	2.4	1.6	.8
3	.8	1.6	2.4	1.2
4	.4	.8	1.2	1.6

The entry .8 in row 3, column 1 indicates that .8 is the expected number of times the player will have $1 if he started with $3.

We can find the expected length of the game from T, since the expected number of games before absorption can be found by adding the entries in each row of T. Thus, if a player starts with $3, the expected number of games before absorption is .8 + 1.6 + 2.4 + 1.2 = 6.0. If a player starts with $1, the expected number of games before absorption is 1.6 + 1.2 + .8 + .4 = 4.0.

● **PROBLEM 24-28**

A and B play the following game. Each tosses a die in turn. If a six is tossed the game is over and whoever tossed the six wins the bets. If a 4 or a 5 is tossed the player tosses again. If a 1, 2 or 3 is tossed the die passes to the other player. Is the game biased in favor of the starting player?

961

Solution: Analyze this game using stochastic matrices. A stochastic matrix is defined to be a square matrix in which every row contains nonnegative elements that sum to one. The rows and columns represent states of the system and element p_{ij} represents the probability of transition from state i to state j.

The matrix of transition probabilities for the given problem is:

	A's Turn	B's Turn	A wins	B wins	
A's Turn	$\frac{1}{3}$	$\frac{1}{2}$	$\frac{1}{6}$	0	
B's Turn	$\frac{1}{2}$	$\frac{1}{3}$	0	$\frac{1}{6}$	= T
A wins	0	0	1	0	
B wins	0	0	0	1	

We will briefly explain how the entries in the transition matrix are found. In row 1, when it is A's turn there are 3 possibilities: (1) A will throw a 4 or 5 and thus have another turn. The probability of this happening is $\frac{2}{6}$ or $\frac{1}{3}$. (2) A will throw a 1, 2 or 3, and thus B will have his turn. The probability of this happening is $\frac{3}{6}$ or $\frac{1}{2}$. (3) A will throw a 6 and wins. The probability of this event is $\frac{1}{6}$. Therefore, since there is no possibility of B winning when A wins, the probability of B winning is 0. Thus the entries in the first row are $\frac{1}{3}$, $\frac{1}{2}$, $\frac{1}{6}$, 0, respectively. The entries of row 2 are found by similar reasoning. In row 3, when A wins, the game has ended. Thus the probabilities of A having another turn, of B having another turn or of B winning are all 0. Thus, the entries in the third row are 0, 0, 1, 0, respectively. The fourth row is found by similar reasoning.

Assume that A starts the game and that the game continues for an indefinite number of moves. Squaring the transition matrix, T, gives the probabilities of transitions between the states occurring in two stages. In general the nth power of T shows the probabilities of transitions between states after n stages. For example, A^4 gives

	A's Turn	B's Turn	A wins	B wins
A's Turn	$\frac{313}{1296}$	$\frac{312}{1296}$	$\frac{148}{1296}$	$\frac{138}{1296}$
B's Turn	$\frac{312}{1296}$	$\frac{313}{1296}$	$\frac{138}{1296}$	$\frac{148}{1296}$
A wins	0	0	$\frac{1296}{1296}$	0
B wins	0	0	0	$\frac{1296}{1296}$

The above matrix is to be interpreted as follows. The entry $\frac{313}{1296}$ indicates that if A starts then after four throws the probability that it will again be his turn is $\frac{313}{1296}$. The entry $\frac{148}{1296}$ gives the probability of A's winning in four moves or less when A starts.

To estimate the respective chances of the starting player we must consider the behaviour of T^n as $n \to \infty$.

To do this, partition the transition matrix into block matrices. Let

$$P = \left[\begin{array}{c|c} Q & R \\ \hline O & I \end{array} \right]$$

where $Q = \begin{bmatrix} \frac{1}{3} & \frac{1}{2} \\ \frac{1}{2} & \frac{1}{3} \end{bmatrix}$, $R = \begin{bmatrix} \frac{1}{6} & 0 \\ 0 & \frac{1}{6} \end{bmatrix}$

and O and I are the remaining 2x2 zero and identity matrices respectively. Then, using the rules of block multiplication:

$$P^2 = \left[\begin{array}{c|c} Q^2 & QR + R \\ \hline O & I \end{array} \right], \quad P^3 = \left[\begin{array}{c|c} Q^3 & (Q^2 + Q + I)R \\ \hline O & I \end{array} \right]$$

and, in general $P^n = \left[\begin{array}{c|c} Q^n & (Q^{n-1}+Q^{n-2}+\ldots+Q^2+ Q + I)R \\ \hline O & I \end{array} \right]$

But $(I + Q + Q^2 + \ldots + Q^{n-1}) (I - Q) = I - Q^n$

or, $(I + Q + Q^2 + \ldots + Q^{n-1}) = (I - Q^n) (I - Q)^{-1}$

assuming that $(I - Q)^{-1}$ exists. To show that it exists consider the behaviour of

$$Q^n = \begin{bmatrix} \frac{1}{3} & \frac{1}{2} \\ \frac{1}{2} & \frac{1}{3} \end{bmatrix} \quad \text{as } n \to \infty.$$

First find a diagonal matrix that is similar to Q. The eigenvalues of Q are $5/6$, $-1/6$ with associated eigenvectors $\begin{bmatrix} 1 \\ 1 \end{bmatrix}$ and $\begin{bmatrix} 1 \\ -1 \end{bmatrix}$ respectively. Hence

$$D = \begin{bmatrix} \frac{5}{6} & 0 \\ 0 & -\frac{1}{6} \end{bmatrix}, \quad P = \begin{bmatrix} 1 & 1 \\ 1 & -1 \end{bmatrix} \quad \text{and } P^{-1} = \begin{bmatrix} \frac{1}{2} & \frac{1}{2} \\ \frac{1}{2} & -\frac{1}{2} \end{bmatrix}.$$

Hence $\quad Q = P^{-1} DP$

and $\quad Q^n = P^{-1} D^n P$.

But as $n \to \infty$ $\quad D^n \to 0$ and hence $Q^n \to 0$.

Since $Q^n \to 0$ as $n \to \infty$, $(I - Q^n) \to I$ and $(I + Q + Q^2 + \ldots + Q^{n-1}) \to (I - Q)^{-1}$. Thus

$$P^n = \left[\begin{array}{c|c} O & (I - Q)^{-1} R \\ \hline O & I \end{array} \right]$$

Applying this to the given problem,

$$Q = \begin{bmatrix} \frac{1}{3} & \frac{1}{2} \\ \frac{1}{2} & \frac{1}{3} \end{bmatrix}, \quad I - Q = \begin{bmatrix} \frac{2}{3} & -\frac{1}{2} \\ -\frac{1}{2} & \frac{2}{3} \end{bmatrix}$$

$$(I - Q)^{-1} = \begin{bmatrix} \frac{24}{7} & \frac{18}{7} \\ \frac{18}{7} & \frac{24}{7} \end{bmatrix}.$$

and, since $R = \begin{bmatrix} \frac{1}{6} & 0 \\ 0 & \frac{1}{6} \end{bmatrix}$,

$$(I - Q)^{-1} R = \begin{bmatrix} \frac{24}{7} & \frac{18}{7} \\ \frac{18}{7} & \frac{24}{7} \end{bmatrix} \begin{bmatrix} \frac{1}{6} & 0 \\ 0 & \frac{1}{6} \end{bmatrix} = \begin{bmatrix} \frac{4}{7} & \frac{3}{7} \\ \frac{3}{7} & \frac{4}{7} \end{bmatrix}$$

But $(I - Q)^{-1}$ R is the part of the matrix P^n which is concerned with the probabilities of making a transition from the various possible starting states to the various possible winning states. Hence, if A starts the probability of his winning provided the game goes on long enough is 4/7. In other words, the game is biased in favour of the starting player.

● PROBLEM 24-29

Bob, Ted, Carol and Alice are throwing a ball to one another. Alice always throws it to Ted; Bob is equally likely to throw it to anybody else; Carol throws it to the boys with equal frequency; Ted throws it to Carol twice as often as to Alice and never throws it to Bob. Construct a stochastic matrix to answer the following question:

What is the probability that the ball will go from

a)   Bob to Carol in two throws

b)   Carol to Alice in three throws?

<u>Solution</u>:   A stochastic matrix is a square matrix in which the sum of the elements in each row is one.   These elements represent transition probabilities.

	A	B	C	T
A	0	0	0	1
B	$\frac{1}{3}$	0	$\frac{1}{3}$	$\frac{1}{3}$
C	0	$\frac{1}{2}$	0	$\frac{1}{2}$
T	$\frac{1}{3}$	0	$\frac{2}{3}$	0

The above table gives the matrix of transition probabilities for the given matrix.   An element p_{ij} represents the probability that one player will throw it to another. For example p_{TC} = 2/3 (the probability that Ted will throw the ball to Carol, and p_{CT} = 1/2).

a)   If we square the matrix, the terms in the squared transition matrix give the probabilities of a transition taking place in two throws.

$$
\begin{bmatrix}
0 & 0 & 0 & 1 \\
\frac{1}{3} & 0 & \frac{1}{3} & \frac{1}{3} \\
0 & \frac{1}{2} & 0 & \frac{1}{2} \\
\frac{1}{3} & 0 & \frac{2}{3} & 0
\end{bmatrix}
\begin{bmatrix}
0 & 0 & 0 & 1 \\
\frac{1}{3} & 0 & \frac{1}{3} & \frac{1}{3} \\
0 & \frac{1}{2} & 0 & \frac{1}{2} \\
\frac{1}{3} & 0 & \frac{2}{3} & 0
\end{bmatrix}
=
\begin{bmatrix}
\frac{1}{3} & 0 & \frac{2}{3} & 0 \\
\frac{1}{9} & \frac{1}{6} & \frac{2}{9} & \frac{1}{2} \\
\frac{1}{3} & 0 & \frac{1}{2} & \frac{1}{6} \\
0 & \frac{1}{3} & 0 & \frac{2}{3}
\end{bmatrix}
\qquad (1)
$$

965

To understand (1) consider the last row of the squared matrix. The entry 1/3 has the following interpretation. The probability that Ted will throw the ball to Alice and that Alice will then throw the ball to Bob is (1/3) (0) = 0. The probability that Ted will throw to Bob and Bob to himself is (0) (0) = 0. The probability that Ted will throw to Carol and then Carol to Bob is (2/3) (1/2) = 1/3. Finally the probability that Ted will throw it to himself and then to Bob is 0 (0) = 0. Adding, yields the probability that the ball will go from Ted to Bob is 1/3.

Thus, observe that the probability of the ball going from Bob to Carol in two throws is (1/3) (0) + 0 (1/3) + (1/3) (0) + (1/3) (2/3) = 2/9.

b) For three throws, find the third power of the original matrix.

$$A^3 = \begin{bmatrix} \frac{1}{3} & 0 & \frac{2}{3} & 0 \\ \frac{1}{9} & \frac{1}{6} & \frac{2}{9} & \frac{1}{2} \\ \frac{1}{3} & 0 & \frac{1}{2} & \frac{1}{6} \\ 0 & \frac{1}{3} & 0 & \frac{2}{3} \end{bmatrix} \begin{bmatrix} 0 & 0 & 0 & 1 \\ \frac{1}{3} & 0 & \frac{1}{3} & \frac{1}{3} \\ 0 & \frac{1}{2} & 0 & \frac{1}{2} \\ \frac{1}{3} & 0 & \frac{2}{3} & 0 \end{bmatrix}$$

$$= \begin{bmatrix} 0 & \frac{1}{3} & 0 & \frac{2}{3} \\ \frac{2}{9} & \frac{1}{9} & \frac{7}{18} & \frac{5}{18} \\ \frac{1}{18} & \frac{1}{4} & \frac{1}{9} & \frac{7}{12} \\ \frac{1}{3} & 0 & \frac{5}{9} & \frac{1}{9} \end{bmatrix}$$

Thus, the probability that the ball will go from Carol to Alice is 1/18.

● PROBLEM 24-30

Suppose that the President of the United States tells person A his intention either to run or not to run in the next election. A relays it to B, B to C and so on. Assume that there is a probability p > 0 that any one person, when he gets the message, will reverse it before passing it on to the next person. What is the probability that the nth man to hear the message will be told that the President will run?

Solution: This is an example of a two state Markov chain, with states indicated by "yes" and "no." The process is in

state "yes" at time n if the nth person to receive the message was told that the President would run. It is in state
"no" if he was told that the President would not run. The
transition matrix is therefore

$$P = \begin{array}{c c} & \begin{array}{c c} \text{yes} & \text{no} \end{array} \\ \begin{array}{c} \text{yes} \\ \text{no} \end{array} & \begin{bmatrix} 1-p & p \\ p & 1-p. \end{bmatrix} \end{array}$$

Since $p > 0$ and $p < 1$ all the entries in the above
matrix are regular. Thus there exists a matrix W with
identical rows ω such that $P^n \to W$. Let $t = [t_1\ t_2]$ be this
fixed probability vector. Then

$$[t_1\ t_2] \begin{bmatrix} 1-p & p \\ p & 1-p \end{bmatrix} = [t_1\ t_2]$$

where $t_1 + t_2 = 1$. Solving the set, $pt_1 = pt_2$. Using the
condition $t_1 + t_2 = 1$, $p(1 - t_2) = pt_2$ and $1 - t_2 = t_2$.
Thus $t_2 = 1/2$, $t_1 = 1/2$ and the fixed probability vector
is $t = [1/2\ \ 1/2]$

Hence the probabilities for the nth man's being told
"yes" or "no" approach 1/2 independently of the initial
decision of the President.

GRAPH THEORY

● **PROBLEM** 24-31

Give examples of the following concepts

a)   Graph

b)   Digraph

c)   Matrix of a digraph

d)   Incidence matrix.

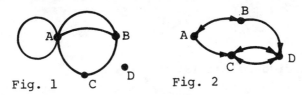

Fig. 1   Fig. 2

Solution: a)   We first define the notion of a graph.   A
graph is a collection of a finite number of vertices P_1, P_2,
...., P_n together with a finite number of edges $P_i P_j$ joining
a pair of vertices P_i and P_j

967

In Fig. 1 there are four vertices A, B, C, D. There are 5 edges, i.e., lines joining some of the vertices. Note that a vertex can be the starting and final point of an edge as the loop from A to A demonstrates. Furthermore, two vertices may be joined by more than one edge (as AB) and a vertex need not have any edge.

b) A digraph is a directed graph. This means that the edges joining vertices have direction.

In Fig. 2, A → B and B → A. Also, A → C. Note that we cannot have a loop in a digraph since a vertex cannot be directed toward itself (a point has no direction). Also, the directed edge AB is different from the directed edge BA.

c) Every digraph has a matrix representation. To determine the matrix representative of a digraph, use a square matrix in which the entry in row A column B is the number of directed edges from A to B. For example, the digraph in Fig. 2 has the matrix

	A	B	C	D
A	0	1	1	0
B	1	0	0	1
C	0	0	0	2
D	0	0	2	0

d) An incidence is the matrix of a digraph that has only the entries 0 or 1. Consider the following digraph.

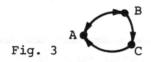

Fig. 3

The matrix representative is

	A	B	C
A	0	1	0
B	1	0	1
C	1	0	0

This is an incidence matrix.

● **PROBLEM 24-32**

The figure below shows a schematic map of the interconnections between the airports in three different countries a, b and c.

The figures beside the links denote the number of choices
along the links. For example, a 4 indicates that four air-
lines fly services along that route.

Construct a matrix showing the number of choices of routes
between the airports in country a and country c.

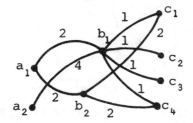

Solution: First tabulate the information given in the
figure.

P:	b_1	b_2		Q:	c_1	c_2	c_3	c_4
a_1	2	2		b_1	1	1	1	1
a_2	4	0		b_2	2	0	0	2

Now, a route from a_1 to c_1 may pass through b_1 or b_2. In
the first case there are 2x1 choices and in the second case
2x2 choices. This means 6 choices in all and the other
entries for the required matrix may be worked out in the
same way, giving

	c_1	c_2	c_3	c_4
a_1	6	2	2	6
a_2	4	4	4	4

But this is simply the matrix product of P and Q:

$$PQ = \begin{bmatrix} 2 & 2 \\ 4 & 0 \end{bmatrix} \begin{bmatrix} 1 & 1 & 1 & 1 \\ 2 & 0 & 0 & 2 \end{bmatrix} = \begin{bmatrix} 6 & 2 & 2 & 6 \\ 4 & 4 & 4 & 4 \end{bmatrix}$$

The natural interpretation of the product matrix is as
follows: the two rows of P represent ways of flying from
a_1 and a_2 to b_1 and b_2 respectively. The columns of Q represent ways in which c_i can be reached from b_1 and b_2. The
product of P and Q therefore gives the number of choices of
route between the airports in country a and country c.

Consider an abstract map giving the routes between cities
as shown in Figure 1.
Show how matrix methods may be used to count the number of
different routes between two points on a map.

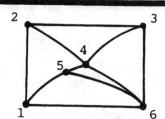

Fig. 1

Solution: The map in Fig. 1 is a graph. The six numbered
dots are called vertices and the connecting lines are called
edges. Two vertices are adjacent if there is an edge directly connecting them. For example, vertex 2 and vertex 4 are
adjacent whereas vertices 3 and 5 are not. A route in a
graph is a sequence of edges such that the terminal point
of one is the initial point of the next edge in the sequence.
For example, $6 \rightarrow 3 \rightarrow 2 \rightarrow 4$ denotes the route starting at
vertex 6 and proceeding to vertex 3, then continuing to ver-
tex 2, and finally ending at vertex 4. The same edge can
appear more than once in a route; for example, the route
$3 \rightarrow 4 \rightarrow 2 \rightarrow 4 \rightarrow 5 \rightarrow 1$ uses the edge joining 2 and 4 twice.

A graph can be numerically described by its incidence
matrix. The size of this nxn matrix is determined by the
number of vertices in the graph. In the present example,
the incidence matrix is a 6x6 matrix where the i,j entry a_{ij}
equals the number of edges joining vertex i to vertex j.
For example, $a_{16} = 1$ since vertices 1 and 6 have one edge
joining them, while $a_{52} = 0$ since there is no edge joining
vertex 5 to vertex 2. Thus, the incidence matrix is:

$$\begin{bmatrix} 0 & 1 & 0 & 0 & 1 & 1 \\ 1 & 0 & 1 & 1 & 0 & 0 \\ 0 & 1 & 0 & 1 & 0 & 1 \\ 0 & 1 & 1 & 0 & 1 & 1 \\ 1 & 0 & 0 & 1 & 0 & 1 \\ 1 & 0 & 1 & 1 & 1 & 0 \end{bmatrix}$$

The length of a route is the number of edges in the
sequence that forms the route. If the same edge is used
more than once, it is counted more than once in determining
the length of the route. For example, the route $4 \rightarrow 2 \rightarrow 3$
$\rightarrow 4 \rightarrow 2 \rightarrow 1$ has length 5. The i,j entry in the incidence
matrix can be interpreted as the number of routes of length

1 from vertex i to vertex j, since there is a route of length
1 from i to j exactly when there is an edge joining i and j.

If we form the product

$$A^2 = AA = \begin{bmatrix} 3 & 0 & 2 & 3 & 1 & 1 \\ 0 & 3 & 1 & 1 & 2 & 3 \\ 2 & 1 & 3 & 2 & 2 & 1 \\ 3 & 1 & 2 & 4 & 1 & 2 \\ 1 & 2 & 2 & 1 & 3 & 2 \\ 1 & 3 & 1 & 2 & 2 & 4 \end{bmatrix}$$

observe that the i,j entry of this matrix gives the number
of routes from i to j of length 2. This follows since

$$\text{i, j entry of } A^2 = a_{i1} a_{1j} + a_{i2} a_{2j} + \ldots + a_{i6} a_{6j}.$$

The first product $a_{i1} a_{1j}$, counts the number of routes
that run from i to 1 to j; the second product $a_{i2} a_{2j}$ counts
the number of routes that run from i to 2 to j and so on.
For example, the entry 3 in the 1, 4 position of A^2 counts
the three routes $1 \rightarrow 2 \rightarrow 4$, $1 \rightarrow 5 \rightarrow 4$, $1 \rightarrow 6 \rightarrow 4$.

In general, if k is any positive integer, the i, j
entry of A^k is the number of routes of length k from vertex
i to vertex j. Using k = 3, we have

$$A^3 = \begin{bmatrix} 2 & 8 & 4 & 4 & 7 & 9 \\ 8 & 2 & 7 & 9 & 4 & 4 \\ 4 & 7 & 4 & 7 & 5 & 9 \\ 4 & 9 & 7 & 6 & 9 & 10 \\ 7 & 4 & 5 & 9 & 4 & 7 \\ 9 & 4 & 9 & 10 & 7 & 6 \end{bmatrix}$$

The entry 9 in the 4, 2 position counts the nine routes
of length 3 from vertex 4 to vertex 2: $4 \rightarrow 2 \rightarrow 1 \rightarrow 2$, $4 \rightarrow 2$
$\rightarrow 3 \rightarrow 2$, $4 \rightarrow 2 \rightarrow 4 \rightarrow 2$, $4 \rightarrow 3 \rightarrow 4 \rightarrow 2$, $4 \rightarrow 5 \rightarrow 1 \rightarrow 2$, $4 \rightarrow 5$
$\rightarrow 4 \rightarrow 2$, $4 \rightarrow 6 \rightarrow 1 \rightarrow 2$, $4 \rightarrow 6 \rightarrow 3 \rightarrow 2$, $4 \rightarrow 6 \rightarrow 4 \rightarrow 2$.

● **PROBLEM** 24-34

The results of a six-person round-robin backgammon tournament
are given in the graph below.

An arrow pointing from player i to player j means that i

971

beat j in their match. For example, player 2 beat player 1 but player 6 beat player 2. Who is the best player?

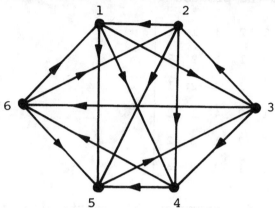

Solution: A matrix can be used to describe the results of the tournament. Let A be the 6x6 matrix where $a_{ij} = 1$ if player i beat player j in their match and $a_{ij} = 0$ if j beat i. Hence,

$$A = \begin{bmatrix} 0 & 0 & 1 & 1 & 1 & 0 \\ 1 & 0 & 0 & 1 & 1 & 0 \\ 0 & 1 & 0 & 1 & 0 & 1 \\ 0 & 0 & 0 & 0 & 1 & 1 \\ 0 & 0 & 1 & 0 & 0 & 0 \\ 1 & 1 & 0 & 0 & 1 & 0 \end{bmatrix}$$

To find how many matches each player won, add across each row. We have

Player #	# wins
1	3
2	3
3	3
4	2
5	1
6	3

and hence the ranking of the players is

Place	Player #
1st	1, 2, 3, ʹ
5th	4
6th	5

972

One method of breaking the four way tie for first place is as follows:

First form the product of A with itself.

$$A^2 = AA = \begin{bmatrix} 0 & 0 & 1 & 1 & 1 & 0 \\ 1 & 0 & 0 & 1 & 1 & 0 \\ 0 & 1 & 0 & 1 & 0 & 1 \\ 0 & 0 & 0 & 0 & 1 & 1 \\ 0 & 0 & 1 & 0 & 0 & 0 \\ 1 & 1 & 0 & 0 & 1 & 0 \end{bmatrix} \begin{bmatrix} 0 & 0 & 1 & 1 & 1 & 0 \\ 1 & 0 & 0 & 1 & 1 & 0 \\ 0 & 1 & 0 & 1 & 0 & 1 \\ 0 & 0 & 0 & 0 & 1 & 1 \\ 0 & 0 & 1 & 0 & 0 & 0 \\ 1 & 1 & 0 & 0 & 1 & 0 \end{bmatrix}$$

$$= \begin{bmatrix} 0 & 1 & 1 & 1 & 1 & 2 \\ 0 & 0 & 2 & 1 & 2 & 1 \\ 2 & 1 & 0 & 1 & 3 & 1 \\ 1 & 1 & 1 & 0 & 1 & 0 \\ 0 & 1 & 0 & 1 & 0 & 1 \\ 1 & 0 & 2 & 2 & 2 & 0 \end{bmatrix} \qquad (1)$$

The matrix (1) gives the two step wins for i over j. To illustrate the meaning of this statement consider the following: Multiplying the first row of A by its second column yields $0 \cdot 0 + 0 \cdot 0 + 1 \cdot 1 + 1 \cdot 0 + 1 \cdot 0 + 0 \cdot 1$

Player 1 who did not play with himself, lost to player 2. But he beat player 3 who beat player 2. Thus player 1 had a two step win over player 2 and hence the (1,2) element of A^2 has 1.

Similarly player 3 had 3 two-step wins over player 5 since he beat player 2 who beat player 5; he also beat player 4 who beat player 5 and finally, beat player 6 who beat player 5.

Now let $B = A + A^2$:

$$B = \begin{bmatrix} 0 & 1 & 2 & 2 & 2 & 2 \\ 1 & 0 & 2 & 2 & 3 & 1 \\ 2 & 2 & 0 & 2 & 3 & 2 \\ 1 & 1 & 1 & 0 & 2 & 1 \\ 0 & 1 & 1 & 1 & 0 & 1 \\ 2 & 1 & 2 & 2 & 3 & 0 \end{bmatrix}$$

The i, j entry of B equals the number of direct wins plus the number of two-step wins that player i has over player j. Adding across rows yields

Player #	# of direct wins + # of 2-step wins
1	9
2	9
3	11
4	6
5	4
6	10

Now the ranking becomes

Place	Player #
1st	3
2nd	6
3rd	1, 2
5th	4
6th	5

Assuming that this method measures a player's strength, observe that player 3 is the strongest player.

Note that there are other ways of measuring relative strength. It is possible to use $2A + A^2$ which gives twice as much weight to a direct win as a two step win.

● **PROBLEM** 24-35

The digraph given below represents relationships between people in a group. An arrow denotes "is friendly to" while a double arrow indicates mutual friendship.

Find the number of cliques and the people belonging to each clique.

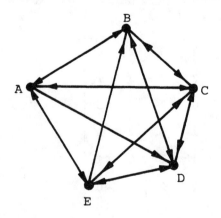

Solution: The incidence matrix of the given digraph is

	A	B	C	D	E
A	0	1	1	1	1
B	1	0	1	1	0
C	1	1	0	1	1
D	0	1	1	0	1
E	1	1	1	1	0.

Next define clique. A clique is the largest collection of three or more individuals with the property that any two of them are mutual friends.

This means that if three people are mutual friends and if each is a mutual friend of a fourth person, the three do not form a clique since they do not form the largest collection of mutual friends.

From the given incidence matrix we can obtain a symmetric matrix by considering only pairs connected by a double arrow, namely

$$A \leftrightarrow B, \ A \leftrightarrow C, \ A \leftrightarrow E, \ C \leftrightarrow E$$

$$B \leftrightarrow C \quad B \leftrightarrow D \quad C \leftrightarrow D \quad D \leftrightarrow E.$$

The resulting symmetric matrix S is

$$S = \quad$$

	A	B	C	D	E
A	0	1	1	0	1
B	1	0	1	1	0
C	1	1	0	1	1
D	0	1	1	0	1
E	1	0	1	1	0

To find cliques in a group examine the entries in the matrix S^3. The reason for this is that the entries along the diagonal of S^3 give the number of three stage relations between a person and himself. This means mutual friendships with at least two other people. We find S^3:

$$S^3 = \quad$$

	A	B	C	D	E
A	4	8	8	4	8
B	8	4	8	8	4
C	8	8	8	8	8
D	4	8	8	4	8
E	8	4	8	8	4

Now use the following facts

1. If $S_{ii}^{(3)}$ is positive, the person P_i belongs to at least one clique.

2. If $S_{ii}^{(3)} = 0$, the person belongs to no clique.

Since all the diagonal entries in S are positive every person in the group belongs to a clique. The size and number of cliques is found by using the theorem below:

Let $S = (s_{ij})$ be the symmetric matrix associated with the incidence matrix of a clique digraph. An individual is a member of exactly one clique with k members if and only if his diagonal entry in S^3 equals $(k - 1)(k - 2)$ i.e., $S_{ii}^{(3)} = (k - 1)(k - 2)$.

For the given digraph the number of possible members in a clique can be 3, 4, or 5. Since none of the diagonal entries is 2, 6, or 12 each person must belong to more than one clique.

Consider A whose diagonal entry is 4. Since 4 = 2 + 2, A belongs to two cliques each with three members. Similarly B, D and E belong to two cliques each containing three people. The diagonal entry 8 in S^3 can only be obtained from 2, 6 and 12 by 2 + 2 + 2 + 2 or 2 + 6. Thus C belongs to four cliques of three persons each or else to two cliques one containing 3 persons and the other containing 4 persons. But since no one else belongs to a 4-person clique, C must belong to 4 cliques of 3 persons each.

Consulting the matrix S we can determine the composition of the 4 cliques. They are: {A, B, C}, {A, C, E}, {B, C, D} and {C, D, E}.

● **PROBLEM 24-36**

Suppose that six individuals have been meeting in group therapy for a long time and their leader, who is not part of the group, has drawn the digraph G below to describe the influence relations among the various individuals.

Who is the leader of the group? Who is the person who influences no one else?

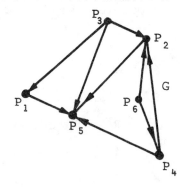

Solution: Every directed graph or digraph has a matrix representation, called the incidence matrix. The incidence matrix for the digraph G is given below:

	P_1	P_2	P_3	P_4	P_5	P_6
P_1	0	0	0	0	1	0
P_2	0	0	0	0	1	0
P_3	1	1	0	0	1	0
P_4	0	1	0	0	1	0
P_5	0	0	0	0	0	0
P_6	0	1	0	1	0	0

A 1 in the i, j position indicates that the ith person dominates (has influence over) the jth person.

Define the leader of the group to be the person who dominates the greatest number of people. Let P(i) denote the number of people that the ith person dominates.

P(1) = 1, P(2) = 1 P(3) = 3 P(4) = 2

P(5) = 0 P(6) = 2.

Note that no persons dominate themselves. P_3 influences three people - more than any other individual. Thus P_3 is called the leader of the group. P_5 is the person who influences no one.

● PROBLEM 24-37

Which of the following matrices can be interpreted as perfect communications matrices? For those that are, find the two stage and three stage communication lines that are feedbacks.

	A	B	C	D
A	0	1	1	2
B	1	0	1	0
C	1	1	0	1
D	2	0	1	0

	A	B	C
A	0	1	1
B	1	0	1
C	1	1	0

Solution: In a perfect communication model we have a group of vertices and a mode of communication between them. These could be telephone lines or highways. In such a model if vertex A communicates with vertex B, then necessarily vertex B communicates with vertex A.

First, draw the digraph of the given matrix.

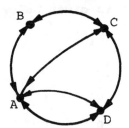

Fig. 1

Observe that the matrix for Fig. 1 is a perfect communication matrix. Examining the matrix itself, observe that it is symmetric. This follows from the fact that if A is connected by an edge to B, then B is connected by an edge to A.

If the original matrix is squared, we obtain the number of lines of communication between two vertices passing through exactly one other vertex.

$$\text{Now, } M^2 = \begin{bmatrix} 6 & 1 & 3 & 1 \\ 1 & 2 & 1 & 3 \\ 3 & 1 & 3 & 2 \\ 1 & 3 & 2 & 5 \end{bmatrix}$$

The entries given in M^2 can be verified from Fig. 1. For example A can communicate with C by the lines $A \to B \to C$, $A \to D \to C$, $A \to D \to C$. Similarly B can communicate with itself through $B \to A \to B$; $B \to C \to B$.

The matrix M^3 gives the number of lines of communication between two vertices passing through exactly two other vertices.

$$M^3 = \begin{bmatrix} 6 & 1 & 3 & 1 \\ 1 & 2 & 1 & 3 \\ 3 & 1 & 3 & 2 \\ 1 & 3 & 2 & 5 \end{bmatrix} \begin{bmatrix} 0 & 1 & 1 & 2 \\ 1 & 0 & 1 & 0 \\ 1 & 1 & 0 & 1 \\ 2 & 0 & 1 & 0 \end{bmatrix} = \begin{bmatrix} 6 & 9 & 8 & 15 \\ 9 & 2 & 6 & 3 \\ 8 & 6 & 6 & 9 \\ 15 & 3 & 9 & 4 \end{bmatrix}$$

The feedback from a vertex to itself is given by the entries along the main diagonal. To find out how many ways A can obtain two-stage or three-stage feedback, it is necessary to only look at the value in the diagonal of $M + M^2 + M^3$.

978

	A	B	C	D
A	12	11	12	18
B	11	4	8	6
C	12	8	9	12
D	18	6	12	9

$M + M^2 + M^3 =$ (matrix above)

Thus, A has 12 ways to get feedback. Similarly B has 4, C has 9 and D has 9.

The digraph for the second matrix is

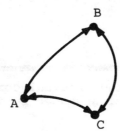

Fig. 2

Thus, the matrix is a perfect communication matrix. Observe that it is symmetric with only ones or zeros as elements. It is an incidence matrix. The square of the matrix is

	A	B	C
A	2	1	1
B	1	2	1
C	1	1	2

$M^2 =$ (matrix above)

This matrix illustrates the number of two stage communication lines between pairs of vertices in the group. The 2's on the diagonal indicate that there are two ways in which each vertex can get information back to itself through one other vertex. The matrix M^3 will give the number of three-stage communication lines between pairs of vertices. The values along the diagonal give the number of ways a vertex can get information back to itself through two other vertices. Here, A has two ways of obtaining feedback information, namely $A \rightarrow B \rightarrow C \rightarrow A$ and $A \rightarrow C \rightarrow B \rightarrow A$.

	A	B	C
A	2	3	3
B	3	2	3
C	3	3	2

$M^3 =$ (matrix above)

Again, to find out how many ways A, B, or C can obtain

two-stage or three-stage feedback, we need only look at the values in the diagonal of $M + M^2 + M^3$

$$
M + M^2 + M^3 = \begin{array}{c} \\ A \\ B \\ C \end{array}
\begin{array}{ccc}
A & B & C \\
4 & 5 & 5 \\
5 & 4 & 5 \\
5 & 5 & 4
\end{array}
$$

Thus, A, B and C have four ways to get feedback. The 5's in row i column j indicate the total number of ways for i to communicate with j using one stage, two stages or three stages.

● **PROBLEM** 24-38

The following figure shows the internal communication digraph as it exists in an organization headed by five officials, A, B, C, D and E:

a) Construct the incidence matrix corresponding to the digraph

b) Is the graph connected?

c) Which officials are liaison officials?

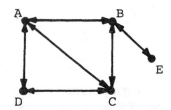

Fig. 1

Solution: The graph in Fig. 1 is an example of a business communication model. This model obeys the following assumptions:

1) If official A communicates with official B then necessarily B communicates with A.

2) No official communicates with himself

3) Two officials may or may not communicate at all

4) There exists at most one line of direct communication between any two officials.

Assumptions 1) and 4) guarantee that the matrix of the digraph will be a symmetric incidence matrix.

a) The incidence matrix is

	A	B	C	D	E
A	0	1	1	1	0
B	1	0	1	0	1
C	1	1	0	1	0
D	1	0	1	0	0
E	0	1	0	0	0

b) We first define the notion of a path. A path between two vertices v_1 and v_2 is a collection of edges and vertices of the form

$$v_1 \rightarrow v_3 \rightarrow v_4 \rightarrow \dots \rightarrow v_2$$

in which no vertex is repeated. For example, in Fig. 1,

$$A \rightarrow B \rightarrow C \rightarrow A \rightarrow D$$

is not a path but $A \rightarrow B \rightarrow C \rightarrow D$ is a path. Also, $E \rightarrow B \rightarrow C \rightarrow B \rightarrow A$ is not a path but $E \rightarrow B \rightarrow A$ is a path from E to A.

A graph is said to be connected if there is a path between every two officials. Observe from Fig. 1 that the given digraph is connected.

c) A liaison official in a connected digraph is an official whose removal from the digraph results in a disconnected graph. In terms of organization theory, the removal of a liaison official will cause a break-down in some line of communication between two other officials.

The following theorem is useful in determining liaison officials:

A business communications graph is disconnected if and only if the corresponding incidence matrix M of the graph has the property that $M + M^2 + M^3 + \dots + M^{n-1}$ has one or more zero entries.

Intuitively, this means that a graph is disconnected when there is no path between two officials (with n officials, any path can have at most n-1 vertices). Recall that the kth power of an incidence matrix corresponds to the kth stage communication between officials.

To determine whether a person P from among n officials is a liaison official, delete from the incidence matrix M the column and the row corresponding to P. Call the deleted matrix N. This matrix represents a disconnected digraph if

$$N + N^2 + \dots + N^{n-2}$$

contains zero entries. In this case P is a liaison official.

Thus, to check whether A is a liaison official, first delete the first row and column from the incidence matrix to obtain

	B	C	D	E
B	0	1	0	1
C	1	0	1	0
D	0	1	0	0
E	1	0	0	0

Now find N^2 and N^3:

$$N^2 = \begin{bmatrix} 2 & 0 & 1 & 0 \\ 0 & 2 & 0 & 1 \\ 1 & 0 & 1 & 0 \\ 0 & 1 & 0 & 1 \end{bmatrix};$$

$$N^3 = \begin{bmatrix} 0 & 3 & 0 & 2 \\ 3 & 0 & 2 & 0 \\ 0 & 2 & 0 & 1 \\ 2 & 0 & 1 & 0 \end{bmatrix}$$

Then $N + N^2 + N^3$

$$= \begin{bmatrix} 2 & 4 & 1 & 3 \\ 4 & 2 & 3 & 1 \\ 1 & 3 & 1 & 1 \\ 3 & 1 & 1 & 1 \end{bmatrix}$$

A is not a liaison official.

Similarly, we find that C, D and E are not liaison officials and that only B is a liaison official. Examining the digraph, observe that if B is removed, no communication is possible between E and the other officials.

● **PROBLEM 24-39**

Recent archeological investigations in Asia Minor have unearthed clay tablets that give details of transactions between merchants from 1900 B.C. Suppose the following hypothetical communication matrix represents tablets between members of a group of merchants. Show how this matrix can help to identify contemporary merchants.

	1	2	3	4	5	6	7	8	9	10
1	0	0	0	0	0	1	0	0	0	1
2	0	0	1	0	1	0	0	0	0	0
3	0	1	0	0	1	0	0	0	0	0
4	0	0	0	0	0	0	0	1	0	1
5	0	1	1	0	0	0	0	0	0	0
6	0	0	0	0	0	0	0	0	0	0
7	0	0	0	0	0	0	0	1	0	0
8	0	0	1	0	0	0	0	0	1	0
9	0	0	0	0	0	0	1	0	0	0
10	1	0	0	0	0	0	0	0	0	0

Solution: An entry of 1 in the ith row and jth column in-dicates that merchant i sent a tablet to merchant j. To solve the problem, note that if we can obtain the equivalence classes for the merchants in the matrix, these equivalence classes will represent contemporary merchants.

The equivalence classes are constructed by taking the intersections of i) the list of merchants the kth merchant could have sent a tablet to and ii) the list of merchants the kth merchant could have received a tablet from.

The send-to list is:

Merchant #	Send-to
1.	1, 10, 6
2.	2, 3, 5
3.	3, 2, 5
4.	4, 8, 10, 3, 9, 1, 2, 5, 7, 6
5.	5, 2, 3
6.	6
7.	7, 8, 3, 9, 2, 5
8.	8, 3, 9, 2, 7, 5
9.	9, 7, 8, 3, 2, 5
10.	10, 1, 6

To construct the Receive-from list, we go down the send-list. If merchant k is on merchant i's send-to list, then merchant i is on merchant k's received-from list. Thus,

Merchant #	Send-to	Receive-from
1.	1, 10, 6	1, 4, 10
2.	2, 3, 5	2, 3, 4, 5, 7, 8, 9
3.	3, 2, 5	2, 3, 4, 5, 7, 8, 9
4.	4,8,10,3,9,1,2,5,7,6	4
5.	5, 2, 3	2, 3, 4, 5, 7, 8, 9
6.	6	1, 4, 6, 10
7.	7, 8, 3, 2, 5, 9	4, 7, 8, 9
8.	8, 3, 9, 2, 7, 5	4, 7, 8, 9
9.	9, 7, 8, 3, 2, 5	4, 8, 9, 7
10.	10, 1, 6	4, 10, 1

Taking the intersections of the send-to and receive-from lists we obtain the equivalence classes {4} {7, 8, 9}, {2, 3, 5}, {1, 10}, {6} . Members of a given equivalence class are contemporary. But is is not clear which of the equivalence classes is earlier, merely from the one-way communication between them. However, further analysis of the content of the messages might help to establish this. For instance, if one of the messages exchanged among merchants 7, 8 and 9 were related to one of the messages exchanged among 2, 3 and 5, then it would be reasonable to assume that they are all contemporaries. This, of course, is an empirical, not a mathematical, matter.

MATRICES

DEFINITION 1. Matrix of order m×n. A *matrix*, $A_{m \times n}$, *of order* $m \times n$ (read "*m* by *n*") is a rectangular array of $m \times n$ numbers (or functions) called *elements* arranged in m rows and n columns subject to a set of rules of operation (to be defined) and is denoted by the symbol

$$A_{m \times n} = \begin{bmatrix} a_{11} & a_{12} & \cdots & a_{1n} \\ a_{21} & a_{22} & \cdots & a_{2n} \\ \cdot & & & \\ \cdot & & & \\ \cdot & & & \\ a_{m1} & a_{m2} & \cdots & a_{mn} \end{bmatrix} \qquad [2.1]$$

where a_{ij} is the element of $A_{m \times n}$ located in the *i*th row and the *j*th column.

We shall use upper-case letters to represent matrices. When the order is immaterial in the discussion, a single letter A, without subscripts, will be used to denote a matrix. We should point out also that when the order is the only information required, expression $[2.1]$ may then be abbreviated as

$$A = [a_{ij}]_{m \times n}$$

which gives essentially the same information as that in Equation $[2.1]$. The element a_{ij} is called the typical element or the (i,j) element. As in the case of a determinant, the first subscript i of a_{ij} indicates its row position, and the second subscript, j, shows its column position.

The reader should note that a determinant is always a square array; whereas, a matrix, say $A_{m \times n}$, is, in general, rectangular in shape unlesss $m = n$; that is, unless the number of rows equal the number of columns. In this case, $A_{m \times n}$ becomes a square matrix and a single-subscripted letter A_n may be used to indicate such information.

DEFINITION 2. Real matrix. A matrix $R = [r_{ij}]_{m \times n}$ is called a *real matrix* if and only if r_{ij} is real for every pair of i and j; that is, R is a real matrix if and only if *all* its elements are real.

In general, the elements of a matrix may be real or complex (or even functions). If one of its elements is complex, the matrix is a "complex" matrix.

DEFINITION 3. Conjugate of a matrix. Let $A = [a_{ij}]_{m \times n}$ be a matrix of complex elements of order mn. If every element $a_{ij} = \alpha_{ij} + \beta_{ij} \sqrt{-1}$ (α, β real) of A is replaced by its complex conjugate $\bar{a}_{ij} = \alpha_{ij} - \beta_{ij} \sqrt{-1}$, the resulting matrix, denoted by $\bar{A}$, is defined as the *conjugate of A*.

DEFINITION 4. Row and column matrices. A *row matrix* of m elements is a matrix of order $1 \times m$; and a *column matrix* of n elements is matrix of order $1 \times n$.

DEFINITION 5. Scalar matrix. A *scalar matrix* is defined as a diagonal matrix such that all the elements in the major diagonal are equal to a scalar In other words, a square matrix $K = [k_{ij}]_n$ of order n is called a diagona matrix if and only if $k_{ii} = \alpha$, a scalar for all i; and $k_{ij} = 0$ for all $i \neq j$.

DEFINITION 6. Submatrix of a matrix. A *submatrix* of a given matrix is the resulting matrix that is obtained by deleting certain rows and/or column of the given matrix.

DEFINITION 7. Rank of a matrix. The *rank* of a matrix A of order $m \times n$ is defined to be the order of the determinant of a highest-order non singular submatrix A_r of A. We denote the rank of A by $R(A) = r$ and say that matrix A is of rank r.

DEFINITION 8. Elementary transformations of a matrix. The following three operations are called the *elementary transformations* of a matrix:

(1) Interchange of any two rows (or columns).

(2) Multiplication of all the elements of any row (or column) by a nonzero constant.

(3) Addition of the elements of a row (or column) multiplied by a nonzero constant to the corresponding elements of another row (or column).

THEOREM 1. The rank of a matrix will remain unaltered under (repeated applications of) any or all of the three operations defined in Definition

DEFINITION 9. Equality of matrices. Two matrices, $A = [a_{ij}]_{m \times n}$ and $B = [b_{ij}]_{p \times q}$, are said to be *equal* if and only if

(a) A and B are of the same order: $m = p$, $n = q$; and

(b) $a_{ij} = b_{ij}$ for every pair of i and j.

In short, A and B are equal if and only if they are of the same order and the corresponding elements are equal one by one.

DEFINITION 10. Conformability for addition. Two matrices A and B ar said to be *conformable* for addition if they both have the same order.

DEFINITION 11. Addition of matrices. If two matrices A and B are co formable for addition and if $A = [a_{ij}]_{m \times n}$ and $B = [b_{ij}]_{m \times n}$, then the *su* $A + B$, is defined as $A + B = [(a_{ij} + b_{ij})]_{m \times n}$.

DEFINITION 12. Negative of a matrix. If $A = [a_{ij}]_{m \times n}$, then the *negative* of A, denoted by $-A$, is defined as $-A = [-a_{ij}]_{m \times n}$; that is, the negative of a matrix is obtained by taking the negative of every element of the matrix.

DEFINITION 13. Zero (null) matrix. A matrix of order $m \times n$ is called a *zero matrix* (or *null matrix*) if all its elements are equal to zero.

DEFINITION 14. Subtraction of matrices. If $A = [a_{ij}]_{m \times n}$ and $B = [b_{ij}]_{m \times n}$, then the *difference*, $A - B$, is defined as $A = B = A + (-B) = [(a_{ij} - b_{ij})]_{m \times n}$.

THEOREM 2. If A, B, and C are three matrices of the same order, then both the commutative and associative laws for addition hold for the three matrices; that is,

$$A + B = B + A \quad \text{(commutative law)}$$

$$A + (B + C) = (A + B) + C \quad \text{(associative law)}$$

DEFINITION 15. Scalar multiple of a matrix. The *product of a matrix* $A = [a_{ij}]_{m \times n}$ *and a scalar* k is the matrix $kA = [ka_{ij}]_{m \times n}$; that is, when a matrix A is multiplied by a scalar k, every element of A is multiplied by k.

DEFINITION 16. Conformability for multiplication. Two matrices, $A = [a_{ij}]_{m \times n}$ and $B = [b_{ij}]_{p \times q}$, are said to be *conformable for multiplication* in the order AB if and only if $n = p$. In other words, A and B are conformable for multiplication in the order AB if and only if the number of columns in A is equal to the number of rows in B.

DEFINITION 17. Product of two matrices. The *product* C *of two matrices*, $A = [a_{ij}]_{m \times n}$ and $B = [b_{ij}]_{n \times q}$, which are conformable for multiplication in the order AB is defined by

$$A \times B = C = [c_{ij}]_{m \times q}$$

where

$$c_{ij} = a_{i1}b_{1j} + a_{i2}b_{2j} + \cdots + a_{in}b_{nj} = \sum_{k=1}^{n} a_{ik}b_{kj}$$

In other words, we say that the (i,j) element c_{ij} of the product matrix $C = A \times B$ is the sum of the products of the elements in the ith row of A and the corresponding elements in the jth column of B.

In general, matrix multiplication is not commutative

$$\mathbf{AB} \neq \mathbf{BA}$$

Matrix multiplication is associative

$$A(BC) = (AB)C$$

The distributive law for multiplication and addition holds as in the case of scalars,

$$(A + B)C = AC + BC$$
$$C(A + B) = CA + CB$$

In some applications, the term-by-term product of two matrices A and B of identical order is defined as

$$C = A * B$$

where

$$c_{ij} = a_{ij}b_{ij}$$

$$(ABC)' = C'B'A'$$
$$(ABC)^H = C^H B^H A^H$$

If both A and B are symmetric, then $(AB)' = BA$. Note that the product of two symmetric matrices is generally not symmetric.

DEFINITION 18. Transpose of a matrix. The matrix A^T of order $n \times m$, obtained by interchanging the rows and columns of an $m \times n$ matrix A, is defined as the *transpose* of A. In symbolic terms, if $A = [a_{ij}]_{m \times n}$, then $A^T = [a_{ji}]_{n \times m}$.

Property 1: The transpose of a matrix always exists.

Property 2: The orders of a matrix and its transpose are the same if and only if the matrix is a square matrix.

Property 3: The transpose of the transpose of a matrix is the matrix itself; that is, $(A^T)^T = A$.

Property 4: The transpose of the scalar multiple of a matrix is equal to the scalar multiple of its transpose; that is, $(kA)^T = kA^T$.

Property 5: The transpose of the sum of two matrices is equal to the sum of their transposes. Thus, if A and B are two matrices of the same order, we have $(A + B)^T = A^T + B^T$.

Property 6: The transpose of the product of two matrices is equal to the product of the two transposes *in the reverse order*. That is, if A and B are conformable for multiplication, we write $(AB)^T = B^T A^T$.

APPENDIX 2

Physical Constants and Conversion Factors

The tables in this chapter supply some of the more commonly needed physical constants and conversion factors.

The International System of Units (SI) established in 1960 by the General Conference of Weights and Measures under the Treaty of the Meter is based upon: the meter (m) for length, defined as 1 650 763.73 wave-lengths in vacuum corresponding to the transition $2p_{10} - 5d_5$ of kypton 86; the kilogram (kg) for mass, defined as the mass of the prototype kilogram at Sevres, France; the second (s) for time, defined as the duration of 9 192 631 770 periods of the radiation corresponding to the transition between the two hyperfine levels of cesium 133; the kelvin (K) for temperature, defined as 1/273.16 of the thermodynamic temperature of the triple point of water; the ampere (A) for electric current, defined as the current which, if flowing in two infinitely long parallel wires *in vacuo* separated by one meter, would produce a force of 2×10^{-7} newtons per meter of length between the wires; and the candela (cd) for luminous intensity, defined as the luminous intensity of 1/600 000 square meter of a perfect radiator at the temperature of freezing platinum.

All other units of SI are derived from these base units by assigning the value unity to the proportionality constants in the defining equations (official symbols for other SI units appear in Tables 2.1 and 2.2). Taking 1/100 of the meter as the unit for length and 1/1000 of the kilogram as the unit for mass gives rise similarly to the cgs system, often used in physics and chemistry.

SI, as it is ordinarily used in electromagnetism, is a rationalized system, i.e., the electromagnetic units of SI relate to the quantities appearing in the so-called rationalized electromagnetic equations. Thus, the force per unit length between two current-carrying parallel wires of infinite length separated by unit distance *in vacuo* is $2f = \mu_0 i_1 i_2/4\pi$, where μ_0 has the value $4\pi \times 10^{-7}$H/m. The force between two electric charges *in vacuo* is correspondingly given by $f = q_1 q_2/4\pi\epsilon_0 r^2$, ϵ_0 having the value $1/\mu_0 c^2$, where c is the speed of light in meters per second. ($\epsilon_0 \sim 8.854 \times 10^{-12}$F/m)

Setting μ_0 equal to unity and deleting 4π from the denominator in the first equation above defines the cgs-emu system. Setting ϵ_0 equal to unity and deleting 4π from the denominator in the second equation correspondingly defines the cgs-esu system. The cgs-emu and the cgs-esu systems are most frequently used in the unrationalized forms.

Table 2.1. Common Units and Conversion Factors, CGS System and SI

Quantity	SI Name	CGS Name	Factor
Force	newton (N)	dyne	10^5
Energy	joule (J)	erg	10^7
Power	watt (W)	....	10^7

Table 2.2. Names and Conversion Factors for Electric and Magnetic Units

Quantity	SI name	emu name	esu name	emu-SI factors	esu-SI factors
Current	ampere (A)	abampere	statampere	10^{-1}	$\sim 3 \times 10^9$
Charge	coulomb (C)	abcoulomb	statcoulomb	10^{-1}	$\sim 3 \times 10^9$
Potential	volt (V)	abvolt	statvolt	10^8	$\sim (1/3) \times 10^{-2}$
Resistance	ohm (Ω)	abohm	statohm	10^9	$\sim (1/9) \times 10^{-11}$
Inductance	henry (H)	centimeter	....	10^9	$\sim (1/9) \times 10^{-11}$
Capacitance	farad (F)	....	centimeter	10^{-9}	$\sim 9 \times 10^{11}$
Magnetizing force	$A \cdot m^{-1}$	oersted	....	$4\pi \times 10^{-3}$	$\sim 3 \times 10^9$
Magnetomotive force	A	gilbert	....	$4\pi \times 10^{-1}$	$\sim 3/10^6$
Magnetic flux	weber (Wb)	maxwell	....	10^8	$\sim (1/3) \times 10^{-2}$
Magnetic flux density	tesla (T)	gauss (G)	....	10^4	$\sim (1/3) \times 10^{-6}$
Electric displacement	....	....	....	10^{-5}	$\sim 3 \times 10^5$

Example: If the value assigned to a current is 100 amperes its value in abamperes is $100 \times 10^{-1} = 10$.

989

The values of constants given in Table 2.3 are based on an adjustment by Taylor, Parker, and Langenberg, Rev. Mod. Phys. 41, p.375 (1969). They are being considered for adoption by the Task Group on Fundamental Constants of the Committee on Data for Science and Technology, International Council of Scientific Unions. The uncertainties given are standard errors estimated from the experimental data included in the adjustment. Where applicable, values are based on the unified scale of atomic masses in which the atomic mass unit (u) is defined as 1/12 o.: the mass of the atom of the ^{12}C nuclide.

Table 2.3. Adjusted Values of Constants

Constant	Symbol	Value	Uncertainty ‡	Systeme International (SI)		Centimeter-gram-second (CGS)	
Speed of light in vacuum	c	2.997 925 0	±10	$\times 10^8$	m/s	$\times 10^{10}$	cm/s
Elementary charge	e	1.602 191 7	70	10^{-19}	C	10^{-20}	$cm^{1/2}g^{1/2}$ *
		4.803 250	21			10^{-10}	$cm^{3/2}g^{1/2}s^{-1}$ †
Avogadro constant	N_A	6.022 169	40	10^{23}	mol^{-1}	10^{23}	mol^{-1}
Atomic mass unit	u	1.660 531	11	10^{-27}	kg	10^{-24}	g
Electron rest mass	m_e	9.109 558	54	10^{-31}	kg	10^{-28}	g
		5.485 930	34	10^{-4}	u	10^{-4}	u
Proton rest mass	m_p	1.672 614	11	10^{-27}	kg	10^{-24}	g
		1.007 276 61	8	10^0	u	10^0	u
Neutron rest mass	m_n	1.674 920	11	10^{-27}	kg	10^{-24}	g
		1.008 665 20	10	10^0	u	10^0	u
Faraday constant	F	9.648 670	54	10^4	C/mol	10^3	$cm^{1/2}g^{1/2}mol^{-1}$ *
		2.892 599	16			10^{14}	$cm^{3/2}g^{1/2}s^{-1}mol^{-1}$ †
Planck constant	h	6.626 196	50	10^{-34}	J $\cdot$ s	10^{-27}	erg $\cdot$ s
	$\hbar$	1.054 591 9	80	10^{-34}	J $\cdot$ s	10^{-27}	erg $\cdot$ s
Fine structure constant	a	7.297 351	11	10^{-3}		10^{-3}	
	$1/a$	1.370 360 2	21	10^2		10^2	
Charge to mass ratio for electron	e/m_e	1.758 802 8	54	10^{11}	C/kg	10^7	$cm^{1/2}g^{1/2}$ *
		5.272 759	16			10^{17}	$cm^{3/2}g^{-1/2}s^{-1}$ †
Quantum-charge ratio	h/e	4.135 708	14	10^{-15}	J $\cdot$ s/C	10^{-7}	$cm^{3/2}g^{1/2}s^{-1}$ *
		1.379 523 4	46			10^{-17}	$cm^{1/2}g^{1/2}$ †
Compton wavelength of electron	λ_C	2.426 309 6	74	10^{-12}	m	10^{-10}	cm
	$\lambda_C/2\pi$	3.861 592	12	10^{-13}	m	10^{-11}	cm
Compton wavelength of proton	$\lambda_{C,p}$	1.321 440 9	90	10^{-15}	m	10^{-13}	cm
	$\lambda_{C,p}/2\pi$	2.103 139	14	10^{-16}	m	10^{-14}	cm
Rydberg constant	$R\infty$	1.097 373 12	11	10^7	m^{-1}	10^5	cm^{-1}
Bohr radius	a_0	5.291 771 5	81	10^{-11}	m	10^{-9}	cm
Electron radius	r_e	2.817 939	13	10^{-15}	m	10^{-13}	cm
Gyromagnetic ratio of proton	γ	2.675 196 5	82	10^8	rad $\cdot$ $s^{-1}T^{-1}$	10^4	rad $\cdot$ $s^{-1}G^{-1}$ *
	$\gamma/2\pi$	4.257 707	13	10^7	Hz/T	10^3	$s^{-1}G^{-1}$ *
(uncorrected for diamagnetism, H_2O)	γ'	2.675 127 0	82	10^8	rad $\cdot$ $s^{-1}T^{-1}$	10^4	rad $\cdot$ $s^{-1}G^{-1}$ *
	$\gamma'/2\pi$	4.257 597	13	10^7	Hz/T	10^3	$s^{-1}G^{-1}$ *
Bohr magneton	μ_B	9.274 096	65	10^{-24}	J/T	10^{-21}	erg/G *
Nuclear magneton	μ_N	5.050 951	50	10^{-27}	J/T	10^{-24}	erg/G *
Proton moment	μ_p	1.410 620 3	99	10^{-26}	J/T	10^{-23}	erg/G *
	μ_p/μ_N	2.792 782	17	10^0		10^0	
(uncorrected for diamagnetism, H_2O)	μ'_p/μ_N	2.792 709	17	10^0		10^0	
Gas constant	R	8.314 34	35	10^0	J $\cdot$ K^{-1} mol^{-1}	10^7	erg $\cdot$ K^{-1} mol^{-1}
Normal volume perfect gas	V_0	2.241 36	39	10^{-2}	m^3/mol	10^4	cm^3/mol
Boltzmann constant	k	1.380 622	59	10^{-23}	J/K	10^{-16}	erg/K
First radiation constant ($8\pi hc$)	c_1	4.992 579	38	10^{-24}	J $\cdot$ m	10^{-15}	erg $\cdot$ cm
Second radiation constant	c_2	1.438 833	61	10^{-2}	m $\cdot$ K	10^0	cm $\cdot$ K
Stefan-Boltzmann constant	σ	5.669 61	96	10^{-8}	W $\cdot$ $m^{-2}K^{-4}$	10^{-5}	erg $\cdot$ $cm^{-2}s^{-1}K^{-4}$
Gravitational constant	G	6.673 2	31	10^{-11}	N $\cdot$ m^2/kg^2	10^{-8}	dyn $\cdot$ cm^2/g^2

‡Based on 1 std. dev; applies to last digits in preceding column.   *Electromagnetic system.   †Electrostatic system.

Table 2.4. Miscellaneous Conversion Factors

Standard gravity, g_0	= 9.806 65 meters per second per second*
Standard atmospheric pressure, P_0	= 1.013 25 $\times$ 10^5 newtons per square meter*
	= 1.013 25 $\times$ 10^6 dynes per square centimeter*
1 thermodynamic calorie,[1] cal_c	= 4.1840 joules*
1 IT calorie[2], cal_s	= 4.1868 joules*
1 liter, l	= 10^{-3} cubic meter*
1 angstrom unit, Å	= 10^{-10} meter*
1 bar	= 10^5 newtons per square meter*
	= 10^6 dynes per square centimeter*
1 gal	= 10^{-2} meter per second per second*
	= 1 centimeter per second per second*
1 astronomical unit, AU	= 1.496 $\times$ 10^{11} meters
1 light year	= 9.46 $\times$ 10^{15} meters
1 parsec	= 3.08 $\times$ 10^{16} meters
	= 3.26 light years

1 curie, the quantity of radioactive material undergoing 3.7×10^{10} disintegrations per second*.

1 roentgen, the exposure of x- or gamma radiation which produces together with its secondaries 2.082×10^9 electron-ion pairs in 0.001 293 gram of air.

The index of refraction of the atmosphere for radio waves of frequency less than 3×10^{10} Hz is given by $(n - 1)10^6 = (77.6/t) (p + 4810e/t)$, where n is the refractive index; t, temperature in kelvins; p, total pressure in millibars; e, water vapor partial pressure in millibars.

Factors for converting the customary United States units to units of the metric system are given in Table 2.5.

Table 2.5. Factors for Converting Customary U.S. Units to SI Units

1 yard	0.914 4 meter*
1 foot	0.304 8 meter*
1 inch	0.025 4 meter*
1 statute mile	1 609.344 meters*
1 nautical mile (international)	1 852 meters*
1 pound (avdp.)	0.453 592 37 kilogram*
1 oz. (avdp.)	0.028 349 52 kilogram
1 pound force	4.448 22 newtons
1 slug	14.593 9 kilograms
1 poundal	0.138 255 newtons
1 foot pound	1.355 82 joules
Temperature (Fahrenheit)	32 + (9/5) Celsius temperature*
1 British thermal unit[3]	1055 joules

Geodetic constants for the international (Hayford) spheroid are given in Table 2.6. The gravity values are on the basis of the revised Potsdam value. They are about 14 parts per million smaller than previous values. They are calculated for the surface of the geoid by the international formula.

Table 2.6. Geodetic Constants

$a = 6\ 378\ 388$ m; $f = 1/297$; $b = 6\ 356\ 912$ m

Latitude	Length of 1' of longitude	Length of 1' of latitude	g
	Meters	*Meters*	*m/s²*
0°	1 855.398	1 842.925	9.780 350
15	1 792.580	1 844.170	9.783 800
30	1 608.174	1 847.580	9.793 238
45	1 314.175	1 852.256	9.806 154
60	930.047	1 856.951	9.819 099
75	481.725	1 860.401	9.828 593
90	0	1 861.666	9.832 072

[1] Used principally by chemists.

[2] Used principally by engineers.

[3] Various definitions are given for the British thermal unit. This represents a rounded mean value differing from none of the more important definitions by more than 3 in 10^4.

* Exact value.

Elementary Analytical Methods

3.1. Binomial Theorem and Binomial Coefficients; Arithmetic and Geometric Progressions; Arithmetic, Geometric, Harmonic and Generalized Means

Binomial Theorem

3.1.1

$$(a+b)^n = a^n + \binom{n}{1}a^{n-1}b + \binom{n}{2}a^{n-2}b^2$$

$$+ \binom{n}{3}a^{n-3}b^3 + \ldots + b^n$$

(n a positive integer)

Binomial Coefficients (see chapter 24)

3.1.2

$$\binom{n}{k} = {}_nC_k = \frac{n(n-1)\ldots(n-k+1)}{k!} = \frac{n!}{(n-k)!k!}$$

3.1.3 $\quad \binom{n}{k} = \binom{n}{n-k} = (-1)^k\binom{k-n-1}{k}$

3.1.4 $\quad \binom{n+1}{k} = \binom{n}{k} + \binom{n}{k-1}$

3.1.5 $\quad \binom{n}{0} = \binom{n}{n} = 1$

3.1.6 $\quad 1 + \binom{n}{1} + \binom{n}{2} + \ldots + \binom{n}{n} = 2^n$

3.1.7 $\quad 1 - \binom{n}{1} + \binom{n}{2} - \ldots + (-1)^n\binom{n}{n} = 0$

Table of Binomial Coefficients $\binom{n}{k}$

3.1.8

n \ k	0	1	2	3	4	5	6	7	8	9	10	11	12
1___	1	1											
2___	1	2	1										
3___	1	3	3	1									
4___	1	4	6	4	1								
5___	1	5	10	10	5	1							
6___	1	6	15	20	15	6	1						
7___	1	7	21	35	35	21	7	1					
8___	1	8	28	56	70	56	28	8	1				
9___	1	9	36	84	126	126	84	36	9	1			
10___	1	10	45	120	210	252	210	120	45	10	1		
11___	1	11	55	165	330	462	462	330	165	55	11	1	
12___	1	12	66	220	495	792	924	792	495	220	66	12	1

3.1.9

Sum of Arithmetic Progression to n Terms

$$a + (a+d) + (a+2d) + \ldots + (a+(n-1)d)$$

$$= na + \frac{1}{2}n(n-1)d = \frac{n}{2}(a+l),$$

last term in series $= l = a + (n-1)d$

Sum of Geometric Progression to n Terms

3.1.10

$$s_n = a + ar + ar^2 + \ldots + ar^{n-1} = \frac{a(1-r^n)}{1-r}$$

$$\lim_{n\to\infty} s_n = a/(1-r) \qquad (-1 < r < 1)$$

Arithmetic Mean of n Quantities A

3.1.11 $\quad A = \dfrac{a_1 + a_2 + \ldots + a_n}{n}$

Geometric Mean of n Quantities G

3.1.12 $\quad G = (a_1 a_2 \ldots a_n)^{1/n} \qquad (a_k > 0, k=1,2,\ldots,n)$

Harmonic Mean of n Quantities H

3.1.13

$$\frac{1}{H} = \frac{1}{n}\left(\frac{1}{a_1} + \frac{1}{a_2} + \ldots + \frac{1}{a_n}\right) \qquad (a_k > 0, k=1,2,\ldots,n)$$

Generalized Mean

3.1.14 $\quad M(t) = \left(\dfrac{1}{n}\sum\limits_{k=1}^{n} a_k^t\right)^{1/t}$

3.1.15 $\quad M(t) = 0 (t < 0, \text{ some } a_k \text{ zero})$

3.1.16 $\quad \lim\limits_{t\to\infty} M(t) = \text{max.} \qquad (a_1, a_2, \ldots, a_n) = \text{max. } a$

3.1.17 $\quad \lim\limits_{t\to-\infty} M(t) = \text{min.} \qquad (a_1, a_2, \ldots, a_n) = \text{min. } a$

3.1.18 $\qquad \lim\limits_{t\to 0} M(t) = G$

3.1.19 $\qquad M(1) = A$

3.1.20 $\qquad M(-1) = H$

3.2. Inequalities

Relation Between Arithmetic, Geometric, Harmonic and Generalized Means

3.2.1

$A \geq G \geq H$, equality if and only if $a_1 = a_2 = \ldots = a_n$

3.2.2 $\qquad \text{min. } a < M(t) < \text{max. } a$

3.2.3 $\qquad$ min. $a < G < $ max. a

equality holds if all a_k are equal, or $t < 0$
and an a_k is zero

3.2.4 $M(t) < M(s)$ if $t < s$ unless all a_k are equal,
or $s < 0$ and an a_k is zero.

Triangle Inequalities

3.2.5 $\qquad |a_1| - |a_2| \le |a_1 + a_2| \le |a_1| + |a_2|$

3.2.6 $\qquad \left| \sum\limits_{k=1}^{n} a_k \right| \le \sum\limits_{k=1}^{n} |a_k|$

Chebyshev's Inequality

If $a_1 \ge a_2 \ge a_3 \ge \ldots \ge a_n$
$\qquad b_1 \ge b_2 \ge b_3 \ge \ldots \ge b_n$

3.2.7 $\qquad n \sum\limits_{k=1}^{n} a_k b_k \ge \left(\sum\limits_{k=1}^{n} a_k \right) \left(\sum\limits_{k=1}^{n} b_k \right)$

Hölder's Inequality for Sums

If $\dfrac{1}{p} + \dfrac{1}{q} = 1, p > 1, q > 1$

3.2.8 $\qquad \sum\limits_{k=1}^{n} |a_k b_k| \le \left(\sum\limits_{k=1}^{n} |a_k|^p \right)^{1/p} \left(\sum\limits_{k=1}^{n} |b_k|^q \right)^{1/q};$

equality holds if and only if $|b_k| = c|a_k|^{p-1}$ $(c = $ constant $> 0)$. If $p = q = 2$ we get

Cauchy's Inequality

3.2.9
$\left[\sum\limits_{k=1}^{n} a_k b_k \right]^2 \le \sum\limits_{k=1}^{n} a_k^2 \sum\limits_{k=1}^{n} b_k^2$ (equality for $a_k = cb_k$,

c constant).

Hölder's Inequality for Integrals

If $\dfrac{1}{p} + \dfrac{1}{q} = 1, p > 1, q > 1$

3.2.10
$\int_a^b |f(x) g(x)| dx \le \left[\int_a^b |f(x)|^p dx \right]^{1/p} \left[\int_a^b |g(x)|^q dx \right]^{1/q}$

equality holds if and only if $|g(x)| = c|f(x)|^{p-1}$
$(c = $ constant $> 0)$.
$\quad$ If $p = q = 2$ we get

Schwarz's Inequality

3.2.11
$\left[\int_a^b f(x) g(x) dx \right]^2 \le \int_a^b [f(x)]^2 dx \int_a^b [g(x)]^2 dx$

Minkowski's Inequality for Sums

If $p > 1$ and $a_k, b_k > 0$ for all k,

3.2.12
$\left(\sum\limits_{k=1}^{n} (a_k + b_k)^p \right)^{1/p} \le \left(\sum\limits_{k=1}^{n} a_k^p \right)^{1/p} + \left(\sum\limits_{k=1}^{n} b_k^p \right)^{1/p},$

equality holds if and only if $b_k = ca_k$ $(c = $ constant $> 0)$.

Minkowski's Inequality for Integrals

If $p > 1$,

3.2.13
$\left(\int_a^b |f(x) + g(x)|^p dx \right)^{1/p} \le \left(\int_a^b |f(x)|^p dx \right)^{1/p}$

$\qquad + \left(\int_a^b |g(x)|^p dx \right)^{1/p}$

equality holds if and only if $g(x) = cf(x)$ $(c = $ constant $> 0)$.

3.3. Rules for Differentiation and Integration
Derivatives

3.3.1 $\qquad \dfrac{d}{dx}(cu) = c\dfrac{du}{dx}, c$ constant

3.3.2 $\qquad \dfrac{d}{dx}(u+v) = \dfrac{du}{dx} + \dfrac{dv}{dx}$

3.3.3 $\qquad \dfrac{d}{dx}(uv) = u\dfrac{dv}{dx} + v\dfrac{du}{dx}$

3.3.4 $\qquad \dfrac{d}{dx}(u/v) = \dfrac{v \, du/dx - u \, dv/dx}{v^2}$

3.3.5 $\qquad \dfrac{d}{dx} u(v) = \dfrac{du}{dv}\dfrac{dv}{dx}$

3.3.6 $\qquad \dfrac{d}{dx}(u^v) = u^v \left(\dfrac{v}{u}\dfrac{du}{dx} + \ln u \dfrac{dv}{dx} \right)$

Leibniz's Theorem for Differentiation of an Integral

3.3.7
$\dfrac{d}{dc} \int_{a(c)}^{b(c)} f(x, c) dx$

$\qquad = \int_{a(c)}^{b(c)} \dfrac{\partial}{\partial c} f(x, c) dx + f(b, c)\dfrac{db}{dc} - f(a, c)\dfrac{da}{dc}$

993

Leibniz's Theorem for Differentiation of a Product

3.3.8

$$\frac{d^n}{dx^n}(uv)=\frac{d^nu}{dx^n}v+\binom{n}{1}\frac{d^{n-1}u}{dx^{n-1}}\frac{dv}{dx}+\binom{n}{2}\frac{d^{n-2}u}{dx^{n-2}}\frac{d^2v}{dx^2}$$

$$+\ldots+\binom{n}{r}\frac{d^{n-r}u}{dx^{n-r}}\frac{d^rv}{dx^r}+\ldots+u\frac{d^nv}{dx^n}$$

3.3.9
$$\frac{dx}{dy}=1\Big/\frac{dy}{dx}$$

3.3.10
$$\frac{d^2x}{dy^2}=\frac{-d^2y}{dx^2}\left(\frac{dy}{dx}\right)^{-3}$$

3.3.11
$$\frac{d^3x}{dy^3}=-\left[\frac{d^3y}{dx^3}\frac{dy}{dx}-3\left(\frac{d^2y}{dx^2}\right)^2\right]\left(\frac{dy}{dx}\right)^{-5}$$

Integration by Parts

3.3.12
$$\int u\,dv=uv-\int v\,du$$

3.3.13
$$\int uv\,dx=\left(\int u\,dx\right)v-\int\left(\int u\,dx\right)\frac{dv}{dx}\,dx$$

Integrals of Rational Algebraic Functions

(Integration constants are omitted)

3.3.14
$$\int(ax+b)^n\,dx=\frac{(ax+b)^{n+1}}{a(n+1)}\qquad(n\neq-1)$$

3.3.15
$$\int\frac{dx}{ax+b}=\frac{1}{a}\ln|ax+b|$$

The following formulas are useful for evaluating $\int\frac{P(x)\,dx}{(ax^2+bx+c)^n}$ where $P(x)$ is a polynomial and $n>1$ is an integer.

3.3.16
$$\int\frac{dx}{(ax^2+bx+c)}=\frac{2}{(4ac-b^2)^{\frac{1}{2}}}\arctan\frac{2ax+b}{(4ac-b^2)^{\frac{1}{2}}}$$
$$(b^2-4ac<0)$$

3.3.17
$$=\frac{1}{(b^2-4ac)^{\frac{1}{2}}}\ln\left|\frac{2ax+b-(b^2-4ac)^{\frac{1}{2}}}{2ax+b+(b^2-4ac)^{\frac{1}{2}}}\right|$$
$$(b^2-4ac>0)$$

3.3.18
$$=\frac{-2}{2ax+b}\qquad(b^2-4ac=0)$$

3.3.19
$$\int\frac{x\,dx}{ax^2+bx+c}=\frac{1}{2a}\ln|ax^2+bx+c|-\frac{b}{2a}\int\frac{dx}{ax^2+bx+c}$$

3.3.20
$$\int\frac{dx}{(a+bx)(c+dx)}=\frac{1}{ad-bc}\ln\left|\frac{c+dx}{a+bx}\right|\qquad(ad\neq bc)$$

3.3.21
$$\int\frac{dx}{a^2+b^2x^2}=\frac{1}{ab}\arctan\frac{bx}{a}$$

3.3.22
$$\int\frac{x\,dx}{a^2+b^2x^2}=\frac{1}{2b^2}\ln|a^2+b^2x^2|$$

3.3.23
$$\int\frac{dx}{a^2-b^2x^2}=\frac{1}{2ab}\ln\left|\frac{a+bx}{a-bx}\right|$$

3.3.24
$$\int\frac{dx}{(x^2+a^2)^2}=\frac{1}{2a^3}\arctan\frac{x}{a}+\frac{x}{2a^2(x^2+a^2)}$$

3.3.25
$$\int\frac{dx}{(x^2-a^2)^2}=\frac{-x}{2a^2(x^2-a^2)}+\frac{1}{4a^3}\ln\left|\frac{a+x}{a-x}\right|$$

Integrals of Irrational Algebraic Functions

3.3.26
$$\int\frac{dx}{[(a+bx)(c+dx)]^{1/2}}=\frac{2}{(-bd)^{1/2}}\arctan\left[\frac{-d(a+bx)}{b(c+dx)}\right]^{1/2}\qquad(bd<0)$$

3.3.27
$$=\frac{-1}{(-bd)^{1/2}}\arcsin\left(\frac{2bdx+ad+bc}{bc-ad}\right)\qquad(b>0,d<0)$$

3.3.28
$$=\frac{2}{(bd)^{1/2}}\ln|[bd(a+bx)]^{1/2}+b(c+dx)^{1/2}|\qquad(bd>0)$$

3.3.29
$$\int\frac{dx}{(a+bx)^{1/2}(c+dx)}=\frac{2}{[d(bc-ad)]^{1/2}}\arctan\left[\frac{d(a+bx)}{(bc-ad)}\right]^{1/2}\qquad(d(ad-bc)<0)$$

3.3.30
$$=\frac{1}{[d(ad-bc)]^{1/2}}\ln\left|\frac{d(a+bx)^{1/2}-[d(ad-bc)]^{1/2}}{d(a+bx)^{1/2}+[d(ad-bc)]^{1/2}}\right|\qquad(d(ad-bc)>0)$$

3.3.31

$$\int [(a+bx)(c+dx)]^{1/2}dx$$
$$=\frac{(ad-bc)+2b(c+dx)}{4bd}[(a+bx)(c+dx)]^{1/2}$$
$$-\frac{(ad-bc)^2}{8bd}\int\frac{dx}{[(a+bx)(c+dx)]^{1/2}}$$

3.3.32

$$\int\left[\frac{c+dx}{a+bx}\right]^{1/2}dx=\frac{1}{b}[(a+bx)(c+dx)]^{1/2}$$
$$-\frac{(ad-bc)}{2b}\int\frac{dx}{[(a+bx)(c+dx)]^{1/2}}$$

3.3.33

$$\int\frac{dx}{(ax^2+bx+c)^{1/2}}$$
$$=a^{-1/2}\ln|2a^{1/2}(ax^2+bx+c)^{1/2}+2ax+b|\,(a>0)$$

3.3.34
$$=a^{-1/2}\operatorname{arcsinh}\frac{(2ax+b)}{(4ac-b^2)^{1/2}}$$
$$(a>0,\ 4ac>b^2)$$

3.3.35
$$=a^{-1/2}\ln|2ax+b|\,(a>0,\ b^2=4ac)$$

3.3.36
$$=-(-a)^{-1/2}\arcsin\frac{(2ax+b)}{(b^2-4ac)^{1/2}}$$
$$(a<0,\ b^2>4ac,\ |2ax+b|<(b^2-4ac)^{1/2})$$

3.3.37

$$\int(ax^2+bx+c)^{1/2}dx=\frac{2ax+b}{4a}(ax^2+bx+c)^{1/2}$$
$$+\frac{4ac-b^2}{8a}\int\frac{dx}{(ax^2+bx+c)^{1/2}}$$

3.3.38

$$\int\frac{dx}{x(ax^2+bx+c)^{1/2}}=-\int\frac{dt}{(a+bt+ct^2)^{1/2}}\ \text{where}\ t=1/x$$

3.3.39

$$\int\frac{xdx}{(ax^2+bx+c)^{1/2}}$$
$$=\frac{1}{a}(ax^2+bx+c)^{1/2}-\frac{b}{2a}\int\frac{dx}{(ax^2+bx+c)^{1/2}}$$

3.3.40
$$\int\frac{dx}{(x^2\pm a^2)^{\frac{1}{2}}}=\ln|x+(x^2\pm a^2)^{\frac{1}{2}}|$$

3.3.41
$$\int(x^2\pm a^2)^{\frac{1}{2}}dx=\frac{x}{2}(x^2\pm a^2)^{\frac{1}{2}}\pm\frac{a^2}{2}\ln|x+(x^2\pm a^2)^{\frac{1}{2}}|$$

3.3.42
$$\int\frac{dx}{x(x^2+a^2)^{\frac{1}{2}}}=-\frac{1}{a}\ln\left|\frac{a+(x^2+a^2)^{\frac{1}{2}}}{x}\right|$$

3.3.43
$$\int\frac{dx}{x(x^2-a^2)^{\frac{1}{2}}}=\frac{1}{a}\arccos\frac{a}{x}$$

3.3.44
$$\int\frac{dx}{(a^2-x^2)^{\frac{1}{2}}}=\arcsin\frac{x}{a}$$

3.3.45
$$\int(a^2-x^2)^{\frac{1}{2}}dx=\frac{x}{2}(a^2-x^2)^{\frac{1}{2}}+\frac{a^2}{2}\arcsin\frac{x}{a}$$

3.3.46
$$\int\frac{dx}{x(a^2-x^2)^{\frac{1}{2}}}=-\frac{1}{a}\ln\left|\frac{a+(a^2-x^2)^{\frac{1}{2}}}{x}\right|$$

3.3.47
$$\int\frac{dx}{(2ax-x^2)^{\frac{1}{2}}}=\arcsin\frac{x-a}{a}$$

3.3.48

$$\int(2ax-x^2)^{\frac{1}{2}}dx=\frac{(x-a)}{2}(2ax-x^2)^{\frac{1}{2}}+\frac{a^2}{2}\arcsin\frac{x-a}{a}$$

3.3.49

$$\int\frac{dx}{(ax^2+b)(cx^2+d)^{\frac{1}{2}}}$$
$$=\frac{1}{[b(ad-bc)]^{\frac{1}{2}}}\arctan\frac{x(ad-bc)^{\frac{1}{2}}}{[b(cx^2+d)]^{\frac{1}{2}}}\qquad(ad>bc)$$

3.3.50

$$=\frac{1}{2[b(bc-ad)]^{\frac{1}{2}}}\ln\left|\frac{[b(cx^2+d)]^{\frac{1}{2}}+x(bc-ad)^{\frac{1}{2}}}{[b(cx^2+d)]^{\frac{1}{2}}-x(bc-ad)^{\frac{1}{2}}}\right|$$
$$(bc>ad)$$

3.4. Limits, Maxima and Minima

Indeterminate Forms (L'Hospital's Rule)

3.4.1 Let $f(x)$ and $g(x)$ be differentiable on an interval $a\le x<b$ for which $g'(x)\ne 0$.

If

$$\lim_{x\to b-}f(x)=0\ \text{and}\ \lim_{x\to b-}g(x)=0$$

or if

$$\lim_{x\to b-}f(x)=\infty\ \text{and}\ \lim_{x\to b-}g(x)=\infty$$

and if

$$\lim_{x\to b-}\frac{f'(x)}{g'(x)}=l\ \text{then}\ \lim_{x\to b-}\frac{f(x)}{g(x)}=l.$$

Both b and l may be finite or infinite.

Maxima and Minima

3.4.2 (1) *Functions of One Variable*

The function $y=f(x)$ has a maximum at $x=x_0$ if $f'(x_0)=0$ and $f''(x_0)<0$, and a minimum at $x=x_0$ if $f'(x_0)=0$ and $f''(x_0)>0$. Points x_0 for which $f'(x_0)=0$ are called stationary points.

3.4.3 (2) *Functions of Two Variables*

The function $f(x, y)$ has a maximum or minimum for those values of (x_0, y_0) for which

$$\frac{\partial f}{\partial x}=0, \frac{\partial f}{\partial y}=0,$$

and for which $\begin{vmatrix} \partial^2 f/\partial x \partial y & \partial^2 f/\partial x^2 \\ \partial^2 f/\partial y^2 & \partial^2 f/\partial x \partial y \end{vmatrix}<0;$

(a) $f(x,y)$ has a maximum

if $\frac{\partial^2 f}{\partial x^2}<0$ and $\frac{\partial^2 f}{\partial y^2}<0$ at (x_0, y_0),

(b) $f(x,y)$ has a minimum

if $\frac{\partial^2 f}{\partial x^2}>0$ and $\frac{\partial^2 f}{\partial y^2}>0$ at (x_0, y_0).

3.5. Absolute and Relative Errors

(1) If x_0 is an approximation to the true value of x, then

3.5.1 (a) the *absolute error* of x_0 is $\Delta x=x_0-x$, $x-x_0$ is the correction to x.

3.5.2 (b) the *relative error* of x_0 is $\delta x=\dfrac{\Delta x}{x} \approx \dfrac{\Delta x}{x_0}$

3.5.3 (c) the *percentage error* is 100 times the relative error.

3.5.4 (2) The absolute error of the sum or difference of several numbers is at most equal to the sum of the absolute errors of the individual numbers.

3.5.5 (3) If $f(x_1, x_2, \ldots, x_n)$ is a function of $x_1, x_2, \ldots, x_n$ and the absolute error in x_i ($i=1, 2, \ldots n$) is Δx_i, then the absolute error in f is

$$\Delta f \approx \frac{\partial f}{\partial x_1} \Delta x_1 + \frac{\partial f}{\partial x_2} \Delta x_2 + \ldots + \frac{\partial f}{\partial x_n} \Delta x_n$$

3.5.6 (4) The relative error of the product or quotient of several factors is at most equal to the sum of the relative errors of the individual factors.

3.5.7

(5) If $y=f(x)$, the relative error $\delta y=\dfrac{\Delta y}{y} \approx \dfrac{f'(x)}{f(x)} \Delta x$

Approximate Values

If $|\epsilon|<<1$, $|\eta|<<1$, $b<<a$,

3.5.8 $$(a+b)^k \approx a^k+ka^{k-1}b$$

3.5.9 $$(1+\epsilon)(1+\eta) \approx 1+\epsilon+\eta$$

3.5.10 $$\frac{1+\epsilon}{1+\eta} \approx 1+\epsilon-\eta$$

3.6. Infinite Series

Taylor's Formula for a Single Variable

3.6.1

$$f(x+h)=f(x)+hf'(x)+\frac{h^2}{2!}f''(x)$$
$$+ \ldots +\frac{h^{n-1}}{(n-1)!}f^{(n-1)}(x)+R_n$$

3.6.2

$$R_n=\frac{h^n}{n!}f^{(n)}(x+\theta_1 h)=\frac{h^n}{(n-1)!}(1-\theta_2)^{n-1}f^{(n)}(x+\theta_2 h)$$
$$(0<\theta_{1,2}(x)<1)$$

3.6.3

$$=\frac{h^n}{(n-1)!}\int_0^1 (1-t)^{n-1}f^{(n)}(x+th)dt$$

3.6.4

$$f(x)=f(a)+\frac{(x-a)}{1!}f'(a)+\frac{(x-a)^2}{2!}f''(a)+$$
$$\ldots +\frac{(x-a)^{n-1}}{(n-1)!}f^{(n-1)}(a)+R_n$$

3.6.5 $$R_n=\frac{(x-a)^n}{n!}f^{(n)}(\xi) \qquad (a<\xi<x)$$

Lagrange's Expansion

If $y=f(x)$, $y_0=f(x_0)$, $f'(x_0) \neq 0$, then

3.6.6

$$x=x_0+\sum_{k=1}^{\infty} \frac{(y-y_0)^k}{k!} [\frac{d^{k-1}}{dx^{k-1}}\{\frac{x-x_0}{f(x)-y_0}\}^k]_{x=x_0}$$

3.6.7

$$g(x)=g(x_0)$$
$$+\sum_{k=1}^{\infty} \frac{(y-y_0)^k}{k!} [\frac{d^{k-1}}{dx^{k-1}}(g'(x)\{\frac{x-x_0}{f(x)-y_0}\}^k)]_{x=x_0}$$

where $g(x)$ is any function indefinitely differentiable.

Binomial Series

3.6.8

$$(1+x)^a=\sum_{k=0}^{\infty} \binom{a}{k} x^k \qquad (-1<x<1)$$

3.6.9

$$(1+x)^\alpha = 1 + \alpha x + \frac{\alpha(\alpha-1)}{2!}x^2 + \frac{\alpha(\alpha-1)(\alpha-2)}{3!}x^3 + \cdots,$$

3.6.10

$$(1+x)^{-1} = 1 - x + x^2 - x^3 + x^4 - \cdots \qquad (-1<x<1)$$

3.6.11

$$(1+x)^{\frac12} = 1 + \frac{x}{2} - \frac{x^2}{8} + \frac{x^3}{16} - \frac{5x^4}{128} + \frac{7x^5}{256} - \frac{21x^6}{1024} + \cdots$$
$$(-1<x<1)$$

3.6.12

$$(1+x)^{-\frac12} = 1 - \frac{x}{2} + \frac{3x^2}{8} - \frac{5x^3}{16} + \frac{35x^4}{128} - \frac{63x^5}{256}$$
$$+ \frac{231x^6}{1024} - \cdots \qquad (-1<x<1)$$

3.6.13

$$(1+x)^{\frac13} = 1 + \frac13 x - \frac19 x^2 + \frac{5}{81}x^3 - \frac{10}{243}x^4$$
$$+ \frac{22}{729}x^5 - \frac{154}{6561}x^6 + \cdots \qquad (-1<x<1)$$

3.6.14

$$(1+x)^{-\frac13} = 1 - \frac13 x + \frac29 x^2 - \frac{14}{81}x^3 + \frac{35}{243}x^4$$
$$- \frac{91}{729}x^5 + \frac{728}{6561}x^6 - \cdots \qquad (-1<x<1)$$

Asymptotic Expansions

3.6.15 A series $\sum_{k=0}^{\infty} a_k x^{-k}$ is said to be an asymptotic expansion of a function $f(x)$ if

$$f(x) - \sum_{k=0}^{n-1} a_k x^{-k} = O(x^{-n}) \text{ as } x \to \infty$$

for every $n=1, 2, \ldots$. We write

$$f(x) \sim \sum_{k=0}^{\infty} a_k x^{-k}.$$

The series itself may be either convergent or divergent.

Operations With Series

$$s_1 = 1 + a_1 x + a_2 x^2 + a_3 x^3 + a_4 x^4 + \cdots$$
$$s_2 = 1 + b_1 x + b_2 x^2 + b_3 x^3 + b_4 x^4 + \cdots$$
$$s_3 = 1 + c_1 x + c_2 x^2 + c_3 x^3 + c_4 x^4 + \cdots$$

	Operation	c_1	c_2	c_3	c_4
3.6.16	$s_3 = s_1^{-1}$	$-a_1$	$a_1^2 - a_2$	$2a_1a_2 - a_3 - a_1^3$	$2a_1a_3 - 3a_1^2a_2 - a_4 + a_2^2 + a_1^4$
3.6.17	$s_3 = s_1^{-2}$	$-2a_1$	$3a_1^2 - 2a_2$	$6a_1a_2 - 2a_3 - 4a_1^3$	$6a_1a_3 + 3a_2^2 - 2a_4 - 12a_1^2a_2 + 5a_1^4$
3.6.18	$s_3 = s_1^{\frac12}$	$\frac12 a_1$	$\frac12 a_2 - \frac18 a_1^2$	$\frac12 a_3 - \frac14 a_1a_2 + \frac{1}{16}a_1^3$	$\frac12 a_4 - \frac14 a_1a_3 - \frac18 a_2^2 + \frac{3}{16}a_1^2a_2 - \frac{5}{128}a_1^4$
3.6.19	$s_3 = s_1^{-\frac12}$	$-\frac12 a_1$	$\frac38 a_1^2 - \frac12 a_2$	$\frac34 a_1a_2 - \frac12 a_3 - \frac{5}{16}a_1^3$	$\frac34 a_1a_3 + \frac38 a_2^2 - \frac12 a_4 - \frac{15}{16}a_1^2a_2 + \frac{35}{128}a_1^4$
3.6.20	$s_3 = s_1^n$	na_1	$\frac12(n-1)c_1a_1 + na_2$	$c_1a_2(n-1) + \frac16 c_1a_1^2(n-1)(n-2) + na_3$	$na_4 + c_1a_3(n-1) + \frac12 n(n-1)a_2^2 + \frac12(n-1)(n-2)c_1a_1a_2 + \frac{1}{24}(n-1)(n-2)(n-3)c_1a_1^3$
3.6.21	$s_3 = s_1 s_2$	$a_1 + b_1$	$b_2 + a_1b_1 + a_2$	$b_3 + a_1b_2 + a_2b_1 + a_3$	$b_4 + a_1b_3 + a_2b_2 + a_3b_1 + a_4$
3.6.22	$s_3 = s_1/s_2$	$a_1 - b_1$	$a_2 - (b_1c_1 + b_2)$	$a_3 - (b_1c_2 + b_2c_1 + b_3)$	$a_4 - (b_1c_3 + b_2c_2 + b_3c_1 + b_4)$
3.6.23	$s_3 = \exp(s_1 - 1)$	a_1	$a_2 + \frac12 a_1^2$	$a_3 + a_1a_2 + \frac16 a_1^3$	$a_4 + a_1a_3 + \frac12 a_2^2 + \frac12 a_2a_1^2 + \frac{1}{24}a_1^4$
3.6.24	$s_3 = 1 + \ln s_1$	a_1	$a_2 - \frac12 a_1c_1$	$a_3 - \frac13(a_2c_1 + 2a_1c_2)$	$a_4 - \frac14(a_3c_1 + 2a_2c_2 + 3a_1c_3)$

Reversion of Series

3.6.25 Given

$$y = ax + bx^2 + cx^3 + dx^4 + ex^5 + fx^6 + gx^7 + \ldots$$

then

$$x = Ay + By^2 + Cy^3 + Dy^4 + Ey^5 + Fy^6 + Gy^7 + \ldots$$

where

$$aA = 1$$

$$a^3 B = -b$$

$$a^5 C = 2b^2 - ac$$

$$a^7 D = 5abc - a^2 d - 5b^3$$

$$a^9 E = 6a^2 bd + 3a^2 c^2 + 14b^4 - a^3 e - 21ab^2 c$$

$$a^{11} F = 7a^3 be + 7a^3 cd + 84ab^3 c - a^4 f$$
$$- 28a^2 bc^2 - 42b^5 - 28a^2 b^2 d$$

$$a^{13} G = 8a^4 bf + 8a^4 ce + 4a^4 d^2 + 120a^2 b^3 d$$
$$+ 180a^2 b^2 c^2 + 132b^6 - a^5 g - 36a^3 b^3 e$$
$$- 72a^3 bcd - 12a^3 c^3 - 330ab^4 c$$

Kummer's Transformation of Series

3.6.26 Let $\sum_{k=0}^{\infty} a_k = s$ be a given convergent series and $\sum_{k=0}^{\infty} c_k = c$ be a given convergent series with known sum c such that $\lim_{k \to \infty} \frac{a_k}{c_k} = \lambda \neq 0$.

Then

$$s = \lambda c + \sum_{k=0}^{\infty} \left(1 - \lambda \frac{c_k}{a_k}\right) a_k.$$

Euler's Transformation of Series

3.6.27 If $\sum_{k=0}^{\infty} (-1)^k a_k = a_0 - a_1 + a_2 - \ldots$ is a convergent series with sum s then

$$s = \sum_{k=0}^{\infty} \frac{(-1)^k \Delta^k a_0}{2^{k+1}}, \quad \Delta^k a_0 = \sum_{m=0}^{k} (-1)^m \binom{k}{m} a_{k-m}$$

Euler-Maclaurin Summation Formula

3.6.28

$$\sum_{k=1}^{n-1} f_k = \int_0^n f(k)\,dk - \frac{1}{2}[f(0) + f(n)] + \frac{1}{12}[f'(n) - f'(0)]$$

$$- \frac{1}{720}[f'''(n) - f'''(0)] + \frac{1}{30240}[f^{(V)}(n) - f^{(V)}(0)]$$

$$- \frac{1}{1209600}[f^{(VII)}(n) - f^{(VII)}(0)] + \ldots$$

3.7. Complex Numbers and Functions

Cartesian Form

3.7.1
$$z = x + iy$$

Polar Form

3.7.2 $\qquad z = re^{i\theta} = r(\cos\theta + i\sin\theta)$

3.7.3 $\qquad \textit{Modulus: } |z| = (x^2 + y^2)^{\frac{1}{2}} = r$

3.7.4 $\textit{Argument: } \arg z = \arctan(y/x) = \theta$ (other notations for arg z are am z and ph z).

3.7.5 $\qquad$ Real Part: $x = \mathscr{R}z = r\cos\theta$

3.7.6 $\qquad$ Imaginary Part: $y = \mathscr{I}z = r\sin\theta$

Complex Conjugate of z

3.7.7 $\qquad\qquad \bar{z} = x - iy$

3.7.8 $\qquad\qquad |\bar{z}| = |z|$

3.7.9 $\qquad\qquad \arg \bar{z} = -\arg z$

Multiplication and Division

If $z_1 = x_1 + iy_1$, $z_2 = x_2 + iy_2$, then

3.7.10 $\qquad z_1 z_2 = x_1 x_2 - y_1 y_2 + i(x_1 y_2 + x_2 y_1)$

3.7.11 $\qquad\qquad |z_1 z_2| = |z_1||z_2|$

3.7.12 $\qquad \arg(z_1 z_2) = \arg z_1 + \arg z_2$

3.7.13 $\quad \dfrac{z_1}{z_2} = \dfrac{z_1 \bar{z}_2}{|z_2|^2} = \dfrac{x_1 x_2 + y_1 y_2 + i(x_2 y_1 - x_1 y_2)}{x_2^2 + y_2^2}$

3.7.14 $\qquad\qquad \left|\dfrac{z_1}{z_2}\right| = \dfrac{|z_1|}{|z_2|}$

3.7.15 $\qquad \arg\left(\dfrac{z_1}{z_2}\right) = \arg z_1 - \arg z_2$

Powers

3.7.16 $\quad z^n = r^n e^{in\theta}$

3.7.17 $\qquad = r^n \cos n\theta + ir^n \sin n\theta$
$$(n = 0, \pm 1, \pm 2, \ldots)$$

3.7.18 $\qquad z^2 = x^2 - y^2 + i(2xy)$

3.7.19 $\qquad z^3 = x^3 - 3xy^2 + i(3x^2 y - y^3)$

3.7.20 $\quad z^4 = x^4 - 6x^2 y^2 + y^4 + i(4x^3 y - 4xy^3)$

3.7.21 $\quad z^5 = x^5 - 10x^3 y^2 + 5xy^4 + i(5x^4 y - 10x^2 y^3 + y^5)$

3.7.22

$$z^n = \left[x^n - \binom{n}{2} x^{n-2} y^2 + \binom{n}{4} x^{n-4} y^4 - \ldots\right]$$

$$+ i\left[\binom{n}{1} x^{n-1} y - \binom{n}{3} x^{n-3} y^3 + \ldots\right],$$

$$(n = 1, 2, \ldots)$$

If $z^n=u_n+iv_n$, then $z^{n+1}=u_{n+1}+iv_{n+1}$ where

3.7.23 $\quad u_{n+1}=xu_n-yv_n;\quad v_{n+1}=xv_n+yu_n$

$\mathscr{R}z^n$ and $\mathscr{I}z^n$ are called harmonic polynomials.

3.7.24 $\qquad \dfrac{1}{z}=\dfrac{\bar{z}}{|z|^2}=\dfrac{x-iy}{x^2+y^2}$

3.7.25 $\qquad \dfrac{1}{z^n}=\dfrac{\bar{z}^n}{|z|^{2n}}=(z^{-1})^n$

Roots

3.7.26 $\quad z^{\frac12}=\sqrt{z}=r^{\frac12}e^{i\theta/2}=r^{\frac12}\cos\frac12\theta+ir^{\frac12}\sin\frac12\theta$

If $-\pi<\theta\leq\pi$ this is the principal root. The other root has the opposite sign. The principal root is given by

3.7.27 $\quad z^{\frac12}=[\frac12(r+x)]^{\frac12}\pm i[\frac12(r-x)]^{\frac12}=u\pm iv$ where $2uv=y$ and where the ambiguous sign is taken to be the same as the sign of y.

3.7.28 $\quad z^{1/n}=r^{1/n}e^{i\theta/n}$, (principal root if $-\pi<\theta\leq\pi$). Other roots are $r^{1/n}e^{i(\theta+2\pi k)/n}$ $(k=1,2,3,\ldots,n-1)$.

Inequalities

3.7.29 $\qquad \Big||z_1|-|z_2|\Big|\leq|z_1\pm z_2|\leq|z_1|+|z_2|$

Complex Functions, Cauchy-Riemann Equations

$f(z)=f(x+iy)=u(x,y)+iv(x,y)$ where $u(x,y)$, $v(x,y)$ are real, is *analytic* at those points $z=x+iy$ at which

3.7.30 $\qquad \dfrac{\partial u}{\partial x}=\dfrac{\partial v}{\partial y},\quad \dfrac{\partial u}{\partial y}=-\dfrac{\partial v}{\partial x}$

If $z=re^{i\theta}$,

3.7.31 $\qquad \dfrac{\partial u}{\partial r}=\dfrac{1}{r}\dfrac{\partial v}{\partial\theta},\quad \dfrac{1}{r}\dfrac{\partial u}{\partial\theta}=-\dfrac{\partial v}{\partial r}$

Laplace's Equation

The functions $u(x,y)$ and $v(x,y)$ are called harmonic functions and satisfy Laplace's equation:

Cartesian Coordinates

3.7.32 $\qquad \dfrac{\partial^2 u}{\partial x^2}+\dfrac{\partial^2 u}{\partial y^2}=\dfrac{\partial^2 v}{\partial x^2}+\dfrac{\partial^2 v}{\partial y^2}=0$

Polar Coordinates

3.7.33 $\quad r\dfrac{\partial}{\partial r}\left(r\dfrac{\partial u}{\partial r}\right)+\dfrac{\partial^2 u}{\partial\theta^2}=r\dfrac{\partial}{\partial r}\left(r\dfrac{\partial v}{\partial r}\right)+\dfrac{\partial^2 v}{\partial\theta^2}=0$

3.8. Algebraic Equations

Solution of Quadratic Equations

3.8.1 Given $az^2+bz+c=0$,

$$z_{1,2}=-\left(\dfrac{b}{2a}\right)\pm\dfrac{1}{2a}\,q^{\frac12},\ q=b^2-4ac,$$

$$z_1+z_2=-b/a,\ z_1z_2=c/a$$

If $q>0$, two real roots,
$q=0$, two equal roots,
$q<0$, pair of complex conjugate roots.

Solution of Cubic Equations

3.8.2 Given $z^3+a_2z^2+a_1z+a_0=0$, let

$$q=\dfrac{1}{3}\,a_1-\dfrac{1}{9}\,a_2^2;\ r=\dfrac{1}{6}\,(a_1a_2-3a_0)-\dfrac{1}{27}\,a_2^3.$$

If $q^3+r^2>0$, one real root and a pair of complex conjugate roots,

$q^3+r^2=0$, all roots real and at least two are equal,

$q^3+r^2<0$, all roots real (irreducible case).

Let

$$s_1=[r+(q^3+r^2)^{\frac12}]^{\frac13},\ s_2=[r-(q^3+r^2)^{\frac12}]^{\frac13}$$

then

$$z_1=(s_1+s_2)-\dfrac{a_2}{3}$$

$$z_2=-\dfrac{1}{2}\,(s_1+s_2)-\dfrac{a_2}{3}+\dfrac{i\sqrt{3}}{2}\,(s_1-s_2)$$

$$z_3=-\dfrac{1}{2}\,(s_1+s_2)-\dfrac{a_2}{3}-\dfrac{i\sqrt{3}}{2}\,(s_1-s_2).$$

If z_1, z_2, z_3 are the roots of the cubic equation

$$z_1+z_2+z_3=-a_2$$

$$z_1z_2+z_1z_3+z_2z_3=a_1$$

$$z_1z_2z_3=-a_0$$

Solution of Quartic Equations

3.8.3 Given $z^4+a_3z^3+a_2z^2+a_1z+a_0=0$, find the real root u_1 of the cubic equation

$$u^3-a_2u^2+(a_1a_3-4a_0)u-(a_1^2+a_0a_3^2-4a_0a_2)=0$$

and determine the four roots of the quartic as solutions of the two quadratic equations

$$v^2+\left[\dfrac{a_3}{2}\mp\left(\dfrac{a_3^2}{4}+u_1-a_2\right)^{\frac12}\right]v+\dfrac{u_1}{2}\mp\left[\left(\dfrac{u_1}{2}\right)^2-a_0\right]^{\frac12}=0$$

If all roots of the cubic equation are real, use the value of u_1 which gives real coefficients in the quadratic equation and select signs so that if

$$z^4+a_3z^3+a_2z^2+a_1z+a_0=(z^2+p_1z+q_1)(z^2+p_2z+q_2),$$

then

$$p_1+p_2=a_3, p_1p_2+q_1+q_2=a_2, p_1q_2+p_2q_1=a_1, q_1q_2=a_0.$$

If z_1, z_2, z_3, z_4 are the roots,

$$\Sigma z_i=-a_3, \ \Sigma z_iz_jz_k=-a_1,$$

$$\Sigma z_iz_j=a_2, \ z_1z_2z_3z_4=a_0.$$

3.9. Successive Approximation Methods

General Comments

3.9.1 Let $x=x_1$ be an approximation to $x=\xi$ where $f(\xi)=0$ and both x_1 and ξ are in the interval $a \leq x \leq b$. We define

$$x_{n+1}=x_n+c_nf(x_n) \qquad (n=1, 2, \ldots).$$

Then, if $f'(x)\geq 0$ and the constants c_n are negative and bounded, the sequence x_n converges monotonically to the root ξ.

If $c_n=c=\text{constant}<0$ and $f'(x)>0$, then the process converges but not necessarily monotonically.

Degree of Convergence of an Approximation Process

3.9.2 Let x_1, x_2, x_3, $\ldots$ be an infinite sequence of approximations to a number ξ. Then, if

$$|x_{n+1}-\xi|<A|x_n-\xi|^k, \qquad (n=1, 2, \ldots)$$

where A and k are independent of n, the sequence is said to have convergence of at most the kth degree (or order or index) to ξ. If $k=1$ and $A<1$ the convergence is linear; if $k=2$ the convergence is quadratic.

Regula Falsi (False Position)

3.9.3 Given $y=f(x)$ to find ξ such that $f(\xi)=0$, choose x_0 and x_1 such that $f(x_0)$ and $f(x_1)$ have opposite signs and compute

$$x_2=x_1-\frac{(x_1-x_0)}{(f_1-f_0)} \ f_1=\frac{f_1x_0-f_0x_1}{f_1-f_0}.$$

Then continue with x_2 and either of x_0 or x_1 for which $f(x_0)$ or $f(x_1)$ is of opposite sign to $f(x_2)$.

Regula falsi is equivalent to inverse linear interpolation.

Method of Iteration (Successive Substitution)

3.9.4 The iteration scheme $x_{k+1}=F(x_k)$ will converge to a zero of $x=F(x)$ if

(1) $|F'(x)|\leq q<1$ for $a\leq x\leq b$,

(2) $a\leq x_0\pm\dfrac{|F(x_0)-x_0|}{1-q}\leq b.$

Newton's Method of Successive Approximations

3.9.5

Newton's Rule

If $x=x_k$ is an approximation to the solution $x=\xi$ of $f(x)=0$ then the sequence

$$x_{k+1}=x_k-\frac{f(x_k)}{f'(x_k)}$$

will converge quadratically to $x=\xi$: (if instead of the condition (2) above),

(1) *Monotonic convergence*, $f(x_0)f''(x_0)>0$ and $f'(x)$, $f''(x)$ do not change sign in the interval (x_0, ξ), or

(2) *Oscillatory convergence*, $f(x_0)f''(x_0)<0$ and $f'(x)$, $f''(x)$ do not change sign in the interval (x_0, x_1), $x_0 \leq \xi \leq x_1$.

Newton's Method Applied to Real nth Roots

3.9.6 Given $x^n=N$, if x_k is an approximation $x=N^{1/n}$ then the sequence

$$x_{k+1}=\frac{1}{n}\left[\frac{N}{x_k^{n-1}}+(n-1)x_k\right]$$

will converge quadratically to x.

If $n=2$, $x_{k+1}=\dfrac{1}{2}\left(\dfrac{N}{x_k}+x_k\right),$

If $n=3$, $x_{k+1}=\dfrac{1}{3}\left(\dfrac{N}{x_k^2}+2x_k\right).$

Aitken's δ^2-Process for Acceleration of Sequences

3.9.7 If x_k, x_{k+1}, x_{k+2} are three successive iterates in a sequence converging with an error which is approximately in geometric progression, then

$$\bar{x}_k=x_k-\frac{(x_k-x_{k+1})^2}{\Delta^2 x_k}=\frac{x_kx_{k+2}-x_{k+1}^2}{\Delta^2 x_k};$$

$$\Delta^2 x_k=x_k-2x_{k+1}+x_{k+2}$$

is an improved estimate of x. In fact, if $x_k=x+O(\lambda^k)$ then $\bar{x}=x+O(\lambda^k)$.

INDEX

REA'S
PROBLEM
SOLVERS®

PROBLEM SOLVERS

Physics

The Perfect Resource for:
- any class
- any exam
- any problem

A Reference for Life

The PROBLEM SOLVERS® are comprehensive supplemental textbooks designed to save time in finding solutions to problems. Each PROBLEM SOLVER® is the first of its kind ever produced in its field. It is the product of a massive effort to illustrate almost any imaginable problem in exceptional depth, detail, and clarity. Each problem is worked out in detail with a step-by-step solution, and the problems are arranged in order of complexity from elementary to advanced. Each book is fully indexed for locating problems rapidly.

Accounting	**Genetics**
Advanced Calculus	**Geometry**
Algebra & Trigonometry	**Linear Algebra**
Automatic Control Systems/Robotics	**Mechanics**
Biology	**Numerical Analysis**
Business, Accounting & Finance	**Operations Research**
Calculus	**Organic Chemistry**
Chemistry	**Physics**
Differential Equations	**Pre-Calculus**
Economics	**Probability**
Electrical Machines	**Psychology**
Electric Circuits	**Statistics**
Electromagnetics	**Technical Design Graphics**
Electronics	**Thermodynamics**
Finite & Discrete Math	**Topology**
Fluid Mechanics/Dynamics	**Transport Phenomena**

If you would like more information about any of these books,
complete the coupon below and return it to us or visit your local bookstore.

Research & Education Association
61 Ethel Road W., Piscataway, NJ 08854
Phone: (732) 819-8880   **website: www.rea.com**

Please send me more information about your Problem Solver® books.

Name _____

Address _____

City _____ State _____ Zip _____